AF781301

THE LIBRARIES OF THE UNIVERSITY OF CALIFORNIA SAN DIEGO

# Lecture Notes in Artificial Intelligence 4529

Edited by J. G. Carbonell and J. Siekmann

Subseries of Lecture Notes in Computer Science

UNIVERSITY OF [redacted] SAN DIEGO
3 1 [redacted] 0928

Patricia Melin Oscar Castillo
Luis T. Aguilar Janusz Kacprzyk
Witold Pedrycz (Eds.)

# Foundations of Fuzzy Logic and Soft Computing

12th International Fuzzy Systems Association World Congress, IFSA 2007
Cancun, Mexico, June 18-21, 2007
Proceedings

Springer

Series Editors

Jaime G. Carbonell, Carnegie Mellon University, Pittsburgh, PA, USA
Jörg Siekmann, University of Saarland, Saarbrücken, Germany

Volume Editors

Patricia Melin
Oscar Castillo
Tijuana Institute of Technology
Department of Computer Science, Tijuana, Mexico
E-mail: {pmelin,ocastillo}@tectijuana.mx

Luis T. Aguilar
CITEDI-IPN
Av. del Parque, Tijuana, Mexico
E-mail: laguilar@citedi.mx

Janusz Kacprzyk
Polish Academy of Sciences
Systems Research Institute
ul. Newelska 6, 01-447 Warsaw, Poland
E-mail: kacprzyk@ibspan.waw.pl

Witold Pedrycz
University of Alberta
Department of Electrical and Computer Engineering
Edmonton, Alberta T6G 2V4, Canada
E-mail: pedrycz@ece.ualberta.ca

Library of Congress Control Number: 2007927820

CR Subject Classification (1998): I.2, F.4.1, J.1, I, H.2

LNCS Sublibrary: SL 7 – Artificial Intelligence

ISSN 0302-9743
ISBN-10 3-540-72917-8 Springer Berlin Heidelberg New York
ISBN-13 978-3-540-72917-4 Springer Berlin Heidelberg New York

This work is subject to copyright. All rights are reserved, whether the whole or part of the material is concerned, specifically the rights of translation, reprinting, re-use of illustrations, recitation, broadcasting, reproduction on microfilms or in any other way, and storage in data banks. Duplication of this publication or parts thereof is permitted only under the provisions of the German Copyright Law of September 9, 1965, in its current version, and permission for use must always be obtained from Springer. Violations are liable to prosecution under the German Copyright Law.

Springer is a part of Springer Science+Business Media

springer.com

© Springer-Verlag Berlin Heidelberg 2007
Printed in Germany

Typesetting: Camera-ready by author, data conversion by Scientific Publishing Services, Chennai, India
Printed on acid-free paper SPIN: 12072392 06/3180 5 4 3 2 1 0

# Preface

This book comprises a selection of papers from IFSA 2007 on new methods and theories that contribute to the foundations of fuzzy logic and soft computing. These papers were selected from over 400 submissions and constitute an important contribution to the theory and applications of fuzzy logic and soft computing methodologies. Soft computing consists of several computing paradigms, including fuzzy logic, neural networks, genetic algorithms, and other techniques, which can be used to produce powerful intelligent systems for solving real-world problems. The papers of IFSA 2007 also make a contribution to this goal.

This book is intended to be a major reference for scientists and engineers interested in applying new computational and mathematical tools to achieve intelligent solutions to complex problems. We consider that this book can also be used to get novel ideas for new lines of research, or to continue the lines of research proposed by the authors of the papers contained in the book.

The book is divided into 14 main parts. Each part contains a set of papers on a common subject, so that the reader can find similar papers grouped together. Some of these parts comprise the papers of organized sessions of IFSA 2007 and we thank the session organizers for their incredible job in forming these sessions with invited and regular paper submissions.

In Part I, we begin with a set of papers on the "Relation Between Interval and Fuzzy Techniques" from a session organized by Vadik Kreinovich and Hung T. Nguyen. This set of papers show important results on the relations between interval and fuzzy techniques, which has become very important lately in part because of the progress of interval type-2 fuzzy logic.

In Part II, we have a set of papers on "Intuitionistic Fuzzy Sets and Their Applications" from a session organized by Eulalia Szmidt and Janusz Kacprzyk. These papers show important theoretical results, as well as novel applications of intuitionistic fuzzy logic. The area of intuitionistic fuzzy logic has also become a potential area of promissory results for the future of fuzzy logic.

In Part III, we have a collection of papers on the topic of "The Application of Fuzzy Logic and Soft Computing in Flexible Querying" from a session organized by Guy DeTre and Slawek Zadrozny. These papers show important theoretical results and applications of fuzzy logic and soft computing in achieving flexible querying for database systems. The area of flexible querying has become an important subject for achieving intelligent interfaces with human users and for managing large databases.

In Part IV, we have a collection of papers on "Philosophical and Human Scientific Aspects of Soft Computing" from a session organized by Vesa A. Niskanen. These papers show the interesting relationships between the philosophical aspects of soft computing and the formal-scientific aspects of soft computing. Papers on this subject are very important because they help in understanding the area of soft computing, and also enable new theories and methods in this area to be proposed.

In Part V, we have a collection of papers on "Search Engine and Information Processing and Retrieval" from a special FLINT session organized by Masoud Nikravesh. These papers describe important contributions on search engines for the Web, summarization, computing with words and granular computing, for information processing and retrieval. Papers on these subjects are very important theoretically as well as for applications because of the importance of Web search for documents and images.

In Part VI, we have a set of papers on "Perception-Based Data Mining and Decision-Making" from a session organized by Ildar Batyrshin, Janusz Kacprzyk, and Ronald R. Yager. These papers constitute an important contribution to data mining and linguistic summarization using fuzzy logic. Papers on these subjects are very important because data mining and building summaries are necessary in managing large amounts of data and information.

In Part VII, we have a collection of papers on "Joint Model-Based and Data-Based Learning: The Fuzzy Logic Approach" from a session organized by Joseph Aguilar-Martin and Julio Waissman Vilanova. These papers describe important contributions to solving the problems of learning in different types of models using fuzzy logic. Also, the new learning methods are applied to different applications. Learning from data and models is very important for solving real-world problems.

In Part VIII, we have a group of papers on "Fuzzy/Possibilistic Optimization" from a session organized by Weldon Lodwick. These papers describe important theoretical results and applications of fuzzy optimization. The optimization problem is considered from the point of view of fuzzy logic, which gives better results than traditional approaches.

In Part IX, we have a group of papers on the subject of "Fuzzy Trees" from a session organized by Ziheng Huang and Masoud Nikravesh. These papers show important theoretical results and applications of fuzzy trees. The use of fuzzy trees is very important as a model of human decision making and for this reason can have many real-world applications.

In Part X, we have a collection of papers on "Fuzzy Logic Theory" that describe different contributions to the theory of fuzzy logic. These papers show mainly theoretical results on fuzzy logic that can help advance the theory and/or provide fundamental tools for possible solutions to real-world problems.

In Part XI, we have a set of papers on "Type-2 Fuzzy Logic" that describe several contributions to the theory and applications of interval type-2 fuzzy logic. The papers represent an important contribution to the state of the art in this area, and also show that real problems can be solved using interval type-2 fuzzy logic.

In Part XII, we have a group of papers on "Fuzzy Logic Applications" that show a wide range of applications of fuzzy logic theory. The papers describe in detail important real-world problems that have been solved satisfactorily with fuzzy systems. Also, the fuzzy solutions are shown to be better than traditional solutions to these problems.

In Part XIII, we have a collection of papers on "Neural Networks and Control" that comprise theoretical contributions on neural networks and intelligent control, as well as real applications of these areas. The papers represent an important contribution to the state of the art in neural network and intelligent control.

In Part XIV, we have a collection of papers on "Intelligent Agents, Knowledge Bases and Ant Colony Optimization" that comprise important contributions in these fields. These papers show theoretical results and important applications of intelligent agents and knowledge-based systems. Also, there are papers on ant colony optimization that show the application of this type of evolutionary methods.

We end this preface by thanking all the people who helped or encouraged us during the writing of this book. We would like to thank our colleagues working in soft computing, who are too many to mention individually. Of course, we thank the supporting agencies in our countries for their help during this project. We would like to thank our institutions, for always supporting our projects.

March 2007

Patricia Melin
Oscar Castillo
Luis T. Aguilar
Janusz Kacprzyk
Witold Pedrycz

# Organization

IFSA 2007 was organized by the International Fuzzy Systems Association (IFSA), the Hispanic American Fuzzy Systems Association (HAFSA), and the Division of Graduate Studies of the Tijuana Institute of Technology, with the support of DGEST and CONACYT of Mexico.

## Executive Committee

| | |
|---|---|
| Honorary Chair | Lotfi Zadeh (University of California at Berkeley) |
| General Chair | Oscar Castillo (Tijuana Institute of Technology) |
| Program Chair | Patricia Melin (Tijuana Institute of Technology) |
| Advisory Committee | Janusz Kacprzyk (Polish Academy of Sciences)<br>Witold Pedrycz (University of Alberta)<br>Ronald R. Yager (Iona College, USA) |
| Panels Chair | Masoud Nikravesh (University of California at Berkeley) |
| Workshop Chair | Eduardo Gomez-Ramirez (La Salle University, Mexico) |

## International Program Committee

Ildar Batyrshin, Mexican Petroleum Institute, Mexico
Z. Zenn Bien, KAIST, Korea
Ulrich Bodenhofer, Johannes Kepler University, Austria
Piero Bonissone, GE Global Research, USA
Patrick Bosc, IRISA/ENSSAT, France
Bernadette Bouchon-Meunier, University Pierre Marie Curie, France
Christer Carlsson, IAMSR/Abo Akademi University, Finland
Joao P. B. Carvalho, INESC, TU Lisbon, Portugal
Oscar Castillo, Tijuana Institute of Technology, Mexico
Carlos Coello, CINVESTAV, Mexico
Oscar Cordon, European Centre for Soft-Computing, Spain
Guoqing Chen, Tsinghua University, China
Bernard De Baets, Ghent University, Belgium
Hai-Bin Duan, Beihang University, China
Pablo Estevez, University of Chile, Chile
Dimitar Filev, Ford Research Center, USA
Takeshi Furuhashi, Nagoya University, Japan
Eduardo Gomez-Ramirez, La Salle University, Mexico
Fernando Gomide, State University of Campinas, Brazil

Madan M. Gupta, University of Saskatchewan, Canada
Mustafa Gunes, King Abdulaziz University, Saudi Arabia
Janos Grantner, Western Michigan University, USA
Kaouru Hirota, Tokyo Institute of Technology, Japan
Hani Hagras, University of Essex, UK
Robert John, De Montfort University, UK
Cengiz Kahraman, Istanbul Technical University, Turkey
Janusz Kacprzyk, Polish Academy of Sciences, Poland
Jim Keller, University of Missouri, USA
Etienne Kerre, Ghent University, Belgium
George Klir, Binghamton University-SUNY, USA
Lazlo Koczy, Budapest University of Technology and Economics, Hungary
Rudolf Kruse, Universitaet Magdeburg, Germany
Vladik Kreinovich, University of Texas, El Paso, USA
Reza Langari, Texas A & M University, USA
Vincenzo Loia, Universita di Salerno, Italy
Luis Magdalena, European Centre for Soft Computing, Spain
E. Mamdani, University of London, UK
Patricia Melin, Tijuana Institute of Technology, Mexico
Jerry Mendel, University of Southern California, USA
Evangelia Micheli-Tzanakou, Rutgers University, USA
K. C. Min, Korea, Korea Fuzzy Logic and Intelligent Society, Korea
Ying Mingsheng, Tsinghua University, China
Masoud Nikravesh, University of California, Berkeley, USA
V. Novak, University of Ostrava, Czech Republic
Jose A. Olivas-Varela, Universidad de Castilla La Mancha, Spain
Gabriella Pasi, Universita degli Studi di Milano, Italy
Witold Pedrycz, University of Alberta, Canada
Frank Chung-Hoon Rhee, Hanyang University, Korea
Rajkumar Roy, Cranfield University, UK
Edgar N. Sanchez, CINVESTAV-Guadalajara, Mexico
Pilar Sobrevilla, Universidad Politecnica de Cataluña, Spain
Joao M. C. Sousa, IDMEC, TU Lisbon, Portugal
Thomas Sudkamp, Wright State University, USA
Michio Sugeno, Doshisha University, Japan
Ricardo Tanscheit, Catholic University of Rio de Janeiro, Brazil
Tsu-Tian Lee, National Taipei University of Technology, Taiwan
Enric Trillas, European Centre for Soft Computing, Spain
I. Burhan Turksen, Economics and Technology University, Turkey, University of Toronto, Canada
Jose L. Verdegay, University of Granada, Spain
Michael Wagenknecht, Univ. of Applied Sciences Zittau, Germany
Ronald R. Yager, Iona College, USA
Liu Yingming, Sichuan University, China
Hans Zimmerman, Aachen University of Technology, Germany
Jacek Zurada, University of Louisville, USA

## Local Organizing Committee

Luis Tupak Aguilar Bustos, CITEDI-IPN, Tijuana
Arnulfo Alanis, Tijuana Institute of Technology, Tijuana
Oscar Castillo, Tijuana Institute of Technology, Tijuana
Carlos A. Coello Coello, CINVESTAV, Mexico
Mario Garcia V., Tijuana Institute of Technology, Tijuana
Pilar Gómez Gil, Universidad de las Americas Puebla, Puebla
Eduardo Gómez Ramírez, Universidad de La Salle, México
Miguel Angel Lopez, Tijuana Institute of Technology Tijuana
Gerardo Maximiliano Méndez, Instituto Tecnológico de Nuevo León, Monterrey
Alejandra Mancilla, Tijuana Institute of Technology, Tijuana
Patricia Melin, Tijuana Institute of Technology, Tijuana
Oscar Montiel Ross, CITEDI-IPN, Tijuana
Antonio Rodríguez Díaz, Univ. Autónoma de Baja California, Tijuana
Edgar N. Sánchez, CINVESTAV, Guadalajara
Roberto Sepúlveda Cruz, CITEDI-IPN, Tijuana

## List of Additional Reviewers

Troels Andreassen, Denmark
Gloria Bordogna, Italy
Jesús Campana, Spain
Eliseo Clementini, Italy
Oscar Cordon, Spain
Fabio Crestani, Switzerland
Guy De Tre, Belgium
Marcin Detyniecki, France
Axel Hallez, Belgium
Enrique Herrera-Viedma, Spain
Janusz Kacprzyk, Poland
Donald Kraft, USA
Zongmin Ma, China
Luis Mancilla, Mexico
Milan Mares, Czech Republic
Nicolas Marin, Spain
Christophe Marsala, France
Maria-Jose Martin-Bautista, Spain
Tom Matthe, Belgium
Oscar Montiel, Mexico
Adam Niewiadomski, Poland
Gabriella Pasi, Italy
Frederick E. Petry, USA
Olivier Pivert, France
Olga Pons, Spain
Ellie Sanchez, France
Roberto Sepúlveda, Mexico
Eulalia Szmidt, Poland
Joerg Verstraete, Belgium
Manolis Wallace, Greece
Slawomir Zadrozny, Poland

# Table of Contents

## I Relation Between Interval and Fuzzy Techniques



## II Intuitionistic Fuzzy Sets and Their Applications

## III The Application of Fuzzy Logic and Soft Computing in Flexible Querying



## IV Philosophical and Human-Scientific Aspects of Soft Computing

## VIII Fuzzy Possibilistic Optimization



## IX Fuzzy Trees



## X Fuzzy Logic Theory

## XI Type-2 Fuzzy Logic

## XII Fuzzy Logic Applications

## XIII Neural Networks and Control



## XIV Intelligent Agents and Knowledge Ant Colony

# Part I

# Relation Between Interval and Fuzzy Techniques

# Estimating Variance Under Interval and Fuzzy Uncertainty: Case of Hierarchical Estimation

Gang Xiang and Vladik Kreinovich

Department of Computer Science, University of Texas at El Paso,
El Paso, TX 79968, USA
gxiang@utep.edu, vladik@utep.edu

## 1 Estimating Variance Under Interval and Fuzzy Uncertainty: Motivations and Known Results

*Computing statistics is important.* Traditional data processing in science and engineering starts with computing the basic statistical characteristics such as the population mean and population variance:

$$E = \frac{1}{n} \cdot \sum_{i=1}^{n} x_i \quad V = \frac{1}{n} \cdot \sum_{i=1}^{n} (x_i - E)^2.$$

*Additional problem.* Traditional engineering statistical formulas assume that we know the *exact* values $x_i$ of the corresponding quantity. In practice, these values come either from measurements or from expert estimates. In both case, we get only *approximations* $\widetilde{x}_i$ to the actual (unknown) values $x_i$.

When we use these approximate values $\widetilde{x}_i \neq x_i$ to compute the desired statistical characteristics such as $E$ and $V$, we only get approximate valued $\widetilde{E}$ and $\widetilde{V}$ for these characteristics. It is desirable to estimate the accuracy of these approximations.

*Case of measurement uncertainty.* Measurements are never 100% accurate. As a result, the result $\widetilde{x}$ of the measurement is, in general, different from the (unknown) actual value $x$ of the desired quantity. The difference $\Delta x \stackrel{\text{def}}{=} \widetilde{x} - x$ between the measured and the actual values is usually called a *measurement error.*

The manufacturers of a measuring device usually provide us with an upper bound $\Delta$ for the (absolute value of) possible errors, i.e., with a bound $\Delta$ for which we guarantee that $|\Delta x| \leq \Delta$. The need for such a bound comes from the very nature of a measurement process: if no such bound is provided, this means that the difference between the (unknown) actual value $x$ and the observed value $\widetilde{x}$ can be as large as possible.

Since the (absolute value of the) measurement error $\Delta x = \tilde{x} - x$ is bounded by the given bound $\Delta$, we can therefore guarantee that the actual (unknown) value of the desired quantity belongs to the interval $[\widetilde{x} - \Delta, \widetilde{x} + \Delta]$.

*Traditional probabilistic approach to describing measurement uncertainty.* In many practical situations, we not only know the interval $[-\Delta, \Delta]$ of possible

P. Melin et al. (Eds.): IFSA 2007, LNAI 4529, pp. 3–12, 2007.
© Springer-Verlag Berlin Heidelberg 2007

values of the measurement error; we also know the probability of different values $\Delta x$ within this interval [7].

In practice, we can determine the desired probabilities of different values of $\Delta x$ by comparing the results of measuring with this instrument with the results of measuring the same quantity by a standard (much more accurate) measuring instrument. Since the standard measuring instrument is much more accurate than the one use, the difference between these two measurement results is practically equal to the measurement error; thus, the empirical distribution of this difference is close to the desired probability distribution for measurement error.

*Interval approach to measurement uncertainty.* As we have mentioned, in many practical situations, we do know the probabilities of different values of the measurement error. There are two cases, however, when this determination is not done:

- First is the case of cutting-edge measurements, e.g., measurements in fundamental science. When a Hubble telescope detects the light from a distant galaxy, there is no "standard" (much more accurate) telescope floating nearby that we can use to calibrate the Hubble: the Hubble telescope is the best we have.
- The second case is the case of measurements on the shop floor. In this case, in principle, every sensor can be thoroughly calibrated, but sensor calibration is so costly – usually costing ten times more than the sensor itself – that manufacturers rarely do it.

In both cases, we have no information about the probabilities of $\Delta x$; the only information we have is the upper bound on the measurement error.

In this case, after performing a measurement and getting a measurement result $\widetilde{x}$, the only information that we have about the actual value $x$ of the measured quantity is that it belongs to the interval $\mathbf{x} = [\widetilde{x} - \Delta, \widetilde{x} + \Delta]$. In this situation, for each $i$, we know the interval $\mathbf{x}_i$ of possible values of $x_i$, and we need to find the ranges $\mathbf{E}$ and $\mathbf{V}$ of the characteristics $E$ and $V$ over all possible tuples $x_i \in \mathbf{x}_i$.

*Case of expert uncertainty.* An expert usually describes his/her uncertainty by using words from the natural language, like "most probably, the value of the quantity is between 6 and 7, but it is somewhat possible to have values between 5 and 8". To formalize this knowledge, it is natural to use *fuzzy set theory*, a formalism specifically designed for describing this type of informal ("fuzzy") knowledge [3,6].

As a result, for every value $x_i$, we have a fuzzy set $\mu_i(x_i)$ which describes the expert's prior knowledge about $x_i$: the number $\mu_i(x_i)$ describes the expert's degree of certainty that $x_i$ is a possible value of the $i$-th quantity.

An alternative user-friendly way to represent a fuzzy set is by using its $\alpha$-cuts $\{x_i \mid \mu_i(x_i) > \alpha\}$ (or $\{x_i \mid \mu_i(x_i) \geq \alpha\}$). For example, the $\alpha$-cut corresponding to $\alpha = 0$ is the set of all the values which are possible at all, the $\alpha$-cut corresponding to $\alpha = 0.1$ is the set of all the values which are possible with degree of certainty at least 0.1, etc. In these terms, a fuzzy set can be viewed as a nested family of intervals $[\underline{x}_i(\alpha), \overline{x}_i(\alpha)]$ corresponding to different level $\alpha$.

*Estimating statistics under fuzzy uncertainty: precise formulation of the problem.* In general, we have fuzzy knowledge $\mu_i(x_i)$ about each value $x_i$; we want to find the fuzzy set corresponding to a given characteristic $y = C(x_1, \ldots, x_n)$. Intuitively, the value $y$ is a reasonable value of the characteristic if $y = f(x_1, \ldots, x_n)$ for some reasonable values $x_i$, i.e., if for some values $x_1, \ldots, x_n$, $x_1$ is reasonable, and $x_2$ is reasonable, ..., and $f = f(x_1 \ldots, x_n)$. If we interpret "and" as min and "for some" ("or") as max, then we conclude that the corresponding degree of certainty $\mu(y)$ in $y$ is equal to $\mu(y) = \max\{\min(\mu_1(x_1), \ldots, \mu_n(x_n)) | C(x_1, \ldots, x_n) = y\}$.

*Reduction to the case of interval uncertainty.* It is known that the above formula (called *extension principle*) can be reformulated as follows: for each $\alpha$, the $\alpha$-cut $\mathbf{y}(\alpha)$ of $y$ is equal to the range of possible values of $C(x_1, \ldots, x_n)$ when $x_i \in \mathbf{x}_i(\alpha)$ for all $i$. Thus, from the computational viewpoint, the problem of computing the statistical characteristic under fuzzy uncertainty can be reduced to the problem of computing this characteristic under interval uncertainty; see, e.g., [2].

In view of this reduction, in the following text, we will consider the case of interval uncertainty.

*Estimating statistics under interval uncertainty: a problem.* In the case of interval uncertainty, instead of the true values $x_1, \ldots, x_n$, we only know the intervals $\mathbf{x}_1 = [\underline{x}_1, \overline{x}_1], \ldots, \mathbf{x}_n = [\underline{x}_n, \overline{x}_n]$ that contain the (unknown) true values of the measured quantities. For different values $x_i \in \mathbf{x}_i$, we get, in general, different values of the corresponding statistical characteristic $C(x_1, \ldots, x_n)$. Since all values $x_i \in \mathbf{x}_i$ are possible, we conclude that all the values $C(x_1, \ldots, x_n)$ corresponding to $x_i \in \mathbf{x}_i$ are possible estimates for the corresponding statistical characteristic. Therefore, for the interval data $\mathbf{x}_1, \ldots, \mathbf{x}_n$, a reasonable estimate for the corresponding statistical characteristic is the range

$$C(\mathbf{x}_1, \ldots, \mathbf{x}_n) \stackrel{\text{def}}{=} \{C(x_1, \ldots, x_n) \,|\, x_1 \in \mathbf{x}_1, \ldots, x_n \in \mathbf{x}_n\}.$$

We must therefore modify the existing statistical algorithms so that they compute, or bound these ranges.

*Estimating mean under interval uncertainty.* The arithmetic average $E$ is a monotonically increasing function of each of its $n$ variables $x_1, \ldots, x_n$, so its smallest possible value $\underline{E}$ is attained when each value $x_i$ is the smallest possible ($x_i = \underline{x}_i$) and its largest possible value is attained when $x_i = \overline{x}_i$ for all $i$. In other words, the range $\mathbf{E}$ of $E$ is equal to $[E(\underline{x}_1, \ldots, x_n), E(\overline{x}_1, \ldots, \overline{x}_n)]$. In other words, $\underline{E} = \frac{1}{n} \cdot (\underline{x}_1 + \ldots + \underline{x}_n)$ and $\overline{E} = \frac{1}{n} \cdot (\overline{x}_1 + \ldots + \overline{x}_n)$.

*Estimating variance under interval uncertainty.* It is known that the problem of computing the exact range $\mathbf{V} = [\underline{V}, \overline{V}]$ for the variance $V$ over interval data $x_i \in [\widetilde{x}_i - \Delta_i, \widetilde{x}_i + \Delta_i]$ is, in general, NP-hard; see, e.g., [4,5]. Specifically, there is a polynomial-time algorithm for computing $\underline{V}$, but computing $\overline{V}$ is, in general, NP-hard.

In many practical situations, there are efficient algorithms for computing $\overline{V}$: e.g., an $O(n \cdot \log(n))$ time algorithm exists when no two narrowed intervals $[x_i^-, x_i^+]$, where $x_i^- \stackrel{\text{def}}{=} \widetilde{x}_i - \dfrac{\Delta_i}{n}$ and $x_i^+ \stackrel{\text{def}}{=} \widetilde{x}_i + \dfrac{\Delta_i}{n}$, are proper subsets of one another, i.e., when $[x_i^-, x_i^+] \not\subseteq (x_j^-, x_j^+)$ for all $i$ and $j$ [1].

## 2 Hierarchical Case: Formulation of the Problem

*Hierarchical case: description.* In some practical situations, we do not know the individual values of the observations $x_i$, we only have average values corresponding to several ($m < n$) groups $I_1, \ldots, I_m$ of observations. Typically, for each group $j$, we know the frequency $p_j$ of this group (i.e., the probability that a randomly selected observation belongs to this group), the average $E_j$ over this group, and the standard deviation $\sigma_j$ within $j$-th group.

*Hierarchical case: analysis.* In this case, the overall average $E$ can be described as

$$E = \frac{1}{n} \cdot \sum_{i=1}^{n} x_i = \frac{1}{n} \cdot \sum_{j=1}^{m} \sum_{i \in I_j} x_i.$$

By definition of the group average $E_j$, we have $E_j = \dfrac{1}{n_j} \cdot \sum\limits_{i \in I_j} x_i$, where $n_j = p_j \cdot n$ denotes the overall number of elements in the $j$-th group. Thus, $\sum\limits_{i \in I_j} x_i = n_j \cdot E_j = p_j \cdot n \cdot E_j$, hence

$$E = \sum_{j=1}^{m} p_j \cdot E_j. \tag{1}$$

Similarly, the overall variance $V = \sigma^2$ can be described as

$$V = \frac{1}{n} \cdot \sum_{i=1}^{n} x_i^2 - E^2 = \frac{1}{n} \cdot \sum_{j=1}^{m} \sum_{i \in I_j} x_i^2 - E^2.$$

For each $j$ and for each $i \in I_j$, we have $x_i = (x_i - E_j) + E_j$, hence $x_i^2 = (x_i - E_j)^2 + E_j^2 + 2(x_i - E_j) \cdot E_j$. Therefore,

$$\sum_{i \in I_j} x_i^2 = \sum_{i \in I_j} (x_i - E_j)^2 + n_j \cdot E_j^2 + 2E_j \cdot \sum_{i \in I_j} (x_i - E_j).$$

The first sum, by definition of population variance $\sigma_j$, is equal to $n_j \cdot \sigma_j^2$; the third sum, by definition of the population mean, is equal to 0. Thus, $\sum\limits_{i \in I_j} x_i^2 = n_j \cdot (\sigma_j^2 + E_j^2)$, where $n_j = p_j \cdot n$, and thus,

$$V = V_E + V_\sigma, \tag{2}$$

where

$$V_E \stackrel{\text{def}}{=} M_E - E^2, \tag{3}$$

$$M_E \stackrel{\text{def}}{=} \sum_{j=1}^{m} p_j \cdot E_j^2, \tag{4}$$

$$V_\sigma \stackrel{\text{def}}{=} \sum_{j=1}^{m} p_j \cdot \sigma_j^2. \tag{5}$$

*Hierarchical case: situation with interval uncertainty.* It is reasonable to consider the situations when we only know the values $E_j$ and $\sigma_j$ with interval uncertainty, i.e., when we only know the intervals $\mathbf{E}_j = [\underline{E}_j, \overline{E}_j]$ and $[\underline{\sigma}_j, \overline{\sigma}_j]$ that contain the actual (unknown) values of $E_j$ and $\sigma_j$. In such situations, we must find the ranges of the possible values for the population mean $E$ (as described by the formula (1)) and for the population variance $V$ (as described by the formula (2)).

*Analysis of the interval problem.* The formula (1) that describes the dependence of $E$ on $E_j$ is monotonic in $E_j$. Thus, we get an explicit formula for the range $[\underline{E}, \overline{E}]$ of the population mean $E$:

$$\underline{E} = \sum_{j-1}^{m} p_j \cdot \underline{E}_j; \quad \overline{E} = \sum_{j-1}^{m} p_j \cdot \overline{E}_j.$$

Since the terms $V_E$ and $V_\sigma$ in the expression for $V$ depend on different variables, the range $[\underline{V}, \overline{V}]$ of the population variance $V$ is equal to the sum of the ranges $[\underline{V}_E, \overline{V}_E]$ and $[\underline{V}_\sigma, \overline{V}_\sigma]$ of the corresponding terms:

$$\underline{V} = \underline{V}_E + \underline{V}_\sigma; \quad \overline{V} = \overline{V}_E + \overline{V}_\sigma.$$

Due to similar monotonicity, we can find an explicit expression for the range $[\underline{V}_\sigma, \overline{V}_\sigma]$ for $V_\sigma$:

$$\underline{V}_\sigma = \sum_{j=1}^{m} p_j \cdot (\underline{\sigma}_j)^2; \quad \overline{V}_\sigma = \sum_{j=1}^{m} p_j \cdot (\overline{\sigma}_j)^2.$$

Thus, to find the range of the population variance $V$, it is sufficient to find the range of the term $V_E$. So, we arrive at the following problem:

*Formulation of the problem in precise terms.*

GIVEN:

- an integer $m \geq 1$;
- $m$ numbers $p_j > 0$ for which $\sum\limits_{j=1}^{m} p_j = 1$; and
- $m$ intervals $\mathbf{E}_j = [\underline{E}_j, \overline{E}_j]$.

COMPUTE the range $\mathbf{V}_E = \{V_E(E_1, \ldots, E_m) \,|\, E_1 \in \mathbf{E}_1, \ldots, E_m \in \mathbf{E}_m\}$, where

$$V_E \stackrel{\text{def}}{=} \sum_{j=1}^{m} p_j \cdot E_j^2 - E^2; \quad E \stackrel{\text{def}}{=} \sum_{j=1}^{m} p_j \cdot E_j.$$

## 3 Main Result

Since the function $V_E$ is convex, we can compute its minimum $\underline{V}_E$ on the box $\mathbf{E}_1 \times \ldots \times \mathbf{E}_m$ by using known polynomial-time algorithms for minimizing convex functions over interval domains; see, e.g., [8].

For computing maximum $\overline{V}_E$, even the particular case when all the values $p_j$ are equal $p_1 = \ldots = p_m = 1/m$, is known to be NP-hard. Thus, the more general problem of computing $\overline{V}_E$ is also NP-hard. Let us show that in a reasonable class of cases, there exists a feasible algorithm for computing $\overline{V}_E$.

For each interval $\mathbf{E}_j$, let us denote its midpoint by $\widetilde{E}_j \stackrel{\text{def}}{=} \dfrac{\underline{E}_j + \overline{E}_j}{2}$, and its half-width by $\Delta_j \stackrel{\text{def}}{=} \dfrac{\overline{E}_j - \underline{E}_j}{2}$. In these terms, the $j$-th interval $\mathbf{E}_j$ takes the form $[\widetilde{E}_j - \Delta_j, \widetilde{E}_j + \Delta_j]$.

In this paper, we consider narrowed intervals $[E_j^-, E_j^+]$, where

$$E_j^- \stackrel{\text{def}}{=} \widetilde{E}_j - p_j \cdot \Delta_j, \quad E_j^+ \stackrel{\text{def}}{=} \widetilde{E}_j + p_j \cdot \Delta_j.$$

We show that there exists an efficient $O(m \cdot \log(m))$ algorithm for computing $\overline{V}_E$ for the case when no two narrowed intervals are proper subsets of each other, i.e., when $[E_j^-, E_j^+] \not\subseteq (E_k^-, E_k^+)$ for all $j$ and $k$.

*Algorithm.*

- First, we sort the midpoints $\widetilde{E}_1, \ldots, \widetilde{E}_m$ into an increasing sequence. Without losing generality, we can assume that

$$\widetilde{E}_1 \le \widetilde{E}_2 \le \ldots \le \widetilde{E}_m.$$

- Then, for every $k$ from 0 to $m$, we compute the value $V_E^{(k)} = M^{(k)} - (E^{(k)})^2$ of the quantity $V_E$ for the vector $E^{(k)} = (\underline{E}_1, \ldots, \underline{E}_k, \overline{E}_{k+1}, \ldots, \overline{E}_m)$.
- Finally, we compute $\overline{V}_E$ as the largest of $m+1$ values $V_E^{(0)}, \ldots, V_E^{(m)}$.

To compute the values $V_E^{(k)}$, first, we explicitly compute $M^{(0)}$, $E^{(0)}$, and $V_E^{(0)} = M^{(0)} - E^{(0)}$. Once we computed the values $M^{(k)}$ and $E^{(k)}$, we can compute

$$M^{(k+1)} = M^{(k)} + p_{k+1} \cdot (\underline{E}_{k+1})^2 - p_{k+1} \cdot (\overline{E}_{k+1})^2$$

and

$$E^{(k+1)} = E^{(k)} + p_{k+1} \cdot \underline{E}_{k+1} - p_{k+1} \cdot \overline{E}_{k+1}.$$

# 4 Proof

*Number of computation steps.*

- It is well known that sorting requires $O(m \cdot \log(m))$ steps.
- Computing the initial values $M^{(0)}$, $E^{(0)}$, and $V_E^{(0)}$ requires linear time $O(m)$.
- For each $k$ from 0 to $m-1$, we need a constant number $O(1)$ of steps to compute the next values $M^{(k+1)}$, $E^{(k+1)}$, and $V_E^{(k+1)}$.
- Finally, finding the largest of $m+1$ values $V_E^{(k)}$ also requires $O(m)$ steps.

Thus, overall, we need

$$O(m \cdot \log(m)) + O(m) + m \cdot O(1) + O(m) = O(m \cdot \log(m)) \text{ steps.}$$

*Proof of correctness.* The function $V_E$ is convex. Thus, its maximum $\overline{V}_E$ on the box $\mathbf{E}_1 \times \ldots \times \mathbf{E}_m$ is attained at one of the vertices of this box, i.e., at a vector $(E_1, \ldots, E_m)$ in which each value $E_j$ is equal to either $\underline{E}_j$ or to $\overline{E}_j$.

To justify our algorithm, we need to prove that this maximum is attained at one of the vectors $E^{(k)}$ in which all the lower bounds $\underline{E}_j$ precede all the upper bounds $\overline{E}_j$. We will prove this by reduction to a contradiction. Indeed, let us assume that the maximum is attained at a vector in which one of the lower bounds follows one of the upper bounds. In each such vector, let $i$ be the largest upper bound index followed by the lower bound; then, in the optimal vector $(E_1, \ldots, E_m)$, we have $E_i = \overline{E}_i$ and $E_{i+1} = \underline{E}_{i+1}$.

Since the maximum is attained for $E_i = \overline{E}_i$, replacing it with $\underline{E}_i = \overline{E}_i - 2\Delta_i$ will either decrease the value of $V_E$ or keep it unchanged. Let us describe how $V_E$ changes under this replacement. Since $V_E$ is defined in terms of $M$ and $E$, let us first describe how $E$ and $M$ change under this replacement. In the sum for $M$, we place $(\overline{E}_i)^2$ with

$$(\underline{E}_i)^2 = (\overline{E}_i - 2\Delta_i)^2 = (\overline{E}_i)^2 - 4 \cdot \Delta_i \cdot \overline{E}_i + 4 \cdot \Delta_i^2.$$

Thus, the value $M$ changes into $M + \Delta_i M$, where

$$\Delta_i M = -4 \cdot p_i \cdot \Delta_i \cdot \overline{E}_i + 4 \cdot p_i \cdot \Delta_i^2. \tag{6}$$

The population mean $E$ changes into $E + \Delta_i E$, where

$$\Delta_i E = -2 \cdot p_i \cdot \Delta_i. \tag{7}$$

Thus, the value $E^2$ changes into $(E + \Delta_i E)^2 = E^2 + \Delta_i(E^2)$, where

$$\Delta_i(E^2) = 2 \cdot E \cdot \Delta_i E + (\Delta_i E)^2 = -4 \cdot p_i \cdot E \cdot \Delta_i + 4 \cdot p_i^2 \cdot \Delta_i^2. \tag{8}$$

So, the variance $V$ changes into $V + \Delta_i V$, where

$$\Delta_i V = \Delta_i M - \Delta_i(E^2) = -4 \cdot p_i \cdot \Delta_i \cdot \overline{E}_i + 4 \cdot p_i \cdot \Delta_i^2 + 4 \cdot p_i \cdot E \cdot \Delta_i - 4 \cdot p_i^2 \cdot \Delta_i^2 =$$

$$4 \cdot p_i \cdot \Delta_i \cdot (-\overline{E}_i + \Delta_i + E - p_i \cdot \Delta_i).$$

By definition, $\overline{E}_i = \widetilde{E}_i + \Delta_i$, hence $-\overline{E}_i + \Delta_i = -\widetilde{E}_i$. Thus, we conclude that

$$\Delta_i V = 4 \cdot p_i \cdot \Delta_i \cdot (-\widetilde{E}_i + E - p_i \cdot \Delta_i). \tag{9}$$

So, the fact that $\Delta_i V \leq 0$ means that

$$E \leq \widetilde{E}_i + p_i \cdot \Delta_i = E_i^+. \tag{10}$$

Similarly, since the maximum of $V_E$ is attained for $E_{i+1} = \underline{E}_{i+1}$, replacing it with $\overline{E}_{i+1} = \underline{E}_{i+1} + 2\Delta_{i+1}$ will either decrease the value of $V_E$ or keep it unchanged. In the sum for $M$, we replace $(\underline{E}_{i+1})^2$ with

$$(\overline{E}_{i+1})^2 = (\underline{E}_{i+1} + 2\Delta_{i+1})^2 = (\underline{E}_{i+1})^2 + 4 \cdot \Delta_{i+1} \cdot \underline{E}_{i+1} + 4 \cdot \Delta_{i+1}^2.$$

Thus, the value $M$ changes into $M + \Delta_{i+1} M$, where

$$\Delta_{i+1} M = 4 \cdot p_{i+1} \cdot \Delta_{i+1} \cdot \underline{E}_{i+1} + 4 \cdot p_{i+1} \cdot \Delta_{i+1}^2. \tag{11}$$

The population mean $E$ changes into $E + \Delta_{i+1} E$, where

$$\Delta_{i+1} E = 2 \cdot p_{i+1} \cdot \Delta_{i+1}. \tag{12}$$

Thus, the value $E^2$ changes into $E^2 + \Delta_{i+1}(E^2)$, where

$$\Delta_{i+1}(E^2) = 2 \cdot E \cdot \Delta_{i+1} E + (\Delta_{i+1} E)^2 = 4 \cdot p_{i+1} \cdot E \cdot \Delta_{i+1} + 4 \cdot p_{i+1}^2 \cdot \Delta_{i+1}^2. \tag{13}$$

So, the term $V_E$ changes into $V_E + \Delta_{i+1} V$, where

$$\Delta_{i+1} V = \Delta_{i+1} M - \Delta_{i+1}(E^2) =$$

$$4 \cdot p_{i+1} \cdot \Delta_{i+1} \cdot \underline{E}_{i+1} + 4 \cdot p_{i+1} \cdot \Delta_{i+1}^2 - 4 \cdot p_{i+1} \cdot E \cdot \Delta_{i+1} - 4 \cdot p_{i+1}^2 \cdot \Delta_{i+1}^2 =$$

$$4 \cdot p_{i+1} \cdot \Delta_{i+1} \cdot (\underline{E}_{i+1} + \Delta_{i+1} - E - p_{i+1} \cdot \Delta_{i+1}).$$

By definition, $\underline{E}_{i+1} = \widetilde{E}_{i+1} - \Delta_{i+1}$, hence $\underline{E}_{i+1} + \Delta_{i+1} = \widetilde{E}_{i+1}$. Thus, we conclude that

$$\Delta_{i+1} V = 4 \cdot p_{i+1} \cdot (\widetilde{E}_{i+1} - E - p_{i+1} \cdot \Delta_{i+1}). \tag{14}$$

Since $V_E$ attains maximum at $(E_1, \ldots, E_i, E_{i+1}, \ldots, E_m)$, we have $\Delta_{i+1} V \leq 0$, hence

$$E \geq \widetilde{E}_{i+1} - p_{i+1} \cdot \Delta_{i+1} = E_{i+1}^-. \tag{15}$$

We can also change both $E_i$ and $E_{i+1}$ at the same time. In this case, from the fact that $V_E$ attains maximum, we conclude that $\Delta V_E \leq 0$.

Here, the change $\Delta M$ in $M$ is simply the sum of the changes coming from $E_i$ and $E_{i+1}$:

$$\Delta M = \Delta_i M + \Delta_{i+1} M, \tag{16}$$

and the change in $E$ is also the sum of the corresponding changes:

$$\Delta E = \Delta_i E + \Delta_{i+1} E. \tag{17}$$

So, for

$$\Delta V = \Delta M - \Delta(E^2),$$

we get

$$\Delta V = \Delta_i M + \Delta_{i+1} M - 2 \cdot E \cdot \Delta_i E - 2 \cdot E \cdot \Delta_{i+1} E - (\Delta_i E)^2 - (\Delta_{i+1} E)^2 -$$
$$2 \cdot \Delta_i E \cdot \Delta_{i+1} E.$$

Hence,

$$\Delta V = (\Delta_i M - 2 \cdot E_i \cdot \Delta_i E - (\Delta_i E)^2) + (\Delta_{i+1} M - 2 \cdot E_{i+1} \cdot \Delta_{i+1} E - (\Delta_{i+1} E)^2) -$$
$$2 \cdot \Delta E_i \cdot \Delta E_{i+1},$$

i.e.,

$$\Delta V = \Delta_i V + \Delta_{i+1} V - 2 \cdot \Delta_i E \cdot \Delta_{i+1} E. \tag{18}$$

We already have expressions for $\Delta_i V$, $\Delta_{i+1} V$, $\Delta_i E$, and $\Delta_{i+1} E$, and we already know that $E_{i+1}^- \le E \le D_i^+$. Thus, we have $D(E) \le 0$ for some $E \in [E_{i+1}^-, E_i^+]$, where

$$D(E) \stackrel{\text{def}}{=} 4 \cdot p_i \cdot \Delta_i \cdot (-E_i^+ + E) + 4 \cdot p_{i+1} \cdot \Delta_{i+1} \cdot (E^- i + 1 - E) + 8 \cdot p_i \cdot \Delta_i \cdot p_{i+1} \cdot \Delta_{i+1}.$$

Since the narrowed intervals are not subsets of each other, we can sort them in lexicographic order; in which order, midpoints are sorted, left endpoints are sorted, and right endpoints are sorted, hence $E_i^- \le E_{i+1}^-$ and $E_i^+ \le E_{i+1}^+$.

For $E = E_{i+1}^-$, we get

$$D(E_{i+1}^-) = 4 \cdot p_i \cdot \Delta_i \cdot (-E_i^+ + E_{i+1}^-) + 8 \cdot p_i \cdot \Delta_i \cdot p_{i+1} \cdot \Delta_{i+1} =$$
$$4 \cdot p_i \cdot \Delta_i \cdot (-E_i^+ + E_{i+1}^- + 2 \cdot p_{i+1} \cdot \Delta_{i+1}).$$

By definition, $E_{i+1}^- = E_{i+1} - p_{i+1} \cdot \Delta_{i+1}$, hence $E_{i+1}^- + 2 \cdot p_{i+1} \cdot \Delta_{i+1} = E_{i+1}^+$, and

$$D(E_{i+1}^-) = 4 \cdot p_i \cdot \Delta_i \cdot (E_{i+1}^+ - E_i^+) \ge 0.$$

Similarly,

$$D(E_i^+) = 4 \cdot p_{i+1} \cdot \Delta_{i+1} \cdot (E_{i+1}^- - E_i^+) \ge 0.$$

The only possibility for both values to be 0 is when interval coincide; in this case, we can easily swap them. In all other cases, all intermediate values $D(E)$ are positive, which contradicts to our conclusion that $D(E) \le 0$. The statement is proven.

**Acknowledgments.** This work was supported in part by NSF grants EAR-0225670 and DMS-0532645 and by the Texas Department of Transportation grant No. 0-5453.

## References

1. E. Dantsin, V. Kreinovich, A. Wolpert, and G. Xiang, "Population Variance under Interval Uncertainty: A New Algorithm", *Reliable Computing*, 2006, Vol. 12, No. 4, pp. 273–280.
2. D. Dubois, H. Fargier, and J. Fortin, "The empirical variance of a set of fuzzy intervals", *Proceedings of the 2005 IEEE International Conference on Fuzzy Systems FUZZ-IEEE'2005*, Reno, Nevada, May 22–25, 2005, pp. 885–890.
3. G. Klir and B. Yuan, *Fuzzy sets and fuzzy logic: theory and applications.* Prentice Hall, Upper Saddle River, New Jersey, 1995.
4. V. Kreinovich, L. Longpré, S. A. Starks, G. Xiang, J. Beck, R. Kandathi, A. Nayak, S. Ferson, and J. Hajagos, "Interval Versions of Statistical Techniques, with Applications to Environmental Analysis, Bioinformatics, and Privacy in Statistical Databases", *Journal of Computational and Applied Mathematics*, 2007, Vol. 199, No. 2, pp. 418–423.
5. V. Kreinovich, G. Xiang, S. A. Starks, L. Longpre, M. Ceberio, R. Araiza, J. Beck, R. Kandathi, A. Nayak, R. Torres, and J. Hajagos, "Towards combining probabilistic and interval uncertainty in engineering calculations: algorithms for computing statistics under interval uncertainty, and their computational complexity", *Reliable Computing*, 2006, Vol. 12, No. 6, pp. 471–501.
6. H. T. Nguyen and E. A. Walker, *A first course in fuzzy logic*, CRC Press, Boca Raton, Florida, 2005.
7. S. Rabinovich, *Measurement Errors and Uncertainties: Theory and Practice*, Springer-Verlag, New York, 2005.
8. S. A. Vavasis, *Nonlinear Optimization: Complexity Issues*, Oxford University Press, New York, 1991.

# Testing Stochastic Arithmetic and CESTAC Method Via Polynomial Computation

René Alt[1], Jean-Luc Lamotte[1], and Svetoslav Markov[2]

[1] CNRS, UMR 7606, LIP6,University Pierre et Marie Curie, 4 place Jussieu, 75252 Paris cedex 05, France
Rene.Alt@lip6.fr, Jean-Luc.Lamotte@lip6.fr
[2] Institute of Mathematics and Informatics, Bulgarian Academy of Sciences, "Acad. G. Bonchev" st., bl. 8, 1113 Sofia, Bulgaria
smarkov@bio.bas.bg

**Abstract.** The CESTAC method and its implementation known as CADNA software have been created to estimate the accuracy of the solution of real life problems when these solutions are obtained from numerical methods implemented on a computer. The method takes into account uncertainties on data and round-off errors. On another hand a theoretical model for this method in which operands are gaussian variables called stochastic numbers has been developed. In this paper numerical examples based on the Lagrange polynomial interpolation and polynomial computation have been constructed in order to demonstrate the consistency between the CESTAC method and the theory of stochastic numbers. Comparisons with the interval approach are visualized.

## 1 Introduction

The CESTAC method is an approach to deal with numerical problems involving uncertainties. It has been created to estimate the accuracy of the solution of real life problems when these solutions are obtained from numerical methods implemented on a computer. Such applications to real life problems can be found in [4], [6], [10] and [11]. This method is of Monte-Carlo-type and consists in performing each arithmetic operation several times using an arithmetic with a random rounding mode, see [2], [12], [13]. In other words, real numbers are considered as random values with some prescribed probabilities. In the simplest case one considers gaussian distributed random values, so-called *stochastic numbers*. Stochastic numbers possess only two probability parameters: mean value and standard deviation, and allow for simple arithmetic operations over them. Working with them can be considered as a particular case of granular computing in the same way as it has been done for intervals [9]. The difference is that here, intervals are confidence intervals and the operations on them are also different. The classical operations on gaussian continuous functions is called *Stochastic Arithmetic* or more precisely *Continuous Stochastic Arithmetic* (*CSA*).

In the CESTAC method a stochastic number is represented by several, say $k$, samples $x_j$, $j = 1, ..., k$, representing a given number $x$. The operations on these

P. Melin et al. (Eds.): IFSA 2007, LNAI 4529, pp. 13–22, 2007.
© Springer-Verlag Berlin Heidelberg 2007

samples are those of the computer in use followed by a random rounding. The samples are randomly generated in a known confidence interval. The mean value $\overline{x}$ is the best approximation of the exact value $x$ and the number of significant digits on $\overline{x}$ is computed by:

$$C_{\overline{x}} = log_{10}\left(\frac{\sqrt{k}\,|\overline{x}|}{\sigma\ \tau_{\eta}}\right), \tag{1}$$

wherein

$$\overline{x} = \frac{1}{k}\sum_{j=1}^{k} x_j, \qquad \sigma^2 = \frac{1}{k-1}\sum_{j=1}^{k}(x_j - \overline{x})^2$$

and $\tau_{\eta}$ is the value of the Student distribution for $k-1$ degrees of freedom and a probability level 0.95. This type of computation on samples approximating the same value is called *Discrete Stochastic Arithmetic* (*DSA*).

Operations on stochastic numbers are used as a model for operations on imprecise numbers, i. e. real numbers containing an unknown error, which is supposed to be centered gaussian with a known standard deviation. Some fundamental properties of stochastic numbers are considered in [3], [14].

This work is part of a more general one, which consists in studying the algebraic structures induced by the operations on stochastic numbers in order to provide a good algebraic understanding of the performance of the CESTAC method [1], [7], [8].

The operations addition and multiplication by scalars are well-defined for stochastic numbers and their properties have been studied in some detail. More specifically, it has been shown that the set of stochastic numbers is a commutative monoid with cancelation law in relation to addition. The operator multiplication by $-1$ (negation) is an automorphism and involution. These properties imply a number of interesting consequences, see, e. g. [7], [8].

In the sequel we first briefly present some algebraic properties of the system of stochastic numbers with respect to the arithmetic operations addition, negation, multiplication by scalars, multiplication between two stochastic numbers and the relation inclusion. This theoretical results are the bases for the numerical experiments presented in the second part of the paper.

## 2 Stochastic Arithmetic Theory (SAT) Approach

A *stochastic number* $a$ is written in the form $a = (a'; a'')$. The first component $a'$ is interpreted as *mean value*, and the second component $a''$ is the *standard deviation*. A stochastic number of the form $(a'; 0)$ has zero standard deviation and represents a (pure) mean value, whereas a stochastic number of the form $(0; a'')$ has zero mean value and represents a (pure) standard deviation. In this work we shall always assume $a'' \geq 0$; however, in some cases it is convenient to consider negative standard deviations. Denote by $\mathbb{S}$ the set of all stochastic numbers, $\mathbb{S} = \{(a'; a'') \mid a' \in \mathbb{R}, a'' \in \mathbb{R}^+\}$.

**Linear operations.** For two stochastic numbers $(m_1; s_1)$, $(m_2; s_2)$, $s_1, s_2 \geq 0$, we define addition by

$$(m_1; s_1) + (m_2; s_2) \stackrel{def}{=} (m_1 + m_2;\ \sqrt{s_1^2 + s_2^2}), \tag{2}$$

Multiplication by real scalars $\gamma \in \mathbb{R}$ is defined by:

$$\gamma * (m_1; s_1) \stackrel{def}{=} (\gamma m_1;\ |\gamma| s_1). \tag{3}$$

In particular multiplication by $-1$ (*negation*) is

$$-1 * (m_1; s_1) = (-m_1;\ s_1), \tag{4}$$

and subtraction of $(m_1; s_1)$, $(m_2; s_2)$ is:

$$(m_1; s_1) - (m_2; s_2) \stackrel{def}{=} (m_1; s_1) + (-1) * (m_2; s_2) = (m_1 - m_2;\ \sqrt{s_1^2 + s_2^2}). \tag{5}$$

**Symmetric stochastic numbers.** A symmetric (centered) stochastic number has the form $(0; s), s \in \mathbb{R}$. The arithmetic operations (2)–(5) show that mean values subordinate to familiar real arithmetic whereas standard deviations induce a special arithmetic structure that deviates from the rules of a linear space. If we denote addition of standard deviations defined by (2) by "$\oplus$" and multiplication by scalars by "$*$", that is:

$$s_1 \oplus s_2 = \sqrt{s_1^2 + s_2^2}, \tag{6}$$

$$\gamma * s_1 = |\gamma| s_1, \tag{7}$$

then we can say that the space of standard deviations is an abelian additive monoid with cancellation, such that for any two standard deviations $s, t \in \mathbb{R}^+$, and real $\alpha, \beta \in \mathbb{R}$:

$$\begin{aligned} \alpha * (s \oplus t) &= \alpha * s \oplus \alpha * t, \\ \alpha * (\beta * s) &= (\alpha\beta) * s, \\ 1 * s &= s, \\ (-1) * s &= s, \\ \sqrt{\alpha^2 + \beta^2} * s &= \alpha * s \oplus \beta * s. \end{aligned}$$

**Examples.** Here are some examples for computing with standard deviations:

$$1 \oplus 1 = \sqrt{2}, \quad 1 \oplus 2 = \sqrt{5}, \quad 3 \oplus 4 = 5, \quad 1 \oplus 2 \oplus 3 = \sqrt{14}.$$

Note that $s_1 \oplus s_2 \oplus ... \oplus s_n = t$ is equivalent to $s_1^2 + ... + s_n^2 = t^2$.

**Multiplication of two stochastic numbers.** The product of two stochastic numbers $(m_1; s_1)$, $(m_2; s_2)$, $s_1, s_2 \geq 0$, is defined as:

$$(m_1; s_1)\ _s\!*\ (m_2; s_2) \stackrel{def}{=} \left( m_1 m_2;\ \sqrt{{m_2}^2 {s_1}^2 + {m_1}^2 {s_2}^2\ + s_1^2\ s_2^2} \right). \tag{8}$$

Some properties of the multiplication of stochastic numbers are the following:

It is easy to show that associativity holds. If $X, Y, Z$ are stochastic numbers then the proof that $X {}_s\!* (Y {}_s\!* Z) = (X {}_s\!* Y) {}_s\!* Z$ is a straightforward calculation.

Concerning the distributivity with addition, it can be seen that it is not true in general. More specifically, the difference $X {}_s\!* (Y + Z) - (X {}_s\!* Y + X {}_s\!* Z)$ has the form $(0; s)$, $s \neq 0$, i. e. it is a symmetric stochastic number. Anyhow, if we denote by $\Omega$ the set of symmetric stochastic numbers and the relation between two stochastic numbers defined by: $X \sim Y \stackrel{def}{=} X - Y \in \Omega$, then it can be shown that relation "$\sim$" is an equivalence and that distributivity holds for the corresponding equivalence classes. A ring structure can thus be obtained for these equivalence classes.

**Inclusion.** Inclusion of stochastic numbers plays important roles in applications. We next discuss two relations for inclusion of stochastic numbers. The so-called *interval inclusion* (briefly: i-inclusion) is defined for $X_1 = (m_1; s_1)$, $X_2 = (m_2; s_2) \in \mathbb{S}$, by:

$$X_1 \subseteq_i X_2 \Longleftrightarrow |m_2 - m_1| \leq s_2 - s_1. \tag{9}$$

Note that addition is i-inclusion isotone, that is: $X_1 \subseteq X_2$ implies $X_1 + Y \subseteq X_2 + Y$ [1]. However, it is easy to see that inverse inclusion isotonicity does not hold, i. e. $X_1 + Y \subseteq X_2 + Y$ does not imply $X_1 \subseteq X_2$. If we want that

$$X_1 \subseteq X_2 \Longleftrightarrow X_1 + Y \subseteq X_2 + Y$$

holds in $\mathbb{S}$, then the inclusion relation "$\subseteq_s$" between two stochastic numbers should be defined by

$$X_1 \subseteq_s X_2 \Longleftrightarrow (m_2 - m_1)^2 \leq s_2^2 - s_1^2. \tag{10}$$

Relation (10) will be called *stochastic inclusion*, briefly: s-inclusion.

**Proposition 1.** Addition and multiplication by scalars are (inverse) inclusion isotone (invariant with respect to s-inclusion).

*Proof.* Denote $X_1 = (m_1; s_1)$, $X_2 = (m_2; s_2)$, $X = (m; s) \in \mathbb{S}$. We shall prove that

$$X_1 \subseteq_s X_2 \Longleftrightarrow X_1 + X \subseteq_s X_2 + X.$$

According to (2)

$$X_1 + X = (m_1; s_1) + (m; s) = (m_1 + m;\ \sqrt{s_1^2 + s^2}),$$

$$X_2 + X = (m_2; s_2) + (m; s) = (m_2 + m;\ \sqrt{s_2^2 + s^2}),$$

and according to (10) $X_1 + X \subseteq_s X_2 + X$ is equivalent to

$$((m_2 + m) - (m_1 + m))^2 \leq (s_2^2 + s^2) - (s_1^2 + s^2),$$

that is $(m_2 - m_1)^2 \leq s_2^2 - s_1^2$, which means that $X_1 \subseteq_s X_2$.

The equivalence $X_1 \subseteq_s X_2 \Longleftrightarrow \gamma * X_1 \subseteq_s \gamma * X_2$ is proved similarly. □

We shall next compare relations (10) and (9). To this end we introduce an end-point presentation.

**End-point presentation.** We shall next look for an end-point presentation for stochastic inclusion. This presentation may be useful when dealing with confidence intervals. The confidence interval corresponding to the stochastic number $(m; s)$ is $[m - \gamma s, m + \gamma s]$, where $\gamma > 0$ is a chosen number (usually $\gamma \approx 2$). For simplicity in the sequel we assume $\gamma = 1$, which corresponds to usual compact intervals on $\mathbb{R}$.

Recall that the relation between the end-point presentation of an interval $A = [a^-, a^+] \subseteq \mathbb{R}$ and its mid-point/radius presentation $A = (a';\ a'')$ is given by:

$$a^- = a' - a'', \qquad a^+ = a' + a'';$$

$$a' = (a^- + a^+)/2, \quad a'' = (a^+ - a^-)/2.$$

Recall also the relation $a^+a^- = a'^2 - a''^2$.

The i-inclusion (9) admits a simple end-point presentation, namely for $A \subseteq_i B$ condition $|b' - a'| \leq b'' - a''$ is presented in end-point form as $b^- \leq a^-$, $a^+ \leq b^+$. We next look for an end-point presentation for the s-inclusion (10): $A \subseteq_s B \Longleftrightarrow (b' - a')^2 \leq b''^2 - a''^2$.

The condition $(b' - a')^2 \leq b''^2 - a''^2$ can be written as $b'^2 - b''^2 + a'^2 + a''^2 \leq 2a'b'$. Replacing $b'^2 - b''^2 = b^+b^-$, $a' = (a^- + a^+)/2$, $a'' = (a^+ - a^-)/2$, etc. we obtain: $2b^+b^- + {a^+}^2 + {a^-}^2 \leq (a^+ + a^-)(b^+ + b^-)$. Thus the end-point condition for s-inclusion obtains the form:

$$A \subseteq_s B \Longleftrightarrow {a^+}^2 + {a^-}^2 + 2b^+b^- \leq (a^+ + a^-)(b^+ + b^-),$$

equivalently: $A \subseteq_s B \Longleftrightarrow 2(b^+b^- - a^+a^-) \leq (a^+ + a^-)(b^+ + b^- - a^+ - a^-)$.

**Proposition 2.** Interval inclusion (9) implies stochastic inclusion (10).

*Proof.* We sketch the proof for proper stochastic numbers. Assume that $A = (a';\ a'')$ is i-included in $B = (b';\ b'')$, $A \subseteq_i B$, which according to (9) means $|b' - a'| \leq b'' - a''$. We have to show that (10) holds true. Note first that from (9) we have $0 \leq a'' \leq b''$. Now from $|b' - a'| \leq b'' - a''$ we have $(b' - a')^2 \leq (b'' - a'')^2 \leq (b'' - a'')(b'' + a'') = b''^2 - a''^2$. □

As a consequence from Proposition 2, stochastic addition is i-inclusion isotone.

## 3 Application: Lagrange Interpolation

The goal of this section is to compare the results obtained with the theory developed in this paper, which is named *Continuous Stochastic Arithmetic (CSA)*, with respective results obtained with the CESTAC method and with interval arithmetic [3], [12]–[14].

As said before, in the CESTAC method, each stochastic variable is represented by a $k$-tuple of gaussian random values with known mean value $m$ and standard deviation $\sigma$. The method also uses a special arithmetic called *Discrete Stochastic Arithmetic (DSA)*, which acts on the above mentioned $k$-tuples.

Within the scope of granular computing [15], as seen above, *CSA* operates on stochastic numbers and is directly derived from operations on independent gaussian random variables. Hence a stochastic number is a granule and continuous stochastic arithmetic is a tool for computing with these granules.

Within the same point of view, in *DSA* which is used in the CESTAC method, a granule is composed by a $k$-tuple representing $k$ samples of the same mathematical result of an arithmetic operator implemented in floating point arithmetic. These samples differ from each other because the data are imprecise and because of different random rounding. The operator acting on these granules is a floating point operator corresponding to the exact arithmetical operator which is performed $k$ times in a synchronous way with random rounding. Thus the result is also a granule. This granule is called a *discrete stochastic number*. It has been shown that *DSA* operating on discrete stochastic numbers possesses many properties (but not all) of real numbers; in particular the notion of stochastic zero has been defined [12]–[14]. The CADNA library merely implements the DSA [2].

To compare the two models, a specific library has been developed which implements both continuous and discrete stochastic arithmetic. The computations are done separately. The *CSA* implements the mathematical rules defined in Section 2.

The comparison has been first done on the Lagrangian interpolation method.

Let $(x_i,\ y_i)$, $i = 1, ..., n$, be a set of $n$ pairs of numbers where all $x_i$ are different. The Lagrangian polynomial $p$ at the point $t$ is:

$$p(t) = y_0 l_0(t) + y_1 l_1(t) + \cdots + y_n l_n(t), \quad l_i(t) = \frac{\prod_{i \neq j}(t - x_j)}{\prod_{i \neq j}(x_i - x_j)}.$$

We consider the situation when the values of $y_i$ are imprecise and $x_i$ are considered exact.

For all examples presented below, we take $n = 11$; the exact $x$-values are defined as $x_i = i$, $i = 1, ..., n$, and the imprecise values $y_i$ are close to 1. This means that in the interval case all intervals $y_i$ have a midpoint 1, whereas in the stochastic case they have a mean value 1.

### 3.1 Interval Approach

Assume first that some guaranteed bounds are given for the $y_i$'s in the form of intervals $Y_i$, that is $y_i \in Y_i$, $i = 1, ..., n$. Then it is well-known that at each $t$

$$p(t) \in P(t) = l_0(t) * Y_0 + l_1(t) * Y_1 + \cdots + l_n(t) * Y_n.$$

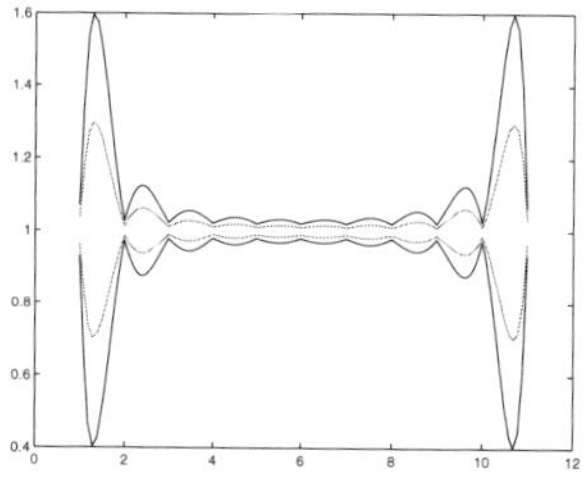

**Fig. 1.** Lagrange, Interval computations

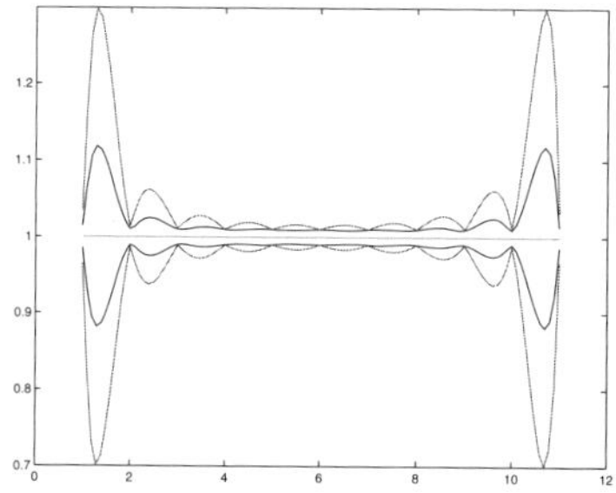

**Fig. 2.** Lagrange, Interval+CSA

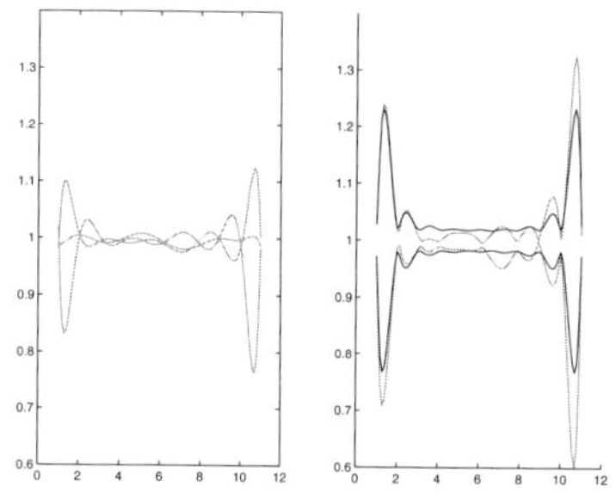

**Fig. 3.** Lagrange DSA 3 samples+CSA

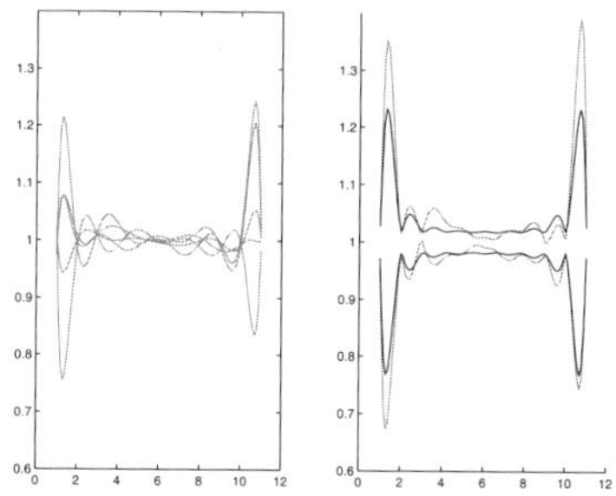

**Fig. 4.** Lagrange DSA 5 samples+CSA

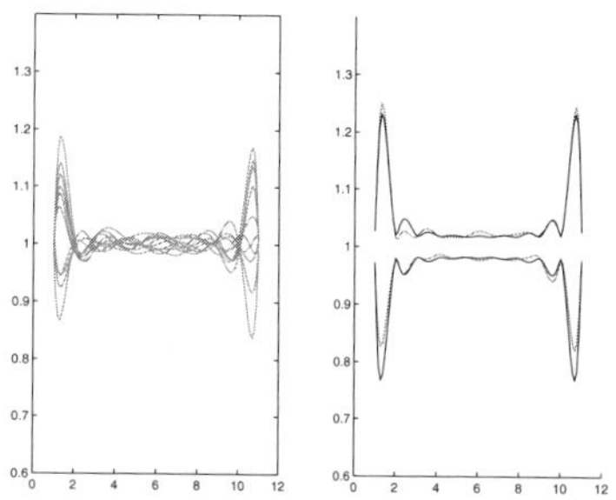

**Fig. 5.** Lagrange DSA 10 samples+CSA

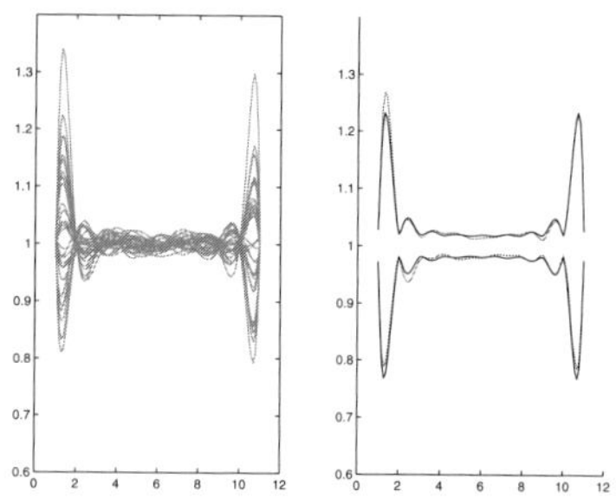

**Fig. 6.** Lagrange DSA 30 samples+CSA

The computation of the interval polynomial $P(t)$ has been performed with the Intlab implementation [5] of interval arithmetic. The maximum error on the $Y_i$ value is $ierr = 0.02$. With the case $Y_i = [1 - ierr; 1 + ierr] =$ constant and $x_i = i,\ i = 1, ..., 11$, the upper and lower bounds of $P$ are shown on Fig. 1. In this example so-called naive interval arithmetic produces exact (sharp) bounds. Normally, naive interval arithmetic produces pessimistic bounds. In most cases, such sharp bounds cannot be obtained by naive interval arithmetic and more sophisticate methods should be used.

### 3.2 The Continuous Stochastic Arithmetic

Corresponding computations are performed on stochastic numbers with the *CSA*. As seen in the preceding sections, this approach is based on operations defined on gaussian random variable $(m; \sigma)$. It is well-known that 95% of the samples of a such variable are inside the interval $[m - 2\sigma, m + 2\sigma]$. Thus, to compare the results with the interval approach, the value of $\sigma$ is taken $\sigma = ierr/2 = 0.01$, so that $(m; \sigma)$ is equal to $(1; 0.01)$.

The computation has been performed with our specific implementation of *CSA*. The gray lower and upper curves in the Fig. 2 represent the results of the *CSA* computation. Each point of the lower curve (respectively the upper curve) is equal to $m - 2\sigma$ (respectively $m + 2\sigma$). More specifically, a set of values $(m_{P(t_i)}; \sigma_{P(t_i)})$ is obtained. Each point of the lower curve (respectively the upper curve) on Fig. 2 is equal to $m_{P(t_i)} - 2\sigma_{P(t_i)}$ (respectively $m_{P(t_i)} + 2\sigma_{P(t_i)}$).

### 3.3 The Discrete Stochastic Arithmetic

The last goal is to compare the results obtained with *CSA* and those obtained with the CESTAC method with $k$ samples, i.e. with *DSA*, $k$ taking successively the values 3, 5, 10, 30. The results obtained for each value of $k$ are reported in figures 3–6 in which the lower and upper curves obtained with the *CSA* are shown. All figures are composed of two sub-figures. The left sub-figure shows the curves obtained as result of the $k$ samples. The right part compares the computed mean value and standard deviation obtained from the $k$-samples to the theoretical mean value and standard deviation obtained with *CSA*.

As observed from the figures, if $\overline{P(t_i)}$ is the mean value of the samples obtained at point $t_i$ with the *DSA* for the computation of $P(t_i)$, then we always have: $m_{P(t_i)} - 2\sigma_{P(t_i)} \leq \overline{P(t_i)} \leq m_{P(t_i)} + 2\sigma_{P(t_i)}$. Thus the numerical experiment shows clearly that the continuous stochastic arithmetic is a good model for the CESTAC method.

## 4 Computation of a Polynomial

In the above section it has been shown experimentally that the theory of stochastic numbers is consistent with the CESTAC method for linear computation. We show now that it is also true in the non-linear case with the computation of the value of a polynomial. Anyhow it must be noted that an hypothesis of the theory is that the stochastic numbers involved in the operations are independent. This hypothesis is clearly not fulfilled in the case of the computation of a polynomial. So one can expect that the order of magnitude of the results are the same for the theoretical and experimental result but that there may be anyhow some differences. In fact these differences may exist but are rather small.

A great number of polynomials have been tested for which the results are always consistent. As an example the results obtained with *DSA* (experimental),

*CSA* (theoretical) and the values provided by the CADNA software for the two simple polynomials:

$$p(x) = x^2 - 2x + 1,$$
$$q(x) = x^3 - 3x^2 + 3x - 1$$

are reported in Table 1 and Table 2. The values of the polynomials have been computed for several values of $x$ with the CADNA software implementing the CESTAC method (i. e. with the *DSA*) and with the *CSA*. In the *DSA* case, the mean value and standard deviation of the result are reported for $k = 3$ and $k = 20$ samples. The values provided by the CADNA software are those obtained with the *DSA* with $k = 3$ which are printed with as many significant digits as computed by the software, i. e. according to formula (1).

When the value is non-significant then the symbol @.0 is printed.

**Table 1.** *Values of* $p(x) = x^2 - 2x + 1$ *computed with DSA and CSA*

| $x$ | DSA 3 samples | DSA 20 samples | CSA | CADNA3 |
|---|---|---|---|---|
| $(2; 0.0001)$ | $(1.000015; 0.000232)$ | $(0.999896; 0.000197)$ | $(1.000000; 0.000346)$ | $0.100E+001$ |
| $(2; 0.001)$ | $(0.999401; 0.003352)$ | $(0.999789; 0.002116)$ | $(1.000000; 0.003464)$ | $0.10E+001$ |
| $(2; 0.01)$ | $(0.999779; 0.018623)$ | $(1.000769; 0.020472)$ | $(1.000000; 0.034641)$ | $0.90E+000$ |
| $(2; 0.1)$ | $(0.853213; 0.104408)$ | $(0.961586; 0.131396)$ | $(1.000000; 0.346411)$ | @.0 |
| $(10; 0.01)$ | $(80.92840; 0.023482)$ | $(81.02938; 0.165771)$ | $(81.00000; 0.142832)$ | $0.81E+002$ |
| $(10; 0.1)$ | $(81.58815; 1.031574)$ | $(81.58087; 1.851930)$ | $(81.00000; 1.428320)$ | $0.8E+002$ |

**Table 2.** *Values of* $q(x) = x^3 - 3x^2 + 3x - 1$ *computed with DSA and CSA*

| $x$ | DSA 3 samples | DSA 20 samples | CSA | CADNA3 |
|---|---|---|---|---|
| $(2; 0.0001)$ | $(0.999796; 0.000196)$ | $(1.000075; 0.000374)$ | $(1.000000; 0.001136)$ | $0.999E+000$ |
| $(2; 0.001)$ | $(1.001783; 0.002189)$ | $(0.999558; 0.002989)$ | $(1.000000; 0.011367)$ | $0.10E+001$ |
| $(2; 0.01)$ | $(1.018297; 0.029464)$ | $(0.989954; 0.028460)$ | $(1.000000; 0.113670)$ | $0.1E+001$ |
| $(2; 0.1)$ | $(1.361246; 0.503563)$ | $(1.013540; 0.313214)$ | $(1.000000; 1.136706)$ | @.0 |
| $(10; 0.01)$ | $(728.7242; 4.028841)$ | $(729.4479; 1.780299)$ | $(729.0000; 1.783594)$ | $0.72E+003$ |
| $(10; 0.1)$ | $(720.1993; 23.03887)$ | $(726.4653; 22.87894)$ | $(729.0000; 17.83594)$ | $0.7E+003$ |

## 5 Conclusion

Starting from a minimal set of empirically known facts related to stochastic numbers, we formally deduce a number of properties and relations. We investigate the set of all stochastic numbers and show that this set possesses nice algebraic properties. We point out to the distinct algebraic nature of the spaces of mean-values and standard deviations. Based on the algebraic properties of the stochastic numbers we propose a natural relation for inclusion, called stochastic inclusion. Numerical examples based on Lagrange interpolation and polynomial computation demonstrate the consistency between the CESTAC method and the presented theory of stochastic numbers. This is one more justification for the practical use of the CADNA software.

## References


1. Alt, R., J.-L. Lamotte, S. Markov, Abstract structures in stochastic arithmetic, In: B. Bouchon-Meunier, R. R. Yager (Eds.), Proc. 11-th Conference on Information Processing and Management of Uncertainties in Knowledge-based Systems (IPMU 2006), Editions EDK, Paris, 2006, 794–801.
2. Alt, R., J. Vignes, Validation of Results of Collocation Methods for ODEs with the CADNA Library. *Appl. Numer. Math.* 20 (1996), 1–21.
3. Chesneaux, J. M., J. Vignes, Les fondements de l'arithmétique stochastique, *C.R. Acad. Sci.*, Paris, Sér. I, Math. 315 (1992), 1435–1440.
4. Delay F.; J.-L. Lamotte. Numerical simulations of geological reservoirs : improving their conditioning through the use of entropy. *Mathematics and Computers in Simulation* 52, 311–320, 2000.
5. INTLAB—INTerval LABoratory V. 5.2., www.ti3.tu-harburg.de/~rump/intlab/
6. Lamotte J.L.; Y. Epelboin. Study of the numerical stability of a X-RAY diffraction model. In *Computational Engineering in Systems Applications, CESA'98 IMACS Multiconference, Nabeul-Hammamet, Tunisia*, volume 1, 916–919, 1998.
7. Markov, S., R. Alt, Stochastic arithmetic: Addition and Multiplication by Scalars, *Appl. Numer. Math.* 50 (2004), 475–488.
8. Markov, S., R. Alt, J.-L. Lamotte, Stochastic Arithmetic: S-spaces and Some Applications, *Numer. Algorithms* 37 (1–4), 275–284, 2004.
9. Rokne, J. G., Interval arithmetic and interval analysis: an introduction. In: Granular computing: an emerging paradigm, *Physica-Verlag GmbH* (2001), 1–22.
10. Toutounian, F., The use of the CADNA library for validating the numerical results of the hybrid GMRES algorithm, *Appl. Numer. Math.* 23 (1997), 275–289.
11. Toutounian, F., The stable $A^TA$-orthogonal $s$-step Orthomin($k$) algorithm with the CADNA library, *Numer. Algo.* 17 (1998), 105–119.
12. Vignes, J., R. Alt, An Efficient Stochastic Method for Round-Off Error Analysis, in: *Accurate Scientific Computations, LNCS 235*, Springer, 1985, 183–205.
13. Vignes, J., Review on Stochastic Approach to Round-Off Error Analysis and its Applications. *Math. and Comp. in Sim.* 30, 6 (1988), 481–491.
14. Vignes, J., A Stochastic Arithmetic for Reliable Scientific Computation, *Math. and Comp. in Sim.* 35 (1993), 233–261.
15. Yao, Y. Y., Granular Computing: basic issues and possible solutions, In: P. P. Wang (Ed.), *Proc. of the 5-th Joint Conference on Information Sciences, Vol. I, Atlantic City, New Jersey, USA, February 27–March 3, 2000*, Association for Intelligent Machinery, 186–189.

# Friction Model by Using Fuzzy Differential Equations

Barnabás Bede[1,2], Imre J. Rudas[2], and János Fodor[2]

[1] Department of Mathematical Sciences
The University of Texas at El Paso
500 West University, El Paso, TX 79968, USA
bbede@utep.edu
[2] Department of Intelligent Engineering Systems
John von Neumann Faculty of Informatics,
Budapest Tech Polytechnical Institution
Doberdó út 96/B , H-1034 Budapest, Hungary
bede.barna@bgk.bmf.hu, rudas@bmf.hu, fodor@bmf.hu

**Abstract.** In the present paper we propose a novel approach for modeling friction, by using fuzzy differential equations under the strongly generalized differentiability concept. The key point is a continuous fuzzyfication of the signum function. The lack of the uniqueness for the solutions of a fuzzy differential equation allows us to choose the solution which better reflects the behavior of the modeled real-world system, so it allows us to incorporate expert's knowledge in our model. Numerical solutions of the fuzzy differential equations modeling dry friction are proposed. In order to show how the expert's knowledge can be incorporated in the system, we study the dry friction equation with different additional assumptions.

## 1 Introduction

The existing models of the friction forces show discontinuous variation at the zero transition of the velocity (see e.g. [5], [19], [17], [11], [12]). The effects of friction at low velocities are due to local properties of the materials and an accurate model of these phenomena is possible by taking into account properties both at the molecular and macroscopic level ([19], [16]). Since the information on the molecular level is usually unavailable we have uncertianties. These uncertainties are usually modeled by considering the friction force as a multivalued function and in this case the equations of motion are considered as differential inclusions. This approach is used in several works (see [1], [15], [8], etc.). The idea behind using differential inclusions is substituting the signum function by a multivalued function. This model often manifests chaotic behavior ([11], [10]).

The above discussion shows that the model of a system with friction is often subject of non-statistical uncertainties. So, in order to model the behavior of a system under the presence of the friction forces we have to take into account these uncertainties. In order to take into account these uncertainties we propose

P. Melin et al. (Eds.): IFSA 2007, LNAI 4529, pp. 23–32, 2007.
© Springer-Verlag Berlin Heidelberg 2007

in the present paper an alternative fuzzy model based on fuzzy differential equations (FDEs). Surely other techniques can be easily imagined (such as interval methods) but these are subject of further research. Also, in our proposed method it is possible to incorporate expert knowledge about the system under study and this property can be turned into an advantage in future studies.

FDEs appear naturally as tools for modeling dynamical systems under uncertainty. Till now, they are rarely used in modeling real-world systems since their theory was developed relatively recently. Also, as it is shown in several recent papers, FDEs are not just an easy extension of the theory of ODEs to the fuzzy case (see e.g. [13], [14], [2]). This fact is also slowing down the extension of the applicability of FDEs. There are several different interpretations of the notion of a FDE (for a discussion about them please refer to [2]). In the present paper we will use the so called strongly generalized differentiability concept introduced recently as a method which solved some problems with the other FDE interpretations (H-derivative (see [18]) or fuzzy differential inclusions (see [7])).

Strongly generalized differentiability was introduced in [3]. The strongly generalized derivative is defined for a larger class of fuzzy-number-valued functions than the H-derivative and fuzzy differential equations can have solutions with decreasing length of their support (this was not the case for the H-derivatve). Also, contrary to the case of differential inclusions, the derivative concept for fuzzy-number-valued function is defined and this makes this method more appropriate for numerical computations. First order linear fuzzy differential equations are investigated in [4] and the behavior of their solutions motivate also the use of the above cited results in the present paper for building a novel friction model.

The key point in our discussion is how to fuzzify the classical model in order to get meaningful conclusions. The key role in this fuzzification is played by the frictional term in the equations of movement with friction. In the present paper, following [7], we fuzzify in a heuristic way the Signum function, by making this term also continuous fuzzy valued function. However we have gained the continuity of the frictional term, since it is a fuzzy one, we obtain a fuzzy solution for our model. The interpretation of this model is the fuzzy set of trajectories, attainable by the system ([7]). The lack of uniqueness of the solution of a fuzzy differential equation under the generalized differentiability concept at first sight could be seen as a disadvantage. But it is turned into an advantage in the present paper since we are able this way to include in our model knowledge based on observations of the modeled system. In the present paper we do not deal with the problem of control, this being subject of further research.

After a preliminary section we propose in Section 3 the heuristic fuzzy model for friction forces with a discussion on dry friction. In Section 4 we present also some preliminary results how friction is modeled by using the proposed approach. We end up with some conclusions and further research topics.

## 2 Preliminaries

We denote by $\mathbf{R}_F$ the space of fuzzy numbers, i.e., fuzzy subsets of the real line $u : \mathbf{R} \to [0,1]$, satisfying the following properties:

(i) $u$ is normal i.e. $\exists x_u \in \mathbf{R}$ with $u(x_u) = 1$;

(ii) $u$ is convex fuzzy set (i.e. $u(tx + (1-t)y) \geq \min\{u(x), u(y)\}, \forall t \in [0,1]$, $x, y \in \mathbf{R}$);

(iii) $u$ is upper semi-continuous on $\mathbf{R}$;

(iv) $\overline{\{x \in \mathbf{R} : u(x) > 0\}}$ is compact, where $\overline{A}$ denotes the closure of $A$.

For $0 < r \leq 1$, denote $[u]^r = \{x \in \mathbf{R}; u(x) \geq r\}$ and $[u]^0 = \overline{\{x \in \mathbf{R};\ u(x) > 0\}}$. Then it is well-known that for any $r \in [0,1]$, $[u]^r$ is a bounded closed interval. For $u, v \in \mathbf{R}_F$, and $\lambda \in \mathbf{R}$, the sum $u + v$ and the product $\lambda \cdot u$ are defined by $[u+v]^r = [u]^r + [v]^r$, $[\lambda \cdot u]^r = \lambda [u]^r$, $\forall r \in [0,1]$.

Let $D : \mathbf{R}_F \times \mathbf{R}_F \to \mathbf{R}_+ \cup \{0\}$, $D(u,v) = \sup_{r \in [0,1]} \max\{|u_-^r - v_-^r|, |u_+^r - v_+^r|\}$, be the Hausdorff distance between fuzzy numbers, where $[u]^r = [u_-^r, u_+^r]$, $[v]^r = [v_-^r, v_+^r]$. In this case $(\mathbf{R}_F, D)$ is a complete metric space. The above operations and the metric space structure allows us to build a mathematical analysis over the space of fuzzy numbers, however some problems appear due to the lack of some properties.

The so called H-difference or Hukuhara difference will play a key role in the present paper. Let us recall its definition.

**Definition 1.** *(see e.g. [18]). Let $x, y \in \mathbf{R}_F$. If there exists $z \in \mathbf{R}_F$ such that $x = y + z$, then $z$ is called the H-difference of $x$ and $y$ and it is denoted by $x \ominus y$.*

In this paper the "$\ominus$" sign stands always for H-difference and let us remark that $x \ominus y \neq x + (-1)y$. We will denote for simplicity $x + (-1)y$ by $x - y$.

Let us recall the definition of strongly generalized differentiability proposed in [3].

**Definition 2.** *Let $f : (a,b) \to \mathbf{R}_F$ and $x_0 \in (a,b)$. We say that $f$ is strongly generalized differentiable at $x_0$, if there exists an element $f'(x_0) \in \mathbf{R}_F$, such that*

*(i) for all $h > 0$ sufficiently small, $\exists f(x_0 + h) \ominus f(x_0)$, $f(x_0) \ominus f(x_0 - h)$ and the limits (in the metric $D$)*

$$\lim_{h \searrow 0} \frac{f(x_0+h) \ominus f(x_0)}{h} = \lim_{h \searrow 0} \frac{f(x_0) \ominus f(x_0-h)}{h} = f'(x_0),$$

*or*

*(ii) for all $h > 0$ sufficiently small, $\exists f(x_0) \ominus f(x_0 + h)$, $f(x_0 - h) \ominus f(x_0)$ and thc limits*

$$\lim_{h \searrow 0} \frac{f(x_0) \ominus f(x_0+h)}{(-h)} = \lim_{h \searrow 0} \frac{f(x_0-h) \ominus f(x_0)}{(-h)} = f'(x_0),$$

*or*

*(iii) for all $h > 0$ sufficiently small, $\exists f(x_0 + h) \ominus f(x_0)$, $f(x_0 - h) \ominus f(x_0)$ and the limits*

$$\lim_{h \searrow 0} \frac{f(x_0+h) \ominus f(x_0)}{h} = \lim_{h \searrow 0} \frac{f(x_0-h) \ominus f(x_0)}{(-h)} = f'(x_0),$$

*or*

*(iv) for all $h > 0$ sufficiently small, $\exists f(x_0) \ominus f(x_0+h)$, $f(x_0) \ominus f(x_0-h)$ and the limits*

$$\lim_{h \searrow 0} \frac{f(x_0) \ominus f(x_0+h)}{(-h)} = \lim_{h \searrow 0} \frac{f(x_0) \ominus f(x_0-h)}{h} = f'(x_0).$$

*(division by $h$ and $(-h)$ is understood as the multiplication of a fuzzy number by the scalars $\frac{1}{h}$ and $-\frac{1}{h}$, respectively).*

We say that a function is (i)-differentiable if it is differentiable as in the previous Definition 2, (i), etc.

Concerning the existence of solutions of a fuzzy initial value problem under generalized differentiability in [3] we have proved that under some relaxed conditions (for which the reader is asked to consult [3]) the fuzzy initial value problem

$$\begin{cases} y' = f(x, y) \\ y(x_0) = y_0 \end{cases}$$

has two solutions (one (i)-differentiable and the other one (ii)- differentiable) $y, \overline{y} : [x_0, x_0 + r] \to B(y_0, q)$ and the successive iterations

$$y_0(x) = y_0$$

$$y_{n+1}(x) = y_0 + \int_{x_0}^{x} f(t, y_n(t))dt, \tag{1}$$

and

$$\overline{y}_0(x) = y_0$$

$$\overline{y}_{n+1}(x) = y_0 \ominus (-1) \cdot \int_{x_0}^{x} f(t, \overline{y}_n(t))dt \tag{2}$$

converge to these two solutions respectively.

The FDEs will have in the present paper will have input data trapezoidal fuzzy numbers. We recall that for $a < b < c < d$, $a, b, c, d \in \mathbf{R}$, the trapezoidal fuzzy number $u = (a, b, c, d)$ determined by $a, b, c$ and $d$ is given such that $u_-^r = a + (b-a)r$ and $u_+^r = d - (d-c)r$ are the endpoints of the $r-$level sets, for all $r \in [0, 1]$.

## 3 The Heuristic Fuzzy Model of Friction

In this section we propose a fuzzy differential equation modeling dry friction, model which is similar to the multivalued models in [1], [15] and the fuzzy model in [7]. In our proposed model we will fuzzify the signum function similarly to, etc. but simultaneously we transform it into a continuous term. As a consequence, the signum function will be in our model continuous fuzzy-valued function and the friction force as well. The velocity and position will be solutions of a system of FDEs and so, these are fuzzy terms.

The fuzzy differential equation modeling dry friction is

$$y'' + \alpha y' + \mu \cdot Sgn(y') + y = u(t), \tag{3}$$

where $\alpha, \mu \in \mathbf{R}$ are positive constants, $u(t)$ is a control signal and the signum function $Sgn(y')$ is given by (4) in our model (for simplicity we do not show the parameters $\varepsilon, \delta$ at each time they occur. The coordinate $y : \mathbf{R} \to \mathbf{R}_F$ is considered to be a trapezoidal fuzzy valued function. The initial conditions are considered to be crisp values.

The trapezoidal-valued signum function is

$$Sgn_{\varepsilon,\delta}(v) = \begin{cases} -1, \text{ if } \overline{v} \leq -\varepsilon \\ (-1, -1+\delta, 1-\delta, 1) \ominus \\ \ominus \left(-\frac{2}{\varepsilon}, -\frac{2+\delta}{\varepsilon}, -\frac{\delta}{\varepsilon}, 0\right) \cdot \overline{v}, \text{ if } |\overline{v}| < \varepsilon \\ 1, \text{ if } \overline{v} > \varepsilon \end{cases}. \tag{4}$$

It is easy to see that

$$\lim_{\varepsilon,\delta \to 0} Sgn_{\varepsilon,\delta}(v) = \begin{cases} -1, \text{ if } \overline{v} < 0 \\ [-1, 1], \text{ if } |\overline{v}| = 0 \\ 1, \text{ if } \overline{v} > 0 \end{cases},$$

which coincides with the interval-valued signum function proposed in [7] (the convergence is understood surely only pointwise).

In order to solve the equation we rewrite it as a system of first order FDEs as follows

$$\begin{cases} y' = v \\ v' = u(t) - \alpha v - \mu \cdot Sgn(v) - y \end{cases},$$

with the initial conditions $y(0) = y_0$ and $v(0) = v_0$. Analogously to the proof of the existence result in [3] a similar theorem can be proved for systems of equations. As a conclusion, the above system may have locally several solutions

$$y_{n+1}(t) = y_0 + \int_{t_0}^{t} v_n dt, \text{ or} \tag{5}$$

$$\overline{y}_{n+1}(t) = y_0 \ominus (-1) \int_{t_0}^{t} v_n dt, \tag{6}$$

and

$$v_{n+1}(t) = v_0 + \int_{t_0}^{t} (u(t) - \alpha v_n - \mu \cdot Sgn(v_n) - y_n) dt \text{ or} \tag{7}$$

$$v_{n+1}(t) = v_0 \ominus (-1) \int_{t_0}^{t} (u(t) - \alpha v_n - \mu \cdot Sgn(v_n) - y_n) dt. \tag{8}$$

In order to solve the problem we will employ a numerical method based on the classical Euler method. We consider the approximation given by this method sufficient for our purposes. Surely theoretical study and implementation of more sophisticated methods is subject of future research.

One step of the Euler's method in our case is given by

$$y(t+h) = y(t) + hv(t), \text{ or} \tag{9}$$
$$y(t+h) = y(t) \ominus (-1)hv(t) \tag{10}$$

and

$$v(t+h) = v(t) + h(u(t) - \alpha v(t) - \mu \cdot Sgn(v(t)) - y(t)) \text{ or} \tag{11}$$
$$v(t+h) = v(t) \ominus (-1)h(u(t) - \alpha v(t) - \mu \cdot Sgn(v(t)) - y(t)), \tag{12}$$

$h \in \mathbf{R}$ being the step size.

Since there may exist locally two solutions, if both of them exist we have to chose locally the one which better reflects the behavior of the real-world system modeled by the given equation. The possibility of this choice, allows us to incorporate further assumptions or observations about the behavior of the system.

## 4 Experimental Results

In the present section we will examine the above proposed model. The lack of uniqueness allows us to introduce is the system additional assumptions and based on these assumptions we chose locally the solution according to a choice function. As a measure of the uncertainty we have used the length of the 0-level set. So, if we say increasing uncertainty we understand increasing length of the 0-level set. Surely several other measures of the uncertainty exist in the literature.

We propose to use and compare experimentally several choice functions in two experimental settings. These are as follows: In the first experimental setting we have put $u_1(t) = \sin(t)$, $\alpha_1 = 1$, $\mu_1 = 0.4$, $\varepsilon_1 = 0.0001$, $\delta_1 = 0.6$ and in the second one

$$u_2(t) = \begin{cases} 5 \text{ if } 0 \leq t < 2 \\ -5 \text{ if } 2 \leq t < 10 \\ 4 \text{ if } 10 \leq t \leq 15 \end{cases},$$

$\alpha_2 = 2$, $\mu_2 = 1.4$, $\varepsilon_2 = 0.01$, and $\delta_2 = 0.7$.

In each of the Figures presented in the present paper, the upper graph represents the coordinate, while the lower graph will represents the velocity.

The choice functions which were tested in these experiments are described as follows.

- First is choosing always the "old" Hukuhara differentiable solution. Surely this is the most inconvenient choice, since uncertainty cannot be decreasing decrease under the Hukuhara differentiability concept ([7]). The experimental results show this behavior expected from the theory.

- Second is choosing solutions with increasing support if the "core", i.e. midpoint of the 1-level set is increasing in absolute value (this choice is based on the

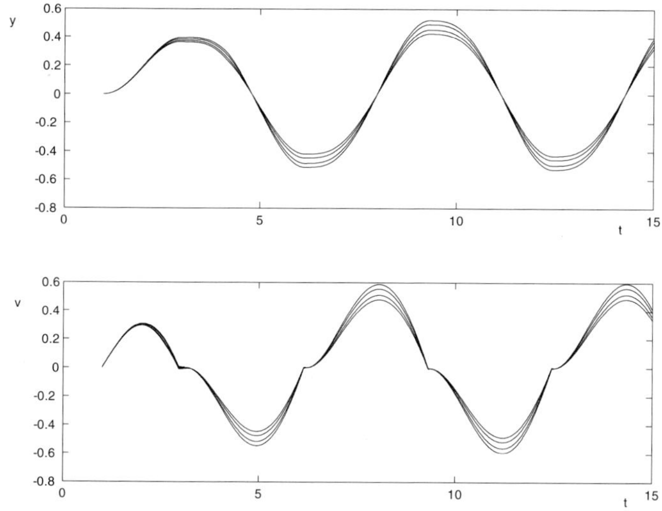


**Fig. 1.** Solution under the second choice function, first experiment

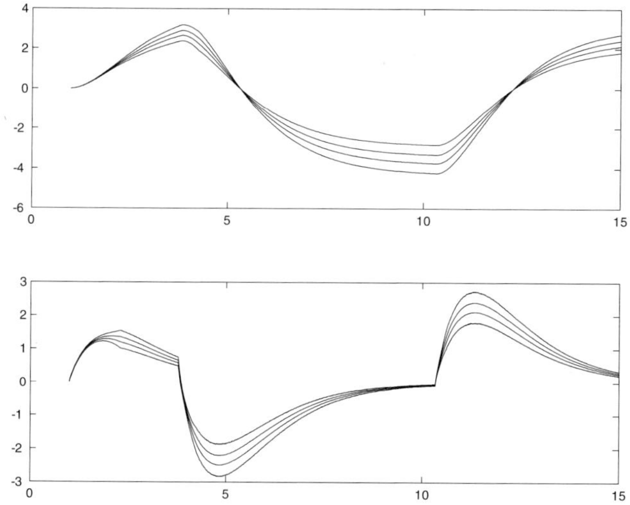


**Fig. 2.** Solution under the second choice function, second experiment

hypothesis that the uncertainty increases together with the value). In our model this is not consistent with the usual real behavior of the velocity. That is the static friction appears at velocity 0 and in this case around zero the uncertainty should increase (see Figs. 1, 2).

- The last choice is based on the expert opinion that when velocity is small the uncertainty is increasing. According to this choice function we set a threshold value for the velocity, under which we assume that the uncertainty increases.

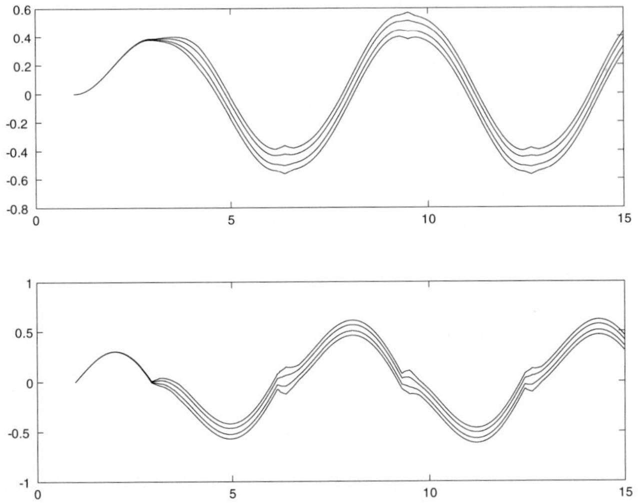


**Fig. 3.** Solution with the assumption that small velocity implies increasing uncertainty, first experiment

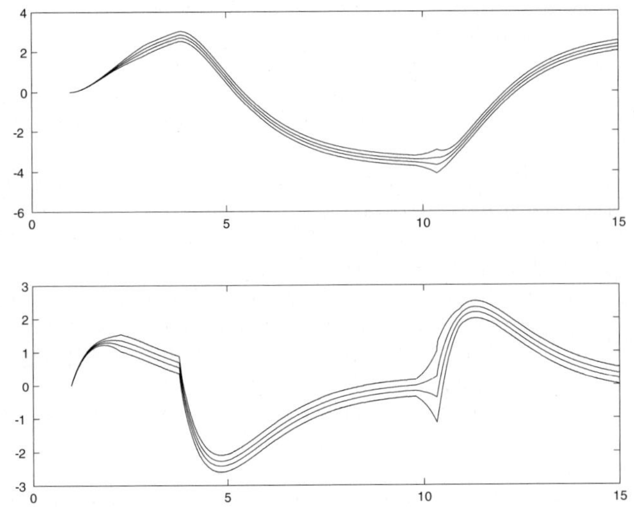


**Fig. 4.** Solution with the assumption that small velocity implies increasing uncertainty, second experiment

Otherwise we allow uncertainty to decrease. This choice is the most well motivated by the physical properties of the system since the principal source of uncertainty is the interaction at low velocities ([5], [12]). See Figs. 3, 4) for numerical results in this case. Surely an experimental comparison will be necessary in order to decide which choice function reflects better the real phenomena, but this is subject of future research.

## 5 Concluding Remarks

We have proposed a fuzzy model for dry friction and we have performed numerical experiments on it. Surely a more accurate comparison with the available experimental data and existing models is a subject of further research.

In the numerical experiments proposed in the present paper we have tested several choice functions based on different assumptions. These assumptions were crisp ones in this paper. As a next step in this research, we propose the use of fuzzy rules in the choice functions together with the fuzzy differential equations to build up a fuzzy model with the expert knowledge incorporated.

## References


1. J.P. Aubin, A. Cellina, Differential Inclusions, Springer-Verlag, Berlin, Heidelberg, New York, Tokyo, 1984.
2. B. Bede, T. Gnana Bhashkar, Perspectives of fuzzy initial value problems, submitted.
3. B. Bede, S.G. Gal, Generalizations of the differentiability of fuzzy-number-valued functions with applications to fuzzy differential equations, Fuzzy Sets and Systems, 151(2005), 581-599.
4. B. Bede, I.J. Rudas, A. Bencsik, First order linear differential equations under generalized differentiability, Information Sciences, to appear.
5. M. Bröcker, M. Lemmen: "Nonlinear Control Methods for Disturbance Rejection on a Hydraulically Driven Flexible Robot", in the Proc. of the Second International Workshop On Robot Motion And Control, RoMoCo'01, October 18-20, 2001, Bukowy Dworek, Poland, pp. 213-217.
6. Y. Chalco-Cano, H. Roman-Flores, On new solutions of fuzzy differential equations, Chaos, Solitons and Fractals, to appear.
7. P. Diamond, Stability and periodicity in fuzzy differential equations, IEEE Transactions on Fuzzy Systems 8 (2000) 583-590.
8. K. Deimling, Multivalued Differential Equations. W. De Gruyter, Berlin, 1992.
9. S.G. Gal, Approximation Theory in Fuzzy Setting, in: G.A. Anastassiou (Ed.), Handbook of Analytic-Computational Methods in Applied Mathematics Chapman & Hall/CRC press, 2000, pp. 617-666.
10. M. Feckan, Chaos in ordinary differential equations with perturbations: Applications to dry friction probelms, Nonlinear Analysis, Theory, Methods & Applications, 30(1997) 1355-1364.
11. B. Feeny, F.C. Moon, Chaos in a forced dry friction oscillator: experiments and numerical modeling, Journal of Sound and Vibration, 170(1994), 303-323.
12. B. Feeny, A. Guran, N. Hinrichs, K. Popp, A historical review on dry friction and stick–slip phenomena, ASME Applied Mechanics Reviews 51 (1998) 321–341.
13. T. Gnana Bhaskar, V. Lakshmikantham, V. Devi, Revisiting fuzzy differential equations, Nonlinear Analysis 58 (2004) 351-358.
14. O. Kaleva, A note on fuzzy differential equations, Nonlinear Analysis 64(2006) 895-900.
15. M. Kunze, Non-Smooth Dynamical Systems, Springer, 2000.
16. V. Lampaert, F. Al-Bender and J. Swevers, Experimental characterization of dry friction at low velocities on a developed tribometer setup for macroscopic measurements, Tribology Letters, 16(2004), 95-105.

17. Y. F. Lim and Kan Chen, Dynamics of dry friction: A numerical investigation, Physical Review E, 58(1998), 5637-5642.
18. M. Puri, D. Ralescu, Differentials of fuzzy functions, J. Math. Anal. Appl., 91(1983), 552-558.
19. W. Sextro, Dynamical contact problems with friction, Models, Methods, Experiments and Applications, Springer, 2002.
20. J.K. Tar, I.J. Rudas, J.F. Bitó: "Group Theoretical Approach in Using Canonical Transformations and Symplectic Geometry in the Control of Approximately Modeled Mechanical Systems Interacting with Unmodelled Environment", Robotica, Vol. 15, pp. 163-179, 1997.

# From Interval Computations to Constraint-Related Set Computations: Towards Faster Estimation of Statistics and ODEs Under Interval, p-Box, and Fuzzy Uncertainty

Martine Ceberio[1], Vladik Kreinovich[1], Andrzej Pownuk[2], and Barnabás Bede[2]

[1] Department of Computer Science, University of Texas at El Paso, El Paso, TX 79968, USA
mceberio@cs.utep.edu, vladik@utep.edu

[2] Department of Mathematical Sciences, University of Texas at El Paso, El Paso, TX 79968, USA
ampownuk@utep.edu, bbede@utep.edu

**Abstract.** In interval computations, at each intermediate stage of the computation, we have intervals of possible values of the corresponding quantities. In our previous papers, we proposed an extension of this technique to *set computations*, where on each stage, in addition to intervals of possible values of the quantities, we also keep sets of possible values of pairs (triples, etc.). In this paper, we show that in several practical problems, such as estimating statistics (variance, correlation, etc.) and solutions to ordinary differential equations (ODEs) with given accuracy, this new formalism enables us to find estimates in feasible (polynomial) time.

## 1 Formulation of the Problem

*Need for data processing.* In many real-life situations, we are interested in the value of a physical quantity $y$ that is difficult or impossible to measure directly. Examples of such quantities are the distance to a star and the amount of oil in a given well. Since we cannot measure $y$ directly, a natural idea is to measure $y$ *indirectly*. Specifically, we find some easier-to-measure quantities $x_1, \ldots, x_n$ which are related to $y$ by a known relation $y = f(x_1, \ldots, x_n)$; this relation may be a simple functional transformation, or complex algorithm (e.g., for the amount of oil, numerical solution to a partial differential equation). Then, to estimate $y$, we first measure or estimate the values of the quantities $x_1, \ldots, x_n$, and then we use the results $\widetilde{x}_1, \ldots, \widetilde{x}_n$ of these measurements (estimations) to compute an estimate $\widetilde{y}$ for $y$ as $\widetilde{y} = f(\widetilde{x}_1, \ldots, \widetilde{x}_n)$

Computing an estimate for $y$ based on the results of direct measurements is called *data processing*; data processing is the main reason why computers were invented in the first place, and data processing is still one of the main uses of computers as number crunching devices.

P. Melin et al. (Eds.): IFSA 2007, LNAI 4529, pp. 33–42, 2007.
© Springer-Verlag Berlin Heidelberg 2007

*Measurement uncertainty: from probabilities to intervals.* Measurement are never 100% accurate, so in reality, the actual value $x_i$ of $i$-th measured quantity can differ from the measurement result $\widetilde{x}_i$. Because of these *measurement errors* $\Delta x_i \stackrel{\text{def}}{=} \widetilde{x}_i - x_i$, the result $\widetilde{y} = f(\widetilde{x}_1, \ldots, \widetilde{x}_n)$ of data processing is, in general, different from the actual value $y = f(x_1, \ldots, x_n)$ of the desired quantity $y$.

It is desirable to describe the error $\Delta y \stackrel{\text{def}}{=} \widetilde{y} - y$ of the result of data processing. To do that, we must have some information about the errors of direct measurements.

What do we know about the errors $\Delta x_i$ of direct measurements? First, the manufacturer of the measuring instrument must supply us with an upper bound $\Delta_i$ on the measurement error. If no such upper bound is supplied, this means that no accuracy is guaranteed, and the corresponding "measuring instrument" is practically useless. In this case, once we performed a measurement and got a measurement result $\widetilde{x}_i$, we know that the actual (unknown) value $x_i$ of the measured quantity belongs to the interval $\mathbf{x}_i = [\underline{x}_i, \overline{x}_i]$, where $\underline{x}_i = \widetilde{x}_i - \Delta_i$ and $\overline{x}_i = \widetilde{x}_i + \Delta_i$.

In many practical situations, we not only know the interval $[-\Delta_i, \Delta_i]$ of possible values of the measurement error; we also know the probability of different values $\Delta x_i$ within this interval. This knowledge underlies the traditional engineering approach to estimating the error of indirect measurement, in which we assume that we know the probability distributions for measurement errors $\Delta x_i$.

In practice, we can determine the desired probabilities of different values of $\Delta x_i$ by comparing the results of measuring with this instrument with the results of measuring the same quantity by a standard (much more accurate) measuring instrument. Since the standard measuring instrument is much more accurate than the one use, the difference between these two measurement results is practically equal to the measurement error; thus, the empirical distribution of this difference is close to the desired probability distribution for measurement error. There are two cases, however, when this determination is not done:

– First is the case of cutting-edge measurements, e.g., measurements in fundamental science. When we use the largest particle accelerator to measure the properties of elementary particles, there is no "standard" (much more accurate) located nearby that we can use for calibration: our accelerator is the best we have.
– The second case is the case of measurements in manufacturing. In principle, every sensor can be thoroughly calibrated, but sensor calibration is so costly – usually costing ten times more than the sensor itself – that manufacturers rarely do it.

In both cases, we have no information about the probabilities of $\Delta x_i$; the only information we have is the upper bound on the measurement error.

In this case, after we performed a measurement and got a measurement result $\widetilde{x}_i$, the only information that we have about the actual value $x_i$ of the measured quantity is that it belongs to the interval $\mathbf{x}_i = [\widetilde{x}_i - \Delta_i, \widetilde{x}_i + \Delta_i]$. In such situations, the only information that we have about the (unknown) actual value of $y =$

$f(x_1, \ldots, x_n)$ is that $y$ belongs to the range $\mathbf{y} = [\underline{y}, \overline{y}]$ of the function $f$ over the box $\mathbf{x}_1 \times \ldots \times \mathbf{x}_n$:

$$\mathbf{y} = [\underline{y}, \overline{y}] = f(\mathbf{x}_1, \ldots, \mathbf{x}_n) \stackrel{\text{def}}{=} \{f(x_1, \ldots, x_n) \,|\, x_1 \in \mathbf{x}_1, \ldots, x_n \in \mathbf{x}_n\}.$$

The process of computing this interval range based on the input intervals $\mathbf{x}_i$ is called *interval computations*; see, e.g., [4].

*Case of fuzzy uncertainty and its reduction to interval uncertainty.* An expert usually describes his/her uncertainty by using words from the natural language, like "most probably, the value of the quantity is between 3 and 4". To formalize this knowledge, it is natural to use *fuzzy set theory*, a formalism specifically designed for describing this type of informal ("fuzzy") knowledge [5].

In fuzzy set theory, the expert's uncertainty about $x_i$ is described by a fuzzy set, i.e., by a function $\mu_i(x_i)$ which assign, to each possible value $x_i$ of the $i$-th quantity, the expert's degree of certainty that $x_i$ is a possible value. A fuzzy set can also be described as a nested family of $\alpha$-cuts $\mathbf{x}_i(\alpha) \stackrel{\text{def}}{=} \{x_i \,|\, \mu_i(x_i) > \alpha\}$.

Zadeh's extension principle can be used to transform the fuzzy sets for $x_i$ into a fuzzy set for $y$. It is known that for continuous functions $f$ on a bounded domain, this principle is equivalent to saying that for every $\alpha$, $\mathbf{y}(\alpha) = f(\mathbf{x}_1(\alpha), \ldots, \mathbf{x}_n(\alpha))$. In other words, fuzzy data processing can be implemented as layer-by-layer interval computations.

In view of this reduction, in the following text, we will mainly concentrate on interval computations.

*Outline.* We start by recalling the basic techniques of interval computations and their drawbacks, then we will describe the new set computation techniques and describe a class of problems for which these techniques are efficient. Finally, we talk about how we can extend these techniques to other types of uncertainty (e.g., classes of probability distributions).

## 2 Interval Computations: Brief Reminder

*Interval computations: main idea.* Historically the first method for computing the enclosure for the range is the method which is sometimes called "straightforward" interval computations. This method is based on the fact that inside the computer, every algorithm consists of elementary operations (arithmetic operations, min, max, etc.). For each elementary operation $f(a, b)$, if we know the intervals $\mathbf{a}$ and $\mathbf{b}$ for $a$ and $b$, we can compute the exact range $f(\mathbf{a}, \mathbf{b})$. The corresponding formulas form the so-called *interval arithmetic*:

$$[\underline{a}, \overline{a}] + [\underline{b}, \overline{b}] = [\underline{a} + \underline{b}, \overline{a} + \overline{b}]; \quad [\underline{a}, \overline{a}] - [\underline{b}, \overline{b}] = [\underline{a} - \overline{b}, \overline{a} - \underline{b}];$$

$$[\underline{a}, \overline{a}] \cdot [\underline{b}, \overline{b}] = [\min(\underline{a} \cdot \underline{b}, \underline{a} \cdot \overline{b}, \overline{a} \cdot \underline{b}, \overline{a} \cdot \overline{b}), \max(\underline{a} \cdot \underline{b}, \underline{a} \cdot \overline{b}, \overline{a} \cdot \underline{b}, \overline{a} \cdot \overline{b})];$$

$$1/[\underline{a}, \overline{a}] = [1/\overline{a}, 1/\underline{a}] \text{ if } 0 \notin [\underline{a}, \overline{a}]; \quad [\underline{a}, \overline{a}]/[\underline{b}, \overline{b}] = [\underline{a}, \overline{a}] \cdot (1/[\underline{b}, \overline{b}]).$$

In straightforward interval computations, we repeat the computations forming the program $f$ step-by-step, replacing each operation with real numbers by the corresponding operation of interval arithmetic. It is known that, as a result, we get an enclosure $\mathbf{Y} \supseteq \mathbf{y}$ for the desired range.

*From main idea to actual computer implementation.* Not every real number can be exactly implemented in a computer; thus, e.g., after implementing an operation of interval arithmetic, we must enclose the result $[r^-, r^+]$ in a computer-representable interval: namely, we must round-off $r^-$ to a smaller computer-representable value $\underline{r}$, and round-off $r^+$ to a larger computer-representable value $\overline{r}$.

*Sometimes, we get excess width.* In some cases, the resulting enclosure is exact; in other cases, the enclosure has excess width. The excess width is inevitable since straightforward interval computations increase the computation time by at most a factor of 4, while computing the exact range is, in general, NP-hard [6], even for computing the population variance $V = \frac{1}{n} \cdot \sum_{i=1}^{n} (x_i - \overline{x})^2$, where $\overline{x} = \frac{1}{n} \cdot \sum_{i=1}^{n} x_i$ [3]. If we get excess width, then we can use more sophisticated techniques to get a better estimate, such as centered form, bisection, etc. [4].

*Reason for excess width.* The main reason for excess width is that intermediate results are dependent on each other, and straightforward interval computations ignore this dependence. For example, the actual range of $f(x_1) = x_1 - x_1^2$ over $\mathbf{x}_1 = [0, 1]$ is $\mathbf{y} = [0, 0.25]$. Computing this $f$ means that we first compute $x_2 := x_1^2$ and then subtract $x_2$ from $x_1$. According to straightforward interval computations, we compute $\mathbf{r} = [0, 1]^2 = [0, 1]$ and then $\mathbf{x}_1 - \mathbf{x}_2 = [0, 1] - [0, 1] = [-1, 1]$. This excess width comes from the fact that the formula for interval subtraction implicitly assumes that both $a$ and $b$ can take arbitrary values within the corresponding intervals $\mathbf{a}$ and $\mathbf{b}$, while in this case, the values of $x_1$ and $x_2$ are clearly not independent: $x_2$ is uniquely determined by $x_1$, as $x_2 = x_1^2$.

## 3 Constraint-Based Set Computations

*Main idea.* The main idea behind constraint-based set computations (see, e.g., [1]) is to remedy the above reason why interval computations lead to excess width. Specifically, at every stage of the computations, in addition to keeping the *intervals* $\mathbf{x}_i$ of possible values of all intermediate quantities $x_i$, we also keep several *sets*:

- sets $\mathbf{x}_{ij}$ of possible values of pairs $(x_i, x_j)$;
- if needed, sets $\mathbf{x}_{ijk}$ of possible values of triples $(x_i, x_j, x_k)$; etc.

In the above example, instead of just keeping two intervals $\mathbf{x}_1 = \mathbf{x}_2 = [0, 1]$, we would then also generate and keep the set $\mathbf{x}_{12} = \{(x_1, x_1^2) \mid x_1 \in [0, 1]\}$. Then, the desired range is computed as the range of $x_1 - x_2$ over this set – which is exactly $[0, 0.25]$.

To the best of our knowledge, in interval computations context, the idea of representing dependence in terms of sets of possible values of tuples was first described by Shary; see, e.g., [7] and references therein.

How can we propagate this set uncertainty via arithmetic operations? Let us describe this on the example of addition, when, in the computation of $f$, we use two previously computed values $x_i$ and $x_j$ to compute a new value $x_k := x_i + x_j$. In this case, we set $\mathbf{x}_{ik} = \{(x_i, x_i + x_j) \mid (x_i, x_j) \in \mathbf{x}_{ij}\}$, $\mathbf{x}_{jk} = \{(x_j, x_i + x_j) \mid (x_i, x_j) \in \mathbf{x}_{ij}\}$, and for every $l \neq i, j$, we take

$$\mathbf{x}_{kl} = \{(x_i + x_j, x_l) \mid (x_i, x_j) \in \mathbf{x}_{ij}, (x_i, x_l) \in \mathbf{x}_{il}, (x_j, x_l) \in \mathbf{x}_{jl}\}.$$

*From main idea to actual computer implementation.* In interval computations, we cannot represent an arbitrary interval inside the computer, we need an enclosure. Similarly, we cannot represent an arbitrary set inside a computer, we need an enclosure.

To describe such enclosures, we fix the number $C$ of granules (e.g., $C = 10$). We divide each interval $\mathbf{x}_i$ into $C$ equal parts $\mathbf{X}_i$; thus each box $\mathbf{x}_i \times \mathbf{x}_j$ is divided into $C^2$ subboxes $\mathbf{X}_i \times \mathbf{X}_j$. We then describe each set $\mathbf{x}_{ij}$ by listing all subboxes $\mathbf{X}_i \times \mathbf{X}_j$ which have common elements with $\mathbf{x}_{ij}$; the union of such subboxes is an enclosure for the desired set $\mathbf{x}_{ij}$.

This implementation enables us to implement all above arithmetic operations. For example, to implement $\mathbf{x}_{ik} = \{(x_i, x_i + x_j) \mid (x_i, x_j) \in \mathbf{x}_{ij}\}$, we take all the subboxes $\mathbf{X}_i \times \mathbf{X}_j$ that form the set $\mathbf{x}_{ij}$; for each of these subboxes, we enclosure the corresponding set of pairs $\{(x_i, x_i + x_j) \mid (x_i, x_j) \in \mathbf{X}_i \times \mathbf{X}_j\}$ into a set $\mathbf{X}_i \times (\mathbf{X}_i + \mathbf{X}_j)$. This set may have non-empty intersection with several subboxes $\mathbf{X}_i \times \mathbf{X}_k$; all these subboxes are added to the computed enclosure for $\mathbf{x}_{ik}$. Once can easily see if we start with the exact range $\mathbf{x}_{ij}$, then the resulting enclosure for $\mathbf{x}_{ik}$ is an $(1/C)$-approximation to the actual set – and so when $C$ increases, we get more and more accurate representations of the desired set.

Similarly, to find an enclosure for

$$\mathbf{x}_{kl} = \{(x_i + x_j, x_l) \mid (x_i, x_j) \in \mathbf{x}_{ij}, (x_i, x_l) \in \mathbf{x}_{il}, (x_j, x_l) \in \mathbf{x}_{jl}\},$$

we consider all the triples of subintervals $(\mathbf{X}_i, \mathbf{X}_j, \mathbf{X}_l)$ for which $\mathbf{X}_i \times \mathbf{X}_j \subseteq \mathbf{x}_{ij}$, $\mathbf{X}_i \times \mathbf{X}_l \subseteq \mathbf{x}_{il}$, and $\mathbf{X}_j \times \mathbf{X}_l \subseteq \mathbf{x}_{jl}$; for each such triple, we compute the box $(\mathbf{X}_i + \mathbf{X}_j) \times \mathbf{X}_l$; then, we add subboxes $\mathbf{X}_k \times \mathbf{X}_l$ which intersect with this box to the enclosure for $\mathbf{x}_{kl}$.

*Limitations of this approach.* The main limitation of this approach is that when we need an accuracy $\varepsilon$, we must use $\sim 1/\varepsilon$ granules; so, if we want to compute the result with $k$ digits of accuracy, i.e., with accuracy $\varepsilon = 10^{-k}$, we must consider exponentially many boxes ($\sim 10^k$). In plain words, this method is only applicable when we want to know the desired quantity with a given accuracy (e.g., 10%).

*Cases when this approach is applicable.* In practice, there are many problems when it is sufficient to compute a quantity with a given accuracy: e.g., when we detect an outlier, we usually do not need to know the variance with a high accuracy, an accuracy of 10% is more than enough.

Let us describe the case when interval computations do not lead to the exact range, but set computations do – of course, the range is "exact" modulo accuracy of the actual computer implementations of these sets.

*Example: estimating variance under interval uncertainty.* Suppose that we know the intervals $\mathbf{x}_1, \ldots, \mathbf{x}_n$ of possible values of $x_1, \ldots, x_n$, and we need to compute the range of the variance $V = \frac{1}{n} \cdot M - \frac{1}{n^2} \cdot E^2$, where $M \stackrel{\text{def}}{=} \sum\limits_{i=1}^{n} x_i^2$ and $E \stackrel{\text{def}}{=} \sum\limits_{i=1}^{n} x_i$.

A natural way to to compute $V$ is to compute the intermediate sums $M_k \stackrel{\text{def}}{=} \sum\limits_{i=1}^{k} x_i^2$ and $E_k \stackrel{\text{def}}{=} \sum\limits_{i=1}^{k} x_i$. We start with $M_0 = E_0 = 0$; once we know the pair $(M_k, E_k)$, we compute $(M_{k+1}, E_{k+1}) = (M_k + x_{k+1}^2, E_k + x_{k+1})$. Since the values of $M_k$ and $E_k$ only depend on $x_1, \ldots, x_k$ and do not depend on $x_{k+1}$, we can conclude that if $(M_k, E_k)$ is a possible value of the pair and $x_{k+1}$ is a possible value of this variable, then $(M_k + x_{k+1}^2, E_k + x_{k+1})$ if a possible value of $(M_{k+1}, E_{k+1})$. So, the set $\mathbf{p}_0$ of possible values of $(M_0, E_0)$ is the single point $(0, 0)$; once we know the set $\mathbf{p}_k$ of possible values of $(M_k, E_k)$, we can compute $\mathbf{p}_{k+1}$ as $\{(M_k + x^2, E_k + x) \,|\, (M_k, E_k) \in \mathbf{p}_k, x \in \mathbf{x}_{k+1}\}$. For $k = n$, we will get the set $\mathbf{p}_n$ of possible values of $(M, E)$; based on this set, we can then find the exact range of the variance $V = \frac{1}{n} \cdot M - \frac{1}{n^2} \cdot E^2$.

What $C$ should we choose to get the results with an accuracy $\varepsilon \cdot \overline{V}$? On each step, we add the uncertainty of $1/C$; to, after $n$ steps, we add the inaccuracy of $n/C$. Thus, to get the accuracy $n/C \approx \varepsilon$, we must choose $C = n/\varepsilon$.

What is the running time of the resulting algorithm? We have $n$ steps; on each step, we need to analyze $C^3$ combinations of subintervals for $E_k$, $M_k$, and $x_{k+1}$. Thus, overall, we need $n \cdot C^3$ steps, i.e., $n^4/\varepsilon^3$ steps. For fixed accuracy $C \sim n$, so we need $O(n^4)$ steps – a polynomial time, and for $\varepsilon = 1/10$, the coefficient at $n^4$ is still $10^3$ – quite feasible.

*Comment.* When the accuracy increases $\varepsilon = 10^{-k}$, we get an exponential increase in running time – but this is OK since, as we have mentioned, the problem of computing variance under interval uncertainty is, in general, NP-hard.

*Other statistical characteristics.* Similar algorithms can be presented for computing many other statistical characteristics. For example, for every integer $d > 2$, the corresponding higher-order central moment $C_d = \frac{1}{n} \cdot \sum\limits_{i=1}^{n} (x_i - \overline{x})^d$ is a linear combination of $d$ moments $M^{(j)} \stackrel{\text{def}}{=} \sum\limits_{i=1}^{n} x_i^j$ for $j = 1, \ldots, d$; thus, to find the exact range for $C_d$, we can keep, for each $k$, the set of possible values of $d$-dimensional tuples $(M_k^{(1)}, \ldots, M_k^{(d)})$, where $M_k^{(j)} \stackrel{\text{def}}{=} \sum_{i=1}^{k} x_i^j$. For these computations, we need $n \cdot C^{d+1} \sim n^{d+2}$ steps – still a polynomial time.

Another example is covariance $C = \frac{1}{n} \cdot \sum_{i=1}^{n} x_i \cdot y_i - \frac{1}{n^2} \cdot \sum_{i=1}^{n} x_i \cdot \sum_{i=1}^{n} y_i$. To compute covariance, we need to keep the values of the triples $(C_k, X_k, Y_k)$, where $C_k \stackrel{\text{def}}{=} \sum_{i=1}^{k} x_i \cdot y_i$, $X_k \stackrel{\text{def}}{=} \sum_{i=1}^{k} x_i$, and $Y_k \stackrel{\text{def}}{=} \sum_{i=1}^{k} y_i$. At each step, to compute the range of

$$(C_{k+1}, X_{k+1}, Y_{k+1}) = (C_k + x_{k+1} \cdot y_{k+1}, X_k + x_{k+1}, Y_k + y_{k+1}),$$

we must consider all possible combinations of subintervals for $C_k$, $X_k$, $Y_k$, $x_{k+1}$, and $y_{k+1}$ – to the total of $C^5$. Thus, we can compute covariance in time $n \cdot C^5 \sim n^6$.

Similarly, to compute correlation $\rho = C/\sqrt{V_x \cdot V_y}$, we can update, for each $k$, the values of $(C_k, X_k, Y_k, X_k^{(2)}, Y_k^{(2)})$, where $X_k^{(2)} = \sum_{i=1}^{k} x_i^2$ and $Y_k^{(2)} = \sum_{i=1}^{k} y_i^2$ are needed to compute the variances $V_x$ and $V_y$. These computations require time $n \cdot C^7 \sim n^8$.

*Systems of ordinary differential equations (ODEs) under interval uncertainty.* A general system of ODEs has the form $\dot{x}_i = f_i(x_1, \ldots, x_m, t)$, $1 \le i \le m$. Interval uncertainty usually means that the exact functions $f_i$ are unknown, we only know the expressions of $f_i$ in terms of parameters, and we have interval bounds on these parameters.

There are two types of interval uncertainty: we may have global parameters whose values are the same for all moments $t$, and we may have noise-like parameters whose values may different at different moments of time – but always within given intervals. In general, we have a system of the type $\dot{x}_i = f_i(x_1, \ldots, x_m, t, a_1, \ldots, a_k, b_1(t), \ldots, b_l(t))$, where $f_i$ is a known function, and we know the intervals $\mathbf{a}_j$ and $\mathbf{b}_j(t)$ of possible values of $a_i$ and $b_j(t)$.

*Example.* For example, the case of a differential inequality when we only know the bounds $\underline{f}_i$ and $\overline{f}_i$ on $f_i$ can be described as $\widetilde{f}_i(x_1, \ldots, x_n, t) + b_1(t) \cdot \Delta(x_1, \ldots, x_n, t)$, where $\widetilde{f}_i \stackrel{\text{def}}{=} (\underline{f}_i + \overline{f}_i)/2$, $\Delta(t) = (\overline{f}_i - \underline{f}_i)/2$, and $\mathbf{b}_1(t) = [-1, 1]$.

*Solving systems of ordinary differential equations (ODEs) under interval uncertainty.* For the general system of ODEs, Euler's equations take the form $x_i(t + \Delta t) = x_i(t) + \Delta t \cdot f_i(x_1(t), \ldots, x_m(t), t, a_1, \ldots, a_k, b_1(t), \ldots, b_l(t))$. Thus, if for every $t$, we keep the set of all possible values of a tuple $(x_1(t), \ldots, x_m(t), a_1, \ldots, a_k)$, then we can use the Euler's equations to get the exact set of possible values of this tuple at the next moment of time.

The reason for exactness is that the values $x_i(t)$ depend only on the previous values $b_j(t - \Delta t)$, $b_j(t - 2\Delta t)$, etc., and not on the current values $b_j(t)$.

To predict the values $x_i(T)$ at a moment $T$, we need $n = T/\Delta t$ iterations.

To update the values, we need to consider all possible combinations of $m+k+l$ variables $x_1(t), \ldots, x_m(t), a_1, \ldots, a_k, b_1(t), \ldots, b_l(t)$; so, to predict the values at moment $T = n \cdot \Delta t$ in the future for a given accuracy $\varepsilon > 0$, we need the running time $n \cdot C^{m+k+l} \sim n^{k+l+m+1}$. This is is still polynomial in $n$.

*Other possible cases when our approach is efficient.* Similar computations can be performed in other cases when we have an iterative process where a fixed finite number of variables is constantly updated.

In such problems, there is an additional factor which speeds up computations. Indeed, in the modern computers, fetching a value from the memory, in general, takes much longer than performing an arithmetic operation. To decrease this time, computers have a hierarchy of memories – from registers from which the access is the fastest, to cash memory (second fastest), etc. Thus, to take full use of the speed of modern processors, we must try our best to keep all the intermediate results in the registers. In the problems in which, at each moment of time, we can only keep (and update) a small current values of the values, we can store all these values in the registers – and thus, get very fast computations (only the input values $x_1, \ldots, x_n$ need to be fetched from slower-to-access memory locations).

*Comment.* The discrete version of the class of problems when we have an iterative process where a fixed finite number of variables is constantly updated is described in [8], where efficient algorithms are proposed for solving these discrete problems – such as propositional satisfiability. The use of this idea for interval computations was first described in Chapter 12 of [6].

*Additional advantage of our technique: possibility to take constraints into account.* Traditional formulations of the interval computation problems assume that we can have arbitrary tuples $(x_1, \ldots, x_n)$ as long as $x_i \in \mathbf{x}_i$ for all $i$. In practice, we may have additional constraints on $x_i$. For example, we may know that $x_i$ are observations of a smoothly changing signal at consequent moments of time; in this case, we know that $|x_i - x_{i+1}| \le \varepsilon$ for some small known $\varepsilon > 0$. Such constraints are easy to take into account in our approach.

For example, if know that $\mathbf{x}_i = [-1, 1]$ for all $i$ and we want to estimate the value of a high-frequency Fourier coefficient $f = x_1 - x_2 + x_3 - x_4 + \ldots - x_{2n}$, then usual interval computations lead to en enclosure $[-2n, 2n]$, while, for small $\varepsilon$, the actual range for the sum $(x_1 - x_2) + (x_3 - x_4) + \ldots$ where each of $n$ differences is bounded by $\varepsilon$, is much narrower: $[-n \cdot \varepsilon, n \cdot \varepsilon]$ (and for $x_i = i \cdot \varepsilon$, these bounds are actually attained).

Computation of $f$ means computing the values $f_k = x_1 - x_2 + \ldots + (-1)^{k+1} \cdot x_k$ for $k = 1, \ldots$ At each stage, we keep the set $\mathbf{s}_k$ of possible values of $(f_k, x_k)$, and use this set to find

$$\mathbf{s}_{k+1} = \{(f_k + (-1)^k \cdot x_{k+1}, x_{k+1}) \mid (f_k, x_k) \in \mathbf{s}_k \,\&\, |x_k - x_{k+1}| \le \varepsilon\}.$$

In this approach, when computing $f_{2k}$, we take into account that the value $x_{2k}$ must be $\varepsilon$-close to the value $x_k$ and thus, that we only add $\le \varepsilon$. Thus, our approach leads to almost exact bounds – modulo implementation accuracy $1/C$.

In this simplified example, the problem is linear, so we could use linear programming to get the exact range, but set computations work for similar nonlinear problems as well.

## 4 Possible Extension to p-Boxes and Classes of Probability Distributions

*Classes of probability distributions and p-boxes: a reminder.* Often, in addition to the interval $\mathbf{x}_i$ of possible values of the inputs $x_i$, we also have partial information about the probabilities of different values $x_i \in \mathbf{x}_i$. An exact probability distribution can be described, e.g., by its cumulative distribution function $F_i(z) = \text{Prob}(x_i \leq z)$. In these terms, a partial information means that instead of a single cdf, we have a *class* $\mathcal{F}$ of possible cdfs.

A practically important particular case of this partial information is when, for each $z$, instead of the exact value $F(z)$, we know an interval $\mathbf{F}(z) = [\underline{F}(z), \overline{F}(z)]$ of possible values of $F(z)$; such an "interval-valued" cdf is called a *probability box*, or a *p-box*, for short; see, e.g., [2].

*Propagating p-box uncertainty via computations: a problem.* Once we know the classes $\mathcal{F}_i$ of possible distributions for $x_i$, and a data processing algorithms $f(x_1, \ldots, x_n)$, we would like to know the class $\mathcal{F}$ of possible resulting distributions for $y = f(x_1, \ldots, x_n)$.

*Idea.* For problems like systems of ODES, it is sufficient to keep, and update, for all $t$, the set of possible joint distributions for the tuple $(x_1(t), \ldots, a_1, \ldots)$.

*From idea to computer implementation.* We would like to estimate the values with some accuracy $\varepsilon \sim 1/C$ and the probabilities with the similar accuracy $1/C$. To describe a distribution with this uncertainty, we divide both the $x$-range and the probability ($p$-) range into $C$ granules, and then describe, for each $x$-granule, which $p$-granules are covered. Thus, we enclose this set into a finite union of p-boxes which assign, to each of $x$-granules, a finite union of $p$-granule intervals.

A general class of distributions can be enclosed in the union of such p-boxes. There are finitely many such assignments, so, for a fixed $C$, we get a finite number of possible elements in the enclosure.

We know how to propagate uncertainty via simple operations with a finite amount of p-boxes [2], so for ODEs we get a polynomial-time algorithm for computing the resulting p-box for $y$.

*For p-boxes, we need further improvements to make this method practical.* Formally, the above method is polynomial-time. However, it is not yet practical beyond very small values of $C$. Indeed, in the case of interval uncertainty, we needed $C^2$ or $C^3$ subboxes. This amount is quite feasible even for $C = 10$.

To describe a p-subbox, we need to attach one of $C$ probability granules to each of $C$ $x$-granules; these are $\sim C^C$ such attachments, so we need $\sim C^C$ subboxes. For $C = 10$, we already get an unrealistic $10^{10}$ increase in computation time.

**Acknowledgments.** Thanks to NSF grants EAR-0225670 and DMS-0532645 and Texas Dept. of Transportation grant No. 0-5453. Many thanks to anonymous referees and to Sergey P. Shary for valuable suggestions.

## References

1. M. Ceberio et al., "How To Take Into Account Dependence Between the Inputs: From Interval Computations to Constraint-Related Set Computations", *Proc. 2nd Int'l Workshop on Reliable Engineering Computing*, Savannah, Georgia, February 22–24, 2006, pp. 127–154; final version in *Journal of Uncertain Systems*, 2007, Vol. 1, No. 1 (to appear).
2. S. Ferson. *RAMAS Risk Calc 4.0.* CRC Press, Boca Raton, Florida, 2002.
3. S. Ferson *et al.*, "Computing Variance for Interval Data is NP-Hard", *ACM SIGACT News*, 2002, Vol. 33, No. 2, pp. 108–118.
4. L. Jaulin, M. Kieffer, O. Didrit, and E. Walter, *Applied Interval Analysis*, Springer, London, 2001.
5. G. Klir and B. Yuan, *Fuzzy sets and fuzzy logic: theory and applications.* Prentice Hall, Upper Saddle River, New Jersey, 1995.
6. V. Kreinovich et al., *Computational complexity and feasibility of data processing and interval computations*, Kluwer, Dordrecht, 1997.
7. S. P. Shary, "Solving tied interval linear systems", *Siberian Journal of Numerical Mathematics*, 2004, Vol. 7, No. 4, pp. 363–376 (in Russian).
8. P. Yu. Suvorov, "On the recognition of the tautological nature of propositional formulas", *J. Sov. Math.*, 1980, Vol. 14, 1556–1562.

# Non-commutative System of Fuzzy Interval Logic Generated by the Checklist Paradigm Measure $m_3$ Containing Early Zadeh Implication


Eunjin Kim[1] and Ladislav J. Kohout[2]

[1] Dept. of Computer Science, University of North Dakota, Grand Forks, ND 58202-9015, USA
ejkim@cs.und.edu

[2] Dept. of Computer Sci., Florida State University, Tallahassee FL 32306-4530, USA
kohout@cs.fsu.edu
http://www.cs.fsu.edu/~kohout



**Abstract.** This paper continues investigation of systems of fuzzy interval logics based on the Checklist Paradigm semantics of Bandler and Kohout [1] [2]. While the early papers dealt with checklist paradigm based interval systems containing commutative AND and OR, this paper is the fifth in the series of papers in which we have been describing the systems in which these connective types are non-commutative. In the present paper we investigate non-commutative interval system generated from implication operators based on the Checklist Paradigm measure $m_3$ of Bandler and Kohout. This system includes the well-known Early Zadeh implication operator (PLY) which is not contrapositive. While the commutative systems can be sufficiently characterized by an 8-element group of transformations, the non-commutative systems require the 16 element group $\mathcal{S}_{2\times2\times2\times2}$.


## 1 Introduction

The major theme of this paper is a non-commutative interval system generated from implication operators based on the measure $m_3$ of Bandler and Kohout by the group transformations provided by the group $\mathcal{S}_{2\times2\times2\times2}$. This system includes the well-known Early Zadeh implication operator (PLY) which is not contrapositive. It generates system of connectives which forms a subgroup $\mathcal{S}_{2\times2\times2}$ of the 16 element group $\mathcal{S}_{2\times2\times2\times2}$. Non-contrapositivity of PLY induces non-commutativity of AND and OR connectives under the group transformations. Section 2.2 deals with group transformations of this system, while section 4 describes the system and also overviews three different classifications of connectives of this interval system.

### 1.1 Interval Logics Generated by the Checklist Paradigm

In 1979 Bandler and Kohout [3] derived five interesting systems from *ab initio* principles based on the Checklist paradigm. The structure of each of these fuzzy


P. Melin et al. (Eds.): IFSA 2007, LNAI 4529, pp. 43–55, 2007.


© Springer-Verlag Berlin Heidelberg 2007

interval systems is generated by a distinct measure that performs the *summarization* of the information contained in certain well-defined binary structures called *fine structures.* See the Appendix below for the definitions of the measures we refer to in this section. The interval produced by a measure $m_i$ pair of connectives of one type can be generically characterized by the following inequality:

$$conbot \leq m_i \leq contop$$

For example, Bandler and Kohout [3] listed the following five inequalities linking the interval bounds for implication operators $\rightarrow_{bot}, \rightarrow_{top}$ with corresponding measures[1] $m_i$, $i = \{1, 2, 3, 4, 5\}$:

1. The Kleene-Dienes implication(KD) and Łukasiewicz implication (Ł) respectively, are attainable lower and upper bounds of $m_1$:
$min(1, 1 - a + b) \geq m_1(\rightarrow) \geq max(1 - a, b)$
2. A certain new function of $(a, b)$ and the Goguen-Gaines (G43) implication (the left-hand side) are respectively attainable lower and upper bounds of $m_2$:
$min(1, b/a) \geq m_2(\rightarrow) \geq max(0, (a + b - 1)/a)$,
3. Another function of $(a, b)$ and the Early Zadeh implication (EZ) are respectively attainable lower and upper bounds of $m_3$:
$max[min(a, b), 1 - a] \geq m_3(\rightarrow) \geq max(a + b - 1, 1 - a)$.
4. Still another function of $(a, b)$ and the Wilmott implication (W) respectively, are attainable lower and upper bounds of $m_4$:
$min[max(a + b - 1, 1 - a), max(b, 1 - a - b)] \leq m_4(\rightarrow) \leq min[max(1 - a, b), (max(a, (1 - b), min(b, 1 - a)))]$
5. Yet another function of $(a, b)$ and one of G43 respectively, are attainable lower and upper bounds of $m_5$:
$max[min(1, b/a), 1 - a] \geq m_5(\rightarrow) \geq max[(a + b - 1)/a, 1 - a]$.

The above quoted paper [3] gave the impetus for more systematic investigation of the systems of connectives that can be generated for above listed implicational intervals by group transformations. The formal semantics for all the interval systems so far described in various papers by Bandler and Kohout, Kohout and Bandler and also by Kohout and Kim are derived by means of an exact mathematical method, which also has a sound ontological and epistemological base. It is based on the **checklist paradigm** introduced by Bandler and Kohout [3],[4],[5],[6]. In order to make this paper self-explanatory, a brief overview of the Checklist Paradigm is provided in the Appendix. The definitions of measures $m_i$, $i = \{1, 2, 3, 4, 5\}$ is also presented in the Appendix.

## 2 The Structure of Some Classes of Fuzzy Interval Logic Systems

The *checklist paradigm* puts ordering on the *pairs of* distinct implication operators and other pairs of connectives. Hence, it provides a theoretical justification of *interval-valued* approximate inference.

[1] See the Appendix below for the definitions of the measures we refer to in this section.

The interval logic given by the pair of TOP and BOT connectives has as its membership function a *fuzzy valued* function $\mu_{Fuzz} : X \longrightarrow \mathcal{F}(\mathbb{R})$, with the rectangular shape. Hence it is a special case of fuzzy sets of the second type. We shall call such a logic system checklist paradigm *fuzzy logic of the second type*, or proper interval fuzzy logic [7]. In this paper we deal only with the interval system based on $m_3$, the conditions of a collapse of the interval system into a point system will be presented elsewhere.

### 2.1 Group Transformations of Logics

Logic transformations are useful in investigating the mutual interrelationships of logic connectives. The global structure of systems of logic connectives can be fruitfully studied by employing the abstract group properties of their group transformations. Transformations are functors that, taking one connective as the argument will produce another connective.

**Definition 1.** *Logic Group Transformations: Symmetric (Piaget)*
*Let $f$ be any one of the 10 two-argument propositional connectives of a logic system, and $\neg$ be an involutive negation. Then we define the following transformations over f:*

1. $I(f) = f(x,y)$; *Identity Transformation;*
2. $D(f) = \neg f(\neg x, \neg y)$; *Dual Transformation;*
3. $C(f) = f(\neg x, \neg y)$; *Contradual Transformation;*
4. $N(f) = \neg f(x,y)$; *Negation Transformation;*

The transformations $\{I, D, C, N\}$ are called *identity, dual, contradual, negation* transformation, respectively. It is well known that for the crisp (2-valued) logic these transformations determine the Piaget group. These transformations are symmetric in their arguments. This group of transformations is a concrete realization of the abstract Klein 4-element group.

In Definition 1 above we have listed the set of the transformations $\mathcal{T}_p = \{I, D, C, N\}$. Adding new non-symmetrical transformations to those defined by Piaget enriches the algebraic structure of logical transformations. In 1979 Kohout and Bandler[8],[9] added the following non-symmetric operations to the above defined four symmetrical transformations:

**Definition 2.** *Logic Group Transformations: non-symmetric (Kohout and Bandler [8],[9]*

1. $LC(f) = f(\neg x, y)$; *Left Contradual;*
2. $RC(f) = f(x, \neg y)$; *Right Contradual;*
3. $LD(f) = \neg f(\neg x, y)$; *Left Dual;*
4. $RD(f) = \neg f(x, \neg y)$; *Right Dual.*

This yields a new 8-element group of transformations. This enlarged set of transformations $\mathcal{T} = \{I, D, C, N, LC, RC, LD, RD\}$ forms 8-element group $\{T, \circ\}$ called $\mathcal{S}_{2\times2\times2}$ group. For its group operations table, see e.g. [12], page 159.

## 2.2 Commutative vs Non-commutative Systems

It is obvious that a non-commutative system of logic has to contain two AND and also two OR connectives. As a consequence of non-commutativity it will also contain two implication operators (denoted by right arrows $\rightarrow$) and two co-implication operators operators $\leftarrow$. As the consequence some other connectives will be also in duplicate. We denote the duplicate connectives by bullets attached to the symbol of the connective. For the purpose of capturing non-commutativity, a new transformation will be introduced, namely a commutator $K$ that is added to the eight already introduced transformations:

This is applied to AND and OR connectives. The following expression is the equational definition of the commutator. It has two parts[2]

**Definition 3.** *Commutator*

1. **syntactic** *For any connective* $*$, *the commutator yields* $K(a * b) = b * a$.
2. **sematic** *If a connective is non-commutative then* $val(a*b) \neq val(b*a)$ *where* $val(a * b), val(b * a) \in [0, 1]$.

**Definition 4.** *Equational Definition of Notation for Non-Commutative Systems of AND and OR:*

$$a\&^{\bullet}b = K(a\&b) = b\&a; \quad a \vee^{\bullet} b = K(a \vee b) = b \vee a.$$

**Definition 5.** *An implication operator is contrapositive, if its valuation satisfies the semantic equality*

$$a \rightarrow b = \neg b \rightarrow \neg a.$$

*Otherwise, it is non-contrapositive.*

We can approach this in a more general way. In order to do this we first shall use definition of logic group transformations from Sec. 2.1.

Commutativity involves restrictions on transformations of connectives, as does the contrapositivity. In the abstract group (see [7]), these restrictions are expressed abstractly as congruences. It is convenient to express such restrictions equationally.

**Theorem 1.** *Link between commutativity and contrapositivity [12]*

1. *For any contrapositive* $\rightarrow$, *the following equality holds:*

$$C[K(a \rightarrow b)] = K[C(a \rightarrow b)] = a \rightarrow b$$

2. *For a non-contrapositive* $\rightarrow$, *the following equality holds:*

$$K(C(K(C(a \rightarrow b)))) = a \rightarrow b$$

*Proof.* *For the proofs of (1) and (2), see Kohout and Kim 1997 [12].*

---

[2] We assume as usual in fuzzy logic [11] that the connectives are truth-functional.

When an implication operator, say $\rightarrow$ is non-contrapositive, the AND and OR connectives generated by the application of the group-compliant logic transformation T to $\rightarrow$ are non-commutative.

The implication operators listed in Sec. 1.1 above have the following properties [3]:

1. The implication operators generated by measures $m_1, m_4, m_5$ are contrapositive.
2. The implication operators generated by measures $m_2, m_3$ lack the contrapositivity.
3. The Kleene-Dienes implication operator (which we shall denote by $PLY\{m_1, Top, KD\}$) giving the upper bound of the interval is obtained in system generated $m_1$ can be obtained by the upper contrapositivization of $PLY\{m_3, Top, EZ\}$.
4. The Wilmott implication operator (which we shall denote by $PLY\{m_4, Top, W\}$) giving the upper bound of the interval is obtained in system generated $m_4$ can be obtained by the lower contrapositivization of $PLY\{m_3, Top, EZ\}$. (For definitions of the upper and lower contrapozitivation see [3]).

## 3 A Brief Overview of the Previous Work

### 3.1 Commutative Interval Systems Generated by the Checklist Paradigm Measures

While the transformation a non-commutative interval system generated from implication operators based on the measure $m_3$ of Bandler and Kohout by the group transformations provided by the group $\mathcal{S}_{2\times2\times2\times2}$. This system includes the well-known Early Zadeh implication operator (PLY) which is not contrapositive. It generates system of connectives which forms a subgroup $\mathcal{S}_{2\times2\times2}$ of the 16 element group $\mathcal{S}_{2\times2\times2\times2}$. Non-contrapositivity of PLY induces non-commutativity of AND and OR connectives under the group transformations.

**System Generated by $m_1$ Measure.** All sixteen interval connective pairs generated by $m_1$ involving the interval pair of Kleene-Dienes and Lukasiewicz implication operators have been first investigated by Bandler and Kohout in detail in [4] as well as in several subsequent papers [5],[6]. The interval logic system based on $m_1$ are globally characterized by the group transformations captured by $\mathcal{S}_{2\times2\times2}$ [9]. Detailed investigation of this system appeared in [1] and [10]. The exhaustive examination of all congruences yielding the subgroups that characterize individual types of connectives of the $m_1$ system was presented in [7]. Applications of the interval system $m1$ have been discussed in [13],[2] and elsewhere.

**Systems Generated by $m_4$ and $m_5$ Measures.** These systems have contrapositive implication operators. As stated above (c.f. Sec.1.1 and Sec. 2.1), the Wilmott implication operator $PLY\{m_4, Top, W\}$ ) giving the upper bound of the interval is obtained by the lower contrapositivization of the Early Zadeh operator $PLY\{m_3, Top, EZ\}$.

On the other hand, $a \rightarrow_{4'} b = max[min(1, b/a), 1 - a]$ operator of the type $PLY\{m_5, Top, G43'\}$ giving the upper bound of the interval in the system $m_5$ is obtained by the lower contrapositivization of the Goguen-Gaines implication operator $PLY\{m_3, Top, G43\}$. (see Sec. 1.1 above and for more details [3].)

**More General Link Between Commutativity of AND, OR and PLY.** As a consequence of Theorem 5 of Sec. 2.1 above we have the following theorem.

**Theorem 2.** *Let $S_{2\times2\times2}$ be represented by the set of transformations $\mathcal{T} = \{I, N, C, D, LC, RC, LD, RD\}$ applied to the set CON= $\{\&, \vee, \downarrow, |, \rightarrow, \leftarrow, \not\rightarrow, \not\leftarrow\}$. Then, if $\rightarrow$ is contrapositive, the corresponding & and $\vee$ in CON must be commutative.*

*Proof.* *See L.J. Kohout and E. Kim [12], p.240, Theorem 2.*

Commutative Interval Systems Generated by the Checklist.

### 3.2 Non-commutative Interval Systems Generated by the Checklist Paradigm

**Characterization by a 16-Element S-Group.** As indicated by Theorem 5 in Sec. 2.1 above, the relationship between implication $\rightarrow$ and co-implication $\leftarrow$ connectives that are not contrapositive is more complex than for contrapositive ones. Indeed, the assumption of semantic equality (c.f Definition 3 in Sec. 2.1) factories the equality of Theorem 5-(2), thus yielding the congruence on group of transformations captured by Theorem 5-(1). This suggests that the group of transformations for systems with non-commutative AND and OR will have higher complexity.

We have seen [10],[7] that for a contrapositive, $\rightarrow$ the corresponding & and $\vee$ in *CON* must be commutative. In this case, the effect of transformations $\mathcal{T} = \{I, N, C, D, LC, RC, LD, RD\}$ applied to the set *CON*= $\{\&, \vee, \downarrow, |, \rightarrow, \leftarrow, \not\rightarrow, \not\leftarrow\}$ is captured algebraically by the 8-element group $S_{2\times2\times2}$.

On the other hand, in the case of non-commutative systems, one needs to introduce the commutator $K$. Then the application of $\mathcal{T}_{+K} = \{I, N, C, D, LC, RC, LD, RD, K\}$ to its system of connectives yields the 16-element group $S_{2\times2\times2\times2}$. The group multiplication table of this group appeared in [12] as well as in some of our subsequent publications.

**System Generated by $m_2$ Measure.** What is indeed the case for the fuzzy interval logic system based on the Checklist paradigm measure $m_3$ was shown in 1997 by Kohout and Kim in [12].

**Theorem 3.** *The closed set of connectives generated by $\{\rightarrow_{G43}, \leftarrow_{G43}, K\}$ is a representation of the 16 element abstract group $S_{2\times2\times2\times2}$.*

*Proof.* *See L.J. Kohout and E. Kim [12].*

This theorem involves the non-contrapositive pair of Goguen-Gaines implication $a \rightarrow_{G43} b = min(1, b/a)$ and co-implication operator $a \leftarrow_{G43} b = min(1, a/b)$ together with the commutator K.

Earlier, in 1992, Kohout and Bandler proved the following theorem:

**Theorem 4 (Kohout & Bandler [9]).** *The closed set of connectives generated from the ply-top implication operator $a \rightarrow_{G43} b = min(1, b/a)$ by the transformation $T$ is listed below. This set of connectives together with $T$ is a realization of the abstract group $\mathcal{S}_{2\times2\times2}$.*

$\rightarrow \; = I(\rightarrow_4) = min(1, b/a)$
$\leftarrow^{\bullet} \; = C(\rightarrow_4) = min(1, 1 - b/1 - a)$
$\not\leftarrow^{\bullet} \; = D(\rightarrow_4) = max(0, b - a/1 - a)$
$\not\rightarrow \; = N(\rightarrow_4) = max(0, a - b/a)$
$\vee \; = LC(\rightarrow_4) = min(1, b/1 - a)$
$\downarrow \; = LD(\rightarrow_4) = max(0, 1 - a - b/1 - a)$
$| \; = RC(\rightarrow_4) = min(1, 1 - b/a)$
$\& \; = RD(\rightarrow_4) = max(0, a + b - 1/a)$.

Although the above transformation of PLY yields an 8 element group, this groups $\mathcal{S}_{2\times2\times2}$ does not subsume the whole system but is only a subgroup of the larger 16 element group $S_{2\times2\times2\times2}$ that characterizes the whole $m_3$ based interval fuzzy logic system. The co-implication operator $\leftarrow_{G43}$ generates another 8-element subgroup (see [14],[15]).

**System Generated by $m_3$ Measure.** Implication operators in this system are non-contrapositive, hence the AND and OR connectives turn out to be non-commutative. The rest of this paper is concerned with investigation of this system.

## 4 Non-commutative System of Fuzzy Interval Logic Generated by $m_3$ Containing Early Zadeh Implication

### 4.1 An Overview of the System $m3$

All the connectives of the non-commutative interval system based on $m_3$ measure are listed in Table 1. The system listed in the table was obtained by the application of the set of transformations
$\mathcal{T} = \{I, N, C, D, LC, RC, LD, RD\}$ to the TOP and BOT implication operators
$\rightarrow \{TOP, m_3, EZ\}$,
$\rightarrow \{BOT, m_3\}$ and the corresponding co-implication operators derived by the checklist paradigm – c.f. Bandler and Kohout [3], Theorem 6.3-(3).

The graph of transformations which displays the efect of transformations on the individual connectives is given in Fig. 1.

The effect of the commutator $K$ can be read directly from this graph. For example, we have $a \rightarrow b = K(a \leftarrow b)$, $a \leftarrow b = K(a \rightarrow b)$, $a\&^{\bullet}b = K(a\&b)$, ... $a \downarrow b = K(a \mid b)$, etc.

The Table 1 clearly indicates which are the TOP and which are the BOT connectives. It also distinguishes *Maxdiag* from *Mindiag* connectives, a distinction

**Table 1.** System of Fuzzy Interval Logic based on $m3$

| $\mathcal{T}$ | Type | Maxdiag | Mindiag | Type |
|---|---|---|---|---|
| **I** | $\rightarrow$ | $max(min(a,b),\ 1-a)$ | $\geq max(a+b-1,\ 1-a)$ | $\rightarrow$ |
| **N** | $\nrightarrow$ | $min(max(1-a,1-b),\ a)$ | $\leq min(2-(a+b),\ a)$ | $\nrightarrow$ |
| **C** | $\leftarrow^\bullet$ | $max(min(1-a,1-b),\ a)$ | $\geq max(1-(a+b),\ a)$ | $\leftarrow^\bullet$ |
| **D** | $\nleftarrow^\bullet$ | $min(max(a,b),\ 1-a)$ | $\leq min(a+b,\ 1-a)$ | $\nleftarrow^\bullet$ |
| **LC** | $\vee$ | $max(min(1-a,b),\ a)$ | $\geq max(b-a,\ a)$ | $\vee$ |
| **RC** | $\mid$ | $max(min(a,1-b),\ 1-a))$ | $\geq max(a-b,\ 1-a)$ | $\mid$ |
| **LD** | $\downarrow$ | $min(max(a,1-b),\ 1-a)$ | $\leq min(1-b+a,\ 1-a)$ | $\downarrow$ |
| **RD** | & | $min(max(1-a,b),\ a)$ | $\leq min(1-a+b,\ a)$ | & |
| $\mathcal{T}$ | Type | Maxdiag | Mindiag | Type |
| **I** | $\leftarrow$ | $max(min(a,b),\ 1-b)$ | $\geq max(a+b-1,\ 1-b)$ | $\leftarrow$ |
| **N** | $\nleftarrow$ | $min(max(1-a,1-b),\ b)$ | $\leq min(2-(a+b),\ b)$ | $\nleftarrow$ |
| **C** | $\rightarrow^\bullet$ | $max(min(1-a,1-b),\ b)$ | $\geq max(1-(a+b),\ b)$ | $\rightarrow^\bullet$ |
| **D** | $\nrightarrow^\bullet$ | $min(max(a,b),\ 1-b)$ | $\leq min(a+b,\ 1-b)$ | $\nrightarrow^\bullet$ |
| **LC** | $\mid^\bullet$ | $max(min(1-a,b),\ 1-b)$ | $\geq max(b-a,\ 1-b)$ | $\mid^\bullet$ |
| **RC** | $\vee^\bullet$ | $max(min(a,1-b),\ b)$ | $\geq max(a-b,\ b)$ | $\vee^\bullet$ |
| **LD** | $\&^\bullet$ | $min(max(a,1-b),\ b)$ | $\leq min(1-b+a,\ b)$ | $\&^\bullet$ |
| **RD** | $\downarrow^\bullet$ | $min(max(1-a,b),\ 1-b)$ | $\leq min(1-a+b,\ 1-b)$ | $\downarrow^\bullet$ |

that stems from the mathematics of the Checklist Paradigm. (See the Appendix for the explanation of these concepts). It is important for classification of properties of individual connectives.

### 4.2 Classification of Connectives

We have 3 kinds of classification of connectives generated by the checklist paradigm semantics, namely,

**Interval-Based.** TOP-BOT interval pair giving the bounds on the values of the interval. See Table 1 in Sec.4

**Group Transformation-Based.** which stems from group transformations (see Sec 2.2) and is performed by examining the subgroups of the transformation group $S_{2\times2\times2\times2}$.

**Constraint-Based.** Maxdiag-Mindiag pair: provides the classification of connectives in Checklist Paradigm based systems of interval fuzzy logics system into 2 classes, namely the Max-diagonal class (MAXD) and the Min-diagonal class (MinD). This classification stems directly from the meta-logical features of the mathematics of the Checklist Paradigm and is induced by the characteristics of the fine structure of the checklist paradigm captured by the Maxdiag and Mindiag constraint tables which determine the interval pairs of the connectives (see the Appendix below).

Both, the Interval based and the Constraint based (i.e. Maxdiag/Mindiag) classification of connectives of $m_3$ system which can be seen in the Table 1

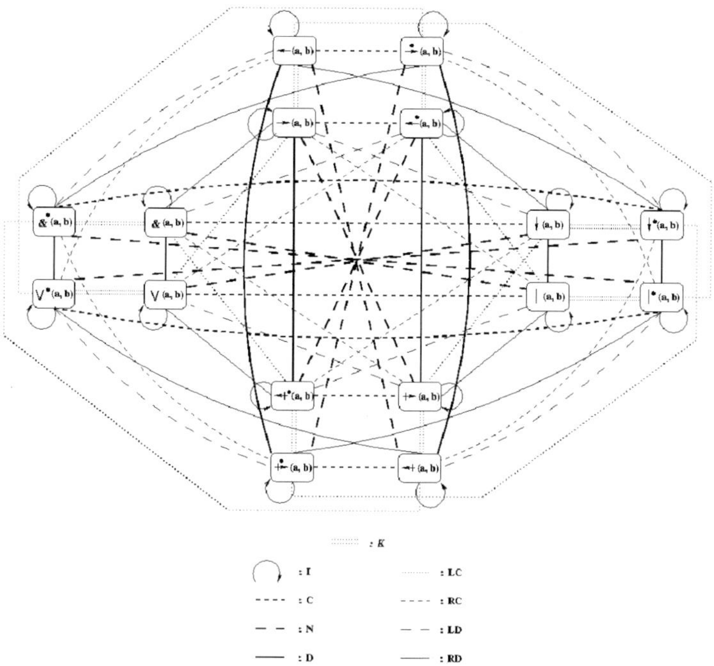


**Fig. 1.** Graph of Group Transformations for Interval System $m_3$

display the classification of the $m_3$ system of interval connectives into 2 classes, namely the Max-diagonal class (MAXD) and the Min-diagonal class (MinD). This classification is derived for MAXDIAG and MINDIAG constraint tables discussed above in section 2.1.

**Closed Systems of Connectives in the Interval System $m_3$.** We shall now recall some useful definitions that appeared explicitly or implicitly in previous papers.

**Definition 6 (Closed Set).** *Bandler and Kohout [9]*

*Let $\mathcal{T}$ be a family of group transformations. Given a logic connective (or a set of connectives) CON, one can generate a new set of connectives $\mathcal{T}(CON)$ by the application of $\mathcal{T}$ to CON. This set is closed, iff after the i-th application $\mathcal{T}(CON)_i = \mathcal{T}(CON)_{i-1}$ where $i \geq 1$. We shall denote this closed set by $Clo(CON, \mathcal{T})$.*

*We say that $Clo(CON, \mathcal{T})$ is the closed set of connectives with respect to $\mathcal{T}$ generated by CON*

**Definition 7 (Elementary Closed Set).** *Kohout and Kim [10],[7]*

*A subset B of a set A of connectives is an elementary closed set iff it is a knot in A. This means that (i) any element in B is reachable from any other element in B, and (ii) no element outside B is reachable from B by any number of* *applications of transformation T. The closure is relative with respect to application of a specific set of transformations $\mathcal{T}$.*

**Definition 8 (Pure Set of Connectives).** *Kohout and Kim [10],[7]*
*A pure set of connectives is a set that contains (i) only the connectives of one classification type (e.g. only TOP, or only BOT, or only Min-diag, or only Max-diag, etc.). (ii) It is an elementary closed set.*

**Theorem 5.** ***Classification of TOP Connectives***

1. *Let* $\mathcal{T}_1 = \{I, C, LC, RC\}$. *Then* $Clo(\rightarrow_{TOP}, \mathcal{T}_1) = \{\rightarrow_{TOP}, \overset{\bullet}{\leftarrow}_{TOP}, \vee_{TOP}, |_{TOP}\}$ *is the Maxdiag set closed with respect to* $\mathcal{T}_1$ *generated by* $\rightarrow_{TOP}$ *.*
2. *Let* $\mathcal{T}_2 = \{I, C, LC, RC\}$. *Then* $Clo(\leftarrow_{TOP}, \mathcal{T}) = \{\leftarrow_{TOP}, \overset{\bullet}{\rightarrow}_{TOP}, \overset{\bullet}{|}_{TOP}, \overset{\bullet}{\vee}_{TOP}\}$ *is the Maxdiag set closed with respect to* $\mathcal{T}$ *generated by* $\leftarrow_{TOP}$ *.*

**Theorem 6.** ***Classification of BOT Connectives***

1. *Let* $\mathcal{T}_1 = \{I, C, LC, RC\}$. *Then* $Clo(\rightarrow_{BOT}, \mathcal{T}) = \{\rightarrow_{BOT}, \overset{\bullet}{\leftarrow}_{BOT}, \vee_{BOT}, |_{BOT}\}$ *is the Mindiag set closed with respect to* $\mathcal{T}$ *generated by* $\rightarrow_{BOT}$ *.*
2. *Let* $\mathcal{T}_2 = \{I, C, LC, RC\}$. *Then* $Clo(\leftarrow_{BOT}, \mathcal{T}_2) = \{\leftarrow_{BOT}, \overset{\bullet}{\rightarrow}_{BOT}, \overset{\bullet}{|}_{BOT}, \overset{\bullet}{\vee}_{BOT}\}$ *is the Mindiag set closed with respect to* $\mathcal{T}_2$ *generated by* $\leftarrow_{BOT}$ *.*

**Theorem 7.** *All closed sets of Theorem 12 and 13 are Pure sets of connectives.*

*Proof.* (i) They are closed (by Th. 12 and 13. (ii) They are elementary closed sets. This follows from the properties of the $S_{2\times2\times2\times2}$ group. The group has one element of order 1 (namely I) and 15 elements of order 2. (iii) Each of the sets contains only the TOP or BOT connectives. Each of these sets contains only Maxdig or Mindiag connectives.

### 4.3 Types of Logic

The logic which has as its membership function a *real valued* function that yields a single point as the value of a logic formula, in the way analogous to the assignment of values to the elements of ordinary fuzzy sets (i.e. fuzzy sets of the first type) is called a **singleton logic**, or checklist *fuzzy logic of the first kind.*

In the logic of 2nd type the atomic object, the basic element of the valuation space is a subset of a rectangular shape. In the logic of the 1st type, the atomic object is a singleton.

## References


1. Kohout, L.J., Bandler, W.: Interval-valued systems for approximate reasoning based on the checklist paradigm. In Wang, Paul, P., ed.: Advances in Fuzzy Theory and Technology, vol. 1. Bookwrights Press, Durham, N.C. (1993) 167–193
2. Kohout, L., Stabile, I.: Interval-valued inference in medical knowledge-based system Clinaid. Interval Computations 2(3) (1993) 88–115

3. Bandler, W., Kohout, L.: Semantics of implication operators and fuzzy relational products. Internat. Journal of Man-Machine Studies 12 (1980) 89116 Reprinted in Mamdani, E.H. and Gaines, B.R. eds. Fuzzy Reasoning and its Applications. Academic Press, London, 1981, pages 219–246.
4. Bandler, W., Kohout, L.: Unified theory of multiple-valued logical operators in the light of the checklist paradigm. In: Proc. of the 1984 IEEE Conference on Systems, Man and Cybernetics, New York, IEEE (1984) 356–364.
5. Bandler, W., Kohout, L.: The interrelations of the principal fuzzy logical operators. In Gupta, M., Kandel, A., Bandler, W., Kiszka, J., eds.: Approximate Reasoning in Expert Systems. North-Holland, Amsterdam (1985) 767–780
6. Bandler, W., Kohout, L.: The use of checklist paradigm in inference systems. In Negoita, C., Prade, H., eds.: Fuzzy Logic in Knowledge Engineering. Verlag TÜV Rheinland, Köln (1986) 95–111.
7. Kohout, L., Kim, E.: Characterization of interval fuzzy logic systems of connectives by group transformations. Reliable Computing 10 (2004) 299–334
8. Kohout, L., Bandler, W.: Checklist paradigm and group transformations. Technical Note EES-MMS-ckl91.2 (1979) Dept. of Electrical Engineering, University of Essex. U.K.
9. Kohout, L., Bandler, W.: How the checklist paradigm elucidates the semantics of fuzzy inference. In: Proc. of the IEEE Internat. Conference on Fuzzy Systems 1992. IEEE, New York (1992) 571–578
10. Kohout, L., Kim, E.: Global characterization of fuzzy logic systems with paraconsistent and grey set features. In Wang, P., ed.: Proc. 3rd Joint Conf. on Information Sciences JCIS97 (5th Int. Conf. on Fuzzy Theory and Technology), Research Triangle Park, NC, Duke University (March 1997) 238–241 In volume 1 Fuzzy Logic, Intelligenr Control and Genetic Algorithms.
11. Hájek, P.: A remark on Bandler-Kohout products of relations. Internat. Journal of General Systems 25(2) (1996) 165–166
12. Kohout, L., Kim, E.: Group transformations of systems of logic connectives. In: Proc. of IEEE-FUZ97, vol. 1,, New York, IEEE (July 1997) pp. 157–162
13. Kohout, L., Bandler, W.: Fuzzy interval inference utilizing the checklist paradigm and BK-relational products. In Kearfott, R., Kreinovich, V., eds.: Applications of Interval Computations. Kluwer, Boston (1996) 291–335
14. Kohout, L., Kim, E.: Non-commutative fuzzy interval logics with approximation semantics based on the checklist paradigm and their group transformations. In: Proc. of FUZZ-IEEE2005, Piscataway, NJ, IEEE Neural Network Council, IEEE (May 2005) CD-ROM.
15. Kohout, L., Kim, E.: Fuzzy interval logic system m2 with non-commutative and and or containing goguen-gaines implication. WSEAS Transactions on Systems 4 (2005).

## Appendix: Mathematics of the Checklist Paradigm

A number of distinct interval systems of fuzzy logics arise when the fuzzy membership function $\mu(S)$ of a fuzzy proposition $S$ is interpreted as a summarization of a two-valued (also called crisp) logical n-tuple that represents a checklist that records *yes-no* answers to $n$ questions concerning the truth status of a logic statement. The structure of the checklist with answers can be conveniently captured by a contingency table. Putting different measures on the contingency tables generate different fuzzy logic systems.

**Definition 9.** *Checklist A checklist template $Q$ is a finite family of properties $\langle Q_1, Q_2, ..., Q_i, ..., Q_n \rangle$; With a template $Q$, and a given proposition $A$, one can associate a specific checklist $Q^A = \langle Q, A \rangle$ which pairs the template $Q$ with a given proposition $A$. A valuation $g_A$ of a checklist $Q_A$ is a function from $Q$ to $\{0, 1\}$.*

**Definition 10.** *Let us denote by symbol $a_Q$ the degree $\delta(A)$ to which the proposition $A$ holds with respect to a template $Q$. This degree is given by the formula*

$$a_Q = \delta(A(Q)) = \sum_{i=1}^{n} q_i^A$$

*where n=card $Q$ and $q_i^A = g_A(P_i)$. Obviously, $q_i^A \in \{0, 1\}$.*

**Definition 11.** *Fine structure*
*A fine valuation structure a pair of propositions $A$, $B$ with respect to the template $Q$ is a function $g^Q_{A,B}$ from $Q$ into$\{0, 1\}$ assigning to each attribute $P_i$ the ordered pair of its values $\langle q_i^A, q_i^B \rangle$. The cardinality of the set of all attributes $P_i$ such that $g^Q_{A,B}(Q_i) = \langle j, k \rangle$ will be denoted by $\alpha_{j,k}$. (Obviously, $q_i^A, q_i^B \in \{0, 1\}$, but $\alpha_{j,k} \in [0, n]$).*

We have the following constraint on the values: $\alpha_{00} + \alpha_{01} + \alpha_{10} + \alpha_{11} = n$. Further, we define $r_0 = \alpha_{00} + \alpha_{01}$, $r_1 = \alpha_{10} + \alpha_{11}$, These entities can be displayed systematically in a *contingency table.* In such a table, the inner fine-summarization structure consists of the four $\alpha_{j,k}$ appropriately arranged, and of margins $c_0, c_1, r_0, r_1$ as shown below:

| | No for B | Yes for B | Row total |
|---|---|---|---|
| No for A | $\alpha_{00}$ | $\alpha_{01}$ | $r_0$ |
| Yes for A | $\alpha_{10}$ | $\alpha_{11}$ | $r_1$ |
| Column Total | $c_0$ | $c_1$ | $n$ |

Now let $F$ be any logical propositional function of propositions $A$ and $B$. For $i, j \in \{0, 1\}$, let $f(i, j)$ be the classical truth value of $F$ for the pair $(i, j)$ of truth values; let $u(i, j) = \alpha_{i,j}/n$ . Then we define the (non-truth-functional) *fuzzy assessment of the truth* of the proposition $F(A, B)$ to be

$$m_{fin}(F(A, B)) = \sum\nolimits_{i,j} f(i, j) \cdot u_{ij}$$

This assessment operator will be called the *value of the fine structure.*

The four interior cells $\alpha_{00}, \alpha_{01}, \alpha_{10}, \alpha_{11}$ of the constraint table constitute its *fine structure*; the margins $r_0, r_1, c_0, c_1$ constitute its *coarse structure* (see Fig. above). As shown elsewhere [3],[4],[5],[6] the *coarse* structure imposes bounds upon the fine structure, without determining it completely. Hence, associated with the various logical connectives between propositions are their *extreme values*. Thus we obtain the inequality restricting the possible values of $m_i(F)$:

$$conbot \leq m_{fin}(F) \leq contop$$

where `con` is the name of connective represented by $f(i, j)$.

When the fine structure is not available, we can still determine the bounds on the possible values from the values of the coarse structure that appear as the margins in the contingency table (see figure above).

**Maxdiag and Mindiag Constraints.** The bounds on the values of the interior of the contingency table are displayed in the two constraint structures, labeled as MINDIAG and MAXDIAG. The first is obtained by minimizing the values of the diagonal of the contingency table, the second by maximizing these.

| | | |
|---|---|---|
| $1-a$ | $0$ | $1-a$ |
| $a-b$ | $b$ | $a$ |
| 1-b | b | 1 |

MAXDIAG if $b \leq a$

| | | |
|---|---|---|
| $1-b$ | $b-a$ | $1-a$ |
| $0$ | $a$ | $a$ |
| $1-b$ | $b$ | $1$ |

MAXDIAG if $b \geq a$

| | | |
|---|---|---|
| $0$ | $1-a$ | $1-a$ |
| $1-b$ | $a+b-1$ | $a$ |
| $1-b$ | $b$ | $1$ |

MINDIAG if $a+b \geq 1$

| | | |
|---|---|---|
| $1-(a+b)$ | $b$ | $1-a$ |
| $a$ | $0$ | $a$ |
| $1-b$ | $b$ | $1$ |

MINDIAG if $a+b \leq 1$

The above seen MAXDIAG and MINDIAG tables that summarize the solutions of constraint inequalities can be conveniently used for direct readout of various pairs of TOP and BOT connectives (as demonstrated e.g. in [7]).

In general, MINDIAG and MAXDIAG constraint structures can be computed by solving the constraint inequalities involving the variables of the fine structure displayed in the contingency table. For the proofs and further explanation see [3],[4],[6].

**A List of Measures Used in Bandler and Kohout [3].** Referring to the contingency table above in this Appendix, let us define $u_{lk} = \alpha_{lk}/n$. This is used to define the following measures that yield implication interval connectives (see Sec. 1.1 above):

1. $m_1(F) = 1 - (\alpha_{10}/n)$
2. If for *con* type we choose an implication again, but only the evaluation "by performance" (that is, we are only concerned with the cases in which the evaluation of A is 1), we obtain $m_2 = u_{11}/(u_{10} + u_{11})$
3. Still another contracting measure which distinguishes the proportion of satisfactions "by performance", $u(1,1)$, and "by default", $u_{00} + u_{01}$, we obtin $m_3 = u_{11} \bigvee (u_{00} + u_{01}))$.
4. The lower contrapositivization of $m_3$ yields $m_4 = (u_{11} \bigvee (u_{00} + u_{01})) \bigwedge (u_{00} \bigvee (u_{01} + u_{11}))$
5. Yet another measure arises by taking for the "performance" part the less conservative $m_2$, giving $m_5 = m_2 \bigvee (u_{00} + u_{11})$.

# Points with Type-2 Operations

Carol L. Walker and Elbert A. Walker

Department of Mathematical Sciences
New Mexico State University
Las Cruces, NM 88003
hardy@nmsu.edu, elbert@nmsu.edu

**Abstract.** The algebra of truth values for fuzzy sets of type-2, due to Zadeh, contains as subalgebras those of type-1 and of interval-valued fuzzy sets. It also contains many other interesting subalgebras, some of which could possibly serve as a basis of a useful fuzzy set theory. This paper is about one such subalgebra which we call the subalgebra of points, and which generalizes type-1. We investigate it as an algebra, and determine its automorphism group. In particular, we show that is it a characteristic subalgebra and that its automorphisms are exactly those induced by automorphisms of the containing truth value algebra of fuzzy sets of type-2.

**Keywords:** Fuzzy truth values, type-2 fuzzy set, type-1 fuzzy set, interval-valued fuzzy set.

## 1 Introduction

Type-2 fuzzy sets—that is, fuzzy sets with fuzzy sets as truth values were introduced by Zadeh [9] in 1975, and have been the subject of many papers. A basic mathematical treatment is given in [6] and it subalgebras and their automorphisms are the subject of [7,8]. There are several subalgebras of special interest, including copies of the truth value algebra of ordinary type-1 fuzzy and sets and of the truth value algebra of interval-valued fuzzy sets. These are not only subalgebras, but are characteristic in the sense that every automorphism of the algebra induces an automorphism of each of these subalgebras, making them very special subalgebras, further testimony to the "correctness" of Zadeh's generalization. Subalgebras in general are of interest because each could serve as the basis of a fuzzy set theory, where a fuzzy set in this theory is a mapping of a universal set into this subalgebra.

This paper is about one special subalgebra. It is a generalization of the truth value algebra of type-1 fuzzy sets. The elements of the truth value algebra of type-2 fuzzy sets are all functions $[0,1] \rightarrow [0,1]$. The elements of the truth value algebra of type-1 fuzzy sets correspond to the characteristic functions of points. That is, they are those functions that are non-zero at exactly one point and have value 1 at that point. The algebra concerned with here is the algebra of those functions which are non-zero at exactly one point but can have any value in

P. Melin et al. (Eds.): IFSA 2007, LNAI 4529, pp. 56–65, 2007.
© Springer-Verlag Berlin Heidelberg 2007

$(0,1]$ at that point. This is indeed a subalgebra of the algebra of truth values of type-2 fuzzy sets, and is more general than that of the type-1 truth values. This subalgebra and its automorphisms are principal topics of this paper. It plays a special role in the development of the general theory of the truth value algebra of type-2 fuzzy sets since obviously every mapping $[0,1] \to [0,1]$ is the pointwise join of elements of this subalgebra. It is especially important to understand this subalgebra and its automorphisms in determining the automorphisms of the larger algebra.

Elements of the algebra of points will be represented by a pair of points from the unit interval which is reminiscent of interval-valued fuzzy sets. But, as you will see, the operations are not the same, and the algebra of points has quite different properties than the algebra of intervals.

The basic mathematical properties of the truth value algebra of type-2 fuzzy sets are given in [6]. We begin with a review of some relevant definitions.

## 2 Type-2 Fuzzy Sets

The algebra of truth values for fuzzy sets of type-2 is the set of all mappings of $[0,1]$ into $[0,1]$ with operations certain convolutions of operations on $[0,1]$. These operations are as follows.

**Definition 1.** *The* ***algebra of truth values for type-2 fuzzy sets*** *is the algebra*

$$\mathbb{M} = \left( [0,1]^{[0,1]}, \sqcup, \sqcap, ^*, \bar{0}, \bar{1} \right) \tag{1}$$

*where the operations are defined by*

1. $(f \sqcup g)(x) = \sup\{f(y) \wedge g(z) : y \vee z = x\}$
2. $(f \sqcap g)(x) = \sup\{f(y) \wedge g(z) : y \wedge z = x\}$
3. $f^*(x) = \sup\{f(y) : 1 - y = x\}$
4. $\bar{1}(x) = 1$ *if* $x = 1$ *and* $\bar{1}(x) = 0$ *if* $x \neq 1$
5. $\bar{0}(x) = 1$ *if* $x = 0$ *and* $\bar{0}(x) = 0$ *if* $x \neq 0$

Note that $f^*(x) = f(1-x)$. A **fuzzy subset of type-2 of a set** $S$ is a mapping $f : S \to [0,1]^{[0,1]}$, and operations on the set $\mathcal{F}_2(S)$ of all such fuzzy subsets are given pointwise from the operations in $\mathbb{M}$. Thus we have the **algebra** $\mathcal{F}_2(S) = (Map(S,[0,1]^{[0,1]}), \sqcup, \sqcap, ^*, \bar{0}, \bar{1})$ **of fuzzy subsets of type-2 of the set** $S$. The same equations hold in $\mathcal{F}_2(S)$ as in $\mathbb{M}$.

Determining the properties of the algebra $\mathbb{M}$ is a bit tedious, but is helped by introducing the following auxiliary operations.

**Definition 2.** *For* $f \in \mathbb{M}$*, let* $f^L$ *and* $f^R$ *be the elements of* $\mathbb{M}$ *defined by*

$$f^L(x) = \sup\{f(y) : y \leq x\}$$
$$f^R(x) = \sup\{f(y) : y \geq x\}$$

Note that $f^L$ is monotone increasing and that $f^R$ is monotone decreasing, and these are the pointwise smallest such functions above $f$. With this definition, the operations $\sqcup$ and $\sqcap$ in $\mathbb{M}$ can be expressed in terms of the pointwise max and min of functions in two ways, as follows.

**Theorem 1.** *The following hold for all $f, g \in \mathbb{M}$.*

$$f \sqcup g = \left(f \wedge g^L\right) \vee \left(f^L \wedge g\right) = (f \vee g) \wedge \left(f^L \wedge g^L\right) \tag{2}$$
$$f \sqcap g = \left(f \wedge g^R\right) \vee \left(f^R \wedge g\right) = (f \vee g) \wedge \left(f^R \wedge g^R\right)$$

Using these auxiliary operations, it is fairly routine to verify the following basic properties of the algebra $\mathbb{M}$. The details may be found in [6].

**Corollary 1.** *The following hold for $f,\ g,\ h \in \mathbb{M}$.*

1. $f \sqcup f = f$ *and* $f \sqcap f = f$
2. $f \sqcup g = g \sqcup f$ *and* $f \sqcap g = g \sqcap f$
3. $\bar{1} \sqcap f = f$ *and* $\bar{0} \sqcup f = f$
4. $f \sqcup (g \sqcup h) = (f \sqcup g) \sqcup h$ *and* $f \sqcap (g \sqcap h) = (f \sqcap g) \sqcap h$
5. $f \sqcup (f \sqcap g) = f \sqcap (f \sqcup g)$
6. $f^{**} = f$
7. $(f \sqcup g)^* = f^* \sqcap g^*$ *and* $(f \sqcap g)^* = f^* \sqcup g^*$

It is not known whether or not every equation satisfied by $\mathbb{M}$ is a consequence of these; that is, whether or not these equations form an equational base for the variety generated by $\mathbb{M}$. It is also not known whether or not the variety generated by $\mathbb{M}$ is generated by a finite algebra. (See [3] for background on varieties.)

## 3 The Subalgebra $\mathbb{P}$ of Points

As indicated in the introduction, this paper is about the subalgebra of those functions, called **points**, which are non-zero at exactly one element of their domain, but can have any value in $(0, 1]$ at that element. This is indeed a subalgebra, and our concern is with its properties, its automorphisms, and its being a characteristic subalgebra of $\mathbb{M}$.

It is a generalization of the truth value algebra of type-1 fuzzy sets—the subalgebra of points with value 1, and seems to be a reasonable candidate for applications. For a function in $\mathbb{P}$, its support could be viewed as degree of membership, and its value as level of confidence. This generalizes type-1 fuzzy sets, where the "level of confidence" is always 1.

One feature of our development is realizing this subalgebra as an algebra of pairs with simple pointwise operations. This is elaborated on below.

**Definition 3.** *The* **points** *of* $\mathbb{M}$ *are those functions that are non-zero at exactly one point of the domain* $[0, 1]$. *We denote by* $\langle a, p\rangle$ *the function that has value* $p$ *at* $a$ *and is* $0$ *elsewhere.*

Note that $\langle 0, 1\rangle$ and $\langle 1, 1\rangle$ are the functions previously denoted by $\bar{0}$ and $\bar{1}$, respectively. Also,

1. $\langle a,p\rangle^L(x) = 0$ if $x < a$, $\langle a,p\rangle^L(x) = p$ if $x \geq a$
2. $\langle a,p\rangle^R(x) = p$ if $x \leq a$, $\langle a,p\rangle^R(x) = 0$ if $x > a$

**Proposition 1.** *Let $P$ denote the set of points of $\mathbb{M}$. Then*

$$\mathbb{P} = (P, \sqcup, \sqcap, ^*, \langle 0,1\rangle, \langle 1,1\rangle) \tag{3}$$

*is a subalgebra of $\mathbb{M}$.*

*Proof.* The constants of $\mathbb{M}$, $\langle 0,1\rangle$ and $\langle 1,1\rangle$, are in $\mathbb{P}$. Also,

$$\begin{aligned}\langle a,p\rangle \sqcup \langle b,q\rangle &= (\langle a,p\rangle \vee \langle b,q\rangle) \wedge \langle a,p\rangle^L \wedge \langle b,q\rangle^L = \langle a \vee b, p \wedge q\rangle \\ \langle a,p\rangle \sqcap \langle b,q\rangle &= (\langle a,p\rangle \vee \langle b,q\rangle) \wedge \langle a,p\rangle^R \wedge \langle b,q\rangle^R = \langle a \wedge b, p \wedge q\rangle \\ \langle a,p\rangle^* &= \langle 1-a, p\rangle\end{aligned}$$

Thus $\mathbb{P}$ is closed under all of the operations of $\mathbb{M}$.

We call the subalgebra $\mathbb{P}$ the **subalgebra of points** of $\mathbb{M}$. From the formulas in the proof above, we see that $\mathbb{P}$ may be regarded as the algebra

$$\mathbb{P} = ([0,1] \times (0,1], \sqcap, \sqcup, ^*, \langle 0,1\rangle, (1,1)) \tag{4}$$

with operations $\sqcup, \sqcap$, and $^*$ given by

$$\begin{aligned}\langle a,p\rangle \sqcup \langle b,q\rangle &= \langle a \vee b, p \wedge q\rangle \\ \langle a,p\rangle \sqcap \langle b,q\rangle &= \langle a \wedge b, p \wedge q\rangle \\ \langle a,p\rangle^* &= \langle 1-a, p\rangle\end{aligned} \tag{5}$$

Viewing the elements of $\mathbb{P}$ as pairs is conceptually simpler, not involving the notions of points being functions and operations being convolutions. We just operate with pairs of elements from the unit interval with the basic operations coming directly from ordinary max and min. Of course $\mathbb{P}$ satisfies the properties of Corollary 1. By [6], since points are convex functions, $\mathbb{P}$ also satisfies the distributive laws

$$\begin{aligned}\langle a,p\rangle \sqcap (\langle b,q\rangle \sqcup \langle c,r\rangle) &= (\langle a,p\rangle \sqcap \langle b,q\rangle) \sqcup (\langle a,p\rangle \sqcap \langle c,r\rangle) \\ \langle a,p\rangle \sqcup (\langle b,q\rangle \sqcap \langle c,r\rangle) &= (\langle a,p\rangle \sqcup \langle b,q\rangle) \sqcap (\langle a,p\rangle \sqcup \langle c,r\rangle)\end{aligned} \tag{6}$$

The algebra $\mathbb{P}$ is not a lattice, because the absorption laws $(x \sqcup y) \sqcap x = x$ and $(x \sqcap y) \sqcup x = x$ fail to hold. For example, if $q < p$,

$$(\langle a,p\rangle \sqcup \langle b,q\rangle) \sqcap \langle a,p\rangle = \langle a \vee b, q\rangle \sqcap \langle a,p\rangle = \langle a,q\rangle \neq \langle a,p\rangle \tag{7}$$

By Corollary 1,

$$(\langle a,p\rangle \sqcap \langle b,q\rangle) \sqcup \langle a,p\rangle = (\langle a,p\rangle \sqcup \langle b,q\rangle) \sqcap \langle a,p\rangle \tag{8}$$

so the other absorption law fails as well. However, the properties that do hold say that $\mathbb{P}$ is a **distributive De Morgan bisemilattice** [1]. The algebra $\mathbb{P}$ inherits the De Morgan laws (Corollary 4, property 7) from $\mathbb{M}$. A **semilattice** is a set with one binary operation that is associative, commutative and idempotent. The two semilattices involved here are monoids (each has an identity)

$$\mathbb{P}_1 = ([0,1] \times (0,1], \sqcap, \langle 1,1 \rangle) \text{ and } \mathbb{P}_2 = ([0,1] \times (0,1], \sqcup, \langle 0,1 \rangle) \tag{9}$$

These two semilattices induce the same partial order on $\mathbb{P}$, namely $\langle a,p \rangle \sqsubseteq \langle b,q \rangle$ if and only if $a \leq b$ and $p \geq q$.

## 4 $\mathbb{P}$ is a Characteristic Subalgebra of $\mathbb{M}$

We have shown in [7,8] that several interesting subalgebras of $\mathbb{M}$ are characteristic subalgebras—that is, they are taken to themselves by every automorphism of $\mathbb{M}$. Intuitively, this means that these subalgebras sit in $\mathbb{M}$ is a very special way. These include the subalgebra of singletons (characteristic functions of one-element sets), of interval-valued fuzzy sets (characteristic functions of closed intervals), of normal functions, of convex functions, and of convex normal functions, among others. In fact, we have shown this to be true in a more general setting, namely in the algebra $\mathbf{M} = \left([0,1]^{[0,1]}, \sqcup, \sqcap, \bar{0}, \bar{1}\right)$ that does not include the negation $^*$. **We will assume until further notice that the operations do not include** $^*$. That is, we will be concerned with the subalgebra $\mathbf{P} = ([0,1] \times (0,1], \sqcup, \sqcap, \langle 0,1 \rangle, \langle 1,1 \rangle)$ with operations given by

$$\langle a,p \rangle \sqcup \langle b,q \rangle = \langle a \vee b, p \wedge q \rangle \text{ and } \langle a,p \rangle \sqcap \langle b,q \rangle = \langle a \wedge b, p \wedge q \rangle \tag{10}$$

Note that if $\mathbb{A}$ is a subalgebra of $\mathbb{M}$ and $\mathbb{A}$ without $^*$ is a characteristic subalgebra of $\mathbf{M}$, then $\mathbb{A}$ is a characteristic subalgebra of the algebra $\mathbb{M}$.

One subalgebra of $\mathbf{P}$ is the algebra $\mathbf{S} = \{[0,1] \times \{1\}, \sqcup, \sqcap, \langle 0,1 \rangle, \langle 1,1 \rangle\}$, that is, the elements of $\mathbf{P}$ with second coordinate 1. This subalgebra is called the subalgebra of **singletons,** and is isomorphic to $\mathbf{I} = ([0,1], \wedge, \vee, 0, 1)$ via $x \to \langle x, 1 \rangle$.

We show now that $\mathbf{P}$ is a characteristic subalgebra of $\mathbf{M}$. It follows, then, that $\mathbb{P}$ is a characteristic subalgebra of $\mathbb{M}$. In the following section, we determine precisely the automorphisms of $\mathbf{P}$, and we show that they are all induces by automorphisms of $\mathbf{M}$.

**Theorem 2.** *Let $\varphi \in Aut(\mathbf{M})$ and $\langle a,p \rangle$ be a point function at $a$. Then $\varphi(\langle a,p \rangle)$ is a point function.*

*Proof.* From [7] we know that automorphisms of $\mathbf{M}$ induce automorphisms of the subalgebra of singletons. Specifically, $\varphi(\langle a,1 \rangle) = \langle \varphi_s(a), 1 \rangle$ for some $\varphi_s \in$ *Aut* $(\mathbf{I})$. Note that

$$\langle a,p \rangle \sqcup \langle a,1 \rangle = \langle a \vee a, p \wedge 1 \rangle = \langle a,p \rangle \text{ and } \langle a,p \rangle \sqcap \langle a,1 \rangle = \langle a \wedge a, p \wedge 1 \rangle = \langle a,p \rangle$$

So we have

$$\begin{aligned}\varphi(\langle a,p\rangle) &= \varphi(\langle a,p\rangle \sqcup \langle a,1\rangle) \\ &= \varphi(\langle a,p\rangle) \sqcup \varphi(\langle a,1\rangle) \\ &= \varphi(\langle a,p\rangle) \sqcup \langle \varphi_s(a),1\rangle \\ &= (\varphi(\langle a,p\rangle) \vee \langle \varphi_s(a),1\rangle) \wedge \varphi(\langle a,p\rangle)^L \wedge \langle \varphi_s(a),1\rangle^L\end{aligned}$$

Therefore, $\varphi(\langle a,p\rangle) = 0$ to the left of $\varphi_s(a)$ since this is the case for $\langle \varphi_s(a),1\rangle^L$. Similarly, using $\langle a,p\rangle \sqcap \langle a,1\rangle = \langle a,p\rangle$, $\varphi(\langle a,p\rangle) = 0$ to the right of $\varphi_s(a)$. Therefore, $\varphi(\langle a,p\rangle)$ is a point function at $\varphi_s(a)$.

**Corollary 2.** *The algebra* $\mathbf{P}$ *is a characteristic subalgebra of* $\mathbf{M}$*, and the algebra* $\mathbb{P}$ *is a characteristic subalgebra of* $\mathbb{M}$.

Since $\varphi_s$ is an automorphism of the unit interval, we also have the following.

**Corollary 3.** *For each* $p \in (0,1]$*, and* $\varphi \in Aut(\mathbf{M})$*,* $\varphi(\langle 0,p\rangle)$ *is a point function at* $0$ *and* $\varphi(\langle 1,p\rangle)$ *is a point function at* $1$.

## 5 Automorphisms of P

The fact that $\mathbf{P}$ is a characteristic subalgebra of $\mathbf{M}$ says that every automorphism of $\mathbf{M}$ induces an automorphism of $\mathbf{P}$. But does $\mathbf{P}$ have any other automorphisms? That is, does every automorphism of $\mathbf{P}$ extend to one of $\mathbf{M}$.

For $\alpha,\beta \in Aut(\mathbf{I})$, and $f \in \mathbf{M}$, define $\alpha_L(f)$ by $\alpha_L(f)(x) = \alpha(f(x))$ and $\beta_R(f)(x) = f(\beta(x))$. Then $\alpha_L$ and $\beta_R$ are both automorphisms of $\mathbf{M}$. (In [8] we showed that every automorphism of $\mathbf{M}$ is of the form $\alpha_L\beta_R$.) If $\varphi$ is an automorphism of $\mathbf{M}$ and $\varphi = \alpha_L\beta_R$, then $\alpha_L\beta_R(\langle a,p\rangle) = \langle \beta^{-1}(a), \alpha(p)\rangle$. To see this, recall that $\langle a,p\rangle$ is the function defined by $\langle a,p\rangle(a) = p$ and $\langle a,p\rangle(x) = 0$ if $x \neq a$. Then for $x \in [0,1]$

$$\begin{aligned}\alpha_L\beta_R(\langle a,p\rangle)(x) &= \alpha(\langle a,p\rangle(\beta(x))) = \begin{cases} 0 & \text{if } \beta(x) \neq a \\ \alpha(p) & \text{if } \beta(x) = a \end{cases} \\ &= \langle \beta^{-1}(a), \alpha(p)\rangle(x)\end{aligned} \tag{11}$$

Thus every pair of automorphisms of the unit interval yields an automorphism of $\mathbf{P}$. We will now show that every automorphism of $\mathbf{P}$ comes about this way.

In order to determine the automorphisms of $\mathbf{P}$, it is convenient to determine the irreducibles elements of $\mathbf{P}$. An element $\langle a,p\rangle$ is **join irreducible** if $\langle a,p\rangle = \langle b,q\rangle \sqcup \langle c,r\rangle$ implies that $\langle a,p\rangle = \langle b,q\rangle$ or $\langle a,p\rangle = \langle c,r\rangle$. Similarly, $\langle a,p\rangle$ is **meet irreducible** if $\langle a,p\rangle = \langle b,q\rangle \sqcap \langle c,r\rangle$ implies that $\langle a,p\rangle = \langle b,q\rangle$. An element $\langle a,p\rangle$ is **irreducible** if it is both join and meet irreducible. It is easy to see that irreducible elements are taken to irreducible elements by automorphisms.

**Theorem 3.** *The irreducibles of* **P** *are exactly the singletons—that is, the points of the form* $\langle a,1\rangle$ *for* $a \in [0,1]$.

*Proof.* For any point $\langle a,p\rangle \in \mathbf{P}$, we have

$$\langle a,p\rangle = \langle a,1\rangle \sqcup \langle 0,p\rangle = \langle a,1\rangle \sqcap \langle 1,p\rangle$$

If $p < 1$, then $\langle a,p\rangle \neq \langle a,1\rangle$. Also at least one of $\langle 0,p\rangle$ and $\langle 1,p\rangle$ is different from $\langle a,p\rangle$. Thus, in this case, $\langle a,p\rangle$ is either join or meet reducible or both. For $p=1$, suppose

$$\langle a,1\rangle = \langle b,q\rangle \sqcup \langle c,r\rangle$$

Then $q \wedge r = 1$ implies that $q = r = 1$, and $a = b \vee c$ implies $a = b$ or $a = c$, so $\langle a,1\rangle$ is join irreducible. And if

$$\langle a,1\rangle = \langle b,q\rangle \sqcap \langle c,r\rangle$$

then $q \wedge r = 1$ implies that $q = r = 1$, and $a = b \wedge c$ implies $a = b$ or $a = c$. Thus $\langle a,1\rangle$ is both join and meet irreducible, and hence irreducible.

Since an automorphism must take irreducible elements to irreducible elements, the singletons are a characteristic subalgebra of **P**. But this subalgebra is isomorphic to the unit interval, so we have the following.

**Corollary 4.** *Every automorphism of* **P** *induces an automorphism of* **I** *by its action on singletons.*

We already had this result for automorphisms $\varphi$ of **M**, and called the induced automorphism on the unit interval $\varphi_s$. We will continue to use that notation for automorphisms $\varphi$ of **P**. Thus for every automorphism $\varphi$ of **P**, $\varphi(\langle a,1\rangle) = \langle \varphi_s(a), 1\rangle$ for any $a \in [0,1]$. Note that if $\alpha \in Aut(\mathbf{I})$, then $\alpha_R^{-1}(\langle a,p\rangle) = \langle \alpha(a), p\rangle$ is an automorphism of **P**, so every such $\alpha$ is realized as a $\varphi_s$, in fact, $\alpha = \left(\alpha_R^{-1}\right)_s$.

Call $p$ the **height** of $\langle a,p\rangle$ for $\langle a,p\rangle \in \mathbf{P}$.

**Theorem 4.** *For any* $a,b \in [0,1]$ *and* $\varphi \in Aut(\mathbf{P})$, *the points* $\varphi(\langle a,p\rangle)$ *and* $\varphi(\langle b,p\rangle)$ *have the same height.*

*Proof.* We may as well assume that $a \leq b$. Then

$$\langle a,p\rangle \sqcup \langle b,p\rangle = \langle a \vee b, p \wedge p\rangle = \langle b,p\rangle \text{ and } \langle a,p\rangle \sqcap \langle b,p\rangle = \langle a \wedge b, p \wedge p\rangle = \langle a,p\rangle$$

Write $\varphi(\langle a,p\rangle) = \langle c,q\rangle$ and $\varphi(\langle b,p\rangle) = \langle d,r\rangle$. Then

$$\begin{aligned}\varphi(\langle a,p\rangle \sqcup \langle b,p\rangle) &= \varphi(\langle a,p\rangle) \sqcup \varphi(\langle b,p\rangle) = \langle c \vee d, q \wedge r\rangle \\ &= \varphi(\langle b,p\rangle) = \langle d,r\rangle\end{aligned}$$

which implies $q \geq r$. And

$$\begin{aligned}\varphi(\langle a,p\rangle \sqcap \langle b,p\rangle) &= \varphi(\langle a,p\rangle) \sqcap \varphi(\langle b,p\rangle) = \langle c \wedge p, q \wedge r\rangle \\ &= \varphi(\langle a,p\rangle) = \langle c,q\rangle\end{aligned}$$

which says $r \geq q$. Thus $r = q$—that is, they have the same height.

Since this height depends only on $p$, it is a function. which we will call $\varphi_k$. We extend this function to the unit interval by defining $\varphi_k(0) = 0$.

**Theorem 5.** *If $\varphi \in Aut(P)$, then $\varphi_k \in Aut(\mathbf{I})$.*

*Proof.* Suppose $p \leq q$. Then $\varphi(\langle a,p\rangle \sqcup \langle a,q\rangle) = \varphi(\langle a, p \wedge q\rangle) = \varphi(\langle a,p\rangle)$ is a point of height $\varphi_k(p)$, and

$$\begin{aligned}\varphi(\langle a,p\rangle \sqcup \langle a,q\rangle) &= \varphi(\langle a,p\rangle) \sqcup \varphi\langle a,q\rangle = \langle b, \varphi_k(p)\rangle \sqcup \langle c, \varphi_k(q)\rangle \\ &= \langle b \vee c, \varphi_k(p) \wedge \varphi_k(q)\rangle\end{aligned}$$

is a point of height $\varphi_k(p) \wedge \varphi_k(q)$. Thus $\varphi_k(p) \wedge \varphi_k(q) = \varphi_k(p)$, so $\varphi_k$ is an increasing function. It is also clearly one-to-one. We have already observed that $\varphi_k(1) = 1$. Thus, extending $\varphi_k$ to the unit interval with the definition $\varphi_k(0) = 0$ yields an automorphism of $\mathbf{I}$.

**Corollary 5.** $\varphi(\langle a,p\rangle) = \langle \varphi_s(a), \varphi_k(p)\rangle$.

Thus every automorphism $\varphi$ of $\mathbf{P}$ induces an automorphism $\varphi_k$ of $\mathbf{I}$ by its action on the heights of points.

**Lemma 1.** *If $\varphi \in Aut(\mathbf{P})$ then $\varphi(\langle 0,p\rangle) = \langle 0, \varphi_k(p)\rangle$.*

*Proof.* Writing $\langle 0,p\rangle = \langle 0,1\rangle \sqcap \langle 0,p\rangle$, we get

$$\begin{aligned}\varphi(\langle 0,p\rangle) &= \varphi(\langle 0,1\rangle) \sqcap \varphi(\langle 0,p\rangle) = \langle 0,1\rangle \sqcap \langle \varphi_s(0), \varphi_k(p)\rangle \\ &= \langle 0, \varphi_k(p)\rangle\end{aligned}$$

**Theorem 6.** *Automorphisms of $\mathbf{P}$ are of the form $\varphi(\langle a,p\rangle) = \langle \alpha(a), \beta(p)\rangle$ for $\alpha, \beta \in Aut(\mathbf{I})$.*

*Proof.* Let $\varphi \in Aut(\mathbf{P})$ and write $\langle a,p\rangle = \langle a,1\rangle \sqcup \langle 0,p\rangle$. Then

$$\begin{aligned}\varphi(\langle a,p\rangle) &= \varphi(\langle a,1\rangle) \sqcup \varphi(\langle 0,p\rangle) = \langle \varphi_s(a), 1\rangle \sqcup \langle 0, \varphi_k(p)\rangle \\ &= \langle \varphi_s(a), \varphi_k(p)\rangle\end{aligned}$$

By the results above, both $\varphi_s$ and $\varphi_k$ are automorphisms of the unit interval, and $\varphi = (\varphi_k)_L \left(\varphi_s^{-1}\right)_R$.

This establishes a one-to-one correspondence between $Aut(\mathbf{P})$ and $Aut(\mathbf{I})^2$. It is routine to show that this correspondence preserves composition of functions. It is also easy to see that $\varphi$ commutes with $*$ if and only if $\varphi_s$ commutes with $'$. This gives the following:

**Corollary 6.** *$Aut(\mathbf{P}) \approx Aut(\mathbf{I})^2$ and $Aut(\mathbb{P}) \approx Aut(\mathbb{I}) \times Aut(\mathbf{I})$.*

## 6 Summary

In this paper we investigated a special subalgebra of the truth value algebra of type-2 fuzzy sets, namely, the algebra $\mathbf{P}$ of points which generalizes type-1 fuzzy sets We showed that this subalgebra of $\mathbf{M}$, and that its automorphism group is isomorphic to $Aut(\mathbf{I}) \times Aut(\mathbf{I})$, where $\mathbf{I}$ is the unit interval with the usual order. These same facts hold for these algebras with the usual negation added to their type.

Computations in this algebra of points are simple compared to general type-2 computations, and there could be some interesting applications of fuzzy sets with $\mathbf{P}$ as the algebra of truth values. The additional parameter, which we called height, is a conservative measure, as the height of the meet or join of finitely many of these functions has the minimum height.

With universal set $\{0, 1, 2, 3, 4, 5\}$, the fuzzy subset "close to 0" might be modeled by the function

$$\begin{bmatrix} 0 & \langle 1,1\rangle \\ 1 & \langle 1,0.8\rangle \\ 2 & \langle 0.8,1\rangle \\ 3 & \langle 0.4,0.8\rangle \\ 4 & \langle 0.2,0.8\rangle \\ 5 & \langle 0,1\rangle \end{bmatrix}$$

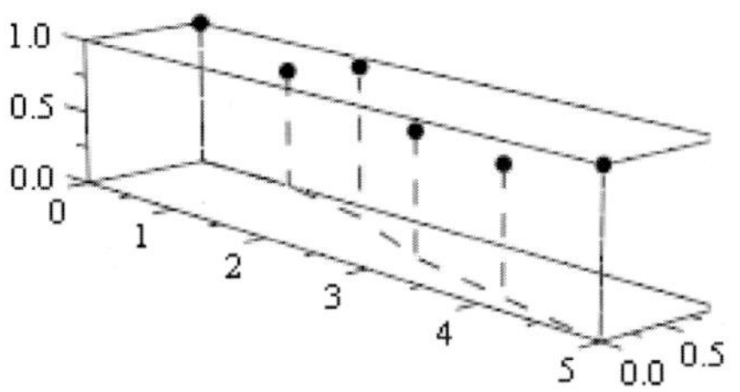


(sometimes written as $\langle 1,1\rangle/0 + \langle 1,0.8\rangle/1 + \langle 0.8,1\rangle/2 + \langle 0.4,0.8\rangle/3 + \langle 0.2,0.8\rangle/4 + \langle 0,1\rangle/5$). Operations between two such fuzzy sets depend only on the elements in the range. So, for example, "close to 0" or "close to 3" might be

$$\begin{bmatrix} 0\ \langle 1,1\rangle \\ 1\ \langle 1,0.8\rangle \\ 2\ \langle 0.8,1\rangle \\ 3\ \langle 0.4,0.8\rangle \\ 4\ \langle 0.2,0.8\rangle \\ 5\ \langle 0,1\rangle \end{bmatrix} \sqcup \begin{bmatrix} 0\ \langle 0,1\rangle \\ 1\ \langle 0.2,0.8\rangle \\ 2\ \langle 0.8,1\rangle \\ 3\ \langle 1,1\rangle \\ 4\ \langle 0.8,1\rangle \\ 5\ \langle 0.2,0.8\rangle \end{bmatrix} = \begin{bmatrix} 0\ \langle 1,1\rangle \\ 1\ \langle 1,0.8\rangle \\ 2\ \langle 0.8,1\rangle \\ 3\ \langle 1,0.8\rangle \\ 4\ \langle 0.8,0.8\rangle \\ 5\ \langle 0.2,0.8\rangle \end{bmatrix}$$

In this toy example, it is not very intuitive why the doubt about the degree to which 3 is close to 0 should affect the confidence in which 3 is close to 3, but it is our hope that practitioners might find situations that this setting models in a useful way.

## References


1. Brzozowski, J.A., De Morgan Bisemilattices, ISMVL, 30th IEEE International Symposium on Multiple-Valued Logic (2000) 173.
2. Gehrke, M., Walker, C., and Walker, E., Some Comments on Interval-Valued Fuzzy Sets, *International Journal of Intelligent Systems* 11(1996) 751-759.
3. McKenzie, R., McNulty, G., Taylor, W., *Algebras, Lattices, Varieties, Vol. I*, Wadsworth, Inc., Belmont, California (1987).
4. Walker, C. and Walker, E., Algebraic Structures on Fuzzy Sets of Type-2, *Proceedings of the International Conference on Fuzzy Information Processing*, Vol. 1, Liu, Chen, Ying, and Cai, eds., (FIP 2003), pp. 97-100.
5. Walker, C. and Walker, E., Algebraic Properties of Type-2 Fuzzy Sets, *Proceedings of the Third International conference on Intelligent Technologies and Third Vietnam-Japan Symposium on Fuzzy Systems and Applications*, Hanoi Vietnam, December 3-5, 2002, Nguyen Hoang Phuong, Hung T. Nguyen, Nguyen Cat Ho, Pratit Santiprabhob, eds., (InTech/VJFuzzy'2002), p. 9.
6. Walker, C., and Walker, E., The Algebra of Fuzzy Truth Values, *Fuzzy Sets and Systems*, 149 (2005) 309-347.
7. Walker, C., and Walker, E., Automorphisms of the Algebra of Fuzzy Truth Values, *International Journal of Uncertainty, Fuzziness and Knowledge-based Systems*, 14(6) (2006) 711-732.
8. Walker, C., and Walker, E., Automorphisms of the Algebra of Fuzzy Truth Values II, preprint
9. Zadeh, L., The Concept of a Linguistic Variable and Its Application to Approximate Reasoning, *Inform Sci.* 8 (1975) 199-249.

# Part II

# Intuitionistic Fuzzy Sets and Their Applications

# Atanassov's Intuitionistic Fuzzy Sets as a Classification Model

J. Montero[1], D. Gómez[2], and H. Bustince[3]

[1] Facultad de Matemáticas, Universidad Complutense, Madrid, Spain
monty@mat.ucm.es
[2] Escuela de Estadística, Universidad Complutense, Madrid, Spain
dagomez@estad.ucm.es
[3] Automática y Computación, Universidad Pública de Navarra, Pamplona, Spain
bustince@unavarra.es

**Abstract.** In this paper we show that Atanassov's Intuitionistic Fuzzy sets can be viewed as a classification model, that can be generalized in order to take into account more classes than the three classes considered by Atanassov's (membership, non-membership and non-determinacy). This approach will imply, on one hand, to change the meaning of these classes, so each one will have a positive definition. On the other hand, this approach implies the possibility of a direct generalization for alternative logics and additional valuation states, being consistent with Atanassov's focuss. From this approach we shall stress the absence of any structure within those three valuation states in Atanassov's model. In particular, we consider this is the main cause of the dispute about Atanassov's model: acknowledging that the name *intuitionistic* is not appropriate, once we consider that a crisp direct graph is defined in the valuation space, formal differences with other three-state models will appear.

**Keywords:** Atanassov's Intuitionistic Fuzzy Sets, Interval Valued Fuzzy Sets, Type-2 Fuzzy Sets, $L$-Fuzzy sets.

## 1 Introduction

The fuzzy scientific community has been attending with great interest the recent dispute about Atanassov's *Intuitionistic* Fuzzy Sets [6,14]. This model was originally proposed by Atannassov [4] as a generalization of Zadeh's Fuzzy Sets [27].

According to Atanassov [4,5], given a set of objects $X$, each object $x \in X$ is being described by means of the degree of membership, $\mu(x) \in [0,1]$, together with the degree of non-membership, $\nu(x) \in [0,1]$, not imposing as Zadeh that these two values should sum up to 1 (i.e., Atanassov does not assumes that the degree of non-membership is the standard negation of the degree of membership, $\nu(x) = 1-\mu(x), \forall x \in X$). On the contrary, a remaining degree on non-determinacy (*hesitation* margin) is allowed by Atanassov, $\pi(x) = 1-\mu(x)-\nu(x) \in [0,1], \forall x \in X$.

Atanassov's model has deserved a serious consideration from theoretical and applied researchers: as pointed out in [7], his papers have more than 1000 references in scientific papers, see also [5]. Nevertheless, we should acknowledge

P. Melin et al. (Eds.): IFSA 2007, LNAI 4529, pp. 69–75, 2007.
© Springer-Verlag Berlin Heidelberg 2007

that the term *intuitionistic* is not appropriate in Atanassov's model (see [9]). A proper *intuitionistic* model was proposed in [23].

Moreover, Atanassov's proposal is not fully clear about the meaning and estimation of the degrees of non-membership and non-determinacy (see, e.g., [14]).

In addition, Atanassov's model, see [13,17], is equivalent to *interval valued fuzzy sets* [22,24], which are defined by the family of mappings

$$\mu : \mathcal{X} \longrightarrow Int([0,1])$$

where $Int([0,1])$ is the set of all closed subintervals in $[0,1]$ and $\mu(x)$ represents the plausible range of the degree to which object $x \in \mathcal{X}$ verifies a certain fuzzy property.

This paper is organized as follows: in section 2 we propose to consider Atanassov's model as a classification model with two main valuation states, plus non-determinacy. In section 3 we propose a direct generalization of his model, allowing more than two main valuation states and providing such a valuation space with a directed graph. In section 4 we stress the consequences of the non-existence of such a graph in Atanassov's model, which we consider the main cause of the confusion with other models. Finally, we conclude with some additional comments about other Atanassov's assumptions that help to explain why such a demanded structure does not seem relevant under his approach.

## 2 Partition-Based Classification Models

Following [2], we consider here a finite valuation space $\mathcal{C}$ of *valuation states*, fuzzy in nature but well defined from a representation point of view, in such a way that for each object $x \in X$ we can evaluate the degree $\mu_x(c) \in [0,1]$ to which such an object verifies properties defining each class $c \in \mathcal{C}$. No restriction is imposed by definition on these degrees of membership, but the possibility of modifying those values through learning (see also [1]).

In order to be properly defined, each one of those valuation spaces (perhaps represented by a linguistic label), should be positively defined. From this point of view, Atanassov's *non-membership* should be changed into a dual or opposite class, different than negation. For example, the opposite of *tall* is *short* and the opposite of *good* is *bad.* Neither negation of *tall* is *short* or the negation of *good* is *bad.*

Let us remind here that the concept of fuzzy partition, introduced by Ruspini [21] in order to generalize the classical crisp partition concept, assumes the existence of a discrete family $\mathcal{C}$ of *classes*, in such a way that

$$\sum_{c \in \mathcal{C}} \mu_c(x) = 1, \forall x \in X$$

holds. Each object $x \in X$ may belong to several classes -to certain degrees-, and the total degree of membership is distributed among all classes (a crisp partition will appear whenever $\mu_c(x) \in \{0,1\}, \forall c \in \mathcal{C}, \forall x \in X$).

Hence, we can characterize Atanassov's model by means of a mapping

$$\mu : \mathcal{X} \longrightarrow [0,1] \times [0,1]$$

meanwhile its two coordinates never sum more that 1. But notice that Atanassov is giving a true role to *non-determinacy*. Hence, Atanassov should be more likely introduced as a mapping

$$\mu : \mathcal{X} \longrightarrow [0,1] \times [0,1] \times [0,1]$$

which defines a Ruspini's partition [21] on $\mathcal{X}$, i.e.,

$$\mu(x) + \nu(x) + \pi(x) = 1, \forall x \in x$$

The introduction of such an extra *non-determinacy* state is from our point of view a key characteristic of Atanassov's approach. We propose here to associate *ignorance* to this state, in such a way that whenever there is no information the degree of membership is concentrated in this state, from which learning will evolve (decreasing the degree of *ignorance* and increasing the degrees of membership to the main valuation states). *Ignorance* should be the initial state of a necessary learning process about degrees of membership.

More important is to realize that those *valuation states* are related between them, defining a particular structure. Quite often the main valuation states define a linear ordering.

In particular, it should be noted that most of us assume as granted a linear order within those two main classes Atanassov considers (*membership* is being associated to 1 which is higher than 0, which is associated to *non-membership*). One may think this is not a relevant issue when only two classes are taken into account, but we are stressing that Atanassov's model refers to three classes, and moreover, what happens if the number of states is bigger? For example, we could introduce a middle state between *membership* and *non-membership*, as in Lukasiewicz logic. The introduction of an extra evaluation state brings a question about where this new state should be allocated with respect to the other evaluation states. This structure must be specified, as the structure of objects (see, e.g., [18,19]).

Therefore (see [20] for more details), we claim here that more attention should be devoted to those type-2 fuzzy sets given by a mapping

$$\mu : \mathcal{X} \to [0,1]^{\mathcal{C}}$$

where

A1 $\mathcal{X}$ is a well-defined non-empty, but finite, set of *objects* such that
- A1.1 There exists a crisp directed graph $(\mathcal{X}, \mathcal{P})$ showing physical immediacy between two distinct objects $x, y \in \mathcal{X}$ ($p_{xy} = 1$ in case there is immediacy between $x, y \in \mathcal{X}$ and $p_{xy} = 0$ otherwise).
- A1.2 There exists a logic on $\mathcal{X}$ allowing a consistent evaluation of questions about objects.

A2 $\mathcal{C}$ is a finite valuation space, with at least three elements, such that

A2.1 There is a crisp directed graph $(\mathcal{C}, \mathcal{R})$, in such a way that $r_{ij} = 1$ in case there is immediacy between $i, j \in \mathcal{C}$ and $r_{ij} = 0$ otherwise.

A2.2 There exists a logic in $\mathcal{C}$, allowing a consistent evaluation of questions about valuation states.

A3 There exists an *ignorance* state $I \in \mathcal{C}$ such that

A3.1 For every $i \in \mathcal{C}$ there exist a path connecting $I$ with $i$.

A3.2 $\mu_x(I) = 1, \forall x \in \mathcal{X}$, and $\mu_x(i) = 0, \forall x \in \mathcal{X}, \forall i \neq I$, when there is no available information (complete ignorance).

Notice that we are not specifying any of the above two logics for objects and classes, although we must point out that results must be consistent. Moreover, according to [1], they should allow the quality evaluation of the classification we have obtained. Of course there is a variety of possible consistent logics supporting a more general notion of partition than the one due to Ruspini [21] (see, e.g., [10,12]), but notice we are imposing almost no restriction in our model. Most standard assumptions can be possible, if existing and desired, only after a more or less long learning process.

The key argument at this stage is to note that no operation is allowed between non related objects (under the recursive arguments of [3,11], for example, we should be considering only chains of immediate values). Most authors use to assume that disjunction and conjunction should be based on a unique t-conorm and a unique t-norm, respectively (see [8,16]), so calculus is commutative and associative. But aggregation based upon OWA operators [26] or uninorms [25] can also be an alternative to be considered.

Since most authors do not refer to any structure for objects we can assume, by default, the following condition on the family of objects.

B1 $\mathcal{X}$ is a well-defined non-empty, but finite, set of *objects* such that its associated crisp directed graph $(\mathcal{X}, \mathcal{P})$ verifies that $p_{xy} = 0, \forall x \neq y$, so reference to the structure of objects can be avoided (no immediacy between two objects).

Analogously, most authors aggregate states without taking care of any structure. So we can assume, by default, the following condition on the family of classes.

B2 $\mathcal{C}$ is a finite valuation space, with at least three elements, such that its associated crisp directed graph $(\mathcal{C}, \mathcal{R})$ verifies that for all $i \neq j$, either $r_{ij} = 1$ holds or $r_{ji} = 1$ holds (or both), so reference to the structure of states can be partially avoided (there is immediacy between every two valuation states).

A common family of structures for valuation states will be the *linear structure*, where $\mathcal{C} - \{I\}$ defines a linear order. We can find in the literature many alternative valuation structures fitting our model. For example, depending on the role we associate to the *average* state in the standard 5-value scale (*None, Poor, Average, Very and Complete*), we may need an additional *ignorance* state in order to fit our model and Atanassov's approach.

Classical 4 epistemic valuation structure with four states (truth "$T$", falsehood "$F$", ignorance "$I$" and contradiction "$C$", see e.g. [15]) is a nice example in order to show that the structure connecting states should not be avoided. There are several possible alternatives on how states are related between them, and our classification problem will change if such a directed graph is modified. Such a directed graph is related to learning, and it is usually needed in order to fix meaning of each state (notice anyway that the *ignorance* state plays a different role than the main valuation states or classes).

## 3 Atanassov Fuzzy Sets

The absence of a directed graph in Atanassov's model does not allow, from our point of view, to fix the exact meaning of his approach, producing as we shall show a natural confusion.

For example, we can find in the literature three relevant models based on a three-stated valuation space: Atanassov's fuzzy sets, interval valued fuzzy sets and Lukasiewicz logic. But the relation between these three states in interval valued fuzzy sets and states in Lukasiewicz model is different than their relation in Atanassov's model. Each one of these three models shows a middle state, but each one of these middle states is "in the middle" of each valuation space in a different way. It will be again important to realize that disjunction (a part of the required logic) between the two non-ignorance states may be not allowed, in case these two states are not connected (in this case the question if the negation of ignorance equals the disjunction of the two main valuation states has no sense, since this second value is not defined). Notice for example that the initial estimation of a degree of membership within interval valued fuzzy sets should be associated to the whole interval, posterior information will allow to declare that certain values are either *too big* or *too small*, but these two extreme states do not show a natural connection in the standard real line representation.

## 4 Final Comments

First of all, note that, on one hand, Atanassov's non-membership has been associated here to duality, so both main valuation states are positively defined (let us remind again that *short* is the dual notion for *tall*, and in no way *short* should be confused with *not tall*).

On the other hand, Atanassov's *indeterminacy* has been also given a particular meaning (ignorance state).

Then, notice that Atanassov does not make any reference to object connectives, so condition B1 applies. Condition B2 also applies (objects are not connected but states are all connected).

Hence, we can maintain Atanassov's approach allowing a more general space of states, and still we should choose a particular directed graph in this space.

Another choice we have to make is about the logic within the space of states. Atanassov assumes Lukasiewicz logic, which is applied once by definition the three degrees of membership sum up to 1. Under this condition it seems that Atanassov does not need to make explicit the underlying structure. But in fact both main (non-ignorance) valuation states are aggregated in his model, so they are assumed to be connected. Otherwise their aggregation should not be evaluated (this is the case in interval valued fuzzy Sets).

If Atanassov's approach is extended into this more general context, we would be concluding that both interval valued fuzzy sets and his own model are different particular cases. Confusion can be therefore justified because of a number of particular choices introduced in his original model. In addition, not specifying the structure of the state of spaces does not allow in our opinion to fix meanings, so depending on the context different intuitions may appear in practical situations.

**Acknowledgments.** This paper has been partially supported by the National Science Foundation of Spain, grant TIN2006-06290.

## References

1. A. Amo, D. Gómez, J. Montero and G. Biging (2001): *Relevance and redundancy in fuzzy classification systems.* Mathware and Soft Computing 8:203–216.
2. A. Amo, J. Montero, G. Biging and V. Cutello (2004): Fuzzy classification systems. *European Journal of Operational Research* 156:459–507.
3. A. Amo, J. Montero, and E. Molina (2001): *Representation of consistent recursive rules.* European Journal of Operational Research 130:29–53.
4. K.T. Atanassov (1983): *Intuitionistic fuzzy sets.* In: V. Sgurev, ed., VII ITKR's Session, Sofia, June 1983 (deposed in Central Science and Technical Library, Bulgarian Academy of Sciences, 1697/84, in Bulgarian).
5. K.T. Atanassov (1999): *Intuitionistic Fuzzy sets*, Physica-Verlag., Heidelberg, New York.
6. K.T. Atanassov (2005): *Answer to D. Dubois, S. Gottwald, P. Hajek, J. Kacprzyk and H. Prade's paper, Terminological difficulties in fuzzy set theory - the case of intuitionistic fuzzy sets.* Fuzzy Sets and Systems 156:496–499.
7. K.T. Atanassov (in press): A personal view on intuitionistic fuzzy sets. In H. Bustince, F. Herrera and J. Montero, eds.: *Fuzzy Sets and Their Extensions: Representation, Aggregation and Models*, Springer Verlag, Berlin.
8. T. Calvo, A. Kolesarova, M. Komornikova and R. Mesiar, Aggregation operators (2002): properties, classes and construction methods. In T. Calvo, G. Mayor and R. Mesiar, Eds.: *Aggregation Operators*, Springer; 3–104.
9. G. Cattaneo and D. Ciucci (in press): *Basic intuitionistic principles in fuzzy set theories and its extensions (a terminological debate on Atanassov IFS).* Fuzzy sets and Systems.
10. P. Cintula (2005): *Basics of a formal theory of fuzzy partitions.* Proceedings EUSFLAT'05 Conference, Technical University of Catalonia, Barcelona; pp. 884–888.
11. V. Cutello and J. Montero (1999): Recursive connective rules. *Int. J. Intelligent Systems* 14:3–20.
12. B. De Baets and R. Mesiar (1998): T-partitions. *Fuzzy Sets and Systems* 97: 211–223.
13. G. Deschrijver and E.E. Kerre (2003): On the relationship between some extensions of fuzzy sets theory. *Fuzzy Sets and Systems* 133:227–235.

14. D. Dubois, S. Gottwald, P. Hajek, J. Kacprzyk and H. Prade (2005): Terminological difficulties in fuzzy set theory - the case of intuitionistic fuzzy sets. Fuzzy Sets and Systems 156:485–491.
15. P. Fortemps and R. Slowinski (2002): *A graded quadrivalent logic for ordinal preference modelling: Loyola-like approach.* Fuzzy Optimization and Decision making 1:93–111.
16. E.P. Klement, R. Mesiar and E. Pap (2002): *Triangular Norms*, Kluwer Academic Publishers, Dordrecht.
17. J.M. Mendel (2007): *Advances in type-2 fuzzy sets and systems.* Information Sciences 117:84–110.
18. J. Montero (1986): Comprehensive fuzziness. *Fuzzy Sets and Systems* 20:89–86.
19. J. Montero (1987): Extensive fuzziness. *Fuzzy Sets and Systems* 21:201–209.
20. J. Montero, D. Gómez and H. Bustince (2006): *On the relevance of some families of fuzzy sets.* Technical report.
21. E.H. Ruspini (1969): A new approach to clustering. *Information and Control* 15: 22–32.
22. R. Sambuc (1975): Function $\Phi$-flous, application a l'aide au diagnostic en pathologie thyroidienne. *These de doctorat en Medicine*, Marseille.
23. G. Takeuti and S. Titani (1984): *Intuitionistic fuzzy logic and intuitionistic fuzzy set theory.* Journal of Symbolic Logic 49:851–866.
24. I.B. Türksen (1986): Interval valued fuzzy sets based on normal forms. *Fuzzy Sets and Systems* 20:191–210.
25. R.R. Yager and A. Rybalov (1996): Uninorm aggregation operators. *Fuzzy Sets and Systems* 80:111–120.
26. R.R. Yager (1988): On ordered averaging aggregation operators in multicriteria decision making. *IEEE Transactions on Systems, Man and Cybernetics* 18: 183–190.
27. L.A. Zadeh (1965): *Fuzzy sets. Information and control* 8:338–353.

# Classification with Nominal Data Using Intuitionistic Fuzzy Sets

Eulalia Szmidt and Janusz Kacprzyk

Systems Research Institute, Polish Academy of Sciences
ul. Newelska 6, 01–447 Warsaw, Poland
{szmidt,kacprzyk}@ibspan.waw.pl

**Abstract.** The classical classification problem with nominal data is considered. First, to make the problem practically tractable, some transformation into a numerical (real) domain is performed using a frequency based analysis. Then, the use of a fuzzy sets based, and – in particular - an intuitionistic fuzzy sets based technique is proposed. To better explain the procedure proposed, the analysis is heavily based on an example. Importance of the results obtained for other areas exemplified by decision making and case based reasoning is mentioned.

## 1 Introduction

We deal with a classification problem, a "meta-problem" in many areas, notably computer science, decision making, etc., with nominal (categorical) data. In nominal data, names (usually belonging to a small set) are assigned to objects as labels. Initially, the only comparison possible is that if the names are the same, the two data items belong to the same category, otherwise they are different, i.e. "equality" or "inequality".

A number of approaches to classification with nominal data have been proposed (cf. Bock and Diday [3]). Basically, they boil down to some trickery to obtain a numeric assessment of nominal values, or maybe rather relations between them. Some solutions can also be found in the context of database queries and linguistic database summarization in Zadrożny [24].

For instance, in Li and Biswas [13] a similarity-based agglomerative clustering (SBAC) is proposed based on a similarity measure proposed by Goodall for biological taxonomy that gives a greater weight to uncommon feature value matches in similarity computations and makes no assumptions of the underlying distributions of the feature values. An agglomerative algorithm is used to derive a dendrogram, and then using a heuristic technique a partition of the data is extracted. Fountoukis, Bekakos and Kontos [9] present an extension of the well known decision tree approach to classification. Cheng et al. [4] propose how to define a good distance (dissimilarity) measure between patterns with nominal attributes by using adaptive dissimilarity matrices for measuring dissimilarities between nominal values. These matrices are learned via optimizing an error function on training samples. This is different than the conventionally

P. Melin et al. (Eds.): IFSA 2007, LNAI 4529, pp. 76–85, 2007.
© Springer-Verlag Berlin Heidelberg 2007

employed value difference metric (VDM) used to define a real-valued distance measure on nominal values. De Carvalho et al. [5], [6], [7] proposed some proximity measures based on histograms. Ichino and Yaguchi [11] used a Minkowski metric, and then extended their analysis in Ichino, Yaguchi and Diday [12] to obtain a fuzzy classifier. Quinlan's [15] ID3 algorithm proved to be effective to handle both numeric and nominal data but it can be viewed to fail to handle a "topological" aspect of knowledge as it does not consider how sure the classification is, what the most typical example is, etc. To deal with these issues one has to resort to numeric analysis, notably via a similarity/proximity measure. Narazaki and Ralescu [14] proposed an alternative model which involves two stages: the configuration stage mapping the symbolic problem into a numerical domain by devising an appropriate distance measure, and then the classification of examples via the distance measure developed.

Here we propose two alternative approaches to the classification of nominal data attempting to involve merits of those approaches above using fuzzy sets (cf. Zadeh [23] ), and intuitionistic fuzzy sets (cf. Atanasov [1], [2]).

## 2 A Brief Introduction to A-IFSs

One of the possible generalizations of a fuzzy set in $X$ (Zadeh [23]), given by

$$A^{'} = \{< x, \mu_{A'}(x) > | x \in X\} \tag{1}$$

where $\mu_{A'}(x) \in [0, 1]$ is the membership function of the fuzzy set $A^{'}$, is an A-IFS, i.e. Atanassov's intuitionistic fuzzy set, (Atanassov [1], [2]) $A$ given by

$$A = \{< x, \mu_A(x), \nu_A(x) > | x \in X\} \tag{2}$$

where: $\mu_A : X \rightarrow [0, 1]$ and $\nu_A : X \rightarrow [0, 1]$ such that $0 \leq \mu_A(x) + \nu_A(x) \leq 1$, and $\mu_A(x)$, $\nu_A(x) \in [0, 1]$ denote a degree of membership and a degree of non-membership of $x \in A$, respectively.

Obviously, each fuzzy set may be represented by the following A-IFS $A = \{< x, \mu_{A'}(x), 1 - \mu_{A'}(x) > | x \in X\}$. For each A-IFS in $X$, we will call

$$\pi_A(x) = 1 - \mu_A(x) - \nu_A(x) \tag{3}$$

an *intuitionistic fuzzy index* (or a *hesitation margin*) of $x \in A$, and it expresses a lack of knowledge of whether $x$ belongs to $A$ or not (cf. Atanassov [2]). It is obvious that $0 \leq \pi_A(x) \leq 1$, for each $x \in X$.

An A-IFS gives us an additional degree of freedom, i.e. a possibility to represent more aspects of imperfect knowledge – cf. Szmidt and Kacprzyk's papers, given in the references, where applications of intutionistic fuzzy sets to group decision making, negotiations, etc. are presented.

Distances are clearly of utmost importance, and to be more specific we will use the normalized Euclidean distance between intuitionistic fuzzy sets $A, B$ in

$X = \{x_1, x_2, \ldots, x_n\}$ (Szmidt and Kacprzyk [17],[20]):

$$q_{IFS}(A,B) = (\frac{1}{2n}\sum_{i=1}^{n}(\mu_A(x_i) - \mu_B(x_i))^2 + (\nu_A(x_i) - \nu_B(x_i))^2 + \\ + (\pi_A(x_i) - \pi_B(x_i))^2)^{\frac{1}{2}} \quad (4)$$

where, clearly, $q_{IFS}(A,B) \in [0,1]$; we can also use different normalized distances (cf. Szmidt and Kacprzyk [17],[20])).

## 3 A Brief Description of the Problem

For clarity we present the problem considered, i.e. classification with nominal data, using the famous Quinlan's [15] example, the so-called "Saturday Morning". We have objects described by attributes. Each attribute measures a feature and takes on discrete, mutually exclusive values. For example, if the objects were "Saturday Mornings" and the classification involved the weather, possible attributes might be [15]:

- **outlook**, with values {sunny, overcast, rain},
- **temperature**, with values {cold, mild, hot},
- **humidity**, with values {high, normal}, and
- **windy**, with values {true, false},

Taken together, the above attributes provide a zeroth-order language for characterizing objects in the universe (the attributes are nominal). A particular Saturday morning, an *example*, might be described as: outlook: overcast; temperature: cold; humidity: normal; windy: false. Each object (example) belongs to one of mutually exclusive classes, $C$. We assume that there are only two classes, i.e., $C = \{P, N\}$, where: $P$ denotes the set of *positive examples*, and $N$ – that of *negative examples*. There are 14 training examples as shown in Table 1. Each training example $e$ is represented by the attribute-value pairs, i.e., $\{(A_i, a_{i,j}); i = 1, \ldots, l_i\}$ where $A_i$ is an attribute, $a_{i,j}$ is its value – one of possible $j$ values (for each $i$-th attribute $j$ can be different, e.g., for *outlook*: $j = 3$, for *humidity*: $j = 2$ etc.).

First, we propose, making use of the frequency description of the problem (cf. Table 2), to express the data in terms of two fuzzy sets - $Pos^+$ (a fuzzy set describing the positive examples $P$) and $Pos^-$ (a fuzzy set describing the negative examples $N$). In its spirit the method proposed is close to that of De Carvalho et al. [5], [6], [7] who use histograms to derive some proximity measures.

The frequency measure (Table 2) used for description of the data (Table 1):

$$f(A_i, a_{i,j}, C) = V(C; A_i = a_{i,j})/p_C \quad (5)$$

where $C = \{P, N\}$; $V(C; A_i = a_{i,j})$ – the number of training examples of $C$ for which $A_i = a_{i,j}$; $p_C$ – the number of the training examples of $C$.

**Table 1.** The "Saturday Morning" data from Quinlan [15]

| No. | Attributes | | | | Class |
|---|---|---|---|---|---|
| | Outlook | Temperature | Humidity | Windy | |
| 1 | sunny | hot | high | false | N |
| 2 | sunny | hot | high | true | N |
| 3 | overcast | hot | high | false | P |
| 4 | rain | mild | high | false | P |
| 5 | rain | cool | normal | false | P |
| 6 | rain | cool | normal | true | N |
| 7 | overcast | cool | normal | true | P |
| 8 | sunny | mild | high | false | N |
| 9 | sunny | cool | normal | false | P |
| 10 | rain | mild | normal | false | P |
| 11 | sunny | mild | normal | true | P |
| 12 | overcast | mild | high | true | P |
| 13 | overcast | hot | normal | false | P |
| 14 | rain | mild | high | true | N |

**Table 2.** The frequencies obtained

| | Outlook | | | Temperature | | | Humidity | | Windy | |
|---|---|---|---|---|---|---|---|---|---|---|
| | S | O | R | H | M | C | H | N | T | F |
| Positive | 2/9 | 4/9 | 3/9 | 2/9 | 4/9 | 3/9 | 3/9 | 6/9 | 3/9 | 6/9 |
| Negative | 3/5 | 0 | 2/5 | 2/5 | 2/5 | 1/5 | 4/5 | 1/5 | 3/5 | 2/5 |

### 3.1 Solution Via Two Fuzzy Sets

Based on the frequency measure (5) - cf. Table 2, we convert relative frequency distribution functions into fuzzy sets (cf. Yamada's [22]). In effect, the data (Table 2) can be expressed in terms of two fuzzy sets: $Pos^+$ - for the data belonging to class $P$, and $Pos^-$ - for the data belonging to class $N$. The results are given in Table 3. Finally, we classify an example either to $P$ or $N$ assigning to each attribute the values: 1, -1, or 0 according to the following rule:

$$a_{i,j} = \begin{cases} 1 & \text{if} \quad Pos^+ > Pos^- \\ -1 & \text{if} \quad Pos^+ < Pos^- \\ 0 & \quad\ \text{otherwise} \end{cases} \tag{6}$$

Applying rule (6) to the data in Table 3, we assign: $S = -1, O = 1, R = 1$; $H = -1, M = 0, C = 1$; $H = -1, N = 1$; $T = -1, F = 1$, and in result we obtain the description of "Saturday Morning" data in the form given in Table 4.To classify the examples, we calculate the *centers of gravity* $CG(P)$ and $CG(N)$ for both classes ($P$ and $N$, respectively) while the $i$-th element of $CG(C)$ for category $C$ is calculated in the following way:

$$CG(C)_i = \sum_{j=1}^{p} e_{i,j}^{C} / p_C \tag{7}$$

**Table 3.** The possibilities obtained

| | Outlook | | | Temperature | | | Humidity | | Windy | |
|---|---|---|---|---|---|---|---|---|---|---|
| | S | O | R | H | M | C | H | N | T | F |
| $Pos^+$: "Positive" possibilities (for $P$) | 0.67 | 1 | 0.89 | 0.67 | 1 | 0.89 | 0.67 | 1 | 0.67 | 1 |
| $Pos^-$: "Negative" possibilities (for $N$) | 1 | 0 | 0.8 | 1 | 1 | 0.6 | 1 | 0.4 | 1 | 0.8 |

**Table 4.** The "Saturday Morning" example derived by the transformation employed

| No. | Attributes | | | | Class |
|---|---|---|---|---|---|
| | Outlook | Temperature | Humidity | Windy | |
| 1 | -1 | -1 | -1 | 1 | N |
| 2 | -1 | -1 | -1 | -1 | N |
| 3 | 1 | -1 | -1 | 1 | P |
| 4 | 1 | 0 | -1 | 1 | P |
| 5 | 1 | 1 | 1 | 1 | P |
| 6 | 1 | 1 | 1 | -1 | N |
| 7 | 1 | 1 | 1 | -1 | P |
| 8 | -1 | 0 | -1 | 1 | N |
| 9 | -1 | 1 | 1 | 1 | P |
| 10 | 1 | 0 | 1 | 1 | P |
| 11 | -1 | 0 | 1 | -1 | P |
| 12 | 1 | 0 | -1 | -1 | P |
| 13 | 1 | -1 | 1 | 1 | P |
| 14 | 1 | 0 | -1 | -1 | N |

where $C = \{P, N\}$; $e_{i,j}^C$ is the attribute value of $A_i$ of the $j$-th training example of $C$ among its $p_C$ positive training examples (for each class, respectively).

We obtain the following centers of gravity (cf. Table 4):

$$CG(P) = (5/9, 1/9, 3/9, 3/9) \quad CG(N) = (-1/5, -1/5, -3/5, -1/5) \tag{8}$$

Knowing $CG(C)$ we can calculate the normalized Euclidean distances $q(e, CG(C))$ between each example $e$ and the centers of gravity, and then to classify example $e$ according to the following rule:

$$e \in C \text{ if } q(e, CG(C)) = \min_{C_j} q(e, CG(C_j)) \tag{9}$$

where $q(e, CG(C_j)) = (\frac{1}{4}\sum_{i=1}^{4}(e_i - CG(C_j)_i)^2)^{\frac{1}{2}}$; $j$ denotes the class ($P$ or $N$), $i$ is the number of an attribute.

The results are given in Table 5. We can see that in several cases some problems occured. First, it was not possible to classify example 3 because the distances to both centers are for this examples the same.

Next, examples 6 and 12 are incorrectly classified which is an obvious result of the model – the same examples (6 and 7, 12 and 14 – Table 4) belong to different classes.

**Table 5.** The "Saturday Morning" data – the Euclidean distances between the examples and the centers of gravity (cf. the approach proposed in Section 3.1)

| No. | Distance | | Class | |
|---|---|---|---|---|
| | from CG(P) | from CG(N) | learned | real |
| 1 | 0.86 | <u>0.6</u> | N | N |
| 2 | 0.95 | <u>0.51</u> | N | N |
| 3 | 0.68 | 0.68 | ? | P |
| 4 | <u>0.55</u> | 0.62 | P | P |
| 5 | <u>0.49</u> | 0.93 | P | P |
| 6 | <u>0.63</u> | 0.87 | P | N |
| 7 | <u>0.63</u> | 0.87 | P | P |
| 8 | 0.76 | <u>0.53</u> | N | N |
| 9 | <u>0.72</u> | 0.87 | P | P |
| 10 | <u>0.37</u> | 0.83 | P | P |
| 11 | 0.76 | 0.7 | ? | P |
| 12 | 0.69 | <u>0.53</u> | N | P |
| 13 | <u>0.54</u> | 0.87 | P | P |
| 14 | 0.69 | <u>0.53</u> | N | N |

To summarize, the proposed method based on fuzzy sets exhibits some inherent deficiencies, and is not sufficient.

### 3.2 Solution Based on Frequency Measures

The use of histograms and frequency measures for dealing with nominal data is quite popular (cf. De Carvalho and Souza [7]). In our context, Narazaki and Ralescu [14] proposed for the frequency measures given in (5) to assign to each attribute either 1, 0 or -1 due to:

$$a_{i,j} = \begin{cases} 1 & \text{if } f(A_i, a_{i,j}, P) > f(A_i, a_{i,j}, N) \\ -1 & \text{if } f(A_i, a_{i,j}, P) < f(A_i, a_{i,j}, N) \\ 0 & \text{otherwise} \end{cases} \tag{10}$$

Using (10) the following assignments were done: $S = -1$, $O = 1$, $\underline{R = -1}$; $H = -1$, $\underline{M = 1}$, $C = 1$; $H = -1$, $N = 1$; $T = -1$, $F = 1$, and we have underlined the attributes the values of which are different from those assigned in Section 3.1 (previously, the values assigned were: $R = 1$, $M = 0$). The counterpart centers of gravity (different from (8)) are:

$$CG(P) = (-1/9, 5/9, 3/9, 3/9) \quad CG(N) = (-1, 1/5, -3/5, -1/5) \tag{11}$$

An example is classified due to the rule

$$e \in C \text{ if } m(e, C) = \max_{C_j} \, m(e, C_j) \tag{12}$$

**Table 6.** The memberships values (12) – cf. Narazaki and Ralescu [14].

| No. | 1 | 2 | 3 | 4 | 5 | 6 | 7 | 8 | 9 | 10 | 11 | 12 | 13 | 14 |
|---|---|---|---|---|---|---|---|---|---|---|---|---|---|---|
| m(e,C) | 0.43 | 0.37 | 0.52 | <u>0.46</u> | 0.61 | <u>0.52</u> | 0.59 | <u>0.46</u> | 0.61 | 0.61 | <u>0.52</u> | <u>0.51</u> | 0.59 | 0.36 |
| Class: learned | N | N | P | <u>N</u> | P | <u>P</u> | P | N | P | P | P | P | P | N |
| Class: real | N | N | P | P | P | N | P | N | P | P | P | P | P | N |

**Table 7.** The counterpart intuitionistic fuzzy model

| | Outlook | | | Temperature | | | Humidity | | Windy | |
|---|---|---|---|---|---|---|---|---|---|---|
| | S | O | R | H | M | C | H | N | T | F |
| Hesitation margins | 0.67 | 0 | 0.69 | 0.67 | 1 | 0.49 | 0.67 | 0.4 | 0.67 | 0.8 |
| membership values | 0 | 1 | 0.2 | 0 | 0 | 0.4 | 0 | 0.6 | 0 | 0.2 |
| non-membership values | 0.33 | 0 | 0.11 | 0.33 | 0 | 0.11 | 0.33 | 0 | 0.33 | 0 |

where $m(e, C) = q(e, CG(C))/\sum_j q(e, CG(C_j))$. The advantage is that all the examples are univocally described. The final results are presented in Table 6. Formally, only examples 4 and 6 are incorrectly classified. But, as a result of a deeper analysis outlined in Narazaki and Ralescu [14], five examples were pointed out (the examples for which the values of $m(e, C)$ are underlined in Table 6) as belonging neither to positive nor to negative ones. Existence of such doubtful "disguised" cases, even if finally classified correctly, speaks against the method e.g., if applied in case based reasoning while only the reliable examples should be considered.

### 3.3 Solution Via Intuitionistic Fuzzy Sets

Now we will use A-IFSs to describe and classify the "Saturday Morning" data. First, we use an algorithm proposed in Szmidt and Baldwin [16] to assign the parameters of an A-IFS model which describes the attributes (with the relative frequency distributions given in Table 2 as the starting point of the algorithm). The results are given in Table 7. A description of attributes in terms of A-IFSs (Table 7) is used for further calculations.

The main idea is to depart from the traditional assignment: 1, 0, -1 as in Sections 3.1 and 3.2, and use the values of all the three functions describing IFSs, i.e. the values of the membership and nonmembership degrees, and of the hesitation margin. So we have a table in which instead of the value "-1" for "sunny" (cf. Table 4), we have a description in terms of intuitionistic fuzzy sets (cf. Table 7), i.e., $(\mu(.), \nu(.), \pi(.))$; for instance, $(0, 0.33, 0.67)$ is for "sunny". After the construction of the table in which the attributes of each example are expressed in terms of the intuitionistic fuzzy sets, we calculate the centers of gravity, $CG(.)$, using the same method as in the previous Sections 3.1 and 3.2.

The centers of gravity, in terms of the A-IFSs, $CG(.) = (Outlook, Temperature, Humidity, Windy)$ are [cf. (7)] are, for the examples from class $P$ and $N$, respectively:

$$CG(P) = ((0.5, 0.1, 0.4), (0.1, 0.1, 0.8), (0.4, 0.1, 0.5), (0.1, 0.1, 0.8)) \tag{13}$$

**Table 8.** The "Saturday Morning" example – results obtained from the intuitionistic fuzzy ( $CG(.)$: (13)–(14))

| No. | Distance | | Class | | Dissimilarity |
|---|---|---|---|---|---|
| | from CG(P) | from CG(N) | learned | real | |
| 1 | 0.151 | 0.067 | N | N | 0.44 |
| 2 | 0.156 | 0.058 | N | N | 0.37 |
| 3 | 0.151 | 0.216 | P | P | 0.7 |
| 4 | 0.130 | 0.079 | N | P | 0.61 |
| 5 | 0.114 | 0.139 | P | P | 0.82 |
| 6 | 0.119 | 0.136 | P | N | 0.88 |
| 7 | 0.145 | 0.245 | P | P | 0.6 |
| 8 | 0.152 | 0.075 | N | N | 0.5 |
| 9 | 0.138 | 0.137 | P | P | 1 |
| 10 | 0.106 | 0.128 | P | P | 0.83 |
| 11 | 0.139 | 0.122 | N | P | 0.9 |
| 12 | 0.16 | 0.22 | P | P | 0.73 |
| 13 | 0.131 | 0.238 | P | P | 0.55 |
| 14 | 0.136 | 0.072 | N | N | 0.53 |

$$CG(N) = ((0.1, 0.2, 0.7), (0.1, 0.1, 0.8), (0.1, 0.3, 0.6), (0.1, 0.2, 0.7)) \quad (14)$$

Comparing the above $CG(P)$ and $CG(N)$ we can notice that attribute 2, (*Temperature*), does not help much in the classification as in both cases the values are the same, i.e. $((0.1, 0.1, 0.8)$ for both $CG(P)$ and $CG(N))$.

The order of the most discriminative attributes (it can be given after calculating the distances (4) between the respective components of the $CG(P)$s and $CG(N)$s) is, from the most to the least discriminative one: *Outlook, Humidity, Windy, Temperature.* This result is fully consistent with the results obtained in [15] while looking for the order of checking the attributes by calculating a statistical property called *information gain.* It can easily be checked that for the centers of gravity obtained in the two previous methods it does not happen.

Having the centers of gravity, we calculate the distances of each example from both the centers of gravity, and classify an example according to the rule:

$$e \in C \text{ if } q(e, CG(C)) = \min_{C_j} q(e, CG(C_j)) \quad (15)$$

where $q(e, CG(C_j))$ is given by (4), $C_j$ denotes the class ($P$ or $N$). The results are given in Table 8 (the distances to the closer $CG(P)$ – significantly closer, i.e. such that the difference between the distances to both $CG(P)$ justifies the classification – are underlined).

Last column of Table 8 contains the measure of *dissimilarity* calculated as:

$$\min_{C_j} q(e, CG(C_j)) / \max_{C_j} q(e, CG(C_j))$$

and a motivation of using this measure is given in [19].

As it can be seen in Table 8, the examples are properly classified if dissimilarity is less than 0.84. This is not fulfilled for examples: 6, 9, and 11 which are incorrectly classified.

The only exception is example 4 (dissimilarity is less than 0.84). The A-IFS approach better renders details concerning small changes of distances to the centers of gravity because it involves three indices.

Notice that both the previous methods had the problem with example 4 which is not "reliable". Even if the fuzzy approach (cf. Table 5) turned out to be successful, i.e. classified the case properly, the distances to both the centers of gravity were almost the same (an "unstable" situation). The frequency based approach (cf. Table 6) gave incorrect classification (but with an advice given in [14] to learn the system to classify the example properly). Only the A-IFS approach definitely pointed out the example as a "troublesome" one – the classification is done incorrectly while using the proper indicators (e.g, the respective distances supporting the incorrect classification). In effect a user obtains a hint not to rely on the example (e.g., in the sense of using it in case based reasoning).

### 3.4 Conclusions

We proposed how to deal with nominal attributes in classification, using – for clarity – a popular example, the so called "Saturday Morning" by Quinlan [15]. We proposed a fuzzy, and then intuitionistic fuzzy approach (cf. Atanassov [2]). We compared the new methods proposed with the one based on frequency analysis due to Narazaki and Ralescu [14].

However, it seems that the effectiveness, measured as the number of correctly classified examples, is not to be considered as the only criterion. The reliability is relevant, and in this respect the use of the intuitionistic fuzzy sets may be justified as we can not only indicate "unstable" cases but to somehow trace sources of that instability. This may be relevant in the context of many applications, notably in case based reasoning when, for instance, correctly classified but unstable cases should not be added to the library of cases.

## References


1. Atanassov K. (1983), Intuitionistic Fuzzy Sets. VII ITKR Session. Sofia (Deposed in Centr. Sci.-Techn. Library of Bulg. Acad. of Sci., 1697/84) (in Bulgarian).
2. Atanassov K. (1999), Intuitionistic Fuzzy Sets: Theory and Applications. Springer.
3. Bock H.-H., Diday E. (2000) Analysis of Symbolic Data. Springer, Heidelberg.
4. Cheng V., Li Ch.-H., Kowk J.T., Li Ch.-K. Dissimilarity learning for nominal data. Pattern Recognition, 37, 1471–1477.
5. De Carvalho F.A.T. (1994), Proximity coefficients between Boolean symbolic objects, in New Approaches in Classification and Data Analysis, Diday et al. (Eds.), Springer, Heidelberg, 387–394.
6. De Carvalho F.A.T. (1998), Extension based proximities between Boolean symbolic objects, in Data Science, Classification and Related Methods, Hayashi, C. et al. (eds.), Springer, Tokyo, 370–378.

7. De Carvalho F.A.T., Souza R.M.C. (1998), Statistical proximity functions of Boolean symbolic objects based on histograms. In : Rizzi, A., Vichi, M., Bock, H.-H. (Eds.) : Advances in Data Science and Classification, Springer, Heidelberg, 391–396.
8. Fisher D., Langley, P. (1986), Conceptual Clustering and its Relation to Numerical Taxonomy. Addison-Wesley Longman, Boston.
9. Fountoukis S.G., Bekasos M.P., Kontos J.P. (2001), Rule extraction from decision trees with complex nominal data. Neural, Parallel & Scient. Comput., 9 , 119–128.
10. Goodall D.W. (1966), A new similarity index based on probability. Biometrics, 22, 882–907.
11. Ichino, M., Yaguchi H. (1994), Generalized Minkowsky metrics for mixed feature type data analysis. IEEE Trans. on Syst., Man and Cybern., 24, 698–708.
12. Ichino M., Yaguchi H., Diday E. (1996), A fuzzy symbolic pattern classifier, In: Ordinal and Symbolic Data Analysis, Diday, E. et al. (Eds.), Springer, 92–102.
13. Li C. and Biswas G. (2002), Unsupervised learning with mixed numeric and nominal data. IEEE Trans. On Knowledge and Data Eng., 14 (4), 673–690.
14. Narazaki H. and Ralescu A. (1994), Iterative induction of a category membership function. Int. J. of Uncert. Fuzziness and Knowledge-Based Systems, 2, (1), 91–100.
15. Quinlan J.R. (1986), Induction of decision trees. Machine Learning, 1, 81–106.
16. Szmidt E. and Baldwin J. (2006), Intuitionistic Fuzzy Set Functions, Mass Assignment Theory, Possibility Theory and Histograms. 2006 IEEE World Congress on Computational Intelligence, 237–243.
17. Szmidt E. and Kacprzyk J. (2000), Distances between intuitionistic fuzzy sets, Fuzzy Sets and Systems, 114 (3), 505–518.
18. Szmidt E., Kacprzyk J. (2001), Entropy for intuitionistic fuzzy sets. Fuzzy Sets and Systems, 118 (3), 467–477.
19. Szmidt E. and Kacprzyk J. (2004), Similarity of intuitionistic fuzzy sets and the Jaccard coefficient. Proc. IPMU 2004, Perugia, 1405–1412.
20. Szmidt E. and Kacprzyk J. (2006), Distances Between Intuitionistic Fuzzy Sets: Straightforward Approaches may not work. 3rd Int. IEEE Conf. "Intelligent Systems", 716–721.
21. Szmidt E. and Kacprzyk J. (2006), An Application of Intuitionistic Fuzzy Set Similarity Measures to a Multi- criteria Decision Making Problem. ICAISC 2006, LNAI 4029, Springer, 314– 323.
22. Yamada K. (2001), Probability–Possibility Transformation Based on Evidence Theory. IFSA–NAFIPS'2001, Vancouver, 70–75.
23. Zadeh L.A. (1965), Fuzzy sets. Information and Control, 8, 338–353.
24. Zadrożny S. (2006), Imprecise queries and linguistic summarisation of the data bases. Exit Publishers, Warsaw (in Polish).

# Intuitionistic Fuzzy Histograms of an Image

Ioannis K. Vlachos and George D. Sergiadis

Aristotle University of Thessaloniki
Faculty of Technology
Department of Electrical & Computer Engineering, Telecommunications Division
University Campus, GR–54124, Thessaloniki, Greece
ivlachos@mri.ee.auth.gr, sergiadi@auth.gr
http://mri.ee.auth.gr

**Abstract.** This paper proposes an automated approach for constructing the intuitionistic fuzzy histograms ($IF$-histograms) of a gray-scale digital image, based on the notion of intuitionistic fuzzy numbers ($IF$-numbers). A method for constructing parametric $IF$-numbers from their fuzzy counterparts using intuitionistic fuzzy generators (IFGs) is also presented, using an entropic optimization criterion. Finally, experimental results demonstrate the ability of the proposed approach to obtain efficiently the $IF$-histograms of gray-scale images.

## 1 Introduction

Since Zadeh introduced fuzzy sets (FSs) theory [1], many theories treating imprecision have been proposed. Among the various extensions of FSs, Atanassov's intuitionistic fuzzy sets (A–IFSs) [2,3,4] provide a flexible and intuitive framework to deal with vagueness originating out of imperfect or/and imprecise information. The sound advantage of A–IFSs is that they are consistent with the human behavior of decision making, expressing the fact that linguistic negation does not always coincides with logical negation.

Digital image processing algorithms can be roughly classified into two main categories; histogram- and pixel-based approaches. Many of the algorithms consider and depend solely on the histogram of the image. Histogram-based techniques are characterized by their simplicity and speed compared to their pixel-based counterparts. Therefore, the concept of histogram, as a descriptor of the underlying statistics of images, is very important in the context of image processing.

In this paper, a novel method for constructing the intuitionistic fuzzy histograms ($IF$-histograms) of digital images is presented. The method is based on the concept of intuitionistic fuzzy numbers ($IF$-numbers) and intuitionistic fuzzy generators (IFGs).

The paper is organized as follows. In Sect. 2 the basic elements of A–IFSs theory are outlined and the concept of intuitionistic fuzzy generators is briefly discussed. Sect. 3 presents the concepts of fuzzy numbers ($F$-numbers) and fuzzy histogram ($F$-histogram), as well as their intuitionistic fuzzy equivalents. An

P. Melin et al. (Eds.): IFSA 2007, LNAI 4529, pp. 86–95, 2007.
© Springer-Verlag Berlin Heidelberg 2007

entropic optimization approach for selecting the optimal $IF$-number to construct the $IF$-histogram is also described. Finally, experimental results are given in Sect. 4, while conclusions are drawn in Sect. 5.

## 2 Atanassov's Intuitionistic Fuzzy Sets

In this section, we briefly describe the basic notions, concepts, and definitions of A–IFSs theory.

**Definition 1.** *An FS $\tilde{A}$ defined on a universe $X$ may be given as [1]*

$$\tilde{A} = \{\langle x, \mu_{\tilde{A}}(x)\rangle | x \in X\} \ , \tag{1}$$

*where $\mu_{\tilde{A}}(x) : X \to [0,1]$ is the membership function of $\tilde{A}$.*

The membership function of $\tilde{A}$ describes the *degree of belongingness* of $x \in X$ in $\tilde{A}$.

**Definition 2.** *An A–IFS $A$ defined on a universe $X$ is given by [2,3,4]*

$$A = \{\langle x, \mu_A(x), \nu_A(x)\rangle | x \in X\} \ , \tag{2}$$

*where*

$$\mu_A(x) : X \to [0,1] \qquad and \qquad \nu_A(x) : X \to [0,1] \ ,$$

*with the condition*

$$0 \leqslant \mu_A(x) + \nu_A(x) \leqslant 1 \ , \tag{3}$$

*for all $x \in X$.*

The values of $\mu_A(x)$ and $\nu_A(x)$ denote the *degree of belongingness* and the *degree of non-belongingness* of $x$ to $A$, respectively. For an A–IFS $A$ in $X$ we call the *intuitionistic index* of an element $x \in X$ in $A$ the following expression

$$\pi_A(x) = 1 - \mu_A(x) - \nu_A(x) \ . \tag{4}$$

We can consider $\pi_A(x)$ as a *hesitancy degree* of $x$ to $A$ [2,3,4]. From (4) it is evident that

$$0 \leqslant \pi_A(x) \leqslant 1 \tag{5}$$

for all $x \in X$.

FSs can also be represented using the notation of A–IFSs. An FS $\tilde{A}$ defined on $X$ can be represented as the following A–IFS

$$A = \{\langle x, \mu_A(x), 1 - \mu_A(x)\rangle | x \in X\} \ , \tag{6}$$

with $\pi_A(x) = 0$ for all $x \in X$.

**Definition 3.** *The complementary set $A^c$ of $A$ is defined as*

$$A^c = \{\langle x, \nu_A(x), \mu_A(x)\rangle | x \in X\} \ . \tag{7}$$

Finally, throughout this paper by $\mathscr{IFS}(X)$ we denote the set of all A–IFSs defined on $X$. Correspondingly, $\mathscr{FS}(X)$ is the set of all FSs on $X$.

### 2.1 Intuitionistic Fuzzy Generators

**Definition 4 (Bustince et al. [5]).** *A function $\varphi : [0,1] \to [0,1]$ will be called intuitionistic fuzzy generator (IFG) if*

$$\varphi(x) \leqslant 1 - x \tag{8}$$

*for all $x \in X$.*

The *first characterization theorem* of IFGs is stated as follows.

**Theorem 1 (Bustince et al. [5]).** *Let $\varphi : [0,1] \to [0,1]$. Then, $\varphi$ is a continuous IFG if and only if there exists a continuous function $f : [0,1] \to [0,1]$ such that*

- *$f(x) \leqslant x$ for all $x \in [0,1]$,*
- *$\varphi(x) = (f \circ N)$ for all $x \in X$,*

*where $N$ denotes the standard negation, $N : [0,1] \to [0,1]$ given by $N(x) = 1 - x$ for all $x \in X$.*

Furthermore, an A–IFS can be constructed from an FS and an IFG according to the following theorem.

**Theorem 2 (Bustince et al. [5]).** *Let $\tilde{A}$ be an FS on the universe $X \neq \emptyset$, and let $\varphi$ be an IFG. Then, the set*

$$A = \{\langle x, \mu_{\tilde{A}}(x), \varphi\left(\mu_{\tilde{A}}(x)\right)\rangle | x_i \in X\} \tag{9}$$

*is an A–IFS on $X$.*

## 3 The Quest for the Intuitionistic Fuzzy Histograms

### 3.1 From Fuzzy Numbers to Fuzzy Histograms

A fuzzy number, hereinafter denoted as $F$-number, $\tilde{g}$ is an FS of the real line that is normal and convex. Different forms of $F$-numbers can be constructed [6,7]. In this paper we consider symmetrical $F$-numbers, which are conceptually suitable to represent the notion of gray level "*approximately* $g$". A symmetrical triangular $F$-number can be defined as

$$\mu_{\tilde{g}}(x) = \max\left\{0, 1 - \frac{|x-g|}{p}\right\}, \tag{10}$$

where the positive real parameter $p$ controls the shape of the $F$-number. Fig. 1(a) illustrates the notion of $F$-number "*approximately* 100" for different values of the parameter $p$.

By modelling gray levels using $F$-numbers, the notion of histogram of a digital image can be extended into a fuzzy setting [8,9]. The fuzzy histogram ($F$-histogram) of a digital image is a sequence $h_A^F(g)$ and is given according to the following definition.

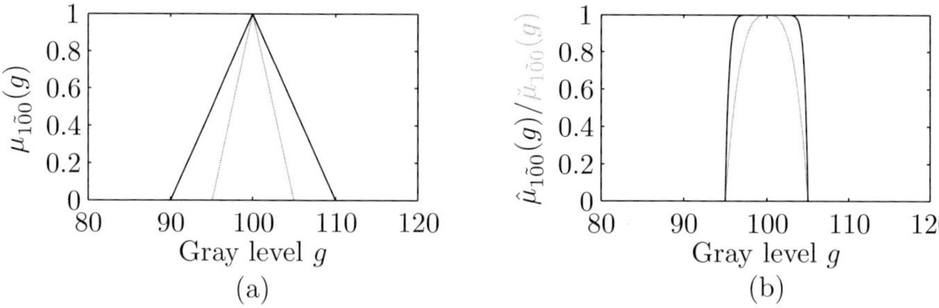


**Fig. 1.** (a) $F$-number "*approximately* 100" with $p = 5$ (gray line) and $p = 10$ (black line) and (b) corresponding $IF$-number for $p = 5$ and $\lambda = 3$. For Fig. 1(b) lower and upper membership functions are denoted by gray and black lines, respectively.

**Definition 5.** *The $F$-histogram of an image $A$ is a sequence $h_A^F(g)$ given by*

$$h_A^F(g) \triangleq \left|\left|\{\langle(i,j), \mu_{\tilde{g}}^A(g_{ij})\rangle | i \in \{1,\ldots,M\},\ j \in \{1,\ldots,N\}\}\right|\right| , \tag{11}$$

*with $g \in \{0,\ldots,L-1\}$, where $||\cdot||$ stands for the cardinality of an FS and $g$ is the gray level of the image.*

Moreover, $h_A^F(g)$ denotes the frequency of occurrence of gray level "*approximately* $g$". It should be stressed out that due to its definition, the $F$-histogram fails to be a probability density function. Therefore, in order for the $F$-histogram to meet the aforementioned requirement the following normalized version is obtained as

$$\bar{h}_A^F(g) = \frac{h_A^F(g)}{\sum_{k=1}^{L-1} h_A^F(k)} , \tag{12}$$

with $g \in \{0,\ldots,L-1\}$, where $L$ is the number of gray levels of the image.

### 3.2 Intuitionistic Fuzzy Numbers

The notion of intuitionistic fuzzy numbers ($IF$-numbers) has been thoroughly studied by many researchers. In this work, we present a method for constructing $IF$-numbers from an $F$-number using the concept of IFGs. A similar approach was employed in [10] for constructing the A–IFS that optimally models the intensity levels of an image.

Let us consider the continuous function $f(x) = x^\lambda$, with $\lambda \geqslant 1$. One can clearly verify that $f(x) \leqslant x$ and $f : [0,1] \to [0,1]$ for all $x \in [0,1]$, under the condition of $\lambda \geqslant 1$. Thus, according to Theorem 1, the following IFG is obtained

$$\varphi(x) = (1-x)^\lambda , \tag{13}$$

where $x \in [0,1]$ and $\lambda \geqslant 1$.

Based on the $F$-number of (10) and considering the IFG of (13), we derive a two-parameter $IF$-number with membership function

$$\mu'_{\tilde{g}}(x) = \max\left\{0, 1 - \frac{|x-g|}{p}\right\} \tag{14}$$

and a non-membership given by

$$\nu'_{\tilde{g}}(x) = \left(1 - \max\left\{0, 1 - \frac{|x-g|}{p}\right\}\right)^{\lambda} . \tag{15}$$

Motivated by the notion of interval-valued fuzzy sets (IVFSs) [11], we employ an alternative, more intuitive, representation of $IF$-numbers, by considering their *upper* and *lower* membership functions, instead of the membership and non-membership ones, described respectively by

$$\check{\mu}'_{\tilde{g}}(x) = \mu'_{\tilde{g}}(x) \tag{16}$$

and

$$\hat{\mu}'_{\tilde{g}}(x) = 1 - \nu'_{\tilde{g}}(x) , \tag{17}$$

for all $x \in X$.

### 3.3 Intuitionistic Fuzzy Histograms

Deriving the optimal $IF$-number for the construction of the $IF$-histogram of the image must encompass optimization of both membership and non-membership functions on the basis of a common parameter. Therefore, taking into account the involution property of fuzzy complements [6,7], the following modified versions of the membership and non-membership functions of (14) and (15) are used, described by

$$\mu_{\tilde{g}}(x;\lambda) = 1 - \left(1 - \max\left\{0, 1 - \frac{|x-g|}{p}\right\}\right)^{\lambda-1} \tag{18}$$

and

$$\nu_{\tilde{g}}(x;\lambda) = \left(1 - \max\left\{0, 1 - \frac{|x-g|}{p}\right\}\right)^{\lambda(\lambda-1)} , \tag{19}$$

with $\lambda \geqslant 1$. Fig. 1(b) depicts the $IF$-number corresponding to the $F$-number "*approximately* 100" of Fig. 1(a) for $p = 5$ and $\lambda = 3$, using the representation involving upper and lower membership functions.

By varying the free parameter $\lambda$ of (18) and (19), different $IF$-numbers can be generated and consequently different $IF$-histograms can be derived. Therefore, an optimization criterion is needed in order to obtain the optimal value of $\lambda$, for which the generated $IF$-histograms model the image under consideration in an optimal way. In the proposed approach we apply a modified version of the *maximum intuitionistic fuzzy entropy principle* introduced in [10].

Among other researchers, Burillo and Bustince [12] were the first to state and propose an axiomatic skeleton of entropy for A–IFS and IVFSs.

**Definition 6 (Burillo and Bustince [12]).** *A real function $E : \mathscr{IFS}(X) \to \mathbb{R}^+$ is called an entropy on $\mathscr{IFS}(X)$, if $E$ has the following properties*

– $E(A) = 0$ *if and only if* $A \in \mathscr{F}\mathscr{S}(X)$,
– $E(A) = Cardinal(X)$ *if and only if* $\mu_A(x) = \nu_A(x) = 0$ *for all* $x \in X$,
– $E(A) = E(A^c)$ *for all* $A \in \mathscr{I}\mathscr{F}\mathscr{S}(X)$,
– $E(A) \geqslant E(B)$ *if* $\mu_A(x) \leqslant \mu_B(x)$ *and* $\nu_A(x) \leqslant \nu_B(x)$ *for all* $x \in X$.

Additionally, they proposed an entropy measure for A–IFSs satisfying their set of axiomatic requirements, given by

$$E(A) = \sum_{x \in X} \pi_A(x) \ , \tag{20}$$

which expresses the *degree of intuitionism* of the set $A$. It should be mentioned that in [13] and [14] different sets of axiomatic requirements were given in order for a measure to be qualified as an entropy in the setting of A–IFSs theory.

Vlachos and Sergiadis [10] introduced the following definition of an image $A$ of size $M \times N$ pixels having $L$ gray levels $g$ ranging between 0 and $L-1$ in the setting of A-IFSs theory, by generalizing its corresponding representation using FSs proposed in [15,16,17].

**Definition 7.** *An image* $A$ *is described by the A–IFS*

$$A = \{\langle g_{ij}, \mu_A(g_{ij}), \nu_A(g_{ij}) \rangle | g_{ij} \in \{0, \ldots, L-1\}\} \ , \tag{21}$$

*with* $i \in \{1, \ldots, M\}$ *and* $j \in \{1, \ldots, N\}$, *where* $\mu_A(g_{ij})$ *and* $\nu_A(g_{ij})$ *denote the degrees of membership and non-membership of the* $(i,j)$*-th pixel to the set* $A$ *associated with an image property.*

Considering the $IF$-numbers of (18) and (19) for modelling the gray levels of the image $A$ and taking also into account that the image can be considered as an array of intuitionistic fuzzy singletons, the entropy of the image can be written as

$$E(A) = \sum_{g=0}^{L-1} h_A(g) E_g(A; g) \ , \tag{22}$$

where

$$E_g(A; g) = \sum_{k=0}^{L-1} h_A(k) \left(1 - \mu_{\tilde{g}}^A(k; g) - \nu_{\tilde{g}}^A(k; g)\right) \ , \tag{23}$$

for each $g \in \{0, \ldots, L-1\}$, with $h_A$ being the crisp histogram of the image.

From (22) and taking into account (23), as well as (18) and (19), we obtain that

$$\begin{aligned} E(A;\lambda) = \sum_{g=0}^{L-1} \Bigg( h_A(g) \sum_{k=0}^{L-1} h_A(k) \Bigg( & \left(1 - \max\left\{0, 1 - \frac{|k-g|}{p}\right\}\right)^{\lambda - 1} \\ & - \left(1 - \max\left\{0, 1 - \frac{|k-g|}{p}\right\}\right)^{\lambda(\lambda-1)} \Bigg) \Bigg) \end{aligned} \ , \tag{24}$$

where the entropy $E(A;\lambda)$ of the image $A$ is also considered as a function of the free parameter $\lambda$. The entropy of (24) serves as the optimization criterion for selecting the optimal parameter $\lambda_{opt}$. This criterion can be formulated as

$$\lambda_{opt} = \arg\max_{\lambda \geqslant 1} \{E(A;\lambda)\} \ . \tag{25}$$

Having obtained $\lambda_{opt}$, the $IF$-histograms of the image are defined as follows.

**Definition 8.** *For an image $A$, the lower (minimum) $IF$-histogram is given by*

$$\check{h}_A^{IF}(g) \triangleq \left|\left|\{\langle (i,j), \mu_{\tilde{g}}^A(g_{ij})\rangle | i \in \{1,\dots,M\},\ j \in \{1,\dots,N\}\}\right|\right| \ , \tag{26}$$

*while the upper (maximum) $IF$-histogram as*

$$\hat{h}_A^{IF}(g) \triangleq \left|\left|\{\langle (i,j), 1-\nu_{\tilde{g}}^A(g_{ij})\rangle | i \in \{1,\dots,M\},\ j \in \{1,\dots,N\}\}\right|\right| \ . \tag{27}$$

Practically, the upper and lower $IF$-histograms are computed using the formulas

$$\check{h}_A^{IF}(g) = \sum_{k=0}^{L-1} h_A(k)\mu_{\tilde{g}}^A(k) \tag{28}$$

and

$$\hat{h}_A^{IF}(g) = \sum_{k=0}^{L-1} h_A(k)\left(1-\nu_{\tilde{g}}^A(k)\right) \ , \tag{29}$$

respectively. It is evident that since $\mu_{\tilde{g}}^A(g) \leqslant 1-\nu_{\tilde{g}}^A(g)$ for all $g \in \{0,\dots,L-1\}$, it follows immediately from (28) and (29) that $\check{h}_A^{IF}(g) \leqslant \hat{h}_A^{IF}(g)$.

Moreover, corresponding normalized $IF$-histograms are defined as

$$\bar{\check{h}}_A^{IF}(g) = \frac{\check{h}_A^{IF}(g)}{\sum_{k=0}^{L-1}\left(\hat{h}_A^{IF}(k) - \check{h}_A^{IF}(k)\right)} \tag{30}$$

and

$$\bar{\hat{h}}_A^{IF}(g) = \frac{\hat{h}_A^{IF}(g)}{\sum_{k=0}^{L-1}\left(\hat{h}_A^{IF}(k) - \check{h}_A^{IF}(k)\right)} \ . \tag{31}$$

As a result, the estimated frequency of occurrence of the gray level "*approximately g*" is given by the interval $[\bar{\check{h}}_A^{IF}(g), \bar{\hat{h}}_A^{IF}(g)]$. We can interpret the above interval as the *possible frequency of occurrence* of intensity level "*approximately g*", with its lower bound corresponding to the *minimum possible frequency of occurrence of gray level "approximately g"* and the upper one to the *maximum possible frequency of occurrence of gray level "approximately g"*. The aforementioned normalization factor was selected in order for the sequence of the lengths of the $[\bar{\check{h}}_A^{IF}(g), \bar{\hat{h}}_A^{IF}(g)]$ intervals, to constitute a probability density function over

**Fig. 2.** Gray-scale test images of size 256 × 256 pixels with 8 bits-per-pixel gray-tone resolution

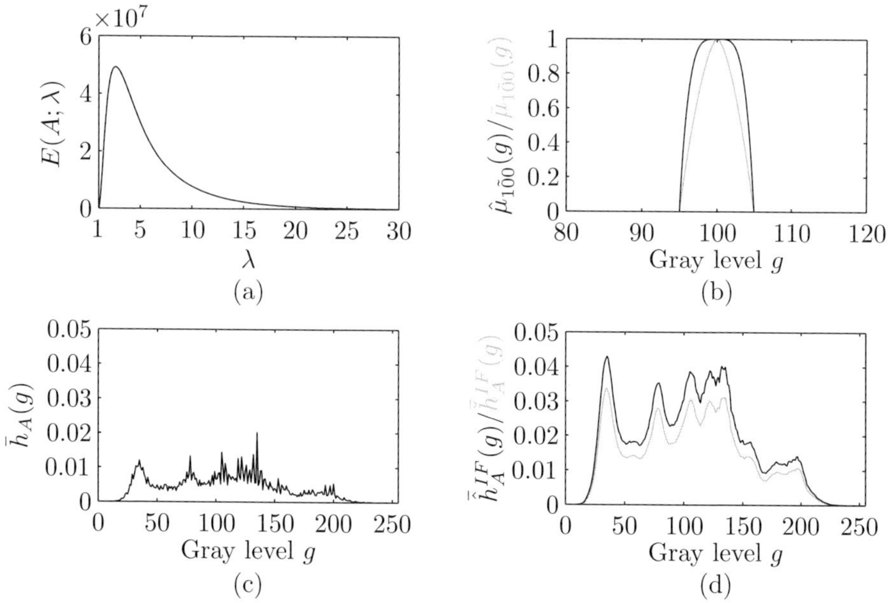


**Fig. 3.** (a) Intuitionistic fuzzy entropy curve for the image of Fig. 2(a), used for obtaining the optimal parameter $\lambda_{opt}$ of the $IF$-numbers for the construction of the $IF$-histograms. (b) $IF$-number expressing the gray level "*approximately* 100" for $p = 5$ and $\lambda_{opt} = 2.62$. Lower and upper membership functions are denoted by gray and black lines, respectively. (c) Normalized crisp histogram for the image of Fig. 2(a). (d) Corresponding lower (gray line) and upper (black line) normalized $IF$-histograms for the aforementioned values of parameters $p$ and $\lambda_{opt}$.

the gray-level range. Finally, it should be mentioned that the difference between $\hat{\bar{h}}_A^{IF}$ and $\check{\bar{h}}_A^{IF}$ allows also for the definition of the *hesitancy histogram* of the image.

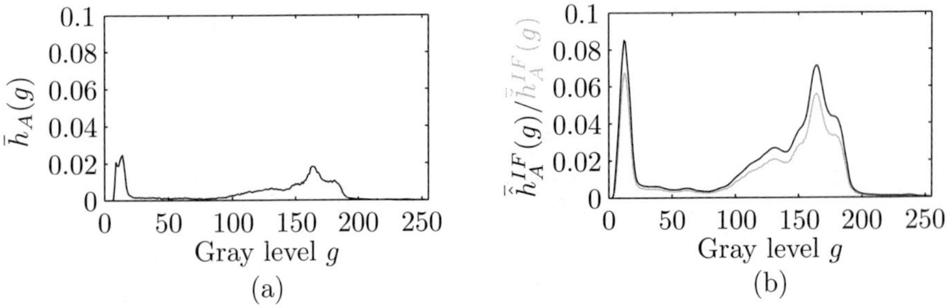


**Fig. 4.** (a) Normalized crisp histogram for the image of Fig. 2(b). (b) Corresponding lower (gray line) and upper (black line) normalized $IF$-histograms for $p = 5$ and $\lambda_{opt} = 2.54$.

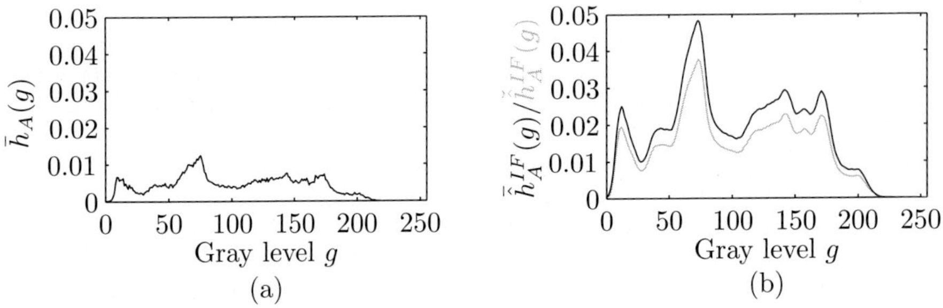


**Fig. 5.** (a) Normalized crisp histogram for the image of Fig. 2(c). (b) Corresponding lower (gray line) and upper (black line) normalized $IF$-histograms for $p = 5$ and $\lambda_{opt} = 2.60$.

## 4 Experimental Results

For the simulations we considered gray-scale images of size $256 \times 256$ pixels with 8 bits-per-pixel gray-tone resolution, shown in Fig. 2. The entropic optimization criterion for obtaining the optimal parameter $\lambda_{opt}$ of the $IF$–numbers is shown in Fig. 3(a). The derived $IF$–number, depicted in Fig. 3(b), is used to model the gray levels, as well as to construct the $IF$–histograms of the image expressing its intuitionistic fuzzy statistics. Finally, Figs. 3(c) and 3(d) illustrate the normalized crisp and $IF$–histograms of the image of Fig. 2(a), while Figs. 4 and 5 the ones corresponding to the images of Figs. 2(b) and 2(c), respectively.

## 5 Conclusions

In this paper we introduced the notion of the $IF$-histograms of a digital image. An entropic optimization method was also proposed for constructing the optimal $IF$-number using the concept of IFGs. Application of the proposed scheme to gray-scale images demonstrated its ability to model efficiently their gray levels, allowing for the extension of image statistics into the intuitionistic fuzzy setting. Since histogram is a key concept in digital image processing, it is expected that its proposed intuitionistic fuzzy extension will find wide application in many diverse image processing tasks. Finally, our future work involves a thorough study of the concepts presented in this paper in view of performing intuitionistic fuzzy histogram equalization.

## References


1. Zadeh, L.A.: Fuzzy sets. Inf. Control **8** (1965) 338–353
2. Atanassov, K.T.: Intuitionistic fuzzy sets. Fuzzy Sets Syst. **20** (1986) 87–96
3. Atanassov, K.T.: More on intuitionistic fuzzy sets. Fuzzy Sets Syst. **33** (1989) 37–45
4. Atanassov, K.T.: Intuitionistic Fuzzy Sets: Theory and Applications. Studies in Fuzziness and Soft Computing. Physica–Verlag, Heidelberg (1999)
5. Bustince, H., Kacprzyk, J., Mohedano, V.: Intuitionistic Fuzzy Generators–Application to Intuitionistic Fuzzy Complementation. Fuzzy Sets Syst. **114** (200) 485–504
6. Klir, G.J., Folger, T.A.: Fuzzy Sets, Uncertainty, and Information. Prentice Hall, Englewood Cliffs, NJ (1988)
7. Klir, G.J., Yuan, B.: Fuzzy Sets and Fuzzy Logic: Theory and Application. Prentice Hall, New Jersey (1995)
8. Jawahar, C.V., Ray, A.K.: Fuzzy statistics of digital images. IEEE Signal Process. Lett. **3** (1996) 225–227
9. Jawahar, C.V., Ray, A.K.: Incorporation of gray-level imprecision in representation and processing of digital images. Pattern Recognit. Lett. **17** (1996) 541–546
10. Vlachos, I.K., Sergiadis, G.D.: Intuitionistic Fuzzy Image Processing. In: Soft Computing in Image Processing: Recent Advances. Volume 210 of Studies in Fuzziness and Soft Computing. Springer (2006) 385–416
11. Zadeh, L.A.: The conept of a linguistic variable and its application to approximate reasoning–I. Inform. Sci. **8** (1975) 199–249
12. Burillo, P., Bustince, H.: Entropy on intuitionistic fuzzy sets and on interval-valued fuzzy sets. Fuzzy Sets Syst. **78** (1996) 305–316
13. Szmidt, E., Kacprzyk, J.: Entropy for intuitionistic fuzzy sets. Fuzzy Sets Syst. **118** (2001) 467–477
14. Szmidt, E., Kacprzyk, J.: Entropy and similarity of intuitionistic fuzzy sets. In: Proc. Information Processing and Management of Uncertainty in Knowledge-Based Systems, Paris, France (2006) 2375–2382
15. Pal, S.K., King, R.A.: Image enhancement using fuzzy set. Electron. Lett. **16** (1980) 376–378
16. Pal, S.K., King, R.A.: Image enhancement using smoothing with fuzzy sets. IEEE Trans. Syst. Man Cybern. **11** (1981) 495–501
17. Pal, S.K., King, R.A.: A note on the quantitative measure of image enhancement through fuzziness. IEEE Trans. Pattern Anal. Mach. Intell. **4** (1982) 204–208

# Image Threshold Using A-IFSs Based on Bounded Histograms

Pedro Couto[1], Humberto Bustince[2], Vitor Filipe[1], Edurne Barrenechea[2], Miguel Pagola[2], and Pedro Melo-Pinto[1]

[1] CETAV – University of Trás-os-Montes e Alto Douro, Ap. 1014, 5001-911 Vila Real, Portugal
[2] Departamento de Automática y Computación - Universidad Pública de Navarra, Campus de Arrosadía, s/n, 31006 Pamplona, Spain

**Abstract.** Atanassov's intuitionistic fuzzy sets (A-IFSs) have been used recently to determine the optimal threshold value for gray-level image segmentation [1]. Atanassov's intuitionistic fuzzy index values are used for representing the unknowledge/ignorance of an expert on determining whether a pixel of the image belongs to the background or the object of the image. This optimal global threshold of the image is computed automatically, regardless of the actual image analysis process.

Although global optimal thresholding techniques give good results under experimental conditions, when dealing with real images having several objects and the segmentation purpose is to point out some application-specific information, one should use heuristic techniques in order to obtain better thresholding results.

This paper introduces an evolution of the above mentioned technique intended for use with such images. The proposed approach takes into account the image and segmentation specificities by using a two-step procedure, with a restricted set of the image gray-levels.

Preliminary experimental results and comparison with other methods are presented.

**Keywords:** Fuzzy Sets Theory Applications, Atanassov's Intuitionistic Fuzzy Sets (A-IFSs), computer Vision, Pattern Recognition, Digital Image Processing.

## 1 Introduction

Many image analysis techniques take as starting point a segmentation of the image, that is, the image is decomposed into meaningful parts for further analysis, resulting in the partition of the set of pixels in the image into a finite set of regions (subsets) according to a certain criterion.

In reality, the segmentation of digital images is the process of dividing an image into disjoined parts, regions or subsets so that each one must satisfy a distinct and well-defined property or attribute.

The most commonly used strategy for segmenting images is global thresholding that refers to the process of partitioning the pixels in an image into object

P. Melin et al. (Eds.): IFSA 2007, LNAI 4529, pp. 96–103, 2007.
© Springer-Verlag Berlin Heidelberg 2007

and background regions on the basis of the different intensity levels of gray of the pixels in the image. This partition is made by establishing a threshold, in such a way that all the pixels with intensity greater or equal than the threshold belong to the background (or to the object) and all the pixels with intensity lower than the threshold belong to the object (or to the background).

Extensive research has been conducted in this research field over the last years, and many types of segmentation techniques have been proposed in the literature, each one of them based on a certain methodology to classify the regions [2,3,4,5,6].

The proposed approach is an evolution/extension of the methodology, based on Atanassov's intuitionistic fuzzy sets (A-IFSs), presented in [1] intended for use with specific images within a particular image analysis process. This approach uses a two-step procedure, applying the methodology presented in [1] first to all the image pixels and then to a restricted set of the original image gray-levels' set.

## 2 Image Threshold Computation by Modelizing Knowledge/Unknowledge by Means of A-IFSs

Being $(x, y)$ the coordinates of each pixel on the image $Q$, and being $q(x, y)$ the gray level of the pixel $(x, y)$ so that $0 \leq q(x, y) \leq L-1$ for each $(x, y) \in Q$ where $L$ is the image grayscale, many methods have been proposed for determining the threshold $t$ of an image considering fuzzy set theory as an efficient tool in order to obtain a good segmentation of the image considered. The most commonly algorithm used to obtain the threshold is the one that uses the concept of fuzzy entropy and its main steps are the following:

(a) Assign $L$ fuzzy sets $Q_t$ to each image $Q$. Each one is associated to a level of intensity $t$, $(t = 0, 1, \cdots, L-1)$, of the grayscale $L$ used.
(b) Calculate the entropy of each one of the $L$ fuzzy sets $Q_t$ associated with $Q$.
(c) Take, as the *best threshold* gray level $t$, associated with the fuzzy set corresponding to the lowest entropy.

The main problem of this algorithm is the step (a). In [1] this problem is solved using A-IFSs in the following way: In order to choose/construct the membership function of each pixel of the image to the associated fuzzy set, three numerical values are assigned to each one of them.

- A value for representing the expert knowledge of the membership of the pixel to the background. A membership function, constructed by the expert using dissimilarity functions, is used to obtain this value (see [7]).
- Dissimilarity functions are also used by the expert to construct a membership function to retrieve a value for representing the expert knowledge of the membership of the pixel to the object.
- The expert knowledge/ignorance, in determining the above mentioned membership functions, is represented by a third value obtained trough Atanassov's intuitionistic index.

The value represented by Atanassov's intuitionistic index indicates the knowledge/ignorance of the expert when assigning a pixel either to the background or the object, so that, when the expert is absolutely sure that a pixel belongs either to the background or the object the Atanassov's intuitionistic index associated with that pixel has the value of zero. This value increases with respect to the unknowledge/ignorance of the expert as to whether the pixel belongs to the background or the object. So, if the expert doesn't know if a pixel belongs to the background or the object, its membership to both must be represented with the value 0.5, and in such conditions, it is said that the expert used the greatest unknowledge/ignorance/intuition allowed in the construction of the membership functions, of the set associated with that pixel, to the background and the object resulting in a Atanassov's Intuitionistic Fuzzy Index maximum value. For this reason, A-IFSs (Atanassov's Intuitionistic Fuzzy Set [9,10]) are used.

In a second stage, the entropy values of each one of the L A-IFSs associated with the image are calculated. In this methodology, entropy on A-IFSs is interpreted as a measure of the degree of a A-IFS that a set has with respect to the fuzzyness of the said set (see [8]). Under these conditions the entropy will be null when the set is a FSs and will be maximum when the set is totally intuitionistic.

Finally, the gray level $t$ associated with the fuzzy set with the lowest entropy is selected for the best threshold.

A possible implementation of this methodology [1], and the one used in this work, is now presented.

***(Step A)*** - Construct $L$ fuzzy sets $Q_{Bt}$ associated with the background and $L$ fuzzy sets $Q_{Ot}$ associated with the object. Each one of these fuzzy sets is associated with a gray level $t$ of the grayscale $L$ used. The membership functions of these sets are defined by means of restricted dissimilarity functions and the expressions are:

$$\mu_{Q_{Bt}}(q) = F\left(d\left(\frac{q}{L-1}, \frac{m_{\mathrm{B}}(t)}{L-1}\right)\right)$$

$$\mu_{Q_{Ot}}(q) = F\left(d\left(\frac{q}{L-1}, \frac{m_{\mathrm{O}}(t)}{L-1}\right)\right)$$

where

$$m_{\mathrm{B}}(t) = \frac{\sum_{q=0}^{t} qh(q)}{\sum_{q=0}^{t} h(q)} \tag{1}$$

$$m_{\mathrm{O}}(t) = \frac{\sum_{q=t+1}^{L-1} qh(q)}{\sum_{q=t+1}^{L-1} h(q)} \tag{2}$$

and

$$F(x) = 1 - 0.5x$$

being $h(q)$ the number of pixels of the image with the gray level q.

Note that $F(x)$ is only one of the possible $F$ functions that could be used (see [1]).

***(Step B)*** - As it has been said before, the unknowledge/ignorance of the expert in the construction of the fuzzy sets (in *Step A*) is represented by means of Atanassov's intuitionistic fuzzy index ($\pi$), meaning that, it is considered that $\mu_{Q_{Bt}}$ ($\mu_{Q_{Ot}}$) indicates the expert's degree of knowledge of the pixel belonging to the background (object).

If the expert is certain of the pixel belonging to the background or the object, then the value of $\pi$ must be zero. The value of $\pi$ increases as the unknowledge/ignorance of the expert grows. However, the unknowledge/ignorance must have the least possible influence on the choice of the membership degree, so, in this implementation, in the worst case, the unknowledge will have a maximum influence of 25 percent.

Under these conditions, the following expression is used to calculate $\pi$:

$$\pi(q) = (1 - \mu_{Q_{Bt}}(q))(1 - \mu_{Q_{Ot}}(q)).$$

Again, this expression is only one of the possible ones (see [1]).

***(Step C)*** - Construct an A-IFS, using $\pi$, with each one of the fuzzy sets $Q_{Bt}$ and $Q_{Ot}$.

$$\begin{aligned}\tilde{Q}_{Bt} = \{(q, \mu_{\tilde{Q}_{Bt}}(q), \nu_{\tilde{Q}_{Bt}}(q)) | q = 0, 1, \cdots, L-1\}, &\text{ given by} \\ \mu_{\tilde{Q}_{Bt}}(q) = \mu_{Q_{Bt}}(q) & \\ \nu_{\tilde{Q}_{Bt}}(q) = 1 - \mu_{\tilde{Q}_{Bt}}(q) - \pi(q) = (1 - \mu_{Q_{Bt}}(q)) \cdot \mu_{Q_{Ot}}(q) &\end{aligned}$$

and

$$\begin{aligned}\tilde{Q}_{Ot} = \{(q, \mu_{\tilde{Q}_{Ot}}(q), \nu_{\tilde{Q}_{Ot}}(q)) | q = 0, 1, \cdots, L-1\}, &\text{ given by} \\ \mu_{\tilde{Q}_{Ot}}(q) = \mu_{Q_{Ot}}(q) & \\ \nu_{\tilde{Q}_{Ot}}(q) = 1 - \mu_{\tilde{Q}_{Ot}}(q) - \pi(q) = (1 - \mu_{Q_{Ot}}(q)) \cdot \mu_{Q_{Bt}}(q) &\end{aligned}$$

***(Step D)*** - Calculate the entropy $IE$ of each one of the $L$ Atanassov's intuitionistic fuzzy sets, using the following expression, so that $0 \leq IE(\tilde{Q}_{Bt}) \leq 0.25$.

$$IE(\tilde{Q}_{Bt}) = \frac{1}{N \times M} \sum_{q=0}^{L-1} h(q)(1 - \mu_{Q_{Bt}}(q))(1 - \mu_{Q_{Ot}}(q)) \tag{3}$$

where $N \times M$ are the image dimensions in pixels.

***(Step E)*** - Finally, the gray level associated with the Atanassov's intuitionistic fuzzy set $\tilde{Q}_{Bt}$ of lowest entropy $IE$ is chosen as the best threshold.

## 3 Materials and Methods

In the image processing system boarded in this work, the main goal is to perform kinematic analysis for the left hindlimb in treadmill walking rats. The method used for the analysis of the hindlimb movement involved the placing of markers on the skin surface overlying joints under analysis. These markers are to be tracked by the system in order to characterize the hindlimb movement [11,12,13].

Image sequences acquired at the usual rate of 25 images per second are insufficient to characterize the rat's hindlimb movement, particularly due to aliasing phenomena's. In order to avoid this aliasing problem, a high-speed digital image camera (Redlake PCI 1000S, San Diego, USA) was used to record the rat gait at 125 frames per second, resulting in images of $480 \times 420$ pixels codified in 8 bits (256 gray levels).

Due to the high speed acquisition, other problems arise in contrast, noise, illumination, resolution, etc., resulting in noisy images with imprecision on the gray levels that conducts to fuzzy boundaries and ill defined regions, which makes the current approach to the segmentation of such images the natural approach.

## 4 Proposed Approach

### 4.1 First Step

In this step the pixels belonging to the background of the image are identified and withdraw from the image. In order to do so, the methodology presented in section 2 is applied to the image.

However, the threshold value $th$, computed at this point, is not used to segment the image, but to bind the original histogram of the image. Thus, all pixels below $th$ value are "extracted" from the original histogram, hence their presence in the image will be ignored for further processing at step two. All other pixels (greater then $th$) will remain with their original gray level.

If we denote by $P$ the number of pixels below threshold $th$, then

$$P = \sum_{q=0}^{th} h(q)$$

and the new image to be processed in step two is a $N \times M - P$ image with $L - th$ gray levels, where $q(x, y)$ is the gray level of the pixel $(x, y)$ so that $th \leq q(x, y) \leq L - 1$ for each $(x, y) \in Q$, and where $[th + 1, \cdots, L - 1]$ is the image grayscale.

### 4.2 Second Step

At this stage, the same methodology is applied to image resulting from the *First Step*. In order to apply the section2 methodology, the following adjustments where made:

In *Step A*, $L - th$ fuzzy sets $Q_{Bt}$ associated with the background and $L - th$ fuzzy sets $Q_{Ot}$ associated with the object are constructed, instead of the original $L$ fuzzy sets. Also, equations 2 and 2 become:

$$m_B(t) = \frac{\sum_{q=th}^{t} qh(q)}{\sum_{q=th}^{t} h(q)} \qquad m_O(t) = \frac{\sum_{q=th+t+1}^{L-1} qh(q)}{\sum_{q=th+t+1}^{L-1} h(q)}$$

respectively.

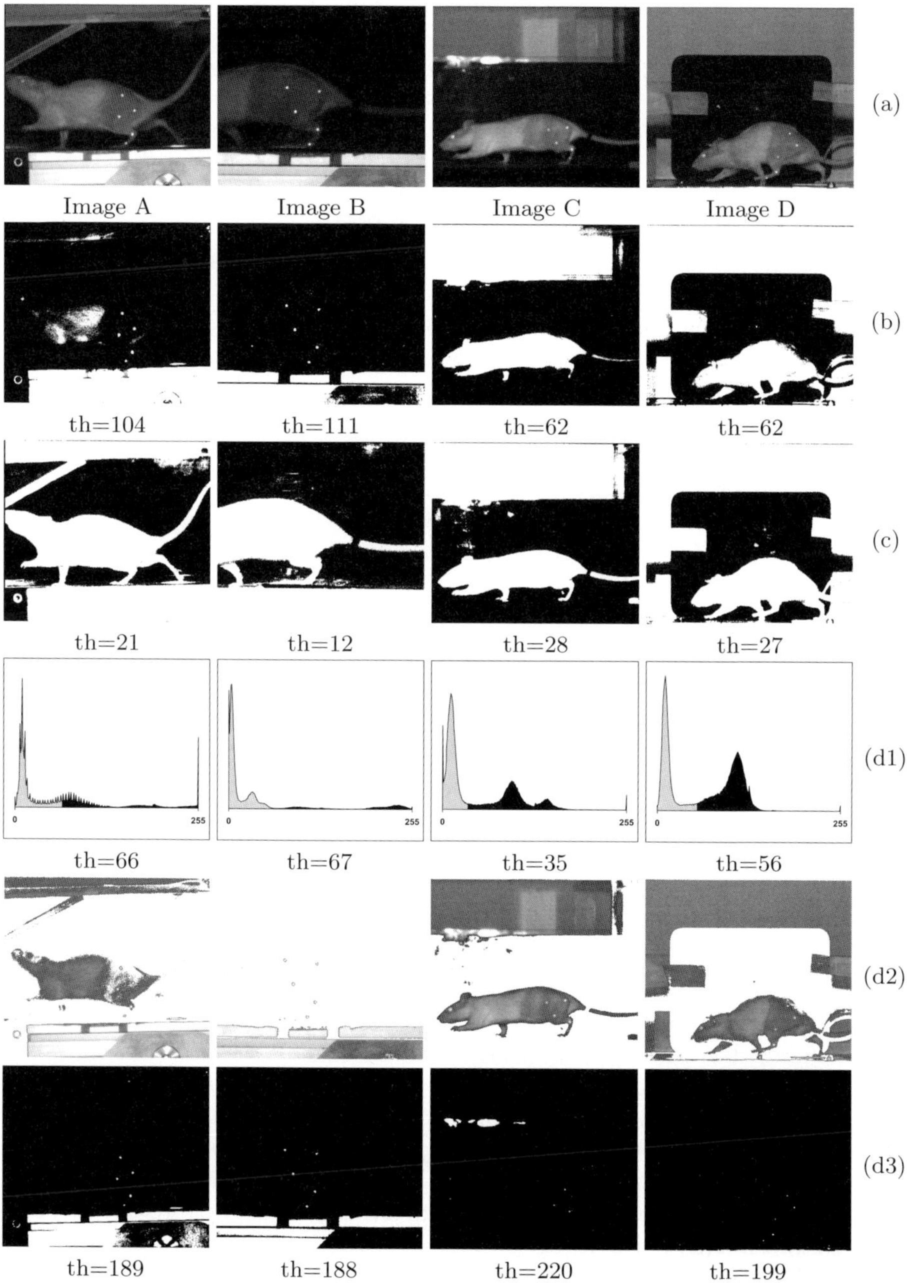


**Fig. 1.** (a) Original image (b) Binary image obtained with the Otsu algorithm (c) Binary image obtained with Kittlers algorithm (d1) Histogram of the original image, where the gray portion represents the gray level intensities that were "removed" from the image (d2) Resulting image after the *First Step* (d3) Binary image obtained with the proposed algorithm after the *Second Step*

In *Step B* and *Step C* although all expressions remain the same, note that $q \in [th+1, \cdots, L-1]$.

In *Step D* equation 3 become

$$IE(\tilde{Q}_{Bt}) = \frac{1}{N \times M} \sum_{q=th+1}^{L-1} h(q)(1 - \mu_{Q_{Bt}}(q))(1 - \mu_{Q_{Ot}}(q))$$

## 5 Experimental Results

In order to test the performance of the proposed approach, four images, presenting contrast problems (more prone to difficulties), from the walking rats' sequences were selected and used as test images. Each one of these images is part of a sequence of images where the rat gait is recorded and is analyzed by tracking the markers placed on the rats in each image. The purpose of the segmentation step in this process is to point out those markers.

We also compare the obtained results with non fuzzy well known methodologies, the Otsu technique [14] and the clustering-based Kittler method [15]. The original images and the results of the used techniques are illustrated in Fig. 1.

The results obtained with the Otsu (Fig.1b) and Kittler (Fig.1c) methodologies do not perform well in identifying the rats markers. Only in one situation (Image B with Otsu method) the markers are clearly identified for further processing. On the contrary, the proposed methodology (Fig.1d3) succeeds in identifying the markers for all the images and, thus, is more reliable for the necessary further processing in order to extract the markers position in the image.

## 6 Conclusions and Future Work

The problem of segmentation in spite of all the work over the last decades, is still an important research field in image processing mostly due to the fact that finding a global optimal threshold is not trivial, and is indeed a very difficult task. One of the most commonly used strategy for segmenting images is global thresholding that refers to the process of partitioning the pixels in an image into regions on the basis of the different intensity levels of gray of the pixels in the regions without distinguishing the pixels within a region, even if their gray values are significantly different in the original image. For this reason, finding an algorithm that can be successfully applied to all kinds of images is a difficult task that, probably, will never be accomplished. Thus, it is suitable to develop new threshold techniques, or new extensions to the existing ones, that can effectively lead us to an optimal threshold within the specificities of one's application.

Although the previous methods presented give good results under experimental conditions, they do not always take into account the specificities of the image analysis process in which it is going to be applied. The new approach presented, successfully intended to endow the algorithm with heuristic techniques that enable to adapt the algorithm with the particular image analysis process.

The preliminary results show that all of the tested images can be properly segmented according to our image analysis process needs and application purpose. Further work is intended, focusing on the adaptation of the proposed algorithm towards a multi-threshold approach and to color image segmentation.[9].

## Acknowledgments

This work has been partially supported by *Acções integradas Luso-Espanholas* 2007 and POCI 2010 (*Programa Operacional Ciência e Investigação* 2010).

## References

1. Bustince, H., Pagola, M., Melo-Pinto, P., Barrenechea, E., Couto, P.: Fuzzy Sets and Their Extensions: Representation, Aggregation and Models, in Studies in Fuzziness and Soft Computing. Image threshold computation by modelizing knowledge/unknowledge by means of A-IFSs. Springer-Verlag, Berlin Heidelberg New York. (next to appear 2007) 225–240
2. Jawahar, C.V., Biswas, P.K., Ray, A.K.: Analysis of fuzzy thresholding schemes. Pattern Recognition **33** (2000) 1339–1349
3. Sezgin, M., Sankur, B.: Survey over image thresholding techniques and quantitative performance evaluation. Journal of Electronic Imaging **13-1** (2004) 146–165
4. Fu, K., Mui, J.: A survey on image segmentation. Pattern Recognition **13** (1981) 3–16
5. Haralick, R., Shapiro, L.: Image segmentation techniques. Computer Vision,Graphics and Image Processing. **29** (1985) 100–132
6. Pal, N., Pal, S.: A review on image segmentation techniques. Pattern Recognition. **26** (1993) 1277–1294
7. Barrenechea, E.: Image Thresholding with Interval-valued Fuzzy Sets. Edge Detection. Contrast. PHD Thesis - Universidad Pública de Navarra.(2005)
8. Burillo, P., Bustince, H.: Entropy on intuitionistic fuzzy sets and on intervalvalued fuzzy sets. Fuzzy Sets and Systems. **78** (1996) 81–103
9. Atanassov, K.: Intuitionistic fuzzy sets. Fuzzy Sets and Systems. **20** (1986) 87–96
10. Atanassov, K.: Intuitionistic fuzzy sets. Theory and applications. Springer-Verlag, Berlin Heidelberg New York. **1** (1999)
11. Pereira, J., Cabrita, A., Filipe, V., Bulas-Cruz, J., Couto, P., Melo-Pinto, P., Costa, L., Geuna, S., Maurcio, A., Varejo, A.: A comparison analysis of hindlimb kinematics during overground and treadmill locomotion in rats. Behavioural Brain Research. **172** (2006) 212–218
12. Filipe, V., Pereira, J., Costa, L., Maurcio, A., Couto, P., Melo-Pinto P., Varejo A.: Effect of skin movement on the analysis of hindlimb kinematics during treadmill locomotion in rats. Journal of Neuroscience Methods. **153** (2006) 55–61
13. Varejo, A., Cabrita, A., Patrcio, J., Bulas-Cruz, J., Gabriel, R., Melo-Pinto, P., Couto, P., Meek, M.: Functional assesment of peripheal nerve recovery in the rat: Gait Kinematics. Microsurgery. **21** (2001) 383–388
14. Otsu, N.: A Threshold Selection Method from Gray-Level Histograms. IEEE Trans. on Systems, Man, and Cybern. **9** (1979) 62–66
15. Kittler, J., Illingworth, J.: Minimum Error Thresholding. Pattern Recognition. **19** (1986)

# The Role of Entropy in Intuitionistic Fuzzy Contrast Enhancement

Ioannis K. Vlachos and George D. Sergiadis

Aristotle University of Thessaloniki
Faculty of Technology
Department of Electrical & Computer Engineering, Telecommunications Division
University Campus, GR–54124, Thessaloniki, Greece
ivlachos@mri.ee.auth.gr, sergiadi@auth.gr
http://mri.ee.auth.gr

**Abstract.** In this paper we study the impact of selecting different entropy measures in the framework of intuitionistic fuzzy image processing and especially in the process of intuitionistic fuzzification of images. Different notions of entropy characterized by different properties are reviewed and their behavior is thoroughly studied under the scope of performing contrast enhancement. Finally, experimental results using gray-scale images reveal the characteristics of the aforementioned measures.

## 1 Introduction

Entropy, a measure of information carried by a system, is a fundamental concept in digital image processing. Therefore, it is not surprising that new theories, such as fuzzy sets (FSs) theory, as well as their extensions, seek intuitive ways to adopt and express the notion of entropy in their particular context.

Intuitionistic fuzzy image processing (IFIP), recently introduced in [1] and [2], provides a flexible, yet solid, mathematical framework for dealing with the vagueness present in a digital image. This is carried out by modelling the hesitancy characterizing image pixels, using Atanassov's intuitionistic fuzzy sets (A–IFSs) theory [3,4]. A–IFSs constitute a generalization of Zadeh's fuzzy sets (FSs) [5], by considering also corresponding degrees of hesitancy. It is this additional degree of freedom that allows for the flexible modelling of imprecise or/and imperfect information often present in images.

In this paper the different concepts of intuitionistic fuzzy entropy are reviewed and their behavior is studied in the context of IFIP for performing contrast enhancement. Evaluation of these measures using real-world images reveal their particular characteristics that are to be exploited for different applications of contrast enhancement in the context of IFIP.

This paper is organized as follows. In Sect. 2 elements of A–IFSs are presented and their geometrical representation of in two- and three-dimensional spaces is discussed. Sect. 3 reviews different concepts of entropy in the intuitionistic fuzzy setting. An overview of the IFIP framework is presented in Sect. 4. Finally, experimental results are given in Sect. 5, while conclusions are drawn in Sect. 6.

P. Melin et al. (Eds.): IFSA 2007, LNAI 4529, pp. 104–113, 2007.
© Springer-Verlag Berlin Heidelberg 2007

# 2 Intuitionistic Fuzzy Sets

In this section, we briefly review the basic notions, concepts, and definitions of A–IFSs, as well as their geometrical representations in two- and three-dimensional spaces.

## 2.1 Elements of Intuitionistic Fuzzy Sets Theory

**Definition 1.** *An FS $\tilde{A}$ defined on a universe $X$ may be given as [5]*

$$\tilde{A} = \{\langle x, \mu_{\tilde{A}}(x)\rangle | x \in X\} \ , \tag{1}$$

*where $\mu_{\tilde{A}}(x) : X \to [0,1]$ is the membership function of $\tilde{A}$.*

The membership function of $\tilde{A}$ describes the *degree of belongingness* of $x \in X$ in $\tilde{A}$.

**Definition 2.** *An A–IFS $A$ defined on a universe $X$ is given by [3,4]*

$$A = \{\langle x, \mu_A(x), \nu_A(x)\rangle | x \in X\} \ , \tag{2}$$

*where*

$$\mu_A(x) : X \to [0,1] \qquad \textit{and} \qquad \nu_A(x) : X \to [0,1] \ ,$$

*with the condition*

$$0 \leqslant \mu_A(x) + \nu_A(x) \leqslant 1 \ , \tag{3}$$

*for all $x \in X$.*

The values of $\mu_A(x)$ and $\nu_A(x)$ denote the *degree of belongingness* and the *degree of non-belongingness* of $x$ to $A$, respectively. For an A–IFS $A$ in $X$ we call the *intuitionistic index* of an element $x \in X$ in $A$ the following expression

$$\pi_A(x) = 1 - \mu_A(x) - \nu_A(x) \ . \tag{4}$$

We can consider $\pi_A(x)$ as a *hesitancy degree* of $x$ to $A$ [3,4]. From (4) it is evident that

$$0 \leqslant \pi_A(x) \leqslant 1 \tag{5}$$

for all $x \in X$.

FSs can also be represented using the notation of A–IFSs. An FS $\tilde{A}$ defined on $X$ can be represented as the following A–IFS

$$A = \{\langle x, \mu_A(x), 1 - \mu_A(x)\rangle | x \in X\} \ , \tag{6}$$

with $\pi_A(x) = 0$ for all $x \in X$.

**Definition 3.** *The complementary set $A^c$ of $A$ is defined as*

$$A^c = \{\langle x, \nu_A(x), \mu_A(x)\rangle | x \in X\} \ . \tag{7}$$

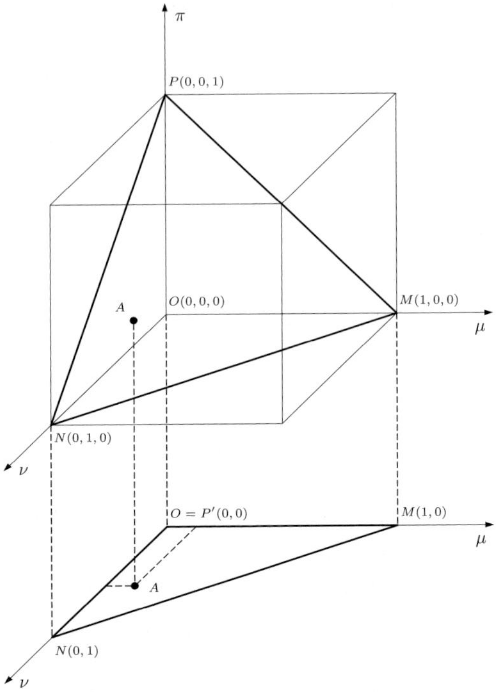


**Fig. 1.** Geometrical representation of an A–IFS $A$ in $X = \{x\}$. (Adopted from [6].)

Throughout this paper by $\mathscr{IFS}(X)$ we denote the set of all A–IFSs defined on $X$. Correspondingly, $\mathscr{FS}(X)$ is the set of all FSs on $X$, while $2^X$ denotes the set of all crisp sets.

Finally, Atanassov [4] proposed an operator, namely the *Atanassov's operator*, to de-construct an A–IFS into an FS.

**Definition 4.** *If* $A \in \mathscr{IFS}(X)$, *then* $D_\alpha : \mathscr{IFS}(X) \to \mathscr{FS}(X)$, *where*

$$D_\alpha(A) = \{\langle x, \mu_A(x) + \alpha\pi_A(x), \nu_A(x) + (1-\alpha)\pi_A(x)\rangle | x \in X\} \ , \qquad (8)$$

*with* $\alpha \in [0, 1]$.

### 2.2 Geometrical Representation of A–IFSs

Generalizing Kosko's [7] geometrical representation of FSs, Atanassov [4] proposed a similar interpretation of A–IFSs in a Euclidean plane with Cartesian coordinates. Szmidt and Kacprzyk [8] extended Atanassov's approach by considering all three parameters of A–IFSs and proposed the geometrical representation of A–IFSs as a mapping into a simplex in the unit cube. Moreover, they demonstrated that Atanassov's interpretation is simply the orthogonal projection of the simplex of their definition into the Euclidean plane. Both representations are illustrated in Fig. 1 for an A–IFS $A$ in $X = \{x\}$.

# 3 Notions of Entropy in the Intuitionistic Fuzzy Setting

Entropy plays an important role in digital image processing. Therefore, it comes as no surprise that the notion of entropy constituted a fundamental aspect from the beginning of the development of FSs theory. De Luca and Termini [9] were the first to introduce an axiomatic skeleton of a nonprobabilistic entropy in the setting of FSs theory that captured our intuition regarding the very essence of fuzzy entropy.

As a natural consequence, the quest for entropy measures in the context of A–IFSs was a very interesting topic that intrigued many researchers working in this field. Burillo and Bustince [10] were the first to state and propose an axiomatic skeleton of entropy for A–IFSs and interval-valued fuzzy sets.

**Definition 5 (Burillo and Bustince [10]).** *A real function* $E : \mathscr{I}\mathscr{F}\mathscr{S}(X) \rightarrow \mathbb{R}^+$ *is called an entropy on* $\mathscr{I}\mathscr{F}\mathscr{S}(X)$, *if* $E$ *has the following properties*

**(E1)** $E(A) = 0$ *if and only if* $A \in \mathscr{F}\mathscr{S}(X)$,
**(E2)** $E(A) = Cardinal(X)$ *if and only if* $\mu_A(x) = \nu_A(x) = 0$ *for all* $x \in X$,
**(E3)** $E(A) = E(A^c)$ *for all* $A \in \mathscr{I}\mathscr{F}\mathscr{S}(X)$,
**(E4)** $E(A) \geqslant E(B)$ *if* $\mu_A(x) \leqslant \mu_B(x)$ *and* $\nu_A(x) \leqslant \nu_B(x)$ *for all* $x \in X$.

Motivated by De Luca and Termini's set of axiomatic requirements, Szmidt and Kacprzyk [8] proposed an alternative interpretation of entropy, accompanied by a different set of axioms.

**Definition 6 (Szmidt and Kacprzyk [8]).** *A real function* $E' : \mathscr{I}\mathscr{F}\mathscr{S}(X) \rightarrow \mathbb{R}^+$ *is called an entropy on* $\mathscr{I}\mathscr{F}\mathscr{S}(X)$, *if* $E$ *has the following properties*

**(E5)** $E'(A) = 0$ *if and only if* $A \in 2^X$,
**(E6)** $E'(A) = 1$ *if and only if* $\mu_A(x) = \nu_A(x)$ *for all* $x \in X$,
**(E7)** $E'(A) = E'(A^c)$ *for all* $A \in \mathscr{I}\mathscr{F}\mathscr{S}(X)$,
**(E8)** $E'(A) \leqslant E'(B)$ *if*
$\mu_A(x) \leqslant \mu_B(x)$ *and* $\nu_A(x) \geqslant \nu_B(x)$ *for* $\mu_B(x) \leqslant \nu_B(x)$
*or*
$\mu_A(x) \geqslant \mu_B(x)$ *and* $\nu_A(x) \leqslant \nu_B(x)$ *for* $\mu_B(x) \geqslant \nu_B(x)$ *for all* $x \in X$.

The aforementioned definition degenerates to De Luca and Termini's definition when FSs are considered. A generalized framework of Definition 6 was introduced in [11]. Finally, it should be mentioned that a connection between the different concepts of entropy for A–IFSs was explored and proved in [12].

## 3.1 Review of Intuitionistic Fuzzy Entropy Measures

Based on the aforementioned notions and definitions of entropy, different entropy measures for A–IFSs were proposed in the literature. Along with their definition of intuitionistic fuzzy entropy, Burillo and Bustince [10] proposed the following entropy

$$E_1(A) = \frac{1}{n} \sum_{i=1}^{n} \pi_A(x_i) \ , \tag{9}$$

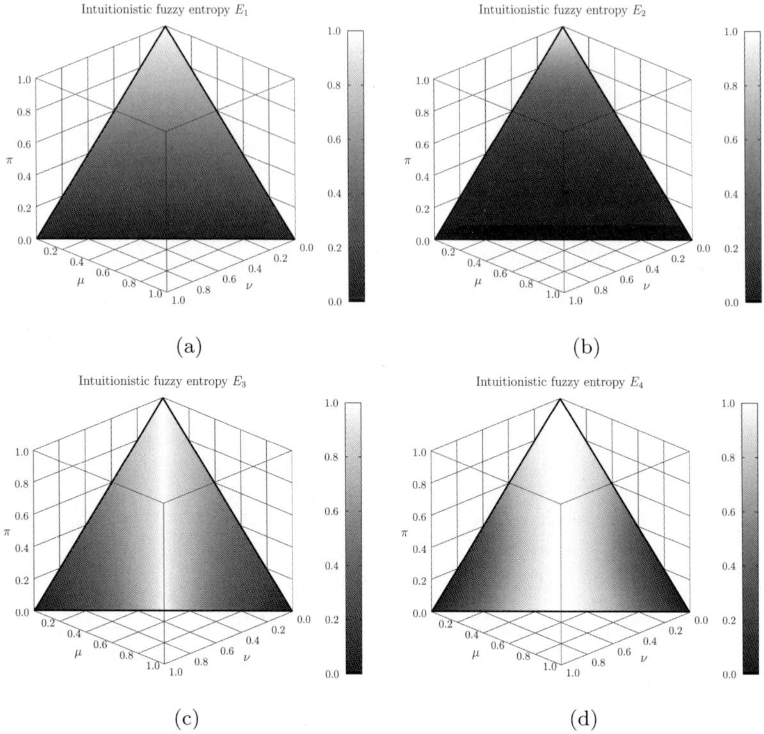


**Fig. 2.** Plots of entropy measures (a) $E_1$, (b) $E_2$, (c) $E_3$, and (d) $E_4$ for an A–IFS defined in $X = \{x\}$

which satisfies the axiomatic requirements **E1**–**E4** and expresses the *degree of intuitionism* of the set $A$. In (9), the normalization factor $\frac{1}{n}$ has been added, in order for the entropy $E_1$ to lie in the $[0,1]$ interval. Additionally, Burillo and Bustince proposed an alternative entropy measure given by

$$E_2(A) = \frac{1}{n}\sum_{i=1}^{n}\left(1 - (\mu_A(x_i) + \nu_A(x_i))\, e^{1-(\mu_A(x_i)+\nu_A(x_i))}\right) . \tag{10}$$

The first measure of entropy satisfying axioms **E5**–**E8** was introduced by Szmidt and Kacprzyk [8] as a ratio of distances between an A–IFS and its nearest and farthest crisp sets, respectively. The aforementioned entropy is given by

$$E_3(A) = \frac{1}{n}\sum_{i=1}^{n}\left(\frac{\max Count\,(A_i \cap A_i^c)}{\max Count\,(A_i \cup A_i^c)}\right) , \tag{11}$$

where $\max Count$ is the *biggest cardinality* of an A–IFS calculated using the following formula

$$\max Count(A) = \sum_{i=1}^{n} (\mu_A(x_i) + \pi_A(x_i)) \tag{12}$$

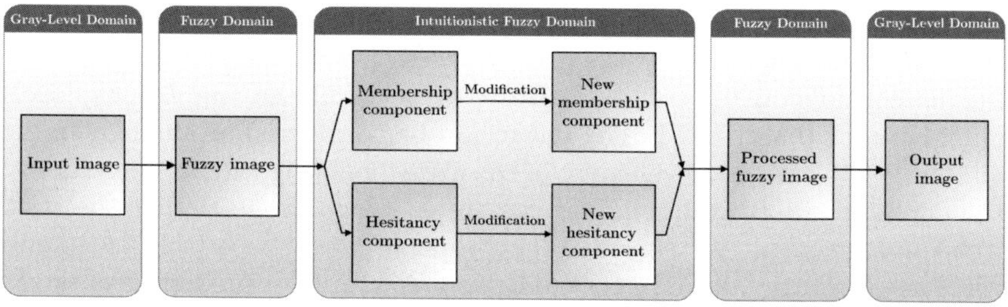


**Fig. 3.** Overview of the IFIP framework

and $A_i$ denotes the single-element A–IFS corresponding to the $i$-the element $x_i$ of the universe $X$, described as $A_i = \{\langle x_i, \mu_A(x_i), \nu_A(x_i)\rangle\}$; i.e. $A_i$ is the $i$-th "component" of $A$.

An inner product-based entropy measure, also satisfying the axiomatic requirements of **E5**–**E8**, was introduced in [13] and is given by

$$E_4(A) = \frac{1}{n}\sum_{i=1}^{n} \frac{2\mu_A(x_i)\nu_A(x_i) + \pi_A^2(x_i)}{\pi_A^2(x_i) + \mu_A^2(x_i) + \nu_A^2(x_i)} . \tag{13}$$

Fig. 2 illustrates the aforementioned entropy measures using the geometrical representation of A–IFSs described in Sect. 2.2 for a singe-element universe $X = \{x\}$. The gray level of each point $(\mu_A(x), \nu_A(x), \pi_A(x))$ on the simplex denotes the entropy value of the set $A = \{\langle x, \mu_A(x), \nu_A(x)\rangle | x \in X\}$ corresponding to that point.

Finally, the entropy measures listed in this section, satisfying and expressing different concepts of intuitionistic fuzzy entropy, will be evaluated in order to assess their behavior, under the scope of performing contrast enhancement, using the IFIP framework.

# 4 From Images to A–IFSs: Entropy Optimization

## 4.1 The Intuitionistic Fuzzy Image Processing Framework

Intuitionistic fuzzy image processing (IFIP) [1,2], involves in general a set of operations carried out using the concepts and elements of A–IFSs theory. Fig. 3 shows an overview of the IFIP framework. In the first stage the image is transferred into the fuzzy domain and sequentially into the intuitionistic fuzzy domain, where the main processing is performed. The inverse procedure is carried out in order to obtain the processed image in the gray-level domain. In this paper we focus on the role of intuitionistic fuzzy entropy measures in the stage of analyzing the image into its intuitionistic fuzzy components; i.e. the stage of "*intuitionistic fuzzification*".

## 4.2 Intuitionistic Fuzzification

In [2], an intuitionistic fuzzification scheme for constructing the A–IFS corresponding to a gray-scale image was proposed, based on the optimization of its intuitionistic fuzzy entropy. In this section, we briefly describe the aforementioned approach.

Let us consider an image $A$ of size $M \times N$ pixels having $L$ gray levels $g$ ranging between 0 and $L-1$. The image can be considered as an array of fuzzy singletons [14,15], with each element of the array denoting the membership value of the corresponding pixel, with respect to an image property. For the task of contrast enhancement we consider the property "*brightness*" of the intensity levels. Therefore, the image in the fuzzy domain can be represented as the FS

$$\tilde{A} = \{\langle g_{ij}, \mu_{\tilde{A}}(g_{ij})\rangle | g_{ij} \in \{0, \ldots, L-1\}\} \ , \tag{14}$$

with $i \in \{1, \ldots, M\}$ and $j \in \{1, \ldots, N\}$.

A basic procedure of IFIP is the derivation of a combination of membership and non-membership functions that model the gray levels of the image in an optimal way. The optimality is considered under the scope of maximizing the intuitionistic fuzzy entropy of the image and thus it is called "*maximum intuitionistic fuzzy entropy principle*" [2]. The family of parametric membership and non-membership functions, used for optimization, is given respectively by

$$\mu_A(g;\lambda) = 1 - (1 - \mu_{\tilde{A}}(g))^{\lambda} \tag{15}$$

and

$$\nu_A(g;\lambda) = (1 - \mu_{\tilde{A}}(g))^{\lambda(\lambda+1)} \ , \tag{16}$$

with $\lambda \geqslant 0$, where the membership function $\mu_{\tilde{A}}(g)$ of the fuzzified image is given by

$$\mu_{\tilde{A}}(g) = \frac{g - g_{min}}{g_{max} - g_{min}} \ . \tag{17}$$

Moreover, the optimization criterion involved can be formulated as follows

$$\lambda_{opt} = \arg\max_{\lambda \geqslant 1} \{E(A;\lambda)\} \ , \tag{18}$$

where $E$ is an entropy measure.

After obtaining the optimal parameter $\lambda_{opt}$, the image is represented as the following A–IFS

$$A_{opt} = \{\langle g, \mu_A(g;\lambda_{opt}), \nu_A(g;\lambda_{opt})\rangle | g \in \{0, \ldots, L-1\}\} \ . \tag{19}$$

By applying Atanassov's operator to the A–IFS $A_{opt}$, we obtain different representations of the image in the fuzzy domain, depending on the parameter $\alpha$ selected. The "*maximum index of fuzziness intuitionistic defuzzification*" procedure was proposed in [12] for selecting the optimal parameter $\alpha_{opt}$, according to the following scheme

$$\alpha_{opt} = \begin{cases} 0 \ , & \text{if } \alpha'_{opt} < 0 \\ \alpha'_{opt} \ , & \text{if } 0 \leqslant \alpha'_{opt} \leqslant 1 \\ 1 \ , & \text{if } \alpha'_{opt} > 1 \end{cases} \ , \tag{20}$$

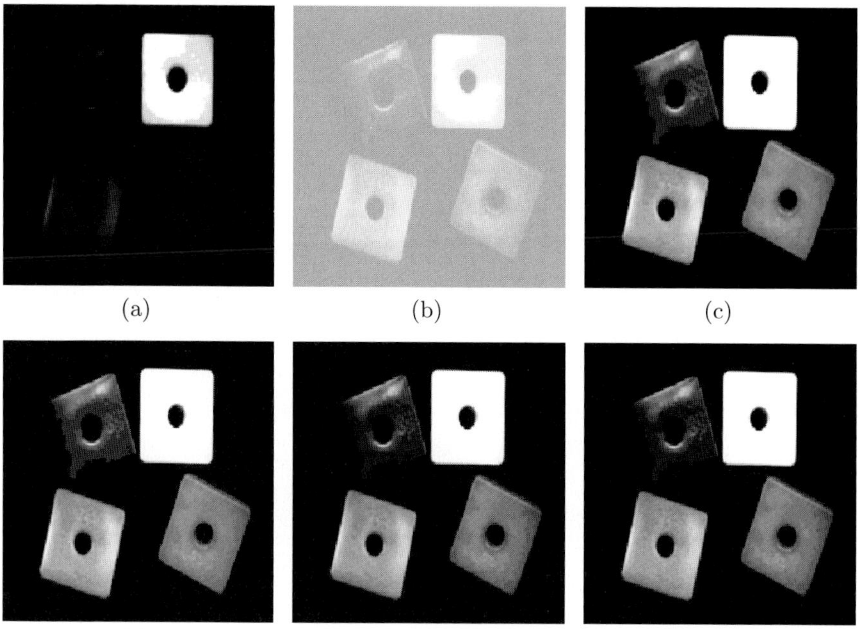

**Fig. 4.** (a) Under-exposed gray-scale image and images obtained using (b) the histogram equalization technique and the IFIP framework employing entropy (c) $E_1$ ($\lambda_{opt} = 6.71$), (d) $E_2$ ($\lambda_{opt} = 6.87$), (e) $E_3$ ($\lambda_{opt} = 11.30$), and (f) $E_4$ ($\lambda_{opt} = 11.05$)

where

$$\alpha'_{opt} = \frac{\sum_{g=0}^{L-1} h_{\tilde{A}}(g)\pi_A(g;\lambda_{opt})\left(1 - 2\mu_A(g;\lambda_{opt})\right)}{2\sum_{g=0}^{L-1} h_{\tilde{A}}(g)\pi_A^2(g;\lambda_{opt})} , \quad (21)$$

with $h_{\tilde{A}}$ being the histogram of the fuzzified image $\tilde{A}$.

Finally, the image in the gray-level domain is obtained as

$$g' = (L-1)\mu_{D_{\alpha_{opt}}(A_{opt})}(g) , \quad (22)$$

where

$$\mu_{D_{\alpha_{opt}}(A_{opt})}(g) = \alpha_{opt} + (1-\alpha_{opt})\mu_A(g;\lambda_{opt}) - \alpha_{opt}\nu_A(g;\lambda_{opt}) , \quad (23)$$

and $g'$, $g$ are the new and initial intensity levels, respectively.

## 5 Experimental Results

The main purpose of this work is to explore the role of intuitionistic fuzzy entropy in the process of intuitionistic fuzzification of images. Therefore, the aforementioned intuitionistic fuzzy entropy measures, were applied to low-contrasted images in order to perform contrast enhancement.

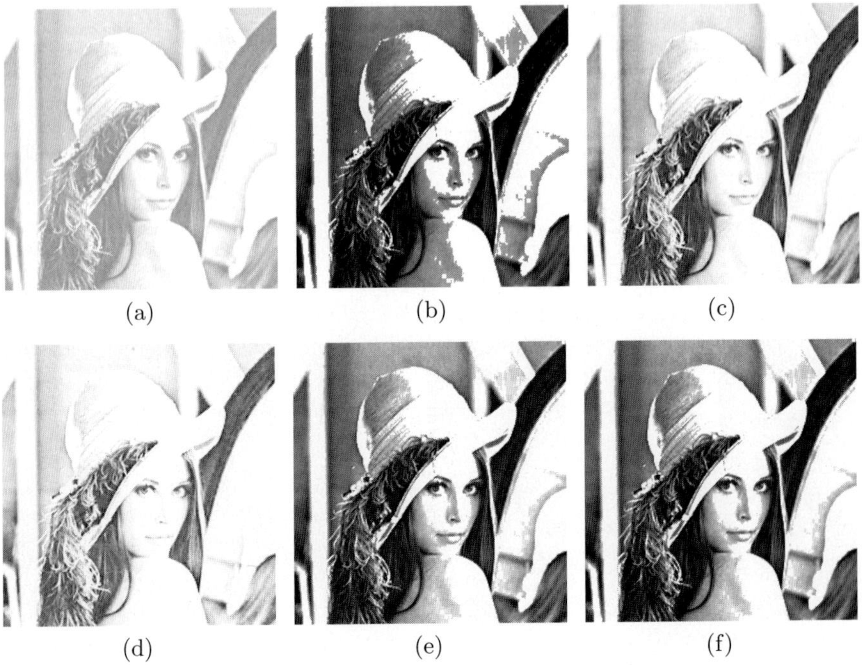

**Fig. 5.** (a) Over-exposed gray-scale image and images obtained using (b) the histogram equalization technique and the IFIP framework employing entropy (c) $E_1$ ($\lambda_{opt} = 1.60$), (d) $E_2$ ($\lambda_{opt} = 2.72$), (e) $E_3$ ($\lambda_{opt} = 0.59$), and (f) $E_4$ ($\lambda_{opt} = 0.45$)

Figs. 4(a) and 4(b) depict an under-exposed image along with its histogram equalized version. Figs. 4(c)–4(f) illustrate images processed using the IFIP framework employing the intuitionistic fuzzy entropy measures $E_1$, $E_2$, $E_3$, and $E_4$, respectively. One may observe that the images obtained using the IFIP framework have been drastically enhanced, revealing high-frequency edges and constant-intensity regions initially not visible due to the low contrast. Moreover, employing the entropy measures $E_1$ and $E_2$, results in a more radical enhancement of the initial image, with $E_2$ exhibiting a slightly better performance. Compared to the histogram-equalized image of Fig. 4(b) the IFIP framework delivers better results for contrast enhancement.

On the other hand, for the over-exposed image of Fig. 5 one may observe that even though entropies $E_1$ and $E_2$ enhance the initial image, the results are not satisfactory compared to the ones obtained using entropy measures $E_3$ and $E_4$ or to the image derived by the histogram equalization technique. However, the IFIP framework equipped with the entropies $E_3$ and $E_4$ yields images exhibiting an overall drastic, yet smooth, enhancement, in contrast with the histogram-equalized image of Fig. 5(b), which appears to be somewhat not natural, possessing regions that have been over-enhanced.

As a final remark, we can outline that the performance of an entropy measure depends more to the set of axioms that it conforms with, than to the form of the measure itself. By examining the corresponding images, as well as the values of $\lambda_{opt}$, entropy $E_3$ performs better for dark low-contrasted images, while $E_4$ for brighter ones. Finally, entropies satisfying properties **E5**–**E8** exhibit in general a better performance for any type of low-contrasted image.

## 6 Conclusions

In this paper we explored the role of entropy in the context of intuitionistic fuzzy image processing. Different entropy measures for A–IFSs with different characteristics were evaluated and their behavior to contrast enhancement of low-contrasted images was examined. Finally, experimental results to real-world images demonstrated that the different notions of intuitionistic fuzzy entropy treat images in different ways, thus making the selection of the appropriate entropy measure to be application-dependent.

## References


1. Vlachos, I.K., Sergiadis, G.D.: Towards intuitionistic fuzzy image processing. In: Proc. International Conference on Computational Intelligence for Modelling, Control and Automation, 2005. Volume 1., Vienna, Austria (2005) 2–7
2. Vlachos, I.K., Sergiadis, G.D.: Intuitionistic Fuzzy Image Processing. In: Soft Computing in Image Processing: Recent Advances. Volume 210 of Studies in Fuzziness and Soft Computing. Springer (2006) 385–416
3. Atanassov, K.T.: Intuitionistic fuzzy sets. Fuzzy Sets Syst. **20** (1986) 87–96
4. Atanassov, K.T.: Intuitionistic Fuzzy Sets: Theory and Applications. Studies in Fuzziness and Soft Computing. Physica–Verlag, Heidelberg (1999)
5. Zadeh, L.A.: Fuzzy sets. Inf. Control **8** (1965) 338–353
6. Szmidt, E., Kacprzyk, J.: Distances between intuitionistic fuzzy sets. Fuzzy Sets Syst. **114** (2000) 505–518
7. Kosko, B.: Fuzziness vs. probability. Int. J. Gen. Syst. **17** (1990) 211–240
8. Szmidt, E., Kacprzyk, J.: Entropy for intuitionistic fuzzy sets. Fuzzy Sets Syst. **118** (2001) 467–477
9. De Luca, A., Termini, S.: A definition of a nonprobabilistic entropy in the setting of fuzzy sets theory. Inf. Control **20** (1972) 301–312
10. Burillo, P., Bustince, H.: Entropy on intuitionistic fuzzy sets and on interval-valued fuzzy sets. Fuzzy Sets Syst. **78** (1996) 305–316
11. Szmidt, E., Kacprzyk, J.: Entropy and similarity of intuitionistic fuzzy sets. In: Proc. Information Processing and Management of Uncertainty in Knowledge-Based Systems, Paris, France (2006) 2375–2382
12. Vlachos, I.K., Sergiadis, G.D.: Intuitionistic fuzzy information–Applications to pattern recognition. Pattern Recognit. Lett. **28** (2006) 197–206
13. Vlachos, I.K., Sergiadis, G.D.: Inner product based entropy in the intuitionistic fuzzy setting. Int. J. Uncertainty Fuzziness Knowl.-Based Syst. **14** (2006) 351–366
14. Pal, S.K., King, R.A.: Image enhancement using fuzzy set. Electron. Lett. **16** (1980) 376–378
15. Pal, S.K., King, R.A.: Image enhancement using smoothing with fuzzy sets. IEEE Trans. Syst. Man Cybern. **11** (1981) 495–501

# Representation of Rough Sets Based on Intuitionistic Fuzzy Special Sets

Zheng Pei, Li Zhang, and Honghua Chen

School of Mathematics & Computer, Xihua University,
Chengdu, Sichuan, 610039, China
pqyz@263.net

**Abstract.** Intuitionistic fuzzy special sets is a special case of intuitionistic fuzzy sets. In this paper, under the framework of information systems, the relationship between intuitionistic fuzzy special sets and rough sets is analyzed. Based on basic intuitionistic fuzzy special sets of information systems, intuitionistic fuzzy special $\sigma$-algebra are generated, and rough sets are embedded in the intuitionistic fuzzy special $\sigma$-algebra. Naturally, distances (*e.g.*, Hamming distance or Euclidean distance) of intuitionistic fuzzy special sets in intuitionistic fuzzy special $\sigma$-algebra can be used to evaluate predication rules of information systems which is an important subject of rough set theory.

## 1 Introduction

As generalization of fuzzy sets, intuitionistic fuzzy sets use the degree of membership and nonmembership of object $x$ [1]-[10]. In some cases, intuitionistic fuzzy sets has more advantages than classical fuzzy sets in describing uncertain concepts. Rough sets theory (RST) proposed by Z. Pawlak is an important theory for data mining [11], [12]. In the process of uncertain information, fuzzy sets theory and rough sets theory both have advantages, respectively.

Formally, intuitionistic fuzzy subset $A$ is $A = \{(x, \mu_A(x), \nu_A(x))|x \in X\}$, in which, $X$ is domain, $\mu_A : X \to [0,1]$ and $\nu_A : X \to [0,1]$ are membership function and nonmembership function of object $x$ in $A$ such that $\forall x \in X$, $0 \leq \mu_A(x)+\nu_A(x) \leq 1$. Intuitionistic fuzzy special subset (IFSS) is $A = \{X, A_1, A_2\}$, in which, $X \neq \emptyset$, $A_1 \subseteq X$, $A_2 \subseteq X$ and $A_1 \cap A_2 = \emptyset$.

1. $\forall B \subset X$, define $B' = \langle X, B, B^c\rangle$ ($B^c$ is the complement of $B$ in $X$), then $B \longleftrightarrow B'$. This means that IFSS is extension of classical subset.
2. Let $A = \{X, A_1, A_2\}$ be IFSS, define

$$\mu_{A_1}(x) = \begin{cases} 1 & \text{if} \quad x \in A_1 \\ 0 & \text{otherwise}, \end{cases} \qquad \nu_{A_2}(x) = \begin{cases} 1 & \text{if} \quad x \in A_2 \\ 0 & \text{otherwise}, \end{cases} \tag{1}$$

   then $\{(x, \mu_{A_1}(x), \nu_{A_2}(x))|x \in X\}$ such that $\forall x \in X$, due to $A_1 \cap A_2 = \emptyset$, $\mu_{A_1}(x)+\nu_{A_2}(x) = 1$ or 0, hence, IFSS is a special case of intuitionistic fuzzy sets.

P. Melin et al. (Eds.): IFSA 2007, LNAI 4529, pp. 114–121, 2007.
© Springer-Verlag Berlin Heidelberg 2007

In IFSS, the following operations can be defined [4]: Let $A = \langle X, A_1, A_2\rangle$, $B = \langle X, B_1, B_2\rangle$ and $\{A_i|i \in J\}$, where $A_i = \langle X, A_i^1, A_i^2\rangle$ be IFSS on $X$, then

1. $A \subset B$ if and only if $A_1 \subset B_1$ and $A_2 \supset B_2$;
2. $A = B$ if and only if $A \subset B$ and $A \supset B$;
3. $\overline{A} = \langle X, A_2, A_1\rangle$;
4. $\emptyset^- = \langle X, \emptyset, X\rangle$, $X^- = \langle X, X, \emptyset\rangle$;
5. $\cup A_i = \langle X, \cup A_i^1, \cap A_i^2\rangle$, $\cap A_i = \langle X, \cap A_i^1, \cup A_i^2\rangle$;
6. $A - B = A \cap \overline{B}$.

Based on the above operations of IFSS, intuitionistic fuzzy special $\sigma$-algebra can be defined as following

**Definition 1.** *[4] Intuitionistic fuzzy special $\sigma$-algebra $\Phi$ on $X$ is such that*

*a)* $X^- \in \Phi$;
*b) If* $A \in \Phi$, *then* $\overline{A} \in \Phi$;
*c) If* $\forall n \in N$, $A_1, \cdots, A_n \in \Phi$, *then* $\cup_{i=1}^n A_i \in \Phi$.

## 2 Representation of Rough Set Based on IFSS

Rough sets are defined on information systems. Formally, an information system is expressed as a quaternion denoted as $(U, A, V, f)$, where $U$ is a non-empty set of objects, $A$ is a non-empty finite set of attributes, $V = \bigcup_{a\in A} V_a$ and $V_a$ is the domain of $a$, $f : U \times A \to V$ is information function. In $(U, A, V, f)$, $\forall a \in A$ and $x_i, x_j \in U$, define $x_i \sim_a x_j$ if and only if $f(x_1, a) = f(x_2, a)$, then $\sim_a$ is an equivalence relation on $U$. As we known, intersection of equivalence relations is also equivalence relation, hence, $\sim_A$, which is intersection of all $\sim_a$ $(a \in A)$, is an equivalence relation on $U$, denotes $U/\sim_A = \{U^k|k = 1, \cdots, n\}$, where $U^k$ is an equivalence class. Based on $U/\sim_A$, $\forall X \subseteq U$, define

$$\underline{X} = \bigcup\{U^k \in U/\sim_A |U^k \subseteq X\}, \quad \overline{X} = \bigcup\{U^k \in U/\sim_A |U^k \cap X \neq \emptyset\}, \quad (2)$$

when $\underline{X} \neq \overline{X}$, $(\underline{X}, \overline{X})$ is $\sim_A$ rough set. In (2), due to $\underline{X} \subseteq \overline{X}$, $\underline{X} \cap (U - \overline{X}) = \emptyset$ is obviously. Hence, $\langle U, \underline{X}, U - \overline{X}\rangle$ is IFSS based on $\sim_A$, this means that in the framework of $U/\sim_A$, $\forall X \subseteq U$, there is the following one to one mapping

$$(\underline{X}, \overline{X}) \longleftrightarrow \langle U, \underline{X}, U - \overline{X}\rangle. \quad (3)$$

**Definition 2.** *Let $(U, A, V, f)$ be an information system, $\forall X \subseteq U$, $\langle U, \underline{X}, U-\overline{X}\rangle$ is called IFSS representation of $(\underline{X}, \overline{X})$.*

As a special case, $(U^i, U^i) \longleftrightarrow \langle U, U_\theta^i, (U_\theta^i)^c\rangle = \langle U, U_\theta^i, \bigcup_{j\neq i} U_\theta^j\rangle$.

**Definition 3.** *Let $(U, A, V, f)$ be an information system, $U/\sim_A = \{U^k|k = 1, \cdots, n\}$. If $\langle U, U^{k_1}, U^{k_2}\rangle$ such that $U^{k_1}, U^{k_2} \in U/\sim_A$ and $k_1 \neq k_2$, then $\langle U, U^{k_1}, U^{k_2}\rangle$ is called basic IFSS based on $\sim_A$. Denote $B_{\sim_A} = \{\langle U, U^{k_1}, U^{k_2}\rangle|U^{k_1}, U^{k_2} \in U/\sim_A, k_1 \neq k_2\}$.*

In the Definition, for a fixed $(U, A, V, f)$, if $|U/\sim_A| = n$, then $|B_{\sim_A}| = n(n-1)$. Let a class of recursive set $\Phi(B_{\sim_A})$ such that

1. $\forall\langle U, U^{k_1}, U^{k_2}\rangle \in B_{\sim_A}, \langle U, U^{k_1}, U^{k_2}\rangle \in \Phi(B_{\sim_A})$;
2. If $\langle U, A_1, B_1\rangle, \langle U, A_2, B_2\rangle \in \Phi(B_{\sim_A})$, then
$$\langle U, A_1, B_1\rangle \cap \langle U, A_2, B_2\rangle = \langle U, A_1 \cap A_2, B_1 \cup B_2\rangle \in \Phi(B_{\sim_A}).$$
3. If $\langle U, A_1, B_1\rangle \in \Phi(B_{\sim_A})$, then $\overline{\langle U, A_1, B_1\rangle} = \langle U, B_1, A_1\rangle \in \Phi(B_{\sim_A})$.

*Property 1.* $\forall\langle U, A_1, B_1\rangle \in \Phi(B_{\sim_A})$, $A_1 \cap B_1 = \emptyset$, *i.e.*, $\langle U, A_1, B_1\rangle$ is IFSS.

*Proof.* According to structure of $\langle U, A_1, B_1\rangle \in \Phi(B_{\sim_A})$ and $\overline{\langle U, U^{k_1}, U^{k_2}\rangle} = \langle U, U^{k_2,}U^{k_1}\rangle \in B_{\sim_A}$

a) If there exists $\langle U, U^{k_1}, U^{k_2}\rangle \in B_{\sim_A}$ such that $\langle U, A_1, B_1\rangle = \langle U, U^{k_1}, U^{k_2}\rangle$, then $A_1 \cap B_1 = \emptyset$, $\langle U, A_1, B_1\rangle$ is IFSS.

b) If there exist $\langle U, U^{k_1}, U^{k'_1}\rangle, \cdots, \langle U, U^{k_m}, U^{k'_m}\rangle \in B_{\sim_A}$ such that $\langle U, A_1, B_1\rangle = \bigcap_{i=1}^{m}\langle U, U^{k_i}, U^{k'_i}\rangle$, $\forall i, k_i \neq k'_i$, then

$$A_1 \cap B_1 = (\bigcap_{i=1}^{m} U^{k_i})\bigcap(\bigcup_{i=1}^{m} U^{k'_i}) = \bigcup_{j=1}^{m}((\bigcap_{i=1}^{m} U^{k_i}) \cap U^{k'_j}),$$

in $(\bigcap_{i=1}^{m} U^{k_i}) \cap U^{k'_j}$, due to $U^{k_j} \cap U^{k'_j} = \emptyset$, hence, every $(\bigcap_{i=1}^{m} U^{k_i}) \cap U^{k'_j} = \emptyset$, $\langle U, A_1, B_1\rangle$ is IFSS.

c) If there exist $\langle U, A_1, B_1\rangle, \langle U, A_2, B_2\rangle \in \Phi(B_{\sim_A})$ such that $\langle U, A_1, B_1\rangle = \bigcap_{i=1}^{m}\langle U, U^{k_i}, U^{k'_i}\rangle$, $\langle U, A_2, B_2\rangle = \bigcap_{j=1}^{r}\langle U, U^{l_j}, U^{l'_j}\rangle$ and $\langle U, A_3, B_3\rangle = \langle U, A_1, B_1\rangle \cap\langle U, A_2, B_2\rangle$, then

$$A_3 \cap B_3 = (A_1 \cap A_2)\bigcap(B_1 \cup B_2) = ((A_1 \cap A_2) \cap B_1)\bigcup((A_1 \cap A_2) \cap B_2),$$

according to b), $A_3 \cap B_3 = \emptyset$, $\langle U, A_3, B_3\rangle$ is IFSS.

d) If there exist $\langle U, A_1, B_1\rangle, \langle U, A_2, B_2\rangle \in \Phi(B_{\sim_A})$ such that $\langle U, A_3, B_3\rangle = \overline{\langle U, A_1, B_1\rangle} \cap \langle U, A_2, B_2\rangle$ or $\langle U, A_3, B_3\rangle = \overline{\langle U, A_1, B_1\rangle} \cap \overline{\langle U, A_2, B_2\rangle}$, then

$$A_3 \cap B_3 = (B_1 \cap A_2)\bigcap(A_1 \cup B_2) = ((B_1 \cap A_2) \cap A_1)\bigcup((B_1 \cap A_2) \cap B_2), \text{ or}$$

$$A_3 \cap B_3 = (B_1 \cap B_2)\bigcap(A_1 \cup A_2) = ((B_1 \cap B_2) \cap A_1)\bigcup((B_1 \cap B_2) \cap A_2),$$

according to c), $A_3 \cap B_3 = \emptyset$, $\langle U, A_3, B_3\rangle$ is IFSS.

**Theorem 1.** *$\Phi(B_{\sim_A})$ is an intuitionistic fuzzy special $\sigma$-algebra generated by $B_{\sim_A}$.*

*Proof.* According to $\Phi(B_{\sim_A})$, a) and c) of Definition 1 need to be proved. Fixed $k_1, k'_1 \in N$, then $\forall k_2(\neq k_1) \in N$ and $\forall k'_2(\neq k'_1) \in N$, $\langle U, U^{k_1}, U^{k_2}\rangle \in B_{\sim_A}$ and $\langle U, U^{k'_1}, U^{k'_2}\rangle \in B_{\sim_A}$, hence,

$$\bigcap_{k_2\neq k_1, k_2=1}^{n}\langle U, U^{k_1}, U^{k_2}\rangle = \langle U, U^{k_1}, \bigcup_{k_2\neq k_1, k_2=1}^{n} U^{k_2}\rangle \in \Phi(B_{\sim_A}),$$
$$\bigcap_{k'_2\neq k'_1, k'_2=1}^{n}\langle U, U^{k'_1}, U^{k'_2}\rangle = \langle U, U^{k'_1}, \bigcup_{k'_2\neq k'_1, k'_2=1}^{n} U^{k'_2}\rangle \in \Phi(B_{\sim_A}).$$

Due to $U^{k_1} \cap U^{k'_1} = \emptyset$ and

$$(\bigcup_{k_2 \neq k_1, k_2=1}^{n} U^{k_2}) \bigcup (\bigcup_{k'_2 \neq k'_1, k'_2=1}^{n} U^{k'_2}) = U,$$

$$\langle U, \emptyset, U\rangle = \langle U, U^{k_1}, \bigcup_{k_2 \neq k_1, k_2=1}^{n} U^{k_2}\rangle \bigcap \langle U, U^{k'_1}, \bigcup_{k'_2 \neq k'_1, k'_2=1}^{n} U^{k'_2}\rangle \in \Phi(B_{\sim_A}),$$

$$U^{-} = \langle U, U, \emptyset\rangle = \overline{\langle U, \emptyset, U\rangle} \in \Phi(B_{\sim_A}).$$

For c) of Definition 1, due to

$$\langle U, U^{k_1}, U^{k_2}\rangle \bigcup \langle U, U^{k'_1}, U^{k'_2}\rangle = \overline{\overline{\langle U, U^{k_1}, U^{k_2}\rangle} \bigcap \overline{\langle U, U^{k'_1}, U^{k'_2}\rangle}} \in \Phi(B_{\sim_A}).$$

*Property 2.* In $\Phi(B_{\sim_A})$,

1. $\langle U, \emptyset, \emptyset\rangle \in \Phi(B_{\sim_A})$;
2. $\forall U^i \in U/\sim_A$, $\langle U, U^i, (U^i)^c\rangle \in \Phi(B_{\sim_A})$;
3. $\forall X \subseteq U$, $\langle U, \emptyset, \underline{X}\rangle, \langle U, \emptyset, \overline{X}\rangle \in \Phi(B_{\sim_A})$;
4. $\forall X \subseteq U$, $\langle U, \underline{X}, U - \overline{X}\rangle \in \Phi(B_{\sim_A})$.

*Proof.* Let $\underline{X} = \{U^{k'_1}, \cdots, U^{k'_m}\}$, then

$$\langle U, \emptyset, \underline{X}\rangle = \bigcap_{i=1}^{m} \langle U, U^{k_i}, U^{k'_i}\rangle = \langle U, \bigcap_{i=1}^{m} U^{k_i}, \bigcup_{i=1}^{m} U^{k'_i}\rangle, \tag{4}$$

in which, $\exists i$ and $j$ such that $k_i \neq k_j$. The others can be proved similarly.

Based on intuitionistic fuzzy special $\sigma$-algebra $\Phi(B_{\sim_A})$, measures on $\Phi(B_{\sim_A})$ can be defined.

**Definition 4.** *[13] $\forall A = \langle U, A^1, A^2\rangle \in \Phi(B_{\sim_A})$, define $\mu : \Phi(BU^{\theta}) \rightarrow [0, \infty)$ as following:*

$$\mu(A) = 1 + \frac{|A^1|}{|U|} - \frac{|A^2|}{|U|}, \tag{5}$$

*then $\mu$ is a measure on $\Phi(B_{\sim_A})$, where $|X|$ is cardinality of $X$.*

For $\langle U, A_1, B_1\rangle, \langle U, A_2, B_2\rangle \in \Phi(B_{\sim_A})$, the distance between $\langle U, A_1, B_1\rangle$ and $\langle U, A_2, B_2\rangle$ can be defined similarly as in intuitionistic fuzzy sets [14]-[20], *e.g.*, Hamming distance and Euclidean distance,

$$d_1 = \sum_{x \in U} |\mu_{A_1}(x) - \mu_{A_2}(x)| + |\nu_{B_1}(x) - \nu_{B_2}(x)| + |\pi_{C_1}(x) - \pi_{C_2}(x)|, \tag{6}$$

$$d_2 = \sqrt{\sum_{x \in U} (\mu_{A_1}(x) - \mu_{A_2}(x))^2 + (\nu_{B_1}(x) - \nu_{B_2}(x))^2 + (\pi_{C_1}(x) - \pi_{C_2}(x))^2} \tag{7}$$

where, $\pi_{C_1}(x) = 1 - \mu_{A_1}(x) - \nu_{B_1}(x)$ and $\pi_{C_2}(x) = 1 - \mu_{A_2}(x) - \nu_{B_2}(x)$. In intuitionistic fuzzy sets, $\mu_*, \mu_*$ and $\pi_*$ are fuzzy sets. In intuitionistic fuzzy special sets, $\mu_*, \mu_*$ and $\pi_*$ are characteristic functions of $*$, *i.e.*, the forms of (1).

# 3 Prediction Based on the Distance of IFSS

As a special case of information system, decision information systems $\langle U, \Omega, V, f\rangle$ are widely used in application. In decision information systems, attributes $\Omega$ are divided by two parts: one is called condition attribute set, denoted by $Q$; the other is called decision attribute set, denoted by $D$. From real world application point of view, the prediction problem is expressed as following [12]

- given a decision attribute $d \in D$, which is the "best" attribute set $C \subseteq Q$ to predict the $d$-value of an object $x \in U$, given the values of $x$ under the features contained in $Q$?

The prediction problem raises two questions:

- Which subsets $C$ of $Q$ are candidates to be such a "best attribute set"?
- What should a metric look like to determine and select the "best attribute set"?

In this paper, the distance of IFSS is used to solve the prediction problem. Suppose that all information of decision information system $I = \langle U, Q \cup D, V, f\rangle$ are known, let $C = \{c_1, \cdots, c_k\} \subseteq Q$ and $d \in D$, decision rule is expressed as

$$R : c_1 \wedge \cdots \wedge c_k \longrightarrow d. \tag{8}$$

According to the above discussions, equivalence relations $\sim_C$, $\sim_d$ and $\sim_{C\cup\{d\}}$ on $U$ can be obtained in $\langle U, C \cup \{d\}, V, f\rangle$, denote

$$U/\sim_C = \{U_C^1, \cdots, U_C^m\}, \quad U/\sim_d = \{U_d^1, \cdots, U_d^n\}, \tag{9}$$

$$U/\sim_{C\cup\{d\}} = \{U_{C\cup\{d\}}^1, \cdots, U_{C\cup\{d\}}^p\}. \tag{10}$$

For every $U_C^{m'} \in U/\sim_C$ and $U_d^{n'} \in U/\sim_d$, based on $U/\sim_{C\cup\{d\}}$, their rough set, *i.e.*, $(\underline{U_C^{m'}}, \overline{U_C^{m'}})$ and $(\underline{U_d^{n'}}, \overline{U_d^{n'}})$, can be obtained, respectively. By (3), IFSS representations of $(\underline{U_C^{m'}}, \overline{U_C^{m'}})$ and $(\underline{U_d^{n'}}, \overline{U_d^{n'}})$, *i.e.*, $\langle U, \underline{U_C^{m'}}, U - \overline{U_C^{m'}}\rangle$ and $\langle U, \underline{U_d^{n'}}, U - \overline{U_d^{n'}}\rangle$, can be obtained, respectively. Based on $U/\sim_{C\cup\{d\}}$, (9) can be rewritten as following

$$U/\sim_C = \{\langle U, \underline{U_C^1}, U - \overline{U_C^1}\rangle, \cdots, \langle U, \underline{U_C^m}, U - \overline{U_C^m}\rangle\}, \tag{11}$$

$$U/\sim_d = \{\langle U, \underline{U_d^1}, U - \overline{U_d^1}\rangle, \cdots, \langle U, \underline{U_d^n}, U - \overline{U_d^n}\rangle\}. \tag{12}$$

For every $\langle U, \underline{U_C^{m'}}, U - \overline{U_C^{m'}}\rangle \in U/\sim_C$ and $\langle U, \underline{U_d^{n'}}, U - \overline{U_d^{n'}}\rangle \in U/\sim_d$, using (6) and (7), their Hamming distance $d_{m'n'}^H$ and Euclidean distance $d_{m'n'}^E$ can be calculated, respectively, denote

$$\overline{d_{C,d}^H} = \frac{\sum_{m'=1}^{m}\sum_{n'=1}^{n} d_{m'n'}^H}{mn}, \tag{13}$$

$$\overline{d_{C,d}^E} = \frac{\sum_{m'=1}^{m}\sum_{n'=1}^{n} d_{m'n'}^E}{mn}. \tag{14}$$

(13) and (14) are average values of Hamming distances and Euclidean distances about rule $R : c_1 \wedge \cdots \wedge c_k \longrightarrow d$, respectively.

From the standpoint of logical systems, $c_1 \wedge \cdots \wedge c_k \longrightarrow d$ is logical proposition. Considering information system $\langle U, C \cup \{d\}, V, f\rangle$, $c_1 \wedge \cdots \wedge c_k \longrightarrow d$ is meaning when all objects, which have attributes $c_1, \cdots, c_{k-1}$ and $c_k$, are asserted to own attribute $d$. From the standpoint of information theory, if $c_1 \wedge \cdots \wedge c_k \longrightarrow d$ is considered as knowledge, then $\langle U, \underline{U_C^{m'}}, U - \overline{U_C^{m'}}\rangle \in U/\sim_C$ and $\langle U, \underline{U_d^{n'}}, U - \overline{U_d^{n'}}\rangle \in U/\sim_d$ can be regarded as certain information about objects have or not attributes $\{c_1, \cdots, c_k\}$ and $\{d\}$ based on the knowledge, respectively. Hence, from objects point of view, $d^H_{m'n'}$ (or $d^E_{m'n'}$) express similarity degree of the certain information between conditions and conclusion of $c_1 \wedge \cdots \wedge c_k \longrightarrow d$. In this paper, $\overline{d^H_{C,d}}$ (or $\overline{d^E_{C,d}}$) are selected as evaluation index of rule $c_1 \wedge \cdots \wedge c_k \longrightarrow d$, *i.e.*, for a fixed decision attribute $d$, the "best" attribute set $C \subseteq Q$ is such that

$$\overline{d^H_{C,d}} = min\{\overline{d^H_{C_1,d}}, \cdots, \overline{d^H_{C_s,d}}\}, \tag{15}$$

$$\overline{d^E_{C,d}} = min\{\overline{d^E_{C_1,d}}, \cdots, \overline{d^E_{C_s,d}}\}, \tag{16}$$

in which, $C_1, \cdots, C_s \subseteq Q$ are candidates to predict $d$.

*Example 1.* Table 1 [12] is a heart disease diagnosis information system, the condition attributes are $S$: smoker and $BMI$: avoirdupois, the decision attribute is $HD$: heart disease.

**Table 1.** Heart disease diagnosis information system.

| No | S | BMI | HD |
|---|---|---|---|
| 1 | no | normal | no |
| 2 | no | obese | no |
| 3 | no | normal | no |
| 4 | no | obese | no |
| 5 | yes | normal | yes |
| 6 | yes | normal | yes |
| 7 | yes | obese | no |
| 8 | yes | obese | yes |
| 9 | no | normal | no |

In Table 1, the following equivalence classes can be obtained: $\{S\} : \{1, 2, 3, 4, 9\}$, $\{5, 6, 7, 8\}$; $\{BMI\} : \{1, 3, 5, 6, 9\}$, $\{2, 4, 7, 8\}$; $\{S, BMI\} : \{1, 3, 9\}$, $\{2, 4\}$, $\{5, 6\}$, $\{7, 8\}$; $\{HD\} : \{1, 2, 3, 4, 7, 9\}$, $\{5, 6, 8\}$.

There are three candidates to predict $HD$, *i.e.*,

$$S \longrightarrow HD, \quad BMI \longrightarrow HD, \quad S \wedge BMI \longrightarrow HD.$$

For $S \longrightarrow HD$, the equivalence classes are $\{1, 2, 3, 4, 9\}$, $\{5, 6, 8\}$ and $\{7\}$, according to (11) and (12), the following can be obtained

$$U/\sim_S = \{\langle U, \{1, 2, 3, 4, 9\}, \{5, 6, 7, 8\}\rangle, \langle U, \{5, 6, 7, 8\}, \{1, 2, 3, 4, 9\}\rangle\},$$

$$U/\sim_{HD} = \{\langle U, \{1, 2, 3, 4, 7, 9\}, \{5, 6, 8\}\rangle, \langle U, \{5, 6, 8\}, \{1, 2, 3, 4, 7, 9\}\rangle\},$$

By (6) and (7),

$$d_{11}^H(\langle U,\{1,2,3,4,9\},\{5,6,7,8\}\rangle,\langle U,\{1,2,3,4,7,9\},\{5,6,8\}\rangle)=2,$$
$$d_{12}^E(\langle U,\{1,2,3,4,9\},\{5,6,7,8\}\rangle,\langle U,\{1,2,3,4,7,9\},\{5,6,8\}\rangle)=\sqrt{2},$$

the others can be calculated similarly,

$$\overline{d_{S,HD}^H}=\frac{d_{11}^H+d_{12}^H+d_{21}^H+d_{22}^H}{4}=9,$$
$$\overline{d_{S,HD}^E}=\frac{d_{11}^H+d_{12}^H+d_{21}^H+d_{22}^H}{4}=\frac{\sqrt{2}+4}{2}.$$

For $BMI \longrightarrow HD$ and $S \wedge BMI \longrightarrow HD$, average values of Hamming distances and Euclidean distances can be obtained similarly, respectively, see Tabel 2.

**Table 2.** Average values of prediction rules

| | Hamming distance | Euclidean distance |
|---|---|---|
| $\{S\}\rightarrow\{HD\}$ | 9 | $\frac{\sqrt{2}+4}{2}$ |
| $\{BMI\}\rightarrow\{HD\}$ | $\frac{21}{2}$ | $\frac{2\sqrt{2}+2\sqrt{10}+\sqrt{14}}{4}$ |
| $\{S,BMI\}\rightarrow\{HD\}$ | 9 | $\frac{3\sqrt{2}+4\sqrt{3}+2\sqrt{6}+\sqrt{10}+4}{8}$ |

According to Table2, if using Hamming distance, then the "best" attribute set, which is used to predict $HD$, is $\{S\}$ or $\{S,BMI\}$. If using Euclidean distance, then the "best" attribute set is $\{S\}$. The conclusion is same as in [12].

## 4 Conclusion

In this paper, under the framework of information systems, intuitionistic fuzzy special $\sigma$-algebra are generated by basic IFSS of information systems, and rough sets are embedded in the intuitionistic fuzzy special $\sigma$-algebra. Hamming distance (or Euclidean distance) of intuitionistic fuzzy special sets in intuitionistic fuzzy special $\sigma$-algebra are used to evaluate predication rules of information systems.

## Acknowledgments

This work was supported by the youth foundation of sichuan province (grant no. 06ZQ026-037) and the education department foundation of sichuan province (grant no. 2005A121, 2006A084).

## References

1. Atanassov, K. T.: Two theorems for intuitionistic fuzzy sets. Fuzzy Sets and Systems. 110 (2000) 267–269
2. De, S. K., Biswas, R., Roy, A. R.: Some operations on intuitionistic fuzzy sets. Fuzzy Sets and Systems. 114 (2000) 477–484
3. Bustince, H., Kacprzyk, J., Mohedano, V.: Fuzzy generators application to intuitionistic fuzzy complementation. Fuzzy Sets and System. 114 485–504
4. Coskun, E.: Systems on intuitionistic fuzzy special sets and intuitionistic fuzzy special measures. Information Sciences. 128 (2000) 105–118
5. Szmidt, E., Kacprzyk, J.: Entropy for intuitionistic fuzzy sets Fuzzy Sets and Systems. 118 (2001) 467–477
6. Davvaz, B., Dudek, W. A., Jun, Y. B.: Intuitionistic fuzzy $H_v$-submodules Information Sciences. 176 (2006) 285–300
7. Mondal, T. K., Samanta, S. K.: On intuitionistic gradation of openness. Fuzzy Sets and Systems. 131 (2002) 323–336
8. Grzegorzewski, P., Mrowka, E.: Some notes on (Atanassov's) intuitionistic fuzzy sets. Fuzzy Sets and Systems. 156 (2005) 492–495
9. Deschrijver, G., Kerre, E. E.: On the relationship between some extensions of fuzzy set theory. Fuzzy Sets and Systems. 133 (2003) 227–235
10. Wang, G. J., Tic, Y. Y.: Intuitionistic fuzzy sets and L-fuzzy sets. Fuzzy Sets and Systems. 110 (2000) 271–278
11. Pawlak, Z.: Rough Sets. International Journal of Computer and Information Science. 11(5) (1982) 341–356
12. Duntsch, I., Gediga, G.: Uncertainty measures of rough set prediction. Artificial Intelligence. 106 (1998) 109–137
13. Pei,Z., Qin, K. Y.: intuitionistic fuzzy special set expression of rough set and its application in reduction of attributes. Pattern Recognition and Artificial Intelligence. 17(3) (2004) 262–268
14. Gregori, V., Romaguera, S., Veeramani, P.: A note on intuitionistic fuzzy metric spaces. Chaos, Solitons and Fractals. 28 (2006) 902–905
15. Szmidt, E., Kacprzyk, J.: Distances between intuitionistic fuzzy sets. Fuzzy Sets and Systems. 114 (2000) 505–518
16. Li, D. F.: Some measures of dissimilarity in intuitionistic fuzzy structures. Journal of Computer and System Sciences. 68 (2004) 115–122
17. Li, D. F.: Multiattribute decision making models and methods using intuitionistic fuzzy sets. Journal of Computer and System Sciences. 70 (2005) 73–85
18. Wang, W., Xin, X. L.: Distance measure between intuitionistic fuzzy setsPattern Recognition Letters. 26 (2005) 2063–2069
19. Liang, Z. Z., Shi, P. F.: Similarity measures on intuitionistic fuzzy sets. Pattern Recognition Letters. 24 (2003) 2687–2693
20. Shu, M. H., Cheng, C. H., Chang, J. R.: Using intuitionistic fuzzy sets for fault-tree analysis on printed circuit board assembly. Microelectronics Reliability. (2006) (in press)

# Part III

# The Application of Fuzzy Logic and Soft Computing in Flexible Querying

# Towards Vague Query Answering in Logic Programming for Logic-Based Information Retrieval

Umberto Straccia

ISTI - CNR, Pisa, Italy
straccia@isti.cnr.it

**Abstract.** We address a novel issue for logic programming, namely the problem of evaluating ranked top-k queries. The problem occurs for instance, when we allow queries such as "find cheap hotels close to the conference location" in which vague predicates like cheap and close occur. Vague predicates have the effect that each tuple in the answer set has now a score in [0,1]. We show how to compute the top-k answers in case the set of facts is huge, without evaluating all the tuples.

**Keywords:** Logic Programming, Fuzzy, Top-k retrieval.

## 1 Introduction

In this paper we address a novel issue for *Logic Programs* (LPs) with a huge set of facts, namely the problem of *evaluating ranked top-k queries*. In classical logic programming, an answer to a query is a set of tuples that satisfy a query. Each tuple may or may not satisfy the predicates in the query. However, very often the information need of a user involves so-called *fuzzy/vague predicates*. For instance, a user may have the following information need: "Find *cheap* hotels *near* to the conference location". Here, *cheap* and *near* are fuzzy predicates. Unlike the classical case, tuples satisfy now these predicates to a score (usually in $[0, 1]$). In the former case the score may depend, e.g., on the price, while in the latter case it may depend e.g. on the distance between the hotel location and the conference location.

Therefore, a major problem we have to face with in such cases is that now an answer is a set of tuples *ranked* according to their *score*. This poses a new challenge in case we have to deal with a huge amount of facts. Indeed, virtually every tuple may satisfy a query with a non-zero score and, thus, has to be ranked. Of course, computing all these scores, ranking them and then selecting the top-$k$ ones is not feasible in practice.

In this work, we address the top-$k$ retrieval problem for Datalog. At the extensional level, each fact may have a score, while at the intentional level rules describe the domain of application. Queries are conjunctive queries in which vague predicates may occur.

P. Melin et al. (Eds.): IFSA 2007, LNAI 4529, pp. 125–134, 2007.
© Springer-Verlag Berlin Heidelberg 2007

## 2 Preliminaries

The formalism we consider is defined as follows. The *score space* (i.e. *truth space*) is $\mathcal{S} = [0, 1]$. We anticipate informally that an interpretation will assign a score (i.e. truth) to a ground atom and that the answers to a query (i.e. ground instances of an atom) will be ranked (in decreasing order) according to their scores.

A *term* is either a *variable* or a *constant.*

Let $\mathcal{V}_E$ and $\mathcal{V}_I$ be disjoint sets of $n$-ary *extensional* and *intentional* predicate symbols, respectively. An *atom* is of the form $p(t_1, \ldots, t_n)$, where $p$ is an $n$-ary predicate symbol and all $t_j$ are terms. An atom is *ground* if no variable occurs in it. A *logic program* $\mathcal{P}$ is made out of an *extensional database* (EDB), $\mathcal{P}_E$, and an *intentional database* (IDB), $\mathcal{P}_I$. The extensional database is a set of *facts* of the form $r(\mathbf{c}) \leftarrow b$, where $r(\mathbf{c})$ is a ground atom, $r$ is an extensional predicate and $b \in \mathcal{S}$ is a score value. The intuition here is that $b$ is the assigned score to tuple $\mathbf{c}$ in relation $r$. For convenience, for each $n$-ary extensional predicate $r$, we represent the facts $r(c_1, \ldots, c_n) \leftarrow b$ in $\mathcal{P}$ by means of a relational $n+1$-ary table $T_r$, containing the records $\langle c_1, \ldots, c_n, b \rangle$. Thus, the table contains all the instances of $r$ together with their scores. We assume that there cannot be two records $\langle c_1, \ldots, c_n, b_1 \rangle$ and $\langle c_1, \ldots, c_n, b_2 \rangle$ in $T_r$ with $b_1 \neq b_2$ (in case their are, we remove the one with the lower score). Each table is sorted in descending order with respect to the scores. Usually, the score of a tuple in a relation has been computed (possibly off-line) by a specific system. For instance, we may have an underlying image retrieval system that for each identified object in an image has $m$ scores, one for each of the $m$ attributes (see, e.g. [8]).

The intentional database is a set of *rules* in which all variables in the head do also appear in the rule body. To facilitate the reading, we first give an example of rule with its intended meaning and then provide the formal definition. Assume we would like to represent the set of good conference hotels $q(x)$, which are hotels close to the conference location. Assume that we have a relational table of hotels, their location and their price, a table of conferences and their location, a distance table reporting the distance among two locations, and two tables for the extensional predicates, *cheap* and *close*, whose instances are scored with respect to the following functions, where the former depends on the price, while the latter depends on the distance: $s_{cheap}(p) = \max(0, 1 - p/200)$, and $s_{close}(d) = \max(0, 1 - d/2000)$. Then following rule may be a candidate rule: for a given conference $c$

$$q(h) \leftarrow \min[hotel(h, hLoc, price), conference(c, cLoc), distance(hLoc, cLoc, d)] \cdot cheap(price) \cdot close(d).$$

Essentially, in the above rule, the score of $q(h)$ is determined by taking the min of the first three atoms and then take the product of it with the last expression. This is similar as it happens in top-$k$ retrieval in the context of relational databases [3,4,6]: the data is represented in relational tables and the SQL query language is extended to allow to express a scoring function, which may use the values occurring in the retrieved records, to compute the final score of the retrieved record.

So, let $\mathcal{F}$ be a set of total *score combination functions*, i.e. computable functions [1] $f\colon [0,1]^n \rightarrow [0,1]$ used to manipulate score values, e.g. $\min, \max, \cdot, +, \ldots$. Score combination functions will have a fixed interpretation, i.e. we will consider them as *built-in* functions. Then an intentional database is a set of rules of form $p(\mathbf{x}) \leftarrow f(A_1, \ldots, A_n)$, where $(i)$ $p$ is an intentional predicate; $(ii)$ $A_i$ is an atom $q(\mathbf{t})$ and $\mathbf{t}$ is a tuple of terms ($q$ is either an intentional or an extensional predicate symbol); and $(iii)$ $f$ is a score combination function, which is assumed to be monotone in its arguments. Note that the extensional predicates do not occur in the head of rules of the intentional database. Essentially, we do not allow that the fact predicates occurring in $\mathcal{P}_E$ can be redefined by $\mathcal{P}_I$. A *classical rule* is one in which $f$ is min.

From the semantics point of view, the *Herbrand universe* $H_\mathcal{P}$ of $\mathcal{P}$ is the set of constants appearing in $\mathcal{P}$. If there is no constant symbol in $\mathcal{P}$ then consider $H_\mathcal{P} = \{c\}$, where $c$ is an arbitrary chosen constant. The *Herbrand base* $B_\mathcal{P}$ of $\mathcal{P}$ is the set of ground instantiations of atoms appearing in $\mathcal{P}$ (ground instantiations are obtained by replacing all variable symbols with constants of the Herbrand universe). Let $\mathcal{P}^*$ be the set of ground rule instantiations obtained from $\mathcal{P}$. Note that $\mathcal{P}^*$ is always *finite*. An *interpretation* $I$ is a partial mapping from intentional and extensional atoms to $[0,1]$ (we recall that for a constant $c$, $I(c) = c$). Note that, as $I$ may be a partial function, some atoms may not have a score. Alternatively, we may assume $I$ to be a total function. We use the former formulation to distinguish the case where a tuple $\mathbf{c}$ may be retrieved, though the score is 0, from the tuples which do not satisfy the query and, thus, would be not retrieved. In particular, if a tuple does not belong to an extensional relation then its score is assumed to be undefined, while if $I$ is total, then the score of this tuple would be 0. We denote with $def(I)$ the set of ground atoms on which $I$ is defined. We say that $I$ is a *model* of $\mathcal{P}$, denoted $I \models \mathcal{P}$, iff for all facts $A \leftarrow b \in \mathcal{P}_E$, $I(A) \geq b$ and for all rules $A \leftarrow f(A_1, \ldots, A_n) \in \mathcal{P}_I$ such that all $I(A_i)$ are defined, $I(A) \geq f(I(A_1), \ldots, I(A_n))$ holds (note that the function $f \in \mathcal{F}$ has a fixed interpretation, which we identify with $f$ itself). We say that an interpretation $I$ is a *minimal model* of $\mathcal{P}$ iff $I \models \mathcal{P}$ and for any other model $J$ of $\mathcal{P}$, $def(I) \subseteq def(J)$ and for all $A \in def(I)$ $I(A) \leq J(A)$ holds. [2] It is not difficult to see that there is an unique minimal model $M_\mathcal{P}$ of $\mathcal{P}$. The proof is based on the existence of a partial monotone immediate consequence operator $T_\mathcal{P}$, whose fixed-points are models of $\mathcal{P}$: for any ground atom $A \in B_\mathcal{P}$ $T_\mathcal{P}(I)(A) = \max\{I(\varphi) \mid A \leftarrow \varphi \in \mathcal{P}^*\}$, where $\max \emptyset$ is undefined ($\mathcal{P}^*$ is finite, so max can be used).

A *query* is an intentional predicate symbol $q$. The *answer set* of $q$ w.r.t. $\mathcal{P}$ is defined as the set $ans(q, \mathcal{P})$ of tuples $\langle \mathbf{c}, s \rangle \in H_\mathcal{P} \times \ldots \times H_\mathcal{P} \times [0,1]$ such that $M_\mathcal{P}(q(\mathbf{c})) = s$ (the score of $\mathbf{c}$ is $s$ in the minimal model).

*Example 1.* Given the logic program $\mathcal{P} = \{(q(x) \leftarrow 0.5 \cdot (p(x) + r(x))), (p(a) \leftarrow 0.9), (p(b) \leftarrow 0.2), (r(b) \leftarrow 0.4)\}$ then $M_\mathcal{P}(p(a)) = 0.9$, $M_\mathcal{P}(p(b)) = 0.2$,

---

[1] With computable we mean that for any input, the value of $f$ can be determined in finite time.

[2] The least interpretation is unique and is $\mathtt{I}_\perp$, where $def(\mathtt{I}_\perp) = \emptyset$, i.e. $\mathtt{I}_\perp$ is undefined everywhere.

$M_{\mathcal{P}}(r(b)) = 0.4$, $M_{\mathcal{P}}(q(b)) = 0.3$ and $ans(q, \mathcal{P}) = \{\langle b, 0.3 \rangle\}$, while $ans(p, \mathcal{P}) = \{\langle a, 0.9 \rangle, \langle b, 0.2 \rangle\}$. If $M_{\mathcal{P}}$ has to be a total interpretation then additionally $M_{\mathcal{P}}(r(a)) = 0$, $M_{\mathcal{P}}(q(a)) = 0.45$ and $\langle a, 0.45 \rangle \in ans(q, \mathcal{P})$.

*Example 2.* Given the logic program $\mathcal{P} = \{(q(x) \leftarrow (q(x)+1)/2), (p(a) \leftarrow 0.4)\}$ then $M_{\mathcal{P}}(p(a)) = 0.4$, $M_{\mathcal{P}}(q(a)) = 1$ and $ans(q, \mathcal{P}) = \{\langle a, 1 \rangle\}$. Note that $\mathcal{P}$ exhibits a well-known behaviour, requiring $\omega$ steps of $T_{\mathcal{P}}$ iterations to obtain the minimal model [5,13].

The basic reasoning service that mainly concerns us is:

**Top-$k$ retrieval:** Given $\mathcal{P}$, retrieve the top-$k$ ranked tuples of the answer set of $q$ w.r.t. the score, denoted $ans_k(q, \mathcal{P}) = \mathtt{Top}_k(ans(q, \mathcal{P}))$.

We note that retrieving the top-$k$ answers of an extensional predicate symbol $r$ is trivial as we have just to retrieve the first $k$ tuples in the relational table $T_r$ associated to $r$. Hence, we restrict top-$k$ retrieval to intentional predicates only.

## 3 Top-$k$ Information Retrieval

We next provide an incremental top-down top-$k$ query answering algorithm. Note that, as Example 2 shows, computing an answer (and, thus, the top-$k$ answers) may not be possible in finite time in general. A usual way to overcome this situation is to rely on *bounded* score combination functions $f$, i.e. for all $i$, $f(x_1, \ldots, x_n) \leq x_i$. In this case it can be shown that the least-fixed point is reached after a finite number of $T_p$ iterations [9].

To start with, we use the usual relation "directly depends on" among predicate symbols, i.e. given $\mathcal{P}$, we say that predicate symbol $p$ *directly depends on* predicate symbol $q$ if there is a rule in $\mathcal{P}$ such that $p$ occurs in the head of it and $q$ occurs in the body of it. The relation *depends on* is the transitive closure of "directly depends on". The *dependency graph* of $\mathcal{P}$ is a directed graph where nodes are predicate symbols and the set of edges is the "directly depends on" relation. The program is *recursive* if there is a cycle in the dependency graph (i.e. there is $p$ depending on $p$). We also say that $\mathcal{P}$ is *deterministic* if for each intentional predicate symbol $p$ there is at most one rule in $\mathcal{P}$ having $p$ in its head.

A practical useful case is when the logic program contains only classical rules, except for the rules having the query in the head. This depicts the scenario when a top-$k$ query involving vague predicates is issued on top of a classical logic program (deductive database), as for the "find cheap hotels" example. We call such programs *classical top-$k$* programs. The top-$k$ retrieval problem for non-recursive classical top-$k$ programs has been addressed in [11], were it has been shown that for a non-recursive classical top-$k$ program $\mathcal{P}$ and query predicate $q$, $ans_k(q, \mathcal{P})$ can be determined in LogSpace w.r.t. the size of $\mathcal{P}_E$.

The procedure is based on a query reformulation step, in which a rule involving the query predicate in the head is reformulated by replacing an atom $A$ in the body by means of the rule body $\phi$, for $A \leftarrow \phi \in \mathcal{P}$, and finally applying a top-$k$ algorithm for relational databases to the obtained query transformations.

For the more general case, this simple strategy is no longer possible. Of course, we always have the possibility to compute all answers (whenever termination is guaranteed), to rank them and to select the top-$k$ ones only. However, this requires to compute the score of all answers. We would like to avoid this in cases in which the extensional database is large and potentially too many tuples would satisfy the query. A distinguishing feature of our query answering procedure is that we do not determine all answers by discovering *all proofs* as e.g. in [5,13], but rather apply a variant of so-called *memoing* techniques developed for classical logic programming –see, e.g. [15] for an overview. Essentially, the basic idea of our procedure is to collect, during the computation, all answers incrementally together in a similar way as it is done for classical Datalog. Hence, for instance, we do not rely on any notion of *atom unification*, but rather iteratively access relational tables using relational algebra.

The presentation of our algorithm proceeds as follows. We present a top-$k$ answering procedure for deterministic logic programs (at most one rule per predicate symbol $p$ in the head). Due to lack of space we are not able to include also the more general case of non-deterministic LPs as well (more than one rule per predicate symbol $p$ in the head), which will be included in an extended version of the paper. For this latter case, we just show the problem introduced and outline the solution for it. For the rest of this paper we will assume that the score combination functions are bounded, to avoid such cases as shown in Example 2.

Given $r : q(\mathbf{x}) \leftarrow \phi \in \mathcal{P}$, with $\mathtt{s}(q, r)$ we denote the set of *sons* of $q$ w.r.t. $r$, i.e. the set of intentional predicate symbols occurring in $\phi$. With $\mathtt{p}(q)$ we denote the set of *parents* of $q$, i.e. the set $\mathtt{p}(q) = \{p_i : q \in \mathtt{s}(p_i, r)\}$ (the set of predicate symbols directly depending on $q$).

*Top-k Query Answering for Deterministic LPs.* The procedure *TopAnswers* is detailed in Table 1. The procedure uses some auxiliary functions and data structures: (*i*) the variable `rankedList` contains, for each intentional predicate $p$, the current top-ranked tuples together with their score. For each $p$, the tuples $\langle \mathbf{c}, s \rangle$ in `rankedList`($p$) are ranked in decreasing order with respect to the score $s$. We do not allow $\langle \mathbf{c}, s \rangle$ and $\langle \mathbf{c}, s' \rangle$ to occur in `rankedList`($p$) with $s \neq s'$ (if so, we remove the tuple with the lower score); (*ii*) the variable `dg` collects the predicate symbols the query predicate $q$ depends on; (*iii*) the array variable `exp` traces the rule bodies that have been "expanded" (the predicate symbols occurring in the rule body are put into the active list); (*iv*) the variable `in` keeps track of the predicate symbols that have been put into the active list so far due to an expansion (to avoid, to put the same predicate symbol multiple times in the active list due to rule body expansion). There are other variables, which however do play a role in the procedure *getNextTuple* only (see Table 1), and are defined in the *TopAnswers* procedure as they act as global variables. We will discuss them in detail once we address the *getNextTuple* procedure later on.

Overall, the procedure works as follows. Assume, we are interested in determining the top-$k$ answers of $q(\mathbf{x})$. We start with putting the predicate symbol $q$

**Table 1.** The top-$k$ query answering procedure

```
Procedure TopAnswers(P, q, k)
Input: Logic program P, query predicate q, k ≥ 1;
Output: Mapping rankedList such that rankedList(q) contains top-k answers of q
Init: δ = 1, for all rules r : p(x) ← φ in P do
          if p intentional then rankedList(p) = ∅, Q(p, r) := ∅;
          if p extensional then rankedList(p) = T_p endfor
1.  loop
2.    A := {q}, dg := {q}, in := ∅, rL' := rankedList, for all rules r : p(x) ← φ do exp(p, r) = false;
3.    while (A ≠ ∅) do
4.      select p ∈ A where r : p(x) ← φ, A := A \ {p}, dg := dg ∪ s(p, r);
5.      Δ_r := getNextTuple(p, r)
6.      if Δ_r ≠ ∅ then rankedList(p) := rankedList(p) ∪ Δ_r, A := A ∪ (p(p) ∩ dg);
7.      if not exp(p, r) then exp(p, r) = true, A := A ∪ (s(p, r) \ in), in := in ∪ s(p, r);
      endwhile
8.    Update threshold δ;
9.  until (rankedList(q) does contain k top-ranked tuples with score above δ) or (rL' = rankedList);
10. return top-k ranked tuples in rankedList(q);
```

```
Procedure getNextTuple(p, r)
Input: Intentional predicate symbol p and rule r : p(x) ← f(A_1, ..., A_n) ∈ P;
Output: Next tuple satisfying the body of the r together with the score
Init: Let p_i be the predicate symbol occurring in A_i;
1. if Q(p, r) ≠ ∅ then
     ⟨t, s⟩ := getTop(Q(p, r)), remove ⟨t, s⟩ from Q(p, r), return {⟨t, s⟩} fi
   loop
2.   Generate the set T of all new join tuples t, using all tuples seen so far in all rankedList(p_i) using symmetric hash join
3.     for all t ∈ T do
4.       s := compute the score of p(t) using f;
5.       if neither ⟨t, s'⟩ ∈ rankedList(p) nor ⟨t, s'⟩ ∈ Q(p, r) with s ≤ s' then insert ⟨t, s⟩ into Q(p) endfor
   until Q(p, r) ≠ ∅ or no new valid join tuple can be generated
6. if Q(p, r) ≠ ∅ then ⟨t, s⟩ := getTop(Q(p, r)), remove ⟨t, s⟩ from Q(p), return {⟨t, s⟩} else return ∅ fi
```

in the *active* list of predicate symbols A. At each iteration step we select a new predicate $p$ from the queue A and get a new tuple ($getNexTuple(p, r)$) satisfying the rule body $r$ whose head contains $p$ with respect to the answers gathered so far. If the evaluation leads to a new answer for $p$ ($\Delta_r \neq \emptyset$), we update the current answer set rankedList($p$) and add all predicates $p_j$ directly depending on $p$ to the queue A. At some point the active list will become empty and we have actually found correct answers of $q(\mathbf{x})$. A threshold will be used to determine when we can stop retrieving tuples. Indeed, the threshold determines when any newly retrieved tuple for $q$ scores lower than the current top-$k$ and, thus, cannot modify the top-$k$ ranking (step 9). So, step 1 loops until we do not have $k$ answers above the threshold or, two successive loops do not modify the current set of answers (step 9). Step 2 initializes the active list of predicates. Step 3. loops until no predicate has to be processed anymore. In step 4, we select a predicate symbol to be processed. In step 5, we retrieve the next answer for $p$. If a new answer has been retrieved (step 6, $\Delta_r \neq \emptyset$) then we update the current answer set rankedList($p$) and add all predicates $p_j$, that directly depend on $p$, to the queue A. In step 7, we put once all intentional predicate symbols appearing in the rule body of $p$ in the active list for further processing.

We next describe the *getNextTuple* procedure (see Table 1). It's main purpose is, given a predicate symbol $p$ and a rule $r : p(\mathbf{x}) \leftarrow \phi$, to get back the next tuple (and its score) satisfying the conditions of the rule $r$. The procedure is a generalization of the analogous *getNext* procedure described in [4] and uses the so-called *symmetric Hash Rank Join* (HRJN) algorithm. This is not surprising as the list of atoms in a rule body may be seen as multiple joins together with a scoring function.

Let us first describe the intuition behind the procedure. For the sake of illustrative purposes assume that the rule $r$ associated to $p$ is $p(x) \leftarrow p_1(x,y) \cdot p_2(y,z) \in \mathcal{P}$. The idea is as follows:

1. We incrementally generate new valid join combinations $\langle x,y,z \rangle$ from the tuples in $\texttt{rankedList}(p_1)$ and $\texttt{rankedList}(p_2)$ using some join strategy. In particular, as [4], we alternatively access first $\texttt{rankedList}(p_1)$ and then $\texttt{rankedList}(p_2)$. We select the next unseen tuple in $\texttt{rankedList}(p_1)$ and then build all join combinations with the tuples seen so far in $\texttt{rankedList}(p_2)$. Then we select the next unseen tuple in $\texttt{rankedList}(p_2)$ and then build all join combinations with the tuples seen so far in $\texttt{rankedList}(p_1)$ and so on until we find some valid join tuples ($\texttt{Q}(p,r) \neq \emptyset$).
2. The join tuples and their scores will be put in the queue $\texttt{Q}(p,r)$ and the top-ranked one is returned.

Specifically, in step 1, whenever we already have some tuples in the queue $\texttt{Q}(p,r)$ of $p$ (obtained by a previous call) then we just return the top-ranked one. Ties are split randomly. In step 2 we generate all candidate joins, involving all seen tuples of the predicates occurring in the rule body of $p$. For each of them we compute its score (step 4). We put the results on the queue $\texttt{Q}(p,r)$ (step 5) and return the top-ranked one. As $\texttt{Q}(p,r)$ may still contain answers for $p$, the next time we ask for a next tuple with respect to $p$ and $r$, we access $\texttt{Q}(p,r)$ directly (step 1).

Finally, threshold $\delta$ is determined as follows. It is computed as in [4]. Suppose that for the query predicate $q$ we have a rule $r : q(x) \leftarrow f(p_1,\ldots,p_n) \in \mathcal{P}$. Let $\mathbf{t}_i^{\perp}$ be the last tuple seen in $\texttt{rankedList}(p_i)$, while let $\mathbf{t}_i^{\top}$ be the top ranked one in $\texttt{rankedList}(p_i)$. With $\mathbf{t}_i.score$ we indicate the tuple's score. Then $\delta$ is the maximum of the following $n$ values:

$$
\begin{aligned}
\delta_1 &= f(\mathbf{t}_1^{\perp}.score, \mathbf{t}_2^{\top}.score, \ldots, \mathbf{t}_n^{\top}.score)\\
\delta_2 &= f(\mathbf{t}_1^{\top}.score, \mathbf{t}_2^{\perp}.score, \ldots, \mathbf{t}_n^{\top}.score)\\
&\;\vdots\\
\delta_n &= f(\mathbf{t}_1^{\top}.score, \mathbf{t}_2^{\top}.score, \ldots, \mathbf{t}_n^{\perp}.score)\ .
\end{aligned}
$$

For instance, for $q(x) \leftarrow p_1(x,y) \cdot p_2(y,z) \in \mathcal{P}$ we have

$$
\begin{aligned}
\delta_1 &= \mathbf{t}_1^{\perp}.score \cdot \mathbf{t}_2^{\top}.score\\
\delta_2 &= \mathbf{t}_1^{\top}.score \cdot \mathbf{t}_2^{\perp}.score\\
\delta &= \max(\delta_1,\delta_2)\ .
\end{aligned}
$$

It is not difficult to see that whenever we consider a new join combination, its score will be below to $\delta$. Indeed, if we consider a new join tuple using the next unseen tuple from $\texttt{rankedList}(p_1)$ and a seen tuple in $\texttt{rankedList}(p_2)$, its score will be below $\delta_1$, while if we consider a new join tuple using the next unseen tuple from $\texttt{rankedList}(p_2)$ and a seen tuple in $\texttt{rankedList}(p_1)$, its score will be below $\delta_2$. Therefore, overall the score will be below $\delta$. It is thus not surprising that whenever we have top-$k$ answers for $q$ with score above $\delta$ we can stop the retrieval process (see step 9 of *TopAnswers*). This property can be generalized to $n$-ary joins (see [4], Theorem 4.2.1). For the sake of illustrative purposes, let us consider the following abstract examples.

*Example 3.* Assume that we have the following query rule $q(x) \leftarrow \min(r_1(x,y), r_2(y,z))$, where $q$ is the query predicate and $r_1, r_2$ are extensional predicates with tables (with millions of tuples)

| *recId* | $r_1$ | $r_2$ |
|---|---|---|
| 1 | $a\ b\ 1.0$ | $m\ h\ 0.95$ |
| 2 | $c\ d\ 0.9$ | $m\ j\ 0.85$ |
| 3 | $e\ f\ 0.8$ | $f\ k\ 0.75$ |
| 4 | $l\ m\ 0.7$ | $m\ n\ 0.65$ |
| 5 | $o\ p\ 0.6$ | $p\ q\ 0.55$ |
| ⋮ | ⋮ ⋮ ⋮ | ⋮ ⋮ ⋮ |

The table below reports a top-2 retrieval computation. The left table reports data related to each loop in the *TopAnswers* procedure, while the other shows at each iteration the execution *getNextTuple* ($r_j(i)$ means that we access the $i$-th tuple in relation $r_j$). The first call of *getNextTuple*$(q)$ requires several alternative accesses to $r_i$ before a tuple can be found ($\langle e,k,0.75\rangle$). In the second call we get immediately two candidate tuples. In the third call, as $\mathtt{Q}(q,r) \neq \emptyset$ we get immediately the next candidate ($\langle l,j,0.7\rangle$). Finally, in the fourth call, we retrieve $\langle l,n,0.65\rangle$. As now `rankedList`$(q)$ contains 2 answers above the threshold of 0.7, we can stop and return $\{\langle e,k,0.75\rangle, \langle l,h,0.7\rangle\}$. Note that no new retrieved answer may have a score above 0.7. Indeed, the next one would be $\langle o,q,0.55\rangle$ and, thus, *not all tuples are processed* (which would be unfeasible in practice).

| *TopAnswers* | | | | | |
|---|---|---|---|---|---|
| *Iter* | `A` | $p$ | $\Delta_r$ | `rankedList`$(p)$ | $\delta$ |
| 1. | $q$ | $q$ | $\langle e,k,0.75\rangle$ | $\langle e,k,0.75\rangle$ | 0.8 |
| 2. | $q$ | $q$ | $\langle l,h,0.7\rangle$ | $\langle e,k,0.75\rangle, \langle l,h,0.7\rangle$ | 0.75 |
| 3. | $q$ | $q$ | $\langle l,j,0.7\rangle$ | $\langle e,k,0.75\rangle, \langle l,h,0.7\rangle, \langle l,j,0.7\rangle$ | 0.75 |
| 4. | $q$ | $q$ | $\langle l,n,0.65\rangle$ | $\langle e,k,0.75\rangle, \langle l,h,0.7\rangle, \langle l,j,0.7\rangle, \langle l,n,0.65\rangle$ | 0.7 |

| *getNextTuple* | | | |
|---|---|---|---|
| *Iter* | $p_i$ | $\langle t_i, s_i\rangle$ | $\mathtt{Q}(p,r)$ |
| 1. | $r_1$ | $r_1(1)$ | – |
| | $r_2$ | $r_2(1)$ | – |
| | $r_1$ | $r_1(2)$ | – |
| | $r_2$ | $r_2(2)$ | – |
| | $r_1$ | $r_1(3)$ | – |
| | $r_2$ | $r_2(3)$ | $\langle e,k,0.75\rangle$ |
| 2. | $r_1$ | $r_1(4)$ | $\langle l,h,0.7\rangle, \langle l,j,0.7\rangle$ |
| 3. | – | – | $\langle l,j,0.7\rangle$ |
| 4. | $r_2$ | $r_2(4)$ | $\langle l,n,0.65\rangle$ |

From computational point of view, by a similar analysis as in [9], it can be shown that *TopAnswer* is exponential with respect to $|\mathcal{P}|$ (combined complexity), but polynomial in $|\mathcal{P}_E|$ (data complexity), and we have:

**Proposition 1.** *Given a deterministic logic program $\mathcal{P}$ in which all scoring functions are bounded, then $TopAnswers(\mathcal{P},q,k)$ terminates with $TopAnswers(\mathcal{P},q,k) = ans_k(q,\mathcal{P})$.*

*Top-k Query Answering for General LPs.* We first illustrate the problem that is introduced in the case a predicate symbol $p$ is in the head of multiple rules and then sketch how we solve it. Our top-$k$ retrieval algorithm is based on the fact that whenever we find a new instance $\langle \mathbf{c}, s\rangle$ for a predicate $p$ occurring in $\mathcal{P}$, any successive retrieved instance $\langle \mathbf{c}', s'\rangle$ for $p$ is scored lower than $\langle \mathbf{c}, s\rangle$, i.e. $s' \leq s$ (the fact that score combination functions are bounded is crucial here), which allows us to apply the stopping criteria based on a threshold. Unfortunately, if

$p$ is in the head of more than one rule this is no longer true. Indeed, clearly two rules $p(\mathbf{x}) \leftarrow \phi_1$ and $p(\mathbf{x}) \leftarrow \phi_2$ are equivalent to the rule $p(\mathbf{x}) \leftarrow \max(\phi_1, \phi_2)$, and max is not a bounded score combination function. So, it is not difficult to find an example where given a retrieved instance $\langle \mathbf{c}, s \rangle$ for $p$, a successive retrieved instance $\langle \mathbf{c}', s' \rangle$ for $p$ may have a score higher than $\langle \mathbf{c}, s \rangle$, i.e. $s' > s$.

*Example 4.* Consider the seven rules $q(x) \leftarrow t_1(x)$, $q(x) \leftarrow p_1(x)$, $p_1(x) \leftarrow t_2(x)$, $t_1(a) \leftarrow 0.4$, $t_1(b) \leftarrow 0.3$, $t_2(c) \leftarrow 0.5$, $t_2(d) \leftarrow 0.2$. A naive extension of our procedure, may retrieve first $\langle a, 0.4 \rangle$ for $q$, second $\langle c, 0.5 \rangle$, third $\langle b, 0.3 \rangle$ and, eventually $\langle d, 0.2 \rangle$.

As we can see in the above example, it may not be guaranteed that any successive retrieved tuple for $q$ is scored lower than the previous one. However, there is still a simple strategy to overcome to this problem. In fact, note that any successively retrieved tuple for rule $r_1 : q(x) \leftarrow t_1(x)$ is scored lower than the one retrieved before for $r_1$. Similarly, any successively retrieved tuple for rule $r_2 : q(x) \leftarrow p_1(x)$ is scored lower than the one retrieved before for $r_2$. Therefore, one strategy may be to gather at least one answer for each of the rules $r_1$ and $r_2$ and only then merge the retrieved answers for $r_1$ and $r_2$ to build the answers for $q$. This will guarantee that successively retrieved answers for $r_1$ and $r_2$ are scored lower than the already retrieved ones for $q$. Of course, the threshold $\delta$ for $q$ is now $\delta = \max(\delta_{r_1}, \delta_{r_2}\}$, where $\delta_{r_1}$ and $\delta_{r_2}$ are computed as previously for rule $r_1$ and $r_2$, respectively. The detailed procedure will be described in more detail in an extended work.

## 4 Conclusions

The problem of top-$k$ retrieval will be an important problem, e.g. in logic-based (multimedia) information retrieval. We have addressed this issue in the context of logic programs. We are unaware of any other work addressing this problem for many-valued (recursive) logic programs (for non-recursive logic programs, see [11,14]), computing the answers iteratively accessing relational tables using relational algebra (and, thus, is not resolution-based).

Major topics for future research include: (*i*) can we refine our strategy in case the score combination functions are not bounded? (*ii*) How can we deal with aggregates (maybe relying on [7])? (*iii*) Can we apply similar ideas to other popular logical formalism, such as *Description Logics* (DLs) [1] (the only work we know about are [12,10]) and their combination with LPs?

## References

1. Franz Baader, Diego Calvanese, Deborah McGuinness, Daniele Nardi, and Peter F. Patel-Schneider, editors. *The Description Logic Handbook: Theory, Implementation, and Applications.* Cambridge University Press, 2003.
2. Didier Dubois and Henri Prade. Possibility theory, probability theory and multiple-valued logics: A clarification. *Annals of Mathematics and Artificial Intelligence*, 32(1-4):35–66, 2001.

3. Ronald Fagin. Combining fuzzy information: an overview. *SIGMOD Rec.*, 31(2):109–118, 2002.
4. Ihab F. Ilyas, Walid G. Aref, and Ahmed K. Elmagarmid. Supporting top-k join queries in relational databases. In VLDB-03, pages 754–765, 2003.
5. Michael Kifer and V.S. Subrahmanian. Theory of generalized annotated logic programming and its applications. *Journal of Logic Programming*, 12:335–367, 1992.
6. Chengkai Li, Kevin Chen-Chuan Chang, Ihab F. Ilyas, and Sumin Song. RankSQL: query algebra and optimization for relational top-k queries. In SIGMOD-05, pages 131–142, New York, NY, USA, 2005. ACM Press.
7. Chengkai Li, Kevin Chen-Chuan Chang, and Ihab F. Ilyas. Supporting ad-hoc ranking aggregates. In SIGMOD-06, USA, 2006. ACM Press.
8. Carlo Meghini, Fabrizio Sebastiani, and Umberto Straccia. A model of multimedia information retrieval. *Journal of the ACM*, 48(5):909–970, 2001.
9. Umberto Straccia. Query answering in normal logic programs under uncertainty. In ECSQARU-05, LNCS 3571, pages 687–700, Barcelona, Spain, 2005. Springer Verlag.
10. Umberto Straccia. Answering vague queries in fuzzy DL-Lite. In *Proceedings of the 11th International Conference on Information Processing and Management of Uncertainty in Knowledge-Based Systems, (IPMU-06)*, 2006.
11. Umberto Straccia. Towards top-k query answering in deductive databases. In *Proceedings of the 2006 IEEE International Conference on Systems, Man and Cybernetics (SMC-06)*. IEEE, 2006.
12. Umberto Straccia. Towards top-k query answering in description logics: the case of DL-Lite. In JELIA-06, LNCS 4160, 2006. Springer Verlag.
13. Peter Vojtáš. Fuzzy logic programming. *Fuzzy Sets and Systems*, 124:361–370, 2001.
14. Peter Vojtáš. Fuzzy logic aggregation for semantic web search for the best (top-$k$) answer. In Elie Sanchez, editor, *Fuzzy Logic and the Semantic Web*, Capturing Intelligence, chapter 17, pages 341–359. Elsevier, 2006.
15. David S. Warren. Memoing for logic programs. *Commun. ACM*, 35(3):93–111, 1992.

# On Browsing Domain Ontologies for Information Base Content

Troels Andreasen and Henrik Bulskov

Department of Computer Science,
Roskilde University,
P.O. Box 260, DK-4000 Roskilde, Denmark
{troels,bulskov}@ruc.dk

**Abstract.** This paper presents a conceptual querying approach with reference to a document collection where the target is the concepts appearing in documents, rather than the documents in the collection. An ontology formalism and a special notion of "instantiated ontology" are introduced. The latter is a structure reflecting the content in the document collection in that; it is a restriction of a general world knowledge ontology to the concepts instantiated in the collection. The notion of ontology-based similarity is briefly described and a language for navigation and retrieval of concepts in the ontology is presented.

## 1 Introduction

In this paper we address an approach to conceptual querying where a set of concepts can be examined. The general idea is to restrict a general world knowledge ontology to the given set of concepts – extending this with relations and related concepts – and thereby providing a structure, a so-called instantiated ontology, for navigation and further investigation of the concepts. Conceptual investigation of a set of documents can be performed by extracting the set of concepts appearing in the documents and by providing means for navigation and retrieval within the set of extracted concepts.

For this purpose a language for navigation and retrieval of concepts in the ontology is presented that include constructs for investigating connected and close concepts to a given concept. Closeness is based on the notion of similarity that is briefly introduced and further described in [1]. Extraction of concepts from text is not covered in this paper and we refer to [2,3] for a discussion on this issue.

One important application aimed is to provide a tool for investigating and describing content of any collection of documents with reference to general world knowledge resources such as WordNet.

## 2 Representation of Ontologies

The purpose of the ontology in this context is to define and relate concepts that may appear in the document collection or in queries to this.

P. Melin et al. (Eds.): IFSA 2007, LNAI 4529, pp. 135–144, 2007.
© Springer-Verlag Berlin Heidelberg 2007

We define a generative ontology framework where a basis ontology situates a set of atomic term concepts $\mathcal{A}$ in a concept inclusion lattice. A concept language (description language) defines a set of well-formed concepts, including both atomic and compound term concepts.

The concept language used here, ONTOLOG[4], is a lattice-algebraic description language. Its basic elements are concepts and binary relations between concepts. The algebra introduces two closed operations on concept expressions $\varphi$ and $\psi$ [4]:

- conceptual *sum* $(\varphi + \psi)$, interpreted as the concept being $\varphi$ or $\psi$
- conceptual *product* $(\varphi \times \psi)$, interpreted as the concept being $\varphi$ and $\psi$

also called *join* and *meet* respectively. Relationships $r$ are introduced algebraically by means of a binary operator (:), known as the Peirce product $(r : \varphi)$, which combines a relation $r$ with an expression $\varphi$. The Peirce product is used as a factor in conceptual products, as in $x \times (r\colon y)$, which can be rewritten to form the feature structure $x[r\colon y]$, where $[r\colon y]$ is an *attribution* of the concept $x$.

Compound concepts can be formed by attribution. Given atomic concepts $\mathcal{A}$ and semantic relations $\mathcal{R}$ , the set of well-formed terms $\mathcal{L}$ is:

$$\mathcal{L} = \{\mathcal{A}\} \cup \{x[r_1 : y_1, ..., r_n : y_n] | x \in \mathcal{A}, r_i \in \mathcal{R}, y_i \in \mathcal{L}\}$$

Compound concepts can thus have multiple as well as nested attributions. For instance with $\mathcal{R}= \{\text{WRT}, \text{CHR}, \text{CBY}, \text{TMP}, \text{LOC}, \ldots\}$[1] and $\mathcal{A}$ ={*entity, physical_entity, abstract_entity, location, town, cathedral, old*} we get:

$$\mathcal{L} = \{entity, physical\_entity, abstract\_entity, location, town, cathedral, old, \\ \ldots, cathedral[\text{LOC}\colon town, \text{CHR}\colon old], cathedral[\text{LOC}\colon town[\text{CHR}\colon old]], \ldots\}$$

## 3 Modelling Ontologies

The objective here is to provide instantiated ontologies as targets for conceptual querying and the approach is to establish a general ontology and derive instantiated ontologies by restriction to sets of concepts.

### 3.1 The General Ontology

Sources for knowledge base ontologies may have various forms. Typically a taxonomy can be supplemented with, for instance, word and term lists as well as dictionaries for definition of vocabularies and for handling of morphology. The well-known and widespread resource WordNet [5] is among the more interesting and useful resources for general ontologies.

We will not go into details on the modeling here but just assume the presence of a taxonomy $\mathcal{T}$ over the set of atomic concepts $\mathcal{A}$. $\mathcal{T}$ and $\mathcal{A}$ expresses the domain and world knowledge provided.

[1] For *with respect to, characterized by, caused by, temporal, location*, respectively.

Based on $\hat{\mathcal{T}}$ the transitive closure of $\mathcal{T}$ we can generalize to an inclusion relation "$\leq$" over all well-formed terms of the language $\mathcal{L}$ by the following [6]:

$$\begin{aligned} \text{``} \leq \text{``} = \hat{\mathcal{T}} & \\ & \cup \{<x[\ldots, r\colon z], y[\ldots]> \mid <x[\ldots], y[\ldots]> \in \hat{\mathcal{T}}\} \\ & \cup \{<x[\ldots, r\colon z], y[\ldots, r\colon z]> \mid <x[\ldots], y[\ldots]> \in \hat{\mathcal{T}}\} \\ & \cup \{<z[\ldots, r\colon x], z[\ldots, r\colon y]> \mid <x, y> \in \hat{\mathcal{T}}\} \end{aligned}$$

where repeated ... denote zero or more attributes of the form $r_i\colon w_i$.

The general ontology $\mathcal{O} = (\mathcal{L}, \leq, \mathcal{R})$ thus encompasses a set of well-formed expressions $\mathcal{L}$ derived in the concept language from a set of atomic concepts $\mathcal{A}$, an inclusion relation generalized from the taxonomy relation in $\mathcal{T}$, and a supplementary set of semantic relations $\mathcal{R}$. For $r \epsilon \mathcal{R}$, we obviously have $x[r\colon y] \leq x$, and that $x[r\colon y]$ is in relation $r$ to $y$. Observe that $\mathcal{O}$ is generative and that $\mathcal{L}$ therefore is potentially infinite.

### 3.2 Instantiated Ontology

Given a general ontology $\mathcal{O} = (\mathcal{L}, \leq, \mathcal{R})$ and a set of concepts $C$ the instantiated ontology $\mathcal{O}_C = (\mathcal{L}_C, \leq_C, \mathcal{R})$ is a restriction of $\mathcal{O}$ to cover only the concepts in $C$ and corresponds to "upper expansion" $\mathcal{L}_C$ of $C$ in $\mathcal{O}$

$$\begin{aligned} \mathcal{L}_C &= C \cup \{x | y \in C, y \leq x\} \\ \text{``} \leq_C \text{``} &= \{(x, y) | x, y \in \mathcal{L}_C, x \leq y\} \end{aligned}$$

Thus $\mathcal{O}_C$ is not generative. "$\leq_C$" may be represented by a minimal set "$\leq'_C$"$\subseteq$"$\leq_C$" such that "$\leq_C$" is derivable from "$\leq'_C$" by means of transitivity of "$\leq$" and monotonicity of attribution:

$$\begin{aligned} \mathit{transitivity} &: x \leq y, y \leq z \Rightarrow x \leq z \\ \mathit{monotonicity} &: x \leq y \Rightarrow z[r\colon x] \leq z[r\colon y] \end{aligned}$$

Figure 1 shows an example of an instantiated ontology. The general ontology is based on (and includes) WordNet and the ontology shown is "instantiated" wrt. the following set of concepts:

$$\begin{aligned} C = \{ & \mathit{cathedral}[\text{LOC}\colon \mathit{town}[\text{CHR}\colon \mathit{old}]], \mathit{abbey}, \\ & \mathit{fortification}[\text{CHR}\colon \mathit{large}, \text{CHR}\colon \mathit{old}], \mathit{stockade}\} \end{aligned}$$

### 3.3 Deriving Similarity

A domain ontology, that reflects a document collection, may provide an excellent means to survey and give perspective to the collection. However as far as access to documents is concerned ontology reasoning is not the most obvious evaluation strategy as it may well entail scaling problems. Applying measures of similarity

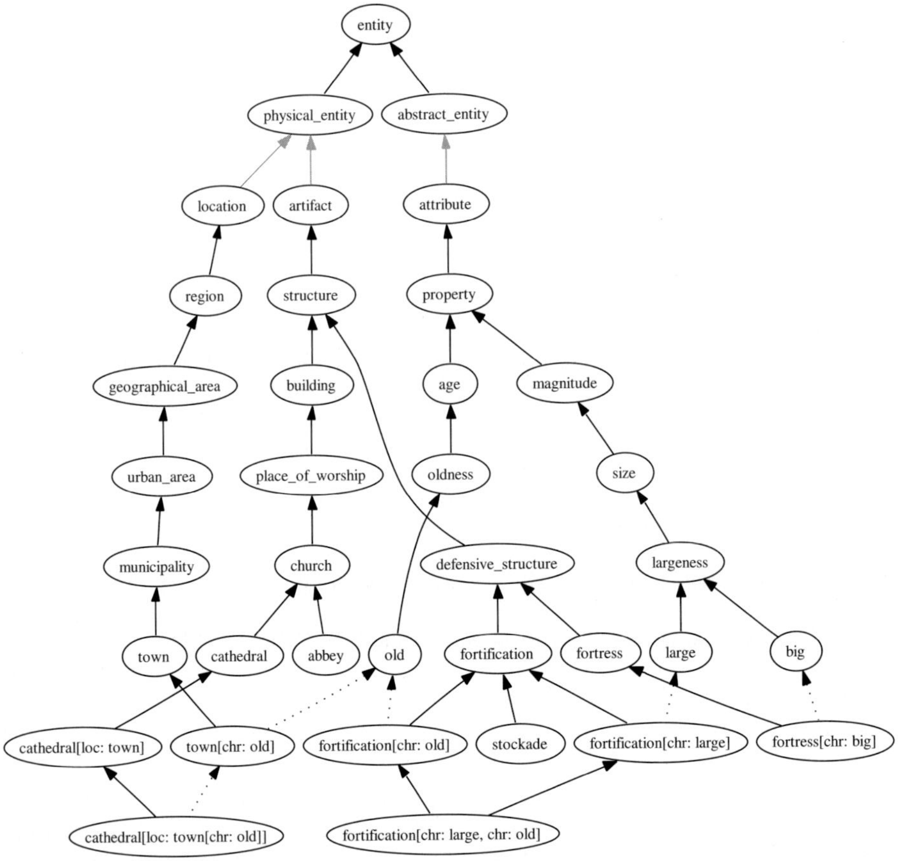


**Fig. 1.** An instantiated ontology based on a Wordnet ontology and the set of instantiated concepts {*cathedral*[LOC: *town*[CHR: *old*]], abbey, *fortification*[CHR: *large*, CHR: *old*], stockade}

derived from the ontology is a way to replace reasoning with simple computation still influenced by the ontology.

One obvious way to measure similarity in ontologies, given the graphical representation, is to evaluate the distance between the concepts being compared, where a shorter distance implies higher similarity and vice versa.

A number of different ontological similarity measures along this line have over the years been proposed. *Shortest Path Length* [7] forms the basis of a group of measures classified as path length approaches. The *Weighted Shortest Path* [8] is a generalization of *Shortest Path Length* where weights are assigned to relations in the ontology. Two different alternatives are *Information Content* [9] and *Weighted Shared Nodes* [10], where the former uses the probability of encountering concepts in a corpus to define the similarity between concepts, and the latter uses the density of concepts shared by the concepts being compared to measure the similarity.

## 4 Querying Ontologies

In a framework where the domain of a database is reflected in a knowledge base, for instance as expressed in an ontology, obviously not only the database, but also the domain ontology may in some cases be the target of interest for queries. Knowledge about existence of concepts, how concepts are related and about similarities between concepts may be relevant to users. In addition knowledge about the actual content of the database can be viewed through the ontology simply by means of revealing only concepts that exist in the database. In other words the instantiated ontology plays a specific role here, since it constitute the means by which the user can obtain a conceptual view of the database content – navigating through all the concepts and relations actually instantiated and only those.

### 4.1 Conceptual Querying

In developing a conceptual querying approach we restrict here to considerations involving explicitly specified concepts and seek appropriate operators to support questions of interest. The main issue is expressivity and observations may (at least to some extend) be generalized to other means of querying such as NL and logic-based approaches.

Questions of interest concerning an ontology typical refer to existence of, and connectivity among concepts. To identify categories of queries we distinguish three aspects that should be covered by the expressivity of the language: basic lattice operations, unified operations involving also connectivity by other relations, and soft closeness operations involving concept similarities.

The basic operations are mainly strict taxonomy and lattice operations concerning connectivity along inclusion only.

Features attached by semantic relations may as well be of interest when querying the ontology and the unified relation operations involves also connectivity through these. For instance querying connectivity on concepts *yellow* and *bird* would possibly lead to nothing (or anything) in the strict inclusion interpretation, but reveal whether a bird with the feature *yellow* can be found in the database when the relation CHR is taken into account.

Soft closeness operations involve similarity measuring. Having introduced ontology based similarity measuring degrees of closeness, we can, apart from crisp (connectivity), also distinguish soft (similarity) questions dealing with sets of similar, rather than sets of connected concepts.

Thus to encircle a suggested set of operators we obviously have numerous possibilities. Below we consider possible operations related to the three mentioned aspects in order.

**Basic Operations.** Keeping in mind that our focus is on instantiated ontologies comprising concepts actually instantiated, the most simple questions should be to the *existence* of concepts. In [11] Trissl and Leser introduces the categories of *reachability, ancestor- or successor set* and *Least common ancestor* of two or more nodes (multiple concepts). Reachability is to verify whether a given concept has another as ancestor, for instance "Is cat a mammal". Ancestor/Successor set for a given node is the set of all ancestors/successors nodes respectively. For instance the successors of "animal" answer the question " What animal do we have". Least common ancestor is of interest when a common origin of two or more nodes should be computed. For instance, we could verify that cat and dog are connected through a least common ancestor mammal, while cat and snake first join in animal. Since we are considering lattice ontologies rather than only strict hierarchical taxonomies, concepts are not only upwards but may also be downwards connected. We should therefore at least include a *greatest common successor*. For instance we could investigate what colors dogs come in by extracting the set of greatest common successors of dog and color.

Expressions to cover the two first categories are straightforward and our ontology formalization, conceptual algebra, comprises the operators $+$ (conceptual sum) and $\times$ (conceptual product) operators for the two latter categories. We therefore only need in addition to cover ancestor/successor set expressions as suggested below.

Assuming an ontology $O_D$ over a concept set $D$ and concepts $c, c_1, c_2$ and concept set $C$ the basic query operations are:

- **Existence** $c$ evaluate existence of $c$
- **Reachability** $c_1 \leq c_2$ evaluate to true if $c_2$ is reachable from (is an ancestor of) $c_1$
- **Ancestor set** $c\ll$ denotes any ancestor of $c$
- **Successor set** $c\gg$ denotes any successor to $c$
- **Ancestor-Successor set** $c*$ denotes any ancestor or successor to $c$
- **Any** $*$ denotes any concept $c$
- **Least common ancestor** $c_1 + c_2$ denotes the conceptual sum (join), that is the least common ancestor to $c_1$ and $c_2$
- **Greatest common successor** $c_1 \times c_2$ denotes the conceptual product (meet), that is the greatest common successor of $c_1$ and $c_2$

Questions involving the operators $\ll, \gg, *$ such as $c \ll, c \gg, c*, *$ are open queries leading to sets of concepts meeting the specifications. With reference to figure 1 simple examples of queries applying these operators and answers are the following.

- $town[\text{CHR}: old] = \{town[\text{CHR}: old]\}$
- $large[\text{CHR}: stockade] = \{\}$
- $building\ll = \{building, structure, artifact, physical\_entity, entity\}$
- $place\_of\_worship\gg = \{place\_of\_worship, church, cathedral, abbey, cathedral[\text{LOC}: town], cathedral[\text{LOC}: town[\text{CHR}: old]]\}$

- $cathedral[\text{LOC}: town\gg] = \{cathedral[\text{LOC}: town], cathedral[\text{LOC}: town[\text{CHR}: old]]\}$
- $church\gg[\text{LOC}: *] = \{cathedral[\text{LOC}: town]\ cathedral[\text{LOC}: town[\text{CHR}: old]]\}$
- $fortification[\text{CHR}: old] + fortress[\text{CHR}: big] = defensive\_structure$

**Unified Relation Operations.** The operations above provide means for navigation directly along the inclusion relation and indirectly along semantic relations attaching features to concepts. We can pose questions like $x[\text{CHR} : *]$ and $x[\text{CHR} : y \gg]$ leading to all concepts $x[\text{CHR} : y] \in L_C$ respectively all concepts $x[\text{CHR} : z] \in L_C$ with $z \leq y$. However to navigate directly along feature connections we need either means for specifying which relations to traverse or means for combining different relations into one unified connectivity. Again the possibilities for language constructs are many. In this context we will only focus on a unified approach. We choose here to introduce two categories:

- **Unified least common ancestor** $c_1 +' c_2$ denotes the conceptual sum of $c_1$ and $c_2$ in the unified lattice where all connections (inclusion as well as features) are combined and considered ordering
- **Unified greatest common successor** $c_1 \times' c_2$ denotes the conceptual product of $c_1$ and $c_2$ in the unified lattice where all connections are combined and considered ordering

Apart from exploiting features through the connectivity they represent, we might also be interested in the specific values they take. If we want, for instance to know about colors of animals we can of course retrieve the set of all the colored animals, but in some cases it would be more convenient rather to receive the set of colors. For this purpose we introduce an unfold construct:

- **Unfold.** The set of concepts appearing as features for a given concept. For instance $xINc[\text{CHR} : x]$ denotes the set of concepts appearing as CHR-features to the concept $c$.

Again with reference to figure 1 simple examples of queries applying the above operators are the following.

- $fortification[\text{CHR}: old] + town[\text{CHR}: old] = physical\_entity$
- $fortification[\text{CHR}: old] +' town[\text{CHR}: old] = old$
- $region \times' age = \{town[\text{CHR}: old]\}$
- $(oldness \times' defensive\_structure)\gg = \{fortification[\text{CHR}: old], fortification[\text{CHR}: large, \text{CHR}: old]\}$
- $fortification[\text{CHR}: *] = \{fortification[\text{CHR}: old], fortification[\text{CHR}: large]\}$
- $x\ IN\ fortification[\text{CHR}: x] = \{old, large\}$

**Soft Closeness Operations.** All operations mentioned above deal with connectedness among concepts. Thus given two concepts our interest is whether these are connected or not – with no regard to the length of the connecting path. Closeness is about involving similarity – typically influenced by length of connecting path. While connectedness is expressed by crisp sets, closeness may involve degree of similarity and is thus more conveniently expressed as fuzzy sets or, alternatively, crisp sets derived as cuts from fuzzy sets.

Closeness operations may simply be supported by one additional operator $sim(c)$, for simplicity introduced here as a cut, assuming a default threshold applied.

- **Similarity** $sim(c)$. The set of concepts similar to a given concept (according to measure in use) $c$.

Examples of closeness operations on the instantiated ontology in figure 1 are the following.

- $sim(\mathit{fortress}[\text{CHR}: \mathit{big}]) = \{\mathit{fortress}[\text{CHR}: \mathit{big}], fortress,$
  $\mathit{fortification}[\text{CHR}: \mathit{large}]\}$
- $\mathit{stockade} \gg [\text{CHR}: sim(\mathit{big})] = \{\mathit{fortification}[\text{CHR}: \mathit{large}]\}$

### 4.2 Summarizing Descriptions

Conceptual querying as introduced with the operations sketched above concerns retrieval of concepts appearing in an instantiated ontology. If $D$ is a document or a set of documents and $C$ is the set of concepts appearing in $D$, then $C$ can be considered as a description of $D$ and through queries to $O_C$ we can investigate in more detail aspects of this description.

Also of interest, when investigating what $D$ is all about, is an entire summary of $D$. Of course $C$ in itself is such a summary, but it may be huge. Among simple alternative summary approaches is to list the most frequent concepts in $C$. However since $O_C$ defines a structure over $C$ and in addition contains more general concepts not included in $C$ obviously we should attempt to apply $O_C$ in summarizing $D$.

To approach an ontology-based summary, first of all observe that if we have two connected concepts as for instance $\{cat, dog\}$ their least common ancestor ($lca$), $lca(\{cat, dog\}) = animal$[2] may be a good summarizing description of these. More generally, our approach to summarizing a description of a set of concepts[3] $C = \{c_1, \ldots, c_n\}$ is to cluster $C$ into $\widehat{C} = \{\widehat{C}_1, \ldots, \widehat{C}_k\} = \{\{c_{11}, \ldots, c_{1n_1}\}, \ldots, \{c_{k1}, \ldots, c_{kn_k}\}\}$ and provide the set of least common ancestors ($lca$) for each cluster as description $\{\widehat{c}_1, \ldots, \widehat{c}_k\} = \{lca(\widehat{C}_1), \ldots, lca(\widehat{C}_k)\}$.

So the challenge, in providing summarizing descriptions, is to find good clustering principles. Based on preliminary investigations we introduce two different clustering approaches; connectivity clustering and similarity clustering to be covered in more detail in a subsequent paper.

**Connectivity Clustering Description** is clustering based solely on the ontology $O_C$. For a set of concepts $C = \{c_1, \ldots, c_n\}$ a description $\delta(C) = \{\widehat{c}_1, \ldots, \widehat{c}_k\}$ is a connectivity clustering description if $\{\widehat{c}_1, \ldots, \widehat{c}_k\} = \{lca(\widehat{C}_1), \ldots, lca(\widehat{C}_k)\}$ where $\{\widehat{C}_1, \ldots, \widehat{C}_k\}$ is a clustering of $C$ such that

[2] $lca(\{c_1, \ldots, c_n\}) = c_1 + \ldots + c_n$.

[3] Perhaps the set of concepts $C$ appearing in a document or a set of documents $D$.

$\forall i,j : \widehat{c}_i \nless \widehat{c}_j$
$\forall i \neg\exists j,k,x : x \in L_C \wedge 1 \leqq j < k \leqq n \wedge 1 \leqq i \leqq k \wedge c_j < x \wedge c_k < x \wedge x < \widehat{c}_i$

thus $\delta(C)$ is a "most specific generalization" of $C$ and to obtain an appropriate description of $C$ we might have to apply $\delta$ several times. At some point $m$ we have that $\delta^m(C) = \top$.

With reference to figure 1 we have for instance (notice that $entity = \top$):

- $C \quad = \{church, fortification[\text{CHR}: old], stockade, fortress[\text{CHR}: big]\}$
- $\delta(C) \; = \{church, fortification, fortress[\text{CHR}: big]\}$
- $\delta^2(C) = \{church, defensive\_structure\}$
- $\delta^3(C) = \{structure\}$
- $\delta^4(C) = \{entity\}$

One alternative option is to provide clusters based on the unified structure considering all connections through all relations in the ontology rather than only inclusion as assumed above.

**Similarity Clustering** is a clustering based on a given similarity measure rather than on direct connectivity in the ontology. Thus any approach to clustering a set applying a distance measure can be used.

Classical clustering techniques divides the concepts in $C$ into crisp clusters, where each concept $c_i$ belongs to exactly one cluster. An alternative approach is to use fuzzy clustering, such as Fuzzy C-Means Clustering [12].

One obvious challenge in introducing fuzzy clustering is how to derive the least common ancestor of a fuzzy cluster. In principle all concepts in $C$ could belong to every fuzzy cluster, and thus the least common ancestor would then be identical for all the clusters. One solution to this would be, for each candidate $lcaL$, to aggregate over the membership grades of the concepts covered (subsumed) by $L$ and to pick the candidate(s) achieving the highest score.

## 5 Conclusion

In this paper we have sketched a query language that extends the conceptual algebra presented as the ontology representation formalism. The main goal is to provide means for querying concepts describing documents, rather than querying documents directly. We present also the notion of instantiated ontology – a conceptual structure reflecting the content of a given document collection and therefore in particular well suited as target for conceptual querying. A special similarity construct in the language can be used to denote sets of concepts and to provide summarizing descriptions of sets of documents or concepts.

The language presented is preliminary and needs to be further refined, however the formalization of most operators is straightforward. Of particular interest, as topic for future research, is investigation and further refinement of similarity measures and of clustering techniques (ontology as well as fuzzy) for summarizing description.

In addition further work should be guided by experiments within a framework that include a realistic general world knowledge resource. Currently the one of the best candidates is probably WordNet because it is quite general with a good coverage of common language. However, since WordNet is more to be considered as a network than as an ontology, more profound domain-ontologies such as those available in bioscience and medicine should also be tested as general world knowledge resources.

## References

1. Knappe, R., Bulskov, H., Andreasen, T.: Perspectives on ontology-based querying. International Journal of Intelligent Systems (2006) to appear.
2. Bulskov Styltsvig, H.: Ontology-based Information Retrieval. PhD thesis, Roskilde University (2006)
3. Andreasen, T., Jensen, P.A., Nilsson, J.F., Paggio, P., Pedersen, B.S., Thomsen, H.E.: Content-based text querying with ontological descriptors. Data Knowl. Eng. **48**(2) (2004) 199–219 ISSN 0169-023X.
4. Nilsson, J.F.: A logico-algebraic framework for ontologies – ONTOLOG. In Jensen, P.A., Skadhauge, P., eds.: Proceedings of the First International OntoQuery Workshop, Department of Business Communication and Information Science, Kolding, Denmark, University of Southern Denmark (2001)
5. Miller, G.A.: Wordnet: a lexical database for english. Commun. ACM **38**(11) (1995) 39–41
6. Bulskov, H., Knappe, R., Andreasen, T.: On querying ontologies and databases. In Christiansen, H., Hacid, M.S., Andreasen, T., Larsen, H.L., eds.: LNAI 3055, 6th International Conference on Flexible Query Answering Systems, Lyon, France, Flexible Query Answering Systems, Springer-Verlag (2004)
7. Rada, R., Mili, H., Bicknell, E., Blettner, M.: Development and application of a metric on semantic nets. IEEE Transactions on Systems, Man, and Cybernetics **19**(1) (January 1989) 17–30
8. Bulskov, H., Knappe, R., Andreasen, T.: On measuring similarity for conceptual querying. In: FQAS '02: Proceedings of the 5th International Conference on Flexible Query Answering Systems, Springer-Verlag (2002) 100–111
9. Resnik, P.: Semantic similarity in a taxonomy: An information-based measure and its application to problems of ambiguity in natural language (1999)
10. Andreasen, T., Knappe, R., Bulskov, H.: Domain-specific similarity and retrieval. In Liu, Y., Chen, G., Ying, M., eds.: 11th International Fuzzy Systems Association World Congress. Volume 1., Beijing, China, IFSA 2005, Tsinghua University Press (2005) 496–502
11. Silke Trissl, U.L.: Querying ontologies in relational database systems. In: 2nd Workshop on Data Integration in the Life Sciences (DILS05). Number 3615 in Lecture Notes in Computer Science, San Diego, USA, Springer (2005) 63 – 79
12. Bezdek, J.C.: Pattern Recognition with Fuzzy Objective Function Algorithms. Kluwer Academic Publishers, Norwell, MA, USA (1981)

# Fuzzy Tree Mining: Go Soft on Your Nodes

Federico Del Razo Lopez[1], Anne Laurent[1], Pascal Poncelet[2], and Maguelonne Teisseire[1]

[1] LIRMM-CNRS UMR5506, Université Montpellier 2, 161 rue Ada, 34392 Montpellier, France
{delrazo,laurent,teisseire}@lirmm.fr
[2] LGI2P - EMA, France
pascal.poncelet@ema.fr

**Abstract.** Tree mining consists in discovering the frequent subtrees from a forest of trees. This problem has many application areas. For instance, a huge volume of data available from the Internet is now described by trees (e.g. XML). Still, for several documents dealing with the same topic, this description is not always the same. It is thus necessary to mine a common structure in order to query these documents. Biology is another field where data may be described by means of trees. The problem of mining trees has now been addressed for several years, leading to well-known algorithms. However, these algorithms can hardly deal with real data in a soft manner. Indeed, they consider a subtree as *fully included* in the super-tree. This means that all the nodes must appear. In this paper, we extend this definition to fuzzy inclusion based on the idea that a tree is included to a certain degree within another one, this fuzzy degree being correlated to the number of *matching nodes.*

## 1 Introduction

Tree mining is a subfield of data mining aiming at discovering automatically all the subtrees that appear frequently in a database of trees. This research area has several applications, including the discovery of mediator schemas. The background in this research is mainly constituted by the work by Asai et al. and Zaki et al.[2,10,13,17,18]. This work address the problem of tree mining considering several ways to define when a tree $S$ is included within another one $T$. Inclusion is then decided depending on the way ancestry and brotherhood are considered. In this respect, the authors distinguish between approaches where (i) either all the pair of connex nodes in the tree $S$ must be found in $T$ with no intermediate node, (ii) or some intermediate nodes are accepted. Figure 1 illustrates this difference.

The work from the literature is then twofolded, considering both:

- the representation of the trees,
- the extraction of frequent subtrees.

P. Melin et al. (Eds.): IFSA 2007, LNAI 4529, pp. 145–154, 2007.
© Springer-Verlag Berlin Heidelberg 2007

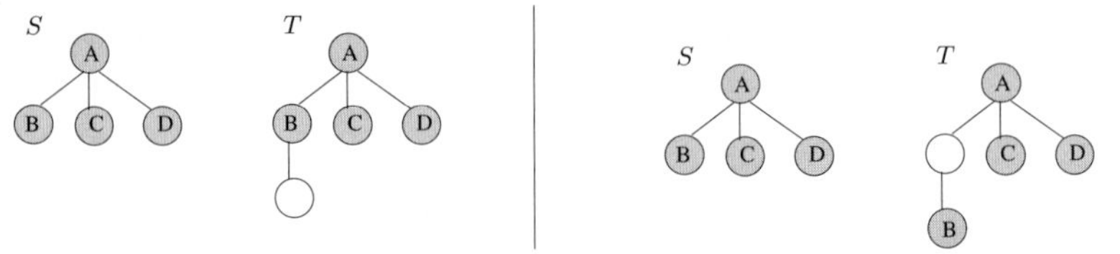


(a) with connex nodes (b) with intermediate nodes

**Fig. 1.** Traditional tree inclusion

It should be noted that designing efficient algorithms to tackle the problem of extracting frequent subtrees is highly correlated to the representation of the trees, as this representation may help scanning the trees. The process of extracting frequent subtrees is based on the Apriori process[1], which is a recursive process. It can be divided into the following steps: for each size of trees (i) generation of candidates and (ii) validation of candidates. A candidate is a tree that is considered as being potentially frequent. The candidates of size $k$ are built based on the frequent subtrees of size $k-1$. This family of methods is well-known and has been applied for tree mining. However, all the existing methods consider that a tree **is** or **is not** included within another one, which is too restrictive to be efficient and relevant. We first propose the concept of fuzzy tree mining, which has been introduced in [11]. This concept has been detailed in [12], where we have defined a fuzzy ancestor-descendant relation (fuzzy vertical path). In this paper, we consider another way to softenize the tree inclusion definition by considering that some nodes may be discarded (partial inclusion). In classical approaches, *all* the nodes of a subtree $S$ must be included in a tree $T$ if $S$ is included in $T$. For instance, Figure 2 shows a tree $S$ that will not be considered as being included in $T$. However, we argue that this is too restrictive when mining data from the real world where imperfections are often present. For instance, in Figure 2 $S$ has 75% of its nodes included in $T$.

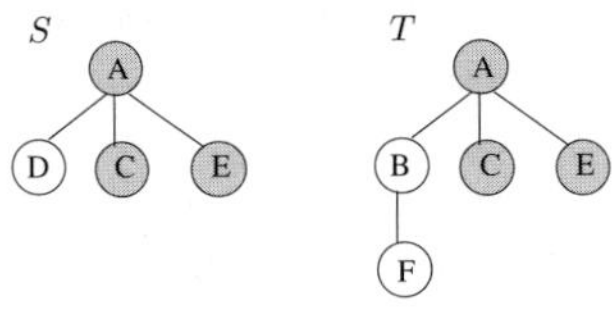


**Fig. 2.** Partial Inclusion

The challenging part of our work is that we want to remain efficient, in the framework of fuzzy data mining. The paper is organized as follows: Section 2 recalls the existing work on tree mining and our previous work on dealing with fuzzy tree mining. Section 3 introduces the necessary definitions for dealing with partial inclusion. Section 4 introduces the algorithms we design for extracting frequent subtrees from a tree database in a soft manner by considering partial inclusion. Finally, Section 5 concludes this work and presents our future working directions.

## 2 Background

In this section, we recall from the literature and from previous work the basic definitions of tree mining and the ways trees can be represented.

### 2.1 Tree Mining

A *tree* is a connected graph containing no cycle. A tree is composed by nodes, which are linked by edges such that their exists a particular node called *root* and such that all the nodes but the root are composed by sub-trees. A tree is said to be an *ordered tree* if the children from a node are ordered. A tree is said to be an *unordered tree* otherwise.

Let $\mathcal{L} = \{a, b, c, ...\}$ be a set of labels. A *labeled ordered tree* is a tree $T = (r, N, B, L, \preceq)$ where: $r$ is the root, $N$ is the set of nodes, $B$ is the set of edges such that $B \subseteq N^2$, $(L : N \to \mathcal{L})$ is a mapping from the set of labels $\mathcal{L}$ to the set of nodes $N$, and $\preceq$ is an ordered relation between brother nodes.

Tree Mining refers to the process of extracting all the subtrees that appear frequently in a database $D$ of trees. The frequency is computed using the notion of support: Given a database $D$, the *support* of a tree $S$ is the proportion of trees from the database where $S$ is embedded:

$$Support(S) = \frac{\#\ of\ trees\ where\ S\ is\ embedded}{\#\ of\ trees\ in\ D}$$

$S$ is said to be frequent if $Support(S) \geq \sigma$ where $\sigma$ is a user-defined minimal support threshold.

Several kinds of tree inclusion can be defined [3], depending on the way ancestors and siblings are considered. For instance, [18] defines the inclusion as follows:

**Definition 1.** *A tree $S$ is* embedded *into a tree $T$ if there exists an injective and total function $\phi : N_S \to N_T$ such as for all $n, m \in N_S$:*

- *$\phi$ keeps the labels: $L_S(n) = L_T(\phi(n))$;*
- *$\phi$ keeps the relations* ancestor-descendant*: $(n, m) \iff (\phi(n), \phi(m))$;*
- *$\phi$ keeps order relations: $(n \preceq_S m) \iff (\phi(n) \preceq_T \phi(m))$.*

As highlighted in [11], fuzzy data mining can help when mining frequent subtrees from a tree database. Four ways to soften classical approaches has been proposed:

- *ancestor-descendant degree*: in classical approaches, a node *is* or *is not* an ancestor of another one. In our approach, we propose to indicate by a degree

between 0 and 1 to which extent a node is an ancestor of another one, meaning that if there are too many nodes between them, then this degree will decrease.
- *sibling ordering degree*: in classical approaches, nodes *are* or *are not* searched in the initial order. In our approach, we propose to indicate by a degree the sibling disorder.
- *partial inclusion*: in classical approaches, all the nodes from the candidate must be in the tree. In our approach, we propose to soften this rule by considering the degree to which the nodes are embedded in the tree.
- *Node similarity*: in classical methods, a node label *is* or *is not* the same as another one. In our approach, we propose to soften this by indicating by a degree to which extent two nodes are similar (*e.g.* based on a taxonomy).

The *ancestor-descendant degree* has been studied in [12]. In the rest of this paper, we focus on the partial inclusion.

**Algorithms.** Several algorithms have been designed to address the problem of tree mining: *TreeMiner* in [18], *FreqT* in [2], *Chopper* [14], *FreeTreeMiner* [5] and *CMTreeMiner* [4]. All of them are based on a process consisting of the following two iterative or recursive steps: generation of candidates and validation of candidates. This process starts from the candidates that contain only one node, to discover the frequent 1-node subtrees, which are used to build the 2-node candidates, and so on.

These two steps have been studied. The generation of candidates is either based on methods that build trees containing $n$ nodes by considering one tree containing $n-1$ nodes and adding another node, or is based on methods that mix two subtrees containing $n$ nodes and sharing $n-1$ nodes to build a new candidate subtree containing $n+1$ nodes.

The validation aims at checking whether a tree is embedded within another one. Several approaches have been proposed. In our previous work, we have defined some algorithms that are based on the idea of *anchoring*: we try to anchor the root of the subtree until we find a node that matches. Then the following nodes are tested until (i) it is no more possible to find some remaining nodes for matching, or (ii) an incompability has been detected or (iii) the subtree fully matches.

### 2.2 Tree Representation

Several ways of representing trees have been proposed to support the algorithms cited above. The representation impacts the two steps discussed above (generation and validation of candidates). However, it may be the case that the representation is too rich and requires too much memory (e.g. representing trees as strings). We have thus proposed in previous work a low-memory representation of trees: RSF. This representation is defined below.

When representing a tree $T$, we keep in mind the following property: all the nodes but the root have one and only one predecessor. We propose thus to use two vectors to represent a tree, as proposed in [15]. The first vector is denoted by $st$. It stores the position of each node predecessor. Nodes are numbered considering a depth-first traversal. The root is numbered as being at position 0, with $st[0] = -1$ since it has no predecessor. The values $st[i], i = 1, 2, ..., k-1$ correspond to all other predecessor positions, as shown on Figure 3.

This representation provides a constant-time method to retrieve the predecessor of a node. Moreover, it allows us to find directly the most right leaf when considering an index $k$. Finally, when visiting the tree, it is possible to build all direct links from predecessors to descendants.

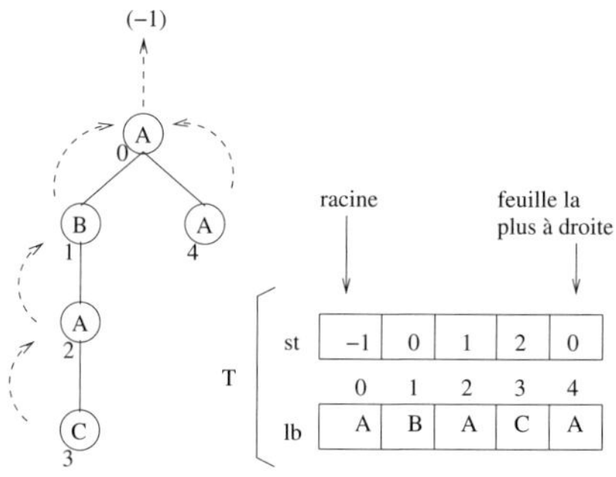


**Fig. 3.** Representation of a Tree

The second vector is denoted by $lb$. It is used to store all the tree labels. $lb[i], i = 0, 1, ..., k-1$ are the labels of each node $n_i \in T$.

The data structure we have chosen needs very low memory since it is reduced to the size of $2|T|$. Moreover, it has good properties when mining frequent subtrees.

As presented in [12], in order to manage trees as efficiently as possible, each tree $T$ is transformed into a binary representation denoted by $T_B$ where each node cannot have more than two children [9]. For this purpose, we propose the following transformation: the first child of a node is put as the left-hand child while the other children are put in the right-hand path, as illustrated in Fig. 4 b).

**Encoding Binary Trees.** Once the tree has been transformed into a binary tree, nodes must be encoded in order to be retrieved. The encoding is then used first in order to identify each node and second in order to determine whether a node is a child or a brother. In order to do so, we consider the Huffman algorithm [8] which we slightly modify in order to fit our needs. The root has address 1. The other node addresses are computed by concatenating the father address with: 1

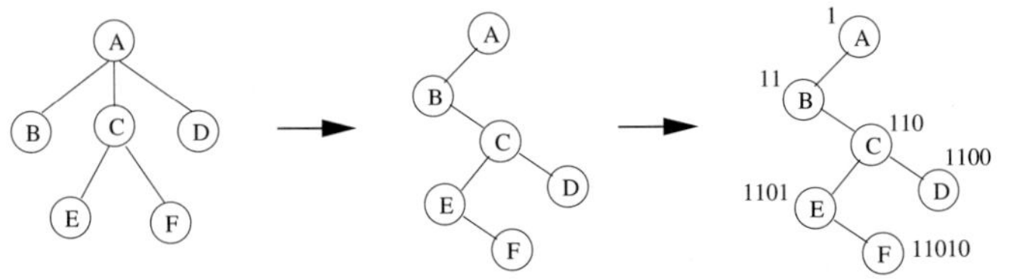


a) $T$ a tree in $D$ | b) $T_B$: binary transformation | c) $T_{BC}$: tree encoding

**Fig. 4.** Example of a Binary Tree Transformation and Node Addressing

if it is a child (left-hand path) and 0 otherwise (right-hand path), as shown on Fig. 4 c).

## 3 FTMnodes: Definitions

In this paper, we formally extend the definition of tree inclusion to partial inclusion based on the number of nodes that are matched. Partial inclusion is defined as follows:

**Definition 2.** *Given a null value* $\perp$, *a tree* $S$ *is* partially embedded *into a tree* $T$ *with a degree* $\delta(S,T)$ *if there exists an injective and total function* $\phi : N_S \rightarrow N_T \cup \perp$ *such that for all* $n, m \in N$*:*

- $\phi$ *keeps the labels:* $L_S(n) = L_T(\phi(n))$ *or* $\phi(n) = \perp$*;*
- $\phi$ *keeps the relations* ancestor-descendant*:* $(n, m) \iff (\phi(n), \phi(m))$ *or* $(\phi(n), \phi(m)) = \perp$*;*
- $\phi$ *keeps the order relations:*
  $(n \preceq_S m) \iff (\phi(n) \preceq_T \phi(m))$ *or* $(\phi(n) \preceq_T \phi(m)) = \perp$*;*
- $\delta(S,T) = \frac{|\{n \in S :\, \phi(n) \neq \perp\}|}{\#\ of\ nodes\ in\ S}$.

From this definition, it is possible to define the support of a subtree, as follows:

**Definition 3.** *Given a database* $D$ *and a tree* $S$*, the support of* $S$ *in* $D$ *is given by:*

$$Support(S) = Agg_{T \in D}(\delta(S,T))$$

*where* $Agg$ *is a function of aggregation.*

For instance, we may use Ordered Weighted Aggregators (also known as OWA) [16]. An OWA operator of dimension $n$ is a mapping

$$F : R^n \rightarrow R$$

that has an associated $n$ vector $W = (w_1, w_2, \ldots, w_n)^T$ such that $w_i \in [0,1]$ and $\sum_{i=1}^{n} w_i = 1$. We have $F(a_1, a_2, \ldots, a_n) = \sum_{j=1}^{n} w_j \cdot b_j$ where $b_j$ is the $j^{th}$ largest value of the $a_i$.

For instance, the average may be applied:

$$Support(S) = \frac{\sum_{T \in D} \delta(S,T)}{\#\ of\ trees\ in\ D}$$

In fact, we consider a thresholded $\Sigma$-count so that:

- a tree cannot be considered as being embedded within another one if the number of embedded nodes is too low,
- the degree to which a tree is embedded within another one is taken into account.

We thus have:

**Definition 4.** *Given a database $D$, a threshold $\tau$ and a tree $S$, the support of $S$ in $D$ is given by:*

$$Support(S) = \sum_{T \in D} (\alpha_\tau(\delta(S,T)))$$

*where*

$$\alpha_\tau(x) = \begin{cases} 0 \text{ if } x > \tau \\ x \text{ otherwise} \end{cases}$$

## 4 FTMnodes: Algorithms

Note that in the classical case, mining totally included trees allows to cut in the database scan since whenever a node cannot be matched, there is not necessary to look for the other ones. In our approach, outliers are accepted, which may be considered as a drawback considering scalability. However, it is still possible to cut off the search when the proportion has been overpassed.

As defined previously, we consider that a tree cannot be considered as being embedded within another one if the number of matching nodes is not greater than a user-defined threshold $\tau$. This definition not only guarantees the quality of the research from a semantic point of view, but it also guarantees the scalability of our approach. Indeed, it is then possible to draw the property of anti-monotonicity which is the basis of levelwise algorithms. We have the following properties:

*Considering that the first $n$ nodes of tree $S$ matched to nodes from $T$, and that $\pi$% of the nodes of $S$ have been matched, then the first $n+1$ nodes of $S$ cannot be embedded in $T$ to a proportion greater than $\pi$.*

This property comes from the fact that if it has not been possible to match $\nu$ nodes among the first $n$ nodes of $S$, then the number of nodes being not matched when going ahead in the process to the first $n+1$ nodes will either be equal or will be greater (equal to $\nu+1$).

As a consequence, whenever the threshold $\tau$ is overpassed, the process can be stopped for this path as it will never be considered in the thresholded $\sum$-count.

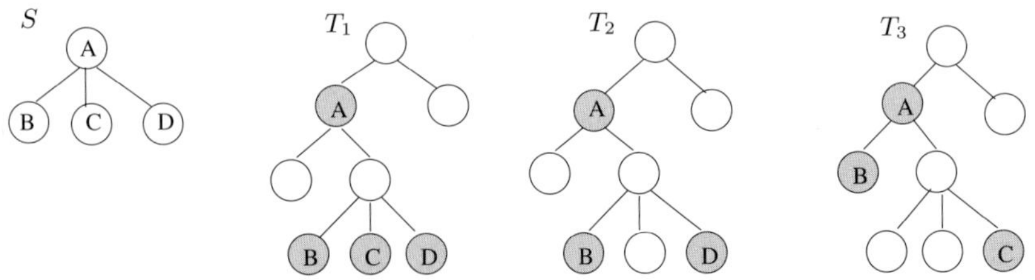


**Fig. 5.** Several ways of including $S$ in $\{T_1, T_2, T_3\}$ with $\tau = 0.75$ (at least 3 nodes out of 4)

Note that it may be the case that a subtree is included within another one in different manners, as illustrated by Figure 5. In this case, the best degree of inclusion will be considered and this best degree is found by maintaining all the possible ways of inclusion until all the solutions have been considered as studied in [7] when considering fuzzy sequential patterns.
The following process is thus considered in our approach (algorithms 1 and 2):

- anchoring
- for each possible anchor, for each node $n$ in $S$ to be matched
  - scan the nodes of $T$ until $n$ is matched, start another way to find the other possible matches,
  - if no match is possible then go to the next node in $n$ and increment the number of mismatched nodes
  - if the number of mismatched nodes is greater than the threshold $\tau$ or if $T$ has been fully scanned, then discard this anchor
- compute the best inclusion from non discarded anchoring paths

```
Data: S //subtree to validate,
          T //tree from database
Result: true //if S is embedded within T
M    // mapping set of S within T;
foreach node m ∈ N_T do
    n ← root(S);
    if L(n) = L(m) then
        PartialInclusionDegree(S, T, n, m, M);
        M ← ∪M;
return the best inclusion {M ∈ M|MIN{M.mismatchedNodes}};
```

**Algorithm 1.** Anchoring

Note that our approach is consistent, meaning that if $\tau = 1$ (i.e. all the nodes must be mapped), then our algorithms are exactly the same as the ones defined in the crisp case [6].

```
Data: S //subtree to validate, T //tree from database,
      n, m //anchoring point, n ∈ S to m ∈ T,
      M //occurrence of S

M[n] ← m;
n ← n + 1;
if n <= |S| then
    P ← {w: w ∈ T such that L(w) = L(n) and m ⪯ w and ancestor(w) =
          M[ancestor(n)]};
    if P ≠ ∅ then
        foreach node w ∈ P do
            PartialInclusionDegree(S, T, n, w, M);
    else
        M.mismatchedNodes ← M.mismatchedNodes + 1;
        if M.mismatchedNodes >= τ then
            exit;
        else
            PartialInclusionDegree(S, T, n, m, M);
return;
```

**Algorithm 2.** PartialInclusionDegree

## 5 Conclusion

In this paper, we have detailed our previous work on fuzzy tree mining by giving the necessary definitions and algorithms in order to address the partial inclusion. Partial inclusion is a big deal in tree mining as it is not possible to consider full matches in real applications. However, it is necessary to remain scalable as the volumes of data being considered in real databases is huge. We thus design solutions based on levelwise algorithms, which consider anti-monotonic properties that guarantee the scalability. The algorithms presented here are currently implemented, and it is possible to conclude that this approach allows the extraction of more frequent subtrees (as fuzziness is introduced) while remaining scalable. Future work include the comparison of the results depending on the choices of the aggregation function. This comparison will be lead both on the quality of frequent subtrees and on runtime, as some aggregation functions are easier to compute than other ones.

## References


1. R. Agrawal and R. Srikant. Fast algorithms for mining association rules in large databases. *In Proceedings of the 20th* VLDB *Conference,* Santiago, Chile, 2002.
2. T. Asai, K. Abe, S. Kawasoe, H. Arimura, and H. Sakamoto. Effcient substructure discovery from large semi-structure data. In *2nd Annual* SIAM *Symposium on Data Mining, SDM2002,* Arlington, VA, USA, 2002. Springer-Verlag.

3. Y. Chi, R. R.Muntz, S. Nijssen, and J. N. Kok. Frequent subtree mining - an overview. *Fundamenta Informaticae XXI,* pages 1001–1038, 2005.
4. Y. Chi, Y. Yang, and R. Muntz. Cmtreeminer: Mining both closed and maximal frequent subtrees. In *The Eighth Pacific-Asia Conference on Knowledge Discovery and Data Mining (PAKDD'04),* 2004.
5. Y. Chi, Y. Yang, and R. R. Muntz. Indexing and mining free trees. In *International Conference on Data Mining 2003 (ICDM2003),* 2003.
6. F. Del Razo, A. Laurent, P. Poncelet, and M. Teisseire. Rsf - a new tree mining approach with an efficient data structure. In *Proceedings of the joint Conference: 4th Conference of the European Society for Fuzzy Logic and Technology (EUSFLAT 2005),* pages 1088–1093, 2005.
7. C. Fiot, A. Laurent, and M. Teisseire. From crispness to fuzziness: Three algorithms for soft sequential pattern mining. *IEEE Transactions on Fuzzy Systems. (to appear),* 2007.
8. D. Huffman. A method for the construction of minimum-redundancy codes. In *Proceedings of the Institute of Radio Engineers,* 1952.
9. D. Knuth. *The Art of Computer Programming, Volume 1: Fundamental Algorithms.* Addison-Wesley, 1973.
10. M. Kuramochi and G. Karypis. Frequent subgraph discovery. In *IEEE International Conference on Data Mining (ICDM),* 2001.
11. A. Laurent, M. Teisseire, and P. Poncelet. *Fuzzy Data Mining for the Semantic Web: Building XML Mediator Schemas,* chapter 12. Elsevier, E. Sanchez(ed.), To appear, 2006.
12. S. Sanchez, A. Laurent, P. Poncelet, and M. Teisseire. Fuzbt: a binary approach for fuzzy tree mining. In *Proceedings of the 11th IPMU International Conference on Information Processing and Management of Uncertainty in Knowledge-Based Systems (IPMU 2006),* 2006.
13. A. Termier, M.-C. Rousset, and M. Sebag. Treefinder, a first step towards XML data mining. In *IEEE Conference on Data Mining (ICDM),* pages 450–457, 2002.
14. C. Wang, Q. Yuan, H. Zhou, W.Wang, and B. Shi. Chopper: An eficient algorithm for tree mining. *Journal of Computer Science and Technology,* 19:309–319, May 2004.
15. M. A. Weiss. *Data Structures And Algorithm Analysis In* C. AddisonWesley, 1998.
16. R. Yager. Families of owa operators. *Fuzzy Sets and Systems,* 57(3):125–148, 1993.
17. X. Yan and J. Han. gspan: Graph-based substructure pattern mining. In *IEEE Conference on Data Mining (ICDM),* 2002. 18.
18. M. J. Zaki.*Efficiently Mining Frequent Trees in a Forest.* In KDD'02, Edmonton, Alberta, Canada, 2002. ACM.

# Numerical Properties of Fuzzy Regions: Surface Area

Jörg Verstraete and Axel Hallez

DDCM, Telin, Gent University, Sint-Pietersnieuwstraat 41, 9000 Ghent, Belgium
{Jorg.Verstraete,Axel.Hallez}@telin.ugent.be
http://telin.ugent.be/ddcm

**Abstract.** This paper concerns the modelling of fuzzy information in geographic databases. In the past, a theoretical model for fuzzy regions has been presented, along with various operations useful in a geographic database: union, intersection, topology, bounding rectangle, etc. and feasible models for implementation based on this theoretical model. Now, the attention is directed at some of the problems that can occur when determining numerical properties of fuzzy regions: what type of result is expected (and desired), and how does this impact the definitions of the operations. As an example, the definition of the surface area of a fuzzy set is studied in more detail.

## 1 Fuzzy Regions

### 1.1 Introduction

Geographic Information Systems (GIS for short) or are complex pieces of software, consisting basically of a database (with optimisations to work with geographic information) and a complex query-engine which allows for graphical input and output of information ([6], [8]). These systems are often used to model geographic features, both static or slow-changing like soil composition, as dynamic and fast changing, like traffic models, flow of water and weather systems. As these systems are used to reflect reality, the currently used crisp data models are sometimes inadequate [5]; consider for example the soil composition: where does one type of soil stops, and another type of soil begins? Several concepts have been developed to improve on the traditional model, by allowing regions with undetermined boundaries to be modelled ([1], [2], [7]). To achieve this, we introduced fuzzy regions, which make use of fuzzy set theory to model uncertainty or imprecission regarding the points belong to it.

### 1.2 Fuzzy Regions

To allow for richer modelling of reality, we presented a model theoretical model that incorporates fuzzy set theory with spatial information. Traditionally, a region in a GIS is modelled by means of its outline, a concept that was maintained in many extensions ([1], [2]). However, a region can also be considered to be a set of locations (namely all the locations that belong to the region). Using this

P. Melin et al. (Eds.): IFSA 2007, LNAI 4529, pp. 155–161, 2007.
© Springer-Verlag Berlin Heidelberg 2007

as the basic concept of a region, a fuzzy region is easily extended as a fuzzy set of locations (i.e. a fuzzy set over $\mathbb{R}^2$).

**Definition 1.** *Let $A \subseteq U$ the set of all the points that will belong to the region (this is a crisp set). The crisp set (or region) $A$ is then generalized to a fuzzy set (or region) $\tilde{A}$, defined as:*

$$\tilde{A} = \{(p, \mu_{\tilde{A}}(p)) | p \in U, \mu_(\tilde{A})(p) > 0\} \tag{1}$$

*where*

$$\mu_{\tilde{A}} : U \rightarrow ]0, 1]$$
$$p \mapsto \mu_{\tilde{A}}(p)$$

*Here, $U$ is the universe of all locations $p$; the membership grade $\mu_{\tilde{A}}(p)$ expresses the extent to which $p$ belongs to the fuzzy region.*

Note that the membership grades of a fuzzy region are interpreted veristic [3]: all points belong to the set, but some to a greater extent than others. A fuzzy region in which the membership grades are interpreted possibilistic [3], can be considered to represent a point at an imprecise or uncertain location, a fuzzy point: each location represented by this fuzzy set is a possible location for a point, with the membership grade indicating the possibility. For this paper, only fuzzy regions are considered.

For fuzzy regions, implementable models have been developed, as well as a number of operations; for both the models and operations we refer to [9], [10]. These operations included intersection, union, bounding rectangle, etc. This paper concerns the surface area of a fuzzy region, and the possible problems that need to be taken into account. To define the surface area, the $\alpha$-cut of a fuzzy region is needed. As a fuzzy region is in essence a fuzzy set over a two dimensional domain, the $\alpha$-cut definitions are straightforward.

**Definition 2.** *The* weak $\alpha$-cut *of a fuzzy region $\tilde{A}$ is defined as:*

$$\tilde{A}_\alpha = \{x | \mu_{\tilde{A}}(x) \geq \alpha, \forall x\}$$

**Definition 3.** *The* strong $\alpha$-cut *of a fuzzy set $\tilde{A}$ is defined as:*

$$\tilde{A}_{\overline{\alpha}} = \{x | \mu_{\tilde{A}}(x) > \alpha, \forall x\}$$

### 1.3 Fuzzy Numbers

A classic concept in fuzzy set theory is the concept of *fuzzy numbers*. In general, a fuzzy number is a fuzzy set over the domain of the real numbers $\mathbb{R}$, that represents numbers close to a given number. By definition ([4]), a fuzzy number $B$ is defined as a fuzzy set over $\mathbb{R}$ that satisfies the following three properties:

- $B$ is normalized, i.e. there is at least one element $x$ for which $\mu_A(x) = 1$
- $\forall \alpha \in ]0, 1] : B_{\overline{\alpha}}$ is a closed interval
- the support $B_0$ of $B$ must be bounded

The first property indicates that the concept of the set of real numbers *close to a given number* is fully satisfied by at least one number: the given number itself. The other two properties allow for the definition of meaningful arithmetic operations in terms of standard arithmetic operations and interval calculus ([4]).

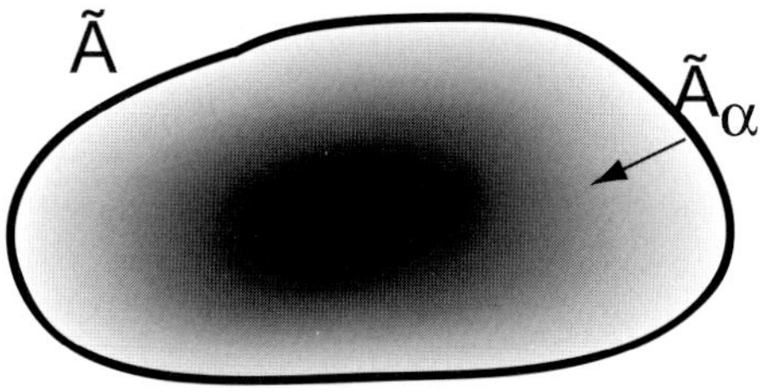


**Fig. 1.** Illustration of a fuzzy region. Grey scales are used to indicate the membership grade: the darker the colour, the higher the membership grade. For clarity, the outline of the region is also shown.

## 2 Numeric Properties: Surface Area

In GIS, a number of operations are possible. Some return a new region (like the set-operations intersection, union, etc. or geo-spatial operations (minimum bounding rectangle, convex hull, etc.), whereas others return numeric values (surface area, distance between two objects, etc.). In this paper, the latter operations are considered, more specifically, the surface of a fuzzy region.

### 2.1 Different Interpretations of the Surface Area

The surface area of a fuzzy region lends itself to two interpretations; depending on the interpretation of the fuzzy region. If the fuzziness in the region used to represent uncertainty regarding the points, this uncertainty should be reflected in the surface area. Consequently, the surface area will be a fuzzy number, where the possibility distribution is dependant of the uncertainty that occurs in the fuzzy region. On the other hand, if the fuzziness is used to represent intrinsic fuzziness, there is no uncertainty regarding the surface area, hence it will yield a crisp number. This crisp number is an extension of the concept of fuzzy cardinality.

### 2.2 Surface Area Yields a Fuzzy Result

**Concept.** The fuzzy surface area $\tilde{S}^f$ of a fuzzy region $\tilde{A}$ in the first interpretation will result in a fuzzy number (indicated by $f$) that represents the possible surface areas. This interpretation is meaningful in a system capable of working with fuzzy numbers and fuzzy arithmetic. Current system don't have this functionality, but this interpretation is just one part of our model for fuzzy regions; in which a GIS with support for fuzzy information is developed. Fuzzy arithmetic allows to work with numbers that are imprecise (or uncertain); each number is represented by a membership function (which associates membership grades in the range $[0, 1]$ with crisp numbers). This membership function is a model for the uncertainty or imprecision of the number: higher membership grades (up to 1) mean a high certainty or precision, lower membership grades (down to 0) indicate a low certainty or precision. Using a fuzzy number to represent the area of a fuzzy region makes sense: any imprecision or uncertainty in the region should be reflected in the number representing this surface area.

To obtain the fuzzy result, first all possible surface areas for the given region must be considered; these are obtained from all the possible $\alpha$-cuts of the fuzzy region.

For this first definition, only the weak $\alpha$-cut (definition 2)is needed. As each weak $\alpha$-cut is a crisp region, the surface area of it can be calculated and is denoted $S(A_\alpha), \forall\alpha$. With each possible surface area, an appropriate membership grade is associated. This leads to the definition.

**Definition 4.**

$$\tilde{S}^f(\tilde{A}) = \{(S(\tilde{A}_\alpha), \mu_{\tilde{S}^f(\tilde{A})}(S(\tilde{A}_\alpha))), \forall\alpha \in ]0,1]\} \tag{2}$$

*where*

$$\begin{aligned} \mu_{S(\tilde{A}_\alpha)} &: \mathbb{R} \to [0,1] \\ &x \mapsto \sup\{\alpha | \alpha \in ]0,1] \wedge S(\tilde{A}_\alpha) = x\} \end{aligned}$$

*and $S$ represents the operation that yields the surface area of a crisp region.*

At every $\alpha$ level, the above definition holds the (crisp) area of the matching $\alpha$ level of the fuzzy region. This is illustrated in fig. 2.

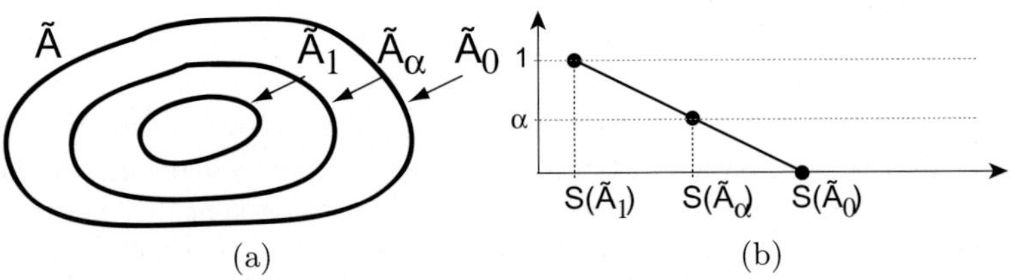


**Fig. 2.** Illustration of the surface area of a fuzzy region: (a) the fuzzy region illustrated using contourlines; the line of only one $\alpha$-cut is shown, (b) the surface area as obtained from the above definition

For future computations, is would be useful to have a fuzzy number as the result of a fuzzy surface area: calculations with different surface areas can be performed using fuzzy arithmetic ([4]). It must now be verified if the result satisfies the properties of a fuzzy number.

– The result is always normalized, even if there are no points $p$ in the fuzzy region $\tilde{A}$ for which $\mu_{\tilde{A}}(p) = 1$ (i.e. it is not a normalized fuzzy region), the result of the surface area using this definition will yield a normalized fuzzy set:

$$\sup\{\alpha | \alpha \in ]0,1] \wedge S(\tilde{A}_\alpha) = x\} \tag{3}$$

equals 1 for for $x = 0$.

- The support is bounded.
- However, not every $\alpha$-cut of the result yields a closed interval; a simple example suffices to illustrate the problems. Consider a fuzzy region $\tilde{A}$ defined by a number of points with membership grade 1 and a number of points with membership grade 0.5 (as illustrated on fig. 3a). The fuzzy surface area using the above definition would yield:

$$\tilde{S}^f(\tilde{A}) = \{(s, 1), (s, 0.5)\} \tag{4}$$

with $s$ the surface area of the square (illustrated on fig.3b). It is now easily verified that:

$$\forall \alpha \in ]0, 0.5[: (\tilde{S}^f(\tilde{A}))_{\overline{\alpha}} \text{ is not a closed interval} \tag{5}$$

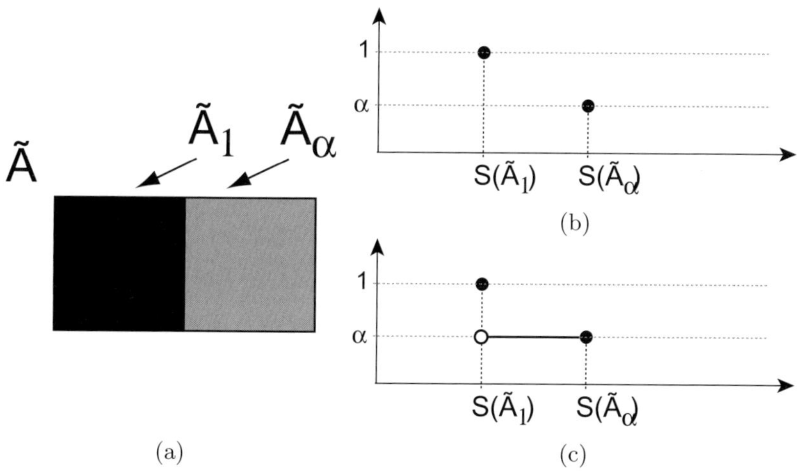


**Fig. 3.** Illustration of the surface of a discontinuous fuzzy region: (a) the fuzzy region, (b) the surface area as obtained from the first definition, (c) the surface area as obtained from the second definition

Note that this problem does not occur if the points at a given membership grade less than 1 form a one dimensional object: i.e. if the membership grades are strictly decreasing from the center outward. In fig. 2a, all the points at any given membership grade in $]0, 1[$ form a closed line. In fig. 3a however, the points at membership grade 0.5 form a two dimensional structure (a filled square in this case).

**Alternative Definition.** Similarly as before, the surface areas of the different $\alpha$-levels are considered; however this time both the strong and the weak $\alpha$-cuts are required (definition 2, 3).

The weak and the strong $\alpha$-cut of a fuzzy region $\tilde{A}$ both yield a crisp region, denoted $\tilde{A}_\alpha$ respectively $\tilde{A}_{\overline{\alpha}}$. For these crisp regions and for every $\alpha \in ]0, 1]$:

$$S(\tilde{A}_{\overline{\alpha}}) \leq S(\tilde{A}_\alpha) \tag{6}$$

where $S$ is the notation for the calculation of the surface area of a crisp region. The equality only occurs if $S(\tilde{A}_\alpha - \tilde{A}_{\overline{\alpha}}) = 0$; this happens if the points $p$ for which $\mu_{\tilde{A}}(p) = \alpha$ form a one dimensional object. This allows us to define the surface area as:

**Definition 5.**

$$\tilde{S}^f(\tilde{A}) = \{(x, \mu_{\tilde{S}^f(\tilde{A})}(x)), \forall\alpha \in ]0,1]\} \tag{7}$$

*where*

$$\mu_{\tilde{S}^f(\tilde{A})}(x) : \mathbb{R} \rightarrow [0,1] \tag{8}$$

$$x \mapsto \begin{cases} 1 & \textit{if } x = S(\tilde{A}_1) \\ \sup\{\alpha | S(\tilde{A}_{\overline{\alpha}}) \leq x \leq S(\tilde{A}_\alpha)\} & \\ 0 & \textit{elsewhere} \end{cases} \tag{9}$$

The main difference with the previous definition is that all surface areas $x$ between $S(\tilde{A}_{\overline{\alpha}})$ and $S(\tilde{A}_\alpha)$ are considered and assigned an appropriate membership grade. This grade is the largest $\alpha$ for which the $x \in [S(\tilde{A}_{\overline{\alpha}}), \leq S(\tilde{A}_\alpha)]$. Again, the properties of fuzzy numbers must be verified.

- The result is always normalized; if there are no points $p$ for which $\mu_{\tilde{A}}(p) = 1$, then $\mu_{S^f(\tilde{A})}(0) = 1$.
- The support is bounded.
- Contrary to the previous case, every $\alpha$-cut for $\alpha \in ]0,1]$ yields a closed interval. To illustrate with the simple example from the previous paragraph, the fuzzy surface area obtained using this alternate definition yields:

$$\tilde{S}^f(\tilde{A}) = \{(s,1)\} \cup \{(x, 0.5) | x \in ]s, 2s]\} \tag{10}$$

  with $s$ the surface area of each of the squares. This membership function is illustrated on fig. 3c.
  For $\alpha \in ]0, 0.5[$, the strong $\alpha$-cut results in the interval $[s, 2s]$.
  For $\alpha \in [0.5, 1]$, the strong $\alpha$-cut yields the (degenerate) interval $[1, 1]$.

## 2.3 Surface Area Yields a Crisp Result

In the second interpretation, the fuzziness is used to indicate the intrinsic vagueness of a region. The surface area therefore becomes a crisp number. This number takes all points into consideration, where the membership grade for each point determines how much is will contribute: a point with a membership grade 0.5 will only contribute half of what a point with membership grade 1 will contribute. In a discrete set, this number resembles the fuzzy cardinality, but for infinite sets this needs to be extended.

The notation for this surface calculation is $\tilde{S}^c$.

$$\tilde{S}^c(\tilde{A}) = \int_{p \in U} (p\ \mu_{\tilde{A}}(p))dp \tag{11}$$

For the simple example in fig. 3a, the result will be $0.5s + 1s = 1.5s$, where $s$ is the crisp surface area of each square.

This notion can be interesting, either depending on the interpretation given to the fuzzy region and/or the surface area, as well as for future calculations. The fact that the result is a crisp number makes it easier to integrate this operation with systems that are unable to work with fuzzy arithmetic.

## 3 Conclusion

When extending a geographic system to work with fuzzy regions, it is interesting to have operations that yield a crisp number for crisp regions yield a fuzzy number for fuzzy regions. In this paper, the intuitive definition for the surface area of a fuzzy region has been considered, and it was shown that this not necessarily yields a fuzzy number. Similar issues occur with other numeric properties, such as the distance to a fuzzy region. To overcome this for the surface area, an alternative definition has been presented. It was also illustrated that this alternative definition yields result which in all cases is a fuzzy number.

## References


1. Clementini E., Di Felice P.; An algebraic model for spatial objects with undetermined boundaries; 1994; GISDATA Specialist Meeting - revised version.
2. Cohn A. G., Gotts N. M.; Spatial regions with undetermined boundaries; 1994; Proceedings of the Second ACM Workshop on Advances in Geographic Information Systems; pages 52–59.
3. Dubois D., Prade H.; 1997; The three semantics of fuzzy sets; 1999; Fuzzy Sets and Systems 90, pages 141-150.
4. Klir G. J., Yuan B.; 1995; Fuzzy sets and fuzzy logic: Theory and applications; New Jersey: Prentice Hall.
5. Morris A.; Why Spatial Databases Need Fuzziness; 2001; Proceedings of Nafips 2001; pages 2446–2451.
6. Rigaux P., Scholl M., Voisard A.; 2002; Spatial Databases with Applications to GIS. Morgan Kaufman Publishers.
7. Schneider M.; ; Modeling Spatial Objects with Undetermined Boundaries Using the Realm/ROSE Approach
8. Shekhar S., Chawla S.; 2003; Spatial Databases: A tour; Pearson Education Inc.
9. Verstraete J, De Tré G., De Caluwe R., Hallez A.; 2005; Field Based Methods for the modelling of Fuzzy Spatial Data; in Fuzzy modeling with Spatial Information for Geographic Problems; eds. Petry F., Robinson v., Cobb M.; Springer-Verlag; pages 41–69.
10. Verstraete J, Hallez A., De Tré G.; 2006; Bitmap Based Structures for the modeling of Fuzzy Entities; in special issue of Contro & Cybernetics; accepted.
11. L.A. Zadeh; 1971; Quantitative fuzzy semantics, Information Sciences **3** 2 (1971); pages 177–200.

# Qualification of Fuzzy Statements Under Fuzzy Certainty

A. González, N. Marín, O. Pons, and M.A. Vila

Dpto. Ciencias de la Computación e I.A.
Universidad de Granada
18071 Granada, Spain
{A.Gonzalez,nicm,opc,vila}@decsai.ugr.es

**Abstract.** In many problems the information can be imprecise and uncertain simultaneously. Linguistic terms can be then used to represent each one of these aspects. In some applications it is desirable to combine imprecision and uncertainty into a single value which appropriately describes the original information. We propose a method to combine imprecision and uncertainty when they are expressed as trapezoidal fuzzy numbers and the final goal is to obtain a normalized fuzzy number. This property is very useful in several applications like flexible querying processes, where the linguistic label used in the query is always normalized.

## 1 Introduction

The aim of this work is to propose a solution to the problem of uncertainty qualification of fuzzy statements [4] when the certainty is expressed as a fuzzy number. In a previous paper [6] we proposed a method to solve this problem when the certainty is expressed as a real number. The main idea of this proposal was the following. Let us suppose we have a fuzzy value A understood as acting as a fuzzy restriction on the possible values of a variable X, and this value is affected by a certainty degree, say $\alpha$. Then, the problem is to represent a qualified statement like "it is $\alpha$-certain that X is A".

This situation can be formulated as a conditional expression, using the generalized modus ponens, in the following terms:

- *if the certainty level is 1, then the value is $A$.*
- *if the certainty level is* $\alpha < 1$, then the value is $T(A)$, where $T(A)$ is a transformation of the original fuzzy set $A$.

In this way, the qualified statements "it is $\alpha$-certain that X is A" is represented as "X is T(A)".

Therefore, a natural way to solve the problem is to consider that the transformation we are handling is $T(A)$ defined as: $\mu_{T(A)}(x) = I(\alpha, \mu_A(x))$ where $I$ is a material implication function which reflects the interpretation given to the compatibility degree.

P. Melin et al. (Eds.): IFSA 2007, LNAI 4529, pp. 162–170, 2007.
© Springer-Verlag Berlin Heidelberg 2007

There exist in the literature two main ways of dealing with imprecise and uncertain data and can be interpreted as follows.

1. To truncate: if the datum is "A with certainty $\alpha$", then $T(A)$ is defined by the membership function $\mu_{T(A)}(x) = min(\alpha, \mu_A(x))$ which directly implies that we are using Mamdani's implication in our reasoning.
2. To expand: if we assume that $\alpha$ is a necessity, then $T(A)$ is given by the membership function $\mu_{T(A)}(x) = max(1 - \alpha, \mu_A(x))$, which corresponds to Kleene-Dienes' implication as foundation of our reasoning.

These proposals can be useful in many applications, but they can also be inappropriate in many others. Thus, Mamdani's implication obliges us to work with non-normalized fuzzy values. Kleene-Dienes' implication obliges to assign the same possibility to all the points of the underlying domain independently from the distance to the support set of the fuzzy value. Therefore, the proposed solutions give rise to a series of inconveniences: the interpretability, in some cases, and those ones derived from the use of non-normalized or non-trapezoidal fuzzy sets.

As an alternative proposal, in [8,6] we proposed a certainty qualification method that consists in increasing the imprecision around the support set of value A depending on an uncertainty value, that is, the imprecision is distributed according to a metric which takes into account the nearness to the original information. This proposal is based on the use of information measures that allow us to transform the uncertainty of the fuzzy statement into imprecision. For example, when we have the information that *"X is black"* with certainty $\alpha$, it is not very convenient to assign a positive possibility to color white but to colors near enough to black depending on value $\alpha$.

Therefore, the process we proposed in [6] was to define $T(A)$ in two steps:

1. First, by considering that the height of a fuzzy number is the certainty degree associated to it [2,5], we use the certainty degree $\alpha$ associated to the fuzzy value $A$ to truncate it at level $\alpha$. After this operation, we obtain a non-normalized fuzzy set $A^\alpha$. Nevertheless, the resulting fuzzy value remains trapezoidal.
2. Since, in many applications, non-normalized fuzzy sets give rise to a series of inconveniences, in a second step we normalize it. To do this, we assume that uncertainty is being translated into imprecision under certain conditions. The most important point to be considered is that the amount of information provided by the fuzzy number remains equal before and after the normalization process. $T_\alpha(A)$ will stand for the obtained normalized fuzzy value, whose imprecision is, obviously, larger than $A^\alpha$ imprecision, as it has been made completely true (its height is 1 again).

In fact, in the fuzzy querying process the linguistic labels used are always normalized what makes it necessary that the stored data are also normalized in order to carry out a semantically coherent matching computation.

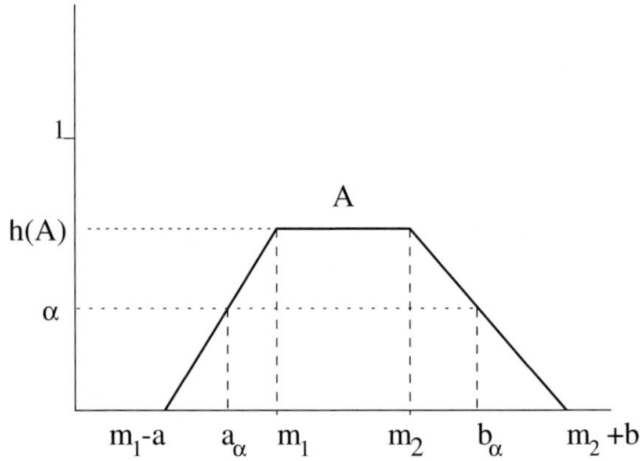


**Fig. 1.** Trapezoidal Fuzzy Number

## 2 Previous Results

A fuzzy value is a fuzzy representation about the real value of a property when it is not precisely known. We will use $\widetilde{I\!R}$ to denote the set of fuzzy numbers.

The interval $[a_\alpha, b_\alpha]$ (see figure 1) is called the $\alpha$-cut of A. Therefore, fuzzy numbers are fuzzy quantities whose $\alpha$-cuts are closed and bounded intervals: $A_\alpha = [a_\alpha, b_\alpha]$ with $\alpha \in (0, 1]$. The set $Supp(A) = \{x \in I\!R \mid A(x) > 0\}$ is called the *support set of* $A$[1] and $h(A)$ denotes the height of the fuzzy number $A$. If there is, at least, one point $x$ verifying $A(x) = 1$ we say that $A$ is a *normalized* fuzzy number.

Usually, a trapezoidal shape is used in order to represent fuzzy numbers. This representation is very useful as the fuzzy number is completely characterized by five parameters $(m_1, m_2, a, b)$ and the height $h(A)$, as figure 1 shows. The interval $[m_1, m_2]$ (i.e, the set $\{x \in Supp(A) \mid \forall\ y \in I\!R,\ A(x) \geq A(y)\}$) will be called *modal set.* The values $a$ and $b$ are called left and right *spreads*, respectively.

The basic idea underlying this work is that when a fuzzy number is not normalized, the situation can be interpreted as a lack of confidence in the information provided by such a number [2,5]. In fact, the height of the fuzzy number could be considered as a certainty degree of the represented value, and this implies that normalized fuzzy numbers represent imprecise quantities on which we have complete certainty.

Since the first step in our proposal is to truncate, we can consider that the truncated fuzzy number represents the imprecise information and moreover it shows a certain level of uncertainty.

In [6], we show how uncertainty can be translated, using a suitable transformation, into imprecision, taking into account that to reduce the uncertainty about a fuzzy number implies to increase the imprecision of such number. This

---

[1] In the rest of the paper $A(x)$ will stand for $\mu_A(x)$.

transformation is made in such a way that the amount of information provided by the fuzzy number is the same before and after the modification.

Our idea is to transform the truncated fuzzy number in order to obtain a completely certain fuzzy number.

As pointed out in the previous section, we are going to translate fuzzy uncertainty into imprecision under given conditions. The most important of these conditions is that the amount of information provided by the fuzzy number remains equal before and after the transformation. Therefore, the first step is to define an information function for fuzzy numbers.

In [6], we propose an axiomatic definition of information, partially inspired in the theory of generalized information given by Kampé de Fériet [7] and that can be related to the precision indexes [3] and the specificity concept introduced by Yager in [11].

**Definition 1.** *Let* $\mathcal{D} \subseteq \widetilde{I\!R} \mid I\!R \subseteq \mathcal{D}$*; we say that* $I : \mathcal{D} \longrightarrow [0, 1]$ *is an* **information function** *on* $\mathcal{D}$ *if it verifies:*

1. $I(A) = 1, \ \forall \ A \in I\!R$
2. $\forall \ A, B \in \mathcal{D} \ \mid \ h(A) = h(B)$ *and* $A \subseteq B \Longrightarrow I(B) \leq I(A)$.

The information about fuzzy numbers may depend on different factors, in particular, on imprecision and certainty. In this work, we focus on general types of information related only to these two factors.

**Definition 2.** *The* **imprecision** *[5] of a fuzzy number is defined as follows:*

$$\forall \ A \in \widetilde{I\!R}, \ imp(A) = \int_0^{h(A)} (b_\alpha - a_\alpha)d\alpha$$

With respect to the height (certainty) and the imprecision of a fuzzy value, we define the following general type of function [5]:

$$\forall \ A \in \widetilde{I\!R}, \ I(A) = \frac{h(A)}{k * imp(A) + 1}$$

where $h(A)$ is A height, imp(A) is the imprecision associated to $A$ and $k \neq 0$ is a parameter which depends on the domain scale. This is the simplest function that verifies the mentioned properties of information functions.

Once we have an information function on fuzzy numbers, we can use it to define transformations which preserve the information amount it provides. The idea is to find an *equivalent* representation of the considered fuzzy number in such a way that we change uncertainty by imprecision keeping constant the relationship between them, which is determined by the information function.

The aim of the transformations we are proposing in this section is, basically, to be able to modify the height of a fuzzy number but keeping the information contained in it.

The definition of transformation will be obtained from the condition of equality in the information but, as a first step, we must establish what we understand for transformation of a fuzzy number on a subset of $\widetilde{I\!R}$.

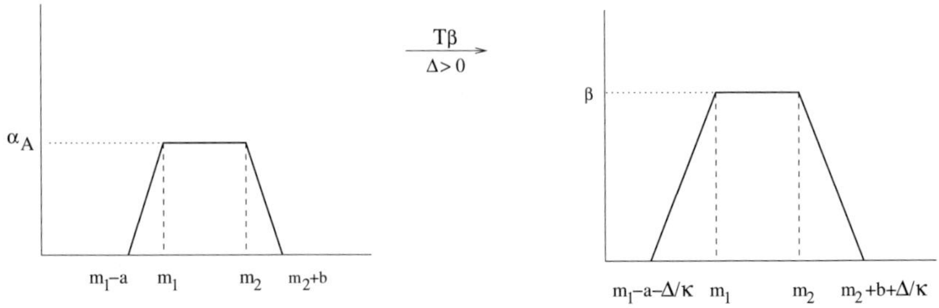


**Fig. 2.** Transformation that increases imprecision

**Definition 3.** *Let us consider* $\alpha \in (0, 1]$ *and the class of fuzzy numbers* $D \subseteq \widetilde{I\!R}$. *We say that*

$$T_\alpha : D \longrightarrow \widetilde{I\!R}$$

*is a* **transformation** *for an information function* $I$ *on* $D$*, if it verifies that:*

1. $T_\alpha(A) \in D$
2. $h(T_\alpha(A)) = \alpha$
3. $I(T_\alpha(A)) = I(A), \ \forall \ A \in D$

We will note by $\tau$ the class of trapezoidal fuzzy numbers on $I\!R$. Given a fuzzy number $A \in \tau$, we are looking for the conditions that another fuzzy number $B$, with fixed height $\alpha \in (0, 1]$, must hold to have the same information amount as $A$. Assuming the following conditions:

1. modal imprecision is preserved,
2. the increase/decrease of imprecision is equally distributed in the right and left sides of the fuzzy number independently from its shape,

we proposed in [6] the following transformation:

**Definition 4.** *Let* $A \in \tau$ *such that*

$$A = \{(m_1, m_2, a, b), \alpha_A\}$$

*where* $m_1, m_2, a$ *and* $b$ *are shown in figure 1 and* $\alpha_A$ *is the height of* $A$.

*Let* $\alpha \in (0, 1]$ *be. We will denote* $\Delta(\alpha_A, \alpha) = \Delta$ *and define*

$$T_\alpha(A) = \{(m_1, m_2, a + \frac{\Delta}{k}, b + \frac{\Delta}{k}), \alpha\}$$

*for those* $\alpha$ *in which the transformation makes sense.*

In figure 2 it is shown how an increment of height produces an increment of imprecision.

In the proposed transformation, the relation between certainty and imprecision is the following:

– An increase of certainty means an increase of imprecision.
– A decrease of imprecision means a decrease of certainty.

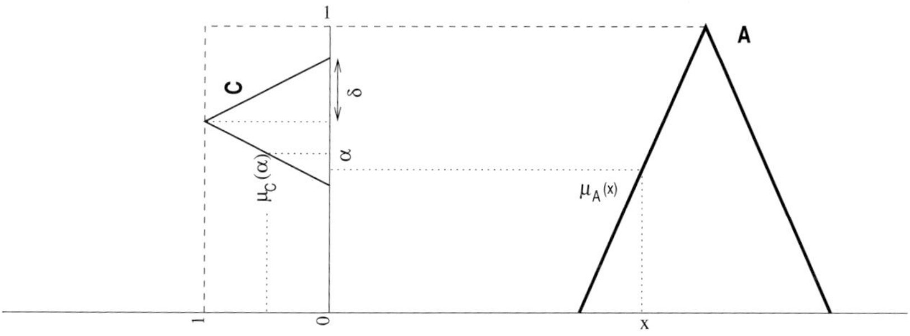


**Fig. 3.** Fuzzy Certainty on a Fuzzy Value: The scale used in both axels is not the same for the sake of clarity

## 3 Fuzzy Uncertainty

Once, we know how to solve the qualification problem when the uncertainty is represented as a real value, now the problem is to extend this process when the uncertainty is represented as a fuzzy value. Thus, we want to translate the information "X is A is C", when C is a fuzzy trapezoidal number, into "X is $T_C(A)$".

The difficulty is now to give a suitable procedure for computing $T_C(A)$ since now C is a trapezoidal fuzzy number. To do this, we will consider that, for any possible truncation level $\alpha$, the membership function of the linguistic label modifies in a certain way the certainty level. In fact we can assume that:

$$(X \ is \ A) \ is \ C \longleftrightarrow \forall \alpha \in [0,1], \ X \ is \ A \ to \ a \ degree \ C(\alpha), \ \alpha \in [0,1]$$

Figure 3 depicts the general problem we are trying to explain.

A possible way to solve this problem is to define $T_C(A)$ in such a way that it summarizes the right side of the above sentence by means of some average. It should be remarked that the membership function $C(.)$ induces two fuzzy measures (possibility/necessity) on the [0,1] interval and that the membership function of any fuzzy number transformed at certainty level $\alpha$ can be considered as a function depending on both $\alpha \in [0,1]$ and $x$, which ranges on another real interval. A method that allows the use of such average is the Sugeno's integral.

In [10], Sugeno introduced the concept of fuzzy integral of a fuzzy measure as a way to compute some kind of average value of a function in terms of the underlying fuzzy measure. Obviously, fuzzy measures formally include possibility/necessity measures as special cases. Fuzzy integrals are interpreted as subjective evaluations of objects where subjectivity is represented by means of fuzzy measures.

The fuzzy integral over a referential set $X$ of a function $f(x)$ with respect to a fuzzy measure $g$ is defined as follows:

$$\int_X f(x) \circ g(.) = sup_{\alpha \in [0,1]} \{\alpha \wedge g(F_\alpha)\}$$

where $F_\alpha = \{x|f(x) \geq \alpha\}$.

In the case that the measure g is a possibility defined by means of the membership function of a fuzzy set $\mu(x)$ with referential X, the Sugeno's integral has the following expression [9]:

$$\int_X f(x) \circ g(.) = sup_{x\in X}(f(x) \wedge \mu(x)).$$

On the other hand, if we assume the considered fuzzy measure g is a necessity induced by the fuzzy set $\mu(x)$, then we have the following expression [9]:

$$\int_X f(x) \circ g(.) = inf_{x\in X}(f(x) \vee (1 - \mu(x))).$$

As we have stated above, the basic idea of our approaches is to use the fuzzy measures (possibility, necessity) induced by the membership function $C(.)$ of the linguistic evaluation of certainty, to compute the *average* of the transformed fuzzy number, by means of Sugeno's integral.

At this point, it is necessary to remark that the transformation process of any fuzzy number $A(.)$ with *crisp certainty* value $\alpha$ has two steps:

(i) Truncating the fuzzy number at the level $\alpha$, obtaining an non-normalized fuzzy number $A^\alpha(.)$.
(ii) Transforming $A^\alpha(.)$ into a normalized fuzzy number $T(A)$.

The idea is the following. In a first step, we apply the Sugeno's integral to the function $f(\alpha, x) = A^\alpha(x)$ with respect to the $\alpha$ variable, obtaining a possibly non-normalized fuzzy number. This fuzzy number will be transformed into a normalized one in the step ii. This process can be done in two different ways depending on wether we use the possibility or the necessity measures to perform the integral.

Thus, let $\Pi_C(.)$ stand for the possibility measure induced by $C$ and $T_p(.)$ stand for the mean of the truncated fuzzy numbers. Then we have:

$$T_p(x) = \int_{[0,1]} A^\alpha(x) \circ \Pi_C(\alpha) = sup_{\alpha\in[0,1]}(A^\alpha(x) \wedge C(\alpha)) =$$
$$= sup_{\alpha\in[0,1]}(A(x) \wedge \alpha \wedge C(\alpha)) = A(x) \wedge sup_{\alpha\in[0,1]}(\alpha \wedge C(\alpha))$$

If $C_p = sup_{\alpha\in[0,1]}(\alpha \wedge C(\alpha))$, then we finally have:

$$T_p(x) = A(x) \wedge C_p$$

which indicates that, in the case of the possibility measure, the mean of truncated values is the result of truncating with an specific value which only depends on the linguistic label $C(.)$.

Alternatively, let $N_C(.)$ stand for the necessity measure induced by $C$ and $T_n(.)$ stand for the mean of the truncated fuzzy numbers. Using expression in section 3, we have:

$$T_n(x) = \int_{[0,1]} A^\alpha(x) \circ N_C(\alpha) = inf_{\alpha\in[0,1]}(A^\alpha(x) \vee (1 - C(\alpha))) =$$
$$= inf_{\alpha\in[0,1]}(A(x) \wedge \alpha \vee (1 - C(\alpha))) = A(x) \wedge inf_{\alpha\in[0,1]}(\alpha \vee (1 - C(\alpha)))$$

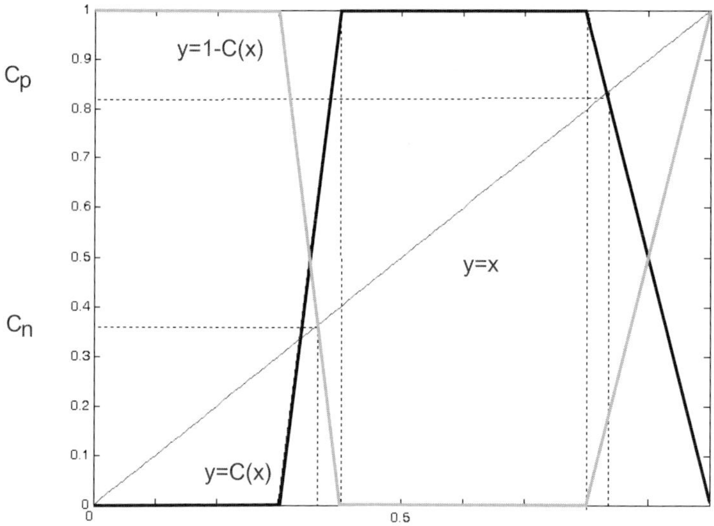


**Fig. 4.** Upper and lower measures

If $C_n = inf_{\alpha \in [0,1]}(\alpha \vee (1 - C(\alpha)))$, then we finally have:

$$T_n(x) = A(x) \wedge C_n$$

which indicates that, also in the case of the necessity measure, the mean of truncated values is the result of truncating with an specific value which only depends on the linguistic label $C(.)$

With the previous expression we have got two proposals for making this truncation or, what is the same, we obtain two different fuzzy values $T_p(x)$ and $T_n(x)$. In this first step we have integrated the fuzzy uncertainty C in the truncation process.

As it happens with all dual measures, the expert can choose either to work with both of them or to decide which one is the most suitable for the purpose of the system. In figure 4 we graphically show the results obtained considering that the linguistic label C has a trapezoidal membership function.

After the truncation, it is necessary to perform the corresponding transformations in order to obtain a normalized fuzzy number. $T_N(.), T_P(.)$ will stand for the transformed $T_n(.)$ and $T_p(.)$, respectively. They can be directly obtained by the process described in section 2.

Moreover, we can conclude that $T_N(.)$ offers us a more imprecise transformed fuzzy number that $T_P(.)$ since

$$T_P(.) \subseteq T_N(.).$$

## 4 Conclusions

We have addressed the problem of dealing with linguistic uncertainty associated with a fuzzy quantity. With the basic idea of transforming uncertainty into imprecision, two possible approaches have been presented; all of them give transformations of the initial fuzzy number that lead to normalized fuzzy numbers. Explicit expressions of such transformed fuzzy numbers have also been obtained. This is a particularly useful property from the storage point of view (e.g. within the databases world or in a data warehousing context), since it provides us with a simple and unified representation for both certain and uncertain fuzzy values.

## References


1. Bordogna G., Pasi G. *Modeling Linguistic Qualifiers of Uncertainty in a Fuzzy Database.* Int. Journ. of Intelligent Systems, Vol. 15, pp. 995-1014. (2000)
2. Dubois, D. *Modeles mathematiques de l'imprecis et de l'incertain en vue d'applications aux techniques d'aide a la decision.* Ph.D Thesis. Université Scientifique et Médicale de Grenoble. (1983)
3. Dubois D., Prade H. *Possibility Theory.* Plenum Press. (1985)
4. Dubois D., Prade H., Yager R. *Merging Fuzzy Information.* In *Fuzzy Sets in Approximate Reasoning and Information Systems*, pp- 335-401. Eds. J.C. Bezdek, D. Dubois, H. Prade. Kluwer Academic Publishers. (1999)
5. Gonzalez, A. *Métodos Subjetivos para la Comparación de Números Difusos.* PhD Thesis. Granada University. (1987)
6. Gonzalez A., Pons O., Vila M.A. *Dealing with uncertainty and imprecision by means of fuzzy numbers.* Int. Journal of Approximate Reasoning, vol. 21, pp. 233-256. (1999)
7. Kampé de Feriét, J. *La theorie generaliseé de l'information et le mesure subjetive de l'information.* Lecture notes in Math: Théories de l'information. Eds. Kampé de Fèriet and C.F. Picard. Springer-Verlag, vol. 398, pp. 1-28. (1973)
8. Pons O., Cubero J.C., Gonzalez A., Vila M.A. *Uncertain Fuzzy Values still in the Framework of First Order Logic.* Proc. of Second Int. Workshop on Logic Programming and Soft Computing: Theory and Applications. Manchester (U.K.). (1989)
9. Suárez García F. and Gil Alvarez P., *Two families of fuzzy integrals.* Fuzzy Sets Syst., vol. 18 (1), pp. 67–81. (1986)
10. Sugeno M. *Theory of fuzzy integrals and its applications.* Thesis. Tokio Inst. of Technology. (1974)
11. Yager, R.R. *Measurement of Properties on Fuzzy Sets and Possibility Distributions.* E.P. Klement Ed. Proc. 3rd Intern. Seminar on Fuzzy Set Theory. Johannes Keplas Univ. Liuz, pp. 211-222. (1981)

# Weighted Conjunctive and Disjunctive Aggregation of Possibilistic Truth Values

Tom Matthé, Guy De Tré, and Axel Hallez

Ghent University, Dept. of Telecommunications and Information Processing,
Sint-Pietersnieuwstraat 41, B-9000 Gent, Belgium
{Tom.MATTHE,Guy.DETRE,Axel.HALLEZ}@UGent.be

**Abstract.** In this paper, weighted aggregation of extended possibilistic truth values will be presented. The impact of the weights on both conjunctive and disjunctive aggregation will be handled, and special care will be given to the case where both conjunctive and disjunctive aggregation are mixed together. The problems that can arise will be illustrated and solutions will be proposed.

## 1 Introduction

Databases are continuously growing in size, and, using traditional querying techniques, users are confronted with difficulties to find the information they are looking for. So, more and more, users are realising the benefits of using flexible querying systems. Flexible querying has many aspects, but this paper will only focus on one of them, namely weighted aggregation. In general, a flexible query will impose several different (flexible) selection criteria or (fuzzy) constraints, which can all have a different importance, indicated by the user. Weighted aggregation will combine all individual satisfaction grades of the respective constraints, taking into account their possible importance value, to produce a global satisfaction grade for the entire query [1].

This paper will not handle about the evaluation of selection criteria, but about the aggregation of the results of the evaluation of the individual selection criteria (which can be flexible or not). It is assumed the satisfaction grades of the individual selection criteria are known or calculated in advance, and as such, no assumption needs to be made about the underlying database. As logical framework, Extended Possibilistic Truth Values will be applied. They can be used to model the satisfaction grades of the flexible constraints and as such, they will be the input (and outcome) of the aggregation.

In Section 2, the basics of Extended Possibilistic Truth Values (EPTVs) will be presented, together with their arithmetic rules and basic aggregation operators. Section 3 will then focus on weighted aggregation, presenting how weights have an impact on EPTVs, what problems can arise when applying the weights and how to solve these problems. Finally, Section 4 presents some conclusions.

P. Melin et al. (Eds.): IFSA 2007, LNAI 4529, pp. 171–180, 2007.
© Springer-Verlag Berlin Heidelberg 2007

## 2 Extended Possibilistic Truth Values

### 2.1 Extended Truth Values

In traditional Boolean logic, with truth values 'True' ($T$) and 'False' ($F$), every proposition is considered to be either true or false. If $I = \{T, F\}$ is the set of truth values, and $P$ the universe of propositions, then we can define the truth value of a proposition by means of a function $t$ as follows:

$$t : P \to I : p \mapsto t(p)$$

$t(p)$ equals to $T$ if $p$ is true, i.e. if $p$ corresponds to reality. Otherwise $t(p)$ equals to $F$.

However, when evaluating propositions it is often the case that the truth value of a proposition cannot be determined as being either true or false. For instance, consider the proposition "Person X scored more than 8/10 on a math test". If it is known for sure that the person did not take the test, the truth value of the proposition is neither true nor false, but the proposition is inapplicable. For these kind of cases the set $I$ will be extended to $I^* = \{T, F, \bot\}$, where the additional truth value, $\bot$, represents inapplicable or undefined. The extended truth value $t^*(p)$ of a proposition $p \in P$ can then be defined as follows:

$$t^* : P \to I^* : p \mapsto t^*(p)$$

where

- $t^*(p) = T$, if $p$ corresponds to reality, i.e. if $p$ is true;
- $t^*(p) = F$, if $p$ does not correspond to reality, i.e. if $p$ is false;
- $t^*(p) = \bot$, if $p$ is (partially) inapplicable, is undefined or does not exist; in these cases it is not meaningful to decide whether or not $p$ corresponds to reality; in this case $p$ is neither true or false, but inapplicable.

The arithmetic rules for extended truth values are as follows:

- *Negation:*

$$\forall\, p \in P : t^*(NOT\ p) = \neg(t^*(p))$$

where $\neg : I^* \to I^* : x \mapsto \neg(x)$ is defined by the truth table

| $x$ | $\neg x$ |
|---|---|
| $T$ | $F$ |
| $F$ | $T$ |
| $\bot$ | $\bot$ |

- *Conjunction:*

$$\forall\, p, q \in P : t^*(p\ AND\ q) = t^*(p) \wedge t^*(q)$$

where $\wedge : I^* \times I^* \to I^* : (x, y) \mapsto x \wedge y$ is defined by the truth table

| $\wedge$ | $T$ | $F$ | $\bot$ |
|---|---|---|---|
| $T$ | $T$ | $F$ | $\bot$ |
| $F$ | $F$ | $F$ | $F$ |
| $\bot$ | $\bot$ | $F$ | $\bot$ |

– *Disjunction:*

$$\forall\, p, q \in P : t^*(p\; OR\;\; q) = t^*(p) \vee t^*(q)$$

where $\vee : I^* \times I^* \rightarrow I^* : (x, y) \mapsto x \vee y$ is defined by the truth table

| $\vee$ | $T$ | $F$ | $\perp$ |
|---|---|---|---|
| $T$ | $T$ | $T$ | $T$ |
| $F$ | $T$ | $F$ | $\perp$ |
| $\perp$ | $T$ | $\perp$ | $\perp$ |

## 2.2 Extended Possibilistic Truth Values

In the real world there are a lot of situations where one cannot unambiguously say that a proposition is either *completely* true, *completely* false or *completely* inapplicable. Often there is some form of uncertainty about the truth value of a proposition. Example are "his/her score on the math test was high", "the house is cheap", etc. To adequately model the uncertainty about the truth value of a proposition, extended possibilistic truth values can be used. The concept 'extended possibilistic truth value' (EPTV) [2] is defined as a (normalized) possibility distribution over the universal set $I^* = \{T, F, \perp\}$ of extended truth values. With the understanding that $P$ represents the universe of all propositions and $\tilde{\wp}(I^*)$ denotes the set of all possible fuzzy sets that can be defined over the set $I^* = \{T, F, \perp\}$, the **Extended Possibilistic Truth Value** $\tilde{t}^*(p)$ of a proposition $p \in P$ can then be defined as follows:

$$\tilde{t}^* : P \rightarrow \tilde{\wp}(I^*) : p \mapsto \tilde{t}^*(p)$$

which associates a fuzzy set $\tilde{t}^*(p)$ with each proposition $p \in P$. The fuzzy set $\tilde{t}^*(p)$ represents a possibility distribution; its membership grades are interpreted as grades of uncertainty:

$$\forall\, x \in I^* : \pi_{t^*(p)}(x) = \mu_{\tilde{t}^*(p)}(x)$$

or

$$\forall\, p \in P : \pi_{t^*(p)} = \tilde{t}^*(p)$$

Generally, an extended possibilistic truth value is a fuzzy set of the form

$$\tilde{t}^*(p) = \{(T, \mu_{\tilde{t}^*(p)}(T)), (F, \mu_{\tilde{t}^*(p)}(F)), (\perp, \mu_{\tilde{t}^*(p)}(\perp))\} \tag{1}$$

where $\mu_{\tilde{t}^*(p)}(T)$ represents the possibility that proposition $p$ is true, $\mu_{\tilde{t}^*(p)}(F)$ represents the possibility that proposition $p$ is false and $\mu_{\tilde{t}^*(p)}(\perp)$ represents the possibility that some parts of $p$ are not applicable, undefined or not supplied.

In this way, EPTVs provide an epistemological representation of the truth of a proposition, which allows to reflect the knowledge about the actual truth and additionally allow to explicitly deal with those cases where the truth value of a proposition is (partly) inapplicable.

An overview of some special values of EPTVs is given in table 1. As an example, consider the modelling of an unknown truth value by the possibility distribution $\{(T,1),(F,1)\}$, which denotes that it is completely possible that the proposition is true ($T$), or it is also completely possible that the proposition is false ($F$).

**Table 1.** Special cases of EPTVs

| $\tilde{t}^*(p)$ | Interpretation |
|---|---|
| $\{(T,1)\}$ | $p$ is true |
| $\{(F,1)\}$ | $p$ is false |
| $\{(T,1),(F,1)\}$ | $p$ is unknown |
| $\{(\perp,1)\}$ | $p$ is undefined |
| $\{(T,1),(F,1),(\perp,1)\}$ | no information |

In the context of flexible database querying, EPTVs can be used to model the satisfaction grade with which a database record satisfies a flexible constraint, imposed by a user query [4]. In general, a flexible query will impose several different selection criteria, interconnected by logical operators for negation ($NOT$), conjunction ($AND$) and disjunction ($OR$). The logical operators for EPTVs are given in [2,3,7]. In this paper we will consider (conjunctive and disjunctive) aggregation based on t-norms and t-conorms. With $(i,u)$ a (t-norm, t-conorm) pair, the operators can be defined as follows:

– *Negation:*

$$\forall\, p \in P : \tilde{t}^*(NOT\ p) = \tilde{\neg}(\tilde{t}^*(p))$$

where $\tilde{\neg} : \tilde{\wp}(I^*) \rightarrow \tilde{\wp}(I^*) : \tilde{V} \mapsto \tilde{\neg}(\tilde{V})$ is calculated as follows:

$$\begin{aligned}\mu_{\tilde{\neg}(\tilde{V})}(T) &= \mu_{\tilde{V}}(F)\\ \mu_{\tilde{\neg}(\tilde{V})}(F) &= \mu_{\tilde{V}}(T)\\ \mu_{\tilde{\neg}(\tilde{V})}(\perp) &= \mu_{\tilde{V}}(\perp)\end{aligned}$$

– *Conjunction:*

$$\forall\, p,q \in P : \tilde{t}^*(p\ AND\ \ q) = \tilde{t}^*(p)\tilde{\wedge}\tilde{t}^*(q)$$

where $\tilde{\wedge} : \tilde{\wp}(I^*) \times \tilde{\wp}(I^*) \rightarrow \tilde{\wp}(I^*) : (\tilde{U},\tilde{V}) \mapsto \tilde{U}\ \tilde{\wedge}\ \tilde{V}$ is calculated as follows:

$$\begin{aligned}\mu_{\tilde{U}\tilde{\wedge}\tilde{V}}(T) &= i(\mu_{\tilde{U}}(T),\mu_{\tilde{V}}(T))\\ \mu_{\tilde{U}\tilde{\wedge}\tilde{V}}(F) &= u(\mu_{\tilde{U}}(F),\mu_{\tilde{V}}(F))\\ \mu_{\tilde{U}\tilde{\wedge}\tilde{V}}(\perp) &= u\left(\begin{array}{c} u\left(\begin{array}{c} i(\mu_{\tilde{U}}(T),\mu_{\tilde{V}}(\perp)),\\ i(\mu_{\tilde{U}}(\perp),\mu_{\tilde{V}}(T))\end{array}\right),\\ i(\mu_{\tilde{U}}(\perp),\mu_{\tilde{V}}(\perp))\end{array}\right)\end{aligned}$$

– *Disjunction:*

$$\forall\, p, q \in P : \tilde{t}^*(p\ OR\ q) = \tilde{t}^*(p)\tilde{\vee}\tilde{t}^*(q)$$

where $\tilde{\vee} : \tilde{\wp}(I^*) \times \tilde{\wp}(I^*) \to \tilde{\wp}(I^*) : (\tilde{U}, \tilde{V}) \mapsto \tilde{U}\ \tilde{\vee}\ \tilde{V}$ is calculated as follows:

$$\mu_{\tilde{U}\tilde{\vee}\tilde{V}}(T) = u(\mu_{\tilde{U}}(T), \mu_{\tilde{V}}(T))$$
$$\mu_{\tilde{U}\tilde{\vee}\tilde{V}}(F) = i(\mu_{\tilde{U}}(F), \mu_{\tilde{V}}(F))$$
$$\mu_{\tilde{U}\tilde{\vee}\tilde{V}}(\bot) = u\left( u\begin{pmatrix} i(\mu_{\tilde{U}}(F), \mu_{\tilde{V}}(\bot)), \\ i(\mu_{\tilde{U}}(\bot), \mu_{\tilde{V}}(F)) \end{pmatrix}, \begin{matrix} \\ \\ i(\mu_{\tilde{U}}(\bot), \mu_{\tilde{V}}(\bot)) \end{matrix} \right)$$

Examples of suitable t-(co)norms are:

– Zadeh t-(co)norm: $\begin{aligned} i_{Za}(x, y) &= \min(x, y) \\ u_{Za}(x, y) &= \max(x, y) \end{aligned}$
– Probabilistic t-(co)norm: $\begin{aligned} i_{Pb}(x, y) &= x \cdot y \\ u_{Pb}(x, y) &= x + y - x \cdot y \end{aligned}$

## 3 Weighted Aggregation of EPTVs

In flexible querying, besides introducing preferences *inside* query conditions, using flexible constraints and resulting in an EPTV, it is also possible to take into account preferences *between* query criteria, using weights to indicate the difference in importance of the different criteria. So weights $w_i$ can be attached to the individual conditions $C_i$, with $w_i \in [0, 1]$. $w_i = 1$ means condition $C_i$ is fully important, while $w_i = 0$ means condition $C_i$ is not important at all and can be forgotten. In order to have an appropriate scaling, it is assumed that $\max_i w_i = 1$.

In what follows, a shortcut notation $(\mu_{\tilde{t}^*(C_i)}(T); \mu_{\tilde{t}^*(C_i)}(F); \mu_{\tilde{t}^*(C_i)}(\bot))$ will be used for an EPTV, in stead of $\left\{(T, \mu_{\tilde{t}^*(C_i)}(T)), (F, \mu_{\tilde{t}^*(C_i)}(F)), (\bot, \mu_{\tilde{t}^*(C_i)}(\bot))\right\}$.

### 3.1 Weight Impact

Suppose that all individual query conditions have been evaluated, resulting each in an EPTV. When aggregating the individual EPTVs, the respective weights have to be taken into account. Let $g$ be the operator that represents the influence of the weights on the individual EPTVs:

$$g : [0, 1] \times \tilde{\wp}(I^*) \to \tilde{\wp}(I^*) : \left(w_i, \tilde{t}^*(C_i)\right) \mapsto g\left(w_i, \tilde{t}^*(C_i)\right)$$

Let $g_T$, $g_F$ and $g_\bot$ represent the impact of a weight on the individual membership grades ($\mu_{\tilde{t}^*(C_i)}(T), \mu_{\tilde{t}^*(C_i)}(F), \mu_{\tilde{t}^*(C_i)}(\bot)$ respectively) of an EPTV:

$$g_T : [0, 1] \times [0, 1] \to [0, 1] : (w_i, t) \mapsto g_T(w_i, t)$$
$$g_F : [0, 1] \times [0, 1] \to [0, 1] : (w_i, f) \mapsto g_F(w_i, f)$$
$$g_\bot : [0, 1] \times [0, 1] \to [0, 1] : (w_i, b) \mapsto g_\bot(w_i, b)$$

with $t = \mu_{\tilde{t}^*(C_i)}(T)$, $f = \mu_{\tilde{t}^*(C_i)}(F)$ and $b = \mu_{\tilde{t}^*(C_i)}(\bot)$.

In order to be a suitable operator, $g$ needs to meet following requirements [5,6]:

- for weight 1, all membership grades must remain unchanged:

$$g\left(1, \tilde{t}^*(C_i)\right) = \tilde{t}^*(C_i)$$

- for weight 0, the EPTV needs to be mapped to the neutral element for the aggregation ($(1;0;0)$ in case of conjunction, but $(0;1;0)$ in case of disjunction:

$$g\left(0, \tilde{t}^*(C_i)\right) = (1;0;0) \text{ in case of conjunction}$$
$$g\left(0, \tilde{t}^*(C_i)\right) = (0;1;0) \text{ in case of disjunction}$$

- the operators $g_T$, $g_F$ and $g_\perp$ need to be monotonic in the membership grades (with $g_\bullet$ either $g_T$, $g_F$ or $g_\perp$ and $x_1, x_2$ membership grades of $T$, $F$ or $\perp$ respectively):

$$\forall w, x_1, x_2 \in [0,1] : x_1 \geq x_2 \Rightarrow g_\bullet(w, x_1) \geq g_\bullet(w, x_2)$$

- the operators $g_T$, $g_F$ and $g_\perp$ need to be monotonic in the weight (with $g_\bullet$ either $g_T$, $g_F$ or $g_\perp$ and $x$ the membership grade of $T$, $F$ or $\perp$ respectively):

$$\forall w_1, w_2, x \in [0,1] : w_1 \geq w_2 \Rightarrow g_\bullet(w_1, x) \geq g_\bullet(w_2, x)$$
$$\text{or}$$
$$\forall w_1, w_2, x \in [0,1] : w_1 \geq w_2 \Rightarrow g_\bullet(w_1, x) \leq g_\bullet(w_2, x)$$

  depending on the kind of the aggregation and the kind of operator (for either $T$, $F$ or $\perp$)

Implicator functions $f_{im}$ and coimplicator functions $f_{im}^{co}$ can be used to model the influence of weights. $f_{im}$ and $f_{im}^{co}$ are $[0,1]$-valued extension of Boolean implication and coimplication, and hence can be rewritten as $f_{im}(x,y) = \neg x \vee y$ and $f_{im}^{co}(x,y) = \neg f_{im}(\neg x, \neg y) = \neg(\neg(\neg x) \vee \neg y) = \neg x \wedge y$. When looking at the extreme points $x = 0$ and $x = 1$, $f_{im}(x,y)$ reduces to 1 ($x = 0$) and $y$ ($x = 1$), while $f_{im}^{co}(x,y)$ reduces to $y$ ($x = 0$) and 0 ($x = 1$). So, the implicator, with $x = w$, can be used in the cases where the membership degree should remain unchanged for $x = 1$ and should be drawn towards 1 for $x = 0$ ($\mu(T)$ in case of conjunction, $\mu(F)$ in case of disjunction, with $x = w$ and $y$ the respective membership degree). The coimplicator on the other hand, with $x = 1 - w$, can be used in the cases where the membership degree should remain unchanged for $x = 0$ and should be drawn towards 0 for $x = 1$ ($\mu(F)$ and $\mu(\perp)$ in case of conjunction, $\mu(T)$ and $\mu(\perp)$ in case of disjunction, with $x = 1 - w$ and $y$ the respective membership degree).

The impact of a weight on an EPTV can then be defined as follows [5]:

- Weight operator for conjunction

$$g^\wedge : [0,1] \times \tilde{\wp}(I^*) \to \tilde{\wp}(I^*)$$
$$(w, \tilde{V}) \mapsto g^\wedge(w, \tilde{V})$$
$$\text{where:}$$
$$\mu_{g^\wedge(w,\tilde{V})}(T) = f_{im}(w, \mu_{\tilde{V}}(T))$$
$$\mu_{g^\wedge(w,\tilde{V})}(F) = f_{im}^{co}(1 - w, \mu_{\tilde{V}}(F))$$
$$\mu_{g^\wedge(w,\tilde{V})}(\perp) = f_{im}^{co}(1 - w, \mu_{\tilde{V}}(\perp))$$

– Weight operator for disjunction

$$\begin{aligned} g^{\vee} : [0,1] \times \tilde{\wp}(I^*) &\rightarrow \tilde{\wp}(I^*) \\ (w, \tilde{V}) &\mapsto g^{\vee}(w, \tilde{V}) \\ &\text{where:} \\ \mu_{g^{\vee}(w,\tilde{V})}(T) &= f_{im}^{co}(1-w, \mu_{\tilde{V}}(T)) \\ \mu_{g^{\vee}(w,\tilde{V})}(F) &= f_{im}(w, \mu_{\tilde{V}}(F)) \\ \mu_{g^{\vee}(w,\tilde{V})}(\bot) &= f_{im}^{co}(1-w, \mu_{\tilde{V}}(\bot)) \end{aligned}$$

Some interesting implicators and coimplicators are:

– Kleene-Dienes: $\begin{aligned} f_{im_{KD}}(x,y) &= \max(1-x, y) \\ f_{im_{KD}}^{co}(x,y) &= \min(1-x, y) \end{aligned}$

– Reichenbach implicator: $\begin{aligned} f_{im_{Rb}}(x,y) &= 1 - x + x \cdot y \\ f_{im_{Rb}}^{co}(x,y) &= (1-x) \cdot y \end{aligned}$

– Gödel implicator: $\begin{aligned} f_{im_{Go}}(x,y) &= \begin{cases} 1 \text{ if } x \leq y \\ y \text{ otherwise} \end{cases} \\ f_{im_{Go}}^{co}(x,y) &= \begin{cases} 0 \text{ if } x \geq y \\ y \text{ otherwise} \end{cases} \end{aligned}$

As an example consider the weight operator for conjunction based on the Kleene-Dienes implicator:

$$g^{\wedge}(w, \tilde{V}) = \{\max(1-w, \mu_{\tilde{V}}(T));\ \min(w, \mu_{\tilde{V}}(F));\ \min(w, \mu_{\tilde{V}}(\bot))\}$$

It is easy to see that this is indeed a monotonic operator, where the result for $w = 1$ will reduce to $\{\mu_{\tilde{V}}(T); \mu_{\tilde{V}}(F); \mu_{\tilde{V}}(\bot)\}$ and for $w = 0$ to $\{1; 0; 0\}$, as was required for a suitable conjunction weight operator.

Using the definitions of the weight operators $g^{\wedge}$ and $g^{\vee}$, and the aggregation operators $\tilde{\wedge}$ and $\tilde{\vee}$, an extended operator for weighted conjunction $\tilde{\wedge}^{w}$ and disjunction $\tilde{\vee}^{w}$ of EPTVs can now be defined:

$$\begin{aligned} &\tilde{\wedge}^{w} : ([0,1] \times \tilde{\wp}(I^*))^2 \rightarrow \tilde{\wp}(I^*) \\ &\quad ((w_1, \tilde{V}_1), (w_2, \tilde{V}_2)) \mapsto g^{\wedge}(w_1, \tilde{V}_1)\ \tilde{\wedge}\ g^{\wedge}(w_2, \tilde{V}_2) \\ &\tilde{\vee}^{w} : ([0,1] \times \tilde{\wp}(I^*))^2 \rightarrow \tilde{\wp}(I^*) \\ &\quad ((w_1, \tilde{V}_1), (w_2, \tilde{V}_2)) \mapsto g^{\vee}(w_1, \tilde{V}_1)\ \tilde{\vee}\ g^{\vee}(w_2, \tilde{V}_2) \end{aligned} \tag{2}$$

## 3.2 Combining Different Types of Aggregation

The operators presented thus far are suitable for the aggregation of weighted selection criteria, as long as one stays within one type of aggregation (either conjunctive or disjunctive). This follows from the fact that the impact of the weights can be calculated before the actual aggregation itself. So, after applying the weight operators, it is like working with '*regular*' EPTVs and all properties (associativity, commutativity) of the weighted aggregation are implied by the underlying aggregation operators $\tilde{\wedge}$ and $\tilde{\vee}$.

However, this is no longer the case when combining conjunctions and disjunctions together. This can be seen in following simple examples ( $(w, \tilde{t}^*(C_i))$ stands for the EPTV resulting from the evaluation of condition $C_i$, with weight $w$).

**Example 1.**

a) $\left[(0, \tilde{t}^*(C_1)) \ \tilde{\wedge}^w \ (0, \tilde{t}^*(C_2))\right] \ \tilde{\vee}^w \ (1, \tilde{t}^*(C_3))$
Intuitively, this should reduce to $\tilde{t}^*(C_3)$, since constraints with weight 0 can be forgotten. However, if we blindly apply the definitions given in 3.1, this leads to (EPTV $(1;0;0)$ is neutral element of conjunction):
$[(1;0;0) \ \tilde{\wedge} \ (1;0;0)] \ \tilde{\vee} \ \tilde{t}^*(C_3)$
Further calculation leads to:
$(1;0;0) \ \tilde{\vee} \ \tilde{t}^*(C_3) = (1;0;0)$, not the expected $\tilde{t}^*(C_3)$

b) $\left[(0, \tilde{t}^*(C_1)) \ \tilde{\vee}^w \ (0, \tilde{t}^*(C_2))\right] \ \tilde{\wedge}^w \ (1, \tilde{t}^*(C_3))$
Intuitively, this should reduce to $\tilde{t}^*(C_3)$, since constraints with weight 0 can be forgotten. However, if we blindly apply the definitions given in 3.1, this leads to (EPTV $(0;1;0)$ is neutral element of disjunction):
$[(0;1;0) \ \tilde{\vee} \ (0;1;0)] \ \tilde{\wedge} \ \tilde{t}^*(C_3)$
Further calculation leads to:
$(0;1;0) \ \tilde{\wedge} \ \tilde{t}^*(C_3) = (0;1;0)$, not the expected $\tilde{t}^*(C_3)$

It is obvious this is not the way to handle weights. The reason for this insufficiency is that the impact of the weights is different when using conjunction than when using disjunction (they work in a different '*direction*', towards their respective neutral elements). The neutral element of the one, is the absorbing element of the other.

To solve this problem, the weights need to be propagated throughout the calculations, i.e. the weights of the intermediate results need to be *remembered* to take into account when aggregating further. For instance, in the first example above, the result of the conjunction $(0, \tilde{t}^*(C_1)) \ \tilde{\vee}^w \ (0, \tilde{t}^*(C_2))$ should also have weight 0. In that case, the resulting $(1;0;0)$ would, in the next step (disjunction), again be transformed to $(0;1;0)$, leading to the final (correct!) result $\tilde{t}^*(C_3)$. A number of rules can be imposed for the calculation of the weight of an intermediate result:

- Constraints with weight 0 are not allowed to have any impact, so both $(0, \tilde{t}^*(C_1)) \ \tilde{\wedge}^w \ (w, \tilde{t}^*(C_2))$ and $(0, \tilde{t}^*(C_1)) \ \tilde{\vee}^w \ (w, \tilde{t}^*(C_2))$ should produce an intermediate result with weight $w$.
- The importance of an intermediate result is always greater than or equal to the greatest of weights from the arguments of the aggregation.

Remark that here, opposed to the actual impact of the weights, the rules are the same for conjunction and disjunction. A t-conorm meets this rules and can thus be chosen to calculate the resulting weight of an aggregation. So, both for conjunction and disjunction: $w_{res} = u(w_1, w_2)$, where $u$ is a t-conorm, $w_{res}$ is the weight resulting from an aggregation of arguments with respective weights $w_1$ and $w_2$.

From the above, it follows that an EPTV cannot be seen separable from its importance (indicated by a weight). Weights are propagated and the result of a weighted aggregation of EPTVs is a new EPTV with again a weight attached to it. Remark also that, when using an appropriate scaling of weights ($\max(w_i) = 1$) and a t-conorm to calculate the weight of an (intermediate) result, the weight of the final result will always be 1.
So, the extended operators (2) for weighted conjunction $\tilde{\wedge}^w$ and disjunction $\tilde{\vee}^w$ of EPTVs, presented above, must be adjusted to also calculate a weight for the result of the aggregation, resulting finally in following operators:

$$\begin{aligned} \tilde{\wedge}^w &: ([0,1] \times \tilde{\wp}(I^*))^2 \rightarrow [0,1] \times \tilde{\wp}(I^*) \\ &((w_1, \tilde{V}_1), (w_2, \tilde{V}_2)) \mapsto (u(w_1, w_2), g^\wedge(w_1, \tilde{V}_1) \ \tilde{\wedge} \ g^\wedge(w_2, \tilde{V}_2)) \\ \tilde{\vee}^w &: ([0,1] \times \tilde{\wp}(I^*))^2 \rightarrow [0,1] \times \tilde{\wp}(I^*) \\ &((w_1, \tilde{V}_1), (w_2, \tilde{V}_2)) \mapsto (u(w_1, w_2), g^\vee(w_1, \tilde{V}_1) \ \tilde{\vee} \ g^\vee(w_2, \tilde{V}_2)) \end{aligned} \tag{3}$$

Also an extended operator for negation can be defined:

$$\begin{aligned} \tilde{\neg}^w &: [0,1] \times \tilde{\wp}(I^*) \rightarrow [0,1] \times \tilde{\wp}(I^*) \\ &(w, \tilde{V}) \mapsto (w, \tilde{\neg}(\tilde{V})) \end{aligned} \tag{4}$$

When using these operators, with propagation of the weights throughout the calculations, the problems arising when combining disjunction and conjunction, as presented above, are solved, as can be seen when looking back at Example 1.a):

**Example 2.**

$$\begin{aligned} &\left[(0, \tilde{t}^*(C_1)) \ \tilde{\wedge}^w \ (0, \tilde{t}^*(C_2))\right] \ \tilde{\vee}^w \ (1, \tilde{t}^*(C_3)) \\ \equiv &\left[u(0,0), ((1;0;0) \ \tilde{\wedge} \ (1;0;0))\right] \ \tilde{\vee}^w \ (1, \tilde{t}^*(C_3)) \\ \equiv &(0, (1;0;0)) \ \tilde{\vee}^w \ (1, \tilde{t}^*(C_3)) \\ \equiv &(u(0,1), ((0;1;0) \ \tilde{\vee} \ \tilde{t}^*(C_3))) \\ \stackrel{!}{\equiv} &(1, \tilde{t}^*(C_3)) \end{aligned}$$

## 4 Conclusion

In flexible queries, weights can be attached to constraints to indicate user preferences between the different constraints. This paper showed how weights have their impact on the satisfaction grades for the individual selection criteria, modeled by Extended Possibilistic Truth Values. The impact of the weights, modeled by means of modification functions based on (co)implicators, is different according to the aggregation type (conjunction or disjunction). It has been shown that the weights need to be propagated throughout the calculations, and that therefore the EPTVs cannot be viewed separately from their associated weight.

## References


1. P. Bosc, D.H. Kraft and F.E. Petry, "Fuzzy sets in database and information systems: Status and opportunities", *Fuzzy Sets and Systems* **156** 3 (2005) 418–426.
2. G. De Tré, "Extended Possibilistic Truth Values", *International Journal of Intelligent Systems* **17** (2002) 427–446.
3. G. De Tré, R. De Caluwe, J. Verstraete and A. Hallez, "Conjunctive Aggregation of Extended Possibilistic Truth Values and Flexible Database Querying", *Lecture Notes in Artificial Intelligence* **2522** (2002) 344–355.
4. G. De Tré and R. De Caluwe, "Modelling Uncertainty in Multimedia Database Systems: An Extended Possibilistic Approach", *International Journal of Uncertainty, Fuzziness and Knowledge-Based Systems* **11** 1 (2003) 5–22.
5. G. De Tré and B. De Baets, "Aggregating Constraint Satisfaction Degrees Expressed by Possibilistic Truth Values", *IEEE Transactions on Fuzzy Systems* **11** 3 (2003) 361–368.
6. D. Dubois, H. Fargier and H. Prade, "Beyond min aggregation in multicriteria decision: (ordered) weighted min, discri-min and leximin", in: *The ordered weighted averaging operators: Theory and applications.*, R.R. Yager en J. Kacprzyk (eds.) (Kluwer Academic Publishers, Boston, USA, 1997) 181–192.
7. H. Prade, "Possibility sets, fuzzy sets and their relation to Lukasiewicz logic", in: *Proc. of the 12th International Symposium on Multiple-Valued Logic, Paris, France* (1982) 223–227.

# Bipolar Queries Using Various Interpretations of Logical Connectives

Sławomir Zadrożny[1,2] and Janusz Kacprzyk[1]

[1] Systems Research Institute, Polish Academy of Sciences, Warszawa, Poland
[2] WIT – Warsaw School of Information Technology, Warszawa, Poland
{Janusz.Kacprzyk,Slawomir.Zadrozny}@ibspan.waw.pl

**Abstract.** In [1,2] we studied various concepts of bipolar queries (cf. Dubois and Prade [3]). We advocated the use of a fuzzified version of the original crisp definition by Lacroix and Lavency [4]. However, the general fuzzification proposed leaves open the choice of a representation of logical connectives and quantifiers. In the present paper we study the influence of the choice some representations that are popular in fuzzy logic on matching degrees of the tuples and their resulting ordering.

## 1 Introduction

One of the most important dimensions of the querying flexibility is the ease with which a user may express his or her requirements as to the data sought. In many scenarios the easiest way is to use in a query some linguistic descriptions such as a "young employee", "high salary", etc. Fuzzy logic provides effective and efficient means to model such linguistic descriptions within queries (cf., e.g., [5]). Much research in this area has resulted in some extensions of such powerful query languages like the relational algebra or SQL.

For some time there has been a growing interest in some special, nonconventional forms of queries that are important from a practical point of view. This interest dates back to a seminal work of Lacroix and Lavency [4]. Their proposal has been recently generalized to a new broader approach. Basically, it attempts to distinguish *regular* requirements the data to be retrieved should satisfy and some *preference-related requirements* of a more subtle, not always obligatory, nature. Thus the name proposed by Chomicki [6] is *queries with preferences*.

The original proposal of Lacroix and Lavency was quickly adopted for the fuzzy case by Bosc and Pivert [7,8]. In 2002 Dubois and Prade [3] introduced an equivalent concept of a *bipolar query* which provides for another interesting interpretation. Thus even if bipolar queries turn out to be a special case of queries with preferences they still are worthwhile to be distinguished due to their clear semantics that is interesting and relevant from a practical point of view.

Lacroix and Lavency's approach deals with crisp conditions only. We proposed their direct "fuzzification" in [1]. In this paper we further analyze this proposal taking into account various fuzzy logic based interpretations of logical connectives and quantifiers. In Section 2 we gather all necessary definitions

P. Melin et al. (Eds.): IFSA 2007, LNAI 4529, pp. 181–190, 2007.
© Springer-Verlag Berlin Heidelberg 2007

of a wide array of operators most often used to model logical connectives and quantifiers in fuzzy logic. Section 3 recalls briefly the basics of the concept of bipolar queries. Section 4 constitutes the main contribution of this paper and lists the properties of bipolar queries under selected combinations of the logical connective representations. Finally Section 5 summarizes the main results and points out the directions of a further research.

## 2 Preliminaries

The fuzzification of the crisp concept of bipolar queries (cf. Section 3) calls first of all for some interpretation of logical connectives and quantifiers. We use for this purpose $t$-norms, $t$-conorms, as well as the negation and implication operators, as is usually done.

A $t$-norm operator $t$ is used to represent the fuzzy logical connective of conjunction and is defined as a function $t : [0,1] \times [0,1] \longrightarrow [0,1]$ such that: (1) $t(x,1) = x$, $\forall x$, (2) $x \leq y \Rightarrow t(x,z) \leq t(y,z)$, $\forall x,y,z$, (3) $t(x,y) = t(y,x)$, $\forall x,y$, and (4) $t(x,t(y,z)) = t(t(x,y),z)$, $\forall x,y,z$.

In particular we will consider the following popular $t$-norm operators:

$$\textit{minimum} \qquad t_{min}(x,y) = x \wedge y = \min(x,y) \tag{1}$$

$$\textit{product} \qquad t_{\Pi}(x,y) = x \cdot y \tag{2}$$

$$\textit{Łukasiewicz t-norm} \qquad t_W(x,y) = \max(0, x+y-1) \tag{3}$$

A $t$-conorm operator $s$ is used to represent the fuzzy logical connective of disjunction and is defined as a function $s : [0,1] \times [0,1] \longrightarrow [0,1]$ such that the same conditions as for a $t$-norm are satisfied except for the boundary condition, $s(x,0) = x$,

Notice that $s(x,1) = 1$, $\forall x$.

In particular we will consider the following popular $t$-conorm operators:

$$\textit{maximum} \qquad s_{max}(x,y) = x \vee y = \max(x,y) \tag{4}$$

$$\textit{probabilistic sum} \qquad s_{\Pi}(x,y) = x + y - x \cdot y \tag{5}$$

$$\textit{Łukasiewicz t-conorm} \qquad s_W(x,y) = \min(1, x+y) \tag{6}$$

The negation operator $n$ is used to represent the fuzzy logical connective of negation and is defined as a function $n : [0,1] \longrightarrow [0,1]$ such that:

$$n(0) = 1, \ n(1) = 0, \ x \leq y \Rightarrow n(x) \geq n(y) \tag{7}$$

For our purposes a reasonable choice is here the most popular negation operator:

$$N(x) = 1 - x \qquad \forall x \tag{8}$$

which additionally possesses the property of *involution*: $N(N(x)) = x$, $\forall x$.

Due to associativity, both $t$-norms and $t$-conorms are naturally generalized to $m$-ary operators, i.e., instead of, e.g., $t(x,t(y,z))$ we will write $t(x,y,z)$.

We will consider combinations of the $t$-norms and $t$-conorms that together with the negation operator $N$ (8) form a De Morgan Triplet, i.e., such a triple $(t, s, n)$ of $t$-norm $t$, $t$-conorm $s$ and negation $n$ that:

$$n(s(x, y)) = t(n(x), n(y)) \tag{9}$$

Due to involution of $N$, (9) is equivalent to $s(x, y) = n(t(n(x), n(y)))$. The triples $(t_{min}, s_{max}, N)$, $(t_{\Pi}, s_{\Pi}, N)$ and $(t_W, s_W, N)$ are the De Morgan Triplets and will be used in what follows.

The formulae to be used to describe the fuzzy set of tuples being an answer to a bipolar query employ the classical quantifiers. Usually they are modelled in fuzzy logic using the inf and sup for the general and existential quantifier, respectively. Thus, the truth values of them are:

$$\text{truth}(\forall x A(x)) = \inf_x \mu_A(x) \tag{10}$$

$$\text{truth}(\exists x A(x)) = \sup_x \mu_A(x) \tag{11}$$

where $A$ denotes both a fuzzy predicate symbol and a fuzzy set being its interpretation. For the finite cas, as in this paper, the min and max replace the inf and sup.

However, a more consistent treatment of the classical quantifiers in the framework of fuzzy logic may be pursued with the use of $t$-quantifiers and $s$-quantifiers (cf., e.g., [9]). The idea of such a type of quantifiers is based on the observation that a formula with the general (existential) quantifier may be identified with a – possibly infinite – conjunction (disjunction):

$$\forall x A(x) \Leftrightarrow A(a_1) \wedge A(a_2) \wedge \ldots \tag{12}$$

$$\exists x A(x) \Leftrightarrow A(a_1) \vee A(a_2) \vee \ldots \tag{13}$$

where $a_i$ are constants corresponding to the elements of the universe of discourse.

In fact we are interested in the finite domains only. Thus the use of a $t$-norm $t$ ($t$-conorm $s$) to model the conjunction (disjunction) implies:

$$\text{truth}(\forall x A(x)) = t(\mu_A(a_1), \mu_A(a_2), \ldots, \mu_A(a_m)) \tag{14}$$

$$\text{truth}(\exists x A(x)) = s(\mu_A(a_1), \mu_A(a_2), \ldots, \mu_A(a_m)) \tag{15}$$

In particular for the $t$-norms (1)-(3) and $t$-conorms (4)-(6) one obtains:

$$\forall_{min} x\ A(x) = \min(x_1, \ldots, x_n) \tag{16}$$

$$\forall_{\Pi} x\ A(x) = \prod_{i=1}^{n} x_i \tag{17}$$

$$\forall_W x\ A(x) = \max(\sum_{i=1}^{n} x_i - n + 1, 0) \tag{18}$$

$$\exists_{max} x\ A(x) = \max(x_1, \ldots, x_n) \tag{19}$$

$$\exists_{\Pi} x\ A(x) = \sum_i x_i - \sum_{i<j} x_i \cdot x_j + \sum_{i<j<k} \prod_{l \in \{i,j,k\}} x_l - \ldots + + (-1)^{n+1} \prod_{1 \le l \le n} x_l \tag{20}$$

$$\exists_W x\ A(x) = \min(\sum_i x_i, 1) \tag{21}$$

where $x \in X$ and $X = \{x_1, \ldots, x_n\}$. In this paper we use the symbols $\forall$ and $\exists$ with subscripts to denote both the logical symbols of the quantifiers in various formulae and the operators defined by the above equations.

The logical connective of implication is represented by an implication operator which is assumed to be a function $i : [0,1] \longrightarrow [0,1]$ such that:

$$x \le u \Rightarrow i(x,y) \ge i(u,y) \qquad \forall x, y, u \tag{22}$$

$$y \le z \Rightarrow i(x,y) \le i(x,z) \qquad \forall x, y, z \tag{23}$$

$$i(0,y) = 1\ i(x,1) = 1\ i(1,0) = 0 \tag{24}$$

We will consider two most popular approaches to defining this operator with respect to the assumed De Morgan Triplet $(t, s, n)$. These are the so-called $S$-implications and $R$-implications defined as follows:

$$R - \text{implication:} \quad x \rightarrow y = \sup\{z : t(x,z) \le y\} \tag{25}$$

$$S - \text{implication:} \quad x \rightarrow y = s(n(x), y) \tag{26}$$

For both the above types of implications another property holds: $i(1,x) = x\ \forall x$.

In particular we will consider the following $R$-implication operators:

$$\text{Gödel's implication} \qquad i_{R-min}(x,y) = \begin{cases} 1 \text{ for } x \le y \\ y \text{ for } x > y \end{cases} \tag{27}$$

$$\text{Goguen's implication} \qquad i_{R-\Pi}(x,y) = \begin{cases} 1 & \text{for } x = 0 \\ \min\{1, \frac{y}{x}\} & \text{for } x \ne 0 \end{cases} \tag{28}$$

$$\text{Łukasiewicz' implication} \qquad i_{R-W}(x,y) = \min(1 - x + y, 1) \tag{29}$$

and the following $S$-implication operators:

$$\text{Kleene–Dienes' implication} \qquad i_{S-max}(x,y) = \max(1 - x, y) \tag{30}$$

$$\text{Reichenbach's implication} \qquad i_{S-\Pi}(x,y) = 1 - x + x \cdot y \tag{31}$$

The $S$-implication operator $i_{S-W}$ is identical with $i_{R-W}$ as given by (29).

## 3 Bipolar Queries

The concept of a bipolar query has been introduced by Dubois and Prade [3] in 2002 (the roots of this concept may be traced back to earlier works of Dubois and Prade as well as other authors, cf., e.g., [10]). The idea is to distinguish in a query two types of conditions: a *required* and a *preferred* one. The former have to be unconditionally met by a tuple, while the latter are to some extent optional and less important. However, the facultative character of the latter is rather subtle and cannot be directly grasped with, e.g., the notion of importance weights. A query of this type may be exemplified with:

$$\textit{Find a cheap house preferably near a railway station} \tag{32}$$

in which the *required*, *strict* condition concerns the price and the *preferred* condition refers to the distance to a railway station.

According to our fuzzification of the Lacroix and Lavency's approach this query is to be interpreted as follows. A house sought has to be *cheap* and *if possible* also *near* the station. The possibility of satisfying both conditions corresponds to the existence of a house that meets both of them. Thus if such a house exists, then only such houses are of interest, i.e., only they belong to the answer of the bipolar query (32). Otherwise it is enough for a house to be cheap to belong to the answer to the query.

Such an interpretation of bipolar queries may be expressed by a logical formula in case the conditions are crisp as in Lacroix and Lavency [4]. In case of fuzzy conditions their formula may be adapted as we did it in [1].

We adopt the following notation: $X = \{x_j\}$ is a set of tuples to be queried; $C(\cdot)$ and $P(\cdot)$ are, *fuzzy* in general, predicates corresponding to the *required* and *preferred* conditions, respectively. We will identify these predicates with fuzzy sets and $C(x)$ and $P(x)$ will denote their membership function values. Then the Lacroix and Lavency's interpretation of bipolar queries may be more formally expressed by the following description of the set of tuples sought ([4]):

$$\{x \in X : C(x) \wedge (\exists y(C(y) \wedge P(y)) \longrightarrow P(x))\} \tag{33}$$

In [1,2] we introduced a specific fuzzy version of (33) denoting the characteristic/membership function of the resulting fuzzy set of tuples as $\gamma(C, P, x, X)$:

$$\gamma(C, P, x, X) = \min(C(x), \max(1 - \max_{y \in X} \min(C(y), P(y)), P(x))) \tag{34}$$

In this formula the $(t_{min}, s_{max}, N)$ De Morgan Triplet is used along with the $i_{S-max}$ (30) implication operator and the existential quantifier $\exists$ modelled via the max operator (cf. (15)).

The characteristic feature of the interpretation represented via (34) is that the value of a matching degree, $\gamma(C, P, x, X)$, for a tuple $x$ depends not only on $x$ but also on the whole set of tuples $X$ (what is appropriately accounted for by the fourth parameter, $X$, of $\gamma$).

## 4 Alternative Interpretations

Now we will reinterpret the concepts of bipolar queries using the three De Morgan Triplets mentioned earlier and their related implication operators.

In Table 1 we have collected various reinterpretations of the formula (33) obtained using the particular De Morgan Triplets, implication operators and quantifiers corresponding to the selected $t$-norm (cf., (16)–(18)). These correspond to the formula (34) we proposed in [1]. The subscripts of $\gamma$ indicate the De Morgan Triplet and implication operator used. For example, $\gamma_{\Pi,S}$ refers to the $(t_{\Pi}, s_{\Pi}, N)$ De Morgan Triplet and associated $S$-implication operator (cf. (1)–(3), (4)-(6), (25)-(29)).

**Table 1.** Various reinterpretations of the concept of a bipolar query

| | |
|---|---|
| $\gamma_{min,S}$ | $\min(C(x), \max(1 - \max_{y \in X} \min(C(y), P(y)), P(x)))$ |
| $\gamma_{min,R}$ | $\begin{cases} C(x) & \text{if } \max_y \min(C(y), P(y)) \leq P(x) \\ \min(C(x), P(x)) & \text{otherwise} \end{cases}$ |
| $\gamma_{\Pi,S}$ | $C(x) \cdot (1 - \exists_{\Pi}(C(y_i) \cdot P(y_i)) \cdot (1 - P(x)) =$ <br> $C(x) \cdot (\prod_i (1 - C(y_i) \cdot P(y_i)) \cdot (1 - P(x)) + P(x))$ |
| $\gamma_{\Pi,R}$ | $\begin{cases} C(x) & \text{if } \exists_{\Pi}(C(y_i) \cdot P(y_i)) = 0 \\ C(x) \cdot \min(\frac{P(x)}{\exists_{\Pi}(C(y_i) \cdot P(y_i))}, 1) & \text{otherwise} \end{cases}$ |
| $\gamma_W$ | $t_W(C(x), i_W(\exists_W\ t_W(C(y), P(y)), P(x)))$ |

Notice that in Table 1 the consecutive rows correspond to columns label by I, II, III, IV and V in Table 2.

Let us compare the particular interpretations of Table 1 on a simple example given in Table 2 (in the latter table, $C(y)$ and $P(y)$ are denoted, due to space limitation, as $C$ and $P$). Let us first compute the truth of $\exists y\ C(y) \wedge P(y)$ for various combinations of the logical connectives:

$$\exists_{min} y\ t_{min}(C(y), P(y)) = 0.8 \tag{35}$$

$$\exists_{\Pi} y\ t_{\Pi}(C(y), P(y)) = 0.96 \tag{36}$$

$$\exists_W y\ t_W(C(y), P(y)) = 1.0 \tag{37}$$

**Notational remark.** As the truth value of the formula $\exists y\ C(y) \wedge P(y)$ is fixed for a given set of tuples $X$ and a chosen De Morgan Triplet and will be important for our further analysis, we will denote it for brevity by $\exists CP$.

It may easily be noticed that various interpretations lead to different matching degrees of particular tuples (i.e., values of the $\gamma$ functions) as well as to different resulting ordering of the tuples (ranking). In what follows we study the differences between particular interpretations. We show some properties and examples. In particular we study the effects of choosing between:

– the standard interpretation (i.e., via the max operator) and a $t$-norm based interpretation of the classical quantifier $\exists$
– the $R$- and $S$-implications,

**Table 2.** A comparison of interpretations: an example

| No | $C$ | $P$ | I | | II | | III | | IV | | V | |
|---|---|---|---|---|---|---|---|---|---|---|---|---|
| | | | $rank$ | $\gamma_{min,S}$ | $rank$ | $\gamma_{min,R}$ | $rank$ | $\gamma_{\Pi,S}$ | $rank$ | $\gamma_{\Pi,R}$ | $rank$ | $\gamma_W$ |
| 1 | 0.9 | 0.7 | 2 | 0.7 | 2 | 0.7 | 3 | 0.64 | 3 | 0.63 | 2 | 0.6 |
| 2 | 0.8 | 0.8 | 1 | 0.8 | 1 | 0.8 | 2 | 0.65 | 2 | 0.67 | 2 | 0.6 |
| 3 | 0.7 | 1.0 | 2 | 0.7 | 2 | 0.7 | 1 | 0.7 | 1 | 0.7 | 1 | 0.7 |
| 4 | 1.0 | 0.0 | 4 | 0.2 | 4 | 0.0 | 4 | 0.04 | 4 | 0.0 | 4 | 0.0 |

**Property 1.** For any combination of a $t$-norm, $t$-conorm and $S$-implication or $R$-implication if there exists a tuple $x$ such that $C(x) = 1$ and $P(x) = 1$, then (33) y turns into $C(x) \wedge P(x)$, where $\wedge$ is represented by a given $t$-norm.

This property stems from the general characteristics of $t$-norms, $t$-conorms and $S$- and $R$-implications. This proves that the use of all of them preserves a basic feature of a bipolar query valid in the crisp case: if there is a tuple satisfying both the required and preferred conditions, then only the tuples satisfying both of them are interesting, i.e., (33) turns into a simple conjunction.

**Property 2.** For any combination of a $t$-norm, $t$-conorm and $S$-implication or $R$-implication and any set of tuples $X$ if for a tuple $x \in X$ $P(x) = 1$, then the formula (33) turns into $C(x)$.

This property stems from the characteristic feature of implication. It is fairly intuitive: if a tuple fully satisfies the preferred condition $P$, then its overall matching degree is equal to its satisfaction of the condition $C$. On the other hand, a rank of such a tuple depends also on the matching degrees of other tuples of the set $X$.

**Property 3.** The use of the usual fuzzy existential quantifier $\exists_{max}$, i.e., the max operator, instead of $\exists_\Pi$ or $\exists_W$: (1) yields greater or equal matching degrees, (2) may change the resulting ordering of the tuples.

Property 3.A is because max is the smallest of all $t$-conorms and because of the monotonicity of any implication and $t$-norm. Property 3.B is illustrated by the following examples. We denote the matching degrees computed with the particular existential quantifiers as $\gamma^{\exists_{\max}}(\ldots)$, $\gamma^{\exists_\Pi}(\ldots)$ and $\gamma^{\exists_W}(\ldots)$.

**Order reversal by changing the $\exists_\Pi$ quantifier to the $\exists_{max}$ quantifier under the De Morgan Triplet** $(t_\Pi, s_\Pi, N)$. Let us consider two tuples $x, y \in X$ such that $C(x) = 0.6, P(x) = 0.82$ and $C(y) = 0.85, P(y) = 0.4$. Then, $\gamma^{\exists_\Pi}_{\Pi,S}(C,P,x,X) = 0.528$, $\gamma^{\exists_\Pi}_{min,S}(C,P,y,X) = 0.3$, $\gamma^{\exists_{max}}_{min,S}(C,P,x,X) = 0.535$, $\gamma^{\exists_{max}}_{min,S}(C,P,y,X) = 0.544$. Thus

$$\gamma^{\exists_\Pi}_{\Pi,S}(C,P,x,X) \geq \gamma^{\exists_\Pi}_{\Pi,S}(C,P,y,X), \text{ but} \tag{38}$$

$$\gamma^{\exists_{max}}_{\Pi,S}(C,P,x,X) \leq \gamma^{\exists_{max}}_{\Pi,S}(C,P,y,X) \tag{39}$$

**Order reversal by changing the $\exists_W$ quantifier to the $\exists_{max}$ quantifier under the De Morgan Triplet $(t_W, s_W, N)$.** Let us consider two tuples $x, y \in X$ such that $C(x) = 1.0, P(x) = 0.5$ and $C(y) = 0.8, P(y) = 0.8$. Then, $\gamma_W^{\exists_W}(C, P, x, X) = 0.5$, $\gamma_W^{\exists_W}(C, P, y, X) = 0.6$, $\gamma_W^{\exists_{max}}(C, P, x, X) = 0.9$, $\gamma_W^{\exists_{max}}(C, P, y, X) = 0.8$. Thus

$$\gamma_W^{\exists_W}(C, P, x, X) \leq \gamma_W^{\exists_W}(C, P, y, X) \text{ but} \tag{40}$$

$$\gamma_W^{\exists_{\max}}(C, P, x, X) \geq \gamma_W^{\exists_{max}}(C, P, y, X) \tag{41}$$

Now let us check what the effect of changing an $S$-implication by an $R$-implication in (33) is while keeping the representation of all other elements fixed.

**Order reversal by changing the $S$-implication to the $R$-implication under the De Morgan Triplet $(t_{min}, s_{max}, N)$.** Let us assume that $\exists CP = 0.5$ and let us consider two tuples $x, y \in X$ such that $C(x) = 0.4, P(x) = 0$ and $C(y) = 0.3, P(y) = 0.2$. Then, $\gamma_{min,S}(C, P, x, X) = 0.4$, $\gamma_{min,S}(C, P, y, X) = 0.3$, $\gamma_{min,R}(C, P, x, X) = 0.0$, $\gamma_{min,R}(C, P, y, X) = 0.2$. Thus

$$\gamma_{min,S}(C, P, x, X) \geq \gamma_{min,S}(C, P, y, X) \text{ but} \tag{42}$$

$$\gamma_{min,R}(C, P, x, X) \leq \gamma_{min,R}(C, P, y, X) \tag{43}$$

The order reversal exemplified above makes the choice between the implication operators an important issue.

A further analysis leads to the following observation.

**Property 4.** Assuming the $(t_{min}, s_{max}, N)$ De Morgan Triplet, for tuples $x$ verifying the conditions:

$$(P(x) \geq \exists CP) \text{ or } ((P(x) \leq \exists CP) \text{ and } (P(x) \geq 1 - \exists CP)) \tag{44}$$

it holds that

1. $\gamma_{min,R}(C, P, x, X) \geq \gamma_{min,S}(C, P, x, X)$
2. replacing the $R$-implication with $S$-implication or vice-versa preserves the resulting order of the tuples, i.e., for $x, y$ verifying (44) it holds:

$$\gamma_{min,S}(C, P, x, X) \geq \gamma_{min,S}(C, P, y, X) \Leftrightarrow \\ \gamma_{min,R}(C, P, x, X) \geq \gamma_{min,R}(C, P, y, X) \tag{45}$$

The validity of this property may be easily proved as follows. In order to show 1., let us observe that for $x$ such that $P(x) \geq \exists CP$ it holds:

$$\gamma_{min,R}(C, P, x, X) = \min(C(x), 1) = C(x) \tag{46}$$

because $i_{R-min}(\exists CP, P(x)) = 1$ due to the fact that $P(X) \geq \exists CP$ while

$$\gamma_{min,S}(C, P, x, X) = \min(C(x), \max(1 - \exists CP, P(x))) \leq C(x) = \gamma_{min,R}(C, P, x, X)$$

On the other hand, for $x$ such that $P(x) \leq \exists CP$ and $P(x) \geq 1 - \exists CP$ it holds:

$$\gamma_{min,S}(C, P, x, X) = \gamma_{min,R}(C, P, x, X) = \min(C(x), P(x)) \qquad (47)$$

so that 1. is trivially verified.

In order to show 2. we will consider 3 cases:

I. both $(P(x) \geq \exists CP)$ and $(P(y) \geq \exists CP)$,
II. both $P(x) \leq \exists CP$ and $P(x) \geq 1 - \exists CP$ as well as $P(y) \leq \exists CP$ and $P(y) \geq 1 - \exists CP$,
III. $P(x) \geq \exists CP$ and $P(y) \leq \exists CP$ and $P(y) \geq 1 - \exists CP$.

*Case I.* Then, $P(x) \geq \exists CP \geq C(x)$ because $\exists CP = \max_y \min(C(y), P(y))$. Moreover (46) holds.

Now, let us assume that the left hand side of (45) holds. It means that:

$$\min(C(x), \max(1 - \exists CP, P(x))) \geq \min(C(y), \max(1 - \exists CP, P(y)))$$

but because $P(x) \geq C(x)$ and $P(y) \geq C(y)$ thus the above reduces to $C(x) \geq C(y)$, which means that $\gamma_{min,R}(C, P, x, X) \geq \gamma_{min,R}(C, P, y, X)$. The proof in the opposite direction of $\Leftrightarrow$ in (45) is obvious.

*Case II.* Then, (47) holds and thus (45) is trivially verified.

*Case III.* Now, due to a similar reasoning as in Case I and due to (47), the left hand side of (45) reduces to: $C(x) \geq \min(C(y), P(y))$. But due to (46) and (47) it is identical to the right hand side of (45).

Thus for the tuples $x$ satisfying (44) their resulting order does not depend on the choice between the $S$-implication and the $R$-implication. The troublesome tuples may appear both for high and low values of $\exists CP$. However in the former case these are less interesting tuples, i.e., such $x$'s that $P(x)$ is small (more precisely $\leq 1 - \exists CP$) while there are tuples well satisfying both $C$ and $P$ (as $\exists CP$ is high). The latter case is worse and should be taken into account while executing bipolar queries with varying logical connectives interpretations.

## 5 Concluding Remarks

We discuss various reinterpretations of previously studied [1,2] definition of bipolar queries. In particular we analyze the effect of the choice of various interpretations of the existential quantifiers and implication operators. The basic conclusion is that such a choice have to be careful as it changes not only the values of matching degree but also the resulting ordering of the tuples. We show also that in some special cases the formula defining bipolar queries reduces to a simpler form under any considered interpretation of logical connectives (cf. Properties 1 and 2).

A further research is needed to classify effects of other choices of logical connectives interpretations. Moreover we plan to extend the similar analysis to the case of the *winnow* operator we introduced in [2] as well as to its relation with bipolar queries.

## References

1. Zadrożny S.: Bipolar queries revisited. In Torra V., Narukawa Y., Miyamoto S., eds.: Proc. of MDAI 2005. LNAI, 3558. Springer, (2005).
2. Zadrożny S., Kacprzyk J.: Bipolar queries and queries with preferences. In: Proc.of DEXA'06, Krakow, Poland, IEEE Computer Soc. (2006) 415–419.
3. Dubois D., Prade H.: Bipolarity in flexible querying. In Andreasen T. et al., ed.: FQAS 2002. LNAI, 2522. Springe (2002) 174–182.
4. Lacroix M., Lavency P.: Preferences: Putting more knowledge into queries. In: Proc. of VLDB'1987, Brighton, UK (1987) 217–225.
5. Rosado A., Ribeiro R., Zadrożny S., Kacprzyk J.: Flexible query languages for relational databases: an overview. In Bordogna G., Psaila G., eds.: Flexible databases supporting imprecision and uncertainty. Springer (2006) 3–53.
6. Chomicki J.: Querying with intrinsic preferences. LNCS, 2287, (2002) 34–51.
7. Bosc P., Pivert O.: Discriminated answers and databases: fuzzy sets as a unifying expression means. In: Proc. FUZZ-IEEE'1992, San Diego, USA (1992) 745–752.
8. Bosc P., Pivert O.: An approach for a hierarchical aggregation of fuzzy predicates. In: Proc, of FUZZ-IEEE'1992, San Francisco, USA (1993) 1231–1236.
9. Mesiar R., Thiele H.: On T-Quantifiers and S-Quantifiers. In Novak V., Perfilieva I., eds.: Discovering the World with Fuzzy Logic. Physica-Verlag (2000) 310–326.
10. Dubois D., Prade H.: Using fuzzy sets in flexible querying: why and how? In Andreasen T., Christiansen H., Larsen H.L., eds.: Flexible Query Answering Systems. Kluwer (1997) 45–60.

# A Hierarchical Approach to Object Comparison

Axel Hallez and Guy De Tré

Ghent university, Dept. of Telecommunications and Information Processing
St.-Pietersnieuwstraat 41
9000 Gent, Belgium
Axel.Hallez@UGent.be, Guy.Detre@UGent.be

**Abstract.** Comparing objects is a very general problem domain. The way objects are compared can differ a lot, in the way objects are related to each other as well as in the meaning that is attributed to the result of a comparison. In this paper we develop a theoretical framework for comparing objects. This framework can be tailored to the specific needs of an application through the choice of appropriate operators and evaluation domain (i.e. the domain in which the comparison results are expressed).

## 1 Introduction

Object comparison is not a trivial problem by any means; several problem domains that are heavily researched topics on their own have to be combined in order to construct a comparison scheme: construction of similarity functions for individual properties, construction of aggregation operators, dealing with preferences, compensation, etc... Most of these topics have been researched quite extensively now and although new approaches are proposed on a regular basis, many solutions are readily available for use in applications. However, one of the biggest problems that remains, one that cannot be automated, is to find out which techniques are best suited for comparing objects in a particular situation, eg. which aggregation operators to use, finding out the suitable weights for the partial results, etc... In order to do that one has to gain a deep understanding of both the application domain and the techniques that are available for comparing objects. The usual practice is to find a domain expert and a technical expert and let them work together to find out how objects compare to each other. It would of course be much more convenient if the domain expert would be able to work with no or less help of a technical expert. Therefor it would be handy to have a (software) tool at hand that allows a domain expert to construct a comparison scheme and evaluate it. The easier this process gets, the more feasible it becomes to test different comparison methods and to compare the results. So we want to build a framework for the construction of comparison schema where a developer (i.e. the domain expert) can easily specify how objects are to be compared, without having to implement everything from scratch.

In this paper we present a theoretical foundation of a generic framework for the construction of comparison schema. The framework is generic in several respects. Firstly, the framework can be used for a wide range of comparison

P. Melin et al. (Eds.): IFSA 2007, LNAI 4529, pp. 191–198, 2007.
© Springer-Verlag Berlin Heidelberg 2007

methods. The framework is independent of the domain in which the results of the comparisons are expressed (i.e. the evaluation domain). It is assumed that this domain has some properties, but apart from that any domain can be used.

Some previous publications already proposed a framework for comparing objects [1] [2]. In these publications the equals operator that is found in many object oriented programming languages is extended to a fuzzy equals operator that allows for non-boolean result of the comparison and also deals with fuzzyness in the data. The method of comparison is similar to the one presented in here: there is a generic comparison method and the parameters (i.e. the type of operators and the weights of the attributes) are integrated in the source code of the data classes using annotations. This approach has many advantages, the main one being that it is very easy for developers to configure the comparison method. However there are also some disadvantages. The fuzzy equality operator is very tight coupled with the object schema of the application and since the configuration is embedded in the source code, the comparison can not be changed at run-time.

The framework presented below is more general, since it is not tightly coupled with the object schema of the data (in fact it is not even assumed that the data is object oriented). The definition of the comparison scheme is external to data scheme. This makes it possible to combine the data of the objects in ways that are not implied by the structure of the data itself. It also allows for the comparison to be adjusted at run-time, eg. to incorporate user preferences.

Object comparisons have many types of applications. However most of these applications use a common scenario: one object is compared to a set of other objects to find out which objects in this set match or are similar to the former object, which results in a set of objects that are sufficiently similar. This process needs to adapt to the specifics of the application: the 'similarity' of objects needs to be expressed in a way that fits the purpose of the comparison, which is discussed in section 2. The individual aspects of the object need to be compared in an appropriate way and these separate results have to be aggregated such that the importance of each aspect is respected. The construction of the way of aggregation and evaluation of the object properties (i.e. the comparison scheme) is discussed in section 3. In section 4 the limitation of the result set is discussed and before coming to a conclusion, we present an example application.

## 2 The Evaluation Domain

Before starting with the construction of a comparison scheme it is necessary to choose a domain in which the results will be expressed. It is important to choose a domain that matches the purpose of the comparison. If for example the purpose of the comparison is to find objects that are to some extent similar to a given object it is perfectly acceptable to use the unit interval as a scale of similarity [6]. If on the other hand the goal is to tell if two object descriptions are in fact describing the same real world object (eg. if two person descriptions are about the same person) it is more suitable to use a domain that allows for the explicit expression of confidence and uncertainty such as possibilistic truth values [3] [4] [5].

In order to be usable in the framework, we will assume that the domain of the comparison results, which we will refer to as $L$ from now on, possesses some properties. We will assume that an order operator $\leq_L$ for this domain $L$ exists, such that $(L, \leq_L)$ is a bounded lattice with $0_L$ and $1_L$ as lower bound and upper bound respectively.

## 3 Comparison Schema

The first thing to specify in a comparison scheme is the domain of objects. This domain will determine the properties of the objects that can be used in the comparison. The domain of objects will be referred to as $O$.

A comparison scheme is constructed in a hierarchical way. Two kinds of operators are used. At the basis of the scheme lay the so called *evaluation operators.* An evaluation operator is an $O^2 \rightarrow L$ mapping that defines how to compare two objects with respect to some aspect (eg. one attribute). There can be many evaluation operators in a comparison scheme, especially if the compared objects are complex. An evaluation operator in itself can be very complex too.

The results of the evaluation operators are then combined by *aggregation operators.* An aggregation operator is a $L^n \rightarrow L$ mapping where $n$ is a natural number larger than 0.

The aggregation can happen in several steps, i.e. attributes that belong together in some respect can be grouped together such that the results from the evaluation operators on these attributes are first combined into a single result. This result in turn can be used in a higher level aggregation. In the end all results are aggregated into a single global comparison result for the objects that are being compared.

Note that many operators, both aggregation and evaluation operators, have arguments that can be considered as configuration information, because they in fact determine how the operator behaves. We consider the values of these arguments as a part of the comparison scheme such that the operator that is put in the scheme is 'preconfigured' with values for these arguments and the only arguments that are left, are the two objects to compare in the case of an evaluation operator or the results of the underlying operators that provide the arguments of an aggregation operator.

Because of the very general definition of evaluation operators one can actually consider a subtree of a comparison scheme as an evaluation operator, since the combination of all operators of a subtree as described above determine a $O^2 \rightarrow L$ mapping. As a result, there is some freedom left in the design of a comparison scheme about what to consider as an evaluation operator. However, in general it will be in the interest of the designer to define the evaluation operators as specialized as possible and to refrain from putting aggregation functionality in them.

From the hierarchical nature of the comparison scheme, it follows that we can represent the scheme by a tree where the leaf nodes of the tree correspond with the evaluation operators and the non-leaf nodes with the aggregation operators, such that the result of an operator serves as an argument for its parent operator in the tree and that the result of the root operator is the global result of the comparison. In a formal way we define a comparison scheme as follows:

**Definition 1 (Comparison Scheme).** *A comparison scheme with object domain $O$ and evaluation domain $L$ is a tuple $(F, h, t)$ where:*

- *$F$ is non-empty finite subset of $(O^2 \to L) \cup (\bigcup_{n \in \mathbb{N}_0} (L^n \to L))$.*
- *$h$ is the partial order relationship on $F$ that defines the hierarchy of the evaluation tree, i.e. $(F, h)$ is a tree.*
- *$t$ is a total order relationship that defines the post order traversal (and thus the order of the child nodes of non-leaf nodes) in the evaluation tree.*
- *A leaf node $f$ of $(F, h)$ is a $O^2 \to L$ mapping.*
- *a non-leaf node $f$ of $(F, h)$ is a $L^n \to L$ mapping, where $n$ is the degree (i.e. the number of child nodes) of $f$ in $(L, h)$*

*The set of all schema with object domain $O$ and evaluation domain $L$ will be denoted as $\mathcal{S}_{(O,L)}$.*

A comparison scheme $(F, h, t) \in \mathcal{S}_{(O,L)}$ is called a *simple comparison scheme* if $F$ is singleton. When $F$ is not a singleton $(F, h, t)$ is a *composite comparison scheme.* Let $(F, h, t)$ be a composite scheme with $f_r$ the root of the evaluation tree (i.e. $h(f, f_r) \Rightarrow f = f_r$) and $n$ the degree of $f_r$. Then we can find $n$ comparison schema $(F_i, h_i, t_i) \in \mathcal{S}_{(O,L)}, i = 1...n$ such that $(F, h, t)$ is the composition of these schema with $f_r$:

$$F = \{f_r\} \cup (\bigcup_{i=1}^{n} F_i)$$

$$h = \{(f_r, f) \mid f \in F\} \cup (\bigcup_{i=1}^{n} h_i)$$

$$t = \{(f_i, f_j) \mid f_i \in F_i \wedge f_j \in F_j \wedge i < j\} \cup \{(f, f_r) \mid f \in F\} \cup (\bigcup_{i=1}^{n} t_i)$$

The schema $(F_i, h_i, t_i)$ are the subschema of $(F, h, t)$. The subschema are implicitly ordered by the order of their respective root operators in $t$.

For compactness of notation we will denote an arbitrary comparison scheme as $S$ with $S = (F_S, h_S, t_S)$. The subschema of $S$ will be referred to as $S_i$, $i = 1, ..., n$ and we will assume that the order of the subschema as implied by the order of their root operators in $t_S$ is reflected by the indices (i.e. if the root operator of $S_i$ is denoted as $f_i$ we assume that $i \leq j \Leftrightarrow t_S(f_i, f_j)$)

The following definition of the evaluation function $E_{(O,L)}$ that compares two objects using a given scheme takes advantage of the subschema.

**Definition 2 (Evaluation function).**

$$E_{(O,L)}: \quad \mathcal{S}_{(O,L)} \times O^2 \to L$$

$$(S, o_1, o_2) \mapsto \begin{cases} f_r(o_1, o_2) & \textit{if } S \textit{ is simple} \\ f_r(\langle E_{(O,L)}(S_i, o_1, o_2) \mid i = 1, ..., n \rangle) & \textit{else} \end{cases}$$

*with* $f_r$ *the root element of the evaluation tree of* $S$ *and* $S_i, i = 1, ..., n$ *the sub-schema of* $S$ *in case of a composite scheme.*

In order for this function to be computable, it is necessary that all operators, both evaluation and aggregation operators, can be evaluated in finite time. A potential problem is when an evaluation operator compares associated objects in such a way that cycles can occure (i.e. it is possible that in order to compare two objects, the evaluation operator calculates the similarity of the same pair of objects). This problem can be dealt with by using a different comparison scheme in the evaluation operator, in order to prevent endless recursion. For a generic method for solving this problem we refer to [1].

## 4 Thresholds

When searching for similar or equal objects in a large set, it is not only important to do the comparison in a meaningful way, it is also very important to process the results appropriately. One aspect is the ordering of the results, in order to present the better matches first. This can be done by applying the natural order of $L$, putting larger values first. Secondly it is also important to decide which comparison results are relevant and which are not. This can be done by specifying a threshold value such that only matches with a result that is better or equal to this threshold are kept and comparison results that are lower are not presented in the result set. This means that a) the threshold has to be an element of $L$ and b) that there needs to be an order on $L$ to decide whether a result is better or equal to the threshold as required.

Several approaches to applying the threshold are possible. One can reject all results that are below the threshold or one can accept only the result that are larger or equal than the threshold. If $(L, \leq_L)$ is a total order, these approaches are equivalent, but if not, it can make a significant difference.

## 5 Example

In this section we provide an example using the framework as described above to illustrate its use. One of the important aspects of maintaining a customer database (or a contacts list in general) is to avoid double entries. This means that before new customers are added to the database some checks have to be performed to see if the new customer already exists in the database. Although simple methods like comparing the names can already help to a great extent (although people can have the same name), things get complicated because of spelling errors and sometimes legitimate differences in how things like street names are written (eg. 'street' can be abbreviated to 'str.'). A good matching procedure tries to estimate whether differences in spelling can be attributed to spelling errors/legitimate spelling different or not, meaning that the entities are different.

It is not possible to draw conclusions from one attribute alone. People can have the same name, share the same address, etc. However email addresses tend to be quite accurate for uniquely identifying individual people (it is rather rare for employees to share email addresses). On the other hand people can have several email addresses.

Since this kind of comparison deals with uncertainty, i.e. the result needs to express how (un)certain it is whether two customer records describe the same person or not, we will use possibilistic truth values as the evaluation domain.

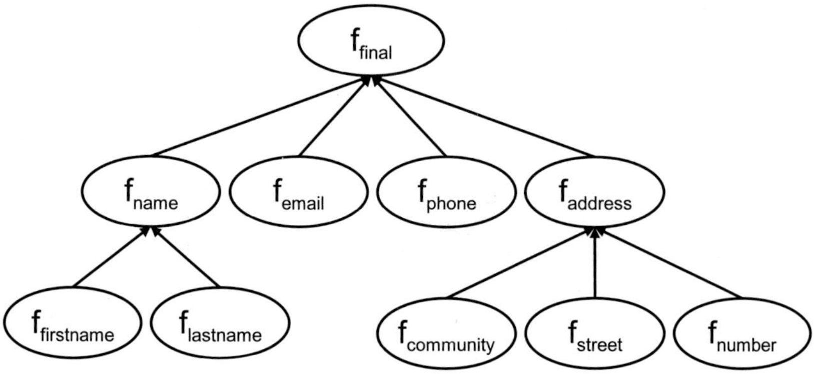


**Fig. 1.** Graphical representation of the comparison scheme

Figure 1 is a graphical representation of the comparison scheme we can use for this application. Note that the evaluation functions calculate how likely it is that the attribute values they compare are equal taking into account possible mistakes/substitutions, rather than indicating how likely it is that the persons are the same based on one attribute. These are the evaluation functions in the scheme:

$f_{firstname}$ The first names can be compared by calculating the editing distance between them. However first names are often substituted (eg. Bill for William) or abbreviated. This needs to be taken into account by using a list of common substitutions.

$f_{lastname}$ The last name too, can be compared by calculating the editing distance. Substitutions although they exist, will be applied less frequently.

$f_{email}$ One person can have multiple email addresses. In a professional context it will be rather rare for people to share their email address. So if two contacts have one or more matching email addresses it is very likely that they are the same person. If they don't share a common email address, it is very likely that they are not the same person, however people can change their email address from time to time. This function compare all of the email addresses of one contact with all of the email addresses of the other contact and takes the best match as a result.

$f_{phone}$ Since it is quite common for the people in one working place to communicate the same phone number to their contacts (eg. the secretary's number) a positive comparison result for the phone number is not conclusive, nor is a negative result, since people can have several phone numbers. This attribute will mainly reinforce the other results.

$f_{community}$ We will take the Belgian address format as a reference here. A Belgian address consists of the street name, the number (of the building), the zip code (a four digit number) and the name of the community. The community name is actually redundant information, since the combination of the zip code and the street name is unique. In this context redundant information is very useful since we can make a decision with more confidence. However, the region of a zip code may correspond to several community names (eg. some communities have both a Dutch and a French name) and some names, like the name of big city, can be used in combination with the several zipcodes. This operator uses primarily the zipcode to match the regions of two addresses, but uses the community name to detect wether a difference in zipcode could be caused by a mistake. This matching is backed by list of accepted combinations.

$f_{street}$ As with the previous attributes, it is possible for the same street name to written in different ways that are acceptable for a human reader, due to common abbreviations or ommitions. So again the comparison needs to take these into account.

$f_{number}$ A house number is what it is, so the comparison needs only to take into account possible errors.

The aggregation operator $f_{name}$ combines the results of the evaluation functions $f_{firstname}$ and $f_{lastname}$ by conjunction to find out whether the name as whole is equal. Three evaluation functions evaluate address attributes. The results of these are combined to find out wether two contacts have the same address by $f_{address}$. In an ideal world, where all addresses are spelled correctly, $f_{address}$ would be a pure conjunction since all three of the comparisons need to be equal for the addresses to be equal. In practice however, the three results will have a different weight depending on quality of the address information.

Finally the results of $f_{name}$, $f_{email}$, $f_{phone}$ and $f_{address}$ are combined by $f_{final}$ to calculate the final result. If we assume that the purpose of the comparison is to provide the user with a list of possible matches, this aggregation does not have to be too strict. We can for example decide that we will confront the user with a match on the condition that at least three of these four results exceed a certain threshold. This can be done by taking the conjuction of the three best results.

## 6 Future Work

In this paper only the basics of the framework have been worked out. This leaves a lot of work to be done on several fields. If one wants to create a framework to be used by domain experts with little technical knowledge, much thought has to be put in the selection of appropriate operators and the way these are presented, so that the user can make an adequate choice.

Another aspect is the creation of an efficient comparison procedure from the comparison scheme that the user creates. In a comparison scheme with many operators, not all operators will be equally efficient with respect resource consumption (i.e. the time it takes a computer to evaluate the function). If one has to evaluate all operators of a comparison scheme before it can be decided if the result will pass the threshold, this information is not of much use, but in many cases one can conclude that the result of an aggregation operator will not pass the threshold without evaluating it, but by checking that one of the arguments that are passed to it are not within a certain range. For example, the result of a conjunction operator (based on a t-norm) will be smaller than or equal to each of its arguments. So if one of the conjunction's arguments does not pass the threshold applied to the result, the other arguments of the conjunction do not have to calculated in order to conclude that the final result will not pass the threshold. This information can be used to optimize the evaluation of a comparison scheme for time performance.

## 7 Conclusion

We developed a theoretical basis for creating schema for object comparison. The separation of structure and semantics makes it a good basis for developing software that allows for the efficient creation of comparison schema. It also allows us to deal with topics such as optimization of comparison speed in a structural manner.

## References

1. N. Marn, J.M. Medina, O. Pons, D. Snchez, M.A Villa (2003), Complex object comparison in a fuzzy context. In: *Information and Software Technology*, volume 45, pages 431-444.
2. F. Berzal, J.C. Cubero, J. Kacprzyk, N. Marn, M.A. Villa, S. Zadrozny (2005), Computing with Words using Metadata-Based Programming, In: *Proc. of the IFSA 2005 World Congress*, Beijing, China, July 2005.
3. G. De Tré, A. Hallez, J. Verstraete and A. Verkeyn (2002), Beyond Conjunctive Aggregation of Possibilistic Truth Values in Database Systems. In: *Proc. of the Eurofuse 2002 Workshop on Information Systems*, pages 137-142, Varenna, Italy, September 2002.
4. G. De Tré and B. De Baets (2003). Aggregating Constraint Satisfaction Degrees Expressed by Possibilistic Truth Values. In *IEEE Transactions on Fuzzy Systems*, volume 11, number 3, pages 354-360.
5. D. Dubois, H. Prade (1988), Possibility theory, New York, Plenum Press.
6. D. Dubois and H. Prade (1997), The three semantics of fuzzy sets,In: *Fuzzy Sets and Systems*, vol. 90, no. 2.

# FuzzyXPath: Using Fuzzy Logic an IR Features to Approximately Query XML Documents

Ernesto Damiani[1], Stefania Marrara[1], and Gabriella Pasi[2]

[1] Università degli Studi di Milano
Dipartimento di Tecnologie dell'Informazione
via Bramante 65 26013 Crema (CR), Italy
{damiani,marrara}@dti.unimi.it
[2] Università degli Studi di Milano Bicocca
DISCO
Via Bicocca degli Arcimboldi, 8 20126 Milano (MI), Italy
pasi@disco.unimib.it

**Abstract.** XML has become a key technology for interoperability, providing a common data model to applications. However, diverse data modeling choices may lead to heterogeneous XML structure and content. In this paper, information retrieval and database-related techniques have been jointly applied to effectively tolerate XML data diversity in the evaluation of flexible queries. Approximate structure and content matching is supported via a straightforward extension to standard XPath syntax. Also, we outline a query execution technique representing a first step toward efficiently addressing structural pattern queries together with predicate support over XML elements content.

## 1 Introduction

Il the last few years, the problem of modeling and querying semi-structured information has been intensively studied by both database and information retrieval research communities, although with a slightly different focus. Database research mostly deals with data which do not conform to a strict database schema, i.e. whose structure is not regular [1]. Information retrieval focuses on documents sharing a basic logical structure, constituted by sub-components (or *sections*). Research on semi-structured information querying also comes in two different flavors: in the context of semi-structured databases, flexible query languages take into account the lack of a rigid schema of the database, thus allowing to enquiry both data and the type/schema [4,8], while in the context of IRSs, modeling flexibility means mainly to take into account the possibility to refer to a non-uniform structure of the documents when formulating queries [16]. De-facto standards for the definition of semi-structured documents such as HTML and XML further enriched this picture by adding some features of hierarchical data models. Today, the XML Infoset is widely employed as a basic data model for semi-structured information, and is now the basic standard for representing semi-structured documents in Information Retrieval. Also, techniques for querying and updating XML data

P. Melin et al. (Eds.): IFSA 2007, LNAI 4529, pp. 199–208, 2007.
© Springer-Verlag Berlin Heidelberg 2007

have been investigated and several standards like XQuery [26] and XUpdate [28] have been proposed. Looking at the amount and diversity of XML data items, it is possible to roughly divide them into 1) *document-centric* items where XML is used for logical markup of text, 2) *data-centric* items where XML is used for exchanging structured data among applications. For the sake of simplicity, in this work we concentrate on the selection aspect of queries. Namely, we extend the idea originally proposed by two of us in [6], and recently re-stated in a more formal garment in [9], in which fuzzy predicates are introduced into a XPath query to express flexible selection conditions and to perform fuzzy subtree matching. In addition to extending structural properties of flexible queries by means of fuzzy operators, we also outline some content-based properties adopting IR features similarly to the proposal of [16]. Our proposal includes modeling of vagueness and imprecision in XML retrieval by means of the following features:

- index term weighting to produce ranked query results,
- specificity-oriented search for retrieving only the most relevant parts of documents,
- use of vague predicates in query formulation to express flexible constraints on stored data,
- structural vagueness, in order to find close matches for structural query conditions.

The aim of the work is to set the foundations for a flexible XML selection language, *FuzzyXPath*, later to be extended to a fully fledged query language[1], which implements imprecision and vagueness for both structural and content-oriented query conditions. The paper is structured as follows: Section 2 motivates the ideas exposed in this work and proposes a running example that will be used in the rest of the paper, Section 3 gives details on query execution and show how it works by means of an example and, finally, Section 4 exposes our conclusions and outlines our future work.

### 1.1 Related Work

Several approaches have been proposed to introduce flexibility in semi-structured information processing and, in particular, in XML querying. An early technique [14] was based on fuzzy encoding of XML data trees. That method did not attempt to define a query language; rather, it supported introducing new nodes in the XML tree structure in order to carry out fuzzy similarity comparison of XML data trees. This approach was later extended in [13] providing some flexibility in content comparison with the concept of XML data *smushing*. A later paper [2] proposed an approach based on XML query rewriting, supporting renaming and deletion of nodes in the query. Hybrid techniques [25] have also been proposed, where XML data are encoded and queries are rewritten. On the one hand, hybrid techniques can

[1] Since the language name originally adopted in [6], *FXPath*, has recently been used with the meaning of *Functional XPath*, in this work we suggest the new name *FuzzyXPath* to overcome confusion.

provide an accurate computation of the query cost; on the other hand, it is very difficult to implement them because they require ad hoc XML data indexing. A recent approach to this problem [20] proposes a dynamic summarization and indexing method, FLUX, based on Bloom filters and $B^+$-trees. Also, the work [12] presents an indexing method to execute approximate queries on XML documents taking into account approximation on both document structure and content. The proposed indexing aims to reduce the complexity of finding approximate query patterns, avoiding sequentially scanning all documents in the collection.

Another recent paper [7] proposes a fuzzy-based XML querying system that performs approximate comparisons between XML query and data trees. This technique supports imprecision on data via possibility distributions. However, while the authors claim that their querying system is fully compatible with XML querying standards since the final rewriting is performed in XQuery, their query rewriting is based on a *mediated architecture* called *MIEL++* that requires several rewriting steps and is unlikely to scale well. In [22] the authors propose an approach for approximate query answering in which, instead of working directly on the data, they interpret the structural component of the query by exploiting a reworking of the documents' schemas by means of a *schema matching* process. The information retrieval angle has been explored in the seminal paper [16], which uses a probabilistic data modeling combined with some IR features in order to query document-centric XML. Of course, several works in the literature had already addressed the problem of defining IR models for structured documents [5,10,8,19,21,27]. They focused on two main aspects: how to index structured documents so as to usefully exploit their structure in their formal representation, and how to define query evaluation mechanisms that can retrieve also document subparts. Passage retrieval is mainly concerned with identify subparts of a text document as retrievable information units. Specifically, passage retrieval aims to identify short blocks of relevant information among irrelevant text [8,17]. In [21] a conceptual model for structured documents has been proposed which supports a query language enabling to retrieve passages based on their context as well as content. Aggregation-based approaches have also been explored for representation and retrieval of structured documents. These approaches estimate the relevance of document subparts based on the aggregation of estimated relevance of their content and of their structurally related parts [3,18,15,23,11]. In order to improve the effectiveness of IRSs some considerable efforts are being spent in trying to define new conceptual models aimed at indexing and querying structured documents [3,24,18,17,27,11]. In [11] the idea of producing a composite representation of structured documents is exploited, enabling focusing retrieval only on some document subparts. In [18] this basic scheme is enriched by modeling uncertainty in content representation. In [10,21,3] other approaches to representation and retrieval of structured documents are presented. Particularly, in [3] a fuzzy model for defining an indexing mechanism is described which exploits users' feedback to create personalized representations of the same structured document. This model exploits the vagueness in the indexing process and enables the system to learn users needs at the indexing level. This pioneering work has been further developed in [5], which is the starting step of this work.

## 2 Motivations and Running Example

Thanks to XML standards several application involving data interchange enjoy interoperability at the level of the data model. Nonetheless, they often have to face the lack of shared semantic models which causes different modeling choices. In turn, these different choices cause differences in XML schemata, leading to heterogeneous XML structure and content. Also, XML data stored in repositories constantly evolve over time. Such evolution is a challenge, as existing applications must continue to work with the evolved data. In our approach, flexible querying of XML data is aimed at tolerating schema-level heterogeneity due either to different modeling choices or to data evolution, allowing for posing the same query to multiple, possibly heterogeneous XML document collections loosely sharing the same semantics. For the sake of concreteness, in the remainder of the paper we shall discuss this notion based on a practical example involving the documents shown in Figure 1.

Both documents contain information about people working at a University. In document (a) `people` is divided into several groups depending on the `department`

```
<university>
 <people>
  <department @name="computer science">
   <technicians>
    <tec> <name>...</name>
     ...
    </tec>
    <tec> ...
    </tec>
   <administrative staff>
     ...
   </administrative staff>
   <research people>
    <professor @qual="full">
       <name> John </name>
       <surname> Smith </surname>
       <CV> Prof. John Smith was born.... His
research interests include Information Retrieval,
security .... He likes reading and horse riding. </cv>
       <office> 11B </office>
    </professor>
    <student @qual="Phd">
       <name> Sarah </name>
       <surname> Plouth </surname>
       <CV> Sara Plouth was born.... Her research
interests include Information Retrieval, security under
the supervision of Professor Smith ....  </cv>
       <office> 12B </office>
    </student>
...
  </department>
 </people>
...

</university>
```

(a)

```
<university>
   <department @name="computer science">
    <employed>
     <administrative staff>
       ...
     </administrative staff>
      <research people>
       <professor @qual="full">
          <name> John Smith</name>
          <CV> Prof. John Smith was born.... His
research interests include Information Retrieval,
security .... He likes reading and horse riding. </cv>
          </professor>
...
     </research people>
    </employed>
     <students>
      <student @qual="Phd">
          <name> Sarah Plouth</name>
          <CV> Sara Plouth was born.... Her research
interests include Information Retrieval, security under
the supervision of Professor Smith ....  </cv>
        </student>
...
 </students>
...
</department>
...
</university>
```

(b)

**Fig. 1.** Two XML documents sharing the same content but with different structure they belong. Then each department is divided into `technicians`, `administrative staff` and `research staff`. Researchers include both professors and Ph.D. students. Inside each `professor` or `student` tag lie their respective `name`, `surname`, `cv` and `office`. The `cv` tag is usually a large text blob. In document (b) the `University` is directly divided into `departments`, each department containing two groups of people: `employees` and `students`. Employees include `administrative staff` and `professors`. Each professor's (or student's) personal details are enclosed in the tag `name` which includes their name, surname, and cv.

## 3 FuzzyXPath

We are now ready to illustrate our approach to fuzzy XML querying in some detail. According to the XML Infoset, XML documents can be represented as a tree of typed nodes. XPath uses a pattern expression to identify nodes in an XML document and retrieves portions of XML documents, namely the set of nodes matching the pattern. Due to space constraints, in this context we focus our attention on a subset of the XPath language informally defined as follows: `XPath*`$:= \varepsilon | l | * | p_1/p_2 | //p_1 | p[q]$ where $p_1$ and $p_2$ are `XPath*` expressions; $\varepsilon$, $l$, $*$ denote the empty path, a label and a wildcard, respectively; / and // stand for child-axis and descendant-or-self-axis; and finally, $q$ is called a qualifier. Following [6] we extend the XPath syntax with three classes of characteristics:

- *Fuzzy Subtree Matching*: namely `NEAR`, `ABOUT`, `BESIDES` and `LIKE`, providing a ranked list of retrieved information items rather than the set oriented one typical of XPath.
- *Fuzzy Predicates*: specifying flexible selection conditions.
- *Fuzzy Quantification*: allowing the specification of linguistic quantifiers as aggregation operators.

XPath 1.0 presents two main features that are relevant to our approach:

- *Path-based selection*: the user formulates a *search path*, in the standard form of XPath expressions, that must be exactly matched against the structure of the target XML documents.
- *Set-oriented query result*: the selection mechanism retrieves the documents sets of nodes that exactly match the user-provided path.

XPath assumes that the user is fully aware of the target schema. This assumption is in itself debatable since most XML documents exist without schemas; even worse, it requires the user to write a different query for each variation of the target schema. In other words, XPath does not tolerate data structure or content diversity. In order to tackle this problem, we extend XPath in order to perform *approximate queries* in which the search path only provides a loose example of the information the user is interested in.

Our approach is based on three steps (Figure 2):

1. in the first step the query is analyzed to extract its keywords, and then these keywords are searched in an inverted file (detailed in Section 3.1).

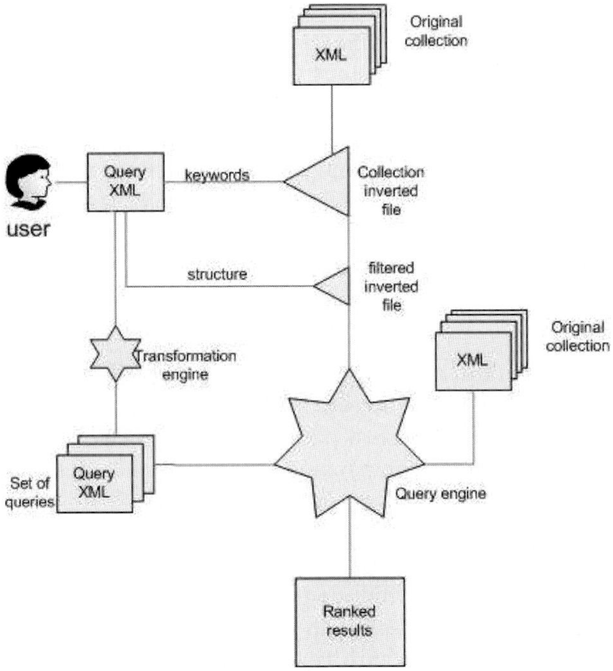


**Fig. 2.** Architecture of the approach

This operation reduces the cardinality of the document collection that will be target of the query execution and extracts a ranked list of candidates on the basis of the *index term weights* contained in the inverted file;

2. then, in order to "easily" process the structural selection conditions, the query is rewritten into a ranked set of queries that approximate the desired information. The ranking of the approximate query is computed on the basis of their distance from the user's query. Each approximate query is computed on the previously computed list of candidate documents by a common query engine and the retrieved results, if any, will be ranked according to the weight of the corresponding query's (*structure weight*).
3. finally retrieved results are shown in a bi-dimensional space in which one dimension is the content-based evaluation retrieval status value (*index term weight*) and the other is the structure-based retrieval status value. The user will choose which dimension he/she prefers to rank the results in the list of retrieved items.

The proposed approach is based on exact tag matching; it would be interesting to address the problem of approximate or semantics-based tag matching. To face this issue, the use of a thesaurus/ontology is needed. At this stage of the work we do not address this problem.

### 3.1 Content Based Constraints

Most of the operating IRSs are based on indexing functions which take into account only marginally the structure of documents, being mainly based on their consideration as a monolithic object, thus providing their overall content synthesis. This allows for the retrieval of a document as a unit of information; in the case of structured documents this retrieval criterion can be in contrast with the real users' needs, as relevant information could be found in just some subparts (fragments) of a document. The work developed in [5] is the basis of the approach presented here. This model is based on the observation that the production of an information surrogate, i.e., a document synthesis used for representing the document content, is strongly subjective and depends on the personal view of the interpreter of this information. Despite of this, most of the existing indexing functions behave as a black box producing the same document representation to all users. To overcome this limitation the indexing model proposed in [3] is able to diversify the document representations, based on the users' view. The proposed indexing model is composed of two main components: a static component, aimed at the pre-calculation of terms occurrence information within the paragraphs (this information has to be stored in the inverted file), i.e., the building bricks of the documents, and an adaptive component that, based on the user's indications computes the index terms weights used by the matching mechanism. In [5] a flexible query language has been also proposed for expressing soft selection conditions on both the documents' structure and contents. The evaluation of these flexible queries makes it possible to select a fuzzy subset of semi-structured documents from a heterogeneous collection. For example, if one is interested in reading scientific papers he/she may filter documents having most of the following *sections: Title, abstract, authors, text references.* The linguistic quantifier most, that specifies the soft constraint on the document structure allows not to disregard potential interesting documents whose structure is not complete with respect to the specifications; for example documents lacking the *abstract* section. Further, users can indicate preferences on the desired sections. The two levels of soft conditions on the structure and on the content of documents are expressed in two distinct phases but are evaluated in a unique step so that the Retrieval Status Value reflects the satisfaction of the global query. In the extension of the XPath query language that we propose in this paper, in order to be able to evaluate content-based query constraints, an file inverted structure for organizing indexes has to be defined. Each term in this structure should point to blocks of information, each block containing: `docId` (document identifier), `field-Id` (textual field corresponding to a leaf node), `path` to reach that field in docId, `n` number of occurrences of that term in that field, a normalization parameter `np` (or two normalization parameters: number of occurrences of the most occurring term in that field and number of total occurrences of words in that field). When a query which specifies both structural and content-based constraints is evaluated, we propose a first pre-processing based on the consideration of terms specified in the query for content based constraints. This pre-processing consists in selecting, based on the information contained in the inverted file, only paths related to the fragments pointed by a given query term (i.e. those which indexed by the term) thus pruning the paths which possibly satisfy the user query. Then on those selected paths the evaluation of structure-based constraints can be performed.

### 3.2 Enforcement of Structure Constraints

Once the document collection has been restricted to a finite set of candidate documents by the content based analysis, the query is analyzed on the basis of its structure. *FuzzyXPath* ([6]) constraints are used to indicate a degree of desired approximation in the query structure. In this step each FuzzyXPath constraint is used to create a set of queries that simulates the degree of approximation specified in the user query. Each of these queries is labeled by a *structure weight* that indicates the distance between the original query structure and the current one. Then, the resulting set of query structures is used to eliminate the documents that do not match at all (not even approximately) the user query from the set of candidate documents. Finally this set of queries is executed by a common query engine (this time the match is computed exactly) and each retrieved result is ranked on the basis of the structure weight of the query that produced it. For example, for the user query:

`university//department{NEAR}/professor[CV cw "Information Retrieval"]/name`

the approach follows these steps:

- We separate the path that carries the desired information (called *info path*) from the path that carries the matching condition (called *match path*). In our example, we obtain the info path `university//department{NEAR}/professor/name` and the match path `university//department{NEAR}/professor[CV cw "Information Retrieval"]`.
- We identify the approximate clauses. In the example, the info path contains the clause `NEAR`, which means that the output node set will be ranked with respect to the number of steps from `department` to `professor`. The matching path contains the construct `cw` (*contains word/s*) that is resolved by means of standard retrieval procedures creating a set of candidate documents by filtering the collection's weighted inverted file. The result of this step is a set of candidate documents represented by their inverted file tuples `(docId, field-Id, path, dw`[2]`)`.
- The info path is then transformed in a set of weighted query paths. This translates our FuzzyXPath NEAR clause into standard XPath constructs. Again in our example we obtain the set of paths $Q$ ={(`university//department/professor/name`,1), (`university//department/*/professor/name`, 0.5),(`university //department/*/*/professor/name`, 0.34), ...}. In our example we compute the structure weight as $w = 1/(1+n)$ where $n$ is the number of steps between `department` and `professor`, but different functions could be used on the basis of the desired approximation ranking.

[2] Note that `dw` $= \frac{n}{np}$.

We can stop the generation of new paths in $Q$ just by selecting a matching threshold $mt$ and stopping when $w < mt$.
- Paths $\in Q$ are used to filter again the candidate documents inverted file eliminating those documents whose `path` does not match any path $\in Q$.
- The match path is used to construct a set of queries $Q'$ where each query $q'$ is composed by the match path and one info path $\in Q$.
- Finally, queries belonging to $Q'$ are executed on the set of candidate documents identified by the filtered inverted file and results are ranked in a bidimensional space that considers both weights `w` and `dw`.

## 4 Conclusions

While fostering application interoperability by providing a common data model, XML Infoset poses unique challenges. Diverse modeling choices may cause differences in XML schemata, leading to heterogeneous XML structure and content. In this paper, information retrieval and database-related techniques have been jointly applied to effectively tolerate this diversity by supporting flexible queries. Fuzzy/approximate matching is supported via a straightforward extension to standard XPath syntax. Also, our indexing technique represents a first step toward efficiently addressing structural pattern queries jointly with predicate support over textual content of XML elements. We plan to address this issue in a future paper.

## References


1. S. Abiteboul. Querying semi-structured data. *LECTURE NOTES IN COMPUTER SCIENCE*, 1186:1–18, 1997.
2. S. Amer-Yahia, S. Cho, and D. Srivastava. Tree pattern relaxation. In Springer, editor, *Proceedings of EDBT*, 2002.
3. G. Bordogna and G. Pasi. Controlling retrieval through a user adaptive representation of documents. *Int. J. of Apporximate Reasoning*, 12:317–339, 1995.
4. G. Bordogna and G. Pasi. Flexible representation and querying of heterogeneous structured documents. *Kibernetika*, 36(6):617–633, 2000.
5. G. Bordogna and G. Pasi. Personalized indexing and retrieval of heterogeneous structured documents. *Information Retrieval*, 8(2):301–318, 2005.
6. D. Braga, A. Campi, E. Damiani, G. Pasi, and PL. Lanzi. FXPath: Flexible querying of xml documents. In *Proceedings of EuroFuse 2002, Varenna, Italy, Sep. 2002.*
7. Patrice Buche, Juliette Dibie-Barthèlemy, and Fanny Wattez. Approximate querying of XML fuzzy data. In Springer, editor, *Proceedings of the 7th International Conference FQAS 2006, Milan, Italy*, 2006.
8. J. Callan. Passage-level evidence in document retrieval. In ACM, editor, *Proceedings of SIGIR 94, Dublin, Ireland*, 1994.
9. A. Campi, S. Guinea, and P. Spoletini. A fuzzy extension for the XPath query language. In Springer, editor, *Proceedings of the 7th International Conference FQAS 2006, Milan, Italy*, 2006.

10. Y. Chiaramella. Information retrieval and structured documents. In F. Crestani M. Agosti and G. Pasi, editors, *Lectures on Information Retrieval*, Lecture Notes in Computer Science. Springer Verlag, 2000.
11. Y. Chiaramella, P. Mulhem, and F. Fourel. A model for multimedia information retrieval. Technical Report Fermi ESPRIT BRA 8134, University of Glasgow, 1996.
12. P. Ciaccia and W. Penzo. The *collection index* to support complex approximate queries. In Springer Verlag, editor, *Proceedings of XSym 2003*, volume 2824, pages 164–179.
13. Ernesto Damiani, Barbara Oliboni, and Letizia Tanca. Fuzzy techniques for XML data smushing. In *Proceedings of the International Conference, 7th Fuzzy Days on Computational Intelligence, Theory and Applications*, pages 637–652, London, UK, 2001. Springer-Verlag.
14. Ernesto Damiani and Letizia Tanca. Blind queries to XML data. In *Database and Expert Systems Applications*, pages 345–356, 2000.
15. M. Frisse. Searching for information in a hypertext medical handbook. *Communication of the ACM*, 31(7):880–886, 1988.
16. N. Fuhr and K. Grobjohann. XIRQL: A query language for information retrieval in xml documents. In ACM, editor, *Proceedings of SIGIR'01, New Orleans, Luisiana, USA*, 2001.
17. M. Kaszkiel and J. Zobel. Passage retrieval revisited. In N.J. Belkin, D. Narasimhalu, and P. Willett, editors, *Proceedings of the 20th SIGIR*, 1994.
18. M. Lalmas. Dempster-shafer's theory of evidence applied to structured documents: modelling uncertainty. In *Proceedings of ACM SIGIR, Philadelphia.*, 1997.
19. M. Lalmas and I. Ruthven. Representing and retrieving structured documents using the dempster-shafer theory of evidence: Modelling and evaluation. *Journal of Documentation*, 54(5):529–565, 1988.
20. Hua-Gang Li, S. Alireza Aghili, Divyakant Agrawal, and Amr El Abbadi. FLUX: Fuzzy content and structure matching of XML range queries. In *Proceedings of WWW 2006, May 23-26, 2006, Edinburgh, Scotland*, 2006.
21. I. Macleod. Storage and retrieval of structured documents. *Information Processing and Management*, 26(2):197–208, 1990.
22. F. Mandreoli, R. Martoglia, and P. Tiberio. Approximate query answering for a heterogeneous XML document base. In Springer, editor, *Proceedings of the 5th Int. Conf on Web Information Systems Engineering*, Brisbane, Australia, November 22-24, 2004.
23. S. Myaeng, D.H. Jang, M.S. Kim, and Z.C. Zhoo. A flexible model for retrieval of sgml documents. In *Proceedings of the 21st ACM SIGIR, Melbourne, Australia.*, pages 138–145, 1998.
24. G. Navarro and R Baeza-Yates. A language for queries on structure and content of textual databases. In *Proceedings of ACM SIGIR, Seattle.*, pages 93–101, 1995.
25. T. Schlieder. Schema-driven evaluation of approximate tree-pattern queries. In Springer, editor, *Proceedings of EDBT*, 2002.
26. W3C. Xquery 1.0: An xml query language, November 2006.
27. R. Wilkinson. Effective retrieval of structured documents. In *Proceedings of the 17th ACM-SIGIR, Dublin.*, pages 311–317, 1994.
28. XML:DB. Xupdate, November 2000. http://xmldb-org.sourceforge.net/xupdate/.

# Part IV

# Philosophical and Human-Scientific Aspects of Soft Computing

# Designing Representative Bodies When the Voter Preferences Are Fuzzy

Hannu Nurmi[1] and Janusz Kacprzyk[2]

[1] Academy of Finland and Department of Political Science
University of Turku
FI-20014 Turku
Finland
hnurmi@utu.fi

[2] Systems Research Institute, Polish Academy of Sciences
ul. Newelska 6, 01-447 Warsaw, Poland
kacprzyk@ibspan.waw.pl

**Abstract.** The theory of fuzzy sets has been applied to social choice primarily in the contexts where one is given a set of individual fuzzy preference relations and the aim is to find a non-fuzzy choice set of winners or best alternatives. We discuss the problem of composing multi-member deliberative bodies starting from a set of individual fuzzy preference relations. We outline methods of aggregating these relations into a measure of how well each candidate represents each voter in terms of the latter's preferences. Our main goal is to show how the considerations discussed in the context of individual non-fuzzy complete and transitive preference relations can be extended into the domain of fuzzy preference relations.[1]

## 1 What Is a Good Representation?

If democracy is ruled by the people, representative democracy must be ruled by representatives of the people. But not all systems in which a group of persons declares themselves the representatives of the others, qualify as representative ones in a deeper sense of the term. To say that a body of persons represents a larger body of people requires that the values, tastes, opinions, attitudes of the latter are somehow reflected in the activities of the former. In the words in Rogowski ([Rogowski 1981], cited by [Chamberlin and Courant 1983, 719]):

> A person, $A$, is represented in some matter by another person, $B$, to the extent that $B$'s actions in the matter reflect what might be called $A$'s ideal preferences – the choices that $A$ would make if $A$ were ideally informed, ideally expert, and ideally clear about his or her own interests.

One might read this as suggesting that a representative body at its best consists of the people itself. This guarantees the presence of every opinion held

[1] An extended version of this paper will appear in New Mathematics and Natural Computation.

P. Melin et al. (Eds.): IFSA 2007, LNAI 4529, pp. 211–219, 2007.
© Springer-Verlag Berlin Heidelberg 2007

by someone among the people to be present in the representative body. But this means that the representative body is no smaller than the body it is supposed to represent, clearly an untractable arrangement.

Chamberlin and Courant impose the following requirements on representation [Chamberlin and Courant 1983]. A committee member represents a voter to the extent that:

1. the committee member "makes present" the voter's opinions in the deliberations that take place within the committee,
2. the committee member is similarly responsive to various kinds of arguments presented in those deliberations as the voter, and
3. the committee member votes in the same way as the voter should the latter be present in the committee.

Clearly, representation of any voter by any member of the committee is a matter of degree and can never be perfect in the sense that all three requirements are satisfied. In fact, the degree of representation for any (member, voter)-pair can be determined only *ex post*, i.e. once the committee's term has expired. This does not solve the problem of electing a representative committee. There are basically two approaches to this election problem. Firstly, one may ask the voters to vote for their favorite committee, that is, to signal their preferences regarding all conceivable committees. The choice of the committee would then be determined on the basis of ballots cast. This is what Chamberlin and Courant call the preferences-over-committees approach. Secondly, the voters could vote for their favorite representative or provide a ranking over candidates. This is the preferences-over-candidates approach. Since representation is mainly about getting one's opinions heard in the committee proceedings rather than influencing the outcomes of the committee decision making, the latter approach seems more appropriate.

Given a preference profile of the voters over candidates and the size of the committee, say $k$, Chamberlin and Courant's proposal for determining the optimal committee composition is equivalent to the following. For each possible committee, compute the number of individuals whose most preferred candidate is present in the committee. Denote this number by $n_1$. It can obviously be any number between 0 and $n$, the total number of voters. Then count the number of voters whose first or second preference candidate is present in the committee and denote this by $n_2$. Continue in this manner until all ranks $1, \ldots, k$ have been considered. Obviously, $n_k = n$. Let now the set of all $k$-member committees be $C^k$ with elements $c_1, c_2, \ldots, c_s$. The value $C(c_i) = \sum_j^k n_j$, for each $i = 1, \ldots, s$ is the indicator of the representativeness of a committee: the higher the value, the better represented are the voters. Clearly, $k \times n$ is the maximum attainable value and is associated with a committee where each voter's first ranked candidate is present. Similarly, 0 is the minimum value of $C(c_i)$. This "worst possible" committee has the distinction that no voter ranks any committee member higher than the $(k+1)$-th in his or her ranking.

It turns out that if one maximizes $C(c_i)$ over all possible $k$-member committees, one – in a specific way to be explained shortly – maximizes the sum of the Borda scores of committee members. The most representative one-member committee is one consisting of the candidate with the maximum Borda score. A maximally representative $k$-member committee, on the other hand, is determined by a modified Borda count. Define each voter's representative as the committee member getting the largest number of Borda points from that voter, i.e. the member ranked highest in the voter's ranking over candidates. Thus, each voter has a representative in each committee. Now, let $B(c_i)$ denote the sum over voters of the Borda points given to their representative in committee $c_i$. The most representative committee is then defined as $c = \arg\max_i B(c_i)$, i.e. the committee where the sum of the Borda points given by each voter to his or her representative is maximal. This is, indeed, a modified Borda count since each voter gives only one score, viz. that of his or her representative.

Chamberlin and Courant show that the method for constructing the most representative committee described above can be given another interpretation, i.e. the most representative committee minimizes the number of objections raised by the voters against various committees. By an objection they refer to a situation where, given a committee $c_i$ not including candidate $x$, voter $j$ would be better represented by a committee including $x$ in the sense that $x$ is preferred by this voter to every member of $c_i$. Now, summing the objections of all voters for each committee gives a index of misrepresentation for the committee. Choosing the committee with the minimum index value amounts to choosing the most representative committee. Thus, the modified Borda count yields the misrepresentation-minimizing committee.

Monroe [Monroe 1995] outlines a similar approach to optimal representation. Its basic concept is also the amount of misrepresentation. However, this concept is applied to pairs consisting of committee members and voters. Consider a committee $C$ and electorate $N$. For each pair $(j, l)$ where $j \in C$ and $l \in N$, let $\mu_{jl}$ be the amount of misrepresentation related to $l$ being represented by $j$. It is reasonable to set $\mu_{jl} = 0$ if $k$ is top-ranked in $l$'s preferences. In searching for the pure fully proportional representation Monroe embarks upon finding a set of $k$ representatives,each representing an equally-sized group of voters (constituency), so that the total misrepresentation – the sum over voters of the misrepresentations of all committee members – is minimal. He suggests a procedure which firstly generates all possible $\binom{n}{k}$ committees of $k$ members. For each committee one then assigns each voter to the representative that represents him or her best. Since this typically leads to committees consisting of members with constituencies of different size, one proceeds by moving voters from one constituency to another so that eventually each constituency has equally many voters. As a criterion in moving voters is the the difference between their misrepresentation in the source and target constituencies: the smaller the difference, the more likely is the voter to be transferred.

For large $n$ and $k$ the procedure is extremely tedious.[2] Potthoff and Brams [Potthoff and Brams 1998] suggest a simplification that essentially turns the

[2] There is also some ambiguity as to how one should proceed in transferring voters from candidates (constituencies) to another.

committee formation problem into an integer programming problem. Let $\mu_{ij}$ be the misrepresentation value of candidate $i$ to voter $j$. Define $x_i$ for $i = 1, \ldots, k$ so that it is 1 if $i$ is present in the committee and 0, otherwise. Furthermore, we define $x_{ij} = 1$ if candidate $i$ is assigned to voter $j$, that is, if $i$ represents $j$ in the committee. Otherwise $x_{ij} = 0$. The objective function we aim at minimizing now becomes:

$$z = \sum_i \sum_j \mu_{ij} \tag{1}$$

In other words, we minimize the sum of misrepresentations associated with the committee members. In the spirit of Monroe, Potthoff and Brams impose the following constraints:

$$\sum_i x_i = k \tag{2}$$

$$\sum_i x_{ij} = 1, \forall i \tag{3}$$

$$\frac{n}{k} x_i + \sum_j x_{ij} = 0, \forall i \tag{4}$$

Condition (2) states that the committee consists on $k$ candidates, condition (3) says that each voter be represented by only one candidate,and condition (4) amounts to the requirement that each committee member represents an equal number of voters. In Monroe's system, $\mu_{ij} = k - 1 - b_{ij}$ where $b_{ij}$ is the number of Borda points given by $j$ to candidate $i$.

## 2 Representation of Fuzzy Preferences

We shall now extend the idea of a fully proportional representation to the domain of fuzzy individual preferences. We assume that the committee to be formed is a group of candidates as in the preceding. In contrast to the preceding, however, the voters have fuzzy preference relations over the candidates. We denote the set of candidates by $K$. With $m$ alternatives these preferences can be represented by $m \times m$ matrices with entry $(i, j) \in [0, 1]$, for $i \neq j$ and $i, j = 1, \ldots, m$. We denote it by $r_{ij}$ which indicates the degree in which alternative $i$ represented by the row is preferred to alternative $j$ represented by the column. In many contexts it is plausible to assume that the fuzzy preferences are reciprocal, that is, $r_{ij} = 1 - r_{ji}$, for all $i$ and $j$. In reciprocal fuzzy relations it is natural to interpret $r_{ij} > 0.5$ as indicating strict preference of $i$ over $j$ with the strength of the preference reaching its maximum at $r_{ij} = 1$. Similarly, $r_{ij} < 0.5$ indicates a preference of $j$ over $i$ and $r_{ij} = r_{ji} = 0.5$ is indifference between the two. We hasten to add that the reciprocal fuzzy relations are by no means universally adopted by the scholarly community (see e.g. [Barrett et al. 1990]). Very little of what is being said in the following assumes reciprocity of the preference relations. Whenever this assumption is made, it will be pointed out.

## 2.1 Maximizing Representation

Consider now the concept of representation in the context of fuzzy individual preference relations. Voter $i$'s preference relation over candidates can be presented as:

$$\begin{array}{cccc} - & r^i_{12} & \dots & r^i_{1k} \\ r^i_{21} & - & \dots & r^i_{2k} \\ \dots & \dots & \dots & \dots \\ r^i_{k1} & r^i_{k2} & \dots & - \end{array}$$

Consider now voter $i$ and a committee $c_t$ consisting of $k$ candidates as required. We are now primarily interested in finding the members of $c_t$ that best represent $i$. Denote the set of these representatives by $B(i, c_t)$. Several plausible ways of finding the best representatives can be envisioned:

1. $B^i_{\text{sum}}(c_t) = \{j \in c_t | \sum_l r_{jl} \geq \sum_l r_{ql}, \forall q \in c_t\}$,
2. $B^i_{\text{min}}(c_t) = \{j \in c_t | \min_l \ r_{jl} \geq \min_l r_{ql}, \forall l \in K, \forall q \in c_t\}$,
3. $B^i_h(c_t) = \{j \in c_t | h(j) \geq h(q), \forall q \in c_t\}$ where $h(j) = p\,(\max_l r_{jl}) + (1 - p)(\min_l r_{jl})$,
4. $B^i_{\text{cop}}(c_t) = \{j \in c_t | cop(j) \geq cop(q), \forall q \in c_t\}$ where $cop(j) = |\{l \in c_t | r_{jl} > r_{lj}, \forall l \in K\}|$.

The first one determines the best representatives on the basis of the sums of the preference degrees obtained by candidates in all pairwise comparisons. This method is very much in the spirit of the Borda count. The second method looks at the minimum preference degree of each candidate when compared with all others and picks the candidate with the largest minimum. It is a variant of the min-max method in social choice theory. The third method is a version of Hurwicz's rule which maximizes the weighted sum of the smallest and largest preference degrees [Milnor 1954]. The fourth method is motivated by Copeland's rule in social choice theory. The Copeland winner is the candidate that defeats more candidates than any other candidate. In the setting of fuzzy preference relation $cop(j)$ is the number of candidates in $c_s$ that are less preferred to $j$ than $j$ is preferred to them. In reciprocal preference matrices, $cop(j)$ is simply the number of entries larger than 0.5 on row $j$.

Each of these methods singles out the best representatives of each voter in any given committee. Since each of the methods is based on a score, we can define a ranking of candidates in accordance with those scores. From the point of view of representation more important is, however, the ranking over committees ensuing from these methods. The most straightforward way to accomplish this is to define the score of committee $c_t$ as follows:

$$S_t = \sum_{i \in N} \sum_{a \in c_t} \sum_{j \in K} r^i_{aj} \tag{5}$$

Thus, the score of a committee is the sum of values given by voters to each of its members. The values, in turn, are the sums of preference degrees in all pairwise comparisons. This method is a variation of the Borda count. The most representative committee $RC^B$ would then be:

$$RC^B = \{c_i \in C^k | S_i \geq S_j, \forall c_j \in C^k\} \tag{6}$$

Although the Chamberlin-Courant approach is very close to the Borda count as well, the above method is not its most plausible fuzzy counterpart. Rather than summing the preference degrees over alternatives and voters, the Chamberlin-Courant approach sums the Borda scores of each voter's representative in any given committee. First we define

$$r_j^i = \sum_{q \in K} r_{jq}^i \tag{7}$$

Then, for each committee $c_t$ we define:

$$V_{it} = max_{j \in c_t} r_j^i \tag{8}$$

This can be viewed as the value of committee $c_t$ to voter $i$ as reflected by the value $i$ assigns to his or her representative in $c_t$.

Now, the most representative committee in the sense of Chamberlin-Courant is:

$$RC_{\text{sum}}^{CC} = \\ = \{c_j \in C^k \mid \sum_i V_{ij} \geq \sum_i V_{iq}, \forall c_q \in C^k, i \in N, j \in K\} \tag{9}$$

The $RC_{Sum}^{CC}$ committee thus defined is based on the summation of preference degrees in individual preference matrices. In analogous manner one can define the most representative committee in the min-max sense. Let $\underline{r_j^i} = min_{q \in K} r_{jq}$.

Now define, for each committee $c_t$ and each voter $i$:

$$V'_{it} = max_{j \in c_t} \underline{r_j^i} \tag{10}$$

Then the most representative committee in the min-max sense is:

$$RC_{\text{min}}^{\prime CC} = \\ = \{c_j \in C^k \mid \sum_i V'ij \geq \sum_i V'_{iq}, \forall c_q \in C^k\} \tag{11}$$

The $RC_{\text{min}}^{\prime CC}$ differs from the previous committee in using the min-max calculus to determine each voter's representative. In a way, $RC_{\text{min}}^{\prime CC}$ mixes two kinds of maximands: the "utilitarian"and "Rawlsian". The former maximizes the average utility, while the latter maximizes the utility of the worst-off individual [Rawls 1971].

A purely Rawlsian committee can also be envisioned. This is obtained as follows:

$$RC^R = \{c_j \in C^k \mid min_i V'_{ij} \geq min_i V'_{iq}, \forall c_q \in C^k\} \tag{12}$$

In a similar vein, one can define the Hurwicz and Copeland committees, $RC^H$ and $RC^{Co}$, respectively. For a fixed value of $p^i \in [0,1]$, let $r_j^{iH} = p^i(\max_q r_{jq}) + (1-p^i)(\min_q r_{jq})$ and $V_{it}^H = max_{j \in c_t} r_j^{iH}$. The set of most representative Hurwicz-type committees is, then:

$$RC^H = \{c_j \in C^k \mid \sum_i V_{ij}^H \geq V_{iq}^H, \forall c_q \in C^k\} \tag{13}$$

Note that the value $p^i$ is a voter specific measure of "optimism", i.e. the weight assigned to $\max_j r_{ij}^i$, the degree of preference assigned to each candidate in the comparison of its weakest competitor. Intuitively speaking the exclusive emphasis on the strongest and weakest pairwise comparisons is somewhat questionable in voting contexts.

To define the Copeland committee, let $RC^{Co}$, in turn, is based on the voters' value function $r_j^{iCo} = \mid \{q \in K \mid r_{jq} > r_{qj}\} \mid$ and the value function $V_{it}^{iCo} = max_{j \in c_t} r_j^{iCo}$. Now,

$$[RC^{Co} = \{c_j \in C^k \mid \sum_i V_{ji}^{iCo} \geq \sum_i V_{qi}^{iCO}, \forall c_q \in C^k\} \tag{14}$$

Of these four types of committees, the Rawlsian and Copeland types utilize the least amount of the voter preference information. The former looks at the minimal level preference of each candidate when compared with all others. The latter uses only the order information of preference degrees. Of course, if the aim is to economize on information usage, the very idea of resorting to fuzzy preference degrees loses much of its appeal.

## 2.2 Committees with Equal-Sized Constituencies

In the preceding we focused on maximally representative committees. Now we approach the committee design problem from the point of view of minimizing misrepresentation, as suggested by Monroe. Again we assume that we are given, for each voter, a matrix of fuzzy preference over all candidates. Our task is to form a committee that minimizes the misrepresentation of voters. It will be recalled that Monroe's procedure has the following elements:

1. every possible committee of $k$ members is considered,
2. for each committee, the voter set $N$ is partitioned into $k$ equal sized groups (constituencies), and
3. each voter is first assigned to the candidate whose election would be accompanied with the smallest degree of misrepresentation to the voter.

The third stage calls typically for some adjustments, i.e. transfers of voters from one candidate to another to obtain equal sized constituencies. With fuzzy preference relations, the first problem to be discussed is how to measure misrepresentation. Consider a situation where voter $i$ prefers one candidate, say $a_w$, to a maximum degree to any other candidate. This would be indicated in $i$'s preference matrix in row $w$ so that it would then consist of straight 1's in all non-diagonal columns. Obviously, then $\sum_j r^i_{wj} = k-1, j \neq w, j = 1,\ldots,k$. A natural measure of misrepresentation for $i$, if candidate $a_v$ is the sole member of the committee is $m_{iv} = k - 1 - \sum_j r^i_{vj}$. In multi-member committees, the same measure is applied to the candidate that represents $i$ in the committee. The best representative, in turn, can be determined as discussed in the preceding subsection.

In the following we shall assume that the $i$'s best representative in $c_t$ is determined as:

$$B^i_{\text{sum}}(c_t) = \{j \in c_t \mid \sum_l r^i_{jl} \geq \sum_l r^i_{ql}, \forall q \in c_t\} = \{j \in c_t \mid r^i_j \geq r^i_q, \forall q \in c_t\}. \tag{15}$$

Let $\max_{j \in c_t} r^i_j = g(i,t)$. The degree of misrepresentation of committee $c_t$ is then:

$$M_t = \sum_i (k-1) - g(i,t) = n(k-1) - \sum_i g(i,t) \tag{16}$$

However, Monroe suggests that the optimal committees be composed of candidates with equal-sized constituencies. This means that committees with the minimum value of $M_t$ are not, in general, acceptable since each committee member is not necessarily the best representative of an equal number of voters. Hence, voters have to be "transferred" from one candidate to another. As a criterion for transfers Monroe suggests that those voters who suffer least from being associated with another committee member be transferred first. For example, suppose that in committee $c_t$ voter 1's best representative is candidate $a_j$ and voter 2's best representative is $a_j$ as well. If $m_{1j} - m_{1l} > m_j - m_{2p}$ where $a_l$ and $a_p$ are the next best representatives of 1 and 2, respectively, then voter 2 is transferred before voter 1. Unfortunately, Monroe does not give full details of the transfer procedure, but, as was pointed out above, Potthoff and Brams have transformed the procedure into an integer programming problem.

It turns out that the Potthoff-Brams procedure can be applied to the fuzzy preference representation problem as well.[3]

The objective function is:

$$\min_t M_t = n(k-1) - \sum_i g(i,t) \tag{17}$$

The constraints, in turn, are exactly the same as those defined by Potthoff and Brams, i.e. (2 – (4).

## 3 Discussion

The problem of optimal representation under fuzzy preferences resembles the problem of electing representative assemblies under various electoral systems. A

[3] In fact, Potthof and Brams extend their analysis to several voting systems including approval voting.

voter's best representative might be one that "defeats" more contestants than any other candidate in the sense of having larger preference degrees in its favor in pairwise contests. This would amount ranking candidates according to their fuzzy Copeland scores. It is, however, questionable whether the notion of defeating has the same unambiguous meaning in individual fuzzy preference relations as in non-fuzzy preference tournaments, especially, if the fuzzy preference relation is non-reciprocal. For this reason, it may make more sense to consider the preference degrees in more detail in defining the degree of various candidates from a voter's point of view. The min-max calculus provides an alternative foundation for such a definition. Similarly, the Hurwicz-type representation calculus takes a closer look at the preference degrees. In our opinion, however, the sum-type definition of representation and misrepresentation is most appropriate to summarize the information contained in fuzzy individual preference relations. It is in the spirit of Borda count but takes advantage of the additional information provided by the degrees of preference.

The sum-type definition of individuals' best representatives has an additional advantage of fitting naturally together with the sum-type definition of a committee's degree of presentation or misrepresentation. We emphasize, however, that the sum-type definition of a committee's degree of representation or misrepresentation is compatible with any method used in aggregating individual fuzzy preference relations into a measure of how well various candidates represent the individual in question. Linear programming provides a useful tool for finding representative committees once the misrepresentation measure is given.

## References


[Barrett et al. 1990] C. R. Barrett, P.Pattanaik and M. Salles, On choosing rationally when preferences are fuzzy, *Fuzzy Sets and Systems.* **34** (1990) 197–212.

[Chamberlin and Courant 1983] J. Chamberlin and P. Courant, Representative deliberations and representative decisions: Proportional representation and the Borda rule, *The American Political Science Review* **77**(1983) 718–733.

[Milnor 1954] J. Milnor, Games against nature, in *Decision Processes*, eds. R. Thrall, C. Coombs and R. Davis, New York: Wiley 1954.

[Monroe 1995] B. Monroe, Fully proportional representation, *The American Political Science Review* **89**(1995) 925–940.

[Potthoff and Brams 1998] R. Potthoff and S. Brams (1998), Proportional representation: Broadening the options, *Journal of Theoretical Politics* **10** (1998) 147–178.

[Rawls 1971] J. Rawls, *A Theory of Justice*, Oxford: Oxford University Press 1971.

[Rogowski 1981] R. Rogowski, Representation in political theory and law, *Ethics* **91** (1981) 395–430.

# Possibility Based Modal Semantics for Graded Modifiers

Jorma K. Mattila

Lappeenranta University of Technology,
Department of Mathematics and Physics
P.O. Box 20, FIN-53851 Lappeenranta, Finland
jorma.mattila@lut.fi

**Abstract.** A brief introduction to basic modifiers is given. Any modifier with its dual and the corresponding negation form a DeMorgan triple similar to that of t-norms, t-conorms, and negation. The lattice structure of the unit interval with the usual partial order is similar to that of the set of all membership functions. This structure has a certain connection to implication, by means of the subsethood of fuzzy sets, and it is possible to create a similar expression for modifiers as the axiom of reflexivity is in modal logic. Also, other connections to modal logics can be found. This motivates to develop a formal semantics to modifier logic by means of that of modal logic. Actually, this kind of logic is so-called *metalogic* concerning either true or false statements about properties of modifiers. This version is based on graded possibility operations. Hence, semantic tools for weakening modifiers are derived. The corresponding things for substantiating modifiers are constructed by means of duality. Finally, some outlines for modifier systems are considered.

**Keywords:** Modality, Modifier, metalogic, Semantics of Modal Logic, Semantics of Modifier Logic, Modifier system.

## 1 Basic Modifiers

So-called *Basic modifiers* are modifiers having a certain effect from two possible properties. A basic modifier either substantiate or weaken the concept being its argument. For example, if the argument of a substantiating modifier is (as usually) a membership function, say $\mu$, then the modifier restricts $\mu$ in similar way at every point $x$ in the domain of the function. What this means, can be seen in Definition 1 below. We consider basic modifiers by giving the definitions and results as a short introduction to modifiers. We consider modifiers as operations in the set of membership functions

$$\mathbb{I}^X = \{\mu \mid \mu : X \longrightarrow \mathbb{I}\} \tag{1}$$

where $\mathbb{I}$ is the lattice $([0,1], \leq)$.

P. Melin et al. (Eds.): IFSA 2007, LNAI 4529, pp. 220–230, 2007.
© Springer-Verlag Berlin Heidelberg 2007

**Definition 1 (Modifier).** We say that a mapping $M : \mathbb{I}^X \to \mathbb{I}^X$ is

(*i*) a *substantiating modifier* if for any fuzzy set $\mu \in \mathbb{I}^X$,

$$\forall x \in X, \quad M(\mu(x)) \leq \mu(x), \tag{2}$$

(*ii*) a *weakening modifier* if for any $\mu \in \mathbb{I}^X$,

$$\forall x \in X, \quad \mu(x) \leq M(\mu(x)), \tag{3}$$

(*iii*) an *identity modifier* if for any $\mu \in \mathbb{I}^X$,

$$\forall x \in X, \quad M(\mu(x)) = \mu(x). \tag{4}$$

Identity modifiers are identity mappings on $\mathbb{I}^X$. They are sometimes needed as links between substantiating and weakening modifiers in some logical structures involving modifier symbols.

Because a modifier $M$ is a map from $\mathbb{I}^X$ to $\mathbb{I}^X$, it has the properties of membership functions. A modifier $M$ is associated with a fuzzy set $\mu$ by means of composition of $M$ and $\mu$ as follows. For any $x \in X$,

$$(M \circ \mu)(x) = M(\mu(x)). \tag{5}$$

A given modifier we can associate with the dual modifier according to the following

**Definition 2 (Dual Modifier).** Let $M$ and $M^*$ be modifiers. We say that $M^*$ is the *dual modifier* associated with $M$, if for any fuzzy set $\mu \in \mathbb{I}^X$,

$$\forall x \in X, \quad M^*(\mu(x)) = \eta(M(\eta(\mu(x)))), \tag{6}$$

where $\eta$ is a strong negation.

**Proposition 1.** *If $M$ is a substantiating modifier then its dual $M^*$ is a* weakening modifier *and vice versa.*

*Proof.* (See also [13].) Suppose $\mu \in \mathbb{I}^X$, and $M$ is a substantiating modifier. Thus $\forall x \in X, M(\mu(x)) \leq \mu(x)$. We have to show that $\forall x \in X, \mu(x) = M^*(\mu(x))$. Let $n$ be a strong negation function. Thus $\forall x \in X,\ M^*((x)) = \eta(M(\eta(\mu(x))))$. Clearly

$$M(\eta(\mu(x))) \leq \eta(\mu(x))$$

by Def.1. From this it follows by the properties of membership functions that

$$\forall x \in X,\ \eta(M(\eta(\mu(x)))) \geq \eta(\eta(\mu(x))) = \mu(x),$$

i.e.,

$$M^*(\mu(x)) \geq \mu(x). \tag{7}$$

Conversely, the result follows in the similar way. □

The condition

$$\forall x \in X, \quad M^*(\mu(x)) = \eta(M(\eta(\mu(x)))) \tag{8}$$

in the previous proof says that the operators $M$, $M^*$, and $\eta$ satisfy DeMorgan's law. Thus dual pairs of modifiers with strong negation form classes called DeMorgan triples of operators ([2]). Originally, a DeMorgan triple consists of a t-norm, corresponding t-conorm, and negation, but the triples involving any modifier, its dual, and negation have the same structure by Definitions 1, 2, and Proposition 1.

We denote $\alpha$-level set of a fuzzy set $\mu$, as usually,

$$\mu_\alpha = \{x \in X \mid \mu(x) \geq \alpha, \ \alpha \in \mathbb{I}\}$$

Thus the $\alpha$-level set of $M(\mu)$ is

$$(M \circ \mu)_\alpha = \{x \in X \mid M(\mu(x)) \geq \alpha, \ \alpha \in \mathbb{I}\}. \tag{9}$$

It is easy to see that modifiers have following properties. Suppose $M$ is a substantiating modifier. Then we have

$$M(\mathbf{0}_X) = \mathbf{0}_X, \tag{10}$$
$$M^*(\mathbf{1}_X) = \mathbf{1}_X, \tag{11}$$
$$(M^*)^*(\mu(x)) = M(\mu(x)), \tag{12}$$

where $\mathbf{0_X}$ and $\mathbf{1}_X$ are the constant functions $\mathbf{0}_X(x) = 0$ and $\mathbf{1}_X(x) = 1$ for all $x \in X$.

The lattice structure of $\mathbb{I}$ gives the natural ordering to a set of modifiers, similar to that of any set of membership functions. This is an implication of the well known fact that the lattice properties of $\mathbb{I}$ can be embedded pointwise to the lattice $\mathbb{I}^X$. Hence, because $\mathbb{I}$ is a distributive and complete lattice, so is $\mathbb{I}^X$, too. The partial ordering relation "$\leq$" involves the principle of implication. This means that if $M$ is a substantiating modifier then for any $x \in X$ and $\mu \in \mathbb{I}^X$, $M(\mu(x)) \leq \mu(x)$, by Definition 1. In fuzzy set theory, this means that a fuzzy set $M(\mu)$ is a subset of the fuzzy set $\mu$ (cf. Zadeh [16]). Hence, if we think that $M$ is the linguistic label of the formal modifier $M$ , i.e., a suitable hedge corresponding to the formal $M$, and $\mu$ is the linguistic interpretation of the formal fuzzy set $\mu$, then the statement

$$M(\mu) \to \mu \tag{13}$$

is an implication being true in any situation where for any $x \in X$ and $\mu \in \mathbb{I}^X$, $M(\mu(x)) \leq \mu(x)$. This means that the statement

$$\mu \to M^*(\mu) \tag{14}$$

is true in some possible worlds. In the same way, we can consider that the statement $\mu \to M^*(\mu)$ is true in the same possible worlds, by Definition 2 and

Proposition 1. If we consider $M$ to be a necessity operator and $M^*$ a possibility operator, then the formula (13) represents the *axiom of reflexivity* in systems where the primitive modal operator is necessity and (14) in modal systems where the primitive modal operator is possibility.

When we consider modifiers as logical operators, we have to notice that they are not truth-functional. This means that the truth values of modified statements do not depend only of the truth value distributions of the primitive symbols in one world, but also of the accessibility to other possible worlds. This is an essential presupposition for the possibility to interpret modifiers as modalities. We will see this in the considerations below.

## 2 Modifier Language

First, consider some fundamental things from history and some motivating things. Lemmon [8], p. 20 - 21, describes Leibniz' basic ideas for motivating the idea of "classical modal logics". He says: "Leibniz' suggestion now becomes: a sentence is necessarily true (in this world) iff that sentence is true in all worlds alternative to this world." See Lemmon [8], p. 20 to check the detailed analysis about what the concept "alternative to this world" means. Lemmon [8], p. 21, continues: "Actually, in many connections it is intuitively simpler to think of world $t$ as accessible from world $u$ rather than alternative to $u$. This at least has the merit of avoiding the temptation to suppose that alternativeness is a symmetric relation between worlds – that if $t$ is alternative to $u$, then $u$ must be alternative to $t$. Indeed, we shall not assume that each world is accessible from itself, or even that to each world there is at least one accessible world: there may be accessibility-isolated worlds. We shall find that to many such assumptions about the accessibility relation between worlds there correspond distinctive modal sentences which come out valid precisely because we have made those assumptions. If necessity means truth in all accessible worlds, then possibility will mean truth in some accessible world. Thus our remarks about the vagueness of the notion of necessity, and the various more precise accounts of it, may be repeated mutantis mutandis for the notion of possibility."

Traditional modal operators are actually modifiers that substantiate or weaken statements or their parts being arguments of the operators. For example, the concepts *possible* and *necessary* make indefinite (i.e. weaken) and make necessary (i.e. substantiate) expressions associated with them, respectively. Thus modal logics can be viewed to be logics of *modifiers*. We define modifiers formally by means of relational structures, or Kripke frames, when they can be interpreted in different ways. We restrict our consideration only to the cases where modifiers are unary operators.

*Propositional modifier language* $\mathcal{L}_{Mod}$ is a language of graded modalities, that can be generally defined by giving its alphabet and rules of formation of formulas as follows.

**Definition 3.** The *alphabet* of a language $\mathcal{L}_{Mod}$ consist of:

(i) propositional letters $p$, $q$, $\ldots (\in \mathit{Prop})$;
(ii) binary truth-functional connectives;
(iii) zero-placed operator $\bot$ (fixes the value *false*);
(iv) non-truth-functional sentence operators $\boldsymbol{F}_1, \boldsymbol{F}_2, \ldots (\in \mathbb{O})$.

**Definition 4.** *Well-formed formulas* (*wff's* or *formulas*, for short) of a language $\mathcal{L}_{Mod}$ are as follows:

(i) propositional letters $p$, $q$, $\ldots (\in \mathit{Prop})$ are formulas;
(ii) $\bot$ is a formula;
(iii) if $\alpha$ and $\beta$ are formulas, and $\blacklozenge$ is any binary truth-functional connective, then $\alpha \blacklozenge \beta$ is a formula;
(iv) if $\alpha$ is a formula, and $\boldsymbol{F}_i \in \mathbb{O}$ then $\boldsymbol{F}_i(\alpha)$ is a formula $(i = 1, 2, \ldots)$.

Formulas of the form $\alpha \to \bot$ are usually abbreviated by $\neg\alpha$.

## 3 Graded Possibility Based Modal Semantics

We consider a relation system

$$\mathbf{R} = \bigcup_{i=0}^{k} \mathbf{R}_i \tag{15}$$

of binary subrelations of a given binary relation $\mathbf{R}$. It is the central part of relational structures we are considering when constructing frames and models for logics of graded modifiers. The relations may have any properties as relations usually have. For example, they are reflexive, symmetric, transitive etc. The only thing we presuppose is that the relations are at least reflexive. Then we can generate systems being similar to aletic modal systems.

**Definition 5.** *frame*, or *relational structure*, is an ordered queue

$$\mathcal{F} = \langle W, \mathbf{R}_0, \mathbf{R}_1, \mathbf{R}_2, \ldots, \mathbf{R}_k \rangle \tag{16}$$

where $W \neq \emptyset$ is a set and $\mathbf{R}_i$ is a binary relation, $i = 1, \ldots, k$. Elements $w$ of $W$ are *points*, *states*, or *possible worlds*.

A *model* corresponding to the frame $\mathcal{F}$ is an ordered queue

$$\mathcal{M} = \langle W, \mathbf{R}_0, \mathbf{R}_1, \mathbf{R}_2, \ldots, \mathbf{R}_k, V \rangle \tag{17}$$

where $V : \mathit{Prop} \to \mathcal{P}(W)$ is a *valuation*.

For every world $w \in W$, $V$ determines a set of those propositional letters, which are true in $w$, namely the set

$$P_w = \{p \mid w \in V(p)\}. \tag{18}$$

The number of relations $k$ may be finite or denumerably infinite.

Every possible world $w$ is a maximal consistent set of formulas, i.e. $\alpha \to \beta \in w$ iff $\alpha \notin w$ or $\beta \in w$, and $\alpha \in w$ iff $\alpha \to \bot \notin w$, i.e. $\neg\alpha \notin w$. Thus a set of worlds $W$ is a set of maximal consistent sets of formulas. We can now define modifiers by means of relations $\mathbf{R}_i$ $(i = 0, 1, 2, \ldots, k)$ of a Kripke frame as follows.

**Definition 6.** Let $\mathcal{F} = \langle W, \mathbf{R}_0, \mathbf{R}_1, \mathbf{R}_2, \ldots, \mathbf{R}_k \rangle$ be a frame, $\alpha$ a formula of $\mathcal{L}_{Mod}$, and $w, t \in W$. Then for every value of the index $i = 0, 1, 2, \ldots, k$, $\boldsymbol{H}_i$ is a *possibility operator*, (i.e., a weakening modifier) iff the condition

$$w\mathbf{R}_i t \Leftrightarrow \{\boldsymbol{H}_i(\alpha) \mid \alpha \in t\} \subseteq w \tag{19}$$

holds, i.e. $w\mathbf{R}_i t$ holds iff for every formula $\alpha \in t$, for which $\boldsymbol{H}_i(\alpha) \in w$. Then we say that $\boldsymbol{H}_i$ is a *possibility operator determined by* $\mathbf{R}_i$. A set of relations $\{\mathbf{R}_0, \mathbf{R}_1, \mathbf{R}_2, \ldots, \mathbf{R}_k\}$ is indicated by $S_{\mathbf{R}}$.

In the sequel, we use the name "modifier" for the modal operators.

**Definition 7.** The *dual* $\boldsymbol{H}_i^*$ of a modifier $\boldsymbol{H}_i$ is defined by the condition

$$\boldsymbol{H}_i^*(\alpha) \equiv \neg\boldsymbol{H}_i(\neg\alpha) \tag{20}$$

($\equiv$ is a symbol of metalanguage presenting equivalence relation between two formulas.)

A set of duals of modifiers defined in Def. 6 is indicated by $\mathbb{O}^*$, i.e.

$$\mathbb{O}^* = \{\boldsymbol{H}_0^*, \boldsymbol{H}_1^*, \ldots, \boldsymbol{H}_k^*\}. \tag{21}$$

Elements of the set (21) are *weakening modifiers*, and we say that $\boldsymbol{H}_i^*$ is a *substantiating modifier determined by* $\mathbf{R}_i$.

**Proposition 2.** *Let* $\mathbf{R}_i \in S_{\mathbf{R}}$, $\boldsymbol{H}_i \in \mathbb{O}$, $(i = 0, 1, \ldots, k)$, *and* $\alpha$ *a formula of* $\mathcal{L}_{Mod}$. *Then the formula*

$$w\mathbf{R}_i t \Leftrightarrow \{\alpha \mid \boldsymbol{F}_i(\alpha) \in w\} \subseteq t \tag{22}$$

*holds.*

*Proof.* Suppose $w\mathbf{R}_i t$, $\boldsymbol{F}_i(\alpha) \in w$, and $\alpha \notin t$. Then $\neg\alpha \in t$, because $t$ is maximal consistent, and further, $\boldsymbol{H}_i(\neg\alpha) \in w$ by (19). From this the condition $\neg\boldsymbol{F}_i(\alpha) \in w$ follows by Def. 7. But this is in contradiction to the supposition, because $w$ is maximally consistent. Thus $w\mathbf{R}_i t \Rightarrow \{\alpha \mid \boldsymbol{F}_i(\alpha) \in w\} \subseteq t$.

In the converse case, suppose $\boldsymbol{F}_i(\alpha) \in w$ for all formulas $\alpha \in t$. From this it follows that $\neg\boldsymbol{F}_i(\alpha) \notin w$ by maximal consistency of $w$, i.e. $\boldsymbol{H}_i(\neg\alpha) \notin w$ by Def. 7, and also $\neg\alpha \notin t$ by maximal consistency of $t$. Thus $\boldsymbol{H}_i(\neg\alpha) \notin w \Rightarrow \neg\alpha \in t$, from which the implication $\alpha \in t \Rightarrow \boldsymbol{H}_i(\alpha) \in w$ by contraposition. Thus $\{\boldsymbol{H}_i(\alpha) \mid \alpha \in t\} \subseteq w$, and thus $w\mathbf{R}_i t$ by (19). This completes the proof.

The truth status of a formula in a given world $w$ follows the principle, according to which $w$ is closed under connectives.

**Definition 8.** A truth value of a formula $\alpha$ of $\mathcal{L}_{Mod}$ in a given world $w$ of a model $\mathcal{M} = \langle W, \mathbf{R}_0, \mathbf{R}_1, \mathbf{R}_2 \ldots, \mathbf{R}_k, V\rangle$ (denote $\models_w^{\mathcal{M}} \alpha$) is defined recursively as follows:

(i) If $p \in Prop$ then $\models_w^{\mathcal{M}} p$ iff $V(p)$.

If $\alpha$ and $\beta$ are formulas and $\boldsymbol{F}_i \in \mathbb{O}$ $(i = 0, 1, 2, \ldots, k)$ then

(ii) $\models_w^{\mathcal{M}} \neg\alpha$ iff $\not\models_w^{\mathcal{M}} \alpha$;
(iii) $\models_w^{\mathcal{M}} \alpha \to \beta$ iff $\not\models_w^{\mathcal{M}} \alpha$, or $\models_w^{\mathcal{M}} \beta$;
(iv) $\models_w^{\mathcal{M}} \boldsymbol{F}_i(\alpha)$ iff for every world $t$, such that if $w\mathbf{R}_i t$ then $\models_t^{\mathcal{M}} \alpha$.

A truth condition for the dual $\boldsymbol{F}_i^*$ of the operator $\boldsymbol{F}_i \in \mathbb{O}(i = 0, 1, 2, \ldots, k)$ can be derived by means of Def. 8 (iv). Let $\alpha$ be a formula of $\mathcal{L}_{Mod}$, then

(v) $\models_w^{\mathcal{M}} \boldsymbol{F}_i^*(\alpha)$ iff there exists $t \in W$, such that $w\mathbf{R}_i t$ and $\models_t^{\mathcal{M}} \alpha$.

A line of argument for the case (v) is analogous to that for $\Diamond$ in Chellas' book [1] on page 7.

For every formula $\alpha$ in $\mathcal{L}_{Mod}$ and for every model $\mathcal{M}$ there exists a set

$$\|\alpha\|^{\mathcal{M}} = \{w \in W \mid \models_w^{\mathcal{M}} \alpha\}. \tag{23}$$

This set is called a *truth set of* $\alpha$ in a model $\mathcal{M}$.

Suppose that a frame $\mathcal{F} = \langle W, \mathbf{R}_0, \mathbf{R}_1, \mathbf{R}_2 \ldots, \mathbf{R}_k\rangle$ is given. We say that a formula $\alpha$ in $\mathcal{L}_{Mod}$ is *valid in* $\mathcal{F}$, denote $\models^{\mathcal{F}} \alpha$ iff $\models_w^{\mathcal{M}} \alpha$ holds in all models $\mathcal{M}$ of $\mathcal{F}$, and $w \in W$. A formula $\alpha$ is *satisfiable in* $\mathcal{F}$ iff $\models_w^{\mathcal{M}} \alpha$ holds in some models $\mathcal{M}$ of $\mathcal{F}$, and $w \in W$. A formula $\alpha$ is *valid* iff $\alpha$ is valin in every $\mathcal{F}$, and *satisfiable* iff $\alpha$ is satisfiable in some $\mathcal{F}$. Equivalently, we can say that $\alpha$ is *valid* iff $\models_w^{\mathcal{M}} \alpha$ holds in all models $\mathcal{M} = \langle W, \mathbf{R}_0, \mathbf{R}_1, \mathbf{R}_2 \ldots, \mathbf{R}_k, V\rangle$, and $w \in W$. It is also very often natural to use the concept "validity of a formula $\alpha$ in a model $\mathcal{M}$", denote $\models^{\mathcal{M}} \alpha$. We say that $\alpha$ is *valid in* $\mathcal{M}$ iff $\models_w^{\mathcal{M}} \alpha$ holds in all $w \in W$. Clearly, $\models^{\mathcal{F}} \alpha$ iff $\models^{\mathcal{M}} \alpha$ for every $\mathcal{M}$ in $\mathcal{F}$, and $\models \alpha$ iff $\models^{\mathcal{M}} \alpha$ for every $\mathcal{M}$.

**Definition 9.** *Semantic entailment* $\alpha \models \beta$ of formulas $\alpha$ and $\beta$ in $\mathcal{L}_{Mod}$ is defined by the condition

$$\alpha \models \beta \text{ iff } \models \alpha \to \beta. \tag{Ent.}$$

Semantic entailment has a special property concerning modifier symbols. It can be proved that the following thing holds:

$$\boldsymbol{H}(\alpha) \models \beta \text{ iff } \models \alpha \to \beta, \tag{Ent$_{WEAK}$}$$

where $\boldsymbol{H}$ is a weakening modifier symbol. The corresponding result for any substantiating modifier symbol is

$$\alpha \models \boldsymbol{F}(\beta) \text{ iff } \models \alpha \to \beta, \tag{Ent$_{SUBST}$}$$

where $\boldsymbol{F}$ is any substantiating modifier symbol.

## 4 Towards Modifier Systems

We consider here some preliminary things we need in creating modifier systems using the semantics considered in Sec. 3. We will have different systems by giving different properties to relations of frames $\mathcal{F} = \langle W, \mathbf{R}_0, \mathbf{R}_1, \mathbf{R}_2 \ldots, \mathbf{R}_k \rangle$. Especially, we fix the relation $\mathbf{R}_0$ for a certain purpose. We give the property

$$w\mathbf{R}_0 t \Leftrightarrow w = t \text{ for all } w, t \in W \tag{24}$$

to it. In this case, the equivalence

$$w\mathbf{R}_0 t \Leftrightarrow \{\alpha \mid \boldsymbol{F}_0(\alpha) \in w\} \subseteq w \Leftrightarrow \{\boldsymbol{F}_0^*(\alpha) \mid \alpha \in t\} \subseteq w \tag{25}$$

follows from the formulas (19) and (22). From this it follows that for such formulas $\alpha$ belonging to the set $\{\alpha \mid \boldsymbol{F}_0(\alpha) \in w\}$, the condition

$$\alpha \in w \Leftrightarrow \boldsymbol{F}_0^*(\alpha) \in w \tag{26}$$

holds, and thus $\|\alpha\|^{\mathcal{M}} = \|\boldsymbol{F}_0^*(\alpha)\|^{\mathcal{M}}$. From the corollary (v) of Def. 8 (vi) we have the condition

(vi) $\models_w^{\mathcal{M}} \boldsymbol{F}_0^*(\alpha) \Leftrightarrow \models_w^{\mathcal{M}} \alpha$

by (24), and this leads to the identity $\|\alpha\|^{\mathcal{M}} = \|\boldsymbol{F}_0^*(\alpha)\|^{\mathcal{M}}$. On the other hand, from Def. 8 (iv) the condition

$$\models_w^{\mathcal{M}} \boldsymbol{F}_0(\alpha) \Leftrightarrow \models_w^{\mathcal{M}} \alpha$$

follows by (24). From this it follows that $\|\alpha\|^{\mathcal{M}} = \|\boldsymbol{F}_0(\alpha)\|^{\mathcal{M}}$. By means of these identities the result

$$\|\boldsymbol{F}_0^*(\alpha)\|^{\mathcal{M}} = \|\boldsymbol{F}_0(\alpha)\|^{\mathcal{M}}. \tag{27}$$

This result can be interpreted by the conclusion that the operator $\boldsymbol{F}_0$ is its own dual. It can be thought that the formulas $\alpha$, $\boldsymbol{F}_0(\alpha)$ and $\boldsymbol{F}_0^*(\alpha)$, whose truth sets are identical, express in each world exactly same states of affairs, and, in this point of view, they are equivalent. When we define the equivalence of two formulas in this way, then especially from the equivalence

$$\alpha \equiv \boldsymbol{F}_0^*(\alpha) \tag{28}$$

we get a result concerning the selfduality of $\boldsymbol{F}_0$ as follows:

$$\begin{aligned} \boldsymbol{F}_0(\alpha) &\overset{(28)}{\equiv} \boldsymbol{F}_0(\boldsymbol{F}_0^*(\alpha)) \equiv \boldsymbol{F}_0(\neg \boldsymbol{F}_0(\neg\alpha)) \equiv \neg \boldsymbol{F}_0^*(\boldsymbol{F}_0(\neg\alpha)) \\ &\overset{(28)}{\equiv} \neg \boldsymbol{F}_0(\neg\alpha) \equiv \boldsymbol{F}_0^*(\alpha). \end{aligned}$$

From this it also follows that the sets in (25) are idetical, i.e.

$$\{\alpha \mid \boldsymbol{F}_0(\alpha) \in w\} = \{\boldsymbol{F}_0^*(\alpha) \mid \alpha \in t\}.$$

Thus the operator $\boldsymbol{F}_0$ is an *identity operator* denoted often by $\Im$.

Next, we consider some additions to the formal semantics considered above. The purpose for this is to give some preliminary things for creating modifier systems, i.e. for choosing a set of axiom scheme and sets of inference rules. These additions concern mainly properties of relations of frames. All the systems use identity operator, and thus the relation $\mathbf{R}_0$ existing in the frames satisfies the condition (24) and determines an identity operator by means of considerations above basing on (19) and (22). The validity of used axioms has been proved in Mattila [11]. Every possible world $w \in W$ mentioned in frames is closed under modus ponens, and thus modus ponens

$$\alpha \to \beta,\ \alpha \vdash_S \beta \tag{MP}$$

appears as an inference rule in every systems. The subscript $S$ in the symbol $\vdash_S$ refers to the system under consideration. In addition to this, according to the basic semantics considered above the (semantical) *rule of possibility* is as follows. Let $\mathbf{H}_i$ be any weakening modifier operator, then

$$\models \boldsymbol{H}_i(\alpha) \Rightarrow \models \alpha \tag{RP$_{SEM}$}$$

holds for all $\boldsymbol{H}_i \in \mathbb{O}^*$, and thus its syntactical counterpart

$$\vdash_S \boldsymbol{H}_i(\alpha) \Rightarrow \vdash_S \alpha \tag{RP}$$

is used as a second inference rule. The duals of the rule of possibility (RP$_{SEM}$) and its syntactical counterpart (RP) are *rule of substantiation*

$$\models \alpha \Rightarrow \models \boldsymbol{F}_i(\alpha) \tag{RS$_{SEM}$}$$

that holds for all $\boldsymbol{F}_i \in \mathbb{O}$, and its syntactical counterpart is

$$\vdash_S \alpha \Rightarrow \vdash_S \boldsymbol{F}_i(\alpha). \tag{RS}$$

The rules (RS$_{SEM}$) and (RS) follow from the rules (RP$_{SEM}$) and (RP), respectively, by the principle of duality. The rule (RS) is a generalization of the *rule of necessitation* (RN) in usual modal logic.

## 5 Concluding Remarks

Modifier logics based on graded modalities are two-valued logics concerning statements about properties and relations of modifiers where the statements are either true or false. Hence, modifier logic can be considered to be so-called *metalogic* considering things in metalevel. Hence, metalogic of basic modifiers is not classical logic but logic of graded modal operators. This means that it is not truth-functional, because modifier operators in modifier logic are not truth-functional.

From the definitions and results in Section 1 concerning modifiers, it seems to be possible to create an alternative formal semantics for aletic modal logic using the structure of modifiers. Some hints to this direction exist already in above considered things.

The set of relations

$$S_{\mathbf{R}} = \{\mathbf{R}_0, \mathbf{R}_1, \mathbf{R}_2, \ldots, \mathbf{R}_k\}$$

is naturally partially ordered by the ordering "$\subseteq$". If the ordering is linear, then we have a nested queue

$$\mathbf{R}_0 \subseteq \mathbf{R}_1 \subseteq \mathbf{R}_2 \subseteq \ldots \subseteq \mathbf{R}_k, \tag{29}$$

where it is possible that $\mathbf{R}_k = \mathbf{R}$, by 15 (or otherwise, $\mathbf{R}_k \subset \mathbf{R}$). If the condition (29) holds with $\mathbf{R}_k = \mathbf{R}$, then it is quite easy to create modifier systems where all the modifiers can be linearly ordered according to the strength of the modifier. Actually, the strength of the modifiers of a system can be defined by means of a nested relation system.

In any case, it is possible to create modifier systems with completeness results similar to graded systems **T**, **S4**, or **S5**, if the relations are reflexive, reflexive and symmetric, or reflexive, symmetric and transitive, respectively. In the nested case, see for example [9] and [12].

Instead of possibility operations, it is also possible to create similar semantics for modifiers using necessity operators, i.e., substantiating modifiers. The basic starting point to that approach is the same as here (cf. [12]).

## References


1. B. F. Chellas, *Modal Logic: An Introduction*, Cambridge University Press, Cambridge, 1980.
2. J. Dombi, *A fuzzy halmazok operátorainak szerkezete a többtényezős döntések szempontjából*, Kandidátusi Értekezés, Hungarian Academy of Sciences, Szeged, 1993.
3. J. Dombi, Theoretical Concepts of Modifiers and Hedges, in: V.A. Niskanen, Jari Kortelainen (eds.), *On the Edge of Fuzziness*, Acta Universitatis Lappeenrantaensis 179, Lappeenranta, 2004.
4. B. Habson, P. Gärdenfors, A guide to intensional semantics, in: *Modality, Morality and Other Problems of Sense and Nonsense: Essays Dedicated to Sören Halldén*, CWK Gleerup Bokförlag, Lund, 1973.
5. G. E. Hughes, M. J. Cresswell, *An Introduction to Modal Logic*, Methuen and Co., reprint, 1985.
6. S. A. Kripke, Semantical considerations on modal logic, *Acta Philosophica Fennica*, Fasc. XVI, 1963.
7. G. Lakoff, Hedges: A study in meaning criteria and the logic of fuzzy concepts, *The Journal of Philosophical Logic*, 2, 1973.
8. E. J. Lemmon, *An Introduction to Modal Logic*, American Philosophical Quarterly, Monograph No. 11 (K. Segerberg (ed.)), Oxford, 1977.
9. J. K. Mattila, Modeling fuzziness by Kripke structures, in: T. Terano, M. Sugeno, M. Mukaidono, K. Shigemasu (eds.), *Fuzzy Engineering toward Human Friendly Systems*, Vol. 2, Proc. of IFES '91, Nov. 13 - 15, 1991, Yokohama, Japan.

10. J. K. Mattila, On modifier logic, in: L. A. Zadeh, J. Kacprzyk (eds.), *Fuzzy Logic for Management of Uncertainty*, John Wiley & Sons, Inc., New York, 1992.
11. J. K. Mattila, Completeness of propositional modifier logic, Research Report 55, Lappeenranta University of Technology, Department of Information Technology, 1995
12. J. K. Mattila, Modifier Logics Based on Graded Modalities, *Journal of Advanced Computational Intelligence and Intelligent Informatics*, Vol. 7 No. 2, 2003
13. J. K. Mattila, Modifiers Based on Some t-norms in Fuzzy Logic, *Soft Computing*, Springer-Verlag, DOI 10.1007/soo5oo-003-0323-x, 2003 (online), 2004 (paper).
14. J.K. Mattila, On Łukasiewicz Modifier Logic, *Journal of Advanced Computational Intelligence and Intelligent Informatics*, Vol. 9 No. 5, 2005
15. G. Resconi, G. J. Klir, U. H. St. Clair, Uncertainty and modal logic, *Proceedings of Fifth International Fuzzy Systems Association World Congress*'93, Vol. I, July 4 - 9, 1993, Seoul, Korea, 1993
16. L.A. Zadeh, Fuzzy Sets, *Information and Control*, 8, 1965.

# New Perspective for Structural Learning Method of Neural Networks

Junzo Watada

Waseda University
Graduate School of Information, Production & Systems
2-7 Hibikino, Wakamatsu, Kitakyushu 808-0135 Japan
junzow@osb.att.ne.jp

**Abstract.** A neural network is developed to mimic a human brain. The neural network consists of units and links that connect between units. Various types of neural networks are categorized into two classes: (1) back-propagation hierarchical neural network and (2) mutual-connected neural network. Generally speaking, it is hard to fix the number of units to build a neural network for solving problems. So the number of units is decided on the basis of experts' experience.

In this paper, we explain a learning method how to decide the structure of a neural network for problems. The learning method is named structural learning. Even if we give a sufficient number of units, the optimal structure will be decided in the process of learning.

The objective of the paper is to explain the structural learning of both hierarchical and mutual connecting neural networks. Both networks obtained and showed the sufficiently good results. In the stock forecast by a general neural network, the operation and the system cost are very large because a lot of numbers of hidden layer units in the network are used. This research tried the optimization of the network by the structured learning, and evaluated the practicality. ...

## 1 Introduction

A neural network is developed to mimic a human brain. The neural network consists of units and links that connect between units. Many types of neural networks are classified into two categories: (1) a back-propagation hierarchical neural network and (2) a connectionist system of a neural network. Generally speaking, it is hard to fix the number of units building a neural network for solving some problem. So the number of units is decided on the basis of experts' experience.

In this paper, a learning method is employed to decide the structure of a neural network in solving a problem. The learning method is named structural learning. Even if we give a sufficient number of units, the optimal structure will be decided in the process of learning.

The author is working on the structural learning in 1990s. First he worked on a hierarchical neural network. He obtained the optimal structure of the hierarchical neural network to solve EXOR logical calculation and classification of iris.

P. Melin et al. (Eds.): IFSA 2007, LNAI 4529, pp. 231–240, 2007.
© Springer-Verlag Berlin Heidelberg 2007

Recently he apply the method to solve the forecasting of the price movement of a stock using a hierarchical neural network. Also he obtained the optimal structure to solve portfolio problems.

Generally, a neural network spends much computation time and cost in forecasting the value and movement of a stock. The reason is because a neural network requires exponential computation time according to the number of units in a hidden layer in pursuing it. In the paper, the hidden layer is optimally structured by structurally learning a neural network. The results enable us to reduce the computational time and cost as well as to understand the structure much easily.

The objective of the paper is to explain the structural learning of both hierarchical and mutual connecting neural networks. Both networks obtained and showed the sufficiently good results. In the stock forecast by a general neural network, the operation and the system cost are very large because a lot of numbers of hidden layer units in the network are used. This research tried the optimization of the network by the structured learning, and evaluated the practicality.

## 2 Structural Learning of Hierarchical Neural Network

Recently, the rapid development of broad-band network accelerates investment on a stock market through inter-networks. Especially, individual investments with small amount are widely prevailed. The survey in 2005 by Kyodo News Service reported that the rate of individual investors is more than 30% in three stock Exchanges in Tokyo, Osaka and Nagoya. The forecast of stock prices has a very high demands in individual investors.

The researches on forecasting stock prices has been pursued for many years. The methods of the forecasting can be classified into two groups; fundamental analysis and technical analysis [4].

Fundaments mean economical basic circumstance for stocks and include economic macro parameters and such micro indices as financial situations of individual companies. Analysis depending on such fundamentals is named fundamentals analysis. The forecasting of each stock price can be achieved by surveying the achievements of many individual companies that are objective to be invested and provides an economical environment The technical analysis stands on that all stock should be a function of a price and time. Therefore, the future of a stock price is analyzed using moving average and stock chart.

In this paper, we build the structural learning of a neural network and employ this model to forecast a stock price. This analysis can be classified into technical analysis.

Generally, a neural network is built using sufficient number of units. This redundancy makes the neural network enable to evaluate the value using huge number of data. On the other hand, the redundancy requires to spend much more computation time and building cost of a system. As well, the complex structure bothers the transparency of a structure.

In this paper, previously employing the structural learning of a neural network and obtaining the optimal structure, it enables us to decrease the computation time and cost keeping the precision of the forecast.

### 2.1 Neural Network of Stock Forecasting

Stock data are time-series data. General neural network cannot deal with a time-series data in the sense of time-series input. It is necessary to input data in a neural network multi-dimensionally or to construct a new structure that can deal with time-series data well. The representative neural network to treat data in time-series is a concurrent neural network. The recurrent neural network is a hierarchical neural network that has feedback of some output from hidden units to input units in order to have a time-series input, although a general hierarchical neural network has the flow of information from input to output. There are Elman type of feedback from hidden units to input and Jordan type of feedback from output [3], [5].

Elman type neural of network feeds back the output from units in a hidden layer and Jordan type of neural network the output from units in an output layer. The information from these units is feedback to the flamed portion. The flamed portion is named a context layer. On the case of the Eman Type of a neural network, a unit in a hidden layer is one to one corresponding to a unit in a context layer. Both the numbers in a hidden layer and context layer are the same. On the case of the Jordan type of a neural network, the number of units in the context layer is the same number of units in the output layer. Therefore, the number of units in the context layer is constrained by the number of units in the output layer defined by the considered problem. On the other hand, the number of units feedback in the Jordan type of a neural network has no constraint depending on the number of units in the output layer, the number of units in the constraint layer can be freely determined. In this research, the Eleman type of a neural network is employed so as to optimize the number of units in the hidden layer by the structural learning.

### 2.2 Structural Learning

In the case of making a hierarchical neural network learn teaching data through buck-propagation method, the numbers of input units and output units are uniquely decided depending on their required and numbers of teaching input and output, respectively. On the other hand, the number of hidden units are depending on a learning method as well as the numbers of the input units and output units. There should be the minimal number of the hidden units for obtaining the results. Generally, such a number is not known previously. Conventionally, the number of units in the hidden layer is decided depending on experiences. On this case, if we can decrease the number of units, the computation speed and system cost can be saved. Also, we experience that a general type of buck-propagation method has much dependence on the initial setting and is hard to forecast an expected value without convergence even if we select the approximately minimal number of units.

In order to overcome such a problem, there are various trials that a structure of a network is changed recursively and gradually to result in reaching to its optimal structure [2], [7]. This process is named structural learning. Such trials of structural learning can be roughly classified into two groups. One group is a generating learning method to start from the small structure of a neural network and expanding recursively and gradually the structure of the neural network up to the optimal one. The other group is an eliminating learning method to start from the sufficiently large structure of a neural network to shrink recursively and gradually the structure of the neural network until the optimal one.

**Generating Learning Method.** The genetic learning method has the following features. On the starting stage, as the structure of a neural network is small, it spends smaller total computation time than the eliminating learning method. Generally, since the learned portion is frozen without being included in the learning process. it is rather rare to stop at the local minimum. Nevertheless, as the partial optimizations are repeated, an unnatural structure is sometimes created. In the paper, we employed the eliminating learning method for structural learning.

**Eliminating Learning Method.** The eliminating learning method requires a little bit large computation time because of starting a large network. The learning can be pursued according to the overall check of the network, it has the possibility to find the optimal solution that the generating learning method cannot reach.

In this research, a goodness factor proposed by Matsunaga [6] is employed in building the optimal structure of a neural network to evaluate the effect of each hidden units in the hidden layer and build the eliminate learning method.

**Elimination of Ineffective Units by Goodness Factor.** The goodness factor defines the total sum of signals propagating in forward direction as follows:

$$G_i^k = \sum_p \sum_j (w_{i,j}^k x_i^k)^2 \tag{1}$$

where $x_i^k$ is an output of unit $i$ of layer $k$, $w_{i,j}^k$ is the link weight from unit $i$ of layer $k$ to unit $j$ of layer $k+1$ , $p$ is summation over whole leaning patterns. As it is easily understood from the definition, the large value of goodness factor produces the large influence on the whole units of level $k+1$ from the unit. Therefore, after calculating goodness factor of all units of layer $k$, the unit with the smallest goodness factor is interpreted as the most useless unit in the layer and named a bad unit.

First, the appropriate number of units in a hidden layer is provided. After buck propagation learning gives the result lower than the error rate given, the learning is taken to terminated and a bad unit is eliminated. When the error rate is greater than arbitrarily defined value then initialize all link weights of a bad unit and then re-learn the teaching patterns. Even if this procedure is executed the result is not improved, then one unit is tacked on to the hidden layer. The above mentioned procedure is pursued repetitively until the smallest number of units is obtained. The structural learning is executed according the procedure.

### 2.3 Example of Determination of Hidden Layer by Stractural Learning

In this sub section, an example is given to show the determination of the number of hidden units. It illustrates and verities how the optimal number of hidden units is determined using Goodness Factor to solve an exclusive logical OR by a neural network. The exclusive logical OR is a kind of logical operations and known a non-linear classification problem. It is also widely known that the minimal structure of a neural network of an exclusive logical OR should have two units for the hidden layer.

**Table 1.** Exclusive Logical OR

| Input Unit 1 | Input Unit 2 | Output |
|---|---|---|
| 0 | 0 | 0 |
| 0 | 1 | 1 |
| 1 | 0 | 1 |
| 1 | 1 | 0 |

In the experiment, we employed an initial neural network with 5 hidden units for structural learning. Figure 1 shows the transit number of hidden units according the structural learning. On the case of the hidden layer with one unit, the learning is not accomplished. Even if starting the learning from the one hidden unit, it could not terminate. So the number of hidden layer was added with one more unit. Therefore, the hidden layer should has two units for succeeded learning. This result is coincident with the widely known result. This result shows our structural learning method works well.

### 2.4 Application of Structural Learning to Stock Price Forecasting

Generally, the structural learning is to build a neural network with the minimal number of units under the consideration of dependency of initial values. In this research, the structural learning is employed to decide the optimal number of units in order that the stock prices are forecasted with high precision.

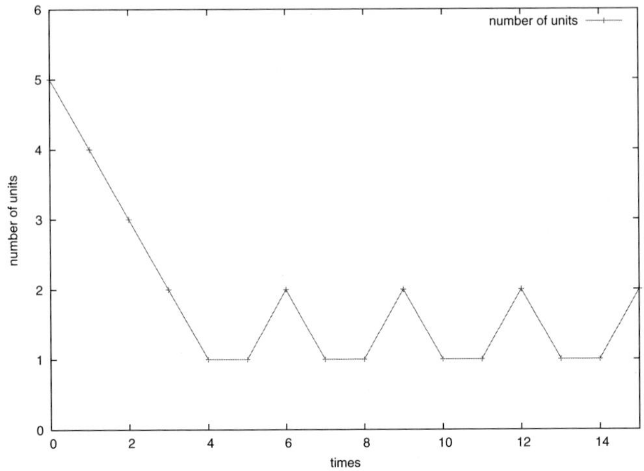


**Fig. 1.** Transit Number of Hidden Units in XOR Problem

The effect of the structural learning in forecasting stock prices by a neural network is verified as follows. First, it is pursued to structural-learn a neural network. Then, depending this procedure, the optimal number of hidden units is determined in this experiment. After then, we compare and verify the effect and efficiency between the neural network with the same number of hidden units and the neural network resulted by the structural learning.

**Input to a Neural Network.** In this paper Plural number of stocks are employed as an input data to a neural networkC Plural stocks are normalized. The plural stocks can remove nose data in learning.

In this experiment, we employed a neural network with 30 units initially and pursue structural learning with 5 input stocks that are selected by SOM. Table 2 shows the resulted forecasting precision in order to verify the effect of the structural learning. The learning is pursued under the same condition as in verification.

From Table 2, in all aspects the results of the structurally learned neural network are better than a neural network with 30 units. In the computational time the structurally learning neural network spent only one third less than a general neural network.

**Table 2.** Forecasting Precision per Stock

| Number of Input Stocks | 1 | 5 |
|---|---|---|
| Forecast Precision (Mean)[%] | 59.6 | 62.7 |
| Forecast Precision (Maximum)[%] | 72.5 | 73.5 |
| Forecast Precision (Minimum)[%] | 52.5 | 55.9 |

**Determination of Input Data.** Before the verification let us discus about input data. In this research we propose to employ plural number of stocks including the focal stock in order that the proposed neural network model can forecast stock prices precisely. Using the plural stocks, the neural network can recognize the tendency of stock movements. Even if focal stock data for forecasting are noisy, the forecasting does not have such an influence.

Also, the effect of plural sets of data is discussed in Appendix 2.4.

In order to select plural stocks relating the focal stock, a large number of stocks from various industries are selected and the time-series data of their stock prices are classified based on a SOM ( Self-Organizing Map ). It is necessary to classify the transit of the stock price instead of classifying the stock prices themselves. In order to do so, we normalize the time-series prices of 50 stocks as a pre-processing.

The normalization is give the following equation:

$$y = \frac{x - \min}{\max - \min} \tag{2}$$

where $y$ is a normalized stock price, $x$ is an original stock price, min andmax are the smallest and the largest of teaching data, respectively.

The stock prices of each of 50 stocks from January 4, 2004 to October 22, 2002 for 197 days are normalized. Figure 2 shows the result classified by SOM using the normalized data. Figure 2 illustrates only the neighboring area adjacent to Hitachi Ltd. [code6501]. The reason why the time-series stock prices is dated because the variation among values measured at stock value.

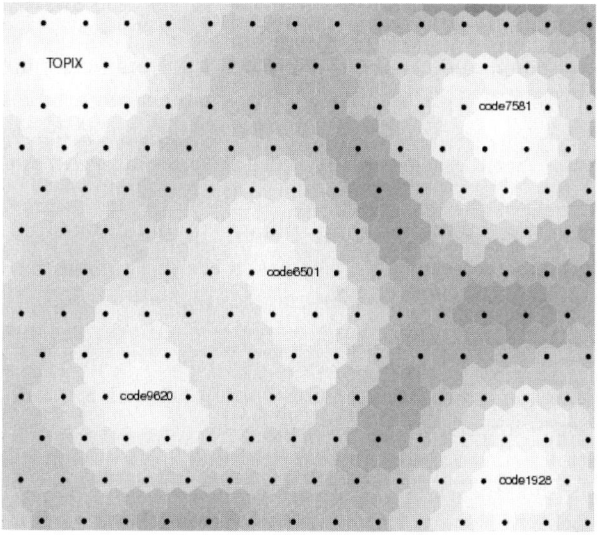


**Fig. 2.** Classification by SOM

Figures 2 shows that input data consists of five stocks.

- Hitachi Ltd [code6501]
- Sekisui House Ltd [code1928]
- Saizeriya Co. Ltd.[code7581]
- Squire Enix [code9620]
- TOPIX

It is necessary in time-series forecasting to input time sequence data that are expanded to space data. In this paper, the time-series movement is realized by giving one bundle consisting of continuous five days of each stock and stock index as input to a neural network and also the bundle of continuous five days regarding one day lag data. Therefore, one input consists of 35 patterns.

**Verification of Structural Learning.** In structural leaning of a neural network for forecasting stock price, each parameter of the neural network and learning condition are defined as follows:

- The initial number of units in hidden layer: 30
- The initial linkage weight and threshold of units: randomly given within a value within -1 to 1.
- Learning rate: 0.1
- maximum number of learning repeatF30000
- Error rate at stopping of learning: less than 0.05

The teaching data for learning of a neural network consist of 192 patterns where one pattern has one bundle of 5 days stock prices for each of 5 stocks and stock index out of 197 days from January 4, 2002 to October 22, 2002. In eliminating a bad hidden unit, as all link weights are maintained from the successful learning, it is not necessary to learn duplicated after elimination. Therefore, as the learning is continuously pursued after learning some length, even if the number of hidden units is 8 and all other parameters and the learning conditions are set to the same, it is not guaranteed to successfully learn the neural network within the given leaning time. So it is necessary to set the maximum learning time larger.

**Application to Forecasting Stock Price.** In the previous section, we obtained the minimal number of hidden units is 8 and the number of input units is 30. In this section we verify the efficiency and effect of forecasting stock price by the structural learning. The same learning condition is employed but the maximum learning time is set to 50,000. The reason is mentioned above. The verification simulation is executed 20 times about the largest, smallest and mean values of forecasting precision and learning time.

The verification is done using the same normalized learning data of 197 days of each of stocks and stock index from January 4, 2002 to October 22, 2002. The forecast of the price of Hitachi Ltd stock for 40 dealt days from October 23, 2002 to December 28, 2002. When the forecasting of price movement is coincident with the real movement, the forecast is counted as success.

Table 3 shows the result of verification.

**Table 3.** Verification Result of Forecasting Stock Movement

| Number of Hidden Units | 8 | 30 |
|---|---|---|
| Forecast Precision (Mean) [%] | 65.1 | 62.7 |
| Forecast Precision (Maximum) [%] | 73.5 | 73.5 |
| Forecast Precision (Minimum) [%] | 58.8 | 55.9 |
| Learning Time (Mean)[sec] | 810.6 | 2289.1 |
| Learning Time (Maximum)[sec] | 1025.3 | 2714.5 |
| Learning Time (Minimum)[sec] | 573.5 | 1849.1 |

### 2.5 Discussions

In structural learning of a neural network, the neural network with 30 hidden units required 1327 seconds as mean learning time. On the other hand, the structurally learned neural network with 8 hidden units can pursue the same results in 413 seconds. So the computational time was shortened into 1/3. Regarding the mean forecasting precision, the neural network with 30 hidden units was 63 %, on the other hand the neural network with 8 hidden units resulted in 67%. Even if the number of hidden units decreases, the forecasting precision was not decreasing but increasing.

Conventionally, it is required to employ a neural network with larger number of hidden units in forecasting data such as stocks using huge number of data. Even if we employed the optimal number of hidden units after structural leaning, it is possible to improve the forecasting in the sense of precision and computation time. In conventional forecasting, although the redundant forecasting is employed considering noise it provided too much fitting. Then the resulted precision of forecasting was not expected good.

The computation cost decreased by eliminating redundant hidden units as well. That is, the elimination realize the decreasing of the memory assignment for the computation and also for the link waits. As a result the computational cost was possible.

### 2.6 Conclusions

The objective of this paper is to remove redundant units from the neural network in forecasting stock prices. The structural learning is employed to decide the optimal number of hidden units of the neural network and its appropriateness is verified. As a result, it was shown in structural learning that the neural network can successfully determined and forecast stock prices in the simulation experiment using real data. The simulation showed to increase forecasting precision and shorten the computational time with better forecasting ability.

### 2.7 Conclusion

In the real investment, considering a limit amount of funds for investment, it can not invest to a large number of stocks. The solution of the portfolio selection problem shows a tendency to increase in the number of invested stocks. This problem is formulated by zero-one mixed integer programming problems. However, it is hard to solve the zero-one mixed integer programming problem because of combinatorial nature. Therefore, it needs an efficient approximate method to solve the large–scale zero-one mixed integer programming problem.

In this paper, we proposed the Meta-controlled Boltzmann machine, based on a Two-layered Boltzmann machine which is proposed by J. Watada et. al [9], to solve the portfolio selection problem which limits the number of invested stocks. The result of numerical examples shows that we can obtain the solution with limited number of selected stocks. Employing the Meta-controlled Boltzmann machine, we can reach the termination within a shorter computing time.

## References


1. Arisawa M., Watada,J.: Enhanced Back-Propagation Learning and Its Application to Business Evaluation Proceeding, *IEEE Interational Conference on Neural Network*, at Orland, Florida, (1994) 1155-1160
2. Asahi,T., Murakami,K., Sagamihara,T.: BP Algorithm for Optimality and highly speeding of Neural Network Structural Learning, *Report of Institute of Electronics, Information*, **NC90-64** (1991)
3. Elman,J.L.: Finding Structure in Time, *Tech.Report CRL-8801*, University of California, San Diego (1988)
4. Hayashi,Y.: *Stock Price Is Forecastable - Introduction to Chart Analysis*, Kanki Publishing (2000) in Japanese
5. Jordan,M.I.: Serial order: A parallel, distributed processing approach, *Tech.Report*, **ICS-8604**, University of California, San Diego (1986)
6. Matsunaga Y., Nakade,Y., Nakagawa,O., Murase,M.: Buck Propagation Algorithm to Auto-Elimitation of hidden units in Higher Neural Network, *Trans of Institute of Electronics, Information*, **J74-D-II(8)** (1991) 1118-1121
7. Matsunaga Y., Nakade,Y., Nakagawa,O., Kawanaka,M.: Buck Propagation Algorithm to auto eliminate redundant hidden units from Competition, *Trans of Institute of Electronics, Information*, **J79-D-II(3)** (1996) 403-412
8. Murata, I., Watada, J., Choi, J: Structural Learning of Neural Network and Its Application to Stock Forecasting, *CD Proceedings, International Conference on Management Engineering (ISME2006)*, Kitakyushu, JAPAN, March 10-12, (2006) R30001-R30006
9. J. Watada and K. Oda:,"Formulation of a Two-layered Boltzmann Machine for Portfolio Selection," *International Journal of Fuzzy Systems*, Vol. 2, No. 1, pp. 39-44., 1999
10. Watada, J.: Keynote Speech on Structural Learning of Neural Network and Its Application, *Proceedings, An International Conference on Intelligent systems (ICIS2005)*, at Kuala Lumpur, Malaysia on Dec. 1-3, (2005)

# Part V

# Search Engine and Information Processing and Retrieval

# Web Usage Mining Via Fuzzy Logic Techniques

Víctor H. Escobar-Jeria[1], María J. Martín-Bautista[2,*], Daniel Sánchez[2], and María-Amparo Vila[2]

[1] Department of Informatics and Computer Science, Metropolitan Technological University of Santiago de Chile, Chile
[2] Department of Computer Science and Artificial Intelligence, University of Granada, Periodista Daniel Saucedo Aranda s/n, 18071, Granada, Spain
mbautis@decsai.ugr.es

**Abstract.** With the increment of users and information on the Web, mining processes inspired in the traditional data mining ones have been developed. This new recent area of investigation is called Web Mining. Within this area, we study the analysis of web log files in what is called Web Usage Mining. Different techniques of mining to discover usage patterns from web data can be applied in Web Usage Mining. We will also study in a more detailed way applications of Fuzzy Logic in this area. Specially, we apply fuzzy association rules to web log files, and we give initial traces about the application of Fuzzy Logic to personalization and user profile construction.

**Keywords:** Web Usage Mining, Fuzzy Logic, Fuzzy Association Rules, Personalization, User Profiles.

## 1 Introduction

In the last decade, new processes to manage huge quantities of data and discover new knowledge have been developed. Knowledge Discovery in Databases is the area that study this pattern discovery, and its main stage is called Data Mining. In traditional Data Mining, non explicit knowledge can be found in data, usually stored in relational databases in a structured form [Agrawal et al., 1993]. When these techniques are applied to other information sources, such as the Web, the particularities of the data, specially the lack of structure, have implied specific features of the processes and techniques. Two new areas are called Text Mining and Web Mining arise to study the applications of mining techniques to documents and Web data, respectively.

The Web grows and changes very rapidly, and its use has been extended not only to the information searching and retrieval but also to make commercial transactions. The competition in the e-commerce makes necessary the application of intelligent methods to store and to examine the information of sessions of Web users or potential customers. For this reason the user's behavior and the user's objectives are elements to obtain. The knowledge about the user is used

* Corresponding author.

P. Melin et al. (Eds.): IFSA 2007, LNAI 4529, pp. 243–252, 2007.
© Springer-Verlag Berlin Heidelberg 2007

not only to characterize to the user but also to discover generic tendencies for marketing purposes and for the web site improvement. These data about the users are collected from the activity of the user in the web site through log files.

The objective is to mine web server logs to find relations among users about navigational aspects. The nature of the data in the log files and the information to predict such as time, user age, cultural level, etc. makes Fuzzy Logic a perfect tool to model this information. From all the techniques with Fuzzy Logic, we will extend our study in fuzzy association rules [Delgado et al., 2003]. Concretely, we present an application of fuzzy association rules in the area of Web Usage Mining.

In this paper, we review some of the main applications of Fuzzy Logic to Web Mining. For this purpose, we first give a general view of the Web Mining area, focusing on Web Usage Mining and the processes of personalization and user profile generation. We also study different aspects of Web Mining with Fuzzy logic applications found in the literature. Concretely, we present some preliminary results of experiments with Fuzzy Association Rules. Finally, we give some conclusions and future trend lines.

## 2 Web Mining

Through Web Mining, mining techniques are used to automatically discover and extract information from web documents and services [Etzioni, 1996]. However, there are some particularities of the Web information that make difficult to get the needed information: the huge amount the data, the diversity of language, the quality of information, the distribution of data at different platforms and finally, one of the most important, the lack of structure in the data. These points, specially the unstructured data and the great heterogeneity, are also the main inconveniences of mining processes in the web.

Cooley distinguishes three forms to understand Web Mining: from the point of view of the content, the structure and the use [Cooley et al., 1997]. Web Content Mining is the automatic discovery of patterns from the content in Web documents [Mitra and Pal., 2002],[Chakrabati, 2000] ; Web Structure Mining consists on studying the structures of link enter or intra documents to discover useful patterns of link structures [Chakrabati, 2000], [Delgado et al., 2002]; and finally, Web Usage Mining, that we will study in this article more detailed. We can define the Web Usage Mining as the process of automatic discovery of access patterns or use of web services, based on the user's behavior when interact with the Web [Delgado et al., 2003].

### 2.1 Web Mining and Fuzzy Logic

In all these types of Web Mining as well as in traditional Data Mining, both from the data or from the technique point of view, optimization tools coming from Soft Computing have been applied such as Fuzzy Logic, Genetic Algorithms, Neural Networks and Rough Sets [Arotaritei and Mitra., 2004]. The Fuzzy Logic is able to represent an user-oriented selection of data, giving flexibility to the system and producing more interpretable solutions [Mitra and Pal., 2002].

Recently, these techniques have been applied to other data areas such as document collections [Justicia et al., 2004] and the Web. In Web Mining, the main techniques utilized in the literature are Fuzzy Clustering and Fuzzy Association Rules. These are the techniques that we will use too to obtain generic navigational tendencies of the user and the construction user profiles.

Fuzzy clustering algorithms such as the fuzzy c-means (FCM), with the fuzzy-c trimmed medoids (FCTM), and fuzzy-c medians (FCLMedS) are used in [Mitra and Pal., 2002] for Web Content and Web Usage Mining. Another application with fuzzy clustering can be looked up in [Nasraoui et al., 1997] for Web Usage and Web Structure Mining. The authors apply an algorithm called CARD (Competitive Agglomeration of Relational Data) to group different sessions of users. For this purpose, not only the entries in log files are considered, but also the structure of the site and the URLs to calculate the similarity between two users' sessions. The objective of this application is to identify users' sessions from the users' accesses to the Web site and its structure. Along with the fuzzy clustering, one of the techniques more utilized in Web Mining is fuzzy association rules. An application of this technique can be found in [Garofalakis et al, 1999], where a refinement of queries from an initial group of documents retrieved from Web is carried out. The textual transactions are constructed also with fuzzy values. The purpose of this work is to provide the system with an ability of reformulation of queries using mining technologies.

Other approach using fuzzy association rules can be found in [Wong et al., 2001], where a proposal with a description of a system architecture to predict Web accesses is presented. The fuzzy association rules and the generation of a fuzzy index tree are utilized to improve the accuracy and the efficiency of predictions of access Web paths.

At the following section, we will comment the Fuzzy Association Rules and experiments related with this technique in the area of Web Usage Mining.

## 3 Web Usage Mining

This area of Web Mining analyze the web log files to obtain user navigational patterns that can evidence user's preferences, the on-line client's behavior and the future directions of improvement of the web site. The log files contain information about the connection server or the user's identity and authentication. This information collects the on-line user's activities and reflects several kinds of behavioral different patterns.

This information is used by the companies from the point of view of marketing, principally, where the assignment of a general profile to a user surfing in the web site based on his/her navigational behavior can be utilized to apply diverse measures, and to identify users with social groups. For the exploitation of different kinds of data, different techniques of pattern discovery can be used. The extracted knowledge can be used to execute tasks like prediction, personalization and improvement of the web site infrastructure.

### 3.1 Web Personalization and User Profiles

The ability of a web site to process the visits in a detailed level and to guide his/her clients or users through useful and pertinent information successfully, is becoming one of the crucial goals for any website nowadays. One of the forms to get this goal is through the personalization of Web.

The personalization of Web can be seen from two points of view: the company and the user. The company's point of view is related to the marketing and identification of demographic classes. The user's point of view is related to navigational recommendations and obtaining of information. This process can be described as a group of actions done by the user in the navigation so that these actions can be processed to improve the web site according to user preferences [Mobasher, 2005]. Part of this information can be stored in what is called user profiles [Martín-Bautista et al, 2002].

User profiles can be defined as a representation of the knowledge about the user's interesting information. In [Martín-Bautista et al, 2002] the authors propose two different types of profiles: the simple profiles, which are represented by data extracted from documents supposedly interesting for the user; and the extended profiles containing additional knowledge about the user such as the age, the language level, the country, among others.

The extended profiles can be described then as a tuple of four variables. One of them is the demographic variable, that is related to social aspects like the user's age range or the education level, among others. Other variable is the identification one, that is related to the information stored in the log files from the user server. Then, the clickstream variables which are related to the page weights, where if one page has a zero value, this implies that the page has not been visited. The last variable is related to the session simple profiles described previously [Martín-Bautista et al, 2002].

For the obtaining of these profiles, clustering and association rules are applicable usually. Through clustering processes, an grouping of clients or data with similar characteristics is initially obtained automatically without having a previous classification. User profiles derived from these groups can be utilized to guide strategies of marketing according to the groups [Nasraoui et al., 1997]. The association rules discover associations and correlations among items where the presence of an item or a group of them in a transaction implies (with a confidence grade) the presence of other items [Carbonell et al., 1998]. One of the most direct applications of association rules to Web Usage Mining comes from the relations among visits of users with a certain navigational pattern to the web site.

The principal inconvenience of handling of profiles in the Web is the lack of knowledge about the identity of the user. Two different situations can rise: first, the unregistered users where users' profile can provide evidence of identity or associating with a social group. A general profile is then assigned to the user. Preferences stored in the profile can be applied to the web site for the user while she/he navigates.

The second situation refers to the registered users. If a user is identified in some way, then the web site can change according to the user's preferences. The system keeps the track of the user in her/his previous visits in the web site together with the users' profile. This information is utilized to personalize the site. To characterize user groups with similar behavior a clustering method can be performed [Martín-Bautista et al, 2002]. In order to have a more ample vision of Web Usage Mining, we will see some previous works in this area.

### 3.2 Related Work

In [Mobasher, 2005], an overview of the process of personalization based in Web Usage Mining is shown. Techniques of data mining such as clustering to discover groups of users are utilized. Furthermore, association rules can be used to find important relations among the articles the users are interested in, based on navigational site patterns. Other different proposal is found in [Martín-Bautista et al, 2002], where the authors propose a scalable clustering methodology, inspired in the natural immunologic system with the facility to learn continuously and to adapt to coming new patterns.

One of the most well-know system developed for personalization can be found in [Cooley et al., 1997]. The system called WebMiner is based on a behavior model of user's navigation. By the grouping of web pages references, the system generate transactions, from which association rules are discovered. Other system related to personalization is presented in [Cernuzzi and Molas, 2004], where a study of Ridier's Web site (http://www.rieder.net.py) is carried out. The log files of the web server are stored and analyzed. From the transactions, behavioral patterns are extracted to describe the users' way of surfing using clustering and association rules. In [Wong et al., 2001] the authors propose a structure for a guided personalization and adaptation in the Web by means of user profiles and the accesses collected through the web log files.

### 3.3 Fuzzy Logic and User Profiles

Sometimes, we do not have explicit information of the users in log files besides the information got from the server. We can complete the user's identity and authentication through another sources or inferring the information through techniques of mining. For example, we can infer in the education level of the user according to her/his navigation or may be according to the information that the users proportionate explicitly.

Therefore, when the extended user profiles are constructed, there is information to manage related to different concepts about the user. Some of these concepts such as the age of the user are imprecise, since the system must approximate the data if the user does not proportionate it, or the patience of the user surfing through the site. These characteristics can be modeled by means of linguistic labels, for example {very low, low, regular, high, very high} [Martín-Bautista et al, 2002].

## 4 Fuzzy Association Rules to Analyze Web Log Files

We have seen different aspects and proposals realized in the area of Web Mining, principally association rules and clustering techniques. In the following, we study the application of fuzzy association rules to Web Usage Mining. The main objective of this application is to search patterns through the fuzzy association rules between the different fields in the web log file. In this way, we can obtained relations between pages to know the page/s that the users visit starting from and initial visited page, or the relations between the navigation time, day of the week, etc. and the visited pages.

Before explaining the experimental stage, we give some theoretical notions about association and fuzzy association rules and their measures, and we show some preliminary results of an experimental example.

### 4.1 Association Rules and Fuzzy Association Rules

The Association Rules look for relations or affinities among of groups of items or fields, generally in a relational database. Let $I$ be a set of elements called "items" and let $T$ be a set of elements called "transactions", each transaction being a set of items.

Let us consider two itemsets (sets of items) $I_1, I_2 \subseteq I$, where $I_1 \cap I_2 = \emptyset$. An association rule [Agrawal et al., 1993] $I_1 \Rightarrow I_2$ is an implication rule meaning that the apparition of itemset $I_1$ in a transaction implies the apparition of itemset $I_2$ in the same transaction. The reciprocal does not have to happen necessarily [Kraft et al., 2003]. $I_1$ and $I_2$ are called antecedent and consequent of the rule, respectively. The measures more utilized to describe the relations among antecedent and consequent of the association rules are the Support, and the Confidence. Support is the percentage of transactions where the rule holds, while confidence measures the strength of the rule as the percentage of transactions containing $I_1$, that contain $I_2$.

Fuzzy Logic [Zadeh, 1975] permits the manipulation and exploitation of incomplete data or with a grade of uncertainty, situation that is very frequent in the data to mine [Delgado et al., 2003]. Several authors have proposed *fuzzy association rules* as a generalization of association rules when data is fuzzy or has been previously fuzzyfied [Lee and Kwang, 1997],[Au and Chan, 1998], [Kuok et al., 1998], [Hong et al., 1999],[Delgado et al., 2003].

Fuzzy association rules can be extracted from a group of fuzzy transactions by means of an algorithm of extraction such as the algorithm APrioriTID [Agrawal et al., 1993].

A fuzzy transaction can be defined as a nonempty subset where $\tilde{\tau} \subseteq I$. For every $i \in I$ we note $\tilde{\tau(i)}$ the membership degree of i in a fuzzy transaction $\tilde{\tau}$ [Etzioni, 1996]. We note$\tau(\tilde{I_0})$ the degree of inclusion of an itemset $I_0 \subseteq I$ in a fuzzy transaction $\tilde{\tau}$, defined in (1):

$$\tilde{\tau}(I_0) = \min_{i \in I} \tilde{\tau}(i) \quad (1)$$

Therefore, fuzzy transactions manage imprecision and give more flexibility because they allow us to deal with intermediate values between 0 and 1 to represent the membership degree of the items to the transaction.

For the evaluation of the performance of the association rules, we employ a semantic approach based on the evaluation of quantified sentenced [Zadeh, 1975].

A quantified sentence is an expression of the form "Q of F are G" where F and G two fuzzy subsets of a finite set X, and Q is a relative fuzzy quantifier. Relative quantifiers are linguistic labels that can be represented by means or fuzzy sets on [0,1], such as "most", "almost", or "many". This way we can define the estimated measures of goodness of rules. We evaluate the sentences by means of the method GD [Delgado et al., 2000], which has been shown to verify good properties with better performance than others. The evaluation of "Q of F are G" by means of GD is defined in (4),

$$GD_Q(\frac{G}{F}) = \sum_{\alpha_i \Delta(\frac{G}{F})} (\alpha_i - \alpha_{i+1}) Q(\frac{|\,(G \cup F)_{\alpha_i}\,|}{F_{\alpha_i}}) \quad (2)$$

Another interesting measure is the certainty factor of a fuzzy association rule [Delgado et al., 2000]. Given a $A \rightarrow C$ , the certainty factor takes values in [1, 1]. It is positive when the dependence between A and C is positive, 0 when there is independence and a negative value when the dependence is negative. We say that a fuzzy association rule is strong when its certainty factor and support are greater than two user-defined thresholds minCF and minSupp, respectively.

### 4.2 Experimental Example

When the users interact with the Web, they leave digital tracks (IP, agents, cookies, etc) that the servers store automatically in a journal of accesses. This activity is principally stored in the log files of the web server, and it is usually completed with other sources of information such as the proxy server and the user's machine.

In our case, the data to analyze is a subset of log files available from the ECML/PKDD Conference 2005 web site [ECML/PKDD 2005]. These files are in a CSV (Comma Separated Value) format. Table 1 shows a line of one of these log files, which is composed by 6 fields (identifier of shopping, date, IP, session, visited page, referenced page).

**Table 1.** Entry line in a log file sample

| **Id shop** | **Date** | **IP** |
|---|---|---|
| 11 | Tue Jan 20 19:00:132004 | 213.235.141.105 |
| **Session** | **Visited Page** | **Referenced Page** |
| 1f75ccd2afbf87dc9abccde23f3 | /dt/?c=11670 | http://www.shop2.cz/ls/index.php |

The sample web log file used in this experiments has 6700 entries. After the preprocessing stage, the number of entries is reduced to 6640. After the transformation of the data set in a transactional form, we can decide the fields to take part of the rule.

For example, if the user chooses the fields of date and visited pages, the extracted knowledge can give us an idea about which pages have been more visited at certain hours. Also, if the user selects the fields of IP and visited pages, we could somehow identify the users that visit those pages.

In order to be able to obtain all this information from the Web log files, we extract the association rules with the algorithm APrioriTID [Agrawal et al., 1993]. Any other Apriori like algorithm can be used, although we have chosen the AprioriTID for its capability of reducing the number of groups considered.

We present an example of results that we can obtain to know the page that the users visit starting from an initial visited page. The form of the rules to extract would be:

**Initial visited page $\longrightarrow$ Referenced page**

1. dt/?c=11670 $\longrightarrow$ http://www.shop2.cz/ls/index.php
   - Support = 0.6
   - Confidence = 1.0
   - FC = 1.0
2. dt/?c=12397 $\longrightarrow$ http://www.shop7.cz/akce/kat=239
   - Support = 0.2
   - Confidence = 1.0
   - FC = 1.0

In these two rules, we can observe that rule 1 appears with a percentage of 60% and rule 2 appears with a percentage of 20%. In both cases, the confidence and the certainty factor are 1, which means that the users visiting the page in the antecedent also visit the page in the consequent.

## 5 Conclusions and Future Work

In this work, we have given a general view of the Web Mining area, focusing on the application of Fuzzy Logic techniques to improve different processes. The two most utilized techniques found in the literature are fuzzy association rules and fuzzy clustering. They have been specially applied to Web Usage Mining for the analysis of web log files.

Fuzzy Logic can be also used to model the user behavior stored in a profile, where most of the elements are imprecise by nature. The future work will be go further in this area principally, continuing with the development of a tool that allows us to integrate other techniques of mining such as clustering and/or fuzzy clustering to group log files and to extract social groups based on their user profiles for personalization purposes.

## References

[Agrawal et al., 1993] Agrawal, R., Imielinski, T., Swami, A.: Mining association rules between sets of items in large databases. In Proceedings of the 1993 ACM SIGMOD Conference. (1993) 207–216

[Arotaritei and Mitra., 2004] Arotaritei, D., Mitra, S.: Web Mining: a survey in the fuzzy framework. Fuzzy Sets and Systems. (2000)

[Au and Chan, 1998] Au, W.H. & Chan, K.C.C.: An effective algorithm for discovering fuzzy rules in relational databases. In Proc. Of IEEE International Conference on Fuzzy Systems, (1998) vol II, 1314–1319.

[Carbonell et al., 1998] Carbonell, J., Carven, M., Fienberg, S., Mitchell T., and Yang, Y.: Report on the conald workshop on learning from text and the web. In CONALD Workshop on Learning from Text and The Web, June, (1998)

[Cernuzzi and Molas, 2004] Cernuzzi, L., Molas, M.L.: Integrando diferentes Tcnicas de Data Mining en procesos de Web Usage Mining. (2003)

[Cooley et al., 1997] Cooley, R., Mobasher, B., and Srivastava, J.: Web mining: Grouping Web Page References into Transactions for Mining World Wide Web Browsing Patterns. (2000) 1–11

[Chakrabati, 2000] Chakrabati, S. Data Mining for hypertext: A tutorial survey. ACM SIGKDD Explorations (2000) 1(2):1–11.

[Delgado et al., 2000] Delgado M., Sánchez D. and Vila M.A.: Fuzzy cardinality based evaluation of quantified sentences. Int. J. Aprox.Reasoning, 3 (2000) 23–

[Delgado et al., 2002] Delgado, M., Martín-Bautista, M.J., Sánchez, D., Vila, M.A.: Mining Text Data: Special Features and Patterns. Pattern Detection and Discovery. Proceedings ESF Exploratory Workshop, D.J. Hand, N. Adams and R. Bolton (Eds.), Lecture Notes in Artificial Intelligence Series (2002) 140–153

[Delgado et al., 2003] Delgado, M., Marín, N., Sánchez, D. & Vila, M.A.,: Fuzzy Association Rules: General Model and Applications. IEEE Transactions on Fuzzy Systems 11 (2003) 214–225.

[ECML/PKDD 2005] ECML/PKDD Conference 2005 Web Site. Porto, Portugal (2005) http://ecmlpkdd05.liacc.up.pt/

[Etzioni, 1996] Etzioni O.: The World Wide Web: Quagmire or gold mine. Comunications of the ACM, 39 (1996) 65–68

[Garofalakis et al, 1999] Garofalakis, M. N., Rastogi, R., Seshadri, S., and Shim, K.:Data Mining and the web: Past, present nad future . In WorkShop on Web Information and Data Managament, (1999) 43–47

[Hong et al., 1999] Hong, T.P., Kuo, C.S. & Chi, S.C. Mining association rules from quantitative data. Intelligent Data Analysis 3 (1999) 363-376

[Justicia et al., 2004] Justicia, C., Martn-Bautista, M. J., Snchez, D.: Minera de textos: Aplicaciones con lgica difusa. Actas del Congreso Espaol de Tecnologas con Lgica Difusa, Jan (2004) (In Spanish)

[Kleinberg et al., 1999] Kleinberg, J. M., Kumar, R. and Raghavan, P: The Web as a graph: measurements, models, and methods. Proceedings of the Fifth Annual International Computing and Combinatorics Conference. (1999)

[Kraft et al., 2003] Kraft, D.H., Martín-Bautista, M.J., Chen, J. & Vila, M.A. Rules and fuzzy rules in text: concept, extraction and usage. International Journal of Approximate Reasoning 34 (2003) 145–161

[Kuok et al., 1998] Kuok, C.-M., Fu, A. & Wong, M. H. Mining fuzzy association rules in databases. SIGMOD Record 27(1) (1998) 41–46

[Lee and Kwang, 1997] Lee, J.H. & Kwang, H.L. An extension of association rules using fuzzy sets. In Proc. of IFSA'97, Prague, Czech Republic (1997)

[Martín-Bautista et al, 2002] Martn-Bautista, M. J., Kraft, D. H., Vila, M. A., Chen, J., Cruz, J: User profiles and fuzzy logic for Web retrieval issues. Soft Computing Journal. Volume 6, Number 5. (2004) 365–372

[Mitra and Pal., 2002] Mitra S., Pal S.K: Data Mining in Soft Computing Framework: A Survey. IEEE Transactions on Neural Networks, (2002) 3–14

[Mobasher, 2005] Mobasher, B.: Web Usage Mining and Personalization. Practical Handbook of Internet Computing M.P.Singh, ed. CRC Press LLC (2005)

[Nasraoui et al., 1997] Nasraoui, O., Frigui, H., Joshi, A., and Krishnappuram, R.: Mining Web accses logs using relational competitive fuzzy clustering. In Proceedings of springs Symposium On Natural Language Proccesing Form the www, Stanford, California. March 1997 (1997)

[Wong et al., 2001] C. Wong, Shiu, S. and Pal, S. :"Mining Fuzzy Association Rules for Web Access Case Adaptation," in Workshop Proceedings of Soft Computing in Case-Based Reasoning Workshop, in conjunction with the 4th International Conference in Case-Based Reasoning, Vancouver, Canada. (2001) 213–220

[Zadeh, 1975] L. Zadeh: The concept of linguistic variable and its application to approximate reasoning I. Information Sciences 8. (1975) 199–251

# Deduction Engine Design for PNL-Based Question Answering System

Zengchang Qin[1,2], Marcus Thint[2], and M.M. Sufyan Beg[1,2]

[1] Berkeley Initiative in Soft Computing
Computer Science Division, EECS Department
University of California, Berkeley CA 94720, USA
{zqin,mmsbeg}@cs.berkeley.edu
[2] Computational Intelligence Group
Intelligent Systems Research Centre
British Telecommunications (BT) Group
marcus.2.thint@bt.com

**Abstract.** In this paper, we present a methodology for designing a Precisiated Natural Language (PNL) based deduction engine for automated Question Answering (QA) systems. QA is one type of information retrieval system, and is regarded as the next advancement beyond keyword-based search engines, as it requires deductive reasoning and use of domain/background knowledge. PNL, as discussed by Zadeh, is one representation of natural language based on constraint-centered semantics, which is convenient for computing with words. We describe a hybrid reasoning engine which supports a "multi-pipe" process flow to handle PNL-based deduction as well as other natural language phrases that do not match PNL protoforms. The resulting process flows in a nested form, from the inner to the outer layers: (a) PNL-based reasoning where all important concepts are pre-defined by fuzzy sets, (b) deduction-based reasoning which enables responses drawn from generated/new knowledge, and (c) key phrase based search when (a) and (b) are not possible. The design allows for two levels of response accuracy improvement over standard search, while retaining a minimum performance level of standard search capabilities.

## 1 Introduction

In general, a Question Answering (QA) system is a type of information retrieval system triggered by an input query. In the particular context of this paper, QA systems should produce direct answers/facts (rather than a ranked list of relevant documents) for a question expressed in natural language. Accordingly, such QA systems are regarded as the next generation search engines, since they require capabilities for query analyzes, recognition of most relevant facts, and deducing new facts and answers including the use of background or domain knowledge where necessary.

Automated QA systems are broadly classified into two categories: open-domain and restricted (or closed) domain systems. Restricted-domain QA deals with

P. Melin et al. (Eds.): IFSA 2007, LNAI 4529, pp. 253–262, 2007.
© Springer-Verlag Berlin Heidelberg 2007

questions about a specific domain (e.g, telecommunications, medicine) or a limited corpus, and is considered an easier task because natural language processing (NLP) systems can exploit domain-specific knowledge frequently formalized in ontologies. Open-domain question answering must deal with virtually any topics, and commonly rely on the World Wide Web (Web) as a knowledge source. Ask.com is an example of an early attempt at open-domain QA system, and Google and MSN (Microsoft) are also integrating question-answering capabilities in their search engines.

Even for restricted-domain QA systems, design complexity depends on the type and size/scale of the knowledge source, input query format, and answer format requirements. QA system based on small, structured databases with structured queries that simply output a list of relevant facts are the least complex, and those based on large, unstructured knowledge sources with natural language queries that require formal/composed answers are the most complex. In this paper, we discuss a problem of intermediate complexity, *i.e.* a QA system based on small, unstructured knowledge sources with natural language queries, where a list of the most relevant facts is an acceptable answer. We introduce the application of PNL-based reasoning for restricted-domain QA systems, which can increase the precision of answers for key topic areas. With a focus on well-rounded system utility, we present a hybrid ("multi-pipe") design that allows integration of PNL-based reasoning, deductive reasoning with natural language phrases, and default keyword based search as a minimal response output.

Although recent QA research investigates open-domain systems since the Web is a vast and readily available knowledge source, restricted-domain QA systems have also been studied. Some restricted-domain QA systems also rely on data redundancy in the Web to validate answers or use the Web as an supplemental knowledge source. For example, Chung *et al.* [4] studied a restricted domain of weather forecasts for Korean cities where weather information is described by well-defined HTML tags from a particular websites. Structured query language (SQL) is used to query the facts of weather conditions, and the *temporal data normalizer* converts temporal expressions such as *today*, *this weekend* and *now* into absolute values that can be used in querying to the database. Tsur *et al.* [11] presented a biographical QA system and empirical studies of machine learning techniques (Ripper, a rule learning algorithm, and support vector machines (SVM) for biography classification). The system first attempts to answer queries by searching a given collection of biographies, and if required information is not found, attempts to find the answer on the Web. Ceusters *et al.* [3] described a commercial QA system package for the health care domain based on domain ontologies. It is a rule-based system where rules are derived from expert knowledge. Benamara [2] presented the WEBCOOP system to generate cooperative responses on the web, integrated with a knowledge base coded in Prolog. Application example domain was restricted to tourism. Doan-Nguyen *et al.* [5] discussed ways to improve precision of answers for restricted-domain QA systems, through re-ranking of candidates proposed by generic Information Retrieval (IR) engine, and heuristic improvements on the IR engine.

## 2 Precisiated Natural Language

We investigate a novel application of Precisiated Natural Language (PNL) enhanced, restricted domain QA system. Vagueness and fuzziness is inherent in natural language, and Zadeh [12] articulates that humans have a remarkable capability to perform a wide variety of physical and mental tasks without any measurements. Humans use perceptions which are intrinsically imprecise. Perceptions partition objects into granules, with a granule being a clump of objects drawn together by indistinguishability, similarity, proximity or function. Perceptions are both fuzzy and granular (referred to as f-granular), and they are difficult for manipulation by computers. Zadeh proposed Precisiated Natural Language (PNL) for reasoning with problems expressed in natural language. The new approaches are not necessarily intended to replace existing NLP tools, but offers new tools for computing with f-granular information in natural language.

In PNL, the meaning of a proposition $p$, can be represented as a generalized constraint on a variable. Schematically, this is represented as:

$$p \rightarrow X \ isr \ R$$

In the generalized constraint expression $X \ isr \ R$, $X$ is the constrained variable, $R$ is the constraining relation, and $r$ is a discrete valued modal variable. The "is" in $isr$ is simply its natural meaning the conjugated verb "to be". Thus, the expression $X \ isu \ R$ means $X$ is usually $R$, and other defined modalities include: possibilistic ($r = blank$); probabilistic ($r = p$); veristic ($r = v$); random set ($r = rs$); fuzzy graph ($r = fg$); bimodal ($r = bm$); and Pawlak set ($r = ps$). It is important to identify the constrained variable, which may also depend on how the information will be consumed/interpreted. For example, a proposition 'Mary is young' can be represented in the two following forms:

$$p \rightarrow Age(Mary) \ is \ young$$

and

$$p \rightarrow Person(young) \ isv \ Mary$$

which answer the questions "How old is Mary?" and "Who is young?" respectively. In Zadeh [14], three basic operations on generalized constraints are proposed (conjunction, projection, and propagation). If fuzzy sets are predefined, these operations can be used to calculate the precisiated implications of propositions.

A protoform (short for "prototypical form") is a key concept that facilitates deduction and reasoning with PNL expressions. Simple NL expressions with a single "to be" verb phrase (in all conjugated forms, tenses, and modalities) can be abstracted into respective "$X \ is \ A$" protoform, such as "Mary is young" or "Price of a golf ball is two dollars". Similarly, "John is a little older than Mary" and "Price of a golf club is much more than the price of a golf ball" can be abstracted to a "$Y \ is \ X + B$" protoform, and "Some balls are blue" can be abstracted to a "$Q_1$ *As are Bs*" protoform. Furthermore, if the first two phrases are recognized as "age(Mary) is young" and "price(golf ball) is two dollars", they can be abstracted to "$f(X)$ *is* $A$" protoform. Details about PNL protoforms can be found in [13] and [14].

Given that a subset of NL expressions are identified according to their protoforms deductions can be applied as follows. Given $X$ *is* $A$, and $Y$ *is* $X + B$ it can be reasoned that $Y$ *is* $A + B$, and given $Q_1 A$'s *are* $B$'s, $Q_2(A$'s&$B$'s) *are* $C$'s, it can be computed that $Q_3 A$'s *are* ($B$'s&$C$'s), where $Q_3 = Q_1 \bullet Q_2$, where $\bullet$ is a product in fuzzy arithmetic. In general, the following computational rules apply to the protoforms:

$$
\begin{array}{lll}
X\ is\ A,\ (X,Y)\ is\ B & \rightarrow Y\ is\ C, & \text{where } \mu_C(v) = \max_u(\mu_A(u) \wedge \mu_B(u,v)) \\
Q_1 A\text{'s } are\ B\text{'s},\ Q_2(A\&B)\text{'s } are\ C\text{'s} \rightarrow & & Q_3 A\text{'s } are\ (B\&C)\text{'s, where } (Q_3 = Q_1 \bullet Q_2) \\
X\ is\ A & \rightarrow g(X)\ is\ B, & \text{where } \mu_B(v) = sup_u(\mu_A(u)),\ v = g(u) \\
f(X)\ is\ A & \rightarrow g(X)\ is\ B, & \text{where } \mu_B(v) = sup_u(\mu_B(f(u))),\ v = g(u)
\end{array}
$$

Each of the forms above can be extended with respect to different modalities (probabilistic, usuality, bimodal interpolation, fuzzy graph interpolation, etc.). The deduction engine is designed to operate offline (i.e. without query input) as well as to be query-driven [14].

## 3 Deduction Module

The deduction module is a major part of an overall PNL-based QA system. A corpus of text documents are processed by the information extraction (IE) module to identify and tag each sentence as a PNL protoform (thus far $X$ *is* $A$, $Y$ *is* $X + B$, $QAs$ *are* $Bs$, $f(X)$ *is* $A$), causal fact (a causes b), if-then fact, procedure, or simply 'fact' (none of the above). Similar techniques are applied to the input question to identify the query-type (what, where, when, how, how much/quantity). Details regarding the precisiation process and its semi-automated implementation in the IE module are described in [1], and the general system design and process flow of the QA system is discussed in [8]. Modules and processes related to the deduction engine are shown in Figure 1 below. The deduction engine has access to all corpus knowledge which were precisiated by the IE module, and represented as a collection of fact-types.

If a sentence is a PNL protoform, it will be processed according to PNL reasoning. Else, phrase-based deductive reasoning is applied. In the presence of an input query, our system analyzes the query key phrases and query-type, selects a subset of relevant facts, and (where possible) combines with background knowledge to generate the most relevant answer set according to a ranking system. If deductive reasoning and concept matching (between query and facts) yield results with low ranking values, a search engine (e.g., Lucene [6]) is invoked for standard keyword-based search, which provides a minimal performance level.

### 3.1 Offline Reasoning

The deduction module is capable of operating in both online and offline mode. In the offline mode, where there is no user input and thus no query, the deduction

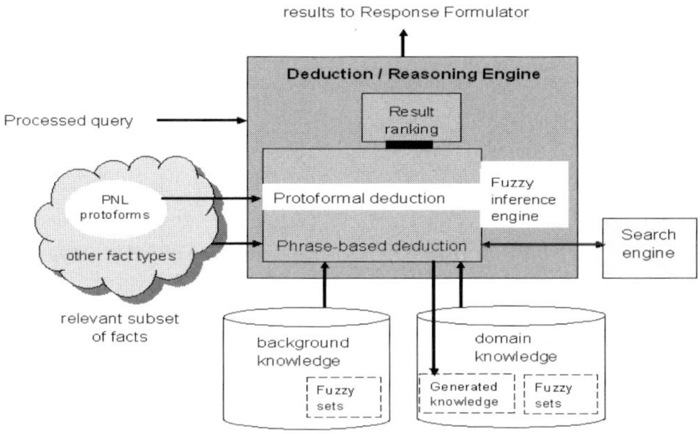


**Fig. 1.** The general structure of deduction engine for the PNL-based QA system

module analyzes existing facts to see if new facts can be generated based on PNL protoforms or phrase-based deductive reasoning. Thus far, standard logic applied to phrases include: negation, transitivity, and chained reasoning.

- Negation: If we have If-Then rules or causal rules like "If A then B" or "A causes B" ($A \rightarrow B$), we can conclude that "if NOT B then NOT A" ($\neg B \rightarrow \neg A$)
- Transitive Reasoning: If we have causal relations $A \rightarrow B$ and $B \rightarrow C$, we then can conclude that $A \rightarrow C$.
- Chaining: chaining is a multi-step transitive reasoning. Given a set of causal relations: $A \rightarrow B$, $B \rightarrow C \cdots E \rightarrow F$, we can conclude that $A \rightarrow F$. Semantic interpretation of chained reasoning results may sometimes appear to be unclear or illogical, hence the deduction module supports an option to show (explain) all intermediate results.

In addition to standard reasoning above, we also combine $X$ *is* $A$ facts with related simple facts. For example, given the facts: (1) Carnivores eat meat. (2) Lions are carnivores. The second fact is in $X$ *is* $A$ form but the first one is simply a fact. The two noun subjects (or subject phrases) are treated as equivalent, and via substitution of 'carnivores' by 'lions' in the first sentence, the deduction module will conclude that "lions eat meat".

As explained in [8] an application manager may provide background knowledge in the forms of simple facts or abstract facts. During offline reasoning, the QA system can also generate new knowledge by combing the facts from a corpus content and background knowledge. For example, in a domain about telecommunication, the background abstract facts: "*wireless *yy is less secure than wired *yy*" and "*wireless *zz simplifies physical installation*" may be provided. The notation '$*yy$' and '$*zz$' are reserved variables, and the deduction module searches for matches of phrase pattern surrounding the reserved words in the corpus text. For example, if another sentence contains information about 'wireless router', then new sentences "wireless router is less secure than wired router" and "wireless router simplifies physical installation" are generated although they do not explicitly appear in the corpus text.

We note that such knowledge augmentation may also generate some meaningless sentences - if a phrase "wireless chat session" is found elsewhere, generated sentence "wireless chat session simplifies physical installation" does not make sense. However, the generated knowledge are passively stored, and when a user inputs a query, the online processing helps to filter sensible results. It is more likely that sentences more relevant to the query will be ranked higher. For instance, if asked, "what type of routers simplify physical installation?" The correct answer "wireless router simplifies physical installation" will be ranked higher than the phrase about wireless chat session.

Offline reasoning improves performance and execution time during online use. The module operates at the phrase-level; each sentence in the corpus is segmented into subject phrase, verb phrase, and object phrase, such that the phrase components can be identified with the $X$, $Y$ or $A$, $B$, $C$ ... components of the PNL or non-PNL protoforms above. Accurate recognition of matching components (e.g. X in one subject phrase with X in another object phrase) depends on the quality of the concept matching module for two input strings. This module removes stop words and checks synonyms of remaining keywords (stemmed & non-stemmed) using WordNet [7] to compare phrases at the concept (rather than string) level. Given a query, the answer generation process flows as described below.

### 3.2 Query-Based Reasoning

The query is analyzed to extract its key phrases and query type (e.g. what, where, when, how, how much). The latter is non-trivial, and requires analysis beyond key word spotting, since "what is the cause...", "what is the method...", "on what occasion...", are actually "why", "how", and "when" type questions respectively, despite the prominent "what" keyword. Similarly, other cases have exceptions, and only cases with higher confidence (e.g. where, whereabouts, location as indicators for 'where'-type questions) are marked. Additional discussion about query typing can be found in [1]. Since this query-typing is used as a secondary factor to refine answer ranking, it is non-critical but helpful when correctly typed, as further explained below.

From the corpus knowledge (comprising facts extracted from a specified corpus), most relevant subset of facts are identified via concept matching with query keywords. If the relevant facts are known PNL protoforms, they are passed to the PNL-based deduction "pipe" and the deduction engine checks for associated fuzzy set definitions. For example, a QA application for an auto sales department may have fuzzy sets defined for "low", "moderate", "powerful", "expensive", "a little more", "a lot more", etc. If fuzzy sets are defined, and the answer to a query is in the form $Y \ is \ A + B$ (e.g. "Horsepower of sports model is a lot more than the horsepower of basic model"), where "Horsepower of basic model is low" our QA system will actually add the fuzzy sets for "low" plus "a lot more" and produce the approximate defuzzified value. If the fuzzy sets are not defined, the answer remains in the NL expression as "Horsepower of sports model is low PLUS a lot more" (i.e. a lot more than "low") which is just a linguistic interpretation of $Y\ is\ A+B$ form. In short, pre-defined fuzzy concepts enable computing more precise responses in the PNL paradigm; otherwise the system will return a linguistic interpretation of PNL protoforms.

As explained in [1] some "f(X)" protoforms (e.g. cost(a flat screen monitor)) can be identified from phrases like "A flat screen monitor is expensive", and the deduction module supports associated computations. Given the following facts:

- A mouse is cheap.
- A computer monitor is two hundred US dollars.
- A keyboard is fifty US dollars.

If asked "what is the cost of a computer monitor plus a keyboard?" the system returns:

> This is a 'compound' type question (e.g. $f(X)$ plus $f(Y)$). Reasoned from sentence: 'A computer monitor is two hundred US dollars.' AND 'A keyboard is fifty US dollars.' We can obtain:
>
> ---
>
> cost(A computer monitor) PLUS cost(A keyboard) IS (two hundred US dollars)+(fifty US dollars)

If we ask the cost for a mouse plus a keyboard, the answer will be 'fifty US dollars' PLUS 'cheap". The defuzzification value will be returned if 'cheap' is predefined by a fuzzy set. Otherwise, only the linguistic answers as above.

### 3.3 Ranking of Answers

If the relevant facts (to a query) are not PNL protoforms, they could be one of the following (as tagged by the IE module): causal fact, if-then fact, procedure, or just "fact". These facts are processed through the phrase-based deduction pipe, and the deduction engine (i) generates primary rankings of facts based on the degree of concept match with the query, (ii) attempts deductive reasoning based on new constraints supplied in the query, and (iii) generates secondary ranking based on the query type (if applicable). Deductive reasoning with new constraints include negation, transitive and chained reasoning as explained above, to produce an answer. In many cases, multiple facts are determined to be relevant to a query, as they include the same concepts/key phrases appearing in the query. Thus, during the secondary ranking process, ranks of causal or if-then facts are incremented for 'why'-type question, ranks of procedure facts would be incremented for 'how'-type questions, and facts containing quantity terms (numeric or fuzzy terms such as "few" or "most") are incremented for how-much type questions. Our system displays $N$ top ranked answers, where $N$ may be specified by the application manager.

A basic ranking algorithm is the count of matched concepts between corpus facts and user query. However, by offline reasoning, we can conclude some implicit relations between sentences and use it to re-rank the sentences. For example, consider the sentences:

1. Up to 16 telephones can be connected to one controller.
2. Up to 6 telephones can be connected to module-Alpha on the wall.
3. The controller is the heart of Beta.

and the query "how many telephones can be connected to Beta?". Typical keyword based systems will rank (1) or (2) as the highest answer, but our system detects that (3) is a $X \ is \ A$ form and "controller" and "the heart of Beta" is equivalent. Therefore, a new generated fact: "Up to 16 telephones can be connected to the heart of Beta" is the highest ranked answer. In some cases where the relevant subset is sparse or the query is ill-formed, and deduction process can not be applied. If concept match results yield low rankings then the highest scores of a standard search engine are returned.

## 4 More Examples

Additional details and output samples below summarize key functionalities of the deduction module. Given example facts:

(i) Horsepower of the basic model is about 150.
(ii) Horsepower of the sports model is a lot more than horsepower of the basic model.

and key concepts defined by fuzzy sets: 'about 150' is defined by a triangular fuzzy set centered on 150 with width of the fuzzy set ± 10% (i.e. ± 15 for this example). The term 'a lot more' is defined by a percentage (e.g. 100%) on the range of the reference fuzzy set 'about 150'. Since the two facts are PNL protoforms $X \ is \ A$ and $Y \ is \ X + B$ respectively, the deduction engine produces the result $Y \ is \ A + B$, and the following fuzzy addition is performed:

$$TriFuzzy(150, 15) + 150 \times 100\% = TriFuzzy(300, 15)$$

where $TriFuzzy(a, b)$ represents a triangular fuzzy set whose center is at $a$ and the width is $b$. The final answer is returned as "about 300" where 300 is the defuzzified value of the composite fuzzy set. QA system built on standard search technology cannot provide this type of response. In practice, all key concepts in a corpus are not likely to be pre-defined with fuzzy set representation, but for a particular subsection(s) where the application manager expects many questions and desires to provide detailed answers, PNL-based computing offers value-added performance.

Examples of phrase-based deductive reasoning are provided below. For example, given two facts:

(i) If over 50 percent of customers complain about new product WRTG54, all types of WRTG routers will be recalled.
(ii) If all types of WRTG routers are recalled, we will start new sale on TRG100 type of router.

If the query is: 'Why did you start new sale of TRG100?' The system returns two answers:

*(1) If all types of WRTG routers are recalled , we will start new sale on TRG100 type of router.*
*(2) If over 50 percent of customers complain about new product WRTG54 then we will start new sale on TRG100 type of router.*

If the query is: 'Why would you not start new sale of TRG100?'. The answers are:

*(3) NOT (we will start on a new sale on TRG100 type of router) IMPLIES NOT (all types of WRTG routers are recalled).*
*(4) NOT (we will start on a new sale on TRG100 type of router) IMPLIES NOT (over 50 percent customers complain about new product WRTG54.*
The second set of answers (3 and 4) result from the transitive implications of facts (i) and (ii).

At this juncture, a result processor has not been implemented to 're-phrase' answers in human-friendly form, but the reasoning logic is correct.

## 5 Summary

We presented an application of PNL-based reasoning, and in particular, discussed its role in a deduction engine for QA systems. While many NL expression can be precisiated manually (i.e. with human interpretations of semantics), a smaller subset is amenable to automated precisiation. As a first attempt at (semi-)automated application of PNL-based reasoning, the input was limited to simple NL sentences (i.e. with single verb phrase) that can be precisiated and abstracted to PNL protoforms. PNL-based processing converts the complex problem of NLP operations and deduction with natural language phrases into simple substitution operations of algebraic expressions. Moreover, if associated fuzzy concepts are pre-defined with fuzzy sets, more accurate answers (within the constraints of f-granularity) can be computed. The limitation (in the context of advanced search or QA applications) however, is that since only a subset of natural language sentence forms can be precisiated in a (semi-)automated sense, PNL-based computing cannot be used as a stand-alone solution, and must be supplemented by standard NLP and information retrieval techniques to realize a complete and practical system. The hybrid, multi-pipe approach described above provides one way to implement an integrated solution. This system remains under development, and further development and testing are planned using TREC's QA data sources [9]. Further research topics also include concept matching, ranking algorithms, different knowledge representation schemes such as RDF and OWL formats [10], and integration of additional NLP tools such as entity extractors for analyzing corpus knowledge and query phrases.

## Acknowledgements

Qin and Beg are British Telecommunications (BT) Research Fellows in BISC Group. This research was funded by BT/BISC Fellowship.

## References

1. M. M. S. Beg, M. Thint, Z. Qin, Precisiating natural language for a question answering system, submitted to *the 11th World Multi Conf. on Systemics, Cybernetics, and Informatics*, 2007.
2. F. Benamara, Cooperative question answering in restricted domains: the WEBCOOP experiment, *Proceedings ACL 2004 Workshop on Question Answering in Restricted Domains*, 2004.
3. W. Ceusters W, B. Smith, M. Van Mol, Using ontology in query answering systems: scenarios, requirements and challenges. In Bernardi R, Moortgat M (Eds.) *Proceedings of the 2nd CoLogNET-ElsNET Symposium*, Amsterdam, pp.5-15, 2003.
4. H. Chung, Y.-I. Song, K.-S. Han, D.-S. Yoon, J.-Y. Lee, H.-C. Rim and S. -H. Kim, A practical QA system in restricted domains, 2004.
5. H. Doan-Nguyen, L. Kosseim, The problem of precision in restricted-domain question-answering: some proposed methods of improvements, *Proceedings ACL 2004 Workshop on Question Answering in Restricted Domains*, 2004.
6. O. Gospodnetic, E. Hatcher, *Lucene in Action*, Manning, 2004.
7. G. Miller, Wordnet: a lexical database, *Communications of the ACM*, 38(11), 1995, pp. 39-41.
8. M. Thint, M. M. S. Beg and Z. Qin, PNL-enhanced restricted domain question answering system, submitted to IEEE-FUZZ, London, UK.
9. Available: http://trec.nist.gov/data/qa.html
10. Available: http://www.w3.org.org/TR/owl-ref
11. O. Tsur, M. de Rijke, and K. Sima'an, BioGrapher: biography questions as a restricted domain question answering task, *Proceedings ACL 2004 Workshop on Question Answering in Restricted Domains*, 2004.
12. L. A. Zadeh, A new direction in AI - toward a computational theory of perceptions, *A.I. Magazine*, Spring 2001.
13. L. A. Zadeh, From computing with numbers to computing with words – from manipulation of measurements to manipulation of perceptions, *Int. J. Appl. Math. Comput. Sci.,* Vol. 12/3: pp. 307-324, 2001.
14. L. A. Zadeh, Toward a generalized theory of uncertainty (GTU) – an outline, *Information Sciences*, Vol. 172: pp. 1-40, 2005

# Granular Computing and Modeling the Human Thoughts in Web Documents

Tsau Young Lin

Department of Computer Science, San Jose State University
San Jose, CA 95192-0249
Berkeley Initiative in Soft Computing
University of California, Berkeley
tylin@cs.sjsu.edu

**Abstract.** The totality of human thoughts in a document set is modeled by a polyhedron. A point represents a THOUGHT, a simplex a CONCEPT, a connected component a COMPLETE CONCEPT, the simplicial structure the whole IDEA. The building block is the simplex; it represents the concept that is carried by a set of high frequency and nearby co-occurring keywords. The simplicial structure of the keywords provides an "informal" formal language about human thoughts in a document set. The model theory of this language gives the desirable model.

**Keyword:** granular computing, neighborhood system, rough set, topology, simplicial complex.

## 1 Introduction

This is our attempt to model the human thinking in a document set. The backbone is concept analysis. First, the collection of concepts in a document set is represented by a geometric structure, called simplicial complex in combinatorial topology [7]. A simplex is simply a high frequency and nearby keywordset that carries a concept. The collection of all these simplexes represents the structure of these concepts. These concepts form an "informal" formal language of the idea in the document. A model of this language is the desirable model.

The concept defined in a document set does satisfy a conditional called Apriori Condition. This is due to we define the concept by frequent keyword set. In general, we may not have the structure of simplicial complex. But the set of concepts is a granular model in the sense of [6,5].

## 2 Granular Computing and Models

Granulation is a natural problem-solving methodology deeply rooted in human thinking; Human body is granulated into head, neck, and etc. The notion is intrinsically fuzzy, vague and imprecise. So the classical strategy Divide and Conquer is actually Granulate and Conquer. Mathematicians idealized it into the notion

P. Melin et al. (Eds.): IFSA 2007, LNAI 4529, pp. 263–270, 2007.
© Springer-Verlag Berlin Heidelberg 2007

of partitions, and developed it into a fundamental problem-solving methodology. Nevertheless, the notion of partitions, which absolutely does not permit any overlapping among its granules, seems to be too restrictive for real world problems. A more general theory seems needed; we have called it Granular Computing [22],[23].

In this paper, we will adopt the idea from topological spaces. A topological space is often defined as a pair of objects that consists of one classical set and one family of open sets that satisfies certain axioms. We will take this scheme and propose to define

**Definition 1.** *A set based granular model is defined to be a pair* $(U, \beta)$*, where* $U$ *is the universe of discourse, called global granular space and* $\beta$ *is a family of subsets* $X_j, j \in J$*, called basic granules or elementary granules, just to be in sync with rough set theory [16], where* $J$ *is an index set.*

We will illustrate the model with a set delicate examples:

**Example 1.** *Let us assume* $U$ *consists of 4 points; see the first tetrahedron in Figure 3 below.* $\beta$ *is organized as follows:*

1. $X_4^1$ *consists of four elements* $\{a, b, c, d\}$*; It can be thought of as a set of linearly independent 4 points in Euclidean space. They determined an open tetrahedron (a open simplex of dimension 3).*
2. *Each one of* $X_3^1$, $X_3^2$, $X_3^3$, *and* $X_3^4$ *consists of three element. Each can be thought of as a set of open triangles (2 dimensional faces) of the tetrahedron.*
3. *Each one of* $X_2^1$, $X_2^2$, $X_2^3$, $X_2^4$ $X_2^5$, *and* $X_2^6$ *consist of two elements. Each is a simplex of dimension 1.*
4. *Each one of* $X_1^1$, $X_1^2$, $X_1^3$, *and* $X_1^4$ *consists of one elements. Each is a simplex of dimension 0.*
5. *This collection of all these simplexes above form a simplicial complex.*

Intuitively $U$ is the data set, the universe of discourse. Each simplex represents some basic knowledge about the data. This granular model has the power set as its granular structure $\beta$.

We can also consider $|U|$= closed Tetrahedron. In this case the $\beta$ is a partition: 1 open tetrahedron, 4 open triangles, 6 open segments and 4 vertices forms a partition. Under this view, each open simplex (tetrahedron, triangles, open segments, vertices) is a basic knowledge - this is a view from rough set theory. If we take the simplicial complex view, each X's is a granule(a combinatorial simplex). It can be realized as an open simplex in Euclidean space, hence the combinatorial simplex should be a basic knowledge, deriving from rough set theory.

## 3 The Structure of Concepts in Documents

In this section, we build the Simplicial Complex of Concepts. A concept is carried by a frequent keywordset; recall the analogous term, frequent itemset in association rules.

The major step is to convert linear-text-representations (documents) into a simplicial-complex-representation. This simplicial-complex is quite foreign to computer scientists, however, it is an old notion in combinatorial topology ( cf Section 3.1). We may schematically summarize our approach as follows: Note that $\Longrightarrow$ denote the "information flow," which not a mathematical mapping yet. The goal of this paper is to have a mathematical model of author's idea, so that $\Longrightarrow$ becomes a mathematical mapping

$$\begin{bmatrix} author's \\ idea \end{bmatrix} \Longrightarrow \begin{bmatrix} linear \\ text \end{bmatrix}$$

$$\Downarrow \text{(summarize)}$$

$$IDEA \equiv \begin{bmatrix} common \\ thoughts \end{bmatrix} \Longrightarrow \begin{bmatrix} simplicial \\ complex \end{bmatrix}$$

### 3.1 Combinatorial Topology

We will review a branch of mathematics, called combinatorial topology. A n-dimensional Euclidean space is a space in which elements can be addressed using the Cartesian product of n sets of real numbers. A unit point is a point which coordinates are all 0 but a single 1, $(0, \ldots, 0,\ 1,\ 0, \ldots,\ 0)$. These unit points will be regarded as vertices. We will use them to illustrate the notion of $n$-simplex.

Let us examine the $n$-simplexes, when $n = 0,\ 1,\ 2,\ 3$. A 0-simplex $\Delta(v_0)$ consists of a vertex $v_0$, which is a point in the Euclidean space. A 1-simplex $\Delta(v_0, v_1)$ consists of two points $\{v_0,\ v_1\}$. These two points can be interpreted as an open segment $(v_0,\ v_1)$ in Euclidean space; note that it does not include its end points. A 2-simplex $\Delta(v_0, v_1, v_2)$ consists of three points $\{v_0,\ v_1,\ v_2\}$. These three points can be interpreted as an open triangle with vertices $v_0$, $v_1$, and $v_2$, which does not include its edges and vertices. A 3-simplex $\Delta(v_0, v_1, v_2, v_3)$ consists of four points $\{v_0,\ v_1,\ v_2,\ v_3\}$ that can be interpreted as an open tetrahedron. Again, it does not includes any of its boundaries.

In general, vertices do not have to be points in Euclidean space, they can be any kind of objects. Formally,

**Definition 2.** *A $n$-simplex, denoted by $\Delta(v_0, \ldots, v_n)$, is a set of independent abstract vertices $\{v_0, \ldots, v_n\}$. A $q$-subset of a $n$-simplex is called a $q$-face; it is a $q$-simplex $\Delta(v_{j_0}, \ldots, v_{j_q})$ whose vertices are a subset of $\{v_0, \ldots,\ v_n\}$ with cardinality $q + 1$.*

More generally, we have ([19], pp. 108).

**Definition 3.** *A simplicial complex $C$ consists of a set $\{v\}$ of vertices and a set $\{s\}$ of finite nonempty subsets of $\{v\}$ called simplexes such that*

- *Any set consisting of one vertex is a simplex.*
- *Closed condition: Any nonempty subset of a simplex is a simplex.*

*Any simplex $s$ containing exactly $q + 1$ vertices is called a $q$-simplex. We also say that the dimension of $s$ is $q$ and write dim $s$=$q$. We will refer to $C$ as a non-closed simplicial complex, if the closed condition is not fulfilled for all its constituting simplexes.*

A simplex is said to be maximal if it is not a face of any other simplex.

### 3.2 Topology of Linear Texts

Let us first recall the notion of Apriori condition.

**Definition 4.** *Apriori condition: Any q-subset of a n-keywordset is a q-keywordset for* $q \leq n$.

where a set containing exactly q elements is abbreviated as a q-subset. If we regard a keyword as vertex, and a keywordset as a simplex, then the Apriori condition is the closed condition of the simplicial complex (cf Section 3.1). With this observation, we have the following theorem:

**Theorem 1.** *The pair* ($V_{text}$, $S_{text}$) *is an Abstract Simplicial Complex, where*

1. $V_{text}$ *is the set of keywords, and is regarded as a set of abstract vertices, called keyword-vertices.*
2. $S_{text}$ *is the set of keywordsets (associations) and is regarded as a set of abstract simplexes, called keyword-simplexes.*

This simplicial complex is called a Keyword simplicial Complex (KSC). Note that this

- KSC is a combinatorial structure that consists, except the vertices, of "invisible objects" hidden in the document set; see Section 3.3.

To see its strength, let us recall a striking theorem: Using the geometry, we can determine if two sets of documents written in different languages are similar, even without translation.

**Corollary 1.** *Let A and B be two document sets, where B is a translation of A into another language then the simplicial complexes of A and the simplicial complexes of B are isomorphic.*

### 3.3 Hidden Semantics - "Reading Between Lines"

Let us start intuitively. Each document describes some idea in author's mind, which may consist of many levels and wide ranges of concepts. We will explore some of it through high frequency keywords and keywordsets. Following the terminology in association mining, we may also use

- an association and a keywordset are synonyms.

Roughly, association is stressed more on the semantics of a keywordset, while the latter one is a phenomena of document set. They will be used interchangeably throughout this paper. First, let us recall some examples.

1. The keywordset "Wall Street" represents a concept that has nothing to do with "Wall" and "Street"
2. The keywordset "White House" represents an object that has very little to do with "White" and "House."

These examples indicate that the strength of this approach is the ability to capture the notion that is defined implicitly by the keywordset, in plain words,

- the capability of "reading between lines."

## 4 The Language of Concepts in Documents

The simplicial complex is a structure of concepts in the document set. For concept analysis, we need a language (a set of notations) to describe concepts, which consist of set of "authors' thoughts" and "their interactions" hidden in the linear texts; all "interactions" are invisible in the text.

**L1)** IDEA is a notation to denote the totality of common thoughts or concepts among all authors of the documents under consideration; it is the universe of discourse.
**L2)** THOUGHT is a notation to denote some important basic ingredient, called element, in IDEA.

Please note that IDEA, at this point, is not a mathematical object yet. Instead, we merely view it as an element of an informal language. Human THOUGHT is not a mathematical concept either. The goal here is to establish

- IDEA is a set of THOUGHTs.

So both are mathematical objects. Loosely speaking, the main goal of this paper is to discover/construct a reasonable mathematical model for IDEA. So the language (a set of notations) will be interpreted into a mathematical model. This model will be the base of our concept analysis.

**L3)** B-concept($k$), read as basic concept, is a notation to denote the MEANING of the keyword $k$.
**L4)** I-concept($\Delta$), read as intermediate concept, is a notation to denote the MEANING of a $q$-simplex $\Delta$.
**L5)** P-concept, read as primitive concept, is a notation to denote the I-concept of a maximal simplex.
**L6)** C-concept, read as complete concept is a notation to denote the MEANING of a connected component.

So we have sufficient notations to conduct concept analysis. However, we do not have adequate facilities in the language, such as a deductive system, etc... to reason about these concepts and understand the interrelationships among them. Therefore, we will set up, in Tarski's style semantics, namely, a mathematical model to interpret this language.

## 5 Polyhedrons- Latent Semantic Model

We may map $V_{text}$ as the unit points of an Euclidean space (cf Section 3.1). By doing so, each keywordset in $S_{text}$ can be interpreted as an open simplex in Euclidean space [19].

**Definition 5.** *The union of all these open simplexes is the Polyhedron* $|S_{text}|$ *of the Linear Texts.*

The Polyhedron is a closed set (due to the closed condition) in the Euclidean space. Next, we will introduce an important notion: star($\Delta$) [19]. We will denote it by $S^*(\Delta)$.

**Definition 6.** $S^*(\Delta) = \cup\{$ *All open simplexes that have* $\Delta$ *as a face* $\}$.

In the case when $\Delta = k$ is a keyword, $S^*(k)$ is an open set in $|S_{text}|$. It is the minimal open neighborhood of $k$ that is definable as simplexes only. It is referred to as the basic neighborhood or basic granule of $k$. Consider it as the first building block of our simplicial complex. For example, we have

**Proposition 1.** $S^*(\Delta) = \cap\{S^*(k_i) \mid k_i \in \Delta\}$.
$S^*(\Delta_{max}) = \Delta + max$ *for a maximal simplex* $\Delta_{max}$.

### 5.1 Latent Semantic Model

Now, we are ready to introduce the model: Let $P = |S_{text}|$ be the polyhedron of KSC.

**Definition 7.** *The Latent Semantic Model consists of a KSC and its Polyhedron:*

- *IDEA is modeled by the polyhedron P.*
- *a human THOUGHT is an element in IDEA. P is referred to as the Latent Semantic Space.*
- *The mapping* $S^* : KSC \longrightarrow P$ *realize the interpretations, for example, a B-concept is mapped to an open set.*

With this interpretation, all language elements, such as concepts, are granules(subsets) in $P$. In particular,

- B-concept($k$) = $S^*(k)$
- I-concept($\Delta$) = $S^*(\Delta) = \cap\{S^*(k_i) \mid k_i \in \Delta\}$
- P-concept is the I-concept of a maximal simplex.
- C-concept is a connected component.

It should be clear that the set of I-concepts forms a semigroup generated by $S^*(k)$ via intersections. This semigroup is a partial order set obtained using the "inclusion" of set theory. On the other hand, the I-concept($\Delta$) is also a partial order set, but this time, due to the face-relation of simplexes $\Delta$).

**Proposition 2.** *The collection of I-concepts has two partial orderings: One by KSC, and another by the set inclusion in the polyhedron. Of importance, these two partial orders do agree.*

We leave it to the reader to verify that the required properties are verified by these sets. This collection is indexed by the KSC. We so we will call it the Knowledge Complex and denote it by $\beta$. It is far more complex than a concept hierarchy. In particular, the pair $(P, \beta)$, where $\beta$=KSC, is a granular model in the sense of [6,5].

## 6 Conclusions

In this paper, we focus on simplicial complex, which is a special form of (set based) granular model. This is because, we use frequency as the criterion for keywordset. If we use generalized TFIDF, then the set of simplexes may not have the closed condition. So the resulting model is a non-closed subset of the simplicial complex. In this case, simplical complexes are two restrictive, we need the full blown set based granular model.

## References

1. Bairamian, S. *Goal Search in Relational Databases*, Thesis, California State University at Northridge, 1989.
2. Brewer, D. C. and Nash, M. J.: "The Chinese Wall Security Policy" IEEE Symposium on Security and Privacy, Oakland, May, 1988, pp 206-214,
3. Chu, W. and Chen, Q. Neighborhood and associative query answering, *Journal of Intelligent Information Systems*, **1**, 355-382, 1992.
4. D. Dubois and H. Prade, "Putting rough sets and fuzzy sets together," in: Decision Support by Experience - Application of the Rough Sets Theory, R. Slowinski (ed.), Kluwer Academic Publishers, 1992, 203-232.
5. T. Y. Lin, "Granular Computing II: Infrastructures for AI-Engineering." In the Proceedings of 2006 International Conference on Granular Computing, May 10-12, Atlanta, Georgia, USA
6. T. Y. Lin."Granular Computing: Examples, Intuitions and Modeling." In: the Proceedings of 2005 IEEE International Conference on Granular Computing," July 25-27, 2005, Beijing China, 40-44.
7. T. Y. Lin and I-Jen Chiang, "A simplicial complex, a hypergraph, structure in the latent semantic space of document clustering", International Journal of Approximate Reasoning, 55-80, vol 44,2005
8. T. Y. Lin, "Granular Computing: Structures, Representations, Applications and Future Directions" in: the Proceedings of 9th International Conference, RSFDGrC 2003, Chongqing, China, May 2003, Lecture Notes on Artificial Intelligence LNAI 2639, Springer-Verlag, 16-24.
9. T. Y. Lin, "Granular Computing on Binary Relations II: Rough Set Representations and Belief Functions." In: Rough Sets In Knowledge Discovery, A. Skowron and L. Polkowski (eds), Springer-Uerlag, 1998, 121-140.
10. T. Y. Lin, "Granular Computing on Binary Relations I: Data Mining and Neighborhood Systems." In: Rough Sets In Knowledge Discovery, A. Skowron and L. Polkowski (eds), Springer-Uerlag, 1998, 107-121.
11. Lin, T. Y."Topological and Fuzzy Rough Sets," in: Decision Support by Experience - Application of the Rough Sets Theory, R. Slowinski (ed.), Kluwer Academic Publishers, 1992, 287-304.
12. T. Y. Lin, K. J. Huang, Q. Liu, and W. Chen, "Rough Sets, Neighborhood Systems and Approximation," Proceedings of the Fifth International Symposium on Methodologies of Intelligent Systems, Selected Papers, Knoxville, Tennessee, October 25-27, 1990, 130-141.
13. Lin, T. Y. Chinese Wall Security Policy–An Aggressive Model, Proceedings of the Fifth Aerospace Computer Security Application Conference, Tuscon, Arizona, Dec 1989, 282-289.

14. Lin, T. Y. Neighborhood Systems and Relational Database. In: Proceedings of 1988 ACM Sixteen Annual Computer Science Conference, Februay 1988, p. 725 (abstract)
15. James Munkres, Topology (2nd Edition), PRENTICE-HALL, 2000
16. Z. Pawlak, Rough sets. Theoretical Aspects of Reasoning about Data, Kluwer Academic Publishers, 1991
17. G. Polya, "How to Solve It", 2nd ed., Princeton University Press, 1957, ISBN 0-691-08097-6.
18. W. Sierpinski and C. Krieger, General Topology, University of Toronto Press 1952.
19. E. Spanier. *Algebric Topology.* McGraw-Hill Book Company, New York, NY, 1966.
20. Zadeh, L.A., "Fuzzy sets and information granularity," in: M. Gupta, R. Ragade, and R. Yager (Eds.), Advances in Fuzzy Set Theory and Applications, North-Holland, Amsterdam, 3-18, 1979.
21. Zadeh, L.A. "Towards a theory of fuzzy information granulation and its centrality in human reasoning and fuzzy logic." *Fuzzy Sets and Systems*, 1997, 19:111–127.
22. Zadeh, L. A. "Some reflections on soft computing, granular computing and their roles in the conception, design and utilization of information/ intelligent systems," Soft Computing, 1998, 2, 23-25.
23. Zadeh, L. A. "Some Reflections on Information Granulation and its Centrality in Granular Computing, Computing with Words, the Computational Theory of Perceptions and Precisiated Natural Language." In: Lin, Yao, Zadeh (eds), Data Mining, Rough Sets, and Granular Computing. 2002, 3-20.

# Part VI

# Perception Based Data Mining and Decision Making

# Extracting Fuzzy Linguistic Summaries Based on Including Degree Theory and FCA

Li Zhang, Zheng Pei, and Honghua Chen

School of Mathematics & Computer, Xihua University,
Chengdu, Sichuan, 610039, China
zzllxz@yahoo.com.cn, pqyz@263.net

**Abstract.** In information systems (or database), generally, attribute values of objects are numeral or symbols, from application point of view, linguistic information or decision rules are widely used. Hence, fuzzy linguistic summaries would be very desirable and human consistent. In this paper, extracting fuzzy linguistic summaries from a continuous information system is discussed. Due to fuzzy linguistic summaries can not be extracted directly in the information system, fuzzy information system is used to discretize the continuous information system, and level cut set is used to obtain classical information system firstly. Then based on including degree theory and formal concept analysis (FCA), simple fuzzy linguistic summaries are extracted. To extract complex linguistic summaries, logical conjunctions $\vee, \wedge$ and $\rightarrow$ are used. An Example of checking quality of sweetened full cream milk powder is also provided.

## 1 Introduction

As we known, in marketing prediction, information management, etc, Knowledge (or rules) are represented by natural languages. On the other hand, all information are stored by database which is also called information systems (or formal context, etc) in formal concept analysis (FCA), and its values are numeric or symbolic. How to extract linguistic knowledge, such as, fuzzy linguistic summaries [1], from database is a problem of knowledge discovery [2]. The methodology of computing with words (CWW) proposed by Zadeh [3]-[6] may be viewed as an attempt to harness the highly expressive power of natural languages by developing ways of CWW or propositions drawn from a natural language. Based on difference background, many important methods and results have been proposed to extract linguistic summaries from numerical database [1], [7]-[11]. In this paper, extracting fuzzy linguistic summaries are concerned on continuous information systems. Then FCA [12] and including degree theory [13] are used to obtain fuzzy linguistic summaries.

## 2 Continuous Information Systems and Its Discretization

Formally, an information system $S$ (or called formal context in FCA) is a quaternion denoted as $(U, A, V, f)$, where $U = \{x_1, x_2, \cdots, x_n\}$ is a non-empty set of objects, $A = \{a_1, a_2, \cdots, a_m\}$ is a non-empty finite set of attributes, $V = \bigcup_{a \in A} V_a$

P. Melin et al. (Eds.): IFSA 2007, LNAI 4529, pp. 273–283, 2007.
© Springer-Verlag Berlin Heidelberg 2007

and $V_a$ is the domain of $a$, $f : U \times A \rightarrow V$ is information function. In which $\forall x \in U$, $a \in A$, $f(x, a) = f_a(x)$. An information system can be expressed by Table 1.

**Table 1.** Information systems $S = (U, A, V, f)$

| $U \setminus A$ | $a_1$ | $a_2$ | $\cdots$ | $a_{m-1}$ | $a_m$ |
|---|---|---|---|---|---|
| $x_1$ | $f_{a_1}(x_1)$ | $f_{a_2}(x_1)$ | $\cdots$ | $f_{a_{(m-1)}}(x_1)$ | $f_{a_m}(x_1)$ |
| $\vdots$ | $\vdots$ | $\vdots$ | $\vdots$ | $\vdots$ | $\vdots$ |
| $x_n$ | $f_{a_1}(x_n)$ | $f_{a_2}(x_n)$ | $\cdots$ | $f_{a_{(m-1)}}(x_n)$ | $f_{a_m}(x_n)$ |

As our knowledge, so called continuous information systems are such that for some $a \in A$, $|V_a| = \infty$ or $\forall x_i \neq x_j \Longrightarrow f_a(x_i) \neq f_a(x_j)$. As pointed out in rough set and FCA, a continuous information system need to be discretized before extracting knowledge (concept, rules, etc) [14], [15]. In this paper, our discussion is aimed at that attributes are fuzzy linguistic variables, *e.g.*, in checking quality of sweetened full cream milk powder by quantities of protein, fat, lactose, etc, for these linguistic variables, fuzzy linguistic values are high, middle, low, etc. From this point of view, the attribute can be discretized by fuzzy linguistic values, and called fuzzy discretization.

Formally, let attribute $a_i$ be fuzzy linguistic variable, its fuzzy linguistic values be $\{v_{a_i}^1, \cdots, v_{a_i}^{r_i}\}$. Generally, $v_{a_j}^i (i = 1, \cdots, r_j)$ can be generated by linguistic hedge and atomic evaluating syntagm [16]-[18]. In real world application, the number of fuzzy linguistic values is decided by user or experts. Due to the domain of fuzzy linguistic variable $a_j$ be known, suppose membership functions of fuzzy linguistic values are as follows

$$\mu_{v_{a_j}^i} : V_{a_j} \longrightarrow [0, 1]. \tag{1}$$

There are many methods to decide $\mu_{v_{a_j}^i}$ [19], [20], here, suppose membership function $\mu_{v_{a_j}^i}$ of every $v_{a_j}^i$ is known. Based on (1), a continuous information system can be transformed as follows, $\forall x_j \in U$ and $a_i \in A$,

$$\begin{aligned} f' : U \times \{a_i\} &\longrightarrow \mu_{v_{a_i}^1} \times \cdots \times \mu_{v_{a_i}^{r_i}}, \\ (x_j, a_i) &\mapsto (\mu_{v_{a_i}^1}(f_{a_i}(x_j)), \cdots, \mu_{v_{a_i}^{r_i}}(f_{a_i}(x_j))). \end{aligned} \tag{2}$$

Hence, $(U, A, V, f')$ is a fuzzy information system [21], see Table 2. When $\alpha$-level value is considered, *i.e.*, $\forall a_k \in A$,

$$\mu_{v_{a_k}^i}(f_{a_k}(x_j)) = \begin{cases} 1, & if\ \mu_{v_{a_k}^i}(f_{a_k}(x_j)) \geq \alpha, \\ 0, & if\ \mu_{v_{a_k}^i}(f_{a_k}(x_j)) < \alpha. \end{cases}$$

Then Table 2 is transformed as a classical information system (or a classical formal context) $(U, A, V, f'_\alpha)$.

**Table 2.** Fuzzy information systems $S = (U, A, V, f')$

| $U \setminus A$ | $v_{a_1}^1$ | $\cdots$ | $v_{a_1}^{r_1}$ | $\cdots$ | $v_{a_m}^1$ | $\cdots$ | $v_{a_m}^{r_m}$ |
|---|---|---|---|---|---|---|---|
| $x_1$ | $\mu_{v_{a_1}^1}(f_{a_1}(x_1))$ | $\cdots$ | $\mu_{v_{a_1}^{r_1}}(f_{a_1}(x_1))$ | $\cdots$ | $\mu_{v_{a_m}^1}(f_{a_m}(x_1))$ | $\cdots$ | $\mu_{v_{a_m}^{r_m}}(f_{a_m}(x_1))$ |
| $\vdots$ | $\vdots$ | $\vdots$ | $\vdots$ | $\vdots$ | $\vdots$ | $\vdots$ | $\vdots$ |
| $x_n$ | $\mu_{v_a^1}(f_{a_1}(x_n))$ | $\cdots$ | $\mu_{v_{a_1}^{r_1}}(f_{a_1}(x_n))$ | $\cdots$ | $\mu_{v_{a_m}^1}(f_{a_m}(x_n))$ | $\cdots$ | $\mu_{v_{a_m}^{r_m}}(f_{a_m}(x_n))$ |

## 3 Fuzzy Linguistic Summary

A fuzzy linguistic summary of information system consists of [1]: a) summarizer $v_a^i$, *e.g.*, high-protein; b) a quantity in agreement $q_{l'}$, *e.g.*, most of, etc; c) truth $t_{k'}$, *e.g.*, true. Hence, a linguistic summary can be formalized by "($q_{l'}$ $x's$ are $v_a^i$) is $t_{k'}$", in which $q_{l'}$ is a fuzzy linguistic quantifier [16], [22], $v_a^i$ is fuzzy linguistic value on $U = \{x_i | i = 1, \cdots, n\}$, $t_{k'}$ is fuzzy truth value. *e.g.*,

$$most\ of\ (q_{l'})\ milk\ powder(x's)\ are\ high-protein\ (v_a^i)\ is\ true(t_{k'}). \quad (3)$$

In this paper, we always assume that $Q = \{q_1, q_2, \cdots, q_l\}$, $P = \{v_{a_1}^1, \cdots, v_{a_1}^{r_1}, \cdots, v_{a_m}^1, \cdots, v_{a_m}^{r_m}\}$, for each $q_{l'} \in Q$, $v_a^i \in P$ are linguistic terms, *e.g.*, *most of*, *young*. For $v_a^i \in P$, it is the fuzzy linguistic value on $V_a$, *i.e.*, $\mu_{v_a^i} : V_a \longrightarrow [0,1]$. In this paper, suppose each summarizer $v_a^i \in P$ has membership function on $V_a$.

According to (3), extracting linguistic summaries are equal to deciding $q_{l'}$, $v_a^i$ and $t_{k'}$. For fuzzy linguistic quantifier $q_{l'} \in Q$, it expresses that how many object $x_i$ is such that (3). Here, membership functions of $q_{l'} \in Q$ and $t_{k'} \in T$ are defined as follows, respectively [10]:

1. Let $R(U) = \{X | X \subseteq U\}$ be the power set of $U$. Define a binary relation on $R(U)$: $X \sim Y$ if and only if $| X |=| Y |$, where $| X |$ is the cardinal number of $X$. Obviously, "$\sim$" is an equivalence relation on $R(U)$. The factor set of $R(U)$ by $\sim$ is denoted by $\overline{R}(U) = R(U)/\sim$.
2. For each fuzzy linguistic quantifier $q_{l'} \in Q$, its fuzzy set is defined by

$$\mu_{q_{l'}} : \overline{R}(U) \longrightarrow [0,1]. \quad (4)$$

3. For each fuzzy linguistic truth degree $t_{k'} \in T$, its fuzzy set is defined by

$$\mu_{t_{k'}} : [0,1] \longrightarrow [0,1]. \quad (5)$$

## 4 Extracting Simple Fuzzy Linguistic Summaries

In information processing based on information systems, including degree theory and FCA are important tools. Including degree theory expresses soft inclusion relation between two sets from quantity point of view [13]. FCA is used to represent relation between objects and attributes [12]. In fuzzy linguistic summary, deciding fuzzy quantifier $q_{l'}$ involves two aspects: one is objects which have fuzzy linguistic value (attribute) $v_a^i$; the other is how many objects has $v_a^i$. In this paper, FCA is used to decide objects which have fuzzy linguistic value (attribute) $v_a^i$, then including degree theory is used to select fuzzy quantifier $q_{l'}$. Formally, including degree theory and FCA are viewed as follows.

For a classical information system $(U, A, V, f'_\alpha)$, operators $^\uparrow$ and $^\downarrow$ on $X \subseteq U$ and $B \subseteq A$ are defined respectively [12]:

$$X^\uparrow = \{v_a^i \in A | \forall x_j \in X, f'_\alpha(x_j, v_a^i) = 1\}, \tag{6}$$

$$B^\downarrow = \{x_j \in U | \forall v_a^i \in B, f'_\alpha(x_j, v_a^i) = 1\}. \tag{7}$$

If a pair $(X, B)$ satisfies $X^\uparrow = B$ and $B^\downarrow = X$, then the pair $(X, B)$ is called as a formal concept. And $X$ is called as the extent of the formal concept, $B$ is called as the intent of the formal concept.

Let $U$ be a finite set of objects, $\forall X,\ Y \subseteq U$, $D(Y/X)$ is called as including degree if it satisfies [13]:

**(1)** $0 \leq D(Y/X) \leq 1$;
**(2)** $X \subseteq Y \Rightarrow D(Y/X) = 1$;
**(3)** $X \subseteq Y \subseteq Z \Rightarrow D(X/Z) \leq D(X/Y)$.

Generally, there are many forms of $D(Y/X)$, *e.g.*,

$$D(Y/X) = \frac{|X \cap Y|}{|X|}. \tag{8}$$

It is easy to prove (8) is including degrees.

Let a simple linguistic summary be "($q_{l'}$ $x's$ are $v_a^i$) is $t_{k'}$", then $q_{l'}$, $v_a^i$ and $t_{k'}$ can be obtained automatically by the following steps [10],

1. In fuzzy information system $(U, A, V, f')$, fixed $\alpha$-level value, $(U, A, V, f')$ is transformed as a classical information system, denoted $(U, A, V, f'_\alpha)$.
2. In $(U, A, V, f'_\alpha)$, $\forall v_a^i \in P$, obtaining $(v_a^i)^\downarrow$. According to (7), $(v_a^i)^\downarrow = \{x_j \in U | f'_\alpha(x_j, v_a^i) = 1\}$.
3. According to (8), $D((v_a^i)^\downarrow/U) = \frac{|(v_a^i)^\downarrow \cap U|}{|U|}$ $(v_a^i \in P)$.
4. Fixing a level (threshold) $\beta$ which can be done by experts or deciders, denote

$$P_\beta = \{v_a^i | D((v_a^i)^\downarrow/U) \geq \beta\}. \tag{9}$$

5. For $v_a^i \in P_\beta$, corresponding the fuzzy linguistic quantifier $q_{l'} \in Q$ can be selected such that

$$\mu_{q_{l'}}(N) = max\{\mu_{q_1}(N), \mu_{q2}(N), \cdots, \mu_{ql}(N)\}, \tag{10}$$

in which $N = |(v_a^i)^\downarrow| (v_a^i \in P_\beta)$. In the real world practice, maybe $\mu_{q_{l'}}(N)$ is not only one, in this case, one $q_{l'}$ can be selected by deciders or the indexes of linguistic terms, *e.g.*, $q_{l'}$ has a maximal index.

**Table 3.** Sweetened full cream milk powder

| Product\ Index | protein(g/100g) | fat(g/100g) | sugar(g/100g) |
|---|---|---|---|
| Yili($x_1$) | 19.8 | 23.40 | 19.74 |
| Haixiang($x_2$) | 20.60 | 24.89 | 17.06 |
| Hongxing($x_3$) | 19.72 | 24.14 | 13.97 |
| Nate($x_4$) | 18.52 | 21.20 | 19.84 |
| Gongnong($x_5$) | 20.06 | 21.20 | 17.94 |
| Dongxing($x_6$) | 17.09 | 24.10 | 19.86 |
| Wondersun($x_7$) | 19.06 | 21.92 | 19.06 |
| Qinyong($x_8$) | 18.62 | 21.18 | 19.50 |
| Gucheng($x_9$) | 18.82 | 20.46 | 19.58 |
| Senyong($x_{10}$) | 19.67 | 21.93 | 19.72 |
| Flying crane($x_{11}$) | 18.45 | 23.01 | 19.84 |
| Mengniu($x_{12}$) | 18.52 | 22.82 | 19.93 |
| Haihe($x_{13}$) | 20.52 | 23.35 | 17.98 |
| Nanshan($x_{14}$) | 19.01 | 22.94 | 19.95 |
| Heshi($x_{15}$) | 18.88 | 23.33 | 19.00 |
| Shenguo($x_{16}$) | 18.74 | 21.96 | 19.26 |
| Nestle($x_{17}$) | 18.82 | 24.18 | 19.62 |
| Sanlu($x_{18}$) | 18.98 | 22.96 | 16.69 |

6. Selecting a fuzzy linguistic truth degree $t_{k'} \in T$. From the viewpoint of logic, the more objects satisfying statement with the quantifier, the higher the truth degree. On the other hand, the bigger $\mu_{q_{l'}}(N)$ is, the more objects is satisfying the statement with the quantifier. Hence, $t_{k'}$ can be selected as:

$$\mu_{t_{k'}}(\mu_{q_{l'}}(N)) = max\{\mu_{t_1}(\mu_{q_{l'}}(N)), \mu_{t_2}(\mu_{q_{l'}}(N)), \cdots, \mu_{t_k}(\mu_{q_{l'}}(N))\}. \quad (11)$$

*Example 1.* In checking quality of sweetened full cream milk powder (noted by CMP) (see Table 3), in this paper, only three of compositions protein, fat and sugar are considered. Let $P$ = {low-protein(lp), middle-protein(mp), high-protein(hp), low-fat(lf), middle-fat(mf), high-fat(hf), low-sugar(ls), middle-sugar (ms), high-sugar(hs)}, $Q$ = {many(ma), most of(mo), nearly all of(na)}, $T$ = {approximately true (at), true(t), very true(vt)}, their membership functions be given as follows,

$$\mu_{lp}(x)=\begin{cases}1, & x\le 17.5;\\ 18.5-x, & 17.5\le x\le 18.5.\end{cases} \quad \mu_{mp}(x)=\begin{cases}x-17.5, & 17.5\le x\le 18.5;\\ 19.5-x, & 18.5\le x\le 19.5.\\ 0, & \text{otherwise}\end{cases}$$

$$\mu_{hp}(x)=\begin{cases}\frac{2}{3}(x-18.5), & 18.5\le x\le 20;\\ 1, & x\ge 20.\end{cases} \quad \mu_{lf}(x)=\begin{cases}1, & x\le 21.5;\\ \frac{2}{3}(23-x), & 21.5\le x\le 23;\end{cases}$$

$$\mu_{mf}(x)=\begin{cases}\frac{2}{3}(x-21.5), & 21.5\le x\le 23;\\ \frac{2}{3}(24.5-x), & 23\le x\le 24.5;\\ 0, & \text{otherwise}\end{cases} \quad \mu_{hf}(x)=\begin{cases}\frac{2}{3}(x-23), & 23\le x\le 24.5;\\ 1, & x\ge 24.5;\end{cases}$$

$$\mu_{ls}(x)=\begin{cases}1, & x\le 16;\\ \frac{1}{2}(18-x), & 16\le x\le 18.\end{cases} \quad \mu_{ms}(x)=\begin{cases}\frac{1}{2}(x-16), & 16\le x\le 18;\\ \frac{2}{3}(19.5-x), & 18\le x\le 19.5;\\ 0, & \text{otherwise}\end{cases}$$

$$\mu_{hs}(x)=\begin{cases}\frac{2}{3}(x-18), & 18\le x\le 19.5;\\ 1, & x\ge 19.5;\end{cases} \quad \mu_{ma}(x)=\begin{cases}\frac{1}{3}x-2, & 6\le x\le 9;\\ -\frac{1}{3}x+4, & 9\le x\le 12.\end{cases}$$

$$\mu_{mo}(x)=\begin{cases}\frac{1}{2}x-5, & 10\le x\le 12;\\ -\frac{1}{2}x+7, & 12\le x\le 14.\end{cases} \quad \mu_{na}(x)=\begin{cases}\frac{1}{3}(x-13), & 13\le x\le 16;\\ 1, & x\ge 16.\end{cases}$$

$$\mu_{at}(x)=\begin{cases}2x, & 0\le x<0.5;\\ 2(1-x), & 0.5\le x\le 1.\end{cases} \quad \mu_{t}(x)=\begin{cases}0, & 0\le x<0.5;\\ \frac{10}{3}(x-0.5), & 0.5\le x<0.8;\\ 5(1-x), & 0.8\le x\le 1.\end{cases}$$

$$\mu_{vt}(x)=\begin{cases}0, & 0\le x<0.8;\\ 5x-4, & 0.8\le x\le 1.\end{cases}$$

Using above membership functions, Table 3 is transformed into Table 4.

Let $\alpha = 0.3$, then Table 4 is transformed into Table 5. In Table 5, $\forall v_a^i \in P$, $(v_a^i)^{\downarrow}$ can be obtained, *e.g.*, $hf^{\downarrow} = \{x_2, x_3, x_6, x_{17}\}$, *etc.* According to (8), $\forall v_a^i \in P$, $D((v_a^i)^{\downarrow}/U)$ can be obtained, *e.g.*, $D(hf^{\downarrow}/U) = \frac{2}{9}$, *etc.* Let $\beta = 0.4$, then $P_\beta = \{v_a^i | D((v_a^i)^{\downarrow}/U) \geq 0.4\} = \{mp, hp, mf, hs\}$. For $v_a^i = mp \in P_\beta$, according to $\mu_{ma}, \mu_{mo}, \mu_{na}$ and (10), $N = |(mp)^{\downarrow}| = 11$, and $max\{\mu_{ma}(N), \mu_{mo}(N), \mu_{na}(N)\}$ $= \{0.33, 0.5, 0\} = 0.5 = \mu_{mo}(N)$. Hence the fuzzy linguistic quantifier corresponding to $mp$ is $mo$. According to $\mu_{at}$, $\mu_t$, $\mu_{vt}$ and (11), $max\{\mu_{at}(\mu_{mo}(N)),$ $\mu_t(\mu_{mo}(N)), \mu_{vt}(\mu_{mo}(N))\} = \{1, 0, 0\} = 1 = \mu_{at}(\mu_{mo}(N))$, hence, the follows fuzzy linguistic summary about $mp$ is obtained,

$$\text{“}Most\ of\ CMP\ are\ middle-protein\ is\ approximately\ true.\text{”}$$

## 5 Extracting Complex Fuzzy Linguistic Summaries

By using logical conjunctions, $\vee, \wedge$ and $\rightarrow$, complex fuzzy linguistic summaries can be extracted. Here, complex fuzzy linguistic summaries are as follows:

$$(\ q_{l'}\ \ x's\ are\ v_{a_e}^i\ or\ v_{a_f}^j)\ is\ t_{k'}; \tag{12}$$

**Table 4.** Fuzzy sweetened full cream milk powder

| $U \setminus A$ | lp | mp | hp | lf | mf | hf | ls | ms | hs |
|---|---|---|---|---|---|---|---|---|---|
| $x_1$ | 0 | 0 | 0.87 | 0 | 0.73 | 0.27 | 0 | 0 | 1.0 |
| $x_2$ | 0 | 0 | 1.0 | 0 | 0 | 1.0 | 0.47 | 0.53 | 0 |
| $x_3$ | 0 | 0 | 0.81 | 0 | 0.24 | 0.76 | 1.0 | 0 | 0 |
| $x_4$ | 0 | 0.98 | 0.01 | 1.0 | 0 | 0 | 0 | 0 | 1.0 |
| $x_5$ | 0 | 0 | 1.0 | 1.0 | 0 | 0 | 0.03 | 0.97 | 0 |
| $x_6$ | 1.0 | 0 | 0 | 0 | 0.27 | 0.73 | 0 | 0 | 1.0 |
| $x_7$ | 0 | 0.44 | 0.37 | 0.72 | 0.28 | 0 | 0 | 0.29 | 0.71 |
| $x_8$ | 0 | 0.88 | 0.16 | 1.0 | 0 | 0 | 0 | 0 | 1.0 |
| $x_9$ | 0 | 0.68 | 0.21 | 1.0 | 0 | 0 | 0 | 0 | 1.0 |
| $x_{10}$ | 0 | 0 | 0.78 | 0.71 | 0.29 | 0 | 0 | 0 | 1.0 |
| $x_{11}$ | 0.05 | 0.95 | 0 | 0 | 0.99 | 0.01 | 0 | 0 | 1.0 |
| $x_{12}$ | 0 | 0.98 | 0.01 | 0.12 | 0.88 | 0 | 0 | 0 | 1.0 |
| $x_{13}$ | 0 | 0 | 1.0 | 0 | 0.77 | 0.23 | 0.01 | 0.99 | 0 |
| $x_{14}$ | 0 | 0.49 | 0.34 | 0.04 | 0.96 | 0 | 0 | 0 | 1.0 |
| $x_{15}$ | 0 | 0.62 | 0.25 | 0 | 0.78 | 0.22 | 0 | 0.33 | 0.67 |
| $x_{16}$ | 0 | 0.76 | 0.16 | 0.69 | 0.31 | 0 | 0 | 0.16 | 0.84 |
| $x_{17}$ | 0 | 0.68 | 0.21 | 0 | 0.21 | 0.79 | 0 | 0 | 1.0 |
| $x_{18}$ | 0 | 0.52 | 0.32 | 0.03 | 0.97 | 0 | 0.66 | 0.35 | 0 |

$$(\ q_{l'} \ \ x's \ are \ v^i_{a_e} \ and \ v^j_{a_f}) \ is \ t_{k'}; \tag{13}$$

$$(\ q_{l'} \ \ (x's \ are \ v^i_{a_e}) \ are \ v^j_{a_f}) \ is \ t_{k'}. \tag{14}$$

For fixing $\alpha$-level value, corresponding the classical information system is $(U, A, V, f'_\alpha)$, $\forall a_e, a_f \in A$, (12), (13) and (14) are obtained as follows, respectively.

1. Extracting (12) is based on logical conjunction " $\vee$ ": From logic point of view, (12) is equal to

$$(\ q_{l'} \ \ x's \text{ are } (v^i_{a_e} \vee \ v^j_{a_f})) \text{ is } t_{k'},$$

**Table 5.** A classical sweetened full cream milk powder

| $U \setminus A$ | lp | mp | hp | lf | mf | hf | ls | ms | hs |
|---|---|---|---|---|---|---|---|---|---|
| $x_1$ | 0 | 0 | 1 | 0 | 1 | 0 | 0 | 0 | 1 |
| $x_2$ | 0 | 0 | 1 | 0 | 0 | 1 | 1 | 1 | 0 |
| $x_3$ | 0 | 0 | 1 | 0 | 0 | 1 | 1 | 0 | 0 |
| $x_4$ | 0 | 1 | 0 | 1 | 0 | 0 | 0 | 0 | 1 |
| $x_5$ | 0 | 0 | 1 | 1 | 0 | 0 | 0 | 1 | 0 |
| $x_6$ | 1 | 0 | 0 | 0 | 0 | 1 | 0 | 0 | 1 |
| $x_7$ | 0 | 1 | 1 | 1 | 0 | 0 | 0 | 0 | 1 |
| $x_8$ | 0 | 1 | 0 | 1 | 0 | 0 | 0 | 0 | 1 |
| $x_9$ | 0 | 1 | 0 | 1 | 0 | 0 | 0 | 0 | 1 |
| $x_{10}$ | 0 | 0 | 1 | 1 | 0 | 0 | 0 | 0 | 1 |
| $x_{11}$ | 0 | 1 | 0 | 0 | 1 | 0 | 0 | 0 | 1 |
| $x_{12}$ | 0 | 1 | 0 | 0 | 1 | 0 | 0 | 0 | 1 |
| $x_{13}$ | 0 | 0 | 1 | 0 | 1 | 0 | 0 | 1 | 0 |
| $x_{14}$ | 0 | 1 | 1 | 0 | 1 | 0 | 0 | 0 | 1 |
| $x_{15}$ | 0 | 1 | 0 | 0 | 1 | 0 | 0 | 1 | 1 |
| $x_{16}$ | 0 | 1 | 0 | 1 | 1 | 0 | 0 | 0 | 1 |
| $x_{17}$ | 0 | 1 | 0 | 0 | 0 | 1 | 0 | 0 | 1 |
| $x_{18}$ | 0 | 1 | 1 | 0 | 1 | 0 | 1 | 1 | 0 |

The parameters $q_{l'}$, $(v^i_{a_e} \vee v^i_{a_f})$ and $t_{k'}$ can be obtained as follows,

(a) For $v^i_{a_e}, v^j_{a_f} \in P$ $(v^i_{a_e} \neq v^j_{a_f})$ such that $D(((v^i_{a_1})^{\downarrow} \cup (v^j_{a_2})^{\downarrow})/U) \geq \beta$, in which, if $a_e = a_f$,

$$\begin{aligned} N &= |(v^i_{a_e})^{\downarrow} \cup (v^j_{a_f})^{\downarrow}| \\ &= |((v^i_{a_e})^{\downarrow} - (\{v^i_{a_e}, v^j_{a_f}\})^{\downarrow}) \cup ((v^j_{a_f})^{\downarrow} - (\{v^i_{a_e}, v^j_{a_f}\})^{\downarrow})|, \end{aligned} \tag{15}$$

this is called as non-intersection-union. If $a_e \neq a_f$,

$$\begin{aligned} N &= |(v^i_e)^{\downarrow} \cup (v^j_{a_f})^{\downarrow}| \\ &= |(v^i_{a_e})^{\downarrow} \cup (v^j_{a_f})^{\downarrow} - (\{v^i_{a_e}, v^j_{a_f}\})^{\downarrow}|. \end{aligned} \tag{16}$$

(b) According to (10) and (11), a fuzzy linguistic quantifier $q_{l'} \in Q$ and a fuzzy linguistic truth degree $t_{k'} \in T$ can be selected.

2. Extracting (13) is based on logical " $\wedge$ ": (13) is equal to

$$( q_{l'} \;\; x's \text{ are } (v^i_{a_e} \wedge \; v^j_{a_f})) \text{ is } t_{k'},$$

The parameters $q_{l'}$, $(v^i_{a_e} \wedge v^i_{a_f})$ and $t_{k'}$ can be obtained as follows,

(a) For $v^i_{a_e}, v^j_{a_f} \in P$ such that $D(((v^i_{a_e})^{\downarrow} \cap (v^j_{a_f})^{\downarrow})/U) = D((\{v^i_{a_e}, v^j_{a_f}\})^{\downarrow}/U)$ $\geq \beta$. In which, $v^i_{a_e} \neq v^j_{a_f}$.

(b) A fuzzy linguistic quantifier $q_{l'} \in Q$ and a fuzzy linguistic truth degree $t_{k'} \in T$ can be selected according to (10) and (11), in which,

$$N = |(v^i_{a_e})^{\downarrow} \cap (v^j_{a_f})^{\downarrow}| = |(\{v^i_{a_e}, v^j_{a_f}\})^{\downarrow}|.$$

3. Extracting (14) is based on logical " $\rightarrow$ ": (14) is equal to

$$(q_{l'} \;\; x's \text{ are } (v^i_{a_e} \;\; \rightarrow \;\; v^j_{a_f})) \text{ is } t_{k'} \Longleftrightarrow (q_{l'} \;\; x's \text{ are } ( \; v^j_{a_f}|v^i_{a_e})) \text{ is } t_{k'},$$

$( \; v^j_{a_f}|v^i_{a_e})$ means that objects which satisfy $v^j_{a_f}$ are to satisfy $v^i_{a_e}$ firstly. The parameters $q_{l'}$, $v^i_{a_e}$, $v^i_{a_f}$ and $t_{k'}$ can be obtained as follows,

(a) Fixing $v^i_{a_e} \in P$, for level (threshold) $\beta$, $v^j_{a_f}$ is such that

$$D((v^j_{a_f})^{\downarrow}/(v^i_{a_e})^{\downarrow}) \geq \beta.$$

(b) According to (10) and (11), $q_{l'} \in Q$ and $t_{k'} \in T$ can be decided, in which,

$$N = |(v^j_{a_f})^{\downarrow} \cap (v^i_{a_e})^{\downarrow}| \times \frac{|U|}{|(v^i_{a_e})^{\downarrow}|}.$$

*Example 2.* Continue Example 1, let $\beta = 0.5$,

1. For $mp$ and $hp$, using (15), $D((mp^{\downarrow} \cup hp^{\downarrow})/U) = \frac{14}{18} \geq 0.5$, according to (10), (11),

$$max\{\mu_{ma}(N), \mu_{mo}(N), \mu_{na}(N)\} = \{0, 0, 0.33\} = 0.33 = \mu_{na}(N),$$

$$max\{\mu_{at}(\mu_{na}(N)), \mu_t(\mu_{na}(N)), \mu_{vt}(\mu_{na}(N))\} = 0.66 = \mu_{at}(\mu_{na}(N)).$$

Hence, we have complex fuzzy linguistic summary as follows

"*Nearly all of CMP are middle-protein or high-protein is approximately true.*"

2. For $lp$ and $hs$, using (16), $D((mp^{\downarrow} \cup hp^{\downarrow})/U) = \frac{12}{18} \geq 0.5$, according to (10), (11), we have complex fuzzy linguistic summary as follows

$$max\{\mu_{ma}(N), \mu_{mo}(N), \mu_{na}(N)\} = \{0, 1, 0\} = 1 = \mu_{mo}(N),$$

$$max\{\mu_{at}(\mu_{mo}(N)), \mu_t(\mu_{mo}(N)), \mu_{vt}(\mu_{mo}(N))\} = 0.67 = \mu_{at}(\mu_{mo}(N)).$$

"*Most of CMP are low-protein or high-sugar is approximately true.*"

3. For $mp$ and $hs$, $D((mp^{\downarrow} \cap hs^{\downarrow})/U) = \frac{10}{18} \geq 0.5$,

$$max\{\mu_{ma}(N), \mu_{mo}(N), \mu_{na}(N)\} = \{0.67, 0, 0\} = 0.67 = \mu_{ma}(N),$$

$$max\{\mu_{at}(\mu_{ma}(N)), \mu_{t}(\mu_{ma}(N)), \mu_{vt}(\mu_{ma}(N))\} = \{0.67, 0.56, 0\} = \mu_{at}(\mu_{ma}(N)).$$

Hence, we have complex fuzzy linguistic summary as follows
"*Many CMP are middle-protein and high-sugar is approximately true.*"
4. For $ls$ and $hf$, $D(hf^{\downarrow}/ls^{\downarrow}) = \frac{2}{3} \geq 0.5$,

$$max\{\mu_{ma}(N), \mu_{mo}(N), \mu_{na}(N)\} = \{0, 1, 0\} = 1 = \mu_{mo}(N),$$

$$max\{\mu_{at}(\mu_{mo}(N)), \mu_{t}(\mu_{mo}(N)), \mu_{vt}(\mu_{mo}(N))\} = \{0, 0, 1\} = \mu_{vt}(\mu_{ma}(N)).$$

Hence, we have complex fuzzy linguistic summary as follows
"*Most of low-sugar CMP are high-fat is very true.*"

## 6 Conclusion

In framework of continuous information systems, based on including degree theory and formal concept analysis (FCA), extracting simple and complex fuzzy linguistic summaries are discussed, provided Example shows that the method of extracting fuzzy linguistic summaries is useful.

## Acknowledgment

This work is supported by the excellent young foundation of sichuan province (grant no. 06ZQ026-037), the National Natural Science Foundation of China (60474022) and a Project Supported by Scientific Reserch Fund of SiChuan Provincial Education Department(grant no. 2005A121, 2006A084).

## References

1. Yager, R. R.:A new approach to the summarization of data. Information Sciences. 28 (1982) 69-86
2. Zadeh, L. A.: Toward a generalized theory of uncertainty (GTU)–an outline. Information Sciences 172 (2005) 1–40
3. Zadeh, L. A.: Fuzzy logic = computing with words. IEEE Trans. Fuzzy Syst 4 (1996) 103–111
4. Zadeh, L. A.: Toward a theory of fuzzy information granulation and its centrality in human reasoning and fuzzy logic. Fuzzy Sets and Systems 90 (1997) 111–127
5. Zadeh, L. A.: From computing with numbers to computing with wordsfrom manipulation of measurements to manipulation of perceptions. IEEE Transactions on Circuits and Systems. 45 (1999) 105–119
6. Zadeh, L. A.: Precisiated natural language (PNL). AI Magazine. 25(3) (2004) 74–91
7. Kacprzyk, J., Zadrożny, S.: Linguistic database summaries and their protoforms: towards natural language based knowledge discovery tools. Information Sciences. 173 (2005) 281-304

8. Kacprzyk, J., Zadrożny, S.: Computing with words in intelligent database querying: standalone and Internet-based applications. Information Sciences. 134 (2001) 71–109
9. Lawry, J.: A framework for linguistic modelling. Artificial Intelligence. 155 (2004) 1-39
10. Pei, Z., Du, Y. J., Yi, L. Z., Xu, Y.: Obtaining a Complex Linguistic Data Summaries from Database Based on a New Linguistic Aggregation Operator. Lecture notes in computer science. 3512 (2005) 771-778
11. Yager, R. R., Kacprzyk, J.: Linguistic data summaries: A perspective. in Proceedings of IFSA'99 Congress (Taipei, Taiwan). 1 (1999) 44-48
12. Ganter, B., Wille, R.: Formal concept analysis. Springer. (1999)
13. Zhang, W. X., Leung, Y.: Theory of including degrees and its applications to uncertainty inferences. Soft Computeing in Intelligent Systems and Information Processing. New York. IEEE. (1996) 496–501
14. Boulle, M.: Khiops: A Statistical Discretization Method of Continuous Attributes. Machine Learning. 55 (2004) 53-69
15. Zighed, D. A., Rabaseda, S., Rakotomalala, R.:Fusinter: a method for discretization of continuous attributes for supervised learning. International Jounal of Uncertainty, Fuzziness and Knowledge-Based Systems. 33 (1998) 307-326
16. Novák V.: Antonyms and linguistic quantifiers in fuzzy logic. 124 (2001) 335-351
17. Ho, N. C., Khang, T. D., Huynh V. N. and Nguyen, H. C.: Hedge algebras, linguistic valued logic and their application to fuzzy reasoning. International Journal of Uncertainty, Fuzziness and Knowledge-Based Systems. 7 (1999) 347–361
18. Pei, Z., Xu, Y.,: Lattice implication algebra model of linguistic variable Truth and its inference. In edited by Da Ruan et al, Applied Computational Intelligence, World Scientific. (2004) 93–98
19. Cerrada, M., Aguilar, J., Colina, E., Titli, A.: Dynamical membership functions: an approach for adaptive fuzzy modelling. Fuzzy Sets and Systems. 152 (2005) 513-533
20. Hu, Y. C.: Determining membership functions and minimum fuzzy support in finding fuzzy association rules for classification problems. Knowledge-Based Systems. 19 (2006) 57-66
21. Pei, Z., Germano, R., Van Der Wal, A. J., Qin, K. Y., Xu, Y.: Interpreting and extracting fuzzy decision rules from fuzzy information system and their inference. Information Sciences. 176 (2006) 1869-1897
22. Liu, Y. X., Etienne E.K.: An overvies of fuzzy quantifiers (I), (II). Interpretations. Fuzzy Sets and Systems. 95 (1998) 1-21, 135-146

# Linguistic Summarization of Time Series by Using the Choquet Integral

Janusz Kacprzyk[1,2], Anna Wilbik[1], and Sławomir Zadrożny[1,2]

[1] Systems Research Institute, Polish Academy of Sciences
ul. Newelska 6, 01-447 Warsaw, Poland
[2] Warsaw Information Technology (WIT)
ul. Newelska 6, 01-447 Warsaw, Poland
{kacprzyk,wilbik,zadrozny}@ibspan.waw.pl

**Abstract.** We further extend a new approach to a linguistic summarization of time series proposed in our previous works (cf. Kacprzyk, Wilbik and Zadrożny [1,2,3,4,5]) in which we put forward the use of a fuzzy linguistic quantifier driven aggregation of trends (partial scores) via the traditional Zadeh calculus of linguistically quantified propositions and the Sugeno integral. Here we use for this purpose the Choquet integral that has been widely advocated for many decision analytic and economic problems. The results are intuitively appealing and the method is effective and efficient.

## 1 Introduction

Time series data are one of the most relevant and widely occurring types of data that play a crucial role in many cases in which variables, quantities and parameters evolve over time. An acute need to find effective and efficient methods for handling times series data has triggered much research, and a notable example are here statistical methods. Recently, other methods have also been proposed exemplified by those based on neural networks, biologically inspired paradigms, cognitive analyses, etc.

Unfortunately, most of those traditional and new approaches are not human consistent enough as that they do not bridge an essential gap between the human being and the computer in that for the human being the only fully natural means of articulation of assessments, intentions, etc. is natural language which is strange to the "machine". This paper is a further step in a new direction to the analysis of time series proposed in our previous papers (Kacprzyk, Wilbik and Zadrożny [1,2,3,4,5]). In these papers a new approach to the capturing of the very essence of time series data has been proposed using natural language descriptions (statements) to describe in a human consistent way how trends in time series evolve over time, how long some types of behavior last, how rapid changes are, how variable they are, etc.

We use the idea of Yager's linguistic data summaries (cf. Yager [6], then advanced in Kacprzyk and Yager [7], and Kacprzyk, Yager and Zadrożny [8]). These summaries can be exemplified by "most trends are short", "most of long

P. Melin et al. (Eds.): IFSA 2007, LNAI 4529, pp. 284–294, 2007.
© Springer-Verlag Berlin Heidelberg 2007

trends are slowly increasing', etc. To derive such linguistic summaries we proposed first to use Zadeh's classic calculus of linguistically quantified propositions (cf. Kacprzyk, Wilbik and Zadrożny [1]). Next, new types of linguistic summaries of trends were proposed in Kacprzyk, Wilbik and Zadrożny [2], and the use of the Sugeno integral was proposed in Kacprzyk, Wilbik and Zadrożny [5].

In this paper, we use the Choquet integral (cf. [9] for the linguistic quantifier driven aggregation of partial scores (trends) that is essential for our purpose. The Choquet integral has attracted a considerable interest in many scientific communities, and for our purposes a positive experience with the Choquet integral reported widely in the decision making and economic community is of a particular importance, and references are abundant (cf. Narukawa, Murofushi and Torra [10].

## 2 Temporal Data and Trend Analysis

We deal with numerical data that vary over time, and a time series is a sequence of data measured at uniformly spaced time moments. We identify trends as linearly increasing, stable or decreasing functions, and therefore represent given time series data as piecewise linear functions. Evidently, the intensity of an increase and decrease (slope) matters, too. These are partial trends and a global trend concerns the entire time span of the time series, and there also may be trends that concern parts of the entire time span, but more than a particular window taken into account while extracting partial trends by using the Sklansky and Gonzalez [11] algorithm.

Function $f$ is a uniform $\varepsilon$-approximation of a time series, or a set of points $\{(x_i, y_i)\}$, if for a given, context dependent $\varepsilon > 0$, there holds

$$\forall i: \ |f(x_i) - y_i| \leq \varepsilon \tag{1}$$

We employ a modification of the widely used Sklansky and Gonzalez [11] algorithm that finds a linear uniform $\varepsilon$-approximation for subsets of points of a time series. The algorithm constructs an intersection of cones starting from point $p_i$ and including the circle of radius $\varepsilon$ around the subsequent points $p_{i+j}$, $j = 1, 2, \ldots$, until the intersection of all cones starting at $p_i$ is empty. If for $p_{i+k}$ the intersection is empty, then we construct a new cone starting at $p_{i+k-1}$. Figure 1 present the idea of the algorithm. The family of possible solutions is indicated as a gray area. Clearly, many other algorithms can also be used.

To present details of the algorithm, let us denote: `p_0` – a point starting the current cone, `p_1` – the last point checked in the current cone, `p_2` – the next point to be checked, `Alpha_01` – a pair of angles $(\gamma_1, \beta_1)$, meant as an interval defining the current cone as in Fig. 1(a), `Alpha_02` – a pair of angles of the cone starting at `p_0` and inscribing the circle of radius $\varepsilon$ around point `p_2` (cf. $(\gamma_2, \beta_2)$ in Fig. 1(a)), function `read_point()` reads a next point of data series, function `find()` finds a pair of angles of the cone starting at the point `p_0` and inscribing the circle of radius $\varepsilon$ around the point `p_2`.

The pseudocode of the procedure that extracts the trends is depicted in Figure 2.

The bounding values of `Alpha_02` $(\gamma_2, \beta_2)$, computed by function `find()`, correspond to the slopes of two lines which: (1) are tangent to the circle of radius $\varepsilon$ around point $p_2 = (x_2, y_2)$, and (2) start at the point $p_0 = (x_0, y_0)$.

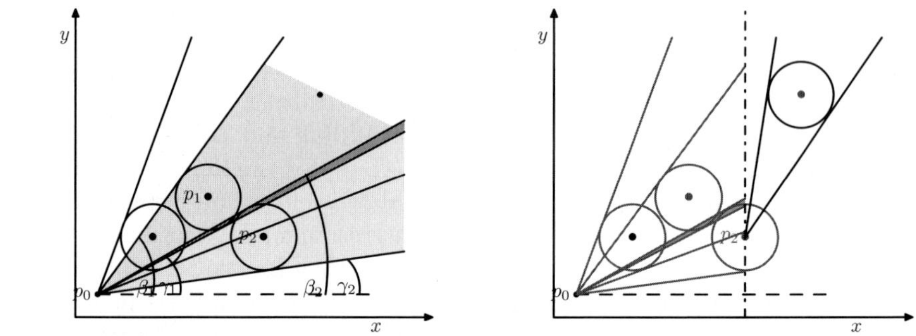


(a) the intersection of the cones is indicated by the dark grey area

(b) a new cone starts in point $p_2$

**Fig. 1.** An illustration of the algorithm for the uniform $\varepsilon$-approximation

Thus

$$\gamma_2, \beta_2 = arctg \left( \frac{\Delta x \cdot \Delta y \pm \varepsilon \sqrt{(\Delta x)^2 + (\Delta y)^2 - \varepsilon^2}}{(\Delta x)^2 - \varepsilon^2} \right)$$

where $\Delta x = x_0 - x_2$ and $\Delta y = y_0 - y_2$.

The resulting $\varepsilon$-approximation of a group of points `p_0, ... ,p_1` is either a single segment, chosen as, e.g., a bisector, or one that minimizes the distance (e.g., the sum of squared errors, SSE) from the approximated points, or the whole family of possible solutions, i.e. rays of the cone. This method is effective and efficient as it requires only a single pass through the data.

## 3 Dynamic Characteristics of Trends

In our approach, while summarizing trends in time series data, we consider the following three aspects: (1) dynamics of change, (2) duration, and (3) variability, and by trends we mean here global trends, concerning the entire time series (or some, probably large, part of it), not partial trends concerning a small time span (window) in the (partial) trend extraction phase via the Sklansky and Gonzales [11] algorithm.

### 3.1 Dynamics of Change

By *dynamics of change* we understand the speed of changes that can be described by the slope of a line representing the trend, (cf. $\alpha$ in Fig. 1(a)). Thus, to quantify dynamics of change we may use the interval of possible angles $\alpha \in \langle -90; 90 \rangle$.

As it might be impractical to use such a scale directly, we may use a fuzzy (linguistic) granulation in order to meet the users' needs and task specificity as, for instance: *quickly decreasing, decreasing, slowly decreasing, constant, slowly*

```
read_point(p_0);
read_point(p_1);
while(1)
{
  p_2=p_1;
  Alpha_02=find();
  Alpha_01=Alpha_02;
  do
  {

       Alpha_01 = Alpha_01 ∩ Alpha_02;

    p_1=p_2;
    read_point(p_2);
    Alpha_02=find();

     } while(Alpha_01 ∩ Alpha_02 ≠ 0);

  save_found_trend();
  p_0=p_1;
  p_1=p_2;
}
```

**Fig. 2.** Pseudocode of the modified Sklansky and Gonzalez [11] procedure for extracting trends

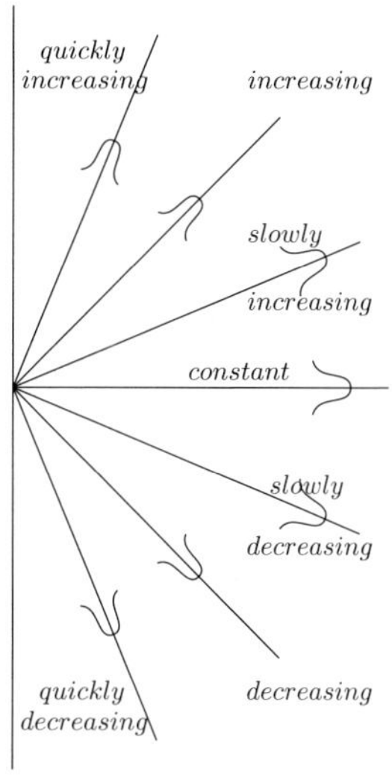


**Fig. 3.** A visual representation of angle granules defining the dynamics of change

*increasing, increasing, quickly increasing,* and Fig. 3 illustrates the lines corresponding to the particular linguistic terms.

In Batyrshin et al. [12,13] there are presented many methods of constructing such a fuzzy granulation. The user may define a membership functions of particular linguistic terms depending on his or her needs.

We map a single value $\alpha$ (or the whole interval of angles corresponding to the gray area in Fig. 1(b)) characterizing the dynamics of change of a trend identified using the algorithm shown as a pseudocode in Fig. 2 into a fuzzy set (linguistic label) best matching a given angle. We can use, for instance, some measure of a distance or similarity, cf. the book by Cross and Sudkamp [14]. Then we say that a given trend is, e.g., "decreasing to a degree 0.8", if $\mu_{decreasing}(\alpha) = 0.8$, where $\mu_{decreasing}$ is the membership function of a fuzzy set representing "decreasing" that is a best match for angle $\alpha$.

### 3.2 Duration

*Duration* describes linguistically the length of a single trend, for instance a "long trend", described by a fuzzy set defined over a time span of time series.

### 3.3 Variability

*Variability* refers to how intensively the consecutive data vary. The following five traditional statistical measures of variability are widely used:

- the range (maximum – minimum);
- the interquartile range (IQR) calculated as the third quartile (the third quartile is the 75th percentile) minus the first quartile (the first quartile is the 25th percentile) that may be interpreted as representing the middle 50% of the data;
- the variance calculated as $\frac{\sum_i (x_i - \bar{x})^2}{n}$, where $\bar{x}$ is the mean value;
- the standard deviation, i.e. the square root of the variance;
- the mean absolute deviation (MAD), calculated as $\frac{\sum_i |x_i - \bar{x}|}{n}$.

We measure the variability of a trend as a distance of the data points covered by this trend from its linear uniform $\varepsilon$-approximation (cf. Section 2). We employ a normalized distance between a point and a family of possible solutions, indicated as a gray cone in Fig. 1(a), and (1) makes it definitely smaller than $\varepsilon$. The normalized distance equals 0 if the point lays in the gray area. In the opposite case it is equal to the distance to the nearest point belonging to the cone, divided by $\varepsilon$.

Similarly as for the dynamics of change, we find for a given value of variability obtained as above a best matching fuzzy set (linguistic label) using, e.g., some measure of a distance or similarity, cf. the book by Cross and Sudkamp [14]. Again the measure of variability is treated as a linguistic variable and expressed using linguistic terms (labels) modeled by fuzzy sets defined by the user.

## 4 Linguistic Data Summaries

A linguistic summary is meant as a (usually short) natural language like sentence (sentences) that subsumes the very essence of a set of data (cf. Kacprzyk and Zadrożny [15], [15]) that is numeric and usually too large to be comprehensible by the human being. Yager's approach (cf. Yager [6], Kacprzyk and Yager [7], and Kacprzyk, Yager and Zadrożny [8]) used here is:

- $Y = \{y_1, \ldots, y_n\}$ is a set of objects (records) in a database, e.g., the set of workers;
- $A = \{A_1, \ldots, A_m\}$ is a set of attributes characterizing $y_i$'s from $Y$, e.g., salary, age, and $A_j(y_i)$ is a value of attribute $A_j$ for object $y_i$.

A linguistic summary of a data set consists of:

- a summarizer $P$, i.e. an attribute with a linguistic value (fuzzy predicate) defined on the domain of $A_j$ (e.g. "low salary" for "salary");
- a quantity in agreement $Q$, i.e. a linguistic quantifier (e.g. most);
- truth (validity) $\mathcal{T}$ of the summary, i.e. a number from $[0, 1]$ assessing the truth (validity) of the summary (e.g. 0.7);
- optionally, a qualifier $R$, i.e. another attribute with a linguistic value (fuzzy predicate) defined on the domain of $A_k$ determining a (fuzzy subset) of $Y$ (e.g. "young" for "age").

Thus, linguistic summaries, a simple one and one with a qualifier, may be exemplified by, respectively:

$$\mathcal{T}(\text{most of employees earn a low salary}) = 0.7 \tag{2}$$

$$\mathcal{T}(\text{most of young employees earn a low salary}) = 0.9 \tag{3}$$

The core of a linguistic summary is a *linguistically quantified proposition* in the sense of Zadeh [16] which, for both types of summaries may be written, respectively, as:

$$Qy\text{'s are } P \tag{4}$$

$$QRy\text{'s are } P \tag{5}$$

and $\mathcal{T}$, i.e., the truth (validity) of them may be calculated by using either original Zadeh's calculus of linguistically quantified propositions (cf. [16]), or other interpretations of linguistic quantifiers, i.e., respectively:

$$T(Qy\text{'s are } P) = \mu_Q\left(\frac{1}{n}\sum_{i=1}^{n}\mu_P(y_i)\right) \tag{6}$$

$$T(QRy\text{'s are } P) = \mu_Q\left(\frac{\sum_{i=1}^{n}(\mu_R(y_i) \wedge \mu_P(y_i))}{\sum_{i=1}^{n}\mu_R(y_i)}\right) \tag{7}$$

## 5 Protoforms of Linguistic Trend Summaries

As advocated by Kacprzyk and Zadrożny [15], and employed here, Zadeh's [17] concept of the protoform is convenient for dealing with linguistic summaries. Basically, a protoform is some prototype (template) of a linguistically quantified proposition. Then, the summaries mentioned above might be represented by two types of the protoforms:

- Summaries based on frequency:
  - a protoform of a short form of linguistic summaries:

    $$Q \text{ trends are } P \tag{8}$$

    and exemplified by "*Most* of trends have a *large variability*",
  - a protoform of an extended form of linguistic summaries:

    $$QR \text{ trends are } P \tag{9}$$

    and exemplified by: "*Most* of *slowly decreasing trends* have a *large variability*".
- Duration based summaries:

- a protoform of a short form of linguistic summaries:

$$\text{The trends that took } Q \text{ time are } P \tag{10}$$

and exemplified by: "The trends that took *most* time have a *large variability*".

- a protoform of an extended form of linguistic summaries:

$$R \text{ trends that took } Q \text{ time are } P \tag{11}$$

and exemplified by: "*Slowly decreasing trends* that took *most* time have a *large variability*".

In this paper we will consider only the simple forms of the summaries.

## 6 Derivation of Linguistic Summaries Via the Choquet Integral

Let $X = \{x_1, \ldots, x_n\}$ be a finite set. Then, (cf., e.g., [18]) a *fuzzy measure* on $X$ is a set function $\mu : \mathcal{P}(X) \longrightarrow$ [0,1], where $\mathcal{P}(X)$ is a family of all susets of $X$, such that:

$$\begin{array}{l} \mu(\emptyset) = 0, \mu(X) = 1; \\ \text{if } A \subseteq B \text{ then } \mu(A) \leq \mu(B), \qquad \forall A, B \in \mathcal{P}(X) \end{array} \tag{12}$$

Let $\mu$ be a fuzzy measure on $X$. The *discrete Choquet integral* of $f : X \longrightarrow [0,1]$, $f(x_i) = a_i$, with respect to $\mu$ is a function $C_\mu : [0,1]^n \longrightarrow [0,1]$ such that

$$C_\mu(a_1, \ldots, a_n) = \sum_{i=1}^{n} a_i \left(\mu(B_i) - \mu(B_{i+1})\right) \tag{13}$$

where $a_i$ is the $i$-th smallest element from among the $a_i$'s, and $B_i = \{x_{k_i}, x_{k_{i+1}} \ldots, x_{k_n}\}$, assuming that $x_{k_i} \leq x_{k_{i+1}} \leq \ldots \leq x_n$.

Note that in our context we view the role of $Q$ in (8) – (11) as a means to attain a linguistic quantifier driven aggregation. A similar role is played by the Choquet integral. For a similar and related point of view, see Bosc et al. [19].

The linguistic quantifier $Q$ is still defined as in Zadeh's calculus as a fuzzy set in $[0, 1]$, exemplified by (16); we assume that $Q$ is monotone and nondecreasing:

$$\mu(0) = 0, \quad \mu(1) = 1 \tag{14}$$

$$x_1 \leq x_2 \Rightarrow \mu_Q(x_1) \leq \mu_Q(x_2) \tag{15}$$

exemplified by

$$\mu_Q(x) = \begin{cases} 1 & \text{for } x \geq 0.65 \\ 4x - 1.6 & \text{for } 0.4 < x < 0.65 \\ 0 & \text{for } x \leq 0.4 \end{cases} \tag{16}$$

Then, for:

- *Simple frequency based summaries defined by (8)*

$$\mathcal{T}(Q \text{ trends are } P) = \sum_{i=1}^{n} \alpha_i \left( \mu_Q \left( \frac{|P_{\alpha_i}|}{|X|} \right) - \mu_Q \left( \frac{|P_{\alpha_{i+1}}|}{|X|} \right) \right) \tag{17}$$

- *Simple duration based summaries defined by (10)*

$$\mathcal{T}(\text{Trends that took } Q \text{ time are } P) = = \sum_{i=1}^{n} \alpha_i \left( \mu_Q \left( \frac{\sum_{j:x_j \in P_{\alpha_i}} time(x_j)}{\sum_{j:x_j \in X} time(x_j)} \right) - \mu_Q \left( \frac{\sum_{j:x_j \in P_{\alpha_{i+1}}} time(x_j)}{\sum_{j:x_j \in X} time(x_j)} \right) \right) \tag{18}$$

## 7 Example

Assume that from some data we extracted trends as in Fig. 4, e.g. via the algorithm shown in Fig. 2, with the granulation of dynamics of change as in Section 3.1.

| id | dynamics of change ($\alpha$ in degrees) | duration (time units) | variability ([0,1]) |
|---|---|---|---|
| 1 | 25 | 15 | 0.2 |
| 2 | -45 | 1 | 0.3 |
| 3 | 75 | 2 | 0.8 |
| 4 | -40 | 1 | 0.1 |
| 5 | -55 | 1 | 0.7 |
| 6 | 50 | 2 | 0.3 |
| 7 | -52 | 1 | 0.5 |
| 8 | -37 | 2 | 0.9 |
| 9 | 15 | 5 | 0.0 |

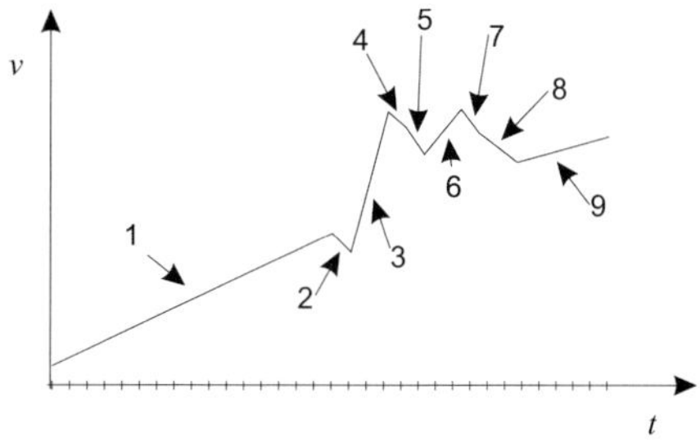


**Fig. 4.** Trends extracted in a tabular and graphical form

We consider the following simple frequency based trend summary:

$$Most\ (Q) \text{ of trends are } decreasing\ (P) \tag{19}$$

where: $Q$ is given as in (16), and $P$ is given as:

$$\mu_P(\alpha) = \begin{cases} 0 & \text{for } \alpha \leq -65 \\ 0.066\alpha + 4.333 & \text{for } -65 < \alpha < -50 \\ 1 & \text{for } -50 \leq \alpha \leq -40 \\ -0.05\alpha - 1 & \text{for } -40 < \alpha < -20 \\ 0 & \text{for } \alpha \geq -20 \end{cases} \tag{20}$$

The truth value of (19) is calculated by (17):

$$\mathcal{T}(\mathit{Most} \text{ trends are } \mathit{decreasing}) = \sum_{i=1}^{n} \alpha_i \left( \mu_Q \left( \frac{|P_{\alpha_i}|}{|X|} \right) - \mu_Q \left( \frac{|P_{\alpha_{i+1}}|}{|X|} \right) \right) = 0.4622$$

In comparison, if we use Zadeh's calculus of linguistically quantified propositions (cf. [1,2]) we obtain: $\mathcal{T}(\mathit{Most}$ trends are $\mathit{decreasing}) = 0.3778$.

If we have the following simple duration based linguistic summary:

$$\text{Trends that took } \mathit{most}\ (Q) \text{ time are } \mathit{slowly\ increasing}\ (P) \tag{21}$$

with $P$ defined as:

$$\mu_P(\alpha) = \begin{cases} 0 & \text{for } \alpha \leq 5 \\ 0.1\alpha - 0.5 & \text{for } 5 < \alpha < 15 \\ 1 & \text{for } 15 \leq \alpha \leq 20 \\ -0.05\alpha + 2 & \text{for } 20 < \alpha < 40 \\ 0 & \text{for } \alpha \geq 40 \end{cases} \tag{22}$$

Then, using (18) we obtain:

$$\begin{aligned} &\mathcal{T}(\text{Trends that took } \mathit{most} \text{ time are } \mathit{slowly\ increasing}) = \\ &= \sum_{i=1}^{n} \alpha_i \left( \mu_Q \left( \frac{\sum_{j:x_j \in P_{\alpha_i}} \mathit{time}(x_j)}{\sum_{j:x_j \in X} \mathit{time}(x_j)} \right) - \mu_Q \left( \frac{\sum_{j:x_j \in P_{\alpha_{i+1}}} \mathit{time}(x_j)}{\sum_{j:x_j \in X} \mathit{time}(x_j)} \right) \right) \\ &= 0.75 \end{aligned}$$

while, using Zadeh's calculus of linguistically quantified propositions (cf. [2,3]) we obtain: $\mathcal{T}$(Trends that took *most* time are *slowly increasing*) $= 0.5667$.

## 8 Concluding Remarks

We extended our approach to the linguistic summarization of time series (cf. Kacprzyk, Wilbik and Zadrożny [1,2,3,4,5]) that linguistically describes how trends in time series evolve over time, how long some types of behavior last, how rapid and variable changes are, etc. We used the Choquet integral. We show that this yields intuitively appealing results, and the method is effective and efficient.

## References


1. Kacprzyk, J., Wilbik, A., Zadrożny, S.: Linguistic summarization of trends: a fuzzy logic based approach. In: Proceedings of the 11th International Conference Information Processing and Management of Uncertainty in Knowledge-based Systems. (2006) 2166–2172 Paris, France, July 2-7, 2006.
2. Kacprzyk, J., Wilbik, A., Zadrożny, S.: On some types of linguistic summaries of time series. In: Proceedings of the 3rd International IEEE Conference "Intelligent Systems", IEEE Press (2006) 373–378 London, UK, September 4-6, 2006.
3. Kacprzyk, J., Wilbik, A., Zadrożny, S.: A linguistic quantifier based aggregation for a human consistent summarization of time series. In Lawry, J., Miranda, E., Bugarin, A., Li, S., Gil, M.A., Grzegorzewski, P., Hryniewicz, O., eds.: Soft Methods for Integrated Uncertainty Modelling. Springer-Verlag, Berlin and Heidelberg (2006) 186–190
4. Kacprzyk, J., Wilbik, A., Zadrożny, S.: Capturing the essence of a dynamic behavior of sequences of numerical data using elements of a quasi-natural language. In: Proceedings of the "2006 IEEE International Conference on Systems, Man, and Cybernetics", IEEE Press (2006) 3365–3370 Taipei, Taiwan, October 8 – 11, 2006.
5. Kacprzyk, J., Wilbik, A., Zadrożny, S.: Linguistic summaries of time series via a quantifier based aggregation using the Sugeno integral. In: Proceedings of 2006 IEEE World Congress on Computational Intelligence, IEEE Press (2006) 3610–3616 Vancouver, BC, Canada, July 16-21, 2006.
6. Yager, R.R.: A new approach to the summarization of data. Information Sciences **28** (1982) 69–86
7. Kacprzyk, J., Yager, R.R.: Linguistic summaries of data using fuzzy logic. International Journal of General Systems **30** (2001) 33–154
8. Kacprzyk, J., Yager, R.R., Zadrożny, S.: A fuzzy logic based approach to linguistic summaries of databases. International Journal of Applied Mathematics and Computer Science **10** (2000) 813–834
9. Narukawa, Y., Murofushi, T.: Choquet integral and Sugeno integral as aggregation function. In Torra, V., ed.: Information Fusion in Data Mining. Springer, Heidelberg (2003) 27–39
10. Narukawa, Y., Murofushi, T., Torra, V.: Decision modelling using the Choquet integral. Lecture Notes in Computer Science **3131** (2004) 183 – 193
11. Sklansky, J., Gonzalez, V.: Fast polygonal approximation of digitized curves. Pattern Recognition **12**(5) (1980) 327–331
12. Batyrshin, I.: On granular derivatives and the solution of a granular initial value problem. International Journal Applied Mathematics and Computer Science **12**(3) (2002) 403–410
13. Batyrshin, I., Sheremetov, L.: Perception based functions in qualitative forecasting. In Batyrshin, I., Kacprzyk, J., Sheremetov, L., Zadeh, L.A., eds.: Perception-based Data Mining and Decision Making in Economics and Finance. Springer-Verlag, Berlin and Heidelberg (2006)
14. Cross, V., Sudkamp, T.: Similarity and Compatibility in Fuzzy Set Theory: Assessment and Applications. Springer-Verlag, Heidelberg and New York (2002)
15. Kacprzyk, J., Zadrożny, S.: Linguistic database summaries and their protoforms: toward natural language based knowledge discovery tools. Information Sciences **173** (2005) 281–304
16. Zadeh, L.A.: A computational approach to fuzzy quantifiers in natural languages. Computers and Mathematics with Applications **9** (1983) 149–184

17. Zadeh, L.A.: A prototype-centered approach to adding deduction capabilities to search engines – the concept of a protoform. In: Proceedings of the Annual Meeting of the North American Fuzzy Information Processing Society (NAFIPS 2002). (2002) 523–525
18. Grabisch, M.: Fuzzy integral as a flexible and interpretable tool of aggregation. In Bouchon-Meunier, B., ed.: Aggregation and Fusion of Imperfect Information. Heidelberg, New York: Physica–Verlag (1998) 51–72
19. Bosc, P., Lietard, L., Pivet, O.: Quantified statements and database fuzzy queries. In Bosc, P., Kacprzyk, J., eds.: Fuzziness in Database Management Systems. Springer-Verlag, Berlin and Heidelberg (1995)

# Visualization of Possibilistic Potentials

Matthias Steinbrecher and Rudolf Kruse

Department of Knowledge Processing and Language Engineering
Otto-von-Guericke University of Magdeburg
Universitätsplatz 2, 39106 Magdeburg, Germany
{msteinbr,kruse}@iws.cs.uni-magdeburg.de

**Abstract.** The constantly increasing capabilities of database storage systems leads to an incremental collection of data by business organizations. The research area of Data Mining has become a paramount requirement in order to cope with the acquired information by locating and extracting patterns from these data volumes. Possibilistic networks comprise one prominent Data Mining technique that is capable of encoding dependence and independence relations between variables as well as dealing with imprecision. It will be argued that the learning of the network structure only provides an overview of the qualitative component, yet the more interesting information is contained inside the network parameters, namely the potential tables. In this paper we introduce a new visualization technique that allows for a detailed inspection of the quantitative component of possibilistic networks.

## 1 Introduction

The ongoing advance in the development of database systems enables today's business organizations to acquire and store huge amounts of data. However, the more data are collected, the stronger is the requirement for sophisticated analyzation methods to extract hidden patterns. The research area of Data Mining addresses these tasks and offers intelligent data analysis techniques such as classification, prediction or concept description, just to name a few.

The latter technique of concept description tries to identify common properties of conspicuous subsets of given samples in the database. For example, an automobile manufacturer may plan to investigate car failures by identifying common properties that are exposed by specific subsets of cars.

Good concept descriptions should have a reasonable length, i. e., they must not be too short in order not to be too general. Then again, long descriptions are too restrictive since they constrict the database samples heavily, resulting in only a few covered sample cases. Since we have to assume that the database entries expose hundreds of attributes, it is essential to employ a feature selection approach that reduces this number to a handy subset of significant attributes.

P. Melin et al. (Eds.): IFSA 2007, LNAI 4529, pp. 295–303, 2007.
© Springer-Verlag Berlin Heidelberg 2007

In this paper, we assume the database entries to have nominal attributes with one distinguished attribute designating the class of each data sample. We will use possibilistic network induction methods to learn a dependence network from the database samples. Further, we only draw our attention to the class attribute and its conditioning attributes, which are its direct parents in the network.

We then show that the network structure alone does not necessarily provide us with a detailed insight into the dependencies between the conditioning attributes and the class attribute. Finally, a new visualization method for these potential tables is presented and evaluated on real-world data.

The remainder of this paper is structured as follows: Section 2 presents a brief review of possibilistic networks. In section 3, arguments for the importance of visualizing the network parameters are produced. This will lead to a concrete application and analysis in section 4. The paper concludes with section 5, giving an outlook of intended further investigations.

## 2 Background

A database $D$, interpreted as a table, shall contain a certain number of tuples (rows) $t_h$ $(1 \leq h \leq N)$, each of which exposes a fixed number of attributes (columns) $\{A_1, \dots, A_n\}$ with respective domains $\text{dom}(A_i) = \{a_{i1}, \dots, a_{ir_i}\}$, i. e. $|\text{dom}(A_i)| = r_i$. We allow $D$ to contain multiple identical tuples which is modeled by a weight function $w : D \to \mathbb{N}^+$ that assigns to each distinct tuple $t \in D$ the number of occurrences in $D$.

In the case of precise tuples, each cell of this table contains exactly one attribute value, i. e. each tuple $t$ assumes one distinct value $a_{ik}$ for each attribute $A_i$: $\forall t \in D : A_i(t) = a_{ik}$, $i = 1, \dots, n$, $1 \leq k \leq r_i$. From such a database (or relation) a joint probability distribution can be estimated for each tuple: $\forall t \in D : p(t) = \frac{w(t)}{N}$. Each attribute can be seen as a random variable:

$$P(A_i = a_{ik}) = \frac{|\{t \in D \mid A_i(t) = a_{ik}\}|}{N}, \quad i = 1, \dots, n, \quad k = 1, \dots, r_i$$

Imprecision now enters through the absence of some of these table entries, i. e. there are tuples that have one or more values missing. Since we do not know the specific value of such cells (usually designated by a '?' or '$*$' in the dataset) we have to take into consideration *all possible* values of the corresponding attribute. Thus, the absence of a specific value of attribute $A$ of tuple $t$ is modeled as $A(t) = \text{dom}(A)$. Of course, this approach can be used as well to model partial ignorance, i. e. we can allow the attribute $A$ to assume any subset of $\text{dom}(A)$. Let us consider the imprecise database depicted in table 1. The first column shows the tuple as it may appear in a data file, the second and third column depict the values of the binary attributes $A$ and $B$, respectively.

Formally, we allow each attribute $A_i$ to be a *random set* [1], rather than a random variable. Let $\Omega$ be the finite set of all possible precise tuples over the Cartesian product of all attributes' domains, i. e. $\Omega = \times_{i=1}^{n} \text{dom}(A_i)$. Then, we can define a mapping $\gamma : D \to 2^{\Omega}$ that assigns to each (possibly imprecise)

**Table 1.** An imprecise example table. Note, that tuples $t_3$, $t_4$ and $t_5$, $t_6$ are identical

| | $A$ | $B$ | $\gamma(t_i)$ |
|---|---|---|---|
| $t_1 = (a_1, *)$ | $\{a_1\}$ | $\{b_1, b_2\}$ | $\{(a_1, b_1), (a_1, b_2)\}$ |
| $t_2 = (a_1, b_2)$ | $\{a_1\}$ | $\{b_2\}$ | $\{(a_1, b_2)\}$ |
| $t_3 = (a_1, b_1)$ | $\{a_1\}$ | $\{b_1\}$ | $\{(a_1, b_1)\}$ |
| $t_4 = (a_1, b_1)$ | $\{a_1\}$ | $\{b_1\}$ | $\{(a_1, b_1)\}$ |
| $t_5 = (*, b_2)$ | $\{a_1, a_2\}$ | $\{b_2\}$ | $\{(a_1, b_2), (a_2, b_2)\}$ |
| $t_6 = (*, b_2)$ | $\{a_1, a_2\}$ | $\{b_2\}$ | $\{(a_1, b_2), (a_2, b_2)\}$ |
| $t_7 = (a_2, b_2)$ | $\{a_2\}$ | $\{b_2\}$ | $\{(a_2, b_2)\}$ |
| $t_8 = (*, *)$ | $\{a_1, a_2\}$ | $\{b_1, b_2\}$ | $\{(a_1, b_1), (a_1, b_2), (a_2, b_1), (a_2, b_2)\}$ |

tuple $t \in D$ all (definitely precise) tuples $\omega \in \Omega$ that are covered by $t$. These sets are shown in the fourth column of table 1.

With this interpretation, each tuple $t \in D$ can be considered a *context*. The example contexts are shown in figure 1. A precise tuple obviously only describes a context that contains itself. Note, that due to the presence of multiple identical tuples ($t_3 \equiv t_4$ and $t_5 \equiv t_6$), we obtain identical contexts as well. The degree of possibility of any precise tuple $\omega \in \Omega$ is the probability of the set of contexts that contain $\omega$:

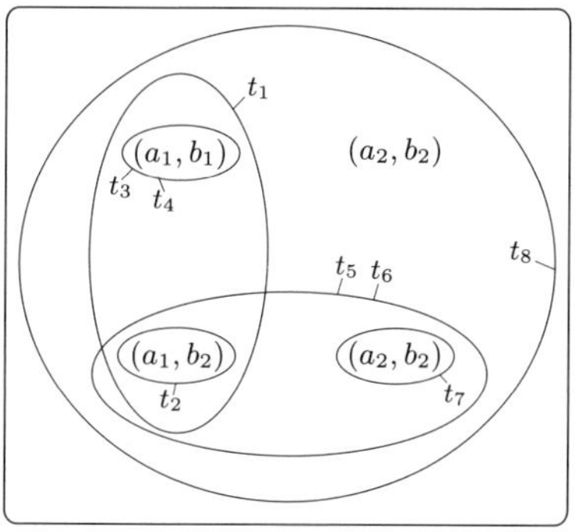


**Fig. 1.** The contexts induced by table 1

$$\pi_D : \Omega \to [0, 1] \quad \text{with}$$
$$\pi_D(\omega) = P_D(\{t \in D \mid \omega \in \gamma(t)\})$$

This coincides with the *one-point coverage* [2] of $\omega$ under $D$. The probability function $P_D$ belongs to the random set and is part of the probability space $(D, 2^D, P_D)$, where in our study each tuple $t \in D$ has the same elementary probability $p(t) = \frac{1}{N}$. In the interpretation from [3] we can derive a possibility measure $\Pi$ from the distribution $\pi_D$ in the following way:

$$\Pi : 2^\Omega \to [0, 1] \quad \text{with} \quad \Pi(E) = \max_{\omega \in E} P_D(\{t \in D \mid \omega \in \gamma(t)\})$$

## 2.1 Possibilistic Networks

Even though the database $D$ will be much smaller than $\Omega$ in practice, we need methods to further reduce the size of the joint possibility distribution induced by $D$. One idea is to exploit certain independency conditions within $\pi_D$ such as the *possibilistic non-interactivity*, which is defined as follows: Let $X = \{A_1, \dots, A_k\}$, $Y = \{B_1, \dots, B_l\}$ and $Z = \{C_1, \dots, C_m\}$ denote three disjoint subsets of attributes, then $X$ and $Y$ are conditionally possibilistically independent given $Z$, if the following equation holds:

$$\begin{aligned}
&\forall a_1 \in \mathrm{dom}(A_1) : \cdots \forall a_k \in \mathrm{dom}(A_k) : \\
&\quad \forall b_1 \in \mathrm{dom}(B_1) : \cdots \forall b_l \in \mathrm{dom}(B_l) : \\
&\quad\quad \forall c_1 \in \mathrm{dom}(C_1) : \cdots \forall c_m \in \mathrm{dom}(C_m) : \\
&\quad\quad\quad \Pi(A_1 = a_1, \ldots, A_k = a_k, B_1 = b_1, \ldots, B_l = b_l \mid C_1 = c_1, \ldots, C_m = c_m) \\
&\quad\quad\quad = \min\{\Pi(A_1 = a_1, \ldots, A_k = a_k \mid C_1 = c_1, \ldots, C_m = c_m), \\
&\quad\quad\quad\quad\quad \Pi(B_1 = b_1, \ldots, B_l = b_l \mid C_1 = c_1, \ldots, C_m = c_m)\}
\end{aligned} \tag{1}$$

where $\Pi(\cdot \mid \cdot)$ denotes the conditional possibility measure defined as follows:

$$\begin{aligned}
&\Pi(A_1 = a_1, \ldots, A_k = a_k \mid B_1 = b_1, \ldots, B_l = b_l) \\
&\quad = \max\{\pi_D(\omega) \mid \omega \in \Omega \wedge \bigwedge_{i=1}^{k} A_i(\omega) = a_i \wedge \bigwedge_{i=1}^{l} B_i(\omega) = b_i\}
\end{aligned} \tag{2}$$

The graph nodes coincide with the attributes. Let parents($A$) denote the set of all nodes that have an edge pointing to node $A$. With these prerequisites we can use a *directed acyclic graph (DAG)* to encode such independencies in the following way: Given an instatiation of the attributes in parents($A$), attribute $A$ is conditional independent of the remaining attributes. Such a DAG is said to carry the *structural* or *global* or *qualitative* information of a possibilistic network.

If a network structure is given, each attribute $A_i$ is assigned a *potential table*, i.e., the set of all conditional distributions, one for each distinct instantiation of the attributes in parents($A_i$). The general layout of such a table is shown in figure 2. Each column (like the one shaded in gray) corresponds to one specific parent attribute instantiation $Q_{ij}$. Each entry $\theta_{ijk}$ is read as

| $A_i$ | $Q_{i1}$ | $\cdots$ | $Q_{ij}$ | $\cdots$ | $Q_{iq_i}$ |
|---|---|---|---|---|---|
| $a_{i1}$ | $\theta_{i11}$ | $\cdots$ | $\theta_{ij1}$ | $\cdots$ | $\theta_{iq_i1}$ |
| $\vdots$ | $\vdots$ | $\ddots$ | $\vdots$ | $\ddots$ | $\vdots$ |
| $a_{ik}$ | $\theta_{i1k}$ | $\cdots$ | $\theta_{ijk}$ | $\cdots$ | $\theta_{iq_ik}$ |
| $\vdots$ | $\vdots$ | $\ddots$ | $\vdots$ | $\ddots$ | $\vdots$ |
| $a_{ir_i}$ | $\theta_{i1r_i}$ | $\cdots$ | $\theta_{ijr_i}$ | $\cdots$ | $\theta_{iq_ir_i}$ |

**Fig. 2.** A general potential table

$$\Pi(A_i = a_{ik} \mid \mathrm{parents}(A_i) = Q_{ij}) = \theta_{ijk}$$

These conditional distributions encode the *parametrical* or *local* or *quantitative* component of the network. The usual learning task of a possibilistic network consists of two components: a search heuristic and an evaluation measure. Examples for the former can be found in [4,5,6], examples for the latter are studied in [7].

## 3 Visualization of Potential Tables

After the learning task for a possibilistic network is completed, we are given a DAG that is encoding the detected (in)dependencies in the above-mentioned manner. A sample network is depicted in figure 3.

Since we are interested in the impact that certain attribute (values) have on the class attribute, we concentrate our attention on the direct ancestors of the class node, i. e., its parent nodes.

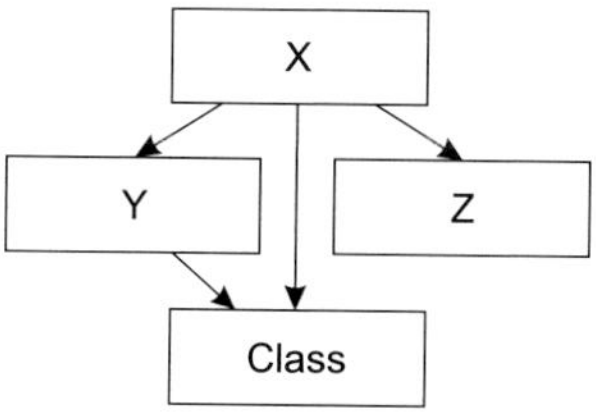


**Fig. 3.** A possibilistic network example

Although such a network conveys valuable information about the underlying data, some important questions remain unanswered. Cut short, it is desirable to know *which combinations* of the conditioning attributes' *values* have *what kind of impact* on *which class values*? The emphasized words in the last sentence mark the entities that carry much more information about the database under consideration. We can use the potential tables — or more specific: the class attribute's potential table — to extract the demanded information. Thus, the goal is to find an intuitive way of representing a potential table graphically, incorporating the entities mentioned above.

In order to represent the entries of a potential table in a chart, we investigate the semantics of these values a little bit further. A value $\theta_{ijk}$ tells us that given the $j$-th instantiation of the parent nodes of attribute $A_i$, then it is possible to a degree of $\theta_{ijk}$ that the attribute $A_i$ assumes the $i$-th value of its domain.

In a probabilistic setting, i. e., if we dealt with Bayesian Networks [8,9], the values $\theta_{ijk}$ would designate probabilities in the following way:

$$P(A_i = a_{ik} \mid \text{parents}(A_i) = Q_{ij}) = \theta_{ijk}$$

For the next considerations, we assume the following abbreviations for the sake of brevity: A subset of sample cases $\sigma_{ijk}$ is defined by the class value $a_{ik}$ and the instantiation of the parent attributes $Q_{ij}$: $\sigma_{ijk} = (Q_{ij}, a_{ik}) := (A, c)$. With this interpretation, each $\sigma_{ijk}$ represents an association rule [10]:

$$\texttt{If}\quad \text{parents}(A_i) = Q_{ij} \quad \texttt{then} \quad A_i = a_{ik} \quad \texttt{with confidence} \quad \theta_{ijk}$$

For each probabilistic entry $\theta_{ijk}$ we would compute three different values of evaluation measures from the domain of association rules,[1] e. g.:

$$m_x = \text{recall}(\sigma_{ijk}), \qquad m_y = \text{lift}(\sigma_{ijk}), \qquad m_z = \text{supp}(\sigma_{ijk})$$

with

$$\text{supp}(\sigma_{ijk}) = P(\text{parents}(A_i) = Q_{ij}, A_i = a_{ik})$$
$$\text{recall}(\sigma_{ijk}) = \frac{\text{supp}(\sigma_{ijk})}{P(A_i = a_{ik})}$$
$$\text{lift}(\sigma_{ijk}) = \frac{\theta_{ijk}}{P(A_i = a_{ik})}$$

[1] For a detailed analysis of such measures we refer to [11].

Finally, we display each $\sigma_{ijk}$ as a circle of size $m_z$ and locate it at position $(m_x, m_y)$ in a two-dimensional chart.

Since we intend to visualize possibilistic values, we interpret the $\sigma_{ijk}$ as possibilistic association rules where the value of $\theta_{ijk}$ represent the degree of possibility rather than the confidence. The presented measures are transformed into their possibilistic counterparts. Of course, we have to check whether the semantics behind these measures remain intact. Since the definition of the conditional possibility is symmetric, i. e., $\forall A, B : \Pi(A \mid B) = \Pi(B \mid A) = \Pi(A, B)$, the definitions for recall, confidence and support would coincide. Therefore, we define them as follows:

$$\mathrm{supp}^{\mathrm{poss}}(\sigma) = \Pi(A, c) \qquad \mathrm{recall}^{\mathrm{poss}}(\sigma) = \frac{\Pi(A, c)}{\Pi(c)}$$

$$\mathrm{conf}^{\mathrm{poss}}(\sigma) = \frac{\Pi(A, c)}{\Pi(A)} \qquad \mathrm{lift}^{\mathrm{poss}}(\sigma) = \frac{\Pi(A, c)}{\Pi(A)\Pi(c)}$$

The justification for this type of definition is as follows: As the degree of possibility for any tuple $t$, we assign the total probability mass of all contexts that contain $t$ [12]. With this interpretation, the term $\Pi(A = a)$ refers to the maximum degree of possibility of all sets of tuples, for which $A(t) = a$ holds, i. e., $\Pi(A = a) = \max\{p(t) = \frac{w(t)}{N} \mid t \in \Omega \wedge A(t) = a\}$. This probabilistic origin allows us to look at the possibility of an event $E$ (i. e., a set of tuples) as an upper bound of elementary events' probablitities contained in $E$ [3].

## 4 Experiments and Evaluation

The visualization technique presented here was introduced during a data mining project in cooperation with an automobile manufacturer. To justify the practical applicability, we intend to present real-world results. Since the underlying datasets are highly confidential, we are not allowed to show any attribute names or values. However, the charts will give a good insight, how suspiciuous subsets of tuples can be identified.

The dataset under analysis contained approximately 50.000 vehicle descriptions, including one class attribute designating, whether the respective car was faulty or not. A network was learned which revealed the class attribute to have two parent nodes, anonymized to $X$ and $Y$. We then chose the three evaluation measures to be *recall*, *lift* and *support*, which resulted in the chart depicted in figure 4. Choosing these measures, a user can apply the following heuristic to identify possible conspicuous tuples:

> "Large circles in the upper right corner are promising candidate subsets of samples that could most likely be suspicuous."

Dark circles represent faulty vehicles, white circles non-faulty ones. In figure 4 we identify a large gray circle on the right side, labeled '1'. The corresponding attribute values for $X$ and $Y$ belonging to this circle were identified by experts as having a causal effect on these 800 faulty vehicles.

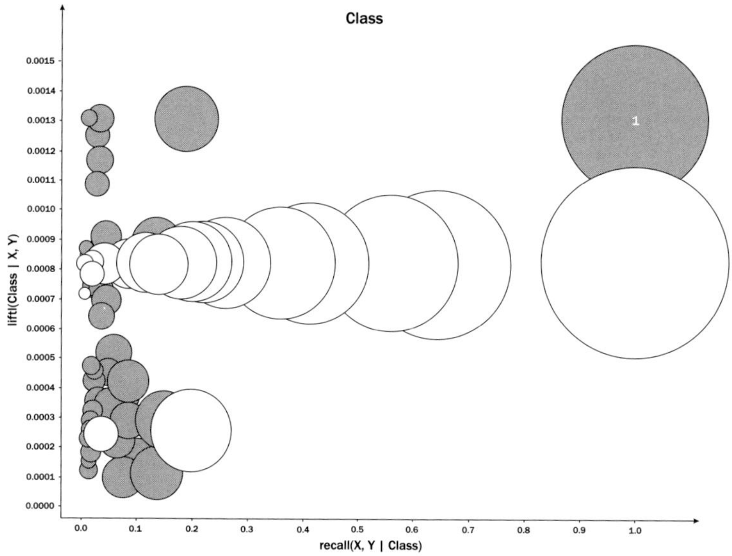


**Fig. 4.** The circle with label '1' represents 800 tuples having a large lift and recall

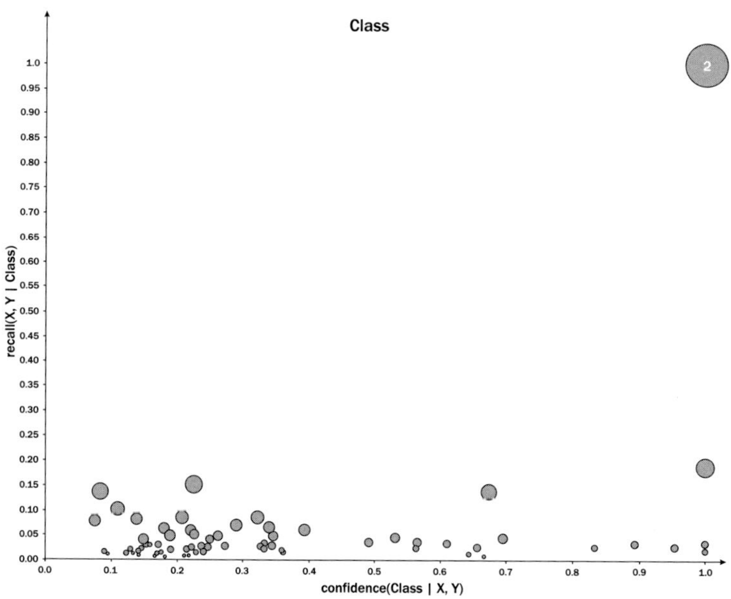


**Fig. 5.** Only faulty vehicles sets are shown. Again, the circle labeled '2' represents eye-catching tuples, that had a causal relationship with the corresponding values for $X$ and $Y$.

Another way of looking at this dataset is by choosing different measures for the x- and y-axis. Motivated by charts from the domain of *Information Retrieval*, we assign the measures *confidence* and *recall* to the x- and y-axis, respectively. Omitting the non-faulty circles for clarity results in the chart of figure 5.

Again, the circle marked as '2' represents (the same) 800 faulty vehicles.

## 5 Conclusion and Future Work

In this paper, we gave a short introduction to possibilistic networks and its ability of handling imprecise data which is becoming more and more a requirement for industrial applications since real-world data often contains missing data. We argued further that the more interesting information is contained inside the quantitative part of a network, namely its potential tables. Then, a new visualization technique was presented that is capable of displaying high-dimensional, nominal potential tables containing possibilistic parameters. This plotting method was evaluated in an industrial setting which produced empirical evidence that the presented visualization method greatly enhances the exploratory data analysis process. Since the presented visualization method aids to find concept descriptions and combines identical tuples (w. r. t. a subset of attributes) it may be promising to try to apply a modified version of this technique in the area of Text Mining, where several documents (again, identical w. r. t. some attribute, e. g. topic or keywords) may be grouped and displayed against other document groups.

## References


1. Nguyen, H.T.: Using Random Sets. Information Science 34 (1984) 265–274
2. Gebhardt, J., Kruse, R.: The Context Model Ů An Integrating View of Vagueness and Uncertainty. Int. Journal of Approximate Reasoning 9 (1993) 283–314
3. Borgelt, C., Kruse, R.: Graphical Models - Methods for Data Analysis and Mining. John Wiley & Sons, United Kingdom (2002)
4. Cooper, G.F., Herskovits, E.: A Bayesian Method for the Induction of Probabilistic Networks from Data (1992)
5. Heckerman, D., Geiger, D., Chickering, D.M.: Learning Bayesian Networks: The Combination of Knowledge and Statistical Data. Technical Report MSR-TR-94-09, Microsoft Research, Advanced Technology Division, Redmond,WA (1994) Revised February 1995.
6. Borgelt, C., Kruse, R.: Some experimental results on learning probabilistic and possibilistic networks with different evaluation measures. In: 1st International Joint Conference on Qualitative and Quantitative Practical Reasoning (ECSQARU/ FAPR'97), Bad Honnef, Germany (1997) 71–85
7. Gebhardt, J., Kruse, R.: Learning Possibilistic Networks from Data (1995)
8. Pearl, J., Russel, S.: Bayesian Networks (1994)
9. Heckerman, D.: A Tutorial on Learning with Bayesian Networks. Technical Report MSR-TR-95-06, Microsoft Research, Advanced Technology Division, Redmond, WA (1995) Revised June 96.

10. Agrawal, R., Imielinski, T., Swami, A.: Mining Association Rules Between Sets of Items in Large Databases (1993).
11. Yao, Y.Y., Zhong, N.: An Analysis of Quantitative Measures Associated with Rules. In: Methodologies for Knowledge Discovery and Data Mining, Springer-Verlag Berlin (1999).
12. Gebhardt, J., Kruse, R.: A Possibilistic Interpretation of Fuzzy Sets by the Context Model. In: IEEE International Conference on Fuzzy Systems, San Diego (1992) 1089–1096.

# Part VII

# Joint Model-Based and Data-Based Learning: The Fuzzy Logic Approach

# Selection Criteria for Fuzzy Unsupervised Learning: Applied to Market Segmentation

Germán Sánchez[1], Núria Agell[1], Juan Carlos Aguado[2], Mónica Sánchez[3], and Francesc Prats[3]

[1] ESADE-URL. Avda. Pedralbes, 60. 08034 Barcelona
{german.sanchez,nuria.agell}@esade.edu
[2] ESAII-UPC. Avda. del Canal Olímpic, s/n. 08860 Castelldefels
juan.carlos.aguado@upc.edu
[3] MA2-UPC. Jordi Girona, 1-3 08034 Barcelona
{monica.sanchez,francesc.prats}@upc.edu

**Abstract.** The use of unsupervised fuzzy learning methods produces a large number of alternative classifications. This paper presents and analyzes a series of criteria to select the most suitable of these classifications. Segmenting the clients' portfolio is important in terms of decision-making in marketing because it allows for the discovery of hidden profiles which would not be detected with other methods and it establishes different strategies for each defined segment. In the case included, classifications have been obtained via the LAMDA algorithm. The use of these criteria reduces remarkably the search space and offers a tool to marketing experts in their decision-making.

**Keywords:** Fuzzy unsupervised learning, criteria for classification selection, marketing applications.

## 1 Introduction

Firms that commercializa their products via other firms ("sell to other firms") are known as B2Bs (or business-to-business companies). One of the main challenges for these firms is knowing and understanding the behavior of the businesses that later sell their products. The distribution of these products is directly affected by the interest and the performance of the points of sale or specialized shops that sell them to the end customer.

For this reason, knowing and understanding how each point of sale behaves is crucial to designing marketing strategies. In this sense, the use of segmentation techniques in the B2B market allows us to determine groups to which the mentioned strategies can be directed. Segmentation in the B2B environment is complex due to the need to consider not only the characteristics of the points of sale but also their relationship with the end customer, the one who ultimately opts for the product.

The use of artificial intelligence techniques in this problem will allow us to determine groups of businesses with different profiles or behaviors. Therefore, it is

P. Melin et al. (Eds.): IFSA 2007, LNAI 4529, pp. 307–317, 2007.
© Springer-Verlag Berlin Heidelberg 2007

necessary to know the variables that determine the differences between the ways these groups act. These variables usually depend on external factors, such as geographical location, the business' area of influence, etc., and also on internal factors, such as the kind of management, the degree of implication by the person responsible, the product display stand in the shop compared to the competitors, etc. [5].

In this paper, an unsupervised learning methodology is used for the automatic generation of segmentations to discover profiles which would not be detected casually. Moreover, a series of criteria for the selection of optimal segmentation is given in order to help marketing professionals make their decisions.

This work is framed within a collaborative agreement between ESADE - Universitat Ramon Llull, the Universitat Politècnica de Catalunya, and the firm, Textil Seu, S.A., specialized in outdoor sporting equipment and clothing.

In this study, we apply a classification technique called LAMDA (Learning Algorithm for Multivariate Data Analysis). Among other advantages, LAMDA allows us to work simultaneously with quantitative and qualitative variables. In order to aggregate the information associated to each pattern, the LAMDA algorithm introduces fuzzy hybrid connectives defined from the convex linear interpolation between a t-norm and its dual t-conorm. Its unsupervised version allows us to obtain a diversity of segmentations from which the marketing team must select the most representative with respect to the posed problem.

This paper begins with the presentation of the LAMDA algorithm and the theoretical concepts on which it is founded. We then analyze and substantiate the criteria to select the optimum segments from among the resulting set of segmentations of an unsupervised learning process. A real B2B case study to which the presented techniques are applied is introduced in Section 3 of this paper. In Section 4, the results obtained are commented on and analyzed. The paper ends with a conclusion and proposals for future work.

## 2 Theoretical Framework: The LAMDA Algorithm

LAMDA is a hybrid connectives-based classification method that combines some of the most interesting capabilities of both purely numeric and purely symbolic algorithms. In order to do so, both employ the interpolation capabilities of logic operators over fuzzy environments [7].

A linearly compensated hybrid connective is an interpolation between a t-norm and its dual t-conorm ($H = \beta T + (1 - \beta)T^*$). It can be noted that for $\beta = 1$, the t-norm is obtained and, for $\beta = 0$, the t-conorm is the result. Taking into account that the t-norms are fuzzy operators associated to an intersection or conjunction and the t-conorms are associated to a union or disjunction, the parameter $\beta$ determines the exigency level of the classification. Obviously, we can define $\lambda = 1 - \beta$ as the tolerance level of a given classification.

In this work, the LAMDA algorithm is employed as a classifier. It was developed by Josep Aguilar in collaboration with a series of authors [1,2,3] as an original technique in which the input data are the values that a set of observations exhibit for a number of descriptors. Each of these descriptors within the database can be either qualitative (a linguistic label) or quantitative (numerical). The basic LAMDA operation is depicted in Figure 1:

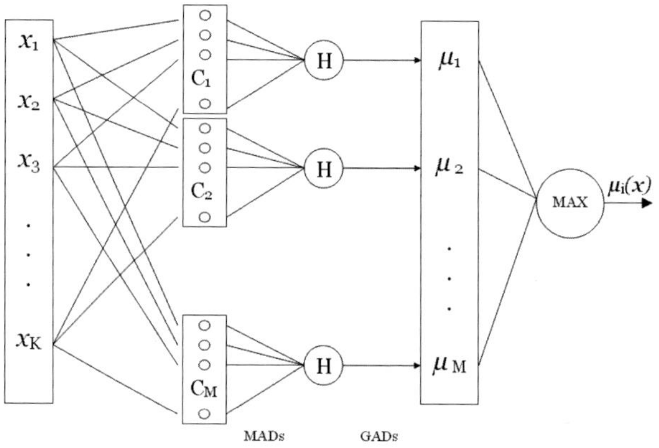


**Fig. 1.** Hybrid Connectives-based Classification

MAD and GAD stand for the Marginal Adequacy Degree and Global Adequacy Degree, respectively, of each individual to a given class. MAD is computed for each descriptor, class and individual, and these partial results are aggregated by means of the hybrid connectives $H$ to supply the GAD of an individual to a class. The visible structural resemblance between the LAMDA algorithm and the Artificial Neural Networks is worth noting, especially for the RBF (Radial Base Functions) type, as can be seen in Figure 2. LAMDA exhibits greater flexibility than neural networks, for example, in its ability to perform either a supervised or an unsupervised learning process indistinctly and its capability to combine pattern recognition with a simple, non-iterative class upgrading.

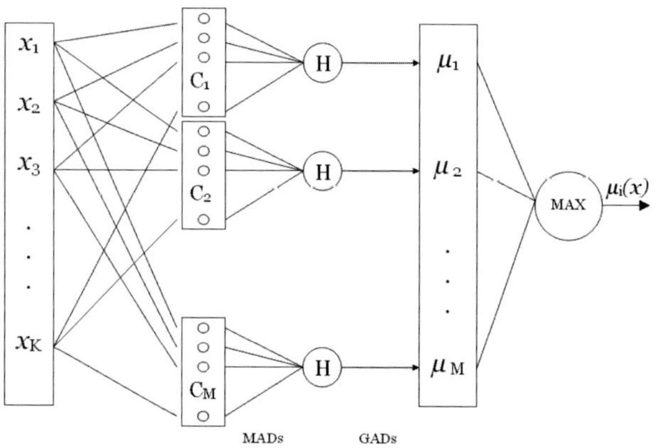


**Fig. 2.** RBF Neural Network Architecture

Whenever the unsupervised learning LAMDA capabilities are employed, the first individual is always placed in a class by itself. Subsequent individuals can then be be assigned to the already existing class(es) or a new one is created. In order to determine when this occurs, the algorithm generates, first of all, a special class called NIC (Non-Informative Class) which represents maximum entropy, with the characteristic of returning the same (low) GAD for every possible individual. As such, the decision-making process consists of comparing the GAD of the individual to the NIC class with the GADs to every other existing class. If one of the real classes returns the maximum GAD, the new individual will be assigned to it and the class will be modified accordingly. But if the NIC is the one with the highest GAD, this means that none of the existing classes are close enough to the individual and so a new class has to be created by the LAMDA algorithm.

The generated classification by unsupervised learning depends, on the one hand, on the hybrid connective employed and, on the other hand, on the selected exigency or tolerance level (also on the assumed distribution for numerical descriptors). Maximum exigency means that an individual will be assigned to a class only if every descriptor points to this, whereas a minimum exigency implies that the individual will be assigned to the class if this is indicated by at least one of the descriptors. Any in-between exigency degree is possible, and our algorithm is able to automatically explore these and generate every possible partition.

Additionally, the obtained results are not only a classical partition. The algorithm returns the Global Adequacy Degree for every individual to each class, that is, a fuzzy partition. These results enable marketing experts to further explore a whole set of possible solutions.

## 3 Selection Criteria for Classifications

The use of unsupervised learning algorithms allows suggesting segmentations that, in principle, are not trivial. In this sense, behavioral patterns or "interesting" profiles can be established by using this type of algorithm and they may reveal new profiles not known by the experts. In an area of application such as marketing, finding new and creative solutions is worthwhile because these allow for the definition of new marketing strategies. However, when a large amount of different segmentations are obtained, how do we choose the one that really adjusts to the proposed objective? Below we define oriented criteria to help solve this problem. A software module based on the outputs obtained by the LAMDA algorithm but adaptable to the outputs of any unsupervised tool is being developed to interpret and decide automatically which is the best segmentation.

**First Criterion: Stable Classification.** The LAMDA algorithm is iterative, that is, it processes all the individuals several times, modifying the classes until reaching a stability in which the classification remains unalterable or until reaching a prefixed maximum number of iterations.

The tool then discards those classifications that have not obtained said stability. Only classifications that remain stable after the maximum number of iterations are considered.

**Second Criterion: Useful and Manageable Number of Classes.** Based on the interpretability and usability of the classification, it is worth interesting examining segmentations which have a sufficient number of classes to generate new knowledge, but small enough to obtain an easily interpretable and usable model. Specifically, in marketing environments in which these classifications are used to extract behavioral patterns to design market strategies, the usual the number of classes distinguished is between 3 and 5 [5]. Designing marketing campaigns with less than 3 or more than 5 different strategies is not very useful.

Therefore, the tool discards classifications with a number of classes lower than a fixed minimum or higher than a fixed maximum.

**Third Criterion: Balanced Classes.** Cases in which one class encompasses most of the individuals are worth avoiding. These segmentations usually do not contribute to create new knowledge. The need for this criterion appears in marketing problems [5]. In this environment, an unbalanced classification supposes a classification effort that does not contribute a proportional benefit.

In order to ensure this, the *Indicator of Balanced Classes* $I_B$ is defined as the variation coefficient for the variable $Y$ = "number of elements of each class in a given segmentation":

$$I_B = CV_Y = \frac{\sigma_Y}{\bar{Y}}, \tag{1}$$

with

$$\sigma_Y = \sqrt{\frac{1}{M}\sum_{i=1}^{M}(Y_i - \bar{Y})^2} \quad \text{and} \quad \bar{Y} = \frac{1}{M}\sum_{i=1}^{M} Y_i, \tag{2}$$

where $\sigma_Y$ is the dispersion of the variable $Y$, $M$ is the number of classes of this segmentation, and $Y_i$ is the number of individuals within the class $i$. It is easily observed that if a classification contains all the classes with an equal number of individuals, this value will be zero. The tool requires that the $I_B$ indicator does not exceed a certain $K_B$ value:

$$I_B < K_B.$$

Because of the fact that the number of classes within the obtained classifications, i.e., the number of variable $Y$ observations, is not fixed, we do not recommend directly comparing the deviations obtained for each one of the classifications. However, the variation coefficient considers the existing proportion between averages and standard deviations, which allows us to compare the different obtained classifications.

**Fourth Criterion: Consistent Segmentation.** A segmentation is considered consistent when the differences between the MADs are small enough for each class and each individual.

As explained in Section 2, in fuzzy classifications, such as those obtained through the LAMDA algorithm, each individual belongs to each class with a certain degree of adequacy. The consistency indicator ensures that the global degrees of adequacy are obtained from similar values of MADs. If we take, for example, $\mu_{ijk}$ as the MAD of the individual $j$ on the description $k$ of class $i$, with $N$ as the number of individuals and $M$ as the number of classes of certain segmentation, the *Consistency Indicator* is defined as follows for each one of the obtained segmentations:

$$I_C = \frac{\sum_i \sum_j \max_{k,k'} |\mu_{ijk} - \mu_{ijk'}|}{M \cdot N}. \tag{3}$$

Alternativa:

$$I_C = \frac{\sum_i \sum_j \sigma_{ij}}{M \cdot N}, \tag{4}$$

where

$$\sigma_{ij} = \sqrt{\frac{1}{V}\sum_{k=1}^{V}(\mu ijk - \bar{\mu ij})^2} \quad \text{and} \quad \bar{\mu ij} = \frac{1}{V}\sum_{i=1}^{V}\mu ijk, \tag{5}$$

where $V$ is the number of variables.
The tool requires that the consistency indicator does not exceed a certain $K_C$ value:

$$I_C < K_C.$$

**Fifth Criterion: Dependency on External Variables.** This criterion uses suitable external variables, called *control variables*, chosen by some experts, in order to evaluate the relevance of the classifications obtained. The dependency degree of a classification on a control variable is defined by means of the value of the statistic $\chi^2$ computed via the contingency table (Table 1), where $C_1...C_i...C_M$ represent the classes of the considered segmentation, $D_1...D_s$ $...D_S$ the values of the external variable and $q_{is}$ the number of observations that take the $D_s$ value in $C_i$ class.

**Table 1.** Contingency Table

| Class | Descriptors or intervals | | | | Total |
|---|---|---|---|---|---|
| | $D_1$ | $D_2$ | ... | $D_S$ | classes |
| $C_1$ | $q_{11}$ | $q_{12}$ | ... | $q_{1S}$ | $M_{1+}$ |
| ... | ... | ... | ... | ... | ... |
| $C_i$ | $q_{i1}$ | $q_{i2}$ | ... | $q_{iS}$ | $M_{i+}$ |
| ... | ... | ... | ... | ... | ... |
| $C_M$ | $q_{M1}$ | $q_{M2}$ | ... | $q_{MS}$ | $M_{M+}$ |
| Total descriptors | $M_{+1}$ | $M_{+2}$ | ... | $M_{+S}$ | $N$ |

If we take, for example, $e_{is}$ as the expected frequency obtained when the variable is independent of the classification, i.e.:

$$e_{is} = \frac{M_{i+} \cdot M_{+s}}{N} \tag{6}$$

the statistic $\chi^2$ is determined through the expression:

$$\chi^2 = \sum_{i=1}^{N} \sum_{s=1}^{S} \frac{(q_{is} - e_{is})^2}{e_{is}} \qquad (7)$$

For each classification, the dependency of each of the control variables with respect to the segmentation is studied, and those classifications that have a high dependency on these external variables will be chosen. For this reason the statistic $\chi^2$ has to have high value.
It is important to note that this criterion can be used directly when the control variables are qualitative. In the case of quantitative control variables, it is advisable to discretize these previously into intervals ($D_s$). For each problem, the discretization criterion will be specifically chosen [6,8].
Because the range of values of $\chi^2$ can change according to the number of classes of the segmentation, Tschuprow's coefficient [10] is used to calculate the *Indicator of Relation with the Control Variable*:

$$I_R = \frac{\chi^2}{N \cdot \sqrt{M-1} \cdot \sqrt{S-1}}. \qquad (8)$$

where $N$ is the number of individuals, $M$ is the number of classes of this segmentation and $S$ is the number of possible control variable values, if it is qualitative, or the number of considered intervals if it is quantitative. The tool requires that the statistic considered exceed a predetermined $K_R$ value:

$$I_R > K_R.$$

# 4 Case Study

## 4.1 Introduction

The problem addressed in this paper belongs to a broader project in which Artificial Learning techniques are applied to a field where there have been few contributions to date: market segmentation [4].

In particular, this paper presents a preliminary study evaluating the efficiency of the LAMDA classifier in designing segmentations for B2B environments. The global objective is to create an application capable of identifying and classifying points of sale which will enable appropriate marketing strategies to be defined. This segmentation is made by using the LAMDA algorithm. In particular, the work is based on data collected using the observations, knowledge, and experience of the sales representatives working for the outdoor sporting equipment firm, Textil Seu, S.A. (Grifone, http://www.grifone.com), a company established in La Seu d'Urgell for more than 25 years. Grifone works in a B2B environment. This paper presents the preliminary results obtained from a database with information about 260 shops that distribute Grifone products.

Each of these points of sale is considered an individual and it is described by the variables defined in Table 2:

**Table 2.** Description of Variables

| Type | Number | Description |
|---|---|---|
| Quantitative | 2 | Duration of commercial relationship<br>Number of full-time sales assistants |
| Qualitative | 5 | Specialist store<br>Geographic location<br>Grifone products in the display window<br>Thermal product display<br>Internet |
| Ordinal | 10 | Assessment by Grifone representatives<br>Level of competition<br>Store size<br>Store maintenance<br>Display window size<br>Communicative quality<br>Aesthetics quality<br>Grifone products' importance<br>Price sensitivity<br>Promotions sensitivity |

Thus, each of the points of sale is described by a vector with 17 variables, 2 of which are quantitative, 5 qualitative, and 10 qualitative ordinal.

It should be noticed that the variable, *sensitivity to promotions*, has not been used in the automatic learning process. From the results obtained by the LAMDA algorithm in its version of unsupervised learning, the criteria defined in the previous section are applied to choose the point of sales' optimal classification using the variable, *sensitivity to promotions*, as a control variable.

## 4.2 Obtained Results

### Parameters

**Fuzzy Hybrid Connectives:** only connectives with $T = \min$ have been considered [9].

**Tolerance:** an automatic tolerance level was fixed by employing the unsupervised learning capability of LAMDA.

**Maximum Variability:** a 0 percent variability was chosen to ensure the stability of the obtained segmentations.

**Maximum Number of Iterations:** a sufficiently large number of iterations was chosen to ensure a maximum number of stable segmentations.

Taking advantage of LAMDA's unsupervised learning capability, 97 classifications were obtained, 65 of which reached stability before the fixed maximum number of iterations (10). Applying the criterion of number of class manageability, those segmentations with less than 3 classes and more than 5 were discarded, reducing the set of classifications to 16, as shown in Table 3.

**Table 3.** Final Classifications after eh First and Second Criteria

| Classif. | n. classes | Classif. | n. classes |
|---|---|---|---|
| Cl. 43 | 3 | Cl. 55 | 4 |
| Cl. 44 | 3 | Cl. 56 | 4 |
| Cl. 48 | 5 | Cl. 57 | 3 |
| Cl. 49 | 5 | Cl. 58 | 4 |
| Cl. 50 | 5 | Cl. 59 | 3 |
| Cl. 52 | 4 | Cl. 60 | 3 |
| Cl. 53 | 4 | Cl. 62 | 3 |
| Cl. 54 | 4 | Cl. 63 | 3 |

**Table 4.** Applying Criterion Three

| Clasification | $I_B$ | Discarded | Clasification | $I_B$ | Discarded |
|---|---|---|---|---|---|
| Cl. 43 | 0.5603 | Yes | Cl. 55 | 0.3750 | No |
| Cl. 44 | 0.4950 | No | Cl. 56 | 0.7917 | Yes |
| Cl. 48 | 0.7011 | Yes | Cl. 57 | 0.4124 | No |
| Cl. 49 | 0.7014 | Yes | Cl. 58 | 0.7123 | Yes |
| Cl. 50 | 0.7937 | Yes | Cl. 59 | 0.4431 | No |
| Cl. 52 | 0.7172 | Yes | Cl. 60 | 0.4377 | No |
| Cl. 53 | 0.7243 | Yes | Cl. 62 | 0.6555 | Yes |
| Cl. 54 | 0.3606 | No | Cl. 63 | 0.6568 | Yes |

**Table 5.** Applying Criterion Four and Five

| Classification | $I_C$ | Discarded | $I_R$ |
|---|---|---|---|
| Classif. 44 | 0.4904 | No | 0.0665 |
| Classif. 54 | 0.4722 | No | 0.0466 |
| Classif. 55 | 0.4598 | No | 0.0635 |
| Classif. 57 | 0.4844 | No | 0.0323 |
| Classif. 59 | 0.4757 | No | 0.5279 |
| Classif. 60 | 0.4683 | No | 0.3634 |

The values shown in Table 4 are obtained by computing the index defined in criterion three, which allows evaluating the balance between class cardinality.

As can be seen in Table 4, several segmentations have been discarded due to their variation coefficient ($I_B$ has exceeded the maximum value previously set at 0.5).

The values shown in Table 5 are obtained by computing the rest of indexes:

The application of the fourth criterion does not discard any segmentation (Table 5). The application of the fifth criterion leads to the selection of classification number 59.

Next, a description of each one of the obtained segments is detailed:

**Class 1:** This segment with 35 points of sale includes shops with a long-standing commercial relationship with Grifone. By and large, these shops are not specialists in mountain gear and competition between them is mostly intense. The points of sale, located in cities or non-mountainous-populations, are medium or large in size and count on many shop assistants. The shops are well-maintained, they have a medium-sized window display, and their aesthetic and communicative qualities and abilities are high.

A lot of the shops in this group have a thermal product display and usually market their goods on the Internet, which is rather unusual for these kinds of shops. The importance of the Grifone brand is secondary, in general, and their clients demonstrate a mid-level sensitivity to the promotions.

In short, Class 1 might correspond to multi-sports shops, with large points of sale not found in mountain locations.

**Class 2:** Most shops in this class (98 shops total) do not have a long-standing commercial relationship with Grifone. Competitiveness is medium, they are small or medium in size, and they have a small sales staff. Their maintenance, esthetic quality, and communicative abilities are generally average or good, and they have a moderately small display window.

Most of the shops in Class 2 do not display Grifone products; the importance of Grifone is minor. Almost none has a thermal product display and grifone representatives give these shops the worst evaluation. These shops' customers demonstrate a low or mid-level sensitivity to promotions.

It seems that, from Grifone's point of view, Class 2 corresponds to the worst shops, those in which its brand is placed worst.

**Class 3:** Shops in the third class, the most extensive with 127 shops, are usually sector specialists, with fairly strong competition between them, and, perhaps because of this factor, they are primarily located in mountainous populations. The shops are not big, nor they have a lot of shop assistants, but they are well-maintained, and have excellent aesthetic and communicative qualities. Normally, they have Grifone clothing in the display windows, although they usually do not usually have a thermal product display.

The importance given to the Grifone brand is usually the highest; Grifone is often the principal product. Their clients are quite sensitive to promotions. In Class 3 we find Grifone's favorite clients: small-sized, elegant, and specialists in mountain gear. In this segment, shops with a long-standing commercial relationship with Grifone products are mixed with others with a shorter relationship‘. The latter could potentially become "favorite shops" and be the first possible target of a marketing campaign.

### 4.3 Conclusions and Future Work

In this study, an unsupervised learning methodology is employed in order to automatically generate a set of classifications, and some criteria are proposed to help marketing experts choose the optimal one. Decision-making can be greatly improved this way.

Employing unsupervised learning permits the generation of hidden profiles which could not be detected casually. Meanwhile, the new criteria introduced in this work drastically reduce the search for the optimal segmentation. Within the framework of the AURA project, it is intended to fully develop the enhanced software module and adapt it to the existing LAMDA implementation.

Further in the future, the acquired knowledge is intended to be applied to fully automate the segmentation analysis just from the input variables.

## Acknowledgments

This work has been supported in part by the AURA Project (Qualitative Reasoning Automatic Learning Systems: Intelligent Tools Applied to Finance and Marketing, which is supported by the Spanish Science and Technology Ministry, TIN2005-08873-C02).

The authors would also like to acknowledge the participation and interest of the firm, Seu, S.A. in the AURA project.

## References

1. J. C. Aguado. *A Mixed Qualitative-Quantitative Self-Learning Classification Technique Applied to Situation assessment in Industrial Process Control.* PhD thesis, Universitat Politècnica de Catalunya, 1998.
2. J. C. Aguado, A. Català, and X. Parra. Comparison of structure and capabilities between a non-standard classificarion technique and the radial basis bunction neural networks. In *Proceedings of the 13th European Simulation Multiconference (ICQFN 99)*, volume II, pages 442–448, Warsaw, Poland, 1999.
3. J. Aguilar and R. López de Mántaras. The process of classification and learning the meaning of linguistic descriptors of concepts. *Approximate Reasoning in Decision Analysis*, pages 165–175, 1982.
4. I. Alon, M. Qi, and R.J. Sadowski. Forecasting aggregate retail sales: a comparisong of artificial networks and traditional methods. *Journal of Retailing and Consumer Services*, 8(3):147–156, 2001.
5. M. Casabayó. *Shopping behaviour forecasts: Experiments based on a fuzzy learning technique in the Spanish food retailing industry.* PhD thesis, Univ. Edinburgh, 2005.
6. J. Dougherty, R. Kohavi, and M. Sahami. Supervised and Unsupervised Discretization of Continuous Features. In *Machine Learning: Proceedings of the 12th International Conference*, San Francisco, CA., 1994.
7. J.G. Klir and B Yuan. *Fuzzy sets and fuzzy logic, Theory and Apps.* Pr. Hall, 1995.
8. L. A. Kurgan, and K. J. Cios. CAIM discretization algorithm. In *Knowledge and Data Engineering, IEEE Transactions on*, volume 16, issue 2, pages 145–153, 2004.
9. C. Olmo, G. Sánchez, F. Prats, and N. Agell. Linearly Compensated Hybrid Connectives for Customers Patterns Segmentation. In *Actas del Taller "Inteligencia Computacional: aplicaciones en marketing y finanzas", Caepia 2005*, pages 19–24, Nov. 2005.
10. A. A. TSchuprow, and M. Kantorowitcsch. Principles of the mathematical theory of correlation. In *Journal of the American Statistical Association*, volume 34, number 208, page 755, December 1939.

# Fuzzy Backpropagation Neural Networks for Nonstationary Data Prediction

Ramon Soto C.

Centro de Investigación y Desarrollo de Ingeniería Avanzada
Real #30 Col. Villa Satélite, Hermosillo, Sonora, Mexico
rsotocc@hotmail.com

**Abstract.** The backpropagation neural network is one of the most widely used connectionist model, especially in the solution of real life problems. The main reasons for the popularity of this model are its conceptual simplicity and its ability to tackle a broad range of problems. But, on the other hand, this architecture shows a well known problem for dealing with nonstationary data. In this paper, a variation of feedforward neural model which uses qualitative data both for feeding the network and for back propagating the error correction is presented. The data are coded by means of a fuzzy concept of local stability.

**Keywords:** Neurofuzzy modelling, data prediction.

## 1 Introduction

Soft computing paradigm is a useful way for tackling hard real problems. The main ingredients in most soft computing applications are artificial neural networks and fuzzy logic, being this the most successful combination of intelligent techniques [10].

An artificial neural network (ANN) is a computational model originally inspired in the bioelectrical networks formed in the brain. Several architectures have been proposed and implemented for realizing this metaphor, being the so called multilayer perceptron the best known and the one most extensively used. The multilayer perceptron or multiperceptron consists of multiple layers of processing units, typically interconnected in a feedforward way, what is called a feedforward neural network. Backpropagation algorithm, on the other hand, is the main technique for adjusting the synaptic weights of feedforward neural networks. In this case, the neural network uses to be called a backpropagation neural network (BNN).

The backpropagation neural network is one of the most widely used connectionist model, especially for the solution of real life problems. The main reasons for the popularity of this model are its conceptual simplicity and its ability to tackle a broad range of problems. But this architecture shows a well known problem for dealing with nonstationary data.

Fuzzy sets, on the other hand, are a useful way for representing imprecise concepts from real world. Opposing to classical sets, objects (or values) can

P. Melin et al. (Eds.): IFSA 2007, LNAI 4529, pp. 318–327, 2007.
© Springer-Verlag Berlin Heidelberg 2007

belong to several fuzzy sets at the same time, even if those fuzzy sets represent contradictory concepts. Fuzzy sets have been shown to be a good tool for representing the qualitative categories describing the behavior of a dynamical system. Standard fuzzy sets have been shown to be a successful tool for modelling nonlinear systems [4], while second order fuzzy sets are a good approach for analyzing data with high noise levels [3]. But standard fuzzy sets give an atemporal description of world and do not represent the dynamical evolution of those categories. Temporal and dynamical fuzzy sets have been proposed for dealing with the dynamical nature of the world [5,9].

In this paper, a variation of feedforward neural model which uses fuzzy coded data, both for feeding the network and for back propagating the error correction, is presented. The coding is realized using soft dynamical fuzzy sets, in order of improve the network efficiency when dealing with nonstationary data.

## 2 Backpropagation Neural Networks

A *feedforward neuronal network* is a system integrated by a set of simple processing devices called nodes (neurons) organized in groups called *layers*, with the information flowing from the input layer to the output layer (figure 1). The neurons in each layer are interconnected only to the neurons of the following layer (from input to output). The output of each neuron is used as the input to the neurons of the next layer with which it is connected to. The output of the $i$-th neuron in any layer $l$ is given by

$$x_i^{(l)} = g\left(a_i^{(l)}\right) \tag{1}$$

where, $a_i^{(l)}$ is called the neuron activity level and describes how much the input data influence the neuron activity. $a_i^{(l)}$ is defined as

$$a_i^{(l)} = \left(\sum_{j=1}^{n^{(l-1)}} w_{ij}^{(l)} x_j^{(l-1)}\right) - \Theta_i^{(l)} \tag{2}$$

$w_{ij}^{(l)}$ is the synaptic weight between the $j$-th neuron in layer $l-1$ and the $i$-th neuron in layer $l$; $n^{(l-1)}$ is the number of neurons in layer $l-1$; $x_j^{(l-1)}$ is the output of the $j$-th neuron in layer $l-1$ and $\Theta_i^{(l)}$ is the bias value of the $i$-th neuron in layer $l$. The output function $g(\cdot)$ used in this work is given by

$$g(a) = \frac{1}{1+e^{-Ga}} \tag{3}$$

being $G$ the network gain.

The neural network training consists of adjusting the synaptic weights between the neurons, in such a way that the output of neurons in the output layer corresponds to a certain known value associated to the input data. Several

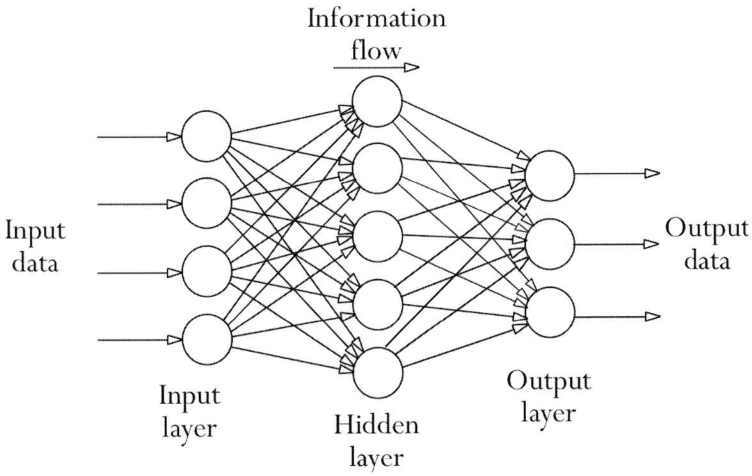


**Fig. 1.** Structure of a feedforward neural network

methods for training a feedforward neural network have been proposed, being the backpropagation algorithm the most extensively used.

In backpropagation algorithm, the output error is computed individually for each neuron in the output layer, and then this error is back propagated towards the first hidden layer.

The synaptic weight $w_{ij}^{(l)}$ that connects the $j$th neuron in layer $l-1$ to the $i$-th neuron in layer $l$, is iteratively updated by means of the equation

$$w_{ij}^{(l)}(k+1) = w_{ij}^{(l)}(k) + \Delta w_{ij}^{(l)}(k) \tag{4}$$

where

$$\Delta w_{ij}^{(l)}(k) = \lambda \delta_i^{(l)}(k) x_j^{(l-1)} + \alpha \Delta w_{ij}^{(l)}(k-1) \tag{5}$$

$\lambda \in (0,1)$ is the learning ratio and $\alpha \in (0,1)$ the momentum. $\delta_i^{(l)}$ is the signal error of the $i$-th neuron in layer $l$ and is defined as

$$\delta_i^{(l)} = \begin{cases} g'\left(a_i^{(l)}\right)\left(y_i - x_i^{(l)}\right) & \text{Neurons at output layer} \\ g'\left(a_i^{(l)}\right)\sum_{m=1}^{n^{(l+1)}} \delta_m^{(l+1)} w_{mi}^{(l+1)} & \text{Neurons at interior layers} \end{cases} \tag{6}$$

The derivative for a sigmoid output function is

$$g'(a) = Gg(a)\left(1 - g(a)\right) \tag{7}$$

A well known problem of backpropagation neural networks is its poor performance when dealing with nonstationary data. This limitation is evident when the behavior of the sigmoid function output is analyzed. From equation 3, can be observed that large activation values $a$ lead to $g(a)$ values very close to the unit, whereas small values of activation lead to $g(a) \approx 0.5$ (see figure 2). This way, large input data values require small synaptic weight values at the input layer, while small input values need large weights, making the synaptic weights tuning a hard task. On the other hand, if the neural network output must be a value in the same range as the input data, then the output function must be attuned to throw values outside the range $[0, 1]$. A common way to solve these problems consists on normalizing the data, but this approach requires to know the complete range of values.

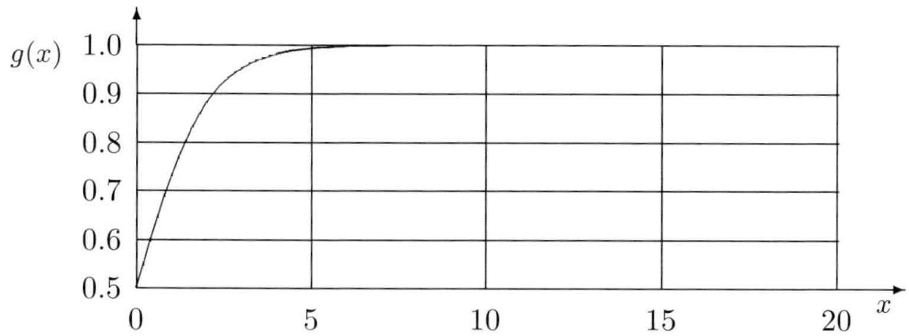


**Fig. 2.** Output function behavior for large activation values

## 3 Fuzzy Local Stability

Given a crisp universe of discourse $U$, a fuzzy set $A$ over $U$ is defined as

$$A = \{(x, \mu_A(x)) \mid x \in U,\ \mu_A : U \rightarrow [0, 1]\} \tag{8}$$

Where $\mu_A(x)$ is the *fixed* degree of membership of $x$ in the fuzzy set $A$.

But, fuzzy categories corresponding to dynamical systems states can show a temporal evolution: The normal values for a given product price, the electric energy demand and the solar activity level, vary along time, in some cases showing some periodicity. If $A$ corresponds to a dynamical category, it is natural to expect that the membership value of $x$ in the fuzzy set $A$ will change along any time interval $T$.

A dynamical fuzzy set [9] $A_D$ is defined as

$$A_D = \left\{(x, \mu^{(x)}(t)) \mid \mu^{(x)}(t) \in [0, 1], t \in T\right\} \tag{9}$$

The behavior of the main variables in many complex dynamical systems can be described in terms of its local stability: the changes on the prices, for example, are say to be raising or dropping. Following this metaphor, the data generated by dynamical systems can be coded in terms of three dynamical fuzzy sets designated as Down, Stable and Up, as is shown in figure 3. Such fuzzy sets describe the local stability of the variable, related to its previous values, and are defined as follows:

– The Stable fuzzy set consist of a pair of increasing-decreasing logistical curves. The membership function for Stable fuzzy set is given by

$$\mu_S(x) = \begin{cases} 0 & x \leq \nu - \varpi \\ 2\left[\frac{\nu-\varpi-x}{\varpi}\right]^2 & \nu - \varpi < x \leq \nu - \varpi/2 \\ 1 - 2\left[\frac{\nu-x}{\varpi}\right]^2 & \nu - \varpi/2 < x \leq \nu + \varpi/2 \\ 2\left[\frac{\nu+\varpi-x}{\varpi}\right]^2 & \nu + \varpi/2 < x < \nu + \varpi \\ 0 & x \geq \nu + \varpi \end{cases} \tag{10}$$

where $\mu_S(x)$ is the membership value of variable $x$ in the Stable fuzzy set; $\nu$ is is the modal value of Stable fuzzy set and represents the reference value for the three fuzzy sets; $\varpi$ is the fuzzy set bandwidth. In this work, the modal value is taken as the last value in the time series previous to the actual value, while the bandwidth value is chosen proportional to the standard deviation value over the adjacent differences on last 10 points in the time series previous to the actual value.

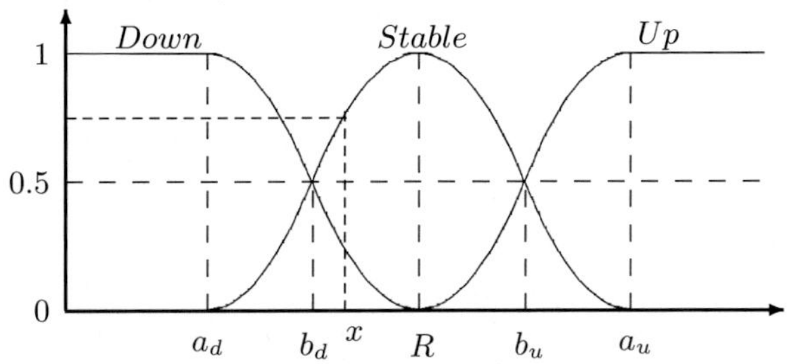


$x \rightarrow (Stable, 0.75, Left)$

**Fig. 3.** Fuzzy sets describing local data stability

– The Down set is chosen as a decreasing logistical curve. The corresponding membership function is given by

$$\mu_D(x) = \begin{cases} 1 & x < \nu - \varpi \\ 1 - 2\left[\frac{\nu-\varpi-x}{\varpi}\right]^2 & \nu - \varpi \leq x < \nu - \varpi/2 \\ 2\left[\frac{\nu-x}{\varpi}\right]^2 & \nu - \varpi/2 \leq x < \nu \\ 0 & x \geq \nu \end{cases} \tag{11}$$

where $\mu_D(x)$, is the membership value of variable $x$ in the Down fuzzy set.

– The Up set is chosen as a growing logistical curve. This set is defined by the fuzzy membership function

$$\mu_U(x) = \begin{cases} 0 & x < \nu \\ 2\left[\frac{\nu-x}{\varpi}\right]^2 & \nu \leq x \leq \nu + \varpi/2 \\ 1 - 2\left[\frac{\nu+\varpi-x}{\varpi}\right]^2 & \nu + \varpi/2 \leq x \leq \nu + \varpi \\ 1 & x \geq \nu + \varpi \end{cases} \tag{12}$$

where $\mu_U(x)$ is the membership value of variable $x$ in the Up fuzzy set.

The fuzzy coding in this paper is based on the qualitative coding method used in the Fuzzy Inductive Reasoning (FIR) technique [1,2]. Such coding uses a triplet formed by a qualitative class value, a numeric membership value and an additional qualitative function side value (see figure 3). The last quantity specifies the membership function side where the corresponding quantitative value is located on. This is a suitable description for local dynamical stability: A variable declared as Stable with a Down tendency, can be coded as $(Stable, \mu, Left)$, for example.

In this coding scheme, a quantitative value is treated as belonging only to the fuzzy set where its membership value is larger. Such coding is equivalent but more economical and readable than the usual approach of coding a quantitative value based on its membership to several fuzzy sets. Two complementary concepts to this representation are those of "neighboring fuzzy sets" and "pointing out to". These concepts are defined as follows:

**Definition 1.** Two fuzzy sets $\alpha$ and $\beta$ are said to be "neighboring fuzzy sets" iff $\alpha \neq \beta$ and $\alpha \cap \beta \neq \phi$, being $\phi$ the empty set. This relations is represented as $\mathcal{N}(\alpha, \beta)$.

**Definition 2.** The record $r$ is said to "point out to" the fuzzy set $\alpha$ iff: 1) $r \in \beta$ and $\mathcal{N}(\alpha, \beta)$, and 2) $\alpha$ is located at the same side of $\beta$ where $r$ is placed on. This relations es represented as $r \mapsto \alpha$.

## 4 Fuzzy Backpropagation Neural Networks

A fuzzy backpropagation neural networks (FBNN) uses the same structure as that of standard feedforward neural network.

The input data for a fuzzy neural network are qualitative triplets $r = (Class, \mu, Side)$. In order to maintain the data coherence, the neural network must operate over the three triplet elements. The output of the $i$-th neuron in layer $l$ is given by

$$X_i^{(l)} = \left(a_i^{(l)}, g(a_i^{(l)}), Side_i^{(l)}\right) \tag{13}$$

- The $\alpha_i^{(l)}$ value is obtained by aggregating the signals coming from the neurons of the previous layer connected to it, $l-1$. This is achieved by computing the next indices

$$I_{\alpha_s} = \sum_{j=1}^{n} w_{i,j}^{(l)} \cdot \chi_j \tag{14}$$

where $\alpha_s \in \{baja, estable, alza\}$, $w_{ij}^{(l)}$ is the synaptic weight between the $j$-th neuron in layer $l-1$ and the $i$-th neuron in the layer $l$. $\chi_j$ is computed by means of the following relation:

$$\chi_j = \begin{cases} g(a_j^{(l-1)}) & X_j^{(l-1)} \in \alpha_s \\ 1 - g(a_j^{(l-1)}) & X_j^{(l-1)} \in \beta, \mathcal{N}(beta, \alpha_s) \text{ y } X_j^{(l-1)} \mapsto \alpha_s \\ 0 & \text{a.o.c.} \end{cases} \tag{15}$$

$\alpha_i^{(l)}$ is chosen as the fuzzy set whose index gets the highest aggregation value.

- $g(\cdot)$ is the output function defined by equation 3, being

$$a_i^{(l)} = \max_s(I_{\alpha_s}) - \vartheta_i^{(l)} \tag{16}$$

  $\vartheta_i^{(l)}$ is the bias value. Since the output membership value always lies in the range $[0.5, 1.0]$, $\vartheta_i^{(l)}$ is set to 0.5 for all neurons.
- The new function side value $Side_i^{(l)}$ is taken as pointing to the fuzzy set whose counter gets the second largest value in the aggregation process.

The error propagation is made using the same process as in the standard backpropagation algorithm, except because now the error must be computed from quelitative data. The error signal for $i$-th neuron in layer $l$, $\delta_i^{(l)}$, is defined as

$$\delta_i^{(l)} = \begin{cases} g'\left(a_i^{(l)}\right) \mathcal{D}(X_i^{(l)}, X_i) & \text{For neurons in the output layer.} \\ g'\left(a_i^{(l)}\right) \sum_{m=1}^{n^{(l+1)}} \delta_m^{(l+1)} w_{mi}^{(l+1)} & \text{For neurons underneath the output layer.} \end{cases} \tag{17}$$

being $\mathcal{D}(\cdot)$ the soft Gower distance function, defined as

**Definition 3.** The soft Gower distance between two records, namely $r_i$ and $r_j$ is

$$\mathcal{D}(r_i, r_j) = \sqrt{\frac{1 - Gow(r_i, r_j)}{Gow(r_i, r_j)}} \tag{18}$$

$X_i$ is the target value for the $i$-th neuron in the output layer and $X_i^{(l)}$ the actual output value at the same node. $Gow(r_i, r_j)$ is given by

$$Gow(r_i, r_j) = \frac{1}{3} \cdot (1 - |\mu_i - \mu_j| + g_c(r_i, r_j) + g_s(r_i, r_j)) \tag{19}$$

where

- $\mu$ is the membership value of the record $r$ as given by the fuzzy triplet $(Class, \mu, Side)$.
- $g_c(r_i, r_j)$ is defined as follows:

$$g_c(r_i, r_j) = \begin{cases} 1 & r_i \in \alpha \text{ and } r_j \in \alpha \\ 0.5 & r_i \in \alpha, r_j \in \beta \text{ and } \mathcal{N}(\alpha, \beta) \\ 0 & \text{Any other case.} \end{cases} \tag{20}$$

  being $\alpha$ and $\beta$ fuzzy sets, and $\phi$ the empty set.

– And $g_s(r_i, r_j)$ is:

$$g_s(r_i, r_j) = \begin{cases} 1 & r_i \in \alpha, r_j \in \alpha \text{ and } S(r_i) = S(r_j) \\ 0.5 & r_i \in \alpha, r_j \in \beta, \mathcal{N}(\alpha, \beta), r_i \mapsto \beta \text{ and } r_j \mapsto \alpha \\ 0 & \text{Any other case.} \end{cases} \quad (21)$$

## 5 Results

The performance of FBNN is compared to that of standard BNN by tackling the following prediction problems:

– The Lorenz attractor, defined by the next differential equations system [8]:

$$\begin{aligned} \frac{dx(t)}{dt} &= \sigma(y - x) \\ \frac{dy(t)}{dt} &= rx - y - xz \\ \frac{dz(t)}{dt} &= xy - bz \end{aligned} \quad (22)$$

where $x$, $y$ and $z$ are state variables and $r$, $\sigma$ and $b$ are positive physical parameters. Following most simulations, the parameter values used in this work are $\sigma = 10$, $r = 28$ and $b = \frac{8}{3}$. The sampling rate is $\tau = 0.01$. A set of 3000 records were generated by numerical integration of equations 22 using a fourth order Runge-Kutta method. The test consist in predicting the $z$ component of Lorenz system. This time series was labeled as "Lorenz 1" in table 1.

– The time series labeled as "Lorenz 2" in table 1 was generated by adding a linear growing tendency and a linear increasing dispersion to the component $z$ of the Lorenz attractor. This series of given by the equation:

$$\breve{z}_k = \frac{k(1 + z_k)}{100}; \quad k = 1 \ldots 3000 \quad (23)$$

– A third time series was generated simulating a quadratic growing tendency and a quadratic increasing dispersion in the $z$ component of the Lorenz attractor. This time series is labeled as "Lorenz 3" in table 1, and is given by the equation:

$$\tilde{z}_k = \frac{k^2(1 + z_k)}{50000}; \quad k = 1 \ldots 3000 \quad (24)$$

– The sunspots number. The data used in this work correspond to the observation range from January 1st 1947 to August 12 1956.

– Foreign currencies market. This problem is a well-known testbed for time series prediction methods. As other financial systems, this one presents many modeling problems [6,7]. The variables to be predicted in this study case are the exchange rates, related to the U.S. dollar, of the next currencies: the Pound Sterling, the Canadian Dollar, the German Mark, the Japanese Yen and the Swiss Franc. The data used in this problem were taken originally from the Monetary Yearbook of the Chicago Mercantile Exchange, and published in Internet by Andreas S. Weigend (http://www.cs.colorado.edu/~andreas/Time-Series/Data/). This database contains 3512 records covering the time period from June 1st, 1973 to May 21st, 1987.

As can be observed in the table, the prediction of stationary time series by the fuzzy neuronal network is slightly better than the prediction obtained with the standard neuronal network. On the other hand, when the stationary conditions disappear, the performance of the fuzzy neuronal network is clearly superior.

In both cases a neuronal network with 15 input nodes, 20 nodes in the only hidden layer, and one unit in the output layer, was used. The training parameters for both neural networks were: gain=1, learning ratio=0.05, momentum=0.5. BNN models used bias=1 while FBNN used bias=0.5.

**Table 1.** NMSE error when predicting different time series with a standard backpropagation neural network and a fuzzy backpropagation neural network

| | **Neural network model** | |
|---|---|---|
| **Time series** | **standard** | **fuzzy** |
| Lorenz 1 | 0.1202752 | 0.0960687 |
| Lorenz 2: linear | 0.3987142 | 0.0933912 |
| Lorenz 3: quadratic | 0.5365449 | 0.0869285 |
| Sunspots number | 0.2661172 | 0.2782048 |
| Pound Sterling | 0.3169981 | 0.0972515 |
| Canadian Dollar | 0.4199859 | 0.1927423 |
| German Mark | 0.2506006 | 0.0495888 |
| Japanese Yen | 0.3608898 | 0.0386783 |
| Swiss Franc | 0.1830959 | 0.0525080 |
| **Average NMSE** | **0.3170246** | **0.1094846** |

## 6 Conclusions

A fuzzy variant on backpropagation neural network has been presented. This model uses dynamical fuzzy sets for describing the local stability in nonstationary data. The performance of fuzzy backpropagation neural networks is compared the performance of standard backpropagation neural networks. The fuzzy variant seems to be clearly superior to the standard one.

## References


1. F. Cellier. General system problem solving paradigm for qualitative modeling. In *Paul A. Fishwick and Paul A. Luker, editors, Qualitative Simulation Modeling and Analysis*, pages 51-71. Springer-Verlag, New York, 1991.
2. Á. de Albornoz. Inductive Reasoning and Reconstruction Analysis: Two Complementary Tools for Qualitative Fault Monitoring of Large-Scale Systems. PhD thesis, Universitat Politcnica de Catalunya, 1996.
3. N. N. Karnik and J. M. Mendel. Applications of type-2 fuzzy logic systems to forecasting of time-series. *Information Sciences*, 120:89–111, 1999.
4. I. Kim and S.-R. Lee. A fuzzy time series prediction method based on consecutive values. *In 1999 IEEE International Fuzzy Systems Conference Proceedings*, pages 703707, Seoul, Korea, 1999.
5. B.R. Kosanović, L.F. Chaparro, and R.J. Sclabassi. Signal analysis in fuzzy information space. *Fuzzy Sets and Systems*, 77(1):49–62, 1996.
6. C. Lee-Giles, S. Lawrence, and A. Chung-Tsoi. Noisy time series prediction using a recurrent neural network and grammatical inference. *Machine Learning*, 44(1/2):161–183, 2001.
7. M. Magdon-Ismail, A. Nicholson, and Y. Abu-Mostafa. Estimating model limitations in financial markets. *Proceedings of the IEEE*, 86(11), 1998.
8. S. Mukherjee, E. Osuna, and F. Girosi. Nonlinear prediction of chaotic time series using support vector machines. In J. Principe, L. Giles, N. Morgan, and E. Wilson, editors, (IEEE) *Workshop on Neural Networks for Signal Processing* (VII), page 511. IEEE Press, 1997.
9. R. Soto and G. Nunez. Dynamical fuzzy sets for time series forecasting. In *IASTED International Conference on Modelling and Simulation* (MS 2003), 2003.
10. A. Tsakonas and G. Dounias. Decision making on noisy time-series data under a neuro-genetic fuzzy rule-based system approach. In *Proceedings of 7th UK Workshop on Fuzzy Systems*, pages 80–89, 2000.

# Fuzzy Model Based Iterative Learning Control for Phenol Biodegradation

Marco Márquez[1,*], Julio Waissman[1], and Olivia Gutú[2]

[1] Centro de Investigación en Tecnologías de la Información y Sistemas
[2] Centro de Investigación en Matemáticas
Unversidad Autónoma del Estado de Hidalgo
Carr. Pachuca-Tulancingo Km. 4.5, Pachuca Hgo. 42084, Mexico
hawk0210@msn.com, {julio,olivia}@uaeh.edu.mx

**Abstract.** In a fedbatch process the operational strategy can consist on controling the influent substrate concentration in the reactor, by means of the input flow manipulation. Due to the repetitive characteristic of the Sequencing Batch Reactor processes, it opens the possibility to explore the information generated in previous cycles to improve the process operation, without having on-line sensors and/or a very precise analytical model. In this work an iterative learning control strategy based on a fuzzy model is proposed. It is assumed that the measurements are analytical and only a few number of them can be obtained. So, an interpolation technique is used to improve the control performance. Simulation results for a phenol biodegradation process are presented.

**Keywords:** Fuzzy Iterative Learning Control, Biotechnological Process.

## 1 Introduction

The Sequencing Batch Reactor (SBR) process operates in a true batch mode with aeration and sludge settling both occurring in the same tank. The major difference between a SBR and a typical activated sludge system is that the SBR tank carries out the functions of equalization, aeration and sedimentation in a time sequence rather than in the conventional space sequence.

The SBR presents some advantages with respect to the continuous activated sludge process. Since SBR is a batch process, the effluent can be held in the reactor until it is treated if the influent can be stored far away. This can minimize the deterioration of effluent quality associated with influent spikes. Also, biomass will not be washed out of a SBR because of flow surges. In addition, settling occurs when there is no inflow or outflow. However, the SBR systems have also some disadvantages, generally related to a higher level of control sophistication.

Identification and real-time control of SBR processes still represent a challenging area of endeavor for control engineers. In particular, the control design is difficult by at least two well-known factors [1]. Firstly, the processes involving

* Scholarship Master Degree Program in Control and Automatization CONACYT.

P. Melin et al. (Eds.): IFSA 2007, LNAI 4529, pp. 328–337, 2007.
© Springer-Verlag Berlin Heidelberg 2007

living microorganisms exhibit large nonlinearities, strongly coupled variables and often poorly understood dynamics. Secondly, the real-time monitoring of many key process variables, which are needed by advanced control algorithms, is hampered by the lack of reliable on-line sensors.

Different trends for the control of fedbatch biotechnological processes have emerged, principally optimal and adaptive approaches [2]. These methods can be applied to the SBR control, since it can be considered as a biotechnological fedbatch process. The main drawback of the model-based optimal control methods, which provides a theorically realizable optimum, is the assumption of a perfectly known model. On the other hand, model-independent adaptive controllers do not guarantee a priori optimality of the control policy results. Finally, the approach based on the concept of minimal modeling of the kinetics has emerged, in order to fill the gap between modeling accuracy and control needs [2]. In this approach, the optimal control of fedbatch processes can be replaced by a common nearly optimal regulation control case. According to this, the control objective is stated in terms of a substrate concentration set point tracking to fix the influent flow rate.

An iterative learning control (ILC) algorithm [3] allows the output-tracking task to be carried out. The choice of an ILC technique is justified by the repetitive nature of the fedbatch cultures. Since ILC uses information from previous executions of the task in order to improve the tracking performance from trial to trial, it does not require any on-line measurement. The ILC differs from the majority of the control methods, as it employs all the possibilities of incorporation of control information from the past process operation, such as the error and input signals, to construct the actual control action.

The ILC control has been proposed for the control of systems that can perform the same task repetitively [4]. Since fedbatch reactors are permanently in a transient regime, the tracking behaviour of the conventional ILC approach deteriorates as the number of off-line measurement samples decreases. The use of ILC for discrete-time systems is a common case, but a very small time period is needed to guarantee the algorithm convergence [5]. In this work, the measurements are considered to be done by a process operator, and so it is desirable to have a sampling period as larger as possible. The effect of a larger sampling period on the ILC control of a SBR process was studied in [6]. The proposed solution consists in generating an estimated output sequence in small intervals, to reduce the tracking error between samples.

In this work, a fuzzy model of the process was developed. The learning law is based in the fuzzy model to improve the convergence rate of the learning algorithm, based in ideas proposed in [7] and [8]. In this work a fuzzy model of the growth rates are coupled with a Takagi-Sugeno type fuzzy model of the process in order to reduce de complexity of the model and reduce the computational effort of the learning law algorithm.

The study is carried out on an analytical model of a SBR system for phenol biodegradation, whose kinetics are characterized by the production and later consumption of an inhibitory metabolic intermediate. The process under study is a pilot plant of the CIQ-UAEH [9]. From this process, an analytical model was previously experimentally validated for fedbatch cultures [10].

## 2 Phenol Biodegradation Model

The mass balance equation for the various constituents of phenol biodegradation is given by the following first order system of differential equations [10]:

$$\begin{aligned}
\dot{X}(t) &= \mu(t)X(t) - \frac{Q_{\mathrm{in}}(t)}{V(t)}X(t) + d_1(t) \\
\dot{S}_1(t) &= -q_{S_1}(t)X(t) + \frac{Q_{\mathrm{in}}(t)}{V(t)}(S_1^{\mathrm{in}} - S_1(t)) + d_2(t) \\
\dot{S}_2(t) &= v_{S_2}(t)X(t) - q_{S_2}(t)X(t) - \frac{Q_{\mathrm{in}}(t)}{V(t)}S_2(t) \\
\dot{V}(t) &= Q_{\mathrm{in}}(t) - Q_{\mathrm{out}}(t) \\
y(t) &= S_1(t)
\end{aligned} \tag{1}$$

where $S_1(t)$ is the phenol concentration, $S_2(t)$ is the main metabolic intermediate concentration, $X(t)$ is the total microbial concentration and $V(t)$ is the volume; $\mu(t)$ is the specific biomass growth rate, $q_{S_1}(t)$ and $q_{S_2}(t)$ are, respectively, the specific consumption rate of phenol and the intermediate; $v_{S_2}(t)$ is the specific intermediate production rate. We consider that the reactor is a fedbatch process and in this case $Q_{\mathrm{out}}(t) = 0$, $d_1(t)$ and $d_2(t)$ are the external disturbances.

The specific biomass growth rate is calculated by

$$\mu(t) = \mu_1(t) + \mu_2(t) \tag{2}$$

where $\mu_1(t)$ is a modified Haldane type equation and $\mu_2(t)$ is a Monod type [10], i.e:

$$\mu_1(t) = \frac{\mu_{\max_1} S_1(t)}{K_{S_1} + S_1 + S_1^2/K_{i_1}} \frac{K_2}{K_2 + S_2(t)} \tag{3}$$

$$\mu_2(t) = \frac{\mu_{\max_2} S_2(t)}{K_{S_2} + S_2(t)} \frac{K_1}{K_1 + S_1(t)} \tag{4}$$

where, $\mu_{\max_1}$ is the maximus growth value due to the phenol concentration, $\mu_{\max_2}$ is the maximus growth value due to the intermediary concentration.

The specific growth and consumption rates are correlated with the constants, biomass to phenol $Y_1$ and biomass to intermediate $Y_2$ as follows:

$$q_{S_1}(t) = \frac{\mu_1(t)}{Y_1} \qquad q_{S_2}(t) = \frac{\mu_2(t)}{Y_2}. \tag{5}$$

The specific production rate of intermediate is linearly correlated to the specific growth rate of biomass on phenol [10] as

$$v_{S_2}(t) = \alpha\mu_1(t). \tag{6}$$

The parameters values are described in [10] in base to a sensitivity analysis, the values used for simulation are given in the Table 1.

**Table 1.** Parameter values

| Symbol | Value | SI units |
|---|---|---|
| $\mu_{\max 1}$ | 0.39 | 1/h |
| $Y_1$ | 0.57 | mg/mg |
| $K_{S_1}$ | 30 | mg/l |
| $K_{i_1}$ | 170 | mg/l |
| $K_2$ | 160 | mg/l |
| $\mu_{\max 2}$ | 0.028 | 1/h |
| $Y_2$ | 0.67 | mg/mg |
| $K_{S_2}$ | 350 | mg/l |
| $K_1$ | 66 | mg/l |
| $\alpha$ | 1.6 | mg/l |

## 3 Fuzzy Model of Microbial Growth Rate

For developing the learning algorithm, it is needed a fuzzy model of the process. In order to reduce the number of possible fuzzy rules a fuzzy model of the microbial growth rate coupled with a fuzzy model of the process is proposed. This approach allows to obtain a reduced number of rules. So, less sub-models are required for the learning law.

There were used five membership functions to evaluate the phenol concentration $\mu_1(t)$, and another five for the metabolic intermediate concentration $\mu_2(t)$ to model both growth rates. So only 25 fuzzy rules were generated.

The consequent parameters of the linear functions can be estimated by least squares from the available data. The condition is that the consequent functions were linear and the number of data items is much greater than the dimension of the regression vector. A diagonal matrix $\Gamma_i \in R^{N\times N}$ is formed with the normalized membership values $\lambda_i(x_k)$ as the $k$-th diagonal element, where $k$ is in $[1, N]$ and $N$ is the number of data items. A $X_e$ matrix was formed by $S_1$, $S_2$ as its columns and a column of ones to determine the parameters $b_i$. I it is so called the extended regressor matrix $X_e = [S_1, S_2, 1]$ [11]. Another matrix $\bar{X} \in R^{N\times nN}$ is formed from the matrices $\Gamma_i$ and $X_e$:

$$\bar{X} = [\Gamma_1 X_e, \Gamma_2 X_e, \ldots, \Gamma_n X_e].$$

If the parameters are defined as a vector $\gamma$, it is computed in the following way:

$$\gamma = \left[\bar{X}^T \bar{X}\right]^{-1} \bar{X}^T y \tag{7}$$

where $y$ is the fuzzy estimation of microbial growth rate $\mu_1(t)$ or $\mu_2(t)$. In this way, $\gamma$ is obtained as the vector:

$$\gamma = [a_1^T b_1, a_2^T b_2, \ldots, a_n^T b_n].$$

The fuzzy estimated values of $\mu_1(t)$ and $\mu_2(t)$ are presented in Fig. 1 and Fig. 2 respectively. Thus, the microbial growth rate is a function of the phenol $S_1$ and the metabolic intermediate $S_2$, with a linguistic interpretation.

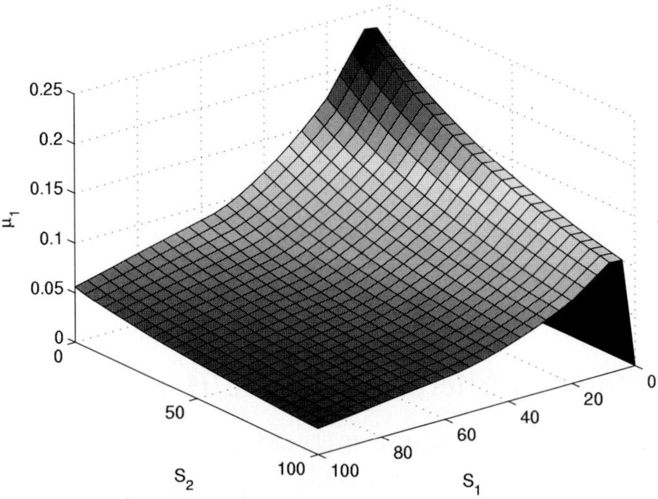


**Fig. 1.** Microbial growth rate due the phenol

Once the fuzzy model of the microbial growth rate is obtained, the number of rules of the process model will be reduced. If there is considered a Takagi-Sugeno fuzzy inference system, the model can be represented by a fuzzy combination of linear systems:

$$R^{(i)} : \text{ If } x_1(t) \text{ is } F_1^{(i)} \text{ and } \cdots \text{ and } x_n(t) \text{ is } F_n^{(i)}, \tag{8}$$

$$\text{Then } \boldsymbol{x}(t+1) \text{ is } A_i(\boldsymbol{x}(t)) + B_i u(t)$$

where $F_j^{(i)}$ is a fuzzy set, $\boldsymbol{x}(t) = [X(t), S_1(t), S_2(t), V(t)]$ denotes the state vector, $u(t)$ denotes the control input, $A_i \in R^{4\times4}$, $B_i^{4\times1}$ are matrices that represent the dynamic of each linear subsystem. The output $y(t)$ is the phenol concentration. A representative prototype of the $i^{th}$ rule is denoted by $\boldsymbol{x}^i = [X^i, S_1^i, S_2^i, V^i]$

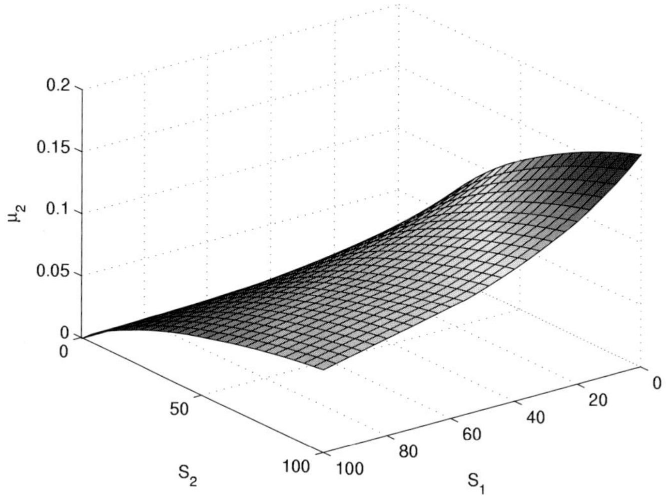


**Fig. 2.** Microbial growth rate due the metabolic intermediate

A normalized membership value can be obtained as:

$$x(t+1) = \sum_{i}^{n} \lambda_i(\boldsymbol{x}(t))(A_i\boldsymbol{x}(t) + B_i u(t)). \tag{9}$$

where

$$\lambda_i(\boldsymbol{x}) = \frac{\prod_j^n \mu_{F_j}(x_j)}{\sum_i^m \prod_j^n \mu_{F_j}(x_j)}. \tag{10}$$

The matrices $A_i$ and $B_i$ for each rule are defined as a linearization of the process model:

$$A_i = \begin{pmatrix} \mu_1^i + \mu_2^i - \frac{Q^i}{V^i} & 0 & 0 & \frac{Q^i X^i}{(V^i)^2} \\ \frac{\mu_1^i}{Y_1} & -\frac{Q^i}{V^i} & 0 & \frac{-Q^i(S_1^{\text{in}} - S_1^i)}{V^i} \\ \alpha\mu_1^i - \frac{\mu_2^i}{Y_2} & 0 & -\frac{Q^i}{V^i} & \frac{Q^i S_2^i}{(V^i)^2} \\ 0 & 0 & 0 & 0 \end{pmatrix} \tag{11}$$

$$B_i = \begin{pmatrix} -\frac{X^i}{V^i} \\ (S_1^{\text{in}} - S_1^i)\frac{1}{V^i} \\ -\frac{S_2^i}{V^i} \\ 1 \end{pmatrix} \tag{12}$$

where $\mu_1^i$ and $\mu_2^i$ are the fuzzy estimation of the growth rates for the prototype of the $i^{th}$ rule. This allows to use a reduced number of rules to represent the process.

## 4 Model Based Fuzzy ILC

The Iterative Learning Control (ILC) is a technique generating an input sequence for a working cycle of a system, by using a learning law and the information of the input and output sequences of the last cycle, as well as the desired output sequence. The learning law has to assure the convergence of the output sequence to the desired one after several working cycles.

A desired output sequence $\{y_d(0), y_d(\Delta t), \ldots, y_d(N\Delta t))\}$ is imposed and it is defined on the interval $[0, T]$. The error is obtained at each point as:

$$e_i(k\Delta t) = y_d(k\Delta t) - y_i(k\Delta t)$$

where $\{y_i(0), y_i(\Delta t), \ldots, y_i(N\Delta t))\}$ is the output sequence of the system for the $i$ iteration. A commonly assumption for ILC is that the initial condition is bounded for each working cycle. The proposed learning law is:

$$u_{i+1}(k\Delta t) = u_i(k\Delta t) + \beta_i(k\Delta t)e_i(k\Delta t) \tag{13}$$

where $\{u_{i+1}(0), u_{i+1}(\Delta t), \ldots, u_{i+1}(N\Delta t))\}$ is the input sequence for the next iteration. The control gain $\beta_i(k\Delta t)$ is obtained by the fuzzy model of the process by:

$$\beta_i(k\Delta t) = \sum_{j=0}^{n} \lambda_j(\boldsymbol{x}(k\Delta t))\beta_i^j \tag{14}$$

$\beta_i^j$ is the learning gain of the $j^{th}$ rule of the model in the $i^{th}$ iteration.

The learning gains $\beta_i^j$ are obtained minimizing a quadratic performance index [12]. In the iteration $i$, the sequence gain is computed for the $j$-th rule as:

$$\beta_i^j = (G_j^T Q G_j + R)^{-1} G_j^T Q \tag{15}$$

where $Q$ and $R$ are symmetric positive definite matrices.

$G_j$ is obtained from the super-vector notation of the system. This notation is in discrete time and is formulated as $y_j(k\Delta t) = G_j u(k\Delta t)$, where

$$G_j = \begin{pmatrix} CB_j & 0 & \cdots & 0 \\ CA_jB_j & CB_j & \cdots & 0 \\ \vdots & \vdots & \ddots & \vdots \\ CA_j^{N-1}B_j & CA_j^{N-2}B_j & \cdots & CB_j \end{pmatrix} \tag{16}$$

where $A_j$ and $B_j$ are the model matrices for the $j^{th}$ rule of the fuzzy system.

## 5 Simulation Results

An application example was developed for the phenol biodegradation process. The main goal was to follow a phenol concentration profile. It was assumed that only a reduced number of analytical samples can be made. To guarantee convergence an interpolation is needed.

As it is shown in [5], to ensure a reduced tracking error for a sampled data non linear system, it is needed a small sample time. Due to the system measurement is assumed to be by the way of operator analysis, the sample time can not be small. In the other hand, a small sampling period implies that the matrices $G_j(k\Delta t)$ will be too big, and the control signal computation can requires a considerable computational effort.

The input signal was computed with a bigger sampling period, and then a finer input sequence was generated using a correct interpolation technique. In [6] is shown that a good choice of the interpolation technique allows to have a bigger sampling period for the output sequence.

The system was simulated for working cycles of 15 hours. In order to ensure a bounded initial condition error for biomass and volume, a purge at the end of each cycle is considered has a realistic assumption. The initial conditions of biomass concentration and volume are $X(0) = 450$ mg/l $\pm$ 3% and $V(0) = 1$ l. Nevertheless, it is not possible to reduce the phenol and the intermediate concentrations from one cycle to other. Then, both substances have to be completely consumed at the end of each working cycle. For this reason, $S_1(0) = S_2(0) = 0$ for the first iteration and for the other iterations, $S_1(0)$ and $S_2(0)$ take the values of the final concentration of the last working cycle.

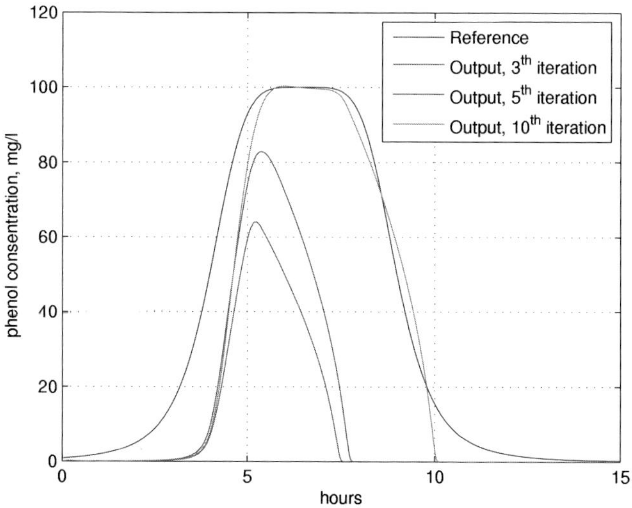


**Fig. 3.** Simulation of the phenol concentration

The signal reference is the concentration profile defined by:

$$y_d(t) = 70 \left| \frac{t - 6.5}{2.5} \right|^{-5} . \tag{17}$$

A white noise was added to states $X$ and $S_1$ in order to simulate the analytical measurement error.

The sample time was 20 minutes, so 46 values are used for each state in each iteration. A cubic spline technique was used to interpolate the control signal in order to have a 1.2 minutes as virtual sample time, and apply a sequence of 751 values to the process.

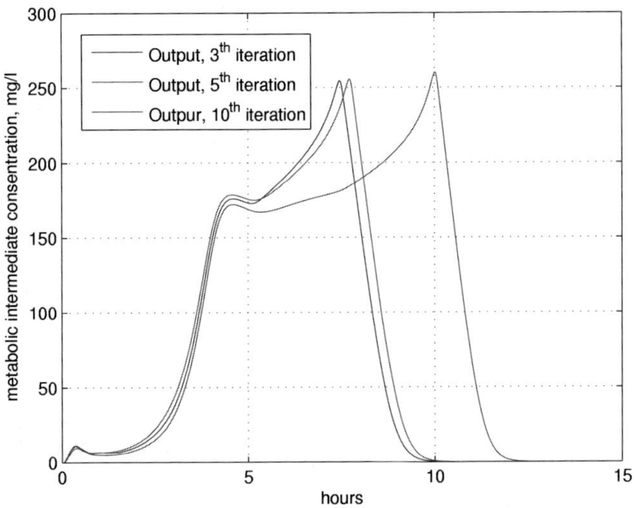


**Fig. 4.** Simulation of the metabolic intermediate

For the fuzzy process model, three membership functions were defined for the phenol and the metabolic intermediate concentration, two functions for the volume, the inlet flow, and the biomass concentration.

In Fig. 3, the simulation of 10 iterations are presented for the phenol concentration. A good performance is presented after 10 iterations. Nevertheless, an important tracking error is presented at the begining of the phenol concentration profile, due to the large sampled time used.

In order to avoid an effect of bioaccumulation of intermediate, it is important to ensure that the intermediate is completely consumed at the end of each working cycle. In Fig. 4 the simulation shows that it is consumed before 15 hours in all iterations.

# 6 Conclusion

The use of a fuzzy model whose consequents are linear systems is a good alternative to control a bioreactor when the measurement can not be made on-line. The number of measurements can be reduced if an interpolation technique is used to have an small virtual sample period with a cost in the performance.

# References


1. Bastin G. and Dochain D.: *On-line estimation and adaptive control of bioreactors.* Elsevier, Ed. Amsterdam: Elsevier, 1990.
2. Van Impe J. and Bastin G.: Optimal adaptive control of fed-batch fermentation processes. *Cont. Eng. Prac.*, **3** (1995) 939–954.
3. Bien Z. and Xu J.: *Iterative learning control analysis, design, integration and applications.* Ed. Boston: Kluwer Academic Publishers, 1998.
4. Arimoto S., Kawamura S. and Miyazaki S.: Bettering operation of robots by learning. *J. Robotics Systems*, **1**(2), 123–140, 1984.
5. Chien C.: A sampled-data iterative learning control using fuzzy network desing. *Int. J. Control*, **73**(10), 902–913, 2000.
6. Márquez M., Ben Youssef C., Vázquez G. and Waissman J.: Iterative learning control of a sbr reactor by using a limited number of samples. In *Proc. of the Int. Meeting on Env. Biotech. and Eng.*, Mexico, 2006.
7. Zhang L. and Yang F.: Fuzzy iterative laerning control design for output traking of discrete-time fuzzy systems. In *Proc. of the Third Conf. on Machine Learning and Cybernetics*, Shanghai, 2004, 678–682.
8. Hätönen J., Owens D. and Moores K.: An algebraic approach to iterative learning control. *Int.J. Control*, **77**(1) (2004) 45-54.
9. Vázquez G., Ben-Youssef C. and Waissman J.: Two step modeling of the biodegradation of phenol by an acclimated activated sludge. *Chem. Eng. J.*, **117** (2006) 245–252.
10. Ben-Youssef C., Waissman J. and Vázquez G.: A new model for phenol biodegradation and activated sludge growth in fedbatch cultures. In *Proc. of the IFAC Symposium on Advanced Control of Chemical Processes*, Grammados, 2006.
11. Babuška R.: *Fuzzy modeling for control.* International series in intelligent technologies, K. Academic, Ed. Boston: Kluwer Academic Publishers, 1998.
12. Kim H. and Shin Y.: Design of adaptive fuzzy sliding mode controller based on fuzzy basis function expansion for ufv depth control. *Int. J. Control Aut. and Syst.*, **3**(2) 2005 217–224.

# Fuzzy Modelling Methodologies for Large Database

Virgilio López Morales[1], Julio Cesar Ramos Fernández[2], Gilles Enea[3], and Jean Duplaix[3]

[1] Universidad Autonoma del Edo. de Hidalgo, CITIS, Carr. Pachuca-Tulancingo Km. 4.5, 42184, Pachuca, Hgo., Mexico
Tel.:(+52)7717172197
virgilio@uaeh.edu.mx

[2] Universidad Tecnologica Tula Tepeji, Carr. Tula-Tepeji, Km. 8, Col. El 61, Ejido El Carmen, Tula de Allende, Hgo., Mexico
Tel.:(+52)7737329100
jramos@uttt.edu.mx

[3] Laboratoire des Sciences de l'Information et des Systemes UMR-CNRS6168, Univ. du Sud Toulon Var, France
Tel.:(+33)494142349
{enea,duplaix}@univ-tln.fr

**Abstract.** In this paper we analyze two recent modelling methodologies: one based on a database preprocessing and then, the application of the fuzzy C-means to highlight useful characteristics used in target selection for direct marketing which is our first study case. The second one is based on fuzzy clustering and cubic splines in the rule consequents. Some examples are given in order to illustrate the advantages and drawbacks of these methods.

**Keywords:** Modelling, Identification, Data Clustering, Cubic Splines.

## 1 Introduction

In control engineering, modelling and identification are important steps in the design of control laws, supervision and fault-detection systems. When poorly understood complex systems are considered, or there is a strong presence of inaccurate values, or the complexity of the resulting model grows up, all these cases are limiting factors to apply this mechanistic modelling approach. Nevertheless, process under study can be approximated by using some general structure, used as a general function approximator. Modelling problem then reduces to postulating an appropriate structure of the approximator, in order to correctly capture the dynamics and the nonlinearities of the system. The identification problem consists of estimating the model parameters. A severe drawback of this approach is that the structure and parameters of these models usually do not have any physical significance.

P. Melin et al. (Eds.): IFSA 2007, LNAI 4529, pp. 338–347, 2007.
© Springer-Verlag Berlin Heidelberg 2007

**Fuzzy modelling.** Modelling approach is based on fuzzy submodels, which describe relationships between variables by means of **If-Then** rules, such as:

$$R_i : \textbf{ If } x_1 \textit{ in } A_1 \text{AND } \cdots \text{AND } x_m \textit{ in } A_m \textbf{Then } \; y = f(x, u) \tag{1}$$

where $R_i$ is the $i^{th}$ rule or submodel. This structure is well known as Takagi-Sugeno-Khan fuzzy model [1]. Function $f(x, u)$ can have different forms, as in the two cases here analyzed: A constant value (singleton) as we have dealed in the first method or a nonlinear function (cubic splines), as in the second method.

In the following we recall some results from the target selection for direct marketing problem. Further details can be found in [2].

## 2 Target Selection from Large Database for Direct Marketing

Target selection is an important data mining problem in direct marketing. The main task in target selection is to determine the potential customers for a product from a large database by identification of the customer profiles that are previously known [3]. Typically, in a marketing database, the data set contains many empty answers, because the clients do not know these characteristics or they are not interested. Clustering must then be applied to characterize the relations in variables involved in the database. In order to get a model-based completion of the data, this can be obtained in two ways: features with missing values can be removed from the data set, or, similarly, data points corresponding to clients with uncompleted records can be removed. These two approaches (removal of features and removal of records) are contradictory, and hence, a trade-off between the conservation of the features and the conservation of records is needed.

### 2.1 Analytical Data Exploration

Our approach (Cf. [2]) provides an analytical methodology to build the representative matrix of the data base, by showing the principal rows (clients) and columns (features) **from a large database**, and it is based on the transformation of Karhunen - Loeve, also known as Hotelling transformation. **This transformation is based in principal components analysis (PCA)**. The main problem to be solved in this Section is on how to obtain the most representative components of a matrix conformed by rows (clients) and columns (features). The answers of each client for each feature is the conformation of the matrix. We will suppose that it is normalized, since in all company databases, it is natural to have the same units of measure.

In order to get a filtered matrix (the pre-processed matrix) let us define $X^c \in \Re^{\mathrm{M}\times\mathrm{N}}$ (respectively $X^f \in \Re^{\mathrm{N}\times\mathrm{M}}$) representing the complete database of the clients (resp. of features), as follows:

$$X^c = \begin{bmatrix} x_1^{1,c} & \cdots & x_N^{1,c} \\ \vdots & \ddots & \vdots \\ x_1^{M,c} & \cdots & x_N^{M,c} \end{bmatrix}; \quad X^f = \begin{bmatrix} x_1^{1,f} & \cdots & x_1^{M,f} \\ \vdots & \ddots & \vdots \\ x_N^{1,f} & \cdots & x_N^{M,f} \end{bmatrix} \tag{2}$$

A first selection of *most records (clients) retained* and *most features allowed* are calculated in the first stage of our algorithm.

### 2.2 Pre-processing the Original Database

The selection of the principal components, is solved separately by using the eigenvalues and eigenvectors of the covariance matrix for each case (clients and features). We then find out the associated components in the original database, by reconstructing via the Hotelling transformation. The procedure is detailed as follows: From equations (2), calculate the average vectors for the clients and features:

$$m^c = \frac{1}{N}\sum_{j=1}^{N} x_j^{k,c}, \forall\, k;\quad m^f = \frac{1}{M}\sum_{k=1}^{M} x_j^{k,f}, \forall\, j \tag{3}$$

Also, calculate the covariance matrix for clients and features:

$$C^c = \frac{1}{N}\sum_{j=1}^{N} x_j^{k,c} \cdot \left(x_j^{k,c}\right)^T - m^c \cdot (m^c)^T ; \tag{4}$$

$$C^f = \frac{1}{M}\sum_{k=1}^{M} x_j^{k,f} \cdot \left(x_j^{k,f}\right)^T - m^f \cdot \left(m^f\right)^T \tag{5}$$

Since $C^c$ and $C^f$ are symmetric and real, then it is always possible to find a set of $n$ - orthonormal eigenvalues. Let $\lambda_j^c$ and $\lambda_k^f$ be the eigenvalues for clients and features respectively, where $j = 1, ..., M$ and $k = 1, ..., N$. The eigenvalues are sorted from the maximum value up to the minimum, with their corresponding eigenvectors. By using equations (2)- (4) we can apply the Hotelling transformation for clients and features. Let $A^c$ be a matrix whose rows, corresponding to eigenvectors of $C^c$ (resp. $C^f$) from (4), are sorted in a way that the first row of $A^c$ (resp. $A^f$) is the eigenvector corresponds to biggest eigenvalue. The rows of $A^c$ and $A^f$ are orthonormal, i.e. $(A^c)^{-1} = (A^c)^T$ and $(A^f)^{-1} = (A^f)^T$. Obtaining the Hotelling transformation,

$$Y^c = A^c\,(X^c - m^c)\,;\quad Y^f = A^f\left(X^f - m^f\right) \tag{6}$$

for clients and features establish a new coordinates system, where the origin is the center of the population, and therefore the elements are not correlated with respect to the original databases (equations (2)). Another important property of the Hotelling matrix is the reconstruction of (2) via (6). The more eigenvector **in the matrix** $A^c$ **and** $A^f$ are taken to reconstruct the matrices (2), the more accurate estimation of the original characteristics is obtained.

Then, by taking only a few eigenvectors (to be defined latter) of $A^c$ and $A^f$ two approximated matrix are obtained and defined by:

$$\hat{X}^c = (A^c)^T\, Y^c + m^c;\quad \hat{X}^f = \left(A^f\right)^T Y^f + m^f \tag{7}$$

In order to select the amount of populations of clients and features, take the error between the approximation of the $X^c$ and $X^f$. We use only the biggest eigenvalue and their corresponding eigenvector to reconstruct the matrix $X^c$, which is approximate by the matrices $\hat{X^c}$ resp. for $X^f$ by $\hat{X^f}$. Our key idea is based on the selection of the eigenvectors (and theirs eigenvalues) within the average value of complete eigenvalues, in order to obtain the principal components of the most representative clients and features. By using this approach, a natural selection of the pre-processed matrix implies to retain the principal clients and features. Then, the amount of *retained records* and *features retained* is automatized and consequently, it can be used to obtain different and growing accurate targets selection model of potential customers. The average value of eigenvalues in clients and features is given by: $\lambda^c_{Aver} = \frac{1}{N}\sum_{k=1}^{N}\lambda^c_k$; $\lambda^f_{Aver} = \frac{1}{M}\sum_{j=1}^{M}\lambda^f_j$.

**Criterion 1: selected eigenvalues.** The eigenvalues are selected in such a way that satisfy the criterion $\lambda^c_{Aver} \leq \lambda^c_{r-1} \leq \cdots \leq \lambda^c_2 \leq \lambda^c_1$ for the clients, and $\lambda^f_{Aver} \leq \lambda^f_{p-1} \leq \cdots \leq \lambda^f_2 \leq \lambda^f_1$ for the features. Where $r$ (resp. $p$) is the amount of eigenvalues selected in clients (resp. in features). In order to keep (w.r.t. the chosen criterion) the amount of the principal components of database in clients (resp. features), take the greater eigenvalue $\lambda^c_1$, (resp. $\lambda^f_1$) with their corresponding eigenvector, and build the matrix $A^c$ (resp. $A^f$) by employing only the first row, and set the other matrix entries to zero. Then, apply equation (6) and (7) which gives the approximation of the principal clients and features (from (2)) in the original database. Consider the approximation error by computing the root-mean-squared errors as $E^{(c,f)}_j = \sqrt{\sum_{k=1}^{M}\left(\hat{X}^{k,(c,f)}_j - X^{k,(c,f)}_j\right)^2 / M}$ $j = 1, \cdots, N$ which represent the approach error between original database and the approximated one, $E^c_j$ for clients and $E^f_j$ for features.

### 2.3 Ordering The Pre-processed Database

**Criterion 2: Components That Leave the Database.** In a second step, it is necessary to sort (16) from the smaller RMSE to the greater RMSE value for clients and features related in *Criterion 1*, to finally characterize the principal components. Choose the first $r$ values of sorted clients (resp. first $p$ values of sorted features). The first $r$ (resp. $p$) elements have the smallest RMSE. Then, the matrix (2) will be reduced, and with the concatenation of $X^c$- reduced and $X^f$- reduced, we obtain a filtered matrix $X^w$, that we call the *work matrix*. In the structure of $X^w$, the amount of rows (clients) are defined by $m^* = M - r$, and the amount of columns (features) defined by $n^* = N - p$ , and $X^w \in \Re^{m^* \times n^*}$. The work matrix can be used to synthesize a learning fuzzy model. In matrix $X^w$ the rows (resp. the columns) are the most representative clients (resp. features).

The resulting matrix of this descendent reordering step is assigned to the matrix $X^{w'}$ and it can be written in the following general form:

$$X^{w'} = \begin{bmatrix} c_{1^*} f_{1^*} & \cdots & c_{1^*} f_{n^*} \\ \vdots & \cdots & \vdots \\ c_{m^*} f_{1^*} & \cdots & c_{m^*} f_{n^*} \end{bmatrix} \tag{8}$$

where $m^*$ and $n^*$ are the work matrix indexes. Finally the $n^*$ features must be analyzed separately to discriminate between the two groups in the data (respondents and non-respondents). This method is widely used by the experts in the marketing and financial world, and it is called a *Gain Chart*. This behavior is preferred in targeted selection, since the goal is to target as many respondents as possible with the minimal number of evaluated clients. A gain chart is defined by the following equation:

$$G_j(\frac{k^*}{m^*}) = \frac{\sum_{i=1}^{k^*} y_i}{\sum_{i=1}^{m^*} y_i}, k^* = 1, 2, ..., m^*, j = 1, 2, ..., n^*, \tag{9}$$

where $y_{k^*} \in \{0, 1\}$ is the corresponding response label.

### 2.4 Construction of the Fuzzy Model Using Clustering C-Means Approach

The final goal of the former methodology is the construction of a fuzzy model which can be obtained for each feature. Each feature model, involves all clients **from the work matrix** participating on the process of learning in that model. **The process of learning, consist in identifying the rules's parameters:**

- **premises using fuzzy C-means clustering**
- **consequents T-S type singleton, based on local least squares approach.**

For the model construction, the method of fuzzy C-means clustering algorithm (FCM) is used.

By taking a suitable fuzzyness parameter value ($m \in (1, \infty)$) we avoid the computing of radius of the volume prototype of each cluster and then we reduce the computational time. We apply the T-S type reasoning, and chose to estimate the rule consequents from the available training data using a local least squares approach (LS), [4] and [5]. Thus for a complete feature model, we have a total of $n^*$ models. Each feature model contains five rules with the following singleton model structure. $R_{ji}$ : **If** feature $f_{j^*}$ *is* $A_{ji}$ **then response** $y_{ji} = c_{ji}$ where $A_{ji}$ are the linguistic variables $i = 1, ..., 5$, (index for each cluster), and $j = 1, ..., n^*$ (index of each feature). Then, the membership functions have a **trapezoid** geometry by choosing the fuzziness parameter (in our case $m = 1.1$). We use a defuzzification method of output weighted average, to test the models obtained. The consequents are ordered automatically in rows and columns, the consequents for the most preferred feature is the first row and the last column contain the conclusion for the higher class.

### 2.5 Iterative Modelling

The reviewed algorithm proposes a direct selection of targets as:

- Data mining with the approach Hotelling's transformation matrix, **in this step, the work matrix $X^w$ is sorted in a descendent way.**
- Ordering of feature vectors and their best clients for each feature. Then make a *Gain Chart* for each feature vector.
- Construction of the model using fuzzy clustering C-means approach, for each feature, (in our case with 5 clusters and $m = 1.1$), and conclusions T-S with constant parameters .

Nevertheless, when we faced to the model identification of some dynamic plants, it is better to use some different techniques an shown in the following section.

## 3 Fuzzy Modelling Via Clustering and Cubic Splines

Our second comparative methodology proposed is based on the rule base structure proposed by T-S, where nonlinear consequents are used. The main advantages in our approach is an important reduction on the rule-base, and consequents fits in a better way the process nonlinearities since they are cubic with linear parameters. Our modelling method is based on a fuzzy clustering algorithm (G-K) in order to classify similar input/output pair data. Further details can be found in [6].

We use classification trees automatically generated by the clustering algorithm and eigenvector-based input partitioning, which allows us both the orientation and the shape of the input space partitioning. By using the well known Least Square Estimation (LSE), the parameter identification is carried out for cubic splines. The key idea behind the use of cubic polynomials is that some standard fuzzy control schemes can be applied in a straightforward way (inversion model [7], LMI with bounded nonlinearities [8,9], etc).

Finally, the antecedents are defined by trapezoidal membership function, and the consequents by cubic equations.

### 3.1 Gustafson-Kessel Algorithm and Knot Selection

The G-K algorithm is a derived from de basic Fuzzy C-Means (FCM), by adapting a *distance norm*, in order to detect clusters with an ellipsoidal geometrical shapes in the analyzed data set [7].

Each cluster has its own norm-inducing matrix $A_i$, which yields the following inner-product norm: $D^2_{ikA_i} = (z_k - v_i)^T A_i (z_k - v_i)$. In the FCM algorithm the matrices $A_i$ are then identity matrix while, in G-K algorithm the matrices $A_i$ employ an adaptive distance norm. This adaptive distance norm, optimizes the cluster's shape while its volume remains constant, i.e. $|A_i| = \rho_i, \quad \rho_i > 0, \quad \forall i$.

Using the Lagrange multiplier method, $A_i$ reads as $A_i = [\rho_i \det(F_i)]^{1/n} F_i^{-1}$, where $F_i$ is the fuzzy covariance matrix of the $i$th cluster defined by:

$$F_i = \left(\sum_{k=1}^{N} (\mu_{ik})^m (z_k - v_i)(z_k - v_i)^T\right) \cdot \left(\sum_{k=1}^{N} (\mu_{ik})^m\right)^{-1}. \qquad (10)$$

Cf. [10], for further details on the G-K Algorithm.

**Remark 1.** The eigen-structure of the cluster covariance matrix (10), provides information about the shape and orientation of the cluster. The axes directions are given by the unitary eigenvectors of $F_i$. The key idea is to take the longest axis for each obtained cluster, and consecutive three clusters can be naturally regrouped in a cubic spline.

**A simple example**

Let us consider the following nonlinear SISO system:

$$y = 0.0001 \cdot \sin(0.001x^2) \cdot x^3 + \varepsilon, \quad x \in [0, 100] \qquad (11)$$

where $\varepsilon \sim N(0, 25)$ is a normally distributed random noise, added to have a more complex system to be modelled .

System is depicted in Figure 1 and is classified with G-K algorithm with five clusters. Each cluster corresponds to a linear model. The slope by each model is obtained by using the eigenvalues and the eigenvectors. The pairs $(k1, k2)$, $(k3, k4)$,...,$(k9, k10)$ are the coordinates corresponding to the eigenvalues and their respective unitary eigenvector, $c_1, c_2, ..., c_5$ are the centers of clusters.

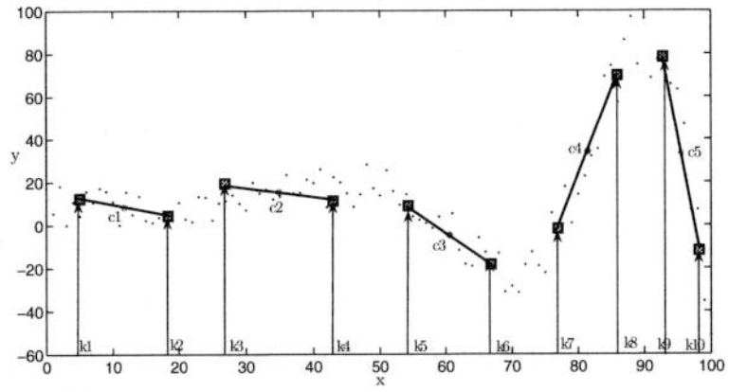


**Fig. 1.** Five local linear models using G-K algorithm

### 3.2 Interpolation Cubic Spline Algorithm for Consequents

By integrating the well known technics of fuzzy clustering, and a synthesis of a cubic spline, our goal is to obtain a structure in the rule base with cubic consequents.

The cubic splines are defined as continuous functions and fitted to the available input/output measured data system (SISO). The input/output vector $Y =$

$[y_1, ..., y_N]^T$, $X = [x_1, ..., x_N]^T$, which are defined by each pair of knot: $[K_1^1, K_2^1]$, $[K_1^2, K_2^2]$,..., $[K_1^{n-1}, K_2^n]$, where $K_1^1 = x_1$; $K_{i+1}^1 = c_{i*3}$, $i = 1, 2, 3, \cdots$; $K_2^j = k_{j*3}$, $j = 1, 2, 3, \cdots$, $K_2^n = x_N$.

Let us define a cubic spline for a knot sequence of cubic polynomials: $x_1 = K_1^1 < K_2^1 < K_1^2 < \cdots < K_2^n = x_N$. By each pair of knots $[K_1^i, K_2^i]$ there is a cubic spline, where $K_1^i$ is the initial knot for the $i$th cubic submodel and $K_2^i$ is the final knot in the $i$th cubic submodel.

**Algorithm 2. (Cubic Spline Approximation)**Let $Z = [[y_{K_1^i}, ..., y_{K_2^i}]^T; [x_{K_1^i}, ..., x_{K_2^i}]^T]$ be the I/O measurement data set, and define $K_1^1 = x_1$ and $K_2^1 = k6$ as the initial and the final knots of the first cluster see Fig. 1. The subsequent knots are defined as $K_1^i = k_j$ and $K_2^i = k_l$, $j = j+2; l = l+4$ with initial values $j = 3, l = 6$, where $j$ is the value on the axis $x$ for the center of cluster $(c_1, c_2, ..., c_5)$, and $l$ are the indices of the coordinates of the eigenvalues with their respective unitary eigenvector. Then, the region of the first cubic spline submodel is in between $[x_1, ..., k6]$, and the second one is in between $[c_3, ..., k_{12}]$, and son on.

By defining in a similar way the next clusters, the cubic spline model to be synthesized has the following structure:

$$S(x) = s_1^p a(x_{K_{1,2}^i}) + s_2^p b(x_{K_{1,2}^i}) + s_1 c(x_{K_{1,2}^i}) + s_2 d(x_{K_{1,2}^i}) \tag{12}$$

where $x_{K_{1,2}^i}$ is the value of $x$ within $K_1^i$ and $K_2^i$.

**Step 1. Calculate the internal coefficients:**

$$h_i = K_2^i - K_1^i \tag{13}$$

**for** $K_1^i \leq x_{K_{1,2}^i} \leq K_2^i$

$$a(x_{K_{1,2}^i}) = \left((K_2^i - x_{K_{1,2}^i})^2 (x_{K_{1,2}^i} - K_1^i)\right) \cdot (h_i)^{-2} \tag{14}$$

$$b(x_{K_{1,2}^i}) = -\left((K_2^i - x_{K_{1,2}^i})(x_{K_{1,2}^i} - K_1^i)^2\right) \cdot (h_i)^{-2} \tag{15}$$

$$c(x_{K_{1,2}^i}) = \left((K_2^i - x_{K_{1,2}^i})^2 (2(x_{K_{1,2}^i} - K_1^i) + h_i)\right) \cdot (h_i)^{-3} \tag{16}$$

$$d(x_{K_{1,2}^i}) = \left((x_{K_{1,2}^i} - K_1^i)^2 (2(K_2^i - x_{K_{1,2}^i}) + h_i)\right) \cdot (h_i)^{-3} \tag{17}$$

**end**

**Step 2. Linear least squares (LLS):** Since each cubic spline model is linear in the parameters, i.e. $\theta = (s_1, s_1^p, s_2, s_2^p)^T$ is a linear parameter vector, then it can be determined by applying the well known LLS.

**Step 3. Cubic splines polynomials:** The cubic polynomials are obtained by (12), former equation, and by some easy algebraic manipulation.

Figure 2 shows the two cubic splines approximating system (11):

Observe that by using **Algorithms G-K and 2**, one gets a smoothed synthesized output respect to the real system output.

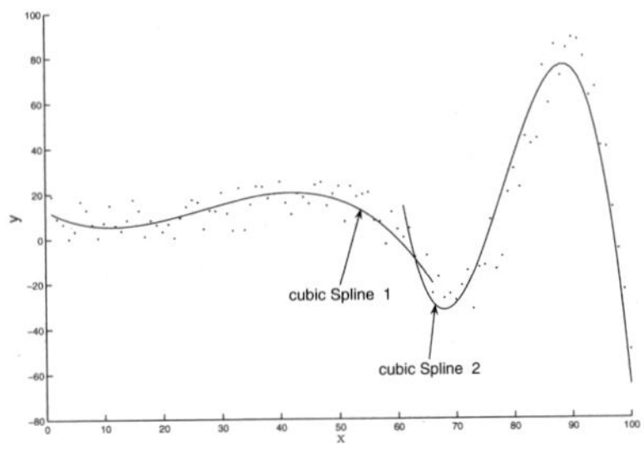


**Fig. 2.** Output approximation of the real system by two cubic splines

### 3.3 The Fuzzy Modelling Approach

By taking the knots, gravity centers and eigenvalues computed by the **G-K Algorithm**, some trapezoidal membership functions are straightforwardly built. The approach of the weighted average of the rule consequents, is well known for the defuzifying a T-S model. One advantage by using weighted average is their computational simplicity in the implementation.

## 4 Concluding Remarks

We have presented the analysis of two modelling method for fuzzy rule-based models from system's SISO measurements data, which provide a good accuracy as well transparency with smoothing trends and low complexity on the resulting rule base. Based on these results we can draw some particular conclusions.

### 4.1 PCA and C-Means Clustering Method

The partition (clusters) in a large database can be predefined by an expert as a function of the available resources. By a principal component analysis (PCA), it seems to be very useful covering the complete data space (data set partitioned in some clusters). This approach provides us a natural method of selection in a large database, for instance for potential customers database.

A simple methodology for data exploration is here reviewed by using the Hotelling transformation properties. The proposed method to filter only the most representative elements using analytical tools is a reliable and clear technique. The proposed method deals with the feature selection in a simple way but showing some kind of robustness. Computational load of the proposed modelling approach was verified to be light.

### 4.2 Clustering and Cubic Splines Consequent Method

The approach has been applied on a model of nonlinear system problem since the cubic consequents catch the nonlinear nature of the phenomena better than linear consequents. The fuzzy modelling approach presented allow us to obtain:Smoothed output from synthesized submodels, transparent rule-based models, and analytical consequents. A drawback in our approach, is that fuzzy linear control techniques can not be straightforwardly used.

**Acknowledgments.** The work of Virgilio López Morales is partially supported by the PAI2006 - UAEH - 55A, SEP - SESIC - PROMEP - UAEH - PTC - 1004 and ANUIES - SEP - CONACyT/ECOS - NORD, M02:M03, research grants. The work of Julio Cesar Ramos Fernández is partially supported by the PROMEP grant and ANUIES - SEP - CONACyT/ECOS NORD, M02:M03.

## References

1. Tanaka, K., Wang, H.: *Fuzzy control systems design and analysis.* John Wiley and Sons, Inc., USA (2001)
2. Ramos-Fernández, J.C., López-Morales, V., Enea, G., Duplaix, J.: Fuzzy model to target selection in direct marketing for large client and preferences database. *Int. Journal Product Lifecycle Management* 1(3) (2006) 249–267
3. Sousa-João, M., Kaymak, U., Madeira, S.: A comparative study of fuzzy target selection methods in direct marketing. In *Proceedings, IEEE World Congress on Computational Intelligence.* (2002) 1251–1256
4. Yen, J., Wang, L., Gillespie, W.C.: Improving the interpretability of TSK fuzzy models by combining global learning and local learning. *IEEE Transactions on Fuzzy Systems* 6(4) (1998) 530–537
5. Setnes, M., Babuka, R., Kaymak, U., Lemke, H.R.V.N.: Similiarity measures in fuzzy rule base simplification. *IEEE Transactions on Systems, Man and Cybernetics* 28(3) (1998) 376386
6. López-Morales, V., Ramos-Fernández, J.C., Enea, G., Duplaix, J.: Fuzzy nonlinear model identification methodology. Submitted to the *Institution of Engineering and Technology* (before IEE Proceedings), Computers & Digital Techniques (February-2007)
7. Babuška, R.: *Fuzzy modelling for control.* Kluwer Academic Publishers, USA (1998)
8. Bernal Reza, M.: *Analysis and Synthesis of fuzzy control systems.* Thesis ph. doct., Czech Technical University in Prague (2005)
9. Zuñiga, E., Santos, O., Paz-Ramos, M., A.: On the robustness of integral time delay systems with pd controllers. In: *2th International Conference on Industrial Electronics, Technology & Automation* (IETA 06), On Line, Springer (2006) CD ROM
10. Foulloy, L., Galichet, S., Titli, A.: *Commande Floue 2, de l'approximation à l'apprentissage.* 1 edn. Hermes Science, France (2003) ISBN: 2-7462-0609-9.

# Part VIII

# Fuzzy Possibilistic Optimization

# On Possibilistic/Fuzzy Optimization

Masahiro Inuiguchi

Department of Systems Innovation
Graduate School of Engineering Science, Osaka University
1-3, Machikaneyama, Toyonaka, Osaka 560-8531, Japan
inuiguti@sys.es.osaka-u.ac.jp

**Abstract.** We focus on possibilistic/fuzzy optimality in the framework of mathematical programming problem with a possibilistic objective function. We observe the interaction between possibilistic objective function values. Two optimality concepts, possible and necessary optimalities are reviewed. The necessary soft optimality is investigated.

## 1 Introduction

Possibilistic programming treats ambiguous parameters in programming problems. Various approaches have been proposed to possibilistic programming problems (see, for example, [1]). As stochastic programming approaches are classified into chance constrained programming approaches, recourse problem (two-stage problem) approaches and distribution problem approaches, possibilistic programming approaches can be classified into modality constrained programming approaches, recourse problem approaches and optimization approaches. Many of possibilistic programming approaches can be regarded as special cases of modality constrained programming approaches. Therefore, modality constrained programming approaches have well-investigated by many authors, so far. Optimization approaches have more or less investigated and recourse problem approaches have little investigated. Many overviews of possibilistic/ fuzzy programming approaches are devoted for modality programming approaches which are more tractable than the others.

In this paper, we review and investigate optimization approaches. In optimization approaches, we analyze the range of optimal solutions with respect to the fluctuation of uncertain parameters. We restrict ourselves to linear programming problems with ambiguous objective function coefficients. First, we observe the induced interaction in the comparison between possibilistic objective function values. Then, possibly and necessarily optimal solutions are defined. Possible and necessary optimality tests are given. Finally, necessarily soft optimal solutions are investigated as a relaxed concept of necessarily optimal solutions.

## 2 Comparison of Possibilistic Objective Function Values

In this paper, we treat the following linear programming problem with ambiguous objective function coefficients:

$$\text{maximize } \boldsymbol{\gamma}^{\mathrm{T}}\boldsymbol{x}, \quad \text{subject to } A\boldsymbol{x} \leq \boldsymbol{b}, \tag{1}$$

P. Melin et al. (Eds.): IFSA 2007, LNAI 4529, pp. 351–360, 2007.
© Springer-Verlag Berlin Heidelberg 2007

where $A = (a_{ij})$ is an $m \times n$ matrix and $\boldsymbol{b} = (b_1, \ldots, b_m)^{\mathrm{T}}$. $\boldsymbol{x} = (x_1, \ldots, x_n)^{\mathrm{T}}$ is a decision variable vector. Moreover, $\boldsymbol{\gamma} = (\gamma_1, \ldots, \gamma_n)^{\mathrm{T}}$ is a vector of ambiguous coefficients. We assume that we know $\boldsymbol{\gamma}$ realizes in a bounded $n$-dimensional fuzzy set $\Gamma$. Such a vector $\boldsymbol{\gamma}$ is said to be a possibilistic variable vector restricted in $\Gamma$. For the sake of simplicity, the feasible region of Problem (1) is denoted by $X$. The membership function $\mu_\Gamma$ of fuzzy set $\Gamma$ is assumed to be upper semi-continuous. The boundedness of fuzzy set $\Gamma$ is characterized by

$$\forall \varepsilon > 0, \exists r \in \mathbf{R}; \{\boldsymbol{c} \in \mathbf{R}^n : \mu_\Gamma(\boldsymbol{c}) \geq \varepsilon\} \subset \{\boldsymbol{c} \in \mathbf{R}^n : ||\boldsymbol{c}|| < r\}. \tag{2}$$

Given a solution $\boldsymbol{x} \neq \mathbf{0}$, by the extension principle, its objective function value is given as a fuzzy set $Y(\boldsymbol{x})$ having the following membership function:

$$\mu_{Y(\boldsymbol{x})}(y) = \sup_{\boldsymbol{c}}\{\mu_\Gamma(\boldsymbol{c}) : \boldsymbol{c}^{\mathrm{T}}\boldsymbol{x} = y\}. \tag{3}$$

There are various ways to compare two fuzzy numbers $Z_1$ and $Z_2$ For example, based on the possibility theory, the following two indices are obtained:

$$\mathrm{POS}(Z_1 \geq Z_2) = \sup_{r_1, r_2}\{\min(\mu_{Z_1}(r_1), \mu_{Z_2}(r_2)) : r_1 \geq r_2\}, \tag{4}$$

$$\mathrm{NES}(Z_1 \geq Z_2) = 1 - \sup_{r_1, r_2}\{\min(\mu_{Z_1}(r_1), \mu_{Z_2}(r_2)) : r_1 < r_2\}. \tag{5}$$

where $\mu_{Z_1}$ and $\mu_{Z_2}$ are membership functions of $Z_1$ and $Z_2$. Possibility degree $\mathrm{POS}(Z_1 \geq Z_2)$ shows to what extent $Z_1$ is possibly larger than or equal to $Z_2$. Similarly, Necessity degree $\mathrm{NES}(Z_1 \geq Z_2)$ shows to what extent $Z_1$ is necessarily larger than or equal to $Z_2$. When $Z_1$ and $Z_2$ are closed intervals $[z_1^{\mathrm{L}}, z_1^{\mathrm{R}}]$ and $[z_2^{\mathrm{L}}, z_2^{\mathrm{R}}]$, respectively, we have the following equivalences which show their meanings and difference remarkably:

$$\mathrm{POS}(Z_1 \geq Z_2) = 1 \Leftrightarrow z_1^{\mathrm{R}} \geq z_2^{\mathrm{L}}, \qquad \mathrm{NES}(Z_1 \geq Z_2) = 0 \Leftrightarrow z_1^{\mathrm{L}} < z_2^{\mathrm{R}}. \tag{6}$$

A comparison index between two fuzzy numbers is often applied to the comparison of possibilistic objective function values discarding their interaction in literature. Next example demonstrates the inadequacy caused by the desertion of the interaction.

**Example 1.** Let $n = 2$ and $\Gamma = [1, 2] \times [-2, -1]$. Namely, we consider a case when each objective function coefficient is given by a closed interval. Consider two feasible solutions $\boldsymbol{x}^1 = (2, 1)^{\mathrm{T}}$ and $\boldsymbol{x}^2 = (3, 1)^{\mathrm{T}}$. We have $Y(\boldsymbol{x}^1) = [0, 3]$ and $Y(\boldsymbol{x}^2) = [1, 5]$. Let us apply (6) discarding the interaction between $Z_1 = Y(\boldsymbol{x}^1)$ and $Z_2 = Y(\boldsymbol{x}^2)$. We obtain $\mathrm{POS}(Z_1 \geq Z_2)$= 1 which implies that the objective function value of $\boldsymbol{x}^1$ can be larger than or equal to that of $\boldsymbol{x}^2$. On the other hand, we have

$$\boldsymbol{c}^{\mathrm{T}}\boldsymbol{x}^1 = 2c_1 + c_2 < 3c_1 + c_2 = \boldsymbol{c}^{\mathrm{T}}\boldsymbol{x}^2, \ \ \forall c_1 \in [1, 2], \ \forall c_2 \in [-2, -1]. \tag{7}$$

This insists that the objective function value of $\boldsymbol{x}^1$ can never be larger than or equal to that of $\boldsymbol{x}^2$. Because the realized values of $c_1$ and $c_2$ are common independent of the selection of a feasible solution to Problem (1), the latter result is correct. Therefore, the direct application of index $\mathrm{POS}(Z_1 \geq Z_2)$ is not adequate for the problem setting.

Similarly, from (6), we obtain $\mathrm{NES}(Z_2 \geq Z_1) = 0$. This implies that there exists $(c_1, c_2)^{\mathrm{T}} \in \Gamma$ such that the objective function value of $\boldsymbol{x}^2$ is less than that of $\boldsymbol{x}^1$. However, this is neither true. As is shown in (7), for all $(c_1, c_2)^{\mathrm{T}} \in \Gamma$, the objective function value of $\boldsymbol{x}^2$ is larger than that of $\boldsymbol{x}^1$.

Now we emphasize the reason why indices defined by (4) and (5) do not work in Example 1. Let $\zeta_1$ and $\zeta_2$ be possibilistic variables restricted by $Z_1$ and $Z_2$. In (4) and (5), it is implicitly assumed that $\zeta_2$ is independent of $\zeta_1$ and vice versa.

In Example 1, we set $Z_1 = Y(\boldsymbol{x}^1)$ and $Z_2 = Y(\boldsymbol{x}^2)$. Namely, they are possible ranges of $\zeta_1 = \boldsymbol{\gamma}^{\mathrm{T}}\boldsymbol{x}^1$ and $\zeta_2 = \boldsymbol{\gamma}^{\mathrm{T}}\boldsymbol{x}^2$, respectively. Both $\zeta_1$ and $\zeta_2$ depend on the possibilistic variable vector $\boldsymbol{\gamma}$ restricted by $\Gamma = [1,2] \times [-2,-1]$. Because of this fact, the implicit assumption in (4) and (5) does not hold. For example, when $\zeta_1 = \boldsymbol{\gamma}^{\mathrm{T}}\boldsymbol{x}^1 = 0$, the possible values of $\gamma \in \Gamma$ are in

$$\{(c_1, c_2)^{\mathrm{T}} \in \mathbf{R}^2 : 2c_1 + c_2 = 0, 1 \leq c_1 \leq 2, -2 \leq c_2 \leq -1\} = \{(1, -2)\}.$$

Namely, from the information $\zeta_1 = 0$, in this case, we know that $\boldsymbol{\gamma}$ uniquely takes $(1,-2)^{\mathrm{T}}$. Therefore, the value $\zeta_2$ takes is also uniquely known, i.e., $\zeta_2 = (1,-2)\boldsymbol{x}^1 = 1$. Generally, when $\zeta_1 = q$, the possible range of $\zeta_2$ is given by

$$\{3c_1 + c_2 : 2c_1 + c_2 = q,\ 1 \leq c_1 \leq 2,\ -2 \leq c_2 \leq -1\}.$$

This range varies depending on $\zeta_1$-value $q$. Therefore, $\zeta_2$ interacts with $\zeta_1$. Similarly, $\zeta_1$ interacts with $\zeta_2$.

Since the implicit assumption of (4) and (5) does not hold, indices defined by (4) and (5) cannot be applied without any modification. For the comparison between possibilistic objective function values, the following modified indices [2] are adequate:

$$\mathrm{POS}(\boldsymbol{\gamma}^{\mathrm{T}}\boldsymbol{x}^1 \geq \boldsymbol{\gamma}^{\mathrm{T}}\boldsymbol{x}^2) = \sup_{\boldsymbol{c}}\{\mu_\Gamma(\boldsymbol{c}) : \boldsymbol{c}^{\mathrm{T}}\boldsymbol{x}^1 \geq \boldsymbol{c}^{\mathrm{T}}\boldsymbol{x}^2\}, \tag{8}$$

$$\mathrm{NES}(\boldsymbol{\gamma}^{\mathrm{T}}\boldsymbol{x}^1 \geq \boldsymbol{\gamma}^{\mathrm{T}}\boldsymbol{x}^2) = 1 - \sup_{\boldsymbol{c}}\{\mu_\Gamma(\boldsymbol{c}) : \boldsymbol{c}^{\mathrm{T}}\boldsymbol{x}^1 < \boldsymbol{c}^{\mathrm{T}}\boldsymbol{x}^2\}. \tag{9}$$

In literature, the desertion exemplified in Example 1 often appears when possibilistic objective function values are compared. Moreover, we note that under other interpretations of fuzzy coefficients, the discussion about the inadequacy is not valid. For example, when a fuzzy objective function is regarded as a collection of objective functions, e.g., a collection of utility functions of many decision makers, the above discussion does not make sense.

## 3 Possibly and Necessarily Optimal Solutions

We define an optimal solution set $S(\boldsymbol{c})$ of Problem (1) with respect to $\boldsymbol{\gamma} = \boldsymbol{c}$ by

$$S(\boldsymbol{c}) = \left\{ \boldsymbol{x} \in X : \boldsymbol{c}^{\mathrm{T}}\boldsymbol{x} = \max_{\boldsymbol{y} \in X} \boldsymbol{c}^{\mathrm{T}}\boldsymbol{y} \right\}. \tag{10}$$

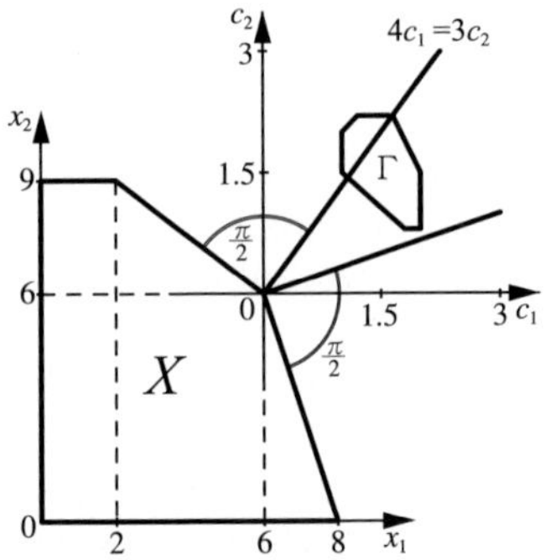


**Fig. 1.** A possibly optimal solution

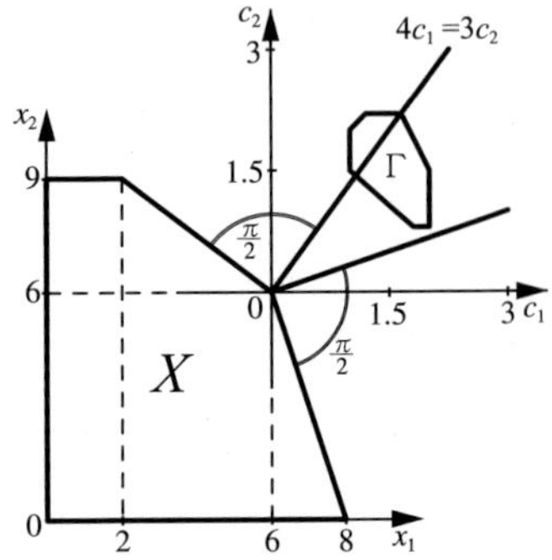


**Fig. 2.** A necessarily optimal solution

Using $S(\boldsymbol{c})$, we can define the following two optimal solution sets $\Pi S$ and $NS$ to Problem (1) when $\Gamma$ is a crisp set:

$$\Pi S = \bigcup_{\boldsymbol{c} \in \Gamma} S(\boldsymbol{c}), \qquad NS = \bigcap_{\boldsymbol{c} \in \Gamma} S(\boldsymbol{c}). \tag{11}$$

$\boldsymbol{x} \in \Pi S$ implies that there exists $\boldsymbol{c} \in \Gamma$ to which $\boldsymbol{x}$ is an optimal solution. Namely, $\boldsymbol{x} \in \Pi S$ can be optimal for at least one possible realization of $\boldsymbol{\gamma}$, and then, it is called a possibly optimal solution. On the other hand, $\boldsymbol{x} \in NS$ implies that for all $\boldsymbol{c} \in \Gamma$, $\boldsymbol{x}$ is optimal. Namely, $\boldsymbol{x} \in NS$ is always optimal for all possible realizations of $\boldsymbol{\gamma}$, and then it is called a necessarily optimal solution. We have $NS \subseteq \Pi S$.

**Example 2.** Consider Problem (1) where $A$ and $\boldsymbol{b}$ are defined by

$$A = \begin{pmatrix} 3 & 3 & 0 & -1 & 0 \\ 4 & 1 & 1 & 0 & -1 \end{pmatrix}^{\mathrm{T}}, \qquad \boldsymbol{b} = (42, 24, 9, 0, 0)^{\mathrm{T}}, \tag{12}$$

and $\Gamma$ is given by

$$\begin{aligned} \Gamma = \{(c_1, c_2)^{\mathrm{T}} : 3.5 \leq 2c_1 + c_2 \leq 5.5,\ 3.4 \leq c_1 + 2c_2 \leq 6, \\ -1 \leq c_1 - c_2 \leq 1.3,\ 1 \leq c_1 \leq 2,\ 0.8 \leq c_2 \leq 2.2\}. \end{aligned} \tag{13}$$

As shown in Fig. 1, $(x_1, x_2)^{\mathrm{T}} = (6, 6)^{\mathrm{T}}$ is optimal for $(c_1, c_2)^{\mathrm{T}} \in \Gamma$ such that $3c_2 \leq 4c_1$, and $(x_1, x_2)^{\mathrm{T}} = (2, 9)^{\mathrm{T}}$ is optimal for $(c_1, c_2)^{\mathrm{T}} \in \Gamma$ such that $3c_2 \geq 4c_1$. Moreover, all solutions on line segment between those solutions are optimal for $(c_1, c_2)^{\mathrm{T}} \in \Gamma$ such that $3c_2 = 4c_1$. Therefore, we have infinitely many possibly optimal solutions on the line segment. However, we have no necessarily optimal solution.

On the other hand, we consider $\Gamma$ defined by

$$\Gamma = \{(c_1, c_2)^{\mathrm{T}} : c_1 + c_2 \geq 3,\ c_1 \geq c_2,\ c_1 \leq 2c_2,\ c_1 \leq 2.5,\ c_2 \leq 2\}. \tag{14}$$

In this case, as shown in Fig. 2, $(x_1, x_2)^{\mathrm{T}} = (6,6)^{\mathrm{T}}$ is optimal for all $\boldsymbol{c} \in \Gamma$. Namely, the solution is a necessarily optimal solution. From $NS \subseteq \Pi S$, $(x_1, x_2)^{\mathrm{T}} = (6,6)^{\mathrm{T}}$ is also a possibly optimal solution.

As shown in Example 2, it is possible that infinitely many possibly optimal solutions exist and that no necessarily optimal solution exists.

In order to show the need of possible and necessary optimality, we continue to treat a case when $\Gamma$ is crisp. In modality constrained programming problems [1], various treatments of possibilistic objective functions are proposed. In crisp case, lower and upper bounds and the center values are often optimized and the width of the possible range of objective function values is minimized. Many approaches may regard a complete optimal solution to the following biobjective programming problem maximizing lower and upper bounds of objective function value as the most reasonable solution, if it exists:

$$\underset{\boldsymbol{x}\in X}{\text{maximize}} \left( \min_{\boldsymbol{c}\in\Gamma} \boldsymbol{c}^{\mathrm{T}}\boldsymbol{x},\ \max_{\boldsymbol{c}\in\Gamma} \boldsymbol{c}^{\mathrm{T}}\boldsymbol{x} \right) \tag{15}$$

Next example shows that, even if a complete optimal solution to Problem (15) exists, it is not always the most reasonable solution to Problem (1).

**Example 3.** Consider Problem (1) with the following $A$ and $\boldsymbol{b}$:

$$A = \begin{pmatrix} 1 & 3 & 0 & -1 & 0 \\ 1 & 1 & 1 & 0 & -1 \end{pmatrix}^{\mathrm{T}}, \qquad \boldsymbol{b} = (12, 24, 9, 0, 0)^{\mathrm{T}} \tag{16}$$

Moreover, $\Gamma$ is defined by

$$\Gamma = \{(c_1, c_2)^{\mathrm{T}} : 7c_1 - 5c_2 \leq 4,\ c_2 \leq 2,\ -3c_1 + 5c_2 \geq 2,\ c_1 \geq 1\}. \tag{17}$$

Consider $(1,1)^{\mathrm{T}}$ and $(3,3)^{\mathrm{T}} \in \Gamma$. For all $\boldsymbol{c} \in \Gamma$, we have $(1,1)^{\mathrm{T}} \leq \boldsymbol{c} \leq (3,3)^{\mathrm{T}}$. Therefore, Problem (15) becomes

$$\underset{\boldsymbol{x}\in X}{\text{maximize}}\ (x_1 + x_2, 3x_1 + 3x_2). \tag{18}$$

As shown in Fig. 3, $\boldsymbol{x}^0 = (6,6)^{\mathrm{T}}$ is a complete optimal solution to Problem (15). However, in Fig. 3, the shaded region of $\Gamma$ where $\boldsymbol{x}^0$ becomes optimal is much smaller than the other region of $\Gamma$. The possibly optimal solution set in this problem is shown as the line segment between points $(6,6)^{\mathrm{T}}$ and $(3,9)^{\mathrm{T}}$. Then, $\boldsymbol{x}^0$ is even extreme in the possibly optimal solution set. From these points of view, $\boldsymbol{x}^0$ is not necessarily the most reasonable solution.

As shown in Example 3, a complete optimal solution to Problem (15) is not always the most reasonable solution. When a necessarily optimal solution exists, it is the most reasonable solution since it is optimal for all possible realizations of $\boldsymbol{\gamma}$. On the other hand, a possibly optimal solution is a solution optimal for at least one possible realization of $\boldsymbol{\gamma}$, therefore, it can be regarded as a solution with minimum rationality. To sum up, possible optimality is the minimum requirement for the optimal solution to Problem (1) while necessary optimality is the ideal.

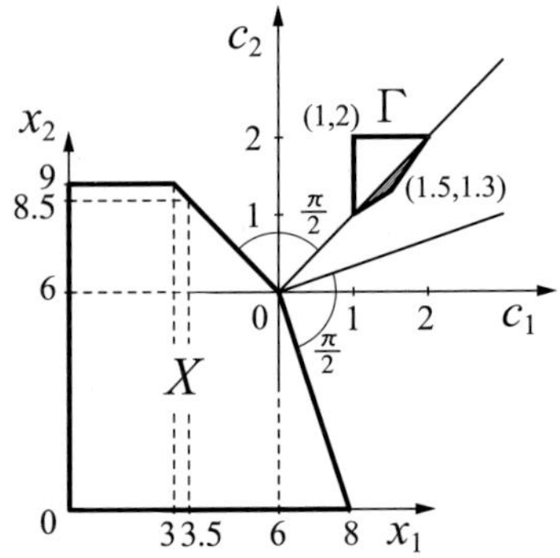


**Fig. 3.** Solution $(6,6)^{\mathrm{T}}$ in Example 3

Now let us describe possibly and necessarily optimal solutions when $\Gamma$ is a fuzzy set. In this case, both possibly optimal solution set $\Pi S$ and necessarily optimal solution set $NS$ become fuzzy sets define by the following membership functions (see [3]):

$$\mu_{\Pi S}(\boldsymbol{x}) = \begin{cases} \sup_{\boldsymbol{c}} \{\mu_\Gamma(\boldsymbol{c}) : \boldsymbol{x} \in S(\boldsymbol{c})\}, & \text{if } \boldsymbol{x} \in X, \\ 0, & \text{if } \boldsymbol{x} \notin X, \end{cases} \tag{19}$$

$$\mu_{NS}(\boldsymbol{x}) = \begin{cases} \inf_{\boldsymbol{c}} \{1 - \mu_\Gamma(\boldsymbol{c}) : \boldsymbol{x} \notin S(\boldsymbol{c})\}, & \text{if } \boldsymbol{x} \in X, \\ 0, & \text{if } \boldsymbol{x} \notin X. \end{cases} \tag{20}$$

Obviously, we have $\mu_{NS}(\boldsymbol{x}) \leq \mu_{\Pi S}(\boldsymbol{x})$, i.e., $NS \subseteq \Pi S$. We also obtain the following stronger relation:

$$\mu_{NS}(\boldsymbol{x}) > 0 \Rightarrow \mu_{\Pi S}(\boldsymbol{x}) = 1. \tag{21}$$

This comes from a relation between possibility and necessity, i.e., if an event is somehow necessary, it should be totally possible.

An optimal solution to a linear programming problem with an objective function $\boldsymbol{c}^{\mathrm{T}}\boldsymbol{x}$ is a feasible solution satisfying $\boldsymbol{c}^{\mathrm{T}}\boldsymbol{x} \geq \boldsymbol{c}^{\mathrm{T}}\boldsymbol{y}$ for all $\boldsymbol{y} \in X$, in other words, there is no feasible solution $\boldsymbol{y} \in X$ such that $\boldsymbol{c}^{\mathrm{T}}\boldsymbol{y} > \boldsymbol{c}^{\mathrm{T}}\boldsymbol{x}$. Applying (8) and (9), for $\boldsymbol{x} \in X$, we have the following properties which corresponds to the property of the optimal solution mentioned above:

$$\mu_{\Pi S}(\boldsymbol{x}) = \inf_{\boldsymbol{y} \in X} \mathrm{POS}(\boldsymbol{\gamma}^{\mathrm{T}}\boldsymbol{x} \geq \boldsymbol{\gamma}^{\mathrm{T}}\boldsymbol{y}) = 1 - \sup_{\boldsymbol{y} \in X} \mathrm{NES}(\boldsymbol{\gamma}^{\mathrm{T}}\boldsymbol{y} > \boldsymbol{\gamma}^{\mathrm{T}}\boldsymbol{x}), \tag{22}$$

$$\mu_{NS}(\boldsymbol{x}) = \inf_{\boldsymbol{y} \in X} \mathrm{NES}(\boldsymbol{\gamma}^{\mathrm{T}}\boldsymbol{x} \geq \boldsymbol{\gamma}^{\mathrm{T}}\boldsymbol{y}) = 1 - \sup_{\boldsymbol{y} \in X} \mathrm{POS}(\boldsymbol{\gamma}^{\mathrm{T}}\boldsymbol{y} > \boldsymbol{\gamma}^{\mathrm{T}}\boldsymbol{x}), \tag{23}$$

where $\mathrm{POS}(\boldsymbol{\gamma}^{\mathrm{T}}\boldsymbol{x}^1 > \boldsymbol{\gamma}^{\mathrm{T}}\boldsymbol{x}^2)$ and $\mathrm{NES}(\boldsymbol{\gamma}^{\mathrm{T}}\boldsymbol{x}^1 > \boldsymbol{\gamma}^{\mathrm{T}}\boldsymbol{x}^2)$ are defined by (8) and (9) with replacements of '$\geq$' and '$\leq$' with '$>$' and '$<$', respectively.

## 4 Possible and Necessary Optimality Tests

Possible and necessary optimalities are fundamental criteria for the solution selection as described earlier. It is worthwhile to calculate degrees of possible and necessary optimalities $\mu_{\Pi S}(\boldsymbol{x})$ and $\mu_{NS}(\boldsymbol{x})$ for justifying the selection of a solution $\boldsymbol{x} \in X$.

Let $A_{j\cdot}$ be the $j$-th row of $A$. Using the optimality condition of the linear programming problem, the necessary and sufficient condition for $\boldsymbol{x} \in X$ to be $\boldsymbol{x} \in S(\boldsymbol{c})$ is given as (see [3])

(a) There exists $j \in \{1, \ldots, m\}$ such that $A_{j\cdot}\boldsymbol{x} - b_j = 0$.
(b) There exists $\boldsymbol{v} \geq \boldsymbol{0}$ such that $A_0^{\mathrm{T}}\boldsymbol{v} = \boldsymbol{c}$, where $A_0$ is a submatrix of $A$ composed all rows $A_{j\cdot}$ such that $A_{j\cdot}\boldsymbol{x} - b_j = 0$.

In what follows, we consider $\boldsymbol{x} \in X$ satisfying (a). From (b), we have

$$\mu_{\Pi S}(\boldsymbol{x}) = \sup_{\boldsymbol{v}}\{\mu_\Gamma\left(A_0^{\mathrm{T}}\boldsymbol{v}\right) : \boldsymbol{v} \geq \boldsymbol{0}\}, \tag{24}$$

$$\mu_{NS}(\boldsymbol{x}) = \inf_{\boldsymbol{c}}\{1 - \mu_\Gamma(\boldsymbol{c}) : \forall \boldsymbol{v} \geq \boldsymbol{0}, A_0^{\mathrm{T}}\boldsymbol{v} \neq \boldsymbol{c}\}. \tag{25}$$

From the boundedness of $\Gamma$, the upper semi-continuity of $\mu_\Gamma$ and (24), $\mu_{\Pi S}(\boldsymbol{x})$ can be obtained as the optimal value of

$$\text{maximize} \quad h, \quad \text{subject to} \quad A_0^{\mathrm{T}}\boldsymbol{v} \in [\Gamma]_h, \ \boldsymbol{v} \geq \boldsymbol{0}. \tag{26}$$

We consider a special case when $\Gamma$ is characterized by membership function,

$$\mu_\Gamma(\boldsymbol{c}) = \min_{k=1,\ldots,p} \varphi(\boldsymbol{d}_k^{\mathrm{T}}\boldsymbol{c}). \tag{27}$$

where $p > n$ and $\varphi : \mathbf{R} \to [0,1]$ is an upper semi-continuous non-increasing function satisfying $\lim_{r \to +\infty} \varphi(r) = 0$. We define a pseudo-inverse of $\varphi$, $\varphi^* : [0,1] \to \mathbf{R}^n \cup \{+\infty\}$ by $\varphi^*(h) = \sup_r\{r : \varphi(r) \geq h\}$. Then, we have

$$\boldsymbol{c} \in [\Gamma]_h \Leftrightarrow \boldsymbol{d}_k^{\mathrm{T}}\boldsymbol{c} \leq \varphi^*(h), \ k = 1, \ldots, p \tag{28}$$

Hence, we obtain the optimal value of Problem (26) as $\varphi(\hat{s})$ by calculating the optimal value $\hat{s}$ of the following linear programming problem:

$$\text{minimize} \quad s, \quad \text{subject to} \quad \boldsymbol{d}_k^{\mathrm{T}}A_0^{\mathrm{T}}\boldsymbol{v} \leq s, \ k = 1, \ldots, p, \ \boldsymbol{v} \geq \boldsymbol{0}. \tag{29}$$

On the other hand, from (25), we have

$$\mu_{NS}(\boldsymbol{x}) \geq h \Leftrightarrow \left(\mu_\Gamma(\boldsymbol{c}) > 1 - h \Rightarrow \exists \boldsymbol{v} \geq \boldsymbol{0}; A_0^{\mathrm{T}}\boldsymbol{v} = \boldsymbol{c}\right). \tag{30}$$

Let $(\Gamma)_{1-h} = \{\boldsymbol{c} : \mu_\Gamma(\boldsymbol{c}) > 1 - h\}$. Then, from (30), we have

$$\begin{aligned}\mu_{NS}(\boldsymbol{x}) &= \sup\left\{h : \sup_{\boldsymbol{c} \in (\Gamma)_{1-h}} \inf_{\boldsymbol{v} \geq \boldsymbol{0}} |A_0^{\mathrm{T}}\boldsymbol{v} - \boldsymbol{c}| \leq 0\right\} \\ &= \max\left\{h : \max_{\boldsymbol{c} \in \mathrm{cl}(\Gamma)_{1-h}} \min_{\boldsymbol{v} \geq \boldsymbol{0}} |A_0^{\mathrm{T}}\boldsymbol{v} - \boldsymbol{c}| \leq 0\right\},\end{aligned} \tag{31}$$

where $\mathrm{cl}(\Gamma)_{1-h}$ is the closure of $(\Gamma)_{1-h}$. From the boundedness of $(\Gamma)_{1-h}$ and the continuity of $|A_0^{\mathrm{T}}\boldsymbol{v}-\boldsymbol{c}|$, we replaced '$(\Gamma)_{1-h}$', 'sup' and 'inf' with '$\mathrm{cl}(\Gamma)_{1-h}$', 'max' and 'min', respectively.

Consider $\Gamma$ defined by (27). We have

$$\boldsymbol{c} \in \mathrm{cl}(\Gamma)_{1-h} \Leftrightarrow \boldsymbol{d}_k^{\mathrm{T}}\boldsymbol{c} \le \bar{\varphi}^*(1-h),\ k=1,\dots,p, \tag{32}$$

where we define $\bar{\varphi}^* : [0,1] \to \mathbf{R}^n \cup \{+\infty\}$ by

$$\bar{\varphi}^*(h) = \begin{cases} \sup_r\{r : \varphi(r) > h\}, & \text{if } h < 1 \\ -\infty & \text{if } h = 1. \end{cases} \tag{33}$$

Since $\Gamma$ is bounded, $\mathrm{cl}(\Gamma)_{1-h}$ becomes a polytope. Let $V(\mathrm{cl}(\Gamma)_{1-h})$ be a set of vertices of $\mathrm{cl}(\Gamma)_{1-h}$. Then any point in $\mathrm{cl}(\Gamma)_{1-h}$ can be expressed as a convex combination of points in $V(\mathrm{cl}(\Gamma)_{1-h})$. Therefore, for any $Q$ such that $V(\mathrm{cl}(\Gamma)_{1-h}) \subseteq Q(h) \subseteq \mathrm{cl}(\Gamma)_{1-h}$, we have

$$\max_{\boldsymbol{c}\in\mathrm{cl}(\Gamma)_{1-h}} \min_{\boldsymbol{v}\ge\boldsymbol{0}} |A_0^{\mathrm{T}}\boldsymbol{v}-\boldsymbol{c}| \le 0 \Leftrightarrow \max_{\boldsymbol{c}\in Q(h)} \min_{\boldsymbol{v}\ge\boldsymbol{0}} |A_0^{\mathrm{T}}\boldsymbol{v}-\boldsymbol{c}| \le 0 \tag{34}$$

If there is a set mapping $Q(h) = \{\boldsymbol{c}^j(h),\ j=1,\dots,u\}$ such that $V(\mathrm{cl}(\Gamma)_{1-h}) \subseteq Q(h) \subseteq \mathrm{cl}(\Gamma)_{1-h}$ for any $h \in (0,1]$, we have $\mu_{NS}(\boldsymbol{x}) = \min_{j=1,\dots,u} h^j$, where $h^j$ is the optimal value of the following linear programming problem:

$$\text{maximize } h, \quad \text{subject to } A_0^{\mathrm{T}}\boldsymbol{v} = \boldsymbol{c}^j(h),\ \boldsymbol{v} \ge \boldsymbol{0}. \tag{35}$$

We may define $\boldsymbol{c}^j(h)$ of $Q(h)$ as the $\boldsymbol{c}$-value of an optimal solution to the following linear programming problem with respect to $P_j$ composed of $n$ elements from $\{1,\dots,p\}$:

$$\text{minimize} \sum_{k\in P_j} s_k, \quad \text{subject to } \boldsymbol{d}_k^{\mathrm{T}}\boldsymbol{c}+s_k = \bar{\varphi}^*(1-h),\ s_k \ge 0,\ k=1,\dots,p. \tag{36}$$

We have $\binom{n}{p}$ $P_j$'s. Therefore, $Q(h)$ is a finite set with at most $u = \binom{n}{p}$ elements $\boldsymbol{c}^j(h)$ for each $h \in (0,1]$.

Since maximizing $h$ is equivalent to maximizing $\bar{\varphi}^*(1-h)$, introducing Problem (36) into Problem (35), we obtain the following two-phase linear programming problem with a sufficiently small positive number $\epsilon$:

$$\text{maximize } -\sum_{k\in P_j} s_k + \epsilon s, \quad \text{subject to } \begin{cases} A_0^{\mathrm{T}}\boldsymbol{v} = \boldsymbol{c},\ \boldsymbol{d}_k^{\mathrm{T}}\boldsymbol{c} + s_k = s, \\ \boldsymbol{v} \ge \boldsymbol{0},\ s_k \ge 0,\ k=1,2,\dots,p \end{cases} \tag{37}$$

Let $s(P_j)$ be $s$-value at an optimal solution to Problem (37). Then we have $\mu_{NS}(\boldsymbol{x}) = \min_{j=1,2,\dots,u}(1-\varphi(s(P_j)))$.

As shown above, $\mu_{NS}(\boldsymbol{x})$ can be obtained by solving multiple linear programming problems. However, it requires to solve $\binom{n}{p}$ problems and thus, it will not be very efficient. A more efficient method can be designed based on global optimization techniques.

We have described computation methods for degrees of possible and necessary optimalities. It is also interesting to obtain all possibly and necessarily optimal solutions. In [4], an enumeration method for possibly optimal extreme pints is proposed.

## 5 Necessarily Soft Optimal Solution

As described before, there is no guarantee that a necessarily optimal solution $\boldsymbol{x}$ such that $\mu_{NS}(\boldsymbol{x}) > 0$ while it is the most reasonable solution. Even if it exists, $\mu_{NS}(\boldsymbol{x})$ is often small. This is because the requirement for the necessary optimality is very strong.

In real world problems, suboptimal solutions are often sufficiently good. Based on this idea, we proposed necessarily soft optimal solutions [5] in which the optimality of necessarily optimal solutions is relaxed to the suboptimality.

Let $\tilde{S}(\boldsymbol{c})$ be a fuzzy optimal solution set to linear programming problem with objective function coefficients $\boldsymbol{c}$. Its membership function can be defined by

$$\mu_{\tilde{S}(\boldsymbol{c})}(\boldsymbol{x}) = \begin{cases} \mu_{Dif}\left(\max_{\boldsymbol{y}\in X} \boldsymbol{c}^{\mathrm{T}}\boldsymbol{y} - \boldsymbol{c}^{\mathrm{T}}\boldsymbol{x}\right), & \text{if } \boldsymbol{x} \in X, \\ 0, & \text{otherwise}, \end{cases} \tag{38}$$

where $\mu_{Dif} : \mathbf{R} \to [0,1]$ is an upper semi-continuous non-increasing function. Equation (38) is based on the difference from the optimal value. We may have a similar approach based on the ratio to the optimal value. When $\forall \boldsymbol{c} \in \Gamma;\ \max_{\boldsymbol{x}\in X} \boldsymbol{c}^{\mathrm{T}}\boldsymbol{x} > 0$, we may define

$$\mu_{\tilde{S}(\boldsymbol{c})}(\boldsymbol{x}) = \begin{cases} \mu_{Rat}\left(\dfrac{\boldsymbol{c}^{\mathrm{T}}\boldsymbol{x}}{\max_{\boldsymbol{y}\in X} \boldsymbol{c}^{\mathrm{T}}\boldsymbol{y}}\right), & \text{if } \boldsymbol{x} \in X, \\ 0, & \text{otherwise}, \end{cases} \tag{39}$$

where $\mu_{Rat} : (-\infty, 1] \to [0,1]$ is upper semi-continuous non-decreasing function.

Using a fuzzy optimal solution set $\tilde{S}(\boldsymbol{c})$, a necessarily soft optimal solution set [5] $\widetilde{NS}$ is defined by the following membership function:

$$\mu_{\widetilde{NS}}(\boldsymbol{x}) = \inf_{\boldsymbol{c}} \max\left(1 - \mu_{\Gamma}(\boldsymbol{c}), \mu_{\tilde{S}(\boldsymbol{c})}(\boldsymbol{x})\right), \tag{40}$$

where when $\tilde{S}(\boldsymbol{c})$ is defined by (39), we assume for all $\boldsymbol{c}$ such that $\mu_{\Gamma}(\boldsymbol{c}) > 0$ $\max_{\boldsymbol{y}\in X} \boldsymbol{c}^{\mathrm{T}}\boldsymbol{y} > 0$.

Based on the necessary soft optimality, the best solution can be an optimal solution to the following problem:

$$\underset{\boldsymbol{x}\in X}{\text{maximize}} \quad \mu_{\widetilde{NS}}(\boldsymbol{x}) \tag{41}$$

The solution is called a best necessarily soft optimal solution.

Now let us consider a case when $\Gamma$ is a crisp set. In this case, for any $\mu_{Dif}$ and $\mu_{Rat}$, Problems (41) with (38) and (39) are reduced to the following problems, respectively:

$$\underset{\boldsymbol{x}\in X}{\text{minimize}}\, R(\boldsymbol{x}) = \max_{\boldsymbol{c}\in\Gamma,\ \boldsymbol{y}\in X} \boldsymbol{c}^{\mathrm{T}}\boldsymbol{y} - \boldsymbol{c}^{\mathrm{T}}\boldsymbol{x}, \quad \underset{\boldsymbol{x}\in X}{\text{maximize}}\, F(\boldsymbol{x}) = \min_{\boldsymbol{c}\in\Gamma} \frac{\boldsymbol{c}^{\mathrm{T}}\boldsymbol{x}}{\max_{\boldsymbol{y}\in X} \boldsymbol{c}^{\mathrm{T}}\boldsymbol{y}}. \tag{42}$$

Those problems are called a minimax regret problem and a maximin regret ratio problem. Those problems have the following good properties:

(a) $R(\boldsymbol{x}) = 0$ ($F(\boldsymbol{x}) = 1$) if and only if $\boldsymbol{x}$ is a necessarily optimal solution.
(b) $R(\boldsymbol{x}) \geq 0$, $\forall \boldsymbol{x} \in X$ ($F(\boldsymbol{x}) \leq 1$, $\forall \boldsymbol{x} \in X$).
(c) Any optimal solution is a possibly optimal solution.

From those properties, optimal solutions to those problems are regarded as possibly optimal solutions minimizing the deviation from necessary optimality.

The minimax regret problem and maximin regret ratio problem include nonconvex programs as their subproblems so that they are not very tractable. However, a solution algorithms based on a relaxation procedure has already proposed. To solve Problem (41), we further introduce the idea of bisection method to a solution method for (42). A solution algorithm converges a relaxation procedure and a bisection method simultaneously has proposed (see [5,6]).

## 6 Concluding Remarks

In this paper, we have reviewed and investigated the optimization approach to possibilistic/fuzzy programming problems. The formulated problems in this approach often include nonconvex subproblems so that applications of global optimization techniques are promising. On the other hand, by the development of interior point method, the range of tractable problems is enlarged. Solution methods for possibilistic/fuzzy optimization problems can also be developed by the introduction of new solution approaches.

The author acknowledges that this work has been partially supported by the Grant-in-Aid for Scientific Research (B) No. 17310098.

## References

1. Inuiguchi, M. and Ramík, J.: Possibilistic linear programming, Fuzzy Sets and Systems, **111**(1), (2000), 3–28.
2. Inuiguchi, M. and Kume, Y.: Extensions of fuzzy relations considering interaction between possibility distributions and an application to fuzzy linear program (in Japanese), Trans. of ISCIE, **3**(4), (1990), 93–102.
3. Inuiguchi, M. and Sakawa, M.: Possible and necessary optimality tests in possibilistic linear programming problems, Fuzzy Sets and Systems, **67**(1), (1994), 29–46.
4. Inuiguchi, M.: Enumeration of all possibly optimal vertices with possible optimality degrees in linear programming problems with a possibilistic objective function, Fuzzy Optimization and Decision Making, **3**(4), (2004), 311–326.
5. Inuiguchi, M. and Sakawa, M.: Robust optimization under softness in a fuzzy linear programming problem, International Journal of Approximate Reasoning, **18**(1-2), (1998), 21–34.
6. Inuiguchi, M., Tanino, T. and Tanaka, H.: Optimization approaches to possibilistic linear programming problems, Proc. of Joint 9th IFSA Congress and 20th NAFIPS International Conference, (2001), 2724–2729.

# The Use of Interval-Valued Probability Measures in Optimization Under Uncertainty for Problems Containing a Mixture of Fuzzy, Possibilisitic, and Interval Uncertainty

Weldon A. Lodwick and K. David Jamison

Department of Mathematics, University of Colorado at Denver
weldon.lodwick@cudenver.edu

**Abstract.** A simple definition of interval-valued probability measure is given and its implications examined. Properties are discussed which allow for the analysis of mixtures of fuzzy, possibilistic, probabilistic, cloud, and interval uncertainty utilizing interval-valued probability theory. It is shown how these properties allow for optimization under uncertainty where the uncertainty is mixed (fuzzy, possibilitic, probabilistic, clouds, and interval ). An example of this type of optimization is given illustrating the usefulness and power of the concepts.

**Keywords:** Imprecise Probability, Fuzzy Set Theory, Possibility Theory, Probability Theory, Interval Analysis, Optimization Under Uncertainty.

## 1 Introduction

A method for optimization under uncertainty where the problems contain a mixture of two or more types of uncertainty - fuzzy, possibilistic, probabilistic, cloud, and/or interval is developed. To solve an optimization problem containing such a mixture, a theory for linking various methods of uncertainty representations is outlined. This is accomplished by showing that each type of the aforementioned uncertainty can be represented within the context of interval-valued probability. This study extends the work of [5] and [11].

In the first section of this paper we define, in a formal way, what we call an interval-valued probability measure as defined by Weichselberger [11]. Weichselberger's definition begins with a set of probability measures and then defines an interval probability, $R-probabilities$ and $F-probabalities$, as a set function providing upper and lower bounds on the probabilities calculated from these measures. It will be seen that $F-probabilities$ are simply the tightest bounds possible for the set of all such probability measures. A discussion of how the various forms of uncertainty representation (possibility, probability, cloud, and interval) may be represented by such measures follows. The next section presents how interval-valued probability measures can be constructed from upper and lower bounding cumulative distribution functions using a result of Kolmogorov

P. Melin et al. (Eds.): IFSA 2007, LNAI 4529, pp. 361–370, 2007.
© Springer-Verlag Berlin Heidelberg 2007

[3]. This construction is particularly useful since it indicates how one may go about setting up a problem in this more general setting. In section three, the extension principle for a function of uncertain variables represented by interval-valued probability measures is given, and integration with respect to functions of interval-valued probability over these measures is presented. Both of these definitions (extension and integration) will be useful in analyzing problems involving uncertainty represented by interval-valued probability measures. The fourth section indicates how optimization is done utilizing interval-valued probability. The conclusions are found in the last section. Throughout this paper we will be primarily interested in interval-valued probability defined on the Borel sets on the real line and real-valued random variables. Much of what is contained in sections 2 and 3 is a synopsis of [2] and [6].

## 2 Interval-Valued Probability Measures - Definitions and Examples

This section begins by defining what is meant by an interval-valued probability measure (IVPM). This generalization of a probability measure includes probability measures, possibility/necessity measures, intervals, and clouds [8]. This set function may be thought of as a method for giving a partial representation for an unknown probability measure. Throughout, arithmetic operations involving set functions are in terms of interval arithmetic [7] and $Int_{[0,1]} \equiv \{[a,b] \mid 0 \leq a \leq b \leq 1\}$.

The basic definitions from Weichselberger (with slight variation in notation) are given next.

**Definition 1.** *(Weichselberger [11]) Given measurable space* $(S, \mathcal{A})$*, an interval valued function* $i_m : \mathcal{A} \to Int_{[0,1]}$ *is called an* ***R-probability*** *if:*
*(a)* $i_m(A) = [a^-(A), a^+(A)] \subseteq [0,1]$ *with* $a^-(A) \leq a^+(A)$
*(b)* $\exists$ *a probability measure* $\Pr$ *on* $\mathcal{A}$ *such that* $\forall A \in \mathcal{A}$, $\Pr(A) \in i_m(A)$.
*By an* ***R-probability field*** *we mean the triple* $(S, \mathcal{A}, i_m)$.
*Given an R-probability field* $\mathcal{R} = (S, \mathcal{A}, i_m)$ *the set* $\mathcal{M}(\mathcal{R}) = \{\Pr \mid \Pr$ *is a probability measure on* $\mathcal{A}$ *such that* $\forall A \in \mathcal{A}, \Pr(A) \in i_m(A)\}$ *is called the* ***structure*** *of* $\mathcal{R}$.
*An R-probability field* $\mathcal{R} = (S, \mathcal{A}, i_m)$ *is called an* ***F-probability field*** *if* $\forall A \in \mathcal{A}$:
*(a)* $a^+(A) = \sup\{\Pr(A) \mid \Pr \in \mathcal{M}(\mathcal{R})\}$
*(b)* $a^-(A) = \inf\{\Pr(A) \mid \Pr \in \mathcal{M}(\mathcal{R})\}$

It is interesting to note that if we have a measurable space $(S, \mathcal{A})$ and a set of probability measures $P$ then defining $a^+(A) = \sup\{\Pr(A) \mid \Pr \in P\}$ and $a^-(A) = \inf\{\Pr(A) \mid \Pr \in P\}$ gives an $F - probability$ and that $P$ is a subset of the structure.

The following examples show how intervals, possibility distributions, clouds and (of course) probability measures can define $R - probability$ fields on $\mathcal{B}$, the Borel sets on the real line.

*Example 1.* (An interval defines an F-probability field): Let $I = [a, b]$ be a non-empty interval on the real line. On the Borel sets define $a^-(A) = \begin{cases} 1 & \text{if } I \subseteq A \\ 0 & \text{otherwise} \end{cases}$ and $a^+(A) = \begin{cases} 1 & \text{if } I \cap A \neq \emptyset \\ 0 & \text{otherwise} \end{cases}$, then $i_m(A) = [a^-(A), a^+(A)]$ defines an F-probability field $\mathcal{R} = (R, \mathcal{B}, i_m)$. To see this simply set $P$ be the set of all probability measures on $\mathcal{B}$ such that $\Pr(I) = 1$.

*Example 2.* (A probability measure is an F-probability field) Let Pr be a probability measure over $(S, \mathcal{A})$. Define $i_m(A) = [\Pr(A), \Pr(A)]$ which is equivalent to having total knowledge about a probability distribution over $S$.

The concept of a cloud was introduced by Neumaier in [8] as follows:

**Definition 2.** *A cloud over set $S$ is a mapping $c$ such that:*
*1) $\forall s \in S$, $c(s) = [\underline{n}(s), \bar{p}(s)]$ with $0 \leq \underline{n}(s) \leq \bar{p}(s) \leq 1$*
*2) $(0, 1) \subseteq \cup_{s \in S} c(s) \subseteq [0, 1]$*
*In addition, random variable $X$ taking values in $S$ is said to belong to cloud $c$ (written $X \in c$) iff*
*3) $\forall \alpha \in [0, 1]$, $\Pr(\underline{n}(X) \geq \alpha) \leq 1 - \alpha \leq \Pr(\bar{p}(X) > \alpha)$*

Clouds are closely related to possibility theory. A function $p : S \to [0, 1]$ is called a regular possibility distribution function if $\sup\{p(x) \mid x \in S\} = 1$. Possibility distribution functions (see [10]) define a possibility measure, $Pos : S \to [0, 1]$ where $Pos(A) = \sup\{p(x) \mid x \in A\}$ and it's dual necessity measure $Nec(A) = 1 - Pos(A^c)$ (we define $\sup\{p(x) \mid x \in \emptyset\} = 0$). We can also define a necessity distribution function $n : S \to [0, 1]$ by setting $n(x) = 1 - p(x)$ and observe that $Nec(A) = \inf\{n(x) \mid x \in A^c\}$ (we define $\inf\{n(x) \mid x \in \emptyset\} = 1$). In [1] we showed that possibility distributions could be constructed which satisfy the following consistency definition.

**Definition 3.** *Let $p : S \to [0, 1]$ be a regular possibility distribution function with associated possibility measure Pos and necessity measure Nec. Then $p$ is said to be **consistent** with random variable $X$ if $\forall$ measurable sets $A$, $Nec(A) \leq \Pr(X \in A) \leq Pos(A)$.*

The concept of a cloud can be stated in terms of certain pairs of consistent possibility distributions which we show in the following proposition.

**Proposition 1.** *(see [2], [6]) Let $\bar{p}$, $\underline{p}$ be a pair of regular possibility distribution functions over set $S$ such that $\forall s \in S$ $\bar{p}(s) + \underline{p}(s) \geq 1$. Then the mapping $c(s) = [\underline{n}(s), \bar{p}(s)]$ where $\underline{n}(s) = 1 - \underline{p}(s)$ (i.e. the dual necessity distribution function) is a cloud. In addition, if $X$ is a random variable taking values in $S$ and the possibility measures associated with $\bar{p}$, $\underline{p}$ are consistent with $X$ then $X$ belongs to cloud $c$. Conversely, every cloud defines such a pair of possibility distribution functions and their associated possibility measures are consistent with every random variable belonging to $c$.*

*Example 3.* **A cloud defines an R-probability field.** Let $c$ be a cloud over the real line. If $Pos^1$,$Nec^1$, $Pos^2$, $Nec^2$ are the possibility measures and their dual necessity measures relating to $\bar{p}(s)$ and $\underline{p}(s)$. Define

$$i_m(A) = \left[\max\left\{Nec^1(A), Nec^2(A)\right\}, \min\left\{Pos^1(A), Pos^2(A)\right\}\right].$$

In [9] Neumaier proved that every cloud contains a random variable $X$. Consistency requires that $\Pr(X \in A) \in i_m(A)$ and thus every cloud defines an R-probability field.

*Example 4.* **A possibility defines an R-probability field.** Let $p : S \to [0,1]$ be a regular possibility distribution function and let $Pos$ be the associated possibility measure and $Nec$ the dual necessity measure. Define $i_m(A) = [Nec(A), Pos(A)]$. If we define a second possibility distribution, $\underline{p}(x) = 1\ \forall x$ then the pair $p$,$\underline{p}$ define a cloud for which $i_m(A)$ defines the R-probability.

## 3 Construction of IVPM from Kolmagorov-Smirnoff Statistics

In this section we construct an $F-$*probability* from upper and lower bounding cumulative distribution functions in a manner allowing practical computation. For example given statistical data we can construct a confidence interval for the underlying cumulative distribution function using the Kolmogorov (see [3]). Then using this confidence interval we can use the following development to construct an interval-valued probability measure. Although the method of setting the interval probability equal to the upper and lower bound over all probability measures contained in the bound is quite simple (this is Weichselberger's approach), it is not clear how to use this definition in practice. The development that follows is more amenable to actual use.

Let $F^u(x) = \Pr(X^u \le x)$ and $F^l(x) = \Pr\left(X^l \le x\right)$ be two cumulative distribution functions for random variables over the Borel sets on the real line, $X^u$ and $X^l$, with the property that $F^u(x) \ge F^l(x)\ \forall x$. Set $\mathcal{M}\left(X^u, X^l\right) = \left\{X \mid \forall x\ F^u(x) \ge \Pr(X \le x) \ge F^l(x)\right\}$ which clearly contains $X^u$ and $X^l$. We will think in terms of an unknown $X \in \mathcal{M}\left(X^u, X^l\right)$. For any Borel set $A$, let $\Pr(A) = \Pr(X \in A)$.

A way to start the construction is with probability bounds for members of the family of sets $\mathcal{I} = \{(a,b], (-\infty,a], (a,\infty), (-\infty,\infty), \emptyset \mid a < b\}$. For $I = (-\infty, b]$, it is clear by definition that $\Pr(I) \in \left[F^l(b), F^u(b)\right]$. For $I = (a,\infty)$, $\Pr(I) \in \left[1 - F^u(a), 1 - F^l(a)\right]$. For $I = (a,b]$, since $I = (-\infty,b] - (-\infty,a]$ and considering minimum and maximum probabilities in each set, $\Pr(I) \in \left[\max\left\{F^l(b) - F^u(a), 0\right\}, F^u(b) - F^l(a)\right]$. Therefore, if the definition of $F^u$, $F^l$ is extended by defining $F^u(-\infty) = F^l(-\infty) = 0$ and $F^u(\infty) = F^l(\infty) = 1$, the following general definition can be made.

**Definition 4.** *For any $I \in \mathcal{I}$, if $I \neq \emptyset$, define $i_m(I) = [a^-(I), a^+(I)] = \left[\max\left\{F^l(b) - F^u(a), 0\right\}, F^u(b) - F^l(a)\right]$ where $a$ and $b$ are the left and right endpoints of $I$ otherwise set $i_m(\emptyset) = [0.0]$.*

*Remark 1.* Note that with this definition $i_m((-\infty,\infty)) = [\max\{F^l(\infty) - F^u(-\infty), 0\}, F^u(\infty) - F^l(-\infty)] = [1,1]$ which matches our intuition and thus it is easy to see that $\Pr(I) \in i_m(I)\ \forall I \in \mathcal{I}$.

This method can be extended to include finite unions of elements of $\mathcal{I}$. For example if $E = I_1 \cup I_2 = (a,b] \cup (c,d]$ with $b < c$, then consider the probabilities $\Pr((a,b]) + \Pr((c,d])$ and $1 - \{\Pr((-\infty,a]) + \Pr((b,c]) + \Pr((d,\infty))\}$ (the probability of the sets that make up $E$ versus one less the probability of the intervals that make up the complement), and the minimum and maximum probability for each case as a function of the minimum and maximum of each set. The minimum for the first sum is $\max\{0, F^l(d) - F^u(c)\} + \max\{0, F^l(b) - F^u(a)$ and the maximum is $F^u(d) - F^l(c) + F^u(b) - F^l(a)$. The minimum for the second is $1 - \left(F^u(\infty) - F^l(d) + F^u(c) - F^l(b) + F^u(a) - F^l(-\infty)\right) = F^l(d) - F^u(c) + F^l(b) - F^u(a)$ and the maximum is $1 - (\max\{0, F^l(\infty) - F^u(d)\} + \max\{0, F^l(c) - F^u(b)\} + \max\{0, F^l(a) - F^u(-\infty)\}) = F^u(d) - \max\{0, F^l(c) - F^u(b)\} - F^l(a)$. This gives

$$\Pr(E) \geq \max\begin{cases} F^l(d) - F^u(c) + F^l(b) - F^u(a) \\ \max\left(0, F^l(d) - F^u(c)\right) + \max\left(0, F^l(b) - F^u(a)\right) \end{cases} \quad \text{and}$$

$$\Pr(E) \leq \min\begin{cases} F^u(d) - \max\left(0, F^l(c) - F^u(b)\right) - F^l(a) \\ F^u(d) - F^l(c) + F^u(b) - F^l(a) \end{cases}$$

so $\Pr(E) \in [\max\{0, F^l(d) - F^u(c)\} + \max\{0, F^l(b) - F^u(a)\}, F^u(d) - \max\{0, F^l(c) - F^u(b)\} - F^l(a)]$ where the final line is arrived at by noting that $\forall x, y\ F^l(x) - F^u(y) \leq \max\{0, F^l(x) - F^u(y)\}$.

*Remark 2.* Note the two extreme cases for $E = (a,b] \cup (c,d]$. For $F^u(x) = F^l(x) = F(x)\ \forall x$, then, as expected, $\Pr(E) = F(d) - F(c) + F(b) - F(a) = \Pr((a,b]) + \Pr((c,d])$,that is, it is the probability measure. Moreover, for $F^l(x) = 0\ \forall x$, $\Pr(E) \in [0, F^u(d)]$. That is, it is a possibility measure for the possibility distribution function $F^u(x)$.

Let $\mathcal{E} = \{\cup_{k=1}^K I_k \mid I_k \in \mathcal{I}\}$, $\mathcal{E}$ is the algebra of sets generated by $I$. Note that every element of $E$ has a unique representation as a union of the minimum number of elements of $\mathcal{I}$ (or, stated differently, as a union of disconnected elements of $\mathcal{I}$). Note also that $R \in \mathcal{E}$ and $\mathcal{E}$ is closed under complements.

Assume $E = \cup_{k=1}^K I_k$ and $E^c = \cup_{j=1}^J M_j$ are the unique representations of $E$ and $E^c$ in $\mathcal{E}$ in terms of elements of $\mathcal{I}$. Then, considering minimum and maximum possible probabilities of each interval it is clear that

$$\Pr(E) \in [\max\{\Sigma_{k=1}^K a^-(I_k), 1 - \Sigma_{j=1}^J a^+(M_j)\}, \\ \min\{\Sigma_{k=1}^K a^+(I_k), 1 - \Sigma_{j=1}^J a^-(M_j)\}].$$

This can be made more concise using the following result.

**Proposition 2.** *(see [2], [6]) If $E = \cup_{k=1}^K I_k$ and $E^c = \cup_{j=1}^J M_j$ are the unique representations of $E$ and $E^c \in \mathcal{E}$ then $\Sigma_{k=1}^K a^-(I_k) \geq 1 - \Sigma_{j=1}^J a^+(M_j)$ and $\Sigma_{k=1}^K a^+(I_k) \geq 1 - \Sigma_{j=1}^J a^-(M_j)$.*

Next $i_m$ is extended to $\mathcal{E}$.

**Proposition 3.** *(see [2], [6]) For any $E \in \mathcal{E}$ let $E = \cup_{k=1}^{K} I_k$ and $E^c = \cup_{j=1}^{J} M_j$ be the unique representations of $E$ and $E^c$ in terms of elements of $\mathcal{I}$ respectively. If*

$$i_m(E) = \left[\Sigma_{k=1}^{K} a^-(I_k), 1 - \Sigma_{j=1}^{J} a^-(M_j)\right]$$

*then $i_m : \mathcal{E} \rightarrow Int_{[0,1]}$ , is an extension of $\mathcal{I}$ to $\mathcal{E}$ and is well defined. In addition, $i_m(E) = [\inf\{\Pr(X) \in E \mid X \in \mathcal{M}(X^u, X^l)\}, \sup\{\Pr(X) \in E \mid X \in \mathcal{M}(X^u, X^l)\}]$.*

The family of sets, $\mathcal{E}$ , is a ring of sets generating the Borel sets $\mathcal{B}$. For an arbitrary Borel set $S$, it is clear that $\Pr(S) \in [\sup\{a^-(E) \mid E \subseteq S, E \in \mathcal{E}\}, \inf\{a^+(F) \mid S \subseteq F, F \in \mathcal{E}\}]$ and this leads to the following proposition.

**Proposition 4.** *(see [2], [6]) Let $i_m : .\mathcal{B} \rightarrow [0,1]$ be defined by*

$$i_m(A) = [\sup\{a^-(E) \mid E \subseteq A, E \in \mathcal{E}\}, \tag{1}$$

$$\inf\{a^+(F) \mid A \subseteq F, F \in \mathcal{E}\}] \tag{2}$$

*The $i_m$ is an extension from $\mathcal{E}$ to $\mathcal{B}$ and is well-defined.*

**Proposition 5.** *The function $i_m : \mathcal{B} \rightarrow Int_{[o,1]}$ defines an F-probability field on the Borel sets and $i_m(B) = [\inf\{\Pr(X \in B) \mid X \in \mathcal{M}(X^u, X^l)\}, \sup\{\Pr(X \in B) \mid X \in \mathcal{M}(X^u, X^l)\}]$. That is, $\mathcal{M}(X^u, X^l)$ defines the structure.*

### 3.1 Interval-Valued Integration, Extension and Independence

Three key concepts needed for the application of IVPMs to mathematical programing problems, integration, extension and independence, are defined.

**Definition 5.** *Given F-probability field $\mathcal{R} = (S, \mathcal{A}, i_m)$ and an integrable function $f : S \rightarrow R$ we define:*

$$\int_A f(x)\, di_m = \left[\inf_{p \in \mathcal{M}(\mathcal{R})} \int_A f(x)\, dp, \sup_{p \in \mathcal{M}(\mathcal{R})} \int_A f(x)\, dp\right] \tag{3}$$

It is easy to see that if $f$ is an $\mathcal{A}-$measurable simple function such that $f(x) = \begin{cases} y & x \in A \\ 0 & x \notin A \end{cases}$ with $A \in \mathcal{A}$, then

$$\int_A f(x)\, di_m = y i_m(A) \tag{4}$$

Further, if $f$ is a simple function taking values $\{y_k \mid k \in K\}$ on an at most countable collection of disjoint measurable sets $\{A_k \mid k \in K\}$ that is, $f(x) = \left\{\begin{array}{ll} y_k & x \in A_k \\ 0 & x \notin A \end{array}\right|$ where $A = \cup_{k \in K} A_k$, then

$$\int_A f(x)\, di_m = \left[a^-\left(\int_A f(x)\, di_m\right), a^+\left(\int_A f(x)\, di_m\right)\right], \tag{5}$$

where

$$a^{+}\left(\int_A f(x)\, di_m\right) = \sup\left\{\Sigma_{k\in K} y_k \Pr(A_k) \mid \Pr \in \mathcal{M}(\mathcal{R})\right\} \tag{6}$$

and

$$a^{-}\left(\int_A f(x)\, di_m\right) = \inf\left\{\Sigma_{k\in K} y_k \Pr(A_k) \mid \Pr \in \mathcal{M}(\mathcal{R})\right\}. \tag{7}$$

Note that these can be evaluated by solving two linear programing problems since $\Pr \in \mathcal{M}(\mathcal{R})$ implies that $\Sigma_{k\in K} \Pr(A) = 1$ and $\Pr(\cup_{l\in L\subset K} A_l) \in i_m(\cup_{l\in L\subset K} A_l)$ so the problem may be tractable. In general, if $f$ is an integrable function and $\{f_k\}$ is a sequence of simple functions converging uniformly to $f$, then we can determine the integral with respect to $f$ by noting that

$$\int_A f(x)\, di_m = \lim_{k\to\infty} \int_A f_k(x)\, di_m,$$

where

$$\lim_{k\to\infty} \int_A f_k(x)\, di_m = \left[\lim_{k\to\infty} a^{-}\left(\int_A f_k(x)\, di_m\right), \lim_{k\to\infty} a^{+}\left(\int_A f_k(x)\, di_m\right)\right]$$

provided the limits exist.

*Example 5.* Consider the IVPM constructed from the interval $[a, b]$. Then $\int_R x di_m = [a, b]$, i.e. the interval-valued expected value is the interval itself.

**Definition 6.** *Let $\mathcal{R} = (S, \mathcal{A}, i_m)$ be an F-probability field and $f : S \to T$ a measurable function from measurable space $(S, \mathcal{A})$ to measurable space $(T, \mathcal{B})$. Then the F-probability $(T, \mathcal{B}, l_m)$ defined by $l_m(B) = [\inf\{\Pr(f^{-1}(B)) \mid \Pr \in \mathcal{M}(\mathcal{R})\}, \sup\{\Pr(f^{-1}(B)) \mid \Pr \in \mathcal{M}(\mathcal{R})\}]$ is called the **extension** of the R-probability field to $(T, \mathcal{B})$.*

That this defines an $F-$*probability* field is clear from our earlier observation. In addition, it's easy to see that this definition is equivalent to setting

$$l_m(A) = i_m\left(f^{-1}(A)\right)$$

which allows for evaluation using the techniques described earlier. The combination of IVPMs when the variables are independent is addressed next. The situation when dependencies may be involved is not discussed here.

Given measurable spaces $(S, \mathcal{A})$ and $(T, \mathcal{B})$ and the product space$(SxT, \mathcal{A} \times \mathcal{B})$ Assume $i_{XxY}$ is an IVPM on $\mathcal{A} \times \mathcal{B}$. Call $i_X$ and $i_Y$ defined by $i_X(A) = i_{XxY}(A \times T)$ and $i_Y(B) = i_{XxY}(S \times B)$ the marginals of $i_{XxY}$. The marginals, $i_X$ and $i_Y$, are IVPMs.

**Definition 7.** *Call the marginal IVPMs* **independent** *iff* $i_{XxY}(AxB) = i_X(A)\, i_Y(B)\ \forall A, B \subseteq S$.

**Definition 8.** *Let* $\mathcal{R} = (S, \mathcal{A}, i_X)$ *and* $\mathcal{Q} = (T, \mathcal{B}, l_Y)$ *be F-probability fields representing uncertain random variables* $X$ *and* $Y$. *Define the F-probability field* $(SxT, \mathcal{A} \times \mathcal{B}, i_{XxY})$ *by*

$$i^+_{XxY}(AxB) = \sup\{\Pr_X(B)\Pr_Y(B) \mid \Pr_X \in \mathcal{M}(\mathcal{R}), \Pr_Y \in \mathcal{M}(\mathcal{Q})\},$$

$$i^-_{XxY}(AxB) = \inf\{\Pr_X(B)\Pr_Y(B) \mid \Pr_X \in \mathcal{M}(\mathcal{R}), \Pr_Y \in \mathcal{M}(\mathcal{Q})\},$$

*where* $(SxT, \mathcal{A} \times \mathcal{B})$ *is the usual product of* $\sigma-$*algebra of sets.*

It is clear from this definition that $i_{XxY}(AxB) \equiv i_X(A)\, i_Y(B)$ for all $A \in \mathcal{A}$ and $B \in \mathcal{B}$. Thus, if several uncertain parameters in a problem are present with the uncertainty characterized by IVPMs, and all independent, an IVPM for the product space can be formed by multiplication and subsequently used as an IVPM.

## 4 Application to Optimization

As an example of possible application of these concepts we will look at a recourse problem. Suppose we wish to optimize $f(x, \boldsymbol{a})$ subject to $g(x, \boldsymbol{b}) = 0$. Assume $\boldsymbol{a}$, and $\boldsymbol{b}$ are vectors of independent uncertain parameters, each with an associated IVPM. Assume the constraint can be violated at a cost so that the problem is to solve (dropping the vector notation):

$$h(x, a, b) = f(x, a) - pg(x, b),$$

where $p$ is the penalty vector.

The IVPM for the product space can be readily formed given the assumption that the uncertainty variables are independent. In this case, the IVPM, $i_{axb}$, is the joint distribution. The interval-valued expected value with respect to this IVPM is:

$$\int_R h(\boldsymbol{x}, a, b)\, di_{axb}$$

To optimize over such a value requires an ordering of intervals. One such ordering is to use the midpoint of the interval on the principle that in the absence of additional data, the midpoint is the best estimate for the true value. Another possible ordering is to use risk/return multi-objective decision making. For example, determine functions $u : R^2 \to R$ and $v : Int_R \to R^2$ by setting, for any interval $I = [a, b]$, $v(I) = \left(\frac{a+b}{2}, b - a\right)$. Thus $v$ gives the midpoint and width of an interval. Then $u$ would measure the decision makes preference for one interval over another considering both its midpoint and width (a risk measure). Then our optimization problem is

$$\max_x u\left(v\left(\int_R h(x, a, b)\, di_{axb}\right)\right).$$

*Example 6.* Consider the problem

$$\begin{aligned} \max f(\boldsymbol{x}, c) =& \ 4x_1 + x_2 \\ & subject\ to \\ g_1(\boldsymbol{x}, \boldsymbol{a}_1, b_1) =& \ x_1 - [1,3]x_2 + 4 = 0 \\ g_2(\boldsymbol{x}, \boldsymbol{a}_2, b_2) =& \ \tilde{2}x_1 - 5x_2 + 1 = 0 \\ & 0 \le x_i \le 2 \end{aligned}$$

where $\tilde{2} = 1/2/3/$, that is, $\tilde{2}$ is the triangular number with support $[1,3]$ and modal value at 2. For $\boldsymbol{p} = (1,1)^T$,

$$h(\boldsymbol{x}, a, b) = 3x_1 + 6x_2 - \tilde{2}x_1 + [1,3]x_2 - 5,$$

so that

$$\begin{aligned} &\int_R h\left(\boldsymbol{x}, a, b\right) di_{axb} \\ &= 3x_1 - [\textstyle\int_0^1 (\alpha - 3) d\alpha, \int_0^1 (-\alpha - 1) d\alpha] x_1 + 6x_2 + [1,3]x_2 - 5 \\ &= 3x_1 - [\frac{3}{2}, \frac{5}{2}]x_1 + 6x_2 + [1,3]x_2 - 5 \\ &= [[\frac{1}{2}, \frac{3}{2}]x_1 + [7,9]x_2 - 5. \end{aligned}$$

Note that $-5$ will not affect the optimization. It will be taken out of the optimization and then re-inserted at the end. Continuing,

$$v(\int_R h\left(\boldsymbol{x}, a, b\right) di_{axb}) = (1,1)x_1 + (8,2)x_2.$$

As an example, let $u(y_1, y_2) = y_1 + y_2$. Then

$$\begin{aligned} \max_{x_i \in [0,2]} u(v(\int_R h\left(\boldsymbol{x}, a, b\right) di_{axb})) &= \max_{x_i \in [0,2]} \{2x_1 + 10x_2\} \\ &= 24. \end{aligned}$$

## 5 Conclusion

The definition of an interval-valued probability measure provides a formal setting in which various representations of uncertainty (for example intervals, probability, possibility and clouds) can be combined. This allows solution methods for problems containing mixed representations. Future research will focus on the theory and applications of such measures to problems in optimization in which uncertainty can not be fully captured by probability alone. We intend to extend the results in [5] which provided an approximation technique for probability measures via inner and outer measures to approximation techniques for functions of cloudy random variables.

## References


1. K. D. Jamison and W. A. Lodwick (2002), "The Construction of Consistent Possibility and Necessity Measures," *Fuzzy Sets and Systems,* **132,** pp. 1-10.
2. K. D. Jamison and W. A. Lodwick (2006), "Interval-valued probability in the analysis of problems that contain a mixture of fuzzy, possibilistic and interval uncertainty" (submitted to *International Journal of Approximate Reasoning* )
3. A. N. Kolmogorov (1941), "Confidence limits for an unknown distribution function," *Annuals of Mathematics Statistics*, **12**, pp. 461-463.
4. A. N. Kolmogorov and S. V. Fomin (1975). *Introductory Real Analysis,*Dover Publications, Inc., New York.
5. W. A. Lodwick and K. D. Jamison (2003), "Estimating and Validating the Cumulative Distribution of a Function of Random Variables: Toward the Development of Distribution Arithmetic", *Reliable Computing* **9**, pp. 127-141.
6. W. A. Lodwick and K. D. Jamison (2006), "Interval-valued probability in the analysis of problems that contain a mixture of fuzzy, possibilistic and interval uncertainty", in K. Demirli and A. Akgunduz (editors), *2006 Conference of the North American Fuzzy Information Processing Society, June 3-6, 2006, Montréal, Canada*, paper 327137.
7. R.E. Moore (1979). *Methods and Applications of Interval Analysis*, SIAM, Philadelphia.
8. A. Neumaier (2004), "Clouds, fuzzy sets and probability intervals" *Reliable Computing* **10**:4, pp. 249-272.
9. A. Neumaier (2005), "On the Structure of Clouds"(submitted - downloadable http://www.mat.univie.ac.at/~neum/papers.html).
10. Z. Wang and G. J. Klir (1992). *Fuzzy Measure Theory,* Plenum Press, New York.
11. K. Weichselberger (2000), "The theory of interval-probability as a unifying concept for uncertainty", *International Journal of Approximate Reasoning,* **24**, pp. 149-170.

# On Selecting an Algorithm for Fuzzy Optimization

Elizabeth Untiedt and Weldon Lodwick

Department of Mathematical Sciences, University of Colorado at Denver
euntiedt@math.cudenver.edu, wlodwick@math.cudenver.edu

**Abstract.** Formulations for fuzzy and possibilistic optimization abound in the literature, but few are implemented in practice. This paper investigates the theory, semantics, and efficacy of a selection of significant fuzzy and possibilistic optimization algorithms via their application to a well-known large-scale problem: the radiation therapy planning problem. The algorithms are compared, critiqued, and organized with the following objective in mind: to guide a decision maker in the selection and implementation of a fuzzy or possibilistic optimization algorithm.

**Keywords:** Fuzzy, Possibilistic, Optimization, Algorithm.

## 1 Introduction

Since much of the decision-making in the real world depends upon constraints, goals, and consequences of potential actions which are not explicitly known, the field of optimization is faced with the challenge of quantifying imprecise data in a meaningful way. The most well-developed branch of optimization under uncertainty is stochastic programming, which utilizes probability distributions for all uncertain data. Not all uncertainty, however, is inherently random. Some uncertainty stems from vagueness about the quantitative meaning of a constraint or goal (as in, "We wish for a return on investment significantly higher than the prime rate.") Some uncertainty stems from a lack of information about, or ambiguity concerning, the value of a goal or constraint (as in, "the body temperature which constitutes a fever is around 100 degrees Fahrenheit"). Known respectively as *fuzziness* and *possibility*, these two forms of uncertainty require quantitative representations which differ from their probabilistic counterparts as well as from each other.

Fuzzy and possibility theory have made contributions to the field of optimization under uncertainty for over 30 years. The literature contains many proposed formulations, but few have been implemented in practice. The desired outcome of this paper is to guide a decision maker in the selection and implementation of a fuzzy or possibilistic optimization algorithm. Section Two introduces several significant fuzzy and possibilistic optimization algorithms. Section Three describes a well-known large-scale problem, the radiation therapy problem, to which the proposed algorithms will be applied. In Section Four, the algorithms are compared, critiqued, and organized for the ease of use for potential decision-makers.

P. Melin et al. (Eds.): IFSA 2007, LNAI 4529, pp. 371–380, 2007.
© Springer-Verlag Berlin Heidelberg 2007

# 2 Background

## 2.1 Optimization Algorithms

Vague parameter values leads to fuzzy programming. When the vagueness represents a willingness on the part of the decision-maker to bend the constraints, fuzzy programming can also be called flexible programming. In these cases, the decision-maker is willing to lower the feasibility requirements in order to obtain a more satisfactory objective function value, or, in some cases, simply in order to reach a feasible solution. Such flexible constraints are commonly referred to as *soft constraints.* In the case of soft constraints, it is the inequality (or equality) itself which is viewed to be fuzzy (i.e. $Ax \tilde{\leq} b$). This is distinct from the case in which the right hand side has vague value, in which case the right hand side is viewed to be fuzzy (i.e. $Ax \leq \tilde{b}$).

In a landmark paper [1], Bellman and Zadeh were the first to propose an approach to mathematical programming with fuzzy components– specifically with soft constraints. Two notable aspects of this approach, which was later formulated for computation by Zimmermann [34], are the reformulation of objective(s) as goal(s), and the drive to maximize the $\alpha$-level of the decision space.

Verdegay [30] published a formulation which yields a fuzzy solution for a fuzzy problem with soft constraints. The optimal solution depends upon the $\alpha$-level of acceptable uncertainty, which must be specified *a priori* by the decision maker.

Recently, Neumaier [20],[13], modeled fuzzy right-hand sides (in contrast to fuzzy inequalities). Neumaier seeks to minimize a surprise function, which is inversely related to membership level associated with a particular outcome.

J.J. Buckley proposes a formulation for possibilistic linear programs in which both the right- and left-hand sides of the constraints and the cost coefficients of the objective function are governed by possibility distributions. Buckley's and Verdegay's approaches are similar in several ways, among them the requirement of an *a priori* $\alpha$-level.

Tanaka, *et al* [26,28], also propose a formulation for possibilistic left-and right-hand sides and cost coefficients. Reframing objective(s) as goal(s), and subtracting the right-hand sides results in a single matrix which the optimization problem seeks to make greater than zero. Tanaka, *et al* introduce the concept of "possibly positive", and the optimal solution is the one for which maximum positivity is attained.

A fuzzy max formulation for possibilistic right and left-hand sides was introduced by Tanaka, *et al.* [28] and by Ramik and Rimanek [22]. The maximum operator is extended to possibilistic numbers and used to define the feasible set of solutions in a possibilistic problem. Because the fuzzy max yields a partial ordering, an acceptable $\alpha$-level is sometime assigned *a priori* to reflect the decision-maker's level of optimism.

Jamison and Lodwick ([7,11]) develop a method for dealing with possibilistic right hand sides that is a possibilistic generalization of the recourse models in stochastic optimization. Violations of constraints are allowable, at a cost determined *a priori* by the decision maker. They choose the utility (that is, valuation) of a given interval of possible values to be its expected average.

A robust optimization formulation can address possibilistic parameters and a fuzzy inequality in the same optimization problem. Robust programming sacrifices some performance in order to guard against excessively variable results. Delgado, *et al.* [16] propose an alternative formulation for the optimization problem with fuzzy inequalities and possibilistic coefficients: they use an *a priori* definition of an acceptable $\alpha$-level, and a fuzzy inequality definition chosen by the implementer.

# 3 Application of the Algorithms

The algorithms were evaluated via the radiation therapy planning problem, a large, well-known application.

## 3.1 The Radiation Therapy Planning Problem

The use of particle beams to treat tumors is called the radiation therapy planning (RTP) problem [5]. Beams of particles are oriented at a variety of angles and with varying intensities to deposit radiation dose (measured as energy/unit mass) to the tumor. The goal is to deposit a tumorcidal dose to the tumor while minimizing damage to surrounding non-tumor tissue.

The process begins with the patient's computed tomography (CT) scan. Each CT image is examined to identify and contour the tumor and normal structures. The image is subsequently vectorized into pixels. Likewise, candidate beams are discretized into beamlets. This study is concerned with a two-dimensional image of a head tumor at a resolution of $64 \times 64$ pixels, and ten beam angles, each divided into 10 beamlets. This is potentially a $64^2 \times 10^2$ problem. However, since not all pixels are in the potential paths of radiation, and not all beamlets have the potential to deliver dose to the tumor, we set these to zero *a priori* and remove them from the analysis. This corresponds to blocking the beam, which is always done in practice.

The RTP in practice proceeds as follows. After the oncologist delineates the tumor and critical structures, a candidate set of beam intensities are obtained by some optimization technique or purely by human choice. These beam intensities are used as inputs to a Federal Drug Administration (FDA) approved dose calculator to produce the graphical depiction of the dose deposited at each pixel. A histogram of the percentage of pixels receiving a particular radiation dose, called the dose volume histogram (DVH), is one such graphical depiction. In general, the radiation oncologist never consults the objective function value; just the final dose distribution is viewed.

A *treatment plan* is the identification of a set of beam angles and weights that provides a lethal dose to the tumor cells while sparing healthy tissue, with a resulting dose distribution acceptable to the radiation oncologist. A *dose transfer matrix* $A$, specific to the patient's geometry, represents how a unit of radiation in beamlet $j$ is deposited in body pixel $i$. The components of $A$ are determined by the fraction of pixel $i$ which intersects with beamlet $j$, attenuated by the distance of the pixel from the place where the beam enters the body. The dose transfer matrix $A$ can be divided into a matrix $T$ which contains dose transfer information to tumor pixels only, matrices $C_1$ through $C_K$ which contain does transfer information to pixel in critical organs 1 though $K$, and body matrix $B$ which contains dose transfer information for all non-tumor and non-critical-organ pixels in the body. The variable vector $x$ represents the beamlet intensities, and the right hand side vector $b$ represents the dosage requirements.

In the RTP literature, there is no agreement on what the objective function function of the RTP problem should be [10]. In this paper, the chosen objective function is minimizing total weighted radiation. Since each element $a_{ij}$ of the attenuation matrix $A$ represents radiation delivered to body pixel $i$ by one unit of radiation from beamlet $j$, the sum of the elements of the column $a_i$ will give the entire amount of radiation delivered to the body by one unit of radiation from beamlet $j$. We therefore select $c_j = \sum_i a_{ij}$ as the objective function coefficient of $x_j$.

The crisp formulation of the RTP is

$$\begin{aligned} \text{minimize } & cx && (1)\\ \text{subject to } B &\le b_{body} \\ C_1 &\le b_{C_1} \\ &\dots \\ C_K &\le b_{C_K} \\ T &\le b_{tumor} \\ -T &\le -b_{tumor}. && (2) \end{aligned}$$

Two inequalities represent the $T = b_{tumor}$ equality constraint. This split representation has a natural interpretation. $T \geq b_{tumor}$ (which is equivalent to $-T \leq -b_{tumor}$) represents the requirement that the dose delivered to tumor pixels is high enough to kill them. $T \leq b_{tumor}$ represents the requirement that no body tissue, tumorous or otherwise, be burned.

Uncertainty might occur in four areas of the mathematical programming problem: the right-hand side, the inequality, the $A$ matrix coefficients, and the cost coefficients. Let us consider what each of these means in the context of the radiation therapy problem:

- Possibilistic uncertainty in the right hand side implies uncertainty about what levels of radiation are tumorcidal and what levels are safe for the body and critical organs. In contrast, fuzzy uncertainty in the right-hand side implies that a dose is safe for organs or tumorcidal to a varying degree (indicated by the membership function.)
- Uncertainty in the inequality is always fuzzy and implies the oncologist's flexibility regarding the satisfaction of the dosage requirements.
- Uncertainty in coefficient $a_{ij}$ implies uncertainty about the degree of radiation beamlet $j$ imparts to pixel $i$. One (possibilistic) source of this type

of uncertainty is patient positioning and breathing [1]. In addition, we have possibilistic uncertainty in the $A$ matrix because the attenuation matrix is a linearization of the non-linear attenuation process. Finally, some uncertainty in the $a_{ij}$s results from the fuzzy nature of the pixels themselves. Biologically, cells are not strictly tumor or non-tumor cells. Some cells take on partial tumor characteristics, which results in a fuzzy membership function.

– Possibilistic uncertainty in cost coefficient $c_j$ implies ambiguity about the impact of the intensity of beamlet $j$ on the overall radiation to the body. Fuzzy uncertainty in the cost coefficient can result from fuzzy coefficients in the attenuation matrix, since the cost coefficients in the this study are defined as the column sums of the attenuation matrix.

### 3.2 Procedure

The Radiation Therapy Planning problem, with a $64 \times 64$ pixel resolution and with the radiation beam discretized into 10 beamlets, was formulated and solved according to each of the solution methods described in Section 2. MATLAB's optimization toolbox was used to produce a solution, which was evaluated via a Dose Volume Histogram. The full run of experiments are evaluated in detail in. The conclusions drawn from these experiments follow in the next section.

## 4 Main Results

As we have seen, fuzzy and possibilistic optimization formulations are abundant. The choice of formulation depends firstly upon the semantics of the problem, secondly upon the desired format of the outcome, and the lastly on the performance of a particular algorithm. What follows are several dichotomies which should aid a decision maker in the selection of an appropriate algorithm.

### 4.1 Hard or Soft Constraints

If the constraints in question are *required* to be met exactly as stated, the decision maker should select a formulation that uses hard constraints, such as Tanaka's h-level technique, the fuzzy max technique, Neumaier's surprise technique, or Buckley's technique. On the other hand, if the constraints indicate preference, are flexible, or can be violated at a cost in order to obtain a more desirable outcome, a formulation with soft constraints is appropriate. Methods we have examined with soft constraints are Zimmermann's, Verdegay's, Jamison and Lodwick's, Luhandjula's, Delgado's, and the fuzzy robust approach. It should be noted that in the case of hard constraints, there may not be feasible solutions in practice.

[1] Ambiguity caused by patient positioning and breathing will not be non-interactive from one body pixel to another, since the body is positioned and breathes as a whole. Since all the fuzzy/possibilistic formulations in this study make the assumption of non-interactivity, we have a less than ideal modeling situation.

### 4.2 Fuzzy or Possibilistic Parameters

Another determinant of the appropriate formulation is the nature of the problem parameters. If the problem contains parameters whose values are known with precision, and the uncertainty lies in the inequality, a problem with crisp parameters, such as Zimmermann's or Verdegay's should be chosen. If the problem contains parameters which are imprecise because there is not sufficient information to fully determine their values, a model that incorporates possibilistic parameters should be chosen, possibly from the following: Luhandjula's technique, the fuzzy max technique, Tanaka's h-level technique, Buckley's technique, fuzzy robust optimization, Jamison and Lodwick's expected average technique, or Delgado's technique. If, however, the value of some parameter is not fully defined because the parameter assumes multiple values to various degrees, the decision maker should choose a formulation with fuzzy parameters. The only model fully examined in this paper which incorporates fuzzy parameters is Neumaier's surprise approach.

### 4.3 Feasibility or Optimality

Model selection will also depend on the decision maker's goal. The problem might be to find a feasible solution (often in these cases, a crisp program is infeasible). In this case, the decision maker should select a model which seeks the most feasible solution, such as Zimmermann's technique, fuzzy robust optimization, surprise, Tanaka's h-level technique, or Delgado's technique. If there is an objective function to be optimized, the decision maker might choose to reformulate the objective function as an additional constraint, as in goal programming. In this case, the problem again becomes one of finding the most feasible solution, and a formulation chosen from those listed above is advised. On the other hand, the decision maker might desire to optimize the objective function for a given feasibility level, choosing a model which seeks an optimal solution such as Buckley's, Luhandjula's, Verdegay's, or the fuzzy max model. The Jamison and Lodwick approach alone among those implemented in this paper affords the decision maker the ability to trade off between optimality and feasibility.

### 4.4 Fuzzy or Crisp Solution

A formulation might also be selected because a crisp solution or a fuzzy/possibilistic solution is desired. In general, algorithms that require $\alpha$ input give fuzzy solutions, and those which do not take an $\alpha$ level *a priori* give crisp solutions. Jamison and Lodwick differ greatly from other formulations, by providing a crisp solution that is not given for a particular $\alpha$ value, but is averaged over all $\alpha$ values.

The question to consider is when to defuzzify. At some point, *all* solutions must become crisp, because one cannot implement a fuzzy or possibilistic solution. For instance, in a transportation problem, the decision maker might send the order to ship $\tilde{3}$ widgets, where $\tilde{3}$ is a triangular fuzzy number whose center is 3, and whose support is $[2, 4]$. In the end, however, the dock-worker will load a crisp 2, a crisp 3, or a crisp 4 on the truck. The trade-off is generally between flexibility and control. A decision maker who selects a formulation which yields a crisp solution knows what solution will be implemented, for there is only one. A decision maker who selects a formulation with a fuzzy or possibilistic solution must at some point choose (or allow another, like the dock-worker in the example above, to choose) a crisp instantiation of the solution to implement. Furthermore, this choice must be made without the aid of an optimization program. Leaving the defuzzifying to a person rather than a program grants flexibility, but relinquishes control over the quality of the solution.

When choosing a fuzzy or crisp solution, it is important to keep the semantics of the problem in mind. As an example, consider Verdegay's fuzzy optimization technique and Buckley's possibilistic approach, in the case that $b$ is possibilistic and $A$ and $c$ are crisp. The two programs give non-crisp solutions that are calculated in exactly the same way, but semantically means very different things. It makes sense to give an *a priori* $\alpha$ level for Buckley, because it corresponds with the decision maker's level of comfort with risk. In Verdegay, the $\alpha$ level corresponds to the decision maker's level of constraint satisfaction. But wouldn't the decision maker always wish to maximize the level of constraint satisfaction, which corresponds to using Zimmermann's method?

Bearing this in mind, the following formulations produce non-crisp solutions: Verdegay's, Buckley's, fuzzy max, and Delgado's. For a crisp solution, good choices are Zimmermann's, surprise, h-level, Luhandjula's, Jamison and Lodwick's, and fuzzy robust optimization.

### 4.5 Complexity and Speed

If multiple formulations fit the semantics of the problem and meet the needs of the decision maker, efficiency may be considered a deciding factor. The following list is a ranking, from fastest to slowest, of the techniques tested in this paper: crisp, Verdegay's, Zimmermann's, Buckley's, robust, $h$-level, fuzzy max, Jamison and Lodwick's.

## 5 Additional Thoughts

The focus of this paper has been the selection of a fuzzy or possibilistic optimization algorithm. The broader question is whether a fuzzy/possibilistic formulation is appropriate in the first place.

### 5.1 Optimization with Both Fuzzy and Possibilistic Parameters

What does the solution to the optimization problem with mixed fuzzy and possibilistic parameters mean?

In both the approaches covered in this section, the same $\alpha$-cuts define the level of ambiguity in the coefficients, *and* the level at which the decision-maker's requirements are satisfied. We interpret the solution to mean that for any value

$\alpha \in (0, 1]$, we have a possibility $\alpha$ of obtaining a solution that satisfies the decision maker to degree $\alpha$. Using the same $\alpha$ value for both the possibilistic and fuzzy components of the problem is convenient, but does not necessarily provide a meaningful model of reality. For example, suppose $\alpha = .8$. If the solution, $x(.8)$ is implemented, there is a .8 possibility that it will provide the decision maker with at least .8 degree of satisfaction. On the other hand, if $x(.3)$ is implemented, there is a .3 possibility that it will provide the decision maker with at least .3 degree of satisfaction. It is apparent that the quality of the solution diverges rapidly. A more useful approach might be to require a higher level of satisfaction for a solution with a smaller possibility of attaining feasibility. For example, if we were to accept a solution that had only .3 possibility of occurring, we might require that it provide us with at least .9 degree of satisfaction. This is akin to the idea of being willing to take greater risks for the possibility of greater rewards.

### 5.2 Optimization Under Uncertainty vs. Crisp Optimization

In the radiation therapy problem, the solution to the problem is multiplied by a crisp attenuation matrix to calculate delivered dose and produce the dose volume histogram. A possibilistic solution, calculated with possibilistic $\hat{A}$, might be feasible–i.e. produce an acceptable dose– when multiplied by the possibilistic $\hat{A}$, but be infeasible when multiplied by a crisp $A$. For an example of this, refer to the Buckley implementation. Since the DVH is the tool used by radiation oncologists to judge the quality of the solution, this is a problem. To accurately display the possibilistic solution, the DVH should be calculated using a possibilistic $\hat{A}$. Even if the radiation oncologist employed a possibilistic DVH, the Federal Drug Administration (FDA) regulates radiation levels with crisp guidelines. A possibilistic solution that produces a feasible DVH based on a possibilistic attenuation matrix might still violate FDA regulations, which are evaluated crisply. We put forth that optimization problems whose solutions will be held to crisp standards should be solved with crisp optimization.

An additional question that must be considered is whether or not the quality of the solution to the fuzzy problem does not significantly surpass the quality of its crisp counterpart. If it does, the marginal quality must be enough to outweigh the computational inefficiency of optimization under uncertainty. In the case of the RTP implementations in this paper, the crisp formulation produced a DVH of quality comparable to the most successful fuzzy and possibilistic techniques in much less time.

## References

1. R. E. Bellman and L. A. Zadeh. Decision-making in a fuzzy environment. *Management Science*, 17(4):141–164, 1970.
2. D. Dubois. Linear programming with fuzzy data. In J. C. Bezdek, editor, *Analysis of Fuzzy Information, vol III: Applications in Engineering and Science*, pages 241–263. CRC Press, 1987.

3. D. Dubois and H. Prade. Systems of linear fuzzy constraints. *Fuzzy Sets and Systems*, 3:37–48, 1980.
4. D. Dubois and H. Prade. Fuzzy numbers: An overview. In J. C. Bezdek, editor, *Analysis of Fuzzy Information, vol. I: Mathematics and Logic*, pages 3–39. CRC Press, 1987.
5. Allen Holder and Bill Salter. A tutorial on radiation oncology and optimization. In H. Greenberg, editor, *Tutorials on Emerging Methodologies and Application in Operations Research*, chapter 4. 2004.
6. M. Inuiguchi, H. Ichihashi, and H. Tanaka. Fuzzy programming: A survey of recent developments. In R. Slowinski and J. Teghem, editors, *Stochastic versus Fuzzy Approaches to Multiobjective Mathematical Programmin under Uncertainty*, pages 45–68. Kluwer Academic Publishers, 1990.
7. K. Jamison and W. Lodwick. Fuzzy linear programming using a penalty method. *Fuzzy Sets and Systems*, 119:97–110, 2001.
8. Y. Lai and C. Hwang. *Fuzzy Mathematical Programming.* Springer-Verlag, Berlin, 1992.
9. W. Lodwick. personal communication, June 2006.
10. W. Lodwick and K. Bachman. Solving large-scale fuzzy and possibilistic optimization problems. *Fuzzy Optimization and Decision Making*, 4:257–278, 2005.
11. W. Lodwick and K. Jamison. A computational method for fuzzy optimization. In B. Ayyub and G. Bilal, editors, *Uncertainty Analysis in Engineering and Sciences: Fuzzy Logic, Statistics, and Neural Network Approach.* Kluwer Academic Publishers, Boston, 1997.
12. W. Lodwick and K. D. Jamison. Theory and semantics for fuzzy and possibilistic optimization. *IFSA*, 2005.
13. W. Lodwick, A. Neumaier, and F. Newman. Optimization under uncertainty: Methods and applications in radiation therapy. *Proc. 10th Intl. Conference on Fuzzy Systems*, 3:1219–1222, 2001.
14. W. A. Lodwick. Analyis of structure in fuzzy linear programs. *Fuzzy Sets and Systems*, 38:15–26, 1990.
15. M.K. Luhandjula. On possibilistic linear programming. *Fuzzy Sets and Systems*, 18:15–30, 1986.
16. M.A. Vila M. Delgado, J.L. Verdegay. A general model for fuzzy linear programming. *Fuzzy Sets and Systems*, pages 21–29, 1989.
17. T. Morrison. personal communication, July 2006.
18. C.V. Negoita. The current interest in fuzzy optimization. *fuzzy Sets and Systems*, 6:261–269, 1981.
19. C.V. Negoita and M. Sularia. On fuzzy mathematical programming and tolerances in planning. *Economic Computation and Economic Cybernetics Studies and Research*, 3(31):3–14, 1976.
20. A. Neumaier. Fuzzy modeling in terms of surprise. *Fuzzy Sets and Systems*, 135:21–35, 2003.
21. D. Ralescu. Inexact solutions for large-scale control problems. *Proceedings of the first international congress on math at the service of man*, 1977.
22. J. Ramik and J. Rimanek. Inequality relation between fuzzy numbers and it's use in fuzzy optimization. *Fuzzy Sets and Systems*, pages 123–138, 1985.
23. A. L. Soyster. Convex programming with set-inclusive constraints and applications to inexact linear programming. *Operations Research*, 21:1154–1157, 1973.
24. A. L. Soyster. A duality theory for convex programming with set-inclusive constraints. *Operations Research*, 22:892–898, 1974.

25. A. L. Soyster. Erratum. *Operations Research*, 22:1279–1280, 1974.
26. H. Tanaka and K. Asai. Fuzzy linear programming with fuzzy numbers. *Fuzzy Sets and Systems*, 13:1–10, 1984.
27. H. Tanaka, H. Ichahashi, and K. Asai. A formulation of fuzzy linear programming problem based on comparison of fuzzy numbers. *Control and Cybernetics*, 13(3), 1984.
28. H. Tanaka, H. Ichihashi, and K. Asai. Fuzzy decision in linear programming with trapezoid fuzzy parameters. In J. Kacpryzk and R.R. Yager, editors, *Management Decision Suppot Systems Using Fuzzy Sets and Possibility Theory*, pages 146–154. Verlag TUV, Koln, 1985.
29. H. Tanaka, T. Okuda, and K. Asai. On fuzzy-mathematical programming. *J. of Cybernetics*, 3(4):37–46, 1974.
30. J.L. Verdegay. *Fuzzy Mathematical Programming*, pages 231–237. North Holland Company, 1982.
31. R.R. Yager. A procedure for odering fuzzy subsets of the unit interval. *Information Sciences*, 24:143–161, 1981.
32. L. A. Zadeh. Fuzzy sets. *Information and Control*, 8:338–353, 1965.
33. L. A. Zadeh. The concept of a linguistic variable and its application to approximate reasoning - i. *Information Sciences*, 8:(199–249), 1975.
34. H-J. Zimmermann. Description and optimization of fuzzy systems. *International J. of General Systems*, 2:209–215, 1976.

# A Risk-Minimizing Model Under Uncertainty in Portfolio

Yuji Yoshida

Faculty of Economics and Business Administration, University of Kitakyushu
4-2-1 Kitagata, Kokuraminami, Kitakyushu 802-8577, Japan
yoshida@kitakyu-u.ac.jp

**Abstract.** A risk-minimizing portfolio model under uncertainty is discussed. In the uncertainty model, the randomness and fuzziness are evaluated respectively by the probabilistic expectation and mean values with evaluation weights and $\lambda$-mean functions. The means, variances and the measurements of fuzziness for fuzzy numbers/fuzzy random variables are applied in the possibility case and the necessity case, and a risk estimation is derived from both random factors and fuzzy factors in the model. By quadratic programming approach, we derive a solution of the risk-minimizing portfolio problem. It is shown that the solution is a tangency portfolio. A numerical example is given to illustrate our idea.

## 1 Introduction and Notations

The portfolio is one of the most important tools for the asset management in finance. Portfolio models have been studied by many authors using mathematical programming on the basis of Markowitz's model (Steinbach 2001, Markowits 1990; Pliska 1997). When we deal with systems like financial markets, fuzzy logic works well since the markets contain the uncertain factors which are different from probabilistic essence and in which it is difficult to specify actual price values exactly (Inuiguchi and Tanino 2000; Tanaka et al. 2000). In this paper, randomness is applied to the uncertainty regarding the belief degree of frequency, and fuzziness is applied to imprecision of data because of a lack of knowledge regarding the current stock market. In this paper, we consider a risk-minimizing portfolio model under uncertainty of randomness and fuzziness.

Estimation of uncertain quantities is important in decision making. To represent uncertainty in a finance model, we use *fuzzy random variables* which have two kinds of uncertainties, i.e. randomness and fuzziness. Recently, (Yoshida 2004, 2006) introduced means, variances and covariance of fuzzy random variables using *evaluation weights* and $\lambda$-mean functions. In this paper, we estimate fuzzy numbers/fuzzy random variables by probabilistic expectation and these evaluations, which are characterized by a *possibility-necessity weight* for subjective estimation and a *pessimistic-optimistic index* for subjective decision. Especially we deal with evaluation weights derived from the possibility measure and the necessity measure for numerical computation in modeling.

P. Melin et al. (Eds.): IFSA 2007, LNAI 4529, pp. 381–391, 2007.
© Springer-Verlag Berlin Heidelberg 2007

In a portfolio model, we use triangle-type fuzzy numbers/fuzzy random variables for simplicity in numerical computation when we apply them to actual models. We discuss a risk-minimizing problem for the portfolio, where the risk is defined by both the random factors and fuzzy factors in the model. By quadratic programming approach, we derive a solution of the risk-minimizing portfolio problem. We show that the solution is a tangency portfolio. A numerical example is given to illustrate our idea.

In this paper, we consider a portfolio model with a bond and $n$ stocks, where $n$ is a positive integer. In the remainder of this section, we describe a bond price process and stock price processes. We deal with a model where an investor's actions do not have any impact on the stock market, so-called *small investors hypothesis* (Pliska 1997). Let $\mathbb{T} := \{0, 1, 2, \cdots, T\}$ be the time space with an expiration date $T$, and $\mathbb{R}$ denotes the set of all real numbers. Let $(\Omega, \mathcal{M}, P)$ be a probability space, where $\mathcal{M}$ is a $\sigma$-field of $\Omega$ and $P$ is a non-atomic probability measure. Take a probability space $\Omega := (\mathbb{R}^{n+1})^{T+1}$ by the product of $\mathbb{R}$. Let a positive number $r_t$ be an *interest rate* of a bond price at time $t$ for $t = 1, 2, \cdots, T$, and put a *bond price process* $\{S_t^0\}_{t=0}^T$ by $S_0^0 = 1$ and

$$S_t^0 := \prod_{s=1}^{t} (1 + r_s) \quad \text{for } t = 1, 2, \cdots, T. \tag{1}$$

For an asset $i = 1, 2, \cdots, n$, a *stock price process* $\{S_t^i\}_{t=0}^T$ is given by *rates of return* $R_t^i$ as follows.

$$S_t^i := S_{t-1}^i (1 + R_t^i) \tag{2}$$

for $t = 1, 2, \cdots, T$, where $\{R_t^i\}_{t=1}^T$ is assumed to be a uniform integrable sequence of independent identically distributed real random variables with values in $[-1, \infty)$. Then we have

$$S_t^i = S_0^i \prod_{s=1}^{t} (1 + R_s^i) \tag{3}$$

for $t = 1, 2, \cdots, T$. In this paper, we present a portfolio model where stock price processes $S_t^i$ take fuzzy values using fuzzy random variables, whose mathematical notations are introduced in the next section.

## 2 A Portfolio Model Under Uncertainty

In this section, we introduce fuzzy numbers/fuzzy random variable and we give a portfolio model under uncertainty. A fuzzy number is denoted by its membership function $\tilde{a} : \mathbb{R} \mapsto [0, 1]$ which is normal, upper-semicontinuous and fuzzy convex and has a compact support (Zadeh 1965). $\mathcal{R}$ denotes the set of all fuzzy numbers. In this paper, we identify fuzzy numbers with their corresponding membership functions. The $\alpha$-cut of a fuzzy number $\tilde{a} (\in \mathcal{R})$ is given by $\tilde{a}_\alpha := \{x \in \mathbb{R} \mid \tilde{a}(x) \geq \alpha\}$ $(\alpha \in (0, 1])$ and $\tilde{a}_0 := \mathrm{cl}\{x \in \mathbb{R} \mid \tilde{a}(x) > 0\}$, where cl denotes the closure of an interval. We write the closed intervals as $\tilde{a}_\alpha := [\tilde{a}_\alpha^-, \tilde{a}_\alpha^+]$ for $\alpha \in [0,1]$. A fuzzy-number-valued map $\tilde{X} : \Omega \mapsto \mathcal{R}$ is called a *fuzzy random variable* if the maps $\omega \mapsto \tilde{X}_\alpha^-(\omega)$ and $\omega \mapsto \tilde{X}_\alpha^+(\omega)$ are measurable for all $\alpha \in (0,1]$, where $\tilde{X}_\alpha(\omega) = [\tilde{X}_\alpha^-(\omega), \tilde{X}_\alpha^+(\omega)] = \{x \in \mathbb{R} \mid \tilde{X}(\omega)(x) \geq \alpha\}$ (Kwakernaak 1987; Puri and Ralescu 1986). Next we need to introduce expectations of fuzzy random variables in order to describe a portfolio model. A fuzzy random variable $\tilde{X}$ is called integrably bounded if both $\omega \mapsto \tilde{X}_\alpha^-(\omega)$ and $\omega \mapsto \tilde{X}_\alpha^+(\omega)$ are integrable for all $\alpha \in (0,1]$. Let $\tilde{X}$ be an integrably bounded fuzzy random variable. The expectation $E(\tilde{X})$ of the fuzzy random variable $\tilde{X}$ is defined by a fuzzy number

$$E(\tilde{X})(x) := \sup_{\alpha \in [0,1]} \min\{\alpha, 1_{E(\tilde{X})_\alpha}(x)\}, \quad x \in \mathbb{R}, \tag{4}$$

where $E(\tilde{X})_\alpha := [\int_\Omega \tilde{X}_\alpha^-(\omega)\,\mathrm{d}P(\omega), \int_\Omega \tilde{X}_\alpha^+(\omega)\,\mathrm{d}P(\omega)]$ for $\alpha \in (0,1]$ (Yager 1981, Puri and Ralescu 1986).

Now we deal with a case when the rate of return $\{R_t^i\}_{t=1}^T$ has some imprecision. We define a *rate of return process with imprecision* $\{\tilde{R}_t^i\}_{t=0}^T$ by a sequence of triangle-type fuzzy random variables

$$\tilde{R}_t^i(\cdot)(x) = \begin{cases} 0 & \text{if } x < R_t^i - c_t^i \\ \frac{x - R_t^i + c_t^i}{c_t^i} & \text{if } R_t^i - c_t^i \leq x < R_t^i \\ \frac{x - R_t^i - c_t^i}{-c_t^i} & \text{if } R_t^i \leq x < R_t^i + c_t^i \\ 0 & \text{if } x \geq R_t^i + c_t^i, \end{cases} \tag{5}$$

where $c_t^i$ is a positive number. We call $c_t^i$ a *fuzzy factor* for asset $i$ at time $t$. Hence we can represent $\tilde{R}_t^i$ by the sum of fuzzy numbers:

$$\tilde{R}_t^i(\omega)(\cdot) := 1_{\{R_t^i(\omega)\}}(\cdot) + \tilde{a}_t^i(\cdot) \tag{6}$$

for $\omega \in \Omega$, where $1_{\{\cdot\}}$ denotes the characteristic function of a singleton and $\tilde{a}_t^i$ is a fuzzy number

$$\tilde{a}_t^i(x) = \begin{cases} 0 & \text{if } x < -c_t^i \\ \frac{x + c_t^i}{c_t^i} & \text{if } -c_t^i \leq x < 0 \\ \frac{x - c_t^i}{-c_t^i} & \text{if } 0 \leq x < c_t^i \\ 0 & \text{if } x \geq c_t^i. \end{cases} \tag{7}$$

For assets $i = 1, 2, \cdots, n$, we define *stock price processes* $\{\tilde{S}_t^i\}_{t=0}^T$ by the *rates of return with imprecision* $\tilde{R}_t^i$ as follows: $\tilde{S}_0^i := S_0^i$ is a constant and

$$\tilde{S}_t^i = \tilde{S}_0^i \prod_{s=1}^{t} (1 + \tilde{R}_s^i) \tag{8}$$

for $t = 1, 2, \cdots, T$ (Yoshida et al. 2003). Hence, we deal with a portfolio with *trading strategies given by portfolio weight vectors* $w = (w^1, w^2, \cdots, w^n)$ such that $w^1 + w^2 + \cdots w^n = 1$ and $w^i \geq 0\,(i = 1, 2, \cdots, n)$. For the trading strategy

$w = (w^1, w^2, \cdots, w^n)$, the rate of return with imprecision for the portfolio is given by

$$\tilde{R}_t := w^1 \tilde{R}_t^1 + w^2 \tilde{R}_t^2 + \cdots w^n \tilde{R}_t^n. \tag{9}$$

This paper discusses a risk-minimizing model regarding (9) under guarantee of rewards.

## 3 Mean, Variance and the Measurement of Fuzziness

There are many research results for the estimation of fuzzy numbers (Campos and Munoz 1989; Fortemps and Roubens 1996; López-Díaz and Gil 1998). In this paper, we use an evaluation method of fuzzy numbers/fuzzy random variables introduced by (Yoshida et al. 2003; Yoshida 2004, 2006) to estimate the rate of return (9) in the portfolio. They studied an evaluation of fuzzy numbers by *evaluation weights* which are induced from fuzzy measures to evaluate a confidence degree that a fuzzy number takes values in an interval. With respect to fuzzy random variables, the randomness is evaluated by the probabilistic expectation and the fuzziness is estimated by evaluation weights and the following function. Let $g^\lambda : \mathcal{I} \mapsto \mathbb{R}$ be a map such that

$$g^\lambda([x, y]) := \lambda x + (1 - \lambda) y, \quad [x, y] \in \mathcal{I}, \tag{10}$$

where $\lambda$ is a constant satisfying $0 \le \lambda \le 1$ and $\mathcal{I}$ denotes the set of all bounded closed intervals. This scalarization is used for the estimation of fuzzy numbers to give a mean value of the interval $[x, y]$ with a weight $\lambda$, and $\lambda$ is called a *pessimistic-optimistic index* and indicates the pessimistic degree in decision making (Fortemps and Roubens 1996). Then, $g^\lambda$ is called a *$\lambda$-mean function.* Let a fuzzy number $\tilde{a} \in \mathcal{R}$. A mean value of the fuzzy number $\tilde{a}$ with respect to $\lambda$-mean functions $g^\lambda$ and an evaluation weights $w(\alpha)$, which depends only on $\tilde{a}$ and $\alpha$, is given as follows (Yoshida 2004, 2006)

$$\tilde{E}^\lambda(\tilde{a}) := \int_0^1 g^\lambda(\tilde{a}_\alpha)\, w(\alpha)\, \mathrm{d}\alpha \Big/ \int_0^1 w(\alpha)\, \mathrm{d}\alpha, \tag{11}$$

where $\tilde{a}_\alpha = [\tilde{a}_\alpha^-, \tilde{a}_\alpha^+]$ is the $\alpha$-cut of the fuzzy number $\tilde{a}$. In (11), $w(\alpha)$ indicates a *confidence degree that the fuzzy number $\tilde{a}$ takes values in the interval $\tilde{a}_\alpha$ at each level $\alpha$*. Hence, an evaluation weight $w(\alpha)$ is called the *possibility evaluation weight $w^P(\alpha)$* and the *necessity evaluation weight $w^N(\alpha)$* induced from the fuzzy number $\tilde{a}$ if they are given respectively by $w^P(\alpha) = 1$ and $w^N(\alpha) = 1 - \alpha$ for $\alpha \in [0, 1]$. Especially, for a fuzzy number $\tilde{a} \in \mathcal{R}$, the mean $\tilde{E}^P(\tilde{a})$ in the possibility case and the mean $\tilde{E}^N(\tilde{a})$ in the necessity case are represented as follows (Yoshida 2004, 2006):

$$\tilde{E}^P(\tilde{a}) = \int_0^1 g^\lambda(\tilde{a}_\alpha)\, \mathrm{d}\alpha \quad \text{and} \quad \tilde{E}^N(\tilde{a}) = \int_0^1 g^\lambda(\tilde{a}_\alpha)\, (2 - 2\alpha)\, \mathrm{d}\alpha. \tag{12}$$

The mean $\tilde{E}^\lambda$ has the following natural properties regarding the linearity and the monotonicity for the fuzzy max order.

**Lemma 1** (Yoshida 2004, 2006). *Let $\lambda \in [0,1]$. Let $\tilde{E}^\lambda = \tilde{E}^P$ or $\tilde{E}^\lambda = \tilde{E}^N$. For fuzzy numbers $\tilde{a}, \tilde{b} \in \mathcal{R}$ and real numbers $\theta, \zeta$, the following (i) – (iv) hold.*

(i) $\tilde{E}^\lambda(\tilde{a} + 1_{\{\theta\}}) = \tilde{E}^\lambda(\tilde{a}) + \theta$, *where* $1_{\{\cdot\}}$ *is the characteristic function of a set.*
(ii) $\tilde{E}^\lambda(\zeta\tilde{a}) = \zeta\tilde{E}^\lambda(\tilde{a})$ *if* $\zeta \geq 0$.
(iii) $\tilde{E}^\lambda(\tilde{a} + \tilde{b}) = \tilde{E}^\lambda(\tilde{a}) + \tilde{E}^\lambda(\tilde{b})$.
(iv) *If* $\tilde{a} \succeq \tilde{b}$, *then* $\tilde{E}^\lambda(\tilde{a}) \geq \tilde{E}^\lambda(\tilde{b})$ *holds, where* $\succeq$ *is the fuzzy max order.*

Next we consider measurements regarding two kinds of uncertainty, i.e. fuzziness and randomness. Fuzziness is based on the imprecision of data and the variance is based on the randomness, and they are given as independent concepts in this paper. Therefore, they should be estimated in different ways. Yoshida 2006) has studied a method to measure the size of fuzziness regarding fuzzy numbers. Let $\tilde{a} \in \mathcal{R}$ be a fuzzy number. A *measurement of fuzziness* $\tilde{F}(\tilde{a})$ of the fuzzy number $\tilde{a}$ is given as follows: Let $\alpha \in [0,1]$. For an interval $\tilde{a}_\alpha = [\tilde{a}_\alpha^-, \tilde{a}_\alpha^+]$, the upper/lower measurements of fuzziness should be given by $m^U(\tilde{a}_\alpha) := \tilde{a}_\alpha^+ - \tilde{a}_\alpha^-$ and $m^L(\tilde{a}_\alpha) := (\tilde{a}_\alpha^+ - \tilde{a}_\alpha^-)/2$. The measurements of fuzziness are related to the imprecision of the data, and they should be defined without the subjective index $\lambda$. Then, for $m = m^U$ or $m = m^L$, a measurement of fuzziness $\tilde{F}(\tilde{a})$ is given by

$$\tilde{F}(\tilde{a}) = \int_0^1 m(\tilde{a}_\alpha)\, w(\alpha)\, \mathrm{d}\alpha \Big/ \int_0^1 w(\alpha)\, \mathrm{d}\alpha, \tag{13}$$

where $\tilde{a}_\alpha = [\tilde{a}_\alpha^-, \tilde{a}_\alpha^+]$ is the $\alpha$-cut of the fuzzy number $\tilde{a} \in \mathcal{R}$.

**Lemma 2.** *Let a fuzzy number $\tilde{a} \in \mathcal{R}$. Then, the measurement of fuzziness in the possibility case and the necessity case are given as follows.*

$$\tilde{F}^P(\tilde{a}) := \int_0^1 m^U(\tilde{a}_\alpha)\, w^P(\alpha)\, \mathrm{d}\alpha \Big/ \int_0^1 w^P(\alpha)\, \mathrm{d}\alpha = \int_0^1 (\tilde{a}_\alpha^+ - \tilde{a}_\alpha^-)\, \mathrm{d}\alpha, \tag{14}$$

$$\tilde{F}^N(\tilde{a}) := \int_0^1 m^N(\tilde{a}_\alpha)\, w^N(\alpha)\, \mathrm{d}\alpha \Big/ \int_0^1 w^N(\alpha)\, \mathrm{d}\alpha = \int_0^1 (\tilde{a}_\alpha^+ - \tilde{a}_\alpha^-)(1-\alpha)\, \mathrm{d}\alpha. \tag{15}$$

Now we have the following natural results about the *possibility fuzziness measure* $\tilde{F}(\cdot) = \tilde{F}^P(\cdot)$ and the *necessity fuzziness measure* $\tilde{F}(\cdot) = \tilde{F}^N(\cdot)$.

**Lemma 3.** *Let $\tilde{F} = \tilde{F}^P$ or $\tilde{F} = \tilde{F}^N$. For fuzzy numbers $\tilde{a}, \tilde{b} \in \mathcal{R}$ and real numbers $\theta, \zeta$, the following (i) – (iv) hold.*

(i) $\tilde{F}(\tilde{a} + 1_{\{\theta\}}) = \tilde{F}(\tilde{a})$.
(ii) $\tilde{F}(\zeta\tilde{a}) = |\zeta|\tilde{F}(\tilde{a})$.
(iii) $\tilde{F}(\tilde{a} \pm \tilde{b}) = \tilde{F}(\tilde{a}) + \tilde{F}(\tilde{b})$.
(iv) *If* $\tilde{a} \supset \tilde{b}$, *then* $\tilde{F}(\tilde{a}) \geq \tilde{F}(\tilde{b})$ *holds, where* $\supset$ *implies the inclusion in the sense of fuzzy sets.*

Let $\tilde{a} \in \mathcal{R}$ be a fuzzy number and let $\nu \in [0,1]$ be a parameter. For applications of the mean values and measurement of fuzziness in actual problems, we introduce a mean value and a measurement of fuzziness with a parameter $\nu$:

$$\tilde{E}^{\lambda,\nu}(\tilde{a}) := \nu \tilde{E}^{P}(\tilde{a}) + (1-\nu)\tilde{E}^{N}(\tilde{a}), \tag{16}$$

$$\tilde{F}^{\nu}(\tilde{a}) := \nu \tilde{F}^{P}(\tilde{a}) + (1-\nu)\tilde{F}^{N}(\tilde{a}). \tag{17}$$

Then, $\nu$ is called a *possibility-necessity weight*, and (16) and (17) are the mean value and the measurement of fuzziness with the possibility-necessity weight $\nu$.

Using evaluation weights, we give means, variances and covariances regarding fuzzy random variables. For the fuzzy random variable $\tilde{X}$, *the mean of the expectation* $E(\tilde{X})$ is a real number

$$E(\tilde{E}^{\lambda}(\tilde{X})) = E\left( \int_0^1 g^{\lambda}(\tilde{X}_{\alpha})\, w(\alpha)\, \mathrm{d}\alpha \Big/ \int_0^1 w(\alpha)\, \mathrm{d}\alpha \right). \tag{18}$$

Then, from Lemma 1, we obtain the following results regarding fuzzy random variables.

**Lemma 4.** *Let $\lambda \in [0,1]$. For a fuzzy number $\tilde{a} \in \mathcal{R}$, integrable fuzzy random variables $\tilde{X}, \tilde{Y}$ and an integrable real random variable $Z$ and a nonnegative real number $\zeta$, the following (i) – (vi) hold.*

(i) $E(\tilde{E}^{\lambda}(\tilde{X})) = \tilde{E}^{\lambda}(E(\tilde{X}))$.
(ii) $E(\tilde{E}^{\lambda}(\tilde{a})) = \tilde{E}^{\lambda}(\tilde{a})$ *and* $E(\tilde{E}^{\lambda}(Z)) = E(Z)$.
(iii) $E(\tilde{E}^{\lambda}(\tilde{X} + \tilde{a})) = E(\tilde{E}^{\lambda}(\tilde{X})) + \tilde{E}^{\lambda}(\tilde{a})$ *and* $E(\tilde{E}^{\lambda}(\tilde{X} + Z)) = E(\tilde{E}^{\lambda}(\tilde{X})) + E(Z)$.
(iv) $E(\tilde{E}^{\lambda}(\zeta\tilde{X})) = \zeta E(\tilde{E}^{\lambda}(\tilde{X}))$.
(v) $E(\tilde{E}^{\lambda}(\tilde{X} + \tilde{Y})) = E(\tilde{E}^{\lambda}(\tilde{X})) + E(\tilde{E}^{\lambda}(\tilde{Y}))$.
(vi) *If* $\tilde{X} \succeq \tilde{Y}$ *almost surely, then* $E(\tilde{E}^{\lambda}(\tilde{X})) \geq E(\tilde{E}^{\lambda}(\tilde{Y}))$ *holds, where* $\succeq$ *is the fuzzy max order.*

Next we introduce variances and covariances of fuzzy random variables from the viewpoint of $\lambda$-mean functions and evaluation weights. For fuzzy random variables $\tilde{X}$ and $\tilde{Y}$, we define variances and covariances as follows (Yoshida 2006).

$$V^{\lambda}(\tilde{X}) := E\left( \int_0^1 (g^{\lambda}(\tilde{X}_{\alpha}) - E(g^{\lambda}(\tilde{X}_{\alpha})))^2\, w(\alpha)\, \mathrm{d}\alpha \Big/ \int_0^1 w(\alpha)\, \mathrm{d}\alpha \right), \tag{19}$$

$$\begin{aligned} &Cov^{\lambda,\gamma}(\tilde{X},\tilde{Y}) \\ &:= E\left( \int_0^1 (g^{\lambda}(\tilde{X}_{\alpha}) - E(g^{\lambda}(\tilde{X}_{\alpha})))(g^{\gamma}(\tilde{Y}_{\alpha}) - E(g^{\gamma}(\tilde{Y}_{\alpha})))\, w(\alpha)\, \mathrm{d}\alpha \Big/ \int_0^1 w(\alpha)\, \mathrm{d}\alpha \right) \end{aligned} \tag{20}$$

for $\lambda, \gamma \in [0,1]$. We can find other approaches in (Carlsson and Fullér 2001; Feng et al. 2001), which discuss the variance of fuzzy numbers by possibility theory.

Hence we obtain the following natural properties about the variance $V^{\lambda}(\cdot)$ and covariance $Cov^{\lambda,\gamma}(\cdot,\cdot)$.

**Lemma 5.** *Let $\lambda, \gamma \in [0,1]$. For fuzzy numbers $\tilde{a}, \tilde{b} \in \mathcal{R}$, integrable fuzzy random variables $\tilde{X}, \tilde{Y}$, integrable real random variables $X, Y$ and a nonnegative real number $\zeta$, the following (i) – (v) hold.*

(i) $V^{\lambda}(\tilde{a}) = 0$ *and* $V^{\lambda}(X) = V(X)$*, where* $V(\cdot)$ *is the variance of real random variables.*
(ii) $V^{\lambda}(\tilde{X} + \tilde{a}) = V^{\lambda}(\tilde{X})$.
(iii) $V^{\lambda}(\zeta\tilde{X}) = \zeta^2 V^{\lambda}(\tilde{X})$.
(iv) $Cov^{\lambda,\gamma}(\tilde{X}, \tilde{a}) = Cov^{\lambda,\gamma}(\tilde{a}, \tilde{X}) = 0$ *and* $Cov^{\lambda,\gamma}(X, Y) = Cov(X, Y)$*, where* $Cov(\cdot,\cdot)$ *is the covariance of real random variables.*
(v) $Cov^{\lambda,\gamma}(\tilde{X} + \tilde{a}, \tilde{Y} + \tilde{b}) = Cov^{\lambda,\gamma}(\tilde{X}, \tilde{Y})$.

## 4 A Risk-Minimizing Model Under Uncertainty in Portfolio

In this section, we discuss a risk-minimizing problem under uncertainty. Let the mean, variance and covariance of the fuzzy random variables $\tilde{R}_t^i$ defined by (5) by $\tilde{\mu}_t^i := E(\tilde{E}^{\lambda}(\tilde{R}_t^i))$, $(\tilde{\sigma}_t^i)^2 := V^{\lambda}(\tilde{R}_t^i)$ and $\tilde{\sigma}_t^{ij} := Cov^{\lambda,\lambda}(\tilde{R}_t^i, \tilde{R}_t^j)$ for $\lambda \in [0,1]$ and $i, j = 1, 2, \cdots, n$. From Lemmas 4 and 5 and (6), we obtain the following results regarding the rates of returns $\tilde{R}_t^i$: Then

(a) $\tilde{\mu}_t^i = \mu_t^i + \tilde{E}^{\lambda}(\tilde{a}^i)$,
(b) $(\tilde{\sigma}_t^i)^2 = (\sigma_t^i)^2$,
(c) $\tilde{\sigma}_t^{ij} = \sigma_t^{ij}$,

where $\mu_t^i := E(R_t^i), (\sigma_t^i)^2 := V(R_t^i) = E((R_t^i - \mu_t^i)^2), \sigma_t^{ij} := Cov(R_t^i, R_t^j) = E((R_t^i - \mu_t^i)(R_t^j - \mu_t^j))$. For the trading strategy $w = (w^1, w^2, \cdots, w^n)$ such that $w^1 + w^2 + \cdots w^n = 1$, from Lemmas 4 and 5 the expectation $\tilde{\mu}_t$ and variance $(\tilde{\sigma}_t)^2$ regarding the *rate of return with imprecision for the portfolio* $\tilde{R}_t := w^1\tilde{R}_t^1 + w^2\tilde{R}_t^2 + \cdots w^n\tilde{R}_t^n$ in (9) is given as follows.

$$\tilde{\mu}_t := \sum_{i=1}^n w^i \tilde{\mu}_t^i = \sum_{i=1}^n w^i(\mu_t^i + \tilde{E}^{\lambda}(\tilde{a}^i)),$$
$$(\tilde{\sigma}_t)^2 := \sum_{i=1}^n (w^i)^2(\sigma_t^i)^2 + 2\sum_{i=1}^n \sum_{j=i+1}^n w^i w^j \sigma_t^{ij} = \sum_{i=1}^n \sum_{j=1}^n w^i w^j \sigma_t^{ij},$$

where $\sigma_t^{ii} := (\sigma_t^i)^2$. Hence, to guarantee the lower bound regarding the expectation $\tilde{\mu}_t$ of the rate of return for the portfolio we estimate $\tilde{\mu}_t$ taking the index $\lambda$ pessimistic ($\lambda = 1$) and the necessity mean $w^N(\alpha) = 1 - \alpha$ ($\alpha \in [0,1]$):

$$\begin{aligned}\tilde{\mu}_t &= \sum_{i=1}^{n} w^i(\mu_t^i + \tilde{E}^\lambda(\tilde{a}_t^i)) \\ &\geq \sum_{i=1}^{n} w^i\left(\mu_t^i - c_t^i\int_0^1 (1-\alpha)(2-2\alpha)d\alpha\right) \\ &= \sum_{i=1}^{n} w^i\left(\mu_t^i - \frac{2}{3}c_t^i\right)\end{aligned}$$

for $\lambda \in [0,1]$, where $\tilde{a}_\alpha^i = [\tilde{a}_\alpha^{i-}, \tilde{a}_\alpha^{i+}] = [-c_t^i(1-\alpha), c_t^i(1-\alpha)]$ from (7). Therefore,

$$\sum_{i=1}^{n} w^i\left(\mu_t^i - \frac{2}{3}c_t^i\right) \tag{21}$$

is the lower bound regarding the expectation (21) of the rate of return for the portfolio. Next, in this model, we deal with the risk derived from the uncertainty which consists of randomness and fuzziness. Since these factors are independent, we define the risk of the portfolio by the combination of the variance $(\tilde{\sigma}_t)^2$ and the measurement of fuzziness $E(\tilde{F}(\tilde{R}_t))$. Then, its upper bound is given by the possibility case $w^P(\alpha) = 1$ ($\alpha \in [0,1]$):

$$\begin{aligned}(\tilde{\sigma}_t)^2 + (E(\tilde{F}(\tilde{R}_t)))^2 &= \sum_{i=1}^{n}\sum_{j=1}^{n} w^i w^j(\sigma_t^{ij} + \tilde{F}(\tilde{a}_t^i)\tilde{F}(\tilde{a}_t^j)) \\ &\leq \sum_{i=1}^{n}\sum_{j=1}^{n} w^i w^j(\sigma_t^{ij} + c_t^i c_t^j).\end{aligned}$$

Therefore,

$$\tilde{\rho} := \sum_{i=1}^{n}\sum_{j=1}^{n} w^i w^j(\sigma_t^{ij} + c_t^i c_t^j) \tag{22}$$

is the upper bound of the risk of randomness and fuzziness. We deal with a risk-minimizing model with a bond. Let $\mu_t^0 = r_t$ and $c_t^0 = 0$. For a given constant $\gamma$ which means the expected rate of return to be guaranteed for the portfolio, we discuss the following problem.

**Risk-minimizing problem (RM):** Minimize the risk

$$\tilde{\rho} := \sum_{i=1}^{n}\sum_{j=1}^{n} w^i w^j(\sigma_t^{ij} + c_t^i c_t^j) \tag{23}$$

with trading strategies $w = (w^0, w^1, w^2, \cdots, w^n)$ satisfying $w^0 + w^1 + w^2 + \cdots w^n = 1$ and $w^i \geq 0\,(i = 1, 2, \cdots, n)$ under the condition

$$\sum_{i=0}^{n} w^i\left(\mu_t^i - \frac{2}{3}c_t^i\right) = \gamma. \tag{24}$$

**Theorem 1.** *Assume $\tilde{\Sigma}^{-1}(\tilde{\mu} - r_t\tilde{m}1) \geq \tilde{m}0$ and $\tilde{\Sigma}^{-1}(\tilde{\mu} - r_t\tilde{m}1) \neq \tilde{m}0$. Then there exists a solution $(\tilde{\rho}^{**}, \tilde{\mu}^{**})$ of Problem (RM), which is given by the following $(\tilde{\rho}^{**}, \tilde{\mu}^{**})$:*

$$(\tilde{\rho}^{**}, \tilde{\mu}^{**}) := \left( \frac{A(r_t)^2 - 2Br_t + C}{(Ar_t - B)^2}, \frac{Br_t - C}{Ar_t - B} \right). \tag{25}$$

*The corresponding trading strategy $w^{**} := (w^0, w^*)$ is given as follows.*

$$\begin{aligned} w^* &= \zeta \tilde{\Sigma}^{-1}(\tilde{\mu} - r_t\tilde{m}1), \\ w^0 &= 1 - (w^*)^{\mathrm{T}}\tilde{m}1, \end{aligned} \tag{26}$$

*with*

$$\zeta := \frac{1}{B - Ar_t}.$$

*where $\tilde{\sigma}^{ij} := \sigma_t^{ij} + c_t^i c_t^j$ for $i, j = 1, 2, \cdots, n$,*

$$\tilde{\mu} := \begin{bmatrix} \mu_t^1 \\ \mu_t^2 \\ \vdots \\ \mu_t^n \end{bmatrix} - \frac{2}{3} \begin{bmatrix} c_t^1 \\ c_t^2 \\ \vdots \\ c_t^n \end{bmatrix}, \quad \tilde{\Sigma} := \begin{bmatrix} \tilde{\sigma}^{11} & \tilde{\sigma}^{12} & \cdots & \tilde{\sigma}^{1n} \\ \tilde{\sigma}^{21} & \tilde{\sigma}^{22} & \cdots & \tilde{\sigma}^{2n} \\ \vdots & \vdots & \ddots & \vdots \\ \tilde{\sigma}^{n1} & \tilde{\sigma}^{n2} & \cdots & \tilde{\sigma}^{nn} \end{bmatrix}, \quad \tilde{m}1 := \begin{bmatrix} 1 \\ 1 \\ \vdots \\ 1 \end{bmatrix},$$

*$A := \tilde{m}1^{\mathrm{T}}\tilde{\Sigma}^{-1}\tilde{m}1$, $B := \tilde{m}1^{\mathrm{T}}\tilde{\Sigma}^{-1}\tilde{\mu}$ and $C := \tilde{\mu}^{\mathrm{T}}\tilde{\Sigma}^{-1}\tilde{\mu}$.*

The solution $(\tilde{\rho}^{**}, \tilde{\mu}^{**})$ is called a *tangency portfolio*, and we can easily check that $(\tilde{\rho}^{**}, \tilde{\mu}^{**})$ maximizes the Sharpe ratio

$$\frac{\tilde{\mu}^{**} - r_t}{\sqrt{\tilde{\rho}^{**}}} \tag{27}$$

with respect to portfolios in the efficient frontier. Tangency portfolios are widely used in the financial market, and it is known that the Sharpe ratio is a measure of risk-adjusted performance of a trading strategy in portfolio theory (Pliska 1997).

## 5 A Numerical Example

Let $n = 3$. Let the interest rate of the bond $r_t = 0.04$. Take a mean and variance-covariance matrix of rate of return and fuzzy factors. From Theorem 1 we obtain the tangency portfolio $(\tilde{\rho}^{**}, \tilde{\mu}^{**}) = (0.0909028, 0.0486702)$ with the trading strategy $w^{**} = (w^0, w^1, w^2, w^3) = (0, 0.022250, 0.623316, 0.354434)$. Hence $\tilde{\mu}^{**} = 0.0486702$ is the mean of the rate of return with imprecision of data and $\tilde{\rho}^{**} = 0.0909028$ is the risk given by uncertainty, which is the sum of variance and the square of the mesurement of fuzziness, for the portfolio. For the tangency portfolio $(\tilde{\rho}^{**}, \tilde{\mu}^{**})$, we can easily calculate that the Sharpe ratio is $\frac{\tilde{\mu}^{**} - r_t}{\sqrt{\tilde{\rho}^{**}}} = 0.0287569$.

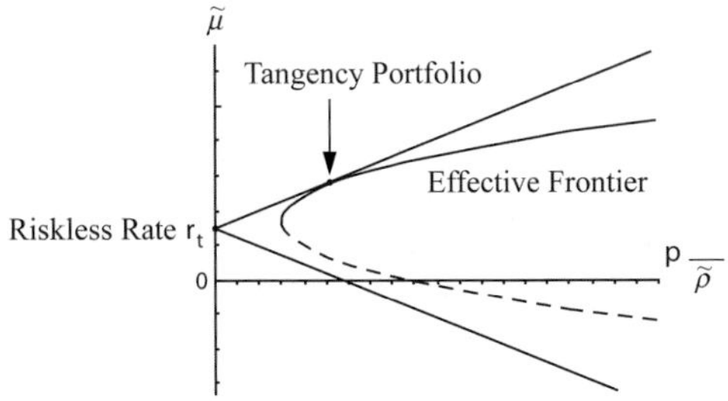


**Fig. 1.** Tangency portfolio

**Table 1.** The mean and variance-covariance matrix of rate of return and fuzzy factors

| Asset $i$ | $\mu_t^i$ |
|---|---|
| 1 | 0.05 |
| 2 | 0.07 |
| 3 | 0.06 |

| $(\sigma_t^i)^2, \sigma_t^{ij}$ | 1 | 2 | 3 |
|---|---|---|---|
| 1 | 0.40 | 0.03 | 0.02 |
| 2 | 0.03 | 0.20 | −0.06 |
| 3 | 0.02 | −0.06 | 0.30 |

| Asset $i$ | $c_t^i$ |
|---|---|
| 1 | 0.01 |
| 2 | 0.03 |
| 3 | 0.02 |

## References


1. Campos, L., Munoz, A.: A subjective approach for ranking fuzzy numbers. Fuzzy Sets and Systems **29** (1989) 145-153
2. Carlsson, C., Fullér, R.: On possibilistic mean value and variance of fuzzy numbers. Fuzzy Sets and Systems **122** (2001) 315-326
3. Feng, Y., Hu L., Su, H.: The variance and covariance of fuzzy random variables. Fuzzy Sets and Systems **120** (2001) 487-497
4. Fortemps, P., Roubens, M.: Ranking and defuzzification methods based on area compensation. Fuzzy Sets and Systems **82** (1996) 319-330
5. Goetshel, R., Voxman, W.: Elementary fuzzy calculus. Fuzzy Sets and Systems **18** (1986) 31-43
6. López-Díaz, M., Gil, M.A.: The $\lambda$-average value and the fuzzy expectation of a fuzzy random variable. Fuzzy Sets and Systems **99** (1998) 347-352
7. Inuiguchi, M., Tanino, T.: Portfolio selection under independent possibilistic information. Fuzzy Sets and Systems **115** (2000) 83-92
8. Kwakernaak, H.: Fuzzy random variables – I. Definitions and theorem. Inform. Sci. **15** (1978) 1-29
9. Markowits, H.: Mean-Variance Analysis in Portfolio Choice and Capital Markets. Blackwell Publ., Oxford (1990)
10. Pliska, S.R.: Introduction to Mathematical Finance: Discrete-Time Models. Blackwell Publ., New York (1997)
11. Puri, M.L., Ralescu, D.A.: Fuzzy random variables. J. Math. Anal. Appl. **114** (1986) 409-422

12. Steinbach, M.C.: Markowitz revisited: Mean-variance model in financial portfolio analysis SIAM Review **43** (2001) 31-85
13. Tanaka, H., Guo, P., Türksen, L.B.: Portfolio selection based on fuzzy probabilities and possibilistic distribution. Fuzzy Sets and Systems **111** (2000) 397-397
14. Yager, R.R.: A procedure for ordering fuzzy subsets of the unit interval. Information Sciences **24** (1981)143-161
15. Yoshida, Y., Yasuda, M., Nakagami, J., Kurano, M.: A discrete-time portfolio selection with uncertainty of stock prices, in: T.Bilgiç, B.DeBaets and O.Kaynac, eds., Fuzzy Sets and Systems - IFSA2003. Lecture Notes in Artificial Intelligence 2715, Springer (2003) 245-252
16. Yoshida, Y.: A mean estimation of fuzzy numbers by evaluation measures, in: M.Ch.Negoita, R.J.Howlett and L.C.Jain, eds., Knowledge-Based Intelligent Information and Engineering Systems, Part II. Lecture Notes in Artificial Intelligence 3214, Springer (2004) 1222-1229
17. Yoshida, Y.: Mean values, measurement of fuzziness and variance of fuzzy random variables for fuzzy optimization, Proceedings of SCIS & ISIS 2006. (2006) 2277-2282
18. Zadeh, L.A.: Fuzzy sets. Inform. and Control **8** (1965) 338-353

# Part IX

# Fuzzy Trees

# Weighted Pattern Trees: A Case Study with Customer Satisfaction Dataset

Zhiheng Huang[1], Masoud Nikravesh[1], Ben Azvine[2], and Tamás D. Gedeon[3]

[1] Electrical Engineering and Computer Science,
University of California at Berkeley, CA 94720, USA
{zhiheng,nikravesh}@cs.berkeley.edu
[2] Computational Intelligence Research Group, Intelligent Systems Research Center
BT Group Chief Technology Office, British Telecom
ben.azvine@bt.com
[3] Department of Computer Science,
The Australian National University, Canberra, ACT 0200, Australia
tom@cs.anu.edu.au

**Abstract.** A pattern tree [1] is a tree which propagates fuzzy terms using different fuzzy aggregations. Each pattern tree represents a structure for an output class in the sense that how the fuzzy terms aggregate to predict such a class. Unlike decision trees, pattern trees explicitly make use of t-norms (i.e., AND) and t-conorms (OR) to build trees, which is essential for applications requiring rules connected with t-conorms explicitly. Pattern trees can not only obtain high accuracy rates in classification applications, but also be robust to over-fitting. This paper further extends pattern trees approach by assigning certain weights to different trees, to reflect the nature that different trees may have different confidences. The concept of weighted pattern trees is important as it offers an option to trade off the complexity and performance of trees. In addition, it enhances the semantic meaning of pattern trees. The experiments on British Telecom (BT) customer satisfaction dataset show that weighted pattern trees can slightly outperform pattern trees, and both of them are slightly better than fuzzy decision trees in terms of prediction accuracy. In addition, the experiments show that (weighted) pattern trees are robust to over-fitting. Finally, a limitation of pattern trees as revealed via BT dataset analysis is discussed and the research direction is outlined.

## 1 Introduction

Most of the existing fuzzy rule induction methods including fuzzy decision trees [9] (the extension of the classic decision tree induction method by Quinlan [6]) focus on searching for rules which only use t-norm operators [7] such as the MIN and algebraic MIN. Disregarding of the t-conorms such as MAX and algebraic MAX is due to the fact that any rule using t-conorms can be represented by several rules which use t-norms only. This is certainly true and it is helpful to simplify the rule induction process by considering t-norms only. However, it may

P. Melin et al. (Eds.): IFSA 2007, LNAI 4529, pp. 395–406, 2007.
© Springer-Verlag Berlin Heidelberg 2007

fail to generate important rules in which fuzzy terms are explicitly connected with t-conorms. Research has been conducted to resolve this problem. For example, Kóczy, Vámos and Biró [3] have proposed fuzzy signatures to model the complex structures of data points using different aggregation operators including MIN, MAX, and average etc. Mendis, Gedeon and Kóczy [4] have investigated different aggregations in fuzzy signatures. Nikravesh [5] has presented evolutionary computation (EC) based multiple aggregator fuzzy decision trees.

Huang and Gedeon [1] have first introduced the concept of pattern trees and proposed a novel pattern tree induction method by means of similarity measures and different aggregations. This paper extends that work to assign certain weights to different pattern trees. As a result, it enhances the semantic meaning of pattern trees and makes them more comprehensible for users. The experiments on BT customer satisfaction dataset show that weighted pattern trees can slightly outperform pattern trees. In addition, this paper shows that pattern trees and weighted pattern trees perform more consistently than fuzzy decision trees. The former are capable of generating classifiers with good generality, while the latter can easily fall into the trap of over-fitting. In fact, weighted pattern trees with only two or three tree levels (depth of tree) are good enough for most experiments carried out in this paper. This provides a very transparent way to model real world applications.

The rest of the paper is arranged as follows: Section 2 provides the definitions for similarity, aggregations and pattern trees, and briefly outlines the pattern tree induction method. Readers may refer to [1][2] for detailed discussion. Section 3 suggests the concept of weighted pattern trees and shows how to use them for classification. Section 4 presents the experimental results over BT customer satisfaction dataset. Finally, Section 5 concludes the paper and points out further research work.

## 2 Definitions and Pattern Tree Induction

Let $A$ and $B$ be two fuzzy sets [10] defined on the universe of discourse $U$. The root mean square error (RMSE) of fuzzy sets $A$ and $B$ can be computed as

$$RMSE(A,B) = \sqrt{\frac{\sum_{j=1}^{m}(\mu_A(x_j) - \mu_B(x_j))^2}{m}}, \quad (1)$$

where $x_j$, $j = 1, \ldots, m$, are the crisp values discretized in the variable domain, and $\mu_A(x_j)$ and $\mu_B(x_j)$ are the fuzzy membership values of $x_j$ for $A$ and $B$. The RMSE based fuzzy set similarity can thus be defined as

$$S(A,B) = 1 - RMSE(A,B). \quad (2)$$

The larger the value of $S(A,B)$, the more similar $A$ and $B$ are. As $\mu_A(x_j), \mu_B(x_j) \in [0,1]$, $0 \leq S(A,B) \leq 1$ holds according to (1) and (2). Note that the pattern tree induction follows the same principle if alternative fuzzy set similarity definitions such as Jaccard are used.

Fuzzy aggregations are logic operators applied to fuzzy membership values or fuzzy sets. They have three sub-categories, namely t-norm, t-conorm, and averaging operators such as weighted averaging (WA) and ordered weighted averaging (OWA) [8].

Triangular norms were introduced by Schweizer and Sklar [7] to model distances in probabilistic metric spaces. In fuzzy sets theory, triangular norms (t-norm) and triangular conorms (t-conorm) are extensively used to model logical operators *and* and *or*. The basic t-norm and t-conorm pairs which operate on two fuzzy membership values $a$ and $b$, $a, b \in [0, 1]$ are shown in Table 1. Although

**Table 1.** Basic t-norms and t-conorms pairs

| Name | t-norm | t-conorm |
|---|---|---|
| MIN/MAX | $min\{a, b\} = a \wedge b$ | $max\{a, b\} = a \vee b$ |
| Algebraic AND/OR | $ab$ | $a + b - ab$ |
| Łukasiewicz | $max\{a + b - 1, 0\}$ | $min\{a + b, 1\}$ |
| EINSTEIN | $\frac{ab}{2-(a+b-ab)}$ | $\frac{a+b}{1+ab}$ |

the aggregations shown only apply to a pair of fuzzy values, they can apply to multiple fuzzy values as they retain associativity. The definition of WA and OWA are shown as follows:

**Definition 1.** *A WA operator of dimension $n$ is a mapping $E : \mathbb{R}^n \rightarrow \mathbb{R}$, that has an associated $n$-elements vector $w = (w_1, w_2, \ldots, w_n)^T$, $w_i \in [0, 1]$, $1 \leq i \leq n$, and $\sum_{i=1}^{n} w_i = 1$ so that $E(a_1, \ldots, a_n) = \sum_{j=1}^{n} w_j a_j$.*

**Definition 2.** *An OWA operator [8] of dimension $n$ is a mapping $F : \mathbb{R}^n \rightarrow \mathbb{R}$, that has an associated $n$-elements vector $w = (w_1, w_2, \ldots, w_n)^T$, $w_i \in [0, 1]$, $1 \leq i \leq n$, and $\sum_{i=1}^{n} w_i = 1$ so that $F(a_1, \ldots, a_n) = \sum_{j=1}^{n} w_j b_j$, where $b_j$ is the $j$th largest element of the collection $\{a_1, \ldots, a_n\}$.*

A fundamental difference of OWA from WA aggregation is that the former does not have a particular weight $w_i$ associated for an element, rather a weight is associated with a particular ordered position of the element.

A pattern tree is a tree which propagates fuzzy terms using different fuzzy aggregations. Each pattern tree represents a structure for an output class in the sense that how the fuzzy terms aggregate to predict such a class. The output class is located at the top as the root of this tree. The fuzzy terms of input variables are on different levels (except the top) of the tree. They use fuzzy aggregations to aggregate from the bottom to the top (root). Assume two fuzzy variables $A$ and $B$ each have two fuzzy linguistic terms $A_i$ and $B_i$, $i = \{1, 2\}$, and the task is to classify the data samples to either class $X$ or $Y$. Fig. 1 shows two example pattern trees, with one for class $X$ and the other for $Y$. It can be seen that pattern trees are built via the aggregation of fuzzy terms. For example, the pattern tree for $X$ is equivalent to fuzzy rule $(B_1 \wedge A_2) \vee A_1 \Rightarrow X$.

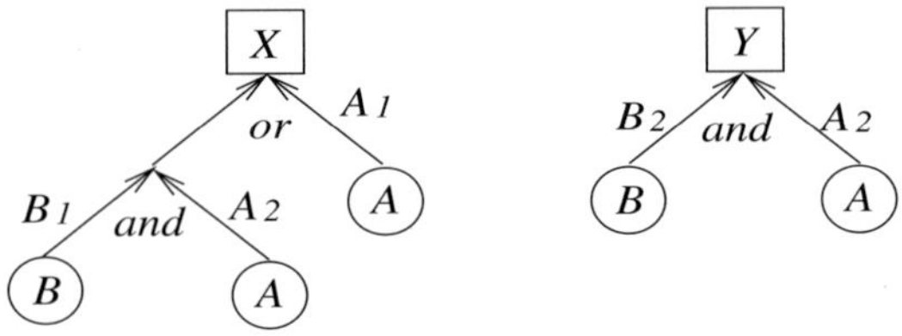


**Fig. 1.** Two example pattern trees

For a classification application which involves several output classes, the worked model should have as many pattern trees as the number of output classes, with each pattern tree representing one class. When a new data sample is tested over a pattern tree, it traverses from the bottom to the top and finishes with a truth value, indicating a degree to which this data sample belongs to the output class of this pattern tree. The output class with the maximal truth value is chosen as the prediction class. For example, consider that a fuzzy data $A_1 = 0.8$, $A_2 = 0.2$, $B_1 = 0$, and $B_2 = 1$ is given for classification. As the truth values of this data over pattern trees for class $X$ and $Y$ are 0.8 and 0.2 respectively, $X$ is chosen as the output class.

The pattern tree induction method as proposed in [1][2] is briefly outlined as follows. Readers may refer to [1][2] for detailed discussion. Without losing generality, assume a dataset has $n$ input variables $A_i$, $i = 1, 2, \ldots, n$ and one output variable $B$. Further assume that input variables have $m$ fuzzy linguistic terms denoted as $A_{ij}$, $i = 1, 2, \ldots, n$, and $j = 1, 2, \ldots, m$, and output variable has $k$ fuzzy or linguistic terms denoted as $B_j$, $j = 1, 2, \ldots, k$. That is, each data point is represented by a fuzzy membership value vector of dimension $(nm + k)$. The task is to build $k$ pattern trees for the $k$ output classes (fuzzy or linguistic terms).

The process of building a pattern tree, say for class $B_0$, is described as follows:

1. From fuzzy term set $\mathcal{S} = \{A_{ij}\}$, $i = 1, 2, \ldots, n$, and $j = 1, 2, \ldots, m$, choose a fuzzy linguistic term $A_{i'j'} \in \mathcal{S}$, which has the highest similarity to the output class $B_0$ as the initial tree. The fuzzy term set is updated as $\mathcal{S} = \mathcal{S} - A_{i'j'}$. The exclusion of fuzzy term $A_{i'j'}$ from $\mathcal{S}$ is to prevent $A_{i'j'}$ from being used more than once in the tree.
2. Try aggregating the current tree with all fuzzy linguistic terms at set $\mathcal{S}$ in turn with different aggregations. Grow the current tree using the term $A_{i'j'}$ from set $\mathcal{S}$ and aggregation which together lead to the highest similarity. The fuzzy term set is updated as $\mathcal{S} = \mathcal{S} - A_{i'j'}$.
3. Keep applying 2 until no fuzzy term and aggregation lead to a higher similarity than the current one.

The above actually presents the induction for *simple pattern trees*. Its extension, the *general pattern trees* induction [2], considers to aggregate not only fuzzy terms, but also other pattern trees. In general, simple pattern trees not only produce high prediction accuracy, but also preserve compact tree structures, while general pattern trees can produce even better accuracy, but as a compromise produce more complex tree structures. Subject to the particular demands (comprehensibility or performance), simple pattern trees and general pattern trees provide an highly effective methodology for real world applications.

## 3 Weighted Pattern Trees

The classification using pattern trees discussed in section 2 is based on the assumption that all pattern trees each have the same confidence on predicting a particular class, though it is not always the case in real world applications. *Weighted trees* are introduced to resolve this problem. For each tree, the similarity of such tree to the output class is served as a degree of confidence, to reflect how confident to use this tree to predict such a class. For example, if the two trees in Fig. 1 have similarities of 0.1 and 0.8 respectively, they can be called weighted pattern trees with weights of 0.1 and 0.8. The prediction using weighted pattern trees is the same as that using pattern trees, except that the final truth values are multiplied by the weights of trees. As an example, let's revise the classification problem in section 2; consider classifying the fuzzy data $A_1 = 0.8$, $A_2 = 0.2$, $B_1 = 0$, and $B_2 = 1$ over pattern trees (with weights of 0.1 and 0.8) in Fig. 1, its truth values over pattern trees for class $X$ and $Y$ change to 0.08 and 0.16 respectively, and $Y$ (rather than $X$) is therefore chosen as the output class. This reflects the fact that, if a tree has a low weight, even an input data has a high firing strength over such pattern tree, the prediction is not confident. Note that this example is merely used to show how weighted pattern trees work. In practice, a pattern tree with weight of 0.1 may not be trusted to predict a class.

The concept of weighted pattern trees is important. It offers an option to trade off the complexity and performance of pattern trees. The pattern tree building process can stop at very compact trees, if it detects that the similarities (weights) of such trees are already larger than a user pre-defined threshold. In addition, it enhances the comprehensibility of pattern trees. For example consider the construction of the pattern tree for class $Y$ in Fig. 1, assume that the tree growing from the primitive tree $B_2 \Rightarrow Y$ to $B_2 \wedge A_2 \Rightarrow Y$ leads to the weight increase from 0.6 to 0.8, this gradual change can be interpreted in a comprehensible way:

$$\text{IF } B = B_2 \text{ THEN it is } \textbf{possible} \text{ that } class = Y, \quad (3)$$

$$\text{IF } B = B_2 \text{ AND } A = A_2 \text{ THEN it is } \textbf{very possible} \text{ that } class = Y, \quad (4)$$

if users pre-define semantic ranges of weights, say *less possible*: $[0, 0.3)$, *possible*: $[0.3, 0.7)$, and *very possible*: $[0.7, 1]$. Thus, the graduate change of confidence of pattern trees can be monitored from the pattern tree induction process. This provides a very transparent way for fuzzy modeling.

# 4 Experimental Results

In this section, different variants of pattern trees, namely simple pattern trees, weighted simple pattern trees, pattern trees, and weighted pattern trees, are applied to a sample customer satisfaction dataset from BT. This dataset has a total of 26 input parameters representing ease of contact, problem understanding, service quality, repair time, and overall event handling. Among the input parameters, 6 are numerical parameters and the rest 20 are category ones, with the number of possible values being from 2 up to 17. The output parameter consists of 7 classes reflecting varying degrees of customer satisfaction.

The BT customer satisfaction dataset has 16698 data points in total. Let *ds*, *ds-odd* and *ds-even* be the datasets which contain the whole, the odd numbered, and the even numbered data points respectively. The number of data per class for these three datasets are shown in Table 2, with $ci$, $i = 0, \ldots, 6$ standing for class $i$. As can be seen, this dataset is not well balanced as the number of data per class varies significantly. The experiments of (weighted) pattern trees are carried out in three combinations of training-test datasets, namely, *odd-even*, *even-odd*, and *ds-ds*. In all experiments, a simple fuzzification method based on three evenly distributed trapezoidal membership functions for each numerical input parameter is used to transform the crisp values into fuzzy values. All aggregations as listed in Table 1 are allowed and the similarity measure as shown in (2) is used.

**Table 2.** Number of data per class for ds, ds-odd and ds-even datasets

| | c0 | c1 | c2 | c3 | c4 | c5 | c6 |
|---|---|---|---|---|---|---|---|
| ds | 1895 | 7289 | 4027 | 382 | 1361 | 853 | 891 |
| ds-odd | 949 | 3659 | 1990 | 197 | 660 | 448 | 446 |
| ds-even | 946 | 3630 | 2037 | 185 | 701 | 405 | 445 |

## 4.1 Prediction Accuracy and Overfitting

The prediction accuracy and rule number of the fuzzy decision trees (FDT) with respect to the number of data points per leaf node (used as criteria to terminate the training), over different combinations of training-test sets are shown in Fig. 2. It reveals that in general the larger number of data points per leaf node, the more compact of the decision trees would be, thus leading to more general trees. The prediction accuracy of pattern trees (PT) and weighted pattern trees (WPT) with respect to different tree levels, over different combinations of training-test sets is shown in Fig. 3. It reveals that (simple) pattern trees maintain good generality even their structure becomes complex.

The experiments show that weighted pattern trees and pattern trees perform roughly the same. In fact, the former slightly outperform the latter. Table 3 shows the highest prediction accuracy of fuzzy decision trees, (weighted) simple pattern trees and (weighted) pattern trees over different combinations of

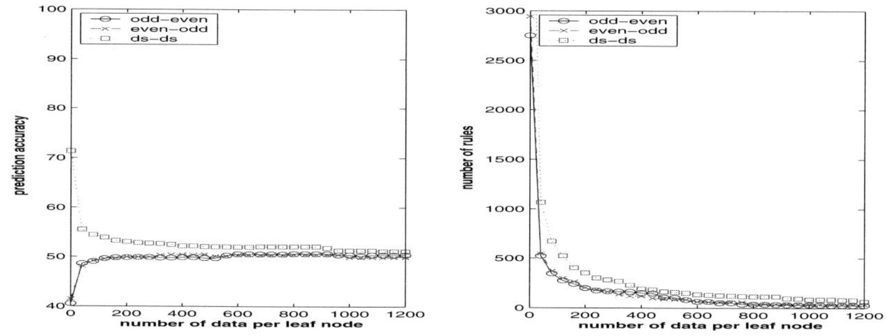


**Fig. 2.** Prediction accuracy and rule number of fuzzy decision trees with different number of data points per leaf node

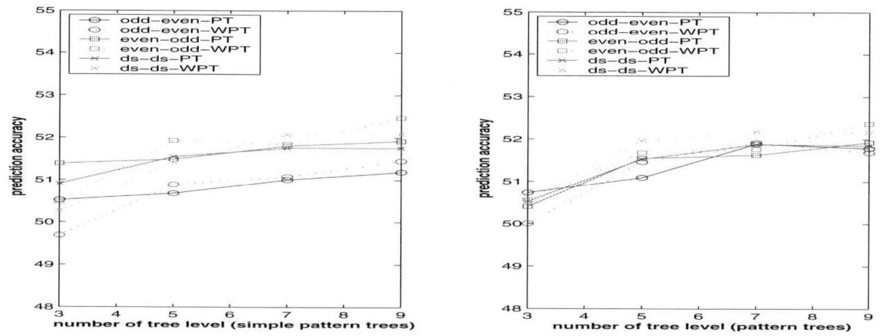


**Fig. 3.** Prediction accuracy of pattern trees and weighted pattern trees with different tree levels

**Table 3.** Highest prediction accuracy of fuzzy decision tree, pattern trees, and weighted pattern trees

| | FDT | SimPT | | PT | |
|---|---|---|---|---|---|
| | | no weight | weight | no weight | weight |
| odd-even | 50.62% | 51.19% | 51.45% | 51.89% | 51.92% |
| even-odd | 50.47% | 51.92% | 52.47% | 51.93% | 52.37% |
| ds-ds | 71.36% | 51.82% | 52.09% | 51.88% | 52.18% |

training-test sets. Both weighted and un-weighed pattern trees can obtain higher prediction accuracy than fuzzy decision trees in *odd-even* and *even-odd* combinations. However, if considering *ds-ds* combination, fuzzy decision trees perform much better. This just reflects the overfitting of fuzzy decision trees, since fuzzy decision trees generate large differences in classification accuracy between the *odd-even*, *even-odd* combinations and *ds-ds* one. The reason is that decision tree induction considers only a portion of the whole training dataset in choosing the branches at low levels of trees. The lack of using the whole training dataset inevitably prevents the method finding generalized tree structures for all the dataset. In contrast, pattern trees make use of the whole data in building each level of the tree, which ensures the tree to keep good generality for classifications. Therefore, even complex pattern trees do not suffer from over-fitting.

In addition, the experiments show that (weighted) pattern trees tend to converge to a accuracy rate when the number of tree level becomes large. It has no trend of overfitting. This property is essential to ensure a stable, compact and effective fuzzy model for the problem at hand. In fact, (weighted) pattern trees with two or three level perform very well for all conducted experiments. That means, pattern trees which consist of maximal $2^3 = 8$ leaf nodes can perform well, in contrast to tens, or even hundred rules used in fuzzy decision trees. This provides a superb solution to achieve a highly effective as well as compact fuzzy model.

### 4.2 Approximate Accuracy

Section 4.1 presented the prediction accuracy of trees in a very strict way. That is, if and only if a data is predicted exactly as its class, this prediction is counted as a correct one. In other words, there is no distinction between "close" errors and "gross" errors. In BT customer dataset, this distinction is necessary as it reflects how far the prediction is away from the actual class. It is much worse if a data of class 0 is mis-predicted to class 5 rather than to class 1. To resolve this problem, three accuracy estimations, namely *accuracy 1*, *accuracy 2*, and *accuracy 3* are employed to estimate prediction accuracy which has no tolerance (the same as the one used in Section 4.1), tolerance of adjacent mis-prediction, and tolerance of mis-prediction within two closest neighbor classes in either direction, respectively. For example in the BT dataset, the mis-prediction of a class 0 data to class 2 is still counted as a correct prediction in the estimation of *accuracy 3*, although it is not counted in either *accuracy 1* or *accuracy 2*.

Table 4 shows the highest prediction accuracy of fuzzy decision trees, (weighted) simple pattern trees and (weighted) pattern trees over odd-even combination of training-test sets (the results on even-odd and ds-ds combinations are similar and thus omitted). Both weighted and unweighted pattern trees can obtain higher prediction accuracy than fuzzy decision trees in estimation of accuracy 1 and 2. In estimation of accuracy 3, weighted pattern trees perform the best, and fuzzy decision trees outperform unweighted pattern trees. Generally, accuracy 2

**Table 4.** Highest prediction accuracy of fuzzy decision trees, pattern trees, and weighted pattern trees over odd-even training-test combination

| | FDT | SimPT | | PT | |
|---|---|---|---|---|---|
| | | no weight | weight | no weight | weight |
| accuracy 1 | 50.62% | 51.19% | 51.45% | 51.89% | 51.92% |
| accuracy 2 | 84.02% | 84.08% | 84.68% | 84.44% | 84.82% |
| accuracy 3 | 92.13% | 91.74% | 92.70% | 91.85% | 92.29% |

and 3 are consistent with accuracy 1. Pattern trees with a high value of accuracy 1 usually have high values of accuracy 2 and 3. This table also shows that both fuzzy decision trees and pattern trees can obtain over 80% prediction accuracy if the closest error can be tolerated.

## 4.3 Interpretation of Pattern Trees

Each pattern tree can be interpreted as a general rule. Considering building level 5 simple pattern trees using odd dataset, 7 simple pattern trees can be obtained, with each representing one output class. Fig. 4 shows the tree for class 0. The ellipses are the input parameters and the rectangle is the output class 0. Over each branch, $i$ and $Fi$, $i = 0, \ldots,$ are category values and fuzzy terms associated with each input parameter. All aggregators as shown in Table 1 are allowed to be used in pattern trees. For example, *A_AND* is algebraic AND, and *WA_0.84* is weighted average with weight vector $w = (0.84, 0.16)$.

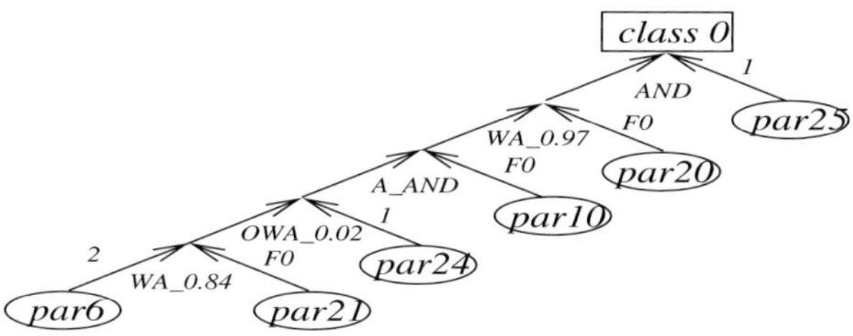


**Fig. 4.** Pattern tree for class 0 using *odd* dataset

Fig. 4 roughly indicates that one example combination yielding highly satisfied customers are: no call re-routing, fast fault reporting time, high technician competence, being well-informed through the repair process, and high satisfaction with company/product in general. Here, we say roughly, as we use different aggregations such as weighted average (WA), ordered weighted average (OWA), algebraic and (A_AND) etc. rather than simple AND.

These 7 pattern trees obtains an accuracy of 51.46%. In particular, the confusion table is shown in Table 5, where $SA$ and $SP$ are number of data for actual and predicted classes respectively.

## 4.4 Limitation

It is a little strange that no prediction is made to *c0* for all test data. From table 5, it can be seen that nearly all data (884 out of 946 in fact) with class 0 are mis-classified to class 1. A first intuition is to raise the weight of pattern tree for class 0. However, this does not work; the raise does not only lead to the data of class 0 to be classified correctly, but also lead to the majority of data of class 1 to be classified as class 0. Considering that there are 3630 data of class 1 and

**Table 5.** Confusion table for pattern tree prediction using odd-even combination

| | | Prediction | | | | | | | |
|---|---|---|---|---|---|---|---|---|---|
| | | c0 | c1 | c2 | c3 | c4 | c5 | c6 | *SA* |
| Actual | c0 | 0 | 884 | 51 | 0 | 3 | 2 | 6 | 946 |
| | c1 | 0 | 3283 | 309 | 0 | 10 | 15 | 13 | 3630 |
| | c2 | 0 | 1122 | 751 | 1 | 53 | 72 | 38 | 2037 |
| | c3 | 0 | 59 | 94 | 0 | 6 | 20 | 6 | 185 |
| | c4 | 0 | 122 | 395 | 1 | 30 | 104 | 49 | 701 |
| | c5 | 0 | 50 | 142 | 0 | 23 | 120 | 70 | 405 |
| | c6 | 0 | 35 | 129 | 0 | 37 | 131 | 113 | 445 |
| | *SP* | 0 | 5555 | 1871 | 2 | 162 | 464 | 295 | 8349 |

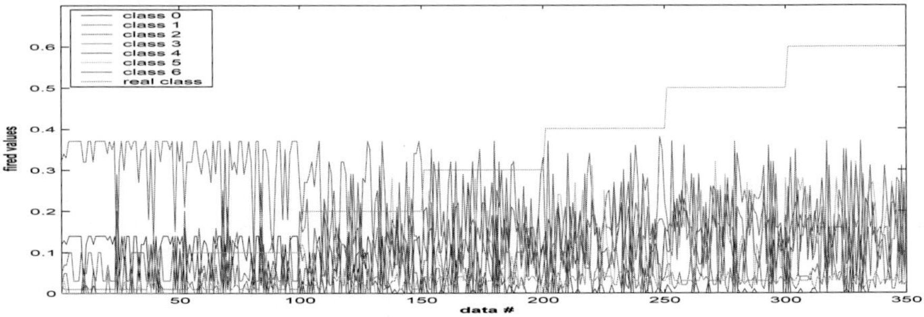


**Fig. 5.** Fired values of first 50 data points per class in *even* dataset over pattern trees constructed from *odd* dataset

only 946 data of class 0 in even dataset, the raise of weight for class 0 tree would therefore cause more mis-classifications. This can be seen in Fig. 5, where the fired values of first 50 data points per class in *even* dataset over pattern trees constructed from *odd* dataset are shown. The real class line indicates the real classes of the data; for example, data numbered from 0 to 49 have class 0, and those from 50 to 99 have class 1.

The phenomena of no prediction on particular classes also occurs in fuzzy decision trees. Considering the highest accuracy of 50.62% which fuzzy decision trees can obtain over odd-even combination, no data is predicted to *c3*, *c4* or *c5*, due to the small fraction of data points in those classes.

**Table 6.** Confusion table for prediction of both fuzzy decision trees and pattern trees using new training and test datasets

| | | Prediction | | |
|---|---|---|---|---|
| | | c0 | c1 | *SA* |
| Actual | c0 | 0 | 884 | 884 |
| | c1 | 1 | 3282 | 3283 |
| | *SP* | 1 | 4166 | 4167 |

An interesting experiment is carried out trying to improve the prediction accuracy for class 0 in Table 5. The data of classes 0 and 1 in odd dataset are selected as a new training dataset, and the data which are of classes 0 and 1 in even dataset and are classified as class 1 in Table 5 are selected as a new test dataset. Both fuzzy decision trees and pattern trees are applied to the new training data and tested over the new test data. Surprisingly, they obtain the same highest accuracy of 78.76%. Table 6 shows the confusion table, which only has one data predicted as class 0 (and it is wrong actually). It can be concluded that the data of class 0 and class 1 can not be separated properly by either fuzzy decision trees or pattern trees.

## 5 Conclusions

This paper further extends pattern trees approach by assigning certain weights to different trees. The concept of weighted pattern trees is important as it not only offers an option to trade off the complexity and performance of trees, but also enhances the semantics of pattern trees.

The experiments on British Telecom customer satisfaction dataset show that weighted pattern trees can slightly outperform pattern trees, and both of them are slightly better than fuzzy decision trees in terms of prediction accuracy. In addition, the experiments show that (weighted) pattern trees are robust to overfitting. In practice, weighted pattern trees with only two or three tree levels are good enough for most experiments carried out in this paper. This of course provides a very transparent way to model the problems at hand.

Further research on assignment of weights to pattern trees is necessary. The current version simply makes use of similarity measures as weights. More sophisticated assignment may be more suitable and can therefore lead to higher accuracy.

## Acknowledgment

The authors would like to thank Marcus Thint for his helpful discussions on the work presented here.

## References

1. Huang, Z. H. and Gedeon, T. D., "Pattern trees," *IEEE International Conference on Fuzzy Systems*, pp. 1784 - 1791, 2006.
2. Huang, Z. H., Gedeon, T. D. and Nikravesh, M., "Pattern trees," submitted to *IEEE Transactions on Fuzzy Systems*, 2006.
3. Kóczy, L. T., Vámos, T. and Biró, G., "Fuzzy signatures," *EUROFUSE-SIC*, pp. 210 – 217, 1999.
4. Mendis, B. S. U., Gedeon T. D., and Kóczy, L. T., "Investigation of aggregation in fuzzy signatures," *3rd International Conference on Computational Intelligence, Robotics and Autonomous Systems*, 2005.

5. Nikravesh, M., "Soft computing for perception-based decision processing and analysis: web-based BISC-DSS," *Studies in Fuzziness and Soft Computing*, vol. 164, pp. 93 -188, Springer Berlin/Heidelberg, 2005.
6. Quinlan, J. R., "Decision trees and decision making," *IEEE Transactions on Systems, Man, and Cybernetics*, vol. 20, no. 2, pp. 339 – 346, 1994.
7. Schweizer, B. and Sklar, A., "Associative functions and abstract semigroups," *Publ. Math. Debrecen*, vol. 10, pp. 69 – 81, 1963.
8. Yager, R. R., "On ordered weighted averaging aggregation operators in multicritera decison making," *IEEE Transactions on Systems, Man and Cybernetics*, vol. 18, pp. 183 – 190, 1988.
9. Yuan, Y. and M. J. Shaw, "Induction of fuzzy decision trees," *Fuzzy Sets and Systems*, vol. 69, no. 2, pp. 125 – 139, 1995.
10. Zadeh, L. A., "Fuzzy sets," *Information and Control*, vol. 8, pp. 338 – 353, 1965.

# Fuzziness and Performance: An Empirical Study with Linguistic Decision Trees

Zengchang Qin[1] and Jonathan Lawry[2]

[1] Berkeley Initiative in Soft Computing
Computer Science Division, EECS Department
University of California, Berkeley CA 94720, USA
zqin@cs.berkeley.edu
[2] Artificial Intelligence Group
Engineering Mathematics Department
University of Bristol, BS8 1TR, UK
j.lawry@bris.ac.uk

**Abstract.** Generally, there are two main streams of theories for studying uncertainties. One is probability theory and the other is fuzzy set theory. One of the basic ideas of fuzzy set theory is how to define and interpret membership functions. In this paper, we will study tree-structured data mining model based on a new interpretation of fuzzy theory. In this new theory, fuzzy labels will be used for modelling. The membership function is interpreted as appropriateness degrees for using labels to describe a fuzzy concept. Each fuzzy concept is modelled by a distribution on the appropriate fuzzy label sets. Previous work has shown that the new model outperforms some well-known data mining models such as Naive Bayes and Decision trees. However, the fuzzy labels used in previous works were predefined. We are interested in study the influences on the performance by using fuzzy labels with different degrees of overlapping. We test a series of UCI datasets and the results show that the performance of the model increased almost monotonically with the increase of the overlapping between fuzzy labels. For this empirical study with the LDT model, we can conclude that more fuzziness implies better performance.

## 1 Introduction

Uncertainty is a nature of our world. Generally, there are two main streams for modelling uncertainties. One is probability theory and the other is fuzzy set theory. Since the first paper published by Zadeh in 1965 [9], fuzzy logic has become an important branch in artificial intelligence as well as some engineering areas such as intelligent control. One of the basic ideas of fuzzy set theory is how to define and interpret membership functions. There are a few different interpretation of fuzziness [8]. In this paper, we will study tree-structured data mining model based on a new interpretation of fuzzy theory. In this new theory, which is referred to as Label Semantics [2], fuzzy labels will be used for modelling.

P. Melin et al. (Eds.): IFSA 2007, LNAI 4529, pp. 407–416, 2007.
© Springer-Verlag Berlin Heidelberg 2007

One inherent disadvantage of classical decision trees is that the model is sensitive to noise. As pointed out by Quinlan [6]: "the results of (traditional) decision trees are categorical and so do not convey potential uncertainties in classification. Small changes in the attribute values of a case being classified may result in sudden and inappropriate changes to the assigned class. Missing or imprecise information may apparently prevent a case being classified at all". This noise is not only due to the lack of precision or errors in measured features but is often present in the model itself since the available features may not be sufficient to provide a complete model of the system. To overcome this problem, some probabilistic or soft decision trees were proposed. The first fuzzy decision tree reference can be back to in 1977. Since then, There are more than forty references on either on fuzzy tree learning or fuzzy rule learning. All these algorithms highlight advantage of using fuzzy rules for classification applications is to maintain transparency as well as a high accuracy rate. According to Olaru and Wehenkel [3]: these fuzzy decision tree algorithms can be roughly divided into two categories:

1. Enable the use of decision trees to manage fuzzy information in the forms of fuzzy inputs, fuzzy classes or fuzzy rules.
2. Using fuzzy logic to improve their predictive accuracy.

Previous work by Lawry and Qin [4] has shown that the LDT model outperforms some well-known data mining models such as Naive Bayes and classical decision trees such as C4.5 [7]. It also can handle fuzzy information and has better transparency comparing to other models. However, the fuzzy labels used in previous works were predefined under some assumptions. We are interested in study the influences of different degrees of overlapping between neighboring fuzzy labels.

## 2 Linguistic Decision Trees

Linguistic decision tree (LDT) [4] is a tree-structured classification model based on label semantics. The information heuristics used for building the tree are modified from Quinlan's ID3 [5] in accordance with label semantics. The nodes of a LDT are linguistic descriptions of variables and leaves are sets of appropriate labels. In such decision trees, the probability estimates for branches across the whole tree is used for classification, instead of the majority class of the single branch into which the examples fall. Linguistic expressions such as *small*, *medium* and *large* are used to learn from data and build a linguistic decision tree guided by information based heuristics. For each branch, instead of labeling it with a certain class (such as positive or negative in binary classification) the probability of members of this branch belonging to a particular class is evaluated from a given training dataset. Unlabeled data is then classified by using probability estimation of classes across the whole decision tree.

### 2.1 Introduction to Label Semantics

Label semantics is a methodology of using linguistic expressions or fuzzy labels to describe numerical values. For a variable $x$ into a domain of discourse $\Omega$ we identify a finite set of fuzzy labels $\mathcal{L} = \{L_1, \cdots, L_n\}$ with which to label the values of $x$. Then for a specific value $x \in \Omega$ an individual $I$ identifies a subset of $\mathcal{L}$, denoted $D_x^I$ to stand for the description of $x$ given by $I$, as the set of labels with which it is appropriate to label $x$. If we allow $I$ to vary across a population $V$ with prior distribution $P_V$, then $D_x^I$ will also vary and generate a random set denoted $D_x$ into the power set of $\mathcal{L}$ denoted by $\mathcal{S}$. We can view the random set $D_x$ as a description of the variable $x$ in terms of the labels in $\mathcal{L}$. The frequency of occurrence of a particular label, say $S$, for $D_x$ across the population then gives a distribution on $D_x$ referred to as a mass assignment on labels. More formally,

**Definition 1 (*Label Description*).** *For $x \in \Omega$ the label description of $x$ is a random set from $V$ into the power set of $\mathcal{L}$, denoted $D_x$, with associated distribution $m_x$, which is referred to as mass assignment:*

$$\forall S \subseteq \mathcal{L}, \quad m_x(S) = P_V(\{I \in V | D_x^I = S\}) \tag{1}$$

*where $m_x(S)$ is called associated mass of $S$ and $\sum_{S \subseteq \mathcal{L}} m_x(S) = 1$. Intuitively mass assignment is a distribution on appropriate label sets and $m_x(S)$ quantifies the evidence that $S$ is the set of appropriate labels for $x$.*

In this framework, *appropriateness degrees* are used to evaluate how appropriate a label is for describing a particular value of variable $x$. Simply, given a particular value $\alpha$ of variable $x$, the appropriateness degree for labeling this value with the label $L$, which is defined by fuzzy set $F$, is the membership value of $\alpha$ in $F$. The reason we use the new term 'appropriateness degrees' is partly because it more accurately reflects the underlying semantics and partly to highlight the

**Algorithm 1.** Linguistic translation

**input** : Given a database $\mathcal{D} = \{\langle x_1(i), \cdots, x_n(i)\rangle | i = 1, \cdots, |\mathcal{D}|\}$ with associated classes $\mathcal{C} = \{C_1, \cdots, C_{|\mathcal{C}|}\}$
**output**: Linguistic dataset $\mathcal{LD}$

**for** $j \leftarrow 1$ **to** $n$ **do**
  **foreach** $x_j$ **do** : Cover the universe of $x_j$ with $N_F$ trapezoidal fuzzy sets with 50% overlap. ;
  **for** $i \leftarrow 1$ **to** $|\mathcal{D}|$ **do**
    **foreach** *Data element* $x_j(i)$ **do** ;
    Read appropriateness degrees for $x_j(i)$ from corresponding fuzzy set. ;
    Calculating corresponding mass assignments:
    $\mathcal{LD}_{i,j} = \langle m_{x(i)}(F_j^1), \cdots, m_{x(i)}(F_j^{h_j})\rangle$ on focal elements from appropriateness degrees. ;
Save dataset $\mathcal{LD}$ where $\mathcal{LD} = \{\mathcal{LD}_{i,j} | i = 1, \cdots, |\mathcal{D}|, j = 1, \cdots, n\}$

quite distinct calculus based on this framework [2]. This definition provides a relationship between mass assignments and appropriateness degrees.

**Definition 2** (***Appropriateness Degrees***)

$$\forall x \in \Omega,\ \forall L \in \mathcal{L} \quad \mu_L(x) = \sum_{S \subseteq \mathcal{L}: L \in S} m_x(S)$$

Based on the underlying semantics, we can translate a set of numerical data into a set of mass assignments on appropriate labels based on the reverse of definition 2 under the following assumptions: consonance mapping, full fuzzy covering and 50% overlapping. These assumptions are fully described in [4] and justified in [2]. These assumptions guarantee that there is unique mapping from appropriate degrees to mass assignments on labels. For example, given $\mu_{middleAged}(30) = 0.3$ and $\mu_{young}(30) = 1$ which are the memberships of being *middleAged* and *young* given a value of 30 (a person's age). The corresponding mass assignment is: $m_{30} = \{young, middleAged\} : 0.3, \{young\} : 0.7$ (More details of mass assignment calculations are available in [2] and [4]). Given a database, we can translate each data element into its mass assignment representation. This process is called *linguistic translation*. The pseudo-code is given in algorithm 1.

### 2.2 Degrees of Overlapping

Through linguistic translation, all numerical data can be represented as mass assignments based on a predefined fuzzy discretization method. In this paper, unless otherwise stated, we will use a percentile-based (or equal points) discretization. The idea is to cover approximately the same number of data points for each fuzzy label. The justification for using this discretization method is given in [4].

Basically, fuzzy discretization provides an interpretation between numerical data and their corresponding linguistic data based on label semantics. We may notice that different fuzzy discretization (fuzzification of a continuous universe) may result in different linguistic data. We introduce a new parameter $PT$ by which to measure the degrees of overlapping between fuzzy labels. As we can see from figure 1, given two fuzzy labels $F$ and $G$, $m$ is the distance between the weighting centers of a fuzzy labels to the meeting point of their membership functions. $a$ is actually the length of the overlapping area. $PT$ is calculated as follows:

$$PT = a/2m \tag{2}$$

$PT = 0.5$ represents 50% of overlapping between each two neighboring fuzzy labels (e.g., figure 1-A). $PT = 0$ represents no overlapping at all (figure 1-C), i.e., the labels are discrete but not fuzzy. Figure 1-B shows a situation that the degree of overlapping is between 0 and 0.5. Figure 1-D also shows the linear relation of parameter $a$ and $PT$.

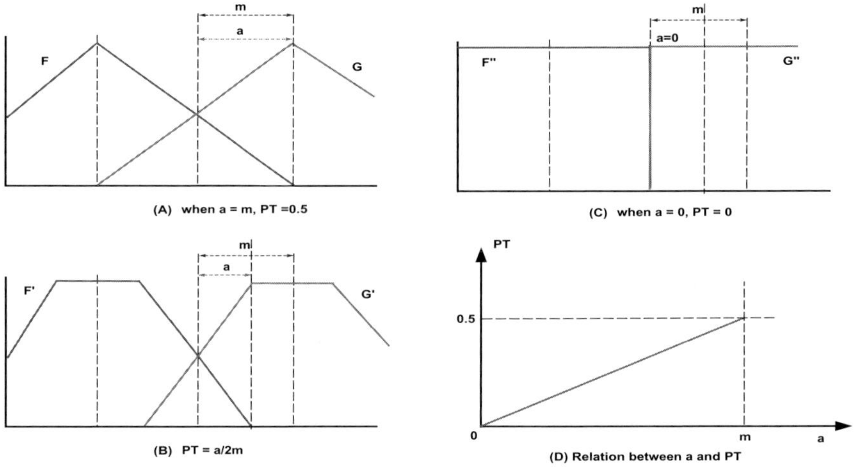


**Fig. 1.** A schematic illustration of calculating the overlap parameter $PT$ given different degrees of overlaps

## 2.3 Classification

Given a database of which each instance is labeled by one of the classes: $\mathcal{C} = \{C_1, \cdots, C_{|\mathcal{C}|}\}$. A linguistic decision tree with $S$ branches built from this database can be defined as follows:

$$T = \{\langle B_1, P(C_1|B_1), \cdots, P(C_{|\mathcal{C}|}|B_1)\rangle, \cdots \langle B_S, P(C_1|B_S), \cdots, P(C_{|\mathcal{C}|}|B_S)\rangle\}$$

where $P(C_k|B)$ is the probability of class $C_k$ given a branch $B$. A branch $B$ with $d$ nodes (i.e., the length of $B$ is $d$) is defined as: $B = \langle F_1, \cdots, F_d \rangle$, where $d \leq n$ and $F_j$ are *focal elements* of attribute $j$. Focal elements are the appropriate label sets with non-zero masses [2]. For example, consider the branch: $\langle\langle\{small_1\}, \{medium_2, large_2\}\rangle, 0.3, 0.7\rangle$. This means the probability of class $C_1$ is 0.3 and $C_2$ is 0.7 given attribute 1 can only be described as *small* and attribute 2 can be described as both *medium* and *large*.

Given a training set $\mathcal{D} = \{\mathbf{x}_1, \cdots, \mathbf{x}_N\}$ where each instance $\mathbf{x}$ has $n$ attributes: $\langle x_1, \cdots, x_n\rangle$. The class probability of $C_k$ given a particular branch $B$ is calculated by the proportion of data covered by this branch and belonging to $C_k$ to all the data covered by this branch:

$$P(C_k|B) = \frac{\sum_{i \in \mathcal{D}_k} P(B|\mathbf{x}_i)}{\sum_{i \in \mathcal{D}} P(B|\mathbf{x}_i)} \tag{3}$$

where $\mathcal{D}_k = \sum_{i:\mathbf{x}_i \rightarrow C_k} \mathbf{x}_i$ is the subset consisting of instances which belong to class $k$. The probability of a branch $B$ given $\mathbf{x}$ can be regarded as the proportion of the data $\mathbf{x}$ covered by branch $B$ and it is evaluated by:

$$P(B|\mathbf{x}) = \prod_{j=1}^{d} m_{x_j}(F_j) \tag{4}$$

where $m_{x_j}(F_j)$ for $j = 1, \cdots, d$ are mass assignments of the single data element $x_j$. Now consider classifying an unlabelled instance in the form of $\mathbf{y} = \langle y_1, \cdots, y_n \rangle$ from the test set. First we apply linguistic translation to $\mathbf{y}$ based on the fuzzy covering of the training data $\mathcal{D}$. According to the Jeffrey's rule the probabilities of class $C_k$ given a LDT with $S$ branches are evaluated as follows:

$$P(C_k|\mathbf{y}) = \sum_{s=1}^{S} P(C_k|B_s)P(B_s|\mathbf{y}) \tag{5}$$

where $P(C_k|B_s)$ and $P(B_s|\mathbf{x})$ are evaluated based on equations 3 and 4.

**Algorithm 2.** Decision Tree Learning

**input** : $\mathcal{LD}$: Linguistic dataset obtained from Algorithm 1.
**output**: $LDT$: Linguistic Decision Tree

Set a maximum depth $M_{dep}$ and a threshold probability $T$.
**for** $l \leftarrow 0$ **to** $M_{dep}$ **do**
    $\mathcal{B} \leftarrow \emptyset$ when $l = 0$
    The set of branches of LDT at depth $l$ is $\mathcal{B}_l = \{B_1, \cdots, B_{|\mathcal{B}_l|}\}$
    **for** $v \leftarrow 1$ **to** $|\mathcal{B}|$ **do**
        **foreach** $B_v$ **do** :
        **for** $t \leftarrow 1$ **to** $|\mathcal{C}|$ **do**
            **foreach** $t$ **do** Calculating conditional probabilities:
            $P(C_t|B_v) = \sum_{i \in \mathcal{D}_t} P(B_v|\mathbf{x}_i) / \sum_{i \in \mathcal{D}} P(B_v|\mathbf{x}_i)$
            **if** $P(C_t|B_v) \geq T$ **then**
                break (step out the loop)
        **if** $\exists\ x_j$: *$x_j$ is free attribute* **then**
            **foreach** $x_j$ **do** : Calculate: $IG(B_v, x_j) = E(B_v) - EE(B_v, x_j)$
            $IG_{max}(B_v) = \max_{x_j}[IG(B_v, x_j)]$
            Expanding $B_v$ with $x_{max}$ where $x_{max}$ is the free attribute we can obtain the maximum $IG$ value $IG_{max}$.
            $\mathcal{B}'_v \leftarrow \bigcup_{F_j \in \mathcal{F}_j} \{B_v \cup F_j\}$.
        **else**
            exit;
    $\mathcal{B}_{l+1} \leftarrow \bigcup_{r=1}^{s} \mathcal{B}'_r$.
$LDT = \mathcal{B}$

### 2.4 LID3 Algorithm

Linguistic ID3 (LID3) is the learning algorithm proposed for building the linguistic decision tree. Similar to the ID3 algorithm [5], search is guided by an information based heuristic, but the information measurements of a LDT are modified in accordance with label semantics. The measure of information defined for a branch $B$ and can be viewed as an extension of the entropy measure used in ID3. The branch entropy of a branch $B$ is given by

$$E(B) = -\sum_{k=1}^{|\mathcal{C}|} P(C_k|B) \log_2(P(C_k|B)) \tag{6}$$

where $|\mathcal{C}|$ is the number of classes. Now, given a particular branch $B$ suppose we want to expand it with the attribute $x_j$. The evaluation of this attribute will be given based on the expected entropy defined as follows:

$$EE(B, x_j) = E[E(x_j|B)] = \sum_{F_j \in \mathcal{F}_j} E(B \cup F_j) P(F_j|B) \tag{7}$$

where $B \cup F_j$ represents the new branch obtained by appending the focal element $F_j$ to the end of branch $B$. The probability of $F_j$ given $B$ can be calculated as follows:

$$P(F_j|B) = \frac{\sum_{i \in \mathcal{D}} (B \cup F_j|\mathbf{x}_i)}{\sum_{i \in \mathcal{D}} (B|\mathbf{x}_i)} \tag{8}$$

We can now define the *Information Gain (IG)* obtained by expanding branch $B$ with attribute $x_j$ as:

$$IG(B, x_j) = E(B) - EE(B, x_j) \tag{9}$$

The pseudo-code of the LID3 algorithm are shown in Algorithm 2.

**Table 1.** Descriptions of the datasets for experiments selected from the UCI machine learning repository [1]

| Dataset | Classes | Size | Attributes | Dataset | Classes | Size | Attributes |
|---|---|---|---|---|---|---|---|
| **Balance** | 3 | 625 | 4 | **Breast-cancer** | 2 | 286 | 9 |
| **Ecoli** | 8 | 336 | 8 | **Glass** | 6 | 214 | 9 |
| **Heart-C** | 2 | 303 | 13 | **Heart-S** | 2 | 270 | 13 |
| **Heptitis** | 2 | 155 | 19 | **Iris** | 3 | 150 | 4 |
| **Liver** | 2 | 345 | 6 | **Pima** | 2 | 768 | 8 |
| **Wcancer** | 2 | 699 | 9 | **Wine** | 3 | 178 | 14 |

## 3 Experiments

In this section, we investigate the influences of overlapping degrees on the accuracy by some empirical studies. First of all, we need to specify the parameter settings for the LDT model. In the following experiments, we use 3 trapezoidal fuzzy sets for discretization (i.e., Alg. 1 line 2: $N_F = 3$). Probability threshold $T = 1$ (Alg. 2 line 1) and we set $M_{dep} = n$ in order to develop a complete LDT

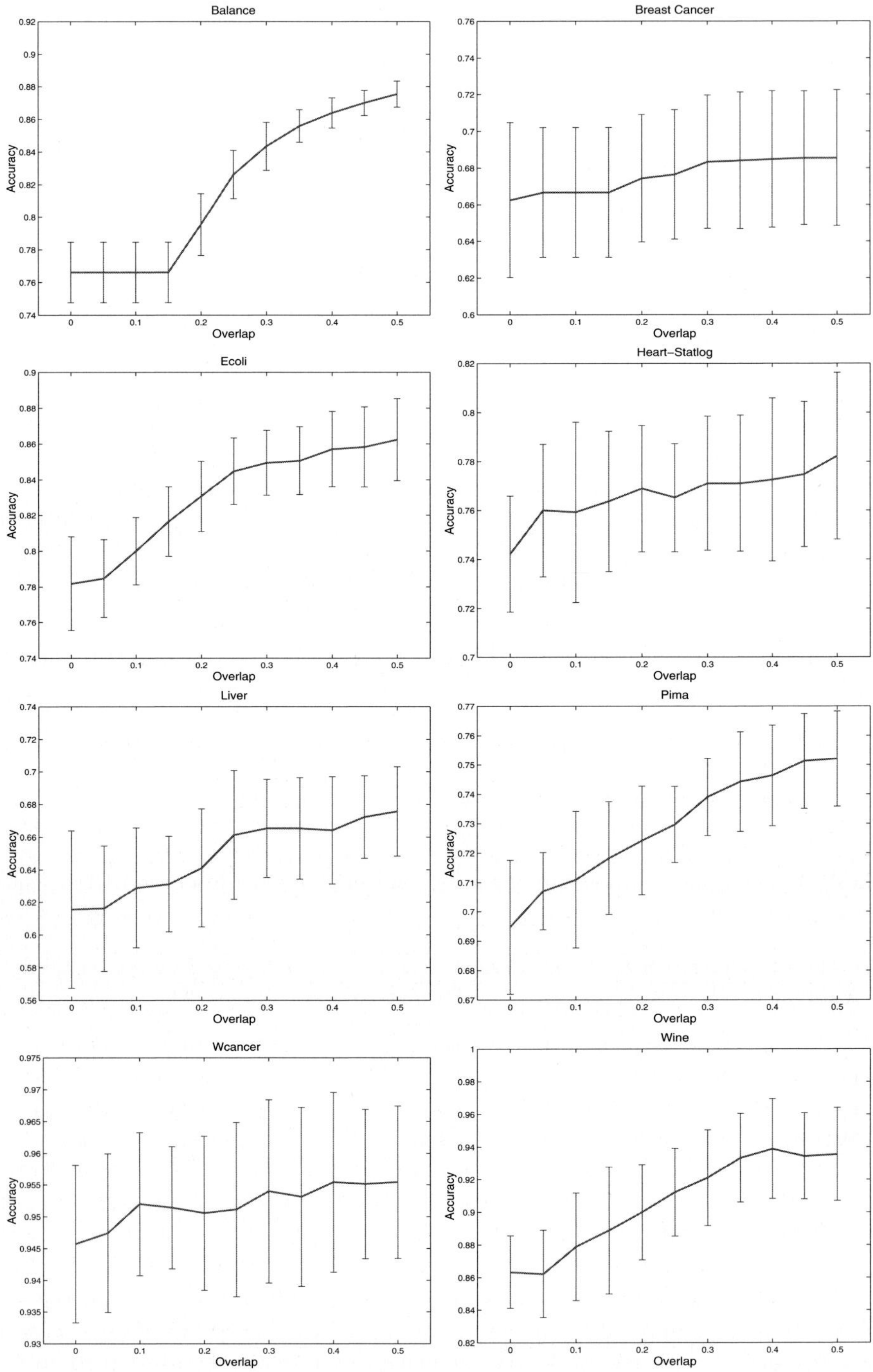


**Fig. 2.** Monotonically increased performance

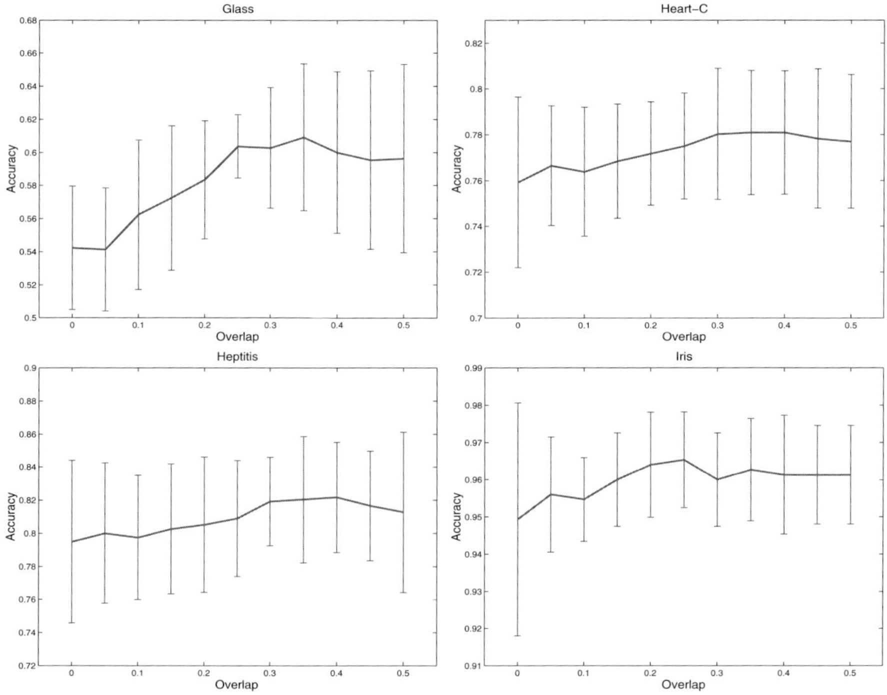


**Fig. 3.** More overlapping does not guarantee the better performance for these datasets

(the growth of LDT will be stopped if all attributes have been used, see Alg 2 line 11). These settings are justified in [4]. We tested 12 datasets taken from UCI [1] machine learning repository. For each experiment, the dataset is partitioned into two parts that the data belonging to the same class are evenly split. One part of the data is for training and the other for test. We will randomly do the split for 10 times and the average results with standard deviation will be calculated. This is referred to as 50-50 split experiments [4]. The experimental results on the given data sets are shown in figures 2 and 3, respectively.

As we can see from these figures, the performance of 8 of the 12 datesets are roughly monotonic increased with the increase of $PT$. It implies that more fuzziness tends to increase the robustness of the LDT model and get better performance. However, from the results in figure 3, we can tell that more overlapping does not guarantee the better performance. For some datasets, 30% of overlapping maybe is enough. More overlapping would not be necessary and it may give worse results sometime. From all the results, we can see that LDTs with fuzzy labels generally outperform the ones with discrete labels (where $PT = 0$). Therefore, in summary, for the case of LDT model, we can say that fuzziness will bring greater performance. The increase is almost monotonically. But the optimal overlapping degrees are depends on the dataset you tested.

## 4 Conclusions

In this paper, we extended the previous work on linguistic decision trees to study the influences on performance by using fuzzy labels with different degrees of overlapping. We tested the LDT model on a series of UCI datasets and the results show that the performance increased almost monotonically with the increase of the overlapping between fuzzy labels. For this empirical study with the LDT model, we can conclude that more fuzziness does imply better performance. However, the optimal overlapping degrees are depends on datasets.

## Acknowledgements

Qin is British Telecommunications (BT) Research Fellow in BISC Group. This research was partly funded by BT/BISC Fellowship.

## References

1. C. Blake and C.J. Merz. UCI machine learning repository, *http://www.ics.uci.edu/~mlearn/ MLRepository.html.*
2. J. Lawry, *Modelling and Reasoning with Vague Concepts*, Springer, 2006.
3. C. Olaru and L. Wehenkel, A complete fuzzy decision tree technique, *Fuzzy Sets and Systems*, Vol. 138: pp.221-254, 2003.
4. Z. Qin and J. Lawry. Decision tree learning with fuzzy labels, *Information Sciences*, Vol. 172/1-2, pp. 91-129, 2005.
5. J. R. Quinlan. Induction of decision trees, *Machine Learning* 1: 81-106. 1986
6. J. R. Quinlan, Decision trees at probabilistic classifiers, *Proceeding of 4th International Workshop on Machine Learning*, pp. 31-37, Morgan Kauffman, 1987.
7. J. R. Quinlan. *C4.5: Programs for Machine Learning.* San Mateo: Morgan Kaufmann, 1993.
8. P. Wang, Interpretations on fuzziness, *IEEE Transactions on Systems, Man and Cybernetics*, Part B, Vol 26(2): 321-326, 1996.
9. L. A. Zadeh, Fuzzy sets, *Information and Control*, Vol 8: pp. 338-353, 1965.

# Part X

# Fuzzy Logic Theory

# Semi–Boolean and Hyper–Archimedean *BL*–Algebras

Esko Turunen

Tampere University of Technology,
P.O. Box 553, 33101 Tampere, Finland
esko.turunen@tut.fi

**Abstract.** We analyze differences between $BL$–algebras and $MV$–algebras. The study has application in mathematical fuzzy logic as the Lindenbaum algebras of Lukasiewicz logic or Hájek's $BL$–logics are $MV$–algebras or $BL$–algebras, respectively. We focus on possible generalizations of Boolean elements of a general $BL$–algebra $L$; we prove that an element $x \in L$ is Boolean iff $x \vee x^* = \mathbf{1}$. $L$ is called semi–Boolean if, for all $x \in L$, $x^*$ is Boolean. We prove that an $MV$–algebra $L$ is semi–Boolean iff $L$ is a Boolean algebra. A $BL$–algebra $L$ is semi–Boolean iff $L$ is a $SBL$–algebra. A $BL$–algebra L is called hyper–Archimedean if, for all $x \in L$, there is an $n \geq 1$ such that $x^n$ is Boolean. We prove that hyper–Archimedean $BL$–algebras are $MV$–algebras. We discuss briefly the applications of our results in mathematical fuzzy logic.

**Keywords:** Mathematical fuzzy logic, $BL$–algrebra, $MV$–algebra.

## 1 Introduction and Preliminaries

The present study is part of a more extensive project to analyse differences between $BL$–algberas and $MV$–algebras (cf. [5,7,8,9,10]). Such studies have direct applications in mathematical fuzzy logic. Indeed, the Lindenbaum algebras of Pavelka–Lukasiewicz logic or Hájek's $BL$–logics are $MV$-algebras or $BL$–algebras, respectively. $MV$-algebras are known to be Hájek's $BL$–algebras satisfying double negation law, for details, see e.g. [4,6].

In previous studies of the project it was proved e.g. (see [7]) that there are no other locally finite $BL$–algebras than locally finite $MV$–algebras, in [9] we demonstated that, for any $BL$–algebra $L$, a subset $MV(L) = \{x^* \mid x \in L\}$ is the largest $MV$–subalgebra of $L$. Then we established several such properties $\Phi$ of a $BL$–algebra $L$ that $L$ has a property $\Phi$ if, and only if the corresponding $MV(L)$ has the property $\Phi$. Such propties $\Phi$ are for example *local, semilocal, quasi–local, perfect, quasi–perfect, bipartite* and *strongly bipartite*, for some of these results, see [2,3], too. Recently, in [5] we demonstrated that states on a $BL$–algebra $L$ are completely determined by states on the corresponding $MV(L)$.

In this paper we focus our concentration on possible generalizations of boolean elements of a general $BL$–algebra. To this end let us recall some basic facts that we will use later. For a start, recall that simple examples of $BL$–algebras are $t$–algebras $\langle [0,1], \wedge, \vee, \odot_t, \rightarrow_t, 0, 1 \rangle$, where $\wedge, \vee$ are *min, max*, respectively,

P. Melin et al. (Eds.): IFSA 2007, LNAI 4529, pp. 419–426, 2007.
© Springer-Verlag Berlin Heidelberg 2007

on the real unit interval $[0,1]$ and $\odot_t$ is a continuous $t$–norm, whereas $\rightarrow_t$ is the corresponding residuum, i.e. $x \odot_t y \leq z$ iff $x \leq y \rightarrow_t z$. The most known $t$–algebras are the following

*Gödel algebra:* $x \odot_t y = \min\{x, y\}$, $x \rightarrow_t y = \begin{cases} 1 & \text{if } x \leq y \\ y & \text{otherwise} \end{cases}$

*Product algebra:* $x \odot_t y = xy$, $x \rightarrow_t y = \begin{cases} 1 & \text{if } x \leq y \\ y/x & \text{otherwise} \end{cases}$

*Lukasiewicz algebra:* $x \odot_t y = \max\{0, x+y-1\}$, $x \rightarrow_t y = \min\{1, 1-x+y\}$.

These three examples are fundamental as any $BL$–algebra on the real unit interval $[0,1]$ can be constructed from them. Moreover, any $BL$–algebra is, up to isomorphism, a subdirect product of linear $BL$–algebras, where linear means that the order relation $\leq$ is a total order (for details, see [4]).

In general, a *$BL$–algebra* is an algebra $L = \langle L, \wedge, \vee, \odot, \rightarrow, \mathbf{0}, \mathbf{1} \rangle$ with four binary operations $\wedge, \vee, \odot, \rightarrow$ and two different constants $\mathbf{0}$, $\mathbf{1}$ such that, for each $x, y, z \in L$, by setting $x \leq y$ iff $x \wedge y = x$ hold:

$$\langle L, \wedge, \vee, \mathbf{0}, \mathbf{1} \rangle \text{ is a distributive lattice with universal bounds } \mathbf{0} \text{ and } \mathbf{1}, \quad (1)$$

$$\odot \text{ is an associative, commutative and isotone operation and } x \odot \mathbf{1} = x, \quad (2)$$

$$x \odot y \leq z \text{ iff } x \leq y \rightarrow z, \quad (3)$$

$$x \wedge y = x \odot (x \rightarrow y), \quad (4)$$

$$(x \rightarrow y) \vee (y \rightarrow x) = \mathbf{1}. \quad (5)$$

In the following text we use the following well–known properties (cf. [6,10]), where $x, y, z$ are elements of a $BL$–algebra.

$$x \rightarrow (y \rightarrow z) = (x \odot y) \rightarrow z = y \rightarrow (x \rightarrow z), \quad (6)$$

in particular, for $z = \mathbf{0}$ and by setting $x^* = x \rightarrow \mathbf{0}$ we get

$$x \rightarrow y^* = (x \odot y)^* = y \rightarrow x^*, \quad (7)$$

$$x \odot x^* = \mathbf{0}, \mathbf{0}^* = \mathbf{1}, \mathbf{1}^* = \mathbf{0}, \quad (8)$$

$$\text{if } x \leq y \text{ then } y^* \leq x^*, \quad (9)$$

$$x \leq x^{**}, \quad (10)$$

$$x^* = x^{***}, \quad (11)$$

$$(x \vee y)^* = x^* \wedge y^* \quad (12)$$

$$x \vee y = [(x \rightarrow y) \rightarrow y] \wedge [(y \rightarrow x) \rightarrow x], \quad (13)$$

$$x \odot \bigvee_{i \in \Gamma} y_i = \bigvee_{i \in \Gamma} (x \odot y_i) \text{ and } \bigvee_{i \in \Gamma} y_i \rightarrow x = \bigwedge_{i \in \Gamma} (y_i \rightarrow x), \quad (14)$$

where $\Gamma$ is a finite set of indices. Moreover, for all $n \geq 1$, set $x^n = \overbrace{x \odot \cdots \odot x}^{\text{n times}}$. Then

$$x^n \leq x, \quad (15)$$

$$(x \vee x^*)^n = x^n \vee (x^*)^n. \quad (16)$$

A $BL$–algebra $L$ is called *local* if, for each element $x \in L$, $x^n = \mathbf{0}$ for some $n \geq 1$ or $(x^*)^m = \mathbf{0}$ for some $m \geq 1$; if such integers exist, then the smallest $n$ and $m$ are denoted by $\mathrm{ord}(x) = n$ and $\mathrm{ord}(x^*) = m$. The case $x \in L$, $x^n > \mathbf{0}$ for all $n \geq 1$ is denoted by $\mathrm{ord}(x) = \infty$.

An *MV–algebra* is a $BL$–algebra $L$ such that $x = x^{**}$ for all $x \in L$. The best known example of $MV$–algebras is Lukasiewicz algebra. In an $MV$–algebra $L$, a binary operation $\oplus$ is defined by $x \oplus y = (x^* \odot y^*)^*$. Moreover, in an $MV$–algebra $L$, for all $x, y \in L$, $x \odot y \leq x \wedge y \leq x, y \leq x \vee y \leq x \oplus y$ and

$$x^* \to y = x \oplus y, \tag{17}$$

$$(x \wedge y)^* = x^* \vee y^*, \tag{18}$$

$$(x \wedge y) \oplus z = (x \oplus z) \wedge (y \oplus z). \tag{19}$$

In [9], we proved that, given a $BL$–algebra $L$, a subset $MV(L) = \{x^* \mid x \in L\}$ of $L$ generates an $MV$–algebra by stipulating $x^* \oplus y^* = x^{**} \to y^*$, which is equivalent to the following

$$x^* \oplus y^* = (x^{**} \odot y)^* = (y \odot x^{**})^* = y \to x^{***} = y \to x^* = (x \odot y)^*. \tag{20}$$

Moreover, $MV(L)$ is the largest $MV$–subalgebra of $L$. In $MV(L)$, the operations $\odot$, $\to$ and $^*$ coincide with those of L and the order relation $\leq$ in $MV(L)$ is that of $L$.

A $BL$–algebra is called *locally finite* if its arbitrary non unit element is of finite order, i.e. if $\mathrm{ord}(x) < \infty$ for all $x < \mathbf{1}$. Locally finite $BL$–algebras, however, are $MV$–algebras and isomorphic to subalgebras of Lukasiewicz algebra.

A *SBL–algebra* is a $BL$–algebra $L$ verifying, for all $x, y \in L$, an equation $(x \odot y)^* = x^* \vee y^*$. Gödel algebra and Product algebra are $SBL$–algebras. Gödel algebra is also an example of a *G–algebra*, a $BL$–algebra with idempotent product $\odot$. All $G$–algebras a known to be $SBL$–algebras. In [9] we characterized $SBL$–algebras by showing that a $BL$–algebra $L$ is a $SBL$–algebra iff $MV(L)$ is a Boolean algebra.

## 2 Hyper–Archimedean $BL$–Algebras

Let $L$ be an $MV$–algebra. For readers comfort, let us recall [1] that a subset

$$B(L) = \{x \in L \mid x \odot x = x\}$$

is the largest subalgbera of $L$ which is at the same time a Boolean algebra. Since $\mathbf{0}, \mathbf{1} \in B(L)$, $B(L)$ is not empty; in $B(L)$, $\odot = \wedge$ and $\oplus = \vee$. Now consider a *Boolean* element $x \in L$, that is

$$x \odot x = x. \tag{21}$$

Then $x \wedge x^* = x \odot (x \to x^*) = x \odot (x \to (x \to \mathbf{0}) = x \odot (x \odot x)^* = x \odot x^* = \mathbf{0}$. Thus,

$$x \wedge x^* = \mathbf{0}. \tag{22}$$

Moreover, if (22) holds then $\mathbf{1} = (x \wedge x^*)^* = x^* \vee x^{**}$, i.e.,

$$x \vee x^* = \mathbf{1}. \tag{23}$$

Conversely, if (23) holds then $\mathbf{0} = (x \vee x^*)^* = x^* \wedge x^{**}$. Hence (22) and (23) are equivalent conditions. Now, (23) holds iff $\mathbf{1} = (x^* \to x) \to x$ iff $x^* \to x \leq x$ iff $x \oplus x \leq x$ iff

$$x \oplus x = x. \tag{24}$$

Moreover, if (23) holds then $x = x \odot \mathbf{1} = x \odot (x \vee x^*) = (x \odot x) \vee (x \odot x^*) = (x \odot x) \vee \mathbf{0} = x \odot x$, so (23) implies (21). We have seen a proof for the well–known fact that equations (21) – (24) are all equivalent conditions in $MV$–algebras.

Next consider a general $BL$–algebra $L$. Then the conditions (21) – (24) are no longer equivalent. For example, (24) is possible only in an $MV$–algebra and (21) holds in all $G$–algebras but does not, in general, imply (23). Moreover, in $G$–algebras (23) imply (22) but not vice versa. Since $MV(L)$ is an $MV$–algebra and $B(MV(L)) \subseteq MV(L) \subseteq L$, it is evident that the Boolean elements of $L$ are precisely those elements $x \in L$ such that (21) and

$$x = x^{**} \tag{25}$$

hold. Trivially, if $x$ is a Boolean element of $L$, then $x^*$ and $x^n, n \geq 1$ are Boolean, too. Moreover, $B(L) = B(MV(L))$. Thus, to analyze Boolean elements of $L$ it is enough to focus on the idempotent elements of the corresponding $MV(L)$.

Now we give another characterization for Boolean elements of a general $BL$–algebra. The result is known while the proof is new and relies on properties of $MV(L)$.

**Proposition 1.** *An element $x$ of a BL-algebra $L$ is Boolean iff (23) holds.*

*Proof.* If $x$ is Boolean then, by (25), $x \in MV(L)$ and in $MV(L)$ the conditions (21) and (23) are equivalent. Conversely, assume $\mathbf{1} = x \vee x^*$. Then

$$\mathbf{1} = (x \to x^*) \to x^* \tag{26}$$

and

$$\mathbf{1} = (x^* \to x) \to x \tag{27}$$

By (26), $(x \odot x)^* = x \to x^* \leq x^*$. Since $x \odot x \leq x$ we have $x^* \leq (x \odot x)^*$. Therefore $x^* = (x \odot x)^*$, and so

$$x^* = x^* \oplus x^*, \tag{28}$$

i.e. $x^*$ is a Boolean element of $MV(L)$. Moreover, by (27), $x^* \to x \leq x$ and as $x \odot x^* = \mathbf{0} \leq x$, we have $x \leq x^* \to x$. Therefore

$$x^* \to x = x. \tag{29}$$

Now, by (29), $x^{**} \leq x = x^* \to x$ iff $x^{**} \odot x^* \leq x$ which holds true as $x^{**} \odot x^* = \mathbf{0}$. We conclude that (25) holds. Then we reason $x \odot x = x^{**} \odot x^{**} = (x^* \oplus x^*)^* = x^{**} = x$, thus (21), too, holds for $x$. The proof is complete.

If $x, y \in L$ are Boolean elements of $L$, then it is easy to see that $x \odot y$ is Boolean, too, while the converse does not, in general, hold. We have, however, the following result.

**Proposition 2.** *If $x \odot y$ is a Boolean element of a BL-algebra $L$ and $x \vee y = \mathbf{1}$, then $x$ and $y$ are Boolean, too.*

*Proof.* Since $x \vee y = \mathbf{1}$, we have $(x \to y) \to y = \mathbf{1}$, thus $(x \to y) \leq y$. By the fact that $y \leq x \to y$ we conclude

$$y = x \to y \text{ and } x = y \to x, \tag{30}$$

where the second equation in proved in a symmetric manner. Moreover,

$$x \wedge y = x \odot (x \to y) = x \odot y. \tag{31}$$

Since $x \odot y$ is Boolean, $x \odot y = (x \odot y)^{**} = (x^* \oplus y^*)^* = x^{**} \odot y^{**}$. Thus, $x^{**} \leq y^{**} \to x \odot y \leq y^{**} \to x \leq y \to x = x$. Similarly $y^{**} \leq y$. We conclude

$$y^{**} = y, x^{**} = x. \tag{32}$$

Therefore $x, y \in MV(L)$ and $\mathbf{1} = x \vee y \leq x \oplus y$, hence

$$x \oplus y = \mathbf{1}. \tag{33}$$

Moreover, $x \odot y \in MV(L)$ and

$$(x \odot y)^* \to x = (x \odot y) \oplus x = (x \wedge y) \oplus x = (x \oplus x) \wedge (y \oplus x) = x \oplus x,$$

that is

$$(x \odot y)^* \to x = x \oplus x. \tag{34}$$

Since $x \odot y$ is Boolean, $(x \odot y) \vee (x \odot y)^* = \mathbf{1}$. Then

$$\begin{aligned} x &= \mathbf{1} \to x \\ &= [(x \odot y) \vee (x \odot y)^*] \to x \\ &= [(x \odot y) \to x] \wedge [(x \odot y)^* \to x] \\ &= \mathbf{1} \wedge (x \oplus x) \\ &= x \oplus x. \end{aligned}$$

In a similar manner we show that $y \oplus y = y$, too. We conclude that $x, y$ are Boolean. The proof is complete.

Our aim is to study various possible generalizations of Boolean elements of a general $BL$–algebra. Thus, first we set the following

**Definition 1.** *A BL–algebra L is called* semi–Boolean *if, for all* $x \in L, x^*$ *is Boolean.*

Then $L$ is semi–Boolean iff for all $x \in L, x^* \vee x^{**} = \mathbf{1}$ iff $MV(L)$ is a Boolean algebra. The following result is obvious.

**Proposition 3.** *An MV–algebra L is semi–Boolean iff L is a Boolean algebra. A BL–algebra L is semi–Boolean iff L is a SBL–algebra.*

Thus, semi–Boolean $BL$–algebras are well–known structures. Cavaccini and Lettieri [1] call an $MV$–algebra $L$ hyper–Archimedean if

$$\forall x \in L : \exists m \geq 1 \text{ such that } x \wedge (mx)^* = \mathbf{0}, \text{ or equivalently, } x \wedge (x^*)^m = \mathbf{0} \quad (35)$$

Notice that, in $MV$–algebras, (35) holds iff $x^{**} \wedge [(x^*)^m]^{**} = \mathbf{0}$, which in turn holds iff $x^* \vee [(x^*)^m]^* = \mathbf{1}$ and, since the complement operation $^*$ is involutive, hyper–Archimedean $MV$–algebras are such that

$$\forall x \in L : \exists m \geq 1 \text{ such that } x \vee (x^m)^* = \mathbf{1}. \quad (36)$$

We generalize the concept of Cavaccini and Lettieri by setting the following

**Definition 2.** *A BL–algebra L is called* hyper–Archimedean *if (36) holds.*

**Proposition 4.** *A BL–algebra L is hyper–Archimedean iff for all* $x \in L$*, there is an* $n \geq 1$ *such that* $x^n$ *is Boolean.*

*Proof.* Assume $x^n$ is Boolean. Then $\mathbf{1} = x^n \vee (x^n)^* \leq x \vee (x^n)^* \leq \mathbf{1}$ for some $n \geq 1$. Thus, $L$ is hyper–Archimedean. Conversely, assume $x \in L$ and, for some $n \geq 1, x \vee (x^n)^* = \mathbf{1}$. Then

$$\begin{aligned}
\mathbf{1} = \mathbf{1} \odot \mathbf{1} &= [x \vee (x^n)^*] \odot [x \vee (x^n)^*] \\
&= \{x \odot [x \vee (x^n)^*]\} \vee \{(x^n)^* \odot [x \vee (x^n)^*]\} \\
&= x^2 \vee [x \odot (x^n)^*] \vee [(x^n)^* \odot x] \vee [(x^n)^* \odot (x^n)^*] \\
&\leq x^2 \vee (x^n)^* \leq \mathbf{1}.
\end{aligned}$$

Hence $x^2 \vee (x^n)^* = \mathbf{1}$. In a similar manner we realize that $\mathbf{1} = \mathbf{1} \odot \mathbf{1} = [x^2 \vee (x^n)^*] \odot [x \vee (x^n)^*] \leq x^3 \vee (x^n)^* = \mathbf{1}$ and, more generally, $x^n \vee (x^n)^* = \mathbf{1}$. Thus, $x^n$ is Boolean. The proof is complete.

**Proposition 5.** *A BL–algebra L is hyper–Archimedean and local iff L is a locally finite MV-algebra.*

*Proof.* Locally finite $BL$–algebras are trivially local. They are hyper–Archimedean, too, as $\text{ord}(x) < \infty$ for all $x < \mathbf{1}$, i.e. $x^n = \mathbf{0} \in B(L)$. Conversely, let $L$ be a hyper–Archimedean and local $BL$–algebra. Assume $x \in L$ is such on element that $\text{ord}(x) = \infty$. Since $L$ is hyper–Archimedean, there is $m \geq 1$ with $x^m \vee (x^m)^* = \mathbf{1}$, where $(x^m)^*$ is of finite order (as $\text{ord}(x^m) = \infty$ and $L$ is local), say $[(x^m)^*]^p = \mathbf{0}$ for some finite $p \geq 1$. We realize that $\mathbf{1} = \mathbf{1}^p = [x^m \vee (x^m)^*]^p = (x^m)^p \vee [(x^m)^*]^p = x^{m+p} \vee \mathbf{0} = x^{m+p} \leq x$. Thus, $x = \mathbf{1}$ and so $L$ is a locally finite $MV$–algebra. The proof is complete.

We are now ready to prove the main results of this study.

**Theorem 1.** *Hyper–Archimedean BL–algebras are MV-algebras.*

*Proof.* Since any $BL$–algebra is a isomorphic to a subalgebra of a direct product $\Pi_{i \in \Gamma} L_i$, where each $L_i, i \in \Gamma$, is linear and, therefore, local $BL$–algebra, it is evident that $L$ is hyper–Archimedean iff each $L_i, i \in \Gamma$, is hyper–Archimedean. By 5, each $L_i, i \in \Gamma$, is a locally finite $MV$–algebra, i.e. for each $x_i \in L_i, i \in \Gamma, x_i^{**} = x_i$. Therefore $x^{**} = x$ for any $x \in L$, and the proof is complete.

**Corollary 1.** *Hyper–Archimedean BL–algebras are, up ito isomorphism, subdirect products of Lukasiewicz algebras.*

## 3 Conclusion

We have studied Boolean elements of a general $BL$–algebra $L$ and given a new proof to the fact that $x \in L$ is Boolean iff $x \vee x^* = \mathbf{1}$. To show its usefulness, we utilized in this and many other prooves the $MV(L)$ construction. Semi–Boolean $BL$–algebras are simple generalizations of Boolean algebras; such algebras are, however, precisely $SBL$–algebras. We introduced hyper–Archimedean $BL$–algebras and proved that $L$ is hyper–Achimedean iff for all $x \in L, x^n$ is Boolean for some finite $n \geq 1$. Alter all, it turns out that hyper–Archimedean $BL$–algebras are $MV$–algebras and isomorphic to subalgebras of direct products of Lukasiewicz algebras.

Our results have direct applications in mathematical fuzzy logic whose Lindenbaum algebras are $BL$–algebras; for example if we wish that any sentence, after a finite repetition, should behave as a Boolean sentence, then we have to approve that all sentences in our logic satisfy the double negation law, too.

## Acknowledgements

These results were obtained when the author visited Academy of Science, Czech Republic, Dept. of Comp. Sciences in Autumn 2006.

## References

1. Cavaccini, V. and Lettieri, A.: Some results on hyper–Archimedean $MV$–algebras. In Barlotti, A. et al. (eds.): Combinatorics '90. Elsevier (1992), 71 79.
2. Di Nola, A., Georgescu, G. and Leustean, L.: Boolean Products of $BL$–algebras. Jour. Math. Anal. Appl. **251**(2000), 106–131.
3. Di Nola, A., Sessa, S., Esteva, F., Godo, L., and Garcia. P.,: The variety generated by perfect $BL$–algebras: an algebraic approach in a fuzzy logic setting. Ann. Math. Artif. Int. **35**(2002), 197–214.
4. Hájek, P.: Metamathematics of Fuzzy Logic, Kluwer (1998).
5. Mertanen, J. and Turunen, E.: Problematic extension of states on $BL$–algebras (submitted).
6. Turunen, E.: Mathematics behind Fuzzy Logic. Springer-Verlag (1999).

7. Turunen, E.: $BL$–algebras of Basic Fuzzy Logic. Mathware and Soft Computing. **6**(1999), 49–61.
8. Turunen, E.: Boolean deductive systems of $BL$–algebras. Archive for Mathematical Logic **40**(2001), 467–473.
9. Turunen, E. and Sessa, S.: Local $BL$–algebras, Multiple Valued Logic **6**(1–2)(2001), 229–249.
10. Turunen, E.: Semilocal $BL$–algeras. IX International IFSA Congress 2005. 28.–31. July 2005, Beiging, China. 252–256.

# A Fuzzy Hahn-Banach Theorem

Wesley Kotzé[1] and Andrew Pinchuck[2]

[1] AIMS, 6 Melrose Rd, Muizenberg, Cape Town, 7945, South Africa
[2] School of Mathematics, University of the Witwatersrand, Private Bag 3, P O WITS 2050, South Africa

**Abstract.** A fuzzy version of the Hahn-Banach theorem is proved based on the classical result. A comparison is also drawn with an earlier published result in this connection.

**Keywords:** Analysis, Topology, $L$-seminormed spaces, Fuzzy extension theorems.

## 1 Introduction

The classical Hahn-Banach Theorem can be stated as in [1] or [2]:

**Theorem 1 (Real case).** *Let $M$ be a subspace of the real vector space $E$ and $p$ a sublinear functional on $E$, i.e. $p(x+y) \leq p(x)+p(y)$ and $p(\alpha x) = \alpha p(x)$ for all $x, y \in E$ and $\alpha \geq 0$, and $f$ a real-valued linear functional on $M$ such that for all $x \in M$, $f(x) \leq p(x)$. Then there exists a linear functional $g$ on $E$, extending $f$ (so $f(x) = g(x)$ on $M$) such that $g(x) \leq p(x)$ for all $x \in E$.*

**Theorem 2 (Complex case).** *Let $M$ be a subspace of the complex vector space $E$ and $p$ a functional on $E$ such that $p(x) \geq 0$, $p(x+y) \leq p(x)+p(y)$, $p(\alpha x) = |\alpha| p(x)$ for all $x, y \in E$ and $\alpha \in \mathbb{C}$, and $f$ is a linear functional on $M$ such that for all $x \in M$, $|f(x)| \leq p(x)$. Then there exists a linear functional $g$ on $E$, extending $f$ such that $|g(x)| \leq p(x)$ for all $x \in E$.*

We'll use these results to prove counterparts for the fuzzy case. The Axiom of Choice is required for the proof of these two theorems, and so is inherent in our results as well.

The idea for this fuzzy version was mooted some eight years ago by J. Chadwick in one of our seminars at Rhodes University. The first author of this paper discussed it afterwards at one of the Linz Seminars on Fuzzy Set Theory. In view of the appearence of [8] it is perhaps appropriate to publish this version and compare it with the result in that paper.

We also had another look at Warren's relationship between neighbourhoods and open sets [10], and in this connection we attached an appendix.

## 2 Preliminaries

We require the following preliminaries. Throughout this section $E$ will denote a vector space over $K$ where $K$ is $\mathbb{R}$ or $\mathbb{C}$ and $(L, \leq, 0_L, 1_L)$ a complete lattice with binary meet being order preserving.

P. Melin et al. (Eds.): IFSA 2007, LNAI 4529, pp. 427–435, 2007.
© Springer-Verlag Berlin Heidelberg 2007

A $L$-subset $\mu$ on $E$ (i.e. $\mu \in L^E$) is *convex* if $\mu(\theta x + (1-\theta)y) \geq \mu(x) \wedge \mu(y)$ whenever $x, y \in E$ and $0 \leq \theta \leq 1$. $\mu$ is called *balanced* (*circled*) if $\mu(\theta x) \geq \mu(x)$ whenever $x \in E$, $\theta \in K$, $|\theta| \leq 1$.

**Note 1.** (i) $\mu$ balanced $\Rightarrow \mu(-x) = \mu(x)$ for all $x$.
(ii) $\mu$ balanced $\Rightarrow \mu(0) = \sup_{x \in E} \mu(x)$.

For $\mu \in L^E, t \in K$ and $x \in E$ $t\mu(x) = \mu(\frac{x}{t})$ for $t \neq 0$ If $t = 0$:

$$t\mu(x) = \begin{cases} 0 & \text{if } x \neq 0 \\ \sup \mu & \text{if } x = 0. \end{cases}$$

**Note 2.** (i) The definition 2 comes from defining $t\mu$ as the Zadeh image operator of $g : E \longrightarrow E$ where $g$ is given by $g(x) = tx$:

$$\begin{aligned} g^{\rightarrow}[\mu](y) &= \begin{cases} \bigvee\{\mu(x) : g(x) = y\} & \\ 0 & \text{if } g(x) \neq y \end{cases} \\ &= \begin{cases} \bigvee\{\mu(x) : tx = y\} & \\ 0 & \text{if } tx \neq y \end{cases} \\ &= \begin{cases} \mu(\frac{y}{t}) & \text{for } t \neq 0 \\ 0 & \text{for } t = 0 \text{ and } y \neq 0 \\ \bigvee \mu(x) & \text{for } t = 0 \text{ and } y = 0. \end{cases} \end{aligned}$$

(ii) $\mu$ convex $\Rightarrow t\mu$ convex for $t \neq 0$:- $t\mu(\theta x + (1-\theta)y) = \mu(\frac{1}{t}\theta x + \frac{1}{t}(1-\theta)y) \geq \mu(\frac{x}{t}) \wedge \mu(\frac{y}{t}) = t\mu(x) \wedge t\mu(y)$.
(iii) It is easy to see that $s(t\mu) = t(s\mu) = (st)\mu$ for all $s, t \in K$ and that $\mu_1 \leq \mu_2 \Rightarrow t\mu_1 \leq t\mu_2$ for all $t \in K$.
(iv) If $\mu$ is balanced, $\frac{t_1}{t_2}\mu(x) = \mu(\frac{t_2}{t_1}x) \geq \mu(x)$ if $|\frac{t_2}{t_1}| \leq 1$. So $t_1\mu \geq t_2\mu$ if $0 < |t_2| \leq |t_1|$.
(v) $\mu$ balanced $\Leftrightarrow t\mu \leq \mu$ whenever $|t| \leq 1$.

$\mu \in L^E$ is *absorbing* if $\sup_{t>0} t\mu = 1_E$ (i.e. $\sup_{t>0} t\mu(x) = 1_L$) i.e. Given any $x \in E$ and $\alpha \in L \backslash \{1_L\}$, then there exists a $t > 0$ such that $(t\mu)(x) > \alpha$, i.e. $\mu(\frac{x}{t}) > \alpha$.

**Note 3.** (i) $\mu$ balanced and absorbing $\Rightarrow \sup_{t>0} t\mu(x) = \mu(0) = 1_L$
(ii) $\mu$ absorbing/balanced/convex $\Rightarrow \mu_\alpha$ and $\mu^\alpha$ absorbing/balanced/convex where $\mu_\alpha = \{x \in E : \mu(x) \geq \alpha\}$, $\mu^\alpha = \{x \in E : \mu(x) > \alpha\}$, $\alpha \in L$, the so-called weak and strong $\alpha$-cuts of $\mu$ respectively.

If $(E, \| \, . \, \|)$ is a normed space and $B$ is the unit ball (open/closed) then $B$ is convex, balanced and absorbing in the classical sense and the $\mathcal{X}_B \in L^E$ (the characteristic function of $B$, i.e. $\mathcal{X}_B(x) = 1_L$ on $B$ and $0_L$ off $B$) has the same properties (as defined above).

Furthermore, if $x \in E, x \neq 0$ then there exists $t > 0$ such that $x \notin tB$ (i.e. $\mathcal{X}_B(\frac{x}{t}) = 0$). This condition distinguishes a norm from a semi-norm.

We therefore define:

[[6]] A convex, balanced and absorbing $\rho \in L^E$ is called a $L$-*seminorm* on $E$. If in addition $\inf_{t>0} t\rho(x) = 0_L$, for all $x \neq 0$, $\rho$ is called a $L$-*norm.* $(E, \rho)$ is called an $L$-*normed* ($L$-*seminormed*) space.

We can now define a $L$-topology on $(E, \rho)$ by means of the following:

The *basic neighbourhoods of* 0 (the zero vector of $E$) are the $L$-subsets $t\rho$ where $t > 0$. A $L$-subset $\sigma$ is called a *neighbourhood of* 0 if there exists a $t > 0$ such that $t\rho \leq \sigma$.

The collection of all neighbourhoods of 0 is denoted by $\mathcal{N}(0)$.

**Proposition 1.** *(a)* $\sigma(0) > 0_L$ *for all* $\sigma \in \mathcal{N}(0)$
*(b)* $\sigma \in \mathcal{N}(0)$ *and* $\sigma \leq \mu \Rightarrow \mu \in \mathcal{N}(0)$
*(c)* $\sigma_1, \sigma_2 \in \mathcal{N}(0) \Rightarrow \sigma_1 \wedge \sigma_2 \in \mathcal{N}(0)$
*(d)* $t\rho \in \mathcal{N}(0)$ *for each* $t > 0$ *(in fact for each* $t \in K, t \neq 0$*)*
*(e)* $\sigma \in \mathcal{N}(0) \Rightarrow \sigma$ *is absorbing*
*(f) If* $\sigma \in \mathcal{N}(0)$*, then there exists a convex* $\sigma^* \in \mathcal{N}(0)$ *such that* $\sigma^* \leq \sigma$.

*Proof.* (a) $\sigma(0) \geq t\rho(0) = \rho(0) = \sup_{x \in E} \rho(x) > 0_L$.
(b) Obvious.
(c) We have $t_1, t_2 > 0$ such that $t_1\rho \leq \sigma_1$ and $t_2\rho \leq \sigma_2$. Thus $t_1\rho \wedge t_2\rho \leq \sigma_1 \wedge \sigma_2$. If $t_2 \leq t_1$, by Note 2 (iv), $t_2\rho \leq \sigma_1 \wedge \sigma_2$ and so $\sigma_1 \wedge \sigma_2 \leq \mathcal{N}(0)$.
(d) Obvious. The remark in parentheses follows for $t \in \mathbb{R}\backslash\{0\}$ since $\rho$ being balanced, $\rho(-x) = \rho(x)$. If $t \in \mathbb{C}\backslash\{0\}$ then by Note 2 (iv), $r\rho \leq t\rho$ for $0 < r \leq |t|$, and so $t\rho \in \mathcal{N}(0)$.
(e) There exists a $t_1 > 0$ such that $t_1\rho \leq \sigma$. So $\sup_{t>0} tt_1\rho \leq \sup_{t>0} t\sigma$, or $1_L = \sup_{s>0} s\rho \leq \sup_{t>0} t\sigma$. Thus $\sup t\sigma = 1_L$.
(f) We have $t\rho \leq \sigma$ for a $t > 0$. $\rho$ is convex, and hence $t\rho$ is convex by Note 2 (ii).

One can define neighbourhoods of an arbitrary point $x \in E$ by translation:

Given $x \in E$ and $\mu \in L^E$, then $x + \mu \in L^E$ is defined as $(x + \mu)(y) = \mu(y - x)$ for all $y \in E$. (cf. $\mathcal{X}_{[a+x,b+x]}(y) = \mathcal{X}_{[a,b]}(y - x)$). The *neighbourhoods of* $x$ are defined as the sets of the form $x + \sigma$ where $\sigma \in \mathcal{N}(0)$. The collection of all the neighbourhoods of $x$ is denoted by $\mathcal{N}(x)$.

**Theorem 3.** *(cf. Appendix) The family* $\mathcal{N}(x)$ *has the following properties:*

*(i)* $\mu \in \mathcal{N}(x) \Rightarrow \mu(x) > 0_L$
*(ii)* $\mu, \nu \in \mathcal{N}(x) \Rightarrow \mu \wedge \nu \in \mathcal{N}(x)$
*(iii)* $\mu = x + \sigma \in \mathcal{N}(x)$ *and* $\mu \leq x + \nu \Rightarrow x + \nu \in \mathcal{N}(x)$
*(iv) If* $\mu \in \mathcal{N}(x)$*, then there exists a* $\omega \in \mathcal{N}(x)$*,* $\omega \leq \mu$ *such that if* $\omega(y) > 0_L$*, then* $\omega \in \mathcal{N}(y)$.

*Proof.* (i) $\mu = x + \sigma$ where $\sigma \in \mathcal{N}(0)$. So $\mu(y) = (x + \sigma)(y) = \sigma(y - x)$ and thus $\mu(x) = \sigma(0) > 0_L$ by Proposition 1 (a).
(ii) $\mu(y) = (x + \sigma_1)(y) = \sigma_1(y - x)$ and $\nu(y) = \sigma_2(y - x)$ with $\sigma_1, \sigma_2 \in \mathcal{N}(0)$. Thus $\mu \wedge \nu(y) = \sigma_1(y - x) \wedge \sigma_2(y - x) = \sigma_3(y - x)$ where $\sigma_3 \in \mathcal{N}(0)$ by Proposition 1 (c) which is equal to $(x + \sigma_3)(y)$.

(iii) $x+\sigma \leq x+\nu \Rightarrow \nu \in \mathcal{N}(0)$ by Proposition 1 (b). Thus $x+\nu \in \mathcal{N}(x)$.
(iv) $\mu = x+\sigma$ and there exists a $t>0$ such that $t\rho \leq \sigma$. So $x+t\rho \leq \mu$. If $\omega(y) = (x+t\rho)(y) > 0_L$, then $t\rho(y-x) > 0_L$, or $y+t\rho(x) > 0_L$ $(\rho(-x) = \rho(x))$ and is a neighbourhood of $y$.

(cf. Appendix) A linear map $T : (E_1, \rho_1) \longrightarrow (E_2, \rho_2)$ between two $L$-(semi) normed spaces is $L$-continuous on $E$ iff $\mu \in \mathcal{N}(Tx_0) \Rightarrow T^{\leftarrow}[\mu] = \mu \circ T \in \mathcal{N}(x_0)$, for every $x_0 \in E_1$.

**Theorem 4.** *A linear map $T : (E_1, \rho_1) \longrightarrow (E_2, \rho_2)$ is $L$-continuous on $E_1$ iff $\mu \in \mathcal{N}(0)$ in $E_2 \Rightarrow T^{\leftarrow}[\mu] = \mu \circ T \in \mathcal{N}(0)$ in $E_1$.*

*Proof.* Let $\mu \in \mathcal{N}(Tx_0)$ in $E_2$. So $\mu = Tx_0 + \sigma$ where $\sigma \in \mathcal{N}(0)$ in $E_2$. Then $\mu \circ T(x) = (Tx_0+\sigma)(Tx) = \sigma(Tx - Tx_0) = \sigma \circ T(x-x_0) = \nu(x-x_0) = x_0 + \nu(x)$ where $\nu = \sigma \circ T \in \mathcal{N}(0)$ in $E_1$ by the assumption. Thus $\mu \circ T = T^{\leftarrow}[\mu]$ is a neighbourhood on $x_0$ in $E_1$.

## 3 The Real Case

Now let $\mathcal{X}$ be the characteristic function of $[-1,1]$ on $\mathbb{R}$, i.e. on $[-1,1]$, $\mathcal{X}(r) = 1_L$. Then $\mathcal{X}$ is a $L$-norm on $\mathbb{R}$. This $L$-norm will be used on $\mathbb{R}$ in the sequel.

### Notation

(1) $E^*$ will denote the set of all $L$-continuous linear functionals from $(E,\rho)$ into $(\mathbb{R}, \mathcal{X})$ where $E$ is a vector space over $\mathbb{R}$.
(2) For $\sigma \in \mathcal{N}(0)$, put $\sigma^0 = \operatorname{supp} \sigma$. Since $\sigma$ is absorbing (Proposition 1 (e), $\sigma^0$ is absorbing, and if $\sigma$ is convex (balanced) so is $\sigma^0$ (Note 3 (ii)).

**Proposition 2.** *Let $f : E \longrightarrow \mathbb{R}$ be linear. Then $f \in E^*$ iff there exists $\sigma \in \mathcal{N}(0)$ in $E$ such that $|f(x)| \leq 1$ for all $x \in \sigma^0$.*

*Proof.* $\Rightarrow$: $f \in E^*$. Now $\mathcal{X}$ is a (basic) neighbourhood of 0 in $\mathbb{R}$ and so $\sigma = f^{\leftarrow}[\mathcal{X}] \in \mathcal{N}(0)$ in $E$. If $x \in \sigma^0$, $\sigma(x) > 0_L$, i.e. $f^{\leftarrow}[\mathcal{X}](x) = \mathcal{X}(f(x)) > 0_L$, so $|f(x)| \leq 1$.
$\Leftarrow$: Assume that $|f(x)| \leq 1$ for $x \in \sigma^0$ where $\sigma \in \mathcal{N}(0)$. Then $\sigma(x) \leq \mathcal{X}(f(x))$ for all $x \in E$. Let $\mu$ be a neighbourhood of 0 in $(\mathbb{R}, \mathcal{X})$. Choose $t_1 > 0$ such that $t_1\mathcal{X} \leq \mu$ and $t_2 > 0$ such that $t_2\rho \leq \sigma$. Then for $x \in E$, $t_1t_2\rho(x) \leq t_1\sigma(x) = \sigma(\frac{x}{t_1}) \leq \mathcal{X}(f(\frac{x}{t_1})) = \mathcal{X}(\frac{1}{t_1}f(x)) = t_1\mathcal{X}(f(x)) \leq \mu(f(x)) = f^{\leftarrow}[\mu](x)$. Hence $t_1t_2\rho \leq f^{\leftarrow}[\mu]$, so $f^{\leftarrow}[\mu] \in \mathcal{N}(0)$. Thus $f$ is $L$-continuous by Theorem 4.

**Corollary 1.** *$E^*$ is a vector space.*

*Proof.* Let $f, g \in E^*$. Choose $\sigma_f, \sigma_g \in \mathcal{N}(0)$ such that $|f(x)| \leq 1$ on $\sigma_f^0$, $|g(x)| \leq 1$ on $\sigma_g^0$. Put $\sigma = \frac{1}{2}\sigma_f \wedge \frac{1}{2}\sigma_g$. So $\sigma \in \mathcal{N}(0)$ (Proposition 1). Now $x \in \sigma^0 \Rightarrow \frac{1}{2}\sigma_f(x) \wedge \frac{1}{2}\sigma_g(x) > 0_L \Rightarrow \frac{1}{2}\sigma_f(x) > 0_L$ and $\frac{1}{2}\sigma_g(x) > 0_L \Rightarrow \sigma_f(2x) > 0_L$ and $\sigma_g(2x) > 0_L \Rightarrow |f(2x)| \leq 1$ and $|g(2x)| \leq 1 \Rightarrow |(f+g)(x)| \leq 1$. So by the preceeding Proposition, $f + g \in E^*$. Likewise we can show that $f \in E^*$, $r$ real $\Rightarrow rf \in E^*$:

The case $r = 0$ is trivially true. So for $r \neq 0$, put $\sigma = \frac{1}{r}\sigma_f$. Then by Proposition 1 (d), $\sigma \in \mathcal{N}(0)$. Then $x \in \sigma^0 \Rightarrow \frac{1}{r}\sigma_f(x) > 0_L \Rightarrow \sigma_f(rx) > 0_L \Rightarrow |f(rx)| \leq 1 \Rightarrow |rf(x)| \leq 1 \Rightarrow rf \in E^*$.

If $(E, \rho)$ is a $L$-(semi) normed space and $M$ is a linear subspace of $E$, then $(M, \rho|_M)$ is a $L$-(semi) normed subspace of $(E, \rho)$. The neighbourhoods of 0 in $(M, \rho|_M)$ are $\sigma|_M$ where $\sigma$ is a neigbourhood of 0 in $(E, \rho)$. Note that $(\sigma|_M)^0 = \{m \in M : \sigma(m) > 0_L\} = M \cap \sigma^0$. $M^*$ has the obvious meaning.

**Theorem 5 (Hahn-Banach - Real case).** *Let $(E, \rho)$ be a $L$-seminormed space over $\mathbb{R}$, $M$ a linear subspace of $E$ and $f \in M^*$. Then there exists a $g \in E^*$ such that $g(m) = f(m)$ for $m \in M$.*

*Proof.* Since $f \in M^*$, there exists a neighbourhood $\zeta$ on 0 in $(M, \rho)$ such that $|f(m)| \leq 1$ for all $m \in \zeta^0$. Now there exists a neighbourhood $\mu$ of 0 in $E$ such that $\zeta = \mu|_M$. By Proposition 1 (f) there exists a convex $\sigma \in \mathcal{N}(0)$ such that $\sigma \leq \mu$. So $|f(m)| \leq 1$ for all $m \in (\sigma|_M)^0 = M \cap \sigma^0$. We can assume $\sigma$ is balanced (or if necessary, replace $\sigma$ by $\sigma \wedge (-\sigma)$. $-\sigma(x) = \sigma(-x)$ as per Definition 2). The convexity and absorption properties are retained - the latter because in a frame arbitrary suprema distribute over finite infima. The set $\sigma^0$ is convex, balanced and absorbing (Note 3). Thus the Minkowski functional ("gauge") $p(x) = \inf\{t > 0 : x \in t\sigma^0\}$ defines a sublinear functional on $E$ as in Theorem 1 (See e.g. [7] or [10]) Now $f(m) \leq p(m)$ for all $m \in M$, $p(m) < 1 \Rightarrow m \in M \cap \sigma^0 \Rightarrow |f(m)| \leq 1$. Now for any $m \in M$, $\theta > 0$, $p(\frac{m}{p(m)+\theta}) = \frac{p(m)}{p(m)+\theta} < 1$. So $|f(\frac{p(m)}{p(m)+\theta})| \leq 1$ or $|f(m)| \leq p(m)+\theta$. Since $\theta > 0$ is arbitrary, $|f(m)| \leq p(m)$, hence $f(m) \leq p(m)$. Apply the classical result Theorem 1 and we obtain a $g$ defined on $E$ such that $g(x) \leq p(x)$ for all $x \in E$ and $g(m) = f(m)$ for all $m \in M$. If $x \in \sigma^0$ then $p(x) \leq 1$, so $g(x) \leq 1$. Also, since $\sigma^0$ is balanced, $g(-x) = -g(x) \leq 1$. Thus $|g(x)| \leq 1$ on $\sigma^0$. It folows that $g \in E^*$.

## 4 The Complex Case

Consider now a vector space $E$ over $\mathbb{C}$ with $E^*$ the set of all $L$-continuous linear functionals from $(E, \rho)$ into $(\mathbb{C}, \mathcal{X})$ where $\mathcal{X}$ is the characteristic function of $B = \{z \in \mathbb{C} : |z| \leq 1\}$, i.e.

$$\mathcal{X}(z) = \begin{cases} 1_L \text{ on } B \\ 0_L \text{ off } B \end{cases}$$

This is a $L$-norm on $\mathbb{C}$. Proposition 2 is also valid for this $E^*$ as is Corollary 1 ($f \in E^*$, $k \in \mathbb{C} \Rightarrow kf \in E^*$ follows again from Proposition 1 (d)).

Theorem 5 then has a counterpart for this case. In the corresponding proof we need $\sigma$ balanced as well. If not, replace it first with $\sigma^* = \sigma \wedge (-\sigma)$. Then for

$\theta$ real and $|\theta| \leq 1$, $\sigma^*(\theta x) \geq \sigma^*(x)$. Then put $\tilde{\sigma}(\theta x) = \sigma^*(|\theta| x)$ for $\theta$ complex. Then $\tilde{\sigma}(\theta x) \geq \sigma^*(x) = \tilde{\sigma}(x)$ (for $|\theta| \leq 1$).

The convexity and absorption properties are carried over from $\sigma$ to $\tilde{\sigma}$. Then $p(x) = \inf\{t > 0 : x \in t\sigma^0\}$ defines a sublinear functional on $E$ (see [7] or [9]) of the type as in Theorem 2 which can be applied to get:

**Theorem 6 (Hahn-Banach - Complex case).** *Let $(E, \rho)$ be a $L$-seminormed space over $\mathbb{C}$, $M$ a linear subspace of $E$ and $f \in M^*$. Then there exists a $g \in E^*$ such that $g(m) = f(m)$ for $m \in M$.*

## 5 Comparison with an Earlier Version

In [8] the following result was proved:

**Theorem 7.** *Let $E$ be a vector space over $\mathbb{R}$, let $\rho$ be $L$-seminorm on $E$ and $M$ a linear subspace of $E$. If $f$ is a linear functional on $M$ such that $\rho \leq \mathcal{X}_{B_f}$ where $\mathcal{X}_{B_f}$ is the characteristic function of $B_f = \{x \in M : |f(x)| \leq 1\}$, then there exists a linear functional $g$ on $E$ such that*

*(1) $f(x) = g(x)$ for all $x \in M$*
*(2) $\rho \leq \mathcal{X}_{B_g}$ where $B_g = \{x \in E : |g(x)| \leq 1\}$.*

Apart from the fact that Theorem 5 is valid for the general $L$-valued case and 7 not, a comparison between these two fuzzy Hahn-Banach theorems in the case of $L = I$ is in order.

The conditions in Theorem 5 and 7 are respectively:

A: $f \in M^*$, i.e. there exists a $\sigma \in \mathcal{N}_E(0)$ such that for all $x \in \sigma^0|_M$, $|f(x)| \leq 1$.

B: On $M$, $\rho \leq \mathcal{X}_{B_f}$ where $B_f = \{x \in M : |f(x)| \leq 1\}$.

Now B $\Rightarrow$ A:

The $L$-seminorm $\rho$ is a neighbourhood on 0. Since $\rho \leq \mathcal{X}_{B_f}$ on $M$, for all $x \in \sigma^0|_M = \rho^0|_M$, $|f(x)| \leq 1$. So $f \in M^*$.

On the other hand A implies :

There exists $t > 0$ such that $t\rho \leq \sigma$ and so on $M$ we have $t\rho \leq \mathcal{X}_{B_f}$.

So Theorem 5 has as corollary:

**Corollary 2.** *If $(E, \rho)$ is a $L$-seminormed space and $M$ a linear subspace of $E$ such that B holds then there exists a $g \in E^*$ such that for all $x \in M$, $g(x) = f(x)$.*

This is exactly the statement of Theorem 7 in view of the comments above.

## 6 Appendix

In order to clarify the relationships between neighbourhoods and open sets, we supply the following. The work by Warren [10] is often quoted in this regard, but his requirement that a neighbourhood $n(x)$ of a point $x$ should contain an open set $g$ such that $\underline{g(x) = n(x)}$, seems to be superfluous (cf. also [3] where a neighbourhood without this condition was called an s-neighbourhood). We dispense here with Greek letters for $L$-subsets in order to emphasize the similarity with the classical (crisp) case.

**The Definition of a $L$-Topology Starting with Neighbourhoods**

A space $X$ is called a *$L$-topological space* if for every $x \in X$ there exists a non-empty family $\mathcal{N}_x$ of $L$-subsets of $X$ called *neighbourhoods of* $x$ satisfying the following axioms:

(i) $U \in \mathcal{N}_x \Rightarrow U(x) > 0_L$
(ii) $U, V \in \mathcal{N}_x \Rightarrow U \wedge V \in \mathcal{N}_x$
(iii) $U \in \mathcal{N}_x,\, U \leq V \Rightarrow V \in \mathcal{N}_x$
(iv) If $U \in \mathcal{N}_x$ then there exists a $W \in \mathcal{N}_x$, $W \leq U$ such that if $W(y) > 0_L$ then $W \in \mathcal{N}_y$.

The family $\mathcal{N} = \{\mathcal{N}_x : x \in X\}$ is called the *neighbourhood system on* $X$.

A non-zero $L$-subset $A$ of a $L$-topological space $X$ is *open* if $A(x) > 0_L$ and there exists a $V \in \mathcal{N}_x$ such that $V \leq A$. (Or we say $A$ is open iff $x$ is an *interior point* of $A$ for all $A(x) > 0_L$ ).

We also take $0_X$ to be open ($0_X(x) = 0_L$ for all $x \in X$). The collection of all the open subsets of $X$ is called *the $L$-topology* on X (generated by $N_x$).

**Proposition 3.** *An open set $A$ is a neighbourhood of $x$ iff $A(x) > 0_L$.*

*Proof.* Follows from Axiom (iii) of neighbourhoods.

**Theorem 8.** *If $X$ is a $L$-topological space, then*

*(i) $1_X$ and $0_X$ are open ($1_X(x) = 1_L$ for all $x \in X$).*
*(ii) If $A$ and $B$ are open then so is $A \wedge B$.*
*(iii) If $A_i$ is open for each $i \in I$, then so is $\bigvee_i A_i$.*

*Proof.* (ii) follows from Axiom (ii) if neighbourhoods.
(iii) follows from Axiom (iii) of neighbourhood.

**Proposition 4.** *Every $U \in \mathcal{N}_x$ contains an open set $A$ such that $A(x) > 0_L$.*

*Proof.* Follows from Axiom (iv) if neighbourhoods.

**The Definition of a $L$-Topology Starting with Open Sets**

A space $X$ is called a *$L$-topological space* if there exists a non-empty family $\tau$ of $L$-subsets of $X$ (i.e. $\tau \subset L^X$) satisfying

(i) $1_X$ and $0_X$ are members of $\tau$.
(ii) $U, V \in \tau \Rightarrow U \wedge V \in \tau$.
(iii) $U_i \in \tau,\, i \in I \Rightarrow \bigvee_i U_i \in \tau$.

$\tau$ is called the *$L$-topology* on $X$ and the members of $\tau$ are called the *open sets.*

A *neighbourhood* of a point $x \in X$ is any $L$-subset $N_x$ which contains an $A \in \tau$ such that $A(x) > 0_L$.

**Proposition 5.** *An open set $A$ is a neighbourhood of $x$ iff $A(x) > 0_L$.*

**Theorem 9.** *If $X$ is a $L$-topological space and $x \in X$, then*

(i) *$U$ is a neighbourhood of $x \Rightarrow U(x) > 0_L$.*
(ii) *$U$ and $V$ neighbourhoods of $x \Rightarrow U \wedge V$ is a neighbourhood of $x$.*
(iii) *$U$ a neighbourhood of $x$ and $U \leq V \Rightarrow V$ is a neighbourhood of $x$.*
(iv) *If $U$ is a neighbourhood of $x$, then there exists a set $W \leq U$ such that $W$ is a neighbourhood of $x$ and which is a neighbourhood of $y$ for every $y$ such that $W(y) > 0_L$.*

*Proof.* Follows from above.

### The Continuity of Maps Between $L$-Topological Spaces

A map $f : (X, \tau_1) \longrightarrow (Y, \tau_2)$ between two $L$-topological spaces is *$L$-continuous* iff for every $A \in \tau_2$, $f^{\leftarrow}[A] \in \tau_1$ where $f^{\leftarrow}[A](x) = Af(x)$.

**Theorem 10 (Characterization of $L$-continuity).** *For a map $f : (X, \tau_1) \longrightarrow (Y, \tau_2)$ between two $L$-topological spaces. the following statements are equivalent:*

(1) *$f$ is $L$-continuous on $X$ (as per definition).*
(2) *$B$ closed in $(Y, \tau_2) \Rightarrow f^{\leftarrow}[B]$ closed in $(X, \tau_1)$.*
(3) *$f^{\rightarrow}[cl_X A] \leq cl_Y(f^{\rightarrow}[A])$ for all $A \in L^X$.*
(4) *For every $x \in X$ and for every neighbourhood $N_{f(x)}$ of $f(x)$ in $Y$, $f^{\leftarrow}[N_{f(x)}]$ is a neighbourhood on $x$ in $X$.*
(5) *For every $x \in X$ and for every neighbourhood $N_{f(x)}$ in $Y$, there exists a neighbourhood $N_x$ of $x$ in $X$ such that $f^{\rightarrow}[N_x] \leq N_{f(x)}$.*

*Proof*

*(1) ⇔ (2): Follows from $f^{\leftarrow}[1_Y - A](x) = (1_Y - A)f(x) = 1_X - f^{\leftarrow}[A]f(x)$.*

*(1) ⇒ (4): $N_{f(x)}$ contains an $A \in \tau_2$ such that $A(f(x)) > 0_L$. So $f^{\leftarrow}[N_{f(x)}](x) = N_{f(x)}f(x) \geq Af(x) = f^{\leftarrow}[A](x)$. By (1), $f^{\leftarrow}[A](x) \in \tau_1$ and is greater than $0_L$. So by definition $f^{\leftarrow}[N_{f(x)}](x)$ is a neighbourhood of $x$.*

*(4) ⇒ (5): Let $N_x = f^{\leftarrow}[N_{f(x)}]$ of (4). Then $f^{\rightarrow}[N_x] = f^{\rightarrow}[f^{\leftarrow}[N_{f(x)}]] \leq N_{f(x)}$.*

*(5) ⇒ (4): We have $f^{\rightarrow}[N_x] \leq N_{f(x)}$. Then $N_x \leq f^{\leftarrow}[f^{\rightarrow}[N_x]] \leq f^{\leftarrow}[N_{f(x)}]$ and hence $f^{\leftarrow}[N_{f(x)}]$ is a neighbourhood on $x$.*

*(4) ⇒ (1): Since $A \in \tau_2$ is a neighbourhood for each $y \in Y$ for which $A(y) > 0_L$, $f^{\leftarrow}[A](x) = Af(x)$ is a neighbourhood for each $x$ for which $A(f(x)) > 0_L$. Thus $f^{\leftarrow}[A] \in \tau_1$.*

*(2) ⇒ (3):* $cl_Y(f^{\rightarrow}[A])$ *is closed in* $Y$ *and so by (2)* $f^{\leftarrow}[cl_Y f^{\rightarrow}[A]]$ *is closed in* $X$*. Now* $A \leq f^{\leftarrow}[f^{\rightarrow}[A]] \leq f^{\leftarrow}[cl_Y f^{\rightarrow}[A]]$*. So* $cl_X A \leq f^{\leftarrow}[cl_Y f^{\rightarrow}[A]]$ *and hence* $f^{\rightarrow}[cl_X A] \leq cl_Y f^{\rightarrow}[A]$*.*

*(3) ⇒ (2):* *Let* $W$ *be closed in* $Y$*. Then by (3)* $f^{\rightarrow}[cl_X f^{\leftarrow}[W]] \leq cl_Y[f^{\rightarrow}[f^{\leftarrow}[W]] \leq cl_Y W = W$*. Hence* $cl_X f^{\leftarrow}[W] \leq f^{\leftarrow}[W]$*. therefore* $f^{\leftarrow}[W]$ *is closed.*

## References


1. G. Bachman & L. Narici: Functional Analysis. Academic Press, N.Y. 1966.
2. N. Dunford & J. T. Schwartz: Linear Operators I. Interscience, N.Y. 1958.
3. M. H. Ghanim & A. S. Mashour: Characterization of fuzzy topologies by s-neighbourhood systems. Fuzzy Sets and Systems 9 (1983) 211-213.
4. A. K. Katsaras & D. B. Liu: Fuzzy Vector Spaces and Fuzzy Topological Vector Spaces. J.Math.Anal.Appl. 58 (1977), 135-146.
5. A. K. Katsaras: Fuzzy Topological Vector Spaces I. Fuzzy Sets and Systems 6 (1981) 85-95.
6. A. K. Katsaras: Fuzzy Topological Vector Spaces II. Fuzzy Sets and Systems 12 (1984) 143-154.
7. J. L. Kelley & I. Namioka: Linear Topological Spaces. Van Nostrand, Princeton, 1963.
8. Gil Seob Rhie & In Ah Hwang: On the fuzzy Hahn-Banach Theorem - an analytic form. Fuzzy Sets and Systems 108 (1999) 117-121.
9. A. E. Taylor: Introduction to Functional Analysis. J. Wiley & Sons, N. Y. 1958.
10. R. H. Warren: Fuzzy topologies characterized by neighbourhood systems. Rocky Mountain Journal of Mathematics 9(4) (1979) 761-764.

# The Algebraic Properties of Linguistic Value "Truth" and Its Reasoning

Zheng Pei

School of Mathematics & Computer, Xihua University,
Chengdu, Sichuan, 610039, China
pqyz@263.net

**Abstract.** From logic and algebra point of view, Computing with words is discussed in this paper. By analyzing the semantically ordering relation of linguistic variable *Truth*, orderings on linguistic hedges $H$ and atomic evaluating syntagm $Tr = \{true, false\}$ are obtained, respectively. Let $H$ be finite chain, then Lukasiewicz product algebra of $T = H \times Tr$ of *Truth* is obtained, and term-set $T(X)$ of *Truth* is embedded into an algebra $\Gamma$ of type $\Delta = \{\vee, \wedge, ', \rightarrow_L\}$. In some cases, $\Gamma$ can be applied in linguistic decision directly, also as truth domain of logic statements. Different with other truth domain, here truth values are linguistic terms rather than numerals (or symbolic).

## 1 Introduction

As it is well known, humans employ words in computing and reasoning, arriving at conclusions expressed as words from premises expressed in a natural language or having the form of mental perceptions. The methodology of computing with words ($CWW$) proposed by Zadeh may be viewed as an attempt to harness the highly expressive power of natural languages by developing ways of $CWW$ or propositions drawn from a natural language[1], [2]. From the point of view of fuzzy logic in broader sense ($FLB$), Novák, et al have proposed the evaluating linguistic predications, the pairs "nominal syntagm-antonym", and the theory of linguistic quantifiers, and developed a theory of natural human reasoning [3]-[5]. In a special fuzzy theory of evaluating syntagms $T^{EV}$, a truth value assigned to a formula is in $L$, where $L = [0, 1]$ with the usual Lukasiewicz operators of conjunction $\otimes$, disjunction $\oplus$, implication $\rightarrow$ and negation $\neg$. In practical cases, however, a linguistic truth value assigned to a formula (or decision), such as, *true, very false, completely true, little false, more or less true*, etc., seems to comply with the intuition. The other work about $CWW$ can be found in [6]-[14].

## 2 Preliminaries

There are special expressions of natural language, which characterize sizes, distances, etc. In general, they characterize a position on an ordered scale. These

P. Melin et al. (Eds.): IFSA 2007, LNAI 4529, pp. 436–444, 2007.
© Springer-Verlag Berlin Heidelberg 2007

special expressions of natural language are called evaluating syntagms [3]-[5]. Among them, simple evaluating syntagms are expressions of the form

$$\langle \text{linguistic hedge} \rangle \langle \text{atomic evaluating syntagm} \rangle.$$

In which, *atomic evaluating syntagm*, such as *small*, *big*, usually forms pairs of antonyms, *i.e.*, the pairs

$$\langle \text{nominal adjective} \rangle \langle \text{antonym} \rangle.$$

There are a lot of pairs of antonyms, for example *young-old*, *ugly-nice*, *stupid-clever*. When completed by the middle term, such as *small-medium-big*, etc., they form the so called basic linguistic trichotomy [4]. Here, we notice that linguistic truth values, *false* and *true*, are also atomic evaluating syntagm. In fact, pair*false-true* can be understood by $0, 1 \in R$, respectively, *i.e.*, We always regard

**Fig. 1.** *false-true*—-position on an ordered scale

that there exists ordering on *atomic evaluating syntagm* (or term-set $T(X)$ of linguistic variable $X$), the ordering is also call the semantically ordering relation [11]-[13], *i.e.*, *true>false*, *very true>true>more or less true*. In this paper, *atomic evaluating syntagm* of *Truth*is denoted by poset

$$(Tr, >) = \{\{true, \quad false\}, >\}. \tag{1}$$

If there is no danger of misunderstanding, $(Tr, >)$ is denoted by $Tr$ simply.

Zadeh has always emphasized two most important characteristics of linguistic variables [15]. The first is the context-independent meaning of linguistic hedges and connectives, whereas the meaning of *atomic evaluating syntagms* is context-dependent. The second is the universality of their structure. Most linguistic variables possess the same basic structure in the sense that their respective linguistic values have the same expressions except for *atomic evaluating syntagms*. Linguistic hedges are decomposed into two classes [3]-[5], one is the linguistic hedges with *narrowing effect* (such as, *very*, *highly*, *more*, etc). The other is the linguistic hedges with *widening effect* (such as, *more or less*, *roughly*, *little*, etc). The set of linguistic hedges is

$$H = \{h_{-k}, \cdots, h_{-1}, h_0, h_1, \cdots, h_k\}. \tag{2}$$

Where, $\{h_1, \cdots, h_k\}$ is the linguistic hedges with *narrowing effect*, $\{h_{-k}, \cdots, h_{-1}\}$ is the linguistic hedges with *widening effect*, $\{h_0\}$ is called *central*. Semantic properties of linguistic hedges can be obtained as follows [11]-[13]:

1. Being similar to *atomic evaluating syntagms*, linguistic terms possesses an intuitive meaning which can be expressed also by a semantically ordering relation, *e.g.*, *very true>more or less true*;

2. linguistic hedges are modifiers or intensifiers and they have their degrees of modification and hence we can compare two hedges, *e.g.*, *very>more or less*;
3. linguistic hedges have the so-called semantic heredity, which stems from the following observation: each linguistic hedge modifies merely a little the meaning of linguistic terms. That is, changing the meaning of a term, it preserves the original essential meaning of this term. Hence, $\forall h_i \in H$, $h_i$*true*>$h_i$*false*.

About semantic heredity of linguistic hedges, "linguistic hedge preserves the original essential meaning of this term" is only one case. Another case is that some linguistic hedges completely change the original essential meaning of a term, *i.e.*, some linguistic hedges express the meaning of antonyms of the term, *e.g.*, **little true** expresses the meaning of **false** with degree, **little false** expresses the meaning of **true** with degree. From logic point of view, there are some linguistic hedges ($H^-$, for short) that are similar to logic operator **Not**, if **Not true** is equal to **false** (or **Not false** is equal to **true**), then for $h \in H^-$, $h$**true**=$h$**Not false**(or $h$**false**=$h$**Not true**) is similar to $\neg\neg a = a$ in logic system. Hence, for $h \in H^-$, $h$ **true**>$h$ **false** can not be always obtained, and $\exists h_i \in H \backslash H^-$ such that $h_i$ **true** and $h$ **false** ($h \in H^-$) are incomparable, the situation can be observed in Chinese especially , and coincides with intuition. In this paper, our studies focus on the case of $H$.

## 3 The Algebra Structure of Term-Set $T(X)$ of Truth

Let linguistic hedges with *widening effect* as $H^-$, *i.e.*,

$$H^- = \{h_{-k}, h_{-(k-1)}, \cdots, h_{-1}\}. \tag{3}$$

and *narrowing effect* as $H^+$, *i.e.*,

$$H^+ = \{h_1, h_2, \cdots, h_k\}. \tag{4}$$

$h_0$ such that $h_0$*true*=*true* and $h_0$*false*=*false*, denotes $H = H^- \cup \{h_0\} \cup H^+$.

**Definition 1.** *$\forall h_i, h_j \in H$ $(i, j \in \{-k, \cdots, k\})$,*

$$h_i \geq h_j \Longleftrightarrow h_i\text{true} \geq h_j\text{true}. \tag{5}$$

*Where, $h_i$true $\geq$ $h_j$true is decided by the semantically ordering relation of $h_i$true and $h_j$true.*

*Example 1.* [11] Let the set of linguistic hedges be $H$ = {**very, more,** $h_0$, **approximately, possibly, more or less, little**}. According to the semantically ordering relation, **very true** > **more true** > $h_0$**true** > **more or less true** > **little true**, or **more true** > $h_0$**true** > **approximately true** > **little true**, or **more true** > $h_0$**true** > **possibly true** > **little true** can be obtained. **more or less true, approximately true** and **possibly true** are incomparable. Hence, **very** > **more** > $h_0$ > **more or less** > **little, more** > $h_0$ > **approximately** > **little, more** > $h_0$ > **possibly** > **little. more or less** , **approximately** and **possibly** are incomparable.

Obviously, by using (5), $(H, \geq)$ is a poset. $(H, \geq)$ is still denoted by $H$ if there is no danger of misunderstanding. Let linguistic hedges $O$ and $I$ such that $\forall h \in H$, $Otrue < htrue < Itrue$.

*Example 2.* Continue Example 1, $H = \{$**very, more,** $h_0$, **approximately, possibly, more or less, little**$\} \cup \{O, I\}$ is a complete lattice (see Fig.2)

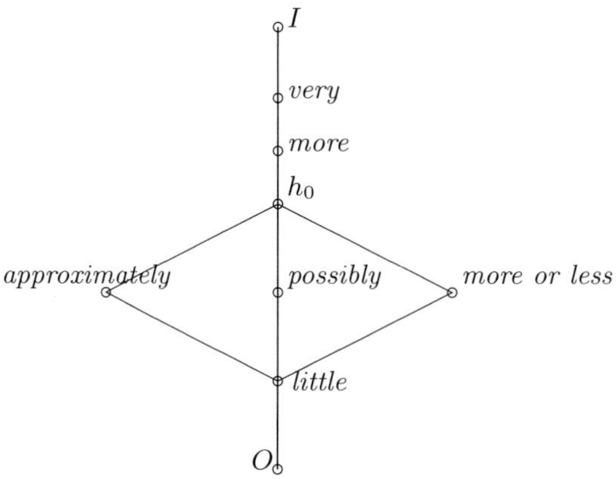


**Fig. 2.** Lattice of $H$

Linguistic hedges $H$ and *atomic evaluating syntagms* $Tr = \{true, false\}$ are foundation to construct term-set $T$ of Truth.

**Lemma 1.** *Let $H$ be finite chain. $T = H \times Tr = \{(h, c)|h \in H, c \in Tr\}$ such that $\forall (h_1, c_1), (h_2, c_2) \in T$, $(h_1, c_1) \leq (h_2, c_2)$ if and only if $h_1 \leq_h h_2$ and $c_1 \leq_c c_2$, then $(T, \leq)$ is a lattice. Where, $\leq_h$ and $\leq_c$ are ordering on $H$ and $Tr$, respectively.*

*Example 3.* Let $H = \{I$, **very, more,** $h_0$, **possibly, little**, $O\}$ and $Tr = \{$**true, false**$\}$, $T = H \times Tr$ (see Fig. 3)

In Example 3, we notice that the ordering of $\{(h, false)|h \in H\}$ doesn't coincide with intuition. The problem will be modified by follows.

**Lemma 2.** *[14] For chain $T^- = H \times \{false\}$, Let $V : T^- \longrightarrow T^-$ be a bijective mapping such that $\forall (h_1, false), (h_2, false) \in T^-$, $V((h_1, false)) \leq_v V((h_2, false))$ if and only if $(h_1, false) \geq (h_2, false)$ in $T$, then $(V(T^-), \leq_v)$ is a chain, and is called the dual of $T^-$.*

**Theorem 1.** *$\forall (h, c) \in (T, \leq)$, define*

$$f : T \longrightarrow T,$$

$$f((h,c)) = \begin{cases} (h, c), & \text{if} \quad c = true, \\ V((h,c)), & \text{if} \quad c = false. \end{cases} \tag{6}$$

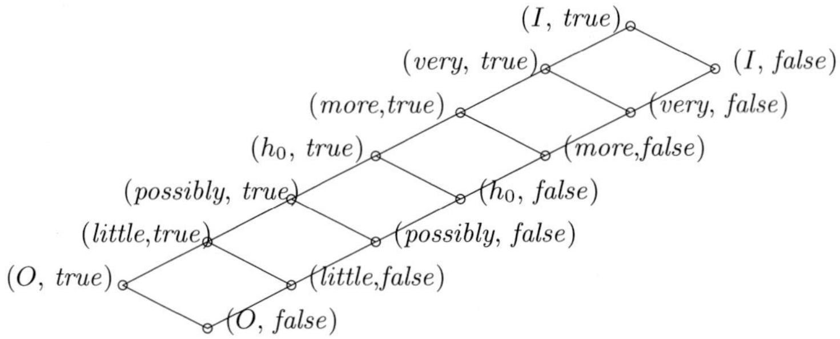


**Fig. 3.** Lattice of $T = H \times Tr$

*and* $\forall (h_j, c_i), (h_k, c_s) \in f(T)$,

$$(h_j, c_i) \vee_f (h_k, c_s) = f(f^{-1}((h_j, c_i)) \vee f^{-1}((h_k, c_s))), \quad (7)$$
$$(h_j, c_i) \wedge_f (h_k, c_s) = f(f^{-1}((h_j, c_i)) \wedge f^{-1}((h_k, c_s))), \quad (8)$$

*then* $(f(T), \vee_f, \wedge_f)$ *is a lattice, denoted by* $(T', \leq_f)$, *and* $(T', \leq_f)$ *is called lattice of term-set* $T$ *of Truth. Mapping* $V$ *is decided by Lemma 2.*

*Example 4.* Continue Example 3, according to Lemma 2 and Theorem 1, Fig. 3 can be modified by Fig. 4

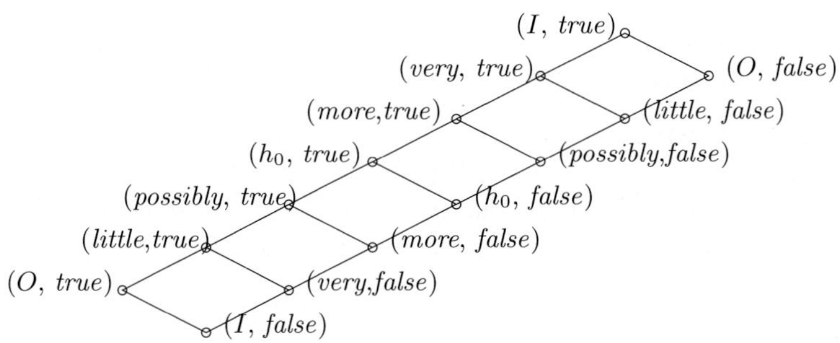


**Fig. 4.** Lattice of $(T', \leq_f)$

## 4 The Algebra Structure of $(T', \leq_f)$

In this Section, $(T', \leq_f)$ is embedded into an algebra of type $\Delta = \{\vee, \wedge, ', \rightarrow_L\}$.

**Lemma 3.** *[3] (Lukasiewicz algebra on finite chains) Let* $L = \{a_i | i = 1, 2, \cdots, n\}$. *For any* $1 \leq j, k \leq n$, *define*

$$a_j \vee a_k = a_{max\{j,k\}}, \quad a_j \wedge a_k = a_{min\{j,k\}}, \quad (9)$$
$$(a_j)' = a_{n-j+1}, a_j \rightarrow_L a_k = a_{min\{n, n-j+k\}}, \quad (10)$$

*then* $(L, \vee, \wedge, ', \rightarrow_L, a_1, a_n)$ *is a Lukasiewicz algebra on finite chains.*

In this paper, $H$ and $Tr$ are finite chains, respectively. Based on (9) and (10), the set of linguistic hedge $H$ and $Tr$ can be embedded into Lukasiewicz algebra on finite chains, respectively, still denoted by $H$ and $Tr$.

**Theorem 2.** *$\forall (h_j, c_i), (h_k, c_s) \in (T, \leq)$, define*

$$(h_j, c_i)' = (h_{n-j+1}, c_{2-i+1}), \quad (h_j, c_i) \vee (h_k, c_s) = (h_{max\{j,k\}}, c_{max\{i,s\}}),$$
$$(h_j, c_i) \wedge (h_k, c_s) = (h_{min\{j,k\}}, c_{min\{i,s\}}),$$
$$(h_j, c_i) \rightarrow_L (h_k, c_s) = (h_{min\{n,n-j+k\}}, c_{min\{2,2-i+s\}}),$$

*then $(T, \vee, \wedge, ', \rightarrow_L, (h_1, \text{false}), (h_n, \text{true}))$ is Lukasiewicz product algebra.*

**Theorem 3.** *$\forall (h_j, c_i), (h_k, c_s) \in (T', \leq_f)$, define*

$$(h_j, c_i) \vee_L (h_k, c_s) = f(f^{-1}((h_j, c_i)) \vee f^{-1}((h_k, c_s))), \tag{11}$$
$$(h_j, c_i) \wedge_L (h_k, c_s) = f(f^{-1}((h_j, c_i)) \wedge f^{-1}((h_k, c_s))), \tag{12}$$
$$(h_j, c_i)' = f((f^{-1}((h_j, c_i)))'), \tag{13}$$
$$(h_j, c_i) \rightarrow_L (h_k, c_s) = f(f^{-1}((h_j, c_i)) \rightarrow_L f^{-1}((h_k, c_s))), \tag{14}$$

*then $\Gamma = (T', \vee, \wedge, ', \rightarrow_L, (h_n, \text{false}), (h_n, \text{true}))$ is an algebra of type $\Delta = \{\vee, \wedge, ', \rightarrow_L\}$. In which $(h_n, \text{false})$ and $(h_n, \text{true})$ are its the least and the greatest elements, respectively. $f$ is decided by (6).*

*Example 5.* Continue Example 4, where $I, v, m, h_0, p, l, O$, standing for $I$, **very, more,** $h_0$, **possibly, little** and $O$, $t$ and $f$ standing for **true**, **false**, we can obtain $\rightarrow_L$ of $\Gamma$ (only ($h$,true) $\rightarrow_L$ ($h$,false) and ($h$,true) $\rightarrow_L$ ($h$,false), see Table 1

**Table 1.** $\rightarrow_L$ of $\Gamma$

| $\rightarrow_L$ | $(O,f)$ | $(l,f)$ | $(p,f)$ | $(h_0,f)$ | $(m,f)$ | $(v,f)$ | $(I,f)$ |
|---|---|---|---|---|---|---|---|
| $(O,t)$ | $(O,f)$ | $(O,f)$ | $(O,f)$ | $(O,f)$ | $(O,f)$ | $(O,f)$ | $(O,f)$ |
| $(l,t)$ | $(l,f)$ | $(O,f)$ | $(O,f)$ | $(O,f)$ | $(O,f)$ | $(O,f)$ | $(O,f)$ |
| $(p,t)$ | $(p,f)$ | $(l,f)$ | $(O,f)$ | $(O,f)$ | $(O,f)$ | $(O,f)$ | $(O,f)$ |
| $(h_0,t)$ | $(h_0,f)$ | $(p,f)$ | $(l,f)$ | $(O,f)$ | $(O,f)$ | $(O,f)$ | $(O,f)$ |
| $(m,t)$ | $(m,f)$ | $(h_0,f)$ | $(p,f)$ | $(l,f)$ | $(O,f)$ | $(O,f)$ | $(O,f)$ |
| $(v,t)$ | $(v,f)$ | $(m,f)$ | $(h_0,f)$ | $(p,f)$ | $(l,f)$ | $(O,f)$ | $(O,f)$ |
| $(I,t)$ | $(I,f)$ | $(v,f)$ | $(m,f)$ | $(h_0,f)$ | $(p,f)$ | $(l,f)$ | $(O,f)$ |

## 5 Reasoning by Directly Handling Linguistic Values of Truth

Normally, a basic element of human knowledge consists of two components: *a vague sentence* and *a truth (belief) degree* which is also expressed in linguistic terms. An elementary vague sentences can be expressed by $p(x; u)$, where $x$ is a variable, $u$ is a vague concept, and $p(.;.)$ is a linguistic analog of classical predicate. An assertion means a pair $A = (p(x;u),(h,c))$, where $p(x;u)$ is a vague sentence and $(h,c)$ is a linguistic truth value. A knowledge base means a finite set $K$ of assertions. From the given knowledge base, new assertions can be deduced by using inference rule. Let $(h,c)$ is a linguistic value of *truth*, linguistic hedges $H$ is finite chain, according to the natural semantic of linguistic sentence, the rule of moving hedges $h$ can be obtained [12]

$$RT_1 : \frac{(p(x;hu),c)}{(p(x;u),(h,c))}, \quad RT_{1'} : \frac{(p(x;u),(h,c))}{(p(x;hu),c)}. \tag{15}$$

Based on rule (15), form $(is(Mary;veryattracive), true)$, we deduce $(is\ (Mary;$ $attracive), (very, true))$, this coincide with our intuition. From propositional calculus point of view, the rules of moving hedges $h$ for $\wedge, \vee$ and $\rightarrow$ can be obtained

$$R_1 : \frac{(p(x;h_1u),c_1) \quad \text{and} \quad (q(y;h_2v),c_2)}{(p(x;u) \wedge q(y,v),(h_1,c_1) \wedge (h_2,c_2))}, \tag{16}$$

$$R_2 : \frac{(p(x;h_1u),c_1) \quad \text{or} \quad (q(y;h_2v),c_2)}{(p(x;u) \vee q(y,v),(h_1,c_1) \vee (h_2,c_2))}, \tag{17}$$

$$R_3 : \frac{(p(x;h_1u),c_1) \rightarrow (q(y;h_2v),c_2)}{(p(x;u) \rightarrow q(y,v),(h_1,c_1) \rightarrow (h_2,c_2)))}, \tag{18}$$

$$R_4 : \frac{(p(x,u) \rightarrow q(y,v),(h_1,c_1)),(p(x,u),(h_2,c_2))}{(q(y,v),(h_1,c_1) \otimes (h_2,c_2))}, \tag{19}$$

$$R_5 : \frac{(p(x,u),(h_1,c_1)) \rightarrow (q(y,v),(h_2,c_2)),(p(x,u),(h_3,c_3))}{(q(y,v),(h_4,c_4))}. \tag{20}$$

Where, operator "$\otimes$" is the lukasiewicz conjunction, *i.e.*, $\forall h_i, h_j \in H$, $h_i \otimes h_j = h_{1 \vee (i+j-n)}$. In (20), $(h_4,c_4) = ((h_1,c_1) \rightarrow (h_2,c_2)) \otimes (h_3,c_3)$, and (20)is extension of (19). Here, $\forall (h_1,c_1),(h_2,c_2) \in T = H \times Tr$,

$$(h_1,c_1) \otimes (h_2,c_2) = (h_1 \otimes h_2, c_1 \otimes c_2), \tag{21}$$

and $\forall (h_1,c_1),(h_2,c_2) \in \Gamma$,

$$(h_1,c_1) \otimes (h_2,c_2) = f(f^{-1}((h_1,c_1)) \otimes f^{-}((h_2,c_2))). \tag{22}$$

*Example 6.* Suppose the following linguistic sentences:

**i)** *If a student studies very hard is true and his university is high-ranking is more true , then he will be a good employee is true .*
**ii)** *The university where Robert studies is high-ranking is little false.*
**iii)** *Robert studies hard is possibly true.*

The question: *how about will Robert be a good employee?*

By formalizing, **i)- iii)** can be rewritten by follows:

**iv)** *(studies(x; very hard), true)* $\wedge$ *(is(Univ(x); high-ranking), (more, true))* $\rightarrow$*(employee(x; good), ($h_0$, true)))*.

**v)** *(is(Univ(Robert); high-ranking), (little, false)).*
**vi)** *(studyies(Robert; hard), (possibly, true)).*

Based on knowledge base **iv)- vi)** and inference rules (15)-(20), we build deduction as follows:

1. *((studying(x; hard), (very, true))∧ (is(Univ(x); high-ranking), (more, true)))* → *(employee(x; good), ($h_0$, true))*, (by **iv)** and (15))
2. *(studying(x; hard) ∧ is(Univ(x); high-ranking), (very, true) ∧ (more, true))* → *(employee(x; good), ($h_0$, true))*, (by 1 and (16))
3. *((studying(x; hard) ∧ is(Univ(x); high-ranking))→ employee(x; good), ((very, true) ∧ (more, true))* → *($h_0$, true))*, (by 2 and (18))
4. *is(Univ(Robert); high-ranking), (little, false))*, (by **v)**)
5. *(studying(Robert; hard), (possibly, true))* (by **vi)**)
6. *(studying(Robert; hard) ∧ is(Univ(Robert); high-ranking), (possibly, true) ∧ (little, false))*, (by 4, 5 and (16))
7. *((studying(Robert; hard) ∧ is(Univ(Robert); high-ranking))→ employee (Robert; good), ((very, true) ∧ (more, true))* → *($h_0$, true))*, ( *Robert* standing for $x$ of 3 )
8. *(employee(Robert; good), ((very, true) ∧ (more, true))* → *($h_0$, true))* ⊗ *((possibly, true) ∧ (little, false)))*. (by 6, 7 and (19))

According to Example 4, we obtain

$$(very, true) \wedge (more, true) = (more, true), \tag{23}$$

$$((very, true) \wedge (more, true)) \to (h_0, true)$$
$$= (more, true) \to (h_0, true) = (very, true), \tag{24}$$

$$(possibly, true) \wedge (little, false) = (more, false), \tag{25}$$

$$(very, true) \otimes (more, false) = (very, false). \tag{26}$$

Hence, the conclusion is *(employee(Robert; good), (very, false))*, *i.e.*, *Robert will be a good employee is very false.*

## 6 Conclusion

In this paper, by analyzing the semantically ordering relation of *Truth*, orderings on linguistic hedges $H$ and atomic evaluating syntagm $Tr = \{true, false\}$ are obtained, respectively. Assuming linguistic hedges $H$ is finite chain, Lukasiewicz product algebra of $T = H \times Tr$ of *Truth* is obtained, and term-set $T(X)$ of *Truth* is embedded into an algebra $\Gamma$ of type $\Delta = \{\vee, \wedge, ', \to_L\}$. By using Lukasiewicz conjunction "$\otimes$", $\Gamma$ can be used to obtain linguistic decision truth.

## Acknowledgments

This work was supported by the youth foundation of sichuan province (grant no. 06ZQ026-037) and the education department foundation of sichuan province (grant no. 2005A121, 2006A084).

## References


1. Zadeh, L. A.: Fuzzy logic = computing with words. Fuzzy Systems. 4 (1996) 103–111
2. Zadeh, L. A.: Toward a theory of fuzzy information granulation and its centrality in houman reasoning and fuzzy logic. Fuzzy Sets and Systems. 90 (1997) 103–111
3. Novák, V., Perfilieva, I., Močkoř, J.: Mathematical principles of fuzzy logic. Kluwer Academic Publishers. (1999)
4. Novák, V.: Antonyms and linguistic quantifiers in fuzzy logic. Fuzzy Sets and Systems. 124 (2001) 335–351
5. Dvořák, A., Novák, V.: Fromal theories and linguistic descriptions. Fuzzy Sets and Systems. 143 (2004) 169–188
6. Wang, P. P.: Computing with words, John Wiley and Sons. (2001)
7. Trillas, E.: On the use of words and fuzzy sets. Information Sciences. 176 (2006) 1463-1487
8. Herrera, F., Herrera-Viedma, E.: Aggregation operators for linguistic weighted information. IEEE Trans.System,Man, Cybernet.-Part A:Systems Humans. 27 (1997) 646–656
9. Herrera, F., Lopez, E., Rodriguez, M. A.: A linguistic decision model for promotion mix management solved with genetic algorithms. Fuzzy Sets and Systems. 131 (2002) 47–61
10. Pei, Z., Du, Y. J., Yi, L. Z., Xu, Y.: Obtaining a complex linguistic data summaries from database based on a new linguistic aggregation operator. Springer-Verlag Berlin Heidelberg. (2005) 771-778
11. Ho, N. C., Khang, T. D., Huynh, V. N.: An algebraic approach to linguistic hedges in Zadehs fuzzy logic. Fuzzy Sets and Systems. 129 (2002) 229–254
12. Ho, N. C., Khang, T. D., Huynh, V. N., Nguyen, H. C.: Hedge algebras, linguistic valued logic and their application to fuzzy reasoning, International Journal of Uncertainty. Fuzziness and Knowledge-Based Systems. 7 (1999) 347–361
13. Ho, N. C., Walfgang, W.: Extended hedge algebras and their application to fuzzy logic. Fuzzy Sets and Systems. 52 (1992) 259–281
14. Pei, Z., Xu, Y.: Lattice implication algebra model of linguistic variable Truth and its inference. in edited by Da Ruan et al, Applied Computational Intelligence, World Scientific. (2004) 93–98
15. Zadeh, L. A.: The concept of linguistic variable and its application to approximate reasoning (I), (II), (III). Information Sciences. 8 (1975) 199–149, 301–357, 9 (1975) 943–80

# Fuzzy Subgroups with Meet Operation in the Connection of Möbius Transformations

Paavo Kukkurainen

Lappeenranta University of Technology, P.O. Box 20
53851 Lappeenranta, Finland
paavo.kukkurainen@lut.fi
http://www.it.lut.fi

**Abstract.** Let $G$ be any subgroup of the group of Möbius transformations and $\mathscr{F}$ a set of stabilizers and their intersections. Taking a fuzzy subgroup of $G$ given by means of stabilizers of $G$ as the mapping $\mathscr{A} : G \longrightarrow \mathscr{F}$, we examine the behaviour of the meet operation $\wedge$ in $\mathscr{F}$.

**Keywords:** Fixed points and multipliers of Möbius transformations, conjugation, stabilizers, fuzzy subgroups, meet operation.

## 1 Preliminaries

We refer to [3] for the concepts of complex analysis.

Let us denote by $\mathbb{C}$ the complex plane and by $\widehat{\mathbb{C}}$ the extended complex plane. A Möbius transformation is a mapping $g : \widehat{\mathbb{C}} \longrightarrow \widehat{\mathbb{C}}$ for which

$$g(z) = \frac{az+b}{cz+d} \quad a,\ b,\ c,\ d \in \mathbb{C}, \quad ad - bc \neq 0$$

and $g(-\frac{d}{c}) = \infty$, $g(\infty) = \frac{a}{c}$. Möbius transformations form a group denoted by $M$. Two transformations $g_1$ and $g_2$ in $M$ are conjugate if $g_1 = \psi g_2 \psi^{-1}$ for some $\psi \in M$. In the same way two groups $G_1$ and $G_2$ of Möbius transformations are conjugate if $G_1 = \psi G_2 \psi^{-1}$. A point $z$ is said to be a fixed point of $g \in M$ if $g(z) = z$. It is known that any $g \in M \setminus \{id\}$ has either one fixed point or two fixed points in $\widehat{\mathbb{C}}$. Let $G$ be a subgroup of $M$. The set $G_z = \{g \in G \mid g(z) = z\}$ is a stabilizer of $G$ at $z \in \widehat{\mathbb{C}}$. Any stabilizer $G_z$ equipped with the composition forms a group.

In [1] and [3], it is presented how to conjugate a Möbius transformation to the standard form: Suppose $g \in M \setminus \{id\}$ has two fixed points $\alpha$ and $\beta$ or only one fixed point $\alpha$. In both cases $g$ may be conjugated by any $\psi \in M$ with $\psi(\alpha) = \infty$, $\psi(\beta) = 0$ and $\psi(g(\beta)) = 1$ if $g(\beta) \neq \beta$. If $g$ has one fixed point $\alpha$, we choose any $\beta \neq \alpha$ in $\mathbb{C}$. This conjugation yields

$$(\psi g \psi^{-1})(\infty) = \infty,$$
$$(\psi g \psi^{-1})(0) = \begin{cases} 0 & \text{if } g(\beta) = \beta \\ 1 & \text{if } g(\beta) \neq \beta \end{cases}$$

P. Melin et al. (Eds.): IFSA 2007, LNAI 4529, pp. 445–450, 2007.
© Springer-Verlag Berlin Heidelberg 2007

Now it is possible to conjugate any Möbius transformation $g \neq id$ to the standard form: There is a Möbius transformation $\psi$ satisfying

$$\begin{aligned}(\psi g \psi^{-1})(z) &= m_1(z) = z+1 \text{ if the only fixed point of } g \text{ is } \alpha,\\ (\psi g \psi^{-1})(z) &= m_k(z) = kz \text{ for some } k \in \mathbb{C},\ k \neq 0,1 \text{ if } \alpha \text{ and } \beta\end{aligned}$$

are the two fixed points of $g$.

In the latter case the number $k$ is independent on the choice of $\psi$.

The above conjugation gives rise to classify Möbius transformations according to the standard forms: Let $g \in M \setminus \{id\}$. Then

$g$ is parabolic iff $g$ is conjugate to $m_1$,
$g$ is loxodromic iff $g$ is conjugate to $m_k$, $|k| \neq 1$,
$g$ is elliptic iff $g$ is conjugate to $m_k$, $|k| = 1$, $k \neq 1$.

It follows that $g$ is parabolic iff it has one fixed point and $g$ is loxodromic or elliptic iff it has two fixed points. Further, conjugate transformations are always of the same type. We may share loxodromic transformations $g$ into the two parts: If $g(D) = D$ for the upper-half plane or the unit disk $D \subset \widehat{\mathbb{C}}$ (more general some upper-half plane or disk), then $g$ is hyperbolic. Otherwise $g$ is strictly loxodromic.

Suppose $g \in M$ has two distinct fixed points $\alpha$ and $\beta$. Then the multiplier of loxodromic or elliptic transformations $g$ is defined as the cross-ratio

$$k(g) = (g(z), z, \alpha, \beta)$$

which holds for every $z \in \widehat{\mathbb{C}} \setminus \{\alpha, \beta\}$. If $g$ is parabolic, we set $k(g) = 1$. Moreover, $k = k(g)$ in the standard forms. The multipliers are invariant in conjugation.

The following concepts can be found from [4]: Let $(P, \leq)$ be a partially ordered set and $A$ a nonempty set. Then a mapping $\mathscr{A} : A \longrightarrow P$ is a $P$-(fuzzy)set on $A$. If $\mathscr{G} = (G, \cdot)$ is a group, then a $P$-set on $G$, $\mathscr{A} : G \longrightarrow P$ is said to be a $P$-(fuzzy) subgroup of $\mathscr{G}$.

## 2 Möbius Transformations Applied to Fuzzy Subgroups with Meet Operation

The following proposition is presented and proved in [2]. The starting point to the proof was the results in [4] and [5]:

**Proposition 1.** *Let $G$ be a subgroup of the group of Möbius transformations and let $G_z = \{g \in G \mid g(z) = z\}$ be a stabilizer of $G$ at $z \in \hat{\mathbb{C}}$. If $\mathscr{F}$ is a set of stabilizers of $G$ and their intersections, and $\mathscr{F} = (\mathscr{F}, \leq)$ is a partially ordered set under $p \leq q$ iff $p \supseteq q$ $(p, q \in \mathscr{F})$, then $\mathscr{A} : G \longrightarrow \mathscr{F}$, as a $\mathscr{F}$-subgroup of $G$, is*

$$\mathscr{A}(g) = \bigcap (G_z \mid g \in G_z\}$$

Denote by $F_g$ (resp. $F_h$) the set of fixed points of $g$ (resp. $h$). Defining the meet operation $\wedge$ as the infimum, we get

**Proposition 2.** *Let $G$ be a subgroup of the group of Möbius transformations and $g, h \in G$.*

1. *If $F_g = F_h$, then*
$$\mathscr{A}(gh) = \mathscr{A}(g) \wedge \mathscr{A}(h),$$
*except in the case where $gh = id$, $g \neq id$ and $h \neq id$ implies*
$$\mathscr{A}(gh) > \mathscr{A}(g) \wedge \mathscr{A}(h).$$

2. *Let $F_g \neq F_h$. If $g$ and $h$ share a fixed point there are three different cases: If $g$ has one fixed point and $h$ two fixed points, then*
$$\mathscr{A}(gh) > \mathscr{A}(g) \wedge \mathscr{A}(h).$$
*If both $g$ and $h$ have two fixed points but only one of them is common to $g$ and $h$, then*
$$\mathscr{A}(gh) \geq \mathscr{A}(g) \wedge \mathscr{A}(h).$$
*The case where $g = id$ and $h \neq id$ implies*
$$\mathscr{A}(gh) = \mathscr{A}(g) \wedge \mathscr{A}(h).$$

3. *Suppose that $g$ and $h$ have no common fixed point. Then*
$$\mathscr{A}(gh) > \mathscr{A}(g) \wedge \mathscr{A}(h).$$

*Proof.* Recall that
$$\mathscr{A}(g) = \bigcap (G_z \mid g \in G_z\}.$$
Assume $g$ and $h$ have both one fixed point $\alpha$ being common to them and let $gh \neq id$. Conjugating with a suitable $\psi \in M$, $\psi(\alpha) = \infty$, we obtain matrix representations for $\psi g \psi^{-1}$ and $\psi h \psi^{-1}$ corresponding to the standard form $z \mapsto z + 1$ and the transformation $z \mapsto z + b$, $b \neq 0, -1$ ($gh \neq id$ implies $b \neq -1$). Using matrix representations, we get

$$\psi(gh)\psi^{-1} = \begin{pmatrix} 1 & 1 \\ 0 & 1 \end{pmatrix} \begin{pmatrix} 1 & b \\ 0 & 1 \end{pmatrix} = \begin{pmatrix} 1 & b+1 \\ 0 & 1 \end{pmatrix},$$
$$\psi(hg)\psi^{-1} = \begin{pmatrix} 1 & b \\ 0 & 1 \end{pmatrix} \begin{pmatrix} 1 & 1 \\ 0 & 1 \end{pmatrix} = \begin{pmatrix} 1 & b+1 \\ 0 & 1 \end{pmatrix}.$$

Now $(\psi(gh)\psi^{-1})(z) = (\psi(hg)\psi^{-1})(z) = z + b + 1$ with one fixed point $\infty$ (b $\neq$ -1). It follows that $gh$ (resp. $hg$) has the only one fixed point $\alpha$. Then

$$\mathscr{A}(gh) = G_\alpha = G_\alpha \wedge G_\alpha = \mathscr{A}(g) \wedge \mathscr{A}(h).$$

Suppose $g$ and $h$ have two common fixed points $\alpha$ and $\beta$. If $gh \neq id$, then

$$\mathscr{A}(gh) = G_\alpha \cap G_\beta = (G_\alpha \cap G_\beta) \wedge (G_\alpha \cap G_\beta) = \mathscr{A}(g) \wedge \mathscr{A}(h).$$

If $g = h = id$, we obtain

$$\mathscr{A}(gh) = \mathscr{A}(id) = (id, \cdot) = (id, \cdot) \wedge (id, \cdot) = \mathscr{A}(g) \wedge \mathscr{A}(h).$$

Let $gh = id$, $g \neq id$ and $h \neq id$. Then $g = h^{-1}$ meaning that $g$ and $h$ have the same fixed points. Therefore $\mathscr{A}(g) = \mathscr{A}(g^{-1}) = \mathscr{A}(h)$ giving

$$\mathscr{A}(gh) = \mathscr{A}(id) = (id, \cdot) > \mathscr{A}(g) = \mathscr{A}(g) \wedge \mathscr{A}(h)$$

This proves $(i)$.

Assume $g$ has one fixed point $\alpha$ and $h$ has two fixed points $\alpha$ and $\beta$, $\beta \neq \alpha$. We choose a suitable $\psi \in M$ with $\psi(\alpha) = \infty$, $\psi(\beta) = 0$ and $\psi(g(\beta)) = 1$ if $g(\beta) \neq \beta$. Conjugation with $\psi$ leads to matrix representations for $\psi g \psi^{-1}$ and $\psi h \psi^{-1}$ corresponding to the standard forms $z \mapsto z + 1$ and $z \mapsto kz$ for some $k \neq 0, 1$. We conclude

$$\begin{aligned} \psi g \psi^{-1} &= \begin{pmatrix} 1 & 1 \\ 0 & 1 \end{pmatrix} \quad \text{and} \quad \psi h \psi^{-1} = \begin{pmatrix} k & 0 \\ 0 & 1 \end{pmatrix}, \\ \psi(gh)\psi^{-1} &= \begin{pmatrix} 1 & 1 \\ 0 & 1 \end{pmatrix} \begin{pmatrix} k & 0 \\ 0 & 1 \end{pmatrix} = \begin{pmatrix} k & 1 \\ 0 & 1 \end{pmatrix}, \\ \psi(hg)\psi^{-1} &= \begin{pmatrix} k & 0 \\ 0 & 1 \end{pmatrix} \begin{pmatrix} 1 & 1 \\ 0 & 1 \end{pmatrix} = \begin{pmatrix} k & k \\ 0 & 1 \end{pmatrix}. \end{aligned}$$

In the former case, $(\psi(gh)\psi^{-1})(z) = kz + 1$ with two distinct fixed points $\infty$ and $\frac{1}{1-k}$ $(k \neq 1)$. In the latter case, $\psi(hg)\psi^{-1}(z) = kz + k$ with fixed points $\infty$ and $\frac{k}{1-k}$. This means that $gh$ has distinct fixed points $\alpha$ and $\psi^{-1}(\frac{1}{1-k}) = \beta_1$ and $hg$ has fixed points $\alpha$ and $\psi^{-1}(\frac{k}{1-k}) = \beta_2$. Since $k \neq 1$ and so $\frac{1}{1-k} < \infty$, then $\beta_1 \neq \alpha$ (also $\beta_2 \neq \alpha$). Therefore

$$\mathscr{A}(gh) = G_\alpha \cap G_{\beta_1} > G_\alpha = G_\alpha \wedge (G_\alpha \cap G_\beta) = \mathscr{A}(g) \wedge \mathscr{A}(h).$$

Correspondingly,

$$\mathscr{A}(hg) > \mathscr{A}(g) \wedge \mathscr{A}(h).$$

Suppose $g$ and $h$ have two fixed points but one common fixed point $\alpha$. Let $g(\alpha) = \alpha$, $g(\beta_1) = \beta_1$, $h(\alpha) = \alpha$ and $h(\beta_2) = \beta_2$, $\beta_1 \neq \beta_2$, $\alpha \neq \beta_1, \beta_2$. By conjugation, we obtain again

$$\begin{aligned} &(\psi g \psi^{-1})(z) = k_1 z \, (k_1 \neq 0, 1) \quad \text{with fixed points} \quad \psi(\alpha) = \infty,\ \psi(\beta_1) = 0, \\ &(\psi h \psi^{-1})(z) = k_2 z + \beta - k_2 \beta \, (k_2 \neq 0, 1) \text{ with fixed points } \psi(\alpha) = \infty,\ \psi(\beta_2) = \beta. \end{aligned}$$

Then

$$\begin{aligned} \psi(gh)\psi^{-1} &= \begin{pmatrix} k_1 & 0 \\ 0 & 1 \end{pmatrix} \begin{pmatrix} k_2 & \beta - k_2\beta \\ 0 & 1 \end{pmatrix} = \begin{pmatrix} k_1 k_2 & k_1\beta - k_1 k_2 \beta \\ 0 & 1 \end{pmatrix}, \\ \psi(hg)\psi^{-1} &= \begin{pmatrix} k_2 & \beta - k_2\beta \\ 0 & 1 \end{pmatrix} \begin{pmatrix} k_1 & 0 \\ 0 & 1 \end{pmatrix} = \begin{pmatrix} k_1 k_2 & \beta - k_2\beta \\ 0 & 1 \end{pmatrix} \end{aligned}$$

leading to

$$(\psi(gh)\psi^{-1})(z) = (k_1k_2)z + k_1\beta - k_1k_2\beta$$

with two distinct fixed points $\infty$ and $\frac{k_1k_2\beta-k_1\beta}{k_1k_2-1}$ if $k_1k_2 \neq 1$ and one fixed point $\infty$ if $k_1k_2 = 1$, and

$$(\psi(hg)\psi^{-1})(z) = (k_1k_2)z + \beta - k_2\beta$$

with two fixed points $\infty$ and $\frac{k_2\beta-\beta}{k_1k_2-1}$, if $k_1k_2 \neq 1$ and one fixed point $\infty$ if $k_1k_2 = 1$.
Then $gh$ has two distinct fixed points $\alpha$ and $\psi^{-1}(\frac{k_1k_2\beta-k_1\beta}{k_1k_2-1}) = \beta_3$ or only one fixed point $\alpha$. Also, $hg$ has fixed points $\alpha$ and $\psi^{-1}(\frac{k_2\beta-\beta}{k_1k_2-1}) = \beta_4$ or only fixed point $\alpha$. In every case, $(G_\alpha \cap G_{\beta_1})$ and $(G_\alpha \cap G_{\beta_2})$ have only two common lower bounds $G_\alpha$ and $G$. Because $G_\alpha \geq G$, the infimum of $(G_\alpha \cap G_{\beta_1})$ and $(G_\alpha \cap G_{\beta_2})$ is $G_\alpha$. Therefore

$$\mathscr{A}(gh) = G_\alpha \cap G_{\beta_3} \geq G_\alpha = (G_\alpha \cap G_{\beta_1}) \wedge (G_\alpha \cap G_{\beta_2}) = \mathscr{A}(g) \wedge \mathscr{A}(h),$$

where the equality holds for if $gh$ has only one fixed point $\alpha$.
If $g = id$ and $h \neq id$, then $gh = h$ and we obtain

$$\mathscr{A}(gh) = \mathscr{A}(h) = (id, \cdot) \wedge \mathscr{A}(h) = \mathscr{A}(g) \wedge \mathscr{A}(h).$$

Also $(ii)$ is proved.

Finally, assume that $g$ and $h$ have no common fixed points. This implies that $G$ is the only common lower bound of $\mathscr{A}(g)$ and $\mathscr{A}(h)$ and therefore $\mathscr{A}(g) \wedge \mathscr{A}(h) = G$. The infimum $G \notin \mathscr{F}$ but $G \in \mathscr{F} \cup \{G\} = \mathscr{F}_1$ which contains $\mathscr{F}$. Also, $(\mathscr{F}_1, \leq)$ is a partially ordered set having the same order as $\mathscr{F}$. On the other hand, $gh$ has at most two fixed points $z_i$, $i = 1, 2$. Consequently

$$\mathscr{A}(gh) = \cap G_{z_i} > G = \mathscr{A}(g) \wedge \mathscr{A}(h).$$

The equality is impossible in any case because both $g$ and $h$ belong to $G$ but they have not the (common) fixed points $z_i$ of $gh$.
This proves $(iii)$. □

**Example 1.** *Consider parabolic transformations* $g(z) = \frac{2z}{z+2}$ *and* $h(z) = \frac{3z}{z+3}$ *both having the fixed point 0. Trivially,* $gh = \frac{6z}{5z+6}$ *is also parabolic with the same fixed point* 0. *Then*

$$\mathscr{A}(gh) = G_0 = G_0 \wedge G_0 = \mathscr{A}(g) \wedge \mathscr{A}(h).$$

*Observe that the consideration of Möbius transformations in Proposition 2 occurs pairwise. It is shown in [4] that in general any group* $G = (G, \cdot)$ *and a lattice-valued fuzzy group* $\mathscr{A} : G \longrightarrow L$ *satisfy*

$$\mathscr{A}(x \cdot y) = \mathscr{A}(x) \wedge \mathscr{A}(y) \quad \text{for all } x \text{ and } y \in G$$

*iff*

$$\mathscr{A}(z) = I \quad \text{for every } z \in G$$

*where $I$ is the supremum of $L$. In this paper it means that $G$ is a trivial group $(id, \cdot)$ (although $\mathscr{F}$ is not a lattice).*

*As stated at the rest of the proof of Proposition 2 the whole group $G$ does not belong to $\mathscr{F}$ which is not therefore a lattice. However, it is the infimum of the complete lattice $\mathscr{F} \cup \{G\}$ or it may be the infimum of some pair of elements in $\mathscr{F}$. The following example illustrates this situation:*

**Example 2.** *Let $g(z) = -z$ with the fixed points $0, \infty$ and $h(z) = \frac{1}{z}$ with the fixed points $-1, 1$. Then $(gh)(z) = -\frac{1}{z}$ with the fixed points $-i, i$.*

$$\mathscr{A}(g) = G_0 \cap G_\infty \subset G,$$
$$\mathscr{A}(h) = G_{-1} \cap G_1 \subset G.$$

*As conclusion, $G$ is the only common lower bound of $\mathscr{A}(g)$ and $\mathscr{A}(h)$. Although $G \notin \mathscr{F}$, it is the infimum of them. This yields*

$$\mathscr{A}(gh) = G_{-i} \cap G_i > G = \mathscr{A}(g) \wedge \mathscr{A}(h).$$

## References


1. A. F. Beardon: The Geometry of Discrete Groups, Springer-Verlag, New York, 1983.
2. P. Kukkurainen: Level Sets as a Topological Base Applied to Vol.9, No.5 (2005), p.511-513.
3. M. Seppälä and T. Sorvali: Geometry of Riemann Surfaces and Teichmüller spaces. In Elsevier Science Publisher B.V.; North Holland, vol. 169, Amsterdam, The Netherland, 1992.
4. B. Ŝeŝelja and A. Tepavĉeviĉ: Fuzzy groups and collections of subgroups, Fuzzy Sets and Systems 83 (1996), p.85-91.
5. B. Ŝeŝelja and A. Tepavĉeviĉ: On a construction on codes by $P$-fuzzy sets. Zb.Rad.Priod.-Mat., Vol.20,pp.71-80,1990.

# A Method for Automatic Membership Function Estimation Based on Fuzzy Measures

Grzegorz Nieradka and Bohdan Butkiewicz

Warsaw University of Technology, Institute of Electronic Systems,
Nowowiejska 15/19, 00 - 665 Warsaw, Poland
nieradka@ise.pw.edu.pl, bb@ise.pw.edu.pl

**Abstract.** Estimation of membership function is one of the most important problems in the application of fuzzy sets. This paper presents one of approaches to this problem. A method for estimation of membership function is proposed, based on fuzzy measures: fuzzy entropy and fuzzy index. Examples of generating membership function in the field of image processing are shown.The method presented in this paper can be used in other fields of computer sciences, where statistical data are available.

## 1 Introduction

Proper choosing of membership function's shape and values is usually not an easy task. There are many methods proposed in the literature also in the image processing field [14]. Some methods adopt an approach based on transformation from a probability distribution to a possibility distribution [3]. Sometimes statistical data describing an image are clustered by C-means algorithm. Membership functions of pixels' brightness are chosen based on these clusters [2]. A method based on optimization objective function is presented in [4]. The application of measures of the fuzzy set such as specificity or consistency is also proposed in [7]. The authors of this paper propose objective function related to entropy measure.

The most similar solution this our proposition is described in [5], but there are numerous discrepancies between the two approaches. Whereas both the works pertain the idea of maximization of entropy, only this paper refers to the measure of entropy other than the idea of probability of fuzzy event as described in [5]. Moreover, this work offers another measure, namely fuzzy index, added for the more exact description of available data image.

In this paper the authors proposed a novel method in which calculated membership function is utilized for modeling linguistic commands used for image processing. Some examples, where such commands are applied prove to be successful approach.

## 2 Membership Function

The shape of S-function is commonly used for the representation of the degree of brightness or whiteness of pixels in grey levels images. This S-function was originally introduced by Zadeh [17].

P. Melin et al. (Eds.): IFSA 2007, LNAI 4529, pp. 451–460, 2007.
© Springer-Verlag Berlin Heidelberg 2007

For flexibility another definition of S-function was proposed [6]:

$$S\left(x;a,b,c\right)=\begin{cases}0, & x\leq a\\ \frac{(x-a)^2}{(b-a)(c-a)}, & a<x\leq b\\ 1-\frac{(x-c)^2}{(c-b)(c-a)}, & b<x\leq c\\ 1, & x\geq c\end{cases} \quad (1)$$

where $x$ is a variable, and $a$, $b$ and $c$ are parameters determining the shape of S-function. In this definition $b$ can be any point between $a$ and $c$. Some examples of possible shapes of this S-function are shown in the Fig. 1

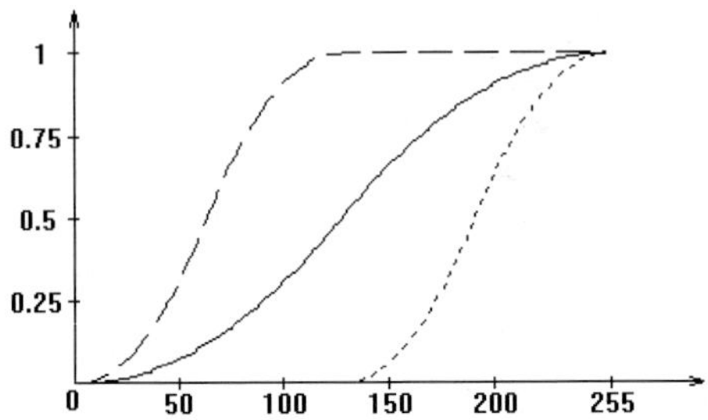


**Fig. 1.** Different shapes of S-function depend on $a$, $b$ and $c$ parameters. Lines denoting parameters: dashed line $a = 0, b = 63.5, c = 127.5$; solid line $a = 0, b = 127.5, c = 255$; dotted line $a = 127.5, b = 191, c = 255$

## 3 Fuzzy Measures

In the literature many fuzzy measures have been proposed [8] [13] as well as measures of fuzzines [15]. This paper incorporates two measures of fuzzy set namely a fuzzy entropy [1] and index of fuzzines introduced by Kaufmann [10].

### 3.1 Fuzzy Entropy

Many definitions of fuzzy entropy [15] [1] exist in the literature. For the purpose of this work authors employ definition of total entropy [1] [16] which is described as follows.

Let $I$ be a set with randomly occuring events $\{x_1, x_2, \ldots, x_n\}$ in an experiment, and $\{p_1, p_2, \ldots, p_n\}$ are respective probabilities of events. Fuzzyfication of set $I$ induces two kinds of uncertainties. Total entropy of the set $F$, being fuzzified set $I$, consists of two parts. The first part of total entropy is a measure deduced from "random" nature of the experiment. Expected value of this uncertainty is computed as Shannon entropy:

$$H\left(p_1, p_2, \ldots, p_n\right) = -\sum_{i=1}^{n} p_i \log\left(p_i\right) \quad (2)$$

The second uncertainty arises from the fuzzines of the fuzzy set $F$ related to the ordinary set. This amount of ambiguity is given by:

$$S(\mu_i) = -\mu_i \log(\mu_i) - (1-\mu_i)\log(1-\mu_i) \tag{3}$$

The statistical average $m$ of the ambiguity for the whole set is given by equation (4):

$$m(\mu, p_1, p_2, \ldots, p_n) = \sum_{i=1}^{n} p_i S(\mu_i) \tag{4}$$

Therefore, the total entropy of the set $F$ is expressed as follows:

$$H_{total} = H(p_1, p_2, \ldots, p_n) + m(\mu, p_1, p_2, \ldots, p_n) \tag{5}$$

### 3.2 Index of Fuzziness

Let $X$ be universum of discourse and $P$ power set of $X$. Kaufmann introduced the index of fuzzines $\gamma$ of fuzzy set $A \subseteq P$:

$$\gamma(A) = \frac{2}{n^k} \times d(A, A^{near}) \tag{6}$$

where $d$ is a suitable metric on the universum $X$, $k$ is positive number and $n$ number of supporting points. $A^{near}$ is the nearest crisp set to $A$ defined as follows:

$$\mu_A^{near}(x) = \begin{cases} 1 \; if \; \mu_A(x) \geq 0.5 \\ 0 \; if \; \mu_A(x) \leq 0.5 \end{cases} \tag{7}$$

Using Minkowski's $q$-norm as a metric, and putting $k = \frac{1}{q}$, the index of fuzzines can be defined as:

$$\gamma(A) = \frac{2}{n^{\frac{1}{q}}} \left\{ \sum_i |\mu_A(x_i) - [1 - \mu_{A^{near}}(x_i)]|^q \right\}^{\frac{1}{q}} \tag{8}$$

In this paper the linear version of this index is used, so the exponential $q = 1$

$$\gamma(A) = \frac{2}{n} \sum_i |\mu_A(x_i) - \mu_{\overline{A}}(x_i)| \tag{9}$$

where $\mu_{\overline{A}}(x_i)$ is complement of set $A$ and $\mu_{\overline{A}}(x_i) = 1 - \mu_A(x_i)$. The difference between the set and nearest ordinal set can be calculated:

$$\gamma(A) = \frac{2}{n} \sum_i [\min\{\mu_A(x_i), \;\; 1 - \mu_A(x_i)\}] \tag{10}$$

Obviously, for an image $O$ of size $M \times N$ with $L$ levels of grey pixels' brightness $g$, and with the histogram $h(g)$ of the image $O$, linear index of fuzzines can be given by:

$$\gamma_{linear}(O) = \frac{2}{MN} \sum_{g=0}^{L-1} h(g) \cdot \min[\mu_O(g), \bar{\mu}_{\overline{O}}(g)] \tag{11}$$

where: $\bar{\mu}_{\overline{O}}$ is complement of $O$ and $\mu_{\overline{O}} = 1 - \mu_O(g)$

## 4 Algorithm of Membership Function Calculation

For estimation of membership function the authors used the measures described above as objective functions for a maximization problem. We would like to find a function which maximizes the information about an image. This condition is reformulated into entropy principle. Taking into consideration that entropy is the measure of information, hence the function which has the maximum entropy is the most informative. However, there still exits need to find a function which describes the fuzzy set in the best way. For this purpose the authors used the index of fuzzines. The result is the function which has the biggest value of index of fuzzines.

Therefore, the problem was reformulated into looking for the function which is optimal with regards to total entropy (5) and index of fuzzines (11).

The objective is find the parameters $a$, $b$ and $c$ of function (1) describing the shape of function, which fulfils conditions of the maximum entropy as well as maximum value of index of fuzzines. This issue is defined as a two criteria problem. The first criterium is founding the set of parameters $a_{Eopt}, b_{Eopt}, c_{Eopt}$ for which the total entropy (5) has the maximal value.

$$H_{total\_\max}\left[S\left(a_{Eopt}, b_{Eopt}, c_{Eopt}\right)\right] = \max_{a,b,c}\left\{H_{total}\left[S\left(a,b,c\right)\right] \;:\; 0 \le a,b,c \le L\right\} \tag{12}$$

The second criterium is founding the set of parameters $a_{\gamma opt}, b_{\gamma opt}, c_{\gamma opt}$ for which the value of the index of fuzzines (11) has the biggest value.

$$\gamma_{linear\_\max}\left[S\left(a_{\gamma opt}, b_{\gamma opt}, c_{\gamma opt}\right)\right] = \max_{a,b,c}\left\{\gamma\left[S\left(a,b,c\right)\right] \;:\; 0 \le a,b,c \le L\right\} \tag{13}$$

After solving eq. (12) and eq. (13) there are two sets of parameters. The S-function described by the average values of the parameters has been chosen as solution.

$$\begin{aligned} a_{opt} &= \tfrac{a_{Eopt}+a_{\gamma opt}}{2} \\ b_{opt} &= \tfrac{b_{Eopt}+b_{\gamma opt}}{2} \\ c_{opt} &= \tfrac{c_{Eopt}+c_{\gamma opt}}{2} \end{aligned} \tag{14}$$

It is assume that S-function (1) described by set of parameters' values given by (14) is the one which is the most informative and describes the fuzzy set in the best way.

Calculations for finding the set of optimal parameters (14) are performed using well known optimization algorithm Particle Swarm Optimization (in short PSO), which is well described in the literature [12] [11].

## 5 Examples of the Algorithm Results

Figures Fig. 2 ... Fig. 6 present the shapes of membership functions which were computed by this algorithm. Five different images with different histograms were chosen for illustration.

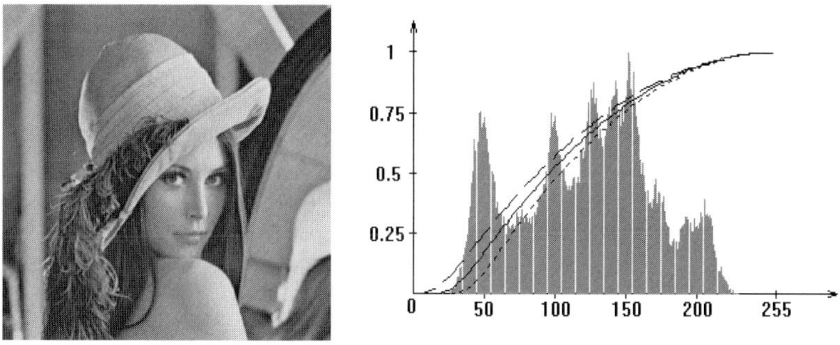


**Fig. 2.** The original image "Lena" (on the left) and histogram with determined membership function (on the right). Significance of lines: dotted – maximum index of fuzzines; dashed – maximum fuzzy entropy; solid – average between maximum entropy and maximum fuzzy index.

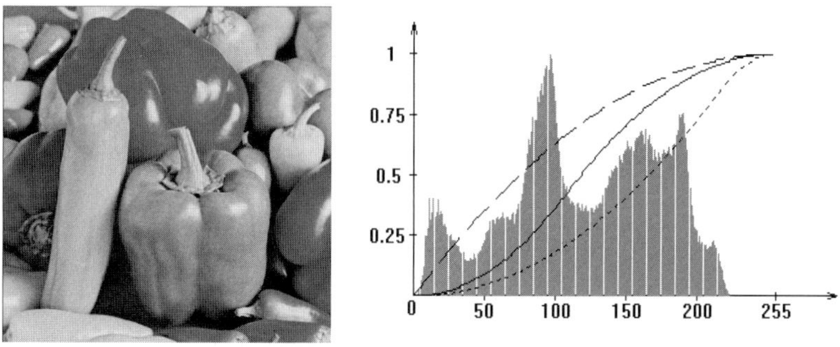


**Fig. 3.** The image "Peppers" (on the left) and histogram with determined membership functions (on the right). Significance of lines: as in the Fig. 2.

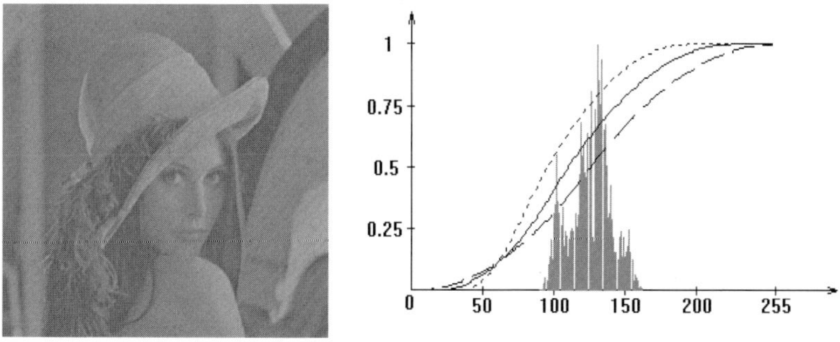


**Fig. 4.** Image "Lena" with very bad contrast (on the left) and histogram with determined membership functions (on the right). Significance of lines: as in the Fig. 2.

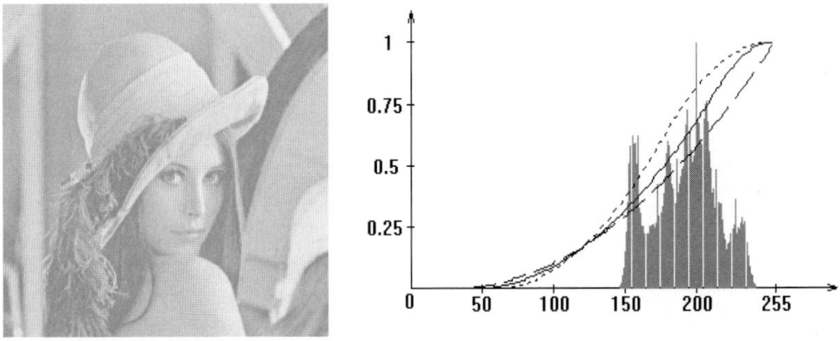


**Fig. 5.** Image "Lena" which is too bright (on the left) and histogram with determined membership functions (in the right). Significance of lines: as in the Fig. 2.

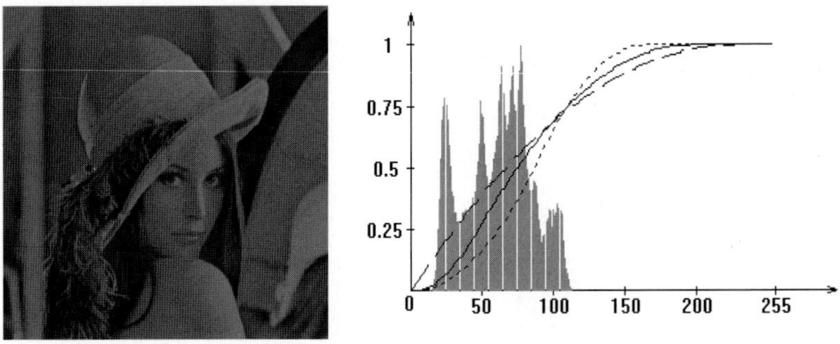


**Fig. 6.** Image "Lena" which is too dark (on the left) and histogram with determined membership functions (on the right). Significance of lines: as in the Fig. 2.

## 6 Linguistic Modifiers for Image Enhancement

For image with estimated membership function linguistic modifiers can be used. The effect of application of different modifiers was measured by *Mean Squared Error* (MSE) (15) calculated as follows:

$$MSE = \frac{\sum_{i=1}^{M_1} \sum_{j=1}^{M_2} \left| O(i,j) - O'(i,j) \right|^2}{M_1 \cdot M_2} \tag{15}$$

where $1 \leq i \leq M_1, 1 \leq j \leq M_2$ and $M_1$ and $M_2$ are respectively heights and widths of the original image $O$ and modified image $O'$ expressed in pixel values. In the presented figures and the tables, there is shown modifiers influence on image.

### 6.1 Modifier *"Increase Contrast"*

For modifier "increase contrast" is applied well known operator of intensification $INT(A)$ of fuzzy set [9]:

$$\mu_{INT(A)}(x) = \begin{cases} 2\left[\mu_A(x)\right]^2 & \forall x : \mu_A(x) < 0.5 \\ 1 - 2\left[1 - \left(\mu_A(x)\right)^2\right] & \forall x : \mu_A(x) \geq 0.5 \end{cases} \quad (16)$$

That application of this operator give the good result as is shown in the Fig.7

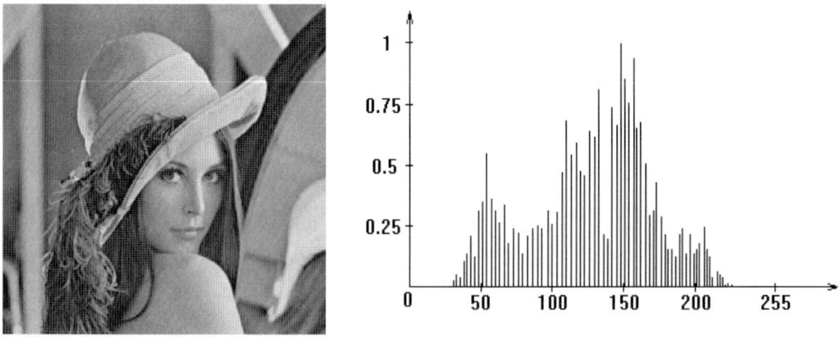


**Fig. 7.** Illustration of using *"increase contrast"* linguistic modifier. The image "Lena" after modification (on the left) and histogram of the modified image (on the right). The original image presented in the Fig. 4 has bad contrast.

**Table 1.** Influence of *"increase contrast"* hedge on the MSE error

| Image | MSE |
|---|---|
| "Lena" image with bad contrast | 1095.5 |
| After using linguistic hedge | 78.7620 |

### 6.2 Modifier "Brighter"

As a representation of linguistic modifier "brighter" the operator of dilation of fuzzy set was chosen. Originally dilation $DIL(A)$ of fuzzy set is defined as [9]:

$$\mu_{DIL(A)}(x) = \left[\mu_A(x)\right]^{0.5} \quad \forall x \in X \quad (17)$$

In this example for better result another power in eq.(17) was chosen. Following equation is used in this example:

$$\mu_{DIL(A)}(x) = \left[\mu_A(x)\right]^{0.25} \quad \forall x \in X \quad (18)$$

The linguistic modifier *"brighter"* as is given by eq.( 18) was used for image with bad dark colors which is presented in the Fig. 6. Application of this modifier gives a very good results. Enhanced image is shown in the Fig. 8. Comparison of $MSE$ error is given in the Tab. 2.

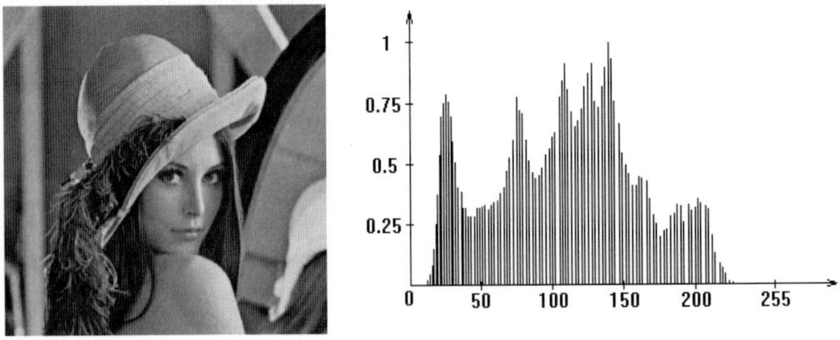


**Fig. 8.** Illustration of using *"brighter"* linguistic hedge. Image "Lena" after brightening (on the left) and histogram of modified image (on the right). Image with bad bright color before modification is presented in Fig. 6.

**Table 2.** Influence of *"brighter"* modifier at the MSE error

| **Image** | **MSE** |
|---|---|
| "Lena" image with bad bright colors | 4261.7 |
| After using linguistic hedge *"brighter"* | 350.5 |

### 6.3 Modifier "Darker"

As a linguistic modifier "darker" concentration of fuzzy set was employed. Originally dilation $CON(A)$ of fuzzy set is defined as [9]:

$$\mu_{CON(A)}(x) = [\mu_A(x)]^2 \quad \forall x \in X \tag{19}$$

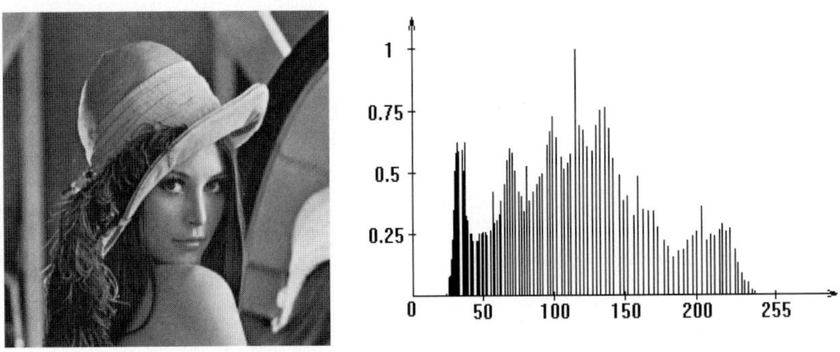


Image "Lena" after darkering Histogram of image "Lena" after darkering

**Fig. 9.** Illustration of using *"darker"* linguistic modifier. Image "Lena" after modification (on the left) and histogram (on the right). Image with bad bright colors as presented in Fig. 5.

However, here the power equal 1.5 for better result is applied in this case. The modifier given by eq.(19) with power 1.5, is utilized to image with bad white colors, as is shown in the Fig. 5. The result after application of this modifier is shown in the Fig. 9. Comparison of $MSE$ errors is given in the Tab. 3

**Table 3.** Influence of *"darker"* hedge at the MSE error

| Image | MSE |
|---|---|
| "Lena" image with bad dark colors | 5419.0 |
| After using linguistic hedge *"darker"* | 505.2 |

## 7 Conclusion

Fuzzy set theory has been successful applied to many tasks in image processing as image filtering or pattern recognition. However, for each usage of fuzzy sets not only in image processing but generally, it is needed to know the membership function's shape and values.

This paper presents the method dealing problem of determination membership function. The well known S-function is used for representing pixels' values which belong to the set of bright pixel. The methodology for choosing parameters of S-function in reasonable way is proposed basing on fuzzy measures. The chosen function is one which compromises the conditions of information and good fuzzines.

The new application area namely, the linguistic modifiers of image is outlined. The way of modelling natural language is shown with the good effects of using linguistic modifiers in image enhancement.

## References


1. Al-sharhan S., Karray F., Basir O.: Fuzzy Entropy: a Brief Survey. IEEE Proceedings of the 10th International Conference on Fuzzy Systems, Melbourne, Vic., Australia **3** (2001) 1135-1138
2. Bezdek J.C.: Pattern Recognition with Fuzzy Objective Function Algorithms. Plenum, New York (1981)
3. Bharathi B., Sarma V.V.S.: Estimation of Fuzzy Membership from Histograms. Information Sciences **35** (1985) 43-59
4. Bloch I., Aurdal L., Bijno D., Müller J.: Estimation of Class Membership Functions for Grey Level Based Image Fusion. Proc. of the Int. Conf. of Image Processing (1997) 268-271
5. Cheng H.D., Cheng J.R.: Automatically Determine the Membership Function Based on the Maximum Entropy Principle **96** (1997) 163-182
6. Cheng H.D., Cheng Y.H.: Thresholding Based on Fuzzy Partition of 2D Histogram. IEEE International Conference on Pattern Recogition, Australia **2** (1998) 1616-1618
7. Civanlar M.R., Trussel H.J.: Construction Membership Functions using Statistical Data. Fuzzy Sets and Systems **18** (1986) 1-13

8. Dubois D., Prade H.: Fuzzy Sets and Systems: Theory and Applications. Academic Press Inc. (1980)
9. Kacprzyk J.: (*in Polish*) Fuzzy Sets in Systems Analysis. Państwowe Wydawnictwa Naukowe (1986)
10. Kaufmann A.: Introduction to the theory of fuzzy subsets. Academic Press (1975)
11. Kennedy J., Eberhart J.: Particle Swarm Optimization. IEEE International Conference on Neural Networks, Perth, Australia (1995) 1942-149
12. Kennedy J., Shi Y..: Swarm Intelligence. Morgan Kaufmann Academic Press (2001)
13. Klir G.J., Folger T.A.: Fuzzy Sets, Uncertainty and Information. Prentice Hall, Englewood Clifs (1988)
14. Medasani S., Kim J., Krishnapuram R.: An Overview of membership function generation techniques for pattern recognition. International Journal of Approximate Reasoning **19** (1998) 391-417
15. Pal N.R., Bezdek J.C.: Measuring Fuzzy Uncertainty. IEEE Transactions of Fuzzy Systems **2(2)** (1994) 107-118
16. Pedrycz W., Czołgała E.: (*in Polish*) Elements and Methods of fuzzy sets theory. Państwowe Wydawnictwa Naukowe (1985)
17. Zadeh L.A.: A fuzzy-algorithmic approach to the definition of complex or imprecise concepts. International Journal Man-Machines Studies **8** (1976) 249-291

# Counting Finite Residuated Lattices*

Radim Belohlavek[1,2] and Vilem Vychodil[2]

[1] Dept. Systems Science and Industrial Engineering, Binghamton University—SUNY
Binghamton, NY 13902, USA
rbelohla@binghamton.edu
[2] Dept. Computer Science, Palacky University, Olomouc
Tomkova 40, CZ-779 00 Olomouc, Czech Republic
vilem.vychodil@upol.cz

**Abstract.** We study finite residuated lattices with up to 11 elements. We present an algorithm for generating all non-isomorphic finite residuated lattices with a given number of elements. Furthermore, we analyze selected properties of all the lattices generated by our algorithm and present summarizing statistics.

## 1 Introduction and Preliminaries

*Problem Setting.* Residuated lattices and particular subclasses of residuated lattices distinguished by identities play a key role in fuzzy logic in both the wide and narrow sense [2,4,6,7,9,12,14,16]. In particular, residuated lattices are considered as structures of truth degrees, i.e., scales of truth degrees equipped with truth functions of (fuzzy) logical connectives. Most common choice of a residuated lattice is a structure defined on the real unit interval $[0,1]$ with multiplication given by a left-continuous t-norm, see [4,9].

In general, non-comparable truth degrees might be of interest for both theoretical and practical reasons (we omit details) which leads us beyond the scope of linearly ordered residuated lattices. Moreover, "small" lattices play an important role. Namely, according to Miller's 7±2 phenomenon well known from psychology [15], humans are able to assign degrees in a consistent manner provided the scale of degrees contains up to 7±2 elements. With more than 7±2 elements, the assignments become inconsistent. Another argument supporting the importance of finite residuated lattices comes from computational considerations. While using $[0,1]$ is satisfactory in many cases, quite a lot of problems leads to infinite structures if $[0,1]$ is used (consider just the simple fact that the set of all fuzzy sets in a finite universe is uncountable when $[0,1]$ is used as a set of truth degrees). Quite often, a natural solution, which is computationally tractable, consists in considering a finite residuated lattice.

These facts bring us to finite residuated lattices with a reasonably small number of elements (7±2, perhaps a bit more). Surprisingly, little has been done in

* Supported by grant No. 1ET101370417 of GA AV ČR, by grant No. 201/05/0079 of the Czech Science Foundation, and by institutional support, research plan MSM 6198959214.

P. Melin et al. (Eds.): IFSA 2007, LNAI 4529, pp. 461–470, 2007.
© Springer-Verlag Berlin Heidelberg 2007

a systematic study in this field. An interesting exception is [1] which studies t-norms on finite chains. This paper is a continuation of [3] where we investigated properties of all non-isomorphic finite lattices up to eleven elements. In this paper, we focus on fundamental questions related to finite *residuated* lattices. For instance, how many non-isomorphic residuated lattices (MTL-algebras, BL-algebras, ...) with $n$ elements are there? Can we generate the structures in an efficient way? Which properties of finite residuated lattices are frequent for small lattices? Our paper answers several questions of this type.

*Preliminaries.* Recall that a residuated lattice is an algebra $\mathbf{L} = \langle L, \wedge, \vee, \otimes, \rightarrow, 0, 1\rangle$ where $\langle L, \wedge, \vee, 0, 1\rangle$ is a bounded lattice, $\langle L, \otimes, 1\rangle$ is a commutative monoid, and $\otimes$ and $\rightarrow$ satisfy $a \otimes b \leq c$ iff $a \leq b \rightarrow c$ for each $a, b, c \in L$ (so-called adjointness). More details about residuated lattices and their role in fuzzy logic can be found in [2,6,7,9].

## 2 Generation of Finite Residuated Lattices

In this section we briefly describe a way to generate residuated lattices of a given size. Given a finite lattice $\mathbf{L} = \langle L, \wedge, \vee, 0, 1\rangle$, we describe an algorithm which generates all pairs $\langle \otimes, \rightarrow\rangle$ of adjoint operations on $\mathbf{L}$. Thus, in order to generate all $n$-element residuated lattices, it suffices to generate all $n$-element lattices and then to generate all the adjoint pairs. Efficient algorithms for generating finite lattices are available, see [3,11].

Let $\mathbf{L} = \langle L, \wedge, \vee, 0, 1\rangle$ be a finite lattice. By definition of a residuated lattice, we are looking for all couples $\langle \otimes, \rightarrow\rangle$ of operations such that $\langle L, \otimes, 1\rangle$ is a commutative monoid, and $\otimes$ and $\rightarrow$ satisfy adjointness. Note that not all finite lattices admit adjoint operations $\otimes$ and $\rightarrow$. The algorithm which will be described later in this section generates only $\otimes$ (multiplication, truth function of "fuzzy conjunction") and tests a condition which is equivalent to the existence of $\rightarrow$ (residuum, truth function of "fuzzy implication") satisfying adjointness with $\otimes$. Namely, we will take advantage of the following assertion:

**Theorem 1.** *Let* $\mathbf{L} = \langle L, \wedge, \vee, 0, 1\rangle$ *be a finite lattice,* $\langle L, \otimes, 1\rangle$ *be a commutative monoid such that* $\otimes$ *is monotone w.r.t.* $\leq$*. The the following are equivalent:*

(i) *there exists (unique)* $\rightarrow$ *satisfying adjointness w.r.t.* $\otimes$*;*
(ii) *for each* $a, b, c \in L$*:* $a \otimes (b \vee c) = (a \otimes b) \vee (a \otimes c)$*;*
(iii) $\rightarrow$ *given by* $a \rightarrow b = \bigvee\{c \in L \mid a \otimes c \leq b\}$ *satisfies adjointness w.r.t.* $\otimes$*.*

*Proof.* Follows from finiteness of $L$ and properties of residuated lattices [2]. We omit the proof due to the limited scope of this paper. □

Due to Theorem 1, it suffices to generate all monotone, commutative, and associative operations $\otimes$ which are neutral with respect to 1 (greatest element of $\mathbf{L}$) and satisfy condition (ii) of Theorem 1. If $\otimes$ satisfies all these conditions, we can use (iii) to compute the residuum $\rightarrow$ of $\otimes$.

The basic idea of our algorithm is that we systematically go through all candidates $\otimes$ which may satisfy assumptions of Theorem 1 and (ii). Multiplication

| $\otimes$ | $0$ | $a_1$ | $a_2$ | $\cdots$ | $a_{n-2}$ | $1$ |
|---|---|---|---|---|---|---|
| $0$ | $0$ | $0$ | $0$ | $\cdots$ | $0$ | $0$ |
| $a_1$ | $0$ | | | | | $a_1$ |
| $a_2$ | $0$ | | | | | $a_2$ |
| $\vdots$ | $\vdots$ | | | | | $\vdots$ |
| $a_{n-2}$ | $0$ | | | | | $a_{n-2}$ |
| $1$ | $0$ | $a_1$ | $a_2$ | $\cdots$ | $a_{n-2}$ | $1$ |

```
procedure fill (i, j):
   if (1)–(3) not satisfied:
      return
   if j ≥ n − 1:
      set i to i + 1
      set j to i
   if i ≥ n − 1:
      store ⊗
      return
   for each b ∈ Bounds(i, j):
      set a_i ⊗ a_j to b
      call fill (i, j + 1)
   set a_i ⊗ a_j to "undefined"
```

**Fig. 1.** Initial assignment of truth degrees to a table for $\otimes$ and procedure *fill*

is a binary operation on $L$. For the sake of simplicity, assume $|L| = n$ and denote the truth degrees by $L = \{0 = a_0, a_1, a_2, \ldots, a_{n-2}, a_{n-1} = 1\}$. We assume that our indexing extends the lattice order, i.e. that $a_i \leq a_j$ implies $i \leq j$. Since for $|L| = n$, there are $n^{n^2}$ distinct binary operations, we cannot generate all such operations before checking the desired properties. Namely, $n^{n^2}$ is a large number even for small $n$. Instead, we take advantage of some of the properties to generate only reasonably small subclass of binary operations.

The task to find a multiplication can be seen as a task to fill a table as the one in Fig. 1 (left) by truth degrees. A table entry given by row $i$ and column $j$ represents value $a_i \otimes a_j$. Since $\otimes$ needs to be commutative, we can focus only on the upper triangle of the table (including the diagonal). Moreover, some truth degrees in the table can be fixed because from properties of residuated lattices we have

$$a \otimes 0 = 0 \otimes a = 0, \qquad a \otimes 1 = 1 \otimes a = a.$$

The other entries in the table can take any values from $L - \{1\}$. Fortunately, we need not go through all possible assignments of degrees from $L - \{1\}$ to blank entries in the table from Fig. 1 (left). We can restrict the set of possible values for each table entry using the following well-known fact:

**Theorem 2.** *Let* $\mathbf{L}$ *be a residuated lattice. Then, for each* $a, b \in L$,
 (i) $a \otimes b \leq a \wedge b$;
 (ii) $\bigvee\{c \otimes d \mid c, d \in L \text{ such that } c \leq a \text{ and } d \leq b\} \leq a \otimes b$. □

As we can see, Theorem 2 provides us with upper and lower bounds for the results of multiplications. In more detail, for each $a, b \in L$, the upper bound is given by (i). The lower bound can be computed using (ii) before each new assignment. Since we assign truth degrees to the table one by one, for a considered pair $a, b \in L$ of truth degrees we can take all $c, d \in L$ such that (i) the value $c \otimes d$ is already assigned, and (ii) $c \leq a$ and $d \leq b$. Then we can compute supremum of values $c \otimes d$ of all such truth degrees which is then the lower bound of $a \otimes b$.

A table for $\otimes$ is filled from its top-left corner to its bottom-right corner. Table entries are traversed in the following order: $a_1 \otimes a_1$, $a_1 \otimes a_2, \ldots, a_1 \otimes a_{n-3}$, $a_1 \otimes a_{n-2}$, $a_2 \otimes a_2$, $a_2 \otimes a_3, \ldots, a_{n-2} \otimes a_{n-2}$. For each new entry being added to the table we check several conditions to see that $\otimes$ represents a "candidate" for multiplication. Namely, for each $a, b, c \in L$ we check

$$a \otimes (b \otimes c) = (a \otimes b) \otimes c, \tag{1}$$

$$a \otimes (b \vee c) = (a \otimes b) \vee (a \otimes c), \tag{2}$$

$$a \leq b \text{ implies } a \otimes c \leq b \otimes c, \tag{3}$$

provided that the expressions (1)–(3) make sense (recall that we deal with a partial operation $\otimes$ which is being constructed, i.e., some values of $x \otimes y$ may not be defined). If the currently assigned value of $a_i \otimes a_j$ violates the conditions above, then we go back and set $a_i \otimes a_j$ to another value.

Otherwise we move to the next blank position in the table and compute possible values of the multiplication result given by Theorem 2. In more detail, for truth degrees $a_i$ and $a_j$ (i.e., degrees corresponding to position given by indices $i$ and $j$ in the table) we consider a set Bounds$(i, j) \subseteq L$ which is an interval

$$\text{Bounds}(i, j) = [b, a_i \wedge a_j]$$

where

$$b = \bigvee\{a_{\min(k,l)} \otimes a_{\max(k,l)} \mid (k = i \text{ and } a_l \prec a_j) \text{ or } (a_k \prec a_i \text{ and } l = j)\}$$

where $a_m \prec a_n$ denotes that $a_m$ is covered by $a_n$, i.e. $a_m \leq a_n$ and $a_m \leq c \leq a_n$ implies $a_m = c$ or $a_n = c$. Then we go through all the values in Bounds$(i, j)$ and set them as the results of $a_i \otimes a_j$. Then we check (1)–(3) for $a_i \otimes a_j$ and the process continues as described above. We finish if we fill the whole table with values satisfying (1)–(3). As we have seen, the algorithm for generating $\otimes$ can be described as a recursive procedure which accepts two parameters: indices of the row and column of table Fig. 1 (left). The procedure is described in Fig. 1 (right). Note that a preliminary version of this method, which was less efficient, was described in [13].

Due to the limited scope of this paper, we postpone proof of soundness of the procedure to a full version of this paper.

## 3 Properties of Generated Residuated Lattices

In this section we present basic characteristics of finite residuated lattices generated by our procedure. We have used our procedure to generate all non-isomorphic residuated lattices with up to 11 elements. Prior to that, we generated all non-isomorphic lattices up to 11 elements, see [3] and studied their properties, see also [11] for a related approach.

We first focus on the numbers of $n$-element residuated lattices. Table 1 contains a basic summary. Columns of the table correspond to sizes of lattices (numbers of their elements). The first row contains numbers of non-isomorphic residuated lattices. The second row contains numbers of non-isomorphic linearly ordered residuated lattices (i.e., lattices with every pair of elements comparable). Note that [1] contains an error since it says that the number for $|L| = 8$ is 2368.

**Table 1.** Numbers of non-isomorphic finite residuated lattices up to 11 elements

| | 1 | 2 | 3 | 4 | 5 | 6 | 7 | 8 | 9 | 10 | 11 |
|---|---|---|---|---|---|---|---|---|---|---|---|
| residuated lattices | 1 | 1 | 2 | 7 | 27 | 142 | 839 | 5803 | 45466 | 406783 | 4134207 |
| linear res. lattices | 1 | 1 | 2 | 6 | 22 | 94 | 451 | 2386 | 13775 | 86417 | 590489 |
| lattices | 1 | 1 | 1 | 2 | 5 | 15 | 53 | 222 | 1078 | 5994 | 37622 |
| res. lattice reducts | 1 | 1 | 1 | 2 | 3 | 7 | 18 | 61 | 239 | 1125 | 6137 |

We can read from the table that small residuated lattices tend to be linear: for $|L| = 5$, 22 residuated lattices out of 27 are linear. On the other hand, with growing sizes of $|L|$, the portion of linear residuated lattices is going down: for $|L| = 11$, only one seventh of all the residuated lattices are linear.

Another interesting thing is the relationship between (numbers of) residuated lattices and (numbers of) their distinct lattice reducts. Recall that if $\mathbf{L} = \langle L, \wedge, \vee, \otimes, \rightarrow, 0, 1\rangle$ is a residuated lattice, its reduct $\langle L, \wedge, \vee, 0, 1\rangle$ resulting by omitting $\otimes$ and $\rightarrow$ is a lattice. Thus, we may ask how many $n$-element lattices are reducts of $n$-element residuated lattices. This is shown in the last two rows of Table 1. The last but one row shows numbers of non-isomorphic $n$-element lattices [3,11]. The last row contains numbers of pairwise distinct non-isomorphic lattice reducts of all non-isomorphic residuated lattices. For instance, the values in column corresponding to $|L| = 11$ mean: there are 37622 non-isomorphic lattices but only 6137 of them can be equipped with $\otimes$ and $\rightarrow$ to form a residuated lattice. An interesting observation here is that even if the number of residuated lattices rapidly grows with growing $|L|$, the number of their lattice reducts compared to the number of all lattices (of that size) is going down. This means that with growing $|L|$, the average number of residuated lattices with the same lattice part is going up. For instance, for $|L| = 8$ the average number of residuated lattices sharing the same lattice part is approximately 95 while for $|L| = 11$ it is 673.

The values in Table 1 may suggest that most residuated lattices can be found on $n$-element chains. This is so for smaller residuated lattices but it is no longer true for large lattices. For instance, if we consider $|L| = 11$, we can depict the numbers of residuated lattices according to their width and height as in Table 2 (see appendices). The rows and columns in Table 2 represent heights and widths

**Table 2.** Numbers of 11-element residuated lattices with given heights and widths

| | 1 | 2 | 3 | 4 | 5 | 6 | 7 | 8 |
|---|---|---|---|---|---|---|---|---|
| 4 | 0 | 0 | 0 | 0 | 0 | 0 | 0 | 1 |
| 5 | 0 | 0 | 0 | 23 | 80 | 64 | 1883 | 0 |
| 6 | 0 | 0 | 684 | 38480 | 31280 | 10470 | 0 | 0 |
| 7 | 0 | 2539 | 113275 | 127288 | 35771 | 0 | 0 | 0 |
| 8 | 0 | 141182 | 428416 | 122677 | 0 | 0 | 0 | 0 |
| 9 | 0 | 825240 | 402523 | 0 | 0 | 0 | 0 | 0 |
| 10 | 0 | 1261842 | 0 | 0 | 0 | 0 | 0 | 0 |
| 11 | 590489 | 0 | 0 | 0 | 0 | 0 | 0 | 0 |

**Table 3.** Numbers of 11-element lattices with given heights and widths

| | 1 | 2 | 3 | 4 | 5 | 6 | 7 | 8 | 9 |
|---|---|---|---|---|---|---|---|---|---|
| 3 | 0 | 0 | 0 | 0 | 0 | 0 | 0 | 0 | 1 |
| 4 | 0 | 0 | 0 | 0 | 123 | 159 | 72 | 15 | 0 |
| 5 | 0 | 0 | 83 | 2212 | 3294 | 1138 | 126 | 0 | 0 |
| 6 | 0 | 0 | 2295 | 8464 | 4387 | 518 | 0 | 0 | 0 |
| 7 | 0 | 164 | 4413 | 5339 | 973 | 0 | 0 | 0 | 0 |
| 8 | 0 | 374 | 2133 | 805 | 0 | 0 | 0 | 0 | 0 |
| 9 | 0 | 217 | 280 | 0 | 0 | 0 | 0 | 0 | 0 |
| 10 | 0 | 36 | 0 | 0 | 0 | 0 | 0 | 0 | 0 |
| 11 | 1 | 0 | 0 | 0 | 0 | 0 | 0 | 0 | 0 |

**Table 4.** Numbers of 11-element lattice reducts with given heights and widths

| | 1 | 2 | 3 | 4 | 5 | 6 | 7 | 8 |
|---|---|---|---|---|---|---|---|---|
| 4 | 0 | 0 | 0 | 0 | 0 | 0 | 0 | 1 |
| 5 | 0 | 0 | 0 | 21 | 75 | 50 | 13 | 0 |
| 6 | 0 | 0 | 157 | 860 | 546 | 86 | 0 | 0 |
| 7 | 0 | 43 | 1021 | 1308 | 257 | 0 | 0 | 0 |
| 8 | 0 | 179 | 865 | 323 | 0 | 0 | 0 | 0 |
| 9 | 0 | 141 | 161 | 0 | 0 | 0 | 0 | 0 |
| 10 | 0 | 29 | 0 | 0 | 0 | 0 | 0 | 0 |
| 11 | 1 | 0 | 0 | 0 | 0 | 0 | 0 | 0 |

**Table 5.** Average characteristics of reducts (legend: ht/wd = avg. height/width of lattices, at = avg. number of (co)atoms, ir/pr = avg. number of irreducible/prime elements, mc/ma = avg. number of maximal chains/antichains

| | 1 | 2 | 3 | 4 | 5 | 6 | 7 | 8 | 9 | 10 | 11 |
|---|---|---|---|---|---|---|---|---|---|---|---|
| ht | 1.00 | 2.00 | 3.00 | 3.50 | 4.33 | 4.86 | 5.44 | 5.87 | 6.31 | 6.68 | 7.01 |
| wd | 1.00 | 1.00 | 1.00 | 1.50 | 1.67 | 2.00 | 2.33 | 2.67 | 3.02 | 3.37 | 3.71 |
| at | 0.00 | 1.00 | 1.00 | 1.50 | 1.33 | 1.71 | 1.89 | 2.10 | 2.23 | 2.38 | 2.50 |
| ir | 1.00 | 2.00 | 3.00 | 3.50 | 4.33 | 5.00 | 5.83 | 6.48 | 7.20 | 7.85 | 8.49 |
| pr | 1.00 | 2.00 | 3.00 | 3.50 | 4.33 | 4.43 | 4.44 | 4.33 | 4.11 | 3.88 | 3.66 |
| mc | 1.00 | 1.00 | 1.00 | 1.50 | 1.67 | 2.14 | 2.56 | 3.13 | 3.65 | 4.27 | 4.95 |
| ma | 1.00 | 2.00 | 3.00 | 3.50 | 4.33 | 5.00 | 5.67 | 6.48 | 7.38 | 8.47 | 9.76 |

of residuated lattices, respectively. By a *height* (or *width*) of a residuated lattice we mean the length of the longest maximal chain (or antichain) contained in its lattice part. The table entries represent numbers of distinct residuated lattices with the dimensions given by the corresponding rows and columns. Table 2 shows that most residuated lattices are defined on "high and thin lattices" but in case of $|L| = 11$, the most frequent residuated lattices are those with width 2 (see second column of Table 2). Let us mention that the distribution of all lattices and all lattice reducts according to their dimensions is quite different from that of

**Table 6.** Numbers of residuated lattices satisfying selected properties

| | 1 | 2 | 3 | 4 | 5 | 6 | 7 | 8 | 9 | 10 | 11 |
|---|---|---|---|---|---|---|---|---|---|---|---|
| all res. lattices | 1 | 1 | 2 | 7 | 27 | 142 | 839 | 5803 | 45466 | 406783 | 4134207 |
| modular | 1 | 1 | 2 | 7 | 27 | 138 | 775 | 4953 | 34833 | 269744 | 2303013 |
| distributive | 1 | 1 | 2 | 7 | 27 | 137 | 748 | 4655 | 31519 | 233186 | 1879285 |
| ($\Pi$1) identity | 1 | 1 | 1 | 4 | 9 | 51 | 290 | 2125 | 18165 | 182390 | 2096717 |
| prelinear | 1 | 1 | 2 | 7 | 23 | 100 | 469 | 2482 | 14256 | 89254 | 608250 |
| ($\Pi$2) identity | 1 | 1 | 1 | 3 | 8 | 31 | 156 | 913 | 6208 | 48054 | 421028 |
| strict | 1 | 1 | 1 | 3 | 7 | 28 | 142 | 842 | 5804 | 45473 | 403000 |
| (WNM) identity | 1 | 1 | 2 | 5 | 11 | 31 | 82 | 253 | 819 | 3064 | 13225 |
| divisible | 1 | 1 | 2 | 5 | 10 | 23 | 49 | 111 | 245 | 547 | 1196 |
| involutive | 1 | 1 | 1 | 3 | 3 | 13 | 17 | 86 | 185 | 779 | 2475 |
| idempotent | 1 | 1 | 1 | 2 | 3 | 5 | 8 | 15 | 26 | 47 | 80 |

**Table 7.** Numbers of selected algebras

| | 1 | 2 | 3 | 4 | 5 | 6 | 7 | 8 | 9 | 10 | 11 |
|---|---|---|---|---|---|---|---|---|---|---|---|
| all res. lattices | 1 | 1 | 2 | 7 | 27 | 142 | 839 | 5803 | 45466 | 406783 | 4134207 |
| MTL-algebras | 1 | 1 | 2 | 7 | 23 | 100 | 469 | 2482 | 14256 | 89254 | 608250 |
| SMTL-algebras | 1 | 1 | 1 | 3 | 7 | 24 | 100 | 472 | 2483 | 14263 | 89254 |
| WNM-algebras | 1 | 1 | 2 | 5 | 9 | 22 | 43 | 98 | 198 | 418 | 842 |
| BL-algebras | 1 | 1 | 2 | 5 | 9 | 20 | 38 | 81 | 161 | 327 | 645 |
| SBL-algebras | 1 | 1 | 1 | 3 | 5 | 10 | 20 | 41 | 82 | 166 | 327 |
| IMTL-algebras | 1 | 1 | 1 | 3 | 3 | 8 | 12 | 36 | 62 | 172 | 339 |
| Heyting algebras | 1 | 1 | 1 | 2 | 3 | 5 | 8 | 15 | 26 | 47 | 80 |
| G-algebras | 1 | 1 | 1 | 2 | 2 | 3 | 3 | 5 | 6 | 8 | 8 |
| NM-algebras | 1 | 1 | 1 | 2 | 1 | 2 | 1 | 5 | 4 | 4 | 3 |
| MV-algebras | 1 | 1 | 1 | 2 | 1 | 2 | 1 | 3 | 2 | 2 | 1 |
| $\Pi$-algebras | 1 | 1 | 0 | 1 | 0 | 0 | 0 | 1 | 0 | 0 | 0 |
| $\Pi$MTL-algebras | 1 | 1 | 0 | 1 | 0 | 0 | 0 | 1 | 0 | 0 | 0 |

residuated lattices. Table 3 and Table 4 show the same characteristics as Table 2 for all 11-element lattices and all lattice reducts of 11-element residuated lattices, respectively. Here we can see that the numbers of lattices follow, more or less, a normal distribution (most of them have average width and height).

Table 5 shows average characteristics of the lattice reducts of residuated lattices. The rows of the table correspond to properties (see the legend in Table 5, for the notions involved we refer to [8]), the columns correspond to sizes of lattices. Table entries are the average values.

We now turn our attention to residuated lattices satisfying additional conditions. We consider the following properties of residuated lattices expressible by identities (see [2,4,5,8,9]):

(MOD) $a \leq c$ implies $a \vee (b \wedge c) = (a \vee b) \wedge c$ *(modularity)*

(DIS) $a \wedge (b \vee c) = (a \wedge b) \vee (a \wedge c)$ *(distributivity)*

| | | |
|---|---|---|
| (MTL) | $(a \to b) \vee (b \to a) = 1$ | (*prelinearity*) |
| ($\Pi$1) | $(c \to 0) \to 0 \leq ((a \otimes c) \to (b \otimes c)) \to (a \to b)$ | ($\Pi$*1-property*) |
| ($\Pi$2) | $a \wedge (a \to 0) = 0$ | ($\Pi$*2-property*) |
| (STR) | $(a \otimes b) \to 0 = (a \to 0) \vee (b \to 0)$ | (*strictness*) |
| (WNM) | $((a \otimes b) \to 0) \vee ((a \wedge b) \to (a \otimes b)) = 1$ | (*weak nilpotent minimum*) |
| (DIV) | $a \wedge b = a \otimes (a \to b)$ | (*divisibility*) |
| (INV) | $a = (a \to 0) \to 0$ | (*involution*) |
| (IDM) | $a = a \otimes a$ | (*idempotency*) |

Table 6 contains numbers of residuated lattices satisfying these conditions. Table 7 summarizes numbers of algebras (particular residuated lattices) which

**Table 8.** Average numbers of idempotent and involutive elements in selected algebras

| | 1 | 2 | 3 | 4 | 5 | 6 | 7 | 8 | 9 | 10 | 11 |
|---|---|---|---|---|---|---|---|---|---|---|---|
| all res. lattices | 1.00 | 2.00 | 2.50 | 3.00 | 3.19 | 3.32 | 3.37 | 3.37 | 3.30 | 3.18 | 3.01 |
| | 1.00 | 2.00 | 2.50 | 3.14 | 3.37 | 3.79 | 4.07 | 4.39 | 4.68 | 4.98 | 5.26 |
| MTL-algebras | 1.00 | 2.00 | 2.50 | 3.00 | 3.17 | 3.42 | 3.58 | 3.74 | 3.87 | 3.97 | 4.01 |
| | 1.00 | 2.00 | 2.50 | 3.14 | 3.30 | 3.71 | 3.97 | 4.26 | 4.51 | 4.77 | 5.03 |
| BL-algebras | 1.00 | 2.00 | 2.50 | 3.20 | 3.67 | 4.25 | 4.74 | 5.31 | 5.80 | 6.33 | 6.82 |
| | 1.00 | 2.00 | 2.50 | 3.00 | 2.78 | 3.10 | 2.89 | 3.14 | 3.02 | 3.03 | 2.97 |
| G-algebras | 1.00 | 2.00 | 3.00 | 4.00 | 5.00 | 6.00 | 7.00 | 8.00 | 9.00 | 10.00 | 11.00 |
| | 1.00 | 2.00 | 2.00 | 3.00 | 2.00 | 2.67 | 2.00 | 3.60 | 2.33 | 2.50 | 2.00 |
| MV-algebras | 1.00 | 2.00 | 2.00 | 3.00 | 2.00 | 3.00 | 2.00 | 4.67 | 3.00 | 3.00 | 2.00 |
| | 1.00 | 2.00 | 3.00 | 4.00 | 5.00 | 6.00 | 7.00 | 8.00 | 9.00 | 10.00 | 11.00 |

**Table 9.** Groups of algebras sharing the same properties

| | MTL-algebra | SMTL-algebra | WNM-algebra | BL-algebra | SBL-algebra | IMTL-algebra | G-algebra | NM-algebra | MV-algebra | $\Pi$-algebra | $\Pi$MTL-algebra |
|---|---|---|---|---|---|---|---|---|---|---|---|
| 3878433 | | | | | | | | | | | |
| 605431 | × | | | | | | | | | | |
| 105952 | × | × | | | | | | | | | |
| 1551 | × | | × | | | | | | | | |
| 617 | × | × | | × | × | | | | | | |
| 603 | × | | | | | × | | | | | |
| 593 | × | | | × | | | | | | | |
| 36 | × | × | × | × | × | | × | | | | |
| 27 | × | | × | × | | | | | | | |
| 18 | × | | × | | | × | | × | | | |
| 10 | × | | | × | | × | | | × | | |
| 4 | × | × | × | × | × | × | × | × | × | × | × |
| 3 | × | | × | × | | × | | × | × | | |

**Table 10.** Groups of algebras sharing the same properties (detail)

| | MOD | DIS | MTL | $\Pi 1$ | $\Pi 2$ | STR | WEA | DIV | INV | IDE | | MOD | DIS | MTL | $\Pi 1$ | $\Pi 2$ | STR | WEA | DIV | INV | IDE |
|---|---|---|---|---|---|---|---|---|---|---|---|---|---|---|---|---|---|---|---|---|---|
| 1195367 | | | | × | | | | | | | 617 | × | × | × | | × | × | | × | | |
| 674576 | × | × | | | | | | | | | 603 | × | × | × | × | | | | | × | |
| 612291 | | | | | | | | | | | 593 | × | × | × | | | | | × | | |
| 586184 | × | × | | × | | | | | | | 543 | × | | | | | | × | | | |
| 429478 | × | × | × | | | | | | | | 444 | × | | | × | | | × | | | |
| 328566 | × | | | × | | | | | | | 442 | × | × | | × | | | × | | | |
| 175953 | × | × | × | × | | | | | | | 355 | × | | | | × | | | | | |
| 163065 | × | × | | | × | × | | | | | 265 | × | × | | | × | × | | × | | |
| 145596 | | | | | × | × | | | | | 227 | × | × | | | | | | × | | |
| 105952 | × | × | × | | × | × | | | | | 221 | × | × | | | × | | | × | | |
| 92801 | × | | | | | | | | | | 146 | × | × | × | × | | | × | | | |
| 39691 | × | | | | × | × | | | | | 76 | × | × | | | × | × | × | × | | × |
| 14040 | | | | | × | | | | | | 73 | × | × | | | × | | × | × | | × |
| 9088 | | | | × | | | × | | | | 38 | × | × | | | | | × | × | | |
| 6413 | × | × | | | × | | | | | | 36 | × | × | × | | × | × | × | × | | × |
| 2750 | | | | | | | × | | | | 27 | × | × | × | | | | × | × | | |
| 2395 | × | × | | | | | × | | | | 18 | × | × | × | × | | | × | | × | |
| 1526 | × | | | × | | | | | × | | 10 | × | × | × | × | | | | × | × | |
| 1405 | × | × | × | | | | × | | | | 6 | × | × | | × | | | × | | × | |
| 742 | × | × | | × | | | | | × | | 4 | × | × | × | × | × | × | × | × | × | × |
| 652 | | | | × | | | | | × | | 3 | × | × | × | × | | | × | × | × | |

are defined by a combination of the above-mentioned properties. The tables show that BL-algebras are very rare among residuated lattices up to 11 elements. The situation for MTL-algebras is better but still, only 15 % of all 11-element residuated lattices are MTL-algebras. An observation which may be surprising is that ($\Pi 1$) is far more frequent a property than prelinearity (for $|L| \leq 11$).

Table 8 shows average numbers of idempotent (upper value in each table entry) and involutive (lower value in each table entry) elements of selected algebras. Here an interesting thing is that the average values for MTL-algebras are approximately the same as for all residuated lattices whereas the average values for BL-algebras are sort of opposite (MTL-algebras have more involutive elements than the idempotent ones in average while BL-algebras have more idempotent than the involutive ones).

Table 6 and Table 7 show the numbers of residuated lattices having each property but do not show, e.g., how many divisible lattices are idempotent. Such information can be found in Table 9 and in more detail in Table 10. Here, the columns denote properties considered in Table 6 and Table 7, the left-most column contains numbers of residuated lattices with given combination of properties. Each row of the tables represents one combination of properties (properties which are present are marked by "×"). Let us note that some combinations of properties are rare and some of them can be found only in larger structures. Just for illustration, the least residuated lattice which satisfies only (MOD) and ($\Pi$2) from all the considered properties has 9 elements.

## 4 Conclusion and Future Research

We have presented a method for generation of finite residuated lattices up to a given size. The generated residuated lattices were used for a preliminary exploration of their quantitative properties. We have focused mainly on the exploration of numbers of various algebras (BL-algebras, MTL-algebras, ...). A database of generated structures can be found at:

`http://vychodil.inf.upol.cz/res/devel/finresl/.`

In our future work we will focus of the following topics:

- incremental algorithms that reuse previously generated structures;
- exploration of further properties of the generated residuated lattices;
- generation of structures with hedges and independent negations [5,10,17].

## References


1. Bartušek T., Navara M.: Program for generating fuzzy logical operations and its use in mathematical proofs. *Kybernetika* **38**, No. 3(2002), 235–244.
2. Belohlavek R.: *Fuzzy Relational Systems: Foundations and Principles.* Kluwer Academic/Plenum Publishers, New York, 2002.
3. Belohlavek R., Vychodil V.: Scales behind computational intelligence: exploring properties of finite lattices (to appear in Proc. 1st IEEE Symposium on Foundations of Comput. Intelligence – FOCI '07, 1–5 Apr 2007, Honolulu, Hawaii, USA).
4. Esteva F., Godo L.: Monoidal t-norm based logic: towards a logic for left-continuous t-norms. *Fuzzy Sets and Systems* **124**(2001), 271–288.
5. Esteva F., Godo L, Hájek P., Navara M.: Residuated fuzzy logics with an involutive negation. *Arch. Math. Logic* **39**(2000), 103–124.
6. Gerla G.: *Fuzzy Logic. Mathematical Tools for Approximate Reasoning.* Kluwer, Dordrecht, 2001.
7. Goguen J. A.: The logic of inexact concepts. *Synthese* **18**(1968-9), 325–373.
8. Gratzer G. A.: *General Lattice Theory.* Birkhauser (1998, 2nd ed.).
9. Hájek P.: *Metamathematics of Fuzzy Logic.* Kluwer, Dordrecht, 1998.
10. Hájek P.: On very true. *Fuzzy Sets and Systems* **124**(2001), 329–333.
11. Heitzig J., Reinhold J.: Counting Finite Lattices. *Algebra Universalis* **48**(1)(2002), 43–53.
12. Höhle U.: On the fundamentals of fuzzy set theory. *J. Math. Anal. Appl.* **201**(1996), 786–826.
13. Kure M.: *Computer-aided study of finite posets.* UP Olomouc (MSc. thesis), 2004.
14. Klir G. J., Yuan B.: *Fuzzy Sets and Fuzzy Logic. Theory and Applications.* Prentice Hall, 1995.
15. Miller G. T.: The magical number seven, plus or minus two: some limits on our capacity for processing information. *The Psychological Review* **63**(1956), 81–97.
16. Pavelka J.: On fuzzy logic I, II, III. *Z. Math. Logik Grundlagen Math.* **25**(1979), 45–52, 119–134, 447–464.
17. Vychodil V.: Truth-depressing hedges and BL-logic. *Fuzzy Sets and Systems* **157**(15)(2006), 2074–2090.

# On Proofs and Rule of Multiplication in Fuzzy Attribute Logic*

Radim Belohlavek[1,2] and Vilem Vychodil[2]

[1] Dept. Systems Science and Industrial Engineering, Binghamton University—SUNY
Binghamton, NY 13902, USA
rbelohla@binghamton.edu

[2] Dept. Computer Science, Palacky University, Olomouc
Tomkova 40, CZ-779 00 Olomouc, Czech Republic
vilem.vychodil@upol.cz

**Abstract.** The paper develops fuzzy attribute logic, i.e. a logic for reasoning about formulas of the form $A \Rightarrow B$ where $A$ and $B$ are fuzzy sets of attributes. A formula $A \Rightarrow B$ represents a dependency which is true in a data table with fuzzy attributes iff each object having all attributes from $A$ has also all attributes from $B$, membership degrees in $A$ and $B$ playing a role of thresholds. We study axiomatic systems of fuzzy attribute logic which result by adding a single deduction rule, called a rule of multiplication, to an ordinary system of deduction rules complete w.r.t. bivalent semantics, i.e. to well-known Armstrong axioms. In this paper, we concentrate on the rule of multiplication and its role in fuzzy attribute logic. We show some advantageous properties of the rule of multiplication. In addition, we show that these properties enable us to reduce selected problems concerning proofs in fuzzy attribute logic to the corresponding problems in the ordinary case. As an example, we discuss the problem of normalization of proofs and present, in the setting of fuzzy attribute logic, a counterpart to a well-known theorem from database theory saying that each proof can be transformed to a so-called RAP-sequence.

## 1 Introduction

If-then rules in their various variants are perhaps the most common way to express our knowledge. Usually, if-then rules are extracted from data to bring up a new knowledge about the data or are formulated by a user/expert to represent a constraint on the data. If-then rules of the form $A \Rightarrow B$, where $A$ and $B$ are collections of attributes, have been used in data mining and in databases. In data mining, rules $A \Rightarrow B$ are called association rules or attribute implications, and have the following basic meaning: If an object has all attributes from $A$ then it has all attributes from $B$. This gives a rise to the first semantics. The rules $A \Rightarrow B$ are interpreted in tables with crisp attributes, i.e., with rows corresponding to objects, columns corresponding to "yes-or-no" attributes, and

* Supported by grant No. 1ET101370417 of GA AV ČR, by grant No. 201/05/0079 of the Czech Science Foundation, and by institutional support, research plan MSM 6198959214.

P. Melin et al. (Eds.): IFSA 2007, LNAI 4529, pp. 471–480, 2007.
© Springer-Verlag Berlin Heidelberg 2007

table entries containing 1 or 0 indicating whether an object does or does not have an attribute. The goal then is to extract "all interesting rules" from data. In databases, rules $A \Rightarrow B$ are called functional dependencies, see [17] for a comprehensive overview, and have the following meaning: If any two objects (items, rows) of a database table agree in their values on all attributes from $A$ then they agree on all attributes from $B$. This gives a rise to the second semantics in which rules $A \Rightarrow B$ are interpreted in database tables.

The two ways of interpreting rules $A \Rightarrow B$ are closely connected. Namely, semantic entailment coincides for both of them. This entailment can also be captured syntactically. This goes back to Armstrong [1] who introduced a set of inference rules, a modified version of which became known as Armstrong axioms. Armstrong axioms are very well known to be complete w.r.t. database semantics and, due to the above-mentioned connection, also to the other semantics. This means that a rule $A \Rightarrow B$ is provable (using Armstrong axioms) from a set $T$ of rules if and only if $A \Rightarrow B$ semantically follows from $T$.

In a series of papers, see e.g. [3,4,5,7,8,9,10] and also an overview paper [6], we started to develop the above-described issues from the point of view of fuzzy logic (note that the first attempt is [18]). We introduced fuzzy attribute implications, i.e., our counterparts to the ordinary if-then rules $A \Rightarrow B$ described above. Among other issues, we studied two types of semantics, the first one given by tables with fuzzy attributes and the second one given by ranked tables over domains with similarities. We proved that these types of semantics have the same semantic entailment. In [4], we introduced a logical calculus, called fuzzy attribute logic, for reasoning with fuzzy attribute implications. We proved its syntactico-semantical completeness, both in the ordinary style ("provable = semantically entailed") and in the graded style ("degree of provability = degree of semantical entailment"). In [5], we presented further results on fuzzy attribute logic. For our present purpose, the most important result of [5] is an invention of a single deduction rule, called a rule of multiplication, which has the following property: Adding the rule of multiplication to an ordinary system of Armstrong axioms (Armstrong axioms are, in fact, deduction rules) yields a syntactico-semantically complete system for reasoning with fuzzy attribute implications. The main aim of the present paper is to focus in more detail on the rule of multiplication and its role in fuzzy attribute logic. As emphasized in [5], the rule of multiplication allows one to consider fuzzy attribute logic as consisting of a system of "ordinary Armstrong rules" plus a single "fuzzy rule" (which is the rule of multiplication). We will show that in addition to this role which is more or less an aesthetic one, the rule of multiplication has also its practical role. Among other things, its properties allow us to almost automatically transfer results known from the ordinary case. As an example, we focus on the problem of normalization of proofs and present, in the setting of fuzzy attribute logic, a counterpart to a well-known theorem from database theory saying that each proof can be transformed to a RAP-sequence.

The paper is organized as follows. Section 2 surveys preliminaries from fuzzy sets and fuzzy logic. Fuzzy attribute logic and its completeness is presented in Section 3. Section 4 deals with the rule of multiplication and related issues. Section 5 presents conclusions.

## 2 Preliminaries

This section surveys preliminaries from fuzzy sets and fuzzy logic. Further details can be found, e.g., in [2,14,16]. As a structure of truth degrees, i.e., a set of truth degrees equipped with truth functions of logical connectives, we use complete residuated lattices with a truth-stressing hedge (shortly, a hedge) [14,15], i.e., algebras $\mathbf{L} = \langle L, \wedge, \vee, \otimes, \rightarrow, {}^*, 0, 1\rangle$ such that $\langle L, \wedge, \vee, 0, 1\rangle$ is a complete lattice with 0 and 1 being the least and greatest element of $L$, respectively; $\langle L, \otimes, 1\rangle$ is a commutative monoid (i.e., $\otimes$ is commutative, associative, and $a \otimes 1 = 1 \otimes a = a$ for each $a \in L$); $\otimes$ and $\rightarrow$ satisfy $a \otimes b \leq c$ iff $a \leq b \rightarrow c$ for each $a, b, c \in L$ (adjointness property); hedge $^*$ satisfies $1^* = 1$, $a^* \leq a$, $(a \rightarrow b)^* \leq a^* \rightarrow b^*$, $a^{**} = a^*$ for each $a, b \in L$. Elements $a$ of $L$ are called truth degrees. $\otimes$ and $\rightarrow$ are (truth functions of) "fuzzy conjunction" and "fuzzy implication". Hedge $^*$ is a (truth function of) logical connective "very true", see [14,15].

A favorite choice of $\mathbf{L}$ is a structure with $L = [0, 1]$ or a subchain of $[0, 1]$, equipped with well-known pairs of $\otimes$ (t-norms or restrictions of t-norms) and the corresponding $\rightarrow$ (residuum to $\otimes$).

Two boundary cases of (truth-stressing) hedges are (i) identity, i.e., $a^* = a$ ($a \in L$); (ii) globalization [19]: $1^* = 1$ and $a^* = 0$ for $a < 1$. A special case of a complete residuated lattice with hedge is the two-element Boolean algebra $\langle\{0, 1\}, \wedge, \vee, \otimes, \rightarrow, {}^*, 0, 1\rangle$, denoted by $\mathbf{2}$, which is the structure of truth degrees of classical logic. That is, the operations $\wedge, \vee, \otimes, \rightarrow$ of $\mathbf{2}$ are the truth functions (interpretations) of the corresponding logical connectives of the classical logic and $0^* = 0$, $1^* = 1$. Note that if we prove an assertion for general $\mathbf{L}$, then, in particular, we obtain a "crisp version" of this assertion for $\mathbf{L}$ being $\mathbf{2}$.

Having $\mathbf{L}$, we define usual notions of an $\mathbf{L}$-set (fuzzy set), $\mathbf{L}$-relation (fuzzy relation), etc. $\mathbf{L}^U$ denotes the collection of all $\mathbf{L}$-sets in $U$. Given $A, B \in \mathbf{L}^U$, we define a subsethood degree

$$S(A, B) = \textstyle\bigwedge_{u \in U} \bigl(A(u) \rightarrow B(u)\bigr), \tag{1}$$

which generalizes the classical subsethood relation $\subseteq$. $S(A, B)$ represents a degree to which $A$ is a subset of $B$. In particular, we write $A \subseteq B$ iff $S(A, B) = 1$. As a consequence, $A \subseteq B$ iff $A(u) \leq B(u)$ for each $u \in U$.

## 3 Fuzzy Attribute Logic and Its Completeness

We now introduce basic concepts of fuzzy attribute logic (FAL). Suppose $Y$ is a finite set of attributes. A *(fuzzy) attribute implication (over attributes $Y$)*, shortly FAI, is an expression $A \Rightarrow B$, where $A, B \in \mathbf{L}^Y$ ($A$ and $B$ are fuzzy sets of attributes). Fuzzy attribute implications are the basic formulas of FAL.

The next step is the semantics of FAL given by interpreting FAIs in data tables with fuzzy attributes. The intended meaning of $A \Rightarrow B$ is: "if it is (very) true that

**Table 1.** Data table with fuzzy attributes

| $I$ | $y_1$ | $y_2$ | $y_3$ | $y_4$ | $y_5$ | $y_6$ |
|---|---|---|---|---|---|---|
| $x_1$ | 1.0 | 1.0 | 0.0 | 1.0 | 1.0 | 0.2 |
| $x_2$ | 1.0 | 0.4 | 0.3 | 0.8 | 0.5 | 1.0 |
| $x_3$ | 0.2 | 0.9 | 0.7 | 0.5 | 1.0 | 0.6 |
| $x_4$ | 1.0 | 1.0 | 0.8 | 1.0 | 1.0 | 0.5 |

$X = \{x_1, \ldots, x_4\}$

$Y = \{y_1, \ldots, y_6\}$

an object has all attributes from $A$, then it has also all attributes from $B$" with the logical connectives being given by $\mathbf{L}$. A *data table with fuzzy attributes* can be seen as a triplet $\langle X, Y, I\rangle$ where $X$ is a set of objects, $Y$ is a finite set of attributes (the same as above in the definition of a fuzzy attribute implication), and $I \in \mathbf{L}^{X\times Y}$ is a binary $\mathbf{L}$-relation between $X$ and $Y$ assigning to each object $x \in X$ and each attribute $y \in Y$ a degree $I(x, y)$ to which $x$ has $y$. $\langle X, Y, I\rangle$ can be thought of as a table with rows and columns corresponding to objects $x \in X$ and attributes $y \in Y$, respectively, and table entries containing degrees $I(x, y)$, see Tab. 1. A row of a table $\langle X, Y, I\rangle$ corresponding to an object $x \in X$ can be seen as a fuzzy set $I_x$ of attributes to which an attribute $y \in Y$ belongs to a degree $I_x(y) = I(x, y)$. Forgetting now for a while about the data table, any fuzzy set $M \in \mathbf{L}^Y$ can be seen as a fuzzy set of attributes of some object with $M(y)$ being a degree to which the object has attribute $y$. For a fuzzy set $M \in \mathbf{L}^Y$ of attributes, we define a *degree* $||A \Rightarrow B||_M \in L$ *to which* $A \Rightarrow B$ *is true in* $M$ by

$$||A \Rightarrow B||_M = S(A, M)^* \to S(B, M). \tag{2}$$

It is easily seen that if $M$ is a fuzzy set of attributes of some object $x$ then $||A \Rightarrow B||_M$ is the degree to which "if it is (very) true that $x$ has all attributes from $A$ then $x$ has all attributes from $B$". For a system $\mathcal{M}$ of $\mathbf{L}$-sets in $Y$, define a degree $||A \Rightarrow B||_{\mathcal{M}}$ to which $A \Rightarrow B$ is true in (each $M$ from) $\mathcal{M}$ by

$$||A \Rightarrow B||_{\mathcal{M}} = \textstyle\bigwedge_{M\in\mathcal{M}} ||A \Rightarrow B||_M. \tag{3}$$

Finally, given a data table $\langle X, Y, I\rangle$ and putting $\mathcal{M} = \{I_x \mid x \in X\}$, $||A \Rightarrow B||_{\mathcal{M}}$ is a degree to which it is true that $A \Rightarrow B$ is true in each row of table $\langle X, Y, I\rangle$, i.e., a degree to which "for each object $x \in X$: if it is (very) true that $x$ has all attributes from $A$, then $x$ has all attributes from $B$". This degree is denoted by $||A \Rightarrow B||_{\langle X,Y,I\rangle}$ and is called a degree to which $A \Rightarrow B$ is true in data table $\langle X, Y, I\rangle$.

*Remark 1.* (1) For a fuzzy attribute implication $A \Rightarrow B$, both $A$ and $B$ are fuzzy sets of attributes. Particularly, both $A$ and $B$ can be crisp (i.e., $A(y) \in \{0, 1\}$ and $B(y) \in \{0, 1\}$ for each $y \in Y$). Ordinary attribute implications (association rules, functional dependencies) are thus a special case of fuzzy attribute implications.

(2) For a fuzzy attribute implication $A \Rightarrow B$, degrees $A(y) \in L$ and $B(y) \in L$ can be seen as thresholds. This is best seen when $^*$ is globalization, i.e., $1^* = 1$ and $a^* = 0$ for $a < 1$. Since for $a, b \in L$ we have $a \leq b$ iff $a \to b = 1$, we have

$$(a \to b)^* = \begin{cases} 1 & \text{iff } a \leq b, \\ 0 & \text{iff } a \nleq b. \end{cases}$$

Therefore, $||A \Rightarrow B||_{\langle X,Y,I\rangle} = 1$ means that a proposition "for each object $x \in X$: if for each attribute $y \in Y$, $x$ has $y$ in degree greater than or equal to (a threshold) $A(y)$, then for each $y \in Y$, $x$ has $y$ in degree at least $B(y)$" has a truth degree 1 (is fully true). In general, $||A \Rightarrow B||_{\langle X,Y,I\rangle}$ is a truth degree of the latter proposition. As a particular example, if $A(y) = a$ for $y \in Y_A \subseteq Y$ (and $A(y) = 0$ for $y \notin Y_A$) $B(y) = b$ for $y \in Y_B \subseteq Y$ (and $B(y) = 0$ for $y \notin Y_B$), the proposition says "for each object $x \in X$: if $x$ has all attributes from $Y_A$ in degree at least $a$, then $x$ has all attributes from $Y_B$ in degree at least $b$", etc. That is, having $A$ and $B$ fuzzy sets allows for a rich expressibility of relationships between attributes which is why we want $A$ and $B$ to be fuzzy sets in general.

We are now coming to the concept of semantic entailment. For simplicity, we consider ordinary sets of FAIs as theories in FAL, i.e., a theory in FAL is a set $T$ of FAIs. Note that more generally, one can consider fuzzy sets of FAIs which is more in the spirit of fuzzy logic, see [5,10]. For a theory $T$, the set $\mathrm{Mod}(T)$ of all *models* of $T$ is defined by

$$\mathrm{Mod}(T) = \{M \in \mathbf{L}^Y \mid \text{for each } A \Rightarrow B \in T\colon ||A \Rightarrow B||_M = 1\}.$$

Therefore, $M \in \mathrm{Mod}(T)$ means that each $A \Rightarrow B$ from $T$ is fully true in $M$. Then, a *degree* $||A \Rightarrow B||_T \in L$ to which $A \Rightarrow B$ *semantically follows* from a set $T$ of attribute implications is defined by

$$||A \Rightarrow B||_T = \textstyle\bigwedge_{M\in\mathrm{Mod}(T)} ||A \Rightarrow B||_M.$$

That is, $||A \Rightarrow B||_T$ can be seen as a truth degree of "$A \Rightarrow B$ is true in each model of $T$".

Consider now the following system of *deduction rules*:

(Ax) infer $A \cup B \Rightarrow A$,
(Cut) from $A \Rightarrow B$ and $B \cup C \Rightarrow D$ infer $A \cup C \Rightarrow D$,
(Mul) from $A \Rightarrow B$ infer $c^* \otimes A \Rightarrow c^* \otimes B$,

for each $A, B, C, D \in \mathbf{L}^Y$, and $c \in L$. Rules (Ax)–(Mul) are to be understood as usual deduction rules: having FAIs which are of the form of FAIs in the input part (the part preceding "infer") of a rule, a rule allows us to infer (in one step) the corresponding fuzzy attribute implication in the output part (the part following "infer") of a rule. (Ax) is a nullary rule (axiom) which says that each $A \cup B \Rightarrow A$ ($A, B \in \mathbf{L}^Y$), i.e., each $C \Rightarrow D$ with $D \subseteq C$, can be inferred in one step.

As usual, a FAI $A \Rightarrow B$ is called *provable* from a set $T$ of fuzzy attribute implications using a set $\mathcal{R}$ of deduction rules, written $T \vdash_{\mathcal{R}} A \Rightarrow B$, if there is a sequence $\varphi_1, \ldots, \varphi_n$ of fuzzy attribute implications such that $\varphi_n$ is $A \Rightarrow B$ and for each $\varphi_i$ we either have $\varphi_i \in T$ or $\varphi_i$ is inferred (in one step) from some of the preceding formulas (i.e., $\varphi_1, \ldots, \varphi_{i-1}$) using some deduction rule from $\mathcal{R}$. If $\mathcal{R}$ consists of (Ax)–(Mul), we say just "provable ..." instead of "provable ... using $\mathcal{R}$" and write just $T \vdash A \Rightarrow B$ instead of $T \vdash_{\mathcal{R}} A \Rightarrow B$. The following theorem was proved in [5].

**Theorem 1 (completeness).** *Let* $\mathbf{L}$ *and* $Y$ *be finite. Let* $T$ *be a set of fuzzy attribute implications. Then*

$$T \vdash A \Rightarrow B \quad \textit{iff} \quad ||A \Rightarrow B||_T = 1.$$

□

Therefore, $A \Rightarrow B$ is provable from $T$ iff $A \Rightarrow B$ semantically follows from $T$ in degree 1.

## 4 Properties and Role of the Rule of Multiplication

Rule (Mul) is called a *rule of multiplication*. System (Ax)–(Mul) improves a previously known complete system presented in [4]. Namely, the system in [4] consists of the following deduction rules:

(Ax') infer $A \Rightarrow S(B, A) \otimes B$,
(Wea') from $A \Rightarrow B$ infer $A \cup C \Rightarrow B$,
(Cut') from $A \Rightarrow e \otimes B$ and $B \cup C \Rightarrow D$ infer $A \cup C \Rightarrow e^* \otimes D$

for each $A, B, C, D \in \mathbf{L}^Y$, and $e \in L$.

Comparing (Ax)–(Mul) to (Ax')–(Cut'), we can see the following distinction. (Ax)–(Mul) results by directly taking two ordinary Armstrong rules, namely (Ax) and (Cut), and by adding a new "fuzzy rule", namely (Mul). In more detail, rules "infer $A \cup B \Rightarrow A$" and "from $A \Rightarrow B$ and $B \cup C \Rightarrow D$ infer $A \cup C \Rightarrow D$" with $A, B, C, D$ being ordinary sets of attributes are well-known ordinary Armstrong rules, see e.g. [17]. Therefore, (Ax) and (Cut) are just these rules with sets replaced by fuzzy sets. Rule (Mul) is a new rule. In the ordinary setting, (Mul) is trivial since it reads "from $A \Rightarrow B$ infer $A \Rightarrow B$" when $c = 1$ and "from $A \Rightarrow B$ infer $\emptyset \Rightarrow \emptyset$" when $c = 0$ (this is easily seen because the only hedge $^*$ in the ordinary setting is the identity mapping). On the other hand, (Ax')–(Cut') result by modifying ordinary rules. For instance, (Cut') results from the above ordinary rule "from $A \Rightarrow B$ and $B \cup C \Rightarrow D$ infer $A \cup C \Rightarrow D$" (with $A, B, C, D$ being ordinary sets) by replacing sets by fuzzy sets and by inserting multiplication by truth degrees $e$ and $e^*$.

The first apparent advantage of system (Ax)–(Mul) is thus aesthetic. System (Ax)–(Mul) can be seen as having two parts, the "ordinary one" consisting of ordinary rules (Ax) and (Cut), and the "fuzzy one" consisting of (Mul). Intuitively, it therefore keeps separated the "ordinary" and the "fuzzy part" with the rule of multiplication "taking care of fuzziness". We are now going to show that there are practical advantages of keeping the ordinary and fuzzy rules separated as well.

The first example is an easily observable fact mentioned already in [5] that if $^*$ is globalization, (Mul) can be omitted. Namely, for $c = 1$, (Mul) becomes "from $A \Rightarrow B$ infer $A \Rightarrow B$" which does not yield anything new. For $c < 1$, (Mul) becomes "from $A \Rightarrow B$ infer $\emptyset \Rightarrow \emptyset$" but $\emptyset \Rightarrow \emptyset$ can be inferred using (Ax) and so (Mul) can be omitted. In the rest of our paper, we are concerned with properties of the rule of multiplication related to the structure of proofs in FAL.

Let us say that (Mul) *commutes* (*backwards*) with a rule (R) of the form "from $A_1 \Rightarrow B_1, \ldots, A_n \Rightarrow B_n$ infer $A \Rightarrow B$" if any FAI $C \Rightarrow D$ which results by first inferring $A \Rightarrow B$ from $A_1 \Rightarrow B_1, \ldots, A_n \Rightarrow B_n$, using (R) and then inferring $C \Rightarrow D$ from $A \Rightarrow B$ using (Mul) can be obtained by first inferring $C_1 \Rightarrow D_1$, $\ldots, C_n \Rightarrow D_n$, from $A_1 \Rightarrow B_1, \ldots, A_n \Rightarrow B_n$, using (Mul), respectively, and then inferring $C \Rightarrow D$ from $C_1 \Rightarrow D_1, \ldots, C_n \Rightarrow D_n$, using (R).

A practical meaning of commutativity of (Mul) with (R) is that in proofs, one can change the order of rules (Mul) and (R) from "first (R), then (Mul)" to "first (Mul), then (R)".

**Lemma 1 (commutativity of (Mul)).** *(Mul) commutes with both* (Ax) *and* (Cut).

*Proof.* Commutativity with (Ax): Let $c^* \otimes (A \cup B) \Rightarrow c^* \otimes A$ result by first inferring $A \cup B \Rightarrow A$ by (Ax) and then inferring $c^* \otimes (A \cup B) \Rightarrow c^* \otimes A$ from $A \cup B \Rightarrow A$ by (Mul). In this case, $c^* \otimes (A \cup B) \Rightarrow c^* \otimes A$ can be obtained by applying (Mul) 0-times (i.e., not applying at all) and then inferring $c^* \otimes (A \cup B) \Rightarrow c^* \otimes A$ by (Ax) since $c^* \otimes (A \cup B) = (c^* \otimes A) \cup (c^* \otimes B)$ and thus, $c^* \otimes (A \cup B) \Rightarrow c^* \otimes A$ is of the form $C \cup D \Rightarrow C$.

Commutativity with (Cut): Let $c^* \otimes (A \cup C) \Rightarrow c^* \otimes D$ result by first inferring $A \cup C \Rightarrow D$ from $A \Rightarrow B$ and $B \cup C \Rightarrow D$, and then inferring $c^* \otimes (A \cup C) \Rightarrow c^* \otimes D$ from $A \cup C \Rightarrow D$ by (Mul). Then, one can first infer $c^* \otimes A \Rightarrow c^* \otimes B$ and $c^* \otimes (B \cup C) \Rightarrow c^* \otimes D$ from $A \Rightarrow B$ and $B \cup C \Rightarrow D$ by (Mul), respectively. Since $c^* \otimes (B \cup C) = (c^* \otimes B) \cup (c^* \otimes C)$, and $c^* \otimes (A \cup C) = (c^* \otimes A) \cup (c^* \otimes C)$, one can infer $c^* \otimes (A \cup C) \Rightarrow c^* \otimes D$ from $c^* \otimes A \Rightarrow c^* \otimes B$ and $c^* \otimes (B \cup C) \Rightarrow c^* \otimes D$ using (Cut). □

Another property of (Mul) is the following one.

**Lemma 2 (idempotency of (Mul)).** *Two (or more) consecutive inferences by* (Mul) *can be replaced by a single inference by* (Mul).

*Proof.* Suppose we start with $A \Rightarrow B$, apply (Mul) to infer $c^* \otimes A \Rightarrow c^* \otimes B$ and then apply (Mul) again to infer $d^* \otimes c^* \otimes A \Rightarrow d^* \otimes c^* \otimes B$. Then, the assertion follows by observing that $d^* \otimes c^* = (d^* \otimes c^*)^*$. Indeed, $d^* \otimes c^* \otimes A \Rightarrow d^* \otimes c^* \otimes B$ is then of the form $a^* \otimes A \Rightarrow a^* \otimes B$, with $a = d^* \otimes c^*$. It remains to prove $d^* \otimes c^* = (d^* \otimes c^*)^*$. "$\geq$" follows directly from subdiagonality of $^*$, i.e., from $b^* \leq b$. "$\leq$" is equivalent to $d^* \leq c^* \rightarrow (d^* \otimes c^*)^*$ which is true. To see this, observe that $d^* \leq c^* \rightarrow (d^* \otimes c^*)$ from which we get

$$d^* = d^{**} \leq (c^* \rightarrow (d^* \otimes c^*))^* \leq c^{**} \rightarrow (d^* \otimes c^*)^* = c^* \rightarrow (d^* \otimes c^*)^*.$$

□

The above-observed commutativity of (Mul) has the following consequence.

**Theorem 2 (normal form of proofs in FAL).** *If $A \Rightarrow B$ is provable from $T$ using* (Ax)–(Mul), *then there exists a proof $A_1 \Rightarrow B_1, \ldots, A_n \Rightarrow B_n$ of $A \Rightarrow B$ from $T$ using* (Ax)–(Mul) *and integers $1 \leq k \leq l \leq n$ such that*

1. *for* $i = 1, \ldots, k$, *we have* $A_i \Rightarrow B_i \in T$,
2. *for* $i = k+1, \ldots, l$, $A_i \Rightarrow B_i$ *results by application of* (Mul) *to some* $A_j \Rightarrow B_j$ *with* $1 \leq j \leq k$,
3. *for* $i = l+1, \ldots, n$, $A_i \Rightarrow B_i$ *results by application of* (Ax) *or* (Cut) *to some* $A_j \Rightarrow B_j$*'s with* $1 \leq j < i$.

*Proof.* The proof follows by induction from Lemma 1 and Lemma 2, we omit details. □

According to Theorem 2, each proof from $T$ in FAL can be transformed to a proof which starts by formulas from $T$, continues by applications of (Mul) to these formulas, and then by applications of (Ax) and (Cut) to preceding formulas. Therefore, the part of the proof which uses "ordinary rules" (Ax) and (Cut) is separated from the part which uses "fuzzy rule" (Mul).

Another way of formulating Theorem 2 is the following. Denote

$$T^* = \{c^* \otimes A \Rightarrow c^* \otimes B \mid A \Rightarrow B \in T, c \in L\}.$$

Then we have

**Theorem 3 (provable from $T$=provable from $T^*$ using ordinary rules).** $A \Rightarrow B$ *is provable from* $T$ *using* (Ax)–(Mul) *iff* $A \Rightarrow B$ *is provable from* $T^*$ *using ordinary rules* (Ax) *and* (Cut).

*Proof.* Easy consequence of Theorem 2 and definitions. □

As an application of the presented results, we now present an analogy of a well-known theorem from relational databases saying that each proof of a functional dependence can be transformed into a RAP-sequence. Due to lack of space, we omit discussion on ramifications of this theorem and refer to [17]. Consider the following deduction rules:

(Ref) infer $A \Rightarrow A$,
(Acc) from $A \Rightarrow B \cup C$ and $C \Rightarrow D \cup E$ infer $A \Rightarrow B \cup C \cup D$,
(Pro) from $A \Rightarrow B \cup C$ infer $A \Rightarrow B$,

for each $A, B, C, D, E \in \mathbf{L}^Y$. The rules result from the well-known ordinary rules of reflexivity, accumulation, and projectivity by replacing sets with fuzzy sets, see [17]. An *MRAP-sequence* for $A \Rightarrow B$ from $T$ is a proof of $A \Rightarrow B$ from $T$ using (Mul), (Ref), (Acc), (Pro), such that

1. the proof starts with FAIs from $T$,
2. continues with FAIs which result by application of (Mul) to formulas from 1.,
3. continues with $A \Rightarrow A$,
4. continues with formulas which result by application of (Acc) to formulas from 1., 2., and 3.,
5. ends with application of (Pro) which results in $A \Rightarrow B$, the last member of the proof.

**Theorem 4 (MRAP-sequence theorem).** *If $A \Rightarrow B$ follows from $T$ in degree 1, then there exists an MRAP-sequence for $A \Rightarrow B$ from $T$.*

*Proof.* Sketch: If $A \Rightarrow B$ follows from $T$ in degree 1, then there exists a proof of $A \Rightarrow B$ from $T$ by Theorem 1. According to Theorem 2, this proof can be transformed into a normal form described in Theorem 2. The transformed proof starts with a sequence $A_1 \Rightarrow B_1, \ldots, A_l \Rightarrow B_l$ which satisfies conditions 1. and 2. of an MRAP-sequence (formulas from $T$ and formulas resulting by application of (Mul)) and continues with a sequence $A_{l+1} \Rightarrow B_{l+1}, \ldots,$ in which only (Ax) and (Cut) are used. It can be shown that the latter sequence can be transformed into a sequence which uses only rules (Ref), (Acc), and (Pro). Since these rules are just "ordinary rules" with ordinary sets replaced by fuzzy sets, one can repeat the ordinary proof, verbatim, see e.g. [17, Theorem 4.2]), showing that sequence $A_{l+1} \Rightarrow B_{l+1}, \ldots,$ can be transformed into a sequence starting with $A \Rightarrow A$, continuing by applications of (Acc), and ending with an application of (Pro) which yields $A \Rightarrow B$. Altogether, this gives an MRAP-sequence for $A \Rightarrow B$. □

## 5 Concluding Remarks

We showed properties of the rule of multiplication in FAL related to the structure of proofs. Proofs in FAL are both of theoretical and practical interest (for instance, proofs are used in algorithms testing redundancy of a set of FAIs, see e.g. [9]). As an application of the properties, we presented theorems concerning normal forms of proofs in FAL. The main benefit of our approach is that separating a system for FAL into an "ordinary part" and a "fuzzy part", which is represented by the rule of multiplication, enables us to use automatically results known from the ordinary case (from the theory of relational databases, formal concept analysis, etc.). Due to the existing relationship to database interpretation of FAIs described above, our results apply to reasoning about functional dependencies in fuzzy setting as well, see e.g. [7] and [11] for comparison of some approaches.

Other issues which we did not present due to the limited scope and issues for further research include

- further properties of the rule of multiplication,
- rule of multiplication in Pavelka-style FAL, see [10],
- further study of the possibility to automatically convey ordinary results to fuzzy setting,
- this includes a more rigorous treatment of issues like the relationship between ordinary deduction rules and their counterparts resulting by replacing sets by fuzzy sets (this was described more or less intuitively in our paper due to lack of space),
- as a long-term goal, a further study of data dependencies in a fuzzy setting.

## References


1. Armstrong W. W.: Dependency structures in data base relationships. *IFIP Congress*, Geneva, Switzerland, 1974, pp. 580–583.
2. Belohlavek R.: *Fuzzy Relational Systems: Foundations and Principles.* Kluwer, Academic/Plenum Publishers, New York, 2002.
3. Belohlavek R., Vychodil V.: Fuzzy attribute logic: attribute implications, their validity, entailment, and non-redundant basis. IFSA 2005, Vol. I, 2005, pp. 622–627. Tsinghua University Press and Springer, ISBN 7–302–11377–7.
4. Belohlavek R., Vychodil V.: Fuzzy attribute logic: syntactic entailment and completeness. Proc. JCIS 2005, Salt Lake City, Utah, 2005, pp. 78–81, ISBN 0–9707890–3–3.
5. Belohlavek R., Vychodil V.: Axiomatizations of fuzzy attribute logic. In: Prasad B. (Ed.): IICAI 2005, Proc. 2nd Indian International Conference on Artificial Intelligence, Pune, India, Dec 20–22, 2005, ISBN 0–9727412–1–6, pp. 2178–2193.
6. Belohlavek R., Vychodil V.: Attribute implications in a fuzzy setting. In: Missaoui R., Schmid J. (Eds.): ICFCA 2006, *Lecture Notes in Artificial Intelligence* **3874**, pp. 45–60, Springer-Verlag, Berlin/Heidelberg, 2006.
7. Belohlavek R., Vychodil V.: Data tables with similarity relations: functional dependencies, complete rules and non-redundant bases. In: M.L. Lee, K.L. Tan, and V. Wuwongse (Eds.): DASFAA 2006, *Lecture Notes in Computer Science* **3882**, pp. 644–658, Springer-Verlag, Berlin/Heidelberg, 2006.
8. Belohlavek R., Vychodil V.: Computing non-redundant bases of if-then rules from data tables with graded attributes. In: Proc. IEEE GrC 2006, 2006 IEEE International Conference on Granular Computing, Atlanta, GA, May 10–12, 2006, pp. 205–210, ISBN 1–4244–0133–X
9. Belohlavek R., Vychodil V.: Properties of models of fuzzy attribute implications. SCIS & ISIS 2006, Int. Conf. Soft Computing and Intelligent Systems & Int. Symposium on Intelligent Systems, Sep 20–24, 2006, Tokyo, Japan, pp. 291–296, ISSN 1880–3741.
10. Belohlavek R., Vychodil V.: Pavelka-style fuzzy logic for attribute implications. Proc. JCIS 2006, Kaohsiung, Taiwan, ROC, pp. 1156–1159.
11. Belohlavek R., Vychodil V.: Codd's relational model of data and fuzzy logic: comparisons, observations, and some new results. In: Mohammadian M. (Ed.): Proc. CIMCA 2006, Sydney, Australia, 2006, 6 pages, ISBN 0–7695–2731–0.
12. Ganter B., Wille R.: *Formal Concept Analysis. Mathematical Foundations.* Springer, Berlin, 1999.
13. Gerla G.: *Fuzzy Logic. Mathematical Tools for Approximate Reasoning.* Kluwer, Dordrecht, 2001.
14. Hájek P.: *Metamathematics of Fuzzy Logic.* Kluwer, Dordrecht, 1998.
15. Hájek P.: On very true. *Fuzzy Sets and Systems* **124**(2001), 329–333.
16. Klir G. J., Yuan B.: *Fuzzy Sets and Fuzzy Logic. Theory and Applications.* Prentice Hall, 1995.
17. Maier D.: *The Theory of Relational Databases.* Computer Science Press, Rockville, 1983.
18. Pollandt S.: *Fuzzy Begriffe.* Springer-Verlag, Berlin/Heidelberg, 1997.
19. Takeuti G., Titani S.: Globalization of intuitionistic set theory. *Annals of Pure and Applied Logic* **33**(1987), 195–211.

# Graded Fuzzy Rules

Martina Daňková

Institute for Research and Applications of Fuzzy Modeling,
University of Ostrava, 30. dubna 22, Czech Republic
martina.dankova@osu.cz
http://irafm.osu.cz

**Abstract.** In this contribution, we will recall graded fuzzy rules introduced in [5] and explain the difference from the classical fuzzy rules. Moreover, properties of formulae, which are used to formalize the graded fuzzy rules, will be recalled.

## 1 Introduction

Let us consider knowledge-based systems of the following two forms:

$$\begin{array}{lll} \texttt{DisK}: & C_1 \ \ \mathsf{OR} \ldots \mathsf{OR} \ \ C_k, \\ \texttt{ConK}: & C_1 \ \mathsf{AND} \ldots \mathsf{AND} \ C_k, \end{array}$$

where AND (OR) relates to conjunction (disjunction), and $C_i$, $i = 1, \ldots, k$, are linguistic expressions. These expressions can have different character. The closest to the practice are knowledge-based systems embodying pieces of knowledge about some complex object or physical system in concern. This system is assumed to be utterable as an relational dependence between two universes of the discourse in the special unified form DisK or ConK, where in each $C_i$, the particular piece of knowledge from the input space is joined with the one from the output space. Since the very beginning [23], the three distinct approaches (dual in some sense) to integration of such pieces of knowledge were pushed ahead:

$$\begin{array}{ll} \texttt{Dis}: & (x \in A_1 \ \mathsf{and}\ y \in B_1) \ \ \mathsf{OR} \ldots \mathsf{OR} \ \ (x \in A_k \ \mathsf{and}\ y \in B_k), \\ \texttt{Con1}: & (x \in A_1 \ \mathsf{or}\ y \in B_1) \ \mathsf{AND} \ldots \mathsf{AND} \ (x \in A_k \ \mathsf{or}\ y \in B_k), \\ \texttt{Con2}: & (\mathsf{If}\ x \in A_1 \ \mathsf{then}\ y \in B_1) \ \mathsf{AND} \ldots \mathsf{AND} \ (\mathsf{If}\ x \in A_k \ \mathsf{then}\ y \in B_k), \end{array}$$

where "$x \in A$" is in the sense of [11], and (or) relates to strong conjunction (strong disjunction). In this special case, each elementary part $C_i$ will be called *fuzzy rule*.

A collection of fuzzy rules incorporates the vagueness of $A_i, B_i$ that comes in degrees and the interpretation of the connectives can carry these degrees in appropriate way. In addition, there may exist a certain amount of dubiousness over the particular fuzzy rules. This is reflected by introducing certain degrees that equip the respective rules. A collection of *graded fuzzy rules* can be visualized in the spirit of [18] (using the same notation as for evaluated formulae)

$$\begin{array}{ll} \texttt{GDis}: & {}^{f_1 \succeq}/(x \in A_1 \ \mathsf{and}\ y \in B_1) \ \ \mathsf{OR} \ldots \mathsf{OR} \ \ {}^{f_k \succeq}/(x \in A_k \ \mathsf{and}\ y \in B_k), \\ \texttt{GCon1}: & {}^{f_1 \preceq}/(x \in A_1 \ \mathsf{or}\ y \in B_1) \ \mathsf{AND} \ldots \mathsf{AND} \ {}^{f_k \preceq}/(x \in A_k \ \mathsf{or}\ y \in B_k), \\ \texttt{GCon2}: & {}^{f_1 \preceq}/(\mathsf{If}\ x \in A_1 \ \mathsf{then}\ y \in B_1) \ \mathsf{AND} \ldots \mathsf{AND} \ {}^{f_k \preceq}/(\mathsf{If}\ x \in A_k \ \mathsf{then}\ y \in B_k), \\ \texttt{GCon3}: & {}^{f_1 \succeq}/(\mathsf{If}\ x \in A_1 \ \mathsf{then}\ y \in B_1) \ \mathsf{AND} \ldots \mathsf{AND} \ {}^{f_k \succeq}/(\mathsf{If}\ x \in A_k \ \mathsf{then}\ y \in B_k), \end{array}$$

P. Melin et al. (Eds.): IFSA 2007, LNAI 4529, pp. 481–490, 2007.
© Springer-Verlag Berlin Heidelberg 2007

where the particular graded rule ${}^{f\succeq}/A$ means "$A$ is valid at most to the degree $f$", analogously, ${}^{f\preceq}/A$ reads as "$A$ is valid at least to the degree $f$", and the degrees $f_1, \ldots, f_k$ belong to some structure for the truth values.

In this contribution, we are going to present appropriate formalization of GDis, GCon1–3 inside a suitable logic, i.e. formulae that fits to the interpretation of the collection of graded fuzzy rules. Then, we will recall the formalization of Dis, Con1–2 and compare them with the formalization of GDis, GCon1–3. In the last part we will concentrate on properties of an approximate reasoning over the collection of graded fuzzy rules using an appropriate rule of inference.

## 2 Background Logic

A suitable logical system should incorporates strong conjunction &, lattice conjunction $\wedge$, implication $\rightarrow$ having the properties specified by schemata of the axioms of MTL, involutive negation $\sim$, i.e. satisfying the following formula:

$$(N1) \qquad (\sim\sim \varphi) \leftrightarrow \varphi,$$

and the truth constant $\bar{0}$. In such a system, the following derived connectives can be introduced

$$\begin{array}{rll}
\neg\varphi & \equiv_{df} \varphi \rightarrow \bar{0} & \textit{negation} \\
\varphi\nabla\psi & \equiv_{df} \sim((\sim\varphi)\&(\sim\psi)) & \textit{strong disjunction} \\
\varphi \Rightarrow \psi & \equiv_{df} (\sim\varphi)\nabla\psi & \textit{S-implication} \\
\varphi \vee \psi & \equiv_{df} [(\varphi \rightarrow \psi) \rightarrow \psi] \wedge [(\psi \rightarrow \varphi) \rightarrow \varphi] & \textit{disjunction} \\
\varphi \leftrightarrow \psi & \equiv_{df} (\varphi \rightarrow \psi)\&(\psi \rightarrow \varphi) & \textit{equivalence}
\end{array}$$

Since MTL contains most of these connectives and schemata of the axioms, it is a good starting point for the additional extension by an involutive negation resulting as $\mathrm{SMTL}_\triangle$ (introduced in [10]) to which we will assign the role of the background logic. This extension is analogous to the one presented in [9].

### 2.1 Syntax and Semantics

We will work with the multi-sorted predicate logic $\mathrm{MTL}_\triangle\forall$. It includes the language $J$ with sorts $s_1, \ldots, s_n$, a non-empty set of predicates of any type, a set of object constants, object variables $x_i, y_i, z_i, \ldots$ of the sorts $s_i$, $i = 1, \ldots, n$, a set of connectives $\{\wedge, \&, \rightarrow, \triangle\}$, truth constants $\bar{0}$ and $\bar{1} \equiv_{df} \bar{0} \rightarrow \bar{0}$, quantifiers $\forall, \exists$ and does not include functional symbols. Terms are object variables and object constants.

$\mathrm{MTL}_\triangle\forall$ consists of the axioms of $\mathrm{MTL}_\triangle$ for connectives together with the axioms on quantifiers. The deduction rules are modus ponens, generalization rule and $\triangle$-necessitation. For the details we refer to [8].

Let us extend the language $J$ of MTL$\forall$ by the new unary connective $\sim$, which will have the highest priority, and the set of axioms of MTL$\forall$ by (N1) and

$$(N2) \qquad \triangle(\varphi \rightarrow \psi) \rightarrow \triangle((\sim\psi) \rightarrow (\sim\varphi)).$$

The resulting extension will be denoted by SMTL$_\triangle\forall$. As above, definable connectives are $\{\neg, \nabla, \Rightarrow, \vee, \leftrightarrow\}$. The above extension within the propositional MTL$_\triangle$ will be denoted by SMTL$_\triangle$. Terms and formulae are build in the usual way.

Interpretation of the connectives $\{\wedge, \&, \rightarrow\}$ is given by the corresponding operations $\{\cap, *, \rightarrow_*\}$, and the constant $\bar{0}$ is interpreted as 0, which together form an MTL-algebra $\mathcal{L} = \langle L, \cap, \cup, *, \rightarrow_*, 0, 1\rangle$. Moreover, the interpretation of $\sim$ will be denoted by $'$ defined as $x' = 1 - x$ for all $x \in L$. Recall that the truth function for $\triangle$ is the following

$$\blacktriangle(x) = \begin{cases} 1, & \text{if } x = 1; \\ 0, & \text{otherwise.} \end{cases}$$

An $\mathcal{L}$-structure for the language $J$ is of the form

$$\mathcal{M} = \langle (D_i)_{for\ s_i}, (r_P)_{P-predicate}, (m_c)_{c-constant}\rangle,$$

where $D_1, \ldots, D_n$ are non-empty sets of objects, $r_P$ is an $L$-fuzzy relation of the respective type and $m_c$ belongs to $D_i$ provided that $c$ is of the type $s_i$.

In analogy with the construction of [9] for SBL$_\triangle$, the completeness of SMTL$_\triangle$ can be shown (see Theorem 3.7 in [10]) and afterwards extended to SMTL$_\triangle\forall$ as in [8] (see Theorem 7 for the completeness of MTL$\forall$).

**Lemma 1.** *In SMTL$_\triangle$, the following formulae are provable:*

$$(\varphi \Rightarrow \psi) \leftrightarrow (\sim \psi \Rightarrow \sim \varphi), \tag{1}$$
$$(\varphi \Rightarrow \psi) \leftrightarrow \sim (\varphi \& \sim \psi), \tag{2}$$
$$(\neg\varphi) \rightarrow (\sim \varphi), \tag{3}$$
$$\sim (\varphi \wedge \psi) \leftrightarrow (\sim \psi \vee \sim \varphi), \tag{4}$$
$$\sim (\varphi \vee \psi) \leftrightarrow (\sim \psi \wedge \sim \varphi), \tag{5}$$
$$((\varphi \& \psi) \rightarrow \chi) \rightarrow (\sim \chi \rightarrow (\varphi \Rightarrow \sim \psi)), \tag{6}$$
$$\psi \rightarrow (\varphi \Rightarrow \sim (\varphi \rightarrow \sim \psi)), \tag{7}$$
$$\psi \rightarrow ((\varphi \rightarrow \sim \psi) \Rightarrow \sim \varphi), \tag{8}$$
$$(\varphi \vee \psi) \rightarrow (\varphi \nabla \psi). \tag{9}$$

*Moreover, in SMTL$_\triangle$ the following is derived inference rule:*

$$(CP) \qquad \frac{\varphi \rightarrow \psi}{\sim \psi \rightarrow \sim \varphi}$$

## 2.2 Formulae as Graded Theorems

In the following, we will prove so called graded theorems (for the first time introduced in [12] as stated in [2]). It means that instead of the usual formulation "If $\vdash \varphi$ then $\vdash \psi$", we are going to find $n \in \mathbb{N}$ for which $\vdash \varphi^n \rightarrow \psi$.

Whenever we do know $\vdash \varphi^n \rightarrow \psi$ and $\vdash \varphi$ then we can derive $\vdash \psi$ easily, however, it is not such a simple case when we proceed the other way round. Hence, the formula $\vdash \varphi^n \rightarrow \psi$ is more general and it can be read as

"Formula $\psi$ is valid at least to the degree of $\varphi^n$."

Of course the previous free reading relates to the notion of truth degrees of $\varphi$ and $\psi$ in the particular model.

Notice that the existence of $n$ follows from the deduction theorem (Theorem 2.2.18 in [13]) and the general method of its estimation is related to the concrete proof (or proofs) – the way of using $\varphi$ as an assumption.

# 3 Logical Insight to Graded Fuzzy Rules

The notion of gradual rule has been introduced in [7] and [6]. Their authors distinguish between gradual rules, certainty rules and their mixture. The gradualness is connected with the properties of implication, certainty with the strong disjunction, and the rules are not equipped with additional truth value as it is in our case. The rules specified by the authors of [7,6] are special cases of our graded rules. We simply take all values $f_i = 0$ (where 0 is associated to the truth constant for the falsity in a structure of truth values) and modify $A_i, B_i$ by means of involutive negation in $\mathtt{GCon1,2}$.

In this work, we interpret $\mathtt{GDis}$ and $\mathtt{GCon1,2}$ by formulae of predicate fuzzy logic. The weight $f$ is obtained from the latter as well. For example, let us consider a rule

$$\mathfrak{R} \equiv (x \in A) \text{ and } (y \in B),$$

which will be interpreted by a formula $A(x)\&B(y)$. Moreover, we can suppose its verity to be supported by examples $D = \{(c_i, d_i) | \, i \in I\}$, where $(c_i, d_i)$ are some object constants of the corresponding language. To each data from $D$ is assigned a weight that impacts the value $f$ in which $\mathfrak{R}$ is satisfied. We may assume that the weight $\omega_i$ of $(c_i, d_i)$ is given as a truth value of $A(c_i)\&B(d_i)$. Note that $\omega_i$ need not be produced necessarily on the basis of definable formula. The final value $f$ of $\mathfrak{R}$ is produced on the basis of knowledge about the data set $D$ with respect to the structure of $\mathfrak{R}$. In this example, we take $f = \max_{i \in I} \omega_i$ and we obtain the graded rule $^{\langle \text{at most} \rangle f} / \mathfrak{R}$.

## 3.1 Formalization of Graded Fuzzy Rules

First, let us recall special formulae (known as normal forms introduced by Perfilieva in [19]), which we will use to formalize collection of graded fuzzy rules.

**Conventions 1.** *For the sake of brevity, let us denote*

$$\begin{array}{ll} [x_1, \ldots, x_n] & \textit{by} \quad \bar{x} \\ \{R_1, \ldots, R_n\} & \textit{by} \quad \bar{R} \\ (R_1(x_1, y_1)\& \cdots \&R_n(x_n, y_n)) & \textit{by}\ R(\bar{x}, \bar{y}) \end{array}$$

*Analogous shortenings we use for any other binary predicate or variable. Moreover, let us suppose $n > 1$, $p \in \mathbb{N}$ such that $1 \leq p \leq n-1$, and $\bar{x}_p = [x_1, \ldots, x_p]$, $\bar{y}_p = [x_{p+1}, \ldots, x_n]$, $\bar{d}_i = [c_{i_1}, \ldots, c_{i_p}]$, and $\bar{e}_i = [c_{i_{p+1}}, \ldots, c_{i_n}]$.*

**Definition 1.** *Let $k \in \mathbb{N}$ and $I_k = \{1, \ldots, k\}$ and let the language $J_k$ be an extension of $J(SMTL_\triangle\forall)$ by*

**(a)** *a finite set of $n$-tuples of object constants $\{\bar{c}_i = [c_{i_1}, \ldots, c_{i_n}] \mid i \in I_k\}$, where each $c_{i_j}$ is of the sort $s_j$ for all $j = 1, \ldots, n$,*
**(b)** *binary predicate symbols $R_1, \ldots, R_n$, each $R_i$ of the type $\langle s_i, s_i \rangle$,*
**(c)** *and an $n$-ary predicate $F$ of the type $\langle s_1, \ldots, s_n \rangle$.*

*We define the following formulae*

$$\mathrm{DNF}_{F,k}(\bar{x}) \equiv_{df} \bigvee_{i \in I_k} (R(\bar{c}_i, \bar{x}) \& F(\bar{c}_i)), \tag{10}$$

$$\mathrm{CNF}^1_{F,k}(\bar{x}) \equiv_{df} \bigwedge_{i \in I_k} (R(\bar{x}, \bar{c}_i) \nabla F(\bar{c}_i)), \tag{11}$$

$$\mathrm{CNF}^2_{F,k}(\bar{x}) \equiv_{df} \bigwedge_{i \in I_k} (R(\bar{x}, \bar{c}_i) \rightarrow F(\bar{c}_i)), \tag{12}$$

$$\mathrm{CNF}^3_{F,k}(\bar{x}) \equiv_{df} \bigwedge_{i \in I_k} (R(\bar{x}, \bar{c}_i) \Rightarrow F(\bar{c}_i)), \tag{13}$$

$$\mathrm{CNF}^4_{F,k}(\bar{x}) \equiv_{df} \bigwedge_{i \in I_k} (F(\bar{c}_i) \rightarrow (R(\bar{x}_p, \bar{d}_i) \rightarrow R(\bar{y}_p, \bar{e}_i)). \tag{14}$$

*Let $\bar{x}$ consist of all free variables of $\varphi$ and $\psi$, and moreover, $\bar{y}$ be substitutable for $\bar{x}$ in $\varphi$ and $\psi$. We define the following formulae:*

$$\begin{array}{rll}
\mathrm{Ext}_{\bar{R}}\,\varphi \equiv_{df} & (\forall \bar{x}, \bar{y})[(R(\bar{x}, \bar{y}) \& \varphi(\bar{x})) \rightarrow \varphi(\bar{y})] & \bar{R}\text{-}\textit{extensionality} \\
\varphi \subseteq \psi \equiv_{df} & (\forall \bar{x})(\varphi(\bar{x}) \rightarrow \psi(\bar{x})) & \textit{inclusion} \\
\varphi \approx \psi \equiv_{df} & (\forall \bar{x})(\varphi(\bar{x}) \leftrightarrow \psi(\bar{x})) & \textit{bi-inclusion} \\
\mathrm{Trans}_R \equiv_{df} & (\forall x, y, z)[(Rxy \& Ryz) \rightarrow Rxz] & \textit{transitivity}
\end{array}$$

Let us assume $n = 2$. Then a partial knowledge formalized by (10) can be interpreted as a collection of $k$ graded fuzzy rules GDis, where $x \in A_i$ is formalized by $R_1(c_{i_1}, x_1)$, $y \in B_i$ by $R_2(c_{i_2}, x_2)$, and the degree $f_i$ by $F(c_{i_1}, c_{i_2})$. Moreover, the part "at most to the degree" is realized using & in DNF. Indeed, since & is in residuated lattice interpreted by $*$, the latter immediately follows from the inequality $x * y \leq y$ for all $x, y$ from the respective support. Hence, the truth value of the $i$'th disjunct in DNF will never exceed the truth value associated to $F(c_{i_1}, c_{i_2})$ for an arbitrary $\mathcal{M}$-valuation of the object variables.

Analogously, the formula (12) formalizes GCon2, where $x \in A_i$ is now formalized by $R_1(x_1, c_{i_1})$, $y \in B_i$ by $R_2(x_2, c_{i_2})$, and the degree $f_i$ by $F(c_{i_1}, c_{i_2})$. The part "at least to the degree" is connected with $\rightarrow$ in $\mathrm{CNF}^1$. Observe that in any residuated lattice $\rightarrow$ is interpreted by $\rightarrow_*$ and the inequality $y \leq x \rightarrow_* y$ holds for all $x, y$. Therefore, the truth value of the $i$'th conjunct in $\mathrm{CNF}^1$ will never fall under the truth value associated to $F(c_{i_1}, c_{i_2})$. Above, we have spoken about the gradualness only w.r.t. fuzzy rules and it needs to be pointed out that on the level of syntax, we deal with (10) and (12), i.e. the formulae in the usual sense.

In any residuated lattice also $x \rightarrow_* y \leq x$ is valid, hence, (14) formalizes GCon3. The remaining formula (11) formalizes GCon1 and (13) formalizes GCon2.

In fact (11) and (13) become equivalent whenever assume different $R$'s in these formulae, moreover, they are dual to (10) by means of $\sim$, and (10) dual to (12) for the extensional $F$ by means of adjunction property but they are not equivalent in general. Because of this fact, we will omit the detailed study of (11) and (13) in the rest of this paper, and we leave it as an exercise to the reader.

Let me add the last comment relating to (14). One may be surprised that there are a good reasons to study this formula. They follow from the study of the "greatest" and "smallest" $\bar{R}$ such that $F$ is extensional w.r.t. $\bar{R}$. A brief introduction is contained in [4].

### 3.2 Formalization of Fuzzy Rules and Its Relationship to Graded Fuzzy Rules

Let us now explain, how are fuzzy rules related to their graded variants. Because of the equivalence explained above, we will focus only on Dis and Con2. First, let us recall formulae interpreting these fuzzy rules in SMTL$_\triangle\forall$.

**Definition 2.** *Under the assumptions of Definition 1, we define the following formulae*

$$\mathrm{DNF}_k(\bar{x}) \equiv_{df} \bigvee_{i \in I_k} (R(\bar{d}_i, \bar{x}_p) \& R(\bar{y}_p, \bar{e}_i)), \tag{15}$$

$$\mathrm{CNF}_k(\bar{x}) \equiv_{df} \bigwedge_{i \in I_k} (R(\bar{x}_p, \bar{d}_i) \to R(\bar{y}_p, \bar{e}_i)). \tag{16}$$

Note that (10) becomes equivalent to (15) simply by taking $F(\bar{x}) \equiv_{df} \bar{1}$. Taking into the consideration (14), we have immediate relationship by taking again $F(\bar{x}) \equiv_{df} \bar{1}$. In the case of (12), the situation is more complicated. Each $R(\bar{y}_p, \bar{e}_i)$ needs to be lifted by $F(\bar{c}_i)$ as follows

$$\vdash \bigwedge_{i \in I_k} (R(\bar{x}, \bar{c}_i) \to F(\bar{c}_i) \leftrightarrow \bigwedge_{i \in I_k} (R(\bar{x}_p, \bar{d}_i) \to (R(\bar{y}_p, \bar{e}_i)) \to F(\bar{c}_i)),$$

where the formula on the righthand-side is in the form of $\mathrm{CNF}_k^2$. Unfortunately, it is not possible to proceed other way round to receive (12). If $\neg$ would satisfy the law of double negation then we could have taken $F(\bar{x}) \equiv_{df} \bar{0}$ to receive one-to-one correspondence between (12) and (16).

## 4 Graded Fuzzy Rules in Practise

### 4.1 Logical Approximation

In this subsection, we will take the results from [19], weaken the requirements and reformulate these results in accordance with a methodology manifested in [2], i.e. in the form of graded theorems. These results are in the scope of the logical approximation, which is a theory aiming at studying properties of a class of formulae in an simplified form relating to some initial formula. There, the significant role is played by a formula of the form $\epsilon \rightarrow (\varphi \leftrightarrow \varphi_S)$ called conditional equivalence (graded theorem of the special form). We may interpret it as a lower bounded or graded equivalence between the given formula $\varphi$ and its simplified version $\varphi_S$. As pointed in [19], if we consider a standard model then the conditional equivalence expresses a precision of approximation of "$\varphi$" by "$\varphi_S$".

**Theorem 1.** *Let* $C_k$ *denote* $(\forall \bar{x}) \bigvee_{i \in I_k} (R(\bar{x}, \bar{c}_i) \& R(\bar{c}_i, \bar{x}))$. *Then the following is provable in* $SMTL_\triangle\forall$*:*

$$\mathrm{Trans}_{\bar{R}} \rightarrow \mathrm{Ext}_{\bar{R}}\, \mathrm{DNF}_{F,k}, \tag{17}$$

$$\mathrm{Trans}_{\bar{R}} \rightarrow \mathrm{Ext}_{\bar{R}}\, \mathrm{CNF}^2_{F,k}, \tag{18}$$

$$\mathrm{Ext}_{\bar{R}}\, F \rightarrow \mathrm{DNF}_{F,k} \subseteq F, \tag{19}$$

$$\mathrm{Ext}_{\bar{R}}\, F \rightarrow F \subseteq \mathrm{CNF}^2_{F,k}, \tag{20}$$

$$\mathrm{Ext}_{\bar{R}}\, F \& C_k \rightarrow \mathrm{DNF}_{F,k} \approx F, \tag{21}$$

$$\mathrm{Ext}_{\bar{R}}\, F \& C_k \rightarrow \mathrm{CNF}^2_{F,k} \approx F. \tag{22}$$

The transitivity of $\bar{R}$ implies the extensionality of DNF as well as $\mathrm{CNF}^2$, see (17), (18). Additionally, the extensionality of $F$ bounds the degree of inclusion of $\mathrm{DNF}_{F,k}$ in $F$ and $F$ in $\mathrm{CNF}^2_{F,k}$. Formulae (21)–(22) show that the truth value of $F \approx \mathrm{D(C)NF}_{F,k}$ is determined by choice of $\bar{c}_i$ and $\bar{R}$ in a particular model, as it can be seen from $C_k$.

### 4.2 Approximate Inference

Approximate inference rule is often considered as a basis for dealing with fuzzy rules and a non-precise input knowledge. A generalized rule of modus ponens as a particular case of compositional rule of inference in the global concept of many-valued logics has been introduced by L. Zadeh in [23]. The analysis of logical aspects of Zadeh's compositional rules of inference was done by P. Hájek in [13] or V. Novák in [16,17,18] (for evaluated syntax). From the other works let us mention e.g. [20,21]. Inference rules were also intensively studied from the algebraical point of view as special operations called compositions (see, e.g., [14]).

In the sequel, we will assume $J_{FC} = J_k \cup \{A^*\}$, where $A^*$ is a predicate of the type $\langle s_1, \ldots, s_p \rangle$. Then, we define

$$B^*_{\mathrm{DNF}}(\bar{y}_p) \equiv_{df} (\exists\, \bar{x}_p)(A^*(\bar{x}_p) \& \,\mathrm{DNF}_{F,k}(\bar{x})), \tag{23}$$

$$B^*_{\mathrm{CNF}^2}(\bar{y}_p) \equiv_{df} (\forall\, \bar{x}_p)(A^*(\bar{x}_p) \rightarrow \mathrm{CNF}^2_{F,k}(\bar{x})), \tag{24}$$

which define $B^*_{\mathrm{DNF}}$ and $B^*_{\mathrm{CNF}^2}$ from $A^*$ using the Zadeh's compositional rule of inference and Bandler-Kouhout's product (BK-product [1]), respectively. Since BK-product can be viewed as a dual to Zadeh's composition, and moreover, $\mathrm{CNF}^2_{F,k}$ is dual to $\mathrm{DNF}_{F,k}$, we conclude that also $B^*_{\mathrm{CNF}^2}$ is in a certain sense dual to $B^*_{\mathrm{DNF}}$.

Approximate inferences based on (23) and (24) can be visualized as inference rules of the following forms

$$(DE)\quad \frac{A^*, \mathrm{DNF}_{F,k}}{B^*_{\mathrm{DNF}}} \quad \text{and} \quad (AB)\quad \frac{A^*, B^*_{\mathrm{CNF}^2}}{\mathrm{CNF}^2_{F,k}}.$$

Here, $F$ represents an ideal situation that can be expressed as $\mathrm{DNF}_F$ and $\mathrm{CNF}^2_F$. Later, we will see that for the extensional $F$, we can prove $F \to \mathrm{CNF}^2_F$ and $\mathrm{DNF}_F \to F$. At this point, we can formulate the following problem: Find $B^*$ such that

$$(a)\quad \begin{array}{l} A^*\&B^* \to F, \\ A^*\&B^* \to \mathrm{CNF}^2_F, \\ \text{and there exists no } B' : \\ (B^* \to B')\&(A^*\&B' \to \mathrm{CNF}^2_F), \end{array} \qquad (b)\quad \begin{array}{l} F \to (A^* \to B^*), \\ \mathrm{DNF}_F \to (A^* \to B^*), \\ \text{and there exists no } B' : \\ (\mathrm{DNF}_F \to (A^* \to B'))\&(B' \to B^*), \end{array}$$

$B'$ is different from $B^*$. In fact, the formulae $B^*_{\mathrm{CNF}^2}$ and $B^*_{\mathrm{DNF}}$ are solutions to the problem (a) and (b), respectively. Since we deal with the extensional $F$ then the first formulae in (a) and (b) can be proved easily. The second ones are obvious by adjunction, and the last requirements are fulfilled by observing

$$(B^*_{\mathrm{CNF}^2} \to B')\&(A^*\&B' \to \mathrm{CNF}^2_F) \to (B^*_{\mathrm{CNF}^2} \leftrightarrow B'),$$
$$(\mathrm{DNF}_F \to (A^* \to B'))\&(B' \to B^*_{\mathrm{DNF}}) \to (B^*_{\mathrm{DNF}} \leftrightarrow B').$$

In the sequel, we select the most interesting formulae from [5] that show properties of approximate reasoning with the graded fuzzy rules formalized by $\mathrm{DNF}_{F,k}$ and $\mathrm{CNF}^2_{F,k}$. In this source the proofs of the following statements can be found.

**Proposition 1.** *In SMTL$_\triangle\forall$, the following is provable:*

$$C_k\&(\exists \bar{x}_p)(A^*(\bar{x}_p))^2 \to B^*_{\mathrm{CNF}^2}(\bar{y}_p) \subseteq B^*_{\mathrm{DNF}}(\bar{y}_p), \tag{25}$$
$$(\mathrm{Ext}_{\bar{R}}\, F)^2 \to B^*_{\mathrm{DNF}}(\bar{y}_p) \subseteq B^*_{\mathrm{CNF}^2}(\bar{y}_p), \tag{26}$$
$$C_k\&(\exists \bar{x}_p)(A^*(\bar{x}_p))^2\&(\mathrm{Ext}_{\bar{R}}\, F)^2 \to B^*_{\mathrm{DNF}}(\bar{y}_p) \approx B^*_{\mathrm{CNF}^2}(\bar{y}_p), \tag{27}$$

*for arbitrary* $p \in \mathbb{N}$, $1 \leq p < n$.

As one would expect from the relationship between D(C)NF and $F$, it is not completely true that $B^*_{\mathrm{DNF}} \subseteq B^*_{\mathrm{CNF}^2}$ (formula (26)) nor $B^*_{\mathrm{CNF}^2} \subseteq B^*_{\mathrm{DNF}}$ (25). From (26), it follows that the extensionality is essential to prove $B^*_{\mathrm{DNF}}(\bar{y}_p) \approx B^*_{\mathrm{CNF}^2}(\bar{y}_p)$.

Now, we will investigate a relationship between the precise value of $F$ and conclusion $B^*$ of the inferences based on the approximate description.

**Theorem 2.** *Let* $P_1 \equiv_{df} \mathrm{Ext}_{\bar{R}}\, F\&C_k$ *and* $P_2 \equiv_{df} (\mathrm{Ext}_{\bar{R}}\, F)^3\&C_k\&A^*(\bar{x}_p)$. *Then SMTL$_\triangle\forall$ prove*

$$P_1 \to F(\bar{x}) \subseteq (A^*(\bar{x}_p) \to B^*_{\mathrm{DNF}}(\bar{y}_p)), \tag{28}$$
$$P_1 \to (A^*(\bar{x}_p)\&B^*_{\mathrm{CNF}^2}(\bar{y}_p)) \subseteq F(\bar{x}), \tag{29}$$
$$P_2 \to [F(\bar{x}) \leftrightarrow B^*_{\mathrm{DNF}}(\bar{y}_p)], \tag{30}$$
$$P_2 \to [F(\bar{x}) \leftrightarrow B^*_{\mathrm{CNF}^2}(\bar{y}_p)]. \tag{31}$$

Since $F$ represents some real situation that is to be approximately described by approximating formulae, we wish to know the relationship w.r.t. $B^*_{\mathrm{D(C)NF}}$ that has been investigated in the above lemma. From $P_{1,2}$ it follows that besides extensionality we need an appropriate partition, i.e. the distribution of input and output fuzzy sets associated to $\bar{R}(\bar{\mathbf{c}}, \bar{x}), \bar{R}(\bar{x}, \bar{\mathbf{c}})$, that lead to the indistinguishability between $B^*_{\mathrm{D(C)NF}}$ and $F$.

**Theorem 3.** *Let us denote* $[c_{i_{p+1}}, \ldots, c_{i_n}]$ *by* $\bar{d}_{i_p}$ *and let*

$$Disj(R) \equiv_{df} (\forall \bar{x}_p) \bigwedge_{i \neq j} \neg[R(\bar{x}_p, \bar{c}_{i_p}) \& R(\bar{x}_p, \bar{c}_{j_p})],$$

$$L_1 \equiv_{df} [R(\bar{c}_{i_p}, \bar{x}_p) \subseteq A^*(\bar{x}_p)] \& (\exists\, \bar{x}_p) R^2(\bar{c}_{i_p}, \bar{x}_p),$$

$$L_2 \equiv_{df} [R(\bar{x}_p, \bar{c}_{i_p}) \subseteq A^*(\bar{x}_p)] \& (\exists\, \bar{x}_p) R^2(\bar{x}_p, \bar{c}_{i_p}),$$

$$L_3 \equiv_{df} [A^*(\bar{x}_p) \subseteq R(\bar{c}_{i_p}, \bar{x}_p)] \& Disj(R),$$

$$L_4 \equiv_{df} [A^*(\bar{x}_p) \subseteq R(\bar{x}_p, \bar{c}_{i_p})] \& Disj(R).$$

*Then* $SMTL_\triangle\forall$ *prove*

$$L_1 \to [R(\bar{d}_{i_p}, \bar{y}_p) \& F(\bar{c}_i)] \subseteq B^*_{\mathrm{DNF}}(\bar{y}_p), \tag{32}$$

$$L_2 \to B^*_{\mathrm{CNF}^2}(\bar{y}_p) \subseteq [R(\bar{y}_p, \bar{d}_{i_p}) \to F(\bar{c}_i)], \tag{33}$$

$$L_3 \to B^*_{\mathrm{DNF}}(\bar{y}_p) \subseteq [R(\bar{d}_{i_p}, \bar{y}_p) \& F(\bar{c}_i)], \tag{34}$$

$$L_4 \to [R(\bar{y}_p, \bar{d}_{i_p}) \to F(\bar{c}_i)] \subseteq B^*_{\mathrm{CNF}^2}(\bar{y}_p). \tag{35}$$

In order to understand the proved relationships (32)–(35), it is worth to pay attention to the following formulae:

$$(\exists\, \bar{x}_p) R^2(\bar{c}_{i_p}, \bar{x}_p), \quad (\exists\, \bar{x}_p) R^2(\bar{x}_p, \bar{c}_{i_p}), \tag{36}$$

$$\bigwedge_{i \neq j} \neg[R(\bar{c}_{i_p}, \bar{x}_p) \& R(\bar{c}_{j_p}, \bar{x}_p)], \quad \bigwedge_{i \neq j} \neg[R(\bar{x}_p, \bar{c}_{i_p}) \& R(\bar{x}_p, \bar{c}_{j_p})]. \tag{37}$$

Let $\langle \{\tilde{R}_j\}_{j \in J}, \{c_i\}_{i \in I} \rangle$ be a model for $\langle \{R_j\}_{j \in J}, \{\bar{c}_i\}_{i \in I} \rangle$, where $j \in \{1, \ldots, n\}$. Then each formula in (36) determines the height of a fuzzy set describing "very close" neighborhood of $c_i$ being a subset of $\tilde{R}(c_i, x)$ or $\tilde{R}(x, c_i)$ that represents this neighborhood. And formula (37) says how much $\tilde{R}_i(c_i, x)$ is disjoint from all the others $\tilde{R}_j(c_j, x)$ (analogously for $\tilde{R}_i(x, c_i)$).

## Acknowledgement

This investigation has been supported by project MSM6198898701 of the MŠMT ČR.

## References


1. W. Bandler, L. Kohout, Special properties, closures and interiors of crisp and fuzzy relations, Fuzzy Sets and Systems 26 (1988) 317–331.
2. L. Běhounek, P. Cintula, From fuzzy logic to fuzzy mathematics: A methodological manifesto, Fuzzy Sets and Systems 157 (2006) 642–646.
3. P. Cintula. From Fuzzy Logic to Fuzzy Mathematics. PhD thesis, Czech Technical University, Prague, 2005.
4. M. Daňková, Short note: On extensionality of fuzzy relations and its applications, Fuzzy Sets and Systems, submitted.
5. M. Daňková, On approximate reasoning with graded rules, Fuzzy Sets and Systems, to appear.
6. D. Dubois, J. Lang, H. Prade, Fuzzy sets in approximate reasoning, Part 2: logical approaches, Fuzzy Sets and Systems 40 (1991) 203–244.
7. D. Dubois , H. Prade, Fuzzy sets in approximate reasoning, part 1: inference with possibility distributions, Fuzzy Sets and Systems 40 (1991) 143–202.
8. F. Esteva, L. Godo, Monoidal t-norm based logic: towards a logic left-continous t-norms, Fuzzy sets and systems 124 (2001) 271-288.
9. F. Esteva, L. Godo, P. Hájek, M. Navara, Residuated fuzzy logics with an involutive negation, Archive for Mathematical Logic 39 (2000) 103–124.
10. T. Flaminio, E. Marchioni, T-norm based logics with an independent involutive negation, Fuzzy Sets and Systems, Vol. 157, Issue 24 (2006) 3125-3144.
11. S. Gottwald, A Treatise on Many-Valued Logics. Studies in Logic and Computation, Vol. 9, Research Studies Press: Baldock, Hertfordshire, England, 2001.
12. S. Gottwald, Fuzzy Sets and Fuzzy Logic: Foundations of Applicationfrom a Mathematical Point of View, Vieweg, Wiesbaden, 1993.
13. P. Hájek, Metamathematics of fuzzy logic, Kluwer, Dordrecht, 1998.
14. F. Klawonn, V. Novák, The relation between inference and interpolation in the framework of fuzzy systems, Fuzzy sets and systems 81 (1996) 331–354.
15. J. Łukasiewicz, Selected works. Studies in Logic and the Foundations of Mathematics, North-Holland Publ. comp., Amsterdam, Warszawa, 1970.
16. V. Novák, S. Lehmke, Logical structure of fuzzy IF-THEN rules, Fuzzy Sets and Systems 157 (2006) 2003–2029.
17. V. Novák, On the logical basis of approximate reasoning, in: V. Novák, J. Ramík, M. Mareš, M. Černý, J. Nekola (Eds.), Fuzzy Approach to Reasoning and Decision Making, Kluwer, Dordrecht, 1992.
18. V. Novák, I. Perfilieva, J. Močkoř, Mathematical Principles of Fuzzy Logic, Kluwer, Boston Dordrecht, 1999.
19. I. Perfilieva, Logical Approximation, Soft Computing 2 (2001) 73–78.
20. I. Perfilieva, Logical Foundations of Rule-Based Systems, Fuzzy Sets and Systems 157 (2006) 615–621.
21. G.-J. Wang, Formalized theory of general fuzzy reasoning, Information SciencesInformatics and Computer Science 160 (2004) 251 – 266.
22. L.A. Zadeh, A theory of approximate reasoning, in: J.E. Hayes, D. Mitchie, L.I. Mikulich (Eds.), Machine Intelligence (9), Wiley, New York, 1979, pp. 149–194.
23. L.A. Zadeh, Outline of a new approach to the analysis of complex systems and decision processes, IEEE Transactions on Systems, Man and Cybernetics 3 (1973) 28–44.

# On External Measures for Validation of Fuzzy Partitions

Alessandro G. Di Nuovo and Vincenzo Catania

Università degli Studi di Catania
Dipartimento di Ingegneria Informatica e delle Telecomunicazioni
Viale Andrea Doria 6, 95125 Catania, Italy
{adinuovo,vcatania}@diit.unict.it

**Abstract.** The procedure of evaluating the results of a clustering algorithm is know under the term cluster validity. In general terms, cluster validity criteria can be classified in three categories: internal, external and relative. In this work we focus on the external criteria, which evaluate the results of a clustering algorithm based on a pre-specified structure $S$, that pertains to the data but which is independent of it. Usually $S$ is a crisp partition (i.e. the data labels), and the most common approach for external validation of fuzzy partitions is to apply measures defined for crisp partitions to fuzzy partitions, using crisp partitions derived (hardened) from them. In this paper we discuss fuzzy generalizations of two well known crisp external measures, which are able to assess the quality of a partition $U$ without the hardening of $U$. We also define a new external validity measure, that we call DNC index, useful for comparing a fuzzy $U$ to a crisp $S$. Numerical examples based on four real world data sets are given, demonstrating the higher reliability of the DNC index.

**Keywords:** Data mining, Fuzzy Clustering, Fuzzy validity index, External Validity Criteria, Fuzzy Rand Index, Partition Assessment, DNC index.

## 1 Introduction

In pattern recognition fuzzy models and algorithms have been widely studied and applied [1,2,3]. In particular one of the major techniques in pattern recognition is fuzzy clustering, that attracts attention because it has been successful in a variety of substantive areas [4,5,6,7] including image recognition, signal processing, business, health, aerospace, and so on. In the fuzzy approach partitions are not really created: elements are associated to each group, with a degree of membership. The result yielded by this algorithm is thus not a simple partitioning (which can, however, still be achieved by assigning each pattern to the group with which it has a higher numerical affinity, for example), but more detailed information on the relations between the patterns and groups.

Even though the clustering methods are viewed as completely unsupervised, this perception is not completely valid. The majority of algorithms has some

P. Melin et al. (Eds.): IFSA 2007, LNAI 4529, pp. 491–501, 2007.
© Springer-Verlag Berlin Heidelberg 2007

parameters to be set before they starts in order to define the clusters in which a data set can be partitioned. Specifically, the objective function, that is to be selected in advance, predefines the shape of the clusters one is interested in finding in the data set. Often the number of clusters is to be set before a clustering algorithm starts. For this reason, given a data set, clustering algorithms will produce as many different partitions as you have time to generate. It is necessary to validate the clustering result. Validation approaches provide an objective measurement of a clustering result and its optimal value is often used to indicate the best possible choice for the values of parameters in the clustering algorithm.

There are three approaches for the evaluation of partitions quality [8]: (i) external criteria; (ii) internal criteria; (iii) relative criteria. Details on validation criteria are given in section 2.

In this paper we focus on the external criteria for validation of fuzzy partitions. Generally the most common approach is to apply measures defined for crisp partitions to fuzzy partitions, using crisp partitions derived (hardened) from them. The main weakness of this approach is that a pattern with degrees of membership which generate uncertainty for its recognition (consider for example, in a partition with two cluster, a pattern with a degree of membership $u_1 = 0.501$ and $u_2 = 0.499$ respectively with cluster 1 and 2), will be evaluated in the same way of an other pattern that has a certain recognition (e.g. $u_1 = 0.999$ and $u_2 = 0.001$). This way the information obtained by the fuzzy algorithm is lost. In order to work around this limitation, in section 4 we present the fuzzy generalization of the two most effective external indices. However in our experiments fuzzy external criteria demonstrate to be too much sensitive to the degree of fuzziness of the partition to be assessed. For this reason, in section 4, we introduce a new external criteria based on fuzzy logic, we call DNC index. In section 5 real world applications are studied, demonstrating the higher reliability of the DNC index. Finally section 6 gives our conclusions.

## 2 Clustering and Validation Criteria

In clustering (also known as exploratory data analysis), a set of patterns, usually vectors in a multi-dimensional space, are organized into coherent and contrasted groups, such as that patterns in the same group are similar in some sense and patterns in different groups are dissimilar in the same sense. Given a data set of $n$ patterns $X = \{\mathbf{x}_1, ..., \mathbf{x}_j, ..., \mathbf{x}_n\}$, the purpose of any clustering technique is to evolve a partition matrix $U(X)$ of the given data set $X$ so as to find a number, say $R$, of clusters $(\{U_1, ..., U_R\})$. The partition matrix $U(X)$ of size $R \times n$ may be represented as $U = [u_{kj}]$, $k = 1, .., R$ and $j = 1, ..., n$, where $u_{kj}$ is the membership of pattern $\mathbf{x}_j$ to cluster $U_k$. In crisp partitioning of the data, the following condition holds: $u_{kj} = 1$ if $\mathbf{x}_j \in U_k$, otherwise $u_{kj} = 0$. In the case of fuzzy clustering, the purpose is to evolve an appropriate partition matrix $U = [u_{kj}]$ where $u_{kj} \in [0, 1]$, such that $u_{kj}$ denotes the degree of membership of the $j$-th pattern to the $k$-th cluster.

The procedure of evaluating the results of a clustering algorithm is know under the term cluster validity. In general terms, cluster validity criteria can be classified in three categories: internal, external and relative. *Relative indices* are used to decide which of two partitions, $U(X)$ or $V(X)$, is a "better choice", where better is defined by the measure that is being used. *Internal criteria* assess the goodness of fit between an algorithmically obtained crisp partition $U(X)$ and the input data using only the data themselves, usually in the form of the distance matrix $D(X) = [\{\delta(x_i, x_j)\}]_{n \times n}$ of the data. *External criteria* are those that match the structure of a partition $U(X)$ computed with $X$ to a partition $S = \{S_1, ..., S_C\}$ of $X$ that pertains to the data but which is independent of it. $C$ is the number of classes. For example, every crisply labeled data set comes with a crisp partition $S$ of $X$. When a measure is a function of $(S,U(X))$, it is called an external criterion. Milligan and Cooper [9] have studied at least 30 external indices for measuring agreement between two partitions in clustering analysis with different number of clusters, and they recommended two statistical criteria, the adjusted Rand index [10] and the Fowlkes-Mallow measure [11], as the indices of choice. More details and fuzzy generalization of these indices are in section 3.

## 3 External Measures for Cluster Validation

In this section, we discuss methods suitable for quantitative evaluation of the clustering results based on a pre-specified structure, known as external cluster validity methods. These methods give an indication of the quality of the resulting partitioning and thus they can only be considered as a tool at the disposal of the experts in order to evaluate the clustering results.

The recognition accuracy and error are widely used as external criteria to assess the quality of partitions, due to their intuitive meaning and the low computational cost. The recognition accuracy, also known as *recognition rate*, measures the number of matches between pairs in $U$ and $S$, i.e. the number of the patterns that are in a cluster of $U(X)$ that is associated with the real class of $S$, while the recognition error measures the number of mismatches, i.e. the number of wrong assignments.

### 3.1 Crisp Statistical Indices

In this subsection we present the adjusted Rand index [10], a more sensitive generalization of the Rand index [12], and the Fowlkes - Mallow index [11]. These validation methods are statistically oriented, and require assumptions about the distribution. In cluster validity the basic hypothesis is to test whether the point of a data set are randomly structured or not.

The Rand indices are based on counting the number of pairwise co-assignments of data items. As a generalization of the Rand Index,the adjusted Rand index additionally introduces a statistically induced normalization in order to yield values close to 0 for random partitions. This normalization removes the bias of the Rand Index with respect to different number of clusters and gives a wider range of values that adjusted Rand index can take on, thus increasing the sensitivity of the index. Let $N(U,S) = N = [n_{lk}] = US^T$, where $n_{lk}$ be the number of patterns that are in both class $s_l$ and cluster $u_k$. Let $n_{l.}$ and $n_{.k}$ be the number of patterns in class $s_l$ and cluster $u_k$ respectively. The notations are illustrated in Table 3.1.

**Table 1.** Notation for the contingency table for comparing two partitions

| Class Cluster | $U_1$ | $U_2$ | ... | $U_R$ | Sums |
|---|---|---|---|---|---|
| $S_1$ | $n_{11}$ | $n_{12}$ | ... | $n_{1R}$ | $n_{1.}$ |
| $S_2$ | $n_{21}$ | $n_{22}$ | ... | $n_{2R}$ | $n_{2.}$ |
| ... | ... | ... | ... | ... | ... |
| $S_C$ | $n_{C1}$ | $n_{C2}$ | ... | $n_{CR}$ | $n_{C.}$ |
| Sums | $n_{.1}$ | $n_{.2}$ | ... | $n_{.R}$ | $n_{..} = n$ |

Under the generalized hypergeometric model, it can be shown that:

$$\Upsilon_R(U,S) = \frac{\sum_{lk}\binom{n_{lk}}{2} - \left[\sum_l \binom{n_{l.}}{2}\sum_k\binom{n_{.k}}{2}\right]/\binom{n}{2}}{\frac{1}{2}\left[\sum_l\binom{n_{l.}}{2} + \sum_k\binom{n_{.k}}{2}\right] - \left[\sum_l\binom{n_{l.}}{2}\sum_k\binom{n_{.k}}{2}\right]/\binom{n}{2}} \tag{1}$$

The Adjusted Rand Index returns values in the interval $[\sim 0, 1]$ and is to be maximized. The expected value of two partitions picked at random is 0. The Fowlkes-Mallows index is the geometric mean of two probabilities: the probability that two randomly chosen observations are in the same cluster given that they are in the same group, and the probability that two randomly chosen observations are in the same group given that they are in the same cluster. Hence a Fowlkes-Mallows index near 1 means that the clusters are good estimates of the groups. To compute the Fowlkes-Mallows index we use the contingency table of the groups and the clusters, as shown in Table 3.1. Then the Fowlkes-Mallows index is given by

$$\Upsilon_{FM} = \frac{\sum\limits_{l,k}\binom{n_{ij}}{2}}{\sqrt{\sum\limits_{i,j}\binom{n_{ij}}{2}\sum\limits_{i,j}\binom{n_{ij}}{2}}} \tag{2}$$

If we define the number T,P and Q:

$$T = \sum_{lk}\binom{n_{lk}}{2}, P = \sum_k \binom{n_{.k}}{2}, Q = \sum_l \binom{n_{l.}}{2} \tag{3}$$

the adjusted Rand index and the Fowlkes-Mallows index could be re-written in a more compact form as follows:

$$\Upsilon_R(U,S) = \frac{T - \frac{(PQ)}{n(n-1)/2}}{\frac{1}{2}(P+Q) - \frac{(PQ)}{n(n-1)/2}} \quad \Upsilon_{FM}(U,S) = \frac{T}{\sqrt{PQ}} \tag{4}$$

### 3.2 Fuzzy Generalization of the Adjusted Rand Index and the Fowlkes-Mallow Index

Fuzzy generalization of the two crisp indices presented in the previous subsection were proposed in [13] and [2]. When the partition matrix $U(X)$ is fuzzy, we could define a fuzzy contingency matrix, $M_f(U,S) = M_f = [m_{f,lk}] = US^T$, where entries are no longer counts of matches and mismatches between pairs in $X \times X$; rather, $m_{f,lk}$ is now interpreted as the similarity between the fuzzy cluster whose membership values are the $i$-th row of $U$ and the fuzzy cluster whose membership values are the $j$-th row of $S$ (which is the $j$ columns of $S^T$). Using the fuzzy contingency matrix $M_f(U,S)$ the numbers T,P and Q can be defined as follows :

$$T_f = \left(\sum_{lk} m^2_{f,lk}\right) - n; P_f = \left(\sum_{k} m^2_{f,.k}\right) - n; Q_f = \left(\sum_{l} m^2_{f,l.}\right) - n \quad (5)$$

Using $T_f, P_f$ and $Q_f$ it is possible to make a direct extension of the adjusted Rand index and the Fowlkes-Mallow index:

$$\Upsilon_{R,f}(U,S) = \frac{T_f - \frac{(P_f Q_f)}{n(n-1)/2}}{\frac{1}{2}(P_f+Q_f) - \frac{(P_f Q_f)}{n(n-1)/2}} \quad \Upsilon_{FM,f}(U,S) = \frac{T_f}{\sqrt{P_f Q_f}} \quad (6)$$

Note that $0 \leq \Upsilon_{FM,f} \leq 1$ while $\Upsilon_{R,f}$ could be less than 0. $\Upsilon_{R,f}$ cannot be used to compare two truly fuzzy partitions because the implication $\Upsilon_{R,f} = 1 \Rightarrow U = S$ is one way; in particular, $\Upsilon_{R,f}(U,U) \neq 1$ if $U$ is a fuzzy partition, so its usefulness lies in validation of a fuzzy $U$ against a crisp $S$. The same statement can be done for $\Upsilon_{FM,f}$. The question is when might this happen? Often, because $S$ is usually just the crisp partition builded using labels of the data, and most of the data sets publicity available are labeled. To compare two truly fuzzy partitions, Back and Hussain [13] propose a measure that they call the MC index for this job.

## 4 DNC Index: A New Fuzzy Validation Index

We say that two degree of membership are close when their distance is below a certain threshold we called $\alpha$. Intuitively, we say that the assignment of a pattern to a real class is uncertain when the pattern's highest degree of membership is close to the one of an other cluster, we say that its assignment to a real class is uncertain. Here we propose an index that take into account this uncertainty that characterize true fuzzy partitions. To measure how much a pattern is confident to be assigned to a cluster we define the degree of confidence ($\Delta_j$). Given a crisp o fuzzy vector of degrees of membership $\mathbf{u}_j = \{u_{1j}, u_{2j}, ..., u_{Rj}\}$ and $a, b \in \{1, 2, ..., R\} : u_{aj} > u_{bj} > u_{Oj}$, with $O = \{1, 2, ..., R\} \cap \{a, b\}$, we define the degree of confidence ($\Delta_j$) of the $j$-th pattern ($\mathbf{x}_j$) as

$$\Delta_j = (u_{aj} - u_{bj}) \quad (7)$$

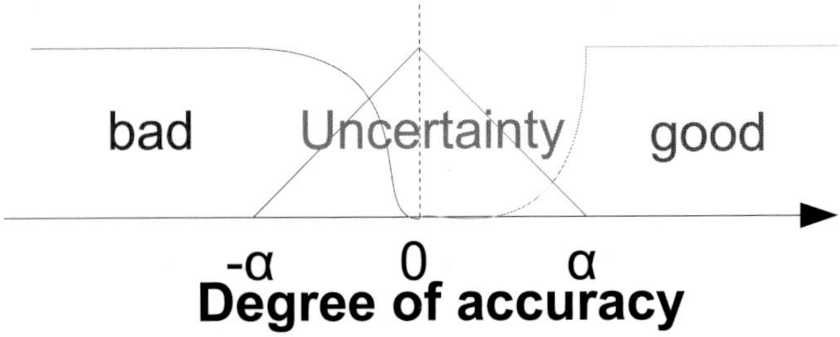


**Fig. 1.** Fuzzy sets: good (match), bad (mismatch) and uncertainty (confidence degree lower than $\alpha$)

To take into account if the pattern $\mathbf{x}_j$ is in a cluster associated with the real group or not, we define the degree of accuracy ($A_j$) as follows:

$$A_j = \begin{cases} \Delta_j \texttt{ if } \texttt{x}_j \texttt{ is in a cluster associated with its real group} \\ -\Delta_j \texttt{ otherwise} \end{cases} \quad (8)$$

Note that $0 \leq \Delta_j \leq 1$ and $-1 \leq A_j \leq 1 \forall j$ by definitions (7) and (8). As shown in figure 1 using $A_j$ we define three fuzzy sets, $\texttt{good}(A_j)$, $\texttt{bad}(A_j)$ and $\texttt{uncertainty}(A_j)$ as follows:

$$\texttt{good}(A_j) = \begin{cases} 1 & A_j \geq \alpha \\ A_j^2/\alpha & 0 < A_j < \alpha \\ 0 & A_j \leq 0 \end{cases} \quad \texttt{bad}(A_j) = \begin{cases} 1 & A_j \leq -\alpha \\ \sqrt{A_j}/\alpha & -\alpha < A_j < 0 \\ 0 & A_j \geq 0 \end{cases}$$
$$\texttt{uncertainty}(A_j) = \begin{cases} 0 & \left|A_j\right| \geq \alpha \\ \left|A_j\right| & \left|A_j\right| < \alpha \end{cases} \quad (9)$$

In the definition of $\texttt{good}$ and $\texttt{bad}$ we wanted to emphasize the loss of certainty of a corrected classification through an elevation to the square and the increase of the certainty of a misclassification with the square root.

Next we define the accuracy index $I_a$ and the uncertainty index $I_u$ as follows

$$I_a = \frac{\sum_j \left(\texttt{good}(A_j) - \texttt{bad}(A_j)\right)}{n} \quad I_u = \frac{\sum_j \texttt{uncertainty}(A_j)}{n} \quad (10)$$

Note that $I_u = 1$ when $|A_j| = 0 \ \forall j$, while $I_u = 0$ when $|A_j| > \alpha \ \forall j$. Accuracy index measures how accurate the partition $U$ is respect to the partition $S$ taking into account the uncertainty in pattern assignment when $U$ is fuzzy. Uncertainty index measures the average uncertainty in pattern assignment.

Finally we define the DNC index ($\Upsilon_{DNC}$) as

$$[h]\Upsilon_{DNC} = I_a \times (1 - I_u). \quad (11)$$

It is easy to show that: $\Upsilon_{DNC} = 1 \Rightarrow U_h = S$ ; $-1 \leq \Upsilon_{DNC} \leq 1 \forall (U,S)$ ; $I_u = 1 \Rightarrow \Upsilon_{DNC} = 0$ ; $I_a = 0 \Rightarrow \Upsilon_{DNC} = 0$ ; $\Upsilon_{DNC} = -1 \Rightarrow$ U and S are completely different. Where $U_h$ is a partition hardened from a fuzzy $U$. However the DNC index can be used also with a crisp $U$ without significant loss. It is useful to compare results obtained with both fuzzy and crisp algorithms.

**Table 2.** Main characteristics of data sets

| Data sets | Patterns | Classes |
|---|---|---|
| Iris | 150 | 3 |
| wine data set | 178 | 3 |
| Wisconsin Breast Cancer | 683 | 2 |
| Pima Indians diabetes | 768 | 2 |

## 5 Numerical Examples

The two fundamental questions that need to be addressed in any typical clustering scenario are: (i) how many clusters are actually present in the data, and (ii) how real is the clustering itself. Here we answer to the first question assuming that number of clusters present in the data is equal to the number of classes. In this section we studied the external criteria to answer to the second question. In order to study the performance of the external indices introduced in previous sections we applied the most famous fuzzy clustering algorithm, the Fuzzy C-Means (FCM) [2], on four benchmark data sets: the Iris data set, the wine data set, the Wisconsin Breast Cancer data set, the Pima Indians diabetes data set. The data sets were obtained from the University of California machine-learning database [14], and their main characteristics are reported in Table 5. FCM uses the principles of fuzzy sets to partition a data set into a fixed number, $R$, of clusters; thereby providing the appropriate $R \times n$ partition matrix. In our analysis we used the Euclidean metric as distance measure and fixed the number of clusters $R$ equal to the number of classes $C$. To get partitions with different degrees of fuzziness we let the weighting exponent of the FCM model (also known as the fuzzyfier, $m$) vary in range (1.0,10.0]. The role of the weighting exponent $m$ was studied for cluster validation indices in [15], where authors discovers that it affects the quality and reliability of the internal indices they studied. They also shown that the higher is the value of $m$ the higher is the partition entropy (i.e. the fuzziness). Our experiments shows that also reliability of external validation indices, both crisp and fuzzy, is affected by this parameter. In particular we see that crisp indices do not consider fuzziness (as expected because they work on hardened partitions) and vice versa fuzzy generalizations are so much related to fuzziness that practically they measure only the grade of the fuzziness. Table 5 lists the outputs of: entropy index, recognition rate, adjusted Rand index $\Upsilon_R$ , Fowlkes-Mallow index $\Upsilon_{FM}$, fuzzy adjusted Rand index $\Upsilon_{R,f}$, fuzzy Fowlkes-Mallow index $\Upsilon_{FM,f}$, and DNC index $\Upsilon_{DNC}$ with confidence level $\alpha = \{0.01; 0.05; 0.10\}$, applied to the Iris data for 19 values of $m$ ranging from 1.01 to 10.0. Crisp indices (recognition rate, $\Upsilon_R$ and $\Upsilon_{FM}$) have higher (i.e. better) values with higher values of $m$, vice versa their fuzzy generalization have better values with lower values of $m$. This means that partition obtained with values of $m$ suggested by crisp indices, have entropy values close to the maximum (1.099), so clusters are significantly overlapped and practically indistinguishable. On the other hand partitions suggested by fuzzy indices are close

**Table 3.** Iris data set, values with $m$ varying between (1,10.0]: Partition entropy, recognition rate, adjusted Rand index, Fowlkes-Mallow index, fuzzy adjusted Rand index, fuzzy Fowlkes-Mallow index, and DNC index with $\alpha = \{0.01; 0.05; 0.10\}$

| $m$ | $U$ entropy | Recognition rate | $\Upsilon_R$ | $\Upsilon_{FM}$ | $\Upsilon_{R,f}$ | $\Upsilon_{FM,f}$ | $\Upsilon_{DNC}$ $\alpha = 0.10$ | $\alpha = 0.05$ | $\alpha = 0.01$ |
|---|---|---|---|---|---|---|---|---|---|
| 1.01 | 0.015 | 0.893 | 0.693 | 0.796 | **0.877** | **0.816** | **0.777** | 0.780 | 0.787 |
| 1.2 | 0.038 | 0.893 | 0.693 | 0.796 | 0.877 | 0.816 | 0.776 | 0.777 | 0.787 |
| 1.4 | 0.104 | 0.886 | 0.678 | 0.786 | 0.874 | 0.810 | 0.773 | 0.773 | 0.773 |
| 1.6 | 0.192 | 0.886 | 0.678 | 0.786 | 0.862 | 0.791 | 0.772 | 0.772 | 0.773 |
| 1.8 | 0.293 | 0.893 | 0.696 | 0.797 | 0.840 | 0.757 | 0.775 | 0.777 | 0.787 |
| 2.0 | 0.396 | 0.893 | 0.696 | 0.797 | 0.812 | 0.715 | 0.773 | 0.783 | 0.787 |
| 2.2 | 0.491 | 0.893 | 0.696 | 0.797 | 0.783 | 0.670 | 0.771 | **0.786** | 0.787 |
| 2.4 | 0.577 | 0.900 | 0.715 | 0.809 | 0.755 | 0.627 | 0.768 | 0.784 | 0.798 |
| 2.6 | 0.650 | 0.900 | 0.715 | 0.809 | 0.728 | 0.588 | 0.763 | 0.783 | 0.800 |
| 2.8 | 0.713 | 0.900 | 0.715 | 0.809 | 0.706 | 0.553 | 0.759 | 0.782 | 0.800 |
| 3.0 | 0.765 | 0.900 | 0.715 | 0.809 | 0.687 | 0.523 | 0.756 | 0.781 | 0.799 |
| 3.5 | 0.862 | 0.906 | 0.733 | 0.822 | 0.649 | 0.466 | 0.747 | 0.776 | 0.800 |
| 4.0 | 0.925 | 0.906 | 0.733 | 0.822 | 0.624 | 0.428 | 0.732 | 0.768 | 0.804 |
| 5.0 | 0.996 | 0.900 | 0.717 | 0.811 | 0.595 | 0.384 | 0.696 | 0.757 | 0.799 |
| 6.0 | 1.032 | 0.913 | 0.754 | 0.835 | 0.580 | 0.361 | 0.651 | 0.743 | 0.803 |
| 7.0 | 1.052 | 0.913 | 0.754 | 0.835 | 0.572 | 0.348 | 0.591 | 0.723 | 0.806 |
| 8.0 | 1.064 | 0.913 | 0.754 | 0.835 | 0.567 | 0.341 | 0.522 | 0.697 | **0.808** |
| 9.0 | 1.072 | 0.920 | 0.773 | **0.848** | 0.563 | 0.336 | 0.480 | 0.67 | 0.808 |
| 10.0 | 1.078 | **0.926** | **0.791** | 0.860 | 0.561 | 0.332 | 0.402 | 0.642 | 0.805 |

**Table 4.** Values of $m$ and entropy for optimal partitions chosen by each index

| Data set | Recognition rate | $\Upsilon_R$ | $\Upsilon_{FM}$ | $\Upsilon_{R,f}$ | $\Upsilon_{FM,f}$ | $\Upsilon_{DNC}$ $\alpha = 0.10$ | $\alpha = 0.05$ | $\alpha = 0.01$ |
|---|---|---|---|---|---|---|---|---|
| | | | | $m$ | | | | |
| Cancer | 7.8-10.0 | 7.8-10.0 | 7.8-10.0 | 1.01 | 1.01 | 1.1 | 1.2 | 5.8 |
| Wine | 8.2-10.0 | 8.3-10.0 | 8.3-10.0 | 1.01 | 1.01 | 1.1 | 1.3 | 8.0 |
| Diabetes | 1.01 | 2.0 | 1.1 | 1.01 | 1.01 | 1.1 | 1.1 | 2.0 |
| | | | | Partition Entropy | | | | |
| Cancer | 0.668-0.679 | 0.668-0.679 | 0.668-0.679 | 0.0161 | 0.0161 | 0.034 | 0.057 | 0.644 |
| Wine | 1.06-1.07 | 1.06-1.07 | 1.06-1.07 | 0.0142 | 0.0142 | 0.038 | 0.067 | 1.056 |
| Diabetes | 0.021 | 0.297 | 0.021 | 0.021 | 0.021 | 0.021 | 0.021 | 0.021 |

to the minimum entropy (0), so these partitions are much more crisp than fuzzy. The DNC index offers three different results, according to the confidence level $\alpha$. The most reliable, obtained with $\alpha = 0.05$, is $m = 2.2$ that it is close to $m = 2$ that is by far the most common choice for this data set because it assures good accuracy with low uncertainty.

Applying FCM to other data sets with $m$ varying in the range $(1, 10]$ and computing the indices, we obtained the results summarized in Table 5 and plotted in Figures 2a, 2b and 2c. We had a confirm that DNC is more reliable than other indices, because their results seems to be not related to the data itself. This can be explained by the fact that classical fuzzy indices measure how much $U$ is similar to $S$, therefore being $S$ crisp they reward the "crispest" partition. Crisp indices, on the other hand, often reward "fuzziest" partition, with entropy close to the maximum, that means that the algorithm is not finding distinct cluster structure. Figure 2c shows that DNC index has different curves depending on the data sets and the value chosen for $\alpha$. With this data sets seems that optimal value for $\alpha$ is in the range $[0.01, 0.05]$, because when $\alpha = 0.10$ results are comparable to other fuzzy index. For diabetes data optimal values of $m$ are very low for all indices because this data set has a significant class overlap which causes a base of uncertainty.

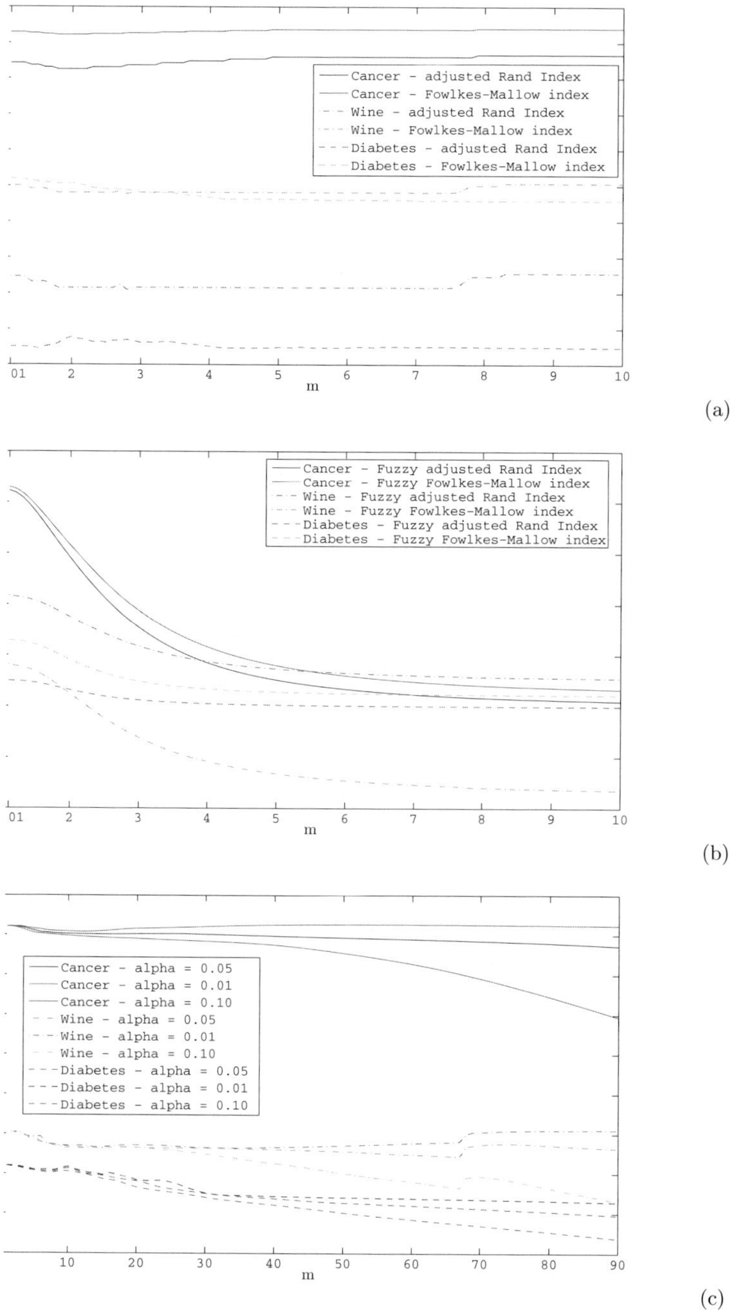


**Fig. 2.** (a) Crisp external indices values with $m$ ranging from 1.01 to 10.;(b) Fuzzy external indices values with $m$ ranging from 1.01 to 10.;(c) DNC index values with $m$ ranging from 1.01 to 10 and $\alpha = 0.05;\ 0.01;\ 0.10$

## 6 Conclusion

External cluster validity is one of the most important issues in cluster analysis. It aims at the evaluation of clustering results based on a-priori information and the selection of the scheme that best fits the underlying data. External cluster validity indices are used in unsupervised learning to assess performances of partitioning algorithms and in partially or fully supervised learning to tune algorithms' parameters. In fuzzy methods often crisp indices are used with the loss of the knowledge stored in fuzzy degree of memberships. In this paper we analyzed the reliability of fuzzy generalization of two well regarded crisp external indices: the fuzzy adjusted Rand index and the fuzzy Fowlkes-Mallow index. In our numerical examples these two indices reward as best partition the "crispest" available, in other words they measure only the uncertainty and always suggest to do not use fuzzy logic. Vice versa when we apply the original crisp indices they measure only the accuracy and do not care about fuzziness, as obvious because they work with hardened partitions. For this reason they often reward partitions without distinct cluster structure. In order to improve these weaknesses we propose a new fuzzy external index, we called DNC index, for validation of fuzzy partitions, that takes into account both accuracy and uncertainty. In real world application the DNC index demonstrated to be more reliable than others.

## References

1. Baraldi, A., Blonda, P.: A survey of fuzzy clustering algorithms for pattern recognition. part I-II. IEEE Transactions on Systems, Man and Cybernetics, Part B (Cybernetics) **29** (1999) 778–801
2. Bezdek, J.C., Keller, J., Krishnapuram, R., Pal, N.R.: Fuzzy Models and algorithms for pattern recognition and image processing. The handbooks of fuzzy sets series. Springer (1999)
3. Mitra, S., Pal, S.K.: Fuzzy sets in pattern recognition and machine intelligence. Fuzzy Sets and Systems **156** (2005) 381–386
4. Hppner, F., Klawonn, F., Kruse, R., Runkler, T.: Fuzzy Cluster Analysis: Methods for Classification Data Analysis and Image Recognition. Wiley, New York (1999)
5. Akay, M.: Nonlinear Biomedical Signal Processing, Fuzzy Logic, Neural Networks, and New Algorithms. Series on Biomedical Engineering. IEEE Press (2000)
6. Jain, L.C., Sato-ILIC, M.: Innovations in Fuzzy Clustering: Theory and Applications Theory And Applications. Studies In Fuzziness And Soft Computing. Springer (2006)
7. de Oliveira, J.V., Pedrycz, W.: Advances in Fuzzy Clustering and its Applications. Wiley, New York (2007)
8. Theodoridis, S., Koutroubas, K.: Pattern recognition. Accademic Press (1999)
9. Milligan, G., Cooper, M.: A study of comparability of external criteria for hierarchical cluster analysis. Multivariate Behavioral Research **21** (1986) 441–458
10. Hubert, L., Arabie, P.: Comparing partitions. Journal of Classification (1985) 193–218
11. Fowlkes, E., Mallows, C.: A method of comparing two hieararchical clusterings. Journal of American Statistical Society **78** (1983) 553–569

12. Rand, W.: Objective criteria for evaluation of clustering methods. Journal of the American Statistical Association **66** (1971) 846–850
13. Back, C., Hussain, M.: Validity measures for fuzzy partitions. In Bock, H., Polasek, W., eds.: Data analysis and information systems. Springer (1995) 114–125
14. (University of california machine learning repository) `http://www.ics.uci.edu/m̃learn/MLRepository.html`
15. Pal, N.R., Bezdek, J.C.: On cluster validity for the fuzzy c-means model. Transactions on Fuzzy Systems **3** (1995) 370–379

# Coherence Index of Radial Conjunctive Fuzzy Systems

David Coufal

Institute of Computer Science AS CR,
Pod Vodárenskou věží 2, 18207 Praha 8, Czech Republic
david.coufal@cs.cas.cz

**Abstract.** The paper introduces the so-called coherence index of a conjunctive radial fuzzy system. The index can be treated as a measure of consistency of knowledge stored in the rule base of the system. Conjunctive fuzzy systems are the systems which employ fuzzy conjunctions for combination of antecedents (IF parts) with consequents (THEN parts) in theirs IF-THEN rules. Radial fuzzy systems are the systems which employ radial functions for representation of membership functions of incorporated fuzzy sets; and exhibiting the radial shape preservation property. Due to this property an effective mathematical analysis of these systems can be carried out.

## 1 Introduction

In area of fuzzy computing the concept of fuzzy system plays the key role. Standard architecture of fuzzy systems, consisting of four building blocks of - fuzzifier, rule base, inference engine and deffuzifier, is notoriously well known [1], [2]. Mathematically, a fuzzy system perform a function from an input space (typically $\mathcal{R}^n$) to an output space (typically $\mathcal{R}$), which corresponds to standard MISO configuration.

The computation of a fuzzy system is mainly affected by the knowledge stored in the rule base of the system. A rule base is canonically formed by a set of $m \in \mathcal{N}$ linguistic IF-THEN rules, and mathematically represents a fuzzy relation on input/output space. From the theory of fuzzy systems it is known that mathematical representation of rule bases admits two main approaches - *conjunctive* and *implicative*. [3], [4].

Under conjunctive approach antecedents of rules (IF parts) are combined with corresponding consequents (THEN parts) by a fuzzy conjunction; and individual rules are then combined by a fuzzy disjunction. Contrary to conjunctive systems, in implicative systems antecedents are combined with consequents by genuine fuzzy implications [1], [5] and individual rules are combined by a fuzzy conjunction.

In connection with fuzzy systems, several theoretical concepts are studied. One of them is the concept of consistency of knowledge stored in the rule base. This question was addressed by several authors [7], [8]. One of the most important works is the paper by Dubois, Prade and Ughetto [9]. In the paper the authors

P. Melin et al. (Eds.): IFSA 2007, LNAI 4529, pp. 502–512, 2007.
© Springer-Verlag Berlin Heidelberg 2007

stress that the question of coherence[1] has good meaning for implicative systems and can be seen as the requirement for normality of fuzzy set which is issued from inference engine, for an arbitrary input to the system. A presence of an non-normal output, i.e., if the fuzzy set undergoing a defuzzification process is not normal, indicates the presence of contradictory rules in the rule base. In the paper there are stated several sufficient conditions for testing the coherence of implicative fuzzy systems.

Concerning conjunctive systems, the question of coherence as formulated for implicative ones has no good meaning, because the rule base is constructed as a list of prototypical points of input/output relation the rule base represents. These prototypes are given by individual rules and an interpolation is performed between the prototypes due to the inference process. However, taking output fuzzy set as a possibilistic distribution of possible crisp action it is reasonable to require the distribution to be unimodal. Unfortunately, as we will see, the strict unimodality cannot be generally reached so we have to adopt the concept of partial unimodality which is measured by the so-called *coherence (or unimodality) index*. This index is a number from interval $[0, 1]$, which is the higher the more unimodal the output fuzzy set is. The equality to 1 indicates strict unimodality.

In the paper we introduce the index mathematically and we present its lower bound which indicates the extent of coherence of knowledge stored in the system. The index will be stated for class of radial conjunctive fuzzy systems. Radial fuzzy systems forms broad class of fuzzy systems. It was shown that for implicative fuzzy systems the question of coherence as presented in [9] can be solved effectively [10]. So this paper can be seen as extension of former research on radial fuzzy systems.

## 2 Radial Conjunctive Fuzzy Systems

This section reviews the concept of radial conjunctive fuzzy systems. We present computational model of these systems. We start by conjunctive systems, then we present radial systems, and finally a fusion of both concepts.

### 2.1 Conjunctive Fuzzy Systems

In conjunctive fuzzy systems antecedents of rules are combined with corresponding consequents by a fuzzy conjunction ($t$-norm). Individual rules are then combined by a fuzzy disjunction ($t$-conorm). The most common choice is to employ the maximum $t$-conorm.

Formally, the antecedent of the $j$-th rule in a conjunctive system (but in implicative too) is given as

$$A_j(\boldsymbol{x}) = A_{j1}(x_1) \star \cdots \star A_{jn}(x_n), \tag{1}$$

[1] From now on we use term *coherence* instead of term consistency to follow the terminology introduced in paper [9].

where $A_{ji}$ are one-dimensional fuzzy sets, $\star$ is a $t$-norm and $\boldsymbol{x} \in \mathcal{R}^n$, $\boldsymbol{x} = (x_1, \dots, x_n)$. Fuzzy relation $R_j$ representing the $j$-th rule has then form

$$R_j(\boldsymbol{x}, y) = A_j(\boldsymbol{x}) \star B_j(y) = A_{j1}(x_1) \star \dots \star A_{jn}(x_n) \star B_j(y), \tag{2}$$

where $B_j$ forms the consequent set of the $j$-th rule. As mentioned, whole rule base is constructed by emploing a fuzzy disjunction:

$$RB(\boldsymbol{x}, y) = \bigvee_j R_j(\boldsymbol{x}, y) = \bigvee_j A_j(\boldsymbol{x}) \star B_j(y). \tag{3}$$

Concerning the most common choice of *singleton fuzzifier* and *CRI inference engine* [1], [2] the output fuzzy set from the engine is obtained as

$$B'(y) = RB(\boldsymbol{x}, y), \tag{4}$$

where $\boldsymbol{x}$ is the actual input into to the system. Expressing the above formula in more details we get the following. Denoting

$$B'_j(y) = R_j(\boldsymbol{x}, y) = A_j(\boldsymbol{x}) \star B_j(y) \tag{5}$$

and employing maximum as fuzzy disjunction we get formula (3) in form

$$B'(y) = \max_j\{B'_j(y)\} = \max_j\{A_j(\boldsymbol{x}) \star B_j(y)\}, \tag{6}$$

which is the most common computational model of conjunctive fuzzy systems. In this model an output fuzzy set from the inference engine is obtained as the maximum from clipped consequents fuzzy sets. The actual crisp output is then determined on the basis of selected defuzzification method.

### 2.2 Radial Fuzzy Systems

Radial fuzzy systems employ radial functions for representation of membership functions of one-dimensional fuzzy sets; and the combination of these radial fuzzy sets given by the selected $t$-norm, preserves the radial shape. This shape preservation property is called *the radial property* [12].

Radial functions are well known in the area of RBF neural networks [11]. Generally, a radial function $f$ has form $f(x) = \Phi(||\boldsymbol{x} - \boldsymbol{a}||)$, where $a \in \mathcal{R}^n$ is the central point of function, $||\cdot||$ is a norm in $\mathcal{R}^n$ space and $\Phi$ is usually monotonic function from $\mathcal{R}_0^+$ to $\mathcal{R}$. Well known example of radial functions are Gaussians curves.

In radial fuzzy systems it is required that both antecedent and consequent fuzzy sets are specified by radial functions in the following form:

$$A_{ji} = act\left(\left|\frac{x_i - a_{ji}}{b_{ji}}\right|\right), \quad B_j = act\left(\frac{\max\{0, |y - c_j| - s_j\}}{d_j}\right), \tag{7}$$

where $a_j, c_j \in \mathcal{R}$ are central points, $b_j, d_j \in \mathcal{R}^+$, i.e., $b_j, d_j > 0$ are (width) scaling parameters and $s_j \in \mathcal{R}_0^+$, i.e., $s_j \geq 0$ is $B_j$'s core (kernel) length driving parameter. The *act* function is a monotonic function from $\mathcal{R}_0^+$ to $[0,1]$ such that $act(0) = 1$ and generally can occur in two versions: 1) Either it is strictly decreasing on its domain and $\lim_{z\to\infty} act(z) = 0$ or 2) there exists $z_0 > 0$ such that *act* is strictly decreasing on $[0, z_0]$ and $act(z) = 0$ for $z \in [z_0, \infty)$.

The other requirement on radial fuzzy systems is the presence of the radial property. The property reflects the preservation of radial shape of one-dimensional fuzzy sets after their combination by the selected $t$-norm in antecedents of rules, equation (1). The next formula states the property mathematically:

$$act\left(\left|\frac{x_1 - a_{j1}}{b_{j1}}\right|\right) \star \cdots \star act\left(\left|\frac{x_n - a_{jn}}{b_{jn}}\right|\right) = act(||\boldsymbol{x} - \boldsymbol{a}_j||_{\boldsymbol{b}_j}), \tag{8}$$

where $\boldsymbol{x} = (x_1, \ldots, x_n)$, $\boldsymbol{x} \in \mathcal{R}^n$, $\boldsymbol{b}_j = (b_{j1}, \ldots, b_{jn})$, $\boldsymbol{b}_j \in \mathcal{R}_+^n$ and $||\cdot||_{\boldsymbol{b}_j}$ is a scaled norm in $\mathcal{R}^n$ space. Employing the radial property, the general formula (1) has form

$$A_j(\boldsymbol{x}) = act(||\boldsymbol{x} - \boldsymbol{a}_j||_{\boldsymbol{b}_j}). \tag{9}$$

Inspecting the above formulas we see that antecedents in radial systems preserve theirs radial shape which is determined by employed *act* function. Actually, due to formula (9) an antecedent is represented by a multi-dimensional radial function in $\mathcal{R}^n$ space with central point $\boldsymbol{a}_j$. The property is not trvial because if the shape (*act* function) of one-dimensional fuzzy sets is selected together with the concrete $t$-norm $\star$ then the form of antecedent is determined (the right part of formula (9)) and need not to be expressible as a multi-dimensional radial fuzzy set of the same shape.

The question of which shapes and $t$-norms can be combined in order to radial property holds is solved in paper [12]. As a by product it is shown that scaled norms arising in representation of antecedents are scaled $\ell_p$ norm. That is, the norm has form $||\boldsymbol{u}||_{\boldsymbol{b}} = (\sum_i |u_i/b_i|^p)^{1/p}$, where $\boldsymbol{b} \in \mathcal{R}_+^n, p \in [1, +\infty]$ are parameters. The most common norms are scaled Euclidean ($p = 2$) and scaled cubic ($p = +\infty$) norms .

### 2.3 Radial Conjunctive Fuzzy Systems

Radial conjunctive fuzzy systems are obtained by fusion of concepts of radial and conjunctive fuzzy systems. Let us go into details. Concerning the $j$-th IF-THEN rule of such a system we get by equations (2), (9) and (7) the following representation:

$$R_j(\boldsymbol{x}, y) = act(||\boldsymbol{x} - \boldsymbol{a}_j||_{\boldsymbol{b}_j}) \star act\left(\frac{\max\{0, |y - c_j| - s_j\}}{d_j}\right), \tag{10}$$

$$B_j'(y) = R_j(\boldsymbol{x}, y) = h_j(\boldsymbol{x}) \star act\left(\frac{\max\{0, |y - c_j| - s_j\}}{d_j}\right). \tag{11}$$

Clearly, for height $h_j$ of $B_j'$ clipped set we have

$$h_j(\boldsymbol{x}) = A_j(\boldsymbol{x}) = act(||\boldsymbol{x} - \boldsymbol{a}_j||_{\boldsymbol{b}_j}). \tag{12}$$

The output fuzzy set from a radial conjunctive fuzzy system is then given by formula (6) as

$$B'(y) = \max_j\{B'_j(y)\} = \max_j\{h_j(\boldsymbol{x}) \star B_j(y)\}, \tag{13}$$

where $\boldsymbol{x}$ is the actual input into the radial fuzzy system.

To finish the specification of presented computational model, an appropriate defuzzification method [1] should be selected so that a crisp output could be specified on the basis of engine's output set $B'$. From the point of view of the research presented in this paper specific defuzzification is not important because we are only interested in properties (namely unimodality) of $B'$ set. So we do not address this question in more details.

## 3 Coherence of Radial Conjunctive Fuzzy Systems

In this section we introduce *coherence index of a radial conjunctive fuzzy system*. This index measures the extent of coherence of knowledge stored in the rule base of such the system.

Let us consider the following well known example. There is a fuzzy system performing a navigation of a car. Its rule base consists of two following rules:

$R_1$ : IF *an obstacle is in front of you* THEN *go left*,

$R_2$ : IF *an obstacle is in front of you* THEN *go right*.

We can easily see that rules in the above rule base are incoherent. Actually, the incoherence is caused by the fact that two rules with identical antecedents have different consequents. Generally, incoherence of knowledge is caused by the fact that similar preconditions have (highly) different conclusions [13]. The question is how to detect such incoherence in a rule base formally.

The above example helps. Expressing lingustic terms *go left*, *go right* by fuzzy sets and making the standard inference the output will be "camel" like multimodal fuzzy set. Using some of common defuzzification methods the crisp output point will be located at the point of local minima of multimodal output fuzzy set. And, this output would navigate the car to crash into the obstacle.

As the membership function of an output fuzzy set can be seen as the possibilistic distribution on an output space, it is highly desirable to this distribution be unimodal, because then output is naturally taken from points making the peak of the distribution.

In the above sense we will treat mathematically coherence of rule bases of conjunctive systems. That is, we say that conjunctive system is coherent if an output fuzzy set is unimodal for arbitrary input. Unfortunately, as we will se later, this strict specification is untenable. So we introduced *coherence index* which measures the extent of unimodality of output fuzzy set. And, we say that the system is the more coherent the higher value has the index.

In the following sections, we report the process which led us to mathematical specification of the index and we investigate some of its properties. As coherence index is related to unimodality of output fuzzy set, we start the process by solving the question of when two clipped radial fuzzy sets combined by maximum yields an unimodal fuzzy set. Then we extent the problem on $m > 2$ fuzzy sets.

### 3.1 Static Unimodality

Let us consider two clipped radial fuzzy sets and theirs union given by maximum. Then the following theorem applies:

**Theorem 1.** *Let $B'_1, B'_2$ be two clipped radial fuzzy sets with fixed heights $h_{1,2} \in [0,1]$*

$$B'_1(y) = h_1 \star act\left(\left|\frac{\max\{0, |y - c_1| - s_1\}}{d_1}\right|\right), \tag{14}$$

$$B'_2(y) = h_1 \star act\left(\left|\frac{\max\{0, |y - c_2| - s_2\}}{d_2}\right|\right), \tag{15}$$

*where $c_1 \leq c_2$ (other parameters given as in radial fuzzy systems). Let*

$$B'(y) = \max\{B'_1(y), B'_2(y)\},$$
$$CS = |c_1 - c_2| - (s_1 + s_2).$$

*If*

- *$CS \leq 0$ then $B'$ is unimodal,*
- *if $CS > 0$ and $h_1 \geq h_2$ then $B'$ is unimodal iff $B'_1(c_2 - s_2) \geq h_2$,*
- *if $CS > 0$ and $h_1 \leq h_2$ then $B'$ is unimodal iff $B'_2(c_1 + s_1) \geq h_1$.*

The above theorem gives us algorithm for checking unimodality of maximum of two clipped radial fuzzy sets on basis of setting of their parameters. We call this algorithm as strict unimodality test. Note that the theorem is stated as equivalence.

In the case of $m > 2$ rules, we are able to test the unimodality of maximum of $m$ clipped radial fuzzy sets on the basis of verifications of pairwise unimodalities.

**Theorem 2.** *Let $B'_1, \ldots, B'_m$ be clipped radial fuzzy sets with fixed heights $h_1, \ldots, h_m \in [0,1]$, i.e., for $j = 1, \ldots, m$,*

$$B'_j(y) = h_j \star act\left(\left|\frac{\max\{0, |y - c_j| - s_j\}}{d_j}\right|\right). \tag{16}$$

*If functions $B_{jk}(y) = \max\{B'_j(y), B'_k(y)\}$ are unimodal for all pairs $j, k \in \{1, \ldots m\}$, then also function $B'(y) = \max\{B'_1(y), \ldots, B'_m(y)\}$ is unimodal.*

The above theorem gives us the sufficient condition for maximum of $m \in \mathcal{N}$ clipped radial fuzzy sets be unimodal. The theorem can be used independently from the fact how the clipped fuzzy sets were obtained. However, our research is aimed on the case when the sets are generated during a computation of a radial conjunctive fuzzy system. In this case, we can check if output of fuzzy system is unimodal for the given concrete input $\boldsymbol{x}$, because the input, by formula (12), generates fixed values of heights $h_j$. But, we would like to say something more for the case of varying input and so varying heights, i.e., we would like to transfer from static to dynamic case of $h_j(\boldsymbol{x}), \boldsymbol{x} \in \mathcal{R}^n$. This question makes the topic of the consequent research.

Unfortunately, it can be shown that even for the simplest case of $m = 2$ rules with a reasonable setting of parameters $c_1 \neq c_2$, $s_1 = s_2 = 0$ there always exists an input making output fuzzy set $B'(y) = \max\{B'_1(y), B'_2(y)\}$ multimodal. This input $\boldsymbol{x}^=$ equalizes heights, i.e., $h_1(\boldsymbol{x}^=) = h_2(\boldsymbol{x}^=)$ and individual coordinates are specified according to formula

$$x_i^= = (a_{1i}b_{2i} + a_{2i}b_{1i})/(b_{1i} + b_{2i}). \tag{17}$$

From this reason we switched from dichotomic notion of unimodality to a measure of unimodality by coherence index $ci$ which is a number from interval $[0, 1]$ with $ci = 1$ iff output fuzzy set is unimodal. The question is how to specify the index mathematically.

### 3.2 Pairwise Incoherence / Coherence Index

In order to mathematically specify the index of coherence we set up natural requirements on how this index should behave in some important cases. We have found out that it is advantanegous to introduce the dual notion of incoherence index $inci$ by formula

$$inci = 1 - ci. \tag{18}$$

Hence the incoherence index is again number from interval $[0, 1]$ and $inci = 0$ iff output fuzzy set is multimodal (i.e., when $ci = 1$).

The following list presents requriments stated on incoherence index of maximum of two clipped radial fuzzy sets:

(1) If one of heights is zero, then incoherence is also zero (output is unimodal).
(2) If one of heights is one, then the results depends on the value of other height linearly
(3) If $h_1 = h_2$ and both heights are increasing, then incoherence is increasing (unimodality is decreasing).
(4) If only one of heights is varying (increasing or decreasing) and the other is constant, then incoherence is decreasing (unimodality is increasing).
(5) The wider is the spread between centers of consequents the higher is incoherence (the higher is multimodality and the less is unimodality).
(6) The wider are fuzzy sets (their supports - if applicable) the less is incoherence (the less is multimodality and the higher is unimodality).

On the basis of these requirements we have introduced the following specification of index of incoherence (multimodality):

**Definition 1.** *Let $B'_j, B'_k$ be two clipped radial fuzzy sets obtained by computation of a conjunctive radial fuzzy system on the basis of actual input $\boldsymbol{x}$. Then their (pairwise) incoherence index is computed according to formula*

$$inci_{jk}(\boldsymbol{x}) = (1 - UMCH_{jk}(\boldsymbol{x})) \cdot K_{jk} \cdot H_{jk}(\boldsymbol{x}), \tag{19}$$

*where*

$$UMCH_{ij}(\boldsymbol{x}) \in \{0,1\} = \textit{strict unimodality test}, \tag{20}$$

$$K \in (0,1) = act\left(-\frac{d_j + d_k}{\max\{0, |c_j - c_k| - (s_j + s_k)\}}\right), \tag{21}$$

$$H_{jk}(\boldsymbol{x}) \in [0,1] = \sqrt{\min\{h_j(\boldsymbol{x}), h_k(\boldsymbol{x})\} \cdot (1 - |h_j(\boldsymbol{x}) - h_k(\boldsymbol{x})|)}. \tag{22}$$

Let us discuss the index. The index is given by the product of three terms.

Term $UMCH_{jk} \in \{0,1\}$ presents the strict unimodality test. The test is based on the procedure presented by Theorem 1. That is, if the output fuzzy set $B'(y) = \max\{B'_j(y), B'_k(y)\}$ is unimodal then $UMCH_{jk} = 1$, term $(1 - UMCH_{jk}) = 0$ and consequentially $inci_{jk}(\boldsymbol{x}) = 0$. So incoherence is minimal and coherence maximal. If $B'(y)$ is multimodal then $UMCH_{jk} = 0$, $(1 - UMCH_{jk}) = 1$ and the value of $inci_{jk}(\boldsymbol{x})$ depends on values of other the two terms.

The specification of other two terms $K_{jk}$ and $H_{jk}(\boldsymbol{x})$ is directly induced by the presented requirements. The main difference between both terms is that $K_{jk}$ is constant with respect to an input $\boldsymbol{x}$ and $H_{jk}(\boldsymbol{x})$ varies with it. Let us discuss the requirements (for the sake of simplicity we omit argument $\boldsymbol{x}$):

(1) $H_{jk} = 0$ due to the minimum term.
(2) $H_{jk} = \sqrt{\min\{1, h_k\} \cdot h_k} = h_k$.
(3) $H_{jk} = \sqrt{\min\{h_j, h_j\}} = \sqrt{h_j}$.
(4) Let $h_j$ be constant and $h_k \geq h_j$ increasing , then $\min\{h_j, h_k\} = h_j$ is constant and the second term of $H_{jk}$ is decreasing due to the increasing difference $|h_j - h_k|$; let $h_j$ be constant, and $h_k \leq h_j$ decreasing, then $\min\{h_j, h_k\} = h_k$ is also decreasing and the second term $H_{jk}$ is decreasing due to the increasing difference $|h_j - h_k|$.
(5) This is assured by employment of $|c_j - c_k|$ term in $K_{jk}$ term; if $|c_j - c_k|$ is increasing, the fraction is decreasing and $K_{jk}$ is increasing due to the properties of *act* function.
(6) This is assured by employment of $d_j + d_k$ term in $K_{jk}$ term; if $d_j + d_k$ is increasing, the fraction is increasing and $K_{jk}$ is decreasing due to the properties of *act* function.

We see that our mathematical specification of incoherence index is in accordance with the list of requirements stated above.

Summarizing the work we have done we can state the following. We have introduced incoherence index of a pair of two fuzzy sets generated by a radial conjunctive fuzzy system for fixed input $\boldsymbol{x}$. The index has value 0 if output is unimodal otherwise it takes value from interval $(0, 1]$. Dually, index of coherence $ci = 1 - inci$ has value 1 if output is unimodal and otherwise it takes values from interval $[0, 1)$. However, still we are in static case. The next section transfers us to dynamic case where we present the main theorem of the paper.

## 4 Dynamic Unimodality Theorem

The value of incoherence/coherence index resides on specification of parameters of the system as well as on the actual value of input. The input drives heights of clipped fuzzy sets. The crucial question is what happens if the input freely moves through the whole input space. How incoherence is varying? By the following theorem we show that we can identify an upper bound of incoherence index.

**Theorem 3.** *Let the incoherence index of two rules $j, k$ of a radial conjunctive fuzzy system be defined as above. Then for any input $\boldsymbol{x} \in \mathcal{R}^n$ the following inequality holds:*

$$inci_{jk}(\boldsymbol{x}) \leq K_{jk} \cdot \sqrt{act(||\boldsymbol{a}_j - \boldsymbol{a}_k||_{2\boldsymbol{b}_M})}, \tag{23}$$

*where* $2\boldsymbol{b}_M = (2b_{M1}, \ldots, 2b_{Mn})$, $b_{Mi} = \max\{b_{ji}, b_{ki}\}$, $i = 1, \ldots, n$.

Due to the lack of space we present the theorem without a proof. We only remark that the proof highly resides on the radial property and that is why we are working with radial conjunctive fuzzy systems (radial C-FSs).

The theorem gives us an upper bound of incoherence index of two rules of a radial C-FS. Clearly, the bound immediately determines a lower bound of coherence index:

$$ci_{jk}(\boldsymbol{x}) \geq 1 - K_{jk} \cdot \sqrt{act(||\boldsymbol{a}_j - \boldsymbol{a}_k||_{2\boldsymbol{b}_M})}. \tag{24}$$

Let us denote the lower bound by $CI_{jk}$, i.e., $CI_{jk} = \sqrt{act(||\boldsymbol{a}_j - \boldsymbol{a}_k||_{2\boldsymbol{b}_M})}$.

Till now we have worked with two rules. Now we transfer to general case of $m > 2$ rules in the rule base. Firstly we define so-called coherence matrix of a radial C-FS as follows:

**Definition 2.** *Coherence matrix $W_{CI}$ of a radial C-FS is the symmetric matrix defined as $W_{CI}(j, k) = CI_{jk}$, i.e.,*

$$W_{CI}(j,k) = \begin{bmatrix} 1 & CI_{12} & \ldots & CI_{1m} \\ CI_{21} & 1 & \ldots & CI_{2m} \\ \vdots & 1 & \ldots & \vdots \\ CI_{m1} & CI_{m2} & \ldots & 1 \end{bmatrix}.$$

Each cell of $W_{CI}$ gives the lower bound of coherence index of corresponding pair of rules. The minimal element of coherence matrix we call *the (overall) coherence index of a radial conjunctive fuzzy system.*

**Definition 3.** *Let $W_{CI}$ be the coherence matrix of a radial C-FS, then the (overall) coherence index of this system is given as*

$$CICFS = ||W_{CI}||_{\min} = \min_{jk}\{W_{CI}(j,k)\}. \tag{25}$$

Hence the coherence index of a radial C-FS is given as the minimum from lower bounds of pairwise coherences of individual rules. Apparently, if for all $j, k$, $W_{CI}(j,k) = 1$, then also $CICFS = 1$ and the system is coherent, which is in accordance with Theorem 2.

## 5 Conclusions

In the paper we have introduced index of coherence of a radial conjuctive fuzzy system. The index is primarily computed pairwisely for pairs of rules and we stated the lower bound for such a pairwise index with respect to arbitrary input to the system. All pairwise lower bounds form so-called coherence matrix of the system and their minimum then states the (overall) coherence index of a radial conjunctive fuzzy system. Dual notion of incoherence index was introduced as well.

The value of index can be seen as a measure of coherence (consistency) of knowledge stored in the rule base of a radial conjunctive fuzzy system. Actually, the greater the index is the more unimodal are output fuzzy sets from the system and defuzzification process is more natural.

In the future research we aim on an incorporation of the notion of coherence (as presented) in the area of radial basis neural networks. This is a tentative research direction because radial C-FSs are computationally almost equivalent to RBF neural networks. So (under slight reconsiderations) the presented index can be used as a measure of coherence (or quality) of learned knowledge and consequently it could be used for discrimination between different learning algorithms.

Concerning practical applications, we plan to use the index for qualitative analysis of neuro-fuzzy classifiers employed in EEG spectrograms signals analysis, which is related to detection of micro-sleeps episodes during car driving [14].

## Acknowledgment

The research presented in the paper was supported by grant ME 701 of The Ministry of Education, Youth and Sports of the Czech Republic.

## References

1. G. J. Klir and B. Yuan. *Fuzzy sets and Fuzzy logic - Theory and Applications.* Prentice Hall, Upper Saddle River, 1995.
2. L. X. Wang. *A Course in Fuzzy Systems and Control.* Prentice Hall, 1997.
3. D. Dubois and H. Prade. What are fuzzy rules and how to use them? *Fuzzy Sets and Systems*, 84(2):169–185, 1996.

4. R. R. Yager and D. P. Filev. *Essentials of Fuzzy Modeling and Control.* John Wiley & Sons, 1994.
5. P. Hájek, *Metamathematics of Fuzzy Logic.* Kluwer Academic Publishers, 1998.
6. E. P. Klement, R. Mesiar, and A. Pap. *Triangular Norms.* Kluwer Academic Publishers, Dordrecht, 2000.
7. K. S. Leung and Y. T. So, "Consistency Checking for Fuzzy Expert Systems," *International Journal of Approximate Reasoning*, vol. 9, pp. 263–282, 1993.
8. S. Viaene, G. Wets, and J. Vanthienen, "A synthesis of fuzzy rule-based system verification," *Fuzzy Sets and Systems*, vol. 113, no. 2, pp. 253–265, 2000.
9. D. Dubois, H. Prade, and L. Ughetto, "Checking the Coherence and Redundancy of Fuzzy Knowledge Bases," *IEEE Transactions on Fuzzy Systems*, vol. 5, no. 3, pp. 398–417, 1997.
10. D. Coufal, "Coherence of Radial Implicative Fuzzy Systems," in *IEEE International Conference on Fuzzy Systems, FUZZ-IEEE 2006, Vancouver, Canada 2006*, 2006, pp. 903–970.
11. S. S. Haykin, *Neural Networks: A Comprehensive Foundation, 2nd Edition.* Upper Saddle River: Prentice Hall, 1998.
12. D. Coufal, "Representation of Continuous Archimedean Radial Fuzzy Systems," in *International Fuzzy Systems Association World Congress IFSA 2005, Beijing, China 2005*, 2005, pp. 1174–1179.
13. D. Driankov, H. Hellendoorn, and M. Reinfrank, *An Introduction to Fuzzy Control.* Berlin Heidelberg: Springer-Verlag, 1993.
14. M. Novák - Ed., *Neurodynamic and Neuroinformatics Studies (Second book on Micro-Sleeps).* Neural Network World - monographs edition, Prague, 2005.

# Topology in Fuzzy Class Theory: Basic Notions

Libor Běhounek[1,⋆] and Tomáš Kroupa[2,⋆⋆]

[1] Institute of Computer Science
Academy of Sciences of the Czech Republic
Pod Vodárenskou věží 2, 182 07 Prague, Czech Republic
behounek@cs.cas.cz, www.cs.cas.cz/hp

[2] Institute of Information Theory and Automation
Academy of Sciences of the Czech Republic
Pod Vodárenskou věží 4, 182 08 Prague, Czech Republic
kroupa@utia.cas.cz

**Abstract.** In the formal and fully graded setting of Fuzzy Class Theory (or higher-order fuzzy logic) we make an initial investigation into basic notions of fuzzy topology. In particular we study graded notions of fuzzy topology regarded as a fuzzy system of open or closed fuzzy sets and as a fuzzy system of fuzzy neighborhoods. We show their basic graded properties and mutual relationships provable in Fuzzy Class Theory and give some links to the traditional notions of fuzzy topology.

## 1 Introduction

Fuzzy topology is among the fundamental disciplines of fuzzy mathematics whose development was stimulated from the very beginning of the invention of fuzzy sets [1]. Following the role of topology in classical mathematics, fuzzy topology should capture the notions of openness, neighborhood, closure, etc., within the setting of fuzzy set theory. The paper [2] by Höhle and Šostak, which is contained in the special issue of Fuzzy Sets and Systems (1995) on fuzzy topology, mentions and classifies a number of conceptual frameworks (lattice-, model-, and category-theoretical) that have arisen during past decades. A detailed and up-to-date exposition of many-valued and fuzzy topologies, mostly based on a categorical viewpoint, is contained in the monograph [3] by Höhle.

This paper follows the footsteps of Ying's attempt [4] to establish fuzzy topology as a non-elementary theory over many-valued logic. We make initial steps towards understanding fuzzy topology as an axiomatic higher-order theory over Hájek-style [5] formal fuzzy logic, following the methodology for formal fuzzy mathematics described in [6]. According to the classification proposed in [2], the models of our theory are closest to "$L$-fuzzy topologies as characteristic morphisms". However, the apparatus of Fuzzy Class Theory, employed in this

⋆ The work of the first author was supported by grant No. B100300502 of GA AV ČR and the Institutional Research Plan AV0Z10300504.

⋆⋆ The work of the second author was supported by grant No. B100300502 of GA AV ČR and grant No. 1M0572 of MŠMT ČR.

P. Melin et al. (Eds.): IFSA 2007, LNAI 4529, pp. 513–522, 2007.
© Springer-Verlag Berlin Heidelberg 2007

paper, makes our notions and the way in which they can be studied quite distinct from (and in some aspects more general than) other approaches to fuzzy topology.

The paper is organized as follows: Section 2 gives a brief exposition of Fuzzy Class Theory and the definitions needed in the paper. Section 3 studies the graded notion of fuzzy topology regarded as a fuzzy system of open (or closed) fuzzy sets. Section 4 then studies graded fuzzy topologies regarded as fuzzy systems of fuzzy neighborhoods.

## 2 Preliminaries

Fuzzy Class Theory FCT, introduced in [7], is an axiomatization of Zadeh's notion of fuzzy set in formal fuzzy logic. Here we use its variant defined over IMTL$_\triangle$ [8], the logic of all left-continuous t-norms whose residual negation is involutive (we shall call them *IMTL t-norms;* the most important example is the Łukasiewicz t-norm $x * y =_{\mathrm{df}} \max(0, x + y - 1)$).

*Remark 1.* We have the following reasons for choosing IMTL$_\triangle$ for the ground logic: the logic MTL$_\triangle$ [8] of all left-continuous t-norms is arguably [9] the weakest fuzzy logic with good inferential properties for fully graded fuzzy mathematics in the framework of formal fuzzy logic [6]. IMTL$_\triangle$ extends it with the law of double negation, which is in fuzzy topology needed for the correspondence between open and closed fuzzy sets. A generalization of fuzzy topology to the logic MTL$_\triangle$ (with independent systems of open and closed fuzzy sets) will be the subject of some future paper.

We assume the reader's familiarity with IMTL$_\triangle$; for details on this logic see [8]. Here we only recapitulate its standard [0, 1] semantics:

| | | |
|---|---|---|
| $\&$ | ... | a left-continuous t-norm $*$ with involutive residual negation |
| $\rightarrow$ | ... | the residuum $\Rightarrow$ of $*$, defined as $x \Rightarrow y =_{\mathrm{df}} \sup\{z \mid z * x \leq y\}$ |
| $\wedge, \vee$ | ... | min, max |
| $\neg$ | ... | $x \Rightarrow 0$; in IMTL$_\triangle$ it is involutive, due to the axiom $\neg\neg\varphi \rightarrow \varphi$ |
| $\veebar$ | ... | the t-conorm dual to $*$ (since $\varphi \veebar \psi$ is defined as $\neg(\neg\varphi \mathbin{\&} \neg\psi)$) |
| $\leftrightarrow$ | ... | the bi-residuum: $\min(x \Rightarrow y, y \Rightarrow x)$ |
| $\triangle$ | ... | $\triangle x = 1 - \mathrm{sgn}(1 - x)$ |
| $\forall, \exists$ | ... | inf, sup; by involutiveness, $(\exists x)\neg\varphi \leftrightarrow \neg(\forall x)\varphi$ |

**Definition 1.** Fuzzy Class Theory FCT *is a formal theory over multi-sorted first-order fuzzy logic (in this paper,* IMTL$_\triangle$*), with the sorts of variables for*

- *atomic objects (lowercase letters $x, y, \ldots$)*
- *fuzzy classes of atomic objects (uppercase letters $A, B, \ldots$)*
- *fuzzy classes of fuzzy classes of atomic objects (Greek letters $\tau, \sigma, \ldots$)*
- *fuzzy classes of the third order (calligraphic letters $\mathcal{A}, \mathcal{B}, \ldots$)*
- *etc., in general for fuzzy classes of the n-th order ($X^{(n)}, Y^{(n)}, \ldots$)*

**Table 1.** Abbreviations used in the formulae of FCT

| | | |
|---|---|---|
| $Ax$ | $\equiv_{\mathrm{df}}$ | $x \in A$ |
| $x_1 \dots x_k$ | $=_{\mathrm{df}}$ | $\langle x_1, \dots, x_k \rangle$ |
| $x \notin A$ | $\equiv_{\mathrm{df}}$ | $\neg(x \in A)$, and similarly for other predicates |
| $(\forall x \in A)\varphi$ | $\equiv_{\mathrm{df}}$ | $(\forall x)(x \in A \rightarrow \varphi)$ |
| $(\exists x \in A)\varphi$ | $\equiv_{\mathrm{df}}$ | $(\exists x)(x \in A \mathbin{\&} \varphi)$ |
| $(\forall x, y \in A)\varphi$ | $\equiv_{\mathrm{df}}$ | $(\forall x \in A)(\forall y \in A)\varphi$, similarly for $\exists$ |
| $\{x \in A \mid \varphi\}$ | $=_{\mathrm{df}}$ | $\{x \mid x \in A \mathbin{\&} \varphi\}$ |
| $\{t(x_1, \dots, x_k) \mid \varphi\}$ | $=_{\mathrm{df}}$ | $\{z \mid z = t(x_1, \dots, x_k) \mathbin{\&} \varphi\}$ |
| $\varphi^n$ | $\equiv_{\mathrm{df}}$ | $\varphi \mathbin{\&} \dots \mathbin{\&} \varphi$ ($n$ times) |

*Besides the crisp identity predicate* $=$, *the language of* FCT *contains:*

- *the membership predicate* $\in$ *between objects of successive sorts*
- *class terms* $\{x \mid \varphi\}$, *for any formula* $\varphi$ *and any variable* $x$ *of any order*
- *symbols* $\langle x_1, \dots, x_k \rangle$ *for* $k$*-tuples of individuals* $x_1, \dots, x_k$ *of any order*

FCT *has the following axioms (for all formulae* $\varphi$ *and variables of all orders):*

- *the logical axioms of multi-sorted first-order logic* $\mathrm{IMTL}_\triangle$
- *the axioms of crisp identity: (i)* $x = x$, *(ii)* $x = y \rightarrow (\varphi(x) \rightarrow \varphi(y))$, *(iii)* $\langle x_1, \dots, x_k \rangle = \langle y_1, \dots, y_k \rangle \rightarrow x_1 = y_1 \mathbin{\&} \dots \mathbin{\&} x_k = y_k$
- *the comprehension axioms:* $y \in \{x \mid \varphi(x)\} \leftrightarrow \varphi(y)$
- *the extensionality axioms:* $(\forall x)\triangle(x \in A \leftrightarrow x \in B) \rightarrow A = B$

*Remark 2.* Notice that in FCT, fuzzy sets are rendered as a *primitive notion* rather than modeled by membership functions. In order to capture this distinction, fuzzy sets are in FCT called *fuzzy classes;* the name *fuzzy set* is reserved for membership functions in the models of the theory.

The models of FCT are systems of fuzzy sets of all orders over a fixed crisp universe of discourse, with truth degrees taking values in an $\mathrm{IMTL}_\triangle$-chain (e.g., the interval $[0, 1]$ equipped with an IMTL t-norm). Thus all theorems on fuzzy classes provable in FCT are true statements about $\mathbf{L}$-valued fuzzy sets, for any $\mathrm{IMTL}_\triangle$-chain $\mathbf{L}$. Notice however that the theorems of FCT have to be derived from its axioms by the rules of the fuzzy logic $\mathrm{IMTL}_\triangle$ rather than classical Boolean logic. For details on proving theorems of FCT see [10] or [11].

**Convention 2.1** *In formulae of* FCT, *we employ usual abbreviations known from classical mathematics, including those listed in Table 1. Usual rules of precedence apply to the connectives of* $\mathrm{IMTL}_\triangle$. *Furthermore we define standard defined notions of* FCT, *summarized in Table 2, for all orders of fuzzy classes.*

*Remark 3.* Notice that in FCT, not only the membership predicate $\in$, but all defined notions are in general fuzzy (unless they are defined as provably crisp). FCT thus presents a fully graded approach to fuzzy mathematics. The importance of full gradedness in fuzzy mathematics is explained in [10,12,13]: its main merit lies in that it allows inferring relevant information even when a property of fuzzy sets is not fully satisfied. Fuzzy topology has a long tradition of attempting full gradedness, cf. graded definitions and theorems e.g. in [3,4].

**Table 2.** Defined notions of FCT

| | | | |
|---|---|---|---|
| $\emptyset$ | $=_{\mathrm{df}}$ | $\{x \mid 0\}$ | empty class |
| $\mathrm{V}$ | $=_{\mathrm{df}}$ | $\{x \mid 1\}$ | universal class |
| $\mathrm{Ker}\, A$ | $=_{\mathrm{df}}$ | $\{x \mid \triangle Ax\}$ | kernel |
| $\alpha A$ | $=_{\mathrm{df}}$ | $\{x \mid \alpha \,\&\, Ax\}$ | $\alpha$-resize |
| $-A$ | $=_{\mathrm{df}}$ | $\{x \mid \neg Ax\}$ | complement |
| $A \cap B$ | $=_{\mathrm{df}}$ | $\{x \mid Ax \,\&\, Bx\}$ | (strong) intersection |
| $A \cup B$ | $=_{\mathrm{df}}$ | $\{x \mid Ax \veebar Bx\}$ | (strong) union |
| $A \times B$ | $=_{\mathrm{df}}$ | $\{xy \mid Ax \,\&\, By\}$ | Cartesian product |
| $\bigcup \tau$ | $=_{\mathrm{df}}$ | $\{x \mid (\exists A \in \tau)(x \in A)\}$ | class union |
| $\bigcap \tau$ | $=_{\mathrm{df}}$ | $\{x \mid (\forall A \in \tau)(x \in A)\}$ | class intersection |
| $\mathrm{Pow}(A)$ | $=_{\mathrm{df}}$ | $\{X \mid X \subseteq A\}$ | power class |
| $\mathrm{Crisp}(A)$ | $\equiv_{\mathrm{df}}$ | $(\forall x)\triangle(Ax \vee \neg Ax)$ | crispness |
| $\mathrm{Ext}_E\, A$ | $\equiv_{\mathrm{df}}$ | $(\forall x, y)(Exy \,\&\, Ax \to Ay)$ | $E$-extensionality |
| $A \subseteq B$ | $\equiv_{\mathrm{df}}$ | $(\forall x)(Ax \to Bx)$ | inclusion |
| $A \approxeq B$ | $\equiv_{\mathrm{df}}$ | $(A \subseteq B) \,\&\, (B \subseteq A)$ | (strong) bi-inclusion |

*Remark 4.* It should be noted that fully graded theories have some peculiar features in which they differ from both classical mathematics and traditional fuzzy mathematics. A detailed account of the unusual features of fully graded theories is given in [14]; some of them can also be found in [10] (available online). Here we only briefly stress the main features of graded mathematics:

- Since $\varphi \to \varphi \,\&\, \varphi$ is not a generally valid law of fuzzy logic, premises may occur several times in theorems. A typical graded theorem has the form $\varphi_1^{k_1} \,\&\, \ldots \,\&\, \varphi_n^{k_n} \to \psi$, where $\varphi^k$ abbreviates $\varphi \,\&\, \ldots \,\&\, \varphi$ ($k$ times, where $\varphi^0$ is 1). The multiplicity $k_i$ of the premise $\varphi_i$ shows how strongly it influences (the lower bound for) the truth of $\psi$ (when only partially true), and depends on how many times the premise is used in the derivation of $\psi$ from $\varphi_1, \ldots, \varphi_k$. The exponent $k$ in $\varphi^k$ can also take the conventional value "$\triangle$", where $\varphi^\triangle$ is understood as $\triangle\varphi$ (recall that $\varphi^\triangle \to \varphi^n$ for all $n$).
- If a complex notion $\Phi$ is defined as a conjunction $\varphi_1 \,\&\ldots\&\, \varphi_n$, then the conjuncts $\varphi_i$ will get different multiplicities in different theorems. It is therefore appropriate to parameterize $\Phi$ by the multiplicities of the components $\varphi_i$ and define it as $\Phi^{k_1,\ldots,k_n} \equiv_{\mathrm{df}} \varphi_1^{k_1} \,\&\ldots\&\, \varphi_n^{k_n}$. (All graded topological notions in the following sections will be defined in this way.) We can write just $\Phi^k$ instead of $\Phi^{k_1,\ldots,k_n}$ if $k_i = k$ for all $i$, and just $\Phi$ if $k_i = 1$ for all $i$.

The following defined predicates will be employed in the next sections.

**Definition 2.** *We define the following (graded) unary predicates:*

| | |
|---|---|
| $\cup$*-closedness:* | $\mathrm{uc}(\tau) \equiv_{\mathrm{df}} (\forall A, B \in \tau)(A \cup B \in \tau)$ |
| $\cap$*-closedness:* | $\mathrm{ic}(\tau) \equiv_{\mathrm{df}} (\forall A, B \in \tau)(A \cap B \in \tau)$ |
| $\bigcup$*-closedness:* | $\mathrm{Uc}(\tau) \equiv_{\mathrm{df}} (\forall \nu \subseteq \tau)\big(\bigcup \nu \in \tau\big)$ |
| $\bigcap$*-closedness:* | $\mathrm{Ic}(\tau) \equiv_{\mathrm{df}} (\forall \nu \subseteq \tau)\big(\bigcap \nu \in \tau\big)$ |
| $\subseteq$*-upperness:* | $\mathrm{Upper}(\tau) \equiv_{\mathrm{df}} (\forall A, B)(A \subseteq B \,\&\, A \in \tau \to B \in \tau)$ |
| *being a filter:* | $\mathrm{Filter}^{v,e,u,i}(\tau) \equiv_{\mathrm{df}} (\mathrm{V} \in \tau)^v \,\&\, (\emptyset \notin \tau)^e \,\&\, \mathrm{Upper}^u(\tau) \,\&\, \mathrm{ic}^i(\tau)$ |

## 3 Topology as a System of Open (Closed) Fuzzy Classes

In classical mathematics, topology can be introduced in several equivalent ways—by open sets, closed sets, neighborhoods, closure, etc. In FCT, however, these approaches yield different concepts. In this paper, we make an initial investigation into two of them, namely the system of open (or closed) classes (in this section) and the system of neighborhoods (in Sect. 4). Due to the limited size of this paper we present only some of the initial results and have to omit all proofs.

The fuzzification of the concept of open (closed) fuzzy topology presented in Def. 3 follows the methodology sketched in [15, §5] and formally elaborated in [7, §7], i.e., reinterpreting the formulae of the classical definition in fuzzy logic.[1]

**Definition 3.** *We define an* (open) $(e, v, i, u)$–fuzzy topology *and a* closed $(e, v, u, i)$–fuzzy topology *respectively by the predicates*

$$\mathsf{OTop}^{e,v,i,u}(\tau) \equiv_{\mathrm{df}} (\emptyset \in \tau)^e \mathbin{\&} (\mathrm{V} \in \tau)^v \mathbin{\&} \mathrm{ic}^i(\tau) \mathbin{\&} \mathrm{Uc}^u(\tau)$$
$$\mathsf{CTop}^{e,v,u,i}(\sigma) \equiv_{\mathrm{df}} (\emptyset \in \sigma)^e \mathbin{\&} (\mathrm{V} \in \sigma)^v \mathbin{\&} \mathrm{uc}^u(\sigma) \mathbin{\&} \mathrm{Ic}^i(\sigma)$$

*(see Remark 4 for the meaning of the parameters $e, v, u, i$).*

Note that this concept of topology is graded, i.e., the predicate $\mathsf{OTop}^{e,v,i,u}$ determines the *degree* to which $\tau$ is an open $(e, v, i, u)$–fuzzy topology.

*Example 1.* Let $*$ be an IMTL t-norm and $\Rightarrow$ its residuum. The $*$-based Zadeh models of open $(1, 1, \triangle, \triangle)$–fuzzy topology, i.e., of the predicate

$$\mathsf{OTop}^{1,1,\triangle,\triangle}(\tau) \equiv \emptyset \in \tau \mathbin{\&} \mathrm{V} \in \tau \mathbin{\&} \triangle\,\mathrm{ic}(\tau) \mathbin{\&} \triangle\,\mathrm{Uc}(\tau)$$

are functions $\tau\colon [0,1]^{\mathrm{V}} \to [0,1]$ satisfying the following conditions:

(i) $\tau(A) * \tau(B) \le \tau(A \cap B)$ for every $A, B \in [0,1]^{\mathrm{V}}$
(ii) $\bigwedge_{A \in [0,1]^{\mathrm{V}}} (\nu(A) \Rightarrow \tau(A)) \le \tau\left(\bigcup \nu\right)$ for every $\nu\colon [0,1]^{\mathrm{V}} \to [0,1]$

where $(A \cap B)(x) = A(x) * B(x)$ and $\left(\bigcup \nu\right)(x) = \bigvee_{A \in [0,1]^{\mathrm{V}}} (\nu(A) * A(x))$. Since both (i) and (ii) are crisp, the degree to which $\tau$ is a $(1, 1, \triangle, \triangle)$-fuzzy topology equals $\tau(\emptyset) * \tau(\mathrm{V})$. These models cover fuzzy topologies studied under the name "*L*-fuzzy topologies of Höhle type" [2].

In $\mathrm{IMTL}_\triangle$, open and closed topologies are interdefinable:

**Definition 4.** *Let* $\tau_c =_{\mathrm{df}} \{A \mid -A \in \tau\}$.

[1] The requirement that both $\emptyset$ and the ground set be open can meaningfully be reinterpreted in fuzzy logic in several ways; here we restrict ourselves to the weakest one, requiring openness just for the two classes $\emptyset$ and V. Stronger notions of topology (e.g., *stratified* topology [3] with the condition $\alpha\mathrm{V} \in \tau$ for all truth degrees $\alpha$) will be studied in subsequent papers.

**Theorem 1.** FCT *proves:* $\mathsf{OTop}(\tau) \leftrightarrow \mathsf{CTop}(\tau_c)$, $\mathsf{CTop}(\sigma) \leftrightarrow \mathsf{OTop}(\sigma_c)$.

**Definition 5.** *Given a class of classes* $\tau$, *we define the* interior *and* closure *in* $\tau$ *as follows:*

$$\mathrm{Int}_\tau(A) =_{\mathrm{df}} \bigcup \{B \in \tau \mid B \subseteq A\}$$
$$\mathrm{Cl}_\tau(A) =_{\mathrm{df}} \bigcap \{B \in \tau_c \mid A \subseteq B\}$$

**Theorem 2.** *It is provable in* FCT*:*

1. $\mathrm{Int}_\tau(A) \subseteq A$
2. $A \subseteq B \rightarrow \mathrm{Int}_\tau(A) \subseteq \mathrm{Int}_\tau(B)$
3. $A \in \tau \rightarrow \mathrm{Int}_\tau(A) \approxeq A$
4. $\mathrm{Int}_\tau(A \cap B) \cap \mathrm{Int}_\tau(A \cap B) \subseteq \mathrm{Int}_\tau(A) \cap \mathrm{Int}_\tau(B)$

**Theorem 3 (OTop and the interior operator).** *It is provable in* FCT*:*

1. $\mathsf{OTop}^{0,0,0,1}(\tau) \rightarrow \mathrm{Int}_\tau(A) \in \tau$
2. $\mathsf{OTop}^{0,0,0,1}(\tau) \rightarrow \mathrm{Int}_\tau(\mathrm{Int}_\tau(A)) \approxeq \mathrm{Int}_\tau(A)$
3. $\mathsf{OTop}^{0,0,1,0}(\tau) \rightarrow \mathrm{Int}_\tau(A) \cap \mathrm{Int}_\tau(B) \subseteq \mathrm{Int}_\tau(A \cap B)$
4. $\mathsf{OTop}^{0,1,0,0}(\tau) \rightarrow \mathrm{Int}_\tau(\mathrm{V}) \approxeq \mathrm{V}$

Since $\mathrm{Cl}_\tau(A) = -\,\mathrm{Int}_\tau(-A)$ is provable in FCT, the next two theorems are just dual counterparts of Th. 2 and 3.

**Theorem 4.** *It is provable in* FCT*:*

1. $A \subseteq \mathrm{Cl}_\tau(A)$
2. $A \subseteq B \rightarrow \mathrm{Cl}_\tau(A) \subseteq \mathrm{Cl}_\tau(B)$
3. $A \in \tau_c \rightarrow \mathrm{Cl}_\tau(A) \approxeq A$
4. $\mathrm{Cl}_\tau(A) \cup \mathrm{Cl}_\tau(B) \subseteq \mathrm{Cl}_\tau(A \cup B) \cup \mathrm{Cl}_\tau(A \cup B)$

**Theorem 5 (OTop and the closure operator).** *It is provable in* FCT*:*

1. $\mathsf{OTop}^{0,0,0,1}(\tau) \rightarrow \mathrm{Cl}_\tau(A) \in \tau_c$
2. $\mathsf{OTop}^{0,0,0,1}(\tau) \rightarrow \mathrm{Cl}_\tau(\mathrm{Cl}_\tau(A)) \approxeq \mathrm{Cl}_\tau(A)$
3. $\mathsf{OTop}^{0,0,1,0}(\tau) \rightarrow \mathrm{Cl}_\tau(A \cup B) \subseteq \mathrm{Cl}_\tau(A) \cup \mathrm{Cl}_\tau(B)$
4. $\mathsf{OTop}^{0,1,0,0}(\tau) \rightarrow \mathrm{Cl}_\tau(\emptyset) \approxeq \emptyset$

**Definition 6.** *A predicate expressing that* $A$ *is a* neighborhood *of* $x$ *in* $\tau$ *is defined as*

$$\mathrm{Nb}_\tau(x, A) \equiv_{\mathrm{df}} (\exists B \in \tau)(B \subseteq A \,\&\, x \in B)$$

*The system of all neighborhoods of* $x$ *will be denoted by* $\nu_x =_{\mathrm{df}} \{A \mid \mathrm{Nb}_\tau(x, A)\}$.

**Theorem 6 (OTop and neighborhoods).** *It is provable in* FCT*:*

1. $x \in \bigcap \nu_x$
2. $\mathrm{Nb}_\tau(x, A) \leftrightarrow x \in \mathrm{Int}_\tau(A)$
3. $\mathsf{OTop}(\tau) \rightarrow \mathrm{Filter}(\nu_x) \,\&\, (\forall A \in \nu_x)(\exists B \in \nu_x)(B \subseteq A \,\&\, (\forall y \in B)\, \mathrm{Nb}_\tau(y, B))$

In general, the system of all open fuzzy topologies is not closed under arbitrary intersections. Nevertheless, the system of all open $\triangle$–fuzzy topologies is at least closed under crisp intersections, which allows introducing the notion of open fuzzy topology generated by a subbase of fuzzy classes:

**Theorem 7.** *Let $\mathcal{X}$ be a fuzzy class of the third order. Then* FCT *proves:*

$$\mathrm{Crisp}(\mathcal{X}) \mathbin{\&} (\forall \tau \in \mathcal{X})(\triangle\, \mathsf{OTop}(\tau)) \to \triangle\, \mathsf{OTop}(\bigcap \mathcal{X})$$

**Definition 7.** *Let $\sigma$ be a fuzzy class of fuzzy classes. Then we define*

$$\tau_\sigma =_{\mathrm{df}} \bigcap \{\tau' \mid \triangle(\mathsf{OTop}(\tau') \mathbin{\&} \sigma \subseteq \tau')\}$$

By Th. 7, FCT proves that $\triangle\, \mathsf{OTop}(\tau_\sigma)$, and obviously also that $\tau_\sigma$ is the least open $\triangle$–fuzzy topology containing $\sigma$.

*Example 2. Interval open fuzzy topology.* Let $\leq$ be a crisp dense ordering (e.g., of real or rational numbers). The fuzzy properties of being an upper resp. lower class in $\leq$ are defined by the predicates

$$\begin{aligned}\mathrm{Upper}_{\leq}(A) &\equiv_{\mathrm{df}} (\forall p, q)(p \leq q \mathbin{\&} Ap \to Aq)\\ \mathrm{Lower}_{\leq}(A) &\equiv_{\mathrm{df}} (\forall p, q)(p \geq q \mathbin{\&} Bp \to Bq)\end{aligned}$$

Fuzzy intervals $[A, B]$ in $\leq$ can be defined [16] as intersections $A \cap B$ of two fuzzy classes $A, B$, where $\triangle\, \mathrm{Upper}_{\leq}(A) \mathbin{\&} \triangle\, \mathrm{Lower}_{\leq}(B)$. An *open* fuzzy interval can be defined by the following fuzzy predicate:[2]

$$\begin{aligned}\mathrm{Op}([A, B]) \equiv_{\mathrm{df}} \triangle(\mathrm{Upper}_{\leq}(A)) \mathbin{\&} (\forall p)(Ap \to (\exists q < p)Aq) \mathbin{\&}\\ \triangle(\mathrm{Lower}_{\leq}(B)) \mathbin{\&} (\forall p)(Bp \to (\exists q > p)Bq)\end{aligned}$$

By Th. 7, the fuzzy system $\sigma = \{[A, B] \mid \mathrm{Op}([A, B])\}$ of open fuzzy intervals generates an open fuzzy topology $\tau_\sigma$—the *interval open fuzzy topology* of $\leq$. It can be shown that $\sigma$ itself is $\cap$-closed; since furthermore $\cap$ distributes over $\bigcup$, FCT proves that $\tau_\sigma = \{\bigcup \nu \mid \nu \subseteq \sigma\}$ (just like in the crisp interval topology).

## 4 Topology Given by a Neighborhood Relation

The following definition of fuzzy topology is an internalization in fuzzy logic of the conditions required from the system of neighborhoods.[3]

[2] Observe that it generalizes the notion of crisp open interval, by the requirement of the appropriate left- or right- continuity of the characteristic function of the interval.

[3] The first condition only determines the type of the neighborhood predicate (i.e., that it is a relation between points and classes), therefore its full validity is required.

**Definition 8.** *We define a* neighborhood $(i, j, k, l)$–fuzzy topology *by the predicate*

$$\begin{aligned}\mathsf{NTop}^{i,j,k,l}(\mathrm{Nb}) \equiv_{\mathrm{df}} &\triangle(\mathrm{Nb} \subseteq \mathrm{V} \times \mathrm{Ker\,Pow}(\mathrm{V})) \;\&\\ &((\forall x, A)(\mathrm{Nb}(x, A) \rightarrow x \in A))^i \;\&\\ &((\forall x, A, B)(\mathrm{Nb}(x, A) \;\&\; A \subseteq B \rightarrow \mathrm{Nb}(x, B)))^j \;\&\\ &((\forall x, A, B)(\mathrm{Nb}(x, A) \;\&\; \mathrm{Nb}(x, B) \rightarrow \mathrm{Nb}(x, A \cap B)))^k \;\&\\ &((\forall x, A)(\mathrm{Nb}(x, A) \rightarrow (\exists B \subseteq A)(\mathrm{Nb}(x, B) \;\&\; (\forall y \in B)\,\mathrm{Nb}(y, B)))^l\end{aligned}$$

**Definition 9.** *Let* $\triangle(\mathrm{Nb} \subseteq \mathrm{V} \times \mathrm{Ker\,Pow}(\mathrm{V}))$. *Then we define (as usual) the system of* Nb-open *classes:*

$$\tau_{\mathrm{Nb}} =_{\mathrm{df}} \{A \mid (\forall x \in A)\,\mathrm{Nb}(x, A)\}$$

It can be shown that even if Nb is a neighborhood fuzzy topology to degree one, $\tau_{\mathrm{Nb}}$ still need not be an open fuzzy topology (in particular, it need not be closed under arbitrary unions). Only the following holds:

**Theorem 8.** FCT *proves:* $\triangle\,\mathsf{NTop}(\mathrm{Nb}) \rightarrow (\forall \sigma \subseteq \tau_{\mathrm{Nb}})\,(\bigcup(\sigma \cap \sigma) \in \tau_{\mathrm{Nb}})$.

This motivates a modified notion of open fuzzy topology:

**Definition 10.** *We define the following predicates:*

$$\mathrm{U_2c}(\tau) \equiv_{\mathrm{df}} (\forall \sigma \subseteq \tau)\Big(\bigcup(\sigma \cap \sigma) \in \tau\Big)$$

$$\mathsf{O_2Top}^{e,v,i,u}(\tau) \equiv_{\mathrm{df}} (\emptyset \in \tau)^e \;\&\; (\mathrm{V} \in \tau)^v \;\&\; \mathrm{ic}^i(\tau) \;\&\; \mathrm{U_2c}^u(\tau)$$

**Theorem 9.** FCT *proves:*

$$(\exists x, A)\,\mathrm{Nb}(x, A) \;\&\; \mathsf{NTop}^{1,3,1,1}(\mathrm{Nb}) \rightarrow \mathsf{O_2Top}(\tau_{\mathrm{Nb}}) \;\&\; (\mathrm{Nb}(x, A) \leftrightarrow \mathrm{Nb}_{\tau_{\mathrm{Nb}}}(x, A))$$

Thus, a "sufficiently monotone" non-empty neighborhood topology determines a corresponding open "topology" which is closed under the operation $\bigcup(\sigma \cap \sigma)$ rather than under usual unions $\bigcup \sigma$. Such systems are met quite often in fully graded fuzzy topology:

*Example 3.* It is well-known from traditional fuzzy mathematics that the system of fuzzy sets fully extensional w.r.t. a fuzzy relation $R$ is closed under unions of arbitrary crisp systems of fuzzy sets and under min-intersections of crisp pairs of fuzzy sets (i.e., it forms a fuzzy topology in the traditional, non-graded sense of [17]). In the *graded* framework of FCT it can be proved that the *fuzzy* system of $R$-extensional classes $\{A \mid \mathrm{Ext}_R\, A\}$ is closed under $\bigcup(\sigma \cap \sigma)$ (but not under arbitrary fuzzy unions), and provided $R \subseteq R \cap R$ (which holds e.g. if $R$ is crisp), it satisfies $\mathsf{O_2Top}$.

Both $\mathsf{OTop}$ and $\mathsf{O_2Top}$ topologies are closed under crisp unions, which leads to a further generalization of the notion of open fuzzy topology:

**Definition 11.** *We define the following predicates:*

$$\mathrm{U}_{\triangle}\mathrm{c}(\tau) \equiv_{\mathrm{df}} (\forall\sigma \subseteq \tau)\big(\mathrm{Crisp}(\sigma) \to \bigcup \sigma \in \tau\big)$$
$$\mathsf{O}_{\triangle}\mathsf{Top}^{e,v,i,u}(\tau) \equiv_{\mathrm{df}} (\emptyset \in \tau)^{e} \;\&\; (\mathrm{V} \in \tau)^{v} \;\&\; \mathrm{ic}^{i}(\tau) \;\&\; \mathrm{U}_{\triangle}\mathrm{c}^{u}(\tau)$$

The models of $\mathsf{O}_{\triangle}\mathsf{Top}$ are among frequently studied fuzzy topological structures called "$L$-fuzzy topologies of Šostak-type" according to the classification proposed in [2].

The definition of the interior operator needs to be modified to have good properties in neighborhood fuzzy topologies:

$$\mathrm{Int}'_{\tau_{\mathrm{Nb}}}(A) =_{\mathrm{df}} \bigcup\{B \mid \triangle(B \in \tau_{\mathrm{Nb}} \;\&\; B \subseteq A)\}$$

**Theorem 10.** *It is provable in* FCT*:*

1. $\mathsf{NTop}^{0,1,0,0}(\mathrm{Nb}) \to \triangle(\mathrm{Int}'_{\tau_{\mathrm{Nb}}}(A) \in \tau_{\mathrm{Nb}})$
2. $A \subseteq B \to \mathrm{Int}'_{\tau_{\mathrm{Nb}}}(A) \subseteq \mathrm{Int}'_{\tau_{\mathrm{Nb}}}(B)$
3. $\triangle(A \in \tau_{\mathrm{Nb}}) \to \mathrm{Int}'_{\tau_{\mathrm{Nb}}}(A) = A$

**Theorem 11 (NTop and interior operator).** *It is provable in* FCT*:*

1. $\triangle(\mathrm{V} \in \tau_{\mathrm{Nb}}) \to \mathrm{Int}'_{\tau_{\mathrm{Nb}}}(\mathrm{V}) = \mathrm{V}$
2. $\mathrm{Int}'_{\tau_{\mathrm{Nb}}}(A) \subseteq A$
3. $\mathsf{NTop}^{0,1,0,0}(\mathrm{Nb}) \to \mathrm{Int}'_{\tau_{\mathrm{Nb}}}(\mathrm{Int}'_{\tau_{\mathrm{Nb}}}(A)) = \mathrm{Int}'_{\tau_{\mathrm{Nb}}}(A)$
4. $\mathsf{NTop}^{0,0,1,0}(\mathrm{Nb}) \to \mathrm{Int}'_{\tau_{\mathrm{Nb}}}(A) \cap \mathrm{Int}'_{\tau_{\mathrm{Nb}}}(B) \subseteq \mathrm{Int}'_{\tau_{\mathrm{Nb}}}(A \cap B)$
5. $\mathrm{Int}'_{\tau_{\mathrm{Nb}}}(A \cap B) \cap \mathrm{Int}'_{\tau_{\mathrm{Nb}}}(A \cap B) \subseteq \mathrm{Int}'_{\tau_{\mathrm{Nb}}}(A) \cap \mathrm{Int}'_{\tau_{\mathrm{Nb}}}(B)$

The following theorem guarantees that neighborhoods defined from a (sufficiently union-closed) open fuzzy topology are exactly the neighborhoods in the sense of predicate $\mathsf{NTop}$.

**Theorem 12 (OTop and NTop).** *It is provable in* FCT*:*

$$\mathsf{OTop}^{1,1,1,2}(\tau) \to \mathsf{NTop}(\mathrm{Nb}_{\tau}) \;\&\; (A \in \tau \leftrightarrow (\forall x \in A)\,\mathrm{Nb}_{\tau}(x, A))$$

*Example 4. Interval neighborhood fuzzy topology.* The (fuzzy) system of open fuzzy intervals of Example 2 allows introducing the *interval neighborhood fuzzy topology* w.r.t. a crisp dense ordering $\leq$, by taking

$$\mathrm{Nb}(x, X) \equiv_{\mathrm{df}} (\exists A, B)\,\triangle\,(\mathrm{Op}([A, B]) \;\&\; [A, B] \subseteq X \;\&\; x \in [A, B])$$

Then it can be shown that FCT proves $\triangle\,\mathsf{NTop}(\mathrm{Nb})$, and in virtue of Th. 9, $\triangle\,\mathsf{O}_2\mathsf{Top}(\tau_{\mathrm{Nb}})$ and $\mathrm{Nb} = \mathrm{Nb}_{\tau_{\mathrm{Nb}}}$. Notice, however, that the interval open topology of Example 2 differs from the interval neighborhood topology introduced here, since in the latter all classes open to degree 1 are crisp.

## 5 Conclusions

We have introduced two notions of fuzzy topology in the graded framework of Fuzzy Class Theory and investigated their basic properties; where appropriate, we gave links to similar notions of fuzzy topology studied previously in traditional fuzzy mathematics. Most of our notions generalize usual concepts of fuzzy topology by allowing full gradedness of all defined predicates and functions. Proofs of the graded theorems, though omitted here due to the limited space, are rather simple and show the strength of the apparatus of higher-order fuzzy logic in fuzzy topology. The results open a wide area of fully graded topological theory and show the possibility of the investigation of more advanced graded topological notions by means of Fuzzy Class Theory.

## References

1. Chang, C.: Fuzzy topological spaces. Journal of Mathematical Analysis and Applications **24** (1968) 182–190
2. Höhle, U., Šostak, A.: A general theory of fuzzy topological spaces. Fuzzy Sets and Systems **73** (1995) 131–149
3. Höhle, U.: Many Valued Topology and Its Applications. Kluwer, Boston (2001)
4. Ying, M.: Fuzzy topology based on residuated lattice-valued logic. Acta Mathematica Sinica (English Series) **17** (2001) 89–102
5. Hájek, P.: Metamathematics of Fuzzy Logic. Kluwer, Dordercht (1998)
6. Běhounek, L., Cintula, P.: From fuzzy logic to fuzzy mathematics: A methodological manifesto. Fuzzy Sets and Systems **157**(5) (2006) 642–646
7. Běhounek, L., Cintula, P.: Fuzzy class theory. Fuzzy Sets and Systems **154**(1) (2005) 34–55
8. Esteva, F., Godo, L.: Monoidal t-norm based logic: Towards a logic for left-continuous t-norms. Fuzzy Sets and Systems **124**(3) (2001) 271–288
9. Běhounek, L.: On the difference between traditional and formal fuzzy logic. Submitted, available at www.cs.cas.cz/hp (2007)
10. Běhounek, L., Cintula, P.: Fuzzy Class Theory: A primer v1.0. Technical Report V-939, Institute of Computer Science, Academy of Sciences of the Czech Republic, Prague (2006) Available at www.cs.cas.cz/research/library/reports_900.shtml.
11. Běhounek, L., Cintula, P.: Fuzzy class theory as foundations for fuzzy mathematics. In: Fuzzy Logic, Soft Computing and Computational Intelligence: 11th IFSA World Congress. Volume 2., Beijing, Tsinghua UP/Springer (2005) 1233–1238
12. Běhounek, L., Bodenhofer, U., Cintula, P.: Relations in Fuzzy Class Theory: Initial steps. Submitted to Fuzzy Sets and Systems (2006)
13. Běhounek, L.: Extensionality in graded properties of fuzzy relations. In: Proceedings of 11th IPMU Conference, Paris, Edition EDK (2006) 1604–1611
14. Běhounek, L., Cintula, P.: Features of mathematical theories in formal fuzzy logic. Submitted to IFSA 2007, available at www.cs.cas.cz/hp (2007)
15. Höhle, U.: Fuzzy real numbers as Dedekind cuts with respect to a multiple-valued logic. Fuzzy Sets and Systems **24**(3) (1987) 263–278
16. Běhounek, L.: Towards a formal theory of fuzzy Dedekind reals. In: Proceedings of Joint EUSFLAT–LFA Conference, Barcelona (2005) 946–954
17. Lowen, R.: Fuzzy topological spaces and fuzzy compactness. Journal of Mathematical Analysis and Applications **56** (1976) 623–633

# Features of Mathematical Theories in Formal Fuzzy Logic

Libor Běhounek* and Petr Cintula**

Institute of Computer Science, Academy of Sciences of the Czech Republic
Pod Vodárenskou věží 2, 187 02 Prague 8, Czech Republic
{behounek,cintula}@cs.cas.cz
http://www.cs.cas.cz/hp

**Abstract.** A genuine fuzzy approach to fuzzy mathematics consists in constructing axiomatic theories over suitable systems of formal fuzzy logic. The features of formal fuzzy logics (esp. the invalidity of the law of contraction) entail certain differences in form between theories axiomatized in fuzzy logic and usual theories known from classical mathematics. This paper summarizes the most important differences and presents guidelines for constructing new theories, defining new notions, and proving new theorems in formal fuzzy mathematics.

**Keywords:** Formal fuzzy logic, axiomatic theories, the law of contraction, fuzzy mathematics, graded properties.

## 1 Introduction

As argued in [1], a genuine fuzzy approach to fuzzy mathematics consists in constructing axiomatic theories over suitable systems of formal fuzzy logic. There are numerous reasons supporting this thesis, let us name just a few: under this approach, there is a strong analogy with classical mathematics; most notions are naturally graded; the connection with real-valued analysis is loosened; a consistent methodology for introducing fuzzy counterparts of crisp notions is provided; hidden crispness can easily be avoided; etc.

The features of formal fuzzy logics (esp. the invalidity in general of the contraction law, see Sect. 2) enforce a specific approach to building axiomatic theories over such logics. Some of the usual practices of classical as well as traditional fuzzy mathematics cease to be useful and need to be adjusted when working in formal fuzzy logic. Examples of such traditional practices are the placement of preconditions in definitions rather than theorems, defining compound notions as conjunctions of several conditions, etc. Furthermore, the properties of the underlying logic entail certain differences in the form as well as strength of theorems that can be proved in theories over formal fuzzy logic as compared to theorems

* The work was supported by the Program "Information Society", Project No. 1ET100300517 and by Institutional Research Plan No. AV0Z10300504.

** The work was supported by grant No. B100300502 of GA AV ČR and by Institutional Research Plan No. AV0Z10300504.

P. Melin et al. (Eds.): IFSA 2007, LNAI 4529, pp. 523–532, 2007.
© Springer-Verlag Berlin Heidelberg 2007

of traditional fuzzy mathematics. In particular, they allow studying graded notions and properties, mostly overlooked by traditional fuzzy mathematics (where usually just the predicate $\in$ is fuzzified).

This paper summarizes the most important differences between theories of classical mathematics or traditional fuzzy mathematics (e.g., [2,3,4]) on the one hand and those axiomatized in formal fuzzy logic (e.g., [5,6,7]) on the other hand, and presents guidelines for introducing new defined notions and formulating meaningful theorems in formal fuzzy mathematics.

## 2 Features of Formal Fuzzy Logics

Many systems of formal fuzzy logic emerged in the last decades. Here we survey their common features relevant to our needs.

Let us start at the propositional level. Formal fuzzy logics share the syntax of classical Boolean logic, only there are usually two different conjunctions—the *residuated* (strong) one & and the *minimum* (weak) one $\wedge$. Although there is a bunch of formal fuzzy logics described in the literature, the deductively well-behaved ones [8] contain some common basic propositional laws (axioms). The shared axioms form the logic MTL [9], the logic of left-continuous t-norms (i.e., the set of truth values is the interval $[0, 1]$, a left-continuous t-norm interprets &, and its residuum interprets implication). In order to enhance its expressive power, one usually adds one more propositional unary connective $\triangle$ with the standard semantics $\triangle x = 1$ if $x = 1$ and $\triangle x = 0$ otherwise. The logic MTL with the connective $\triangle$ will be denoted by $\mathrm{MTL}_\triangle$ further on.

It can be argued [8] that formal fuzzy logics suitable for axiomatizing mathematical theories extend the logic $\mathrm{MTL}_\triangle$; following [8], we shall call them *deductive* fuzzy logics. The most prominent examples of such logics are Łukasiewicz logic, Hájek's BL, involutive MTL, product logic, the logic ŁΠ, etc. (all of them with $\triangle$). The main distinction between classical logic and deductive fuzzy logics is the invalidity in general of the law of contraction $(\varphi \,\&\, \varphi) \leftrightarrow \varphi$ in the latter. Non-contractivity has a huge impact on the axiomatic mathematical theories over deductive fuzzy logics: see Sect. 6–8 for details.

Propositional fuzzy logic is not expressive enough to support mathematical theories; at least first-order fuzzy logic is needed for fuzzy mathematics. For a recent survey of first-order fuzzy logics see [10]; for higher-order fuzzy logics see [11,12,13]. Unless stated otherwise, our background fuzzy logic is supposed to be the first-order logic $\mathrm{MTL}_\triangle$.[1]

Axiomatic mathematical theories are given by a set of formulae, called the axioms of the theory. The theorems of a theory are proved by formal deductions

[1] First-order $\mathrm{MTL}_\triangle$ retains the completeness w.r.t. semantics based on left-continuous t-norms (although this is, in general, not the case of stronger fuzzy logics like BL or Łukasiewicz). This allows us to transfer some results of traditional fuzzy mathematics proven for all left-continuous t-norms automatically into $\mathrm{MTL}_\triangle$. However, these results are usually much weaker than those achievable directly in the axiomatic theory (see Sect. 3 and 5 for more details).

from its axioms by the deduction rules of the underlying formal fuzzy logic. For details on axiomatic theories over fuzzy logics see [14].

Further on, we shall adopt the following useful conventions for formulae of formal fuzzy logic:

**Convention 21.** *In order to save some parentheses, we assume that* $\rightarrow$ *and* $\leftrightarrow$ *have less priority than other binary connectives, and that unary connectives have the highest priority. A chain of implications* $\varphi_1 \rightarrow \varphi_2, \ldots, \varphi_{n-1} \rightarrow \varphi_n$ *will be written as* $\varphi_1 \longrightarrow \varphi_2 \longrightarrow \cdots \longrightarrow \varphi_n$*, and similarly for the equivalence connective.*

## 3 Graded Notions

Traditional fuzzy set theory fuzzifies (at least) the membership predicate: the membership degree $Ax$ of an element $x$ in a set $A$ can take intermediate values between 1 and 0. Fuzzy sets are identified with their membership functions $x \mapsto Ax$; their properties thus can be studied by means of usual methods of classical mathematics (which uses the laws of classical Boolean logic for reasoning), since membership functions are after all crisp objects of real-valued or lattice-valued analysis. Traditional properties of fuzzy sets are therefore *bivalent*: they either hold or not (e.g., a fuzzy relation either is or is not reflexive).

Only some properties of fuzzy sets are sometimes considered *graded* (i.e., with truth values in $[0, 1]$ or a lattice $L$) rather than bivalent (i.e., either true or false): most often the graded inclusion predicate $A \subseteq_{\mathrm{gr}} B$ defined as $\inf_x(Ax \Rightarrow Bx)$, where $\Rightarrow$ is a suitable fuzzy implication (compare it with the non-graded inclusion of fuzzy sets, defined by the condition $Ax \leq Bx$ for all $x$).

Formal fuzzy mathematics, on the other hand, uses formal fuzzy logic rather than classical Boolean logic for reasoning about fuzzy sets or other fuzzy notions, and therefore *all* formulae take truth values in $L$; thus *all* defined notions and *all* statements in general are graded and can be just partially true (unless they are defined as provably crisp).

Consequently, even such properties of fuzzy relations as reflexivity, which in traditional fuzzy mathematics is usually defined as bivalent (by requiring that $Rxx = 1$ for all $x$), are in formal fuzzy logic graded (defined as the truth value in $L$ of the formula $(\forall x)Rxx$, i.e., $\inf_x Rxx$). In principle, all properties in formal fuzzy logic are of a similar kind as the property of *height* of a fuzzy set, which even in traditional fuzzy mathematics naturally takes values in $L$.

Graded properties of fuzzy relations have for the first time been systematically studied in Gottwald's monograph [15], and more recently elaborated in Gottwald's [16] and Bělohlávek's [17]. Graded notions also have a long tradition in fuzzy topology, see e.g. [18]. The graded approach is important for several reasons. First, graded notions generalize the traditional (non-graded) ones, as the latter are definable (by means of $\triangle$) in terms of the former, but not vice versa. Second, graded notions are more informative—they allow inferring relevant information even when traditional notions are simply false (see Sect. 5). Third, graded notions take the idea of fuzziness seriously, as there is no reason to assume that properties of fuzzy sets should only be crisp. Moreover, graded notions can easily be handled within the framework of formal fuzzy logic, so their gradedness does not present too much additional difficulty.

## 4 Natural Fuzzification of Classical Notions

One of the main motivations of formal fuzzy logic is the generalization of classical logic to non-crisp predicates: thus it is natural to fuzzify classical mathematical notions just by re-interpreting them in a suitable formal fuzzy logic. This methodology has been foreshadowed already in Höhle's 1987 paper [19, §5]:

> "It is the opinion of the author that from a mathematical viewpoint the important feature of fuzzy set theory is the replacement of the two-valued logic by a multiple-valued logic. [...I]t is now clear how we can find for every mathematical notion its 'fuzzy counterpart'. Since every mathematical notion can be written as a formula in a formal language, we have only to internalize, i.e. to interpret these expressions by the given multiple-valued logic."

Much later the principle was formalized in [11, §7], and proposed as an important guideline for formal fuzzy mathematics in [1].

Nevertheless, although an important guideline, the method cannot be applied mechanically, as some classically equivalent definitions may no longer be equivalent in the (weaker than classical) fuzzy logic. In some cases, one can select the most suitable version of the definition, by the criteria of fruitfulness, applicability, and the practice of traditional fuzzy mathematics. In other cases, a notion of classical mathematics splits into several meaningful fuzzy notions. This can be exemplified by the notion of equality of fuzzy sets. Besides the crisp identity = of fuzzy sets, at least two graded notions of natural fuzzy equality are defined and used in the literature (e.g., the first one is used in [17] and the second one in [16]):

$$A \approx B \equiv_{\mathrm{df}} (\forall x)[(Ax \to Bx) \mathbin{\&} (Bx \to Ax)] \tag{1}$$

$$A \approxeq B \equiv_{\mathrm{df}} (\forall x)(Ax \to Bx) \mathbin{\&} (\forall x)(Bx \to Ax) \tag{2}$$

These notions are not equivalent (even in rather strong fuzzy logics, e.g., Łukasiewicz), as shown by the following counter-example:

*Example 1.* Let $A, B$ be interpreted in a model over the standard MV-algebra (see [14]) by the following assignment of truth values: $Aa = Bb = 1$ and $Ab = Ba = 0.5$ for some individuals $a$ and $b$, and $Ax = Bx = 0$ otherwise. Then the truth value of $A \approx B$ is 0.5, while the truth value of $A \approxeq B$ is 0.

Notice that in traditional fuzzy mathematics these two notions of graded equality coincide, since $\triangle(A \approx B) \longleftrightarrow \triangle(A \approxeq B) \longleftrightarrow A = B$ (see Prop. 2 in Sect. 8).

## 5 Theorems in the Form of Provable Implications

The general gradedness of all notions in formal fuzzy logic allows proving more general theorems that are not available for non-graded notions in traditional fuzzy mathematics. A typical non-graded theorem of traditional fuzzy mathematics has the following form:

> *If* some (non-graded) assumption is true (i.e., fully true),
> *then* some (non-graded) conclusion is (fully) true.

With graded notions we can formulate and prove much stronger theorems of the following form:

> *The more* some (graded) assumption is true (even if partially),
> *the more* some (graded) conclusion is true (i.e., at least as true as the assumption).

The latter can be expressed in formal fuzzy logic by means of implication $\varphi \rightarrow \psi$, where $\varphi$ is the formula which expresses the assumption and $\psi$ is the formula which expresses the conclusion. In deductive fuzzy logics, if $\varphi \rightarrow \psi$ is provable, then the truth value of $\psi$ is at least as large as the truth value of $\varphi$ in any model. Provable implications thus express exactly the graded theorems of the above form. Since the full truth of $\chi$ is expressed by $\triangle\chi$, the former non-graded theorem of traditional fuzzy mathematics is expressed by the formula $\triangle\varphi \rightarrow \triangle\psi$. The graded theorem $\varphi \rightarrow \psi$ is generally stronger than the non-graded theorem $\triangle\varphi \rightarrow \triangle\psi$, since the latter is an immediate consequence of the former in $\mathrm{MTL}_\triangle$, but not vice versa.

*Example 2.* If we set $Ixy = \begin{cases} 1 \text{ if } x = y \\ 0 \text{ otherwise,} \end{cases}$ then:

- Traditional fuzzy mathematics proves that if a fuzzy relation $R$ is reflexive (in the traditional sense), then $I$ is a fuzzy subset of $R$; i.e., if $Rxx = 1$ for each $x$, then $Ixy \leq Rxy$ for each $x, y$.
- In formal fuzzy logic we can easily prove that the more a fuzzy relation $R$ is reflexive (in the graded sense), the more $I$ is a fuzzy subset of $R$; in symbols, $(\forall x)Rxx \rightarrow (\forall xy)(Ixy \rightarrow Rxy)$. Thus for any left-continuous t-norm $T$ we get $\inf_x Rxx \leq \inf_{x,y} \overrightarrow{T}(Ixy, Rxy)$.

Notice that the latter result is indeed more general than the former one: if $R$ is 0.999-reflexive, the traditional theorem asserts nothing (as $R$ is *not* reflexive in the traditional sense), while the graded theorem ensures that $I$ is a fuzzy subset of $R$ at least to degree 0.999. (Much more complex examples of this kind can be found in [5].)

By the above considerations, it is preferable to prove theorems in the form of implication $\varphi \rightarrow \psi$, rather than traditional non-graded theorems, which in formal fuzzy logic can be formalized as $\triangle\varphi \rightarrow \triangle\psi$.

## 6 Exponents

As stressed in Sect. 2, the law of contraction $(\varphi \,\&\, \varphi) \leftrightarrow \varphi$ is not generally valid in deductive fuzzy logics. Therefore, repeated occurrences of a premise $\varphi_i$ in a theorem of the form

$$\varphi_1 \,\&\, \ldots \,\&\, \varphi_n \rightarrow \psi \tag{3}$$

cannot be contracted into a single occurrence, as usual in classical mathematics. For convenience, the $k$ occurrences of $\varphi_i$ in (3) can be written as $\varphi_i^k$. Thus a typical form of a graded theorem is actually

$$\varphi_1^{k_1} \mathbin{\&} \ldots \mathbin{\&} \varphi_n^{k_n} \to \psi \tag{4}$$

Semantically, since the truth value of $\varphi \mathbin{\&} \varphi$ is in general smaller than $\varphi$ in usual fuzzy logics, the larger the exponent $k_i$ in (4) is, the truer $\varphi_i$ must be to ensure a large truth degree of the conclusion $\psi$. In other words, the conclusion of a theorem depends more on the truth degree of the premises with larger exponents than on those with smaller exponents.

Syntactically, the exponent $k_i$ in a theorem of the form (4) expresses how many times the premise $\varphi_i$ was used in an MTL-proof of $\psi$. This can be seen from the proof of the Local Deduction Theorem for propositional MTL (see [20]), or from the following proposition which justifies proving a conjunction by proving the conjuncts separately:

**Proposition 1.** *Propositional fuzzy logic* MTL *proves (see [9]):*

$$[(\varphi_1 \to \psi_1) \mathbin{\&} (\varphi_2 \to \psi_2)] \to [(\varphi_1 \mathbin{\&} \varphi_2) \to (\psi_1 \mathbin{\&} \psi_2)] \tag{5}$$

$$[(\varphi \to \psi_1) \mathbin{\&} (\varphi \to \psi_2)] \to [\varphi \to (\psi_1 \wedge \psi_2)] \tag{6}$$

Thus if we can prove

$$\varphi_1^{k_1} \mathbin{\&} \ldots \mathbin{\&} \varphi_n^{k_n} \to \psi_1 \quad \text{and} \quad \varphi_1^{l_1} \mathbin{\&} \ldots \mathbin{\&} \varphi_n^{l_n} \to \psi_2 \tag{7}$$

then we also have

$$\varphi_1^{k_1+l_1} \mathbin{\&} \ldots \mathbin{\&} \varphi_n^{k_n+l_n} \to \psi_1 \mathbin{\&} \psi_2 \tag{8}$$

$$\varphi_1^{\max(k_1,l_1)} \mathbin{\&} \ldots \mathbin{\&} \varphi_n^{\max(k_n,l_n)} \to \psi_1 \wedge \psi_2 \tag{9}$$

Notice two different ways of "counting the premises" based on whether we prove conjunction or min-conjunction of conclusions.

Since $\varphi^k \to \psi$ is weaker for larger $k$, one should actually find a counterexample against $\varphi^{k-1} \to \psi$ whenever one proves a theorem of the form $\varphi^k \to \psi$, to show that it cannot be strengthened. This may, however, be quite difficult for more complex theorems. Also if the exponents in a theorem grow too large, it may in some cases be preferable to weaken the theorem and use $\triangle\varphi$ as a premise instead of $\varphi^k$ (for $k \gg 0$).

## 7 Preconditions and Compound Notions

The fact that assumptions get variable exponents in theorems leads to two important guidelines for defining new notions in formal fuzzy mathematics.

In classical mathematics, definitions often have preconditions under which the defined notions are meaningful (e.g., "let $R$ be an ordering"). In formal fuzzy mathematics, such preconditions are in general fuzzy (the notion of ordering is graded). In proofs of graded theorems, such preconditions will be used various numbers of times, and so they will get various exponents.

A notion defined with a fuzzy precondition is therefore of little interest, since only such graded properties are provable about the notion that use the precondition at most *once*; more complex properties will need the precondition several times. Thus it is better to state the definition of the notion without the precondition, and add the precondition with the required multiplicity in a theorem of the form (4). Only such preconditions $\varphi$ can meaningfully be required in definitions that provably satisfy $(\varphi \& \varphi) \leftrightarrow \varphi$ and so they do not acquire differing exponents in theorems. (In particular, *crisp* preconditions satisfy the latter and therefore can meaningfully be used in definitions.)

A similar effect of variable exponents can be seen in notions defined as conjunctions of two or more conditions. We exemplify the effect on the notions of fuzzy preordering and similarity.

*Example 3.* In traditional fuzzy mathematics we say that a fuzzy relation $R$ is a preordering iff $R$ is reflexive and transitive (where $R$ is transitive iff $Rxy * Ryz \leq Rxz$ for all $x, y, z$ and reflexive iff $Rxx = 1$ for all $x$); it is a similarity iff it is reflexive, transitive, and symmetric (where $R$ is symmetric iff $Rxy \leq Ryx$ for all $x, y$). In formal fuzzy logic, *graded* reflexivity, symmetry, and transitivity are defined by the following formulae:

$$\mathrm{Refl}\, R \equiv_{\mathrm{df}} (\forall x) Rxx \tag{10}$$

$$\mathrm{Sym}\, R \equiv_{\mathrm{df}} (\forall xy)(Rxy \to Ryx) \tag{11}$$

$$\mathrm{Trans}\, R \equiv_{\mathrm{df}} (\forall xyz)(Rxy \,\&\, Ryz \to Rxz) \tag{12}$$

The traditional notions of preordering and similarity are then expressed by the formulae $\triangle\, \mathrm{Refl}\, R \,\&\, \triangle\, \mathrm{Trans}\, R$ and $\triangle\, \mathrm{Refl}\, R \,\&\, \triangle\, \mathrm{Sym}\, R \,\&\, \triangle\, \mathrm{Trans}\, R$, respectively. The definition of *graded* preordering or similarity first needs to distinguish which conjunction is used between the conjuncts $\mathrm{Refl}\, R, \mathrm{Sym}\, R, \mathrm{Trans}\, R$. (Notice that in the traditional definition it is immaterial which one is used, since both conjunctions are 1-true under the same conditions.) The default choice is the strong conjunction &, since it allows using *all* conjuncts in proofs, while $\wedge$ only allows using any *one* of them (see [13]). Nevertheless, the definitions

$$\mathrm{Preord}\, R \equiv_{\mathrm{df}} \mathrm{Refl}\, R \,\&\, \mathrm{Trans}\, R \tag{13}$$

$$\mathrm{Sim}\, R \equiv_{\mathrm{df}} \mathrm{Refl}\, R \,\&\, \mathrm{Sym}\, R \,\&\, \mathrm{Trans}\, R \tag{14}$$

still allow using each of the conjuncts just *once* in the proofs. However (cf. Sect. 6), the assumptions $\mathrm{Refl}\, R$, $\mathrm{Sym}\, R$, or $\mathrm{Trans}\, R$ are needed variable times in proofs of various theorems, and thus get variable exponents, independent of each other. Thus, rather than defining preorders and similarities by (13)–(14), it is more meaningful to define parameterized notions of $(r, t)$-preorders and $(r, s, t)$-similarities as follows:

$$\mathrm{Preord}^{r,t}\, R \equiv_{\mathrm{df}} \mathrm{Refl}^{r}\, R \,\&\, \mathrm{Trans}^{t}\, R \tag{15}$$

$$\mathrm{Sim}^{r,s,t}\, R \equiv_{\mathrm{df}} \mathrm{Refl}^{r}\, R \,\&\, \mathrm{Sym}^{s}\, R \,\&\, \mathrm{Trans}^{t}\, R \tag{16}$$

Typical graded theorems on fuzzy preorders or similarities then have the form $\mathrm{Preord}^{r,t}\, R \to \varphi$ resp. $\mathrm{Sim}^{r,s,t}\, R \to \varphi$ (for some $r, s, t$), and thus they are actually theorems on $(r,t)$-preorders and $(r,s,t)$-similarities. Recall from Sect. 6 that the parameters measure the strictness of requiring a large truth value of the respective conjunct; thus $(2,5)$-preorders are more sensitive to imperfections in transitivity than in reflexivity, while $(10,1)$-preorders are much more sensitive to flaws in reflexivity than transitivity.

## 8 Equivalences and Bounds

Many theorems of traditional fuzzy logic have the form of equivalence between two conditions, which in formal fuzzy logic is expressed by a formula of the form $\triangle\varphi \leftrightarrow \triangle\psi$. The graded version of such a theorem, $\varphi \leftrightarrow \psi$, is sometimes provable in formal fuzzy logic; if so, it expresses the fact that the truth degree of $\varphi$ equals the truth degree of $\psi$. (Observe that again the traditional non-graded version of the theorem, which expresses only the fact that $\varphi$ is 1-true iff $\psi$ is 1-true, follows immediately from the graded version.)

Often, however, the graded version of a theorem $\triangle\varphi \leftrightarrow \triangle\psi$ is more complicated than the simple equivalence $\varphi \leftrightarrow \psi$. It can be exemplified by the relationship between the two notions of graded equality (1)–(2) (for a proof, see [5]):

**Proposition 2.** *The following theorems are provable in first order* MTL*:*

1. $A \approx^2 B \longrightarrow A \approxeq B \longrightarrow A \approx B$
2. $\triangle(A \approx B) \longleftrightarrow \triangle(A \approxeq B) \longleftrightarrow A = B$

Observe that the first statement says that the truth value of $A \approxeq B$ is bounded by the truth values of $A \approx^2 B$ (a lower bound) and $A \approx B$ (an upper bound). In traditional non-graded fuzzy mathematics both notions coincide, since they are 1-true under the same conditions, as shown by the second statement of Prop. 2.

The situation that a theorem $\triangle\varphi \leftrightarrow \triangle\psi$ has a graded version of the form $\varphi^n \longrightarrow \psi^m \longrightarrow \varphi^k$ for some $n \geq m \geq k$ occurs regularly under some conditions:

**Theorem 1.** *Let $\varphi$ and $\psi$ be formulae of the first-order logic* MTL *(i.e., they contain no $\triangle$) such that $\triangle\varphi \leftrightarrow \triangle\psi$ is provable in a theory $T$ over (first-order)* $\mathrm{MTL}_\triangle$. *Then there exist $n, m$ such that $\varphi^n \to \psi$ and $\psi^m \to \varphi$ are provable in $T$.*

*Proof.* Follows directly from the $\triangle$–Deduction Theorem and Local Deduction Theorem for the first-order logic $\mathrm{MTL}_\triangle$ resp. MTL (see [10]). If $\varphi, \psi$ are not closed formulae (to which the Deduction Theorems apply), first replace free variables by new constant symbols, which is harmless for provability in $T$. □

**Corollary 1.** *Under the conditions of Th. 1, we get the following mutual estimates for the truth degrees of $\varphi$ and $\psi$ (for $m, n$ from Th. 1):*

$$\varphi^{m \cdot n} \longrightarrow \psi^m \longrightarrow \varphi \tag{17}$$

$$\psi^{m \cdot n} \longrightarrow \varphi^n \longrightarrow \psi \tag{18}$$

It is worth noting that graded theorems of this form have occurred in the traditional fuzzy literature, see e.g. [21, L.16].

We conclude this section by an illustrative example which can be viewed as a graded generalization of (a certain variant of) the well-known Valverde representation theorem for preorders (see [22] for its non-graded version).

**Proposition 3.** [23] *The following graded characterizations are provable in first-order* MTL*:*

$$\mathrm{Refl}\, R \leftrightarrow (\forall xy)[(\forall z)(Rzx \to Rzy) \to Rxy] \tag{19}$$

$$\mathrm{Trans}\, R \leftrightarrow (\forall xy)[Rxy \to (\forall z)(Rzx \to Rzy)] \tag{20}$$

Recall from [9] that the following implications are provable in first-order MTL:

$$((\forall u)(\psi \,\&\, \chi))^2 \longrightarrow (\forall u)\psi \,\&\, (\forall u)\chi \longrightarrow (\forall u)(\psi \,\&\, \chi), \tag{21}$$

and it cannot be improved as the converse implication $(\forall u)(\psi \,\&\, \chi) \to (\forall u)\psi \,\&\, (\forall u)\chi$ does not generally hold in fuzzy logics. As $\mathrm{Preord}^{r,t}\, R \equiv_{\mathrm{df}} \mathrm{Refl}^r\, R \,\&\, \mathrm{Trans}^t\, R$, we obtain just the following graded variant of Valverde representation:

**Corollary 2.** [5] *Define* $\varphi(R)$ *as* $(\forall xy)[Rxy \leftrightarrow (\forall z)(Rzx \to Rzy)]$*. Then*

$$\varphi^2(R) \longrightarrow \mathrm{Preord}^{1,1}\, R \longrightarrow \varphi(R), \tag{22}$$

*i.e.,* $\varphi^2(R)$ *and* $\varphi(R)$ *give respectively the lower and upper bounds for the truth value of* $\mathrm{Preord}^{1,1}\, R$*. Considering only* 1*-truth of both conditions, we get a non-graded characterization* $\triangle\, \mathrm{Preord}\, R \leftrightarrow \triangle\, \varphi(R)$*.*

## 9 Conclusion

As can be seen from the previous sections, in that part of fuzzy mathematics that can be formalized in formal fuzzy logic the apparatus of the latter allows deriving more general (graded) theorems than traditional methods. In order to utilize the strength of the apparatus to the full extent, however, the guidelines sketched in this paper have to be observed, namely:

- Defining new notions graded (§3), by formulae analogical to definitions in classical mathematics (§4); parameterizing definitions of compound notions by (variable) exponents and giving preconditions with variable exponents in theorems rather than definitions (§7)
- Proving theorems in the form of fuzzy implication (§5) rather than crisp consequence of fully true premises, using the laws of formal fuzzy logic (§2) and counting the exponents of premises properly (§6)

This leads to stronger, even though sometimes more complicated (§8) theorems than traditional methods. Failing to respect these unusual features when building graded fuzzy theories would unnecessarily weaken the theorems obtained.

## References


1. Běhounek, L., Cintula, P.: From fuzzy logic to fuzzy mathematics: A methodological manifesto. Fuzzy Sets and Systems **157**(5) (2006) 642–646
2. Dubois, D., Prade, H.: Fuzzy Sets and Systems: Theory and Applications. Academic Press, New York (1980)
3. Dubois, D., Prade, H., eds.: Fundamentals of Fuzzy Sets. Volume 7 of The Handbooks of Fuzzy Sets. Kluwer Academic Publishers, Boston (2000)
4. Klir, G.J., Yuan, B.: Fuzzy Sets and Fuzzy Logic: Theory and Applications. Prentice Hall (1995)
5. Běhounek, L., Bodenhofer, U., Cintula, P.: Relations in Fuzzy Class Theory: Initial steps. Submitted to Fuzzy Sets and Systems (2006)
6. Kroupa, T.: Filters in fuzzy class theory. Submitted (2006)
7. Běhounek, L.: Extensionality in graded properties of fuzzy relations. In: Proceedings of 11th IPMU Conference, Paris, Edition EDK (2006) 1604–1611
8. Běhounek, L.: On the difference between traditional and formal fuzzy logic. Submitted (2007)
9. Esteva, F., Godo, L.: Monoidal t-norm based logic: Towards a logic for left-continuous t-norms. Fuzzy Sets and Systems **124**(3) (2001) 271–288
10. Hájek, P., Cintula, P.: Triangular norm predicate fuzzy logics. To appear in Proceedings of Linz Seminar 2005 (2006)
11. Běhounek, L., Cintula, P.: Fuzzy class theory. Fuzzy Sets and Systems **154**(1) (2005) 34–55
12. Novák, V.: On fuzzy type theory. Fuzzy Sets and Systems **149**(2) (2004) 235–273
13. Běhounek, L., Cintula, P.: Fuzzy Class Theory: A primer v1.0. Technical Report V-939, Institute of Computer Science, Academy of Sciences of the Czech Republic, Prague (2006) Available at www.cs.cas.cz/research/library/reports_900.shtml.
14. Hájek, P.: Metamathematics of Fuzzy Logic. Volume 4 of Trends in Logic. Kluwer, Dordercht (1998)
15. Gottwald, S.: Fuzzy Sets and Fuzzy Logic: Foundations of Application—from a Mathematical Point of View. Vieweg, Wiesbaden (1993)
16. Gottwald, S.: A Treatise on Many-Valued Logics. Volume 9 of Studies in Logic and Computation. Research Studies Press, Baldock (2001)
17. Bělohlávek, R.: Fuzzy Relational Systems: Foundations and Principles. Volume 20 of IFSR International Series on Systems Science and Engineering. Kluwer Academic/Plenum Press, New York (2002)
18. Höhle, U.: Many Valued Topology and Its Applications. Kluwer, Boston, MA (2001)
19. Höhle, U.: Fuzzy real numbers as Dedekind cuts with respect to a multiple-valued logic. Fuzzy Sets and Systems **24**(3) (1987) 263–278
20. Cintula, P.: Weakly implicative (fuzzy) logics I: Basic properties. Archive for Mathematical Logic **45**(6) (2006) 673–704
21. Ying, M.: Fuzzy topology based on residuated lattice-valued logic. Acta Mathematica Sinica (English Series) **17** (2001) 89–102
22. Valverde, L.: On the structure of $F$-indistinguishability operators. Fuzzy Sets and Systems **17**(3) (1985) 313–328
23. Bandler, W., Kohout, L.J.: On the universality of the triangle superproduct and the square product of relations. International Journal of General Systems **25** (1997) 399–403

# A New Method to Compare Dynamical Systems

Juan Moreno-Garcia[1], Jose Jesus Castro-Schez[2], and Luis Jimenez[2]

[1] E. U. de Ingenieria Tecnica Industrial
Universidad de Castilla-La Mancha
Avda. Carlos III s/n, Toledo, Spain
Juan.Moreno@uclm.es
[2] Escuela Superior de Informatica
Universidad de Castilla-La Mancha
Paseo de la Univ. 4, Ciudad Real, Spain
JoseJesus.Castro@uclm.es, Luis.Jimenez@uclm.es

**Abstract.** This work presents a new method to compare two Temporal Fuzzy Chains (TFCs). The TFCs comparison is used to compare the temporal evolution of two dynamical systems (TFCs). Firstly, the two TFCs that represent the two DSs are obtained. After that, the two obtained TFCs are compared, so the *similarity* of the two TFCs is obtained. This *"similarity"* is done by using linguistic labels.

## 1 Introduction

*Dynamical Systems* [1], from now on *DS*, are *systems that changes over time*, understanding *system* as a set of elements with some relations between them. We have worked with MISO DSs (Multiple In, Simple Out). The values of the system variables at a time $t_i$ depend on the values of variables in times $t_{i-1}, \ldots, t_1$. DSs verify the *continuity* feature, i.e., the variable evolution is continued in time, at a time $t_{i+1}$ the variable value $v_{t_{i+1}}$ is *similar* to the variable value $v_{t_i}$ at a time $t_i$. We suppose this hypothesis when we define the TFCs [2].

DSs are modelled by using different traditional techniques [1]. They are modelled to explain and to predict its behavior, and to analyze its dynamical properties. There are some previous works that propose the use of the fuzzy logic to model DSs, for example [3,4,5]. TFCs are used to represent the evolution of the DS. TFCs simulate the DS behavior by using the input variables. In this work, we use the TFCs to compare the *dynamism* between to DSs. For this purpose, the TFC of the two DSs are induced, and later on, the two obtained models are compared to obtain their differences. The aim of this work is to present a method to compare TFCs that represent two DSs. We use linguistic labels to show the similarity between TFCs during consecutive time intervals all through the time. This comparison is more comprehensible by using linguistic labels.

The remainder of the paper is organized as follows. Section 2 presents an introduction to TFCs. Afterwards, the proposed method is shown. A real example of the behavior of the algorithm is explained (Section 4). Finally, the obtained conclusions are given.

P. Melin et al. (Eds.): IFSA 2007, LNAI 4529, pp. 533–542, 2007.
© Springer-Verlag Berlin Heidelberg 2007

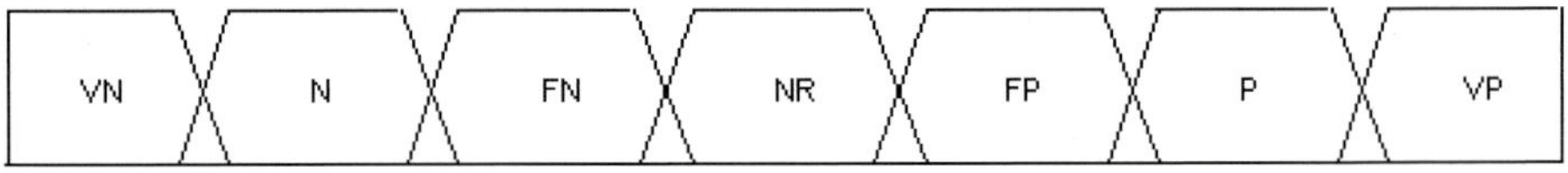


**Fig. 1.** Ordered set of labels $SA_j$

## 2 An Introduction to the TFCs

### 2.1 Some Previous Concepts

A time series is a sequence of data that is ordered in the time. A DS could be represented by using a time series $E$. Each example $E = \{e_1 \dots e_n\}$ of the time series contains the variables values in a time, being $e_i = (x_1^i \dots x_m^i, s^i, t_i)$, where $x_j^i \in X_j$, $s^i \in S$ and $t_i$ is the time when $e_i$ occurs. An ordered set of labels is defined for each input variable $X_j$ (Figure 1). Its structure is $SA_j = \{SA_j^1 \dots SA_j^{i_j}\}$. An ordered set of labels $SC = \{SC^1 \dots SC^{i_y}\}$ is defined for the output variable. The superindex $i$ is the position of the label in these sets, $i_j = |SA_j|$ and $i_y = |SC|$. The order of the labels is based on the defuzzification method named *mean of maximum* (MOM) [6]. These sets of labels are defined a priori.

TFCs work with linguistic variables, named *continuous linguistic variables* (from now on variables), with some restrictions over its domain. These variables can take values from an ordered set of labels. Linguistic labels of the variables, or simply labels, are defined before the TFC is obtained. $X_j$ is a continuous linguistic variable if it has associated a semantic to each $SA_j^i$ defined by itself that verifies: (1) Each label $SA_j^i$ is defined over an ordered domain; (2) The labels defined on $X_j$ are sorted by means of the MOM [6]; (3) The sum of the membership grade of a value $x$ to all labels defined in a continuous variable must be 1, $\sum_{SA_j^i \in SA_j} \mu_{SA_j^i}(x) = 1$.

These variables take a linguistic interval as value. *A linguistic interval* $LI_{j,p}^c$, from now on interval, is a sequence of $c$ consecutive labels defined on $SA_j$ that begins in the label $p$. It is represented as $LI_{j,p}^c = \{SA_j^p \dots SA_j^{p+(c-1)}\}$, where $SA_j = \{SA_j^1 \dots SA_j^{i_j}\}$ is an ordered set of $i_j$ labels for the input variable $X_j$, $p$ is the position in $SA_j$ of the first label of the interval and $c$ is the labels number of the interval. Its membership function is the sum of the membership grade of $a_j$ to each label of the interval (Equation 1 with $z \in [p..c-1]$).

$$\mu_{LI_{j,p}^c}(a_j) = \sum_{SA_j^z \epsilon LI_{j,p}^c} \mu_{SA_j^z}(a_j) \tag{1}$$

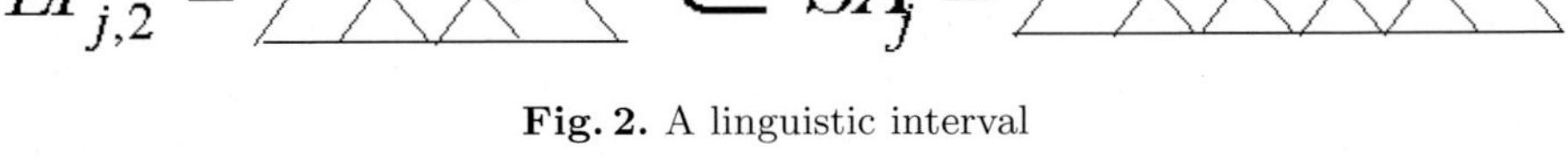


**Fig. 2.** A linguistic interval

An ordered set of $m$ labels is a sequence of $m$ intervals defined over $m$ variables, and it is represented as $SLI_m = \{LI^{c_1}_{1,p_1}, \ldots LI^{c_m}_{m,p_m}\}$. The membership grade of a $SLI_m$ is calculated applying a t-norm between the membership grade of $SLI_m$ intervals (Equation 2).

$$\mu_{SLI_m}(e_i) = *(\mu_{LI^{c_j}_{j,p_j}}(x^i_j)) \tag{2}$$

where $e_i = (x^i_1 \ldots x^i_m, s^i, t_i)$ is an example of $E$, $j\epsilon[1..m]$ and $*$ is a t-norm.

## 2.2 An Introduction to the TFCs

We suggest embodying the DS temporal side by making use of the TFCs. Informally, a TFC is a collection of linguistic states related to each other by means of linguistic transitions. A linguistic state represents a set of consecutive examples in time with a similar behavior, that is, the output variables take the same value for all of them. A linguistic transition is used to relate two consecutive linguistic states: it establishes the necessary conditions that must take place in the system inputs to cause a change in the system output behavior for it to enter into the next linguistic state. The change of state is described in a linguistic way by using fuzzy linguistic values.

A **linguistic state** $i$, or simply state, is defined a tuple $sta_i =< A^i_m, SE_i >$ where $A^i_m$ is the $SLI_m$ of the state $i$ corresponding to the $m$ input variables, and $SE_i$ is the output of the state $i$ corresponding to the output variable. A **linguistic transition** $i$, or simply transition, is a tuple $trans_i =< T^i_m, ST_i >$ where $T^i_m$ is the $SLI_m$ of the transition $i$, and $ST_i$ is the output label of the transition $i$. A **TFC** is a tuple $CHAIN =< STA, TRANS >$ where $STA = \{sta_1 \ldots sta_{ns}\}$ is an ordered set of $ns$ states, and $TRANS = \{trans_1 \ldots trans_{ns-1}\}$ is an ordered set of $ns - 1$ transitions. The transition $i$ represents the conditions to change from $sta_i$ to $sta_{i+1}$.

**Algorithm 1.** TFCs inference process

```
cur ← 1
for i = 1 to |E| do
  if μ_{A_m^cur}(e_i) > μ_{T_m^cur}(e_i) then
    s ← SE_cur
  else
    s ← ST_cur
    cur ← cur + 1
  end if
end for
```

To reproduce the behavior of the DS the TFCs inference method is offered (Algorithm 1). In this algorithm, $cur$ indicates the current state and transition, $E$ is the set of examples, $e_i$ is the example $i$ of $E$ and $s$ is the output label. The algorithm uses a set of examples $E$ as input, and it is based in the definition of the *current state* $sta_{cur}$. $sta_{cur}$ indicates the state in that the DS is in, and it is used to calculate the output for this time. The inference method begins by selecting $sta_1$ as the first current state. After that, it is calculated $\mu_{A_m^{cur}}(e_i)$ and $\mu_{T_m^{cur}}(e_i)$, if $\mu_{A_m^{cur}}(e_i)$ is greater than $\mu_{T_m^{cur}}(e_i)$ then the obtained output is the label output label $SE_{cur}$ (corresponding to $sta_{cur}$), and there is not change of state. In other case, if $\mu_{A_m^{cur}}(e_i)$ is less or equal than $\mu_{T_m^{cur}}(e_i)$, then there is a change of state, the obtained output label is $ST_{cur}$ (corresponding to $trans_{cur}$), and the new current state is the next state to $sta_{cur}$, i.e., $sta_{cur+1}$. This process is repeated for each example in $E$.

**Algorithm 2.** Proposed method

**1.** Defining the phases.
**2.** Generating the TFCs.
**3.** Comparing the TFCs.

## 3 Proposed Method

Algorithm 2 describes the proposed method. As you can see, it consists in three phases. In the first phase, the phases of the studied DS are defined. This is done because the human movements are our usual research application line (human walk, shot putting,...), and the biomechanics study of the movement is made by means of phases. For this reason, the presented algorithm defines a set of phases to study the DS. This is not a limitation of the algorithm, if anybody does not want an study by phases, it must define one single phase.

The two TFCs of the two studied DSs are induced in the next phase of the algorithm. To do that, it is used the algorithm presented in [2]. We must to add a temporal component to indicate the state duration. This suppose a change in the original algorithm (presented in [2]), to add the calculation of the duration of all the states that have been united to obtain the final state in the second phase of the algorithm [2]. To do that, it is used the temporal component $t_i$ of each example $e_i$ used to obtain the state.

Finally, in the last phase of the algorithm, the comparison of the DSs is made by using the Algorithm 3.

The main idea of Algorithm 3 consists of comparing twoTFCs throughout the time. The two TFCs begins in its first state, and it is compared the present state of both TFCs for every time. The current state of each TFC is changed to the following state when the comparison time, denoted by $t$ in the algorithm, is less than the duration of the current state added to the all previous states duration (this is controlled by using the $t_0$ and $t_1$ for CDT0 and CDT1 respecti-vely). The comparison time $t$ is initialized to 0 and it is incremented with the constant *inc*. The indexes $i_0$ and $i_1$ are used to control the current states of $TFC_0$ and $TFC_1$. The list named *phases* is used to handle the phases in the algorithm, for example, if $phases = [0, 0.6, 1.0, 1.2]$ then, the DS has three phases, the first one begins in the time 0 and it finishes in 0.6, the second phases begins in 0.6 and finished in 1.0, and the last one stays from 1.0 to 1.2.

**Algorithm 3.** Comparison of TFCs

$inc \leftarrow CTE$ {CTE indicates the increment of the time}
$phases \leftarrow [0 \ldots t_{max}]$ {*phases* contains the defined phases. It is a lists with the principle and the end values of each one of the phases. It begins in 0 and it finishes with the maximum possible value of the variable time.}
$s \leftarrow 1$ {$s$ indicates the phase by which the algorithm goes}
$t \leftarrow phases[0]$ {$t$ indicates the comparison time}
$i_0 \leftarrow 1$ {$i_0$ points at the current state in which the CDT0 is being compared}
$t_0 \leftarrow$ Duration of the first state of CDT0 {$t_0$ takes the value of the duration of the first state of CDT0}
$i_1 \leftarrow 1$ {$i_1$ points at the current state in which the CDT1 is being compared}
$t_1 \leftarrow$ Duration of the first state of CDT1 {$t_1$ takes the value of the duration of the first state of CDT1}
$P_{phase} \leftarrow \emptyset$ {$P_{phase}$ is a list containing the values of the similarity of the current phase}
$P_{total} \leftarrow \emptyset$ {$P_{total}$ is a list of list. Each list will contain the similarity of each one of the phases. If there are $n$ phases, will be composed by $n$ lists}
**while** $t < t_{max}$ **do**
  {It controls if it has been arrived at a phase end, and changes to the following one}
  **if** $t \geq phases(s)$ **then**
    $P_{total} \leftarrow P_{total} + P_{phase}$ {Add the lists $P_{phase}$ to the list of lists $P_{total}$}
    $s \leftarrow s + 1$ {Make the change of phase}
    $P_{phase} \leftarrow \emptyset$ {the list $P_{phase}$ it is initialized to $\emptyset$ due to the phase change}
  **end if**
  {It controls if a change of state is due to do in CDT0}
  **if** $t \geq t_0$ **then**
    $i_0 \leftarrow i_0 + 1$ {The current state is the following}
    $t_0 \leftarrow t_0 +$ State duration $i_0$ of CDT0 {The duration of the state is incremented}
  **end if**
  {It controls if a change of state is due to do in CDT1}
  **if** $t \geq t_1$ **then**
    $i_1 \leftarrow i_1 + 1$ {The current state is the following}
    $t_1 \leftarrow t_1 +$ State duration $i_1$ of CDT1
  **end if**
  $p \leftarrow Similarity(TFC_0(i_0), TFC_1(i_1))$ {It calculates the similarity of the current states of CDT0 and CDT1}
  $P_{phase} \leftarrow P_{phase} + p$ {Adds the obtained similarity to $P_{phase}$}
  $t \leftarrow t + inc$ {Increments the time with $inc$}
**end while**
$P_{total} \leftarrow P_{total} + P_{phase}$

This first part of the algorithm obtains a list called $P_{total}$ formed by lists containing the numerical values of the similarity for each one of the compared times in each one of the phases, that is, a list of $P_{total}$ has the similarities of each compared time for a phase. For example, if $phases = [0, 0.6, 1.0, 1.2]$ and $inc = 0.05$, then $P_{total}$ will be formed by 3 lists, the first one contains 12 elements (0.6/0.05), the second one has 8 elements (0.4/0.05) and the last one contains 4 elements (0.2/0.05), i.e., $P_{total} = [\ [p_0, p_{0.05}, p_{0.1} \cdots p_{0.6}],\ [p_{0.65}, p_{0.7}, p_{0.75} \cdots p_{1.0}],\ [p_{1.05}, p_{1.1}, p_{1.1}, p_{1.15}, p_{1.2}]\ ]$ where $p_n$ represents the similarity in the time $n$.

To finish this section, we explain how to obtain the similarity of two states. Algorithm 3 uses the function $Similarity(TFC_0(i_0), TFC_1(i_1))$ to do this, where $TFC_0$ and $TFC_1$ are the states to compare. Both states have the same variables, with the same ordered set of labels defined over on them. Equation 3 calculates the similarity between two states, where $\otimes$ is a t-norm, $P_{SLI_m}$ is the similarity between the $SLI_m$ of the states to compare, and $P_{Loutput}$ is the similarity between the output labels of the states.

$$\otimes\left(P_{SLI_m}, P_{Loutput}\right) \tag{3}$$

Equation 4 is used to calculate $P_{SLI_m}$.

$$\otimes\left(S(LI^{f^1}_{j,p_1}, LI^{f^2}_{j,p_2})\right) \tag{4}$$

where $\otimes$ is a t-norm and $S(LI^{f^1}_{j,p_1}, LI^{f^2}_{j,p_2})$ calculates the similarity between two intervals (Equation 5).

$$S(LI^{f^1}_{j,p_1}, LI^{f^2}_{j,p_2}) = 1 - \left|\frac{MOM(LI^{f^1}_{j,p_1}) - MOM(LI^{f^2}_{j,p_2})}{Max_j - Min_j}\right| \tag{5}$$

where $MOM(I^{f^j}_{j,p_j})$ is the *Mean of the Maximum* (MOM) [6], $Min_j$ and $Max_j$ are the minimum and maximum value of the support of $X_j$.

Equation 6 is used to compare the output labels ($P_{Loutput}$), being $A$ and $B$ labels.

$$R(A, B) = 1 - \left|\frac{MOM(A) - M(B)}{MOM(SC^{i_y}) - MOM(SC^1)}\right| \tag{6}$$

To describe the obtained output in a linguistic way, the similarity value $\rho_i$ is calculated for each phase $i$ defined in the beginning of the algorithm. To do that, Equation 7 is used for each one of the $P_{total}$ components, where $P^i_{total}$ is the component $i$ of $P_{total}$, $|P^i_{total}|$ is the element number of $P^i_{total}$ and $P^{i,j}_{total}$ is the value in the position $j$ of $P^i_{total}$.

$$\rho_i = \frac{\sum_{j=1}^{|P^i_{total}|} P^{i,j}_{total}}{|P^i_{total}|} \tag{7}$$

Finally, a ordered set of labels called *similarities* is defined, and the label with more membership grade to each similarity value $\rho_i$ is calculated, this label is considered the similarity for each one of the phases.

## 4 An Example of Proposed Method

This section presents an example of the behavior of the proposed method. Two DSs corresponding at two *jumps with feet together* are used in order to prove the

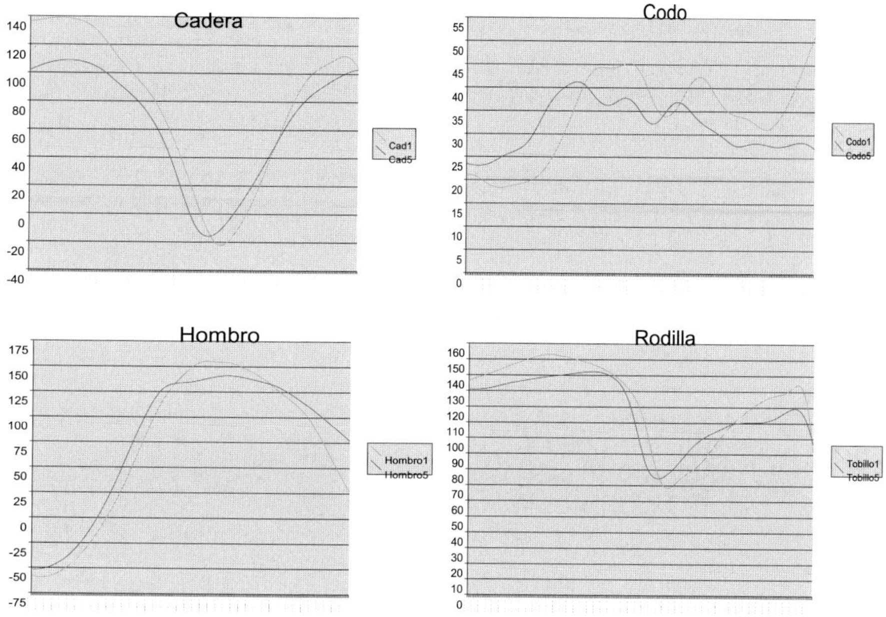


**Fig. 3.** Data graphs

presented algorithms. *Jumps with feet together* consist in to jump the greater possible distance without previous race. To study the jumps, we consider four input variables and one output variable. The input variables correspond to the hip angle ($HA$ variable), the elbow angle ($EA$ variable), the shoulder angle ($SA$ variable), and the knee angle ($KA$ variable), whereas for the output variable we use the ankle angle ($AA$ variable). A sample every 0.005 seconds are obtained in the data capture, that is, 200 samples per second, and a manual digitalization is done. Data correspond to students men of the Faculty of sciences of sport of the Universidad de Castilla-La Mancha. The duration of the two compared jumps are 1.02 seconds (Figure 3). We need an ordered set of labels for each variable

**Table 1.** Ordered set of linguistic labels used in the proof

| Name | a | b | c | d |
|---|---|---|---|---|
| $VN$ | 0 | 0 | 0.15 | 0.18 |
| $N$ | 0.15 | 0.18 | 0.29 | 0.32 |
| $FN$ | 0.29 | 0.32 | 0.43 | 0.46 |
| $NR$ | 0.43 | 0.46 | 0.57 | 0.60 |
| $FP$ | 0.57 | 0.60 | 0.71 | 0.74 |
| $P$ | 0.71 | 0.74 | 0.85 | 0.88 |
| $VP$ | 0.85 | 0.88 | 1.00 | 1.00 |

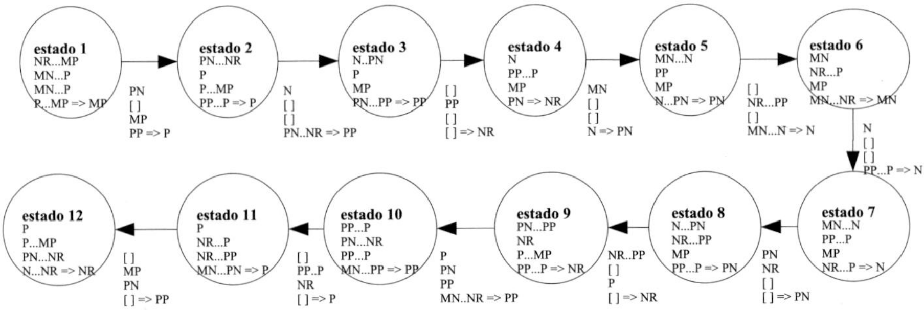


**Fig. 4.** Temporal Fuzzy Chain 0

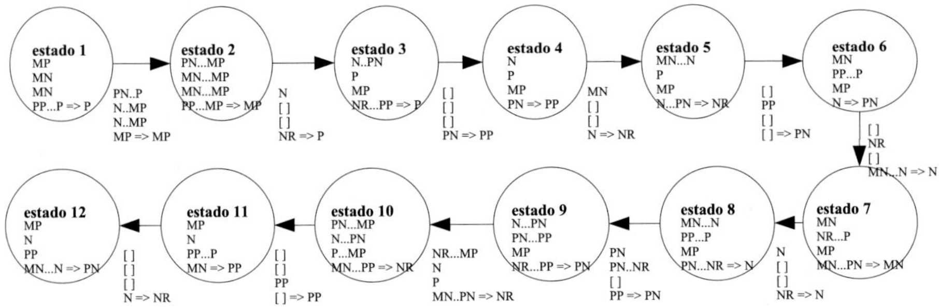


**Fig. 5.** Temporal Fuzzy Chain 1

of the system to induce the TFCs [2], these sets are named $SA_1$, $SA_2$, $SA_3$, $SA_4$ and $SC$ for the variables $HA$, $EA$, $SA$, $KA$ and $AA$ respectively. Each one contains 7 trapezoidal labels, its name and normalized support is shown in the Table 1, being $VN$ very negative, $N$ negative, $FN$ few negative, $NR$ norm, $FP$ few positive, $P$ positive and $VP$ very positive.

Now, we define the phases of the studied DS (Section 3). Each phase has a duration of 0.2 second except the last one that has a duration between the time 0.8 to the time 1.02, i.e., $phases = [0, 0.2, 0.4, 0.6, 0.8, 1.02]$. The increment of the time, $inc$ in the algorithm 3, is assigned to the value 0.005 that matches with the time interval of capture of samples, that is, $inc = 0.005$. The t-norm uses is the *"minimum"* for the Equations 3 and 4.

The induction method [2] obtains the two TFCs (Figures 4 and 5) by using the input data.

A set of 5 linguistic labels called *similarities* is defined to calculate the similarity of each section. Its labels are trapezoidal, and Table 2 shows the values that define each linguistic label. By using *similarities* for each phase we obtain the following results:

**Table 2.** *similarities* Set of labels

| *Name* | *a* | *b* | *c* | *d* |
|---|---|---|---|---|
| *NOTHING* | 0 | 0 | 0.10 | 0.15 |
| *LITTLE* | 0.10 | 0.15 | 0.35 | 0.40 |
| *MEDIUM* | 0.35 | 0.40 | 0.60 | 0.65 |
| *MUCH* | 0.60 | 0.65 | 0.85 | 0.90 |
| *EQUAL* | 0.85 | 0.90 | 1.00 | 1.00 |

**Phase 1:** The two TFCs look like *"MUCH"*, with a membership grade 0.76.
**Phase 2:** Its membership grade is 0.93, and its similarity is *"EQUAL"*
**Phase 3:** The similarity is *"MUCH"* (membership grade 0.78).
**Phase 4:** Its membership grade is 0.76, and its similarity is *"MUCH"*
**Phase 5:** The two TFCs look like *"MUCH"*, with a membership grade 0.70.

To finish this section, we want to emphasize that, by using the similarity values and the set *similarities*, it is possible to be described linguistically [7] [8] [9] the *similarities of the two TFCs* as follows:

> The jumps of the first student ($TFC_1$) and the second student ($TFC_2$) look like MUCH in the phase 1, are EQUAL in the second phase and they have MUCH similar in the phases 3, 4 and 5.

We one to emphasize that the present method allows to automate the obtaining of text in natural language.

## 5 Conclusions

A new method to compare two DSs is presented. This method is based in to compare two TFCs, that model two DSs. In addition, this comparison can be offered of a linguistic way, as we show in Section 4.

We believe that is a good tool to compare DSs linguistically. This method can be used, for example, in automatic sport trainers, or in detection of pathologies (for example, in the human walk). Also, it could be used in other fields, like in economic systems.

As future works, we have thought the use of our method in a concrete practical application to prove its efficiency, its adaptations and improvements to that practical application.

## Acknowledgments

This work is supported in part by the Junta de Comunidades de Castilla-La Mancha PBC06-0064-4504 and PAC-06-0141 grants.

## References


1. Luenberger, D.G.: Introduction to Dynamic Systems: Theory, Models and Applications. John Wiley and sons, 1979.
2. Moreno-Garcia, J., Castro-Schez, J.J., Jimenez, L.: A fuzzy inductive algorithm for modeling dynamical systems in a comprehensible way. To be published in IEEE Transactions on Fuzzy Systems.
3. Moreno-Garcia, J., Jimenez, L., Castro-Schez, J.J. ,Rodriguez, L.: A direct linguistic induction method for systems. Fuzzy Sets and Systems, 146, 79-96, 2004.
4. Castro, J.L., Castro-Schez, J.J., Zurita, J.M.: Learning maximal structure rules in fuzzy logic for knowledge acquisition in expert systems, Fuzzy Sets and Systems, 101, 331-342, 1999.
5. Gonzalez, A.: A learning methodology in uncertain and impresise enviroments. International Journal of Intelligence Systems, 10, 357-371, 1995
6. Leekwijck, W.V., Kerre, E.E.: Defuzzification: criteria and classification. Fuzzy Sets and Systems, 108(2), 159-178, 1999.
7. Zadeh, L.A.: The concept of a linguistic variable and its applications to approximate reasoning, Part I. Information Science, 8, 199-249, 1975.
8. Zadeh, L.A.: The concept of a linguistic variable and its applications to approximate reasoning, II. Information Science, 8, 301-357, 1975.
9. Zadeh, L.A.: The concept of a linguistic variable and its applications to approximate reasoning, Part III. Information Science, 9, 43-80, 1975.

# Advances in the Geometrical Study of Rotation-Invariant T-Norms

Koen C. Maes and Bernard De Baets

Department of Applied Mathematics, Biometrics and Process Control
Ghent University, Coupure links 653, B-9000 Gent, Belgium
{Koen.Maes,Bernard.DeBaets}@UGent.be

**Abstract.** T-norm properties for left-continuous, increasing $[0,1]^2 \to [0,1]$ functions can be fully described in terms of contour lines. For a left-continuous t-norm $T$, the rotation-invariance property comes down to the continuity of its contour line $C_0$. However, contour lines are inadequate to investigate the geometrical structure of these rotation-invariant t-norms. Enforced with the companion and zooms it is possible to totally reconstruct $T$ by means of its contour line $C_0$ and its $\beta$-zoom, with $\beta$ the unique fixpoint of $C_0$.

**Keywords:** Rotation-invariant t-norm, contour line, companion, zoom, associativity.

## 1 Introduction

Originally, triangular norms were introduced in order to generalize the triangle inequality towards probabilistic metric spaces [15]. Nowadays, they are widely used in fuzzy set theory.

**Definition 1.** *A* triangular norm *or* t-norm $T$ *is an associative, commutative, increasing* $[0,1]^2 \to [0,1]$ *function that has neutral element* 1.

So far only the class of continuous t-norms has been fully characterized (see e.g. [8]). In particular, this class comprises the three prototypical t-norms: the *minimum operator* $T_{\mathbf{M}}(x,y) = \min(x,y)$, the *algebraic product* $T_{\mathbf{P}}(x,y) = x\,y$ and the *Łukasiewicz t-norm* $T_{\mathbf{L}}(x,y) = \max(x+y-1,0)$.

A t-norm $T$ is called left-continuous if all its partial functions $T(x,\bullet)$ (and hence also $T(\bullet,x)$) are left-continuous [8]. In most studies dealing with t-norms, it is required that the t-norms in question should be left-continuous. In monoidal t-norm based logic (MTL logic) for example, where the implication is defined as the residuum of the conjunction, left-continuous t-norms ensure the definability of the t-norm-based residual implicator [2].

The rotation-invariance of a left-continuous t-norm $T$ is equivalent with the continuity and with the involutivity of its contour line $C_0$ that determines the intersection of $T$ with the plane containing its domain $[0,1]^2$. In particular, this contour line coincides with the residual negator of $T$ and, therefore, rotation-invariant t-norms are of great interest to people working on involutive monoidal t-norm based logic (IMTL logic) [1,10] and fuzzy type theory [14].

P. Melin et al. (Eds.): IFSA 2007, LNAI 4529, pp. 543–552, 2007.
© Springer-Verlag Berlin Heidelberg 2007

# 2 Tools

Studying the structure of a (left-continuous) increasing $[0,1]^2 \to [0,1]$ function $T$, it is often worthwhile to observe this function from a different point of view. We present here three functions that describe $T$ in an alternative way. They will prove to be indispensable for the decomposition and construction of rotation-invariant t-norms.

## 2.1 Contour Lines

Contour lines of an increasing $[0,1]^2 \to [0,1]$ function $T$ are defined as the upper, lower, right or left limits of its horizontal cuts, *i.e.* the intersections of its graph by planes parallel to the domain $[0,1]^2$. Although there are four different types of contour lines, those determined by the upper limits of the horizontal cuts are of particular interest for the study of rotation-invariant t-norms [12].

**Definition 2.** [11] *Let* $a \in [0,1]$. *The* contour line $C_a$ *of an increasing* $[0,1]^2 \to [0,1]$ *function* $T$ *is the* $[0,1] \to [0,1]$ *function defined by*

$$C_a(x) = \sup\{t \in [0,1] \mid T(x,t) \le a\}\,. \tag{1}$$

For a left-continuous t-norm $T$, the contour line $C_a$ equals the partial function $I_T(\bullet, a)$ of the *residual implicator* $I_T$ (see e.g. [4]). In particular, the contour line $C_0$ coincides with the *residual negator* $N_T$, defined by $N_T = I_T(\bullet, 0)$. Contour lines of a continuous t-norm $T$ are also called *level functions* [9].

*Property 1.* [11,12] A contour line $C_a$, with $a \in [0,1]$, of an increasing $[0,1]^2 \to [0,1]$ function $T$ satisfies the following properties:

1. $C_a$ is decreasing.
2. $C_a \le C_b$, for every $b \in [a,1]$.
3. If $T$ is left-continuous, then $C_a$ is left-continuous.

The greatest merit of contour lines is that they can be used to express all t-norm properties in an alternative way. Further on, this will allow us to provide a geometrical interpretation of the associativity property. Dealing with contour lines of the type $C_a$ the left-continuity of $T$ is required.

**Theorem 1.** [11] *For a left-continuous, increasing* $[0,1]^2 \to [0,1]$ *function* $T$ *having absorbing element* 0 *the following characterizations hold:*

1. *$T$ has neutral element $e \in\, ]0,1]$ if and only if $e \le C_a(x) \Leftrightarrow x \le a$ and $C_a(e) = a$ hold for every $(x,a) \in [0,1]^2$.*
2. *$T$ is commutative if and only if $C_a(x) < y \Leftrightarrow C_a(y) < x$ holds for every $(x,y,a) \in [0,1]^3$.*
3. *$T$ is associative if and only if $C_a(T(x,y)) = C_{C_a(x)}(y)$ holds for every $(x,y,a) \in [0,1]^3$.*

The characterization of the commutativity of $T$ comes down to the **id**-orthosymmetry of its contour lines [13]. Taking into account the tight correspondence between contour lines and the residual implicator of a left-continuous t-norm $T$, the above theorem expresses the associativity of $T$ by means of the *portation law* (*i.e.* $I_T(T(x, y), z) = I_T(x, I_T(y, z))$, for every $(x, y, z) \in [0, 1]^3$) [5].

**Corollary 1.** [12] *For a left-continuous t-norm $T$ it holds for every $(x, y, z, a) \in [0, 1]^4$ that*

$$T(x, y) \leq C_a(z) \Leftrightarrow T(x, z) \leq C_a(y) . \tag{2}$$

Jenei [7] has recently shown that, for a commutative, left-continuous, increasing $[0, 1]^2 \to [0, 1]$ function $T$ that has absorbing element 0, Eq. (2) is equivalent with the associativity of $T$. Note that for his characterization the commutativity of $T$ is required, this in contrast to our characterization in Theorem 1. His result can also be easily retrieved from the last two characterizations in Theorem 1.

## 2.2 The Companion

A second useful tool to study an increasing $[0, 1]^2 \to [0, 1]$ function $T$ is its companion $Q$.

**Definition 3.** [12] *The* companion $Q$ *of an increasing* $[0, 1]^2 \to [0, 1]$ *function* $T$ *is the* $[0, 1]^2 \to [0, 1]$ *function defined by*

$$Q(x, y) = \sup\{t \in [0, 1] \mid C_t(x) \leq y\} .$$

The following properties provide better insight into the geometrical structure of $Q$.

*Property 2.* [13] The companion $Q$ of an increasing $[0, 1]^2 \to [0, 1]$ function $T$ satisfies the following properties:

1. $Q$ is increasing in both arguments.
2. $Q(x, y) = \inf\{T(x, u) \mid u \in ]y, 1]\}$, with $\inf \emptyset = 1$.
3. $T(x, y) \leq Q(x, y)$, for every $(x, y) \in [0, 1]^2$.
4. $Q(x, \bullet)$ is right-continuous for every $x \in [0, 1]$.
5. If $T$ has neutral element 1, then $Q(x, y) \leq T_{\mathbf{M}}(x, y)$, for every $(x, y) \in [0, 1] \times [0, 1[$.

The second property allows to straightforwardly construct the graph of $Q$ (*i.e.* $\{(x, y, Q(x, y)) \mid (x, y) \in [0, 1]^2\}$) from the graph of $T$ (*i.e.* $\{(x, y, T(x, y)) \mid (x, y) \in [0, 1]^2\}$). It suffices to convert the partial functions $T(x, \bullet)$ into right-continuous functions and to replace the set $\{(x, 1, x) \mid x \in [0, 1]\}$ by $\{(x, 1, 1) \mid x \in [0, 1]\}$ as $Q(x, 1) = 1$ must hold for every $x \in [0, 1]$. Clearly, $Q(x, y) = T(x, y)$ whenever $T(x, \bullet)$ is right-continuous in $y \in [0, 1[$. Every left-continuous increasing binary function $T$ that has absorbing element 0 is totally determined by its companion $Q$. Note also that $Q(x, 1) = 1$ and $Q(1, x) = x$ prevent $Q$ from being commutative.

### 2.3 Zooms

Finally, every increasing $[0,1]^2 \to [0,1]$ function $T$ is trivially described by its associated set of $(a,b)$-zooms.

**Definition 4.** *Let $T$ be an increasing $[0,1]^2 \to [0,1]$ function and take $(a,b) \in [0,1]^2$ such that $a < b$ and $T(b,b) \leq b$. Consider an $[a,b] \to [0,1]$ isomorphism $\sigma$. The $(a,b)$-zoom $T^{(a,b)}$ of $T$ is the $[0,1]^2 \to [0,1]$ function defined by*

$$T^{(a,b)}(x,y) = \sigma\left[\max\left(a, T(\sigma^{-1}[x], \sigma^{-1}[y])\right)\right] .$$

*If $b = 1$ we simply talk about the $a$-zoom $T^a$ of $T$.*

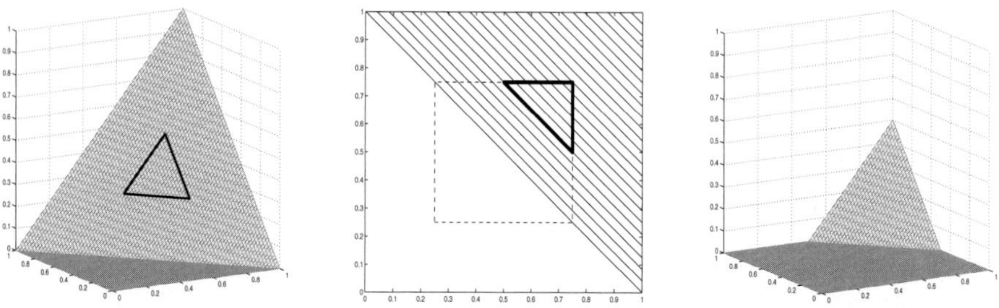

$T_{\mathbf{L}}$ Contour plot of $T_{\mathbf{L}}$ $T_{\mathbf{L}}^{(\frac{1}{4},\frac{3}{4})}$

**Fig. 1.** The $(\frac{1}{4},\frac{3}{4})$-zoom of $T_{\mathbf{L}}$

The graph of $T^{(a,b)}$ is determined by rescaling the set $\{(x,y,T(x,y)) \mid (x,y) \in [a,b]^2 \wedge a < T(x,y)\}$ (zoom in) into the unit cube (zoom out). Figure 1 illustrates this procedure for the Łukasiewicz t-norm $T_{\mathbf{L}}$, with $a = \frac{1}{4}$, $b = \frac{3}{4}$ and $\sigma = \varsigma$, where $\varsigma$ is the *linear rescaling* of $[a,b]$ into $[0,1]$ (*i.e.* $\varsigma(x) = (x-a)/(b-a)$, for every $x \in [a,b]$). In our examples we will always use this linear rescaling function.

Whenever $T(b,b) \leq a$, the function $T^{(a,b)}$ is trivially constant: $T^{(a,b)}(x,y) = a$, for every $(x,y) \in [0,1]^2$. For $b = 1$ the boundary condition $T(1,1) \leq 1$ is always true such that the $a$-zoom of $T$ is defined for every $a < 1$. Note that $T^0 = T_{\sigma^{-1}}$, where $T_{\sigma^{-1}}$ denotes the $\sigma^{-1}$-transform of $T$ (*i.e.* $T_{\sigma^{-1}}(x,y) := \sigma[T(\sigma^{-1}[x], \sigma^{-1}[y])]$).

Since the $(a,b)$-zoom $T^{(a,b)}$ of an arbitrary increasing function $T$ is totally determined by $T|_{[a,b]^2}$, its contour lines and companion can be computed from the contour lines and companion of $T$. In case $T^{(a,b)}$ has neutral element 1, we obtain a straightforward relationship between its contour lines and those of the original function $T$.

*Property 3.* Consider an increasing $[0,1]^2 \to [0,1]$ function $T$. Take $(a,b) \in [0,1]^2$, such that $a < b$ and $T(b,b) \leq b$. Let $\sigma$ be an arbitrary $[a,b] \to [0,1]$ isomorphism. If the $(a,b)$-zoom $T^{(a,b)}$ has contour lines $C_d^{(a,b)}$ and companion $Q^{(a,b)}$ then the following properties hold:

1. $T^{(a,b)}$ is increasing in both arguments.
2. $Q^{(a,b)}(x,y) = \sigma[Q(\sigma^{-1}[x], \sigma^{-1}[y])]$, for every $(x,y) \in [0,1]^2$ s.t. $C_0^{(a,b)}(x) \leq y < 1$.
3. If $T$ is left-continuous, then $T^{(a,b)}$ is left-continuous.
4. $C_d^{(a,b)}(x) = \sigma[C_{\sigma^{-1}[d]}(\sigma^{-1}[x])]$ holds if
   (a) $b = 1$, $T(1,a) \leq a$ and $(x,d) \in [0,1]^2$;
   (b) $T^{(a,b)}$ has neutral element 1 and $(x,d) \in [0,1]^2$ s.t. $d < x$.
5. If $T$ is associative and $\max(T(a,b), T(b,a)) \leq a$, then $T^{(a,b)}$ is also associative.

In accordance to Definition 4 we will usually denote the contour lines of $T^a (= T^{(a,1)})$ as $C^a (= C^{(a,1)})$ and its companion as $Q^a (= Q^{(a,1)})$. Zooms are extremely suited to study an increasing function $T$ that satisfies $T \leq T_{\mathbf{M}}$. The restrictions $T(b,b) \leq b$ (Definition 4), $T(1,a) \leq a$ and $\max(T(a,b), T(b,a)) \leq a$ (Property 3) then trivially hold.

**Definition 5.** [6] *A* t-subnorm $T$ *is an associative, commutative, increasing* $[0,1]^2 \to [0,1]$ *function that satisfies* $T \leq T_{\mathbf{M}}$.

Clearly, all t-norms are t-subnorms. Due to its boundary condition we can construct all $(a,b)$-zooms $(a < b)$ of every t-subnorm. Moreover, all these $(a,b)$-zooms are t-subnorms ass well.

**Corollary 2.** *Consider* $(a,b) \in [0,1]^2$ *such that* $a < b$. *Then the* $(a,b)$*-zoom of a t-subnorm is a t-subnorm and the a-zoom of a t-norm is a t-norm.*

The $(\frac{1}{4}, \frac{3}{4})$-zoom in Fig. 1 is a t-subnorm but not a t-norm. No $(a,b)$-zoom, with $b < 1$, of the Łukasiewicz t-norm $T_{\mathbf{L}}$ can be a t-norm. The latter follows from the observation that $T^{(a,b)}$ has neutral element 1 whenever $T(x,b) = T(b,x) = x$, for every $x \in ]a,b]$. Dealing with $T_{\mathbf{L}}$ this only occurs for $b = 1$. Otherwise, every $(a,b)$-zoom of the minimum operator $T_{\mathbf{M}}$ equals $T_{\mathbf{M}}$ itself.

## 3 Rotation-Invariant T-Norms

### 3.1 A Continuous Contour Line

**Definition 6.** [5] *Let* $N$ *be an involutive negator (*i.e. *an involutive decreasing* $[0,1] \to [0,1]$ *function). An increasing* $[0,1]^2 \to [0,1]$ *function* $T$ *is called* rotation invariant w.r.t. an involutive negator $N$ *if for every* $(x,y,z) \in [0,1]^3$ *it holds that*

$$T(x,y) \leq z \Leftrightarrow T(y, z^N) \leq x^N . \tag{3}$$

This property was first described by Fodor [3]. Jenei [5] emphasized its geometrical interpretation by referring to it as the rotation-invariance of $T$ w.r.t. $N$. Recently, Jenei [7] used Eq. (2) to define the (algebraical) rotation invariance property. However, as pointed out before, Eq. (2) merely expresses the associativity of $T$. As will be shown later on, a geometrical notion of rotation can be attributed to Eq. (3) but in general not to Eq. (2)

**Theorem 2.** [12] *For a left-continuous t-norm $T$, the following assertions are equivalent:*

1. $C_a$ *is continuous.*
2. $C_a$ *is involutive on* $[a,1]$.
3. $T(x,y) = C_a(C_{C_a(x)}(y))$, *for every* $(x,y) \in [0,1]^2$ *s.t.* $C_a(x) < y$.
4. $T(x,y) \leq z \Leftrightarrow T(x, C_a(z)) \leq C_a(y)$, *for every* $(x,y,z) \in [a,1]^3$.
5. $Q(x,y) < C_a(z) \Leftrightarrow Q(x,z) < C_a(y)$, *for every* $(x,y,z) \in [a,1] \times [a,1[^2$.

The third assertion can be seen as an adjustment of the portation law. The fourth and fifth assertion are closely related to Eq. (2). If $a = 0$ then the additional restriction $C_0(x) < y$ in assertion 3 can be omitted. Furthermore, taking into account that $C_0 = N_T$, one can then recognize in assertion 4 the rotation-invariance of $T$ w.r.t. its residual negator $N_T$. Jenei has proven that every t-norm $T$ that is rotation-invariant w.r.t. an involutive negator $N$ is necessarily left-continuous and $N_T = N$ [5]. Therefore, it becomes superfluous to mention the negator $N$ explicitly. For a left-continuous t-norm $T$, its rotation-invariance is also equivalent with the continuity of its contour line $C_0$ (Theorem 2). Herein lies the true meaning of the rotation-invariance property. We briefly call a t-norm *rotation invariant* if it is left-continuous and has a continuous contour line $C_0$. Note that the continuity of $C_0$ does not necessarily imply the left-continuity of $T$ [12].

**Theorem 3.** *Consider a left-continuous t-norm $T$ and take $a \in [0,1]$ such that $a < \alpha := \inf\{t \in [0,1] \mid C_a(t) = a\}$. Then the following assertions are equivalent:*

1. $C_a$ *is continuous on* $]a,1]$.
2. $C_a$ *is involutive on* $]a,\alpha[$.
3. $C_a(]a,\alpha[) = ]a,\alpha[$.
4. $T^{(a,\alpha)}$ *is a rotation-invariant t-norm.*

*In particular, if $C_a$ is continuous then $\alpha = 1$.*

To better comprehend the structure of t-norms that have a (partially) continuous contour line $C_a$ we thus need to focus first on the structure of rotation-invariant t-norms. Studying these t-norms, Jenei provided a real breakthrough by introducing his rotation and rotation-annihilation construction [6]. Unfortunately, his decompositions and constructions were not able to describe all rotation-invariant t-norms [12,13]. The Łukasiewicz t-norm $T_{\mathbf{L}}$, for example, did not fit into his framework. We will present an alternative approach.

### 3.2 Decomposition Revisited

Let $T$ be a rotation-invariant t-norm and $\beta$ be the unique fixpoint of $C_0$. As depicted in Fig. 2, we partition area $\mathcal{D} = \{(x,y) \in [0,1]^2 \mid C_0(x) < y\}$ into four parts:

$$\begin{aligned}
\mathcal{D}_{\mathrm{I}} &= \{(x,y) \in ]\beta,1]^2 \mid C_\beta(x) < y\}\,,\\
\mathcal{D}_{\mathrm{II}} &= \{(x,y) \in ]0,\beta] \times ]\beta,1] \mid C_0(x) < y\}\,,\\
\mathcal{D}_{\mathrm{III}} &= \{(x,y) \in ]\beta,1] \times ]0,\beta] \mid C_0(x) < y\}\,,\\
\mathcal{D}_{\mathrm{IV}} &= \{(x,y) \in ]\beta,1[^2 \mid y \leq C_\beta(x)\}\,.
\end{aligned}$$

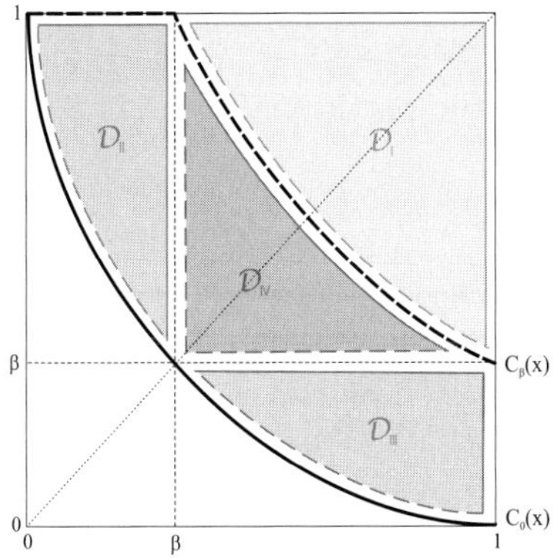


**Fig. 2.** The partition $\mathcal{D} = \mathcal{D}_{\mathrm{I}} \cup \mathcal{D}_{\mathrm{II}} \cup \mathcal{D}_{\mathrm{III}} \cup \mathcal{D}_{\mathrm{IV}}$

Due to the left-continuity of $T$ it is obvious that $T(x, y) = 0$ holds for every $(x, y) \notin \mathcal{D}$.

**Theorem 4.** [12] *Consider a rotation-invariant t-norm $T$. Let $\sigma$ be an arbitrary $[\beta, 1] \to [0, 1]$ isomorphism with $\beta$ the fixpoint of $C_0$. Then there exists a left-continuous t-norm $\widehat{T}$ (with contour lines $\widehat{C}_a$) such that*

$$T(x,y) = \begin{cases} \sigma^{-1}\left[\widehat{T}(\sigma[x], \sigma[y])\right], & \text{if } (x,y) \in \mathcal{D}_{\mathrm{I}}\,, \\ C_0\left(\sigma^{-1}\left[\widehat{C}_{\sigma[C_0(x)]}(\sigma[y])\right]\right), & \text{if } (x,y) \in \mathcal{D}_{\mathrm{II}}\,, \\ C_0\left(\sigma^{-1}\left[\widehat{C}_{\sigma[C_0(y)]}(\sigma[x])\right]\right), & \text{if } (x,y) \in \mathcal{D}_{\mathrm{III}}\,, \\ 0, & \text{if } (x,y) \notin \mathcal{D}\,. \end{cases} \quad (4)$$

*In particular, $\widehat{T} = T^\beta$.*

Note that the isomorphism $\sigma$ is used to compute the $\beta$-zoom $T^\beta$ of $T$. Geometrically, $T|_{\mathcal{D}_{\mathrm{I}}}$ is a rescaled version of $T^\beta|_{\mathcal{D}^\beta}$, where $\mathcal{D}^\beta = \{(x,y) \in [0,1]^2 \mid 0 < T^\beta(x,y)\}$. $T|_{\mathcal{D}_{\mathrm{II}}}$ is obtained by rotating $T|_{\mathcal{D}_{\mathrm{I}}}$ 120 degrees to the left around the axis $\{(x,y,z) \in [0,1]^2 \mid y = x \,\wedge\, z = 1 - x\}$. Similarly, rotating $T|_{\mathcal{D}_{\mathrm{I}}}$ 120 degrees to the right around this axis determines $T|_{\mathcal{D}_{\mathrm{III}}}$. As illustrated in [12], these rotations sometimes have to be reshaped to fit into the areas $\mathcal{D}_{\mathrm{II}}$ and $\mathcal{D}_{\mathrm{III}}$, respectively. The contour lines $C_0$ and $C_\beta$ cause this reshaping. Solely the continuity of the contour line $C_0$ is responsible for the existence of the geometrical (transformed) rotations. T-norms such as the minimum operator $T_{\mathbf{M}}$ that do not have a continuous contour line do not have such geometrical symmetries. Therefore, only Eq. (3) and not Eq. (2) (see [7]) can be understood as the rotation-invariance property.

If $T^\beta$ has no zero divisors, then $\mathcal{D}_{\mathrm{IV}}$ is empty and Eq. (4) totally determines $T$. These particular t-norms have also been (alternatively) described by Jenei [6].

Figure 3 depicts the decomposition of the *nilpotent minimum* $T^{\mathbf{nM}}$ ($T^{\mathbf{nM}}(x,y) = 0$ whenever $x+y \leq 1$ and $T^{\mathbf{nM}}(x,y) = \min(x,y)$ elsewhere). The bold black lines in the figures indicate the partition $\mathcal{D} = \mathcal{D}_{\mathrm{I}} \cup \mathcal{D}_{\mathrm{II}} \cup \mathcal{D}_{\mathrm{III}}$ (for the nilpotent minimum $\mathcal{D}_{\mathrm{IV}} = \emptyset$).

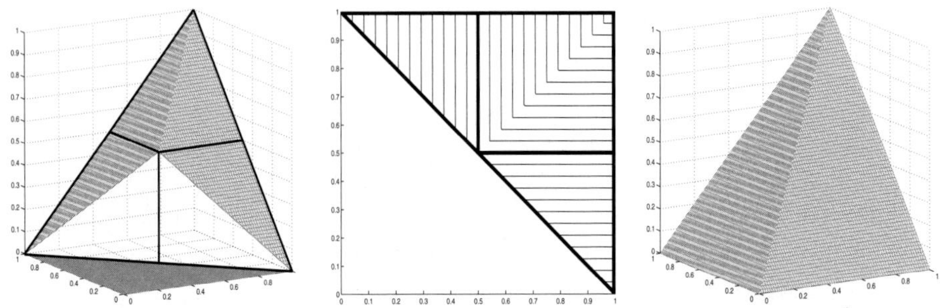

(a) $T^{\mathbf{nM}}$ (b) Contour plot of $T^{\mathbf{nM}}$ (c) $(T^{\mathbf{nM}})^{\frac{1}{2}}$

**Fig. 3.** Decomposition of the nilpotent minimum $T^{\mathbf{nM}}$

As illustrated in [12], $T|_{\mathcal{D}_{\mathrm{IV}}}$ is in general not uniquely determined by $C_0$ and $T^{\beta}$. Examining numerous examples, we noticed that the filling-in of area $\mathcal{D}_{\mathrm{IV}}$ is uniquely fixed whenever both $C_0$ and $C_\beta$ are continuous. Invoking Theorem 3 we can generalize our decomposition from [12] in the following way.

**Theorem 5.** *Consider a rotation-invariant t-norm $T$ for which $C_\beta$ is continuous on $]\beta,1]$, with $\beta$ the unique fixpoint of $C_0$. Let $\sigma$ be an arbitrary $[\beta,1] \to [0,1]$ isomorphism. Then there exists a left-continuous t-norm $\widehat{T}$ (with contour lines $\widehat{C}_a$ and companion $\widehat{Q}$) such that $\widehat{C}_0$ is continuous on $]0,1]$ and*

$$T(x,y) = \begin{cases} \sigma^{-1}\left[\widehat{T}(\sigma[x],\sigma[y])\right], & \text{if } (x,y) \in \mathcal{D}_{\mathrm{I}}\,, \\ C_0\left(\sigma^{-1}\left[\widehat{C}_{\sigma[C_0(x)]}(\sigma[y])\right]\right), & \text{if } (x,y) \in \mathcal{D}_{\mathrm{II}}\,, \\ C_0\left(\sigma^{-1}\left[\widehat{C}_{\sigma[C_0(y)]}(\sigma[x])\right]\right), & \text{if } (x,y) \in \mathcal{D}_{\mathrm{III}}\,, \\ C_0\left(\sigma^{-1}\left[\widehat{Q}(\widehat{C}_0(\sigma[x]),\widehat{C}_0(\sigma[y]))\right]\right), & \text{if } (x,y) \in \mathcal{D}_{\mathrm{IV}}\,, \\ 0, & \text{if } (x,y) \notin \mathcal{D}\,. \end{cases} \quad (5)$$

*In particular, $\widehat{T} = T^{\beta}$ and $\widehat{Q}$ must be commutative on $[0,\hat{\alpha}[^2$, with $\hat{\alpha} = \inf\{t \in [0,1] \mid \widehat{C}_0(t) = 0\}$.*

Geometrically, the filling-in of area $\mathcal{D}_{\mathrm{IV}}$ is obtained by rotating $T|_{\mathcal{D}_{\mathrm{I}} \cap [\beta,\sigma^{-1}(\hat{\alpha})]^2}$ 180 degrees to the front around the axis $\{(x,y,z) \in [0,1]^3 \mid x+y = \beta + \sigma^{-1}[\hat{\alpha}] \wedge z = \beta\}$. In case $C_\beta$ is continuous it holds that $\hat{\alpha} = 1$ and the latter comes down to a 180 degree front-rotation of $T|_{\mathcal{D}_{\mathrm{I}}}$ around the axis $\{(x, y, z) \in [0,1]^3 \mid x + y = \beta + 1 \ \wedge\ z = \beta\}$. Again, the contour lines $C_0$ and $C_\beta$ can cause some additional reshaping.

Figure 4 depicts the decomposition of the *Jenei t-norm* $T^{\mathbf{J}}_{1/4}$ and the Łukasiewicz t-norm $T_{\mathbf{L}}$. $T^{\mathbf{J}}_{1/4}$ can be created from the nilpotent minimum by lowering its values on $[\frac{1}{4}, \frac{3}{4}]^2$ in such a way that its $(\frac{1}{4}, \frac{3}{4})$-zoom equals $T_{\mathbf{L}}$. The bold black lines in the figures indicate the partition $\mathcal{D} = \mathcal{D}_{\mathrm{I}} \cup \mathcal{D}_{\mathrm{II}} \cup \mathcal{D}_{\mathrm{III}} \cup \mathcal{D}_{\mathrm{IV}}$.

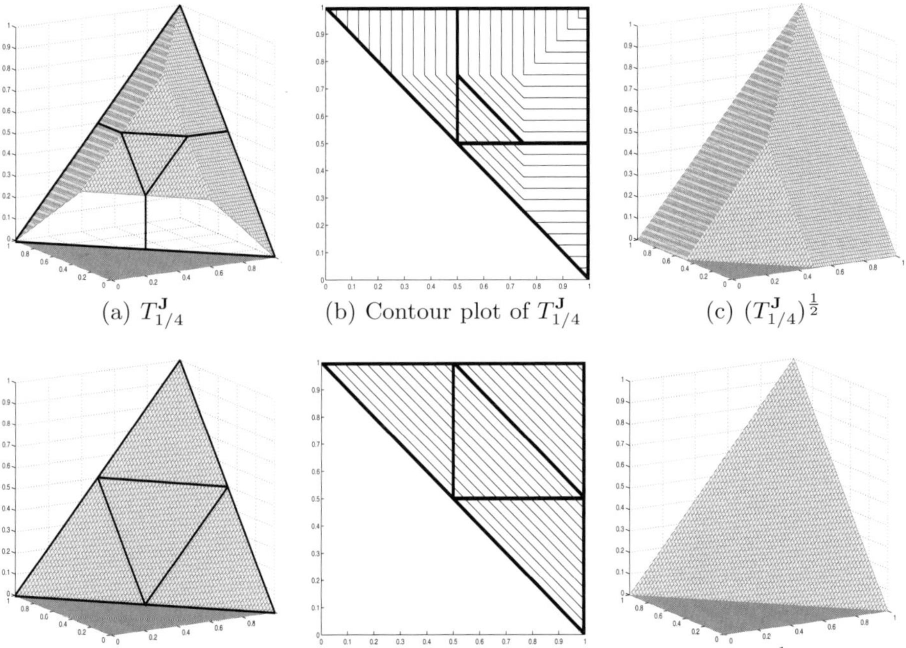

(a) $T^{\mathbf{J}}_{1/4}$ (b) Contour plot of $T^{\mathbf{J}}_{1/4}$ (c) $(T^{\mathbf{J}}_{1/4})^{\frac{1}{2}}$

(d) $T_{\mathbf{L}}$ (e) Contour plot of $T_{\mathbf{L}}$ (f) $T_{\mathbf{L}}^{\frac{1}{2}}$

**Fig. 4.** Decomposition of the Jenei t-norm $T^{\mathbf{J}}_{1/4}$ and the Łukasiewicz t-norm $T_{\mathbf{L}}$

The geometrical symmetries of a rotation-invariant t-norm $T$ establish in fact its associativity. In this respect Eq. (5) can also be used to construct rotation-invariant t-norms. Inspired by the geometrical interpretation of Eq. (5), we have called this construction the *triple rotation method* [13]. As the construction of t-norms falls outside the scope of this paper, we will not go into detail here.

## References


1. F. Esteva, J. Gispert, L. Godo, and F. Montagna, *On the standard and rational completeness of some axiomatic extensions of the monoidal t-norm logic*, Studia Logica **71** (2002), 199–226.

2. F. Esteva and L. Godo, *Monoidal t-norm based fuzzy logic: towards a logic for left-continuous t-norms*, Fuzzy Sets and Systems **124** (2001), 271–288.
3. J. Fodor, *A new look at fuzzy connectives*, Fuzzy Sets and Systems **57** (1993), 141–148.
4. J. Fodor and M. Roubens, *Fuzzy preference modelling and multicriteria decision support*, Kluwer Academic Publishers, 1994.
5. S. Jenei, *Geometry of left-continuous t-norms with strong induced negations*, Belg. J. Oper. Res. Statist. Comput. Sci. **38** (1998), 5–16.
6. S. Jenei, *How to construct left-continuous triangular norms - state of the art*, Fuzzy Sets and Systems **143** (2004), 27–45.
7. S. Jenei, *On the geometry of associativity*, Semigroup Forum, to appear.
8. E.P. Klement, R. Mesiar, and E. Pap, *Triangular norms*, Trends in Logic, Vol. 8, Kluwer Academic Publishers, 2000.
9. E.P. Klement, R. Mesiar, and E. Pap, *Different types of continuity of triangular norms revisited*, New Mathematics and Natural Computation **1** (2005), 195–211.
10. L. Lianzhen and L. Kaitai, *Involutive monoidal t-norm based logic and $R_0$ logic*, Internat. J. Intell. Systems **19** (2004), 491–497.
11. K.C. Maes and B. De Baets, *A contour view on uninorm properties*, Kybernetika **42** (3) (2006), 303–318.
12. K.C. Maes and B. De Baets, *On the structure of left-continuous t-norms that have a continuous contour line*, Fuzzy Sets and Systems, in press (DOI:10.1016/j.fss.2006.09.016).
13. K.C. Maes and B. De Baets, *The triple rotation method for constructing rotation-invariant t-norms*, Fuzzy Sets and Systems, submitted.
14. V. Novák, *On fuzzy type theory*, Fuzzy Sets and Systems **149** (2005), 235–273.
15. B. Schweizer and A. Sklar, *Probabilistic metric spaces*, Elsevier Science, New York, 1983.

# Fuzzy Reversed Posynomial Geometric Programming and Its Dual Form

Bing-yuan Cao

School of Math. and Inform. Sci., Guangzhou University, Guangzhou 510006, China
bycao@gzhu.edu.cn

**Abstract.** With an exponent-type membership function serving as conduct fuzzy value, the author builds a model in fuzzy reversed posynomial geometric programming with its dual programming developed. In the application process of fuzzy reversed posynomial geometric programming , he consequently discovers that the former can be changed into a convex parameter geometric programming with respect to $\alpha$,$\beta$. Therefore, he can get many nice properties like fuzzy posynomial geometric programming. Besides,he advances a dual theory and algorithm in fuzzy reversed posynomial geometric programming. And finally he uses numerical examples to testify the built model and its method effectively.

## 1 Introduction

In 1987, the author extended a classical posynomial geometric programming (PGP) into a fuzzy PGP by using a fuzzy valued-set theory [1,2], and built a fuzzy dual theory and a strong one [2] in 1989 and 1993, respectively. So far, this model with the method has already got the extensive application [3,4,5,6]. In this paper, he first gives out some basic concept in PGP, discusses properties and dual theorem in fuzzy reversed posynomial GP (FRPGP), and advances an exponent-type membership function, which is fitter for the determination of fuzzy reversed GP than the membership function mentioned in Paper [2], playing an important role in convexification of nonconvex FRPGP. Finally he constructs numerical examples.

## 2 FRPGP and Its Dual Form

We call

$$(\tilde{P}) \qquad \begin{aligned} &\widetilde{\min}\ \tilde{g}_0(x) \\ &\text{s.t. } \tilde{g}_i(x) \lesssim 1 (1 \leqslant i \leqslant p'), \\ &\quad\ \ \tilde{g}_i(x) \gtrsim 1, (p'+1 \leqslant i \leqslant p), \quad x > 0 \end{aligned}$$

an FRPGP, where $\tilde{g}_i(x) = \sum_{k=1}^{J_i} \tilde{v}_{ik}(x) (0 \leqslant i \leqslant p)$ are fuzzy posynomials of $x$,

$$\tilde{v}_{ik}(x) = \begin{cases} \tilde{c}_{ik} \prod\limits_{l=1}^{m} x_l^{\tilde{\gamma}_{ikl}}, & (1 \leqslant k \leqslant J_i; 0 \leqslant i \leqslant p'), \\ \tilde{c}_{ik} \prod\limits_{l=1}^{m} x_l^{-\tilde{\gamma}_{ikl}}, & (1 \leqslant k \leqslant J_i; p'+1 \leqslant i \leqslant p), \end{cases}$$

P. Melin et al. (Eds.): IFSA 2007, LNAI 4529, pp. 553–562, 2007.
© Springer-Verlag Berlin Heidelberg 2007

and coefficients $\tilde{c}_{ik} > 0$ and exponents $\tilde{\gamma}_{ikl}$ may be freely fixed in the closed interval $[c_{ik}^-, c_{ik}^+](c_{ik}^- < c_{ik}^+, c_{ik}^-, c_{ik}^+)$ and $[\tilde{\gamma}_{ikl}^-, \tilde{\gamma}_{ikl}^+](\gamma_{ikl}^- < \gamma_{ikl}^+)$ and degree of accomplishment [2] is determined by

$$\tilde{c}_{ik}(c_{ik}) = \begin{cases} 0, & \text{if} \quad c_{ik} < c_{ik}^-, \\ \left(\frac{c_{ik}-c_{ik}^-}{c_{ik}^+ - c_{ik}^-}\right)^r, & \text{if} \quad c_{ik}^- \leqslant c_{ik} \leqslant c_{ik}^+, \\ 1, & \text{if} \quad c_{ik} > c_{ik}^+, \end{cases} \tag{1}$$

$$\tilde{\gamma}_{ikl}(\gamma_{ikl}) = \begin{cases} 0, & \text{if} \ \gamma_{ikl} < \gamma_{ikl}^-, \\ \left(\frac{\gamma_{ikl}-\gamma_{ikl}^-}{\gamma_{ikl}^+ - \gamma_{ikl}^-}\right)^r, & \text{if} \ \gamma_{ikl}^- \leqslant \gamma_{ikl} \leqslant \gamma_{ikl}^+, \\ 1, & \text{if} \ \gamma_{ikl} > \gamma_{ikl}^+, \end{cases} \tag{2}$$

respectively. Here $c_{ik}^-, c_{ik}^+; \gamma_{ikl}^-, \gamma_{ikl}^+$ is left and right endpoints in the intervals, and they and $r$ are arbitrary rational numbers.

For each item $\widetilde{v}_{ik}(x)$ $(0 \leqslant k \leqslant J_i; p'+1 \leqslant i \leqslant p)$ in the reversed inequality $\tilde{g}_i(x) \gtrsim 1$, $x_l$ acts as an exponent in the item by $-\tilde{\gamma}_{ikl}$ instead of by $\tilde{\gamma}_{ikl}$, where symbol '$\lesssim$', or '$\gtrsim$'denotes the fuzzified version of $\leqslant$, or $\geqslant$ and has the linguistic interpretation "essentially smaller than or equal", or "essentially larger than or equal", and $\widetilde{\min}$ is an extension of min operation.

If $\tilde{A}_0 = \{x | \tilde{g}_0(x) \lesssim Z_0, x > 0\}$ is a fuzzy object set, $Z_0$ is an expectation value of $\tilde{g}_0(x)$, then its membership function is defined by

$$\tilde{A}_0(x) = \begin{cases} 1, & \text{if} \quad \bar{g}_0(x) \leqslant Z_0 \\ e^{-\frac{1}{d_0}\left(\bar{g}_0(x) - Z_0\right)}, & \text{if} \quad Z_0 < \bar{g}_0(x) \leqslant Z_0 + d_0. \end{cases} \tag{3}$$

If $\tilde{A}_i^1 = \{x | \tilde{g}_i(x) \lesssim 1, x > 0\}, \tilde{A}_i^2 = \{x | \tilde{g}_i(x) \gtrsim 1, x > 0\}$ are fuzzy feasible solution sets, the membership functions of constraints $\tilde{g}_i(x)$ defined by

$$\tilde{A}_i(x) = \begin{cases} 1, & \text{if} \quad \bar{g}_i(x) \leqslant 1 \\ e^{-\frac{1}{d_i}\left(\bar{g}_i(x) - 1\right)}, & \text{if} \quad 1 < \bar{g}_i(x) \leqslant 1 + d_i, \ \text{at } 1 \leqslant i \leqslant p' \end{cases} \tag{4}$$

$$\tilde{A}_i(x) = \begin{cases} 0, & \text{if} \quad \bar{g}_i(x) \leqslant 1 \\ e^{-\frac{1}{d_i}\left(1 - \bar{g}_i(x)\right)}, & \text{if} \quad 1 < \bar{g}_i(x) \leqslant 1 + d_i, \ \text{at } p'+1 \leqslant i \leqslant p \end{cases} \tag{5}$$

respectively, $d_i \geqslant 0 (0 \leqslant i \leqslant p)$ is a flexible index of $i$-th fuzzy function $\tilde{g}_i(x)$.

If we define $\widetilde{\min}\ \tilde{g}_0(x) \leftarrow \tilde{g}_0(x) \lesssim Z_0$, $Z_0$ might have to be written down as a minimizing goal in order to consider $\tilde{g}_0(x)$ as an upper bound, then $(\tilde{P})$ can be rewritten down as

$$\begin{cases} \tilde{g}_0(x) \lesssim Z_0, \\ \tilde{g}_i(x) \lesssim 1, (1 \leqslant i \leqslant p'), \ \tilde{g}_i(x) \gtrsim 1, (p'+1 \leqslant i \leqslant p), x > 0. \end{cases} \tag{6}$$

**Definition 1.** *Let $\tilde{A}_0(x)$ be a fuzzy object function defined on $X \subset R^m$ and $\tilde{A}_i^1(x), \tilde{A}_i^2(x)$ is a fuzzy feasible solution set defined on $X \subset R^m$. We call $\tilde{Y} = \tilde{A}_0 \cap \tilde{A}^1 \cap \tilde{A}^2 = \tilde{A}_0 \cap \bigcap\limits_{1 \leqslant i \leqslant p'} \tilde{A}_i^1 \cap \bigcap\limits_{p'+1 \leqslant i \leqslant p} \tilde{A}_i^2$ a fuzzy decision for $(\tilde{P})$, satisfying*

$$\tilde{Y}(x) = \tilde{A}_0(x) \wedge \min_{1 \leqslant i \leqslant p'} \tilde{A}_i^1(x) \wedge \min_{p'+1 \leqslant i \leqslant p} \tilde{A}_i^2(x), x > 0, \tag{7}$$

*calling a point $x^*$ a fuzzy optimal solution to $(\tilde{P})$, satisfying*

$$\tilde{Y}(x^*) = \max_{x>0}\{\tilde{Y}(x) = \min\{\tilde{A}_0(x), \min_{1\leqslant i\leqslant p'} \tilde{A}_i^1(x), \min_{p'+1\leqslant i\leqslant p} \tilde{A}_i^2(x)\}\}. \tag{8}$$

*Suppose $\tilde{A}_0$ is a fuzzy optimal point set of $\tilde{g}_0(x)$, calling (8) an FRPGP for $\tilde{g}_0(x)$ with respect to $\tilde{Y}$.*

Here we only involve discussing the equal level for both constraint and optimalization of $x^*$. In particular, $x^*$ is a solution to classical GP $(P)$ at $Y(x^*) = 1$.

**Theorem 1.** *If $\tilde{A}_i(x), \tilde{B}_i(\cdot), \tilde{c}_{ik}$ and $\tilde{\gamma}_{ikl}(1 \leqslant k \leqslant J_i, 0 \leqslant i \leqslant p, 1 \leqslant l \leqslant m)$ are all fuzzy continuous and strictly monotonous function (CSMF), the maximizing of $\tilde{Y}(x)$ is equivalent to*

$$(\bar{P}) \quad \max\ \alpha \tag{9}$$

$$\begin{aligned} s.t.\ \tilde{A}_0(x) &= \{\sum_{k=1}^{J_0} \tilde{c}_{0k}^{-1}(\beta) \prod_{l=1}^{m} x_l^{\tilde{\gamma}_{0kl}^{-1}(\beta)}\}/(\tilde{B}_0^{-1}(\alpha)) \leqslant 1,\\ \tilde{A}_i^1(x) &= \{\sum_{k=1}^{J_i} \tilde{c}_{ik}^{-1}(\beta) \prod_{l=1}^{m} x_l^{\tilde{\gamma}_{ikl}^{-1}(\beta)}\}/(\tilde{B}_i^{-1}(\alpha)) \leqslant 1, (1 \leqslant i \leqslant p'), \\ \tilde{A}_i^2(x) &= \{\sum_{k=1}^{J_i} \tilde{c}_{ik}^{-1}(\beta) \prod_{l=1}^{m} x_l^{\tilde{\gamma}_{ikl}^{-1}(\beta)}\}/(\tilde{B}_i^{-1}(\alpha)) \geqslant 1, (p'+1 \leqslant i \leqslant p),\\ & x > 0, \alpha, \beta \in [0,1]. \end{aligned} \tag{10}$$

*where $\tilde{B}_i^{-1}(\alpha)(0 \leqslant i \leqslant p)$ from (1)-(3) are defined as*

$$\tilde{B}_0^{-1}(\alpha) = Z_0 - d_0 \log\alpha; \tilde{B}_i^{-1}(\alpha) = \begin{cases} 1 - d_i \log\alpha, & (1 \leqslant i \leqslant p'),\\ 1 - d_i \log(1-\alpha), & (p'+1 \leqslant i \leqslant p). \end{cases}$$

*Proof.* Substituting (1)(2)(3)(4)(5) into (9), when $\tilde{A}_0(x) = Z_0 - d_0$, and $\tilde{A}_i(x) = 1 + d_i$, after some rearrangements [2], then (9) arrive at

$$\begin{aligned} \tilde{Y}(x) = {} & \tilde{A}_0(x) \wedge \min_{1\leqslant i\leqslant p'} \tilde{A}_i^1(x) \wedge \min_{p'+1\leqslant i\leqslant p} \tilde{A}_i^2(x) = \exp\{-(\bar{g}_0(x) - Z_0)/d_0\} \wedge \\ & \min_{1\leqslant i\leqslant p'} \exp\{(\bar{g}_i(x) - 1)/d_i\} \wedge \min_{p'+1\leqslant i\leqslant p} \{1 - \exp\{-(\bar{g}_i(x) - 1)/d_i \\ = {} & \exp\Big\{-\frac{\sum_{k=1}^{J_0} \tilde{c}_{0k}^{-1}(\beta) \prod_{l=1}^{m} x_l^{\tilde{\gamma}_{0kl}^{-1}}(\beta) - Z_0}{d_0}\Big\} \wedge \min_{1\leqslant i\leqslant p'} \exp\Big\{-\frac{\sum_{k=1}^{J_i} \tilde{c}_{ik}^{-1}(\beta) \prod_{l=1}^{m} x_l^{\tilde{\gamma}_{ikl}^{-1}(\beta)} - 1}{d_i}\Big\} \\ & \wedge \min_{p'+1\leqslant i\leqslant p} \Big\{\exp\Big\{-\frac{1 - \sum_{k=1}^{J_i} \tilde{c}_{ik}^{-1}(\beta) \prod_{l=1}^{m} x_l^{\tilde{\gamma}_{ikl}^{-1}(\beta)}}{d_i}\Big\}\Big\}. \end{aligned}$$

The maximization decision of $(\tilde{P})$ can be turned into a solution to $x$ such that $\max_{x>0}\{\tilde{Y}(x)\}$. Introducing a new variable $\alpha$, let $\alpha = \tilde{Y}(x)$, then $\max_{x>0} \alpha$ and $\tilde{Y}(x) \geqslant \alpha$. When $0 \leqslant i \leqslant p'$, we have

$$\exp\Big\{-(\sum_{k=1}^{J_i}\tilde{c}_{ik}^{-1}(\beta)\prod_{l=1}^{m}x_l^{\tilde{\gamma}_{ikl}^{-1}(\beta)}-b_i)/d_i\Big\}\geqslant\alpha \tag{11}$$

$$\Longleftrightarrow \bar{g}_i(x)=\sum_{k=1}^{J_i}\tilde{c}_{ik}^{-1}(\beta)\prod_{l=1}^{m}x_l^{\tilde{\gamma}_{ikl}^{-1}(\beta)}\leqslant b_i-d_i\log\alpha. \tag{12}$$

It is equivalent to $\bar{g}_i(x)/(b_i-d_i\log\alpha)\leqslant 1$ because $b_i-d_i\log\alpha>0$ is obvious larger than zero for $\alpha\in[0,1]$, where $b_i=\begin{cases}Z_0, & \text{when } (i=0),\\ 1, & \text{when } (1\leqslant i\leqslant p');\end{cases}$ when $p'+1\leqslant i\leqslant p$, we have

$$\exp\Big\{-(1-\sum_{k=1}^{J_i}\tilde{c}_{ik}^{-1}(\beta)\prod_{l=1}^{m}x_l^{\tilde{\gamma}_{ikl}^{-1}(\beta)})/d_i\Big\}\geqslant\alpha$$

$$\Longleftrightarrow \bar{g}_i(x)=\sum_{k=1}^{J_i}\tilde{c}_{ik}^{-1}(\beta)\prod_{l=1}^{m}x_l^{\tilde{\gamma}_{ikl}^{-1}(\beta)}\geqslant 1+d_i\log\alpha. \tag{13}$$

It is equivalent to $\bar{g}_i(x)/(1+d_i\log\alpha)\geqslant 1$, here $1+d_i\log\alpha>0$ is supposed to represent the sign inequality.

If converse $\tilde{B}_i^{-1}(\alpha)(0\leqslant i\leqslant p)$ in (12) (13) is written down as

$$\tilde{B}_i^{-1}(\alpha)=\begin{cases}b_i-d_i\log\alpha, & (0\leqslant i\leqslant p'),\\ 1+d_i\log\alpha, & (p'+1\leqslant i\leqslant p),\end{cases}$$

then (10) is got. It is obvious that they are greater than 0 for $\alpha\in[0,1]$.

Therefore, the maximizing of $\tilde{Y}(x)$ is equivalent to the truth of (9)-(10).

**Theorem 2.** *Suppose $(x_1^*,x_2^*,\ldots,x_m^*;\alpha,\beta)^T$ to be an optimal solution to (9)-(10), then $(x_1^*,x_2^*,\ldots,x_m^*)^T$ is a fuzzy optimal solution to $(\tilde{P})$.*

*Proof.* Obviously it is easy to prove.

**Theorem 3.** *Suppose that fuzzy-valued functions $\tilde{A}_i=\tilde{B}_i\circ\tilde{\phi}_i$ are CSMF, where $\tilde{B}_i$ and $\tilde{\phi}_i=\tilde{c}_{ik}\circ\tilde{\gamma}_{ikl}(1\leqslant k\leqslant J_i, 0\leqslant i\leqslant p, 1\leqslant l\leqslant m)$ are CSMF. A dual programming of $(\tilde{P})$ is*

$$(\tilde{D})\qquad \widetilde{\max}\tilde{d}'(w)=(\tilde{a}_{0k}/w_{00})^{w_{00}}\prod_{i=0}^{p'}\prod_{k=1}^{J_i}(\tilde{c}_{ik}/\tilde{a}_{ik}w_{ik})^{w_{ik}}$$
$$\times\prod_{i=p'+1}^{p}\prod_{k=1}^{J_i}(\tilde{c}_{ik}/\tilde{a}_{ik}w_{ik})^{-w_{ik}}\prod_{i=1}^{p'}w_{i0}^{w_{i0}}\prod_{i=p'+1}^{p}w_{i0}^{-w_{i0}}$$
$$s.t.\ w_{00}=\sum_{k=1}^{J_0}w_{0k}=1, \tilde{\Gamma}'^{T}w=0, w\geqslant 0,$$

*where $w=(w_{00},w_{01},\ldots,w_{0J_0},w_{p1},\ldots,w_{pJ_p})^T$ is a $J'-$dimensional fuzzy variable vector $(J'=1+J_0+\cdots+J_p)$, and $w_{i0}=w_{i1}+w_{i2}+\cdots+w_{iJ_i}(1\leqslant i\leqslant p)$ is the sum of each set of dual variables corresponding to an objective function $\tilde{g}_0(x)(i=0)$ or constraints function $\tilde{g}_i^*(x)(\tilde{g}_i^*(x)=\tilde{g}_i(x)-1)(1<i\leqslant p)$; $-w_{ik}$ and $-w_{i0}$ denote a reversed direction inequality $\tilde{g}_i(x)\gtrsim 1$ corresponding to factors $(\frac{\tilde{c}_{ik}}{w_{ik}})^{-w_{ik}}$ and $w_{i0}^{-w_{i0}}$ in the upper-right-corner exponent. $\tilde{\Gamma}'$ represents a fuzzy exponent matrix $(\sigma_i\tilde{\gamma}_{ikl}), i=0,\ldots,p; k=1,\ldots,J_i; l=1,\ldots,m; \sigma_i=1, i=1,\ldots,p'; \sigma_i=-1, i=p'+1,\ldots,p$.*

*Proof.* From Theorem 1, $(\tilde{P})$ is turned into (9)-(10).

Similarly, as to the proof in Theorem 4.1 in [2], we deduce the dual programming of $(\bar{P})$ as follows:

$$(\bar{D}) \quad \max\, (-\alpha/\bar{w}_{00})^{\bar{w}_{00}} \prod_{i=0}^{p'} \prod_{k=1}^{J_i} (\tilde{c}_{ik}^{-1}(\beta)/\tilde{B}_i^{-1}(\alpha)\bar{w}_{ik})^{\bar{w}_{ik}}$$

$$\times \prod_{i=p'+1}^{p} \prod_{k=1}^{J_i} (\tilde{c}_{ik}^{-1}(\beta)/\tilde{B}_i^{-1}(\alpha)\bar{w}_{ik})^{-\bar{w}_{ik}} \prod_{i=1}^{p'} \bar{w}_{i0}^{\bar{w}_{i0}} \prod_{i=p'+1}^{p} \bar{w}_{i0}^{-\bar{w}_{i0}}$$

$$\text{s.t.} \quad \bar{w}_{00} = \sum_{k=1}^{J_0} \bar{w}_{0k} = 1, \bar{\Gamma}'^{\mathrm{T}}\bar{w} = 0, \alpha, \beta \in [0,1], \bar{w} \geqslant 0,$$

where $\bar{w} = (w, \alpha, \beta)$, and $\bar{\Gamma}' = (\sigma_i \bar{\gamma}_{ikl}(\beta)), i = 0, \ldots, p; k = 1, \ldots, J_i; l = 1, \ldots, m; \sigma_i = 1, i = 1, \ldots, p'; \sigma_i = -1, i = p'+1, \ldots, p$.

Again, $(\bar{D}), (\tilde{D})$ denotes dual programming of$(\bar{P}), (\tilde{P})$, respectively, therefore, the theorem holds by transitivity of an equivalence relation.

## 3 Properties of Prime FRPGP

**Definition 2.** *Let $\tilde{Y}$ be a fuzzy set defined on $E \subset \Re^m$, (6) is called a problem of FRPGP in $\tilde{g}_0(x)$ with respect to $\tilde{Y}$. If $E \subset \Re^m$ is a convex set, and $-\tilde{g}_i(x)(0 \leqslant i \leqslant p')$ and $\tilde{g}_i(x)(p'+1 \leqslant i \leqslant p)$ are fuzzy (or strongly fuzzy) convex functions [2] with respect to $\tilde{A}_0$ and $\tilde{A}_i$, respectively, (6) is called a problem of fuzzy (or strongly fuzzy) convex reversed PGP with respect to $\tilde{g}_0(x)$.*

**Theorem 4.** *The FRPGP $(\tilde{P})$ is a convex fuzzy set, that is, equivalent to the same with (6), while the fact that (6) is a convex fuzzy set equivalent that* (6) *is a convex reversed PGP for any $\alpha, \beta \in [0,1]$.*

*Proof.* Because of (6) equivalence to $(\tilde{P})$, then (6), a convex fuzzy set, is equivalent that $(\tilde{P})$ is convex fuzzy set obviously.

Because of $Y_{\alpha,\beta} = \{\bar{x}|\tilde{Y}(\bar{x}) \geqslant \alpha\}$, and $\forall \bar{x}_1, \bar{x}_2 \in Y_{\alpha,\beta}$ and $\forall \alpha, \beta \in [0,1]$ in the event of $\tilde{Y}$ being a convex fuzzy set, there is $\tilde{Y}(\lambda\bar{x}_1 + (1-\lambda)\bar{x}_2) \geqslant \tilde{Y}(\bar{x}_1) \bigwedge \tilde{Y}(\bar{x}_2) \geqslant \alpha$, that is, $\lambda\bar{x}_1 + (1-\lambda)\bar{x}_2 \in Y_{\alpha,\beta}$, where $\bar{x} = x(\alpha,\beta), \bar{x}_1 = x_1(\alpha,\beta), \bar{x}_2 = x_2(\alpha,\beta)$. Therefore $Y_{\alpha,\beta}$ is a convex set, such that $\tilde{Y}$ is convex reversed PGP.

Whereas, if $\forall \alpha, \beta \in [0,1]$, then $Y_{\alpha,\beta}$ is a convex set. $\forall\ \bar{x}_1, \bar{x}_2 \in E$, if let $\alpha = \tilde{Y}(\bar{x}_1) \bigwedge \tilde{Y}(\bar{x}_2)$, then $\tilde{Y}(\bar{x}_1) \geqslant \alpha, \tilde{Y}(\bar{x}_2) \geqslant \alpha$, therefore $\bar{x}_1, \bar{x}_2 \in Y_{\alpha,\beta}$. From the convexity of $Y_{\alpha,\beta}$, such that $\lambda\bar{x}_1 + (1-\lambda)\bar{x}_2 \in Y_{\alpha,\beta}$ for any $\lambda \in [0,1]$, that is, $\tilde{Y}(\lambda\bar{x}_1 + (1-\lambda)\bar{x}_2) = \tilde{Y}(\bar{x}_1) \bigwedge \tilde{Y}(\bar{x}_2)$. It is known by the definition of a convex fuzzy set that $\tilde{Y}$ is a convex fuzzy set and so is (6).

Generally, a reversed PGP is non-convex, so is the deformed substitution by $x_i = e^{y_i}$. Therefore, it does not keep nice properties [7] as a PGP. But we find that many results can be obtained in an FRPGP like a fuzzy PGP after fuzzy GP is nonfuzzified by adopting the membership function (3)-(5) mentioned in the paper.

**Theorem 5.** *If* $\tilde{B}_i$ *is* $\begin{cases} \text{a non-creasing function of CSMF for } 1 \leqslant i \leqslant p'; \\ \text{a non-decreasing function of CSMF for } p'+1 \leqslant i \leqslant p; \end{cases}$ *for all* $i \in (0,p)$, $\tilde{c}_{ik}$ *and* $\tilde{\gamma}_{ikl}(1 \leqslant k \leqslant J_i, 0 \leqslant i \leqslant p, 1 \leqslant l \leqslant m)$ *are CSMF, then FRPGP* $(\tilde{P})$ *can be turned into a convex FRPGP.*

*Proof.* Because $(\tilde{P})$ can be changed into $(\bar{P})$ by Formal (1)-(5), with $g_i(\bar{x}) = \sum\limits_{k=1}^{J_i} \tilde{c}_{ik}^{-1}(\beta) \prod\limits_{l=1}^{m} x_l^{\tilde{\gamma}_{ikl}^{-1}(\beta)} / \tilde{B}_i^{-1}(\alpha)$ in $(\bar{P})$, and if writing $f_{ik}(\bar{x}) = \prod\limits_{l=1}^{m} x_l^{\tilde{\gamma}_{ikl}^{-1}(\beta)} / \tilde{B}_i^{-1}(\alpha)$, then $g_i(\bar{x}) = \sum\limits_{k=1}^{J_i} \tilde{c}_{ik}^{-1}(\beta) f_{ik}(\bar{x})$. Again because $\tilde{c}_{ik}^{-1}(\beta) \geqslant 0$, we might as well suppose $f_{ik}(\bar{x})$ to nondecrease, so does $g_i(\bar{x})$. Therefore for all $x^1, x^2 \in X, \lambda, \beta \in [0,1]$, we have $g_i[\lambda \bar{x}_1 + (1-\lambda)\bar{x}_2] \leqslant \lambda g_i(\bar{x}_1) + (1-\lambda) g_i(\bar{x}_2)$, such that as for nondecrease of $\tilde{B}_i$, we get

$$\tilde{B}_i\big(g_i(\lambda \bar{x}^1 + (1-\lambda)\bar{x}^2)\big) \leqslant \tilde{B}_i\big(\lambda g_i(\bar{x}_1) + (1-\lambda) g_i(\bar{x}_2)\big) \leqslant \max\{\tilde{B}_i\big(g_i(\bar{x}^1)\big), \tilde{B}_i\big(g_i(\bar{x}^2)\big)\}.$$

Therefore, $\tilde{A}_i(\bar{x}) = \tilde{B}_i(g_i(\bar{x}))$ is a quasi-convex function valued on [0,1], such that $g_i(\bar{x})$ is a fuzzy convex function for $\tilde{A}_i(\bar{x})$, hence $(\bar{P})$ is a convex reversed PGP with respect to $x$. The theorem holds from arbitrarily of $\alpha, \beta$ in [0,1].

**Theorem 6.** *Let* $\tilde{A}_i(0 \leqslant i \leqslant p)$ *be a CSMF. Then the fuzzy local minimum solution (LMS) to* $(\tilde{P})$ *is also its fuzzy global minimum solution (GMS).*

*Proof.* From Theorem 5, we know $(\tilde{P})$ is fuzzy convex, then, $\forall \alpha, \beta \in [0,1]$,

$$\tilde{Y}(\bar{x}) = \tilde{A}_0(\bar{x}) \wedge \Big(\wedge_{1 \leqslant i \leq p'} \tilde{A}_i^1(\bar{x})\Big) \wedge \Big(\wedge_{p'+1 \leqslant i \leqslant p} \tilde{A}_i^2(\bar{x})\Big)$$

is a convex fuzzy valued function [2]. Therefore, the LMS of $\tilde{Y}(\bar{x})$ is its GMS.

**Corollary 1.** *Let* $\tilde{A}_i(0 \leqslant i \leqslant p)$ *be a CSMF. Then the strictly fuzzy LMS to* $(\tilde{P})$ *is its strictly fuzzy GMS.*

**Corollary 2.** *Let* $(\tilde{P})$ *be a strongly convex FRPGP. Then any fuzzy LMS to* $(\tilde{P})$ *is its unique fuzzy GMS.*

**Theorem 7.** *Let* $\tilde{A}_i(0 \leqslant i \leqslant p)$ *be a CSMF. Then any fuzzy LMS to* $(\tilde{P})$ *is its fuzzy GMS.*

The theorem and corollaries above can be easily proved by means of [2].

## 4 Dual Theorem of the FRPGP

**Definition 3.** *If* $x > 0$, *such that the groups of condition*

$$V(\tilde{g}_i(x) < 1) = 1, (1 \leqslant i \leqslant p'), V(\tilde{g}_i(x) > 1) = 1, (p'+1 \leqslant i \leqslant p),$$

*or* $$V(\tilde{g}_i(x) \lesssim 1) = 1, (1 \leqslant i \leqslant p') V(\tilde{g}_i(x) \gtrsim 1) = 1, (p'+1 \leqslant i \leqslant p)$$

*holds, then we call them fuzzy super-consistence or fuzzy consistence in the FRPGP* $(\tilde{P})$, *respectively, where* $v(\tilde{g}_i(x)\tilde{\circledcirc} 1)$ *denotes a possibility degree of* $\tilde{g}_i(x)\tilde{\circledcirc}1$ *with*

$$v(\tilde{g}_i(x)\tilde{\circledcirc}1) = \sup_{x,y|x\tilde{\circledcirc}y} \min\{\mu_{\tilde{g}_i}(x), \mu_1(y)\}(1 \leqslant i \leqslant p),$$

*here* $\tilde{\circledcirc}$ *denotes one of operators* $<, \lesssim, >$ *and* $\gtrsim$.

**Lemma 1.** *Let* $\tilde{B}_i(0 \leqslant i \leqslant p)$ *and* $\tilde{\phi}_i = \tilde{c}_{ik} \circ \tilde{\gamma}_{ikl}(1 \leqslant k \leqslant J_i, 0 \leqslant i \leqslant p, 1 \leqslant l \leqslant m)$ *be CSMF. Then to any fuzzy feasible solution* $x$ *and feasible solution* $w$ *to prime FRPGP* $(\tilde{P})$ *and its dual program* $(\tilde{D})$, *respectively, we have*

$$\tilde{g}_0(x) \gtrsim \tilde{g}_0(x) \prod_{i=1}^{p'} (\tilde{g}_i(x))^{w_{i0}} \prod_{i=p'+1}^{p} (\tilde{g}_i(x))^{-w_{i0}} \gtrsim \tilde{d}'(w) \tag{14}$$

*and*

$$\tilde{g}_0(x) = \tilde{d}'(w) \tag{15}$$

*iff*

$$w_{ik} = \begin{cases} \tilde{v}_{0k}(x)/\tilde{g}_0(x), & (i = 0; 1 \leqslant k \leqslant J_0) \\ w_{i0}\tilde{v}_{ik}(x), & (i \neq 0; 1 \leqslant k \leqslant J_i). \end{cases} \tag{16}$$

*Proof.* Since $x$ denotes a fuzzy feasible solution to $(\tilde{P})$, then $\bar{x}$ is a feasible one to $(\bar{P})$, that is, for $\tilde{B}_i, \tilde{c}_{ik}$ and $\tilde{\gamma}_{ikl}$, $g_i(\bar{x}) = g_i(x')/\tilde{B}_i^{-1}(\alpha) \leqslant 1(0 \leqslant i \leqslant p')$, and $g_i'(\bar{x}) = \tilde{B}_i^{-1}(\alpha)/g_i(x') \leq 1(p'+1 \leqslant i \leqslant p)$. Again, $\bar{w}_{i0} \geqslant 0$, hence $g_0(\bar{x}) \geqslant g_0(\bar{x}) \prod_{i=1}^{p'} (g_i(\bar{x}))^{\bar{w}_{i0}} \prod_{i=p'+1}^{p} (g_i'(\bar{x}))^{\bar{w}_{i0}}$. And because $g_i(\bar{x}) = \sum_{k=1}^{J_i} v_{ik}(\bar{x}), g_i'(\bar{x}) = 1/g_i(\bar{x})$, from the ordinary geometric inequality, then $(g_i(\bar{x}))^{\bar{w}_{i0}} \geqslant \prod_{k=1}^{J_i} \left(\frac{v_{ik}(\bar{x})}{\bar{w}_{ik}}\right)^{\bar{w}_{ik}} \bar{w}_{i0}^{\bar{w}_{i0}} (0 \leqslant i \leqslant p')$, whereas there must exist $\alpha, \beta \in [0,1]$ for the feasible solution $\bar{x}$ and $\bar{w}$ to $(\bar{P})$ and $(\bar{D})$, respectively, such that $\bar{w}_{00} = 1$ and $\bar{\Gamma}'^{\mathrm{T}}\bar{w} = 0$, then

$$-\alpha \geq (-\alpha) \prod_{i=0}^{p'} (g_i(\bar{x}))^{\bar{w}_{ik}} \prod_{i=p'+1}^{p} (g_i'(\bar{x}))^{\bar{w}_{ik}} \geq \bar{d}'(\bar{w}), \tag{17}$$

where $v_{ik}(\bar{x}) = v_{ik}(x')/\tilde{B}_i^{-1}(\alpha), \bar{w} = w(\alpha, \beta)$. Therefore, $g_0(\bar{x}) \geqslant d'(\bar{w}) = \tilde{B}_0^{-1}(\alpha)\bar{d}'(\bar{w})/(-\alpha)$, and

$$g_0(\bar{x}) = \bar{d}'(\bar{w}) \tag{18}$$

iff

$$\bar{w}_{ik} = \begin{cases} v_{00}(x')/(-\alpha), & (i = 0; k = 0, \alpha \neq 0) \\ \bar{w}_{i0} v_{ik}(x'), & (0 \leqslant i \leqslant p; 1 \leqslant k \leqslant J_i) \end{cases} \tag{19}$$

holds. Since (14)–(16) holds in the light of arbitration of $\alpha, \beta$ in $[0,1]$, and (17)⇔ (14), (18)⇔ (15), (19)⇔ (16), hence the lemma holds.

**Theorem 8.** *(The first dual theorem) Let the FRPGP* $(\tilde{P})$ *be fuzzy super-consistence with fuzzy optimal solution* $x^*$ *existing. Then there must exist a Lagrange multiplier* $\mu^* = (\mu_1^*, \mu_2^*, \ldots, \mu_{p'}^*, \mu_{p'+1}^*, \cdots, \mu_p^*)^T \geqslant 0$, *and CSMF* $\tilde{A}_i(0 \leqslant i \leqslant p)$, *such that*

$$\nabla \tilde{g}_0(x^*) + \sum_{i=1}^{p'} \mu_i^* \nabla \tilde{g}_i(x^*) + \sum_{i=p'+1}^{p} \mu_i^* \nabla(-\tilde{g}_i(x^*)) = 0, \tag{20}$$

$$\mu_i^*(\tilde{g}_i(x^*) - 1) = 0, \quad (1 \leqslant i \leqslant p), \tag{21}$$

*while* $w^*$ *defined by*

$$w_{ik}^* = \begin{cases} \tilde{v}_{0k}(x^*)/\tilde{g}_0(x^*), & (i = 0; 1 \leqslant k \leqslant J_0) \\ \mu_i^* \tilde{v}_{ik}(x^*)/\tilde{g}_0(x^*), & (i \neq 0; 1 \leqslant k \leqslant J_i) \end{cases} \tag{22}$$

*is an optimal solution to dual program* $(\tilde{D})$, *and the optimal value of these two programs is equality, i.e.,*

$$\tilde{g}_0(x^*) = \tilde{d}'(w^*). \tag{23}$$

*Proof.* As for all the CSMF$\tilde{A}_i(0 \leqslant i \leqslant p)$, $(\tilde{P})$ fuzzy super-consistence having a fuzzy optimal solution $x^*$ is equivalent to $(\bar{P})$ being super-consistent with optimal solution $\bar{x}^*$ for all $\alpha, \beta \in [0, 1]$. Again, because $(\bar{P})$ is convex with respect to $x$ for $\alpha, \beta$, satisfying the Slater condition, there exists an optimal parametric solution $\bar{x}^*$. Therefore, there exists a Lagrange multiplier $\mu^* = (\mu_1^*, \mu_2^*, \ldots, \mu_{p'}^*, \mu_{p'+1}^*, \cdots, \mu_p^*)^{\mathrm{T}} \geqslant 0$ from Theorem 1.6.3 in [7], such that

$$\nabla g_0(\bar{x}^*) + \sum_{i=1}^{p'} \mu_i^* \nabla g_i(\bar{x}^*) + \sum_{i=p'+1}^{p} \mu_i^* \nabla(-g_i(\bar{x}^*)) = 0, \tag{24}$$

$$\mu_i^*(g_i(\bar{x}^*) - 1) = 0 \quad (1 \leqslant i \leqslant p), \tag{25}$$

while $\bar{w}_{ik}^*$ defined by

$$\bar{w}_{ik}^* = \begin{cases} v_{00}(x'^*)/(-\alpha) & (i = 0; k = 0, \alpha \neq 0) \\ \mu_i^* v_{ik}(x'^*)/g_0(\bar{x}^*) & (1 \leqslant k \leqslant J_i; 0 \leqslant i \leqslant p) \end{cases} \tag{26}$$

is an optimal parametric solution to dual program $(\bar{D})$, where $\bar{x}^* = x^*(\alpha, \beta), x'^* = x^*(\beta)$, and there exists the determined $\alpha, \beta \in [0, 1]$, such that

$$g_0(\bar{x}^*) = d(\bar{w})\tilde{B}_i^{-1}(\alpha)/(-\alpha) = d'(\bar{w}^*). \tag{27}$$

Again, (24)–(27) hold by the arbitrariness of $\alpha, \beta$ in [0,1], then (20)–(23) hold. It follows that the theorem holds.

**Corollary 3.** *Let a general FRPGP be* $(\tilde{P}_I)$ *and its dual programming be* $(\tilde{D}_S)$. *If, to the CSMF* $\tilde{A}_i(0 \leqslant i \leqslant p)$, $(\tilde{P}_I)$ *is fuzzy super-consistence, with fuzzy constraint infimum [2]* $M_{\tilde{P}_I} > 0$, *there must exist a Lagrange multiplier* $\mu^* = (\mu_1^*, \mu_2^*, \ldots, \mu_{p'}^*, \mu_{p'+1}^*, \cdots, \mu_p^*)^T \geqslant 0$, *such that (17)-(19) hold, and* $w^*$ *is an optimal solution to a dual program* $(\tilde{D}_S)$ *with* $M_{\tilde{P}_I} = \tilde{d}'(w^*)$.

*Proof.* Because $(\tilde{P})$ has a minimum in Theorem 8, correspondingly $(\tilde{P}_I)$ has a constrained minimum which must be positive. Therefore the theorem holds, which is proved in the same way of Theorem 6.

All of the above conclusions assume that there exists a fuzzy optimal solution to FRPGP $(\tilde{P})$. When is a fuzzy optimal one obtained in $(\tilde{P})$?

**Theorem 9.** *(The second dual theorem) Let $(\tilde{P})$ be fuzzily consistent and $(\tilde{D})$ have a feasible solution with components being positive. Then to CSMF $\tilde{A}_i = \tilde{B}_i \circ \tilde{\phi}_i$, where $\tilde{\phi}_i = \tilde{c}_{ik} \circ \tilde{\gamma}_{ikl}(1 \leqslant k \leqslant J_i, 0 \leqslant i \leqslant p, 1 \leqslant l \leqslant m)$, there exists a fuzzy optimal solution in $(\tilde{P})$.*

*Proof.* For CSMF $\tilde{A}_i(0 \leqslant i \leqslant p)$, $(\tilde{P})$ can be changed into a convex GP $(\bar{P})$, while the dual form of $(\bar{P})$ is $(\bar{D})$. Because $(\tilde{P})$ is fuzzy consistency $\Leftrightarrow$ $(\bar{P})$ consistency; again, the constraint condition of $(\tilde{D})$ is equivalent to that of $(\bar{D})$, namely, $(\tilde{D})$ contains a feasible solution with components being positive $\Leftrightarrow$ $(\bar{D})$ has a parameter feasible solution with components being positive, we can get a reformed prime programming of $(\bar{P})$ by replacing $x_l = e^{z_l}(1 \leqslant l \leqslant m)$. It also consists with point range $\{z^k\}$ existing and satisfying

$$G_i(\bar{z}^k) \leqslant 1(1 \leqslant i \leqslant p'), \quad G_i^{-1}(\bar{z}^k) \leqslant 1(p'+1 \leqslant i \leqslant p),$$

such that $\lim_{k\to\infty} G_0(\bar{z}^k) = M_{\bar{P}}$, in similar way of the Theorem 1.8.1 in [7].
Accordingly there must exist $\alpha, \beta \in [0,1]$, and then

$$G_i(\bar{z}^*) = \begin{cases} \sum_{k=1}^{J_i} \tilde{c}_{ik}^{-1}(\beta)/\tilde{B}_i^{-1}(\alpha) \prod_{l=1}^{m} x_l^{*\tilde{\gamma}_{ikl}^{-1}(\beta)} = \begin{cases} = M_{\bar{P}}, \ (i=0), \\ \leqslant 1, \quad (1 \leqslant i \leqslant p'), \end{cases} \\ \sum_{k=1}^{J_i} \tilde{c}_{ik}^{-1}(\beta)/\tilde{B}_i^{-1}(\alpha) \prod_{l=1}^{m} x_l^{*-\tilde{\gamma}_{ikl}^{-1}(\beta)} \geqslant 1, \quad (p'+1 \leqslant i \leqslant p), \end{cases}$$

where $\bar{z}_l^* = z_l^*(\alpha,\beta)$ is a limit point of $z_l^k = z_l^k(\alpha,\beta)$, $\bar{x}_l^* = x_l^*(\alpha,\beta)$ is a limit point of $x_l^k = x_l^k(\alpha,\beta)$, that is, $\bar{x}_l^*(1 \leqslant l \leqslant m)$ is an optimal solution to $(\bar{P})$.

## 5 Numerical Examples

*Example 1.* Find an FRPGP

$$\begin{aligned} &\widetilde{\min}\ g_0(x) = 2x_1 + 3x_2 \\ &\text{s.t. } g_1(x) = x_1^2 + x_2^2 \gtrsim 1, x_1, x_2 > 0. \end{aligned} \tag{28}$$

*Solution 1.* The classical GP in (28) is non-convex. Given $x_1 = e^u, x_2 = e^v$, its deformed programming

$$\begin{aligned} &\min\ 2e^u + 3e^v \\ &\text{s.t. } e^{2u} + e^{2v} \geqslant 1 \end{aligned}$$

is nonconvex [7]. This prevents much perfect relations from fuzzy PGP, such that difficulties appear in the solution to FRPGP. We might as well take $d_0 = 1$,$d_1 = 1/2$ for the sake of convenience and let $g_0(x) = d(w) = 3$. Then (28) can be turned into a determined reversed PGP, its dual solution is $\bar{w}_{00} = 1, \bar{w}_{01} = 0, \bar{w}_{02} = 1, \bar{w}_{11} = 0, \bar{w}_{12} = 1/2, \alpha = 0.7575 \in (0.60653, 1]$. By means of (16), then $(x_1, x_2, \alpha) = (0, 2/3, 0.7575)$ is an optimal solution to prime problem (28); the optimal value of prime problem is equal to its dual optimal value 2.

Let us consider the reformed prime parameter programming as

$$\begin{aligned} &\min(-\alpha) \\ &\text{s.t. } \frac{2e^u + 3e^v}{3 - \log\alpha} \leqslant 1, \ \frac{e^{2u} + e^{2v}}{1 + 2\log\alpha} \geqslant 1, \\ &\quad\ 1 + 2\log\alpha > 0, \alpha \in (0, 1). \end{aligned}$$

$(0, -1, 0.7575)$,$(-1, 0, 0.7575)$ and $(-1/2, -1/2, 0.7575)$ are selected for points $(u, v, \alpha)$, respectively. Therefore, it can be changed into a convex one with respect to $u, v$ and $\alpha$. This outcome differs from a classical GP.

*Example 2.* If we use membership functions $\tilde{A}_0(x) = (1 + \exp\{2x_1 + 3x_2 - 3\})^{-1}, \tilde{A}_1(x) = (1 + \exp\{-(x_1^2 + x_2^2 - 1) - 1/2\})^{-1}$, (28) is solved again.

When $\alpha = (1 + e^{0.5-4/9})^{-1} \approx 0.486 (\in (1 + e^{0.5})^{-1}, 1])$ is chosen, the optimal solution is the same as the one to Example 1.

## 6 Conclusion

This paper changes an FRPGP into a convex parameter GP with respect to $\alpha$,$\beta$, so that the result of the FRPGP differs from that in the reversed PGP. Meanwhile, policymakers in application decide an optimal distribution scheme with respect to $\alpha$,$\beta$ in terms of an objective value, which leaves flexible room for them to decide in.

This research is supported by the National Natural Science Foundation of China and by the Foundation of Guangzhou University.

## References

1. Cao, B.Y.: Solution and theory of question for a kind of fuzzy positive geometric program. Proc. of 2nd IFSA Congress, Tokyo, Vol.1(1987) 205-208
2. Cao, B.Y.: Fuzzy geometric programming (I). Fuzzy Sets and Systems 53(1993) 135-153
3. Cao, B.Y.: Fuzzy geometric programming optimum seeking of scheme for waste water disposal in power plant. Proc. of FUZZY-IEEE/IFES' 95, Yokohama 793-798
4. Cao, B.Y.: Fuzzy geometric programming optimum seeking in power supply radius of transformer substation. IEEE Int. Fuzzy Systems Conference Proceedings, Korea. 3(1999) 1749-1753
5. Mandal, N.K., Roy, T.K., Maiti, M.: Multi-objective fuzzy inventory model with three constraints: A geometric programming approach. Fuzzy Sets and Systems 150(2005) 87-106
6. Liu, S.T.: Fuzzy geometric programming approach to a fuzzy machining economics model. Int. J. Prod. Res., Vol.42,no.16(2004) 3253-3269
7. Wu, F., Yuan, Y.Y.: Geometric programming. Mathematics in Practice and Theory 2(1987) 68-77

# Posynomial Fuzzy Relation Geometric Programming

Ji-hui Yang[1] and Bing-yuan Cao[2]

[1] Department of Mathematics, Shantou University
Guangdong, ZIP 515063, P.R.China
yangjihui@163.com

[2] School of Mathematics and Information Science, Guangzhou University,
Guangdong, ZIP 510006, P.R.China
caobingy@163.com

**Abstract.** In this paper, the concept and type of posynomial fuzzy relation geometric programming is introduced, some basic theories of posynomial fuzzy relation geometric programming is presented, and then a solution procedure is expatiated to solving such a programming based on structure of feasible region. And finally, two practical examples are given for illustration purpose.

**Keywords:** Fuzzy relation equations, Geometric programming, Posynomial fuzzy relation geometric programming, Dual theory, Difficult degree, Optimal solution.

## 1 Introduction

Fuzzy optimization is an important branch of fuzzy mathematics, although it is not accurate to say its specific happening time of the subject branch. However, it is can be sure that its seeds should be as long as the initial era of fuzzy set theory. 1965, The fuzzy set theory was introduced for the first time by Prof.Zadeh, it is the direct starting point of fuzzy set theory father to achieve optimal control to complex systems. This is, in essence, an optimization problem, of course, that time has not yet formed the systemic fuzzy optimization theory. 1970, Bellman and Zadeh clearly given the general solution of fuzzy optimization problems in the literature[1], it became popular after 1970s and were used mainly by researchers in information, engineering and decision-making. The advantages of applying fuzzy optimization model to solving practical problems as follows: on the one hand, it can avoid rigidity and stiffness arising from conventional optimization model deal with some practical optimization problems, on the other hand, it can also efficiently reduce information loss arising from the conventional optimization model operation to data. So fuzzy optimization theory has good development prospect. This paper presents an optimization model as posynomial fuzzy relation geometric programming, let us first review some content of conventional optimization before discussing the model.

P. Melin et al. (Eds.): IFSA 2007, LNAI 4529, pp. 563–572, 2007.
© Springer-Verlag Berlin Heidelberg 2007

Conventional optimization problem can be denoted as follows:

$$\begin{array}{c} \min \ f(x) \\ s.\ t.\ x \in S. \end{array} \tag{1}$$

where $S$ is subset in $n$ dimension space $R^n$, $f(x)$ is real function on $S$, $f(x)$ is called the objective function of optimization (1), $S$ is called feasible region of (1). $S$ can be given by a variety of ways, such as direct statements, by linear or non-linear equations, by linear or nonlinear inequalities, by linear or nonlinear equations and inequalities, and so on. Function that can form $S$ is called constraint function, different types of optimization can be formed by combining the different objective and constraint functions. For instance, linear objective function and linear constraint functions is called linear programming, secondary objective function and linear constraint functions is called quadratic programming, if the objective and constraint functions have the following shape:

$$f_k(x) = \sum_{l=1}^{L_k} f_{kl}(x) = \sum_{l=1}^{L_k} c_{kl} \prod_{j=1}^{n} x_j^{\gamma_{jkl}} \, (0 \le k \le p) \tag{2}$$

then such optimization is called geometric programming. Where function $f_k(x)(0 \le k \le p)$ is called posynomial function, $f_0(x)$ is objective function, $f_k(x)(1 \le k \le p)$ is constraint functions, respectively. In particular, $f_{kl}(x)$ is call monomial function of $x$, coefficient $c_{kl} > 0$, variable $x = (x_1, x_2, \cdots, x_n)^T > 0$, exponent $\gamma_{jkl}$ $(1 \le l \le L_k, 0 \le k \le p, 1 \le j \le n)$ is arbitrary real number.

E. Sanchez, a famous French scholars, who presented fuzzy relation equations theory when he studied the questions coming from medical diagnosis in 1976[2], this is an inverse problem to fuzzy comprehensive evaluation, fuzzy relation equations play an important role in the fuzzy set theory and application, the successful application has been done in control, medicine, agriculture and other related fields[3].

The research of optimization problem with fuzzy relation equations constraint happened in eighties of the last century, but the main research direction is focus on the linear programming with fuzzy relation equations constraint[4][5]. S. C. Fang, an American scholar, who originally presented non-linear programming with fuzzy relation equations constraint in 2001, and then he gave a genetic algorithm procedure for solving such an optimization problem[6]. However, after all, genetic algorithms are a heuristic method, to a large extent, the quality to optimal solution depend on the characteristic of the objective function. Genetic algorithms can easy run into local minimum solution for solving a large class constraint optimization problem with non-linear objective function, moreover, genetic algorithms can only get approximate optimal solution when it solve some optimization problems that conventional optimization algorithms can easy obtain the optimal solution, all those show that there is many insurmountable weaknesses about genetic algorithms in solving the above optimization. It should be said that up to now we still have not found a universally valid algorithm for solving non-linear programming with fuzzy relation equations constraint. Most scholars who work in field of optimization believe it is very difficult and not necessary to trying to find a effective algorithm that fit all nonlinear programming, it is a good studying nonlinear programming method that proposed suitable algorithm based on specific type of the function. For example, the lemke algorithm is proposed to solving quadratic programming problem, the dual algorithm is proposed to solving geometric programming problem. These algorithms are playing an important role in solving practical problems. In this paper, we present posynomial geometric programming with fuzzy relation equations constraint, it is called " posynomial fuzzy relation geometric programming", and then some relevant definitions and algorithms are given to solving posynomial fuzzy relation geometric programming. The reason we raise this issue to a special discussion is: posynomial function has always been considered as a important and most widely used function in real living, it includes not only all linear function, but also the important polynomial function. The issue of many field can be good characterized by such a function. Just like geometric programming have solved a lot of practical problems at that time, we believe the same posynomial fuzzy geometric programming will also solve a lot of practical problems.

## 2 Definition of Posynomial Fuzzy Relation Geometric Programming

**Definition 2.1.** We call

$$A \circ x = b \tag{3}$$

a fuzzy relation equation, where, $x = (x_1, x_2, ..., x_n)^T$, $0 \le x_j \le 1 (1 \le j \le n)$ *is* $n$ dimension fuzzy vector, $A = [a_{ij}]$, $0 \le a_{ij} \le 1$ is $m \times n$ dimension fuzzy matrix, $b = (b_1, b_2, ..., b_m)^T$, $0 \le b_i \le 1 (1 \le i \le m)$ is $m$ dimension fuzzy vector, matrix compose operation " $\circ$ " is $\vee - \wedge$ operator, that is,

$$\bigvee_{j=1}^{n} (x_j \wedge a_{ij}) = b_i \quad (1 \le i \le m)$$

.

**Definition 2.2.** We call the following optimization model

$$\begin{aligned} &\min\ f(x) \\ &\text{s.t}\ \ A \circ x = b, \\ &\qquad 0 \le x_j \le 1 (1 \le j \le n), \end{aligned} \tag{4}$$

posynomial fuzzy relation geometric programming, where $f(x)$ is posynomial function of $x$ the same as (2), that is,

$$f(x) = \sum_{l=1}^{L} f_l(x) = \sum_{l=1}^{L} c_l \prod_{j=1}^{n} x_j^{\gamma_{jl}} \tag{5}$$

where, each $f_l(x)$ is called a monomial function of $x$, coefficient $c_l > 0$, variable $x = (x_1, x_2, \cdots, x_n)^T > 0$, exponent $\gamma_{jl}$ $(0 \le l \le L, 1 \le j \le n)$ is arbitrary real number.

In practical problems, the coefficients and exponents of objective function are sometimes difficult to accurately given. Under the circumstances, these parameters can often be expressed by Fuzzification method. It is a simple approach to using fuzzy numbers to denote these parameters that can not be accurately determined. By and large, the optimal value of the objective function should be fuzzy when the parameters are fuzzy, we call such an optimization as posynomial fuzzy relation geometric programming with fuzzy objective. The corresponding definition (4) is called a posynomial fuzzy relation geometric programming with clear objective.

To facilitate the description, suppose that $1 \geq b_1 > b_2 > ... > b_m > 0$. On the general situation, we can rank $b_i (1 \leq i \leq m)$ from large to small, the corresponding component of $x$, the corresponding row of $A$, the corresponding every term of $f(x)$ can be adjusted based on the ranking of $b_i (1 \leq i \leq m)$.

## 3 Structure of Solution Set on Fuzzy Relation Equations

Since the feasible region of optimization problem (4) is solution set to (3). Solving Equation (3) is very important to optimize (4). Next, we make some summarizer to solution set structure of Equation (3).

**Definition 3.1.** If there exists a solution in Equation (3), it is called compatible[7].

Suppose $X(A,b) = \{x = (x_1, x_2, ..., x_n)^T \in R^n | A \circ x = b, \ \ 0 \leq x_j \leq 1\}$ is the solution set of Equation (3). We define $\forall x^1, x^2 \in X(A,b)$, $x^1 \leq x^2 \Leftrightarrow x_j^1 \leq x_j^2 \ (1 \leq j \leq n)$, such a definition " $\leq$ " is a partial order relation on $X(A,b)$.

**Definition 3.2.** If $\exists \hat{x} \in X(A,b)$, such that $x \leq \hat{x}$, $\forall x \in X(A,b)$, then $\hat{x}$ is called a maximal solution to Equation (3). If $\exists \breve{x} \in X(A,b)$, such that $\breve{x} \leq x$, $\forall x \in X(A,b)$, then $\breve{x}$ is called a minimal solution to Equation (3). And if $\exists \breve{x} \in X(A,b)$, such that $x \leq \breve{x}$, then $x = \breve{x}$, $\breve{x}$ is called a minimum solution to Equation (3). let

$$\hat{x}_j = \wedge\{b_i | b_i < a_{ij}\} \ \ (1 \leq i \leq m), \ \ 1 \leq j \leq n, \tag{6}$$

suppose that $\{\wedge \emptyset = 1\}$.

If $\hat{x} = (\hat{x}_1, \hat{x}_2, ..., \hat{x}_n)$ is a solution to Equation (3), we can easily prove that $\hat{x}$ must be a maximal solution to one. For a maximal solution to Equation (3), we have the following lemma:

**Lemma 3.1.** $A \circ x = b$ is compatible if and only if there exists a maximal solution $\hat{x}$.

**Proof:** The sufficiency is evident. Now we prove necessity.

If $x$ is a solution to $A \circ x = b$, and $\bigvee_{j=1}^{n} (a_{ij} \wedge x_j) = b_i (1 \leq i \leq m)$ then $\forall i, j$, there is $a_{ij} \wedge x_j \leq b_i$. Let i be fixed, and when $a_{ij} \leq b_i$, then $0 \leq x_j \leq 1$, when $a_{ij} > b_i$, then $0 \leq x_j \leq b_i$. According to $\{\wedge \emptyset = 1\}$, we have $x_j \leq \wedge\{b_i | b_i < a_{ij}\} = \hat{x}_j$, that is $x \leq \hat{x}$.

Step forward. Suppose that $b_i < a_{ij}$, since $\hat{x}_j = \wedge\{b_i | b_i < a_{ij}\} \leq b_i$, then $a_{ij} \wedge \hat{x}_j \leq b_i$, suppose that $b_i \geq a_{ij}$, then $a_{ij} \wedge \hat{x}_j \leq a_{ij} \leq b_i$, so we have $\bigvee_{j=1}^{n} (a_{ij} \wedge \hat{x}_j) \leq b_i$, that is, $A \circ \hat{x} \leq b$. Since $x \leq \hat{x}$, then $b = A \circ x \leq A \circ \hat{x} \leq b$. Hence, $A \circ \hat{x} = b$, and $\hat{x}$ is a maximal solution, the proof is complete.

**Corollary 3.1.** If $X(A, b) \neq \emptyset$, then $\hat{x} \in X(A, b)$[8].

If a minimal solution exists in Equation (3), then solution set of (3) can be easily confirmed. However, that is not necessarily the case. The minimal solution does not often exist in Equation (3). Even under the situation of $X(A, b) \neq \emptyset$, we have not found an effective method to confirm whether $X(A, b)$ have minimum solution at present, which makes solution of Equation (3) more complicated. In order to discuss the question conveniently, the paper always assumes that every solution to Equation (3) contains a minimum solution that is less than or equal to it, and the number of minimum solution is finite[9]. If we denote all minimum solution to Equation (3) by $\breve{X}(A, b)$, then solution set of Equation (3) can be denoted as follows.

$$X(A, b) = \bigcup_{\breve{x} \in \breve{X}(A,b)} \{x | \breve{x} \leq x \leq \hat{x}, \quad x \in R^n\}. \tag{7}$$

We can clearly see by Formula (7), solution set structure of Equation (3) can be ascertained by $\breve{X}(A, b)$, solving $\breve{X}(A, b)$ means $X(A, b)$ is known. Although $X(A, b)$ is not convex set, it is composed of union of several $n$-dimensional sup-rectangular solid with every sup-rectangular solid being a closed convex set.

## 4 Solving Method on Posynomial Fuzzy Relation Geometric Programming with Clear Objective

Suppose that solution set of the fuzzy relation equations $X(A, b)$ is gained through conservative path method[10]. Without lost of generality, assume that cardinal number of $\breve{X}(A, b)$ is $| \breve{X}(A, b) |= p$, now let $\breve{x}^k \in \breve{X}(A, b)(1 \leq k \leq p)$, and then the $p$ conventional geometric programming can be stated as follows:

$$\begin{aligned} GP_k \quad & \min f(x) \\ & \text{s.t } \breve{x_j}^k \leq x_j \leq \hat{x_j}(1 \leq j \leq n), \end{aligned} \tag{8}$$

Let $x^{k^*}(1 \leq k \leq p)$ is optimal solution of optimization $GP_k$, take

$$x^\star \in \{x^{l^*} \mid f(x^{l^*}) = \min\{f(x^{k^*})(1 \leq k \leq p)\}\} \tag{9}$$

then $x^\star$ is a optimal solution to Optimization (4).

**Definition 4.1.** If $X(A, b) \neq \emptyset$, then Optimization (4) is said to super-compatible.

**Definition 4.2.** $DD = L + n - 1$ is said to difficult degree of posynomial fuzzy relation geometric programming.

Numerical $D$ reflected the extent of the difficulties in solving posynomial fuzzy relation geometric programming (4). When $D = 0$, the dual programming of posynomial geometric programming (8) has only one solution, this is the simplest case, when $D = 1$, the dual programming is equivalent to a single variable function optimization problems, it can be solved using the method of linear search, in general, if the value of $D$ became more large, the solving optimal solutions of (8) is more difficult[11].

The following give an algorithm procedure to posynomial fuzzy relation geometric programming with clear objective:

**Algorithm 4.1.**

Step 1. According to the order of components of $b$ from large to small, $b$ is rearranged, and $A$, $x$ and $f(x)$ are adjusted corresponding to changed $b$.

Step 2. By Formula (6), solve $\hat{x}$. If $\hat{x}$ is not a solution to Equation (3), then turn to Step 7. Otherwise, turn to Step 3.

Step 3. Solving $\breve{X}(A, b)$ by conservative path method.

Step 4.Solving the optimal solution $x^{k^*}(1 \leq k \leq p)$ of (8) by conventional geometric programming algorithm.

Step 5. Solving optimal solution $x^*$ of the Optimization (4) by (9).

Step 6. Print $f(x^*)$, stop.

Step 7. Print "have no solution", stop.

## 5 Solving Method on Posynomial Fuzzy Relation Geometric Programming with Fuzzy Objective

**Definition 5.1.** Fuzzy set $\widetilde{A}$ is called a fuzzy number on real number field $R$, if satisfied:

1)$\exists x_0 \in R$, suffice to $\widetilde{A}(x_0) = 1$;

2)$\forall \lambda \in [0, 1], \widetilde{A}_\lambda = \{x|\widetilde{A}(x) \geq \lambda\} = [A_\lambda^L, A_\lambda^R]$ is a close interval.

Suppose that $F(R)$ denotes all fuzzy number on $R$.

If the coefficients and exponents of objective function are fuzzy in posynomial fuzzy geometric Optimization (4), then objective value should also be fuzzy, the Optimization (4) has become posynomial fuzzy relation geometric programming with fuzzy objective. Suppose that coefficient $c_l$ and exponent $\gamma_{jl}$ are denoted by fuzzy number $\widetilde{C_l}$, $\widetilde{\Gamma_{jl}}$, respectively, and let $\mu_{\widetilde{C_l}}$, $\mu_{\widetilde{\Gamma_{jl}}}$ denote Membership Function of $\widetilde{C_l}$, $\widetilde{\Gamma_{jl}}$, respectively, then (4) can be denoted as follows:

$$\begin{aligned} \min\ & \widetilde{f(x)} = \sum_{l=1}^{L} \widetilde{C_l} \prod_{j=1}^{n} x_j^{\widetilde{\Gamma_{jl}}} \\ \text{s.t}\ & A \circ x = b, \\ & 0 \leq x_j \leq 1 (1 \leq j \leq n), \end{aligned} \qquad (10)$$

where objective function $\widetilde{f(x)}$ is fuzzy number.

The $\alpha$–cut set of $\widetilde{C_l}$ and $\widetilde{\Gamma_{jl}}$ is denoted by $[(\widetilde{C_l})_\alpha^L, (\widetilde{C_l})_\alpha^U]$ and $[(\widetilde{\Gamma_{jl}})_\alpha^L, (\widetilde{\Gamma_{jl}})_\alpha^U]$, let $c_l \in S(\widetilde{C_l}), \gamma_{jl} \in S(\widetilde{\Gamma_{jl}})$, where $S(\widetilde{C_l})$ and $S(\widetilde{\Gamma_{jl}})$ is support set of fuzzy set $\mu_{\widetilde{C_l}}$ and $\mu_{\widetilde{\Gamma_{jl}}}$, respectively. According to extension principle [12], we can definition

$$\mu_{\widetilde{f(x)}}(y) = \sup_{c_l, \gamma_{jl}} \min\{\mu_{\widetilde{C_l}}(c_l), \mu_{\widetilde{\Gamma_{jl}}}(\gamma_{jl}) \mid \forall\, l, j, y = f(c_l, \gamma_{jl})\},$$

in order to solving $\mu_{\widetilde{f(x)}}(y)$, suppose that

$$\mu^U_{\widetilde{f(x)}_\alpha} = \max\{\, f(c_l, \gamma_{jl}) \mid (\widetilde{C_l})^L_\alpha \leq c_l \leq (\widetilde{C_l})^U_\alpha, (\widetilde{\Gamma_{jl}})^L_\alpha \leq \gamma_{jl} \leq (\widetilde{\Gamma_{jl}})^U_\alpha\},$$
$$\mu^L_{\widetilde{f(x)}_\alpha} = \max\{\, f(c_l, \gamma_{jl}) \mid (\widetilde{C_l})^L_\alpha \leq c_l \leq (\widetilde{C_l})^U_\alpha, (\widetilde{\Gamma_{jl}})^L_\alpha \leq \gamma_{jl} \leq (\widetilde{\Gamma_{jl}})^U_\alpha\},$$

then two level programming can be gained as follows:

$$(U) \quad \max_{\substack{(\widetilde{C_l})^L_\alpha \leq c_l \leq (\widetilde{C_l})^U_\alpha, \\ (\widetilde{\Gamma_{jl}})^L_\alpha \leq \gamma_{jl} \leq (\widetilde{\Gamma_{jl}})^U_\alpha, \\ \forall\, l, j}} \quad \begin{array}{l} \min \sum_{l=1}^{L} c_l \prod_{j=1}^{n} x_j^{\gamma_{jl}} \\ \text{s.t } A \circ x = b, \\ 0 \leq x_j \leq 1 (1 \leq j \leq n), \end{array} \qquad (11)$$

$$(L) \quad \min_{\substack{(\widetilde{C_l})^L_\alpha \leq c_l \leq (\widetilde{C_l})^U_\alpha, \\ (\widetilde{\Gamma_{jl}})^L_\alpha \leq \gamma_{jl} \leq (\widetilde{\Gamma_{jl}})^U_\alpha, \\ \forall\, l, j}} \quad \begin{array}{l} \min \sum_{l=1}^{L} c_l \prod_{j=1}^{n} x_j^{\gamma_{jl}} \\ \text{s.t } A \circ x = b, \\ 0 \leq x_j \leq 1 (1 \leq j \leq n), \end{array} \qquad (12)$$

Optimization (11) and (12) are posynomial fuzzy relation geometric programming with clear objective and parameter $\alpha$ the same as (4), when $0 < \alpha_1 < \alpha_2 \leq 1$, there is

$$\mu^L_{\widetilde{f(x)}_{\alpha_1}} \geq \mu^L_{\widetilde{f(x)}_{\alpha_2}}, \mu^U_{\widetilde{f(x)}_{\alpha_1}} \leq \mu^U_{\widetilde{f(x)}_{\alpha_2}},$$

if $\alpha$ take every value in [0,1], according to decompose theorem of fuzzy set, the Membership Function $\widetilde{f(x)}$can be gotten[12].

$$\mu_{\widetilde{f(x)}}(y) = \begin{cases} L(y), \ \mu^L_{\widetilde{f(x)}_{\alpha_0}} \leq y \leq \mu^L_{\widetilde{f(x)}_{\alpha_1}} \\ 1, \quad\ \ \mu^L_{\widetilde{f(x)}_{\alpha_1}} \leq y \leq \mu^U_{\widetilde{f(x)}_{\alpha_1}} \\ R(y), \ \mu^U_{\widetilde{f(x)}_{\alpha_1}} \leq y \leq \mu^U_{\widetilde{f(x)}_{\alpha_0}} \end{cases}$$

where, $L(y)$ is non-decreasing function on $y$, $R(y)$ is non-increasing function on $y$, if take finite value of $\alpha$, for example, $\alpha$ take

$$0, 0.1, 0.2, 0.3, 0.4, 0.5, 0.6, 0.7, 0.8, 0.9, 1.0,$$

then a better approximation to $\mu_{\widetilde{f(x)}}$ can be gained. In practical problems, the approximate result can often be good utilized. The following give an algorithm procedure to posynomial fuzzy relation geometric programming with fuzzy objective as follows:

**Algorithm 5.1.**

Step 1. According to the order of components of $b$ from large to small, $b$ is rearranged, and $A$, $x$ and $f(x)$ are adjusted corresponding to changed $b$.

Step 2. By Formula (6), solve $\hat{x}$. If $\hat{x}$ is not a solution to Equation (3), then turn to Step 6. Otherwise, turn to Step 3.

Step 3. Solving $\breve{X}(A, b)$ by conservative path method.

Step 4. Solving optimal solution $x^*$ of Optimization model (11) and (12) by conventional geometric programming.

Step 5. Print $f(x^*)$, stop.

Step 6. Print "have no solution", stop.

## 6 Numerical Example

Example 1: Solving

$$\begin{aligned} \min\ & f(x) = 3x_1^{-0.5}x_2^{-0.2} + 4x_1x_3^{-0.4}x_4^2 + x_1x_2x_3x_4 \\ \text{s.t}\ & A \circ x = b, \\ & 0 \le x_j \le 1, \quad (1 \le j \le 4) \end{aligned}$$

$$\text{where } A = \begin{pmatrix} 0.3\ 0.2\ 0.7\ 0.8 \\ 0.5\ 0.4\ 0.4\ 0.9 \\ 0.7\ 0.3\ 0.2\ 0.7 \\ 0.9\ 0.6\ 0.1\ 0.2 \\ 0.8\ 0.5\ 0.6\ 0.4 \end{pmatrix}, \ b = (0.7, 0.4, 0.4, 0.3, 0.6)^T.$$

**Notes:** If $x_j = 0$, we can definition $\frac{1}{x_j} = \infty$.

Solving: According to (3), $\hat{x} = (0.3, 0.3, 1, 0.4)^T$ can be gotten, it is easy to confirm $A \circ \hat{x} = b$, then $\hat{x}$ is maximal solution of $A \circ x = b$. Two solutions of fuzzy relation equation can be obtained by conservative path method, the two solutions are

$$\breve{x} = (0.3, 0, 0.7, 0.4)^T \text{ and } \breve{x} = (0, 0.3, 0.7, 0.4)^T,$$

The solution set of fuzzy relation equation is

$$S = (0.3, [0, 0.3], [0.7, 1], 0.4) \cup ([0, 0.3], 0.3, [0.7, 1], 0.4).$$

Then two conventional posynomial geometric programming can be formulated as follows:

$$\begin{aligned} GP_1 \quad \min\ & f(x) = 3 \cdot 0.3^{-0.5}x_2^{-0.2} + 0.192x_3^{-0.4} + 0.12x_2x_3 \\ \text{s.t}\ & \tfrac{10}{3}x_2 \le 1, \\ & 0.7x_3^{-1} \le 1, \\ & x_3 \le 1, \end{aligned}$$

$$\begin{aligned} GP_2 \quad \min\ & f(x) = 3 \cdot 0.3^{-0.2}x_1^{-0.5} + 0.64x_1x_3^{-0.4} + 0.12x_1x_3 \\ \text{s.t}\ & \tfrac{10}{3}x_1 \le 1, \\ & 0.7x_3^{-1} \le 1, \\ & x_3 \le 1, \end{aligned}$$

the optimal solutions of conventional geometric programming $GP_1$ and $GP_2$ are $(0.3, 0.3, 1, 0.4)$ and $(0.3, 0.3, 1, 0.4)$, respectively, the optimal values are 7.19 and 7.19, respectively. It is easy to see, $\min\{7.19, 7.19\} = 7.19$, therefore, the optimal solution is $(0.3, 0.3, 1, 0.4)$, and the optimal value is 7.19.

Example 2: Soving

$$\begin{aligned} \min\ & \widetilde{f(x)} = (1.2, 1.4, 1.6)x_1^{(0.5,0.6,0.7)}x_2^2 + (0.8, 1.0, 1.2)x_3^{-0.5}x_4^{-2} \\ \text{s.t}\ & A \circ x = b, \\ & 0 \le x_j \le 1, \quad (1 \le j \le 4) \end{aligned}$$

where $A$, $b$ is the same as Example 1.

The following two bilevel programming can be formulated by Optimization model (11) and (12):

$$(U) \quad \max_{\substack{1.2+0.2\alpha \leq c_1 \leq 1.6-0.2\alpha, \\ 0.8+0.2\alpha \leq c_2 \leq 1.2-0.2\alpha, \\ 0.5+0.1\alpha \leq \gamma_{11} \leq 0.7-0.1\alpha}} \begin{array}{l} \min\ c_1 x_1^{\gamma_{11}} x_2^2 + c_2 x_3^{-0.5} x_4^{-2} \\ \text{s.t } A \circ x = b, \\ \quad 0 \leq x_j \leq 1 (1 \leq j \leq 4), \end{array}$$

$$(L) \quad \min_{\substack{1.2+0.2\alpha \leq c_1 \leq 1.6-0.2\alpha, \\ 0.8+0.2\alpha \leq c_2 \leq 1.2-0.2\alpha, \\ 0.5+0.1\alpha \leq \gamma_{11} \leq 0.7-0.1\alpha}} \begin{array}{l} \min\ c_1 x_1^{\gamma_{11}} x_2^2 + c_2 x_3^{-0.5} x_4^{-2} \\ \text{s.t } A \circ x = b, \\ \quad 0 \leq x_j \leq 1 (1 \leq j \leq 4), \end{array}$$

Using the Algorithm 5.1 discussed in previous section, we can solve the above two bilevel posynomial fuzzy relation geometric programming with parameter $\alpha$, the optimal solution of $(U)$ and $(L)$ can be denoted by

$$7.5 - 1.25\alpha \text{ and } 5.0 + 1.25\alpha,$$

respectively. That is

$$\widetilde{f(x)}_\alpha = [5.0 + 1.25\alpha, 7.5 - 1.25\alpha] \ (0 \leq \alpha \leq 1)$$

Corresponding fuzzy set can be gotten from the decomposition theorem as follows:

$$\mu_{\widetilde{f(x)}}(y) = (5.0, 6.25, 7.5).$$

This is a triangular fuzzy numbers, the function image and diagram of different $\alpha$-level value can be depicted as follows:

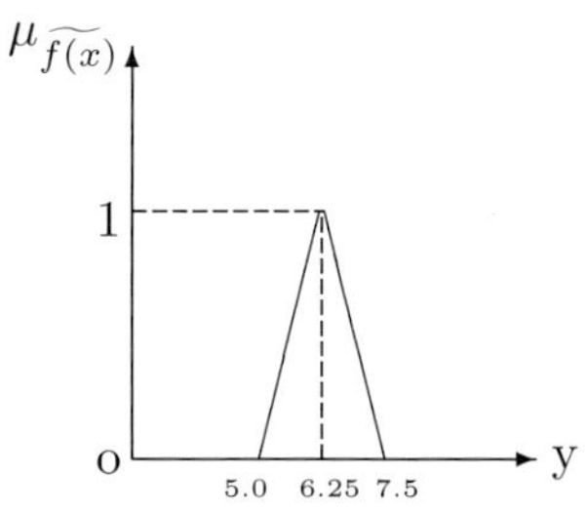


**The range of optimal solution with different $\alpha$-level value**

| $\alpha$ | 0 | 0.2 | 0.4 | 0.6 | 0.8 | 1.0 |
|---|---|---|---|---|---|---|
| Optimal Solutions Range | [5.0,7.5] | [5.25,7.25] | [5.5,7.0] | [5.75,6.75] | [6.0,6.5] | 6.25 |

## 7 Conclusion

So far, we have researched posynomial fuzzy relation geometric programming as (4). When the objective function are clear numbers, we transform the posynomial fuzzy relation geometric programming problem into a conventional geometric programming, and then we get the optimal solution. When the exponents and coefficients of objective function are fuzzy numbers, the paper develops a method that is able to find the membership function of the fuzzy objective optimal value. The idea is based on Zadeh's extension principle to transform the posynomial fuzzy relation geometric programming problem into a pair of bilevel mathematical programming. Based on duality theory and a simple algorithm, the pair of bilevel mathematical programming is transformed into a pair of conventional geometric programming. Solving the pair of geometric programming produces the upper bound and lower bound of the optimal value at specific $\alpha$ level. At last, membership function of objective function optimal value is gotten. Of course, the algorithms are not effective to all posynomial fuzzy relation geometric programming problems(PFRGPs), when the problem scale is very large, at present, there is no an effective numerical algorithms that suit to all PFRGPs, how to looking for a simple and effective procedure which can solve such a special programming is always an worthy attention problem.

## Acknowledgments

I am grateful to the referees for their valuable comments and suggestions. This work supported by the National Natural Science Foundation of China (NSFC) (No. 70271047 and No. 79670012) and Project Science Foundation of Guangzhou University.

## References

1. Bellmann. R. and L. A. Zadeh(1970) Decision Making in Fuzzy Environment . Management Science 17(4):141-164
2. E. Sanchez(1976) Resolution of composite fuzzy relation equations. Inform.and Control 30:38-48
3. Wang J Y(1991) Fuzzy Relation Equations and Their applications . Changsha, Hunan Science and Technology Press
4. S.C.Fang and G.Z.Li(1999) Solving fuzzy relation with a linear objective function. Fuzzy Sets and Systems 103: 107-113
5. Amin Ghodousian(2006) Esmaile Khorram. An algorithm for optimizing the linear function with fuzzy relation equation constraints regarding max-prod composition. Applied Mathematics and Computation 178: 502-509
6. J.J.Lu and S.C.Fang(2001) Solving nonlinear optimization problems with fuzzy relation equation constraints. Fuzzy Sets and Systems 119: 1-20
7. Higashi M and Klir GJ(1984) Resolution of finite fuzzy relation equations. Fuzzy Sets and Systems 13: 65-82
8. Zhang W X(1991) The introduction of fuzzy mathematics. Xi'an, Xi'an Jiaotong University Publishing House
9. Wang X P(2002)Conditions Under Which a Fuzzy Relational Equation has Minimal Solutions in a Complete Brouwerian Lattice. Advances in Mathematics(China)31(3): 220-228
10. P.Z.Wang, D.Z.Zhang, E.Sanchez and E.S.Lee(1991) Latticized linear programming and fuzzy relation inequalities. Journal of mathematical analysis and applications 159(1):72-87
11. M. Avriel,Ed(1980) Advances in Geometric Programming, volume 21 of Mathematical Concept and Methods in Science and Engineering.New York, Plenum Press
12. H. J. Zimmermann(1996) Fuzzy Set Theory and Its applications. Boston, Kluwer-Nijhoff

# Part XI

# Type-2 Fuzzy Logic

# A Vector Similarity Measure for Type-1 Fuzzy Sets

Dongrui Wu and Jerry M. Mendel

Signal and Image Processing Institute, Ming Hsieh Department of Electrical Engineering, University of Southern California, Los Angeles, CA 90089-2564
dongruiw@usc.edu, mendel@sipi.usc.edu

**Abstract.** Comparing the similarity between two fuzzy sets (FSs) is needed in many applications. The focus herein is linguistic approximation using type-1 (T1) FSs, i.e. associating a T1 FS $A$ with a linguistic label from a vocabulary. Because each label is represented by an T1 FS $B_i$, there is a need to compare the similarity of $A$ and $B_i$ to find the $B_i$ most similar to $A$. In this paper, a vector similarity measure (VSM) is proposed for T1 FSs, whose two elements measure the similarity in shape and proximity, respectively. A comparative study shows that the VSM gives best results. Additionally, the VSM can be easily extended to interval type-2 FSs.

## 1 Introduction

Fuzzy sets (FSs), which handle uncertainties in a natural way, have been used in numerous applications. The application of particular interest in this paper is the linguistic approximation problem [1,2] using type-1 (T1) FSs [1], i.e. we have a system whose inputs are linguistic labels modeled by T1 FSs, and after some operations it outputs another T1 FS $A$, and, we want to map $A$ to a linguistic label in a vocabulary so that it can be understood linguistically. Because each label in the vocabulary is represented by a T1 FS $B_i$, there is a need to compare the similarity of $A$ and $B_i$ to find the $B_i$ most similar to $A$.

Many similarity measures for T1 FSs have been introduced. According to Cross and Sudkamp [4], they can be classified into four categories: (1) Set-Theoretic Measures, (2) Proximity-Based Measures, (3) Logic-Based Measures, and (4) Fuzzy-Valued Measures. Two similarity measures proposed particularly for the linguistic approximation problem are Bonissone's method [1,2] and Wenstøp's method [8]. In this paper, a vector similarity measure (VSM) for T1 FSs is proposed. It is simpler than either of these two methods, and has better performance on T1 FSs. Additionally, it can be easily extended to interval T2 FSs [9].

The rest of this paper is organized as follows: Section 2 reviews Bonissone's method and Wenstøp's method for linguistic approximation using T1 FSs.

[1] In this paper we call the original FSs introduced by Zadeh [10] in 1965 *T1 FSs* to distinguish them from their extension, *type-2 FSs*, which were also introduced by Zadeh [11] in 1975 to model more uncertainties.

P. Melin et al. (Eds.): IFSA 2007, LNAI 4529, pp. 575–583, 2007.
© Springer-Verlag Berlin Heidelberg 2007

Section 3 proposes a VSM for the linguistic approximation problem. Section 4 compares the VSM with Bonissone's method and Wenstøp's method. Section 5 draws conclusions. Proofs of the theorems are given in Appendix A.

## 2 Existing Similarity Measures for Linguistic Approximation

The literature on similarity measures for T1 FSs is quite extensive [4]. Two similarity measures, Bonissone's method and Wenstøp's method, which are proposed particularly for linguistic approximation, will be reviewed in this section.

### 2.1 Bonissone's Linguistic Approximation Distance Measure

As mentioned in the Introduction, Bonissone's [1, 2] linguistic approximation distance measure was proposed to identify the linguistic label $B_i$ which most closely resembles a given FS $A$.

The first step of Bonissone's method eliminates from further consideration those linguistic labels determined to be very far away from $A$. For a given T1 FS $A$, the distances between $A$ and $B_i$, $d_1(A, B_i)$, are computed to identify $M$ $B_i$ that are close to $A$ (according to some tolerance parameter). Bonissone [2] first computed four T1 FS features, *centroid*, *cardinality*, *fuzziness* and *skewness*, for $A$ and $B_i$, and then defined $d_1(A, B_i)$ as the weighted Euclidean distance between the two four-dimensional points $[(p_A^1, p_A^2, p_A^3, p_A^4)^T$ and $(p_{B_i}^1, p_{B_i}^2, p_{B_i}^3, p_{B_i}^4)^T]$ represented by the values of the four features for each T1 FS, i.e.,

$$d_1(A, B_i) = \left[\sum_{j=1}^{4} w_j^2 (p_A^j - p_{B_i}^j)^2\right]^{1/2}. \tag{1}$$

The weights[2] $w_j$ $(j = 1, 2, 3, 4)$ have to be pre-specified.

After pre-screening linguistic labels far away from $A$, Bonissone's second step uses the *modified Bhattacharya distance* [6] to discriminate between the $M$ linguistic labels close to $A$, i.e.,

$$d_2(A, B_k) = \left[1 - \int_X \left(\frac{\mu_A(x)\mu_{B_k}(x)}{card(A) \cdot card(B_k)}\right)^{1/2} dx\right]^{1/2} \qquad k = 1, \ldots, M \tag{2}$$

The linguistic label corresponding to the smallest $d_2(A, B_k)$ is considered most similar to $A$.

### 2.2 Wenstøp's Linguistic Approximation Method

Wenstøp [8], who considered the same problem as Bonissone, states: "*a linguistic approximation routine is a function from the set of fuzzy subsets to a set of*

[2] We show $w_j^2$ in (1), because this is the way the equation is stated in [2].

*linguistic values.*" Wenstøp used two parameters of a T1 FS, its *imprecision* (cardinality) and its *location* (centroid). The imprecision ($p^1$) was defined as the sum of membership values, whereas the location ($p^2$) was defined as the center of gravity. He then computed

$$d_W(A, B_i) = \left[(p_A^1 - p_{B_i}^1)^2 + (p_A^2 - p_{B_i}^2)^2\right]^{1/2} \qquad i = 1, \ldots, N \tag{3}$$

and chose $B_i$ with the smallest $d_W(A, B_i)$ as the one most similar to $A$. Observe that Wenstøp's method is a simplified version of Bonissone's first step.

## 3 The VSM for T1 FSs

In this section a VSM for T1 FSs is proposed. Four desirable properties a similarity measure should possess are introduced first.

### 3.1 Four Desirable Properties of a Similarity Measure

The following four properties are proposed for a reasonable similarity measure for T1 FSs.

1) The similarity between two T1 FSs is 1 if and only if they are exactly the same.
2) If two T1 FSs intersect, there should be some similarity between them.
3) If two T1 FSs become more distant from each other, similarity between them should decrease.
4) The similarity between two T1 FSs should be a constant regardless of the order in which they are compared, i.e. $s(A, B) = s(B, A)$.

Next a VSM which possesses these properties is proposed.

### 3.2 The VSM for T1 FSs

When the similarity of two T1 FSs $A$ and $B$ are compared, it is necessary to compare their shapes as well as proximity; hence, a VSM, $\mathbf{s}_v(A, B)$, with two components is proposed,

$$\mathbf{s}_v(A, B) = (s_1(A, B), s_2(A, B))^T, \tag{4}$$

where $s_1(A, B) \in [0, 1]$ is a similarity measure on the shapes of $A$ and $B$, and $s_2(A, B) \in [0, 1]$ is a similarity measure on the proximity of $A$ and $B$. To define $\mathbf{s}_v(A, B)$, $s_1(A, B)$ and $s_2(A, B)$ must first be defined.

### 3.3 Definition of $s_1(A, B)$

Because the proximity of $A$ and $B$ is considered in $s_2(A, B)$, when computing $s_1(A, B)$ $A$ and $B$ are "aligned" so that their shapes can be compared. A reasonable alignment method is to move one or both of $A$ and $B$ so that their centroids, $c(A)$ and $c(B)$, coincide (see Fig. 1). The two T1 FSs can be moved to any location as long as $c(A)$ and $c(B)$ coincide; this will not affect the value of $s_1(A, B)$. In this paper $B$ is moved to $A$ and called $B'$, as shown in Fig. 1.

Once the two T1 FSs are "aligned," $s_1(A, B)$ is computed by Jaccard's *unparameterized ratio model of similarity* [3] [5]:

$$s_1(A, B) = \frac{card(A \cap B')}{card(A \cup B')} = \frac{\int_X \min(\mu_A(x), \mu_{B'}(x))dx}{\int_X \max(\mu_A(x), \mu_{B'}(x))dx}. \quad (5)$$

Observe that $s_1(A, B)$ is a set-theoretic measure [4].

**Theorem 1.** *(a)* $0 \leq s_1(A, B) \leq 1$; *(b)* $s_1(A, B) = 1 \Leftrightarrow A = B'$; *and, (c)* $s_1(A, B) = s_1(B, A)$.

*Proof:* See Appendix A.1. ■

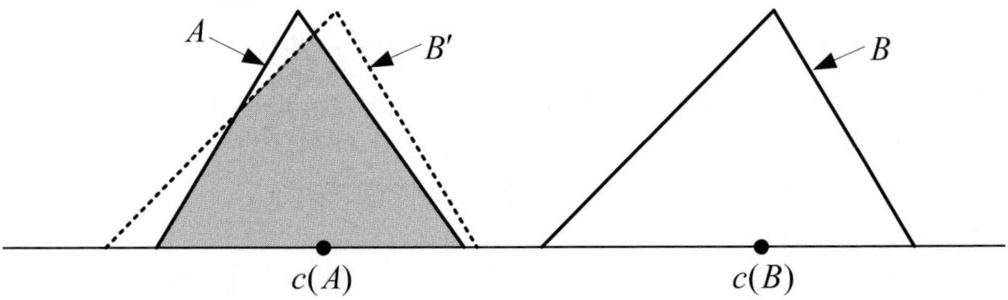


**Fig. 1.** An example of the VSM for T1 FSs. $c(A)$ and $c(B)$ are the centroids of $A$ and $B$, respectively. $B'$ is obtained by moving $B$ so that $c(B)$ coincides with $c(A)$. Note that the shaded region can also be obtained by moving $c(A)$ to $c(B)$.

### 3.4 Definition of $s_2(A, B)$

$s_2(A, B)$ measures the proximity of $A$ and $B$, and is defined as

$$s_2(A, B) \equiv h(d(A, B)) \quad (6)$$

where $d(A, B) = |c(A) - c(B)|$ is the Euclidean distance between the centers of the centroids of $A$ and $B$ (see Fig. 1), and $h$ can be any function satisfying: (1) $\lim_{x\to\infty} h(x) = 0$; (2) $h(x) = 1$ if and only if $x = 0$; and, (3) $h(x)$ decreases monotonically as $x$ increases.

**Theorem 2.** $s_2(A, B) \in [0, 1]$, *and* $s_2(A, B) = 1$ *if and only if* $c(A) = c(B)$.

*Proof:* Theorem 2 is obvious from (6) and the above constraints on $h(x)$. ■

An example of $s_2(A, B)$ is

$$s_2(A, B) = e^{-rd(A,B)}, \quad (7)$$

[3] It is called *coefficient of similarity* by Sneath in [7]. The term *index of communality* has also been used [4].

where $r$ is a positive constant. $s_2(A, B)$ is chosen as an exponential function because we believe the similarity between two FSs should decrease rapidly as the distance between them increases.

### 3.5 On Converting $\mathbf{s}_v(A, B)$ to a Scalar Similarity Measure $s_s(A, B)$

$\mathbf{s}_v(A, B)$ enables us to separately quantify the similarity of two features, shape and proximity. In linguistic approximation $\mathbf{s}_v(A, B_i)$ $(i = 1, 2, \ldots, N)$ need to be ranked to find the $B_i$ most similar to $A$. This can be achieved by first converting the vector $\mathbf{s}_v(A, B_i)$ to a scalar similarity measure $s_s(A, B_i)$ and then ranking $s_s(A, B_i)$ $(i = 1, 2, \ldots, N)$.

In this paper, the scalar similarity between two T1 FSs $A$ and $B$ is computed as the product of their similarities in shape and proximity [4], i.e.

$$s_s(A, B) = s_1(A, B) \times s_2(A, B) \tag{8}$$

Properties of $s_s(A, B)$ include:

**Theorem 3.** *(a)* $A = B \Leftrightarrow s_s(A, B) = 1$; *(b)* $s_s(A, B) > 0$; *(c)* $s_s(A, B) > s_s(A, C)$ *if* $B$ *and* $C$ *have the same shape and* $C$ *is further away from* $A$ *than* $B$ *is; and, (d)* $s_s(A, B) = s_s(B, A)$.

*Proof:* See Appendix A.2. ■

Theorem 3 shows that $s_s(A, B)$ satisfies the four properties stated in Section 3.1.

## 4 Comparisons

### 4.1 Comparison with Bonissone's Linguistic Approximation Distance Measure

Both $\mathbf{s}_v(A, B)$ and Bonissone's method consider the shapes and proximity of $A$ and $B$. The main differences between them are:

(1) $\mathbf{s}_v(A, B)$ is a one-step method, whereas Bonissone's method is a two-step method.
(2) $\mathbf{s}_v(A, B)$ considers two features of $A$ and $B$ (shape and proximity). In Bonissone's first step, four features (centroid, cardinality, fuzziness and skewness) are considered, and in his second step, only one feature is considered (the modified Bhattacharya distance).
(3) $\mathbf{s}_v(A, B)$ measures the similarity between $A$ and $B$, i.e. a larger $\mathbf{s}_v(A, B)$ means $A$ and $B$ are more similar. On the other hand, Bonissone's method measures the distance (or difference) between $A$ and $B$, i.e. a larger $d_2(A, B)$ means $A$ and $B$ are less similar.

---

[4] Recently, Bonissone, et al. [3] defined a similarity measure as a weighted minimum of several sub-similarity measures. Although similar to our idea, their objective is quite different from our objective; hence, their similarity measure is not used in this paper.

### 4.2 Comparison with Wenstøp's Linguistic Approximation Method

Wenstøp's linguistic approximation method is quite similar to the VSM method in that both of them use the centroid and cardinality. The differences are:

(1) The VSM computes the similarity between two T1 FSs, whereas Wenstøp's method computes the difference between two T1 FSs.
(2) The VSM first aligns $A$ and $B$ and then computes the cardinalities of $A \cap B$ and $A \cup B$, whereas Wenstøp's method computes cardinalities of $A$ and $B$ directly.
(3) The VSM can be used for T1 FSs of any shapes, whereas, as shown in [8], the two parameters in Wenstøp's method are insufficient criteria for satisfactory linguistic approximation. As a further refinement, he includes other characteristics of FSs, e.g. non-normality, multi-modality, fuzziness and dilation [8].

### 4.3 Examples

For T1 FSs shown in Fig. 2, the results of Bonissone's linguistic approximation distance measure, Wenstøp's linguistic approximation measure and the VSM are shown in Table 1. The domain of $x$ was discretized into 201 equally-spaced points in all three methods, and $r \equiv 4/|X|$ ($|X|$ is the length of the support of $A \cup B$) in the VSM [see (7)]. Note that all $B_k$ ($k = 1, 2, 3, 4$) are assumed to survive Bonissone's first step, hence (2) was used to compute Bonissone's distance measure. Observe that all methods indicate $B_2$ is more similar to $A$ than is $B_1$, which seems reasonable. When $B_3$ and $B_4$ are considered, Bonissone's measure indicates that they have the same similarity to $A$ [5], and Wenstøp's measure indicates that $B_4$ is more similar to $A$ than $B_3$ is. Both results seem counter-intuitive, because $B_3$ should be more similar to $A$ than $B_4$ is, as indicated by the VSMs.

**Table 1.** Comparisons of similarity measures for T1 FSs $A$ and $B_k$ ($k = 1, \ldots, 4$) shown in Fig. 2

| Measure | $k = 1$ | $k = 2$ | $k = 3$ | $k = 4$ |
|---|---|---|---|---|
| $d_2(A, B_k)$ | 0.2472 | 0.1617 | 1 | 1 |
| $d_W(A, B_k)$ | 28.5679 | 16.6650 | 38.6805 | 37.5736 |
| $s_s(A, B_k)$ | 0.6368 | 0.7208 | 0.0086 | 0.0013 |

[5] If one FS must be chosen from $B_k$ ($k = 1, 2, 3, 4$) so that it is most similar to $A$, then $B_3$ and $B_4$ may be removed during Bonissone's first step because they are too far away from $A$; however, if only $B_3$ and $B_4$ are available and one of them must be chosen so that it is more similar to $A$, Bonissone's method will have a problem because both $B_3$ and $B_4$ survive in the first step, and in the second step $d_2(A, B_3) = d_2(A, B_4)$.

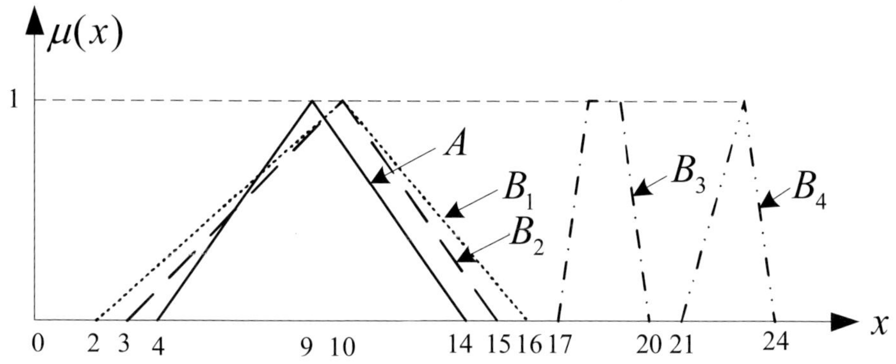


**Fig. 2.** T1 FSs used in the comparative study

## 5 Conclusions

A vector similarity measure for T1 FSs has been proposed in this paper. It is easy to understand, and its two components enable us to consider the similarity between shapes and proximity separately and explicitly. The VSM is simpler than two existing linguistic approximation methods, and yet a comparative study showed that it has better performance. Additionally, the VSM can be easily extended to interval T2 FSs [9].

## References


1. Bonissone, P.P.: A pattern recognition approach to the problem of linguistic approximation. In: Proc. IEEE Int'l Conf. on Cybernetics and Society, Denver, CO (1979) 793–798
2. Bonissone, P.P.: A fuzzy sets based linguistic approach: Theory and applications. In: Proc. 12th Winter Simulation Conference, Orlando, FL (1980) 99–111
3. Bonissone, P.P., Varma, A., Aggour, K.S., Xue, F.: Design of local fuzzy models using evolutionary algorithms. *to be published in* Journal of Computational Statistics and Decision Analysis (2006)
4. Cross, V.V., Sudkamp, T.A.: Similarity and Compatibility in Fuzzy Set Theory: Assessment and Applications. Physica-Verlag, Heidelberg, NY (2002)
5. Jaccard, P.: Nouvelles recherches sur la distribution florale. Bulletin de la Societe de Vaud des Sciences Naturelles **44** (1908) 223
6. Kailath, T.: The divergence and bhattacharyaa distance measure in signal detection. IEEE Trans. on Communication Technology **15** (1967) 609–637
7. Sneath, P.H.A., Sokal, R.R.: Numerical Taxonomy. W.H. Freeman and Company, San Francisco, CA (1973)
8. Wenstøp, F.: Quantitative analysis with linguistic values. Fuzzy Sets and Systems **4** (1980) 99–115

9. Wu, D., Mendel, J.M.: A vector similarity measure for interval type-2 fuzzy sets and type-1 fuzzy sets. *submitted to* Information Sciences (2006)
10. Zadeh, L.A.: Fuzzy sets. Information and Control **8** (1965) 338–353
11. Zadeh, L.A.: The concept of a linguistic variable and its application to approximate reasoning-1. Information Sciences **8** (1975) 199–249

# A Proof of Theorems

## A.1 Proof of Theorem 1

**Proof of (a).** Because

$$0 \leq \min(\mu_A(x), \mu_{B'}(x)) \leq \max(\mu_A(x), \mu_{B'}(x)) \tag{9}$$

it follows that

$$0 \leq \int_X \min(\mu_A(x), \mu_{B'}(x))dx \leq \int_X \max(\mu_A(x), \mu_{B'}(x))dx \tag{10}$$

Consequently,

$$s_1(A, B) = \frac{\int_X \min(\mu_A(x), \mu_{B'}(x))dx}{\int_X \max(\mu_A(x), \mu_{B'}(x))dx} \in [0, 1]. \tag{11}$$

**Proof of (b).** $A = B'$ means $\mu_A(x) = \mu_{B'}(x)$. Substituting these two equations into (5),

$$s_1(A, B) = \frac{\int_X \mu_A(x)dx}{\int_X \mu_A(x)dx} = 1, \tag{12}$$

which proves the necessity of Theorem 1(b).

To prove the sufficiency of the result, observe that $s_1(A, B) = 1$ means

$$\int_X \min(\mu_A(x), \mu_{B'}(x))dx = \int_X \max(\mu_A(x), \mu_{B'}(x))dx \tag{13}$$

(13) holds if and only if

$$\mu_A(x) = \mu_{B'}(x) \qquad \forall x \in X. \tag{14}$$

(14) means $A = B'$.

**Proof of (c).** $s_1(A, B) = s_1(B, A)$ is obvious because the min and max operators in (5) do not concern the order of $\mu_A(x)$ and $\mu_{B'}(x)$, i.e. $\min(\mu_A(x), \mu_{B'}(x)) = \min(\mu_{B'}(x), \mu_A(x))$ and $\max(\mu_A(x), \mu_{B'}(x)) = \max(\mu_{B'}(x), \mu_A(x))$. ■

## A.2 Proof of Theorem 3

**Proof of (a).** Sufficiency: $A = B$ means $s_1(A, B) = 1$ and $s_2(A, B) = 1$; hence, $s_s(A, B) = 1$.

Necessity: $s_s(A, B) = 1$ if and only if $s_1(A, B) = 1$ and $s_2(A, B) = 1$. $s_1(A, B) = 1$ means the shapes of $A$ and $B$ are the same, and $s_2(A, B) = 1$ means the distance between $A$ and $B$ is zero. Consequently, $A = B$.

**Proof of (b).** Observe that $s_1(A,B) > 0$ and $s_2(A,B) > 0$. Consequently, $s_s(A,B) > 0$.

**Proof of (c).** $B$ and $C$ have the same shape means

$$s_1(A,B) = s_1(A,C). \tag{15}$$

$C$ is further away from $A$ than $B$ means

$$s_2(A,B) > s_2(A,C). \tag{16}$$

Hence,

$$s_1(A,B) \times s_2(A,B) > s_1(A,C) \times s_2(A,C), \tag{17}$$

i.e. $s_s(A,B) > s_s(A,C)$.

**Proof of (d).** Because neither $s_1(A,B)$ nor $s_2(A,B)$ concern the order of $A$ and $B$, i.e. $s_1(A,B) = s_1(B,A)$ and $s_2(A,B) = s_2(B,A)$, it follows that $s_s(A,B) = s_s(B,A)$. ■

# On Approximate Representation of Type-2 Fuzzy Sets Using Triangulated Irregular Network

Long Thanh Ngo[1], Long Pham The[1], Phuong Hoang Nguyen[2], and Kaoru Hirota[3]

[1] Center of Simulation Technology, Le Quy Don Technical Universiy, Hanoi, Vietnam
ngotlong@gmail.com, thelong@math.ac.vn
[2] Center of Health Information Technology, Hanoi, Vietnam
nhphuong@fmail.vnn.vn
[3] Department of Computational Intelligence and Systems Science, Graduate School of Science and Engineering, Tokyo Institute of Technology, G3-49, 4259 Nagatsuta, Modori-ku, Yokohama 226-8502, Japan
hirota@hrt.dis.titech.ac.jp

**Abstract.** The type-2 fuzzy sets was extended of original fuzzy sets and there are many researches related to type-2 fuzzy sets and fuzzy logic systems (FLS), recently. The paper deals with the approximate presentation of type-2 fuzzy sets and operations by using triangulated irregular networks (TIN). TIN usually used to represent approximate 3-D surfaces with applications in computer graphics. By using TINs and half-edge structure, authors proposed representation and operations of type-2 fuzzy sets based on this approach. Some results compared with original approaches are shown to confirm the approach be good and correctly.

**Keywords:** type-2 fuzzy sets, TIN, Delaunay algorithm.

## 1 Introduction

The type-2 fuzzy sets was introduced by L. Zadeh as an extension of ordinary fuzzy sets. So the concept of type-2 fuzzy sets is also extended from type-1 fuzzy sets. If $A$ is a type-1 fuzzy set and membership grade of $x \in X$ in $A$ is $\mu_A(x)$, which is a crisp number in [0, 1]. A type-2 fuzzy set in $X$ is $\tilde{A}$, and the membership grade of $x \in X$ in $\tilde{A}$ is $\mu_{\tilde{A}}(x)$, which is a type-1 fuzzy set in [0, 1]. The elements of the domain of $\mu_{\tilde{A}}(x)$ are called *primary memberships* of $x$ in $\tilde{A}$ and the memberships of the primary memberships in $\mu_{\tilde{A}}(x)$ are called *secondary memberships* of $x$ in $\tilde{A}$.

Recenly, there are many researches and applications related to type-2 fuzzy sets becase of the advandcing in uncertainty management. Karnik et al [3] have studied practical algorithms of operations on type-2 fuzzy sets as union, intersection, complement. Karnik and Mendel[4] proposed the method of type-reduction of type-2 fuzzy sets based on centroid defuzzification. J. Mendel and R.John [7] have developed new representation of type-2 fuzzy sets based on embedded type-2 fuzzy sets. This representation obtains the design of type-2 fuzzy logic system

P. Melin et al. (Eds.): IFSA 2007, LNAI 4529, pp. 584–593, 2007.
© Springer-Verlag Berlin Heidelberg 2007

is easy to use and understand. Mendel [8] proposed some practical algorithms in implementing and storing data to speed-up the computing rate of type-2 fuzzy logic system. Coupland et al [9-10] proposed representation type-1 and interval type-2 fuzzy sets and fuzzy logic system by using 2-D computational geometry, the approach is better in computing than analytic approaches. TIN is a method of representation of curved surface in 3-D space with many applications in computer graphics and simulation. Many approaches [11-13] are use to generate TIN from set of points based Delaunay algorithms.

The paper deals with the new representation of type-2 fuzzy sets using TIN. The membership grades of type-2 fuzzy sets being 3-D surfaces are discreted into triangular faces with planar equations. Size of triangle is difference depending on slope of the surface. Authors proposed practical algorithms to implement operations on type-2 fuzzy sets by building computational geometry algorithms on TIN. The result is shown and tested for robustness of the approach.

## 2 Background

### 2.1 Type-2 Fuzzy Sets and Operations

A type-2 fuzzy set in $X$ is $\tilde{A}$, and the membership grade of $x \in X$ in $A$ is $\mu_{\tilde{A}}(x,u)$, $u \in J_x \subseteq [0,1]$, which is a type-1 fuzzy set in [0, 1]. The elements of the domain of $\mu_{\tilde{A}}(x,u)$ are called primary memberships of $x$ in $\tilde{A}$ and the memberships of the primary memberships in $\mu_{\tilde{A}}(x,u)$ are called secondary memberships of x in $\tilde{A}$.

**Definition 1.** *A type $-$ 2 fuzzy set, denoted $\tilde{A}$, is characterized by a type-2 membership function $\mu_{\tilde{A}}(x,u)$ where $x \in X$ and $u \in J_x \subseteq [0,1]$, i.e.,*

$$\tilde{A} = \{((x,u), \mu_{\tilde{A}}(x,u)) | \forall x \in X, \forall u \in J_x \subseteq [0,1]\} \tag{1}$$

*or*

$$\tilde{A} = \int_{x \in X} \int_{u \in J_x} \mu_{\tilde{A}}(x,u))/(x,u), J_x \subseteq [0,1] \tag{2}$$

*in which* $0 \leq \mu_{\tilde{A}}(x,u) \leq 1$.

At each value of $x$, say $x = x'$, the 2-D plane whose axes are $u$ and $\mu_{\tilde{A}}(x',u)$ is called a *vertical slice* of $\mu_{\tilde{A}}(x,u)$. A *secondary membership function* is a vertical slice of $\mu_{\tilde{A}}(x,u)$. It is $\mu_{\tilde{A}}(x = x', u)$ for $x \in X$ and $\forall u \in J_{x'} \subseteq [0,1]$, ie.,

$$\mu_{\tilde{A}}(x = x', u) \equiv \mu_{\tilde{A}}(x') = \int_{u \in J_{x'}} f_{x'}(u)/u, J_{x'} \subseteq [0,1] \tag{3}$$

in which $0 \leq f_{x'}(u) \leq 1$.

Theoric operations of type-2 fuzzy ses as union, intersection and complement is mentioned in [2]. Karnik et al [3] proposed algorithms to compute join ($\sqcup$ - union), meet($\sqcap$-intersection) and negation ($\neg$-complement) of type-2 fuzzy sets. Type-2 fuzzy sets are called an interval type-2 fuzzy sets if the secondary membership function $f_{x'}(u) = 1$ $\forall u \in J_x$.

### 2.2 TIN and Delaunay Triangulation

A *topographic surface* $\sigma$ is the image of a real bivariate function $f$ defined over a domain $D$ in the Euclidean plane, as

$$\sigma = \{(x, u, f(x, u)) \big| (x, u) \in D\} \tag{4}$$

A polyhedral model is the image of a piecewise-line function $f$ being described on a partition of $D$ into polygonal regions $\{T_1, ..., T_k\}$ and the image of $f$ over each region $T_i (i = 1, ..., k)$ is a planar patch. If all of $T_i$s $(i = 1, .., k)$ are triangles then the polyhedral model is called a *Triangulated Irregular Network* (TIN). Hence, $\sigma$ may be represented approximately by a TIN, as

$$\sigma \tilde{=} \sum_{i=1}^{k} \{(x, u, f_i(x, u)) \big| (x, u) \in T_i\}, \bigcup_{i=1}^{k} T_i \equiv D \tag{5}$$

where $f_i$s $(i = 1, ..., k)$ are planar equations.

The *Delaunay triangulation* of a set $V$ of points in $IR^2$ is a subdivision of the convex hull of V into triangles having their vertices at points of $V$, and such that triangles are as much aquiangular as possible. More formally, a triangulation $\tau$ of $V$ is a Delaunay triangulation if and only if, for any triangle $t$ of $\tau$, the circumcircle of $t$ does not contain any point of $V$ in its interior. This property is called the *empty circle property* of the Delaunay triangulation.

The usual input for two-dimensional mesh generation is not merely a set of vertices. Most theoretical treatments of meshing take as their input a planar straight line graph (PSLG). A PSLG is a set of vertices and segments that satisfies two constraints. First, for each segment contained in a PSLG, the PSLG must also contain the two vertices that serve as endpoints for that segment. Second, segments are permitted to intersect only at their endpoints.

The *constrained Delaunay triangulation* (CDT) of a PSLG $X$ is similar to the Delaunay triangulation, but every input segment appears as an edge of the triangulation. An edge or triangle is said to be *constrained Delaunay* if it satisfies the following two conditions. First, its vertices are *visible* to each other. Here, visibility is deemed to be obstructed if a segment of $X$ lies between two vertices. Second, there exists a circle that passes through the vertices of the edge or triangle in question, and the circle contains no vertices of $X$ that are visible from the interior of the edge or triangle.

## 3 Approximate Representation of Type-2 Fuzzy Sets

Extending the concept of interval type-2 sets of upper MF and lower MF, we define a membeship grade of type-2 fuzzy sets by dividing it into subsets: upper (lower) surface and normal surface as follows:

**Definition 2 (Upper, lower and normal surface of T2FS).** *$\tilde{A}_T$ is called a sub-set of type-2 fuzzy set $\tilde{A}$, is defined as follows:*

$$\tilde{A}_T = \int_{x \in X} [\int_{u \in J_x^k} f_x(u)/u]/x \tag{6}$$

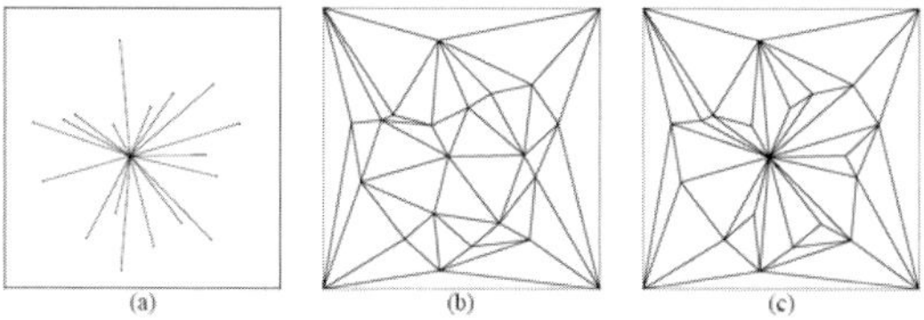


**Fig. 1.** (a) A planar straight line graph. (b) Delaunay triangulation of the vertices of the PSLG. (c) Constrained Delaunay triangulation of the PSLG.

*Upper surface ($\tilde{A}_{US}$), lower surface ($\tilde{A}_{LS}$) and normal surface ($\tilde{A}_{NS}$) are subsets of $\tilde{A}$ in which $J_x^k$ of (6) is $J_x^+$, $J_x^-$, $J_x^*$, respectively, at vertical-slice $x = x'$ are computed as $J_{x'}^+ \subseteq [u_x^{*+}, 1]$, $J_{x'}^- \subseteq [0, u_x^{*-}]$ and $J_{x'}^* \subseteq [u_x^{*-}, u_x^{*+}]$ which $u_{x'}^{*+} = sup\{u|\mu_{\tilde{A}}(x', u) = 1\}$, $u_{x'}^{*-} = inf\{u|\mu_{\tilde{A}}(x', u) = 1\}$.*

So a type-2 fuzzy set $\tilde{A}$ is re-written as $\tilde{A} = \tilde{A}_{US} \cup \tilde{A}_{NS} \cup \tilde{A}_{LS}$ or $\mu_{\tilde{A}}(x, u) = \mu_{\tilde{A}_{US}}(x, u) \cup \mu_{\tilde{A}_{NS}}(x, u) \cup \mu_{\tilde{A}_{LS}}(x, u)$. Fig. 2 is an example of type-2 fuzzy set with three subsets: upper surface, normal surface and lower surface.

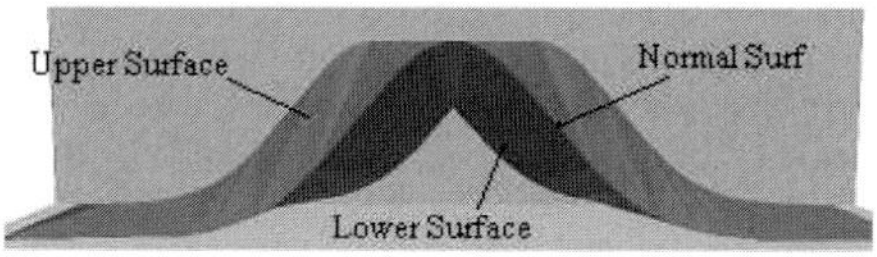


**Fig. 2.** Example of surfaces of type-2 fuzzy sets

A proximate representation of type-2 fuzzy sets is proposed by using a TIN to represent the 3-d membership grade, is expressed as the following theorem.

**Theorem 1 (Approxiamate Theorem).** *Let $\tilde{A}$ be type-2 fuzzy set with membership grade $\mu_{\tilde{A}}(x, u)$ in domain $D(x, u)$. A TIN $M_{\tilde{A}}$ represent approximately $\tilde{A}$ with membership grade $\mu_{\tilde{A}}^M(x, u)$, satisfying:*

$$\|\mu_{\tilde{A}}(x, u) - \mu_{\tilde{A}}^M(x, u)\| < \epsilon, \quad \forall (x, u) \in D. \tag{7}$$

*Proof.* If membership grade of type-2 fuzzy sets only use triangular and trapezoid MF, then $M_{\tilde{A}}$ represent faithfully $\tilde{A}$ satisfying (7). We prove the theorem with non-linear membership grades.

Let $D^*(x, z) \subseteq D(x, z)$ so that $\mu_{\tilde{A}}(x, u)$ is continuous. We prove (7) is true with $\forall (x, u)$ *in* $D^*$ with $T^*$ is a TIN representing $\tilde{A}$. Union of $T^*$s is a TIN satisfying constrained Delaunay that represent approximately $\tilde{A}$.

Let $T_k$ is a TIN that represent $\tilde{A}$ in $D^*$. Suppose that $\exists (x_k, u_k) \in D^*(x, u)$ so that $d_k = |\mu_{\tilde{A}}(x_k, u_k) - \mu_{\tilde{A}}^M(x_k, u_k)| \geq \epsilon$.

We modify $T$ by inserting new vertices and arranging triangular faces, as follows:

1. Find the triangle $t_j(j = 1,\overline{N_k})$, $N_k$ is number of triangles, of $T_k$ that $(x_k, u_k)$ is in $t_j$.
2. Partitioning the $t_j$ into sub-triangles depending on position $(x_k, u_k)$ on $t_j$.
   + If $(x_k, u_k)$ lies on edge $e_k$ of $t_j$. Partitioning $t_j$, $t_i$ be adjacent triangle sharing the edge $e_k$ with $t_j$, into four sub-triangles(Fig. 3a).
   + If $(x_k, u_k)$ is in $t_j$. Partitioning $t_j$ into three sub-triangles as Fig. 3b.
3. Verifying new triangles satisfying Delaunay. Operation may be re-arrange triangles by using *flip* operation of two adjacent triangles.

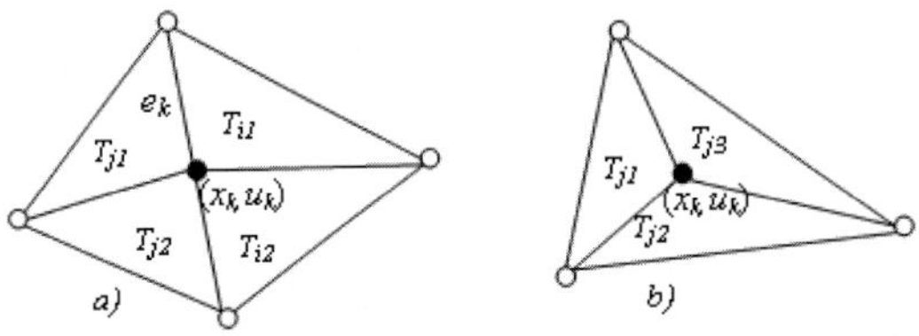


**Fig. 3.** Partitioning the $t_j$ triangle

After number of the limited steps implementing above algorithm, $d_k < \epsilon$, because of the continuous of $\mu_{\tilde{A}}(x, u)$ in $D^*$.

**Definition 3.** *A base-line of a TIN representing a type-2 fuzzy set is a polyline $v_i(i = 1, .., N)$ satisfying $v_i.u = 0$ and $v_iv_{i+1}$ is a edge of triangle of TIN.*

Fig.4 is the TIN that represent approximately of Gaussian type-2 fuzzy sets with $\epsilon = 0.1$. The primary MF is a Gaussian with fixed deviation and mean $m_k \in [m_1, m_2]$ and the secondary MF is a triangular MF. The dask-line is a base-line of TIN.

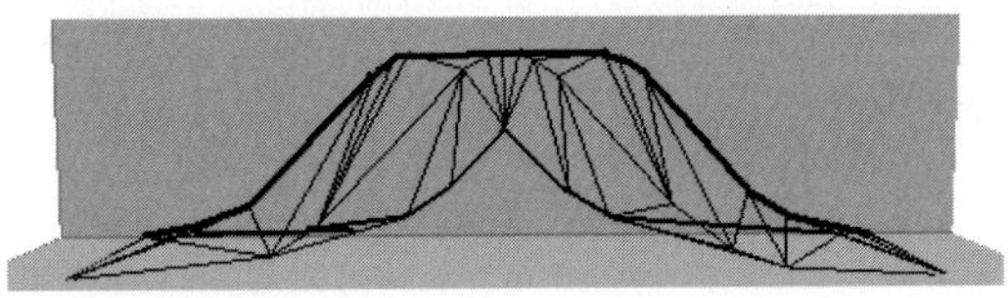

**Fig. 4.** Example of representation of a type-2 Gaussian fuzzy sets

# 4 Operations of TIN

## 4.1 Data

Data storing a TIN includes vertices, indices of faces and relations of them. Data of vertices is a list of 3-d vectors with x, y and z components. Indices of faces are three indices of vertices of triangle.

Relations between vertices and faces is used to speed up algorithms on TIN such as searching or computing algorithms. In [13], authors used the $half-edge$ data structure to store relation between a vertex and neighbouring vertices or faces. A half-edge includes the end point and the adjacent face being on right of edge. List of half-edges of a vertex is arranged clockwise based coordinate of the end point. Fig. 5 shows the storing list of half-egdes of a vertex.

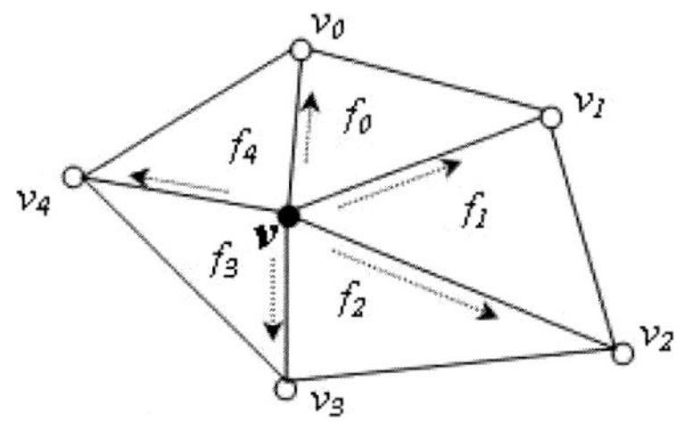


**Fig. 5.** List of half-edges of a vertex

### 4.2 Algorithms Computing on TIN

The section introduces some algorithms operating on TIN such as: algorithm computing intersection of two TINs or algorithm computing minimum or maximum of two TINs. Algorithm computing intersection is to create a polyline being intersection of two TINs and the polyline is a break-line of TINs. Algorithm computing maximum/minimum is to generate new TIN $T_0$ from two TINs $T_1, T_2$ satisfying $\forall(x,u)|\mu^{T_0}(x,u) = min(\mu^{T_1}(x,u), \mu^{T_2}(x,u))$ or $\mu^{T_0}(x,u) = max(\mu^{T_1}(x,u), \mu^{T_2}$ $(x,u))$. The following is the detailed descriptions of algorithms.

**Algorithm 1** (*Algorithm computing intersection*)

*Input: $T_1, T_2$ are two TIN representing two type-2 fuzzy sets. Outputs is $T_1, T_2$ being modified with some new vertices and edges on intersection polylines.*

1. *Computing $L_1, L_2$ are base-lines of $T_1, T_2$, respectively.*
2. *Find $v_k^*(k = 1,.., M)$ are the intersection points of $L_1, L_2$.*
3. *If $M = 0$ or set of intersection points is empty then return.*
4. *For each $v_k^*(k = 1, ..., M)$*
   *$v^* \leftarrow v_k^*$. Init queue $Q_k$.*
   *While not find $v^*$*
   (a) *$v \leftarrow v^*$. Insert $v$ into queue $Q_k$.*
   (b) *Insert $v$ into each of $T_1, T_2$, become $v_{T_1}, v_{T_2}$.*
   (c) *Find adjacent triangle $t_1^*, t_2^*$ of $v_{T_1} and v_{T_2}$, respectively, so that $t_1^*, t_2^*$ are intersected by a segment in $t_1^*$ and $t_2^*$.*
   (d) *If existing new $v^*$ point so that $vv^*$ is a intersecting segment of $t_1^*$ and $t_2^*$ then*
      *$v \leftarrow v^*$*
      *Come back step a).*
      *Else*
      *Come back step 2).*

**Algorithm 2** (*Algorithm computing maximum/minimum*)
*Input: $T_1, T_2$ are two TIN representing two type-2 fuzzy sets. Output is $T_0$ being a result of minimum/maximum operation.*

1. *Computing intersection of $T_1, T_2$ (call algorithm computing intersection).*
2. *Init queue $Q$.*
3. *for each triangle t of $T_1$ or $T_2$.*
   (a) *With maximum algorithm:*
   *if t is triangle of $T_1 (T_2)$ and be upper than $T_2 (T_1)$ then push t into $Q$.*
   (b) *With minimum algorithm:*
   *if t is triangle of $T_1 (T_2)$ and be lower than $T_2 (T_1)$ then push t into $Q$.*
   (c) *Generating TIN from triangles in $Q$.*

**Algorithm 3** (*Join, Meet Operation*)
*Input: $\tilde{A}$, $\tilde{B}$ are two type-2 fuzzy sets with TINs $T_{\tilde{A}}$, $T_{\tilde{B}}$. Output is $\tilde{C}$ being result of join (meet) operation.*

1. *Computing result of upper surface as follows (Call algorithm 3):*
   $T_{\tilde{C}_{US}} = max(T_{\tilde{A}_{US}}, T_{\tilde{B}_{US}})$ *(for join operation).*
   $T_{\tilde{C}_{US}} = min(T_{\tilde{A}_{US}}, T_{\tilde{B}_{US}})$ *(for meet operation).*
2. *Computing result of lower surface as follows (Call algorithm 3):*
   $T_{\tilde{C}_{LS}} = max(T_{\tilde{A}_{LS}}, T_{\tilde{B}_{LS}})$ *(for join operation).*
   $T_{\tilde{C}_{LS}} = min(T_{\tilde{A}_{LS}}, T_{\tilde{B}_{LS}})$ *(for meet operation).*
3. *Computing $T_{\tilde{C}_{NS}}$ from $T_{\tilde{C}_{US}}$ and $T_{\tilde{C}_{US}}$.*

# 5 Applications

This section introduces some applications of new representation of type-2 fuzzy sets using TIN such as computing meet, join, negation operations.

## 5.1 Join Operation and Meet Operation Under Min

Based-on theoric join and meet operation, we proposed algorithms to compute result of operations by using algorithms on TIN. Algorithm for join or meet under min operator use algorithms computing intersection and algorithm computing min/max of 2 TINs in section 4.2.

**Fig. 6.** Example of two fuzzy sets for operations

Fig. 6 is two type-2 fuzzy sets with primary MF is Gaussian and secondary MF is triangular MF. Fig. 7 (Fig. 8) is the result of algorithm for computing join operation (meet operation under min).

**Fig. 7.** Meet operation under min

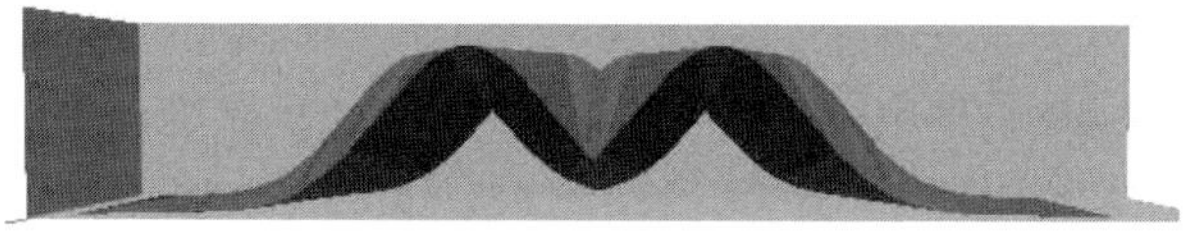

**Fig. 8.** Join operation

**Algorithm 4** (*Meet operation under product*)
*Input: $\tilde{A}$, $\tilde{B}$ are two type-2 fuzzy sets. Output is $\tilde{C}$ being result of meet operation under product.*

1. *Call $\tilde{D}$ is a clone of $\tilde{B}$.*
2. *For each vertices $v_{1k}$ of $T_{US}$, $v_{2k}$ of $T_{LS}$ of $\tilde{D}$.*
   *Computing y-component of $v_k$ as follows:*
   $v_{1k}.y = \mu^T_{\tilde{A}_{US}}(v_{1k}.x, v_{1k}.u) * \mu^T_{\tilde{B}_{US}}(v_{1k}.x, v_{1k}.u)$
   $v_{2k}.y = \mu^T_{\tilde{A}_{LS}}(v_{2k}.x, v_{2k}.u) * \mu^T_{\tilde{B}_{LS}}(v_{2k}.x, v_{2k}.u)$
3. *For each triangle $t_k$ of $T_{US}$ or $T_{LS}$ of $\tilde{D}$.*
   (a) $t \leftarrow t_k$
   (b) *Computing $G$ being the gravity of $t$.*
   (c) $dy = \mu^T_{\tilde{A}_{US}}(G.x, G.u) * \mu^T_{\tilde{B}_{US}}(G.x, G.u)$ *or* $dy = \mu^T_{\tilde{A}_{LS}}(G.x, G.u) * \mu^T_{\tilde{B}_{LS}}(G.x, G.u)$
   (d) *If* $\mid G.y - dy \mid > \epsilon$ *Then*
   - *Insert $G$ into $T_{US}$ (or $T_{LS}$ if $t_k$ in $T_{LS}$), $t_1$, $t_2$, $t_3$ are three new sub-triangles. And set to $G.y = dy$.*
   - *Set $t \leftarrow t_1$, $t \leftarrow t_2$, $t \leftarrow t_3$ and come back step a).*

**Algorithm 5** (*Negation Operation*)
*Input: $\tilde{A}$ is a type-2 fuzzy set. Output is result of negation operation.*

1. *For each vetex $v_k$ of $T_{US}$ or $T_{LS}$ of $\tilde{B}$.*
   $v_k.y = 1.0 - v_k.y$
2. *Set* $T'_{US} \leftarrow T_{LS}$, $T'_{LS} \leftarrow T_{US}$.
3. *Set* $\tilde{B} = \{T'_{US}, T_{NS}, T'_{LS}\}$.

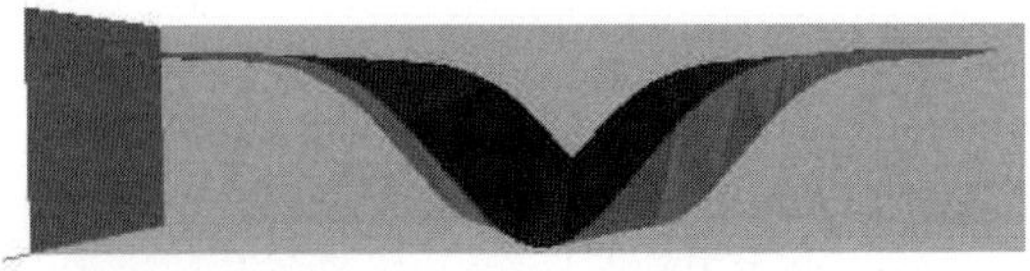

**Fig. 9.** Negation operation

### 5.2 Meet Operation Under Product

### 5.3 Negation Operation

### 5.4 Performance

The new approach uses memory and computations less than previous approaches. If the TIN has N vertices, M faces then it takes N*12 bytes for vertices, M*6 bytes for faces and M*6 for relations between vertices and faces. For examples, triangular or tapezoid takes about 720 bytes with $N \cong M \cong 30$. Gaussian membership grades take about 200 vertices and 300 faces with accuracy $\epsilon = 0.01$, i.e 6000 bytes. Beside, the memoty using for traditional approach takes about 100 000 bytes with step is 0.01 and $x$ takes value in [0, 10].

We also tested the performace of algorithms with different membership grades. We implemented operations in 1000 times for each operation and sumaried run-time (in ms )in table 1.

**Table 1.** The run-time of operations

| | Join | Meet(min) | Meet(product) | Negation |
|---|---|---|---|---|
| Triangular-Triangular | 1 | 1 | 1 | 1 |
| Gaussian - Triangular | 474 | 290 | 312 | 1 |
| Interval Gaussian | 114 | 86 | 91 | 1 |

## 6 Conclusion

The paper introduces the new approach to represent a type-2 fuzzy sets using triangular irreular network. TIN has used to represent 3-D surfaces by partitioning domain $D$ into sub-triangle satisfying Delaunay or constrained Delaunay. So we use this representation for membership grades of type-2 fuzzy sets being 3-D surfaces. W also proposed approach using half-edge to operate TIN with real-time. Based-on this result, we developed new computations to implement operations of type-2 fuzzy sets such as join, meet, negation. These operations is the base to develop computings for type-2 fuzzy logic system.

The next goals is to continue improving geometry algorithms descrease computational time. The second is to apply computings for developping type-2 fuzzy logic system using geometry algorithms.

## References


1. L.A. Zadeh (1975), The concept of linguistic variable and its application to approximate reasoning, J.Inform. Sci. 8: 199 - 249.
2. M. Mizumoto, K. Tanaka (1976) Some properties of fuzzy sets of type-2, J. Inform. and Control 31: 312 34.
3. N. Karnik, J.M. Mendel(2001), Operations on Type-2 Fuzzy Sets, Fuzzy Sets and Systems, vol. 122: 327-348.
4. N. Karnik, J.M. Mendel(2001). Centroid of a type-2 fuzzy set, Information Sciences, 132 : 195-220.
5. N.N Karmil, J.M. Mendel, Q. Liang (1999) Type-2 Fuzzy Logic Systems, IEEE Trans. on Fuzzy Systems, Vol.7, No.6, : 643-658.
6. Q. Liang, J.M. Mendel (2000) Interval Type-2 Fuzzy Logic Systems: Theory and Design, IEEE Trans. on Fuzzy Systems, Vol.8, No.5, 635-650.
7. J.M Mendel, R.B. John (2002) Type-2 Fuzzy Sets Made Simple, Fuzzy Sets and Systems, vol. 10, no.2, 117-127.
8. Jerry M. Mendel (2004) On Computing the Centroid of a Symmetrical Interval Type-2 Fuzzy Set, Proc. Conf. on Information Processing and Management of Uncertainty in Knowledge-Based Systems (IPMU), Perugia, Italy, July 4-9.
9. S. Coupland, R. John (2004) Fuzzy Logic and Computational Geometry, In 5th International Conference on Recent Advances in Soft Computing, 3 - 8.
10. S. Coupland, R. John (2006) Geometric Type-1 and Type-2 Fuzzy Logic Systems, Accepted for publication in IEEE Transactions on Fuzzy Systems.
11. J. R. Shewchuck (2002) Delaunay Refinement Algorithms for Triangular Mesh Generation, Computational Geometry: Theory and Applications 22(1-3):21-74.
12. J. Ruppert (1995) A Delaunay refinement algorithm for Quality 2-Dimensional Mesh Generation, Journal of Algorithms 18(3): 548-585.
13. L.T. Ngo, L.T.Pham (2005) Approach in Generating 3-D Digital Terrain Model based on Refinement Delaunay Algorithm (in Vietnamese), Proc. of Symposium in IT, 248-258.

# Hybrid Control for an Autonomous Wheeled Mobile Robot Under Perturbed Torques

Leslie Astudillo[1], Oscar Castillo[1], Luis T. Aguilar[2], and Ricardo Martínez[1]

[1] Department of Computer Science, Tijuana Institute of Technology, Tijuana, México
Leslie.Astudillo@suntroncorp.com, ocastillo@tectijuana.mx
[2] Instituto Politecnico Nacional,
2498 Roll Dr., #757 Otay Mesa, San Diego CA 92154
luis.aguilar@ieee.org

**Abstract.** This paper focuses on the control of wheeled mobile robot under bounded torque disturbances. Hybrid tracking controller for the mobile robot was developed by considering its kinematic model and Euler-Lagrange dynamics. The procedure consist in minimizing the stabilization error of the kinematic model through genetic algorithm approach while attenuation to perturbed torques is made through type-2 Fuzzy Logic Control (FLC) via backstepping methodology. Type-2 fuzzy logic is proposed to synthesize the controller for the overall system which is claimed to be a robust tool for related applications. The theoretical results are illustrated through computer simulations of the closed-loop system.

## 1 Introduction

Mobile robots have attracted considerable interest in the robotics and control research community, because they posses nonholonomic properties caused by nonintegrable differential constraints. The motion of a nonholonomic mechanical systems [2] is constrained by its own kinematics, so the control laws are not derivable in a straightforward manner (Brockett condition [3]).

Fuzzy logic control [18] has been recognized for its effectiveness in the control of wheeled mobile robots in order to perform missions in uncertain environments where robustness properties must be considered in the control design (see e.g., [4,7,8,14], and the references therein). In parallel with studies devoted to intelligent control of wheeled mobile robots, there has recently been a growing interest in the robustification of the method motivated by attractive features of the type-2 FLC illustrated in [15]. In particular, the result in [8] shows a reactive control architecture for autonomous mobile robots that is based on type-2 FLC to implement the basic navigation behaviors and the coordination between these behaviors to produce a type-2 hierarchical FLC whose result show how the type-2 FLCs can deal in real-time with the uncertainties facing mobile robots in changing and dynamic unstructured environments and that they resulted in good real time control responses that had outperformed the type-1 FLCs.

P. Melin et al. (Eds.): IFSA 2007, LNAI 4529, pp. 594–603, 2007.
© Springer-Verlag Berlin Heidelberg 2007

In this paper is introduced a type-2 fuzzy logic controller to force the wheeled mobile robot to follow a desired trajectory in spite of the existence of unknown but bounded disturbances. Currently, many research papers consider only the kinematic model (steering system) to solve the tracking control problem, where the velocity, used as input control, is assumed to be supplied by the mobile robot whose dynamic of the actuators is neglected (see e.g., [11,13,16,17], and references therein). However, real prototype mobile robots have actuated wheels whose slip rate, rolling, inertia moments, and mass distribution contribute to the forces exerted on the structure of the vehicle thus affecting the accuracy and full manoeuvrability of the robot. Motivated by this, the vehicle dynamics, represented by the Euler-Lagrange equations, is considered to convert a steering system into control inputs for the actual vehicle ([5,6]).

Roughly speaking, the strategy of the type-2 fuzzy logic controller constructed is to drive the kinematic model to a desired trajectory in finite-time taking into account that torque is the real input. Backstepping methodology has long been recognized as a tool for solving the tracking control problem for mobile robots, including both kinematic and Euler-Lagrange models [6]. The backstepping approach consists of two steps: 1) finding the velocities that stabilize the kinematic model and 2) finding a control law such that ensure the converge of real velocities to those values.

In this case, a type-2 membership functions were considered for the error based on Mamdani reasoning approach. In particular, triangle- and trapezoid-shaped membership functions were used in the control design, with three fuzzy partitions and nine fuzzy rules.

This paper is organized as follows: Section 2 presents the problem statement and the kinematic and dynamic model of the unicycle mobile robot. Section 3 introduces the fuzzy logic control system using a Mamdani-type model where the wheel input torques, linear velocity, and angular velocity will be considered as linguistic variables. Section 4 provides a simulation study of the unicycle mobile robot using the controller described in Section 3. Finally, Section 5 presents the conclusions.

## 2 Problem Statement

In this paper, a perturbed unicycle mobile robot is considered as a case study. A unicycle mobile robot is an autonomous, wheeled vehicle capable of performing missions in fixed or uncertain environments. The robot body is symmetrical around the perpendicular axis and the center of mass is at the geometric center of the body. It has two driving wheels fixed to the axis that passes through $C$ and one passive orientable wheel that is placed in front of the axis and normal to it. The two fixed wheels are controlled independently by motors, and the passive wheel prevents the robot from tipping over as it moves on a plane. In what follows, it is assumed that the motion of passive wheel can be ignored in the dynamics of the mobile robot represented by the following set of equations [10]:

$$M(q)\dot{v} + C(q,\dot{q})v + Dv = \tau + p(t) \tag{1}$$

$$\dot{q} = \underbrace{\begin{bmatrix} \cos\theta & 0 \\ \sin\theta & 0 \\ 0 & 1 \end{bmatrix}}_{S(q)} \underbrace{\begin{bmatrix} \nu \\ \omega \end{bmatrix}}_{v} \tag{2}$$

where $q = (x, y, \theta)^T$ is the vector of the configuration coordinates; $v = (\nu, \omega)^T$ is the vector of velocities; $\tau = (\tau_1, \tau_2)$ is the vector of torques applied to the wheels of the robot; $p(t)$ is the $2 \times 1$ uniformly bounded disturbance vector; $M(q)$ is a $2 \times 2$ positive-definite inertia matrix; $C(q,\dot{q})v$ is the vector of centripetal and Coriolis forces; and $D$ is a $2 \times 2$ diagonal positive-definite matrix. Equation (2) represents the kinematics of the system, where $(x, y)$ is the position in the $X - Y$ (world) reference frame; $\theta$ is the angle between the heading direction and the $x$-axis; $\nu$ and $\omega$ are the linear and angular velocities, respectively; and $\tau_1$ and $\tau_2$ denote the torques of the right and left wheel, respectively (see Fig. 1). Furthermore, the system (1)-(2) has the following nonholonomic constraint:

$$\dot{y}\cos\theta - \dot{x}\sin\theta = 0, \tag{3}$$

which corresponds to a no-slip wheel condition preventing the robot from moving sideways [12]. The system (2) fails to meet Brockett's necessary condition for feedback stabilization [3] which implies that no continuous static state-feedback controller exists such that globally stabilizes the closed-loop system around the equilibrium point.

The control objective is established formally as follows: design a fuzzy logic controller $\tau$ such that the posture $q(t)$ reach a reference continuously differentiable, bounded trajectory and orientation $q_d(t) \in \mathbb{R}^3$, that is

$$\lim_{t\to\infty} \|q_d(t) - q(t)\| = 0, \tag{4}$$

while attenuating external disturbances.

## 3 Type-2 Fuzzy Control Design

This section will illustrate the framework to achieve stabilization of a unicycle mobile robot around a desired path. A systematic procedure is given based on the backstepping approach.

The system (1)-(2) is in cascade interconnection; that is, the kinematic subsystem (2) is controlled only indirectly through the velocity vector $v$ (see Fig. 2). Stabilizing control laws for systems in such a hierarchical form can be designed using the method of backstepping [9], consisting in two steps:

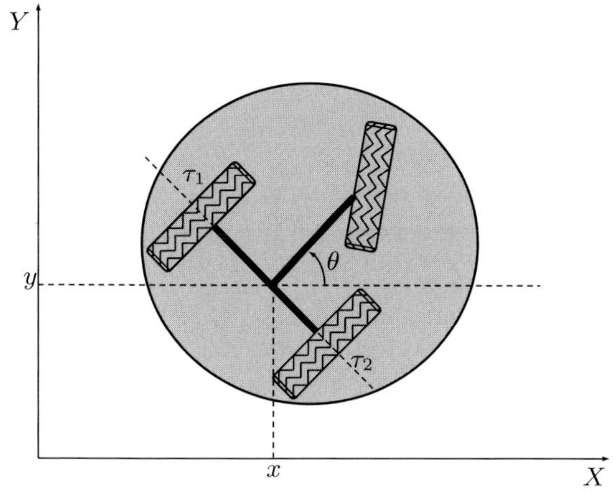


**Fig. 1.** Wheeled mobile robot

- Find the ideal velocity vector $v_r = v$ such that the kinematic system (2) be asymptotically stable, then
- Enforce $\tau$ by using FLC such that

$$\lim_{t \to t_s} \|v_r(t) - v(t)\| = 0 \tag{5}$$

be achieved in finite-time ($t_s < \infty$).

In (5), it is considered that real mobile robots have actuated wheels, so the control input is $\tau$ that must be designed to stabilize the system (1) without destabilizing the system (2) by forcing $v \to v_r$ in finite-time. Roughly speaking, if (5) is satisfied in infinite time ($t_s = \infty$) then $v_r$ will be different from $v$ along $t < \infty$, consequently the mobile robot will be neither positioned nor oriented at the desired point. Figure 2 illustrates the feedback connection which involves the fuzzy controller.

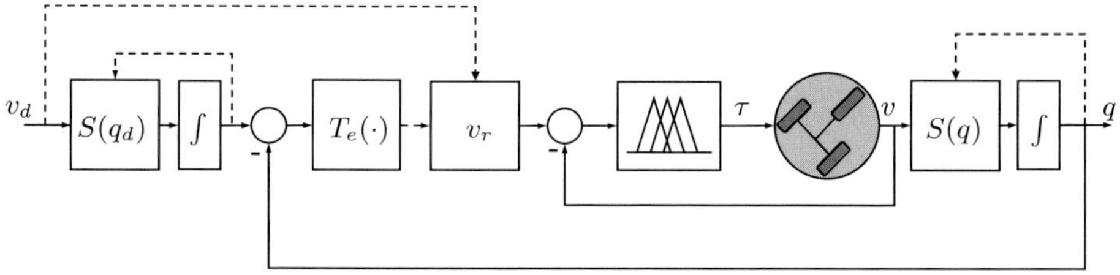


**Fig. 2.** Closed-loop system for the wheeled mobile robot

### 3.1 Finding the Velocity Vector $v_r$

In this subsection is summarized the procedure given in [6] and [10] to derive $v_r$. To begin with, suppose that the reference trajectory $q_d(t)$ satisfies

$$\dot{q}_d = \begin{pmatrix} \cos\theta_d & 0 \\ \sin\theta_d & 0 \\ 0 & 1 \end{pmatrix} \begin{pmatrix} v_{1d} \\ v_{2d} \end{pmatrix} \tag{6}$$

where $\theta_d(t)$ is the desired orientation, and $v_{1d}(t)$ and $v_{2d}(t)$ denote the linear and angular desired velocities, respectively. In the robot's local frame, the error coordinates can be defined as

$$\begin{pmatrix} e_1 \\ e_2 \\ e_3 \end{pmatrix} = \underbrace{\begin{pmatrix} \cos\theta & \sin\theta & 0 \\ -\sin\theta & \cos\theta & 0 \\ 0 & 0 & 1 \end{pmatrix}}_{T_e(\theta)} \begin{pmatrix} x_d - x \\ y_d - y \\ \theta_d - \theta \end{pmatrix}, \tag{7}$$

where $(x_d(t), y_d(t))$ is the desired position in the world $X-Y$ coordinate system, $e_1$ and $e_2$ are the coordinates of the position error vector, and $e_3$ is the orientation error. The associated tracking error model

$$\begin{pmatrix} \dot{e}_1 \\ \dot{e}_2 \\ \dot{e}_3 \end{pmatrix} = \begin{pmatrix} v_2 e_2 - v_1 + v_{d1}\cos e_3 \\ -v_2 e_1 + v_{d1}\sin e_3 \\ v_{d2} - v_2 \end{pmatrix} \tag{8}$$

which is in terms of the corresponding real and desired velocities is then obtained by differentiating (7) with respect to time. Let us recall the result given in [6] for the kinematic model stabilization (2) involving the tracking error model (7) and desired velocities ($v_d \in \mathbb{R}^2$).

**Theorem 1 ([6]).** *Let the tracking error equations (8) be driven by the control law*

$$\begin{aligned} v_{r1} &= v_{d1}\cos e_3 + k_1 e_1 \\ v_{r2} &= v_{d2} + v_{d1} k_2 e_2 + k_3 \sin e_3 \end{aligned} \tag{9}$$

*where $k_1$, $k_2$, and $k_3$ are strictly positive constants. If $v_1 = v_{r1}$ and $v_2 = v_{r2}$ in (2), then the origin of the closed-loop system [(8)-(9)] is asymptotically stable.*

Genetic algorithms are adopted for tuning the kinematic control (9) where it is required to find the gains $k_i$ $(i = 1, 2, 3)$ such that the error $e \in \mathbb{R}^3$ be minimized. The minimization of the performance index

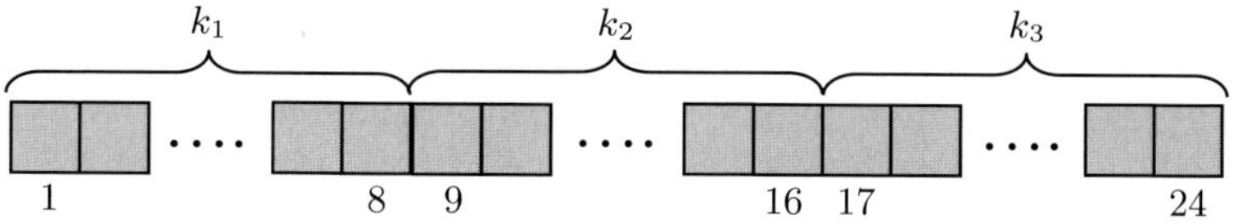


**Fig. 3.** Binary chromosome for gain selection

$$\mathcal{I}(e) = \sqrt{\frac{1}{n}\sum_{i=1}^{n}(|e_1(i)| + |e_2(i)| + |e_3(i)|)} \tag{10}$$

was considered to obtain an objective criterion in the selection of the gains $k_i$ $(i = 1, 2, 3)$.

The genetic algorithm was codified with a chromosome of 24 bits in total, eight bits for each of the gains. Figure 3 shows the binary chromosome representation of the individuals in the population. Different experiments were performed, changing the parameters of the genetic algorithm and the best results were obtained by comparing the corresponding simulations. Changing the crossover rate and number of points used did not affect the results. Also, changing the mutation rate did not affect the optimal results. The advantage of using the genetic algorithm to find the gains is that time-consuming manual search of these parameters was avoided.

### 3.2 Velocity Fuzzy Control Synthesis

In this subsection is derived a fuzzy logic controller designed to force the real velocities of the mobile robot (1)-(2) to match those required in equations (9) of theorem 1 in order to satisfy the control objective (4). Toward this end, the Mamdani Fuzzy model is adopted where the fuzzy rules are presented as a mapping from the linear and angular velocity errors, named linguistic input variables,

$$e_\nu = v_{r1} - v_1 \tag{11}$$

$$e_\omega = v_{r2} - v_2 \tag{12}$$

to the required torque $(\tau_1, \tau_2)$, named linguistic output variables. The membership functions, depicted in Fig. 4, are shaped like triangles and trapezoids with three fuzzy partitions denoted as *negative* (N), *zero* (Z), and *positive* (P). The nine *fuzzy rules* are:

$R_1$ : IF $e_\nu$ is $Z$ and $e_\omega$ is $Z$ THEN $\tau_1$ is $Z$ and $\tau_2$ is $Z$,
$R_2$ : IF $e_\nu$ is $Z$ and $e_\omega$ is $P$ THEN $\tau_1$ is $Z$ and $\tau_2$ is $P$,
$R_3$ : IF $e_\nu$ is $Z$ and $e_\omega$ is $N$ THEN $\tau_1$ is $Z$ and $\tau_2$ is $N$,
$R_4$ : IF $e_\nu$ is $P$ and $e_\omega$ is $P$ THEN $\tau_1$ is $P$ and $\tau_2$ is $P$,
$R_5$ : IF $e_\nu$ is $P$ and $e_\omega$ is $N$ THEN $\tau_1$ is $P$ and $\tau_2$ is $N$,

$R_6$ : IF $e_\nu$ is $P$ and $e_\omega$ is $Z$ THEN $\tau_1$ is $P$ and $\tau_2$ is $Z$,
$R_7$ : IF $e_\nu$ is $N$ and $e_\omega$ is $N$ THEN $\tau_1$ is $N$ and $\tau_2$ is $N$,
$R_8$ : IF $e_\nu$ is $N$ and $e_\omega$ is $P$ THEN $\tau_1$ is $N$ and $\tau_2$ is $P$,
$R_9$ : IF $e_\nu$ is $N$ and $e_\omega$ is $Z$ THEN $\tau_1$ is $N$ and $\tau_2$ is $Z$.

The Centroid of Area

$$z_{COA} = \frac{\int_Z \mu_A(z) z dz}{\int_Z \mu_A(z) dz} \tag{13}$$

was used as the defuzzification method, where $\mu_A(z)$ is the aggregated output of membership function, and $\int$ denotes the union of $(z, \mu(z))$ pairs. Figure 5 shows the input-output curves.

It should be pointed out that a fuzzy logic system (FLS) described using at least one type-2 fuzzy set is called a type-2 FLS. Type-1 FLSs are unable to directly handle rule uncertainties, because they use type-1 fuzzy sets that are certain. On the other hand, type-2 FLSs, are very useful in circumstances where it is difficult to determine an exact, and measurement uncertainties [15]. It is known that type-2 fuzzy set let us to model and to minimize the effects of uncertainties in rule-based FLS. Unfortunately, type-2 fuzzy sets are more difficult to use and understand that type-1 fuzzy sets; hence, their use is not widespread yet. Similar to a type-1 FLS, a type-2 FLS includes *type-2 fuzzyfier*, *rule-base*, *inference engine* and substitutes the defuzzifier by the output processor. The output processor includes a *type-reducer* and a *type-2 defuzzyfier*; it generates a type-1 fuzzy set output (from the type reducer) or a crisp number (from the

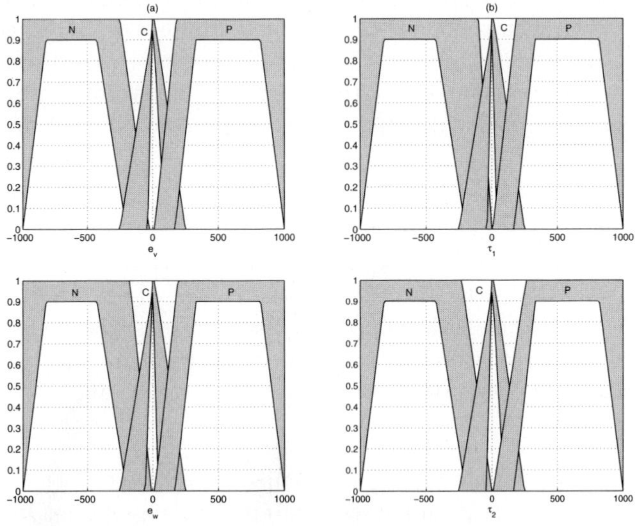


**Fig. 4.** *a*) Fuzzy membership functions of the input variables $(e_\nu, e_\omega)$, and *b*) fuzzy membership functions of output variables $(\tau_1, \tau_2)$

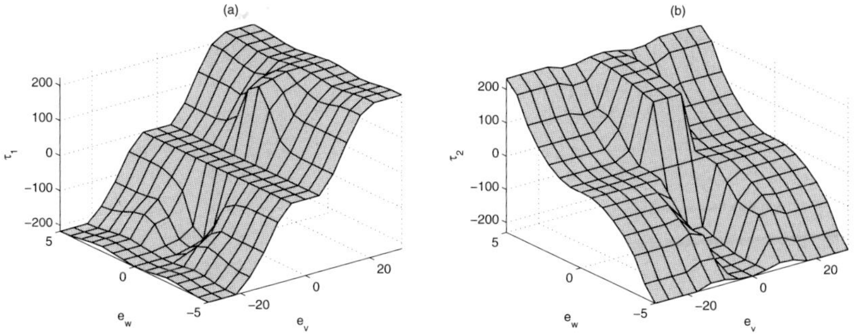


**Fig. 5.** Graphs of the input-output surfaces

defuzzyfier). A type-2 FLS is again characterized by IF-THEN rules, but its antecedent of consequent sets are now type-2. Type-2 FLSs, can be used when the circumstances are too uncertain to determine exact membership grades.

## 4 Simulation Results

In this section, we evaluate, through computer simulation performed in MATLAB® and SIMULINK®, the ability of the proposed controller to stabilize the unicycle mobile robot, defined by (1), (2). The following matrix values

$$M(q) = \begin{bmatrix} 0.3749 & -0.0202 \\ -0.0202 & 0.3739 \end{bmatrix}, \quad C(q,\dot{q}) = \begin{bmatrix} 0 & 0.1350\dot{\theta} \\ -0.1350\dot{\theta} & 0 \end{bmatrix}, \quad D = \begin{bmatrix} 10 & 0 \\ 0 & 10 \end{bmatrix}$$

were taken from [5]. In the simulations, the trajectory

$$v_d(t) = \begin{cases} v_{d1}(t) = 0.25 - 0.25\cos(\frac{2\pi t}{5}) \\ v_{d2}(t) = 0 \end{cases} \tag{14}$$

was chosen in terms of its corresponding desired linear $v_{d1}$ and angular velocities $v_{d2}$, subject to the initial conditions $q(0) = (0.1, 0.1, 0)^T$ and $v(0) = 0 \in \mathbb{R}^2$. The gains $k_i$ $(i = 1, 2, 3)$ of the kinematic model (9) were tuned by using genetic algorithm approach resulting in $k_1 = 43$, $k_2 = 493$, and $k_3 = 195$.

Figure 6 shows the posture errors, $X - Y$ path, velocity errors, and input torques for the closed-loop system for the FLC presented in Section 3 under a periodic disturbance torque defined as

$$p_i(t) = 10\cos(t) \qquad i = 1, 2.$$

It should be noted that the horizontal and vertical displacements are brought to zero in 0.5 s and 1.0 s, respectively, while the orientation converge at $t = 1.2$ s. The controller brings the velocity errors to zero at $t_s = 0.25$ s. Figure 6 also demonstrates the fast switching of the input control. Clearly, the proposed

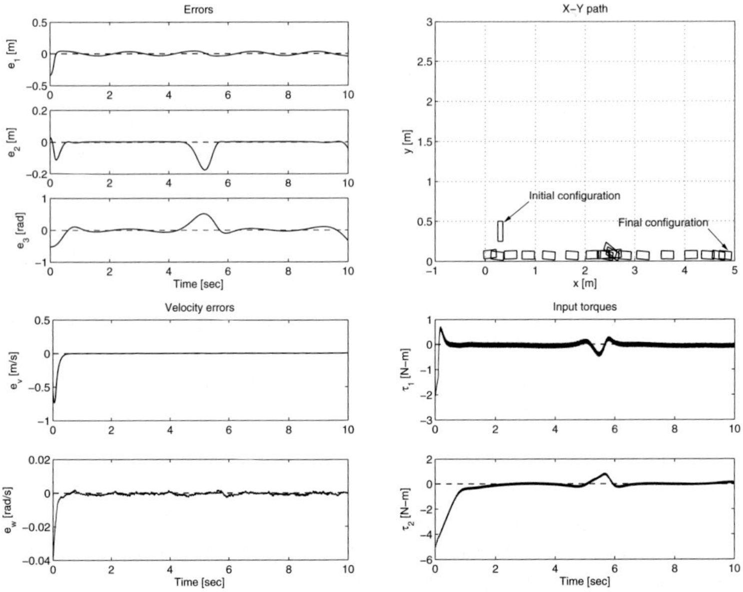


**Fig. 6.** Stabilization of the perturbed mobile robot

controller achieves regulation of the velocity errors in finite-time thus satisfying the control objective while the position and orientation errors remain stable, in spite of the external disturbances.

## 5 Conclusions

This paper addressed the problem of tracking control around a desired position and orientation, taking into account the dynamics of the wheeled mobile robots. The proposed solution is based on the backstepping approach, with an internal loop governed by a Mamdani-based fuzzy logic controller (FLC). In particular, attention have been focused in the Type-2 FLC design forcing the real velocities towards the values required to achieve the control objective while the disturbances forces were attenuated. Two inputs (the linear and angular velocity errors) and two outputs (the torques) were used to create nine IF-THEN fuzzy rules, resulting in minimal software complexity. Genetic algorithm were used for tuning the kinematic control thus minimizing the tracking error. We pointed out that FLC does not require information of the parameters of the Euler-Lagrange equation (masses, inertias, damping, etc.), thus avoiding an extra work on its identification. Capabilities of the derived type-2 fuzzy logic controller are illustrated by simulations. A comparative study of the proposed approach with type-1 fuzzy logic controller has been given in [1].

## Acknowledgment

Authors gratefully acknowledge the financial support given for this work by the Dirección General de Educación Superior Tecnológica under grant no. 314.06-P. Work partially supported by CONACYT under Grant 52246.

## References

1. Astudillo, L.: *Control difuso de un robot autónomo movil en ambientes inciertos.* MSc Thesis, Instituto Tecnológico de Tijuana, 2006.
2. Bloch, A.M.: *Nonholonomic mechanics and control.* Springer Verlag, New York, 2003.
3. Brockett, R.W.: Asymptotic stability and feedback stabilization. In "*Differential Geometric Control Theory*," (eds. R.S. Millman and H.J. Sussman), Birkhauser, Boston, (1983) 181–191.
4. Das, T., Kar, I.N.: Design and implementation of an adaptive fuzzy logic-based controller for wheeled mobile robots. IEEE Trans. Contr. Syst. Technol. **14** (2006) 501–510.
5. Duc Do, K., Zhong-Ping, J., Pan, J.: A global output-feedback controller for simultaneous tracking and stabilization of unicycle-type mobile robots. IEEE Trans. Automat. Contr. **30** (2004) 589–594.
6. Fierro, R., Lewis, F.L.: Control of a nonholonomic mobile robot: backstepping into dynamics. Proc. 34th IEEE Conf. on Decision and Control, New Orleans USA, (1995) 3805–3810.
7. Fukao, T., Nakagawa, H., Adachi, N.: Adaptive tracking control of a nonholonomic mobile robot. IEEE Trans. Robot. Automat. **16** (2000) 609–615.
8. Hagras, H.A.: A hierarchical type-2 fuzzy logic control architecture for autonomous mobile robots. IEEE Trans. Fuzzy Syst. **12** (2004) 524–539.
9. Kristic, M., Kanellakopoulos, I., Kokotovic, P.: *Nonlinear and adaptive control design.* Wiley-Interscience, 1995.
10. Lee, T-C., Song, K-T., Lee, C-H., Teng, C-C.: Tracking control of unicycle-modeled mobile robot using a saturation feedback controller. IEEE Trans. Contr. Syst. Technol. **9** (2001) 305–318.
11. Li, T.H., Chang, S.J., Tong, W.: Fuzzy target tracking control of autonomous mobile robots by using infrared sensors. IEEE Trans. Fuzzy Syst. **12** (2004) 491–501.
12. Liberzon D.: *Switching in Systems and Control.* Birhauser, 2003.
13. Lo Bianco, C.G., Piazzi, A. Romano, M.: Smooth motion generation for unicycle mobile robots via dynamic path inversion. IEEE Trans. Robot. **20** (2004) 884–898.
14. Maeda, M., Maeda, Y., Murakami, S.: Fuzzy drive control of an autonomous mobile robot. Fuzzy Sets and Systems **39** (1991) 195–204.
15. Mendel, J.M.: *Uncertain Rule-Based Fuzzy Logic: Introduction and new directions.* Prentice Hall, USA, 2000.
16. Samson, C.: Control of chained systems applications to path following and time-varying point stabilization of mobile robots. IEEE Trans. Automat. Contr., **40** (1995) 64–77.
17. Yim, H., Butler, A.C.: Motion planning using fuzzy logic control with minimum sensors. In *Proc. of the 1995 IEEE International Symposium on Intelligent Control* (1995), 558–564.
18. Zadeh, L.A.: Fuzzy sets. *Information and control* **8** (1965) 338–353.

# Type-2 Fuzzy Logic for Improving Training Data and Response Integration in Modular Neural Networks for Image Recognition

Olivia Mendoza[1], Patricia Melin[2], Oscar Castillo[2], and Guillermo Licea[1]

[1] Universidad Autonoma de Baja California, Tijuana, Mexico
omendoza@uabc.mx, glicea@uabc.mx
[2] Tijuana Institute of Technology, Tijuana, Mexico
pmelin@tectijuana.mx, ocastillo@tectijuana.mx

**Abstract.** The combination of Soft Computing techniques allows the improvement of intelligent systems with different hybrid approaches. In this work we consider two parts of a Modular Neural Network for image recognition, where a Type-2 Fuzzy Inference System (FIS 2) makes a great difference. The first FIS 2 is used for feature extraction in training data, and the second one to find the ideal parameters for the integration method of the modular neural network. Once again Fuzzy Logic is shown to be a tool that can help improve the results of a neural system, when facilitating the representation of the human perception.

## 1 Methods for Image Recognition

At the moment, many methods for image recognition are available. But most of them include a phase of feature extraction or another type of preprocessing closely related to the type of image to recognize[1][2][3][4][5][6]. The method proposed in this work can be applied to any type of images, because the preprocessing phase does not need specific data about the type of image[7][8][9]. Even if the method was not designed only for face recognition, we have made the tests with the ORL face database [10] composed of 400 images of size 112x92. There are 40 persons, with 10 images of each person. The images are taken at different times, lighting and facial expressions. The faces are in up-right position of frontal view, with slight left-right rotation. To explain the proposed steps of the method, we need to separate it them in two phases: the training phase in Fig.1 and the recognition phase in Fig.2.

## 2 Type-2 Fuzzy Inference System as Edges Detector

In previous work we presented an efficient Fuzzy Inference System for edges detection, in order to use the output image like input data for modular neural networks [11]. In the proposed technique, it is necessary to apply Sobel operators to the original images, and then use a Type-2 Fuzzy Inference System to

P. Melin et al. (Eds.): IFSA 2007, LNAI 4529, pp. 604–612, 2007.
© Springer-Verlag Berlin Heidelberg 2007

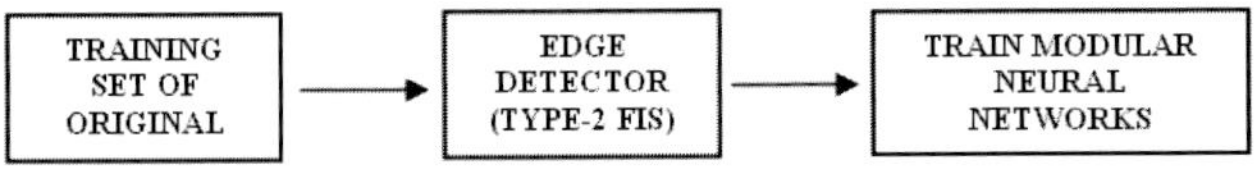


**Fig. 1.** Steps in Training Phase

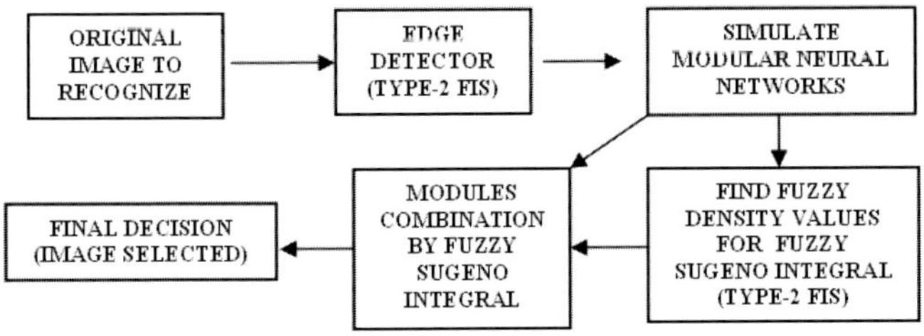


**Fig. 2.** Steps in Recognition Phase

generate the vector of edges that would serve as input data in a neural network. Type-2 Fuzzy Logic enables us to handle uncertainties in decision making and recognition in a more convenient way and for this reason was proposed [12][13]. For the Type-2 Fuzzy Inference System, 3 inputs are required, 2 of them are the gradients with respect to x-axis and y-axis, calculated with (1) and y-axis, calculated with (2), which we call $DH$ and $DV$ respectively. The Sobel edges detector uses a pair of 3x3 convolution masks, one estimating the gradient in the x-direction (columns) and the other estimating the gradient in the y-direction (rows).

$$Sobel_x = \begin{bmatrix} -1\ 0\ 1 \\ -2\ 0\ 2 \\ -1\ 0\ 1 \end{bmatrix} \tag{1}$$

$$Sobel_y = \begin{bmatrix} 1 & 2 & 1 \\ 0 & 0 & 0 \\ -1 & -2 & -1 \end{bmatrix} \tag{2}$$

where $Sobel_x$ y $Sobel_y$ are the Sobel Operators throughout x-axis and y-axis. If we define $I$ as the source image, $g_x$ and $g_y$ are two images, which at each point contain the horizontal and vertical derivative approximations, the latter are computed as (3) and (4).

$$g_x = \sum_{i=1}^{3}\sum_{j=1}^{3}(Sobel_{x,i,j}) * I_{r+i-2,c+j-2} \tag{3}$$

$$g_y = \sum_{i=1}^{3}\sum_{j=1}^{3}(Sobel_{y,i,j}) * I_{r+i-2,c+j-2} \tag{4}$$

where $g_x$ and $g_y$ are the gradients along axis-x and axis-y, and * represents the convolution operator. The other input is a filter that calculates when applying a mask by convolution to the original image. The low-pass filter $hMF$ of (5) allow us to detect image pixels belonging to regions of the input were the mean gray level is lower. These regions are proportionally affected more by noise, but we suppose it is uniformly distributed over the whole image. The goal here is to design a system, which makes it easier to include edges in low contrast regions, but which does not favor false edges by effect of noise[14].

$$hMF = \frac{1}{25} * \begin{bmatrix} 1\,1\,1\,1\,1 \\ 1\,1\,1\,1\,1 \\ 1\,1\,1\,1\,1 \\ 1\,1\,1\,1\,1 \\ 1\,1\,1\,1\,1 \end{bmatrix} \tag{5}$$

The inference rules and membership function parameters allow us to calculate a gray value between -4.5 and 1.5 for each pixel, as shown in Fig.3, where the most negative values corresponds to the dark tone in the edges of the image. Then if we see the rules, only when the increment value of the inputs $DH$ and $DV$ are low the output is HIGH or clear (the background), in the rest of rules the output is LOW or dark (the edges). The complete set of fuzzy rules is given as follows [15]:

```
1.    If (DH is LOW) and (DV is LOW) then (EDGES is HIGH) (1)
2.    If (DH is MEDIUM) and (DV is MEDIUM) then (EDGES is LOW) (1)
3.    If (DH is HIGH) and (DV is HIGH) then (EDGES is LOW) (1)
4.    If (M is LOW) and (DV is MEDIUM) then (EDGES is LOW) (1)
5.    If (M is LOW) and (DH is MEDIUM) then (EDGES is LOW) (1)
```

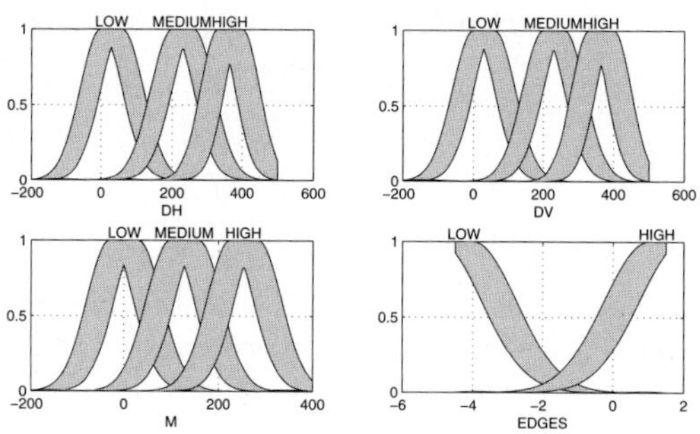


**Fig. 3.** Membership Functions for the Type-2 FIS Edges Detector

Then the inputs for Type-2 FIS are: $DH = g_x$, $DV = g_y$, $M = hMF * I$, where $*$ is the convolution operator, and the output is a column vector contains the values of the image edges, and we can represent that in graphics shown in Fig.4. The Edges Image is smaller than the original because the result of

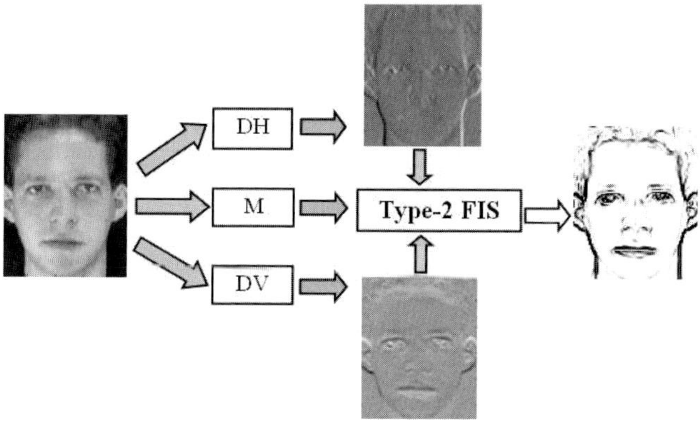


**Fig. 4.** Inputs and output in the Type-2 FIS for Edges Detection

the convolution operation is a central matrix where the convolution has a value. Then in our example, each image with dimension 112x92 was reduced to 108x88.

The edges detector allow us to ignore the background color. We can see in this database of faces, different tones present for the same or another person. Then we eliminate a possible influence of a bad classification by the neural network, without losing detail in the image. Another advantage of edges detector is that the values can be normalized to a homogenous value range, independently the light, contrast or background tone in each image. In the examples in Fig.5, all the edges images have a minimum value of -3.8 and a maximum value of 0.84. In particular for neural network training, we find these values to make the training faster: the mean of the values is near 0 and the standard deviation is near 1 for all the images.

**Fig. 5.** Examples of Edges Detection with the Type-2 FIS method

## 3 The Modular Structure

The design of the Modular Neuronal Network consists of 3 monolithic feedforward neural networks [16], each one trained with a supervised method with the first 7 samples of the 40 images. Then the edges vector column was accumulated

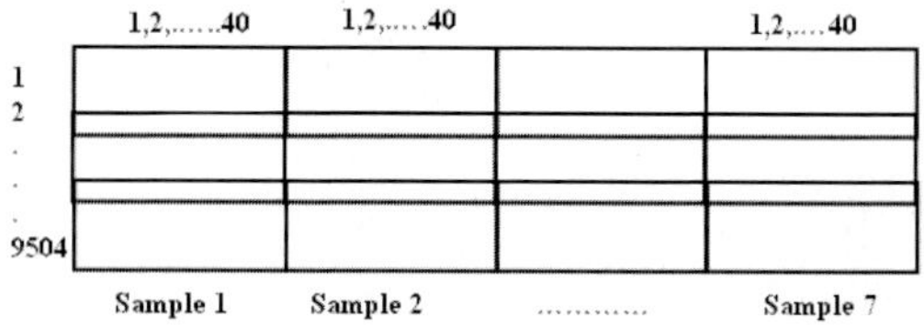


**Fig. 6.** Input: 7 samples for each person

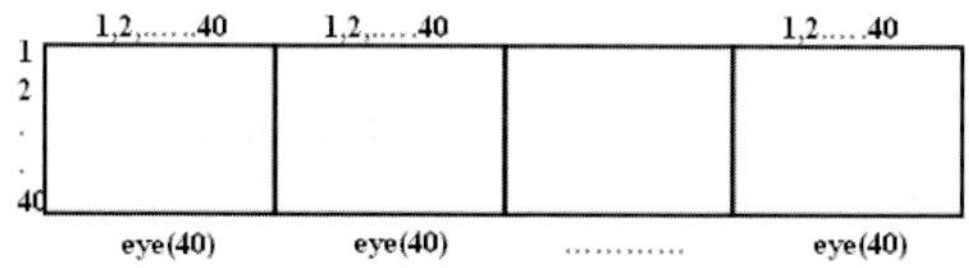


**Fig. 7.** Target: 7 identity matrix with dimension 40, for each sample

until the number of samples to build the input matrix for the neural networks as it is in Fig.6. Once the complete matrix of images was divided in 3 parts, each module was trained with the corresponding part, with some rows of overlapping.

The target to the supervised training method consist of one identity matrix for each sample, building one matrix with dimensions $40x(40 * numberofsamples)$, as shown in Fig.7.

Each Monolithic Neural Network [17] has the same structure and was trained under the same conditions, like we can see in the next code segment:

```
layer1=200; layer2=200; layer3=number_of_subjects;
net=newff(minmax(p),[layer1,layer2,layer3],
{'tansig','tansig','logsig'},'traingdx');
net.trainParam.goal=1e-5;
net.trainParam.epochs=1000;
```

The average number of epochs to meet the goal in each module was 240, and the time for training was 160 seconds.

## 4 Simulation Results

A program was developed in Matlab that simulates each module with the 400 images of the ORL database, building a matrix with the results of the simulation of each module, as it is shown in Fig.8. These matrices are stored in the file "mod.mat" to be analyzed later for the combination of results.

We can observe that in the columns corresponding to the training data, the position with a value near one corresponds the image selected correctly. However in the columns that correspond to the test data this doesn't always happens, reason why it is very important to have a good combination method to be able to recognize more images.

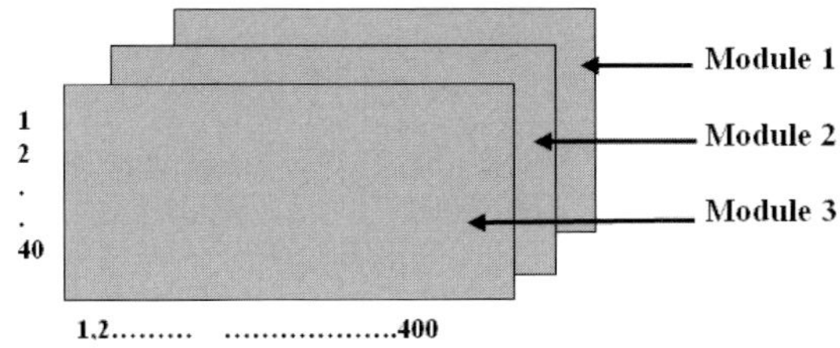


**Fig. 8.** Scheme of the simulation matrices of the 3 modules

### 4.1 Type-2 Fuzzy Inference System to Determine Fuzzy Densities

According to exhaustive tests made in the simulation matrices, we know that recognition of the images that were used for the training of the neural networks is of 100%. Therefore the interest is focused on the recognition of the samples that do not belong to the training set, is to say samples 8,9 and 10. The parameters that will be inferred with the Type-2 Fuzzy Inference System are the Fuzzy Densities for the Sugeno Fuzzy Integral, a value between 0 and 1 for each module, which determines the rate for each module. The parameter *lambda*, according to the theory of fuzzy measures depends on the values of the fuzzy densities, and is calculated by searching for the roots of a polynomial [18].

### 4.2 Inputs and Outputs for the Type-2 Fuzzy Inference System

After the simulation of an image in the Neural Network, the simulation value is the only known parameter to make a decision, then to determine the fuzzy density for each module this is the unique available information. For this reason we analyze the values in many simulations in the matrix and decide that each input to the Type-2 FIS corresponds to the maximum value of each column corresponding to the simulation of each module for each of the 400 images. The process to recognize each one of the images is shown in Fig.9.

Each output corresponds to one fuzzy density, to be applied for each module to perform the fusion of results later with the Fuzzy Sugeno Integral approach. The inference rules find fuzzy densities near 1 when de maximum value in the simulation is between 0.5 and 1, and near 0 when the maximum value in the simulation is near 0. The fuzzy rules are shown below and membership functions in Fig.10 [15].

```
1. If (max1 is LOW) then (d1 is LOW) (1)
2. If (max2 is LOW) then (d2 is LOW) (1)
3. If (max3 is LOW) then (d3 is LOW) (1)
4. If (max1 is MEDIUM) then (d1 is HIGH) (1)
5. If (max2 is MEDIUM) then (d2 is HIGH) (1)
6. If (max3 is MEDIUM) then (d3 is HIGH) (1)
7. If (max1 is HIGH) then (d1 is HIGH) (1)
8. If (max2 is HIGH) then (d2 is HIGH) (1)
9. If (max3 is HIGH) then (d3 is HIGH) (1)
```

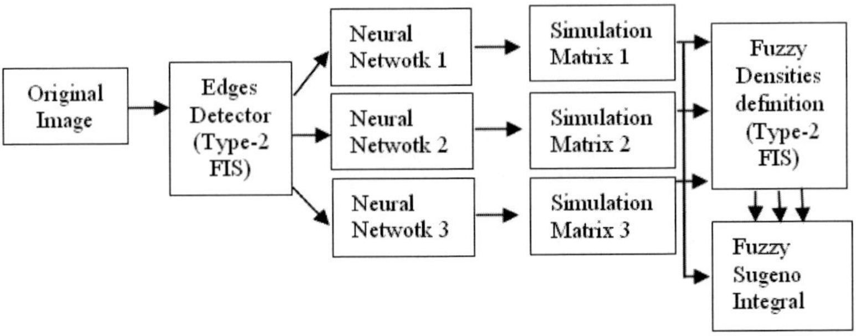


**Fig. 9.** Phases of the Hybrid Type-2 FIS/Modular Neural Network pattern recognition method

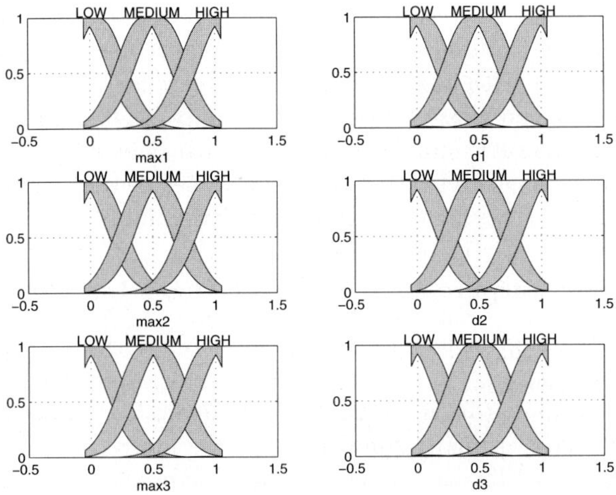


**Fig. 10.** Membership functions for the Type-2 FIS to find fuzzy densities

Although the rules are very simple, they allows to model the fuzziness in evaluating the modules when the simulation results don't reach the maximum value 1. However some of the images don't reach a sufficient value in the simulation of the three modules, in these cases, there are not enough information to select an image in the phase of combination of the modules, and the image is wrongly selected.

### 4.3 Results

In order to measure in an objective form the final results, we developed a method of random permutation, which rearranges the samples of each person before the training. Once a permutation is made, the modular neural networks are trained and combined four times to obtain the sufficient information to validate the

**Table 1.** Summary of the simulation results with the hybrid approach

| Permutation | Train 1 | Train 2 | Train 3 | Train 4 | Average | Maximum |
|---|---|---|---|---|---|---|
| 1 | 92.75 % | 95.00 % | 92.20 % | 93.25 % | 93.30 % | 95.00 % |
| 2 | 96.50 % | 95.25 % | 94.25 % | 95.50 % | 95.37 % | 96.50 % |
| 3 | 91.50 % | 92.00 % | 93.75 % | 95.25 % | 93.12 % | 95.25 % |
| 4 | 94.50 % | 94.50 % | 93.25 % | 94.00 % | 94.06 % | 94.50 % |
| 5 | 93.75 % | 93.50 % | 94.00 % | 96.00 % | 94.31 % | 96.00 % |
| | | | | | 94.04 % | 96.50 % |

results. We show in Table 1 the summary of simulation results for each of the modules and the average and maximum results of the modular network (after fusion or combination of the results).

## 5 Conclusions

We have shown in this paper that the combination of Soft Computing techniques allows the improvement of intelligent systems with different hybrid approaches. In this work we considered two parts of a Modular Neural Network for image recognition, where a Type-2 Fuzzy Inference System (FIS 2) help us improves the performance results in image recognition. The first FIS 2 was used for feature extraction in training data, and the second one to find the ideal parameters for the integration method of the modular neural network. Once again Fuzzy Logic is shown to be a tool that can help improve the results of a neural system, when facilitating the representation of the human perception. In this case, the hybrid fuzzy neural approach can be considered a good alternative for improving the performance of the modular neural model.

## References


1. Renato Salinas, Luis Larraguibel, 2002, *Red Neuronal de Arquitectura Paramtrica en Reconocimiento de Rostros*, Divulgacion Electronica de las Ciencias, Vol.17, http://cabierta.uchile.cl/revista/17/articulos/articulos.html
2. Patricia Melin, Oscar Castillo, 2005, *Hybrid Intelligent Systems for Pattern Recognition*, Springer-Verlag, Heidelberg, Germany.
3. Peng Yang, Bo Du, Shiguang Shan, Wen Gao, 2004, *A Novel Pupil Localization Method Based On Gaboreye Model And Radial Symmetry Operator*, ICT-ISVISION Joint R&D Laboratory for Face Recognition, Beijing, China.
4. V.V. Starovoitov1, D.I Samal, D.V. Briliuk, 2002, *Three Approaches for Face Recognition*, The 6-th International Conference on Pattern Recognition and Image Analysis, Velikiy Novgorod, Russia, pp. 707-711.
5. J. B. Dowdalla, I. Pavlidisa, G. Bebisb, 2003, *Face Detection in the Near-IR Spectrum, Dept. of Computer Science*, University of Nevada, Reno, NV.
6. Mao-Meng Chuang, Ruey-Feng Chang And Yu-Len Huang Automatic, 2000, *Automatic Facial Feature Extraction in Model-based Coding*, Journal of Information Science and Engineering, Vol.16, pp. 447-458.

7. Olivia Mendoza, Patricia Meln, 2004, *Sistema para el Modelado Diseo y Pruebas de Arquitecturas Neuronales Multi-Red*, International Seminar on Computational Intelligence, Tijuana, Mxico., pp. 88-98.
8. Patricia Melin, Alejandra Mancilla, Miguel Lopez,Olivia Mendoza, 2007, *A hybrid modular neural network architecture with fuzzy Sugeno integration for time series forecasting*, Journal of Applied Soft Computing, In Press, Corrected Proof.
9. Olivia Mendoza, Patricia Melin, 2007 *The Fuzzy Sugeno Integral as a Decision Operator in the Recognition of Images with Modular Neural Networks*, In: Hybrid Intelligent Systems:Design and Analysis, Springer-Verlag, Germany, pp. 299-310.
10. AT&T Laboratories Cambridge, *The ORL database of faces*, http://www.cl.cam.ac.uk/research/dtg/attarchive/facedatabase.html
11. Olivia Mendoza, Patricia Meln, 2006, *Sistemas de Inferencia Difusos Tipo-1 y Tipo-2 Aplicados a la Deteccin de Bordes en Imgenes Digitales*, International Seminar on Computational Intelligence 2006 IEEE-CIS chapter Mexico, Tijuana, Mexico, pp. 117-123.
12. Jerry Mendel, 2001, *Uncertain Rule-Based Fuzzy Logic Systems: Introduction and New Directions*, Prentice-Hall, USA.
13. Oscar Castillo, Patricia Melin, Janusz Kacprzyk, Witold Pedrycz, 2007, *Hybrid Intelligent Systems: Design and Analysis*, Springer-Verlag, Germany.
14. Cristiano Jacques Miosso, Adolfo Bauchspiess, 2001, *Fuzzy Inference System Applied to Edges Detection in Digital Images*, V Brazilian Conference on Neural Networks, Brasil.
15. Juan Ramn Castro-Rodrguez, Oscar Castillo-Lpez, Luis Guillermo Martnez-Mndez, 2006, *Tool Box para Lgica Difusa Tipo-2 por Intervalos*, International Seminar on Computational Intelligence 2006 IEEE - CIS Mexico Chapter, Tijuana, Mexico, pp. 100-108.
16. Amanda J.C. Sharkey, *Ensemble and Modular Multi-Net Systems*, 1999, In: Amanda J.C. Sharkey, Combining Artificial Neural Nets, Springer, Germany, pp. 1-25.
17. The Math Works, Inc., 1994-2006 *Neural Network Toolbox*, available on the web page: http://www.mathworks.com/products/neuralnet/
18. Marcin Detyniecki, 2000, *Mathematical Aggregation Operators and Their Application to Video Querying*,PHD Artificial Intelligence, Pierre and Marie Curie University, Paris.

# Part XII

# Fuzzy Logic Applications

# A Fuzzy Model for Olive Oil Sensory Evaluation

Luis Martínez[1], Luis G. Pérez[1], Jun Liu[2], and Macarena Espinilla[1]

[1] Computer Sciences Department
University of Jaen
Campus Las Lagunillas s/n 23071- Jaén, Spain
{martin,lgonzaga}@ujaen.es
[2] School of Computing and Mathematics
University of Ulster at Jordanstown
Newtownabbey, Co. Antrim BT37 0QB, Northern Ireland, UK
j.liu@ulster.ac.uk

**Abstract.** The evaluation is a process that analyzes elements to achieve different objectives such as quality inspection, design, marketing exploitation and other fields in industrial companies. In many of these fields the items, products, designs, etc., are evaluated according to the knowledge acquired via human senses (sight, taste, touch, smell and hearing), in such cases, the process is called *Sensory Evaluation*. In this type of evaluation process, an important problem arises as it is the modelling and management of uncertain knowledge, because the information acquired by our senses throughout human perceptions involves uncertainty, vagueness and imprecision.

The sensory evaluation of Olive oil plays a relevant role for the quality and properties of the commercialized product. In this contribution, we shall present a new evaluation model for Olive oil sensory evaluation based on a decision analysis scheme that will use the Fuzzy Linguistic Approach to facilitate the modelling and managing of the uncertainty and vagueness of the information acquired through the human perceptions in the sensory evaluation process.

**Keywords:** Decision Analysis, Sensory Evaluation, Linguistic variables.

## 1 Introduction

The evaluation is a complex cognitive process that involves different mechanisms in which it is necessary to define the elements to evaluate, fix the evaluation framework, gather the information and obtain an evaluation assessment by means of an evaluation process. The aim of any evaluation process is to obtain information about the worth of an item (product, service, material, etc.), a complete description about different aspects, indicators, criteria in order to improve it or to compare with other items to know which ones are the best. The information gathered in this kind of processes is usually provided by a group of individuals, called panel of experts, where each expert expresses their opinions about the item according to their knowledge and their own perceptions.

P. Melin et al. (Eds.): IFSA 2007, LNAI 4529, pp. 615–624, 2007.
© Springer-Verlag Berlin Heidelberg 2007

This contribution is focused on *Sensory Evaluation* processes [5,12,13] that is an evaluation discipline whose information, provided by the panel of experts, is perceived by the human senses of *sight, smell, taste, touch and hearing.* A suitable mathematical formulation is not easy in this type of problems because human perceptions are subjective and not objective, therefore the assessments provided by the individuals are vague and uncertain. In such a case, linguistic descriptors are direct provided by the experts to express their knowledge about the evaluated element. The Fuzzy Linguistic Approach [15] provides a systematic way to represent linguistic variables in an evaluation procedure.

In decision theory before making a decision is carried out a decision analysis approach that allows people to make decisions more consistently, i.e., it helps people to deal with difficult decisions. The decision analysis is a suitable approach for evaluation processes because it helps to analyze the alternatives, aspects, indicators of the element/s under study that it is the objective of the evaluation processes.

Nowadays, the quality of the olive oil plays a key role in its production and final price. The evaluation of the quality of the olive oil is not an easy task and is usually accomplished by olive oil Tasting Panel, which will evaluate, by means of their perceptions acquired via their senses, the features that describe the samples of olive oil. The aim of this contribution is to propose a linguistic sensory evaluation model based on a decision analysis scheme that uses the Fuzzy Linguistic Approach and the 2-tuple fuzzy representation model [6] to represent the experts' assessments.

This paper is structured as follows, in Section 2 we present and review in short the necessary concepts and processes to develop the linguistic sensory evaluation. In Section 3 we present our proposal of linguistic sensory evaluation model, and in Section 4 we expound an application of this evaluation model. Finally, this paper is concluded in Section 5.

## 2 Background

Our evaluation model is based on the scheme of the Decision Analysis that we present in this section. Moreover, we shall make a brief review of the Fuzzy Linguistic Approach and the Linguistic 2-tuple representation model that will be used to facilitate the computation of the linguistic information in the evaluation process.

### 2.1 Decision Analysis Steps

The Decision Analysis is a discipline, which belongs to Decision Making Theory, whose purpose is to help the decision makers to reach a consistent decision in a decision making problem. Here, we model the evaluation process as a Multi-Expert Decision Making (MEDM) problem. A classical decision analysis scheme is composed by the following phases (see figure 1):

- *Identify decision and objectives.*
- *Identify alternatives.*

- *Model:* For example, a decision problem is modelled as a MEDM [7] model that deals with a type of information.
- *Gathering information:* decision makers provide their information.
- *Rating alternatives:* This phase is also known as "aggregation phase" [11] due to the fact in this phase, the individual preferences are aggregated in order to obtain a collective value for each alternative.
- *Choosing best alternatives:* or "exploitation phase" [11] selects the solution from the set of alternatives applying a choice degree [1,10] to the collective values computed in the previous phase.
- *Sensitive analysis:* in this step the information obtained is analyzed in order to know if it is good enough to make a decision, or otherwise, to go back to initial phases to improve the quantity or/and the quality of the information obtained.
- *Make a decision.*

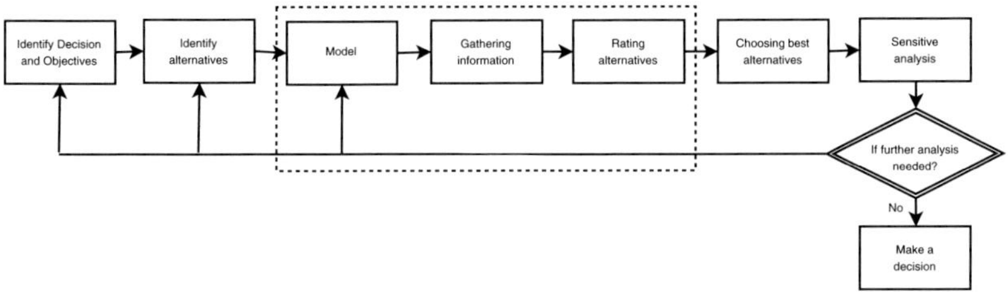


**Fig. 1.** Decision Analysis Scheme

The application of the decision analysis to an evaluation process does not imply the eight phases. The essential phases regarding an evaluation problem are dashed in a rectangle of the Fig. 1.

### 2.2 Fuzzy Linguistic Approach

Although we usually work in quantitative settings where the information is expressed by numerical values, sometimes we shall need to describe activities of the real world that cannot be assessed in a quantitative form, but rather in a qualitative one, i.e., with vague or imprecise knowledge. In that case, a better approach may be to use linguistic assessments instead of numerical values. The variables which participate in these problems are assessed by means of linguistic terms [15]. The fuzzy linguistic approach represents qualitative aspects as linguistic values by means of linguistic variables [15]. We have to choose the appropriate linguistic descriptors for the term set and their semantics. In order to accomplish this objective, an important aspect to analyze is the *"granularity of uncertainty"*, i.e., the level of discrimination among different counts of uncertainty. The universe of the discourse over which the term set is defined can be arbitrary, in this paper we shall use linguistic term sets in the interval $[0, 1]$. In [2] the use of term sets with an odd cardinal was studied, representing the mid term by an assessment of "approximately 0.5", with the rest of the terms being placed symmetrically around it and with typical values of cardinality, such as 7 or 9.

One possibility of generating the linguistic term set consists of directly supplying the term set by considering all terms distributed on a scale on which total order is defined [14]. For example, a set of seven terms $S$, could be given as:

$$S = \{s_0 : none, s_1 : verylow, s_2 : low, s_3 : medium, s_4 : high, s_5 : veryhigh, s_6 : perfect\}$$

In these cases, it is required that in the linguistic term set there exist:

1. A negation operator $\text{Neg}(s_i) = s_j$ such that $j = g$-i (g+1 is the cardinality).
2. A max operator: $\max(s_i, s_j) = s_i \; if \; s_i \geq s_j$.
3. A min operator: $\min(s_i, s_j) = s_i \; if \; s_i \leq s_j$

The semantics of the terms is given by fuzzy numbers. A computationally efficient way to characterize a fuzzy number is to use a representation based on parameters of its membership function [2]. The linguistic assessments given by the users are just approximate ones, some authors consider that linear trapezoidal membership functions are good enough to capture the vagueness of those linguistic assessments. The parametric representation is achieved by the 4-tuple $(a, b, d, c)$, where $b$ and $d$ indicate the interval in which the membership value is 1, with $a$ and $c$ indicating the left and right limits of the definition domain of the trapezoidal membership function [2]. A particular case of this type of representation are the linguistic assessments whose membership functions are triangular, i.e., $b = d$, then we represent this type of membership functions by a 3-tuple $(a, b, c)$. An example may be the following:

$$P = (.83, 1, 1) \quad VH = (.67, .83, 1) \; H = (.5, .67, .83) \; M = (.33, .5, .67)$$
$$L = (.17, .33, .5) \; VL = (0, .17, .33) \;\; N = (0, 0, .17),$$

The use of linguistic variables implies processes of computing with words such as their fusion, aggregation, comparison, etc. To perform these computations there are different models in the literature, such as, the semantic one [3], the symbolic one [4] or the 2-tuple representation model [6]. In the following subsection we shall review the 2-tuple model due to the fact, that it will be the computational model used in our evaluation proposal.

### 2.3 The 2-Tuple Fuzzy Linguistic Representation Model

This model has been presented in [6] and has showed itself as useful to deal with evaluation problems similar to the one we are facing in this paper [9].

This linguistic model takes as basis the symbolic aggregation model [4] and in addition defines the concept of Symbolic Translation and uses it to represent the linguistic information by means of a pair of values called linguistic 2-tuple, $(s, \alpha)$, where $s$ is a linguistic term and $\alpha$ is a numeric value representing the symbolic translation.

**Definition 1.** *Let $\beta$ be the result of an aggregation of the indexes of a set of labels assessed in a linguistic term set $S = \{s_0, ..., s_g\}$, i.e., the result of a symbolic aggregation operation. $\beta \in [0, g]$, being $g + 1$ the cardinality of $S$. Let $i = round(\beta)$ and $\alpha = \beta - i$ be two values, such that, $i \in [0, g]$ and $\alpha \in [-.5, .5)$ then $\alpha$ is called a Symbolic Translation.*

**Definition 2 [6].** *Let $S = \{s_0, ..., s_g\}$ be a linguistic term set and $\beta \in [0, g]$ a value supporting the result of a symbolic aggregation operation, then the 2-tuple that expresses the equivalent information to $\beta$ is obtained with the following function:*

$$\Delta : [0, g] \longrightarrow S \times [-0.5, 0.5)$$

$$\Delta(\beta) = \begin{cases} s_i & i = round(\beta) \\ \alpha = \beta - i & \alpha \in [-.5, .5) \end{cases} \quad (1)$$

*where round is the usual* round *operation, $s_i$ has the closest index label to "$\beta$" and "$\alpha$" is the value of the symbolic translation.*

**Proposition 1 [6].** *Let $S = \{s_0, ..., s_g\}$ be a linguistic term set and $(s_i, \alpha)$ be a 2-tuple. There is a $\Delta^{-1}$ function, such that, from a 2-tuple it returns its equivalent numerical value $\beta \in [0, g] \subset \mathcal{R}$.*

**Proof.** It is trivial, we consider the following function:

$$\Delta^{-1} : S \times [-.5, .5) \longrightarrow [0, g] \quad (2)$$

$$\Delta^{-1}(s_i, \alpha) = i + \alpha = \beta$$

**Remark 1:** *From definitions 2 and 3 and from proposition 1, it is obvious that the conversion of a linguistic term into a linguistic 2-tuple consist of adding a value* 0 *as symbolic translation:* $s_i \in S \Longrightarrow (s_i, 0)$

This representation model has associated a computational model that was presented in [6].

## 3 Linguistic Sensory Evaluation Model Based on Decision Analysis

We must keep in mind that the information provided by the experts in sensory evaluation has been perceived by the senses of sight, touch, smell, taste and hearing, and therefore, those requirements are subjective and involves uncertainty, vagueness and imprecision.

Our aim is to propose a Sensory Evaluation model based on the linguistic decision analysis whose mathematical formalism will be the linguistic 2-tuple model in order to obtain accurate and reliable evaluation results. This proposal consists of the following evaluation phases that are graphically showed in Fig.2.

- *Identify Evaluated Objects.* This phase is problem-dependent and each problem identifies its objects of interest.
- *Model:* this phase defines the evaluation framework that establishes the evaluation context in which the information is assessed and the problem solved.
- *Gathering information:* the experts express their sensory knowledge about the objects by means of linguistic assessments.
- *Rating objects:* we propose to use of the 2-tuple computational model to obtain a rate for every object.

– *Evaluation results:* it consists of analyzing the results obtained in the previous phase with the purpose of achieving the evaluation process. These results can be used in different ways, such as:
  - To learn which element is better considered by the experts.
  - To know which features are better in the evaluated element.
  - To identify which aspects of an element should be improve in order to enhance its quality.
  - Etc.

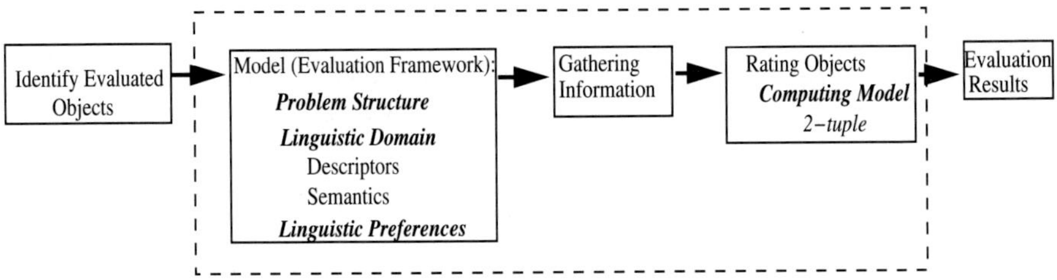


**Fig. 2.** A Linguistic Sensory Evaluation Scheme based on decision analysis

In the following subsections we shall present in further detail the main phases of our sensory evaluation model.

### 3.1 Model

This phase models the evaluation problem defining its evaluation framework, such that, the problem structure is defined and the linguistic descriptors and semantics that will be used by the experts to provide the information about the sensory features of the evaluated objects are chosen.

First of all, we must analyze which sensory features will be evaluated that depend on the evaluated object and which linguistic term set will be used to assess those features. The linguistic term set will be chosen according to:

1. *The accuracy of the evaluations:* since our senses could recognize and assess some features better than others, the granularity of the linguistic term set must be chosen according to the accuracy of our perceptions.
2. *The experience of the experts:* Some of the senses need to be trained. Therefore, the granularity of the linguistic term set used by an expert should be also chosen according to the expert's experience.

In this contribution we deal with an evaluation framework such that the different experts provide their sensory perceptions about item features by means of a linguistic label assessed in a fixed term set according to the above conditions.

### 3.2 Gathering Information

Due to the fact that the linguistic decision analysis we propose is based on the MEDM problems. The experts provide their knowledge by means of utility vectors that contain a linguistic assessment for each evaluated feature.

$\{e_1, ..., e_n\}$, group of experts
$O = \{o_1, ...., o_m\}$, set of evaluated objects
$F = \{f_1, ...., f_h\}$, set of evaluated features for each object
$S = \{s_0, ...., s_g\}$, Linguistic term set
$e_i$ provides his/her preferences in $S$ by means of a utility vector:

$$U_i = \{u^i_{11}, ...., u^i_{1h}, u^i_{21}, ..., u^i_{2h}, ..., u^i_{m1}, ..., u^i_{mh}\}$$

where $u^i_{jk} \in S$ is the assessment provided to the feature $f_k$ of the object $o_j$ by the expert $e_i$. Consequently in the gathering process every expert $e_i$ will provide his/her utility vector $U_i$ expressed by linguistic labels in the linguistic term set $S$ fixed in the evaluation framework.

### 3.3 Rating Objects

In this phase the linguistic utility vectors provided by the experts will be used in processes of Computing with Words in order to rate each evaluated object. To do so, the information gathered will be aggregated.

This proposal will use the linguistic 2-tuple computational model, to operate with the uncertain information provided by the experts it must be remarked that several aggregation operators have been introduced for this computational model [6]. The rating process of this proposal consists of two steps:

1. *Computing collective evaluations for each feature*: in the gathering process each expert, $e_i$ provides his/her preferences for every feature $f_k$ of the object $o_j$ by means of a utility assessment , $u^i_{jk}$. Then, the rating process in first place will compute a collective value for each feature, $u_{jk}$, using an aggregation operator, $AG$, on the assessments provided by the experts:

$$u_{jk} = AG_1(u^1_{jk}, ...., u^n_{jk}) \tag{3}$$

2. *Computing a collective evaluation for each object*: the final aim of the rating process is to obtain a global evaluation, $u_j$, of each evaluated object according to all the experts and features that take part in the evaluation process. To do so, this process will aggregate the collective features values $u_{jk}$ for each object, $o_j$:

$$u_j = AG_2(u_{j1}, ...., u_{jh}) \tag{4}$$

   The aggregation operators, $AG_1$ and $AG_2$, will depend on each evaluation problem taking into account if all experts or features are equally important or there are experts or features more important than the others.

   The collective evaluation obtained will be the score obtained by the evaluated object in the sensory evaluation problem.

**Table 1.** Olive Oil Tasting Panel's utility vectors for the feature *sweetness*

| | $e_1$ | $e_2$ | $e_3$ | $e_4$ | $e_5$ | $e_6$ | $e_7$ | $e_8$ |
|---|---|---|---|---|---|---|---|---|
| $o_1$ | $s_4$ | $s_2$ | $s_5$ | $s_3$ | $s_4$ | $s_5$ | $s_2$ | $s_7$ |
| $o_2$ | $s_4$ | $s_3$ | $s_4$ | $s_2$ | $s_2$ | $s_4$ | $s_5$ | $s_3$ |
| $o_3$ | $s_3$ | $s_3$ | $s_5$ | $s_4$ | $s_3$ | $s_2$ | $s_4$ | $s_2$ |
| $o_4$ | $s_5$ | $s_4$ | $s_4$ | $s_5$ | $s_6$ | $s_3$ | $s_7$ | $s_3$ |

# 4 Evaluating Sweetness of Olive Oil

The quality of the olive oil plays a key role in its production and final price. This quality depends on several aspects such as the condition of olives when enter the factory, the extraction processes and their sedimentation, or their storage.

The evaluation of the quality of the olive oil is usually accomplished by a testing panel that evaluate the features that describe the samples of olive oil, by means of their perceptions acquired via their senses.

The combination of smell and taste is known as flavor and defines the organoleptic properties of the olive oil. These properties, with acidity grade of the olive oil, are essential to obtain their quality. While it is easy to obtain the acidity grade of a sample of olive oil by means of chemical processes, the organoleptic properties need to be evaluated by a Tasting Panel that uses their perceptions to catch different aspects of its flavor such as fruity, bitter, pungent, etc.

Here, we shall show a simple example of how to evaluate four samples of olive oil, in order to find out the values of the organoleptic property of sweetness. These values can be used in order to decide which batches should be mixed to obtain a given flavor.

## 4.1 Evaluation Framework

An Olive oil Tasting Panel of eight connoisseurs $E = \{e_1, ..., e_8\}$ will evaluate the sensory feature $F = \{sweetness\}$ of four samples of Olive Oil $O = \{o_1, ...., o_4\}$. To do so, a linguistic term set $S$ of nine terms is chosen according to conditions presented in subsection 3.1 to assess the sweetness. Its syntax and semantics are the following ones:

$s_8$ : *Very sweet* : $(.88, 1, 1)$ $s_7$ : *Rather sweet* : $(.75, .88, 1)$ $s_6$ : *Sweet* : $(.62, .75, .88)$
$s_5$ : *A bit sweet* : $(.5, .62, .75)$ $s_4$ : *Average* : $(.38, .5, .62)$ $s_3$ : *A bit bitter* : $(.25, .38, .5)$
$s_2$ : *Bitter* : $(.12, .25, .38)$ $s_1$ : *Rather bitter* : $(0, .12, .25)$ $s_0$ : *Very bitter* : $(0, 0, .12)$

## 4.2 Gathering Process

The preferences of our Tasting Panel for sweetness are showed in Table 1.

Now, we shall transform their preferences into 2-tuple representation model (Table 2) to manage easily this information.

**Table 2.** Olive Oil Tasting Panel's utility vectors for the feature *sweetness* over the 2-tuple representation model

| | $e_1$ | $e_2$ | $e_3$ | $e_4$ | $e_5$ | $e_6$ | $e_7$ | $e_8$ |
|---|---|---|---|---|---|---|---|---|
| $o_1$ | $(s_3, 0)$ | $(s_3, 0)$ | $(s_6, 0)$ | $(s_4, 0)$ | $(s_6, 0)$ | $(s_6, 0)$ | $(s_4, 0)$ | $(s_7, 0)$ |
| $o_2$ | $(s_4, 0)$ | $(s_3, 0)$ | $(s_4, 0)$ | $(s_2, 0)$ | $(s_2, 0)$ | $(s_4, 0)$ | $(s_5, 0)$ | $(s_3, 0)$ |
| $o_3$ | $(s_3, 0)$ | $(s_3, 0)$ | $(s_5, 0)$ | $(s_4, 0)$ | $(s_3, 0)$ | $(s_3, 0)$ | $(s_4, 0)$ | $(s_2, 0)$ |
| $o_4$ | $(s_4, 0)$ | $(s_3, 0)$ | $(s_4, 0)$ | $(s_4, 0)$ | $(s_5, 0)$ | $(s_3, 0)$ | $(s_7, 0)$ | $(s_3, 0)$ |

### 4.3 Rating Objects

In this phase we shall carry out the following steps:

1. *Computing collective values for each feature:* In order to simplify the example we have considered that all the experts are equally important. Therefore, we have used the arithmetic mean for 2-tuples [6] for aggregating the information provided by the experts (Table 3) obtaining a collective value for sweetness for each sample according to all the connoisseurs:

**Table 3.** Olive Oil Tasting Panel's collective utility vector for the *sweetness*

| $o_1$ | $o_2$ | $o_3$ | $o_4$ |
|---|---|---|---|
| $(s_5 = A\ b\ sw, -.125)$ | $(s_3 = A\ b\ bit, .375)$ | $(s_3 = A\ b\ bit, .375)$ | $(s_4 = Av, .25)$ |

2. *Computing a collective evaluation for each object:* In this example the objective is to obtain the evaluation of the organoleptic feature. So it is not necessary to obtain a global evaluation of each olive batch according to the property analyzed. However, it is important to point out that if it would be necessary to obtain this global evaluation value we should use an aggregation method able to manage linguistic information assessed in different linguistic term sets as the methods showed in [7,8].

### 4.4 Evaluation Results

The purpose of this evaluation process was to find out the values of different samples of olive oil regarding their sweetness property . If we analyze the aforesaid results (Table 3), the sample $o_1$ obtains the highest score for it.

## 5 Concluding Remarks

When we face a sensory evaluation problem we must realize that we are going to work with knowledge that has been acquired via the human senses sight, taste, touch, smell and hearing. This knowledge is better expressed using words instead of numbers.

In this contribution, we have presented a sensory evaluation model based on the linguistic decision analysis and the 2-tuple computational model.

Finally we have applied this model to a specific sensory evaluation problem, the evaluation of olive oil.

## Acknowledgements

This work has been partially supported by the Research Project TIN2006-02121.

## References


1. K.J. Arrow. *Social Choice and Individual Values.* Yale University Press, New Haven CT, 1963.
2. P.P. Bonissone and K.S. Decker. *Selecting Uncertainty Calculi and Granularity: An Experiment in Trading-Off Precision and Complexity.* In L.H. Kanal and J.F. Lemmer, Editors., Uncertainty in Artificial Intelligence. North-Holland, 1986.
3. R. Degani and G. Bortolan. The problem of linguistic approximation in clinical decision making. *International Journal of Approximate Reasoning*, 2:143–162, 1988.
4. M. Delgado, J.L. Verdegay, and M.A Vila. On aggregation operations of linguistic labels. *International Journal of Intelligent Systems*, 8:351–370, 1993.
5. G.B. Dijksterhuis. *Multivariate Data Analysis in Sensory and Consumer Science, Food and Nutrition.* Press Inc. Trumbull, Connecticut, USA, 1997.
6. F. Herrera and L. Martínez. A 2-tuple fuzzy linguistic representation model for computing with words. *IEEE Transactions on Fuzzy Systems*, 8(6):746–752, 2000.
7. F. Herrera and L. Martínez. A model based on linguistic 2-tuples for dealing with multigranularity hierarchical linguistic contexts in multiexpert decision-making. *IEEE Transactions on Systems, Man and Cybernetics. Part B: Cybernetics*, 31(2):227–234, 2001.
8. F. Herrera, L. Martínez, and P.J. Sánchez. Managing non-homogeneous information in group decision making. *European Journal of Operational Research*, 166(1):115–132, 2005.
9. L. Martínez, J. Liu, J.B. Yang, and F. Herrera. A multi-granular hierarchical linguistic model for design evaluation based on safety and cost analysis. *International Journal of Intelligent Systems.*, 20(12):1161–1194, 2005.
10. S.A. Orlovsky. Decision-making with a fuzzy preference relation. *Fuzzy Sets Systems*, 1:155–167, 1978.
11. M. Roubens. Fuzzy sets and decision analysis. *Fuzzy Sets and Systems*, 90:199–206, 1997.
12. D. Ruan and X. Zeng (Eds.). *Sensory Evaluation: Methodologies and Applications.* Springer, 2004.
13. H. Stone and J.L. Sidel. *Sensory Evaluation Practice.* Academic Press Inc., San Diego, CA, 1993.
14. R.R. Yager. An approach to ordinal decision making. *International Journal of Approximate Reasoning*, 12:237–261, 1995.
15. L.A. Zadeh. The concept of a linguistic variable and its applications to approximate reasoning. *Information Sciences, Part I, II, III*, 8,8,9:199–249,301–357,43–80, 1975.

# An Interval-Based Index Structure for Structure Elucidation in Chemical Databases

Sven Helmer

University of London, Birkbeck College,
London, United Kingdom
sven@dcs.bbk.ac.uk

**Abstract.** We propose to adapt an interval-based index structure, Relational Interval trees, to support the process of determining the structure of an unknown chemical compound. Important information for retrieving relevant substructures that make up a compound can only be described in an imprecise way, resulting in interval-based values specifying chemical shifts. The access method was implemented on top of a commercial database system and evaluated experimentally. The results of these experiments show that Relational Interval trees are an efficient way of indexing data needed for structure elucidation.

## 1 Introduction

Elucidation is the process of determining the chemical structure of an unknown substance. This plays an important role in chemistry and biochemistry (e.g. in such areas as pharmaceutical research, development of technical synthesis, or quality control), since many characteristics of a compound can be identified by analyzing its structure. Consequently, many different systems have been developed for structure elucidation: NMRAnalyst [2], SIGNATURE [1], and CHEMICS [3] just to name a few. We are going to have a closer look at SpecSolv, developed by Will et al. [9], which is based on Nuclear Magnetic Resonance (NMR) spectroscopy (more details on this later). One step in the elucidation process of SpecSolv involves filtering a database of about 700,000 identified substructures for possible candidates from which to construct the complete compound. Substructures are matched with a compound by comparing the location of peaks in spectrograms. Due to the nature of the NMR method, peaks in a compound may be shifted when compared to the peaks of a substructure leading to an imprecise description of the peak locations. Nevertheless, we are able to determine limits of these shifts, thus describing intervals in which a compound's peak must lie to match with one of a substructure. We show how an interval-based index structure devised for relational database management systems (RDBMSs), the Relational Interval (RI) tree, can be adapted and employed for speeding up this filtering step. A prototype was implemented on top of a commercial DBMS and compared to the technique used by SpecSolv before.

The remainder of this paper is structured as follows. The following section gives a brief introduction to NMR spectroscopy, while Section 3 formalizes the

P. Melin et al. (Eds.): IFSA 2007, LNAI 4529, pp. 625–634, 2007.
© Springer-Verlag Berlin Heidelberg 2007

problem we are trying to solve. Section 4 briefly reviews the technique that is currently used. In Section 5 we introduce RI-trees and show how they can be adapted to indexing chemical data. The results of our experiments and their interpretation can be found in Section 6. We conclude the paper with a short summary and an outlook.

## 2 NMR Spectroscopy and SpecSolv

Atomic nuclei (e.g. those of $^{1}H$ and $^{13}C$) resonate at a characteristic frequency when exposed to a magnetic field. Depending on the bonds that an atom forms with neighboring atoms in a compound, it resonates at slightly different frequencies (this is called the *chemical shift*). During spectroscopy all the different resonating frequencies of a compound are measured. The results of this measurement can be plotted as peaks at certain frequencies (see Figure 1). These signals or peaks can then be assigned to atoms or atom groups, gaining knowledge about a compound's structure. The intensity of a signal indicates the number of certain groups that can be found in a molecule. (For an introduction to NMR spectroscopy see [8].)

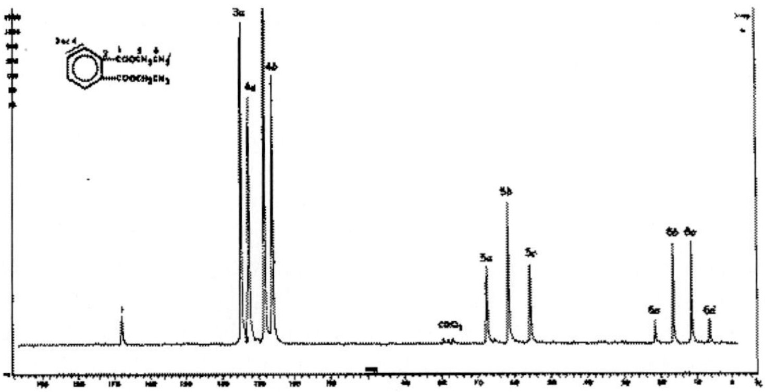

**Fig. 1.** NMR spectroscopy result

In the SpecSolv system the most important signals are those for single carbon atoms (C) and groups consisting of a single carbon and a single/two/three hydrogen atoms (CH, $CH_2$, and $CH_3$, respectively). After obtaining the chemical shifts of an (unknown) substance these are matched to those of known substructures to determine possible building blocks for the unknown, investigated molecule.

Here we focus on filtering out substructures that cannot possibly be contained in the investigated molecule. This filtering is very important: feeding all 700,000 substructures in the database into SpecSolv for a full structural analysis can easily result in a run time of several months for structure elucidation. The filtering step speeds this up considerably; we turn an intractable problem into a feasible one.

## 3 Formalization of the Problem

The peak frequencies of the investigated molecule are stored in four multisets (one for each $CH_x$ group): $M_C$, $M_{CH}$, $M_{CH_2}$, and $M_{CH_3}$, so the search for matching substructures is done four times: a substructure is matched for each $CH_x$ group. As already mentioned, the intensity of a peak indicates the number of groups with a certain shift. Therefore certain frequencies may show up multiple times. We denote the frequency of the $k$-th peak in the multiset $M_{CH_x}$ by $p_k^{M_{CH_x}}$ with $x \in \{0, 1, 2, 3\}$.

For each substructure in the database we know the exact chemical formula and the exact structure. The characteristics of a known substructure $j$ in terms of the $CH_x$ groups it contains are also described by four multisets: $S_C^j$, $S_{CH}^j$, $S_{CH_2}^j$, and $S_{CH_3}^j$. Rather than peak data these multisets contain intervals which describe the range of possible frequencies where peaks can appear for this substructure. As already mentioned, depending on how a substructure is embedded into a molecule, it can have different chemical shifts. $I_m^{S_{CH_x}^j}$ denotes the $m$-th interval of the multiset $S_{CH_x}^j$ (with $x \in \{0, 1, 2, 3\}$). Each interval in turn is described by its lower bound $l_m^{S_{CH_x}^j}$ and its upper bound $u_m^{S_{CH_x}^j}$.

Formally speaking, the filtering step boils down to a bipartite matching between the peaks of the investigated molecule and the intervals describing the known substructures. An interval $I_m^{S_{CH_x}^j}$ matches a peak $p_k^{M_{CH_x}}$, denoted by $I_m^{S_{CH_x}^j} \doteq p_k^{M_{CH_x}}$, if $l_m^{S_{CH_x}^j} \leq p_k^{M_{CH_x}} \leq u_m^{S_{CH_x}^j}$. A substructure $j$ may be a candidate (may be contained in an unknown molecule), if

$$\forall I_m^{S_{CH_x}^j} \in S_{CH_x}^j : \exists p_k^{M_{CH_x}} \in M_{CH_x} : I_m^{S_{CH_x}^j} \doteq p_k^{M_{CH_x}}$$

In addition to this, no two intervals may be matched to the same peak, i.e.

$$\forall I_m^{S_{CH_x}^j}, I_n^{S_{CH_x}^j} \in S_{CH_x}^j \text{with } m \neq n : I_m^{S_{CH_x}^j} \doteq p_k^{M_{CH_x}} \wedge I_n^{S_{CH_x}^j} \doteq p_l^{M_{CH_x}} \Rightarrow k \neq l$$

Only if both of the above conditions are satisfied for all $CH_x$ groups of a substructure, then it is a candidate. Otherwise, this substructure cannot possibly be found in the investigated molecule and is filtered out.

## 4 State of the Art

Implementing the bipartite matching as described in Section 3 directly on a RDBMS in SQL is very awkward and slow. Database administrators at BASF, where SpecSolv is being used, have come up with the following filtering step before checking the bipartite matching after retrieving the relevant information from the database.

In this filtering step the intervals of a substructure are described with the help of a bit vector, each bit representing a fixed interval in the domain of frequency

**Table 1.** Fixed intervals for bit vectors

| Group | Bit 1 | Bit 2 | Bit 3 | Bit 4 | Group | Bit 1 | Bit 2 | Bit 3 |
|---|---|---|---|---|---|---|---|---|
| C | [0,60) | [60,90) | [90,186) | [186,∞) | $CH_2$ | [0,56) | [56,90) | [90,∞) |
| CH | [0,63) | [63,90) | [90,145) | [145,∞) | $CH_3$ | [0,5) | [5,45) | [45,∞) |

ranges. Table 1 gives an overview of the fixed intervals (the groups C and CH use 4 bits, the groups $CH_2$ and $CH_3$ use 3 bits[1])

Whenever an interval $I_m^{S^j_{CH_x}}$ intersects with one of the fixed intervals of group $CH_x$, the bit for this fixed interval will be set to 1. We proceed likewise for the peak frequencies of the investigated molecule. The bits for all intervals in which peaks can be found are set to 1. During querying the bit vector entries of all substructures are scanned and if the bit vectors of the intervals form a subset of the query molecule bit vectors (i.e. all bits in a substructure bit vector are also set in the query bit vector), then we retrieve this entry for checking the bipartite matching. The bit vector index described above is basically a signature file using superimposed coding [7].

Although this technique speeds up the retrieval, we still have to scan through all 700,000 entries and compare their bit vectors to the query bit vector. As we will see later in the section on experimental evaluation, we can do better by employing an index structure made specifically for interval-based data.

## 5 Relational Interval Trees

We are now going to present an alternative to the bit vector index: the relational interval tree (or RI-tree) by Kriegel et al. [4]. It is based on the interval tree by Edelsbrunner [6]. For better understandability we are first going to give a brief introduction to interval trees, which are an efficient way of managing intervals in main memory.

### 5.1 Interval Trees

An interval tree $T$ consists of a *primary* and a *secondary* structure. The primary structure divides up a (discrete) set of points, which w.l.o.g. we denote by $Y = \{y_1, y_2, \ldots, y_{2^n-1}\}$. Note that the set $Y$ is totally ordered. $T$ is a binary tree with the following properties:

1. The root $w$ of $T$ has a discriminant $d(w) = 2^{(n-1)}$ and pointers to two lists $L(w)$ and $R(w)$ that store the (left) lower bounds and (right) upper bounds, respectively, of all intervals which contain $d(w)$. The lists $L$ and $R$ make up the secondary structure of the interval tree.

[1] There are additional bits used for describing other chemical properties of the substructures. However, they are unrelated to the bipartite matching, so we are not going to look at them here.

2. The left subtree of $w$ contains all intervals whose upper bound is smaller than $d(w)$, while the right subtree contains all those whose lower bound is greater than $d(w)$. The left subtree is responsible for managing the subset $y_1, y_2, \ldots, y_{2^{(n-1)}-1}$ while the right subtree manages the other half $y_{2^{(n-1)}+1}$, $y_{2^{(n-1)}+2}, \ldots, y_{2^n-1}$. This continues recursively until a subtree manages only a single point. Each node in a subtree also has a discriminant which splits the managed subset in half.
3. Every node of the primary structure is either *active* or *inactive*. A node is active if its secondary structure (the lists $L$ and $R$) are non-empty or if its subtree contains active nodes. Otherwise a node is inactive. We only store information on active nodes in the data structure, as the primary structure of the tree can be computed (the discriminants follow a very regular structure).

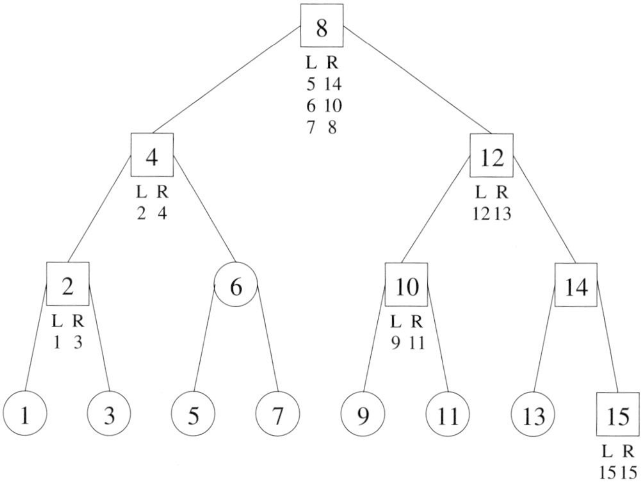


**Fig. 2.** Example of an interval tree

Figure 2 shows an example of an interval tree for the set $Y = \{1, 2, \ldots, 15\}$ into which the intervals [1,3], [2,4], [5,8], [6,14], [7,10], [9,11], [12,13], and [15,15] have been inserted. The square boxes show active nodes, which have been labeled with their discriminant (inactive nodes are represented by circles). The secondary lists $L$ and $R$ are sorted ($L$ ascending and $R$ descending) for optimizing the search. When sorting the lists, the corresponding ends of the intervals have to be linked with each other so that they can be reconstructed.

**Insertion of Intervals.** In order to insert an interval $[s, e]$, we have to find the top-most node $j$ whose discriminant is contained in the interval, i.e. $s \leq d(j) \leq e$. We start the insertion procedure at the root $w$ of the interval tree and work our way downwards. For each node $i$ we encounter we do the following:

1. Check if $s \leq d(i) \leq e$. If this is the case, we insert $s$ into $L(i)$ and $e$ into $R(i)$ and we stop.

2. If $e < d(i)$, then choose the left subtree of $i$ as new current node $i$. Go to 1.
3. If $d(i) < s$, then choose the right subtree of $i$ as new current node $i$. Go to 1.

During the insertion all nodes, i.e. active and inactive nodes, are relevant. An insertion may result in nodes changing from an inactive state to an active one.

**Deletion of Intervals.** Deleting an interval is very simple: we just remove its end points from the $L$ and $R$ lists of the node where it is stored. If the secondary lists of a node are empty afterwards (and all its descendants are inactive), this node changes from an active state to an inactive one. This may also result in some of the ancestor nodes of this node changing from an active state to an inactive one.

**Searching for Intervals.** Here we look at point queries (also called stabbing queries), which are relevant for our application. This means, that we query with a search value $v \in Y$, and we want to find all intervals which contain this value. Basically, we traverse the tree from the root $w$ until we find a node $i$ for which $d(i) = v$ or until we reach an (active) leaf node.

Let $i$ denote the current node; we start the search at the root node $w$:

1. If $d(i) = v$, then output all intervals in the lists $L(i)$ and $R(i)$ and stop the search.
2. If $v < d(i)$, then output all intervals stored at node $i$ for which the lower bound in $L(i)$ is smaller than $v$. If $i$ has an active left subtree, continue the search there (go to 1.), otherwise stop the search.
3. If $d(i) < v$, then output all intervals stored at node $i$ for which the upper bound in $R(i)$ is greater than $v$. If $i$ has an active right subtree, continue the search there (go to 1.), otherwise stop the search.

## 5.2 Adapting Interval Trees to an RDBMS

Now we are going to have a look at how to implement an interval tree in a relational schema (see also [4]). Since the primary structure of a tree can be computed, we only need to store active nodes in the RDBMS. Obviously we need a relation in which to store the intervals, this is done via the relation *Intervals* shown in Figure 3(a). This table contains the lower and upper bounds of an interval, a key to identify the interval, and the number of the node where the interval is stored. In addition to this we create two indexes on the table*Intervals*: *Lindex* and *Rindex* (see Figure 3(b)). These two indexes represent the secondary structure of an interval tree and contain the $L$ and $R$ lists of all nodes.

**Insertion and Deletion of Intervals.** Before inserting an interval into the database, we have to compute the number of the node where it should be stored. Figure 4 shows the insertion algorithm (`root` is the number of the root node and is equal to $2^{(n-1)}$). After computing the node number, an interval can be inserted into the relation *Intervals*. For deleting an interval we simply delete it from the relation *Intervals*.

```
CREATE TABLE Intervals (
  node  integer,
  id    integer primary key,
  lower integer,
  upper integer);
```

(a)

```
CREATE INDEX Lindex
ON TABLE Intervals(node, lower);

CREATE INDEX Rindex
ON TABLE Intervals(node, upper);
```

(b)

**Fig. 3.** Relational schema for an interval tree

```
int findNode(int lower, int upper, int root)
  int node = root;
  for(step = node/2; step >= 1; step /=2) {
    if(upper < node) { node -= step; }
    else if (node < lower) { node += step;}
    else break;
  }
  return node;
```

**Fig. 4.** Insertion algorithm

**Searching for Intervals.** When given a point query with the value $v$, instead of scanning the $L$ and $R$ lists immediately (corresponding to *Lindex* and *Rindex* in the relational schema), we traverse down the tree by computing the relevant node numbers noting whether the $L$, $R$, or both lists have to be scanned. The node numbers are inserted into two (transient) relations called *leftNodes* and *rightNodes*, respectively. These two relations are then used during querying (see Figure 5).

```
SELECT id
FROM   leftNodes l, Intervals i, rightNodes r
WHERE  (i.node = l.node AND i.upper >= v)
OR     (i.node = r.node AND i.lower <= v);
```

**Fig. 5.** SQL query

### 5.3 Using RI-Trees to Index Molecules

We now need to integrate RI-trees into our querying schema for finding substructures of molecules. As there are four different types of intervals, depending on the $CH_x$ group we look at, we allocate a table for each of these groups (see Figure 6(a)). In addition to the interval id, we need an id identifying the substructure an interval belongs to.[2] Also, RI-trees are only suited for integers, while the intervals we want to index are bounded by real numbers. Therefore we round the lower bounds of each interval to the next smaller integer and the upper bounds to the next larger

[2] As a substructure can have more than one interval, we have a composite key.

integer. In doing so, we are on the safe side, i.e. we will not lose any qualifying intervals (but may introduce some more false positives). The attribute *noVal* denotes the number of intervals for a substructure. This is needed later during querying to find out if all intervals have matched with a query peak.

```
CREATE TABLE CH_x (
  node  integer,
  subid integer,
  intid integer,
  lower integer,
  upper integer
  noVal integer,
  primary key
   (subid,intid));
```

(a)

```
SELECT c.subid
FROM   CH_x c, leftPeaks l, rightPeaks r
WHERE  ((c.node = l.node AND c.upper >= l.v)
     OR (c.node = r.node AND c.lower <= r.v))
AND    n.subid = c.subid
GROUP BY c.subid
HAVING COUNT(*) >= c.noVal;
```

(b)

**Fig. 6.** Managing molecule substructures with RI-trees

We can check that each interval matches with at least one peak in SQL (see Figure 6(b)). The relations *leftPeaks* and *rightPeaks* contain the relevant node numbers when traversing down the tree and values for all the peaks of the query molecule. The main problem with this SQL query is that we cannot guarantee that all intervals match with different peaks. This is done outside of the SQL in a C-program using a standard algorithm for determining bipartite matching [5].

## 6 Experimental Evaluation

The RI-tree index for the substructures was realized on top of an IBM DB2 UDB database system running on a Linux machine. The boundaries of the intervals were rounded, the data was loaded into the tables, and the statistics were updated by running the `runstats` command. We used seven different queries, ranging from small to large molecules (containing from 4 to 43 peaks), to test the index. Each query was formulated in three different ways: an SQL query using the indexes (and a C-program to do the bipartite matching), the bit vector approach as described in Section 4 (with the bit vector comparison built into a user-defined function), and a pure SQL query without indexes (that was not able to do the actual bipartite matching) as a reference.

For each of the queries we investigated its runtime (in seconds) and its filtering capabilities (in the number of substructures returned by the query). The results for the run time can be found in Table 2(a), those for the answer set cardinality can be found in Table 2(b).

Let us first turn to the runtime results. Except for one case (the query with four peaks) the bit vector approach is the fastest. Basically it scans the whole table comparing the bit vectors of the substructures with the query bit vector. The comparison can be done using very fast bit vector operations. Furthermore,

**Table 2.** Experimental results

| Query | # of peaks | RI-tree | bit vector | pure SQL |
|---|---|---|---|---|
| 1 | 4 | 390.0s | 483.7s | 2373.7s |
| 2 | 14 | 1302.3s | 703.1s | 2634.7s |
| 3 | 22 | 1558.1s | 729.6s | 2503.5.s |
| 4 | 30 | 1886.7s | 734.1s | 2670.0s |
| 5 | 33 | 2117.6s | 745.0s | 2858.3s |
| 6 | 38 | 2394.9s | 772.6s | 3034.5s |
| 7 | 43 | 2498.2s | 781.7s | 3651.0s |

(a) Runtime

| Query | RI-tree | bit vector | pure SQL |
|---|---|---|---|
| 1 | 416 | 18,349 | 520 |
| 2 | 2156 | 213,136 | 2713 |
| 3 | 9395 | 255,931 | 12,564 |
| 4 | 13,225 | 324,508 | 17,836 |
| 5 | 30,144 | 372,376 | 41,477 |
| 6 | 44,003 | 476,153 | 55,864 |
| 7 | 53,929 | 495,430 | 66,961 |

(b) Result set cardinality

these comparisons are not influenced by the number of peaks in the query. The slight increase in runtime is due to the growing cardinality of the result set that has to be returned. The runtime of the RI-tree and pure SQL queries also increases with larger query molecules. For the pure SQL query this is also mostly due to the growing result set cardinality. The disproportionate increase for the RI-trees can be explained by the higher number of tuples that have to be joined, which makes the join operations costlier. Nevertheless, the RI-tree index is still faster than the pure SQL query. At first glance the bit vector approach seems to be the clear winner. However, it filters very badly for large molecules as we will see in a moment.

In terms of the filtering capabilities the RI-tree index is the clear winner (see Table 2(b)). It always returns the smallest number of candidate substructures. While the differences between the RI-tree index and the pure SQL approach are not very large (RI-trees return about 25% fewer substructures), the differences between RI-trees and the bit vector approach are tremendous. Remember that the database contains about 700,000 substructures which means that with query 5 and upwards more than half of the substructures in the database are returned. This is clearly unacceptable, since the next step, doing the actual structure elucidation, is an NP-complete problem. Keeping the input small is absolutely crucial to the runtime of the SpecSolv algorithm.

## 7 Conclusion and Outlook

An important application depending on chemical databases is structure elucidation, i.e. determining the structure of an unknown molecule. This is relevant for identifying chemical properties of molecules. The process of elucidation involves imprecise data in the form of chemical shifts, which can be described by intervals. We proposed to adapt a Relational Interval tree to index this data to be able to efficiently filter out substructures that cannot possibly be found in the investigated molecule. We implemented an RI-tree on top of a commercial DBMS and demonstrated experimentally that it shows the best overall behavior when compared to a bit vector approach and pure SQL queries. Unlike other specialized index structures, RI-trees do not need to be integrated into the core engine of a DBMS to work, which usually is not possible to do for users of DBMSs.

One of our next goals will be to improve this index structure even further. Since molecular databases for structure elucidation are relatively static (data in the database is rarely changed; the only updates that happen from time to time are insertions of new substructures), it may pay off to materialize even more information in the index, e.g. about other properties of the substructures, or to buffer data intelligently in main memory for faster retrieval.

## Acknowledgments

We want to give special thanks to Dirk Schiebeler for implementing and benchmarking the index and to Martin Will from BASF for providing us with the chemical data.

## References

1. M.S. Diallo, J.L. Faulon, W.A. Goddard, and J.H. Johnson. Humic Substances: Structures, Models and Functions, chapter Binding of Hydrophobic Organic Compounds to Dissolved Humic Substances: A Predictive Approach Based on Computer Assisted Structure Elucidation, Atomistic Simulations and Flory-Huggins Solution Theory. Royal Society of Chemistry, 2001.
2. R. Dunkel, C.L. Mayne, R.J. Pugmire, and D.M. Grant. Computerized analysis of two-dimensional double quantum NMR spectra. In Symposium on Advances in Computer Simulation and Artificial Intelligence in Analytical Chemistry, April 1990.
3. K. Funatsu, Y. Susuta, and S.I. Sasaki. Application of the automated structure elucidation system (CHEMICS) to the chemistry of natural products. Pure and Appl. Chem., 61(3):609–612, 1989.
4. H.P. Kriegel, M. Potke, and T. Seidl. Managing intervals efficienty in objectrelational databases. In Proc. of 26th Int. Conf. on Very Large Data Bases (VLDB), pages 407–418, 2000.
5. U. Manber. Introduction to Algorithms: A Creative Approach. Addison Wesley, New York, 1989.
6. F.P. Preparata and M.I. Shamos. Computational Geometry: An Introduction. Springer, Heidelberg, 1993.
7. C.S. Roberts. Partial-match retrieval via the method of superimposed codes. Proc. of the IEEE, 67(12):1624–1642, December 1979.
8. Wikipedia. NMR spectroscopy. http://en.wikipedia.org/wiki/NMR Spectroscopy.
9. M. Will, W. Fachinger, and J. Richert. Fully automated structured elucidation - a spectroscopists dream comes true. J. Chem. Inf. Comput. Science, 36(2):221–227, 1996.

# Fuzzy Cognitive Layer in RoboCupSoccer

Susana Muñoz-Hernandez and Wiratna Sari Wiguna

Facultad de Informática
Universidad Politécnica de Madrid
susana@fi.upm.es,wiratna@gmail.com

**Abstract.** RoboCupSoccer domain has several leagues which varies in the rule of play such as specification of players, number of players, field size, and time length. Nevertheless, each RoboCup league is a variant of a soccer league and therefore they are based on some basic rules of soccer. A layered design of agents system presented in [1] shows a modular approach to build control for a team of robots participating in RoboCup-Soccer E-League. Based on this design, we propose a generalized architecture offering flexibility to switch between leagues and programming language while maintaining Prolog as cognitive layer. Prolog is a perfect tool to design strategies for soccer players using simple rules close to human reasoning. Sometimes this reasoning needs to deal with uncertainty, fuzziness or incompleteness of the information. In these cases it is useful Fuzzy Prolog [11,8,10,9]. In this paper we propose to use a combination of Prolog (that is crisp) and Fuzzy Prolog to implement the cognitive layer in RoboCupSoccer, which has the advantage of incorporating as conventional logic as fuzzy logic in this layer. A prototype of a team based on this architecture has been build for RoboCup Soccer Simulator, and we show that this approach provides a convenient way of incorporating a team strategy in high level (human-like) manner, where technical details are encapsulated and fuzzy information is represented.

**Keywords:** Logic Programming, Constraint Logic Programming Implementation, Fuzzy Reasoning, Prolog Application, RoboCupSoccer, Cognitive Layer.

## 1 Introduction

The idea of robot playing soccer has been developed since early 90s [7]. Soccer environment is a dynamically changing environment which requires individual skill as well as team skill and therefore is an interesting research field on Artificial Intelligence and robotics. Prolog is a programming language that represent logic reasoning. Is is a perfect tool to represent human reasoning, so it seems to be a good choice for implementing the cognitive layer of soccer players that is a simulation of human behaviour related to this game. For example, applying the rule "if the goal keeper is not at the goal then kick to ball". But many of the most important decisions that are made by soccer players deal with non-crisp issues. They are related to fuzziness (e.g. "if other player of my team is FAR

P. Melin et al. (Eds.): IFSA 2007, LNAI 4529, pp. 635–645, 2007.
© Springer-Verlag Berlin Heidelberg 2007

from me then don't pass him/her the ball"), uncertainty (e.g. "if I CAN get the goal then kick the ball"), or incompleteness (e.g. "if I cannot see the position of a player, by default I'm not going to pass him the ball"). Fuzzy Prolog is an attempt to introduce fuzzy reasoning into logic programming that also deals with uncertainty and incompleteness. It is aimed of this proposal to combine the advantages of these different types of programming and to show how to handle this combination.

There are many works that have been done on this research area related to RoboCup [1,6,5] and to Fuzzy Prolog [11,8,10,9]. This work is the continuation of the research line of the project [2].

The rest of the paper is organized as follow. Next section gives brief overview on RoboCupSoccer and section 3 describes Fuzzy Prolog. Section 4 and 5 talks about our approach and its evaluation. Section 6 concludes this paper and mentions some further works.

## 2 RoboCupSoccer

RoboCup is an international annual event promoting research on Artificial Intelligence, robotics, and related field. The original motivation of RoboCup is RoboCupSoccer. As the nature of soccer game, autonomous robots participating in RoboCupSoccer should have individual ability such as moving and kicking the ball, cooperative ability such as coordinating with teammates, and of course, the ability to deal with dynamic environment.

RoboCupSoccer consists of several leagues, providing test beds for various research scale: Simulation League, Small Size Robot League (F-180), Middle Size Robot League (f-2000), Four-Legged Robot League, Humanoid League, E-League and RoboCup Commentator Exhibition.

Our work is part of a joint research project [2] on RoboCupSoccer E-League with the National University of Comahue (Argentina). However as a preliminary work, we employ RoboCupSoccer Simulation League for the sake of simplicity.

## 3 Fuzzy Prolog

The Ciao Prolog System offers a complete Prolog system supporting ISO-Prolog. Its modular design allows restriction and extension of the language both syntactically and semantically. The Ciao Prolog Development System provides many libraries including a constraint logic programming system and interfaces to some programming languages. In Ciao Prolog terminology, a library is implemented as either a module or a package. Fuzzy Prolog described in [11] and [8,9] is implemented as the package "fuzzy.pl", a syntactic extension of the CLP($\mathcal{R}$) system in the Ciao Prolog System. This is a continuous variant of Fuzzy Prolog.

We use in this work the discrete variant of Fuzzy Prolog that is implemented using CLP($\mathcal{FD}$) as described in [10]. It offers an implementation of a Fuzzy Prolog system with discrete (versus continuous) truth values. The next subsection summarizes the basic formal concepts that are described by the semantics of this language.

### 3.1 Discrete Fuzzy Prolog Language

The set of continuous subintervals on [0,1] is denoted by $\mathcal{E}([0,1])$. The Borel Algebra, $\mathcal{B}([0,1])$, is the power set of $\mathcal{E}([0,1])$. We talk about discrete instead of continuous interval when it is compound by a finite set of elements included in the corresponding continuous interval. The set of discrete subintervals on [0,1] is denoted by $\mathcal{E}_d([0,1])$. We call $\mathcal{B}_d([0,1])$ the Discrete Borel Algebra over the interval [0,1] for representing the set of finite unions of discrete subintervals on [0,1].

As defined in [10], truth values in discrete Fuzzy Prolog are elements of Discrete Borel Algebra over the interval [0,1]. Fuzzy sets are defined by functions of the form $A : X \rightarrow \mathcal{B}_d([0,1])$. The Fuzzy Prolog system with Borel Algebra as in [11] and [8,9] is often referred as continuous Fuzzy Prolog system. Figure 1 illustrate the difference between Borel Algebra and Discrete Borel Algebra.

Notice that the truth value representation of Fuzzy Prolog and discrete Fuzzy Prolog is very general (union of intervals of real numbers) and it can seem to be little intuitive. The applications of this generality are disscussed in [11] and [8,9]. Simple truth values (as a unique interval or a plain real number) can be more adequate to be used in RoboCup game. These simple values are particular cases of the general framework.

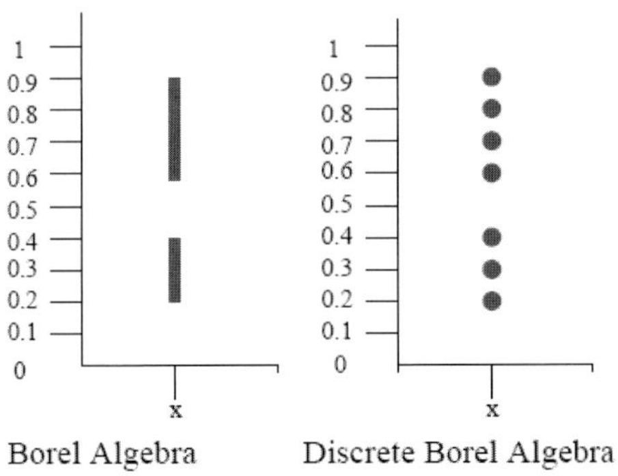


**Fig. 1.** Truth Value: Borel Algebra versus Discrete Borel Algebra [10]

**Definition 1 (discrete-interval).** *A discrete-interval* $[X_1, X_N]_\epsilon$ *is a finite set of values,* $\{X_1, X_2, \ldots, X_{N-1}, X_N\}$, $0 \leq X_1 \leq X_N \leq 1$ *such that* $\exists\ 0 < \epsilon < 1$. $X_i = X_{i-1} + \epsilon, i \in \{2..N\}$.

As indicated in the definition, the set of values of a discrete-interval depends on the choice of $\epsilon$. Smaller $\epsilon$ value represents more precision. For example, an interval $[0.3, 0.5]_{0.1}$ with $\epsilon = 0.1$ is a finite set {0.3, 0.4, 0.5} while the interval $[0.30, 0.50]_{0.01}$ with $\epsilon = 0.01$ is a finite set {0.30, 0.31, 0.32,. . . , 0.48, 0.49, 0.50}.

Therefore we denote the algebras $\mathcal{E}_d([0,1])$ and $\mathcal{B}_d([0,1])$ incorporating the granularity to the notation substituting $d$ by the particular $\epsilon$ (e.g. $\mathcal{B}_{0.1}([0,1])$).

**Definition 2 (discrete-aggregation).** *Discrete-aggregation (in fuzzy sets) is the application of a numeric discrete-aggregation operator (or discrete-aggregation) of type* $f : [0,1]^n \rightarrow [0,1]$. *If it satisfies* $f(0,\ldots,0) = 0$ *and* $f(1,\ldots,1) = 1$, *and in addition it is monotonic.*

Notice that the operator is only monotonic by definition, not continuous (that is why it is called "discrete").

**Definition 3 (discrete-interval-aggregation).** *Given a discrete-aggregation* $f : [0,1]^n \rightarrow [0,1]$, *a discrete-interval-aggregation* $F : \mathcal{E}_\epsilon([0,1])^n \rightarrow \mathcal{E}_\epsilon([0,1])$ *is defined as follows:*

$$F([x_1^l, x_1^u]_\epsilon, \ldots, [x_n^l, x_n^u]_\epsilon) = [f(x_1^l, \ldots, x_n^l), f(x_1^u, \ldots, x_n^u)]_\epsilon$$

*where* $0 < \epsilon < 1$.

Intuitively we can say that $F$ provide a discrete-interval from the aggregation of $n$ discrete-intervals.

**Definition 4 (discrete-union-aggregation).** *Given a discrete-interval-aggregation* $F : \mathcal{E}_\epsilon([0,1])^n \rightarrow \mathcal{E}_\epsilon([0,1])$ *defined over discrete-intervals, a discrete-union-aggregation* $\mathcal{F} : \mathcal{B}_\epsilon([0,1])^n \rightarrow \mathcal{B}_\epsilon([0,1])$ *is defined over union of discrete-intervals as follows:*

$$\mathcal{F}(B_1, \ldots, B_n) = \cup\{F(\mathcal{E}_{1\epsilon}, \ldots, \mathcal{E}_{n\epsilon}) | \mathcal{E}_{i\epsilon} \in B_i\}.$$

The alphabet of the fuzzy language consists of variables, constants, function symbols, and predicate symbols. A term is defined inductively as follows:

1. A variable is a term.
2. A constant is a term.
3. If $f$ is an n-ary function symbol and $t_1, \ldots, t_n$ are terms then $f(t_1, \ldots, t_n)$ is a term.

An atom or atomic formula is defined as the following:
If $p$ is an n-ary predicate symbol and $t_1, \ldots, t_n$ are terms, then $p(t_1, \ldots, t_n)$ is an atom.
A fuzzy program is a finite set of fuzzy facts and fuzzy clauses. Information is obtained from the fuzzy program through a fuzzy query.

**Definition 5 (fuzzy fact).** *If* $A$ *is an atom,* $A \leftarrow v$ *is a fuzzy fact, where* $v$, *a truth value, is an element in* $\mathcal{B}_\epsilon([0,1])$ *and* $0 < \epsilon < 1$.

**Definition 6 (fuzzy clause).** *Let* $A, B_1, \ldots, B_n$ *be atoms,* $A \leftarrow_F B_1, \ldots, B_n$ *is a fuzzy clause where* $F$ *is a discrete-interval-aggregation operator of truth values in* $\mathcal{B}_\epsilon([0,1])$, $0 < \epsilon < 1$, *and* $F$ *induces a discrete-union-aggregation as by definition 4.*

**Definition 7 (fuzzy query).** *A fuzzy query is a tuple* $v \leftarrow A?$ *where* $A$ *is an atom, and* $v$ *is a variable (possibly instantiated) that represents a truth value in* $\mathcal{B}_\epsilon([0,1])$, *where* $0 < \epsilon < 1$.

### 3.2 Discrete Fuzzy Prolog Syntax

Constraint Logic Programming, CLP, is one of the most promising extensions of Logic Programming from the implementation point of view. There are many Prolog systems that implement it [4]. One of the most popular extensions is CLP($\mathcal{FD}$), Constraint Logic Programming over Finite Domains. It is related to constrain the set of possible values of the variables for efficiency. So, the possible values can be obtained according to a set of constraints that should be satisfied by each variable. In Ciao Prolog, CLP($\mathcal{FD}$) works on integer domain. We use it to represent the set of truth values of a fuzzy variable by a set of integer numbers. Therefore, given the $\epsilon$ as in definition 1, truth value is interpreted as a finite union of discrete subintervals on $[0,1]_\epsilon$. Hence, the discrete subinterval is a set of integers.

The interval $[X_1, X_N]_{1/k}$ is interpreted for CLP($\mathcal{FD}$) as the discrete interval $[X_1 * K * 10, X_N * K * 10]_1$

For example, the interval $[0.4, 0.6]_{0.01}$ is interpreted as the set {400, 401, 402,..., 598, 599, 600} while $[0.4, 0.6]_{0.1}$ is interpreted as the set {40, 41,..., 59, 60}. Table 1 shows the syntax for fuzzy facts. We use 0.1, 0.01, etc for simplicity, but any other value (e.g. 0.2, 0.34, ...) can be used also.

**Table 1.** Discrete Fuzzy Prolog syntax for fuzzy fact

| Fuzzy Fact | Discrete Fuzzy Prolog Syntax |
|---|---|
| $p(john) \leftarrow 0.70$ | `p(john,70) ::~.` |
| $p(peter) \leftarrow [0.40, 0.60]_{0.01}$ | `p(peter,V) ::~ { V in 400 .. 600}.` |
| $p(joan) \leftarrow$ $[0.20, 0.50]_{0.1} \cup [0.80, 1]_{0.1}$ | `p(joan,V) ::~ { V in 20 .. 50}.` `p(joan,V) ::~ { V in 80 .. 100}.` |

The fuzzy clause is defined as `Head ::~ Aggregator Body`. For example, the syntax for fuzzy clause

$slow\_dash\ (Distance, Power) \leftarrow_{min}$
$\quad near(Distance),$
$\quad low\_dash\_power(Power)$

is in discrete Fuzzy Prolog

```
slow_dash (Distance,Power,V) ::~ min
          near(Distance,V1),
          dash_power(Power,V2).
```

In Fuzzy Prolog syntax, the query is formulated as an atom $A$ with $v$, the truth value, as additional parameter. The fuzzy query is defined similar with the ones in continuous Fuzzy Prolog. For example, the syntax for the fuzzy query that consults if running slowly is a good option taking into account the distance that the player want to cover and the power of his/her dash $v \leftarrow slow\_dash(Distance, Power)$? is written in Fuzzy Prolog:

```
?-slow_dash(Distance,Power,V).
```

The value of $V$ is obtained from the aggregation (using the operator $min$) of the truth values $V1$ and $V2$. The value of $V$ in Fuzzy Prolog with CLP($\mathcal{FD}$) ranges (values or intervals) between 0 and 100, 0 and 1000, ... (depending on the precision of the program and representing the corresponding truth values between 0 and 1 obtained dividing by 100, 1000, ... respectively). As CLP($\mathcal{FD}$) supports `labeling`, $V$ could be instantiated to one or more values satisfying the constraints. This is the reason why discrete Fuzzy Prolog is more useful for this application than the continuous variant, because it provides constructive answers as values instead of providing constraints (as in the continuous variant).

Note that when more precision is needed, a suitable $\epsilon$ in the definition 1 could be chosen, which results in a larger set of integer between 0 and 1 for truth values.

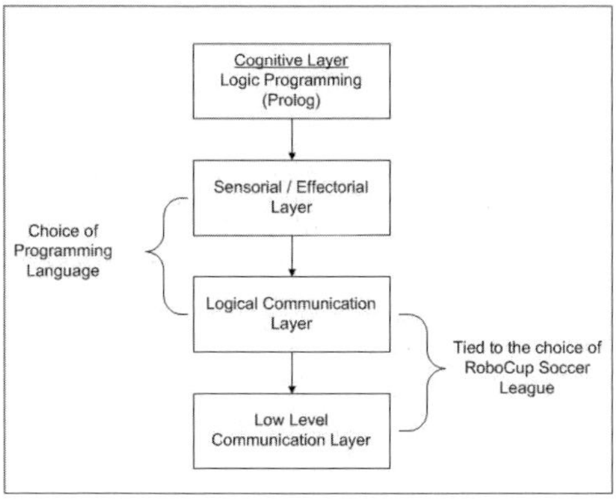


**Fig. 2.** Generic System Architecture

## 4 Architecture and Implementation Details

Based on agent system architecture proposed by [1], we propose a generic system architecture for RoboCup offering flexibility on choice of programming language and minimal modification to switch between leagues. This architecture is shown in figure 2. Prolog is proposed for cognitive layer, and in our work we use Fuzzy Prolog for implementing the cognitive layer. The system architecture of our implementation for RoboCupSoccer Simulation League is shown in figure 3. As it can be seen in the figures and as it is going to be described in this section, the generic architecture is customized in our implementation for Simulation League.

### 4.1 Low Level Communication Layer

As the name suggests, this is the lowest layer of our architecture. This layer includes all hardwares and softwares provided by the league. The robots, infrared transmitter, video camera, communication network, and vision systems belong to this layer. Different leagues in RoboCupSoccer are represented by different

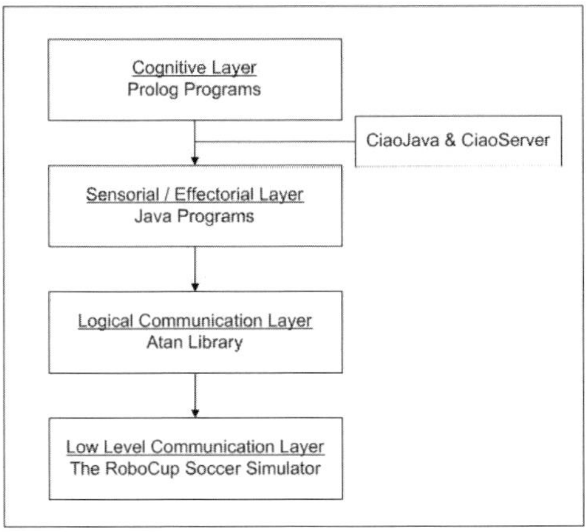


**Fig. 3.** System Architecture for RoboCup Soccer Server

Low Level Communication Layer. E-League has the robots, Doraemon vision package, and communication server as part of this layer, whereas Simulation League has only The RoboCup Soccer Simulator as part of this layer.

### 4.2 Logical Communication Layer

This layer acts as the interface between low level communication layer and the upper layers. It is intended to hide physical structure of the environment from the upper layer. As long as the interface of the services offered by this layer remain unchanged, then the rest of the upper layer can also remain unchanged [1]. Basic services that should be offered for E-league are:

- Reading the packets generated by video server.
- Establishing communication with the communication server.
- Continuous sensing for the referee decision.

In our implementation for Simulation League, this layer is represented by a Java library called Atan [3] which provides following services :

- Connection to the simulation server via UDP.
- Side conversion to support the internal representation of the state of the world.
- Parsing of the output string from simulation server.
- Generation of a command string that can be understood by the simulation server.

### 4.3 Sensorial/Effectorial Layer

This layer serves as a bridging layer between logical communication layer and cognitive layer. It translates visual information into the representation needed by cognitive layer, and also translates output from cognitive layer into basic action to be performed by the robots. In our implementation for Simulation League which use Prolog programs as cognitive layer and Java library as logical communication layer, this means translating visual information into prolog predicates and interpreting prolog query result. Let us show an example of the java translation code used when a player see a ball:

```
public void infoSeeBall
        (double distance, double direction) {
            infobuf.
            append("distance").
            append(PrologConnection.ATOM_SPLITTER).
            append("ball").
            append(PrologConnection.ATOM_SPLITTER).
            append(distance).
            append(PrologConnection.FACT_SPLITTER).
            append("direction").
            append(PrologConnection.ATOM_SPLITTER).
            append("ball").
            append(PrologConnection.ATOM_SPLITTER).
            append(direction).
            append(PrologConnection.FACT_SPLITTER);
}
```

If the player see a ball in distance $x$ and direction $y$, then this information is translated into prolog predicates as `distance(ball,x)` and `direction(ball,y)`.

### 4.4 Cognitive Layer

Cognitive layer is where the strategy is implemented. It is the highest level layer. Our work is focused in this layer where we employ The Ciao Prolog System [4], and in particulas the Fuzzy Prolog library, to do reasoning over provided information. Our approach is providing the capability of handling fuzzy, uncertain and incomplete information to the cognitive layer. This information is very close to the human reasoning, so this framework is improving the human-like control of this layer. A strategy can be easily implemented on this layer without having to put effort on low level technical details more related to the machine than to the human mind.

## 5 Evaluation

For testing our architecture (Figure 3) we have implemented a prototype using Ciao Prolog and its library that provide Fuzzy Prolog, a Java interface programs to connect to the Atan library and a RoboCupSoccer Simulator.

Some simple scenarios have been prepared for observation on difference between fuzzy and crisp approach on similar strategy.

For example, if the strategy is implemented as a prolog program, the program with crisp strategy takes only the best action and fails when there is none. The program with fuzzy strategy proposes the almost best action when there is no best action.

The program with crisp strategy (using classical prolog) looks like the following:

```
get_command (Info,Command) :-
            update_info(Info),
            best_command(Command).
```

And the fuzzy approach for similar strategy:

```
get_command (Info,Command) :-
            update_info(Info), !,
            best_command(Command,100).
get_command (_,Command) :-
            best_command(Command,V),
            V .>. 80.
```

The strategy is coded into the predicate best_command/2. For example, to determine the next action when a player has the ball (it is close to him) the best_command/2 predicate decide if shooting the ball or dribble to goal or pass the ball to a teammate. Below is the code of the first choice:

```
best_command(Command,V) ::~ min
                play_mode(play_on),
                \+ player_role(goalie),
                ball_in_possesion(V1),
                goal_position(Dist,Dir),
                good_to_shoot(Dist,Dir,V2),
                shoot(Dist,Dir,Command).

best_command(Command,V) ::~ min
                ...
                dribble(Dist,Dir,Command).

best_command(Command,V) ::~ min
                ...
                pass_to_teammate(Command).
```

In the above example there are many fuzzy concepts. Depending on the distance to the goal position the player will evaluate if it is a good for shooting to goal. In case it is, he will do it. Otherwise, he will evaluate if his possesion of the ball is save enaugh (no players of the other team are close to the ball). If not, he will dribble with the ball to skip the danger. Finally, if the player is not close to shoot to goal and is is not possible to dribble and then shoot to goal, then he will pass the ball to another player of his team.

This is only an example for representing the necesity of using fuzzy rules and fuzzy concepts in soccer control.

The evaluation includes power calculation using set of rules, rule based decision, and team play. Fuzzy rules enable fuzzy control implementation for power calculation. In rule based decision, there are certain cases where fuzzy approach offers better solution than crisp approach. These small differences in power calculation and decision leads to a slight better performance of fuzzy approach on team play with respect to the crisp approach. However, fuzzy approach requires more processing time than crisp approach and this could lead to poor performance if it is not managed properly. We will evaluate the efficiency of the two approaches (crisp and fuzzy, Prolog and Fuzzy Prolog) in further work.

We have realized that none of the two alternatives (crisp versus fuzzy) are good for all scenarios. Indeed, it seems that a combination of fuzzy reasoning with crisp predicates could be the perfect combination and this is a promising result because Fuzzy Prolog [9] is able to deal with this combination.

## 6 Conclusion

We choose RoboCupSoccer domain as our case problem to employs Fuzzy Prolog system (an approach to incorporate fuzzy reasoning into logic programming). This work is an initial step toward series of research in this area.

On the other side, we believe that logic programming is a perfect environment for dealing with the cognitive layer at RoboCupSoccer league as it is in general to implement cognitive and control issues in robotics.

Our goal is to provide a framework to employ Prolog in general and Fuzzy Prolog in particular for RoboCupSoccer. A generic architecture for RoboCupSoccer is given, with flexibility in changing leagues and programming languages while maintaining a combination of Prolog and Fuzzy Prolog as cognitive layer. For the implementation, we apply the framework to work on the RoboCup Soccer Simulator, by implementing a prototype player with Fuzzy Prolog as cognitive layer and adapting properly the sensorial/effectorial layer. Considering time and technical constraints, we choose the discrete Fuzzy Prolog system over the continuous approach, for this application, providing constructive answers as values (instead of constraints) is translated into more efficiency for the strategy (measurements will be studied in further work).

We observe that the fuzzy program is slower than crisp program for several reasons:

- Fuzzy program needs to be translated into CLP($\mathcal{FD}$)
- Instead of failing immediately when a predicate in a body clause is not satisfied, the evaluation in a fuzzy rule continues and the truth values are aggregated with the truth value 0.
- When there is no best action to be done, fuzzy program attempts to find an alternative action.

Therefore, the fuzzy program should be designed carefully by taking into account the different procedural semantic between a crisp prolog program and a Fuzzy Prolog program.

This work establishes a preliminary groundwork towards a series of research on employing Fuzzy Prolog in RoboCupSoccer E-League. The results of this work are:

- A prototype of discrete Fuzzy Prolog system, implemented as `dfuzzy` package in Ciao Prolog.
- A framework to employ Fuzzy Prolog in RoboCup Soccer, including a prototype of cognitive layer for RoboCup Soccer.
- A prototype of player client as application of the framework for the RoboCup Soccer Simulator League (sensorial/effectorial layer).
- Simple scenarios to demonstrate the utility of fuzzy reasoning for soccer players control.

There is room for improvement of this work. With regards to the cognitive layer, more advanced strategy and various proper aggregation operator can be applied, using either continuous or discrete Fuzzy Prolog system. We intend to distinguish at the players control, when is it better to use fuzzy reasoning (Fuzzy Prolog), and when is it faster to use crisp reasoning (Prolog). Our future work is to improve and employ the cognitive layer for RoboCup Soccer E-League. Another possible work in this research area is to use Fuzzy Prolog as cognitive layer in different RoboCup domain, for example in RoboCup Rescue or RoboCup@Home.

## References


1. A.J.García, G.I.Simari, and T.Delladio. Designing an Agent System for Controlling a Robotic Soccer Team. Argentine Conference on Computer Science (CACIC 2004), 2004.
2. AL05 PID 0040 project. http://faea.uncoma.edu.ar/materias/ia/Robotica.
3. Atan library. http://atan1.sourceforge.net.
4. M. Hermenegildo, F. Bueno, D. Cabeza, M. García de la Banda, P. Lṕez, and G. Puebla. The CIAO Multi-Dialect Compiler and System: An Experimentation Workbench for Future (C)LP Systems. *In Parallelism and Implementation of Logic and Constraint Logic Programming*. Nova Science, Commack, NY, USA, April 1999. http://www.cliplab.org/Software/Ciao.
5. J.M.Santos, H.D.Scolnik, I.Laplagne, S.Daicz, F.Scarpettini, H.Fassi, and C.Castelo. Uba-sot: An approach to control and team strategy in robot soccer. *International Journal of Control, Automation, and Systems*, 1(1):149155, 2003.
6. J.Anderson, J.Baltes, D.Livingston, E.Sklar, and J.Tower. Toward an Undergraduate League for Robocup. In *Proceedings of Seventh International RoboCup Syposium* (RoboCup-2003), 2003.
7. M.Chen, K.Dorer, and E.Foroughi. *Users Manual RoboCup Soccer Server*, 2003.
8. S.Muñoz-Hernandez and C.Vaucheret, editors. *Extending Prolog with Incomplete Fuzzy Information*, Proceedings of the 15th International Workshop on Logic Programming Environments, 2005.
9. S.Muñoz-Hernandez and C.Vaucheret, editors. *Default values to handel Incomplete Fuzzy Information*, volume 14 of IEEE Computational Intelligence Society Electronic Letter, ISSN 0-7803-9489-5. IEEE, 2006.
10. S.Muñoz-Hernandez and J.M.Gomez-Perez, editors. *Solving Collaborative Fuzzy Agents Problems with CLP($\mathcal{FD}$)*, Proceedings of International Symposium on Practical Aspects of Declarative Languages 2005, California, 2005.
11. S.Guadarrama, S.Muñoz, and C.Vaucheret. Fuzzy prolog: A new approach using soft constraints propagation. *Fuzzy Sets and Systems*, 144(1):127150, 2004.

# An Approach to Theory of Fuzzy Discrete Signals

Bohdan S. Butkiewicz

Warsaw University of Technology, Warsaw, Poland,
bb@ise.pw.edu.pl
http://www.ise.pw.edu.pl/~bb/index.html

**Abstract.** The paper presents an approach to description of fuzzy discrete functions and Fourier transform of such functions taking in consideration their uncertainty. Conventional approach to uncertainty employs probabilistic description. Here, fuzzy logic theory is applied to describe this uncertainty. A definition of transform, called later Discrete Fuzzy Fourier Transform and definition of Inverse Discrete Fuzzy Fourier Transform are proposed. Some properties of such transformations and examples of applications and comparison with conventional approach are shown.

## 1 Introduction

Discrete signals play an important part in the theory of many branches of science. Such signals contain fuzziness included in the function itself or in the parameters. For example sampled and quantified signal contains fuzziness itself because exact value of a sample is unknown after quantization process. Such situation occurs also in image transformations by optical systems. Original image is crisp but image on the photographic plate is blurred. Dubois and Prade [8] call this type of ill-known functions fuzzifying functions. Fuzzifying functions have been studied by Sugeno [12] under the name of fuzzy correspondences. These concepts of fuzzy function is mathematically equivalent to fuzzy relation. Conventional approach in such cases of imprecise data uses probability description.

Very often there are no possibility to gather many experimental data and information about situation concerning similar circumstances, occurrences or phenomena. Moreover, we use sometimes our intuition to build the model of phenomenon. For such kind of data, fuzzy description is more justifiable then probabilistic description.

In the paper, an approach based on uncertainty description using fuzzy logic is proposed. Uncertainty can be introduced in two ways. Function can be treated as function of discrete crisp variables and fuzzy parameters or fuzziness can be included in the function itself not in the parameters.

The idea of fuzzy approach to signal processing was proposed by Kosko [9], but he considered signals as crisp with random impulsive noise similarly as in conventional approach. Only during signal processing fuzzy algorithms was used. The concept leads to fuzzy filtering. The concept of fuzzy filters and other fuzzy techniques was developed also by other researchers especially for image processing [1],[10].

P. Melin et al. (Eds.): IFSA 2007, LNAI 4529, pp. 646–655, 2007.
© Springer-Verlag Berlin Heidelberg 2007

The name of fuzzy transform was used also in [11]. However the approach is quite different. It is based on discrete partition of time space on intervals and fuzzy description by membership functions (called basic functions) defined on each interval. It can be used also for description of fuzzy discrete signal.

Notions of fuzzy signal, fuzzy Fourier transform and fuzzy correlation proposed by the author are based on the concept of norms and scalar products as in conventional definitions to preserve similar properties. The first approach to analog fuzzy signals and fuzzy Fourier transform was published by author in [2]. After the definition was enlarged also on fuzzifying functions [3]. Two-dimensional fuzzy Fourier transform and fuzzy convolution were proposed in [4] and [6] with application to image processing.

In this paper an approach to discrete fuzzy function is proposed.

## 2 Concept of Fuzzy Discrete Signal

In theoretical conventional approach the space of discrete signals is defined as follows.

**Definition 1.** *(Class $l^2$) Discrete function $x[n]$ of integer argument $n$ belongs to the class $l^2(n_1, n_2)$ where integers $n_1, n_2 \in (-\infty, \infty)$ if, and only if the sum*

$$E_x = \sum_{n=n_1}^{n_2} |x[n]|^2 \tag{1}$$

*is finite.*

The value $E_x$ represents energy of signal $x[n]$ and $l^2$ is called *class of discrete signals with finite energy.* Square root of $E_x$ plays role of a norm in the space $l^2$, $\| x[n] \| = \sqrt{E_x}$.

This conventional definition can be enlarged for fuzzy discrete signals in following way.

**Definition 2.** *(Class $l^2_f$) Fuzzy discrete function $\mathsf{x}[n, \alpha]$ where $n$ is integer and $\alpha$ is a fuzzy argument, being a fuzzy normal set or vector of fuzzy arguments, belongs to the class $l^2_f(n_1, n_2)$ where $n_1, n_2 \in (-\infty, \infty)$ are integers if, and only if, for any $a \in supp(\alpha)$ the sum*

$$E_x(\alpha) = \sum_{n=n_1}^{n_2} |x[n, a]|^2 \tag{2}$$

*is finite.*

It will be shown that $\sqrt{E_x(\alpha)}$ is a norm $\|\mathsf{x}[n, \alpha]\|$ of $\mathsf{x}[n, \alpha]$ in $l^2_f$. It satisfies the axioms:

(1) $\|\mathsf{x}[n, \alpha]\| \geq 0$ for all $x \in l^2_f$
(2) $\|\mathsf{x}[n, \alpha]\| = 0 \Leftrightarrow \mathsf{x}[n, \alpha] = 0$, i.e. all $x[n, a] = 0$ where $a \in supp(\alpha)$
(3) $\|\lambda \mathsf{x}[n, \alpha]\| = |\lambda| \|\mathsf{x}[n, \alpha]\|$ where $\lambda$ is real constant
(4) $\|\mathsf{x}_1[n, \alpha] + \mathsf{x}_2[n, \beta]\|^2 \leq \|\mathsf{x}_1[n, \alpha]\| + \|\mathsf{x}_2[n, \beta]\|$

The first three properties are obvious. The triangle inequality (4) is proved using inequality for functions in conventional space $l^p(n_1, n_2)$

$$\{\sum_{n=n_1}^{n_2}(x[n]+y[n])^p\}^{1/p} \leq \{\sum_{n=n_1}^{n_2}(x[n]]^p\}^{1/p} + \{\sum_{n=n_1}^{n_2}(y[n])^p dt\}^{1/p}$$

where $p \geq 1$. The inequality is applied for any $a \in supp(\alpha)$ and $b \in supp(\beta)$

*Example 1* (Fuzzy discrete signal). Let function takes fuzzy values $x[n] = 2k + \alpha$ where $k = -4, -3, ..., 4$ and $\alpha$ is fuzzy number described by membership functions of trapezoidal shape $trapeze(-1.5, -0.5, 0.5, 1.5)$. Signal $x[n]$ is shown in the Fig. 1 (left).

*Example 2* (Fuzzy linguistic signal). In one of his papers [6] the author introduced a concept of linguistic signal. It is a discrete fuzzy function $x[n, \alpha]$ taking discrete values $x_l[n]$ where for any discrete level $l$ is attached a linguistic term $v_l$. Let discrete levels $x_l$ for any $n$ can take only one of $l = 1, ..., 9$ different linguistic values $v_l$ called NV, NL,..., PV. Let uncertainty of each value $x_l[n]$ is described by the similar fuzzy set $\alpha$ with membership function $\mu_\alpha(a)$ with trapezoidal shape. The linguistic signal corresponding to fuzzy signal from Example 1 is presented in the Fig. 1 (right).

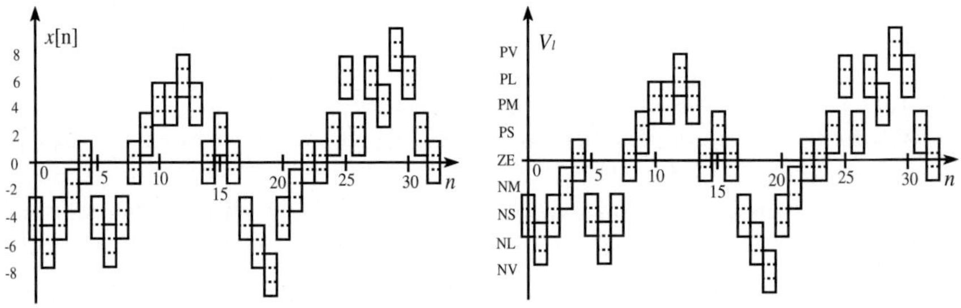


**Fig. 1.** Example of discrete fuzzy signal (left) and linguistic signal (right)

Such situation occurs in practice where analog-digital (A/D) converter can introduce an error, for example the value of last bit is not sure.

## 3 Concept of Fuzzy Discrete Fourier Transform

In conventional approach Fourier transform of discrete set of real or complex numbers $x[n]$, $n \in Z$ (integers) is defined as follows

$$X(f) = \sum_{n=-\infty}^{\infty} x[n]\, e^{-i2\pi n f} \qquad (3)$$

For practical reasons a finite-length sequence $x[n]$ is obviously needed, $n = 0, ..., N-1$. The transform is periodic. Moreover, $X(f)$ is evaluated at an arbitrary number $M$ of uniformly-spaced frequencies $f_m$ across one period $f_m = m/M$, $m = 0, ..., M-1$. Hence, discrete Fourier transform is calculated as following

$$X[m] = \sum_{n=0}^{N-1} x[n]\, e^{-i2\pi nm/M} \tag{4}$$

In Discrete Fourier Transform (DFT) procedures used for calculation the number $M = N$ is assumed. Thus follow definition is used

$$X[m] = \sum_{n=0}^{N-1} x[n]\, e^{-i2\pi nm/N} \tag{5}$$

Inverse transform has a form

$$x[n] = \frac{1}{N} \sum_{m=0}^{N-1} X[m]\, e^{-i2\pi nm/N} \tag{6}$$

Sometimes in (5) and (6) the term $\sqrt{N}$ is used for symmetry.

Fuzzy approach, proposed here, is quite different. Firstly it can be noticed that fuzziness is very often included in the function itself not in the parameters. Therefore, the parameter $\alpha$ in $\mathsf{x}[n, \alpha]$ will be omitted for generality. Next, the concept of discrete pseudo $\alpha$-level curves is introduced. The term "discrete curves" is used here for simplicity and understand as series of values $x[n]$ for $n = 0, ..., N-1$.

**Definition 3.** *(Discrete pseudo $\alpha$-level curves) Let value of fuzzifying function $\mathsf{x}[n]$ of crisp discrete integer variable $n$ for any $n \in (-\infty, \infty)$ is a fuzzy number (convex normal fuzzy set). Let $\mu_{\mathsf{x}}(x[n])$ be membership function of $\mathsf{x}[n]$. Let for any $\alpha \in (0, 1)$ the equation $\mu_{\mathsf{x}}(y[n]) = \alpha$ has two and only two solutions $y[n] = x_\alpha^{\mp}[n]$ and $y[n] = x_\alpha^{\pm}[n]$, and only one solution $y[n] = x[n]$ for $\alpha = 1$. The solutions $x_\alpha^{\mp}[n]$ $x_\alpha^{\pm}[n]$, will be called* pseudo $\alpha$-level curves *of the* $\mathsf{x}[n]$.

This concept is something similar to $\alpha$-level curves $x_\alpha^-(t)$ $x_\alpha^+(t)$ introduced by Dubois and Prade [8] and after in [5]. However, here functions are discrete and the order of the $x_\alpha^{\mp}[n]$ $x_\alpha^{\pm}[n]$ can change with $n$ and they may do not fulfill inequality as $x_\alpha^-[n] \le x_\alpha^+[n]$. Sometimes $x_\alpha^{\mp}[n]$ can be grater than $x_\alpha^{\pm}[n]$. For instance, such situation occurs for fuzzy function $\Pi(t)\, cos(2\pi f_0 t)$ where $\Pi(t)$ is crisp function with rectangular shape and $f_0$ is fuzzy number with membership function with triangular shape $\mu_{f_0}(f) = triangle(0.9, 1, 1.1)$.

**Definition 4.** *(Discrete Fuzzy Fourier Transform) Let value of discrete fuzzifying function $\mathsf{x}[n]$ of crisp integer variable $n$ for any $n \in [0, N-1]$ is a fuzzy number (convex normal fuzzy set). Let pseudo $\alpha$-level curves $x_\alpha^{\pm}[n]$ $x_\alpha^{\mp}[n]$ of the $\mathsf{x}[n]$, are continuous with respect to $\alpha$ and for $\alpha \to 1$ limit $x_1^{\mp}[n] \to x[n]$ and $x_1^{\pm}[n] \to x[n]$. Let all the pseudo $\alpha$-level curves, $\alpha \in (0, 1]$, fulfill condition (1). Then, it is possible to introduce two conventional sums*

$$X_\alpha^{\mp}[m] = \sum_{n=0}^{N-1} x_\alpha^{\mp}[n]\, e^{-i2\pi nm/N} \tag{7}$$

$$X_\alpha^{\pm}[m] = \sum_{n=0}^{N-1} x_\alpha^{\pm}[n]\, e^{-i2\pi nm/N} \tag{8}$$

*which will be called* pseudo $\alpha$-level curves *of DFuzFT. The set* $\mathsf{X}[m]$ *of all pseudo $\alpha$-level curves, $\alpha \in (0,1]$, will be called* Discrete Fuzzy Fourier Transform *(DFuzFT) and written in the form of sum*

$$\mathsf{X}[m] = \sum_{n=0}^{N-1} \mathsf{x}[n]\, e^{-i2\pi nm/N} \tag{9}$$

**Definition 5.** *(Inverse Discrete Fuzzy Fourier Transform) Let $X_\alpha^{\mp}[m]$ and $X_\alpha^{\pm}[m]$ be pseudo $\alpha$-level curves describing DFuzFT $\mathsf{X}[m]$. Then set of sums*

$$x_\alpha^{\mp}[n] = \frac{1}{N}\sum_{m=0}^{N-1} X_\alpha^{\mp}[m]\, e^{i2\pi nm/N} \tag{10}$$

$$x_\alpha^{\pm}[n] = \frac{1}{N}\sum_{m=0}^{N-1} X_\alpha^{\pm}[m]\, e^{i2\pi nm/N} \tag{11}$$

*for all values $\alpha \in (0,1]$ will be called* Inverse Discrete Fuzzy Fourier Transform *(IDFuzFT) and write in the form of sum*

$$\mathsf{x}[n] = \frac{1}{N}\sum_{m=0}^{N-1} \mathsf{X}[m]\, e^{i2\pi nm/N} \tag{12}$$

If $\alpha$ is a parameter with membership function $\mu_\alpha(a)$ and $supp(\alpha) = [a_1, a_2]$, then fuzzy signal can be written in the form of union

$$\mathsf{x}[n] = \int \mu_\alpha(a)/x[n] \tag{13}$$

Let and both transforms DFuzFT and IDFuzFT exist. Transformation $x_\alpha^{\mp}[n] \Leftrightarrow X_\alpha^{\mp}[m]$ is one to one. Thus, $X_\alpha^{\mp}[m]$ conserves the same $\alpha$-level as $x_\alpha^{\mp}[n]$. Therefore, DFuzFT can be written in the form of union

$$\mathsf{X}[m] = \int \mu_\alpha(a)/X[m] \tag{14}$$

Now arise a question what membership function has DFuzFT? Real value of membership can be found as set of all values

$$X_\alpha^{-}([m]) = \min(X_\alpha^{\mp}[m], X_\alpha^{\pm}[m]) \tag{15}$$

$$X_\alpha^{+}([m]) = \max(X_\alpha^{\mp}[m], X_\alpha^{\pm}[m]) \tag{16}$$

Sometimes it will be useful to obtain crisp result for transform. It is possible to find weighted average of $\mathsf{X}[m]$

$$\overline{\mathsf{X}[m]} = \frac{\int_{a_1}^{a_2} \mu_\alpha(a) X[m,a] da}{\int_{a_1}^{a_2} \mu_\alpha(a) da} \tag{17}$$

It will be called *Discrete Fourier Transform of Fuzzy Function* (DFTFF).

*Comments.* One remark must be added. Both operations of transformation, i.e. DFuzFT and DFTFF, are not fuzzy itself. They are only performed on fuzzy functions. The name "fuzzy Fourier" is used to accentuate that result of DFuzFT is fuzzy function in contradiction to DFTFF where result is crisp function.

## 4 Some Properties of Discrete Fuzzy Fourier Transform

It can be shown that many properties of DFuzFT are similar as for conventional DFT. Let $\mathsf{X}[m]$ be DFuzFT of $\mathsf{x}[n]$ written as $\mathsf{x}[n] \leftrightarrow \mathsf{X}[m]$. For instance consider shift, addition, and modulation property.

*Shift.* Let $\mathsf{x}[n] \leftrightarrow \mathsf{X}[m]$ then

$$\mathsf{x}[n-k] \leftrightarrow \mathsf{X}[m] e^{i2\pi km/N} \tag{18}$$

*Proof.* Any $x_\alpha^\mp[n]$ is crisp. Therefore, from conventional shift property it follows $x_\alpha^\mp[n-k] \leftrightarrow X[m]_\alpha^\mp e^{i2\pi km/N}$. Similarly for $x_\alpha^\pm[n]$. Thus, (18) is true.

*Addition.* Let $\mathsf{x}[n] \leftrightarrow \mathsf{X}[m]$ and $\mathsf{y}[n] \leftrightarrow \mathsf{Y}[m]$ then

$$\mathsf{x}[n] + \mathsf{y}[n] \leftrightarrow \mathsf{X}[m] + \mathsf{Y}[m] \tag{19}$$

*Proof.* It was shown in [8] that sum of $\alpha$-level curves conserves $\alpha$-level. Of course, it is true also for discrete case. Then sum $x_\alpha^\mp[n] + y_\alpha^\mp[n]$ has also the same pseudo $\alpha$-level. Thus (19) is true.

*Modulation.* Let $\mathsf{x}[n] \leftrightarrow \mathsf{X}[m]$ then

$$\mathsf{x}[n]\, cos(2\pi k/N) \leftrightarrow \frac{1}{2}(\mathsf{X}[m-k] + \mathsf{X}[m+k]) \tag{20}$$

*Proof.* From conventional modulation property it follows that $x_\alpha^\mp[n] e^{i2\pi nk/N} \leftrightarrow X_\alpha^\mp[m-k]$. Similarly for $x_\alpha^\pm[n]$. Thus, from Euler formula $e^{i\vartheta} = cos\vartheta + isin\vartheta$ it follows that (20) is true.

It is known that Parseval's theorem play an important part in signal theory. Here we start from Carleman's formulation [7] of Parseval's theorem. The theorem can be proved also for discrete signals.

**Theorem 1 (discrete Carleman).** *If the sum*

$$\sum_{n=0}^{N-1} |x[n]|^2 \tag{21}$$

*is finite then following sum with discrete Fourier transforms of* $x[n]$

$$\frac{1}{N}\sum_{m=0}^{N-1} |X[m]|^2 \tag{22}$$

*is finite and they are equals*

$$\sum_{n=0}^{N-1} |x[n]|^2 = \frac{1}{N}\sum_{m=0}^{N-1} |X[m]|^2 \tag{23}$$

Fuzzy version of this theorem is presented below.

**Theorem 2 (fuzzy signal energy).** *Let* $\mathsf{x}[n]$ *belongs to* $l_f^2$ *and* $\mathsf{X}[m]$ *be discrete fuzzy Fourier transform of* $\mathsf{x}[n]$*. Then*

$$\sum_{n=0}^{N-1} |x_\alpha^\mp[n]|^2 = \frac{1}{N}\sum_{m=0}^{N-1} |X_\alpha^\mp[m]|^2 \tag{24}$$

*and*

$$\sum_{n=0}^{N-1} |x_\alpha^\pm[n]|^2 = \frac{1}{N}\sum_{m=0}^{N-1} |X_\alpha^\pm[m]|^2 \tag{25}$$

*It will be written as*

$$\sum_{n=0}^{N-1} |\mathsf{x}[n]|^2 = \frac{1}{N}\sum_{m=0}^{N-1} |\mathsf{X}[m]|^2 \tag{26}$$

*Proof.* Functions $x_\alpha^\mp[n]$, $x_\alpha^\pm[n]$ and their Fourier transforms $X_\alpha^\mp[m]$, $X_\alpha^\pm[m]$ are crisp. Assumption of the theorem assures that $x_\alpha^\mp[n]$ and $x_\alpha^\pm[n]$ belong to class $l^2$. Thus, conventional theorem can be applied to these functions for any $a \in supp(\alpha)$. Therefore, the equality (26) is consistent.

It must be noted that equivalence (26) concerns fuzzy functions. Such definition of signal energy leads to concept of fuzzy energy. Of course, crisp value of energy can be obtain as weighted average

$$E_{\mathsf{x}} = \frac{\int_{a_1}^{a_2} \mu_\alpha(a) \sum_{n=0}^{N-1} |x[n,a]|^2 da}{\int_{a_1}^{a_2} \mu_\alpha(a) da} \tag{27}$$

*Example 3* (DFuzFT). Let vector of values equals $x[n]$=[-2 -3 -2 -1 0 -2 -4 -2 0 1 3 2 5 2 0 1 0 -2 -3 -5 -2 -1 0 0 1 3 1 3 2 1 3 1 0 ], but this values are not certain. Let uncertainty of $\mathsf{x}[n]$ is described by the set of membership functions $\mu_{\mathsf{x}}(x[n])$

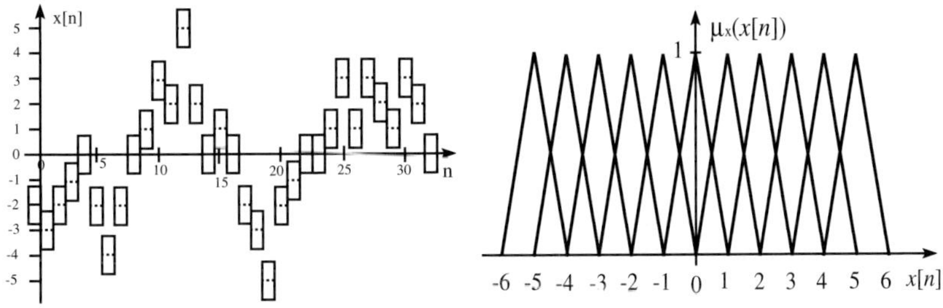


**Fig. 2.** Example of discrete fuzzy signal (left) and its membership function (right)

(Fig. 2). The sets (7) (8) of all pseudo $\alpha$-level curves for DFuzFT were calculated. The result, three series of points $X_0^-[m]$, $X_0^+[m]$ found using (16)(16) and $X_1[m]$, was shown in the Fig. 3 (left) as dashed and continuous lines, for DFuzFT with $\alpha = 0$ and 1. Of course, such result can be presented in similar way as in the Fig. 2 with rectangles showing possible changes of spectrum $X[m]$. Here lines are used for simplicity. In the Fig. 4 conventional FFT and DFTFF are compared.

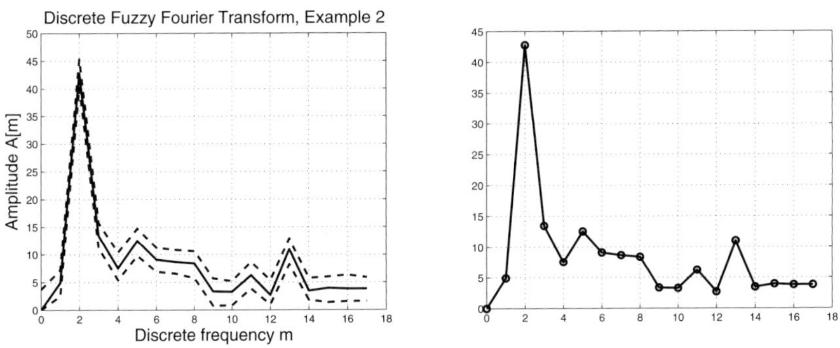


**Fig. 3.** Example 2. Discrete fuzzy Fourier transform (left). Solid line shows $X_1[m]$ where $\alpha = 1$, dashed lines show $X_\alpha^-[m]$ and $X_\alpha^+[m]$ for $\alpha = 0$. Comparison of $X_1[m]$ and conventional FFT (right), $X_1[m]$ - solid line, FFT - points.

*Remark.* Example 2 shows that values of $X_1[m]$ are identical with the result obtained by conventional FFT, see Fig. 3 (right). However, the idea of DFuzFT can be enlarged on the cases where the values of $x_1[n]$ are not unique. It is a case when membership functions of $x[n]$ have for example trapezoidal shapes. In such cases the value $x_1[n]$ and $X_1[m]$ are not unique. Therefore, DFuzFT concept must not be trivial extension of the FFT procedure. Such modification of the DFuzFT concept does not shown in this paper. The author in [6] presents one of possible solution proposed for such discrete fuzzy functions, called linguistic signals, where a concept of linguistic Fourier transform is introduced. In this case fuzzy signal as well as its Fourier transform have linguistic forms.

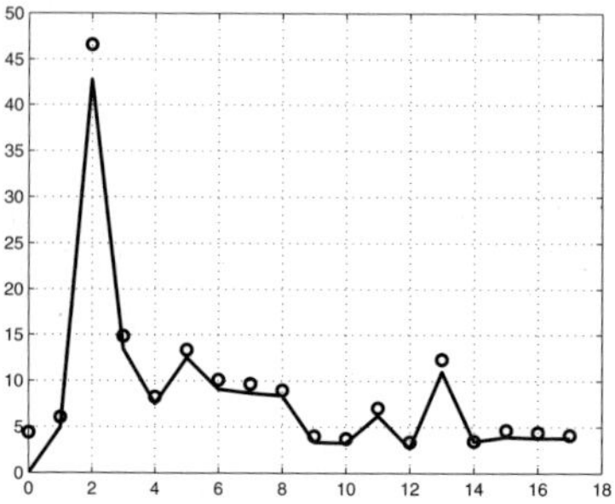


**Fig. 4.** Example 2. Comparison of DFTFF and conventional FFT, DFTFF - points, FFT - solid line.

## 5 Conclusions

In the paper new approach to Fourier transform is proposed based on fuzzy logic. It was shown that fuzzy transforms contain information about uncertainty. It is not the case in the conventional probabilistic approach where expected values are used. This information can be obtained in sufficiently easy way using conventional discrete Fourier transform. Effective numerical procedure can be build using popular programming tools as FFT. New definitions were introduced using concept of norms, i.e. in such a way to conserve general properties of conventional continuous and discrete Fourier transforms. It was shown that principal properties as shift, addition, modulation and many others, and also Parseval's theorem can be generalized for fuzzy case. Crisp result for transform (DFTFF) can be obtained also using average mean or other defuzzification procedure.

## References

1. Bezdek, J.C., Pal, S.K.: Fuzzy Models for Pattern Recognition. IEEE Press, New York (1992)
2. Butkiewicz, B.S.: Fuzzy approach to Fourier transform. Proceedings of SPIE. **6159** (2006) 1045–1050
3. Butkiewicz, B.S.: Fuzzy approach to correlation function. Lecture Notes in Artif. Intellig. LNAI **4029** (2006) 202–211.
4. Butkiewicz, B.S.: Fuzzy description of image processing. Proceedings of SPIE (to appear)
5. Butkiewicz, B.S.: Towards Fuzzy Fourier Transform. Eleventh Int. Conf. Information Processing and Management of Uncertainty in Knowledge-based Systems. Paris (2006) 2560-2565
6. Butkiewicz, B.S.: Linguistic Fourier Transform and Fuzzy Image Description. Zittau Fuzzy Colloquium, Zittau, Germany (2006) 73–78
7. Carleman T.: L'Intégrale de Fourier et questions qui s'y rattachent. Almqvist and Wiksells, Uppsala, Sweden (1944)
8. Dubois, D., Prade, H.: Fuzzy Sets and Systems. Academic Press, New York (1980)

9. Kosko, B.: Fuzzy engineering. Prentice Hall International, Inc., London, New Jersey (1997)
10. Nachtegael, M., Van der Weken, D., Van De Ville, D., Kerre, E.E, Eds. Fuzzy Filters for Image Processing. Springer, Berlin, Germany (2003)
11. Haldeeva, E., Perfilieva, I,: Fuzzy Transformation. Joint 9th IFSA World Congress and 20th NAFIP Int. Conf., Vancouver (2001) 1946-1948
12. Sugeno, M.: Fuzzy Systems with Undelying Deterministic Systems. (Towards Fuzzy descriptions of Complex Systems). Res. Rep. L.A.A.S., Toulouse, France (1977)

# Using Gradual Numbers for Solving Fuzzy-Valued Combinatorial Optimization Problems*

Adam Kasperski[1] and Paweł Zieliński[2]

[1] Institute of Industrial Engineering and Management,
Wrocław University of Technology, Wybrzeże Wyspiańskiego 27,
50-370 Wrocław, Poland
adam.kasperski@pwr.wroc.pl

[2] Institute of Mathematics and Computer Science, Wrocław University of Technology,
Wybrzeże Wyspiańskiego 27, 50-370 Wrocław, Poland
Pawel.Zielinski@pwr.wroc.pl

**Abstract.** In this paper a general approach to combinatorial optimization problems with fuzzy weights is discussed. The results, valid for the interval-valued problems, are extended to the fuzzy-valued ones by exploiting the very recent notion of a gradual number. Some methods for determining the exact degrees of possible and necessary optimality and the possibility distributions of deviations of solutions and elements are proposed. The introduced notions are illustrated by practical examples.

## 1 Introduction

In a combinatorial optimization problem we are given a finite set of elements $E$ and a weight $w_e$ is associated with each element $e \in E$. We seek an object composed of the elements of $E$ whose total weight is minimal or maximal. In this case the solutions and elements can be divided into two groups: optimal and non-optimal ones. One may also evaluate the optimality from the point of view of a *deviation*, that describes how far a solution (an element) is from being optimal. In this paper, we consider the case in which the element weights are ill-known and they are modeled by means of intervals or fuzzy intervals. In the interval-valued case solutions and elements form three groups: those that are optimal for sure (*necessarily optimal*), those that are not optimal for sure and those, whose optimality is unknown (*possibly optimal*). Now, to each solution and element an interval of possible deviations from optimum may be associated. In the fuzzy-valued case the notions of optimality can be extended and every solution (element) can be characterized by *degrees* of possible and necessary optimality and a *possibility distribution* of its deviation.

There exists a wide class of combinatorial optimization problems for which the characterization of optimality of solutions and elements can be efficiently done in

* This work was partially supported by Polish Committee for Scientific Research, grant 3T11C05430.

P. Melin et al. (Eds.): IFSA 2007, LNAI 4529, pp. 656–665, 2007.
© Springer-Verlag Berlin Heidelberg 2007

the interval model. In this paper we extend these results to the fuzzy-valued case by exploiting a very recent notion of a gradual number, which provides a new outlook on fuzzy intervals. This allows to apply interval methods to problems with fuzzy weights. In consequence, we propose some methods for determining the exact degrees of possible and necessary optimality and the possibility distributions of deviations of solutions and elements for some fuzzy-valued combinatorial optimization problems.

## 2 Preliminaries

Let $E = \{e_1, \ldots, e_n\}$ be a finite ground set and let $\Phi \subseteq 2^E$ be a set of subsets of $E$ called the set of the *feasible solutions*. In a deterministic case, for every element $e \in E$ there is given a nonnegative real weight $w_e$. A *combinatorial optimization problem* $\mathcal{P}$ with a linear objective function consists in finding a feasible solution $X \in \Phi$ whose total weight is minimal, namely:

$$\mathcal{P}: \quad F^* = \min_{X \in \Phi} F(X) = \min_{X \in \Phi} \sum_{e \in X} w_e. \tag{1}$$

Formulation (1) encompasses a large variety of deterministic combinatorial optimization problems, for instance shortest path, minimum spanning tree, minimum assignment, 0-1 knapsack etc. Some of them are polynomially solvable while the other ones are NP-hard. For a wide review we refer the reader to [11].

A solution that minimizes (1) is said to be *optimal*. We also call an element $e \in E$ *optimal* if it is a part of an optimal solution. Thus, in the deterministic case, the set of solutions $\Phi$ and the set of elements $E$ can be divided into two groups: the optimal and non-optimal ones. In the case when solution $X \in \Phi$ or element $f \in E$ are not optimal, a natural question arises: how far they are from optimality. One can define a *deviation* $\delta_X$ of solution $X$ in the following way:

$$\delta_X = F(X) - F^* = \sum_{e \in X} w_e - \min_{X \in \Phi} \sum_{e \in X} w_e.$$

Similarly, a *deviation* $\delta_f$ of element $f$ is defined as follows:

$$\delta_f = F_f^* - F^* = \min_{X \in \Phi_f} \sum_{e \in X} w_e - \min_{X \in \Phi} \sum_{e \in X} w_e,$$

where $\Phi_f \subseteq \Phi$ is the set of all solutions containing $f$. In other words, deviation $\delta_X$ (resp. $\delta_f$) describes how far solution $X$ (resp. element $f$) is from being optimal. Obviously, solution $X$ (element $f$) is optimal if and only if $\delta_X = 0$ ($\delta_f = 0$).

## 3 Interval-Valued Combinatorial Optimization Problem

Suppose that the values of the element weights are not precisely known, but they are known to belong to intervals $W_e = [w_e^-, w_e^+]$, $e \in E$. It means that the actual value of a weight will take some value within an interval, but it is not possible to predict at present which one. A *configuration* is a precise instantiation of the weights of each element $\boldsymbol{w} = (w_e)_{e\in E}$, $w_e \in W_e$. Thus, every configuration expresses a realization of the weights which may occur. We denote by $\Gamma$ the set of all configurations, i.e. $\Gamma = \times_{e\in E}[w_e^-, w_e^+]$. We use $w_e(\boldsymbol{w})$ to denote the weight of element $e \in E$ in a specified configuration $\boldsymbol{w} \in \Gamma$.

Among the configurations of $\Gamma$ we distinguish the *extreme* ones, which belong to $\times_{e\in E}\{w_e^-, w_e^+\}$. Let $A$ be a given subset of the element set $E$. In extreme configuration $\boldsymbol{w}_A^+$ all elements $e \in A$ have weights $w_e^+$ and all the remaining elements have weights $w_e^-$. Similarly, in extreme configuration $\boldsymbol{w}_A^-$ all elements $e \in A$ have weights $w_e^-$ and all the remaining elements have weights $w_e^+$. These extreme configurations will play a crucial role in further considerations.

For a given solution $X \in \Phi$, we define its weight under a fixed configuration $\boldsymbol{w} \in \Gamma$ as $F(X, \boldsymbol{w}) = \sum_{e\in X} w_e(\boldsymbol{w})$. We will denote by $F^*(\boldsymbol{w})$ the value of the weight of an optimal solution under $\boldsymbol{w} \in \Gamma$, that is

$$F^*(\boldsymbol{w}) = \min_{X\in\Phi} F(X, \boldsymbol{w}). \tag{2}$$

Notice that if configuration $\boldsymbol{w}$ is fixed, then (2) is problem $\mathcal{P}$. Hence the optimality of a given solution or element now depends on configuration $\boldsymbol{w}$ and the following characterization can be provided: a solution $X \in \Phi$ (element $f \in E$) is said to be *possibly optimal* if there exists a weight configuration $\boldsymbol{w} \in \Gamma$ for which it is optimal; a solution $X$ (element $f$) is said to be *necessarily optimal* if it is optimal for all weights configurations $\boldsymbol{w} \in \Gamma$.

As in the deterministic case, we can obtain an additional information about the optimality of a given solution or element using the concept of a deviation. Denote by $\delta_X(\boldsymbol{w})$ and $\delta_f(\boldsymbol{w})$ a deviation of solution $X$, element $f$, in configuration $\boldsymbol{w}$. We can now define the widest interval $\Delta_X = [\delta_X^-, \delta_X^+]$ of possible values of deviations of solution $X$ in the following way:

$$\delta_X^- = \min_{\boldsymbol{w}\in\Gamma} \delta_X(\boldsymbol{w}) \text{ and } \delta_X^+ = \max_{\boldsymbol{w}\in\Gamma} \delta_X(\boldsymbol{w}). \tag{3}$$

In the same way, we define the widest interval $\Delta_f = [\delta_f^-, \delta_f^+]$ of possible values of deviation of element $f$. Intervals $\Delta_X$ and $\Delta_f$ measure how far $X$ and $f$ are from being possibly (resp. necessarily) optimal. It is clear that solution $X$ is possibly optimal if and only if $\delta_X^- = 0$ and it is necessarily optimal if and only if $\delta_X^+ = 0$. The same relations hold for elements. The following proposition shows the crucial role of the extreme configurations:

**Proposition 1.** *The maximum and minimum of $\delta_X(\boldsymbol{w})$ and $\delta_f(\boldsymbol{w})$ over all configurations $\boldsymbol{w} \in \Gamma$ are attained in extreme configurations from set $\times_{e\in E}\{w_e^-, w_e^+\}$. Moreover, the following equalities hold for every solution $X$:*

$$\delta_X^- = \delta_X(\boldsymbol{w}_X^-) = F(X, \boldsymbol{w}_X^-) - F^*(\boldsymbol{w}_X^-), \tag{4}$$
$$\delta_X^+ = \delta_X(\boldsymbol{w}_X^+) = F(X, \boldsymbol{w}_X^+) - F^*(\boldsymbol{w}_X^+). \tag{5}$$

*Proof.* The fact that the minimum and maximum are attained in extreme configurations follows from the results obtained in [10] and the property that functions $\delta_X(\boldsymbol{w})$ and $\delta_f(\boldsymbol{w})$ are locally monotonic with respect to each variable $w_e \in [w_e^-, w_e^+]$, $e \in E$. In order to prove (4) suppose that $\boldsymbol{w}$ is a configuration that minimizes $\delta_X(\boldsymbol{w})$ and denote by $X^*$ the optimal solution in configuration $\boldsymbol{w}_X^-$. It holds $\delta_X^-(\boldsymbol{w}) = F(X,\boldsymbol{w}) - F^*(\boldsymbol{w}) \geq F(X,\boldsymbol{w}) - F(X^*,\boldsymbol{w}) \geq F(X,\boldsymbol{w}_X^-) - F^*(X^*,\boldsymbol{w}_X^-) = F(X,\boldsymbol{w}_X^-) - F^*(\boldsymbol{w}_X^-)$. Hence, configuration $\boldsymbol{w}_X^-$ also minimizes $\delta_X(\boldsymbol{w})$. The proof of equality (5) is similar. □

From Proposition 1 it follows that if problem $\mathcal{P}$ is polynomially solvable, then interval $\Delta_X$ for a given solution $X$ can be computed in polynomial time. Hence, the possible and necessary optimality of $X$ can be characterized in polynomial time as well. Contrary to the optimality of solutions, there is no an easy characterization of the optimality of the elements. We know that configuration $\boldsymbol{w}$ that minimizes or maximizes $\delta_f(\boldsymbol{w})$ is an extreme one, but it may be hard to compute (observe that there are up to $2^n$ extreme configurations). The complexity of computing interval $\Delta_f$, for a given element $f \in E$, strongly depends on a particular problem $\mathcal{P}$ and it may be NP-hard even if $\mathcal{P}$ is polynomially solvable. For instance, for Shortest Path, determining a configuration that minimizes $\delta_f(\boldsymbol{w})$ is strongly NP-hard [1], but determining a configuration that maximizes $\delta_f(\boldsymbol{w})$ is polynomially solvable in acyclic digraphs [7].

# 4 Fuzzy-Valued Combinatorial Optimization Problem

In this section we extend the notion of a deviation of a solution and element to the more general fuzzy case. In consequence, we also generalize the notions of possible and necessary optimality of solutions and elements. In order to compute the fuzzy deviations we apply a recent concept of a gradual number [8].

## 4.1 Gradual Numbers and Fuzzy Intervals

The classical intervals model uncertainty in a Boolean way: a value in the interval is possible; a value outside is impossible. The idea of fuzziness is to move from the Boolean way to a gradual one. Hence fuzziness makes the boundaries of the interval softer and thus making uncertainty gradual. In order to model the essence of graduality without uncertainty the concept of a gradual number has been recently proposed. Following the notation of [8] a *gradual real number* (or *gradual number* for short) $\tilde{r}$ is defined by an assignment function $\mathcal{A}_{\tilde{r}}$ from $(0,1]$ to $\mathbb{R}$. It can be seen as a number parametrized by a value of $\lambda \in (0,1]$.

Making use of the notion of a gradual number, one can describe a *fuzzy interval* $\widetilde{W}$ by an ordered pair of two gradual numbers $[\tilde{w}^-, \tilde{w}^+]$, where $\tilde{w}^-$ is a *gradual lower bound* of $\widetilde{W}$ and $\tilde{w}^+$ is a *gradual upper bound* of $\widetilde{W}$. In order to ensure the well known shape of a fuzzy interval, $\tilde{w}^-$ and $\tilde{w}^+$ must satisfy the following properties: $\mathcal{A}_{\tilde{w}^-}$ is an increasing function; $\mathcal{A}_{\tilde{w}^+}$ is a decreasing function and $\mathcal{A}_{\tilde{w}^-}(1) \leq \mathcal{A}_{\tilde{w}^+}(1)$. In this paper we will additionally assume that the assignment functions of gradual numbers are continuous and their domains are extended

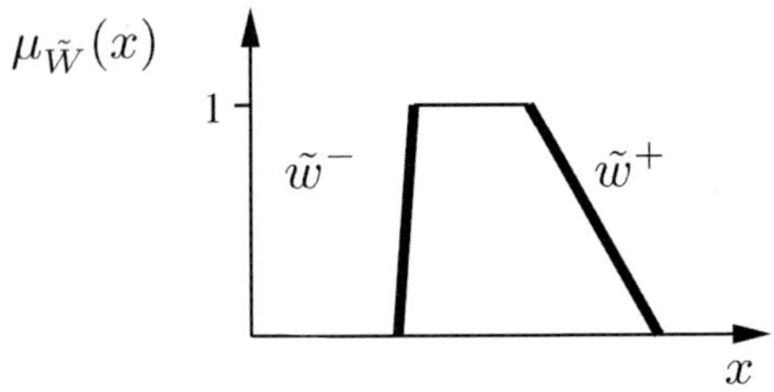


**Fig. 1.** The left and right bounds of fuzzy interval $\tilde{W}$ (in bold)

to interval $[0,1]$. In consequence, the corresponding membership functions of fuzzy intervals are continuous and have a compact support. Observe that a fuzzy interval can be now viewed as an interval of gradual numbers bounded by $\tilde{w}^-$ and $\tilde{w}^+$ (see Fig. 1). For a deeper discussion on gradual numbers and their relationships with fuzzy intervals, we refer the reader to [4,5,8].

The classical arithmetic operations on gradual numbers are defined by operations on their assignment functions. Let $\tilde{r}$ and $\tilde{s}$ be two gradual numbers. The sum of $\tilde{r}$ and $\tilde{s}$ is defined by summing their assignment functions, that is $\mathcal{A}_{\tilde{r}+\tilde{s}}(\lambda) = \mathcal{A}_{\tilde{r}}(\lambda) + \mathcal{A}_{\tilde{s}}(\lambda)$ for all $\lambda \in (0,1]$. The subtraction, maximum and minimum of gradual numbers can be defined in a similar manner. Observe that the minimum and maximum operations on gradual numbers are not selective, that is in general case $\max(\mathcal{A}_{\tilde{r}}, \mathcal{A}_{\tilde{s}}) \neq \mathcal{A}_{\tilde{r}}$ or $\mathcal{A}_{\tilde{s}}$. There are subintervals of $(0,1]$ where $\max(\mathcal{A}_{\tilde{r}}, \mathcal{A}_{\tilde{s}}) = \mathcal{A}_{\tilde{r}}$ and it is $\mathcal{A}_{\tilde{s}}$ in the complementary subinterval. It is worth pointing out that most algebraic properties of real numbers are preserved for gradual real numbers, contrary to the case of fuzzy intervals.

A concept of a *fuzzy interval of the L-R type* (see, e.g. [2]) is very popular and convenient in applications. A fuzzy interval of the L-R type is a fuzzy set in the space of real numbers, whose membership function is of the following form:

$$\mu_{\tilde{W}}(x) = \begin{cases} 1 & \text{for } x \in [w^-, w^+], \\ L\left(\frac{w^- - x}{\alpha_W}\right) & \text{for } x \leq w^-, \\ R\left(\frac{x - w^+}{\beta_W}\right) & \text{for } x \geq w^+, \end{cases}$$

where $L$ and $R$ are continuous non-increasing functions, defined on $[0,+\infty)$, strictly decreasing to zero in those subintervals of the interval $[0,+\infty)$ in which they are positive, and fulfilling the conditions $L(0) = R(0) = 1$. The parameters $\alpha_W$ and $\beta_W$ are non-negative real numbers. A special case of a fuzzy interval is *a triangular fuzzy interval* in which $L(x) = R(x) = \max\{0, 1-x\}$ and $w^- = w^+$. It is denoted by triple $(w, \alpha_W, \beta_W)$. A fuzzy interval $\tilde{W}$ of the L-R type can be described by an ordered pair of gradual numbers $[\tilde{w}^-, \tilde{w}^+]$ with the following assignment functions:

$$\mathcal{A}_{\tilde{w}^-}(\lambda) = w^- - L^{-1}(\lambda)\alpha_W \text{ and } \mathcal{A}_{\tilde{w}^+}(\lambda) = w^+ + R^{-1}(\lambda)\beta_W, \tag{6}$$

where $L^{-1}$ (similarly $R^{-1}$) denotes the inverse function to $L$ in this part of its domain in which it is positive.

### 4.2 A Possibilistic Formulation of the Problem

Assume that the element weights are ill-known, uncontrollable and unrelated parameters $\mathfrak{w}_e$, $e \in E$, with fuzzy sets of more or less possible values. We say that the assertion of the form "$\mathfrak{w}_e$ is $\widetilde{W}_e$", where $\widetilde{W}_e$ is a fuzzy interval associated with $\mathfrak{w}_e$, generates the possibility distribution of $\mathfrak{w}_e$ with respect to the formula $\Pi(\mathfrak{w}_e = x) = \mu_{\widetilde{W}_e}(x)$. For the interpretation of the possibility distribution we refer the reader to [2]. A *configuration* $\boldsymbol{w} = (w_e)_{e\in E}$ represents a state of the world, where $\mathfrak{w}_e = w_e$ for every $e \in E$. The joint possibility distribution over all configurations is as follows:

$$\pi(\boldsymbol{w}) = \Pi(\wedge_{e\in E}(\mathfrak{w}_e = w_e)) = \min_{e\in E} \Pi(\mathfrak{w}_e = w_e) = \min_{e\in E} \mu_{\widetilde{W}_e}(w_e).$$

We can now fuzzyfy intervals $\Delta_X$ and $\Delta_f$ obtaining fuzzy intervals $\tilde{\Delta}_X$ and $\tilde{\Delta}_f$, whose membership functions express the possibility distributions for deviations $\delta_X$ and $\delta_f$. This can be done in the following way:

$$\mu_{\tilde{\Delta}_X}(\delta) = \Pi(\delta_X = \delta) = \sup_{\{\boldsymbol{w}\,:\,\delta_X(\boldsymbol{w})=\delta\}} \pi(\boldsymbol{w}),$$
$$\mu_{\tilde{\Delta}_f}(\delta) = \Pi(\delta_f = \delta) = \sup_{\{\boldsymbol{w}\,:\,\delta_f(\boldsymbol{w})=\delta\}} \pi(\boldsymbol{w}).$$

Having the possibility distribution for $\tilde{\Delta}_X$ we can calculate the possibility and necessity of an event $\delta_X \in [\delta^-, \delta^+]$ in the following way:

$$\Pi(\delta_X \in [\delta^-, \delta^+]) = \sup_{\delta\in[\delta^-,\delta^+]} \mu_{\tilde{\Delta}_X}(\delta),$$
$$\mathrm{N}(\delta_X \in [\delta^-, \delta^+]) = 1 - \Pi(\delta_X \notin [\delta^-, \delta^+]) = 1 - \sup_{\delta\notin[\delta^-,\delta^+]} \mu_{\tilde{\Delta}_X}(\delta).$$

The same formulae hold for an element deviation $\delta_f$. Considering the particular event $\delta_X = 0$ we can compute the *degrees of possible* and *necessary optimality* of solution $X$ in the following way:

$$\Pi(X \text{ is optimal}) = \Pi(\delta_X = 0) = \mu_{\tilde{\Delta}_X}(0), \tag{7}$$
$$\mathrm{N}(X \text{ is optimal}) = \mathrm{N}(\delta_X = 0) = 1 - \Pi(\delta_X > 0) = 1 - \sup_{\delta>0} \mu_{\tilde{\Delta}_X}(\delta). \tag{8}$$

In the same way we define the degrees of possible and necessary optimality of element $f$, using $\tilde{\Delta}_f$.

### 4.3 Application of Gradual Numbers to Computing $\tilde{\Delta}_X$ and $\tilde{\Delta}_f$

In this section we show that the concept of a gradual number allows to extend naturally all the results from Section 3 to the fuzzy-valued case. In consequence we obtain methods for computing the fuzzy deviations $\tilde{\Delta}_X$ and $\tilde{\Delta}_f$. We start with introducing the notion of a *fuzzy configuration*, which is a vector $(\tilde{\boldsymbol{w}})_{e\in E}$

of gradual numbers specified for the weight of every element $e \in E$. Now, the gradual weight of a solution $X \in \Phi$ under a fuzzy configuration $\tilde{\boldsymbol{w}}$ is $\tilde{F}(X, \tilde{\boldsymbol{w}}) = \sum_{e \in X} \tilde{w}_e(\tilde{\boldsymbol{w}})$ and $\tilde{F}^*(\tilde{\boldsymbol{w}}) = \min_{X \in \Phi} \tilde{F}(X, \tilde{\boldsymbol{w}})$. Hence $\tilde{\delta}_X(\tilde{\boldsymbol{w}}) = \tilde{F}(X, \tilde{\boldsymbol{w}}) - \tilde{F}^*(\tilde{\boldsymbol{w}})$ and $\tilde{\delta}_f(\tilde{\boldsymbol{w}}) = \tilde{F}^*_f(\tilde{\boldsymbol{w}}) - \tilde{F}^*(\tilde{\boldsymbol{w}}) = \min_{X \in \Phi_f} \sum_{e \in X} \tilde{w}_e(\tilde{\boldsymbol{w}}) - \tilde{F}^*(\tilde{\boldsymbol{w}})$. Observe, that $\tilde{\delta}_X(\tilde{\boldsymbol{w}})$ and $\tilde{\delta}_f(\tilde{\boldsymbol{w}})$ are now gradual numbers and they are computed by means of operations of the sum, the subtraction and the minimum in the space of gradual numbers.

Suppose now that for every weight $\mathfrak{w}_e$, $e \in E$, there is given a fuzzy interval $\tilde{W}_e = [\tilde{w}_e^-, \tilde{w}_e^+]$, where $\tilde{w}_e^-$ is a gradual lower bound and $\tilde{w}_e^+$ is a gradual upper bound of $\tilde{W}_e$. Now the fuzzy deviations are fuzzy intervals which can be also described by pairs of gradual numbers, namely $\tilde{\Delta}_X = [\tilde{\delta}_X^-, \tilde{\delta}_X^+]$ and $\tilde{\Delta}_f = [\tilde{\delta}_f^-, \tilde{\delta}_f^+]$.

In order to apply the interval methods, given in Section 3, to the fuzzy interval computations, we need to extend extreme configurations to the *fuzzy extreme* ones. Let $A \subseteq E$ be a given subset of elements. In the fuzzy extreme configuration $\tilde{\boldsymbol{w}}_A^+$ all elements $e \in A$ have gradual weights $\tilde{w}_e^+$ and all the remaining ones have gradual weights $\tilde{w}_e^-$. Similarly, in the fuzzy fuzzy extreme configuration $\tilde{\boldsymbol{w}}_A^-$ all elements $e \in A$ have gradual weights $\tilde{w}_e^-$ and all the remaining ones have gradual weights $\tilde{w}_e^+$. Now $\tilde{w}_e(\tilde{\boldsymbol{w}}_A^+)$ and $\tilde{w}_e(\tilde{\boldsymbol{w}}_A^-)$ are gradual weights in fuzzy extreme configurations $\tilde{\boldsymbol{w}}_A^+$ and $\tilde{\boldsymbol{w}}_A^-$, respectively. A fuzzy counterpart of Proposition 1 is the following one:

**Proposition 2.** *The following equalities hold:*

$$\tilde{\delta}_X^- = \min_{\tilde{\boldsymbol{w}} \in \times_{e \in E}\{\tilde{w}_e^-, \tilde{w}_e^+\}} \tilde{\delta}_X(\tilde{\boldsymbol{w}}) = \tilde{\delta}_X(\tilde{\boldsymbol{w}}_X^-),\ \tilde{\delta}_X^+ = \max_{\tilde{\boldsymbol{w}} \in \times_{e \in E}\{\tilde{w}_e^-, \tilde{w}_e^+\}} \tilde{\delta}_X(\tilde{\boldsymbol{w}}) = \tilde{\delta}_X(\tilde{\boldsymbol{w}}_X^+),$$

$$\tilde{\delta}_f^- = \min_{\tilde{\boldsymbol{w}} \in \times_{e \in E}\{\tilde{w}_e^-, \tilde{w}_e^+\}} \tilde{\delta}_f(\tilde{\boldsymbol{w}}),\ \tilde{\delta}_f^+ = \max_{\tilde{\boldsymbol{w}} \in \times_{e \in E}\{\tilde{w}_e^-, \tilde{w}_e^+\}} \tilde{\delta}_f(\tilde{\boldsymbol{w}}),$$

*Proof.* The proof that the minimum and maximum are attained in fuzzy extreme configurations follows from [8] and the fact that functions $\delta_X$ and $\delta_f$ are locally monotonic with respect to each variable. The proof that $\tilde{\boldsymbol{w}}_X^-$ ($\tilde{\boldsymbol{w}}_X^+$) minimizes (maximizes) $\tilde{\delta}_X(\tilde{\boldsymbol{w}})$ follows from the existence of a link between the interval model and the fuzzy one. Observe that $\tilde{W}_e = [\tilde{w}_e^-, \tilde{w}_e^+] = [\mathcal{A}_{\tilde{w}_e^-}(\lambda), \mathcal{A}_{\tilde{w}_e^+}(\lambda)]$ for $\lambda \in [0, 1]$. Hence, from Proposition 1 we conclude that for every fixed $\lambda$ the maximum and minimum are attained either at $\mathcal{A}_{\tilde{w}_e^-}(\lambda)$ or $\mathcal{A}_{\tilde{w}_e^+}(\lambda)$. □

Notice that, similarly to the interval-valued case, we can identify the fuzzy extreme configurations that minimize and maximize the fuzzy deviation for a solution but it may be a hard task for an element.

**Computing $\tilde{\Delta}_X$.** From Proposition 2, we immediately obtain algorithms for determining the gradual lower and upper bounds of fuzzy interval $\tilde{\Delta}_X$ for a given solution $X \in \Phi$. Algorithm 1 is the one for determining the gradual lower

**Algorithm 1.** Determining the gradual lower bound of $\tilde{\Delta}_X$

**Input**: Solution $X \in \Phi$, fuzzy weights $\tilde{W}_e = [\tilde{w}_e^-, \tilde{w}_e^+]$, $e \in E$.
**Output**: The gradual lower bound $\tilde{\delta}_X^-$ of $\tilde{\Delta}_X$.

1 **foreach** $e \in E$ **do**
2     **if** $e \in X$ **then** $\tilde{w}_e(\tilde{\boldsymbol{w}}_X^-) \leftarrow \tilde{w}_e^-$ **else** $\tilde{w}_e(\tilde{\boldsymbol{w}}_X^-) \leftarrow \tilde{w}_e^+$
3 **end**
4 $\tilde{F}(X, \tilde{\boldsymbol{w}}_X^-) \leftarrow \sum_{e \in X} \tilde{w}_e(\tilde{\boldsymbol{w}}_X^-)$ `/* the sum of gradual numbers */`
5 Compute $\tilde{F}^*(\tilde{\boldsymbol{w}}_X^-)$ by solving $\min_{X \in \Phi} \tilde{F}(X, \tilde{\boldsymbol{w}}_X^-) = \min_{X \in \Phi} \sum_{e \in X} \tilde{w}_e(\tilde{\boldsymbol{w}}_X^-)$
6 $\tilde{\delta}_X^- \leftarrow \tilde{F}(X, \tilde{\boldsymbol{w}}_X^-) - \tilde{F}^*(\tilde{\boldsymbol{w}}_X^-)$ `/* the subtraction of gradual numbers */`
7 **return** $\tilde{\delta}_X^-$

bound $\tilde{\delta}_X^-$ (an algorithm for the gradual upper bound $\tilde{\delta}_X^+$ is similar). A key line in Algorithm 1 is line 1, in which we compute the value of $\tilde{F}^*(\tilde{\boldsymbol{w}}_X^-)$. This is a gradual number expressing the fuzzy value of the weight of an optimal solution in fuzzy configuration$\tilde{\boldsymbol{w}}_X^-$. From the technical point of view, $\tilde{F}^*(\tilde{\boldsymbol{w}}_X^-)$ is a function from $[0, 1]$ to $\mathbb{R}$ and it can be obtained by solving a parametric version of the combinatorial optimization problem $\mathcal{P}$. Recall that in a parametric problem every element weight is specified as a function $w(\lambda)$, $\lambda \in \mathbb{R}$, and we wish to compute function $F^*(\lambda)$, so that $F^*(\lambda)$ is the weight of an optimal solution if the element weights are set to $w(\lambda)$. In our case the weights are given as the assignment functions$\mathcal{A}_{\tilde{w}_e(\tilde{\boldsymbol{w}}_X^-)}(\lambda)$ for all $e \in E$. In particular, if we apply fuzzy intervals of the L-R type, then we can use assignment functions of the form (6). In order to compute the value of $\tilde{F}^*(\tilde{\boldsymbol{w}}_X^-)$ some known methods for solving parametric problems with linearly varying weights can be applied (see e.g. [6,12,13]). These algorithms can be directly applied if the fuzzy intervals are trapezoidal or triangular ones (their shape functions are linear). They can be also applied if the fuzzy intervals are of the L-L type (their right and left shape functions are the same) since function $L$ can be then easily linearized. From the knowledge of fuzzy deviation $\tilde{\Delta}_X$, we can also obtain the degrees of possibility and necessity that a given solution $X$ is optimal (see (7) and (8)).

Let us illustrate our algorithms by an example. Consider SHORTEST PATH problem shown in Fig. 2. The arc weights are given as triangular fuzzy intervals: $\tilde{W}_{a_1} = (2, 2, 2)$, $\tilde{W}_{a_2} = (2, 1, 4)$, $\tilde{W}_{a_3} = (6, 2, 1)$, $\tilde{W}_{a_4} = (2, 1, 4)$, $\tilde{W}_{a_5} = (1, 1, 1)$. We wish to compute the fuzzy deviation for the path composed of arcs $a_1$ and $a_3$, that is $\tilde{\Delta}_{\{a_1,a_3\}}$. In Fig. 2a the fuzzy configuration $\tilde{\boldsymbol{w}}^-_{\{a_1,a_3\}}$ and in Fig. 2b the fuzzy configuration $\tilde{\boldsymbol{w}}^+_{\{a_1,a_3\}}$ are shown. Observe, that these configurations induce two parametric shortest path problems in which the parametric weights are computed by means of formula (6). Solving the parametric problems we obtain the gradual numbers $\tilde{F}^*(\tilde{\boldsymbol{w}}^-_{\{a_1,a_3\}})$ and $\tilde{F}^*(\tilde{\boldsymbol{w}}^+_{\{a_1,a_3\}})$. Comparing them to $\tilde{F}(\{a_1, a_3\}, \tilde{\boldsymbol{w}}^-_{\{a_1,a_3\}})$ and $\tilde{F}(\{a_1, a_3\}, \tilde{\boldsymbol{w}}^+_{\{a_1,a_3\}})$ we obtain the gradual lower and the gradual upper bounds of $\tilde{\Delta}_{\{a_1,a_3\}}$ (see Fig. 2c).

We can see that $\Pi(\{a_1, a_3\}$ is optimal$) = 4/7$ and $\mathrm{N}(\{a_1, a_3\}$ is optimal$) = 0$. We can obtain also an additional information, for instance $\Pi(\delta_{\{a_1,a_3\}} \leq 1\frac{3}{5}) = 4/5$.

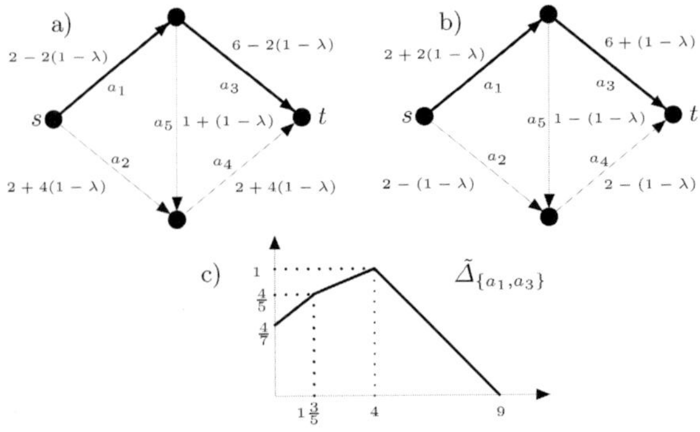


**Fig. 2.** An example of fuzzy-valued SHORTEST PATH PROBLEM

**Computing $\tilde{\Delta}_f$ for Some Practical Problems.** Contrary to the optimality of solutions, there is no an easy characterization of the optimality of the elements and the computational complexity of determining $\tilde{\Delta}_f$ for a given element $f \in E$ strongly depends on a particular problem $\mathcal{P}$.

Recall that $\tilde{\delta}_f(\tilde{\boldsymbol{w}}) = \tilde{F}^*_f(\tilde{\boldsymbol{w}}) - \tilde{F}^*(\tilde{\boldsymbol{w}})$ and we wish to determine $\tilde{\delta}^-_f$ and $\tilde{\delta}^+_f$, that is the gradual numbers that minimize and maximize $\tilde{\delta}_f(\tilde{\boldsymbol{w}})$. While computing these gradual bounds for a given element $f \in E$ two problems arises. The first one is computing the extreme configurations that minimize and maximize $\tilde{\delta}_f(\tilde{\boldsymbol{w}})$. It turns out that the problem of computing the proper extreme configurations is strongly NP-hard for some well-known, polynomially solvable problems like SHORTEST PATH, MINIMUM ASSIGNMENT and MINIMUM CUT. These problems remain strongly NP-hard even in the interval-valued case. The second problem is computing a gradual minimum over all solutions that contain element $f$, that is the value of $\tilde{F}^*_f(\tilde{\boldsymbol{w}})$.

In this section we briefly present some particular problems for which $\tilde{\Delta}_f$ can be efficiently determined. The first problem is MINIMUM SPANNING TREE. Given is an undirected graph $G = (V, E)$ with edge weights specified as fuzzy intervals. Set $\Phi$ consists of all spanning trees of $G$. We wish to determine interval $\tilde{\Delta}_f$ (more precisely the possibility distribution $\mu_{\tilde{\Delta}_f}(x)$) for a given edge $f \in E$. The following proposition immediately follows from the result obtained in [9]:

**Proposition 3.** *The fuzzy configuration $\tilde{\boldsymbol{w}}^-_{\{f\}}$ minimizes and the fuzzy configuration $\tilde{\boldsymbol{w}}^+_{\{f\}}$ maximizes gradual deviation $\tilde{\delta}_f(\tilde{\boldsymbol{w}})$.*

Applying the parametric approach proposed in [6] we can compute the values of $\tilde{F}^*(\tilde{\boldsymbol{w}}^-_{\{f\}})$ and $\tilde{F}^*(\tilde{\boldsymbol{w}}^+_{\{f\}})$. Applying a slightly modified parametric approach we can also compute $\tilde{F}^*_f(\tilde{\boldsymbol{w}}^-_{\{f\}})$ and $\tilde{F}^*_f(\tilde{\boldsymbol{w}}^+_{\{f\}})$. In consequence we obtain $\tilde{\delta}^-_f$ and $\tilde{\delta}^+_f$ for the problem.

The second problem for which $\tilde{\Delta}_f$ can be efficiently computed is SHORTEST PATH when the input graph is restricted to be edge series-parallel digraph (see [14] for a description of this class of graphs). In this problem there is given an edge series parallel digraph $G = (V, A)$ with two distinguished nodes

$s$ and $t$ and arc weights specified as fuzzy intervals. Set $\Phi$ consists of all paths from $s$ to $t$ in $G$. Denote by $Pred(f)$ the set of all arcs that precede arc $f$ and by $Succ(f)$ the set of all arcs that succeed arc $f$ on a path from $s$ to $t$ in $G$. The following proposition is a consequence of the results from [3]:

**Proposition 4.** *The fuzzy configuration* $\tilde{\boldsymbol{w}}^-_{\{Pred(f)\cup\{f\}\cup Succ(f)\}}$ *minimizes and the fuzzy configuration* $\tilde{\boldsymbol{w}}^+_{\{Pred(f)\cup\{f\}\cup Succ(f)\}})$ *maximizes gradual deviation* $\tilde{\delta}_f(\tilde{\boldsymbol{w}})$.

Applying a parametric approach proposed in [12] we can compute the value of $\tilde{F}^*(\tilde{\boldsymbol{w}})$, where $\tilde{\boldsymbol{w}}$ is one of the two fuzzy configurations specified in Proposition 4. The value of $\tilde{F}^*_f(\tilde{\boldsymbol{w}})$ for arc $f = (k,l)$ can be computed using the fact that $\tilde{F}^*_f(\tilde{\boldsymbol{w}}) = \tilde{F}^*_{s-k}(\tilde{\boldsymbol{w}}) + \tilde{w}_f(\tilde{\boldsymbol{w}}) + \tilde{F}^*_{l-t}(\tilde{\boldsymbol{w}})$, where $\tilde{F}^*_{s-k}(\tilde{\boldsymbol{w}})$ is the gradual weight of a shortest path from $s$ to $k$ and $\tilde{F}^*_{l-t}(\tilde{\boldsymbol{w}})$ is the gradual weight of a shortest path from $l$ to $t$ in configuration $\tilde{\boldsymbol{w}}$. Both gradual numbers can be also computed by means of the parametric approach proposed in [12].

## References


1. Chanas, S., Zieliński, P.: The computational complexity of the criticality problems in a network with interval activity times. Eur. J. Oper. Res. 136 (2002) 541–550.
2. Dubois, D., Prade, H.: Possibility theory: an approach to computerized processing of uncertainty. Plenum Press, New York, 1988.
3. Dubois, D., Fargier, H., Galvagnon, V.: On latest starting times and floats in activity networks with ill-known durations. Eur. J. Oper. Res. 147 (2003) 266–280.
4. Dubois, D., Fargier, H., Fortin, J.: A generalized vertex method for computing with fuzzy intervals. In: Fuzz'IEEE (2004) 541–546.
5. Dubois, D., Prade, H.: Gradual Elements in a Fuzzy Set. Soft Comput. To appear.
6. Fernandez-Baca, D., Slutzki, G., Eppstein, D.: Using sparsification for parametric minimum spanning tree problems. In: SWAT 1996 LNCS 1097, pp. 149–160, 1996.
7. Fortin, J., Zieliński, P., Dubois, D., Fargier, H.: Interval analysis in scheduling. In: CP 2005, LNCS 3709, pp. 226–240, 2005
8. Fortin, J., Dubois, D., Fargier, H.: Gradual Numbers and their Application to Fuzzy Interval Analysis. IEEE T. Fuzzy Syst. To appear.
9. Kasperski, A., Zieliński, P.: On combinatorial optimization problems on matroids with uncertain weights. Eur. J. Oper. Res. 177 (2007) 851–864.
10. Moore, R.: Methods and applications of interval analysis. SIAM Studies in Applied Mathematics, 1979.
11. Papadimitriou, C., Steiglitz, K.: Combinatorial optimization, algorithms and complexity. Dover Publications Inc., Mineola, New York., 1998.
12. Raghavan, S., Ball, M. O., Trichur, V. S.: Bicriteria product design optimization: An efficient solution procedure using AND/OR trees. Nav. Res. Log. 49 (2002) 574–592.
13. Young, N., Tarjan, R., Orlin, J.: Faster Parametric Shortest Path and Minimum Balance Algorithms. Networks 21 (1991) 205–221.
14. Valdes, J., Tarjan, R. E., Lawler, E. L.: The recognition of series parallel digraphs. SIAM J. Comput. 11(2) 298–313, 1982.

# Fuzzy Classifier with Probabilistic IF-THEN Rules

Hexin Lv[1], Bin Zhu[1], and Yongchuan Tang[2,*]

[1] College of Information Science and Technology, Zhejiang Shuren University
Hangzhou, Zhejiang Province, 310015, P.R. China
[2] College of Computer Science, Zhejiang University
Hangzhou, Zhejiang Province, 310027, P.R. China
Tel(Fax): 86-571-87953695
yongchuan@263.net

**Abstract.** The typical fuzzy classifier consists of rules each one describing one of the classes. This paper presents a new fuzzy classifier with probabilistic IF-THEN rules. A learning algorithm based on the gradient descent method is proposed to identify the probabilistic IF-THEN rules from the training data set. This new fuzzy classifier is finally applied to the well-known Wisconsin breast cancer classification problem, and a compact, interpretable and accurate probabilistic IF-THEN rule base is achieved.

## 1 Introduction

Fuzzy production rules can deal with the imprecise knowledge and uncertainty information and strengthen the knowledge representation power. However, how to get proper fuzzy rules is still the most important task for a fuzzy model. There are mainly two approaches to tackle this problem. One is directly summarizing the operators' or experts' experiences and translating their knowledge into fuzzy rules. The knowledge acquisition and verification process, however, are difficult and time-consuming. Another approach is obtaining fuzzy rules through machine learning, with which automatically generate or extract knowledge from sample cases or examples.

We focus on the problem of obtaining a compact and accurate fuzzy rule-based classification system from the observation data. Typical fuzzy classifiers consist of interpretable if-then rules with fuzzy antecedents and class labels in the consequent part. The antecedents (if-parts) of the rules partition the input space into a number of fuzzy regions by fuzzy sets, while the consequents (then-parts) describe the output of the classifier in these regions. In order to determine the fuzzy model automatically, many different techniques have been presented. [1] constructed a self-adaptive neuro-fuzzy classification system, this system is capable of self-adapting and self-organizing its internal structure to acquire a parsimonious rule-base for interpreting the embedded knowledge of a

* Corresponding author.

P. Melin et al. (Eds.): IFSA 2007, LNAI 4529, pp. 666–676, 2007.
© Springer-Verlag Berlin Heidelberg 2007

system from the given training data set. [2] discussed the GA-fuzzy modelling from the observation data. In this model, First fuzzy clustering is applied to obtain a compact initial rule-based model. Then this model is optimized by a real-coded GA subjected to constraints that maintain the semantic properties of the rules. A special designed fuzzy-genetic classification system is applied to the Wisconsin breast cancer data [3]. In order to reduce the complexity of the fuzzy model and achieve a compact and interpretable fuzzy classification system, much work based on fuzzy clustering algorithms is tried to initialize the fuzzy model. [4] presented a fuzzy relation-based classifier trained by fuzzy c-mean clustering. However, the obtained membership values have to be projected onto the input variables and approximated by parameterized membership functions that deteriorates the performance of the classifier. To avoid the projection error and maintain the interpretability of the model, the Gath-Geva (GG) clustering algorithm [5] is proposed to identify the fuzzy model. Since GG clustering algorithm doesn't utilize the class labels, [6] proposed a modified GG clustering algorithm to identify the fuzzy classification system.

In this paper we mainly solve the identification problem of fuzzy classification rules from data. But the proposed fuzzy classification rule is different from the classical fuzzy rule. This kind of fuzzy classification rule is similar to that in [6], and is a generalization of that in [15]. That is, the consequent of each fuzzy rule is defined as the probabilities that a given rule represents all classes. The novelty of this new model is that one rule can represent more than one classes with different probabilities. Hence, the same antecedent (fuzzy subregion in the input domain) can infer different classes with different class conditional probabilities. The decision strategy of this new classification system is the Bayes decision rule. According to this kind of decision rule, an objective function based on posterior probabilities of classes given the predictive features is defined, and a gradient-based method is proposed to optimize the parameters in the proposed fuzzy model and improve the classification accuracy.

The paper is organized as follows. Section 2 presents new fuzzy classification rules and its reasoning process. Section 3 describes the identification process based on gradient descent method. Section 4 is our test results in the well-known Wisconsin breast cancer data set. The final section offers the conclusions.

## 2 Structure of the Fuzzy Rule-Based Classifier

The identification of a classifier system means the construction of a model that predicts the class $y_k = \{c_1, c_2, \ldots, c_C\}$ to which pattern $\mathbf{x_k} = [x_{1k}, x_{2k}, \ldots, x_{nk}]$ should be assigned. The classic approach for this problem $C$ classes is based on Bayes' rule. The probability of making an error when classifying an example $\mathbf{x_k}$ is minimized by Bayes' decision rule of assigning it to the class with the largest a posteriori probability:

$$\mathbf{x} \text{ is assigned to } c_i \Longleftrightarrow p(c_i \mid \mathbf{x}) \geq p(c_j \mid \mathbf{x}) \;\; \forall j \neq i. \tag{1}$$

### 2.1 Classical Fuzzy Classifier

Fuzzy classification is a decision process based on the fuzzy logic. Fuzzy classification system consists of classification rules. This system determines a mapping from a given input to an output representing the class, and each rule represents a local mapping from a given input to an output.

Taking into account a classification problem which has $n$ decision variables (attributes or features) $X_i, i = 1, 2, \ldots, n$ and one class variable $y$. Assume $U_i$ is the universe of discourse of variable $X_i$, then the classical fuzzy classification system consists of $C$ fuzzy if-then rules of the forms:

$$\text{If } (X_1 \text{ is } A_{r1}) \text{ and } (X_2 \text{ is } A_{r2}) \ldots \text{ and } (X_n \text{ is } A_{rn}) \\ \text{then } y \text{ is } c_r \text{ with certainty factor } \alpha_r; \quad (2)$$

where $A_{ij}$ is the fuzzy set of $U_j$ and $c_r \in \{c_1, c_2, \ldots, c_C\}$ is the class label. The parameter $\alpha_r$ is the certainty factor which means how certain the relationship between the antecedent and the consequent of this rule is. Furthermore, the certainty factor is within the range of $[0, 1]$.

When taking into account the certainty factor of each rule, given a case which has $n$ attribute values $(x_1, x_2, \ldots, x_n)$, the reasoning method is described as follows:

1. calculate the degree of activation of each rule. For each rule $R(r)$ (see (2)), $A_{r1}(x_1)$, $A_{r2}(x_2), \ldots$, $A_{rn}(x_n)$ are the membership degrees in the antecedent propositions. Assume that the *and* connective is modelled by the product operator allowing for the interaction between the propositions in the antecedent, hence, the degree of activation of the $i$th rule is calculated as:

$$w_r(\mathbf{x}) = \alpha_r \prod_{k=1}^{n} A_{rk}(x_k). \quad (3)$$

2. Determine the output. The output of the classical fuzzy classifier is determined by the *winner takes all* strategy, i.e.,

$$\hat{y} = c_{i^*}, i^* = \arg\max_{1 \le i \le C} w_i(\mathbf{x}). \quad (4)$$

Though any type of membership functions (e.g., the triangle-shaped, trapezoid-shaped and bell-shaped) can be used for fuzzy sets, we employ the gauss-shaped fuzzy sets $A_{rj}$, with the following membership function:

$$A_{rj}(x_j) = \exp\left[-\frac{1}{2}\left(\frac{x_j - a_r^j}{\sigma_r^j}\right)^2\right], \quad (5)$$

where $a_r^j$ represents the center and $\sigma_r^j$ stands for the variance of the Gaussian function. The use of Gaussian function allows for the compaction formulation of (3),

$$\begin{aligned} w_r(\mathbf{x}) &= \alpha_r A_r(\mathbf{x}) \\ &= \alpha_r \exp\left(-\frac{1}{2}(\mathbf{x} - \mathbf{a_r})^T (F_r)^{-1} (\mathbf{x} - \mathbf{a_r})\right), \end{aligned} \quad (6)$$

where $\mathbf{a_r}$ denotes the center of the $r$th multivariate Gaussian and $F_r$ stands for a diagonal matrix that contains the variances $(\sigma_r^j)^2$.

If we assume that the class conditional probability in the classical fuzzy classifier is defined as follows,

$$p(c_r \mid \mathbf{x}) = \frac{w_r(\mathbf{x})}{\sum_{i=1}^{C} w_i(\mathbf{x})}, \tag{7}$$

then the classical fuzzy classifier is a special case of Bayes classifier.

### 2.2 Fuzzy Classifier with Conditional Probabilities

In most situation, the Bayes rule is adopted to make classification decision. In this paper, we propose a new fuzzy classifier. This new fuzzy classifier consists of $m$ fuzzy classification rule, each fuzzy rule can represents more than one class with different probabilities. That is, the $i$th fuzzy classification rule has the form as follows,

$$\begin{aligned} &\text{If } (X_1 \text{ is } A_{r1}) \text{ and } (X_2 \text{ is } A_{r2}) \ \ldots \text{ and } (X_n \text{ is } A_{rn}) \\ &\text{then } y \text{ is } c_1 \ \text{ with } p(c_1 \mid R_r) \ldots y \text{ is } c_C \text{ with } p(c_C \mid R_r) \end{aligned} \tag{8}$$

satisfying

$$0 \leq p(c_i \mid R_r) \leq 1, \sum_{i=1}^{m} p(c_i \mid R_r) = 1.$$

The above formulation shows that the same antecedent can infer different classes with different probabilities. In the classical fuzzy model, the same antecedent can just infer one definite class with some degree of certainty. So we can think the new fuzzy model is an extended fuzzy classifier. This fuzzy model is actual a mixture of classical fuzzy classifiers since the Bayes decision rule in new fuzzy classifier is defined as follows,

$$p(c_r \mid \mathbf{x}) = \sum_{i=1}^{m} p(R_i \mid \mathbf{x}) p(c_r \mid R_i), \tag{9}$$

where

$$p(R_i \mid \mathbf{x}) = \frac{A_i(\mathbf{x})}{\sum_{i=1}^{m} A_i(\mathbf{x})} \tag{10}$$

In the reference [6], authors assign a weight to the fuzzy rule defined as (8) and present an identification method from data based on fuzzy clustering. In our proposed fuzzy classifier structure, the weight assigned to each fuzzy rule is deleted, or fixed as constant one, since we think the fuzzy rule with conditional probabilities has enough represent ability and has well transparency.

The remaining paper focuses on the identification method of new fuzzy classifier from data. In general, the number $m$ of fuzzy rules is predefined. Hence, the parameters required to be identified are the mean and covariance of each gaussian membership function in the antecedent of each fuzzy rule, and conditional probabilities associated with each fuzzy rule. We adopt gradient-based method to work out the parameter identification.

## 3 Identification of Fuzzy Classifier

Assume that the training data set $D$ consists of $N$ training cases $[\mathbf{x_j}; y_j] = [x_{1j}, x_{2j}, \ldots, x_{nj}; y_j]$, where $y_j \in \{c_1, c_2 \ldots, c_C\}$, $j = 1, \ldots, N$. The object to identify the parameters in the fuzzy classifier is to find the parameters to best model the data set. Since we make the classification decision by maximizing posteriori class probability, the notion of "best" means that maximizing the objection function

$$Q(D) = \prod_{j=1}^{N} p(y_j \mid \mathbf{x_j}). \tag{11}$$

We view $Q(D)$ as the function of the parameters in the fuzzy classifier. In fact, it turns out to be easies to maximize the log-function $\ln Q(D)$. Since two functions are monotonically related, maximizing one is equivalent to the other. By Jensen inequality, we have

$$\ln Q(D) \geq \sum_{j=1}^{N} \sum_{i=1}^{n} \mu_{ij} \ln P(y_j|R_i) \tag{12}$$

$$= H(D), \tag{13}$$

where

$$\mu_{ij} = \frac{A_i(\mathbf{x_j})}{\sum_{i=1}^{m} A_i(\mathbf{x_j})}. \tag{14}$$

Since log-function $\ln Q(D)$ is lower bounded by function $H(D)$, we can maximize $H(D)$. In fact, this maximization problem is a constrained optimization problem. That is, for every $i$, we must have that $\sum_{h=1}^{m} p(c_h|R_i) = 1$. Introducing Lagrange multipliers for these constraints, we conclude that our solutions must satisfy the following formulas,

$$\frac{\partial}{\partial p(c_k|R_i)} \left( H(D) + \lambda \left( \sum_{h=1}^{m} p(c_h|R_i) - 1 \right) \right) = 0, \tag{15}$$

for all $i, k$.

Hence we have the following formula,

$$p(c_k|R_i) = \frac{\sum\limits_{j:y_j=c_k} \mu_{ij}}{\sum\limits_{j=1}^{N} \mu_{ij}}, \tag{16}$$

for all $i, k$.

In order to determine the shapes of fuzzy sets in the antecedent of each fuzzy rule, we adopt the gradient-based method to resolve this problem.

By the Eq. (12) and Eq. (14), we can view $H(D)$ as the function of the parameter vector $\mathbf{w}$ which consists of the centers $a_r^k$ and the widths $\sigma_r^k$ of the fuzzy subsets $A_{rk}$. This reduces the maximizing $H(D)$ problem to one of finding a maximum of a multivariate nonlinear function. Algorithms for solving this problem typically take a small steps on the surface whose "coordinates" are the parameters of the function and whose "height" is the value of the function, trying to get to the "highest" point on the surface.

The simplest of this approach, and the one we use, is *gradient descent* (also known as "hill-climbing"). At each point $\mathbf{w}$, it computes $\nabla_{\mathbf{w}}$, the partial derivatives vector with respect to the centers and the widths of the fuzzy subsets. The algorithm then take a small step in the direction of the gradient. Naively, this would be to the point $\mathbf{w}+\alpha\nabla_{\mathbf{w}}$, where $\alpha$ is a step-size parameter. The algorithm terminates when a local maximum is reached, that is, when the gradient is zero.

The partial derivative with respect to the parameter $a_r^k$ can be obtained as follows,

$$\frac{\partial H(D)}{\partial a_r^k} = \sum_{j=1}^{N} \frac{x_{kj} - a_r^k}{(\sigma_r^k)^2} \mu_{rk} \left[ \ln p(y_j \mid R_r) - \sum_{h=1}^{m} \mu_{hj} \ln p(y_j \mid R_h) \right] \tag{17}$$

Similarly, we can obtain the partial derivative with respect to the parameter $\sigma_r^k$,

$$\frac{\partial H(D)}{\partial \sigma_r^k} = \sum_{j=1}^{N} \frac{(x_{kj} - a_r^k)^2}{(\sigma_r^k)^3} \mu_{rk} \left[ \ln p(y_j \mid R_r) - \sum_{h=1}^{m} \mu_{hj} \ln p(y_j \mid R_h) \right] \tag{18}$$

We can now summarize the above discussion in the form of a basic algorithm for learning the fuzzy classifier with conditional probabilities from the training data set. For sake of clarity, we show this basic algorithm in the Fig. 1.

In order to initialize the parameters of the fuzzy classifier, we generate a random matrix $U$ such that

$$0 < \mu_{rj} < 1, \sum_{r=1}^{m} u_{rj} = 1, r = 1, \ldots, m, j = 1, \ldots, N \tag{19}$$

then we use the following formulas to estimate the initial mean values and widths of gaussian membership functions in the fuzzy classifier,

$$a_r^k = \frac{\sum_{j=1}^{N} \mu_{rj} x_{kj}}{\sum_{j=1}^{N} \mu_{rj}}, \tag{20}$$

$$(\sigma_r^k)^2 = \frac{\sum_{j=1}^{N} \mu_{rj} (x_{kj} - a_r^k)^2}{\sum_{j=1}^{N} \mu_{rj}}. \tag{21}$$

---

function F-Algorithm($\mathbf{F}$, $D$) returns an optimized fuzzy classifier
  inputs: $\mathbf{F}$, a fuzzy classifier with parameters $\mathbf{w} = (a_r^k, \sigma_r^k)$
    $D$, a set of training examples

  repeat until $\Delta\mathbf{w} \approx 0$
    for each class $c_k$ and the $r$th rule
      compute probability $p(c_k \mid R_r)$ in $\mathbf{F}$ according to the Eq. (16)
    for each variable $\widetilde{X}_k$, the $i$th rule
      compute $\frac{\partial H(D)}{\partial a_r^k}$ according to the Eq. (17)
      compute $\frac{\partial H(D)}{\partial \sigma_r^k}$ according to the Eq. (18)
      $a_r^k \leftarrow a_r^k + \alpha \frac{\partial H(D)}{\partial a_r^k}$
      $\sigma_r^k \leftarrow \sigma_r^k + \alpha \frac{\partial H(D)}{\partial \sigma_r^k}$
      $\Delta\mathbf{w} = (\frac{\partial H(D)}{\partial a_r^k}, \frac{\partial H(D)}{\partial \sigma_r^k})$
      $\mathbf{w} = (a_r^k, \sigma_r^k)$
  return $\mathbf{F}$

---

**Fig. 1.** A basic algorithm

# 4 Performance Evaluation

In order to examine the performance of the fuzzy classifier one well-known multidimensional classification benchmark problem is presented in this section. The studied Wisconsin breast cancer data comes from the UCI Repository of Machine Learning Database (http://www.ics.uci.edu).

The performance of the obtained classifier measured by 10–fold cross validation. The data is divided into ten sub-sets of cases that have similar size and class distributions. Each sub-set is left out once, while the other nine are applied for the construction of the classifier which is subsequently validated for the unseen cases in the left-out sub-set.

Wisconsin breast cancer data is widely used to test the effectiveness of classification and rule extraction algorithms. The data samples are taken from fine needle aspirates from human breast tissue. The aim of the classification is to distinguish between benign and malignant cancers based on the available nine measurements: $X_1$ clump thickness, $X_2$ uniformity of cell size, $X_3$ uniformity of cell shape, $X_4$ marginal adhesion, $X_5$ single epithelial cell size, $X_6$ bare nuclei, $X_7$ bland chromatin, $X_8$ normal nuclei, and $X_9$ mitosis. The measurements are assigned an integer value between 1 and 10, with 1 being the closest to benign and 10 the most anaplastic. Associated with each sample is its class label, which is either benign or malignant. The original database contains 699 instances however 16 of these are omitted because these are incomplete, which is common with other studies. The class distribution is 65.5% benign and 34.5% malignant, respectively.

**Table 1.** The classification results of the fuzzy classifier constructed for the Wisconsin breast cancer problem

| Number of rules | Min Acc. | Max Acc. | Mean Acc. |
|---|---|---|---|
| $m = 2$ | 95.59% | 100% | 97.16% |
| $m = 3$ | 98.53% | 100% | 98.56% |

Partial previous work on the classification problem for the Wisconsin breast cancer is listed in the reference [6]. The advanced version of C4.5 gives misclassification of 5.26% on 10-fold cross validation (94.74% correct classification) with tree size 25±0.5 [7]. [8] use the decision tree to initialize the fuzzy classifier and GAs to improve the classification accuracy. In this method, the decision tree is used for the selection of the relevant attributes and effective initial partitioning of the input domains of the fuzzy system. The classification accuracy of this fuzzy system with two rules on the Wisconsin breast cancer data is 96.87%. [9] developed a constrained-syntax genetic programming system for discovering classification rules in the Wisconsin breast cancer data, the classification accuracy is compatible with the advanced version of C4.5 [7]. [10] combined neuro-fuzzy techniques with interactive strategies for rule pruning to obtain a fuzzy classifier. An initial rule-base was made by applying two sets for each input, resulting in $2^9 = 512$ rules which was reduced to 135 by deleting the non-firing rules. A heuristic data-driven learning method was applied instead of gradient descent learning, which is not applicable for triangular membership functions. Semantic properties were taken into account by constraining the search space. The final fuzzy classifier could be reduced to two rules with 5-6 features only, with a misclassification of 4.94% on 10-fold validation (95.06% classification accuracy). Rule-generating methods that combine GA and fuzzy logic were also applied to this problem [3]. In this method the number of rules to be generated needs to be determined a priori. This method constructs a fuzzy model that has four membership functions and one rule with an additional else part. [11] developed a fuzzy classification system with FeatureSelector and modified threshold accepting to test the Wisconsin breast cancer data. This classification system can reduce the dimension of the feature space in classification problems of high dimensions without sacrificing the classification power considerably. [12] also developed a fuzzy classifier with the ability of feature selection based on fuzzy entropy measure. The application of this system in the Wisconsin breast cancer data shows that this method can reduce the dimensionality of the problem. [13] has generated similar compact classifier by a two-step rule extraction from a feedforward neural network trained on preprocessed data. [14] presented a hybrid system combining the numerical and linguistical knowledge representation to deal with the medical diagnosis. This model is a hierarchical integration of an incremental learning fuzzy neural network and a linguistic model, i.e., fuzzy expert system, optimized via the genetic algorithm.

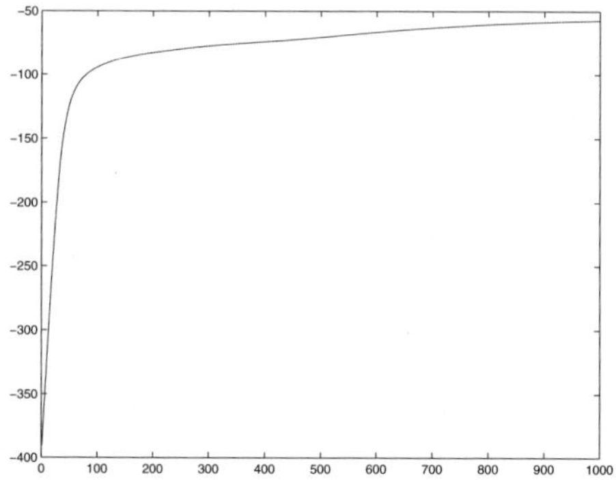


**Fig. 2.** The convergence of the objective function trained by gradient-based method

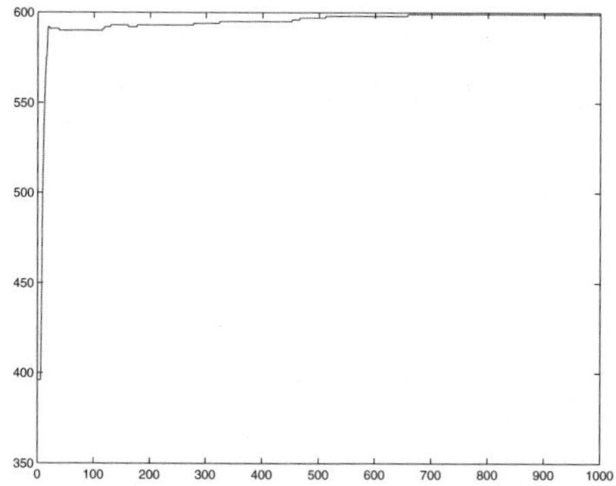


**Fig. 3.** The convergence of the classification accuracy trained by gradient-based method

In Table 1, we list the experimental results on the Wisconsin breast cancer database by using the fuzzy classifier with conditional probabilities. This table shows that our fuzzy classifier can achieve very high accuracy with simple model structure. The fuzzy classifier with 2 fuzzy classification rules achieves 97.16% mean accuracy, and the fuzzy classifier with 3 fuzzy classification rules achieves 98.56% mean accuracy. In this experiment, the learning rate $\alpha = 0.001$ and iteration number is 1000.

Figure 2 illustrates the changing trend of the objective function $H(D)$ of the Wisconsin breast cancer data in one of cross-validation phase. Figure 3 illustrates the corresponding classification accuracy of the fuzzy system. These two figures show that optimizing the objective function can improve the classification accuracy, and the objective function and classification accuracy reach the local maxima almost at the same time very quickly.

## 5 Conclusions

In this paper we provide a gradient-based method to identify a new fuzzy classification system from observation data. This fuzzy classifier consists of rules each one represents more than one class with different conditional probabilities. And this fuzzy classifier can be represented as a mixture model of classical fuzzy classifiers. For the identification of the proposed fuzzy classifier a gradient-based method has been worked out by optimizing an objective function based on the posterior probabilities of classes given the predictive features.

In the proposed identification method, an alternating optimization process is adopted. The conditional class probabilities associated with each fuzzy rule is calculated from the training data and parameters of the fuzzy sets in the antecedents, and the gradient of the parameters of the fuzzy sets in the fuzzy model is computed out from the training data and class conditional probabilities. Moreover, the computation involved in the optimization is easy and direct.

A compact and interpretable fuzzy classifier is achieved by applying the proposed method to the well-known Wisconsin breast cancer data. The test results show that this fuzzy classification system has strong power of knowledge representation and the gradient-based identification method is very effective.

### Acknowledgements

This work has been supported by National Natural Science Foundation of China (No. 60604034), Science and Technology Program of Zhejiang Province (No. 2006C23062, 2007C23G2010148).

## References

1. J.-S. Wang, C. S. G. Lee, Self-adaptive neuro-fuzzy inference systems for classification applications, IEEE Transactions on Fuzzy Systems 10 (6) (2002) 790–802.
2. M. Setnes, H. Roubos, Ga-fuzzy modeling and classification: Complexity and performance, IEEE Transactions on Fuzzy Systems 8 (5) (2000) 509–522.
3. C. A. P. na Reyes, M. Sipper, A fuzzy-genetic approach to breast cancer diagnosis, Artificial Intelligence in Medicine 17 (1999) 131–155.
4. M. Setnes, R. Babuška, Fuzzy relational classifier trained by fuzzy clustering, IEEE Transactions on Systems, Man, and CyberneticsPart B: Cybernetics 29 (5) (1999) 619–625.
5. I. Gath, A. B. Geva, Unsupervised optimal fuzzy clustering, IEEE Transactions on Pattern Analysis and Machine Intelligence 11 (7) (1989) 773–781.
6. J. Abonyi, F. Szeifert, Supervised fuzzy clustering for the identification of fuzzy classifiers, Pattern Recognition Letters 24 (2003) 2195–2207.
7. J. R. Quinlan, Improved use of continuous attributes in c4.5, J. Artificial Intell. Res. 4 (1996) 77–90.
8. J. A. Janos Abonyi, F. Szeifert, Data-driven generation of compact, accurate,and linguistically sound fuzzy classifiers based on a decision-tree initialization, International Journal of Approximate Reasoning 32 (2003) 1–21.

9. A. A. F. Celia C. Bojarczuk, Heitor S. Lopes, E. L. Michalkiewicz, A constrained-syntax genetic programming system for discovering classification rules: application to medical data sets, Artificial Intelligence in Medicine 30 (2004) 27–48.
10. D. Nauck, R. Kruse, Obtaining interpretable fuzzy classification rules from medical data, Artificial Intell. Med. 16 (1999) 149–169.
11. V. Ravi, H.-J. Zimmermann, Fuzzy rule based classification with featureselector and modified threshold accepting, European Journal of Operational Research 123 (2000) 16–28.
12. J.-M. C. Hahn-Ming Lee, Chih-Ming Chen, Y.-L. Jou, An efficient fuzzy classifier with feature selection based on fuzzy entropy, IEEE Transactions on Systems, Man, and CyberneticsPart B: Cybernetics 31 (3) (2001) 426–432.
13. R. Setiono, Generating concise and accurate classification rules for breast cancer diagnosis, Artificial Intelligence in Medicine 18 (2000) 205–219.
14. P. Meesad, G. G. Yen, Combined numerical and linguistic knowledge representation and its application to medical diagnosis, IEEE Transactions on Systems, Man, and CyberneticsPart A: Systems and Humans 33 (2) (2003) 206–222.
15. Y. Tang, S. Sun, A Mixture Model of Classical Fuzzy Classifiers, Q. Li, G. Wang, and L. Feng (Eds.):WAIM 2004, LNCS 3129, pp. 616C621, 2004.

# Fuzzy Adaptive Search Method for Parallel Genetic Algorithm Tuned by Evolution Degree Based on Diversity Measure

Yoichiro Maeda and Qiang Li

Dept. of Human and Artificial Intelligent Systems, Graduate School of Engineering, Univ. of Fukui, 3-9-1 Bunkyo, Fukui 910-8507 Japan
{maeda,qli}@ir.his.fukui-u.ac.jp

**Abstract.** Generally, as for Genetic Algorithms (GAs), it is not always optimal search efficiency, because genetic parameters (crossover rate, mutation rate and so on) are fixed. For this problem, we have already proposed Fuzzy Adaptive Search Method for GA (FASGA) that is able to tune the genetic parameters according to the search stage by the fuzzy reasoning. On the other hand, in order to improve the solution quality of GA, Parallel Genetic Algorithm (PGA) based on the local evolution in plural sub-populations (islands) and the migration of individuals between islands has been researched.

In this research, Fuzzy Adaptive Search method for Parallel GA (FASPGA) combined FASGA with PGA is proposed. Moreover as the improvement method for FASPGA, Diversity Measure based Fuzzy Adaptive Search method for Parallel GA (DM-FASPGA) is also proposed. Computer simulation was carried out to confirm the efficiency of the proposed method and the simulation results are also reported in this paper.

## 1 Introduction

Some tools for monitoring the exploitation/exploration relationship (EER) [1] have been proposed in order to avoid the premature convergence problem and improve GA performance [2,3]. These tools include modified selection and crossover operators and the optimization of parameter setting. Some of these tools are based on the fuzzy logic based techniques [4,5,6]. In our laboratory, we have already proposed Fuzzy Adaptive Search method for Genetic Algorithm (FASGA) and for Genetic Programming (FASGP) [7,8]. FASGA and FASGP are able to realize the efficient search by describing of fuzzy rules to tune GA and GP parameters (crossover rate, mutation rate and so on) based on the maximum and average fitness values according to the search stage. In other words, the GA parameters are not fixed and varying with the search stage. So FASGA has a quick search ability to obtain the best solution than GAs. But FASGA has not very good performance to obtain the higher quality solution, because it could not avoid the premature convergence and fall in the local solution completely.

P. Melin et al. (Eds.): IFSA 2007, LNAI 4529, pp. 677–687, 2007.
© Springer-Verlag Berlin Heidelberg 2007

On the other hand, some GAs in parallel methods were already proposed as the effective method for finding high quality solutions [9,10,11,12,13]. In parallel methods, the total population is divided into independent subpopulations called islands. There are three kinds of distribution models: 1) master-slave model; 2) coarse-grained model (island model); 3) fine-grained model (cellular model). In this research, we use the island model for avoiding to propagation of local optimum solutions in whole population and obtaining high quality solution. After a predetermined number of generations (the migration interval), some genes are moved to the another island at a predetermined migration rate (defined as the number of genes migrating per migration event). In this paper, GA based on parallel methods is called Parallel Genetic Algorithm (PGA). Because the existence of islands and the operation of migration, the variety of solutions is kept in PGA and higher quality solutions are able to be obtained than GAs. However, PGA has also the disadvantage itself. It is that PGA is not always effective in using parallel processing because the migration rate of PGA is a constant. In fact, the migration of individuals is not necessarily performed only in case of necessity. To overcome this problem, many researchers have proposed the solved method, such as, a distributed genetic algorithm with the randomized migration rate [14], a parallel genetic algorithm with the distributed environment scheme [15].

Therefore, we have already proposed Fuzzy Adaptive Search method for Parallel Genetic Algorithm (FASPGA) combined FASGA with PGA. FASPGA is a genetic algorithm in parallel method with both the quick search and high-quality solutions ability. In this method, parameters tuned by fuzzy rules are not only the crossover rate and mutation rate but also the migration rate. So an efficient migration can be realized in FASPGA. We has proved that FASPGA has better performance than PGA and FASGA in [16,17,18]. In this research, we have adopted the phenotypic parameters, maximum and average fitness as the inputs of fuzzy rules, but it could be considered that it is not enough to describe the search stage. Therefore, as an improvement method for FASPGA, Diversity Measure based Fuzzy Adaptive Search method for Parallel GA (DM-FASPGA) is also proposed in this paper. In this method, we introduce the genotypic parameters, such as the hamming distance, into the fuzzy rule. As combining with phenotype parameters, there are many combinations as the input of fuzzy rules. For finding out optimum combination, many simulations are executed in this research. The results of simulations are also reported in this paper.

## 2 FASPGA

We have already proposed Fuzzy Adaptive Search method for Parallel Genetic Algorithm (FASPGA) combined FASGA with quick search ability and Island model Parallel GA with high-quality solutions ability in this research. General concept of FASPGA is introduced in this section.

### 2.1 Migration in FASPGA

In fact, the migration is an operation that some individuals are selected to move from one island to another. By the migration, the better individuals could be spread in all population quickly, and enhance the precision of the solution. The migration of individuals from one island to another is controlled by these parameters: (a) a migration rate; (b) a migration interval; and (c) the topology that defines the connections between islands. In FASPGA, the migration rate is tuned by the fuzzy rule and we just use the random ring model for the topology.

To our knowledge, it is difficult to decide the migration rate properly, but it affects the performance of FASPGA directly. Generally, the individuals of migration are almost the best individuals of each subpopulation. So if the migration rate is larger, it is more advantage to spread the advance individuals to all population and accelerate convergence. However, at the same time it causes the decrease of population diversity and is the disadvantage to explore different regions of the search space.

Generally, the migration rate is a constant in PGA. In the other words, the size of individuals migrated are fixed for each island, regardless of the island with the delayed evolutionary condition or with the advanced evolutionary condition. It is obviously disadvantages to convergence of PGA as spreading of individuals in island with the delayed evolutionary condition.

Therefore, for FASPGA proposed in this paper, the migration rate is not constant and is tuned in a range by fuzzy rule according to states of each island. So by the migration, some individuals in the island with the advanced evolutionary condition are easy to be spread to whole population. On the contrary, some individuals in the island with the delayed evolutionary condition are difficult to be spread to whole population. In there, the fuzzy rule plays a good role in guiding the evolutional direction for improving the search efficiency.

We used the roulette wheel selection as the selection method to select migration individuals in FASPGA. Probability of the roulette wheel selection for selecting individuals with high fitness value is used high in the sender island and low in the receiver island as shown in the following equations. In these equation, $p$ means the populations size of island.

$$IndividualSelectionProbabilityinSenderIsland: \quad \frac{f_i}{\sum_{i=0}^{p} f_i} \tag{1}$$

$$IndividualSelectionProbabilityinReceiverIsland: \frac{f_{m_j} - f_j}{\sum_{j=1}^{p}(f_{m_j} - f_j)} \tag{2}$$

### 2.2 Fuzzy Reasoning of FASPGA

In FASPGA, the crossover rate $r_{c_i}$, the mutation rate $r_{m_i}$ and the migration rate $r_{e_i}$ of each island are not fixed (i : island number) and are tuned by the fuzzy rule. The $IF$ part of the membership function is same as FASGA. It is composed of the average fitness value $f_{a_i}$ and the difference between the maximum and average fitness value ($f_{m_i}$-$f_{a_i}$). But, FASPGA adds a parameter called the migration

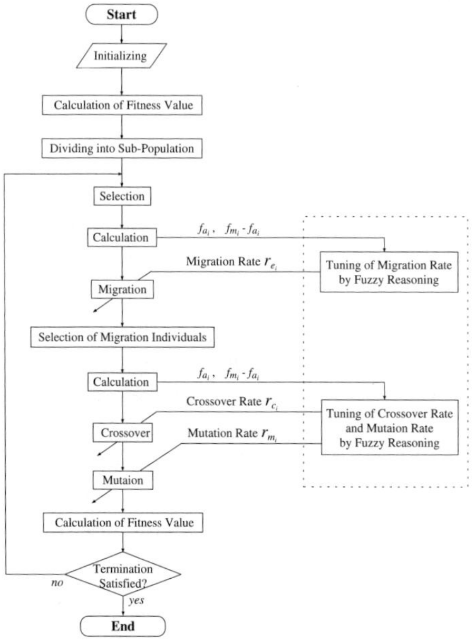


Fig. 1. Algorithm Flow of FASPGA

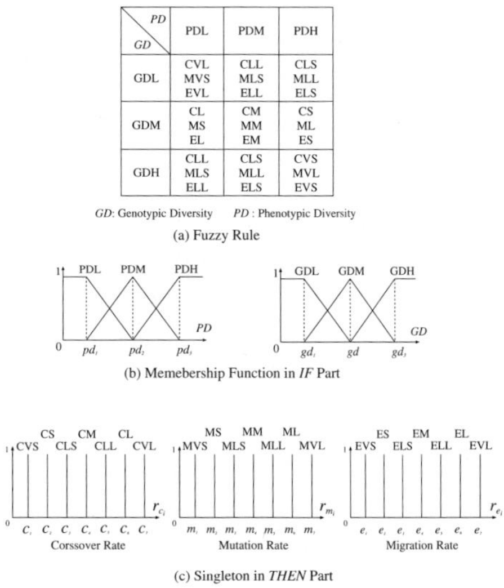


Fig. 2. Fuzzy Reasoning of DM-FASPGA

rate in $THEN$ part. The result of output is calculated by the weighted average based on the firing strength. The fuzzy rule and membership functions in $IF$ part and singletons in $THEN$ part are shown in [18].

## 2.3 Algorithm Flow of FASPGA

At first, an initial population is generated in random. Then the fitness value of each individual is calculated. Next, the initial population is divided into $n$ sub-populations (islands). After the selection by using the roulette wheel selection method, the average fitness value $f_{a_i}$ and the maximum fitness value $f_{m_i}$ are calculated in each island ($i = 1, 2, ..., n$). By estimating a progress degree of the evolution with the average fitness value $f_{a_i}$ and the difference between the maximum and average fitness value ($f_{m_i}$-$f_{a_i}$), the migration rate $r_{e_i}$ in each island are decided by fuzzy rule. The migration rate $r_{e_i}$ is larger, the size of individuals of migration is larger. And the migration is executed with the random ring model. After the operation of migration and before the operation of crossover and mutation, the average fitness value $f_{a_i}$ and the maximum fitness value $f_{m_i}$ of each island are calculated one more time. Because the fuzzy rule depends on current average fitness value $f_{a_i}$ and difference between the maximum and average fitness value ($f_{m_i}$-$f_{a_i}$) to tune crossover rate $r_{c_i}$ and mutation rate $r_{m_i}$. Finally, the terminate condition is checked. If it is contented then terminate the evolution, else return to the operation of selection, and execute once more in the same step. We can regard the flow of FASPGA as putting FASGA algorithm applying to each island. The algorithm flowchart of FASPGA proposed in this paper is shown in Fig. 1. Tuning processes of the crossover rate $r_{c_i}$, the mutation rate $r_{m_i}$ and the migration rate $r_{e_i}$ in each island by the fuzzy reasoning are executed inside the dotted line area.

# 3 DM-FASPGA

For FASPGA, it was confirmed the better search efficiency and higher solution quality than GA and PGA in the multi-peaks function. But as for the single peak function, FASPGA was not confirmed the better than PGA, perhaps because the diversity measure by the phenotypic distance is not efficient for the fitness evaluation in the function with small change of curvature. Therefore, in this paper, we propose an improvement method for FASPGA; Diversity Measure based Fuzzy Adaptive Search method for Parallel Genetic Algorithms (DM-FASPGA). The fuzzy reasoning in DM-FASPGA is almost same to FASPGA except the point of using the diversity measure.

## 3.1 Diversity Measure

There are many kinds of the diversity measure methods proposed by the other researchers [1,6]. Before introduce of the diversity measure methods used in this research, we must explain the following of calculations for the diversity measure executed in an island, not in whole population.

At first, a genotypic diversity measure method based on the Hamming Distances (HD) of the chromosomes between the individual and one individual with average fitness is adopted. Its definition is

$$HD = \frac{\sum_{k=1}^{n} d(I_{ave}, I_k)}{n-1} \tag{3}$$

where $I_{ave}$ shows a chromosome with the closest fitness to average one in the island, $I_k$ shows any one chromosome in the island and d($I_{ave}$, $I_k$) means the hamming distance between $I_{ave}$ and $I_k$. The $n$ means the population size in the island. The range of HD is from 0 to the chromosome's length. If HD is lower, it means more individuals converge on an individual with the average fitness. On the contrary, if HD is higher, it means more individuals diverge from an individual with the average fitness. There are several variant definitions for the hamming distance shown as follows.

$$HD\ II = \frac{\sum_{k=1}^{n} d(I_{best}, I_k)}{n-1} \tag{4}$$

$$HD\ III = \frac{HD\ I - min\{d(I_{ave}, I_k)\}}{max\{d(I_{ave}, I_k)\} - min\{d(I_{ave}, I_k)\}} \tag{5}$$

$$HD\ IV = \frac{HD\ II - min\{d(I_{best}, I_k)\}}{max\{d(I_{best}, I_k)\} - min\{d(I_{best}, I_k)\}} \tag{6}$$

where $I_{best}$ shows a chromosome with the closest fitness to best one.

In the previous section, we introduced FASPGA that adopt the Average Fitness (AF) and the Different Fitness between the best fitness and average fitness (DF), which both are the phenotypic diversity measures. These are shown as in Equations (7) and (8).

$$AF = f_{I_{ave}} \tag{7}$$

$$DF = f_{I_{best}} - f_{I_{ave}} \tag{8}$$

where $f_{I_{best}}$ and $f_{I_{ave}}$ show the best fitness and average fitness value in the island. If DF is samller, it means more individuals converge. On the contrary, if DF is larger, it means more individuals diverge.

There are also several variant definitions for the phenotypic diversity shown as follows.

$$DF\ II = f_{I_{ave}} - f_{I_{min}} \tag{9}$$

$$DF\ III = \frac{\sum_{k=1}^{n}(f_{I_{ave}} - f_{I_k})}{n-1} \tag{10}$$

$$DF\ IV = \frac{\sum_{k=1}^{n}(f_{I_{best}} - f_{I_k})}{n-1} \tag{11}$$

where $f_{I_k}$ show the fitness value of any individual in the island.

At last, we introduce a Relative Phenotypic diversity measure (RP) defined as following. RP values belong to the interval [0,1]. If the value of RP is near to 0, it means the population is convergence and in early search stage, whereas if it is near to 1, it means that the population is divergence and in final search stage.

$$RP = \frac{f_{I_{best}}}{f_{I_{ave}}} \tag{12}$$

### 3.2 Fuzzy Reasoning of DM-FASPGA

In this method, the inputs of the fuzzy rule are PD and GD, which the meaning of the linguistic terms PD and GD are Phenotypic Diversity measure and Genotypic Diversity measure. The outputs of the fuzzy rule are same to FASPGA composed of the crossover rate $r_c$, the mutation rate $r_m$ and the migration rate $r_e$. Fig.2 shows the fuzzy rule, membership functions and singletons of DM-FASPGA. The algorithm flow of DM-FASPGA is same to FASPGA.

## 4 Simulation

To confirm the efficiency of DM-FASPGA proposed in this paper, the computer simulation was executed as compared with PGA. In this simulation, we still used the Rastrigin function and Rosenbrock function as the test function and the elitist strategy is same to FASPGA. In this section, we will introduce the parameters setting. The results and remarks are also reported.

### 4.1 Test Function

In this simulation, we used the Rastrigin and Rosenbrock function as a test function to confirm the efficiency of methods proposed in this paper. Here we briefly describe each of them.

The Rastrigin Function is a typical non-linear multimodal function. It is a $n$-dimensional function with multiple peaks as shown in Equation (13), which has lattice-shaped semi-optimum solutions around an optimum solution, and has no dependence between design parameters. This function is a fairly difficult problem due to its large search space and its large number of local minima.

Rosenbrock's function, also known as Banana function, is a classic optimization problem. It is a single peak function as shown in Equation (14) and has dependence between design parameters. The global optimum is inside along, narrow, parabolic shaped flat valley. The valley is trivial, but the convergence to the global optimum is difficult and hence this problem has been repeatedly used in the performance evaluation of optimization algorithms.

$$F_{Rastr}(x) = 10n + \sum_{i=1}^{n} \{x_i^2 - 10\cos(2\pi x_i)\} \tag{13}$$
$$(-5.12 \leq x_i < 5.12)$$
$$min(F_{Rastr}(x)) = F(0,0,\ldots,0) = 0$$

$$F_{Rosen}(x) = \sum_{i=1}^{n-1} \{100(x_{i+1} - x_i^2)^2 + (1 - x_i)^2\} \tag{14}$$
$$(-2.048 \leq x_i < 2.048)$$
$$min(F_{Rosen}(x)) = F(1,1,\ldots,1) = 0$$

### 4.2 Parameters Setting

In this simulation, we execute the optimization simulations using Rastrigin and Rosecbrock function with 20 variables by the binary coding. The result of simulations based on maximum fitness value is disscussed in this section. All of the

**Table 1.** Parameters setting

| GA Parameters | PGA | FASPGA |
|---|---|---|
| Generations | 1000 | 1000 |
| Chromosome Length | 200(L) | 200 |
| Total Population Size | 500 | 500 |
| Island Size | 10 | 10 |
| Selection Method | Roulette Wheel | Roulette Wheel |
| Crossover Rate | 0.6 (Single Point) | Tuned by Fuzzy Rule |
| Mutation Rate | 1/L | Tuned by Fuzzy Rule |
| Migration Rate | 0.5 | Tuned by Fuzzy Rule |
| Migration Interval | 5 (generations) | Checking Everytime |

**Table 2.** Inputs of Fuzzy Rule

| Simulation | Inputs of Fuzzy Rule | | Coding |
|---|---|---|---|
| PP1 | AF | DF | Binary |
| PP2 | AF | DF II | |
| PP3 | AF | DF III | |
| PP4 | AF | DF IV | |
| PG1 | RP | HD | |
| PG2 | RP | HD II | |
| PG3 | RP | HD III | |
| PG4 | RP | HD IV | |
| PG5 | AF | HD | |
| Gray-PG1 | RP | HD | Gray |
| Gray-PG2 | RP | HD II | |
| Gray-PG3 | RP | HD III | |
| Gray-PG4 | RP | HD IV | |
| Gray-PG5 | AF | HD | |

figures in this paper display the maximum fitness on the $y$-axis, and generations on the $x$-axis. For the convenience of writing computer programs, we increase one minus sign before the test function like as $f_{Rastr}(x) = - F_{Rastr}(x)$. The optimum solution value is 0.

Here we introduce how to decide the value of parameters of membership functions. At first, an initial population of individuals is produced in random. The population size and island size are decided based on the parameter setting. Next we calculate the average fitness $f_{ai}$ and the difference between the maximum fitness and the average fitness $(f_{mi} - f_{ai})$ and so on.

In this computer simulation, the parameter setting is shown in Table 1. The inputs of the fuzzy rule of DM-FASPGA are shown in Table 2. The way to decide the membership functions is same to FASPGA and parameters of the membership functions in Fig. 2 were omitted for want of space.

### 4.3 Simulation Results and Remarks

In this simulation, we compared the simulation results of DM-FASPGA with PGA based on the maximum fitness value. In order to obtain stable results, all of the simulation results are shown in the average value in 10 times. Fig. 3 to Fig. 7 show the simulation results excuted by Rastrigin function. Fig. 4 to Fig. 8 are shown the simulation results excuted by Rosenbrock function.

By Fig. 3 to 7, we could confirm that PGA has almost the worst performacne, as the Rastrigin function is test function. However, in the Fig. 5, we could find that PGA has better performance than PG2 in Rastrigin function. By Fig. 4 to 8, we could confirm that FASPGA has almost better performance than PGA. And it was observed that the total performance of Gray-PG has the best in both Rastrigin and Rosenbrock function. Totally, PP1, PG1 and Gray-PG5 displayed the best performance in case of Rastrigin function, but almost these parameters performed the worst or bad behavior in Rosenbrock function.

From the above-mentioned results, we could find some rules for diversity measure. When the change of curvature of the test function is big, such as Rastrigin function, the phenotypic diversity measure (such as AF, DF) will display good performance than the genotypic diversity measure. On the contrary, when the change of curvature of the test function is small, such as Rosenbrock function, the genotypic diversity measure (such as HD) will display good performance than the phenotypic diversity measure. As this reason, we consider that the change of curvature is bigger, the phenotype distance is bigger, then phenotypic diversity measure is useful to tune genetic parameters by fuzzy reasoning. On the other hand, the change of curvature is smaller, the phenotype distance is smaller, then the phenotypic diversity measure is unutilizable to tune genetic parameters by fuzzy reasoning. But, in this time, the genotypic diversity measure is useful to tune genetic parameters by fuzzy reasoning.

In addition, we confirmed that when the inputs of the fuzzy rule include the genotypic diversity measure, the gray coding will show the better performance. As this reason, we consider that the genotypic diversity measure defines

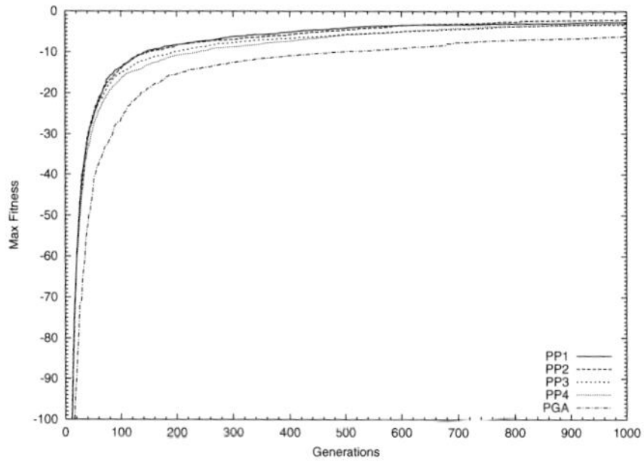


**Fig. 3.** Simulation (PP in Rastrigin)

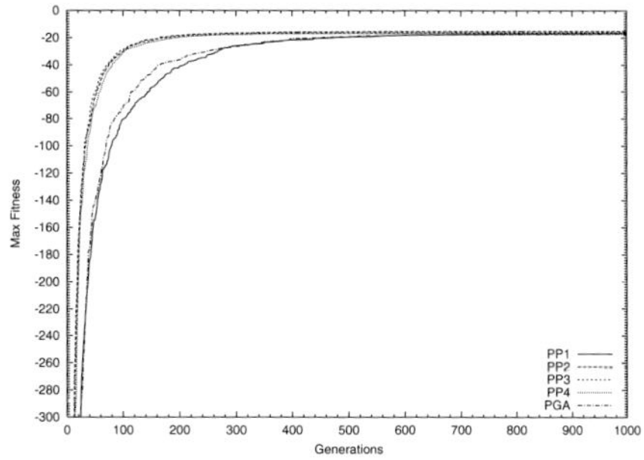


**Fig. 4.** Simulation (PP in Rosenbrock)

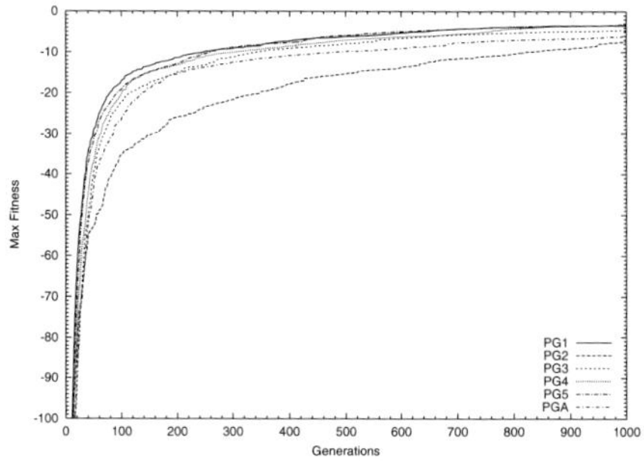


**Fig. 5.** Simulation (PG in Rastrigin)

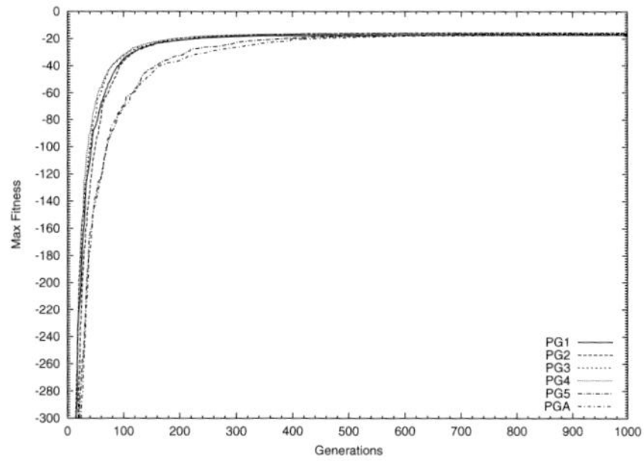


**Fig. 6.** Simulation (PG in Rosenbrock)

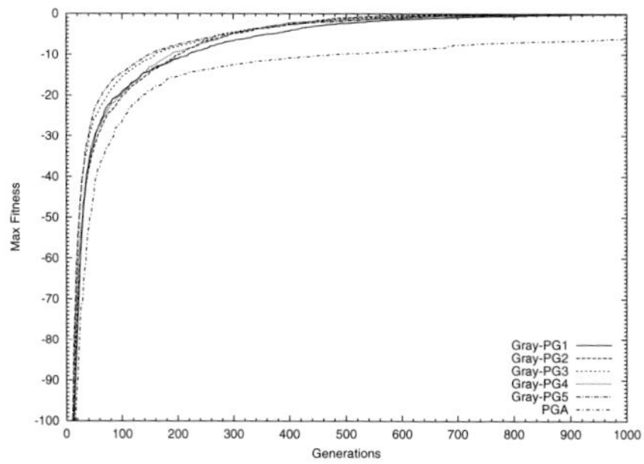


**Fig. 7.** Simulation (Gray-PG in Rastrigin)

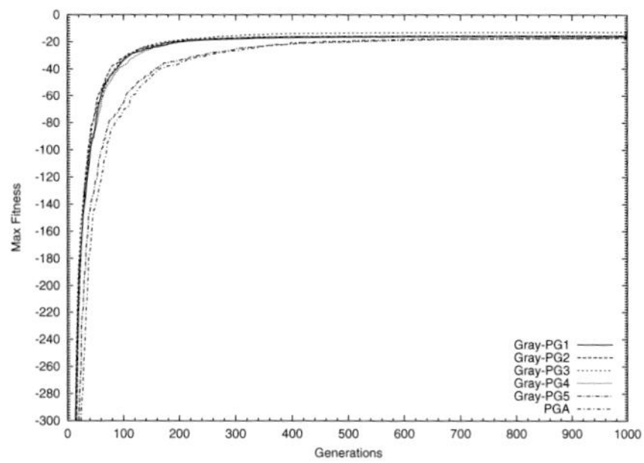


**Fig. 8.** Simulation (Gray-PG in Rosenbrock)

the distance in the bit string, and the genotypic distance of gray coding correspond to the phenotypic distance, so gray coding is useful for the evolution of solution, when the inputs of the fuzzy rule include the genotypic diversity measure.

## 5 Conclusions

In this paper, we proposed the Fuzzy Adaptive Search method for Parallel Genetic Algorithm (FASPGA) combined FASGA with a quick search ability and Island model Parallel GA with a high-quality solutions ability. By using the fuzzy reasoning to tune the genetic parameters, FASPGA displayed a good performance in Rastrigin function. However, FASPGA also displayed a bad performance in Rosenbrock function. To improve the performance of FASPGA, we proposed the Diversity Measure based Fuzzy Adaptive Search method for Parallel Genetic Algorithms (DM-FASPGA). A lots of combinations were adopted as the inputs of the fuzzy rule in this paper and the computer simulation was also excuted. From the simulation results, we confirmed that Hamming Distance (HD) and Relative Phenotypic diversity (RP) were the best combination.

About the research plan for the future, we would like to introduce another geneotypic diversity measure into the inputs of the fuzzy rule. To confirm the usefulness of the proposed method, the application of the FASPGA method in motion learning experiment of robot manipulator is also considered.

## References

1. R. Subbu and P. Bonissone, "A Retrospective View of Fuzzy Control of Evolutionary Algorithm Resources", Proc. FUZZ-IEEE 2003, pp. 143–148 (2003)
2. J.H. Holland, *Adaptation in Netural and Artifical System*, University of Mechigan Press (1992)
3. D.E. Goldberg, *Genetic Algorithms in Search, Optimization, and Machine Learning*, Addison-Wesley Publishing Company (1989)
4. H.Y. Xu and G. Vukovich, "A Fuzzy Genetic Algorithm with Effective Search and Optimization," Int'l J. Conf. on Neural Networks (IJCNN'93), pp. 2967–2970 (1993)
5. M.A. Lee and H. Takagi, "Dynamic Control of Genetic Algorithms using Fuzzy Logic Techniques," Proc. of 5th International Conference on Genetic Algorithms (ICGA'93), pp. 76–83 (1993)
6. F. Herrera and M. Lozamo. "Adaptive Genetic Algorithms Based on Fuzzy Tecniques," In Proc. Sixth Int'l Conf. on Information Processing and Management of Uncertainty in Knowledge Based System (IPMU'96), pp. 775–780 (1996)
7. Y. Maeda, "A Method for Improving Search performance of GA with Fuzzy Rules," Proc. of the 6th Intelligent System symposium, Vol. 3, pp. 27–30 (1996) [In Japanese]
8. Y. Maeda, "Fuzzy Adaptive Search Method for Genetic Programming," International Journal of Advanced Computational Intelligence, Vol.3, No.2, pp. 131–135 (1999)
9. J. Nang, K. Matsuo, "A Survey on the Parallel Genetic Algorithms," Journal of the Society of Instrument and Control Engineering, Vol. 33, No. 6, pp. 500–509 (1994)
10. T. Starkweather, D. Whitley, K. Mathisa, "Optimization Using Distributed Genetic Algorithms," Parallel Problem Solving from Nature, Springer Verlag (1991)
11. R. Tanese, "Distributed Genetic Algorithms," Proc 3rd International Conf on Genetic Algorithms, Morgan-Kaufmann, pp. 434–439 (1989)

12. E. Cant'u-Paz, "A Survey on the Parallel Genetic Algorithms," Calculateurs Paralleles (1998)
13. H. Mühlenbein, "Evolution in Time and Space: The Parallel Genetic Algorithm," In G. Rawlins, editor, FOGA-1, pp. 316–337, Morgan Kaufmann (1991)
14. T. Hiroyasu, M. Miki, M. Negami, "Distributed Genetic Algorithms with Randomized Migration Rate," IEEE Proceedings of Systems, Man and Cybernetics Conference (SMC'99), Vol.1, pp. 689–694 (1999)
15. M. Miki, T. Hiroyasu, O. Kaneco and K. Hatanaka, "A Parallel Genetic Algorithm with Distributed Environment Scheme," IEEE Proceedings of Systems, Man and Cybernetics Conference (SMC'99), pp. 695–700 (1999)
16. Q. Li and Y. Maeda, "Adaptive Search Method for Parallel Genetic Algorithms Used Fuzzy Reasoning," The 23rd Annual Conference of the Robotics Society of Japan, 2B15 (2004)
17. Q. Li and Y. Maeda, "Parallel Genetic Algorithms with Adaptive Migration Rate Tuned by Fuzzy Reasoning," Proceedings of the Fourth International Symposium on Human and Artificial Intelligence Systems (HART 2004), pp. 259–264 (2004)
18. Y. Maeda and Q. Li, "Parallel Genetic Algorithm with Adaptive Genetic Parameters Tuned by Fuzzy Reasoning," International Journal of Innovating Computing, Information and Control, Vol.1, No.1, pp. 95–107 (2005)

# Fuzzy Controller for Robot Manipulators

Basil M. Al-Hadithi[1,2], Fernando Matía[2], and Agustín Jiménez[2]

[1] Area de tecnologías de Información y Comunicaciones, Universidad Alfonso X El Sabio, Spain
bmal@uax.es

[2] Intelligent Control Group, Universidad Politécnica de Madrid, Spain
www.disam.upm.es/control

**Abstract.** This paper deals with the design of a fuzzy (FC) based on pole assignment method for the control of a robotic system. The main idea is to design a supervisory fuzzy controller capable to adjust the controller parameters in order to obtain the desired axes positions under variationes of the robot parameters and payload variations. The behaviour of the closed loop fuzzy system controlled by the fuzzy controller is identical to the linear system whose state transition matrix is the desired one.

The main objective of this paper is the use of the affine Takagi-Sugeno (T-S) fuzzy model to represent both the controlled system and the fuzzy controller taking into consideration the effect of the constant term [1]. Most of the research works analyzed the T-S model assuming that the non-linear system is linearized with respect to the origin in each IF-THEN rule, which means that the consequent part of each rule is a linear function with zero constant term. This will in turn reduce the accuracy of approximating non-linear systems.

## 1 Introduction

In the advanced control of robotic manipulators, it is important for manipulators to track trajectories in a wide range of work place [5], [6], [7], [8]. If speed and accuracy is required, the control using conventional methods is difficult to realize because of the high nonlinearity of the robot system.

Up to now, the fuzzy controller has been the most successful application field for fuzzy logic. Many applications show that the fuzzy controllers yield results superior to those obtained by the conventional control algorithms.

In the past, the design of controllers based on a linearized model of real control systems. In many cases a good response of complex and highly non-linear real process is difficult to obtain by applying conventional control techniques which often employ linear mathematical models of the process. One reason for this lack of a satisfactory performance is the fact that linearization of a non-linear system might be valid only as an approximation to the real system around a determined operating point.

However, fuzzy controllers are basically non-linear, and effective enough to provide the desired non-linear control actions by carefully adjusting their parameters.

P. Melin et al. (Eds.): IFSA 2007, LNAI 4529, pp. 688–697, 2007.
© Springer-Verlag Berlin Heidelberg 2007

In this paper, the non-linear system is represented by affine T-S model, where the consequent part of each rule represents an affine model of the original system in a certain operating point. The final fuzzy system can be obtained by blending of these affine models. The control is carried out based on the fuzzy model via the so-called parallel distributed compensation scheme . The idea is that for each local affine model, an affine linear feedback control is designed. the resulting overall controller, which is also a non-linear one, is again a blending of each individual affine linear controller.

In this study, a design of a pole assignment fuzzy control is presented based on T-S model taking into account the effect of the constant term in both the fuzzy system and controller.

## 2 Modelling of Robot Manipulators

The robot under study is characterized by having six rotational joints driven by hydraulic actuators(motors for the first joint and the robot wrist, and cylinders for other axes).

The main problem in controlling such processes is the nonlinearity. This makes it very difficult the use of conventional control techniques to implement the control job.

In this paper, the robot which is a highly non-linear system is represented by affine T-S model, where the consequent part of each rule represents an affine model of the original system in a certain operating point. The final fuzzy system can be obtained by blending of these affine models. The control is carried out based on the fuzzy model via the so-called parallel distributed compensation scheme. The idea is that for each local affine model, an affine linear feedback control is designed. the resulting overall controller, which is also a non-linear one, is again a blending of each individual affine linear controller.

The behaviour of the robot depends upon the robot working conditions, in particular the axes positions and the payload which are considered as the premise part of the fuzzy rule [5], [6], [7], [8].

The suggested fuzzy control considers every axis as a system whose control variables has to be tuned. It is necessary to establish differences between the first axis, which implies a rotation in the horizontal plane, and the axes 2,3 and 4, which imply rotations in the vertical plane. In the case of the latter two axes, which drive the robot wrist, it is not necessary to adjust the control parameters in real time, and they are automatically adjusted when the robot payload changes. for the latter two axes, due to the short length of the driven links and the robot kinematic configuration, their angular position doesnot have a significant amount of influence on their dynamic behaviour, which is mainly determined by the payload.

We should mention that not all the robot joints will influence the dynamic behaviour. The first axis position does not influence the others.

The angular values of the vertical joints that are placed behined the joint we are considering along the robot kinematic chain, and which influence the dynamic

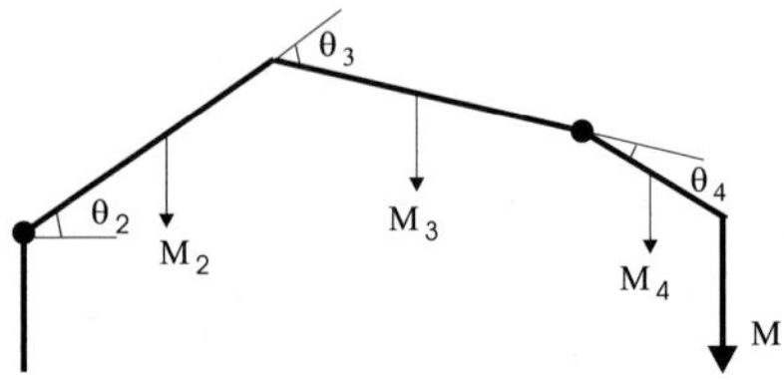


**Fig. 1.** Scheme for the fuzzy input variable for axes 2, 3 and 4

behaviour, can be combined in one fuzzy variable. Denoting the angular value for the joint j by $\theta_j$, the effective angular value $\theta_{ia}$ to be considered as a fuzzy input variable for axes 2, 3 and 4 is:

$$\theta_{ia} = \sum_{j=2}^{i} \theta_j, \quad i = 2, 3, 4$$

Similarly, considering one particular axis, the angular axis, the angular values of the vertical joints that ar placed in front of it, as well as the robot payload, can be combined in the other fuzzy input variable, namely the effective moment of inertia from the considered axis $J_i$. This can be represented as:

$$J_i = f(\theta_{j>i}, M_{j>i}, M)$$

Where

- $J_i$ represents the effective moment of inertia from axis i
- $\theta_{j>i}$ represents the angular values of the axes after i
- $M_{j>i}$ represents the mass of the link j including its actuator
- M represents the mass of payload.

Figure 1 shows the scheme for the fuzzy input variable for axes 2, 3 and 4.

## 3 Takagi and Sugeno's Fuzzy Model

The continuous fuzzy dynamic model proposed by Takagi and Sugeno is described by fuzzy IF-THEN rules [9], [10], where the consequent part is an affine input-output relation. The main feature of this model is to express the local dynamics of each fuzzy implication by an affine model. The final fuzzy system is resulted from blending of the affine linear system models.

The IF-THEN rules are as follows:

$$R^{(i_1 \cdots\cdots i_n)} : \textit{If } x \textit{ is } M_1^{i_1} \textit{ and } \dot{x} \textit{ is } M_2^{i_2} \textit{ and} \ldots \textit{ and } x^{(n-1)} \textit{ is } M_n^{i_n}$$

$$\textit{then } \dot{\mathbf{x}} = \mathbf{a}_o^{(i_1 \ldots i_n)} + \mathbf{A}^{(i_1 \ldots i_n)}\mathbf{x} + \mathbf{B}^{(i_1 \ldots i_n)}\mathbf{u} \qquad (1)$$

where $M_1^{i_1}$ $(i_1 = 1, 2, \ldots, r_1)$ are fuzzy sets for $x$, $M_2^{i_2}$ $(i_2 = 1, 2, \ldots, r_2)$ are fuzzy sets for $\dot{x}$ and $M_n^{i_n}$ $(i_n = 1, 2, \ldots, r_n)$ are fuzzy sets for $x^{(n-1)}$. $\mathbf{x}(t)$ is the

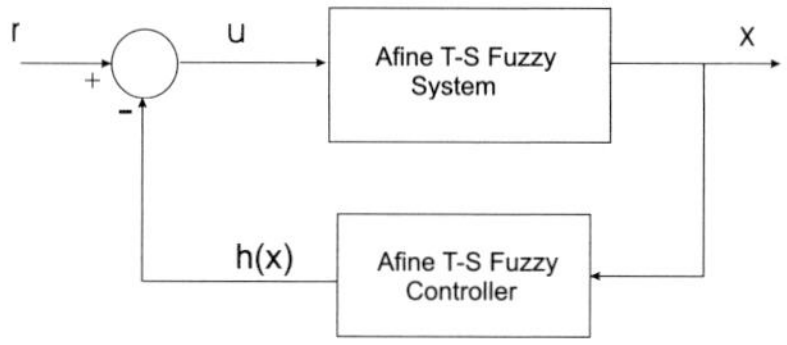


**Fig. 2.** Feedback Fuzzy control system

state vector, $\mathbf{u}(t)$ is the input vector. Therefore the whole fuzzy system consists of $r_1.r_2.\dots.r_n$ rules. The expression for $\mathbf{A}^{(i_1\dots i_n)}$ is

$$\begin{bmatrix} 0 & 1 & 0 & \dots & 0 \\ 0 & 0 & 1 & \ddots & \vdots \\ \vdots & \vdots & \ddots & \ddots & 0 \\ 0 & 0 & \dots & 0 & 1 \\ -a_1^{(i_1\dots i_n)} & -a_2^{(i_1\dots i_n)} & \dots & -a_{n-1}^{(i_1\dots i_n)} & -a_n^{(i_1\dots i_n)} \end{bmatrix}, \ \mathbf{a}_0^{(i_1\dots i_n)} = \begin{bmatrix} 0 \\ \vdots \\ 0 \\ a_0^{(i_1\dots i_n)} \end{bmatrix},$$

$$\mathbf{x}^T = \left[x \; \dot{x} \dots x^{(n-1)}\right]$$

The final output of the fuzzy system can be represented as

$$\dot{\mathbf{x}} = \mathbf{a}_0(\mathbf{x}) + \mathbf{A}(\mathbf{x})\mathbf{x} + \mathbf{B}(\mathbf{x})\mathbf{u}(\mathbf{x})$$

## 4 Fuzzy Controller Based on Pole Assignment Method

Let us consider the problem of designing a FC using the pole assignment method [1], [2]. The fuzzy model is described by T-S model [9]. Figure 2 represents the fuzzy control system.

Let the $(i_1 \dots i_n)^{th}$ rule of T-S fuzzy model be represented by:

$$S^{(i_1\dots i_n)} : \mathit{If}\ x\ \mathit{is}\ M_1^{i_1}\ \mathit{and}\ \dot{x}\ \mathit{is}\ M_2^{i_2}\ \mathit{and}\dots\ \mathit{and}\ x^{(n-1)}\ \mathit{is}\ M_n^{i_n}$$

$$\mathit{then}\ \ \dot{\mathbf{x}} = \mathbf{a}_o^{(i_1\dots i_n)} + \mathbf{A}^{(i_1\dots i_n)}\mathbf{x} + \mathbf{B}^{(i_1\dots i_n)}\mathbf{u} \qquad (2)$$

where the membership functions are given by $\mu_{M_1^{i_1}}(x)$, $\mu_{M_2^{i_2}}(\dot{x})$ and $\mu_{M_n^{i_n}}(x^{(n-1)})$, respectively as shown in figure 3.

The $(j_1 \dots j_n)^{th}$ fuzzy controller rule is:

$$C^{(j_1\dots j_n)} : \mathit{If}\ x\ \mathit{is}\ N_1^{j_1}\ \mathit{and}\ \dot{x}\ \mathit{is}\ N_2^{j_2}\ \mathit{and}\dots\ \mathit{and}\ x^{(n-1)}\ \mathit{is}\ N_n^{j_n}$$

$$\mathit{then}\ \ \mathbf{u} = \mathbf{r} - [\mathbf{k}_o^{(j_1\dots j_n)} + \mathbf{K}^{(j_1\dots j_n)}\mathbf{x}] \qquad (3)$$

where $N_1^{j_1}$ $(j_1 = 1, 2, \dots, s_1)$ are fuzzy sets for $x$, $N_2^{j_2}$ $(j_2 = 1, 2, \dots, s_2)$ are fuzzy sets for $\dot{x}$ and $N_n^{j_n}$ $(j_n = 1, 2, \dots, s_n)$ are fuzzy sets for $x^{(n-1)}$ as shown in

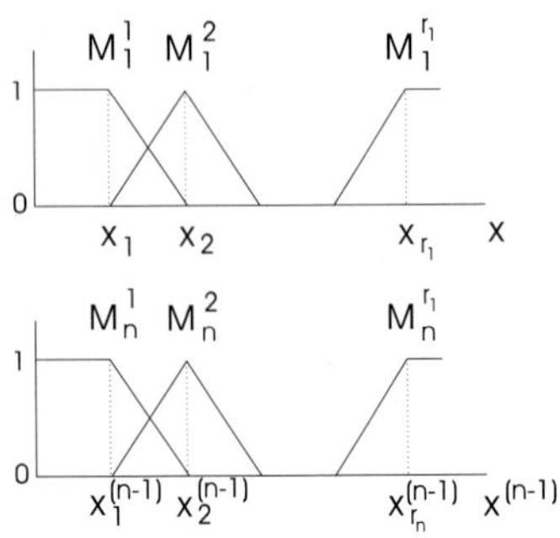

Fig. 3. Membership functions of the fuzzy system

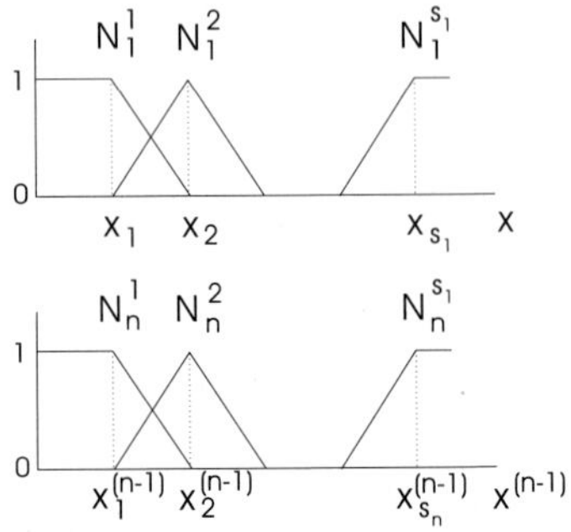

Fig. 4. Membership functions of the fuzzy controller

figure 4. $\mathbf{h}^{(j_1 \cdots j_n)} = \mathbf{k_0}^{(j_1 \cdots j_n)} + \mathbf{K}^{(j_1 \cdots j_n)}\mathbf{x}$ is $\mathbf{r}$-dimensional feedback vector, the reference input vector and $\mathbf{r}$ is r-dimensional and $\mathbf{K}^{(j_1 \cdots j_n)}$ is $r \times n$ matrix. The fuzzy controller will be

$$\mathbf{u} = \mathbf{r} - \mathbf{h}(\mathbf{x}) \tag{4}$$

Substituting (3) in (2), we get the feedback system.

$$SC^{(i_1 \ldots i_n, j_1 \ldots j_n)} : \textit{If } x \textit{ is } (M_1^{i_1} \textit{ and } N_1^{j_1}) \textit{ and } \dot{x} \textit{ is } (M_2^{i_2} \textit{ and } N_2^{j_2}) \ldots \textit{ and } x^{(n-1)}$$

$$\textit{is } (M_n^{i_n} \textit{ and } N_n^{j_n})$$

$$\textit{then } \ \dot{\mathbf{x}} = \mathbf{a_0}^{(i_1 \ldots i_n)} + \mathbf{A}^{(i_1 \ldots i_n)}\mathbf{x} + \mathbf{B}^{(i_1 \ldots i_n)}\Big(\mathbf{r} - [\mathbf{k_0}^{(j_1 \cdots j_n)} + \mathbf{K}^{(j_1 \cdots j_n)}\mathbf{x}]\Big) \tag{5}$$

Suppose that we choose $\mathbf{k_0}^{(j_1 \cdots j_n)}$ such that

$$\mathbf{a_0}^{(i_1 \ldots i_n)} - \mathbf{B}^{i_1 \ldots i_n}\mathbf{k_0}^{j_1 \cdots j_n} = 0 \tag{6}$$

then (5) will be:

$$SC^{(i_1 \ldots i_n, j_1 \ldots j_n)} : \textit{If } x \textit{ is } (M_1^{i_1} \textit{ and } N_1^{j_1}) \textit{ and } \dot{x} \textit{ is } (M_2^{i_2} \textit{ and } N_2^{j_2}) \ldots \textit{ and } x^{(n-1)}$$

$$\textit{is } (M_n^{i_n} \textit{ and } N_n^{j_n})$$

$$\textit{then } \ \dot{\mathbf{x}} = \mathbf{A}^{(i_1 \ldots i_n)}\mathbf{x} + \mathbf{B}^{(i_1 \ldots i_n)}\mathbf{r} - \mathbf{B}^{(i_1 \ldots i_n)}\mathbf{K}^{(j_1 \cdots j_n)}\mathbf{x} \tag{7}$$

Then the final output of the fuzzy closed loop system is

$$\dot{\mathbf{x}} = \frac{\sum_{i_1=1}^{r_1} \cdots \sum_{i_n=1}^{r_n} \sum_{j_1=1}^{s_1} \cdots \sum_{j_n=1}^{s_n} w^{(i_1 \cdots i_n, j_1 \cdots j_n)}(\mathbf{x})\mathbf{A}^{(i_1 \cdots i_n)}\mathbf{x}}{\sum_{i_1=1}^{r_1} \cdots \sum_{i_n=1}^{r_n} \sum_{j_1=1}^{s_1} \cdots \sum_{j_n=1}^{s_n} w^{(i_1 \cdots i_n, j_1 \cdots j_n)}(\mathbf{x})} - \frac{\sum_{i_1=1}^{r_1} \cdots \sum_{i_n=1}^{r_n} \sum_{j_1=1}^{s_1} \cdots \sum_{j_n=1}^{s_n} w^{(i_1 \cdots i_n, j_1 \cdots j_n)}(\mathbf{x})\mathbf{B}^{(i_1 \cdots i_n)}\mathbf{K}^{(j_1 \cdots j_n)}\mathbf{x}}{\sum_{i_1=1}^{r_1} \cdots \sum_{i_n=1}^{r_n} \sum_{j_1=1}^{s_1} \cdots \sum_{j_n=1}^{s_n} w^{(i_1 \cdots i_n, j_1 \cdots j_n)}(\mathbf{x})} \tag{8}$$

where the elements $(k_{i1}^{(j_1 \ldots j_n)}\ k_{i2}^{(j_1 \ldots j_n)}\ \ldots k_{in}^{(j_1 \ldots j_n)})$ of the feedback vector $\mathbf{h}^{(j_1 \ldots j_n)} = \mathbf{k_0}^{(j_1 \ldots j_n)} + \mathbf{K}^{(j_1 \ldots j_n)}\mathbf{x}$ are computed from the following matrix, where its eigen values are the desired closed loop poles:

$$\mathbf{A}^{(i_1 \ldots i_n)} - \mathbf{B}^{(i_1 \ldots i_n)}\mathbf{K}^{(j_1 \ldots j_n)} \tag{9}$$

and the constant term $\mathbf{k_0}^{(j_1 \ldots j_n)}$ can be calculated from (6)

The behaviour of the fuzzy system represented by (2) and controlled by fuzzy controller in (3) is the same as the linear system which contains the desired closed loop poles (9).

The fuzzy controller implemented in this paper is the one shown in (3) modified by introducing the condition given by (6) which can be written as,

$$\tilde{\mathbf{u}}(\mathbf{x}) = \mathbf{r} - \frac{\sum_{j_1=1}^{s_1} \cdots \sum_{j_n=1}^{s_n} w^{(j_1 \ldots j_n)}(\mathbf{x})\mathbf{K}^{(j_1 \ldots j_n)}\mathbf{x}}{\sum_{j_1=1}^{s_1} \cdots \sum_{j_n=1}^{s_n} w^{(j_1 \ldots j_n)}(\mathbf{x})} \tag{10}$$

with

$$-\mathbf{K}^{(j_1 \ldots j_n)}\mathbf{x} = -\left[k_{i1}^{(j_1 \ldots j_n)} \ldots k_{in}^{(j_1 \ldots j_n)}\right] \begin{bmatrix} x \\ \dot{x} \\ \vdots \\ x^{(n-1)} \end{bmatrix} \tag{11}$$

It can be also written in another form,

$$\tilde{\mathbf{u}}^{(j_1 \ldots j_n)}(\mathbf{x}) = \begin{bmatrix} \tilde{u}_1^{(j_1 \ldots j_n)}(\mathbf{x}) \\ \vdots \\ \tilde{u}_r^{(j_1 \ldots j_n)}(\mathbf{x}) \end{bmatrix} = \mathbf{r} - \mathbf{K}^{(j_1 \ldots j_n)}\mathbf{x}$$

$$\tilde{u}_i^{(j_1 \ldots j_n)}(\mathbf{x}) = r - \left[k_{i1}^{(j_1 \ldots j_n)} \ldots k_{in}^{(j_1 \ldots j_n)}\right] \begin{bmatrix} x \\ \dot{x} \\ \vdots \\ x^{(n-1)} \end{bmatrix} \tag{12}$$

## 5 Fuzzy Control of Robotic System

The definition of the fuzzy input is shown in table 1. The input fuzzy variable which represent the axis position is linearized in three operating points. The moment of inertia is linearized in three operating points [4]. The results were obtained from several tens of experiments of the real system [3]. The system has been approximated in each operating point by a linearized mathematical model looking for a suitable model that coincide with the non-linear system. A fuzzy controller is designed which meets the requirements of small overshoot in the transient response and a well damped oscilations with no steady state error.

Firstly, a T-S affine fuzzy model is used to represent the controlled process. The applied T-S model used to represent the robotic system has been modified

**Table 1.** Input fuzzy variables

| Variable | Universe | Label |
|---|---|---|
| $\theta_{1a}$ | $[-180°, 180°]$ | $\{M\}$ |
| $\theta_{2a}$ | $[0°, 115°]$ | $\{S\ M\ B\}$ |
| $\theta_{3a}$ | $[-120°, 90°]$ | $\{S\ M\ B\}$ |
| $\theta_{4a}$ | $[-240°, 90°]$ | $\{S\ M\ B\}$ |
| $J_1$ | [7392, 56729] | $\{S\ M\ B\}$ |
| $J_2$ | [5000, 51540] | $\{S\ M\ B\}$ |
| $J_3$ | [1500, 18564] | $\{S\ M\ B\}$ |
| $J_4$ | [140, 5093] | $\{S\ M\ B\}$ |

in its premise part with respect to the original one explained in section 3. This modification has been made due to the nature of the robotic system. As it will be seen that the premise part of each rule of the T-S model of the robot depends on both the axis position and the moment of inertia, instead of a certain variable and its derivatives as the original one. The consequent part is an affine model which represents the linearized model obtained from the identification of the nonlinear system in a certain operating point, which coincides with the consequent part of th original T-S model. In the original T-S model as in the modified one, the consequent part of each rule depends on the weight of the membership functions of the premise part. In order to simplify the simulation, it is assumed that the membership functions of the fuzzy controller are the same as those of the fuzzy system.

$$R^{11} : If \ \left(\theta_{ia} \ is \ M_1^1\right) \ and \ ( \ J_i \ is \ M_2^1 \ ) \ then$$

$$\ddot{\theta}(t) = -77.4\dot{\theta}(t) - 3947.5\theta(t) + 6615000u(t)$$

$$C^{11} : If \ \left(\theta_{ia} \ is \ M_1^1\right) \ and( \ J_i \ is \ M_2^1 \ ) \ then$$

$$u(t) = r - [k_o^1 + \mathbf{K}^1\theta(t) + \mathbf{K}^2\dot{\theta}(t)]$$

$$R^{12} : If \ \left(\theta_{ia} \ is \ M_1^1\right) \ and \ ( \ J_i \ is \ M_2^2 \ ) \ then$$

$$\ddot{\theta}(t) = -43.8\dot{\theta}(t) - 3276.4\theta(t) + 4839100u(t)$$

$$C^{12} : If \ \left(\theta_{ia} \ is \ M_1^1\right) \ and( \ J_i \ is \ M_2^2 \ ) \ then$$

$$u(t) = r - [k_o^1 + \mathbf{K}^1\theta(t) + \mathbf{K}^2\dot{\theta}(t)]$$

$$R^{13} : If \ \left(\theta_{ia} \ is \ M_1^1\right) \ and \ ( \ J_i \ is \ M_2^3 \ ) \ then$$

$$\ddot{\theta}(t) = -49.2\dot{\theta}(t) - 1754.5\theta(t) + 2496452u(t)$$

$$C^{13} : If \ \left(\theta_{ia} \ is \ M_1^1\right) \ and( \ J_i \ is \ M_2^3 \ ) \ then$$

$$u(t) = r - [k_o^1 + \mathbf{K}^1\theta(t) + \mathbf{K}^2\dot{\theta}(t)]$$

$$R^{21} : If \ \left(\theta_{ia} \ is \ M_1^2\right) \ and \ ( \ J_i \ is \ M_2^1 \ ) \ then$$

$$\ddot{\theta}(t) = -74.4\dot{\theta}(t) - 3452.4\theta(t) + 5952500u(t)$$

$$C^{21} : If \ \left(\theta_{ia} \ is \ M_1^2\right) \ and( \ J_i \ is \ M_2^1 \ ) \ then$$

$$u(t) = r - [k_o^1 + \mathbf{K}^1\theta(t) + \mathbf{K}^2\dot{\theta}(t)]$$

$$R^{22} : If \ \left(\theta_{ia} \ is \ M_1^2\right) \ and \ ( \ J_i \ is \ M_2^2 \ ) \ then$$

$$\ddot{\theta}(t) = -41.7\dot{\theta}(t) - 3007.6\theta(t) + 1.65 + 4590700u(t)$$

$$C^{22} : If \ \left(\theta_{ia} \ is \ M_1^2\right) \ and( \ J_i \ is \ M_2^2 \ ) \ then$$

$$u(t) = r - [k_o^1 + \mathbf{K}^1\theta(t) + \mathbf{K}^2\dot{\theta}(t)]$$

$$R^{23} : If \ \left(\theta_{ia} \ is \ M_1^2\right) \ and \ ( \ J_i \ is \ M_2^3 \ ) \ then$$

$$\ddot{\theta}(t) = -51.1\dot{\theta}(t) - 1832.8\theta(t) + 3.3 + 2647176u(t)$$

$$C^{23} : If \ \left(\theta_{ia} \ is \ M_1^2\right) \ and( \ J_i \ is \ M_2^3 \ ) \ then$$

$$u(t) = r - [k_o^1 + \mathbf{K}^1\theta(t) + \mathbf{K}^2\dot{\theta}(t)]$$

$$R^{31} : If \ \left(\theta_{ia} \ is \ M_1^3\right) \ and \ ( \ J_i \ is \ M_2^1 \ ) \ then$$

$$\ddot{\theta}(t) = -74.1\dot{\theta}(t) - 3540.3\theta(t) + 6399500u(t)$$

$$C^{31} : If \ \left(\theta_{ia} \ is \ M_1^3\right) \ and( \ J_i \ is \ M_2^1 \ ) \ then$$

$$u(t) = r - [k_o^1 + \mathbf{K}^1\theta(t) + \mathbf{K}^2\dot{\theta}(t)]$$

$$R^{32} : If \ \left(\theta_{ia} \ is \ M_1^3\right) \ and \ ( \ J_i \ is \ M_2^2 \ ) \ then$$

$$\ddot{\theta}(t) = -33.4\dot{\theta}(t) - 2379\theta(t) + 11.71 + 3964700u(t)$$

$$C^{32} : If \ \left(\theta_{ia} \ is \ M_1^3\right) \ and( \ J_i \ is \ M_2^2 \ ) \ then$$

$$u(t) = r - [k_o^1 + \mathbf{K}^1\theta(t) + \mathbf{K}^2\dot{\theta}(t)]$$

$$R^{33} : If \ \left(\theta_{ia} \ is \ M_1^3\right) \ and \ ( \ J_i \ is \ M_2^3 \ ) \ then$$

$$\ddot{\theta}(t) = -50.7\dot{\theta}(t) - 1777.6\theta(t) + 23.43 + 2813090u(t)$$

$$C^{33} : If \ \left(\theta_{ia} \ is \ M_1^3\right) \ and( \ J \ is \ M_2^3 \ ) \ then$$

$$u(t) = r - [k_o^1 + \mathbf{K}^1\theta(t) + \mathbf{K}^2\dot{\theta}(t)]$$

Using the pole asignment design method explained above and by choosing the desired closed loop poles, the feedback controller parameters given by the feedback vector $\mathbf{K}^{(j_1 \cdots j_n)}$ can be calcualted from (9) and the constant parameters $\mathbf{k_0}^{(j_1 \cdots j_n)}$ can be determined from (6). Thus the controller parameters are:

$$k_0^1 = 0, \quad k_{11}^1 = 0.013, \quad k_{11}^2 = 0.015$$

$$k_0^1 = 0, \quad k_{12}^1 = 0.16, \quad k_{12}^2 = 0.019$$

$$k_0^1 = 0, \quad k_{13}^1 = 0.13, \quad k_{13}^2 = 0.018$$

$$k_0^1 = 0, \quad k_{21}^1 = 0.22, \quad k_{21}^2 = 0.03$$

$$k_0^1 = .00054, \quad k_{22}^1 = 0.33, \quad k_{22}^2 = 0.03$$

$$k_0^1 = 0.0018, \quad k_{23}^1 = 0.68, \quad k_{23}^2 = 0.04$$
$$k_0^1 = 0, \quad k_{31}^1 = 1.28, \quad k_{31}^2 = 0.05$$
$$k_0^1 = 0.0049, \quad k_{32}^1 = 1.18, \quad k_{32}^2 = 0.048$$
$$k_0^1 = 0.0132, \quad k_{33}^1 = 1.25, \quad k_{33}^2 = 0.05$$

The initial condition applied are: $\theta_{20} = 69°$, $\theta_3 = -60°$, $\theta_4 = -10°$, $J_{20} = 13.3\ Kgm^2$. The transient response of axis 2 using the proposed fuzzy controller is shown in figure 5.

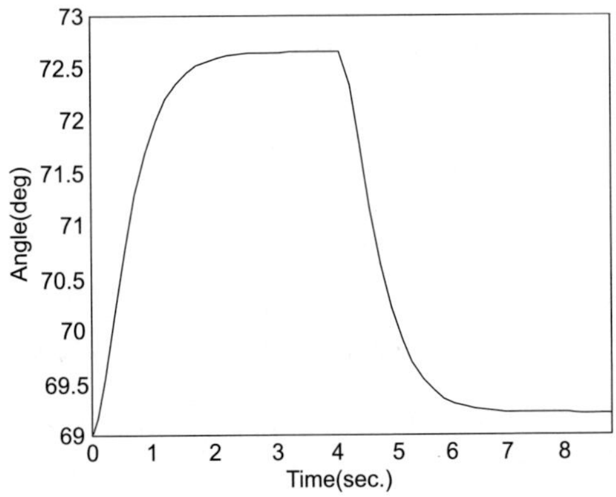


**Fig. 5.** Transient Response of of axis 2 using the proposed fuzzy controller

## 6 Conclusion

The paper presented the design of a fuzzy controller for the control of a robotic system represented by the affine T-S fuzzy model. The robot system has been linearized in various operating points in each fuzzy rule. A fuzzy controller has been designed based on pole assignment method. The main advantage of using the pole assignment method is that behaviour of the closed loop fuzzy system controlled by the suggested fuzzy controller is identical to the linear system whose closed loop poles are the desired ones and that the fuzzy controller designed with the pole assignment method guarantees the stability of the controlled system.

The results obtained in this paper have shown that the proposed controller is capale to deal with the nonlinearities of the robot and the changing of its parameters.

## Acknowledgements

This work is funded by Spanish Ministry of Science and Technology (ROBINT project DPI-2004-07907-C02).

## References


1. Al-Hadithi, B. M. : Global Stability Analysis and Design of Fuzzy Control Systems. Universidad Politcnica de Madrid, Doctorate thesis (in spanish)(2002).
2. Freund, E.: Fast Nonlinear Control with Arbitrary Pole-placement for Industrial Robots and Manipulators. Robotics Research. pages 65–78 **1** (1982).
3. Gamboa, E.,: Control de robots con accionamiento hidráulicos. Aplicación a robots de grandes dimensiones para la construcción. Tesis doctoral. Universidad Politécnica de Madrid. (1996)
4. Ishikawa, T.: A study on fuzzy control of an arm robot. B. E. thesis, Dept. Contr. Eng., Tokyo Inst. Technol(1988).
5. Llama, M. A.: Rafael Kelly, Santibaez V.: A stable motion control system for manipulators via fuzzy self-tuning. In *Fuzzy Sets ans Systems*, volume 124, pages 133–154, 2001.
6. Moudgad, V. G., Kwong, W. A., Passino, K. M., Yurkovich S.: Fuzzy learning control for a flexible link robot. IEEE Trans. Fuzzy Syst. pages 199–210 **3** (1995).
7. Purwar, S., Kar I. N., Jha A. N.: Adaptive control of robot manipulators using fuzzy logic systems under actuator constraints. In *Fuzzy Sets ans Systems*, volume 152, pages 651–664, 2005.
8. Song, Z., Yi J., Zhao D., Li X.: A computed torque controller for uncertain robotic manipulator systems: Fuzzy approach. In *Fuzzy Sets ans Systems*, volume 154, pages 208–226, 2006.
9. Takagi, T., Sugeno M.: Fuzzy Identification of Systems and Its Applications to Modeling and Control. In *IEEE Transactions on Systems, Man and Cybernetics*, volume SMC-15, pages 116–132, 1985.
10. Yoneyama, J., Nishikawa M., Katayama H.: Akira Ichikawa. Design of output feedback controllers for Takagi-Sugeno fuzzy systems. In *Fuzzy Sets ans Systems*, volume 121, pages 127–148, 2001.

# Collaboration Between Hyperheuristics to Solve Strip-Packing Problems*

Pablo Garrido and María Cristina Riff

Universidad Federico Santa María, Departamento de Informática
Av. España No. 1680, Valparaíso, Chile
{pgarrido,mcriff}@inf.utfsm.cl

**Abstract.** In this paper we introduce a collaboration framework for hyperheuristics to solve hard strip packing problems. We have designed a genetic based hyperheuristic to cooperate with a hill-climbing based hyperheuristic. Both of them use the most recently proposed low-level heuristics in the literature. REVAC, which has recently been proposed for tuning [18], has been used to find the best operators parameter values. The results obtained are very encouraging and have improved the results from both the single heuristics and the single hyperheuristics' tests. Thus, we conclude that the collaboration among hyperheuristics is a good way to solve hard strip packing problems.

**Keywords:** Hyperheuristic, Strip Packing, Heuristic Search, Metaheuristics, Parameter Control.

## 1 Introduction

In this paper we focus our attention on methods to solve the two-dimensional strip packing problem, where a set of rectangles (objects) must be positioned on a container (a rectangular space area). This container has a fixed width dimension and a variable height size. The goal is, when possible, to introduce all the objects into the container without overlapping, using a minimum height dimension for the container. Many approaches have been proposed in the literature. In our understanding a more complete revision has been presented in E. Hopper's Thesis [10]. However, in the last few years the interest in this subject has increased, as has the interest in the number of research papers presenting new approaches and improvements to the existing strategies. These approaches are, in general, single heuristics or heuristics incorporated into metaheuristics methods. Recently, the concept of hyperheuristics has been introduced and tested successfully in different problems, [6]. The key idea is to tackle problems using various low-level heuristics and develop a framework that controls the applications of the heuristics. Using this framework, the time consuming task of designing an algorithm with special components for a specific algorithm is reduced. This kind of approach is useful to obtain a good solution for a problem in a reasonable

* Partially Supported by the Fondecyt Project 106377.

P. Melin et al. (Eds.): IFSA 2007, LNAI 4529, pp. 698–707, 2007.
© Springer-Verlag Berlin Heidelberg 2007

amount of time. It emphasizes a trade-off between the quality of the solution and the invested time for designing the algorithm. Our goal is to show that the cooperation between hyperheuristics can be applied to solve Strip Packing Problems providing effective solutions in a time efficient manner. Our collaborative approach is compared to other approaches using well known benchmarks. This paper is organized as follows: First we present an overview of methods based on heuristics to solve the strip packing problem, which are included in our hyperheuristics approach. Next we introduce our framework. We will then present the results obtained using the benchmarks. Finally, our conclusions and future trends in this research area are presented.

## 2 Heuristics Based Methods

In this section, we present a revision of the most recently published heuristics for strip packing problems. Lesh et al. in [14], [15] concentrate their research on improving BLD heuristic. This heuristic is inspired in Baker's work [3]. He introduced Bottom-Left heuristics, which orders the objects according to their area. The objects are then located on the most bottom left coordinate possible. Hopper [11] presented BLD which is an improved strategy of BL, where the objects are ordered using various criteria (height, width, perimeter, area) and the algorithm selects the best result obtained. Lesh. et al's. call their new heuristics $BLD^*$. In $BLD^*$ the objects are randomly ordered according to the Kendall-tau distance from all of the possible fixed orders. This strategy is called Bubble Search, [15] and can be applied to any constructive algorithm as a GRASP algorithm in order to do dynamic ordering. The key of the algorithm is the order of the objects to be placed. In their approach they also include the rotation capability. The results reported indicate that the top-right corner is the most suitable decision, and the most effective order is from their minimal length. Finally, Lesh et al. concluded that the method using Bubble search is the best for solving the most known benchmarks, including Hopper benchmarks. However, in 2006 Bortfeldt [4] introduced a Genetic Algorithm called SPGAL and claimed that it obtained the best results known in the literature. The algorithm generates an initial population using a $BFDH^*$ heuristic which is an improvement on the BFDH heuristic initially proposed in [17]. This heuristic works as follows: The objects are oriented such that their width is no lower than their height, and they are ordered from highest to lowest. Each object is packed in a rectangular sub-area of the container in the bottom left corner. The width of the sub-area is given by the container, and the height is given by the first object packed in this sub-area. In some cases when it is possible to include the current object to be placed on some sub-areas, it is positioned in the sub-area having the least available area. In other cases the algorithm opens a new sub-area above the existing sub-areas positioning the current object in the bottom left corner as the first object of this sub-area. As we mentioned before $BFDH^*$ seeks to improve this heuristic by doing the following: It allows object rotations, so that when the algorithm searches to include the current object into a sub-area it tests both orientations and selects the best. Prior to creating a new sub-area the algorithm searches the holes produced to the right of the sub-areas, dividing the available holes on guillotinable holes. It then tries to include the larger object in the hole farthest left of the available area.

Zhang et al. [20] propose the heuristic $HR$, introducing a recursive algorithm which locates the objects on the bottom left corner. When the first object is positioned in the container it identifies the two remaining areas. It recursively continues placing objects from the largest area to the smallest area. The algorithm gives priority to the objects with largest areas. The authors claim that their algorithm quickly obtains the best results on Hopper's benchmarks.

It seems that the key idea is to find a good order of the objects for any positioning heuristic. In [19] they present a genetic algorithm and a simulated annealing algorithm, both of which try to find the best order for the objects to be placed in the container using the BLF strategy. For our hyperheuristics we have selected $HR$, $BF$, $BLF$, $BFDH^*$ as the low-level heuristics, because they are shown to be individually competitive. However, some small adaptations are required for the heuristics designed for guillotinable problems.

## 3 Collaborative Framework

From the analysis of the four low-level heuristics we can remark the following:

- Performance changes according to the order of the list of the objects, their rotation, and their location (i.e. right or left on the floor).
- The data structure required to obtain a good implementation code is not always the same for all of these heuristics.

Taking into account these remarks we have designed two hyperheuristic approaches which allow us to include a good individual implementation for each heuristic when considering them as black boxes. They communicate following a protocol for both interchanging and cooperation of the current state of the search. Our representation for both hyperheuristic approaches includes the following components: Heuristic $H$, Number of objects to be placed using $H$, $n_H$. The type of ordering of the list of the $n_H$ objects assigned, and finally if $H$ must consider the objects rotated or no. In this paper, we are interested in evaluating a cooperation between a Hill-climbing based hyperheuristic (H-SP) and a genetic inspired hyperheuristic (G-SP). In the following section we briefly describe H-SP.

### 3.1 The Hill-Climbing Based Hyperheuristic: H-SP

In a previous research, we proposed H-SP [2] which is a hyperheuristic based on a Hill-climbing method that has shown encouraging results. In H-SP an operation is accepted only if it can obtain a new pre-solution equal to or better than the current one. The moves that we have defined are: Add, Delete and Replace Heuristics. In the beginning, all of them have the same probability value to be applied. The hyperheuristic initializes its representation considering each heuristic once. It assigns $n_H = \frac{n}{4}$ number of objects to be placed for each heuristic, where $n$ is the total number of objects to be placed. Both, the order of the list and the rotation option are set using the best known strategies reported for the heuristic. It is a very simple hyperheuristic implementation that was able to find quite good solutions quickly.

### 3.2 The Genetic Inspired Hyperheuristic: G-SP

Here we propose a new hyperheuristic which is based on genetic algorithms. There is some genetic inspired hyperheuristic in the literature to solve combinatorial problems, [8], [7], [9]. However, in most of the cases, they use a representation that just corresponds to a simple sequence of low-level heuristics to be applied.

**Representation.** In our approach, we have defined a representation that is able to manage and to exploit more information. We have divided the low-level heuristics according to their functionality. Thus, we distinguish among greedy, ordering and rotation heuristics. This kind of representation allows the algorithm to have a wider combination between low-level heuristics. The chromosome has also included the number of objects to be positioned using each low-level heuristics combination. The chromosome structure is shown in figure 1. In this chromosome we can identify that the algorithm must use the first greedy heuristic using the second ordering heuristic applying the fourth rotation heuristic to locate the first five objects. Note that the chromosome does not have a fixed size.

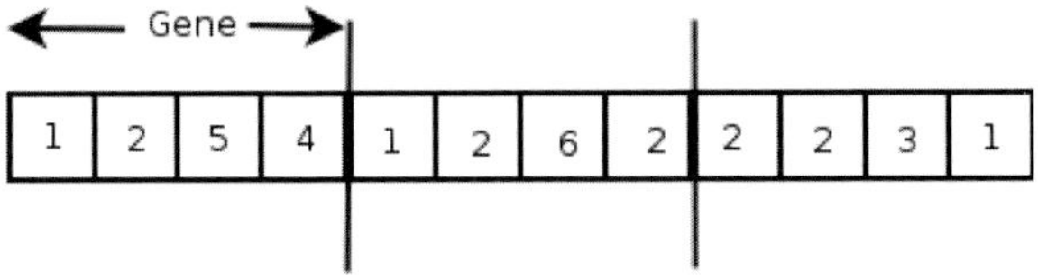


**Fig. 1.** Chromosome Structure

**Specialized Genetic Operators.** The algorithm has four operators. One recombination operator and three mutation-like operators.

- Recombination Operator: Cross-OP In our approach the recombination operator is a one-point crossover. The cross-point is selected such that a cut inside on the gen single structure is forbidden. It takes two parents to generate two offsprings. This operator must do a post-crossing procedure in order to respect the number of objects to be placed by each individual. Either the lack of or the excess of the number of objects is distributed evenly among the genes in the chromosome structure. The goal of this operator, in our approach, is to explore the search space of the low-level heuristics.
- Asexual Operators: Each operator has a special rôle.
  - Add-OP: The algorithm randomly selects a heuristic from the representation of a selected chromosome. A new heuristic is included after the selected heuristic. The new heuristic is required to position $n_1$ of the

objects previously assigned to the selected heuristic and it is randomly selected. The key idea of this operation is to include new heuristics in a different step of the algorithm in order to obtain better cooperation among them.

- Delete-OP: The algorithm randomly selects a heuristic from the selected chromosome. The heuristic is then deleted and the number of objects previously assigned to be located by it are added to the objects of the previous heuristic. The idea here is to allow the algorithm to discard some heuristics that are not helping the algorithm obtain better results.
- Replace-OP: The algorithm randomly selects both a heuristic to be replaced and the heuristic to be included. The new heuristic included inherits the number of objects to be placed. The other components of its representation are randomly generated. The idea of this operation is to give more exploration capability to the algorithm.

**Evaluation Function and Selection.** Our approach uses the traditional fitness function for strip-packing [11], which is to minimize the container's height used. It is supposed that the container's width is fixed. A minimization Roulette Wheel selection is implemented in order to increase the probability of choosing an individual with low height values.

**Parameters Values.** Each operator has its own probability to be applied. In order to find the better values for the operator's probability values, we have implemented the very recently proposed approach [18], named REVAC, for tuning. REVAC is a genetic algorithm that uses some statistical properties to determine the better parameter values and also to discard some genetic operators which, with a statistical significance, do not really improve the algorithm to be tuned. It is based on the shannon entropy to measure the diversity of solutions. The method has shown to be effective, but it is a time consuming task for finding the better values by evaluating many runs of the all the problem instances. For this, we have selected the most difficult instances that really seem to require the investment in this additional computational effort. The operator's parameter values are reported in the Tests section.

### 3.3 Framework H-SP/G-SP

Because H-SP is able to quickly find quite good solutions, our cooperation is implemented in the initial population generation for G-SP. G-SP receives a set of chromosomes from the pre-solutions found by H-SP. In our approach, both algorithms use the same low-level heuristics, thus the chromosome migration is easy to be integrated in G-SP. This kind of cooperation maintains both, the properties and the concepts of the hyperheuristics, as it is easy to implement and does not require specific hard domain knowledge. Our goal is to evaluate if we can improve the results obtained by their single versions using this framework.

# 4 Tests

We have done two kinds of tests. The first one is to compare the results obtained using single low-level heuristics with our hyperheuristics approaches. We report the quality of the solution found and the percentage used of each single low-level heuristics in the hyperheuristics. The second test compares H-SP, G-SP and H-SP/G-SP with the better reported results from the state of the art strip-packing. Both tests use the Hopper's instances [11] for problems $C_1, \ldots, C_7$ as benchmarks.

## 4.1 Hardware

The hardware platform for the experiments was a PC Pentium IV Dual Core, 3.4Ghz with 512 MB RAM under the Mandriva 2006 operating system. The algorithm has been implemented in C++.

## 4.2 Comparison with Low-Level Heuristics

First of all we have used REVAC to obtain the better parameter values for our approaches. As we mentioned before, REVAC is a time consuming algorithm. To determine our parameter values, we have selected the six hardest instances from the 21 problem instances. The running time for instance has been fixed to 30 seconds (3 minutes for each instance set). The number of iterations done by REVAC, as it has been recommended by the authors, was 1000 iterations. Thus, the calibration required was around 48 hours. After tuning, REVAC has determined the following parameter values for each operator probability: 0.640, 0.803, 0.722, 0.493, for Cross-OP, Add-OP, Delete-OP, Replace-OP, respectively. According to these results, we can conclude that the four operators are significant for G-SP. Note that the lower probability values are for the operators which have made more exploration of the search space: Cross-OP and Replace-OP, as we can expect.

In order to obtain significant results, each hyperheuristic has been executed 10 times for each problem category with various initial populations. We limit the running time to 60 seconds. In H-SP/G-SP, these 60 seconds are distributed over 20 seconds for H-SP to generate the 10 individuals of the population, and the remaining 40 seconds for G-SP to obtain a near-optimal solution.

**Gap to the Solution:** Table 1 shows the percentage from both the optimal solution to the best solution found (gap % $= \frac{(best_{found} - opt)}{opt}$) and the average for each single heuristic and for the hyperheuristics H-SP and G-SP.

The quality of the solution found by the single heuristic has been strongly improved using our framework. Furthermore, the hyperheuristic allows both a division of the task and cooperation among the heuristics for positioning of the objects.

**Low-level Heuristics Runs:** In table 2, we report the percentage of the number of times that each heuristic has been applied for each type of problem in our hyperheuristic collaborative framework for the best heuristics combination.

**Table 1.** Gap to the solution for: low-level heuristics, hyperheuristics and collaborative approach

| Algorithms | BLF | HR | *BFDH*$^*$ | BF | H-SP | G-SP | H-SP/G-SP |
|---|---|---|---|---|---|---|---|
| C1 | 6.6 | 6.6 | 6.6 | 5 | 0.0 | 0.0 | 0.0 |
| C2 | 13.3 | 8.8 | 8.8 | 8.8 | 3.22 | 4.00 | 2.67 |
| C3 | 11.1 | 6.6 | 6.6 | 6.6 | 2.22 | 3.89 | 1.56 |
| C4 | 4.4 | 3.8 | 3.8 | 3.3 | 1.72 | 1.94 | 1.72 |
| C5 | 2.6 | 2.6 | 2.6 | 2.6 | 1.15 | 1.63 | 1.15 |
| C6 | 3.1 | 2.7 | 2.7 | 2.5 | 1.33 | 1.86 | 1.36 |
| C7 | 2.6 | 2.6 | 2.6 | 2.2 | 1.17 | 1.67 | 1.28 |
| Average | 6.24 | 4.81 | 4.81 | 4.42 | 1.54 | 2.14 | 1.39 |

**Table 2.** Average use of low-level heuristics in the collaborative approach

| Heuristic | C1 | C2 | C3 | C4 | C5 | C6 | C7 | Average |
|---|---|---|---|---|---|---|---|---|
| BLF | 22.45 | 32.80 | 26.00 | 43.47 | 44.29 | 34.97 | 43.63 | 35.37 |
| HR | 21.84 | 25.73 | 28.13 | 16.43 | 12.74 | 14.33 | 7.38 | 18.08 |
| BFDH$^*$ | 2.08 | 21.20 | 2.65 | 1.87 | 7.60 | 1.39 | 0.64 | 5.35 |
| BF | 53.62 | 20.27 | 43.22 | 38.23 | 35.37 | 49.31 | 48.34 | 41.19 |

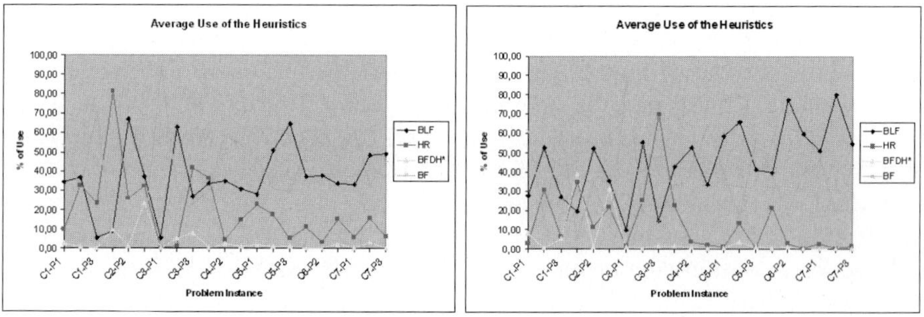


**Fig. 2.** Percentage of low-level heuristics used for the single hyperheuristics: left H-SP, right G-SP

This table can be interpreted as the number of the objects (in percentage) that each heuristic is located on the floor. We can appreciate that each problem requires a different combination of the low-level heuristics. This is the advantage of the implicit natural adaptation of the hyperheuristic framework. A more detailed comparison of the use of the low-level heuristics is shown in figure 2 and 3. Both figures show that BFDH$^*$ tends to be less applied as the size of the problem increases. While BLF shows the exact contrary behaviour. A pattern can not be identified for both BF and HR heuristics. Note however that BF has been used more frequently than HR. In addition, HR is more useful in solving smaller problem categories. Thus, the application percentage of the low-level heuristics

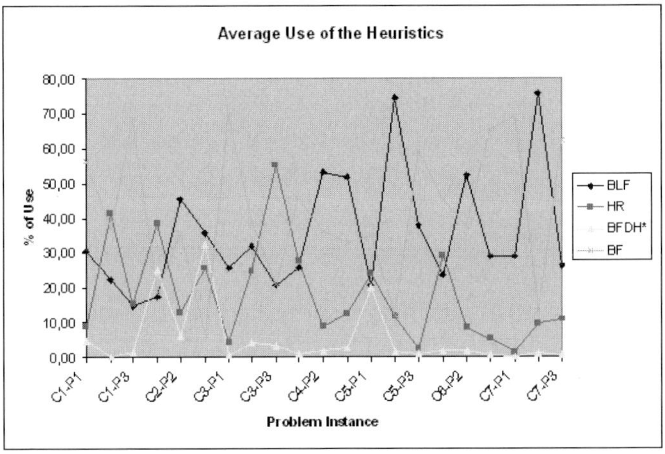


**Fig. 3.** Percentage of low-level heuristics used for the collaborative framework

**Table 3.** Gap to the solution for: state-of-the-art algorithms, H-SP, G-SP and for the collaborative approach

| | Category | | | | | | | |
|---|---|---|---|---|---|---|---|---|
| Average % of Gaps | C1 | C2 | C3 | C4 | C5 | C6 | C7 | Average |
| GA + BLF, [11] | 4 | 7 | 5 | 3 | 4 | 4 | 5 | 4.57 |
| SA + BLF, [11] | 4 | 6 | 5 | 3 | 3 | 3 | 4 | 4 |
| Iori, [12] | 1.59 | 2.08 | 2.15 | 4.75 | 3.92 | 4.00 | - | 3.98* |
| HR, [20] | 8.33 | 4.45 | 6.67 | 2.22 | 1.85 | 2.5 | 1.8 | 3.97 |
| SPGAL-R, [5] | 1.7 | 0.0 | 2.2 | 0.0 | 0.0 | 0.3 | 0.3 | 0.6 |
| SPGAL, [4] | 1.59 | 2.08 | 3.16 | 2.70 | 1.46 | 1.64 | 1.23 | 1.98 |
| BLD*, [13] | - | - | - | - | 2 | 2.4 | - | 2.2* |
| R-GRASP, [1] | 0.0 | 0.0 | 1.08 | 1.64 | 1.10 | 0.83 | 1.23 | 0.84 |
| Martello B&B, [16] | 0.0 | 0.0 | 2.15 | - | - | - | - | 0.71* |
| H-SP | 0.0 | 3.22 | 2.22 | 1.72 | 1.15 | 1.33 | 1.17 | 1.54 |
| G-SP | 0.0 | 4.00 | 3.89 | 1.94 | 1.63 | 1.86 | 1.67 | 2.14 |
| H-SP/G-SP | 0.0 | 2.67 | 1.56 | 1.72 | 1.15 | 1.36 | 1.28 | 1.39 |

depends on the problem instance to be solved. Furthermore, the algorithm is able to self-adapt to the problem at hand.

### 4.3 Comparison with State-of-the-Art Algorithms

Table 3 summarizes the best results found in the literature [12], [11], [20], [4,5], [1], [16], [13,14] and the results obtained by our single and collaborative hyperheuristics for the Hopper's instances. Results show that the hyperheuristics give good quality solutions and even better solutions than various problem-made algorithms (meta heuristics and heuristics) except for the SPGAL-R and

R-GRASP algorithms that present the best solutions. These algorithms are especially designed for these benchmarks. The above demonstrates that our approaches are very competitive ones. Note that the values for HR in this section are not the same as in the previous section. In the previous tests, we have fixed the running time to be 60 seconds. Here the results are the best reported for this technique without imposing any time constraint.

## 5 Conclusions

Our research allows us to conclude that using a collaborative hyperheuristic approach we can improve the performance of both the single heuristics and the single hyperheuristics with regard to the results obtained in the literature. This indicates that both the single and the collaborative framework are very promising approaches in solving hard strip packing problems. Both G-SP and H-SP/G-SP have the following characteristics: flexibles, cheap and easy to be implemented and at the same time are able to obtain quite good solutions. Moreover, the hyperheuristics are able to adapt themselves to the problem by selecting the best combination of the low-level heuristics. We note that the selection of suitable low-level heuristics is a major task when designing hyperheuristics. In order to obtain competitive state-of-the-art solutions, we require selecting efficient low-level heuristics. The key idea is to allow cooperation among them improving their single behaviours.

## References

1. R. Alvarez-Valdes, F. Parre no, and J.M. Tamarit. Reactive grasp for the strip packing problem. *Proceedings Metaheuristic Conference MIC*, 1, 2005.
2. I. Araya, M-C. Riff, and B. Neveu. Towards an efficient hyperheuristic for strip-packing problems. *Proceedings of the 7th EU-Meeting, Málaga, Spain*, 2006.
3. B.S. Baker, E.G. Coffman, and R.L. Rivest. Orthogonal packings in two dimensions. *SIAM Journal on Computing*, 9:846–855, 1980.
4. A. Bortfeldt. A genetic algorithm for the two-dimensional strip packing problem with rectangular pieces. *European Journal of Operational Research*, 172:814–837, 2006.
5. A. Bortfeldt and H. Gehring. New large benchmarks for the two-dimensional strip packing problem with rectangular pieces. *IEEE Proceedings of the* $39^{th}$ *Hawaii International Conference on Systems Sciences*, page 30.2, 2006.
6. E. Burke, G. Kendall, J. Newall, E. Hart, P. Ross, and S. Schulenburg. Hyper-heuristics: an emerging direction in modern search technology. *Handbook of Metaheuristics*, 16:457–474, 2003.
7. P. Cowling, G. Kendall, and L. Han. An adaptive length chromosome hyperheuristic genetic algorithm for a trainer scheduling problem. *Proceedings SEAL*, 2002.
8. P. Cowling, G. Kendall, and L. Han. An investigation of a hyperheuristic genetic algorithm applied to a trainer scheduling problem. *Proceedings CEC*, 2002.
9. L. Han and G. Kendall. Guided operators for a hyper-heuristic genetic algorithm. *Proceedings of AI-2003: Advances in Artificial Intelligence. The 16th Australian Conference on Artificial Intelligence*, pages 807–820, 2003.

10. E. Hopper. *Two-Dimensional Packing Utilising Evolutionary Algorithms and other Meta-Heuristic Methods.* PhD. Thesis Cardiff University, 2000.
11. E. Hopper and B.C.H. Turton. An empirical investigation on metaheuristic and heuristic algorithms for a 2d packing problem. *European Journal of Operational Research*, 128:34–57, 2001.
12. M. Iori, S. Martello, and M. Monaci. *Metaheuristic algorithms for the strip packing problem*, pages 159–179. Kluwer Academic Publishers, 2003.
13. N. Lesh, J. Marks, A. Mc. Mahon, and M. Mitzenmacher. Exhaustive approaches to 2d rectangular perfect packings. *Information Processing Letters*, 90:7–14, 2004.
14. N. Lesh, J. Marks, A. Mc. Mahon, and M. Mitzenmacher. New heuristic and interactive approaches to 2d rectangular strip packing. *ACM Journal of Experimental Algorithmics*, 10:1–18, 2005.
15. N. Lesh and M. Mitzenmacher. Bubble search: A simple heuristic for improving priority-based greedy algorithms. *Information Processing Letters*, 97:161–169, 2006.
16. S. Martello, M. Monaci, and D. Vigo. An exact approach to the strip-packing problem. *INFORMS Journal of Computing*, 15:310–319, 2003.
17. C. Mumford-Valenzuela, J. Vick, and P.Y. Wang. *Heuristics for large strip packing problems with guillotine patterns:An empirical study*, pages 501–522. Kluwer Academic Publishers, 2003.
18. V. Nannen and A.E. Eiben. Relevance estimation and value calibration of evolutionary algorithm parameters. *Proceedings of Joint International Conference for Artificial Intelligence (IJCAI)*, 2006.
19. A. Soke and Z. Bingul. Hybrid genetic algorithm and simulated annealing for two-dimensional non-guillotine rectangular packing problems. *Engineering Applications of Artificial Intelligence*, 19:557–567, 2006.
20. D. Zhang, Y. Kang, and A. Deng. A new heuristic recursive algorithm for the strip rectangular packing problem. *Computers and Operations Research*, 33:2209–2217, 2006.

# Part XIII

# Neural Networks and Control

# Discrete-Time Recurrent High Order Neural Observer for Induction Motors

Edgar N. Sanchez[1], Alma Y. Alanis[2], and Alexander G. Loukianov[2]

[1] CINVESTAV, Unidad Guadalajara, on sabbatical leave at CUCEI, Universidad de Guadalajara, Mexico
[2] CINVESTAV, Unidad Guadalajara, Apartado Postal 31-438, Plaza La Luna, Guadalajara, Jalisco, C.P. 45091, Mexico
sanchez@gdl.cinvestav.mx

**Abstract.** A nonlinear discrete-time neural observer for the state estimation of a discrete-time induction motor model, in presence of external and internal uncertainties is presented. The observer is based on a discrete-time recurrent high order neural network (RHONN) trained with an extended Kalman filter (EKF)-based algorithm. This observer estimates the state of the unknown discrete-time nonlinear system, using a parallel configuration. The paper also includes the stability proof on the basis of the Lyapunov approach. To illustrate the applicability simulation results are included.

## 1 Introduction

Induction motors are widely used in industrial applications due to their reliability, simpler construction and reduced cost with respect, for example, to d.c. motors. On the other hand its model is much more complicated than other machines. Moreover, uncertainties such as load torque and rotor resistance are usually unknown and may have a large degree of variations [4], [8].

Literature in induction motors control is extensive and includes field oriented controller, VSC sliding mode controller, passivity-based controllers, and more recently dynamic feedback linearization method. Most of these methods assume that all the states are available and all motor parameters are known; additionally those works were developed for continuous-time systems ([4] and references therein).

Due to the assumption of system state complete accessibility, the state estimation problem has received much attention by many authors, who have obtained many interesting results in different directions. Most of the existing results need the use of a special nonlinear transformation [9] or a linearization technique [2], [5]; such approaches can be considered as a relatively simple method to construct nonlinear observers. However, they do not consider uncertainties. Other kind of observers which have a good performance even in presence of model and disturbance uncertainties, are called robust; their design process is too complex [8].

All the approaches mentioned above need the previous knowledge of the plant dynamics. Recently, other kind of observer has emerged: neural observers [6], [8],

P. Melin et al. (Eds.): IFSA 2007, LNAI 4529, pp. 711–721, 2007.
© Springer-Verlag Berlin Heidelberg 2007

[10], [12]. For linear unknown systems, observer design has been widely investigated [12]. Instead of if nonlinear systems are considered, there are few results. Besides, most of the exiting results were developed for continuous-time systems; the nonlinear discrete-time case have not been discussed to the same degree.

In this paper, we consider a class of MIMO discrete-time nonlinear system, for which we develop a Luenberger-like observer [10]; then this observer is applied to a discrete-time induction motor model [7]. This observer is based on a recurrent high order neural network (RHONN) [13], which estimates the state vector of the unknown plant dynamics. It deals with the so-called mixed uncertainties (the presence of simultaneous external and internal uncertainties) [10]. The learning algorithm for the RHONN is based on an extended Kalman filter (EKF) ([11] and references therein). This paper also includes the respective stability analysis, on the basis of the Lyapunov approach, for the neural observer trained with the EKF.

## 2 Preliminaries

### 2.1 Stability Definitions

Through this paper, we use $k$ as the step sampling, $k \in 0 \cup \mathbb{Z}^+$ , $|\bullet|$ as the absolute value and, $\|\bullet\|$ as the Euclidian norm for vectors and as any adequate norm for matrices. Consider a MIMO nonlinear system

$$\chi(k+1) = F(\chi(k), u(k)) \tag{1}$$

where $\chi \in \Re^n$, $u \in \Re^m$ and $F \in \Re^n \times \Re^m \to \Re^n$ is nonlinear function.

**Definition 1.** *The system (1) is said to be forced, or to have input. In contrast the system described by an equation without explicit presence of an input $u$, that is*

$$\chi(k+1) = F(\chi(k))$$

*is said to be unforced. It can be obtained after selecting the input $u$ as a feedback function of the state*

$$u(k) = \xi(\chi(k)) \tag{2}$$

*Such substitution eliminates $u$:*

$$\chi(k+1) = F(\chi(k), \xi(\chi(k))) \tag{3}$$

*and yields an unforced system (3).*

**Definition 2.** *The solution of (1) – (2) is semiglobally uniformly ultimately bounded (SGUUB), if for any $\Omega$, a compact subset of $\Re^n$ and all $\chi(k_0) \in \Omega$, there exists an $\varepsilon > 0$ and a number $N(\varepsilon, \chi(k_0))$ such that $\|\chi(k)\| < \varepsilon$ for all $k \geq k_0 + N$.*

In other words, the solution of (1) is said to be SGUUB if, for any a priory given (arbitrarily large) bounded set $\Omega$ and any a priory given (arbitrarily small) set $\Omega_0$, which contains $(0,0)$ as an interior point, there exists a control (2), such that every trajectory of the closed loop system starting from $\Omega$ enters the set $\Omega_0 = \{\chi(k) \mid \|\chi(k)\| < \varepsilon\}$, in a finite time and remains in it thereafter.

*Theorem 1 [1].* Let $V(\chi(k))$ be a Lyapunov function for the discrete-time system (1), which satisfies the following properties:

$$\gamma_1(\|\chi(k)\|) \leq V(\chi(k)) \leq \gamma_2(\|\chi(k)\|)$$
$$V(\chi(k+1)) - V(\chi(k)) = \Delta V(\chi(k)) \leq -\gamma_3(\|\chi(k)\|) + \gamma_3(\zeta)$$

where $\zeta$ is a positive constant, $\gamma_1(\bullet)$ and $\gamma_2(\bullet)$ are strictly increasing functions, and $\gamma_3(\bullet)$ is a continuous, nondecreasing function. Thus if

$$\Delta V(\chi) < 0 \quad \text{for} \quad \|\chi(k)\| > \zeta$$

then $\chi(k)$ is uniformly ultimately bounded, i.e. there is a time instant $k_T$, such that $\|\chi(k)\| < \zeta, \forall\ k < k_T$.

**Definition 3.** *A subset $S \in \Re^n$ is bounded if there is $r > 0$ such that $\|\chi\| \leq r$ for all $\chi \in S$ .*

## 2.2 Discrete-Time Recurrent High Order Neural Networks

Consider the following discrete-time recurrent high order neural network (RHONN):

$$x_i(k+1) = w_i^\top z_i(x(k), u(k)), \quad i = 1, \cdots, n \tag{4}$$

where $x_i$ $(i = 1, 2, \cdots, n)$ is the state of the $i$th neuron, $L_i$ is the respective number of higer-order connections, $\{I_1, I_2, \cdots, I_{L_i}\}$ is a collection of non-ordered subsets of $\{1, 2, \cdots, n\}$, $n$ is the state dimension, $w_i$ $(i = 1, 2, \cdots, n)$ is the respective on-line adapted weight vector, and $z_i(x(k), u(k))$ is given by

$$z_i(x(k), u(k)) = \begin{bmatrix} z_{i_1} \\ z_{i_2} \\ \vdots \\ z_{i_{L_i}} \end{bmatrix} = \begin{bmatrix} \Pi_{j \in I_1} y_{i_j}^{d_{ij}(1)} \\ \Pi_{j \in I_2} y_{i_j}^{d_{ij}(2)} \\ \vdots \\ \Pi_{j \in I_{L_i}} y_{i_j}^{d_{ij}(L_i)} \end{bmatrix} \tag{5}$$

with $d_{j_i}(k)$ being a nonnegative integers, and $y_i$ is defined as follows:

$$y_i = \begin{bmatrix} y_{i_1} \\ \vdots \\ y_{i_1} \\ y_{i_{n+1}} \\ \vdots \\ y_{i_{n+m}} \end{bmatrix} = \begin{bmatrix} S(x_1) \\ \vdots \\ S(x_n) \\ u_1 \\ \vdots \\ u_m \end{bmatrix} \tag{6}$$

In (6), $u = [u_1, u_2, \ldots, u_m]^\top$ is the input vector to the neural network, and $S(\bullet)$ is defined by

$$S(x) = \frac{1}{1 + \exp(-\beta x)} \tag{7}$$

Consider the problem to approximate the general discrete-time nonlinear system (1), by the following discrete-time RHONN [13]:

$$\chi_i\left(k+1\right) = w_i^{*\top} z_i\left(x(k), u(k)\right) + \epsilon_{z_i}, \quad i = 1, \cdots, n \tag{8}$$

where $\chi_i$ is the $i$th plant state, $\epsilon_{z_i}$ is a bounded approximation error, which can be reduced by increasing the number of the adjustable weights [13]. Assume that there exists ideal weights vector $w_i^*$ such that $\|\epsilon_{z_i}\|$ can be minimized on a compact set $\Omega_{z_i} \subset \Re^{L_i}$ The ideal weight vector $w_i^*$ is an artificial quantity required for analytical purpose [13]. In general, it is assumed that this vector exists and is constant but unknown. Let us define its estimate as $w_i$ and the estimation error as

$$\widetilde{w}_i\left(k\right) = w_i^* - w_i\left(k\right) \tag{9}$$

The estimate $w_i$ is used for stability analysis which will be discussed later. Since $w_i^*$ is constant, then $\widetilde{w}_i\left(k+1\right) - \widetilde{w}_i\left(k\right) = w_i\left(k+1\right) - w_i\left(k\right)$, $\forall k \in 0 \cup \mathbb{Z}^+$.

### 2.3 The EKF Training Algorithm

It is known, that Kalman filtering (KF) estimates the state of a linear system with additive state and output white noises [2], [14]. For KF-based neural network training, the network weights become the states to be estimated. In this case the error between the neural network output and the measured plant output can be considered as additive white noise. Due to the fact that the neural network mapping is nonlinear, an EKF-type is required (see [11] and references therein). The training goal is to find the optimal weight values which minimize the prediction error. In this paper, we use a EKF-based training algorithm described by

$$\begin{aligned} w_i\left(k+1\right) &= w_i\left(k\right) + \eta_i K_i\left(k\right) e_i\left(k\right) \\ K_i\left(k\right) &= P_i\left(k\right) H_i\left(k\right) M_i\left(k\right) \\ P_i\left(k+1\right) &= P_i\left(k\right) - K_i\left(k\right) H_i^\top\left(k\right) P_i\left(k\right) + Q_i\left(k\right), \quad i = 1, \cdots, n \end{aligned} \tag{10}$$

with

$$M_i\left(k\right) = \left[R_i\left(k\right) + H_i^\top\left(k\right) P_i\left(k\right) H_i\left(k\right)\right]^{-1} \tag{11}$$

$$e_i\left(k\right) = y\left(k\right) - \widehat{y}\left(k\right) \tag{12}$$

where $e\left(k\right) \in \Re^p$ is the observation error and $P_i\left(k\right) \in \Re^{L_i \times L_i}$ is the weight estimation error covariance matrix at step $k$, $w_i \in \Re^{L_i}$ is the weight (state) vector, $L_i$ is the respective number of neural network weights, $y \in \Re^p$ is the plant output, $\widehat{y} \in \Re^p$ is the NN output, $n$ is the number of states, $K_i \in \Re^{L_i \times p}$ is the Kalman gain matrix, $Q_i \in \Re^{L_i \times L_i}$ is the NN weight estimation noise covariance matrix, $R_i \in \Re^{p \times p}$ is the error noise covariance, and $H_i \in \Re^{L_i \times p}$ is a matrix, in which each entry $(H_{ij})$ is the derivative of the $i$-th neural output with respect to $ij$-th neural network weight, $(w_{ij})$, given as follows:

$$H_{ij}(k) = \left[ \frac{\partial x_i(k)}{\partial w_{i_j}(k)} \right]^\top \tag{13}$$

where $i = 1, ..., n$ and $j = 1, ..., L_i$. Usually $P_i$ and $Q_i$ are initialized as diagonal matrices, with entries $P_i(0)$ and $Q_i(0)$, respectively. It is important to remark that $H_i(k)$, $K_i(k)$ and $P_i(k)$ for the EKF are bounded; for a detailed explanation of this fact see [14].

## 3 Discrete-Time Neural Observers

In this section, we consider to estimate the state of a discrete-time nonlinear system, which is assumed to be observable, given by

$$\begin{aligned} x(k+1) &= F(x(k), u(k)) + d(k) \\ y(k) &= Cx(k) \end{aligned} \tag{14}$$

where $x \in \Re^n$ is the state vector of the system, $u(k) \in \Re^m$ is the input vector, $y(k) \in \Re^p$ is the output vector, $C \in \Re^{p \times n}$ is a known output matrix, $d(k) \in \Re^n$ is a disturbance vector and $F(\bullet)$ is a smooth vector field and $F_i(\bullet)$ its entries; hence (14) can be rewritten as:

$$\begin{aligned} x(k) &= \begin{bmatrix} x_1(k) \dots x_i(k) \dots x_n(k) \end{bmatrix}^\top \\ d(k) &= \begin{bmatrix} d_1(k) \dots d_i(k) \dots d_n(k) \end{bmatrix}^\top \\ x_i(k+1) &= F_i(x(k), u(k)) + d_i(k) \quad , \quad i = 1, \cdots, n \\ y(k) &= Cx(k) \end{aligned} \tag{15}$$

For system (15), we propose a Luenberger neural observer (RHONO) with the following structure:

$$\begin{aligned} \widehat{x}(k) &= \begin{bmatrix} \widehat{x}_1(k) \dots \widehat{x}_i(k) \dots \widehat{x}_n(k) \end{bmatrix}^\top \\ \widehat{x}_i(k+1) &= w_i^\top z_i(\widehat{x}(k), u(k)) + L_i e(k) \\ \widehat{y}(k) &= C\widehat{x}(k), \qquad i = 1, \cdots, n \end{aligned} \tag{16}$$

with $L_i \in \Re^p$, $w_i$ and $z_i$ as in (4); the weight vectors are updated on-line with a decoupled EKF (10) – (13) and the output error is defined by

$$e(k) = y(k) - \widehat{y}(k) \tag{17}$$

and the state estimation error as:

$$\widetilde{x}(k) = x(k) - \widehat{x}(k) \tag{18}$$

Then the dynamic of (18) can be expressed as

$$\begin{aligned}\widetilde{x}_i(k+1) &= x_i(k+1) - \widehat{x}_i(k+1)\\ &= w_i^{*\top} z_i(x(k),u(k)) + \epsilon_{z_i} + d_i(k) - w_i^\top z_i(\widehat{x}(k),u(k)) - L_i e(k)\\ &= \widetilde{w}_i(k)\, z_i(x(k),u(k)) + \epsilon_{z_i} + d_i(k) - L_i e(k) \end{aligned} \tag{19}$$

Considering (16) and (17)

$$e(k) = C\widetilde{x}(k) \tag{20}$$

Then (12) can be rewritten as

$$\widetilde{x}_i(k+1) = \widetilde{w}_i(k)\, z_i(x(k),u(k)) + \epsilon'_{z_i} - L_i C\widetilde{x}(k) \tag{21}$$

with $\epsilon'_{z_i} = \epsilon_{z_i} + d_i(k)$. On the other hand the dynamics of (11) is

$$\widetilde{w}_i(k+1) = w_i^* - w_i(k+1) = \widetilde{w}_i(k) - \eta_i K_i(k)\, e(k) \tag{22}$$

Considering (10) – (22), we establish the main result as the following theorem.

*Theorem 2.* For the system (15), the RHONO (16) trained with the EKF-based algorithm (10), ensures that the output error (17) and the estimation error (18) are semiglobally uniformly ultimately bounded (SGUUB); moreover, the RHONO weights remain bounded.

**Proof:** Consider first the Lyapunov function candidate

$$V_i(k) = \widetilde{w}_i(k)\, P_i(k)\, \widetilde{w}_i(k) + \widetilde{x}_i(k)\, P_i(k)\, \widetilde{x}_i(k)$$

Then

$$\Delta V_i(k) = V(k+1) - V(k) \tag{23}$$

Using (9) and (10) in (23)

$$\begin{aligned}\Delta V_i(k) &= [\widetilde{w}_i(k) - \eta_i K_i(k)\, e(k)]^T [A_i(k)] [\widetilde{w}_i(k) - \eta_i K_i(k)\, e(k)]\\ &\quad + [f(k) - L_i C\widetilde{x}(k)]^T [A_i(k)] [f(k) - L_i C\widetilde{x}(k)]\\ &\quad - \widetilde{w}_i(k)\, P_i(k)\, \widetilde{w}_i(k) - \widetilde{x}_i(k)\, P_i(k)\, \widetilde{x}_i(k) \end{aligned} \tag{24}$$

with $A_i(k) = P_i(k) - D_i(k) + Q_i$ $D_i(k) = K_i(k)\, H_i^\top(k)\, P_i(k)$ and $f(k) = \widetilde{w}_i(k)\, z_i(x(k),u(k)) + \epsilon'_{z_i}$, (24) can be expressed as

$$\begin{aligned}\Delta V_i(k) &= \widetilde{w}_i^T(k)\, P_i(k)\, \widetilde{w}_i(k) - \widetilde{w}_i^T(k)\, [B_i(k)]\, \widetilde{w}_i(k)\\ &\quad + \eta^2 \widetilde{x}^T(k)\, C^T K^T [A_i(k)]\, K_i(k)\, C\widetilde{x}(k) + f^T(k)\, [A_i(k)]\, f(k)\\ &\quad + \widetilde{x}^T(k)\, C^T L_i^T [A_i(k)]\, L_i C\widetilde{x}(k) - \widetilde{w}_i^T(k)\, P_i(k)\, \widetilde{w}_i(k)\\ &\quad - \widetilde{x}_i^T(k)\, P_i(k)\, \widetilde{x}_i(k)\\ &\le \|\widetilde{x}(k)\|^2 \|\eta K_i C\|^2 \|A_i(k)\| - \|\widetilde{x}(k)\|^2 \|L_i C\|^2 \|A_i(k)\|\\ &\quad - \|\widetilde{x}(k)\|^2 P_i(k) - \|\widetilde{w}_i(k)\|^2 \|B_i(k)\| + \left|\epsilon'_{z_i}\right|^2 \|A_i(k)\|\\ &\quad + 2\|\widetilde{w}_i(k)\|\, \|z_i(x(k),u(k))\| \left|\epsilon'_{z_i}\right| \|A_i(k)\|\\ &\quad + \|\widetilde{w}_i(k)\|^2 \|z_i(x(k),u(k))\|^2 \|A_i(k)\| \end{aligned}$$

with $B_i(k) = D_i(k) - Q_i$

$$\Delta V_i(k) \le -\|\widetilde{x}(k)\|^2 E_i(k) - \|\widetilde{w}_i(k)\|^2 F_i(k) + \left|\epsilon'_{z_i}\right|^2 \|A_i(k)\| + 2G_i(k)$$

where

$$\begin{aligned} E_i(k) &= P_i(k) - \|\eta K_i C\|^2 \|A_i(k)\| - \|L_i C\|^2 \|A_i(k)\| \\ F_i(k) &= \|B_i(k)\| - \|z_i(x(k), u(k))\|^2 \|A_i(k)\| \\ G_i(k) &= \|w_i^* - w_{i\max}\| \|z_i(x(k), u(k))\| \left|\epsilon'_{z_i}\right| \|A_i(k)\| \end{aligned}$$

Then $\Delta V_i(k) \le 0$ when

$$\|\widetilde{x}(k)\| > \sqrt{\frac{\left|\epsilon'_{z_i}\right|^2 \|A_i(k)\| + 2G_i(k)}{E_i(k)}} \equiv \kappa_1$$

or

$$\|\widetilde{w}_i(k)\| > \sqrt{\frac{\left|\epsilon'_{z_i}\right|^2 \|A_i(k)\| + 2G_i(k)}{F_i(k)}} \equiv \kappa_2$$

Therefore the solution of (21) and (22) is stable. Hence, the estimation error and the RHONO weights are SGUUB [3]. Considering (16) and (12) it is easy too see that the output error has an algebraic relation with $\widetilde{x}(k)$ for that reason if $\widetilde{x}(k)$ is bounded $e(k)$ is bounded too.

$$\begin{aligned} e(k) &= C\widetilde{x}(k) \\ \|e(k)\| &= \|C\| \|\widetilde{x}(k)\| \end{aligned}$$

■

## 4 Induction Motor Application

### 4.1 Discrete-Time Model

The six-order discrete-time induction motor model in the stator fixed reference frame $(\alpha, \beta)$ under the assumptions of equal mutual inductances and linear magnetic circuit is given by [7]

$$\begin{aligned} \omega(k+1) &= \omega(k) + \frac{\mu}{\alpha}(1-\alpha) - \left(\frac{T}{J}\right) T_L(k) M \left(i^\beta(k)\psi^\alpha(k) - i^\alpha(k)\psi^\beta(k)\right) \\ \psi^\alpha(k+1) &= \cos(n_p\theta(k+1))\rho_1(k) - \sin(n_p\theta(k+1))\rho_2(k) \\ \psi^\beta(k+1) &= \sin(n_p\theta(k+1))\rho_1(k) + \cos(n_p\theta(k+1))\rho_2(k) \\ i^\alpha(k+1) &= \varphi^\alpha(k) + \frac{T}{\sigma}u^\alpha(k) \\ i^\beta(k+1) &= \varphi^\beta(k) + \frac{T}{\sigma}u^\beta(k) \\ \theta(k+1) &= \theta(k) + \omega(k)T + \frac{\mu}{\alpha}\left[T - \frac{(1-a)}{\alpha}\right] \\ &\quad \times M\left(i^\beta(k)\psi^\alpha(k) - i^\alpha(k)\psi^\beta(k)\right) - \frac{T_L(k)}{J}T^2 \end{aligned} \qquad (25)$$

with

$$\begin{aligned}\rho_1(k) &= a\left(\cos\left(n_p\theta(k)\right)\psi^\alpha(k) + \sin\left(n_p\theta(k)\right)\psi^\beta(k)\right)\\ &\quad + b\left(\cos\left(n_p\theta(k)\right)i^\alpha(k) + \sin\left(n_p\theta(k)\right)i^\beta(k)\right)\\ \rho_2(k) &= a\left(\cos\left(n_p\theta(k)\right)\psi^\alpha(k) - \sin\left(n_p\theta(k)\right)\psi^\beta(k)\right)\\ &\quad + b\left(\cos\left(n_p\theta(k)\right)i^\alpha(k) - \sin\left(n_p\theta(k)\right)i^\beta(k)\right)\end{aligned}$$

$$\begin{aligned}\varphi^\alpha(k) &= i^\alpha(k) + \alpha\beta T\psi^\alpha(k) + n_p\beta T\omega(k)\psi^\alpha(k) - \gamma T i^\alpha(k)\\ \varphi^\beta(k) &= i^\beta(k) + \alpha\beta T\psi^\beta(k) + n_p\beta T\omega(k)\psi^\beta(k) - \gamma T i^\beta(k)\end{aligned}$$

where $b = (1-a)M$, $\alpha = \frac{R_r}{L_r}$, $\gamma = \frac{M^2 R_r}{\sigma L_r^2} + \frac{R_s}{\sigma}$, $\sigma = L_s - \frac{M^2}{L_r}$, $\beta = \frac{M}{\sigma L_r}$, $a = e^{-\alpha T}$, $\mu = \frac{M n_p}{J L_r}$ with $Ls$, $L_r$ and $M$ are the stator, rotor and mutual inductance, respectively; $R_s$ and $R_r$ are the stator and rotor resistances, respectively; $n_p$ is the number of pole pairs;$\omega$ represents the angular speed; $i^\alpha$ and $i^\beta$ represent the currents in the $\alpha$ and $\beta$ phases, respectively; $\psi^\alpha$ and $\psi^\beta$ represent the fluxes in the $\alpha$ and $\beta$ phases, respectively and $\theta$ is the rotor angular displacement.

### 4.2 Simulation Results

In this section we apply the above developed scheme to estimate the state of a three-phase induction motor (25). Simulations are performed for the system (25), using the following parameters: $R_s = 14\Omega$; $L_s = 400mH$; $M = 377mH$; $R_r = 10.1\Omega$; $L_r = 412.8mH$; $n_p = 2$; $J = 0.01Kgm^2$; $T = 0.001s$. To estimate the state of system (25) we use the RHONO (16) with $n = 6$ trained with the EKF (10).

$$\begin{aligned}\widehat{x}_1(k+1) &= w_{11}(k)S(\widehat{x}_1(k)) + w_{12}(k)S(\widehat{x}_1)S(\widehat{x}_3(k))\widehat{x}_4(k)\\ &\quad + w_{13}(k)S(\widehat{x}_1)S(\widehat{x}_2(k))\widehat{x}_5(k)\\ \widehat{x}_2(k+1) &= w_{21}(k)S(\widehat{x}_1(k))S(\widehat{x}_3(k)) + w_{22}(k)\widehat{x}_5(k)\\ \widehat{x}_3(k+1) &= w_{31}(k)S(\widehat{x}_1(k))S(\widehat{x}_2(k)) + w_{32}(k)\widehat{x}_4(k)\\ \widehat{x}_4(k+1) &= w_{41}(k)S(\widehat{x}_2(k)) + w_{42}(k)S(\widehat{x}_3(k)) + w_{43}(k)S(\widehat{x}_4(k))\\ &\quad + w_{44}(k)u^\alpha(k)\\ \widehat{x}_5(k+1) &= w_{51}(k)S(\widehat{x}_2(k)) + w_{52}(k)S(\widehat{x}_3(k)) + w_{53}(k)S(\widehat{x}_5(k))\\ &\quad + w_{54}(k)u^\beta(k)\\ \widehat{x}_6(k+1) &= w_{61}(k)S(\widehat{x}_2(k)) + w_{62}(k)S(\widehat{x}_3(k)) + w_{63}(k)S(\widehat{x}_6(k))\end{aligned}$$

where $\widehat{x}_1$ estimates the angular speed $\omega$; $\widehat{x}_2$ and $\widehat{x}_3$ estimate the fluxes $\psi^\alpha$ and $\psi^\beta$, respectively; $\widehat{x}_4$ and $\widehat{x}_5$ estimate the currents $i^\alpha$ and $i^\beta$, respectively; finally $\widehat{x}_6$ estimates the angular displacement $\theta$. The inputs $u^\alpha$ and $u^\beta$ are selected as chirp functions.

The training is performed on-line, using a parallel configuration. All the NN states are initialized randomly. The covariances matrices are initialized as diagonals, and the nonzero elements are: $P_i(0) = 10000$; $Q_i(0) = 500$ and

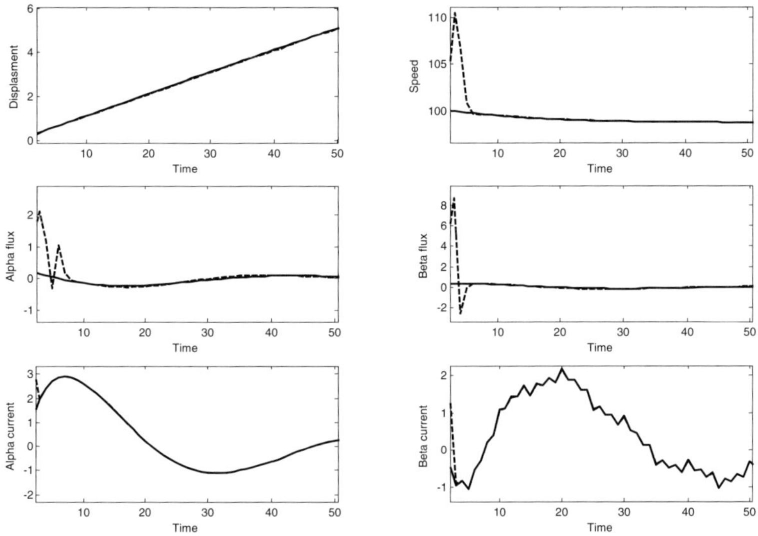


**Fig. 1.** Motor states and its estimates

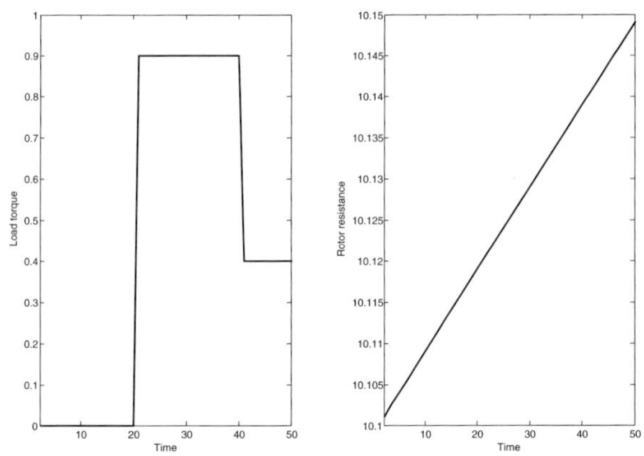


**Fig. 2.** External disturbance (load torque) and parametric variation (rotor resistance)

$R_i(0) = 10000$, $(i = 1, \cdots, 6)$, respectively. Simulation results are presented as: Fig.1 displays the time evolution of the estimated states $\widehat{x}_i(k)$, $(i = 1, \cdots, 6)$, respectively., and Fig. 2 portrays the load torque applied as an external disturbance and the parametric variation introduced in the rotor resistance $(R_r)$ as a variation of 1 Ohm per second.

**Remark 1.** The purpose of this paper is to develop a discrete-time nonlinear observer for a class of MIMO nonlinear systems in discrete-time, by means of the use of the EKF as the neural network learning algorithm without the knowledge of a nominal plant model; this approach is validated by the simulation results presented above.

**Remark 2.** Even if the EKF is not an easy learning algorithm, it presents an excellent performance and has proven to be reliable and practical for many applications over the past ten years ([11] and references therein).

## 5 Conclusions

A RHONN is used to design a Luenberger-like observer for a class of MIMO discrete-time nonlinear system; this observer is trained with an EKF-based algorithm, which is implemented on-line as a parallel configuration. The boundedness of the output and estimation errors is established on the basis of the Lyapunov approach. Simulation results shows the effectiveness of the proposed RHONO, as applied to a discrete-time induction motor model.

*Acknowledgement.* The authors thank the support of CONACYT Mexico, through Projects 39866Y and 39811Y.

## References

1. S. S. Ge, J. Zhang and T.H. Lee,"Adaptive neural network control for a class of MIMO nonlinear systems with disturbances in discrete-time", *IEEE Transactions on Systems, Man and Cybernetics*, Part B, Vol. 34, No. 4, August, 2004.
2. R. Grover and P. Y. C. Hwang, *Introduction to Random Signals and Applied Kalman Filtering*, 2nd ed., John Wiley and Sons, New York, USA, 1992.
3. Y. H. Kim and F. L. Lewis, *High-Level Feedback Control with Neural Networks, World Scientific*, Singapore, 1998.
4. F. Khorrami, P. Krishnamurthy and H. Melkote, *Modeling and Adaptive Nonlinear Control of Electric Motors*, Springer-Verlag, New York, USA, 2003.
5. A. J. Krener and A. Isidori, "Linearization by output injection and nonlinear observers", *System and Control Letters*, vol. 3, pp. 47-52, 1983.
6. A. U. Levin and K. S. Narendra, "Control of nonlinear dynamical systems using neural networks - part II: observability, identification and control", *IEEE transactions on neural networks*, vol. 7, no. 1, pp. 30-42, January, 1996.
7. A. G. Loukianov, J. Rivera and J. M. Cañedo, "Discrete-time sliding mode control of an induction motor", *Proceedings IFAC'02*, Barcelone, Spain, July, 2002.
8. R. Marino, "Observers for single output nonlinear systems", *IEEE Transactions on Automatic Control*, vol. 35, pp. 1054-1058, September,1990.
9. S. Nicosia and A. Tornambe, "High-gain observers in the state and parameter estimation of robots having elastic joins", *System and Control Letters*, vol. 13, pp. 331-337, 1989.
10. A. S. Poznyak, E. N. Sanchez and W. Yu, *Differential Neural Networks for Robust Nonlinear Control*, World Scientific, Singapore, 2001.
11. E. N. Sanchez, A. Y. Alanis and G. Chen, "Recurrent neural networks trained with Kalman filtering for discrete chaos reconstruction", *Proceedings of Asian-Pacific Workshop on Chaos Control and Synchronization'04*, Melbourne, Australia, July, 2004.

12. L. J. Ricalde and E. N. Sanchez, "Inverse optimal nonlinear high order recurrent neural observer", *International Joint Conference on Neural Networks IJCNN 05*, Montreal, Canada, August, 2005.
13. G. A. Rovithakis and M. A. Chistodoulou, *Adaptive Control with Recurrent High -Order Neural Networks*, Springer Verlag, New York, USA, 2000.
14. Y. Song and J. W. Grizzle, *The extended Kalman Filter as Local Asymptotic Observer for Discrete-Time Nonlinear Systems*, Journal of Mathematical systems, Estimation and Control,vol. 5, No. 1, pp. 59-78, Birkhauser-Boston, 1995.

# Strict Generalization in Multilayered Perceptron Networks

Debrup Chakraborty[1] and Nikhil R. Pal[2]

[1] Computer Science Department, CINVESTAV-IPN
Av. IPN No. 2508, Col. San Pedro Zacatenco
Mexico, D.F. 07360, Mexico
[2] Electronics and Communication Sciences Unit
Indian Statistical Institute, 203 B.T. Road
Calcutta 700108, India

**Abstract.** Typically the response of a multilayered perceptron (MLP) network on points which are far away from the boundary of its training data is not very reliable. When test data points are far away from the boundary of its training data, the network should not make any decision on these points. We propose a training scheme for MLPs which tries to achieve this. Our methodology trains a composite network consisting of two subnetworks : a mapping network and a vigilance network. The mapping network learns the usual input-output relation present in the data and the vigilance network learns a decision boundary and decides on which points the mapping network should respond. Though here we propose the methodology for multilayered perceptrons, the philosophy is quite general and can be used with other learning machines also.

## 1 Introduction

Multilayered perceptrons (MLP) are widely used to realize nonlinear mappings between input-output training data. It is known that MLPs can generalize on unknown data with reasonable accuracy. In [3] we demonstrated that the generalization capability of MLPs is generally over estimated and they can generalize well only on test points which are in the vicinity of the training data. The output of an MLP for points which lie far away from the boundary of its training sample is never reliable. This fact though known is seldom considered while training or using neural networks. An user who gets a trained neural network may (usually will) not have the training data with him (her), thus it is not possible for the user to know about the domain in which the network can perform meaningful generalizations. Some experiments reported in [3] clearly demonstrate that for classification problems, a trained MLP can produce very high response for a test point which is far away from the boundary of the training data. And in most cases such responses are *useless*. Ideally, a trained network must not respond to test points which lie far away from its training sample. We call this kind of generalization as "strict generalization". In [3] we proposed a scheme which does so only for classification problems. Also the method in [3] depends on a technique

P. Melin et al. (Eds.): IFSA 2007, LNAI 4529, pp. 722–731, 2007.
© Springer-Verlag Berlin Heidelberg 2007

to generate additional training points to detect the boundary of the training data. This method of generating new points becomes computationally expensive for reasonably high dimensional data. In this paper we address the same problem but with a different methodology which do not have the limitations of the method in [3]. This method is well suited for function approximation problems also and it does not require generation of additional points as in [3].

Our method involves building a composite network consisting of two subnetworks, each for a different task: (a) to learn the input-output mapping present in the training set, and (b) to learn the boundary of the training set. The composite network not only performs the main task of function approximation/classification, but also it learns a decision boundary as in classification problems. We call the first network which learns the input-output mapping as the *mapping network* and the other network which learns the decision boundary as the *vigilance network*. We propose a novel method to train the vigilance network which does not require generation of additional points as in [3], but it involves decomposing the training sample into small subsets, and making the vigilance net learn the boundary of such sets. The vigilance network is then combined with a mapping network to realize strict generalization for both function approximation (FA) and classification tasks.

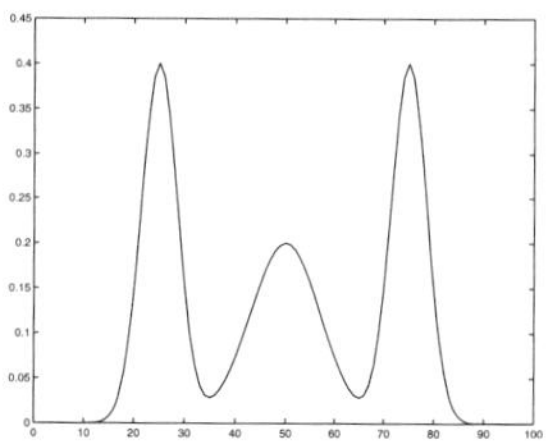
(a)

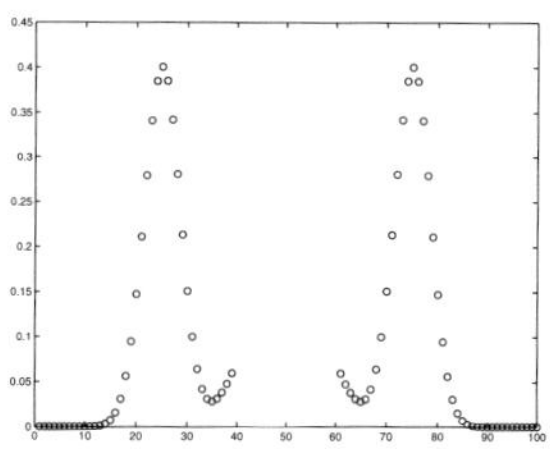
(b)

**Fig. 1.** (a)Plot of 3-Peaks (b)The points in 3-Peaks used for training

## 2 The Motivation

Let us consider the function:

$$y = 0.2e^{-\left(\frac{x-50}{10}\right)^2} + 0.4e^{-\left(\frac{x-25}{5}\right)^2} + 0.4e^{-\left(\frac{x-75}{5}\right)^2}. \tag{1}$$

We call this function as 3-Peaks. Fig. 1(a) depicts the function 3-Peaks. We sample a few points from the function in eq. (1) to train an MLP. Intentionally we sample points in such a manner that there remains a gap in the input space. Figure 1(b) shows the sampled points, we call this set of points as $PT_1$. The MLP trained with these sampled points are tested on a data set which contains 1000 equispaced points generated in [0,100]. Figure 2 shows the generalization done on the test data by four MLPs trained with different initializations. From Fig. 1(b) it is clear that the interval [40, 60] is not represented by any training data, so the MLP is not expected to perform well over this interval. Figure 2 shows some queer generalizations. Specially the generalization shown in Fig. 2(b).

When training data are collected from a live process then there may remain areas in the input space which are not well represented by the training data or are not *at all* represented by the training data. For test points which lie in those areas, ideally, an MLP should not respond at all. But an MLP, as shown in Fig.2, will always produce some output. We device a mechanism here which can take care of this problem.

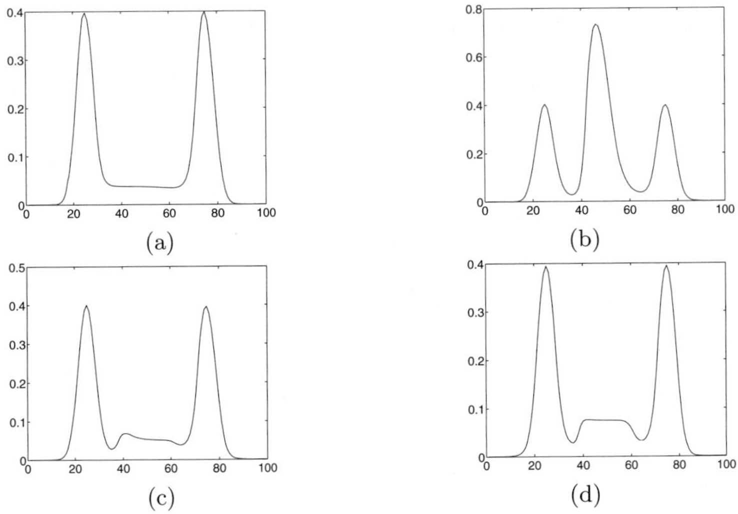

Fig. 2. Generalization an ordinary MLP trained with $PT_1$ for 4 different initializations

## 3 Training Scheme

Let $T = \{(\boldsymbol{x}_1, \boldsymbol{y}_1), (\boldsymbol{x}_2, \boldsymbol{y}_2), \ldots, (\boldsymbol{x}_N, \boldsymbol{y}_N)\}$ be our training set with $N$ training samples where $\boldsymbol{x}_i \in \Re^s$ be an input vector and $\boldsymbol{y}_i \in \Re^t$ be the corresponding output vector. Let $X = \{\boldsymbol{x}_1, \boldsymbol{x}_2, \ldots, \boldsymbol{x}_N\}$ be the set of input vectors in the training set $T$ and $Y = \{\boldsymbol{y}_1, \boldsymbol{y}_2, \cdots, \boldsymbol{y}_N\}$ be the set of output vectors in $T$. The task here is to learn the unknown input-output mapping that exists between $\boldsymbol{x}$ and $\boldsymbol{y}$. An ordinary MLP trained with conventional backpropagation or any other method can accomplish the task with a reasonable accuracy for almost all kinds of data. But we have an additional objective. We want to train an MLP in such a manner that it does not respond to test points which are away from the "boundary" of $X$. This can be realized if we can make the MLP learn the boundary of $X$ along with the input-output mapping between $\boldsymbol{x}$ and $\boldsymbol{y}$. Thus, we want our network to learn a decision boundary as in case of classification problems. To realize this we use two networks. The first one is an usual MLP which learns the input-output mapping, we call this as the *mapping* network. The second network is called the *vigilance net* which decides whether the MLP should respond to a point or not. The final output for a test point is obtained by suitably combining the outputs of both networks.

### 3.1 Training the Vigilance Network with Receptive Fields Around Data Points

We call the vigilance network as the receptive field vigilance network (RVN), because it uses Gaussian receptive fields around clusters of data points.

We can assume that the input vectors of the training set $X$ can be divided into a number of hyperspherical clusters $X_i, i = 1, 2, \ldots, n$, such that $\cup_{i=1}^{n} X_i = X$ and $X_i \cap X_j = \phi, \ \forall i, j; i \neq j$. Such a decomposition into hyperspherical clusters can be done using any conventional clustering algorithm like the $k$-means [4], or the Fuzzy c-means [2]. The vigilance net is trained in such a manner that it can detect whether a test point falls in any of these clusters or not.

This RVN is a three layered network. It has $s$ nodes in the input layer (if $X \subset \Re^s$), $k$ nodes in the hidden layer and one node in the output layer. Each node in the hidden layer has two parameters $\boldsymbol{\mu}_i \in \Re^s$ and $\sigma_i \in \Re$ associated with it. For a input vector $\boldsymbol{x}$, the $i^{th}$ hidden node computes

$$z_i = exp\left(-\frac{||\boldsymbol{x} - \boldsymbol{\mu}_i||^2}{\sigma_i^2}\right), \quad \forall i = 1, 2, \ldots, k. \tag{2}$$

The single output node in the third layer aggregates the outputs of the $k$ hidden nodes to give a single response. Let $b$ be the output of the third layer node :

$$b = \max_{i=1,2,\ldots,k} \{z_i\}. \tag{3}$$

Each node in the hidden layer represents a cluster in the data set $X$. The parameters $\boldsymbol{\mu}_i$ and $\sigma_i$ are decided using the FCM algorithm. If we decide $k$ as the number of hidden nodes then, we find out $k$ clusters from $X$ and denote $\boldsymbol{\mu}_i, \ \ i = 1, 2, \ldots, k$ as the $i^{th}$ cluster center. FCM produces a set of centroids $\boldsymbol{V} = \{\boldsymbol{v}_1, \boldsymbol{v}_2, \ldots, \boldsymbol{v}_k\}$, and a partition matrix $U = [u_{ij}]_{k \times N}$, where $u_{ij}$ denotes the degree to which $\boldsymbol{x}_j$ belongs to the $i^{th}$ cluster and $\boldsymbol{v}_i$ is the centroid of the $i^{th}$ cluster. Here we take $\boldsymbol{\mu}_i = \boldsymbol{v}_i$. The fuzzy partition matrix obtained from FCM can be hardened using the maximum membership rule [2]. In other words, we can consider that a point $\boldsymbol{x}_i \in X$ belongs to cluster $c, 1 \leq c \leq k$, if $u_{ci} = \max_j\{u_{ji}\}$.

So, the clustering output can be used to partition $X$ into $k$ disjoint sets $X_1, X_2, \ldots, X_k$. The $\sigma_i$ is chosen as :

$$\sigma_i = \max_{\boldsymbol{x}_j \in X_i} \{||\boldsymbol{x}_j - \boldsymbol{\mu}_i||\}, \forall i = 1, 2, \ldots, k. \tag{4}$$

For a test point $\boldsymbol{x} \in \Re^s$ each hidden node in the vigilance network gives an output related to the distance of $\boldsymbol{x}$ from the cluster center that the node represents. Thus, if a test point lies in or around the boundary of the cluster that a hidden node represents, then the output of that hidden node will be high. Therefore, for a test point $\boldsymbol{x} \in \Re^s$, if $b$ takes a high value then we conclude that $\boldsymbol{x}$ lies within or around some cluster of $X$; otherwise, it lies far from all $k$

clusters of $X$. So, $b$ can be used as an indicator of whether $\boldsymbol{x}$ lies in or around the boundary of $X$.

The structure of the RVN is similar to a Radial Basis Function (RBF) network but its function is quite different from that of an RBF.

### 3.2 The Composite Network

Another network is trained along with the vigilance network. This second network is an ordinary MLP, which is trained with the points in $X$ along with its associated output, i.e., with $T$. This network is called the *mapping* network (maps input to output). The vigilance network and the mapping network are combined together to a composite network which makes the final decision. Denoting the trained vigilance network as $\mathcal{N}_v$ and the mapping network as $\mathcal{N}_m$, the final trained network $\mathcal{N}$ is represented by the tuple $\mathcal{N} = (\mathcal{N}_m, \mathcal{N}_v)$. If the output dimension of the data is $t$, then the composite network will have $t+1$ output nodes. The first $t$ output nodes correspond to the output of the mapping network ($\mathcal{N}_m$) and the $(t+1)^{th}$ node corresponds to the output of the vigilance network ($\mathcal{N}_v$). We call the output of $\mathcal{N}_v$ as the *boundary indicator component* (BIC)(please refer to Fig. 3). A test point is fed to the composite network, and if the BIC gives a value greater than a threshold $th$, then the output of the test point corresponds to the output of the remaining $t$ nodes. If the BIC bears a value lower than $th$, the network infers that the point is away from the boundary of the training set and hence the net may not produce a correct output (decision) for it. The threshold $th$ is generally user defined. In our simulations we use $th = e^{-1}$. The reason for such a choice is that in case of our vigilance network, $\sigma_i$ is the largest distance of a training point that belongs to the cluster associated with the $i^{th}$ receptive field. So, it is reasonable to assume that the receptive field of a node is extended up to a distance equal to its $\sigma$ or a little beyond that.

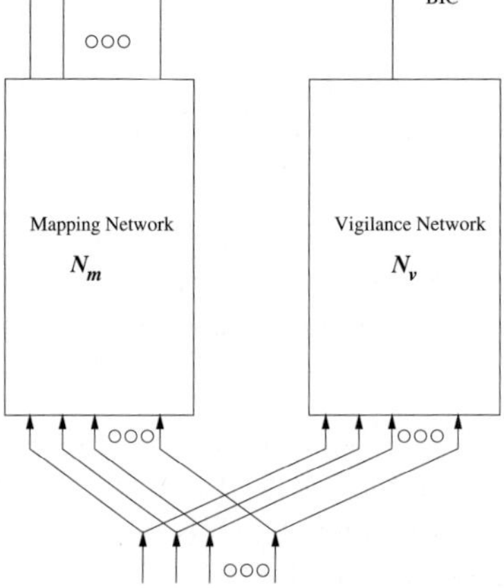


**Fig. 3.** The composite network $\mathcal{N} = (\mathcal{N}_m, \mathcal{N}_v)$

Based on this idea we choose $th$ equal to the response of a node at a distance $\sigma$, which is equal to $e^{-1}$.

# 4 Simulation Results

We use two function approximation and two classification data sets for demonstrating the effectiveness of our network. The function approximation data sets are 3-Peaks and Boston-Housing. For the data set 3-Peaks we can show the generalization properties pictorially and conclude that our network does a good job. But, for the real life data set, Boston-Housing, such a pictorial representation is not possible as this data set is in high dimension. For this data set we define some measures which help us to evaluate the performance of our network. Let $T = \{(\boldsymbol{x}_i, \boldsymbol{y}_i) : i = 1, 2, \ldots, N\}$ be the training set and $X = \{\boldsymbol{x}_i : i = 1, 2, \ldots, N\}$ be the input vectors of the training set $T$. Let $X_{Te} = \{\boldsymbol{x}'_i : i = 1, 2, \ldots, M\}$ be the input vectors of the test set. A trained composite network $\mathcal{N} = (\mathcal{N}_m, \mathcal{N}_v)$, will either respond to a test point $\boldsymbol{x}'_i$ or will not respond to it. Thus, the set $X_{Te}$ can be partitioned into two disjoint sets $X^A_{Te}$ and $X^R_{Te}$. $X^A_{Te}$ contains the points for which the composite network produces a response and $X^R_{Te}$ includes the points for which the composite network does not produce any response. Now, for each test point $\mathbf{x}'_i$ we define a function $\Delta$ as:

$$\Delta(\boldsymbol{x}'_i) = \min_{\boldsymbol{x}_j \in X} ||\boldsymbol{x}'_i - \boldsymbol{x}_j||. \quad (5)$$

Hence, $\Delta(\boldsymbol{x}'_i)$ represents the distance of $\boldsymbol{x}'_i$ from its nearest neighbor in $X$. Let $\mu_{\Delta A}$ and $\mu_{\Delta R}$ respectively denote the mean $\Delta$ for points which are accepted by the vigilance network (i.e., points in $X^A_{Te}$) and the points which are rejected by the vigilance network (i.e., points in $X^R_{Te}$) respectively. Thus,

$$\mu_{\Delta A} = \frac{1}{|X^A_{Te}|} \sum_{\boldsymbol{x}'_i \in X^A_{Te}} \Delta(\boldsymbol{x}'_i), \quad (6)$$

and

$$\mu_{\Delta R} = \frac{1}{|X^R_{Te}|} \sum_{\boldsymbol{x}'_i \in X^R_{Te}} \Delta(\boldsymbol{x}'_i). \quad (7)$$

For a test set $X_{Te}$ if $\mu_{\Delta A} < \mu_{\Delta R}$ then it is reasonable to assume that the network serves the intended purpose. Because $\mu_{\Delta A} < \mu_{\Delta R}$ implies that the points for which the composite network responds are more close to the training data than those for which the network does not respond.

**3-Peaks:** The 3-Peaks data set has been discussed in Section 2. We sample 80 points uniformly from the interval [0,100]-[40,60] and call them $PT_1$. We test the generalization capabilities of trained networks on a test set of 1000 equispaced points generated in the interval [0,100].

As $PT_1$ does not contain any point in the interval [40,60] (refer Fig. 1(b)), an ordinary MLP is not expected to produce meaningful response for test points which lie in the interval [40,60]. In Fig. 2 we have already shown that this is indeed the case.

A composite network $NP_1 = (NP_{m1}, NP_{v1})$ trained with $PT_1$ produces better generalizations. Figure 4 shows the generalizations of 4 different composite networks. Figure 4 reveals that the composite networks do not respond to test points which fall in the area not represented in the training set. Note, the response is plotted only when BIC $\geq 0.368$.

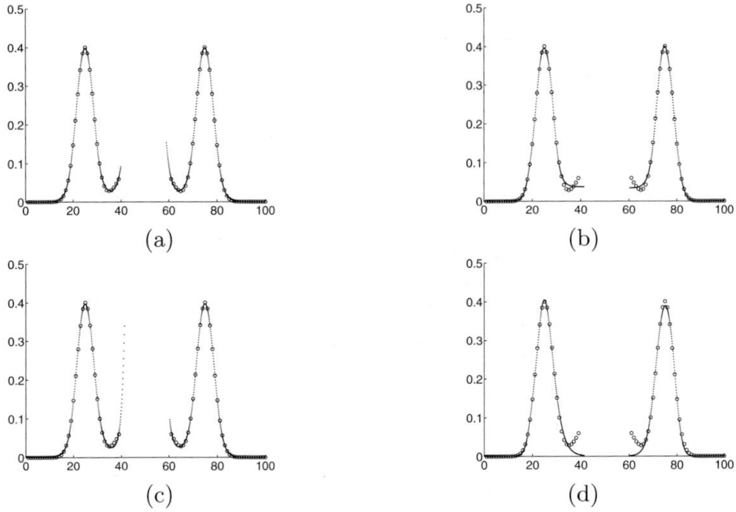


**Fig. 4.** Generalizations produced by $NP_1 = (NP_{m1}, NP_{v1})$, (using RVN) when trained with $PT_1$ for various initializations(the large dots denotes the training points)

**Boston-Housing:** Boston-Housing data set [1] contains 506 samples in 13 dimension and it contains only one output. We use a normalized version of this data set. We divide each input feature and the output by the respective maximum value so that they lie between 0 and 1. For Boston-Housing data, we create a random training-test partition so that the the training and test set contains equal number of data points. We train 10 different composite networks with different initializations. For each run we use a mapping network with 10 hidden nodes and a RVN with 10 receptive fields. Table 1 shows the results on the test sets for this data set. In Table 1, column 2 shows the number of points for which the composite network makes a decision, while column 3 gives the number of cases for which the network refuses to produce an output. Comparing column 4 with column 5 we see that for all cases $\mu_{\Delta A}$ is significantly lower than $\mu_{\Delta R}$, indicating that the points for which the network makes predictions are in the vicinity of the training points. Columns 6 and 7 show the mean test error (the absolute deviation of the network response from the true output) for accepted and the rejected points. Comparing columns 6 and 7, we find that in all cases the network rejects those points which produces more error. Note that, the composite network does not respond to the rejected points (the points in $X_{Te}^R$), but in column 7 of Table 1 we report the deviations of the outputs of the mapping network for the rejected points ignoring the values of the BIC produced by the vigilance network. It is not expected that a trained network will produce good results for test points which are away from the training set, and comparing columns 6 and 7 we see that this is true for all the runs with Boston-Housing data.

**Table 1.** Run statistics for Boston-Housing on 50% Training-Test partition

| Run No. | $\lvert X_{Te}^A \rvert$ | $\lvert X_{Te}^R \rvert$ | $\mu_{\Delta A}$ | $\mu_{\Delta R}$ | Mean Test error for accepted points | Mean test error for rejected points |
|---|---|---|---|---|---|---|
| 1 | 245 | 8 | 0.155 | 0.324 | 0.090 | 0.149 |
| 2 | 246 | 7 | 0.152 | 0.369 | 0.092 | 0.212 |
| 3 | 249 | 4 | 0.158 | 0.395 | 0.068 | 0.386 |
| 4 | 245 | 8 | 0.153 | 0.439 | 0.078 | 0.187 |
| 5 | 251 | 2 | 0.152 | 0.280 | 0.068 | 0.304 |
| 6 | 246 | 7 | 0.157 | 0.330 | 0.115 | 0.140 |
| 7 | 243 | 10 | 0.160 | 0.317 | 0.098 | 0.397 |
| 8 | 246 | 7 | 0.158 | 0.454 | 0.445 | 0.602 |
| 9 | 247 | 6 | 0.161 | 0.425 | 0.438 | 0.585 |
| 10 | 248 | 5 | 0.158 | 0.246 | 0.153 | 0.219 |

To validate that in average $\mu_{\Delta A} < \mu_{\Delta R}$, we perform another experiment. In this experiment, we use all 506 points as training examples and test the networks with 1000 additional points generated in the 10% inflated hyperbox containing the training data. Here too we train 10 different networks and test with different test sets each containing 1000 points. Table 2 shows the results for the 10 networks. From columns 2 and 3 of Table 2 we see that the number of points rejected is much more than the number of points accepted by the composite network. This is due to the fact that the input vectors are 13 dimensional, and we have only 506 training points. So, the training points occupy only a small

**Table 2.** Run statistics for Boston-Housing on artificially generated test data

| Run No. | $\lvert X_{Te}^A \rvert$ | $\lvert X_{Te}^R \rvert$ | $\mu_{\Delta A}$ | $\mu_{\Delta R}$ |
|---|---|---|---|---|
| 1 | 86 | 914 | 0.810 | 1.242 |
| 2 | 155 | 845 | 0.862 | 1.243 |
| 3 | 103 | 897 | 0.804 | 1.242 |
| 4 | 135 | 865 | 0.864 | 1.230 |
| 5 | 108 | 892 | 0.818 | 1.231 |
| 6 | 88 | 912 | 0.799 | 1.233 |
| 7 | 56 | 944 | 0.812 | 1.218 |
| 8 | 143 | 857 | 0.873 | 1.235 |
| 9 | 118 | 882 | 0.796 | 1.241 |
| 10 | 139 | 861 | 0.851 | 1.233 |

part of the total hyperbox bounded by the data. And most of the artificially generated points fall outside the boundary of the training sample. The scenario was different in case of Table 1 as there it is expected that the test points follow the same probability distribution as that of the training points, hence in Table 1 only a few points got rejected. Comparing columns 4 and 5 of Table 2 we see that for all cases $\mu_{\Delta A} < \mu_{\Delta R}$, which shows that the networks respond only to points which are in the vicinity of the training points. As in this case the test data are artificially generated, we cannot measure the deviation of the network output from the true output.

### 4.1 Results on Classification

We report results on two classification data sets : Breast Cancer [1] and Wine [1]. Tables 3 and 4 summarize the run statistics of 10 networks trained and tested for Wine and Breast-Cancer data respectively. For both these data sets we used equal number of points in the training and test sets. Also we used 10 hidden nodes in the mapping network and 10 receptive fields in the RVN. Tables 3 and 4 clearly show that for all networks $\mu_{\Delta R}$ is significantly greater than $\mu_{\Delta A}$.

**Table 3.** Run statistics for Wine

| Run No. | $\|X^A_{Te}\|$ | $\|X^R_{Te}\|$ | $\mu_{\Delta A}$ | $\mu_{\Delta R}$ |
|---|---|---|---|---|
| 1 | 86 | 3 | 0.390 | 0.560 |
| 2 | 87 | 2 | 0.388 | 0.733 |
| 3 | 87 | 2 | 0.387 | 0.814 |
| 4 | 87 | 2 | 0.398 | 0.618 |
| 5 | 84 | 5 | 0.377 | 0.606 |
| 6 | 87 | 2 | 0.406 | 0.767 |
| 7 | 88 | 1 | 0.391 | 0.912 |
| 8 | 88 | 1 | 0.402 | 0.781 |
| 9 | 82 | 7 | 0.394 | 0.723 |
| 10 | 88 | 1 | 0.399 | 0.912 |

**Table 4.** Run statistics for Breast-Cancer

| Run No. | $\|X^A_{Te}\|$ | $\|X^R_{Te}\|$ | $\mu_{\Delta A}$ | $\mu_{\Delta R}$ |
|---|---|---|---|---|
| 1 | 341 | 1 | 2.160 | 8.307 |
| 2 | 341 | 1 | 1.999 | 5.916 |
| 3 | 341 | 1 | 2.054 | 9.165 |
| 4 | 340 | 2 | 2.016 | 7.083 |
| 5 | 337 | 5 | 2.081 | 8.517 |
| 6 | 334 | 8 | 2.016 | 7.058 |
| 7 | 338 | 4 | 1.923 | 7.854 |
| 8 | 341 | 1 | 2.131 | 5.916 |
| 9 | 341 | 1 | 2.083 | 9.165 |
| 10 | 338 | 4 | 2.119 | 6.730 |

## 5 Conclusion

We proposed a training scheme for MLPs which can equip an MLP with the property of strict generalization. Our method uses a composite network that judiciously integrates two subnetworks, a mapping network and a vigilance network. The simulation results demonstrate that our training scheme serves the purpose quite satisfactorily both for function approximation and classification tasks. The basic philosophy of vigilance network is quite general in nature and can be used with other machine learning tools like radial basis function networks.

## References

1. C.L. Blake and C.J. Merz, UCI Repository of machine learning databases [http://www.ics.uci.edu/~mlearn/MLRepository.html]. Irvine, CA: University of California, Department of Information and Computer Science, 1998.
2. J. C. Bezdek, J. Keller, R. Krishnapuram and N. R. Pal, *Fuzzy Models and Algorithms for Pattern Recognition and Image Processing* Kluwer, Massachusetts, 1999.
3. D. Chakraborty, N.R. Pal, "A novel training scheme for multilayered perceptrons to realize proper generalization and incremental learning", *IEEE Trans. Neural Networks*, vol 14, no 1, pp. 1-14, 2003.
4. R.O. Duda, P.E. Hart, D.G. Stork, *Pattern Classification*, John Wiley, New York, 2000.
5. M. T. Hagan and M. B. Menhaj, "Training Feedforward Networks with the Marquardt Algorithm," *IEEE Trans. on Neural Networks*, vol. 5, no. 6, pp. 989–993, 1994.
6. M. Sugeno, T. Yasukawa, "A fuzzy-logic based approach to qualitative modeling," *IEEE Transactions on Fuzzy Systems*, vol. 1, pp. 7–31, 1993.

# Fault Tolerant Control of a Three Tank Benchmark Using Weighted Predictive Control

L.F. Mendonça[1,2], J.M.C. Sousa[1], and J.M.G. Sá da Costa[1]

[1] Center of Intelligent Systems/IDMEC,
Instituto Superior Técnico
Technical University of Lisbon
Av. Rovisco Pais, 1049-001 Lisbon, Portugal
{mendonca,jsousa,sadacosta}@dem.ist.utl.pt
[2] Escola Náutica Infante D. Henrique
Dept. of Marine Engineering
Av. Eng. Bonneville Franco, Lisbon, Portugal
mendonca@enautica.pt

**Abstract.** This paper proposes the application of fault-tolerant control (FTC) using weighted fuzzy predictive control. The FTC approach is based on two steps, fault detection and isolation (FDI) and fault accommodation. Fault detection is performed by a model-based approach using fuzzy modeling. Fault isolation uses a fuzzy decision making approach. The model of the isolated fault is used in fault accommodation with a model predictive control (MPC) scheme. This paper uses a weighted fuzzy predictive control scheme, where fuzzy goals and fuzzy constraints are described in a fuzzy objective function. The criteria (goals or constraints) have an associated weight factor, which are chosen by the decision-maker. Two faults were simulated in a three tank benchmark and the respective fuzzy models were identified. The fuzzy FTC scheme proposed in this paper was able to accommodate the simulated faults.

## 1 Introduction

With the increase of technical processes complexity, safety and reliability become important system requirements. Considering that industrial processes are more and more complex, the probability of occurring a fault increases. Thus, control systems must include automatic supervision of process control to detect and isolate faults as early as possible, and to tolerate some component malfunctions. FTC can be performed by passive methods or by active methods. Passive methods use robust control techniques to ensure that a closed-loop system remains insensitive to certain faults. In active methods, a new control system is redesigned by using the desired properties of performance and robustness in the system without faults. Fault accommodation involves the detection and isolation of faults, and taking appropriate control actions that eliminate or reduce the effect of faults and maintains the control. The use of MPC to deal with the fault accommodation is relatively natural and straightforward, considering the representation of both faults and control objectives [6]. The use of fuzzy goals and

P. Melin et al. (Eds.): IFSA 2007, LNAI 4529, pp. 732–742, 2007.
© Springer-Verlag Berlin Heidelberg 2007

fuzzy constraints in MPC provides ways of representing and dealing with flexible or soft criteria. The optimal trade-off amongst fuzzy goals and fuzzy constraints is determined by maximizing simultaneously the satisfaction of the optimization objectives and the constraints [12]. In [7] a method for satisfying the problem constraints and the goals is proposed, where preference for different constraints and goals can be specified by the decision maker. This paper proposes the uses of weights in the fuzzy MPC scheme to optimize the control performance when a fault occurs.

The FDI approach presented in this paper uses one fuzzy model representing the normal state of the system and one fuzzy model for each fault that can occur in a given system. The faults are detected and isolated based on these fuzzy models. A fuzzy decision making approach is used to isolate the faults. When a fault is isolated, fault accommodation is performed by using the respective faulty model. This paper proposes a fault tolerant control scheme, where the faulty model is used in a fuzzy MPC scheme. A simulation of a three tank benchmark is used in this paper to illustrate the advantages of the proposed approach, and to obtain two faults in the system behavior.

This paper is organized as follows. Next section presents a brief overview of fault detection and isolation, and fault tolerant control. Fuzzy predictive control is presented in Section 3. The architecture for fault tolerant control is described in Section 4. The application example is presented in Section 5 and finally some conclusions are drawn in Section 6.

## 2 Fault Tolerant Control

A system that includes the capacity of detecting, isolating and identifying faults is called a fault detection and isolation system [2]. During the years, many research has been carried out using analytical approaches and model-based approaches. The idea is to generate signals that reflect inconsistencies between normal and faulty system operation and detect and isolate the faults.

The use of FDI in fault tolerant control is very important in the active way of achieving fault-tolerance, by detecting and isolating the faults. After the fault indication by FDI, the system can then be reconfigured or restructured. In some cases, a pre-calculated controller will be activated or the parameters of the controller will be changed according to the real time diagnostic provided by FDI. The active fault tolerant control approach uses the FDI information to make the on-line controller reconfiguration or model selection. Another possible approach is to use all the information given by FDI to improve the ability of on-line controller reconfiguration.

When MPC is used in FTC, some faults can be validated by modifying the constraints in the MPC problem definition. The use of MPC with internal model modification is used also to accommodate the faults behavior. The use of MPC increases the degree of fault tolerance under certain conditions, when the fault is not detected [6]. Thus, MPC in fault tolerant control provides a suitable implementation architecture and increases the system capability to accommodate the faults. Fault accommodation yields taking appropriate control actions that eliminate or reduce the effect of the faults and maintains the control.

A fuzzy logic approach in FTC is used in [5] where Takagi-Sugeno (TS) fuzzy models are used in fault tolerant control of non-linear systems. This paper proposes a fuzzy MPC scheme to perform fault accommodation.

# 3 Fuzzy Predictive Control

## 3.1 Classical Objective Functions

In predictive control of multivariable systems, the output values $\hat{\mathbf{y}}(k+i)$, $i = 1, \ldots, H_p$, depend on the states of the process at the current time $k$ and on the future control signals $\mathbf{u}(k+j), j = 1, \ldots, H_c$, where $H_c$ is the *control horizon.* For multivariable systems the objective function can be represented by $J = \mathbf{e}^T R\mathbf{e} + \Delta\mathbf{u}^T Q \Delta\mathbf{u}$, where the first term accounts for the minimization of the output errors, the second term represents the minimization of the control effort, and $R$ and $Q$ are weighting matrices. Note that these parameters have two functions: they normalize the different outputs and inputs of the system and weight the importance of the two different terms in the objective function over the time steps.

## 3.2 Fuzzy Objective Functions

When fuzzy criteria is used in the objective function, the criteria has some flexibility that can be exploited for improving the optimization objective. Predictive control using fuzzy goals and fuzzy constraints can be defined as a fuzzy decision making problem [11]. Using a process model, a fuzzy decision making algorithm selects the control actions that best meet the specifications. Hence, a control strategy can be obtained that is able to push the process closer to the constraints, and that is able to force the process to a better performance, based on the goals and the constraints set by the operator together with the known conditions provided by the system's designers. The resulting optimization problem is a multistage decision making problem. Multistage decision making has been applied to control by several authors [4].

## 3.3 Weight Selection in Fuzzy Aggregation

Weighted aggregation has been used quite extensively especially in fuzzy decision making, where the weights are used to represent the relative importance that the decision maker attaches to different decision criteria. Almost always an averaging operator has been used for the weighted aggregation, such as the generalized means, fuzzy integrals or the ordered weighted average (OWA) operators. Weighted aggregation of fuzzy sets by using t-norms has been considered first by Yager in [13]. He proposed to modify the membership functions with the associated weight factors before the fuzzy aggregation. The weighted aggregation is then the aggregation of the modified membership functions. A generalized form of this idea leads to the weighted aggregation function.

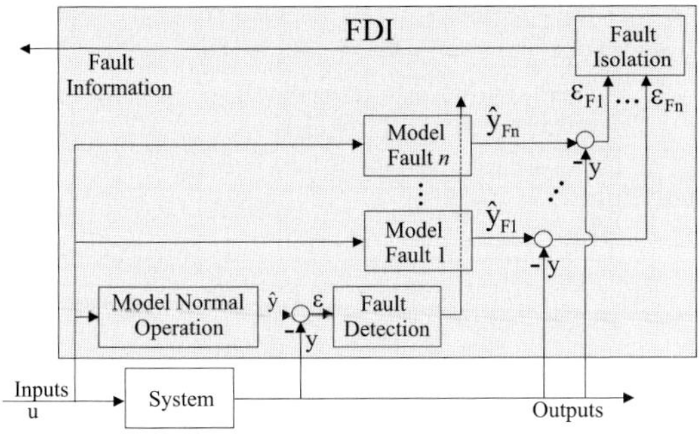


**Fig. 1.** Fault detection and isolation scheme

The weight factors represent the relative importance of various constraints and objectives with respect to one another. The general assumption is that the higher is the weight of a particular constraint, the larger is its importance on the aggregation result. Hence, final optimization result will be closer to the more important constraints. If the objective is more important, the constraints will be relaxed to a larger degree in order to increase the objective function. In this paper, a heuristic is used to select the weight factors, as described in [7], and is summarized as follows:

1. *Initialize all the weight factors to one, and evaluate the control performance using the corresponding objective function.*
2. *Decrease each of the $\tilde{T}$ weight factors to* 0.5 *one by one. Evaluate the performance, and order the criteria, where the first is the one that improved the performance of the system most.*
   *When the number of criteria $\tilde{T}$ is very high, a simplification can be made. In this case, reduce simultaneously a certain criterion for the entire prediction horizon $H_p$. The number of iterations is then reduced from $\tilde{T} = T \times H_p$ to $T$. Thus, instead of evaluating each weight associated with the criterion $\zeta_{ij}$, the same weight is assumed for the criterion $\zeta_j$, i.e. the criterion is considered constant for the entire prediction horizon.*
3. *For each criterion, $\zeta_{ij}$ or $\zeta_j$ depending on the choice in Step 2, reduce the weight factor to* 0.25 *and check if the control performance is better. If this is the case, reduce further the weight to* 0.125*. The weight that yields the best performance is chosen as the weight factor for that criterion.*
4. *When all the criteria have been evaluated, the best combination of weights is determined, and should be used for the system.*

## 4 Architecture for Fault Tolerant Control

This paper uses a simple architecture for fault tolerant control [8]. This approach is based on two steps: the first performs fault detection and isolation, and the second performs fault accommodation.

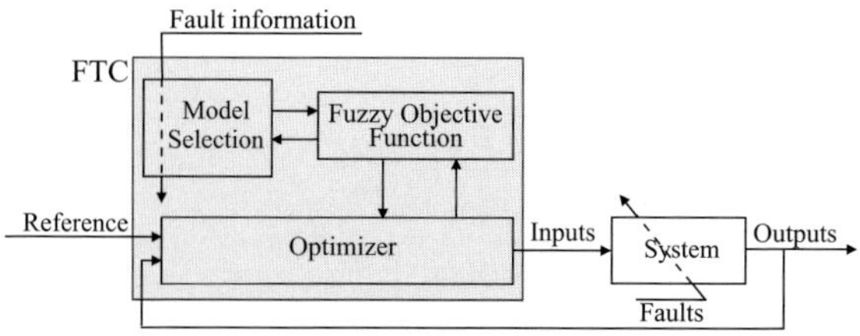


**Fig. 2.** FTC structure

### 4.1 Fault Detection and Isolation

The fault detection and isolation approach is presented in Fig. 1. In the presented FDI approach, the multidimensional input $\mathbf{u}$, enters both the process and a model (observer) in normal operation. The vector of residuals $\boldsymbol{\varepsilon}$ is defined as, $\varepsilon = \mathbf{y} - \hat{\mathbf{y}}$, where $\mathbf{y}$ is the output of the system and $\hat{\mathbf{y}}$ is the output of the model in normal operation. When any component of $\boldsymbol{\varepsilon}$ is bigger than a certain threshold, the system detects faults. In this case, $n$ observers (models), one for each fault, are activated, and $n$ vectors of residuals are computed. Each residual $i$, with $i = 1, \ldots, n$, is computed as, $\varepsilon_{\mathrm{F_i}} = \mathbf{y} - \hat{\mathbf{y}}_{\mathrm{F}_i}$, where $\hat{\mathbf{y}}_{\mathrm{F}_i}$ is the output of the observer for the fault $i$. The residuals $\varepsilon_{\mathrm{F}_1}, \ldots, \varepsilon_{\mathrm{F}_n}$ are evaluated, and the fault or faults detected are the outputs of the FDI system. The fault isolation is performed by evaluating *fuzzy decision factors*, which are built based on residuals. The fuzzy fault isolation used in this paper is based on FDM [10]. In this approach, a membership function $\mu_{\varepsilon_{ij}}$ is derived for each residual $\varepsilon_{ij}$. Note that this method to derive membership functions is common in various fuzzy approaches [9].

The $m$ membership functions $\mu_{\varepsilon_{i1}}, \ldots, \mu_{\varepsilon_{im}}$ must be aggregated using a conjunction operator, which assures that a fault is isolated only when all the residuals $\varepsilon_{ij}$ are close to zero. The aggregation can be given by $\gamma_i = t(\mu_{\varepsilon_{i1}}, \ldots, \mu_{\varepsilon_{im}})$, where $t$ is a triangular norm, as e.g. the minimum operator. An example of $\gamma_i$ for two outputs is shown in [10]. Let $\gamma_i(k) \in [0,1]$, $i = 1, \ldots, n$, be called a *fuzzy decision factor*. These values are computed at each time instant $k$. A vector of fuzzy decision factors can be computed as $\Gamma(k) = [\gamma_1(k) \quad \gamma_2(k) \quad \cdots \quad \gamma_n(k)]$, i.e., one fuzzy decision factor for each fault. A fuzzy decision factor $\gamma_i(k)$ is high only if all the residuals are close to zero.

In order to isolate a fault $i$, the value of $\gamma_i(k)$ must be higher than a *threshold* $R$, which must be close to one. Note that the threshold $T$ is equal for all the faults, because the fuzzy decision factors are already normalized in the interval $[0, 1]$. The threshold is obtained experimentally and defines the regions of fault and no fault. In practice, the definition of this value revealed to be relatively easy, and a value around $R = 0.7$ isolated the faults properly. This value can suffer a slight change in others processes. Note that several $\gamma_i(k)$ can be above the threshold at a certain time $k$. Therefore, a fault $i$ is isolated only when the remaining faults are below $R$. However, even if only one fault is above the threshold at a certain time instant, this can occur due to noise or model errors.

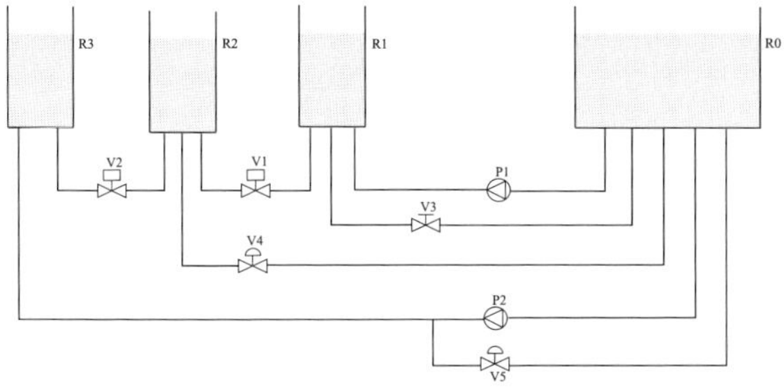


**Fig. 3.** Flowsheet of the three tank benchmark process

Therefore, our approach considers that a fault $i \in \{1, \ldots, n\}$ is only isolated when $\gamma_i$ is above the threshold $R$ and the remaining $\gamma_l$ decision factors are below the same threshold for $t_k$ consecutive time instants. The fuzzy isolation scheme used in this paper is presented in [10].

### 4.2 Fuzzy Fault Tolerant Control

The FTC structure proposed in this paper is shown in Fig. 2. FTC uses a multiple model selection approach, where a fuzzy model for the process running in normal operation and one model for each one of the faults are used.

In MPC, the user can specify requirements like minimal energy use or fast control by changing the cost function. This feature is very useful in FTC, because it allows different control specification for faulty models, in order to have minimal losses when the system is working in a faulty mode. Furthermore, the control action can take into account a time interval (prediction horizon). Also the receding horizon principle allows at each time instant to assess the situation by taking into account any change in the fault status to apply the best control action. The use of nonlinear models results in a non-convex optimization problem. In this case, iterative optimization techniques are usually considered. However, these methods have generally high computational costs and converge often to local minima. By discretizing the control actions, the efficient B&B algorithm can be used to search the discrete space for the best solution. This has proven to give better results than iterative optimization techniques [9].

## 5 Application Example

### 5.1 Description of the System

The process used in this paper is a benchmark which presents the transport of fluids behavior in chemical processes. Many of the faults of chemical processes occur in the level of transport of fluids and raw materials; leaks, clogs, valve blockages and sensor faults are only a few of them. Figure 3 presents the flowsheet of the used benchmark process [3]. It consists of three tanks, R1, R2 and R3 connected with flow paths which serve to supply water from the reservoir R0,

**Table 1.** VAF of fuzzy models with faults and without faults (in %)

| Outputs | Without faults | Faults $F1$ | Faults $F4$ |
|---|---|---|---|
| $h1$ | 99.8 | 99.8 | 99.8 |
| $h2$ | 99.8 | 99.8 | 98.5 |
| $h3$ | 99.9 | 93.7 | 68.6 |

P1 and P2 are pumps driven with DC motors with permanent magnet. There are two configurations to active flow paths available. In the first one, flow is generated by varying the angular speed of the pump P1. In the second case, pump P2 works at constant speed. Flow is then varied by manipulating the valve V5. There is one servo-valve V5 in the plant, driven by a DC motor. Valves V1 and V2 are on-off while V3 and V4 are manual valves. The purpose of the valve V3 is mainly to implement a real fault, i.e. a leakage of the tank R1. The capacity of the reservoir R0 is much larger than the capacity of the tanks, so that its level is practically constant during operation [1].

### 5.2 FDI Results

The faults considered in this paper are: $F1$, leak in R1 with 50% of intensity, and $F4$ clog in branch containing $V4$, both with 50% of intensity. This fault description is presented in [3]. In order to measure modeling accuracy, this paper uses the Variance Accounted For (VAF). The identification data used to build the three tank system model in normal operation contains 2000 samples. The same number of samples was used to identify each fault used in the simulation. A fuzzy model was identified for the model in normal operation. The model inputs when the system is in normal operation and when the system is with faults are: $w_1$, speed of rotation of pump P1, and $S5$, position of the valve V5. The model outputs are: $h1$ level in the tank R1, $h2$ level in the tank R2, and $h3$ level in the tank R3.

The accuracy of the identified fuzzy models,with or without faults is presented in Table 1. In general, the fuzzy models present good accuracy when the system is with or without faults. The FDI step is made considering the scheme presented in [10]. Fault $F1$ occurs at 50 s and fault $F4$ at 90 s. The two faults $F1$ and $F4$ are correctly detected and isolated. The fault isolation time is determined when the residuals are close to zero. The detection of both faults is made in the first instant (time instant 51s, when the fault $F1$ occurs, and time instant 91, when the fault $F4$ occurs). The isolation time of fault $F1$, is 52 s, i.e. only one time instant after the detection time. Fault $F4$ is isolated 2 s after the detection time.

### 5.3 FTC Results

The fault accommodation is made considering the two controlled variables of the three tank benchmark test: the level in the tank R1, $h1$, and the level in the

**Table 2.** Normalized errors using classical objective functions and weighted fuzzy objective functions (Faults F1 and F4)

| | F1 | | | | F4 | | | |
|---|---|---|---|---|---|---|---|---|
| Classical | $R_{h1}$ | $R_{h3}$ | $e_{h1}$ | $e_{h3}$ | $R_{h1}$ | $R_{h3}$ | $e_{h1}$ | $e_{h3}$ |
| | 1.0 | 1.0 | 1.0 | 1.0 | 1.0 | 1.0 | 1.0 | 1.0 |
| Fuzzy | $w_1$ | $w_3$ | $e_{h1}$ | $e_{h3}$ | $w_1$ | $w_3$ | $e_{h1}$ | $e_{h3}$ |
| Step 1 | 1.0 | 1.0 | 1.01 | 0.27 | 1.0 | 1.0 | 0.18 | 2.54 |
| Step 2 | **1.0** | **0.5** | **0.9** | **0.29** | **1.0** | **0.5** | **0.13** | **2.53** |
| Step 3 | 0.5 | 1.0 | 1.37 | 0.19 | 0.5 | 1.0 | 0.16 | 2.54 |
| Step 4 | 1.0 | 0.25 | 1.06 | 0.24 | 1.0 | 0.25 | 0.18 | 2.54 |
| Step 5 | 0.25 | 0.5 | 1.37 | 0.29 | 0.25 | 1.0 | 0.18 | 2.54 |

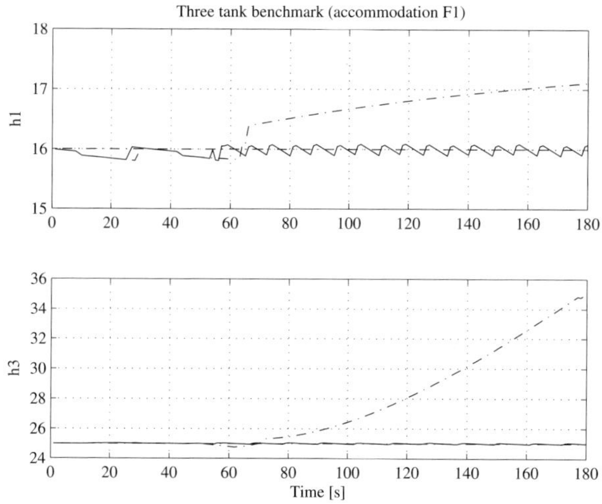


**Fig. 4.** Fault $F1$ accommodation. Solid - fault with weighted accommodation, dash-dotted - fault without accommodation.

tank R3, $h3$. From the possible aggregation methods presented in [7] the best results were obtained with Yager's $t$-norm. In this paper, the Yager's $t$-norm is also used to accommodate the faults. The sum squared error (SSE) is used to evaluate the control performance. Table 2 presents the fault accommodation results with classical, fuzzy and weighted fuzzy objective functions. The heuristic to select the weight factors, which is presented in Section 3.3, is used. The Steps in Table 2 follow the order of this heuristic. Considering the fault F1, the error using the weights $R_{h1} = R_{h3} = 1.0$ of classical objective function is taken as 1 (100%), and it serves as the normalization to be compared with the other errors. The absolute error values obtained are $e_{h1} = 43.2 \times 10^{-5}$ m for the level in the tank R1 and $e_{h3} = 0.26 \times 10^{-5}$ m the level in the tank R3. The best result of fault F1 accommodation is obtained at Step 2 when the fuzzy weighted approach is used. The use of weights in fuzzy objective functions decrease the

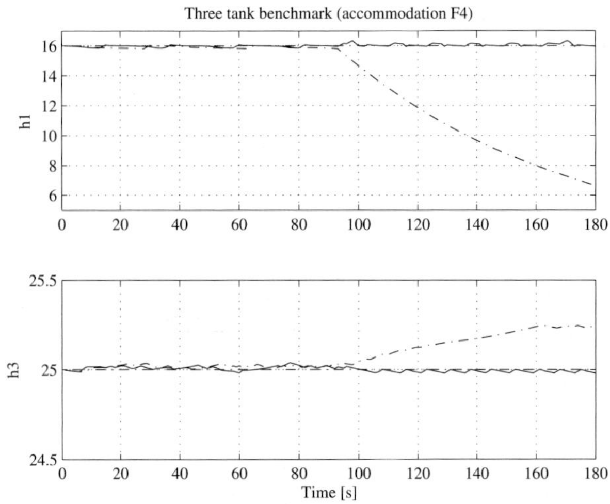


**Fig. 5.** Fault $F4$ accommodation. Solid - fault with weighted accommodation, dash-dotted - fault without accommodation.

errors of the two controlled variables $h1$ and $h3$. The accommodation results for fault F4 are also presented in Table 2. The absolute error values using the weights $R_{h1} = R_{h3} = 1.0$ of classical objective function are $e_{h1} = 8.1 \times 10^{-5}$ m for the level in the tank R1 and $e_{h3} = 0.67 \times 10^{-5}$ m for the level in the tank R3. The weights combination used in Step 2 leads to smaller errors in the level of tank R1. The error $e_{h3}$ is larger when the fuzzy controller is used, but the error $e_{h1}$ is much smaller. As the error $e_{h3}$ was 10 times smaller than $e_{h1}$, the fuzzy controller elegantly improve the overall performance of the controlled system. Recall that fault $F1$ was simulated to start 50 seconds after the beginning of the trajectory. The simulation results at Step 2 in Table 2 and without fault accommodation are depicted in Fig. 4. The fuzzy FTC scheme proposed in this paper was able to detect, isolate and accommodate correctly the fault $F1$. The behavior of fault $F1$ can be observed in all the controlled variables (level $h1$ and level $h3$). Note that when fault $F1$ is isolated the fuzzy model in normal operation is substituted by the fuzzy model considering that fault $F1$ is active. This faulty fuzzy model is used in the weighted fuzzy MPC scheme to derive the proper control actions. When fault $F1$ is active, the MPC controller still presents good performance for the two controlled variables. The level of the tank R1, $h1$, presents small oscillations. Figure 5 presents the results obtained with weighted fuzzy objective functions (Step 2 in Table 2) and without fault accommodation. The proposed FTC approach was also able to detect, isolate and accommodate correctly the fault $F4$. The behavior of fault $F4$ can be observed in the controlled variables, level $h1$ and level $h3$. The controlled output level $h3$ presents small oscillations. The proposed FTC scheme presents good performance when applied to the faults considered in this paper.

## 6 Conclusions

This paper presents the application of a weighted fuzzy FTC scheme to accommodate several faults in a three tank benchmark system. The FTC approach is based on two steps: fault detection and isolation, and fault accommodation. In the first step the FDI scheme is based on fuzzy models for both normal operation and for faulty operation, and on a fuzzy decision making approach. Fault isolation is performed by evaluating fuzzy decision factors that are built based on residuals. In the second step, the fault accommodation is made using weighted fuzzy MPC. Fault accommodation show that the performance of fuzzy predictive controllers can be improved by weighting the control objectives and the constraints. To achieve this, a weighted extension of Yager $t$-norm is used for the aggregation. This operator can model the simultaneous satisfaction of the goals and the constraints, while taking the difference in the importance into account. An algorithm is used for selecting the weights affecting the fuzzy goals and the fuzzy constraints. The application of this approach to a three tank benchmark system shown its ability to detect, isolate and accommodate two faults.

Future research will consider the extension of the proposed FTC scheme to a larger number of faults with different intensities.

## Acknowledgments

This work is partially supported by the "Programa do FSE-UE, PRODEP III, acção 5.3, no âmbito do III Quadro Comunitário de apoio" and by the project POCI/EME/59522/2004, co-sponsored by FEDER, Programa Operacional Ciência e Inovação (POCI) 2010, FCT, Portugal.

## References

1. L. Bahir and M. Kinnaert. Fault detection and isolation for a three tank system based on a bilinear model of the supervised process. *Proc. Int. Conf. Control*'98: UKACC, pages 1486–1491, 1998.
2. R. Chen and R. Patton. *Robust Model-Based Fault Diagnosis for Dynamic Systems.* Kluwer Academic Publishers, Boston, MA, 1999.
3. G. Dolanc, D. Juricic, A. Rakar, J. Petrovcic, and D. Vrancic. Three-tank benchmark test. Technical Report COPL007R, Copernicus project CT94-0237, Jozef Stefan Institute, 1997.
4. J. Kacprzyk. *Multistage Fuzzy Control.* John Wiley & Sons, 1997.
5. C. J. Lopez-Toribio, R. J. Patton, and S. Daley. Takagi-Sugeno fuzzy fault-tolerant control of an induction motor. *Neural Computing & Applications*, 9:19–28, 2000.
6. J. M. Maciejowski and C. N. Jones. MPC fault-tolerant flight control case study: flight 1862. *IFAC Symp. SAFEPROCESS'2003*, pages 121–125, 2003.
7. L. F. Mendonça, J. M. C. Sousa, U. Kaymak, and J. M. G. Sá da Costa. Weighted goals and constraints in fuzzy predictive control. *Journal of Int. & Fuzzy Syst.*, 17(5):517–532, 2006.

8. L. F. Mendonça, J. M. C. Sousa, and J. M. G. Sá da Costa. Fault tolerant control using fuzzy MPC. *Proc. of Safeprocess 2006, 6th IFAC Symposium on Fault Detection, Supervision and Safety of Technical Processes*, pages 1501–1506, September.
9. L. F. Mendonça, J. M. C. Sousa, and J. M. G. Sá da Costa. Optimization problems in multivariable fuzzy predictive control. *Int. Journal Appr. Reasoning*, 36(3):199–221, July 2004.
10. L. F. Mendonça, J. M. C. Sousa, and J. M. G. Sá da Costa. Fault isolation using fuzzy model-based observers. *Proceedings of Safeprocess 2006, 6th IFAC Symposium on Fault Detection, Supervision and Safety of Technical Processes*, pages 781–786, September 2006.
11. J. M. C. Sousa and U. Kaymak. Model predictive control using fuzzy decision functions. *IEEE Trans. on Syst., Man, and Cyb. Part B: Cybernetics*, 31(1):54–65, February 2001.
12. J. M. C. Sousa and U. Kaymak. *Fuzzy Decision Making in Modeling and Control.* World Scientific Pub. Co., Singapore, 2002.
13. R. Yager. Fuzzy decision making including unequal objectives. *International Journal of Man-Machine Studies*, 21:389–400, 1978.

# Synchronization in Arrays of Chaotic Neural Networks

C. Posadas-Castillo[1], C. Cruz-Hernández[2,⋆], and R.M. López-Gutiérrez[1]

[1] FIE, Baja California Autonomous University (UABC)
and FIME, University Autonomous of Nuevo León (UANL), México
[2] DET, Scientific Research and Advanced Studies of Ensenada (CICESE), México
Electronics and Telecommunications Department, P.O. Box 434944, San Diego, CA 92143-4944, USA
Phone: +52.646.1750500; Fax: +52.646.1750554
ccruz@cicese.mx

**Abstract.** In this paper, synchronization in coupled arrays of Cellular Neural Networks (*CNN*) is presented. In particular, synchronization of chaotic neural networks is obtained from complex systems theory. We consider two complex networks composed by second-order $3 \times 4$ *CNN* array, and a *CNN* with delay, the information interactions are defined via coupling law through of the first state of each cell. We impose the dynamics of a *master cell* to *multiple slave cells* of a complex network. We obtain synchronization in the complex network when the *CNN's* cells are globally coupled.

**Keywords:** Synchronization, Cellular Neural Networks (CNNs), Chaos, Complex Systems.

## 1 Introduction

Complex network structures have been observed in physics, biology, economics, ecology, electronics, and computer science. In particular, **Cellular Neural Networks *(CNNs)*** constitute an important example in such cases. CNN is a nonlinear system defined by coupling only identical simple dynamical systems called **cells** (Chua 1998). CNN has broad applications in image and video signal processing, robotic and biological visions (Werblin *et al.*, 1994), etc.

On the other hand, recently chaotic and hyperchaotic synchronization has become a field of active research, see e.g. (Pecora and Carroll 1990; Nijmeijer and Mareels 1997; Fradkov *et al.*, 1998; Cruz-Hernández and Nijmeijer 1999; 2000; Sira-Ramírez and Cruz-Hernández 2000; 2001; Pikovsky *et al.*, 2001; Cruz-Hernández 2004; López-Mancilla and Cruz-Hernández 2005a; 2005b; Aguilar-Bustos and Cruz-Henández 2006) and references therein. However, some applications of great interest require extend this synchronization to synchronizing multiple oscillators that compose complex dynamical networks, see e.g. (Pogromsky *et al.* 2001; Wang

⋆ Corresponding author.

P. Melin et al. (Eds.): IFSA 2007, LNAI 4529, pp. 743–754, 2007.
© Springer-Verlag Berlin Heidelberg 2007

and Chen 2002; Wang 2002; Posadas-Castillo *et al.* 2005; 2006b; 2006c; 2007) with the objective to make a common task, achieve collective behavior, etc.

Natural and artificial systems, many times, are composed by thousands of unities (cells), where we can know (or not) their individual behaviors. For example, the neurons and another brain's neurocells, components of an electronic circuit, the ants that conform an anthill, or a set of pages web of the network connected through link.

The main goal of this paper is to synchronize the dynamics of multiple chaotic neural networks (as cells) operating in different dynamics including chaotic motion, that conform a complex dynamical network. This objective is achieved by appealing to results from complex systems theory (Wang and Chen 2002; Wang 2002). We show that the adopted approach is indeed suitable to synchronize multiple chaotic neural networks. In particular, we will arrange the cells that composed the dynamical complex network in a topology of global coupling.

This paper is arranged as follows: In Section 2, we give a brief review on chaos synchronization of dynamical networks, synchronization conditions, and globally coupled networks. In Section 3, we apply this approach to synchronize multiple chaotic neural networks by using two illustrative examples, reported in (Posadas-Castillo *et al.*, 2006a) for two cells; a second-order $3 \times 4$ $CNN$ array and a $CNN$ with delay. Finally, in Section 4, we give some concluding remarks.

# 2 Brief Review on Synchronization of Dynamical Networks

## 2.1 Complex Dynamical Networks

As in (Posadas-Castillo *et al.*, 2006b; 2006c; 2007), we consider a **complex dynamical network** composes of $N$ identical cells, linearly and diffusively coupled through the first state variable of each cell. In this dynamical network, each cell constitutes a **n-dimensional dynamical system**, described as follows

$$\dot{\mathbf{x}}_i = f(\mathbf{x}_i) + u_i, \qquad i = 1, 2, \ldots, N, \tag{1}$$

where $\mathbf{x}_i = (x_{i1},\ x_{i2},\ \ldots,\ x_{in})^T \in \mathbb{R}^n$ are the **state** variables of the cell or node $i$, $u_i = u_{i1} \in \mathbb{R}$ is the **input** signal of the cell $i$, and is defined by

$$u_{i1} = c \sum_{j=1}^{N} a_{ij} \Gamma \mathbf{x}_j, \qquad i = 1, 2, \ldots, N, \tag{2}$$

the positive constant $c$ represents the **coupling strength** of (1), and $\Gamma \in \mathbb{R}^{n \times n}$ is a constant 0-1 matrix linking coupled state variables. For simplicity, assume that $\Gamma = \mathrm{diag}\,(r_1,\ r_2, \ldots, r_n)$ is a diagonal matrix with $r_i = 1$ for a particular $i$ and $r_j = 0$ for $j \neq i$. This means that two coupled cells are linked through their $i - th$ state variables. Whereas, $\mathbf{A} = (a_{ij}) \in \mathbb{R}^{N \times N}$ is the **coupling matrix**, which represents the coupling configuration of the dynamical network. If there is a connection between cell $i$ and cell $j$, then $a_{ij} = 1$; otherwise, $a_{ij} = 0$ for $i \neq j$. The diagonal elements of coupling matrix $\mathbf{A}$ are defined as

$$a_{ii} = - \sum_{j=1,\ j\neq i}^{N} a_{ij} = - \sum_{j=1,\ j\neq i}^{N} a_{ji}, \qquad i = 1, 2, \ldots, N. \tag{3}$$

If the degree of cell $i$ is $d_i$, then

$$a_{ii} = -d_i, \qquad i = 1, 2, \ldots, N. \tag{4}$$

Besides, suppose that the dynamical network (1) is connected in the sense that there are no isolated clusters. Then, the coupling matrix $\mathbf{A}$ is a symmetric irreducible matrix. In this case, zero is an eigenvalue of $\mathbf{A}$ with multiplicity 1 and all the other eigenvalues of $\mathbf{A}$ are strictly negative (Wang and Chen 2002; Wang 2002).

An **isolated cell** has a single input $u_{i1}$ and single output $y_{i1}$; this output can be defined as $y_{i1} = g_i(\mathbf{x}_i)$, with $g_i(\cdot)$ a nonlinear function of the state vector. Nevertheless, in many cases, the output signal of interest coincides with one of the state variables. In this work, we consider as output of interest to $y_{i1} = x_{i1}$, i.e. the first state of each cell $i$.

Synchronization state of cells in complex systems, can be characterized by the nonzero eigenvalues of the coupling matrix $\mathbf{A}$. The dynamical network (1) is said to achieve (asymptotically) **synchronization**, if (Wang 2002):

$$\mathbf{x}_1(t) = \mathbf{x}_2(t) = \cdots = \mathbf{x}_N(t), \quad \text{as} \quad t \to \infty. \tag{5}$$

The diffusive coupling condition (3) guarantees that the synchronization state is a solution, $\mathbf{s}(t) \in \mathbb{R}^n$, of an isolated cell, that is

$$\dot{\mathbf{s}}(t) = f\left(\mathbf{s}(t)\right), \tag{6}$$

where $\mathbf{s}(t)$ can be an equilibrium point, a periodic orbit, or a chaotic attractor. Thus, stability of the synchronization state,

$$\mathbf{x}_1(t) = \mathbf{x}_2(t) = \cdots = \mathbf{x}_N(t) = \mathbf{s}(t), \tag{7}$$

of dynamical network (1) is determined by the dynamics of an isolated cell.

## 2.2 Synchronization

**Theorem 1** (Wang and Chen 2002; Wang 2002)**.** *Consider the dynamical network (1). Let*

$$0 = \lambda_1 > \lambda_2 \geq \lambda_3 \geq \cdots \geq \lambda_N \tag{8}$$

*be the eigenvalues of its coupling matrix $A$. Suppose that there exists an $n \times n$ diagonal matrix $D > 0$ and two constants $\bar{d} < 0$ and $\tau > 0$, such that*

$$\left[Df(\mathbf{s}(t)) + d\Gamma\right]^T \mathbf{D} + \mathbf{D}\left[Df(\mathbf{s}(t)) + d\Gamma\right] \leq -\tau \mathbf{I}_n \tag{9}$$

*for all $d \leq \bar{d}$, where $I_n \in \mathbb{R}^{n\times n}$ is an unit matrix. If, moreover,*

$$c\lambda_2 \leq \bar{d}, \tag{10}$$

*then, the synchronization state (7) of dynamical network (1) is exponentially stable.*

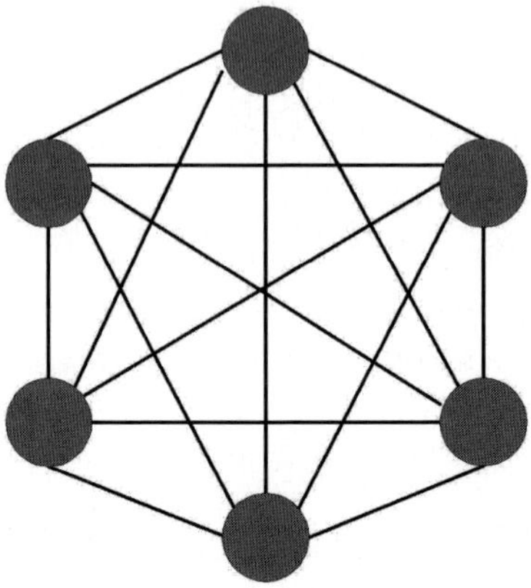

**Fig. 1.** Network of identical cells globally coupled

Since $\lambda_2 < 0$ and $\bar{d} < 0$, inequality (10) is equivalent to

$$c \geq \left| \frac{\bar{d}}{\lambda_2} \right| . \tag{11}$$

A small value of $\lambda_2$ corresponds to a large value of $|\lambda_2|$, which implies that dynamical network (1) can synchronize with a small coupling strength $c$. Therefore, synchronizability of dynamical network (1) with respect to a specific coupling configuration can be characterized by the second-largest eigenvalue of the corresponding coupling matrix **A**.

### 2.3 Globally Coupled Networks

The coupling configurations commonly studied in synchronization of complex networks are the so-called: globally coupled networks, nearest-neighbor coupled networks, and star coupled networks. In this work, we consider only complex networks of **identical cells globally coupled**.

The **globally coupled** configuration means that any two different cells are connected directly, this is shown in Figure 1 for $N = 6$. The corresponding coupling matrix is

$$A_{gc} = \begin{bmatrix} -N+1 & 1 & 1 \cdots & 1 \\ 1 & -N+1 & 1 \cdots & 1 \\ \vdots & \ddots & \ddots & \vdots \\ 1 & 1 & 1 \cdots & 1 \\ 1 & 1 & 1 \cdots & -N+1 \end{bmatrix}, \tag{12}$$

this matrix has a single eigenvalue at 0 and all the others equal to $-N$. Hence, the second largest eigenvalue $\lambda_{2gc} = -N$ decreases to $-\infty$ as $N \to \infty$,

$$\lim_{N \to \infty} \lambda_{2gc} = -\infty. \tag{13}$$

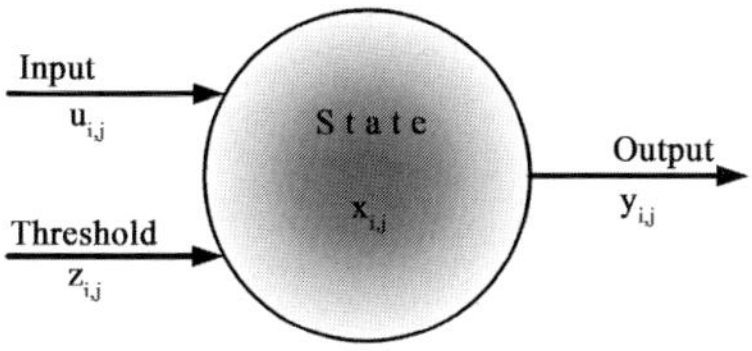


**Fig. 2.** Isolated cell: input $u_{ij}$, threshold $z_{ij}$, state $x_{ij} \in \mathbb{R}^x$, and output $y_{ij}$ for a two-dimensional CNN

## 3 Synchronization of Chaotic Neural Networks: Examples

In this section, we present two illustrative examples of synchronization of multiple chaotic neural networks, to this purpose, let us first briefly give a suitable material on CNN.

**Definition (CNN).** A CNN is any spatial arrangement of **locally coupled cells**, where each cell is a dynamical system which has an **input**, and a **state** evolving according to some prescribed dynamical laws (Chua 1998).
In three-dimensional lattice CNN architecture, mathematically each cell $C_{ijk}$ at location $(i, j, k)$ is a dynamical system whose states evolve according to some prescribed **state equations**, whose dynamics are **coupled** only among the neighboring cells lying within some prescribed **sphere of influence** $S_{ijk}$, centered at $(i, j, k)$. In two-dimensional case, using a double subscript, the variables for an **isolated** cell are: input $u_{ij}(t) \in \mathbb{R}^u$, threshold $z_{ij}(t) \in \mathbb{R}^z$, state $x_{ij}(t) \in \mathbb{R}^x$, and output $y_{ij}(t) \in \mathbb{R}^y$. A CNN cell is said to be isolated if it is not coupled to any other cell (Figure 2).

In this work, we assume that all isolated cells $C_{ij}$ are identical, and that for simplicity we have that $z_{ij}(t)$ is a constant scalar. Besides, we assume that for any $x_{ij}(t_0)$ at $t = t_0$, any threshold $z_{ij}(t)$, and any input $u_{ij}(t)$, the state of each isolated cell $C_{ij}$ is assumed to evolve for all $t \geq t_0$ as a nonautonomous set of ordinary differential equations

$$\begin{aligned} \dot{x}_{ij} &= f(x_{ij}, z_{ij}, u_{ij}), \quad i = 1, 2, \ldots, M; \quad j = 1, 2, \ldots, N, \\ y_{ij} &= g_{ij}(x_{ij}), \end{aligned}$$

where $g_{ij}(\cdot)$ is a nonlinear function of the state. However, in many cases the output of interest often coincides with the state, $y_{ij}(t) = x_{ij}(t)$.

The **standard CNN equations** used most widely in the literature, proposed in (Chua and Yang 1988) for a $M \times N$ CNN array

$$\dot{x}_{ij} = -x_{ij} + z_{ij} + \sum_{kl \in S_{ij}(r)} a_{kl}\, y_{kl} + \sum_{kl \in S_{ij}(r)} b_{kl}\, u_{kl}, \quad i = 1, 2, \ldots, M;\ j = 1, 2, \ldots, N, \tag{14}$$

$$y_{ij} = f(x_{ij}), \tag{15}$$

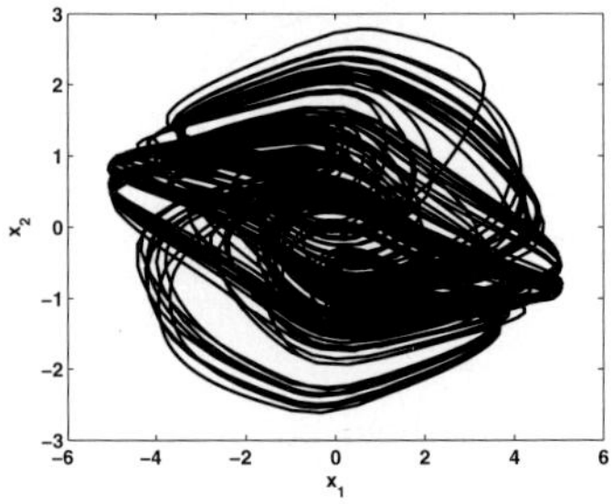


**Fig. 3.** Projection of the chaotic attractor of $3 \times 4$ $CNN$ onto the $(x_1, x_2)$ plane

where $S_{ij}(r)$ is the sphere of influence of radius $r$; $\sum_{kl \in S_{ij}(r)} a_{kl}\ y_{kl}$ and $\sum_{kl \in S_{ij}(r)} b_{kl}\ u_{kl}$ are the local coupling, and

$$f(x_{ij}) = \frac{1}{2}\left(|x_{ij}+1| - |x_{ij}-1|\right) = \begin{cases} 1, & x_{ij} \geq 1 \\ x_{ij}, & |x_{ij}| < 1 \\ -1, & x_{ij} \leq -1 \end{cases}$$

For the particular case where $M = 3$ and $N = 4$, the Eqs. (14)-(15) assume the simpler form $3 \times 4$ $CNN$ array

$$\begin{aligned} \dot{x}_1 &= -x_1 + a_{00} f(x_1) + a_{01} f(x_2) + b_{00} u_1(t), \\ \dot{x}_2 &= -x_2 + a_{0,-1} f(x_1) + a_{00} f(x_2) + b_{00} u_2(t), \\ y_1 &= f(x_1), \\ y_2 &= f(x_2). \end{aligned} \tag{16}$$

**Example 1** (Chua 1998)**.** Consider the second-order nonautonomous CNN. If $a_{00} = 2$, $a_{0,-1} = -a_{0,1} = 1.2$, $b_{00} = 1$, $u_1(t) = 4.04 \sin\left(\frac{\pi}{2}t\right)$, and $u_2(t) = 0$; then Eq. (16) becomes

$$\begin{aligned} \dot{x}_1 &= -x_1 + 2f(x_1) - 1.2f(x_2) + 4.04 \sin\left(\frac{\pi}{2}t\right), \\ \dot{x}_2 &= -x_2 + 1.2f(x_1) + 2f(x_2), \end{aligned} \tag{17}$$

with nonlinear function

$$f(x) = \frac{1}{2}\left(|x+1| - |x-1|\right) = \begin{cases} 1, & x \geq 1 \\ x, & |x| < 1 \\ -1, & x \leq -1 \end{cases} \tag{18}$$

Figure 3 shows a projection of the chaotic attractor of $3 \times 4$ $CNN$ (17)-(18). The waveforms of $(x_1(t), x_2(t))$ corresponding to the $(x_1(0), x_2(0)) = (0.1, -0.1)$.

The **complex network** of $N$ CNN's designed in this work, takes the following form,

$$
\begin{aligned}
\dot{x}_{i1} &= -x_{i1} + 2f(x_{i1}) - 1.2f(x_{i2}) + 4.04\sin\left(\frac{\pi}{2}t\right) + u_{i1}, \\
\dot{x}_{i2} &= -x_{i2} + 1.2f(x_{i1}) + 2f(x_{i2}), \\
f(x_{i1}) &= \frac{1}{2}\left(|x_{i1}+1| - |x_{i1}-1|\right), \\
f(x_{i2}) &= \frac{1}{2}\left(|x_{i2}+1| - |x_{i2}-1|\right), \\
u_{i1} &= c\sum_{j=1}^{N} a_{ij}x_{j1}, \qquad i = 1, 2, \ldots, N.
\end{aligned}
\tag{19}
$$

In particular, we consider $N = 5$, i.e. we have 5 cells constituting the complex network to be synchronized.

**Case 1.** *All the cells have chaotic behavior*, the coupling matrix (12) is given by

$$
A_{gc} = \begin{bmatrix} -4 & 1 & 1 & 1 & 1 \\ 1 & -4 & 1 & 1 & 1 \\ 1 & 1 & -4 & 1 & 1 \\ 1 & 1 & 1 & -4 & 1 \\ 1 & 1 & 1 & 1 & -4 \end{bmatrix}, \tag{20}
$$

the corresponding eigenvalues are $\lambda_1 = \lambda_2 = \lambda_3 = \lambda_4 = -5$, $\lambda_5 = 0$, with a coupling value $c = 500$ obtained from (11), with initial conditions: $x_{11}(0) = 0.1$, $x_{21}(0) = 0.21$, $x_{31}(0) = 0.31$, $x_{41}(0) = 0.4$, and $x_{51}(0) = 0.5$. With these values the Theorem 1 guarantees synchronization of the complex network. Figure 4 shows synchronization of first state of the five cells, $x_{i1}(t)$, $i = 1, 2, \ldots, 5$. Also, is shown the collective behavior of the cells in the complex network, through phase portrait of $x_{11}$ versus $x_{12}$.

**Case 2.** *An isolated cell (master cell) with chaotic behavior*, and the *rest* of the cells *have periodic behavior*, in this case the coupling matrix (12) is given by

$$
A_{gc} = \begin{bmatrix} 0 & 0 & 0 & 0 & 0 \\ 1 & -4 & 1 & 1 & 1 \\ 1 & 1 & -4 & 1 & 1 \\ 1 & 1 & 1 & -4 & 1 \\ 1 & 1 & 1 & 1 & -4 \end{bmatrix}, \tag{21}
$$

the corresponding eigenvalues are $\lambda_1 = \lambda_2 = \lambda_3 = -5$, $\lambda_4 = -1$, and $\lambda_5 = 0$, with a coupling value $c = 500$, with initial conditions: $x_{11}(0) = 0.1$, $x_{21}(0) = 0.21$, $x_{31}(0) = 0.31$, $x_{41}(0) = 0.4$, and $x_{51}(0) = 0.5$. With these values the Theorem 1 guarantees synchronization of the complex network. Figure 5 illustrates synchronization of first state of the five cells, $x_{i1}(t)$, $i = 1, 2, \ldots, 5$. Also, is shown the collective behavior of the cells in the complex network, through phase portrait of $x_{11}$ versus $x_{12}$.

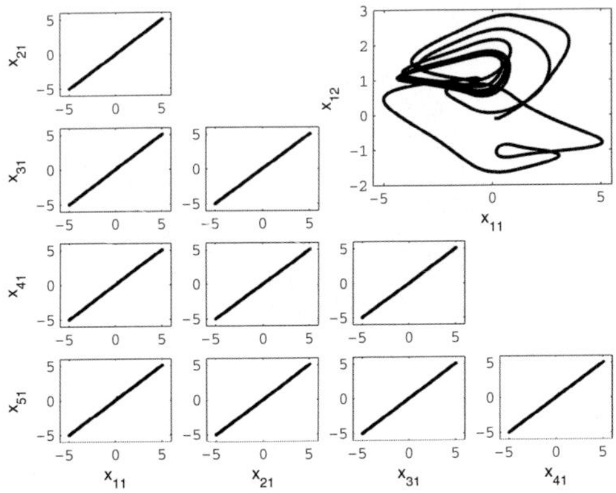

**Fig. 4.** Synchronization of first state of the five cells, $x_{i1}(t)$, $i = 1, 2, \ldots, 5$, and the behavior of the complex network projected onto the $(x_{11}, x_{12})$-plane

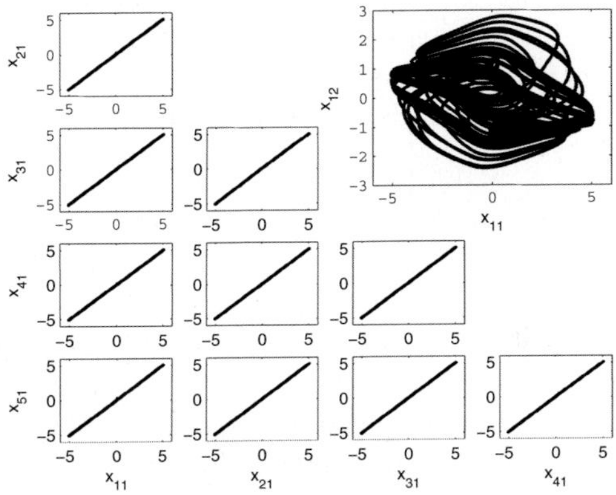

**Fig. 5.** Synchronization of first state of the five cells, $x_{i1}(t)$, $i = 1, 2, \ldots, 5$, and the behavior of the complex network projected onto the $(x_{11}, x_{12})$-plane

**Remark:** We mention that for the second state of the five cells of the complex network ($x_{i2}(t)$, $i = 1, 2, \ldots, 5$) is not possible to obtain exact synchronization. Nevertheless, the corresponding synchronization errors, i.e. $\mathbf{e}_i(t) = \mathbf{x}_i(t) - \mathbf{x}_{i+1}(t)$, $i = 1, 2, \ldots, 4$ remain bounded.

**Example 2.** Time-delay oscillators represent examples of high-dimensional chaos generators. Now, the system considered is a equation for each cell in Cellular Neural Networks with delay (Lu *et al.*, 1998), as follows,

$$\dot{x}(t) = 0.001x(t) - 3.8\left(|x_\tau + 1| - |x_\tau - 1|\right) + 2.85\left(\left|x_\tau + \frac{4}{3}\right| - \left|x_\tau - \frac{4}{3}\right|\right) \quad (22)$$

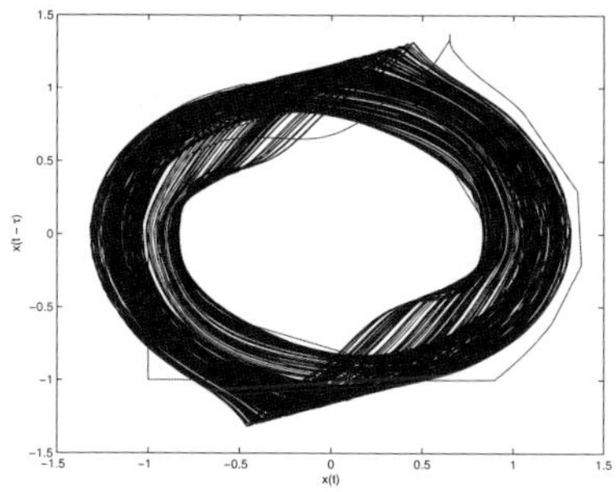


**Fig. 6.** Phase space dynamics for the Cellular Networks with delay projected onto the $(x, x_\tau)$ plane

where $x_\tau = x(t-\tau)$. Its solution is infinite-dimensional, with initial condition as any continuous function defined on the closed interval $[-\tau, 0]$. By considering $\tau = 1$ and initial condition as a constant function equal to 0.5 on $[-1, 0]$, and initial state $x(0) = -1$. Figure 6 shows a projection of the chaotic attractor of the cellular neural network with delay in the $(x, x_\tau)$-plane.

The complex network of $N$ CNN's designed in this example, takes the following form,

$$\dot{x}_i(t) = 0.001 x_i(t) - 3.8\left(|x_{\tau i} + 1| - |x_{\tau i} - 1|\right) + 2.85\left(\left|x_{\tau i} + \frac{4}{3}\right| - \left|x_{\tau i} - \frac{4}{3}\right|\right) + u_i, \tag{23}$$

for $i = 1, 2, \ldots, 5$, and with the state $x_{\tau i} = x_i(t-\tau)$, and input defined by

$$u_i = c\sum_{j=1}^{N} a_{ij} x_{j1}. \tag{24}$$

We consider in particular $N = 5$, i.e. we have 5 cells composing the complex network. For this case *all the cells* in the complex network *have chaotic behavior*, the coupling matrix (12) is given by

$$A_{gc} = \begin{bmatrix} -4 & 1 & 1 & 1 & 1 \\ 1 & -4 & 1 & 1 & 1 \\ 1 & 1 & -4 & 1 & 1 \\ 1 & 1 & 1 & -4 & 1 \\ 1 & 1 & 1 & 1 & -4 \end{bmatrix}, \tag{25}$$

the corresponding eigenvalues are $\lambda_1 = \lambda_2 = \lambda_3 = -5, \lambda_4 = -1$, and $\lambda_5 = 0$, with a coupling value $c = 1$ obtained from (11), with initial conditions: $x_{11}(0) = -1$, $x_{21}(0) = -1.1$, $x_{31}(0) = -1.2$, $x_{41}(0) = 1$, and $x_{51}(0) = 1.1$. With these values the Theorem 1 guarantees synchronization of the complex network. Figure 7 shows that the synchronization of first state of the five cells, $x_{i2}(t)$, $i = 1, 2, \ldots, 5$. Also, is shown the collective behavior of the cells in the complex network, through phase portrait of $x_{11}$ versus $x_{12}$.

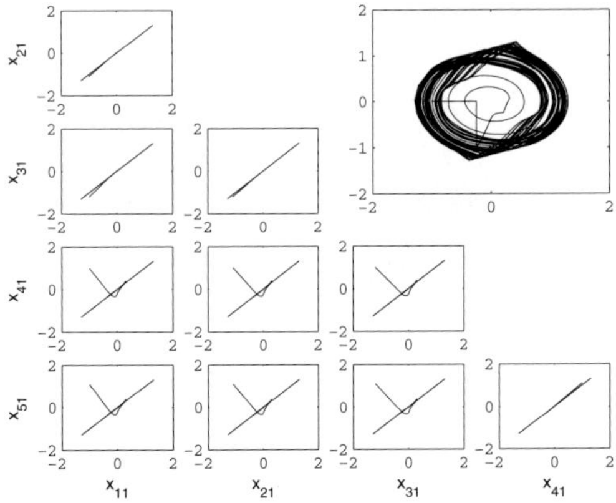


**Fig. 7.** Synchronization of state of the five cells, $x_{i1}(t)$, $i = 1, 2, \ldots, 5$, and the behavior of the complex network projected onto the $(x_{11}, x_{12})$-plane

# 4 Concluding Remarks

In this paper, we have presented synchronization in coupled arrays of Cellular Neural Networks. The synchronization of chaotic neural networks was achieved by using results from complex systems theory. We applied this approach to synchronize multiple chaotic neural networks, through two illustrative examples, reported in (Posadas-Castillo *et al.*, 2006a) for two cells. In particular, we achieved synchronization considered cells in a complex dynamical network by using a topology of global coupling. In conclusion, we have shown that synchronization of complex networks of chaotic CNNs is possible from this viewpoint.

In a forthcoming work we will be concerned with a physical implementation of networks of CNN with electronic circuits, and the synchronization of large chaotic neural networks and possible applications, in particular in communications and image processing.

# Acknowledgments

This work was supported by CONACYT, México under Research Grants No. J49593-Y and No. P50051-Y. And by UABC, México under Research Grant No. 0460.

# References

[1] Aguliar-Bustos, A. Y. and Cruz-Hernández, C. (2006) "Synchronization of discrete-time hyperchaotic systems through extended Kalman filtering," *Nonlinear Dynamics and Systems Theory* **6**(4), pp. 319-336.

[2] Chua, L.O. and Yang, L. (1988) "Cellular neural networks: Theory and applications," *IEEE Trans. Circuits Syst.* **35**, pp. 1257-1290.

[3] Chua, L.O. (1998) *CNN: A Paradigm for Complexity*, World Scientific, Singapore.

[4] Cruz-Hernández, C. and Nijmeijer, H. (1999) "Synchronization through extended Kalman filtering," *New Trends in Nonlinear Observer Design* eds. Nijmeijer & Fossen, Lecture Notes in Control and Information Sciences **244** Springer-Verlag, pp. 469-490.

[5] Cruz-Hernández, C. and Nijmeijer, H. (2000) "Synchronization through filtering," *Int. J. Bifurc. Chaos* **10**(4), pp. 763-775.

[6] Cruz-Hernández, C., Posadas, C., and Sira-Ramírez, H. (2002) "Synchronization of two hyperchaotic Chua circuits: A generalized Hamiltonian systems approach," *Procs. of 15th IFAC World Congress*, July 21-26, Barcelona, Spain.

[7] Cruz-Hernández, C. (2004) "Synchronization of time-delay Chua's oscillators with applications to secure communication," *Nonlinear Dynamics and Systems Theory* **4**(1), pp. 1-13.

[8] Fradkov, A.L., and Pogromsky, A. Yu. (1998) *Introduction to control of oscillations and chaos*, World Scientific Publishing, Singapore.

[9] López-Mancilla, D. and Cruz-Hernández, C. (2005a) "Output synchronization of chaotic systems: model-matching approach with application to secure communication," *Nonlinear Dynamics and Systems Theory.* **5**(2), pp. 141-156.

[10] López-Mancilla, D. and Cruz-Hernández, C. (2005b) "Output synchronization of chaotic oscillators and private communication," *Procs. of 16th IFAC World Congress*, July 3-8, Prague, Czech Republic.

[11] Nijmeijer, H. and Mareels, I.M.Y. (1997) "An observer looks at synchronization," *IEEE Trans. Circuits Syst. I* **44**(10), pp. 882-890.

[12] Pecora, L.M. and Carroll, T.L. (1990) "Synchronization in chaotic systems," *Phys. Rev. Lett.* **64**, pp. 821-824.

[13] Pikovsky, A., Rosenblum, M., and Kurths, J. (2001) *Synchronization: A Universal Concept in Nonlinear Sciences*, Cambridge University Press.

[14] Pogromsky A. Yu and Nijmeijer H. (2001), "Cooperative oscillatory behavior of mutually coupled dynamical systems," *IEEE Trans. Circuits Syst.-I* **48**(2), pp. 152-162.

[15] Posadas-Castillo C., Cruz-Henández C., López-Mancilla D. (2006a) "Synchronization of chaotic neural networks: A generalized Hamiltonian Systems approach," *Hybrid Intelligent Systems Analysis and Design*, Springer Verlag. And in *Procs. of International Conference on Fuzzy Systems, Neural Networks and Genetic Algorithms*, (FNG'2005), Tijuana, México, pp 332-344.

[16] Posadas-Castillo C., Cruz-Henández C., López-Mancilla D. (2005) "Sincronización de múltiples osciladores de Rössler," *Congreso Anual de la Asociación de Control Automático de México*, Cuernavaca Morelos, México.

[17] Posadas-Castillo C., Cruz-Henández C. and López-Gutiérrez R.M. (2006b) "Synchronization in a network of Chua's circuits", *Procs. of 4th IASTED, International Conference on Circuits, Signals, and Systems*, San Francisco, California, USA, pp 236-241.

[18] Posadas-Castillo C., Cruz-Henández C. and López-Gutiérrez R.M. (2006c) "Red compleja de circuitos de Chua: construcción y sincronización," *Procs. 2da conferencia Internacional de Diseño Electrónico* (ICED'2006), Veracruz, México, pp. 130-135.

[19] Posadas-Castillo C., López-Gutiérrez R.M. and Cruz-Henández C. (2007) "Synchronization of chaotic solid-state Nd:YAG lasers: Application to secure communication," to be published in *Communications in Nonlinear Science and Numerical Simulations.*

[20] Sira-Ramírez, H. and Cruz-Hernández, C. (2001) "Synchronization of Chaotic Systems: A Generalized Hamiltonian Systems Approach," *Int. J. Bifurc. Chaos* **11**(5), pp. 1381-1395. And in *Procs. of American Control Conference* (ACC'2000), Chicago, USA, pp. 769-773.

[21] Wang XF, Chen G. (2002) "Synchronization in small-world dynamical networks," *Int J Bifurct Chaos*; **12**(1), pp. 187-92.

[22] Wang XF. (2002) "Complex networks: Topology, dynamics and synchronization," *Int J Bifurct Chaos* 2002; **12**(5), pp.885-916.

[23] Werblin, F., Roska, T., and Chua, L.O. (1994) "The analogic cellular neural network as a bionic eye," *Int. J. Circuit Theor. Appl.* **23**, pp. 541-569.

# Part XIV

# Intelligent Agents and Knowledge Ant Colony

# On Fuzzy Projection-Based Utility Decomposition in Compound Multi-agent Negotiations

Jakub Brzostowski and Ryszard Kowalczyk

Faculty of Information and Communication Technologies
Swinburne University of Technology
Hawthorn, Victoria 3122, Australia
phone: +61392145623
fax: +61398190823
{jbrzostowski,rkowalczyk}@it.swin.edu.au

**Abstract.** In the process of compound multi-agent negotiation a number of agents concurrently negotiate with one or more counterparts in order to satisfy the individual preferences that lead to the collective maximization of the overall utility function imposed on the compound service. In order to perform this task the overall utility function has to be decomposed into individual single-service utility functions. This problem is not trivial, especially in compound multi-agent negotiations involving more complex aggregation patters of negotiated issues. In this paper we propose an approach for derivation of the individual utility functions based of the principles of fuzzy set projection. We also propose a way of modifying the initially generated utility functions in the case where the agreement was not reached with those functions, what allows for reaching an agreement in repeated negotiation.

## 1 Introduction

In automated negotiations a decision maker needs the specification of preferences. Its preferences are typically encoded by an utility function that assigns to each potential alternative agreement a level of satisfaction gained from consuming a product or service. The aim of the negotiation is to find an agreement maximizing the utilities of the negotiating parties [9] [14]. The notion of utility has also widely been used in multi-agent negotiations that involve a number of software agents negotiating on behalf of their users [2] [8] [10] [11] [15]. A number of real-world applications including e-commerce, e-business, planning, resource allocation, scheduling [12] have explored the agent-based negotiations. Recently the automated negotiations has been applied in Web service compositions, especially the negotiations of quality-of-service (QoS) of compound services [3]. A compound service may consist of different atomic services composed according to the various composition patterns. An example of a composition pattern is shown

P. Melin et al. (Eds.): IFSA 2007, LNAI 4529, pp. 757–766, 2007.
© Springer-Verlag Berlin Heidelberg 2007

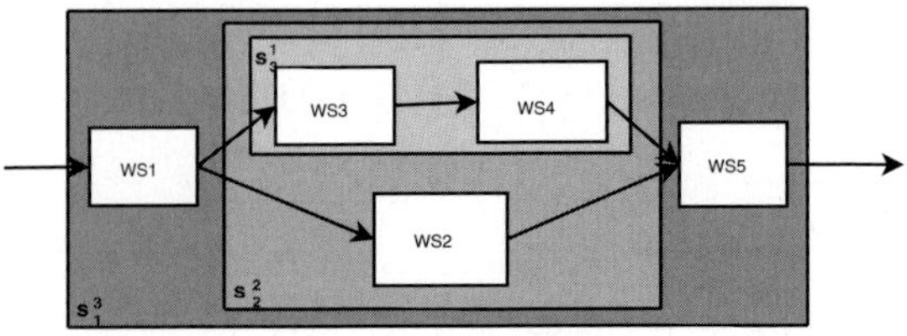


**Fig. 1.** Example of a composition pattern

in Figure 1. The negotiation of QoS typically involves a number of different attributes (negotiation issues) such as price, time, availability etc. The different negotiation issues of the compound service involve different types of aggregation of the attributes negotiated for the atomic services such as sum, max and sum or min.

The negotiation agents have beliefs about their preferences and desired outcomes, represented by the utility functions. Typically the user provides directly the utility function to an agent or the utility function is derived through the process of preference elicitation [14]. However, in compound negotiations a number of agents need to negotiate simultaneously with their counterparts representing the atomic services, and the user provides the overall utility function specifying preferences for the compound service. This means that the user is interested in the end-to-end QoS of the compound service and therefore she can only specify the overall preferences rather than the individual utility functions corresponding to the atomic services. However, the individual agents have to know these utility functions, in order to negotiate the atomic services. Therefore, the individual utility functions have to be derived from the overall utility function specified for the compound service.

In our negotiation scenario, this task is performed by a coordinating agent. The coordinator decomposes the overall utility function into the single-service utility functions and assigns these functions to the individual agents (Figure 2). After the negotiation agents have been provided with the utility functions they negotiate with the agents representing atomic services, in order to find the best compound service according to the individual utilities. If no agreement is found the whole process of assigning individual utility and negotiation is repeated with updated utility functions.

In this paper we focus on decomposition of the overall utility function corresponding to a compound service into individual utility functions corresponding to atomic services that form the compound service. However, the initially decomposed utility functions may not result in agreement and therefore we also propose ways of updating them, so the agreement can be reached in repeated negotiations.

The problem of utility decomposition is weakly related to the problem of preference elicitation [14]. However the elicitation usually focuses on obtaining the

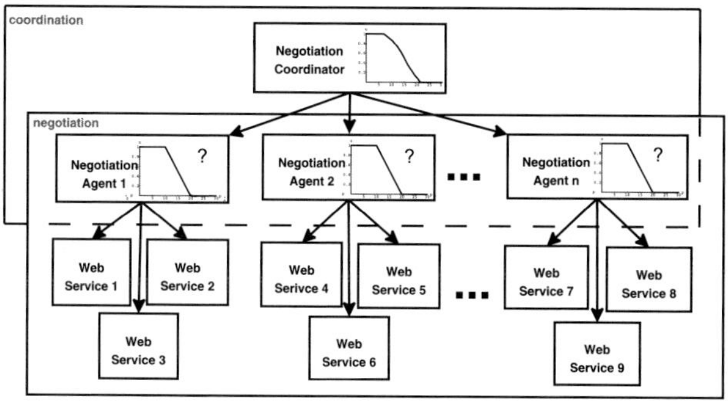


**Fig. 2.** Example of composition pattern with the aggregation of attribute values

utility function from the user and in our case we have to extract the single-service utility functions from the overall utility function that collectively correspond to the overall utility function.

The problem of bundling in combinatorial auctions [5], where the utilities are specified over the possible bundles of atomic services, is also different from the problem of utility decomposition. In our problem the user does not know in advance the atomic services involved and their composition (bundling) patterns. Therefore our utility decomposition may be regarded as an inverse problem in relation to bundling in combinatorial auctions. This problem has received a very little attention in the literature so far. In [19] a specific solution for decomposing time/utility function in utility accrual scheduling of the "distributable threads" in real-time distributed systems is presented. There is a similarity to our problem of utility decomposition, however it considers only one type of attribute (execution time) and the decomposition is tailored to the specific requirements of a different application. The authors present five decomposition techniques that are specific to different classes of utility accrual scheduling algorithms. In our problem we consider multiple QoS attributes and more complex aggregation patterns required in Web service compositions.

In this paper we propose an approach for the extraction of individual utility functions from the overall utility based on the concept of fuzzy set projection. Moreover, because the initial utility function may be too tolerant we also propose ways of modifying the initial function so the agreement can be reached in repeated negotiations. The approach is presented in the context of multi-agent negotiation of compound services consisting of different composition patters in which the agent coordinating the negotiation assigns the single-service utility functions to the individual negotiation agents. Due to space limits but without loosing generality we focus on a selected QoS attribute, namely price. Section 2 introduces the idea of generating the initial utility function as a fuzzy projection. In section 3 we show how to obtain the fuzzy projection of the overall utility function. In the fourth section we describe the way of modifying the initial utility function for the repeated negotiations. The fifth section describes the simulation of multi-agent system that was used to validate the approach. In the sixth section we present the results and finally in the section 7 we present the conclusions and the future work.

## 2 Single-Service Utility as Fuzzy Projection

The utility theory plays an important role in modelling and solving multi-attribute decision-making problems including negotiation [13][18]. It provides a general framework for assessing, comparing and ordering alternative solutions based on their utility values. The relationship between elements of utility theory and fuzzy set theory has been investigated in the literature (e.g. [4][6][16][17]) and incorporated in the framework of agent-based negotiation [10].

In general, a utility function encodes the preferences used in decision making, representing a level of satisfaction with different alternatives. On the other hand, the membership function of a fuzzy set can be regarded a fuzzy constraint [1] assigning a level of satisfaction to each alternative into an interval [0, 1]. Therefore the possible utilities can be considered in the terms of a fuzzy set, with the utility function corresponding to the membership function [4][6][10][16][17]. Considering the fuzzy set theoretical framework the utility set can be defined as follows:

$$U = \{(a, u(a)) \quad | \quad a \in D\} \quad Pow(U) = \sum_{a \in D} u(a)$$

where $U$ is the utility set with a utility function $u$ defined over the space of alternatives $D$. The utility set $U_2 = \{(a, u_2(a))|a \in D\}$ can be regarded as more tolerant than utility set $U_1 = \{(a, u_2(a))|a \in D\}$ if the $U_2$ contains $U_1$:

$$U_1 \subset U_2 \Rightarrow \forall a \in D \quad | \quad u_1(a) < u_2(a)$$

Now we describe the specification of preferences imposed on the compound service and give the definition of fuzzy set projection which will be used to derive the single-service utility function. The space of potential solutions (agreements) $S$ is defined in the following way:

$$S = S_1 \times S_2 \times \cdots \times S_n$$

where the set $S_i$ is a set of potential attribute values corresponding to the $i$-th atomic service. The overall utility function of the compound service requester is defined over the space $S$ in the following way:

$$\begin{aligned} & f : S \to [0, 1] \\ & f\ (s_1, s_2, \dots, s_n) = u(O(s_1, s_2, \dots, s_n)) \end{aligned} \tag{1}$$

where the function $u$ is the single-attribute utility function imposing the preferences of the service requester on the aggregate value of an attribute $O(s_1, s_2, \dots, s_n)$ and the operator $O$ constitutes the aggregation of the values of attributes $s_i$ corresponding to the atomic services (in the case of price it is sum $\sum_{i=1}^{n} s_i$). We have to determine the utility function for each negotiation agent negotiating with the atomic service. This single-service utility function must be extracted from the overall utility function imposed on the compound service. If we can estimate values of the attribute that can be potentially obtained from all the atomic services except $A_i$ then we can easily construct the single-service utility function needed for potential negotiation with the agent $A_i$. This utility function can be obtained by fixing all the values of attribute in the formula 1 except the one corresponding to the agent $A_i$:

$$u_i(a)=f(s_1^0, s_2^0, \ldots, s_{i-1}^0, a, s_{i+1}^0, \ldots, s_n^0)=u(O(s_1^0, s_2^0, \ldots, s_{i-1}^0, a, s_{i+1}^0, \ldots, s_n^0))$$

The sequence of the fixed values $s_1^0, s_2^0, \ldots, s_{i-1}^0, s_{i+1}^0, \ldots, s_n^0$ corresponds to the sequence of estimated values of attributes. Let us consider the set of all utility sets $\mathbb{U}$ obtained as the intersections of the function $f$ with $n-1$-dimensional subspace (all values of attributes were fixed except one in the position $i$):

$$\mathbb{U} = \{U_{\bar{c}}: \quad U_{\bar{c}} = \{(a, u_{\bar{c}}(a)) \quad | \quad a \in S_i\} \quad | \quad \bar{c} = (s_1, s_2, \ldots, s_{i-1}, s_{i+1}, \ldots, s_n) \\ , \quad u_{\bar{c}}(a) = f(s_1, s_2, \ldots, s_{i-1}, a, s_{i+1}, \ldots, s_n)\} \qquad (2)$$

Assuming that the space of all possible intersections $C$ has the following form:

$$C = S_1 \times S_2, \times \ldots, \times S_{i-1} \times S_i \times \cdots \times S_n$$

we can derive the most tolerant utility set by calculating the fuzzy union $P$ of all possible utility sets obtained by different intersections. This fuzzy set $P$ will be called the projection of the overall utility set onto the subspace $S_i$:

$$P = \bigcup_{\bar{c} \in C} U_{\bar{c}} = \{(a, max_{\bar{c} \in C} u_{\bar{c}}(a)) \quad | \quad a \in S_i\}$$

The set $P$ is the smallest set containing all the intersections $U_{\bar{c}}$:

$$\forall \bar{c} \in C \quad | \quad U_{\bar{c}} \subseteq P \Leftrightarrow \forall \bar{c} \in C \quad \forall a \in S_i \quad | \quad u_{\bar{c}}(a) < max_{\bar{c} \in C} u_c(a)$$

The projection $P$ is the smallest utility set that is more tolerant than any utility set $U_c$ obtained by the intersection $c$. Therefore projection my be proposed as the most tolerant single-service utility set. Such a highly tolerant utility set is safe because it gives high chance of reaching an agreement with the individual utility function. However, the overall preferences may not be satisfied or the overall utility value may small, and therefore it has to be modified into less tolerant utility function so the agreement satisfying the overall preferences of bothe parties may be reached in one of the next renegotiations.

## 3 The Projection Function as the Boundary Function

In this section we propose how to derive the single-service utility function in a form of projection from the compound service utility function. We show that the boundary function that can be obtained very easily is equal to the projection function.

In the next theorem we will state that the boundary function is equal to the projection function (the proof will be presented in another paper).

Let us assume that the point $\bar{c}_0(a)$ is the border point of the space $C_i$ ($\bar{c}_0(a) \in C_i$) reduced by the subspace $S_i$: $C_i = S_1 \times S_2 \times \cdots \times S_{i-1} \times S_{i+1} \times \cdots \times S_n$

$$\bar{c}_0(a) = (\underset{1,}{0}, \underset{2,}{0}, \ldots, \underset{i-1,}{0}, \underset{i,}{a}\ \underset{i+1,}{0}, \ldots, \underset{n}{0})$$

where $a$ is any point in the set $S_i$. The border function $u_{\bar{c}_0}(a)$ is obtained by mapping the border point $\bar{c}_0(a)$ by the overall utility function:

$$u_{\bar{c}_0}(a) = f(\bar{c}_0(a)) = \mathrm{f}(\underset{1,}{0}, \underset{2,}{0}, \ldots, \underset{i-1,}{0}, \underset{i,}{a}\ \underset{i+1,}{0}, \ldots, \underset{n}{0})$$

The projection is obtained by the use of fuzzy disjunction of all the cut functions $u_{\bar{c}}(a) = f(\bar{c}(a))$ (where $\bar{c} = (s_1, s_2, \ldots, s_{i-1}, s_{i+1}, \ldots, s_n)$ is any point in $C_i$) in a following way:

$$u_p(a) = max_{\bar{c} \in C_i} u_{\bar{c}}(a)$$

**Theorem 1.** *The border function $u_{\bar{c}_0}$ is equal to the projection $u_p$:*

$$\forall a \in S_i \quad | \quad u_{\bar{c}_0}(a) = u_p(a)$$

## 4 The Modification of Initial Utility Function for Repeated Negotiations

The initial utility function obtained in the form of fuzzy projection of the overall utility function may be too tolerant. It means that the negotiation may end with an agreement (satisfying the preferences of both parties in terms of the utilities of individual agents) but the overall preferences of the client side may not be satisfied. This means that the negotiation has to be repeated with modified (less tolerant) utility functions. We propose two methods for the modification of the initial utility function: the constant shift and bisection algorithm.

### 4.1 The Constant Shift

One way to modify the utility function corresponding to the client agent is to shift it into the direction of the client's best position (left in the case of price attribute). In the terms of fuzzy arithmetic, this shift corresponds to the fuzzy subtraction. If the initial utility function is treated as a fuzzy number (fuzzy preferred alternative) we subtract a crisp number from it, what results in the shift left along the x-axis. After this operation we obtain more restrictive (less tolerant) utility function and the negotiation is repeated with the modified utility function. When after the negotiation the overall client's preferences are still not satisfied then the process of modification and negotiation is repeated multiple times until the agreement satisfying preferences of both parties is met.

The negotiation shift is determined as a percentage of the maximal shift. If the maximal number of repeated negotiations is set to $n$ then the partial shift (for one negotiation) is the $\frac{1}{n}$ of the maximal shift $s_m$. The maximal shift $s_m$ is determined as the distance between the last negotiation result $\bar{o}_k$ (in the $k$-th repeated negotiation) and the lowest indifference curve $I_\epsilon$ of the client's overall utility function:

$$I_\epsilon = \{\bar{x} \mid u(\bar{x}) = \epsilon\}$$

where the $\epsilon$ is a value close to 0.

$$s_m = d(I_\epsilon, \bar{o}_k) = min_{\bar{x} \in I_\epsilon} d(\bar{x}, \bar{o}_k)$$

where the $d$ function is the Euclidean distance. After each repeated $k$-th negotiation the value of $s_m$ is updated using new value of negotiation outcome $\bar{o}_k$. If there are $n - k$ remaining negotiations then the partial shift is calculated in the following way:

$$s_p = \frac{1}{n-k} s_m = \frac{1}{n-k} min_{\bar{x} \in I_\epsilon} d(\bar{x}, \bar{o}_k)$$

### 4.2 The Bisection Algorithm

If the utility function is too tolerant (shifted too far right in the case of price attribute) then the resulting negotiation may give solution satisfying the preferences of the opponent but not satisfying our preferences. If the utility function is too restrictive (shifted too far left in the case of price attribute) than there will be no solution satisfying the preferences of our opponent. The task is to find some middle point (not too large and too small shift) that would lead to the solution satisfying the preferences of both negotiation parties. If we know that the solution (optimal shift $o_s$) lies in an interval $o_s \in [a, b]$ then we can apply the bisection algorithm to determine it. The algorithm starts by evaluating the midpoint $\frac{(a+b)}{2}$ - the utility is shifted by the midpoint and the negotiation is performed. Then depending on the negotiation result one of the subintervals $[a, \frac{(a+b)}{2}]$, $[\frac{(a+b)}{2}, b]$ is chosen. If the shift was too far than the second interval is chosen and if the shift was too close than the first interval is chosen. The procedure is then repeated with the new interval as often as needed to locate the solution.

## 5 Simulation of a Multi-agent System

In order to validate the proposed approach we simulate the multi-agent system consisting of some number of client-agents that are coordinated by the coordinator-agent and a number of provider-agents. Each client-agent may negotiate with some number of candidates (provider-agents) representing the one type of atomic service. In our case each of 3 client agents looking for atomic services of different types negotiates with 3 candidates (provider-agents) as shown in Figure 1. First the coordinator agent assigns the initial utility functions to all negotiation agents. The negotiation agents negotiate with all potential candidates representing atomic services. Next, they forward the results to the coordinator agent that checks if the overall preferences are satisfied. When the overall preferences are not satisfied the coordinator agent assigns to the negotiation agents the modified utility function and all the negotiations are repeated. This procedure is repeated until the agreement is found. Since, each negotiation agents has three options (three potential partners providing the atomic service) and there are three components negotiated by the corresponding negotiation agents, there are 27 possible contracts ($3 \times 3 \times 3 = 27$). The coordinator chooses one contract that maximizes the overall utility. The agents negotiate using the positional bargaining, e.g. the agents start at their best positions according to the utility functions and concede in the utility spaces until the offers of the client-agents exceed the last offers of the provider-agents. The agents concede according the the negotiation strategy described by the $\beta$ parameter indicating the level of convexity or concavity of the concession curve [7]. During the simulation the preferences of the client-agents are updated and the negotiation strategy does not change. The preferences of the provider-agents are randomly chosen (according to the uniform distribution) in the beginning and do not change during the simulation. However, the negotiation strategy is changing after each encounter, e.g. it is randomly chosen after each negotiation (according to the uniform distribution).

## 6 Results and Discussion

The simulation was run 100 times for each of 9 client negotiation strategies ($\beta \in \{0.2, 0.4, 0.6, 0.8, 1.0, 1.25, 1.66, 2.5, 5.0\}$). For each strategy the average utility gain was calculated over the 100 experiments. The average number of negotiations needed to reach agreement and the percentage of successful encounters were calculated.

The negotiation pay-off was calculated as the difference between the utility of negotiation outcome and the utility of the Nash equilibrium point (the point maximizing the product of utilities of both negotiation parties). From the Figures 3 and 4 we can see that the bisection algorithm gives better results in term of pay-off than the constant shift algorithm. That is because the first approach

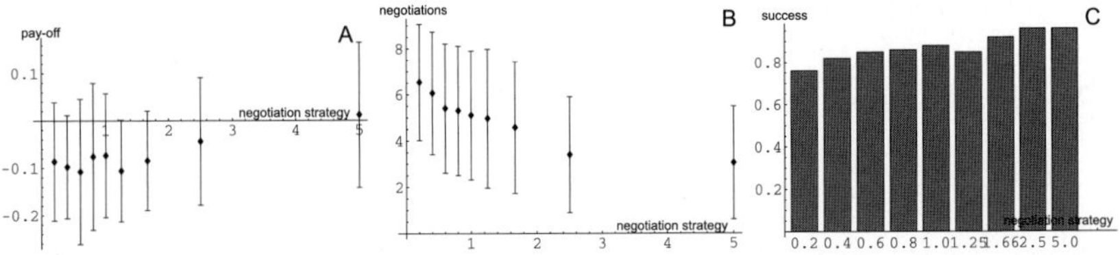


**Fig. 3.** The firs approach: constant shift. A) average pay-off over 100 experiments with the standard deviation of this value for different negotiation strategies used by the client-agents. B) average number of negotiations needed to reach agreement in one simulation averaged over 100 experiments for different negotiation strategies. C) the percentage of successful simulations for different negotiation strategies.

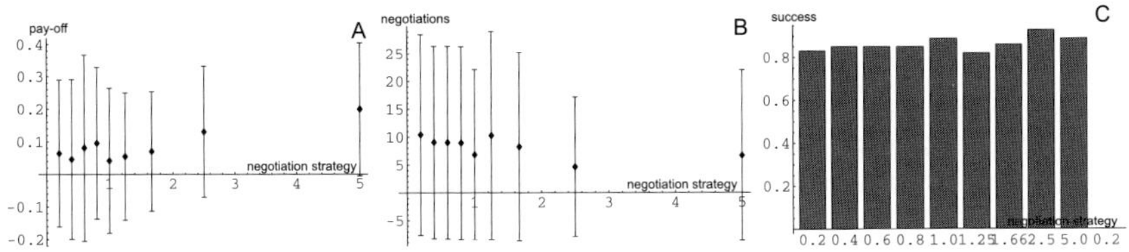


**Fig. 4.** The second approach: bisection. A, B and C show the results according to Fig above.

start with tolerant utility function and changes it gradually into more and more restrictive utility function. The algorithm stops as soon as any solution is found (alternative satisfying preferences of both parties). Therefore this algorithm does not result in high pay-off. The bisection algorithm gives better pay-offs because it always looks for the solution in the middle of divided interval what gives higher chance of reaching good utilities. Although, usually the bisection algorithm gives the solution faster than the constant shift algorithm in the Figure 4 the average number of negotiations is higher than in the case of constant shift (Figure 3). The reason for that is the failed encounters also contribute to the overall average value of number of negotiations. The bisection algorithm stopped after 50 repeated negotiations when no solution was found and this value increased the average number of negotiations.

## 7 Conclusions and Future Work

The extraction of the single-service utility function from the overall preferences imposed on the compound service is an important task of compound multi-agent negotiation. We propose to use fuzzy projection of the overall utility function as the initial single-service utility function or the reference utility function. We also propose two alternative approaches for the modification of the initial utility function, namely the constant shift algorithm and the bisection algorithm. Both algorithms give good solutions. However, the results of the bisection algorithm are better in terms of the negotiation pay-off. In the future work we will consider different ways of modyfing the utility function based on fuzzy arithmetic.

## Acknowledgments

This work is part of the Adaptive Service Agreement and Process Management (ASAPM) in Services Grid project (AU-DEST-CG060081) in collaboration with the EU FP6 Integrated Project on Adaptive Services Grid (EU-IST-004617). The ASAPM project is proudly supported by the *Innovation Access Programme - International Science and Technology* established under the Australian Governments innovation statement, *Backing Australia's Ability*.

## References


1. R.E. Bellman and L.A. Zadeh. Decision-making in a fuzzy environment. *Management Sceince*, 17(4):141–164, 1970.
2. P. Braun, J. Brzostowski, G. Kersten, J. Kim, R. Kowalczyk, S. Strecker, and R. Vahidov. *E-Negotiation Systems and Software Agents Methods, Models, and Applications*. Springer-Verlag London, 2005.
3. M. Chhetri, J. Lin, S. Goh, J. Yan, J.Y. Zhang, and R. Kowalczyk. A coordinated architecture for the agent-based service level agreement negotiation of web service compositions. In J. Han and M. Staples, editors, *The proceedings of The Australian Software Engineering Conference*, page in press, Sydney, Australia, 2006. Springer.
4. F. Chiclana, F. Herrera, and E. Herrera-Viedma. Integrating three representation models in fuzzy multipurpose decision making based on fuzzy preference relations. *Fuzzy Sets and Systems*, 97:33–48, 1998.
5. Sven de Vries and Rakesh V. Vohra. Combinatorial auctions: A survey. *INFORMS Journal on Computing*, 15(3):284–309, 2003.
6. D. Dubois and H. Prade. Possibility theory as a basis for qualitative decision theory. In Chris Mellish, editor, *Proceedings of the Fourteenth International Joint Conference on Artificial Intelligence*, pages 1924–1930, San Francisco, 1995. Morgan Kaufmann.
7. P. Faratin, C. Sierra, and N. Jennings. Negotiation decision functions for autonomous agents. *Int. Journal of Robotics and Autonomous Systems*, 24(3-4): 159–182, 1998.
8. N. R. Jennings, P. Faratin, A. Lomuscio, S. Parson, C. Sierra, and M. Wooldridge. Automated negotiation: Prospects, methods and challenges. *International Journal of Group Decision and Negotiation*, 10(2):199–215, 2001.
9. R. L. Keeny and H. Raiffa. *Decisions with Multiple Objectives: Preferences and Value Tradeoffs*. John Wiley and Sons, 1976.
10. R. Kowalczyk. Fuzzy enegotiation agents. *Journal of Soft Computing, Special Issue on Fuzzy Logic and the Internet*, 6(5):337–347, 2002.
11. S. Kraus. *Strategic Negotiation in Multiagent Environments*. MIT Press, 2001.
12. Guoming Lai, Cuihong Li, Katia Sycara, and Joseph Andrew Giampapa. Literature review on multi-attribute negotiations. Technical Report CMU-RI-TR-04-66, Robotics Institute, Carnegie Mellon University, Pittsburgh, PA, December 2004.
13. R. Lewicki, D. Saunders, , and J. Minton. *Essentials of Negotiation*. Irwin, 1997.
14. H. Raiffa. *The Art and Science of Negotiation*. Belknap Press, 1985.
15. J. Rosenschein and G. Zlotkin. *Rules of Encounter: Designing Conventions for Automated Negotiation among Computers*. MIT Press, 1994.
16. F. Seo. Construction of fuzzy utility functions in group decision making. *Consensus Under Fuziness*, pages 211–230, 1995.
17. F. Seo and M. Sakawa. Fuzzy multiattribute utility analysis for collective choice. *IEEE Transactions on Systems*, 15(1):45–53, 1985.
18. K. Sycara. The persuader. *The Encyclopedia of Artificial Intelligence*, 1992.
19. H. Wu, B. Ravindran, E. D. Jensen, and P. Li. Time/utility function decomposition techniques for utility accrual scheduling algorithms in real-time distributed systems. *IEEE Trans. Computers*, 54(9):1138–1153, 2005.

# Conditional Dempster-Shafer Theory for Uncertain Knowledge Updating

Hexin Lv[1], Bin Zhu[1], and Yongchuan Tang[2,⋆]

[1] College of Information Science and Technology, Zhejiang Shuren University
Hangzhou, Zhejiang Province, 310015, P.R. China
[2] College of Computer Science, Zhejiang University
Hangzhou, Zhejiang Province, 310027, P.R. China
Tel(Fax): 86-571-87953695
yongchuan@263.net

**Abstract.** This paper presents a theory called conditional Dempster-Shafer theory (CDS) for uncertain knowledge updating. In this theory, *a priori* knowledge about the value attained by an uncertain variable is modeled by a fuzzy measure and the evidence about the underlying uncertain variable is modeled by the Dempster-Shafer belief measure. Two operations in CDS are discussed in this paper, the conditioned combination rule and conditioning rule, which deal with evidence combining and knowledge updating, respectively. We show that conditioned combination rule and conditioning rule in CDS satisfy the property of Bayesian parallel combination.

## 1 Introduction

A unified framework to model the knowledge about an uncertain variable is the class of monotonic nonadditive measures called fuzzy measures[13,16] which have properties very suitable for representation and management of uncertain information. When we compile the knowledge about an uncertain variable, we assume the real knowledge can be approximated by a fuzzy measure since of insufficient analysis, that is, we use a fuzzy measure to model *a priori* knowledge. When new evidence about the uncertain variable is available, we should update the knowledge about the underlying uncertain variable from *a priori* knowledge.

We focus on the knowledge updating problem in evidence theory involving *a priori* knowledge in this paper. Mathematical theory of evidence was first introduced by Dempster in the 1960s [4], and later extended by Shafer [10]. This theory, which allows to represent both imprecision and uncertainty, appears as a more flexible and general approach than the Bayesian one. One of its advantages is its ability to consider not only single (or individual) values of the uncertain variable, but also unions of values. Applications were developed in image processing [1,15], signal detection [2], target identification [3], remote sensing classification [7], multiple-attribute decision making [18], and some intelligent

⋆ Corresponding author.

P. Melin et al. (Eds.): IFSA 2007, LNAI 4529, pp. 767–777, 2007.
© Springer-Verlag Berlin Heidelberg 2007

systems [5,6,20]. Other important contributions to Dempster-Shafer theory of evidence are its fuzzy versions [17,19] and its variation–the transferable belief model [11,12].

In this paper we assume that *a priori* knowledge about the uncertain variable is modeled by a fuzzy measure, and the evidence is modeled by the Dempster-Shafer belief measure. We propose two rules called the rule of conditioned Dempster-Shafer combination and the conditioning rule. One is used as the evidence combining rule, the other is used as the knowledge updating rule. When two or more pieces of evidence are available, all available bodies of evidence are first combined by using the rule of conditioned Dempster-Shafer combination, then *a priori* knowledge is updated by using the conditioning rule. The final updated knowledge can be represented by a conditional fuzzy measure conditioned on the combined evidence. One important property of these two rules is that the Möbius inversion formula of the updated fuzzy measure satisfies the property of Bayesian parallel combination.

In fact, the work in this paper is a generalization of R. Mahler's work [8,9] where *a priori* knowledge and evidence about the uncertain variable are all modeled by the Dempster-Shafer belief measures. Since there exists different types of uncertainties, such as randomness, lack of specificity, and imprecision, we adopt a more general framework—fuzzy measure to represent *a priori* information about the uncertain variable [14]. It is possible to construct a more general theory to deal with the problem of knowledge updating wherein both *a priori* knowledge and evidence are modeled by lower (upper) previsions. At present time, however, this is not the case: the theory presented in this paper is adopted because of its intriguing Bayesian-like behavior.

The paper is organized as follows. Section 2 describes some basic elements from conventional Dempster-Shafer theory and fuzzy measure. Section 3 presents conditional Dempster-Shafer (CDS) theory, in which the conditioned combination rule and conditoning rule are discussed in detail. Section 4 shows that the conditioned combination rule and the conditioning rule in CDS satisfy the property of Bayesian parallel combination. The last section is our conclusions.

## 2 Theory of Evidence and Fuzzy Measure

In this section we introduce some preliminary ideas of conventional Dempster-Shafer theory of evidence [4,10] and fuzzy measure [13] needed in this paper.

### 2.1 Dempster-Shafer Theory of Evidence

In Dempster-Shafer theory of evidence, one piece of evidence concerning the value of an uncertain variable is represented by so called "body of evidence" which is a normalized random set $\Sigma$ having the form of

$$\Sigma = \sum_{S \subseteq U} m_{\Sigma}(S) S,$$

such that $m_\Sigma(\emptyset) = 0$, $m_\Sigma(S) \geq 0$ and $\sum_{S \subseteq U} m_\Sigma(S) = 1$. Here, the "focal" subset $S$ of finite universe $U$ such that $m_\Sigma(S) > 0$ represents the hypothesis that the value of the underlying variable is in $S$. The corresponding "mass" $m_\Sigma(S) > 0$ is the confidence which attributes exactly to $S$ but to no smaller subset of $S$.

The belief measure $\beta_\Sigma(\cdot)$ and commonality measure $\delta_\Sigma(\cdot)$ associated with $\Sigma$ are

$$\beta_\Sigma(S) \triangleq \sum_{E \subseteq S} m_\Sigma(E), \delta_\Sigma(S) \triangleq \sum_{E \supseteq S} m_\Sigma(E).$$

The plausibility measure associated with $\Sigma$ is defined by $Pl_\Sigma(S) = 1 - \beta_\Sigma(S^c)$.

If $\Gamma$ is another body of evidence then $\Sigma$ and $\Gamma$ can be fused into a combined body of evidence $\Sigma * \Gamma$ using the Dempster-Shafer rule of combination. That is, $\Sigma * \Gamma = \sum_{E \subseteq U} m_{\Sigma * \Gamma}(E)E$. Here, for any $E \neq \emptyset$,

$$m_{\Sigma * \Gamma}(E) = \frac{\sum_{S \cap T = E} m_\Sigma(S) m_\Gamma(T)}{1 - K}$$

with

$$K = \sum_{S \cap T = \emptyset} m_\Sigma(S) m_\Gamma(T),$$

where $K$ is a normalization constant, called conflict because it measures the degree of conflict between $\Sigma$ and $\Gamma$. $K = 0$ corresponds to the absence of conflict between $\Sigma$ and $\Gamma$, whereas $K = 1$ implies the complete contradiction between $\Sigma$ and $\Gamma$.

### 2.2 Fuzzy Measure

Assume $V$ is a variable which attains its value in the space $U$. In situations in which the exact value of the variable $V$ is unknown, the best we can do is to try to formulate the knowledge about $V$ in a useful mathematical structure—fuzzy measure[16]. One useful feature of this measure is to represent in a unified way different types of characterizations of uncertainty.

Formally a fuzzy measure $f$ on a space $U$ is a mapping from subsets of $U$ into the unit interval $f : 2^U \to [0, 1]$ satisfying the following conditions:

$$f(\emptyset) = 0, \;\; f(U) = 1, f(A) \leq f(B), \text{ if } \;\; A \subset B \subset U.$$

Within the framework of using the fuzzy measures to represent information about an uncertain variable, $f(E)$ can be interpreted as a measure associated with our belief that the value of $V$ is contained in the subset $E$.

It is noticeable that we can get another measure $\nu_f$ associated with fuzzy measure $f$ using the Möbius inversion formulas,

$$\nu_f(S) = \sum_{T \subseteq S} (-1)^{\#(S-T)} f(T), \tag{1}$$

for all $S \subseteq U$, where, for all $X \subseteq U$, $\#X$ denotes the number of elements in $X$. So the fuzzy measure can be recovered from $\nu_f$:

$$f(S) = \sum_{T \subseteq S} \nu_f(T). \tag{2}$$

If $f$ is a probability measure on $U$ then the Möbius transform $\nu_f(\{u\}) = f(\{u\})$ for all $u \in U$ and $\nu_f(S) = 0$ for all $S \subseteq U$ such that $\#S > 1$.

It is easy to verify that both belief and plausibility measures associated with a piece of evidence are fuzzy measures, but a commonality measure isn't a fuzzy measure. So, for a piece of evidence $\Sigma$ on $U$, we have $\nu_{\beta_\Sigma}(S) = m_\Sigma(\mathrm{S})$ for all $S \subseteq U$.

The conditional fuzzy measure of $S \subseteq U$ conditioned on $T \subseteq U$ is defined as follows:

$$f(S \mid T) = \frac{f(S \cap T)}{f(T)} \tag{3}$$

provided that $f(T) \neq 0$; and $f(S \mid T) = 0$ otherwise.

## 3 Combining and Conditioning in the Conditional Dempster-Shafer

In this paper, a fuzzy measure $f$ is used to model *a priori* knowledge about the value of the uncertain variable. When two or more bodies of evidence about the underlying uncertain variable are available, they should be fused using some rule of combination with respect to *a priori* knowledge. In this section, conditional Dempster-Shafer theory (CDS) is developed. In subsection 3.1, the pure algebraic definition of conditioned combination in CDS is introduced. In subsection 3.2, the conditioning rule in CDS is proposed and analyzed.

### 3.1 Combining in CDS

In subsection 2.1, we denote a Dempster-Shafer body of evidence $\Gamma$ as a formal weighted sum $\Gamma = \sum m_\Gamma(S)S$, where $m_\Gamma(S) \geq 0$ is the "mass" allocated to the subset $S$. In CDS discussed in this subsection, for a formal weighted sum $\mathcal{B} = \sum \mathcal{B}_S S$, $\mathcal{B}_S$ is allowed to be a real number. When $\mathcal{B}$ is a piece of evidence, we have $\mathcal{B}_S = m_\mathcal{B}(S)$ for any $S \subseteq U$. This algebraic notation has been employed by Mahler[8].

In this subsection, the conditioned combination rule in CDS is proposed. Using the conditioned combination rule defined in this subsection, when any two bodies of evidence $\mathcal{B}$ and $\mathcal{C}$ are available, we can achieve a fused evidence $\mathcal{B} *_f \mathcal{C}$ with respect to *a priori* knowledge representable by a fuzzy measure $f$ on $U$.

**Definition 1.** *a) Let $R[U]$ denote the vector space generated over the real numbers $R$ by the subsets of $U$. That is, it consists of the vectors of the form $\mathcal{B} = \sum_{S \subseteq U} \mathcal{B}_S S$ for $\mathcal{B}_S \in R$ and where the collection $\{S\}$ for $S \subseteq U$ is assumed to be a set of basic vectors linearly independent over $R$.*

*b) Let $f$ be a fuzzy measure over $U$. Then $R[U; f]$ denotes the subspace of $R[U]$ generated by all of subsets $S \subseteq U$ such that $f(S) > 0$.*

According to the above definition, we know that the set of all pieces of evidence on $U$ is a subset of $R[U]$. Now we are in a position to define the conditioned combination rule in CDS.

**Definition 2 (Conditioned Combination Rule).** *Let $f$ be a fuzzy measure over $U$ and $\mathcal{B}, \mathcal{C} \in R[U]$ where $\mathcal{B} = \sum_{S \subseteq U} \mathcal{B}_S S$ and also $\mathcal{C} = \sum_{T \subseteq U} \mathcal{C}_T T$. Then*
*a) The conditioned agreement of $\mathcal{B}, \mathcal{C}$ with respect to $f$ is defined by*

$$\alpha_f(\mathcal{B}, \mathcal{C}) \triangleq \sum_{S,T \subseteq U} \mathcal{B}_S \mathcal{C}_T \alpha_f(S, T)$$

*where*

$$\alpha_f(S, T) \triangleq \frac{f(S \cap T)}{f(S) f(T)}$$

*provided that $f(S) \neq 0 \neq f(T)$; and $\alpha_f(S, T) = 0$ otherwise.*
*b) The conditioned product of $\mathcal{B}, \mathcal{C} \in R[U]$ with respect to $f$ is defined by*

$$\mathcal{B} \cdot_f \mathcal{C} \triangleq \sum_{S,T \subseteq U} \mathcal{B}_S \mathcal{C}_T \alpha_f(S, T)(S \cap T).$$

*c) The conditioned Dempster-Shafer combination of $\mathcal{B}, \mathcal{C}$ with respect to $f$ is defined by*

$$\mathcal{B} *_f \mathcal{C} \triangleq \frac{\mathcal{B} \cdot_f \mathcal{C}}{\alpha_f(\mathcal{B}, \mathcal{C})}$$

*whenever $\alpha_f(\mathcal{B}, \mathcal{C}) \neq 0$.*

If fuzzy measure $f(S) = 1$ for all non-null subsets $S \subseteq U$, both $\mathcal{B}$ and $\mathcal{C}$ are two pieces of evidence, then CDS combination degenerates to the conventional Dempster-Shafer combination. Furthermore, if fuzzy measure $f$ is a Dempster-Shafer belief measure then CDS reduces to CDS combination discussed by Mahler[8]. The combination rule presented here considers not only all available evidence but also *a priori* knowledge, we may anticipate that this feature augments the representation power of CDS.

Since $\mathcal{B} *_f \mathcal{C}$ is in $R[U]$, it can be written as a linear combination of the basic vectors of $R[U]$. That is, $\mathcal{B} *_f \mathcal{C} = \sum_{E \subseteq U} (\mathcal{B} *_f \mathcal{C})_E E$ where

$$(\mathcal{B} *_f \mathcal{C})_E = \sum_{S \cap T = E} \mathcal{B}_S \mathcal{C}_T \frac{\alpha_f(S, T)}{\alpha_f(\mathcal{B}, \mathcal{C})}.$$

So CDS combination has the same general form as the conventional Dempster-Shafer combination, differing only in that conditioned agreement is redefined so as to reflect the influence of *a priori* knowledge. Especially, when both $\mathcal{B}$ and $\mathcal{C}$ are two bodies of evidence, then $\mathcal{B} *_f \mathcal{C}$ is also a body of evidence.

### 3.2 Conditioning in CDS

When CDS is applied to knowledge updating, *a priori* knowledge represented by a fuzzy measure $f$ should be updated based on the evidence $\mathcal{B}$ at hand. In this subsection, a conditioning rule in CDS is introduced, which carries out this knowledge updating process. The basic idea to introduce this updating rule is to define the posterior fuzzy measure conditioned on the body of evidence $\mathcal{B}$.

For purpose of simplicity, in what follows, we assume that $f(S) \neq 0$ for all $\emptyset \neq S \subseteq U$.

**Proposition 1 (Conditioning Rule).** *Let $\mathcal{B}$ be a body of evidence on $U$. Then $f(S|\mathcal{B})$ is a fuzzy measure for the subset $S \subseteq U$, where $f(S|\mathcal{B}) = \sum_{E \subseteq U} f(S \mid E)\mathcal{B}_E$.*

Proof. It is clear that $f(\emptyset|\mathcal{B}) = 0$, and $f(U|\mathcal{B}) = 1$ since of the following assumption,$f(E) \neq 0$ for all $\emptyset \neq E \subseteq U$. For any $S \subseteq T \subseteq U$, we have

$$f(S|\mathcal{B}) = \sum_{E \subseteq U} \frac{f(E \cap S)\mathcal{B}_E}{f(E)} \leq \sum_{E \subseteq U} \frac{f(E \cap T)\mathcal{B}_E}{f(E)} = f(T|\mathcal{B}).$$

So the result follows. ■

Notice that the (conditional) fuzzy measure, $f(S|\mathcal{B}) = \sum_{E \subseteq U} f(S|E)\mathcal{B}_E$, is a weighted summation of posterior fuzzy measures of $S$ conditioned on all constant subsets of $U$, and the weights are the "mass" of the evidence $\mathcal{B}$. In addition, we have $f(S \mid \mathcal{B}) = \alpha_f(\mathcal{B}, f(S)S)$.

It is very easy to verify that if fuzzy measure $f$ is a probability measure then posterior fuzzy measure $f(\cdot|\mathcal{B})$ conditioned on a piece of evidence $\mathcal{B}$ is still a probability measure. We can also verify that if fuzzy measure $f$ is a belief (plausibility) measure then posterior fuzzy measure $f(\cdot|\mathcal{B})$ conditioned on a piece of evidence $\mathcal{B}$ is still a belief (plausibility) measure.

Especially, when fuzzy measure $f$ over $U$ degenerates to a plausibility measure such that for any non-null subset $E \subseteq U$, $f(E) = 1$, then CDS conditioning rule is consistent with conventional evidence theory.

**Corollary 1.** *Let $f$ over $U$ be a plausibility measure such that for any non-null subset $E \subseteq U$, $f(E) = 1$, and $\Sigma$ be a body of evidence on $U$, then*

$$f(S \mid \Sigma) = Pl_{\Sigma}(S) \tag{4}$$

*for any $S \subseteq U$.*

Usually the total ignorance is modeled by a fuzzy measure such that for any non-null subset $E \subseteq U$, $f(E) = 1$. The above conclusion shows that when one's prior knowledge is in total ignorance, the information represented by new evidence is totally consistent with the updated knowledge. So we can say that the conventional evidence theory is a special CDS where one's beliefs on *a priori* knowledge is in total ignorance.

## 4 Property of Bayesian Parallel Combination in CDS

We have shown that CDS combination rule has some Bayesian-like behavior in previous section. That is, the updated knowledge can be regarded as the conditional fuzzy measure given the evidence. Moreover, we show that CDS combination rule and conditioning rule has the property of Bayesian parallel combination. To see the Bayesian parallel combination in probability theory, assume the following strong independence assumption: events $R, R^{'}$ are conditionally independent in the sense that $p(R, R^{'}|u) = p(R|u)p(R^{'}|u)$ for all $u \in U$. Then Bayes' rule allows us to write the posterior distribution $p(u|R, R^{'})$ in terms of the posterior distributions $p(u|R)$ and $p(u|R^{'})$ as follows:

$$p(u|R, R^{'}) = \frac{p(u|R)p(u|R^{'})p(u)^{-1}}{\sum_{v \in U} p(v|R)p(v|R^{'})p(v)^{-1}}.$$

In CDS, when two bodies of evidence $\mathcal{B}$ and $\mathcal{C}$ are available, the final updated knowledge is represented by a conditional fuzzy measure $f(\cdot \mid \mathcal{B} *_f \mathcal{C})$. One may ask that what relation exists between $f(\cdot \mid \mathcal{B} *_f \mathcal{C})$, $f(\cdot \mid \mathcal{B})$ and $f(\cdot \mid \mathcal{C})$.

In order to illustrate the property of Bayesian parallel combination in CDS, we discuss the relation between $\nu_f(\cdot \mid \mathcal{B} *_f \mathcal{C})$, $\nu_f(\cdot \mid \mathcal{B})$ and $\nu_f(\cdot \mid \mathcal{C})$, where $\nu_f(\cdot \mid \mathcal{B} *_f \mathcal{C})$, $\nu_f(\cdot \mid \mathcal{B})$ and $\nu_f(\cdot \mid \mathcal{C})$ are the möbius transforms of $f(\cdot \mid \mathcal{B} *_f \mathcal{C})$, $f(\cdot \mid \mathcal{B})$ and $f(\cdot \mid \mathcal{C})$, respectively. The property of Bayesian parallel combination in CDS is shown in Theorem 1. Before proving this theorem, we first introduce a new basis $\{e_S\}$ for the vector space $R[U; f]$.

**Definition 3.** *Let $f$ be a fuzzy measure for the subsets of $U$. For each $S \subseteq U$, an element $e_S \in R[U; f]$ is defined by*

$$e_S \triangleq \sum_{T \subseteq S} (-1)^{\#(S-T)} f(T)T.$$

The following proposition shows that the non-vanishing elements of the collection $\{e_S\}$ form a basis for the vector space $R[U; f]$.

**Proposition 2.** *a) The non-vanishing elements of the collection $\{e_S\}$ form a set of orthogonal idempotents for the multiplication operator '$\cdot_f$'.(That is, $e_X \cdot_f e_Y = \delta_{X,Y} e_X$ for all $X, Y \subseteq U$. Here, $\delta_{X,Y}$ is a Kronecker delta: $\delta_{X,Y} = 1$ if $X = Y$ and $\delta_{X,Y} = 0$ otherwise.) Moreover, this collection spans the subspace $R[U; f]$.*

*b) $\alpha_f(e_X, e_Y) = \delta_{X,Y} \nu_f(X)$ for all $X, Y \subseteq U$.*

*c) Let $S \subseteq U$. If $f(S) = 0$ then $\alpha_f(S, e_T) = 0$. If $f(S) \neq 0$ then*

$$\alpha_f(S, e_T) = \frac{\nu_f(T)}{f(S)} \quad (\textit{if } T \subseteq S),$$

$$\alpha_f(S, e_T) = 0 \qquad (\textit{otherwise}).$$

Proof. a) Note that the vector space $R[U]$ is just the semigroup algebra over the real numbers $R$ induced by the semigroup operator '$\cdot_{f_0}$', where $f_0(S) = 1$ for all $S \subseteq U$ (here, $f_0$ isn't a fuzzy measure since $f(\emptyset) = 1$). Abbreviate:

$$d_S \triangleq \sum_{T \subseteq S} (-1)^{\#(S-T)} T.$$

We find that the $d_S$ form a system of orthogonal idempotents under the operator '$\cdot_{f_0}$'. Furthermore, the $d_S$ span the vector space $R[U]$. Next, define the map $\zeta_f : R[U] \to R[U; f]$ by

$$\zeta_f : \sum_{S \subseteq U} \mathcal{B}_S S \to \sum_{S \subseteq U} \mathcal{B}_S f(S) S.$$

Then $\zeta_f$ is a surjective homomorphism of $R-$algebra. Clearly, it is surjective. As for the fact that it is a homomorphism, it is enough to prove this for the generators of $R[U]$. Assume that $f(S) \neq 0 \neq f(T)$. Then

$$\begin{aligned} \zeta_f(S \cdot_{f_0} T) &= f(S \cap T) S \cap T \\ &= \alpha_f(S, T) f(S) f(T) S \cap T \\ &= f(S) f(T) S \cdot_f T = \zeta_f(S) \cdot_f \zeta_f(T). \end{aligned}$$

Consequently,

$$e_X \cdot_f e_Y = \zeta_f(d_X \cdot_{f_0} d_Y) = \delta_{X,Y} \zeta_f(d_X) = \delta_{X,Y} e_X$$

as desired. To prove that the $e_X$ span $R[U; f]$ note that for any $S \subseteq U$ the Möbius inversion formula gives $S = \sum_{T \subseteq S} d_T$. Therefore,

$$f(S) S = \zeta_f(S) = \sum_{X \subseteq S} \zeta_f(d_X) = \sum_{X \subseteq S} e_X.$$

b) From part a) we know that

$$\begin{aligned} \alpha_f(e_X, e_Y) &= \delta_{X,Y} tr(e_X) = \delta_{X,Y} \sum_{E \subseteq X} (-1)^{\#(X-E)} f(E) \\ &= \delta_{X,Y} \nu_f(X) \end{aligned}$$

where $tr(\sum_{S \subseteq U} \mathcal{B}_S S) \triangleq \sum_{S \subseteq U} \mathcal{B}_S$ and where the rightmost equation results from the Möbius inversion formula for fuzzy measure $f$.

c) From the proof for Part b) we know that $f(S) S = \sum_{E \subseteq S} e_E$. Thus

$$f(S) \alpha_f(S, e_T) = \sum_{E \subseteq S} \alpha_f(e_E, e_T) = \sum_{E \subseteq S} \delta_{E,T} \nu_f(E).$$

If $T$ is not a subset of $S$ then the last quantity vanishes; otherwise it is equal to $\nu_f(T)$. If we assume that $f(S) \neq 0$ then the result follows. ■

**Proposition 3.** *Let* $\mathcal{B}, \mathcal{C} \in R[U]$. *Then*

$$\alpha_f(\mathcal{B} \cdot_f \mathcal{C}, e_S) = \alpha_f(\mathcal{B}, e_S)\alpha_f(\mathcal{C}, e_S)\nu_f(S)^{-1},$$

*for* $\nu_f(S) \neq 0$.

Proof. It is enough to prove

$$\alpha_f(X \cdot_f Y, e_S) = \alpha_f(X, e_S)\alpha_f(Y, e_S)\nu_f(S)^{-1}$$

for all $X, Y \subseteq U$. Note that if $f(X) = 0$ or $f(Y) = 0$ then both sides of this equation vanish identically. So, we may assume that $f(X) \neq 0$ and $f(Y) \neq 0$. Hence

$$\begin{aligned}\alpha_f(X \cdot_f Y, e_S) &= \alpha_f(X, Y)\alpha_f(X \cap Y, e_S)\\ &= \frac{f(X \cap Y)}{f(X)f(Y)}\alpha_f(X \cap Y, e_S).\end{aligned}$$

From Proposition 2, we know that $\alpha_f(X \cap Y, e_S) = f(X \cap Y)^{-1}\nu_f(S)$ if $S \subseteq X \cap Y$ and $f(X \cap Y) \neq 0$ and that $\alpha_f(X \cap Y, e_S) = 0$ otherwise. The same is true for $\alpha_f(X, e_S)$ and $\alpha_f(Y, e_S)$. So, we may assume that $S \subseteq X \cap Y$ and $f(X \cap Y) \neq 0$. Thus we get

$$\begin{aligned}\alpha_f(X \cdot_f Y, e_S) &= f(X)^{-1}f(Y)^{-1}\nu_f(S)\\ &= f(X)^{-1}\nu_f(S)f(Y)^{-1}\nu_f(S)\nu_f(S)^{-1}\\ &= \alpha_f(X, e_S)\alpha_f(Y, e_S)\nu_f(S)^{-1}.\end{aligned}$$

■

**Theorem 1.** *Let* $\mathcal{B}, \mathcal{C} \in R[U]$. *Then*

$$\nu_f(S \mid \mathcal{B} *_f \mathcal{C}) = \frac{\nu_f(S \mid \mathcal{B})\nu_f(S \mid \mathcal{C})\nu_f(S)^{-1}}{\sum_{T \subseteq U} \nu_f(T \mid \mathcal{B})\nu_f(T \mid \mathcal{C})\nu_f(T)^{-1}}$$

*for* $\nu_f(S) \neq 0$ *and* $\nu_f(S \mid \mathcal{B} *_f \mathcal{C}) = 0$ *otherwise.*

Proof. Notice that for any body of evidence $\mathcal{B}$, the conditional fuzzy measure $f(S|\mathcal{B}) = \alpha_f(\mathcal{B}, f(S)S) = \sum_{T \subseteq S} \alpha_f(\mathcal{B}, e_T)$, that is, $\alpha_f(\mathcal{B}, e_S)$ is the Möbius transform of $f(S|\mathcal{B})$, or $\nu_f(S \mid \mathcal{B}) = \alpha_f(\mathcal{B}, e_S)$. So this theorem follows immediately from Proposition 3. In other words, we can say that the combination rule and conditioning rule in CDS satisfy the property of Bayesian parallel combination. ■

Notice that if $f$ is a probability measure then we have the following corollary.

**Corollary 2.** *Assume* $f$ *be a probability measure and* $\mathcal{B}, \mathcal{C}$ *be two bodies of evidence, then*

$$f(\{u\}|\mathcal{B} *_f \mathcal{C}) = \frac{f(\{u\}|\mathcal{B})f(\{u\}|\mathcal{C})f(\{u\})^{-1}}{\sum_{v \in U} f(\{v\}|\mathcal{B})f(\{v\}|\mathcal{C})f(\{v\})^{-1}}.$$

## 5 Conclusions

When evaluating the value of an uncertain variable, all available evidence should be fused as well as *a priori* knowledge about the value of the uncertain variable. In this paper, we assume that *a priori* knowledge about the uncertain variable can be modeled by a fuzzy measure—a unified framework to model different kinds of uncertain information. Under this assumption, we present a theory of knowledge updating called conditional Dempster-Shafer (CDS) which is a generalization of conventional evidence theory.

Based on CDS theory, a generalized evidence combination method and a conditioning rule involving *a priori* knowledge are introduced. This combination rule in CDS is called the conditioned combination rule. When Two or more bodies of evidence are available, they are first fused by conditioned combination rule, then *a priori* knowledge is updated by conditioning rule based on fused evidence. We show that these two rules satisfy the property of Bayesian parallel combination. In addition, the posterior knowledge achieved from CDS can be represented by a conditional fuzzy measure, this process is similar to the process of the probabilistic knowledge updating.

## Acknowledgements

This work has been supported by National Natural Science Foundation of China (No. 60604034), Science and Technology Program of Zhejiang Province (No. 2006C23062, 2007C23G2010148).

## References

1. A. Bendjebbour, Y. Delignon, L. Fouque, V. Samson, and W. Pieczynski, Multisensor image segmentation using Dempster-Shafer fusion in markov fields context, IEEE Trans. Geoscience and Remote Sensing 39(8) (2001) 1789–1798.
2. J. R. Boston, A signal detection system based on Dempster-Shafer theory and comparison to fuzzy detection, IEEE Trans. Systems Man and Cybernetics–Part C: Applications and Reviews 30(1) (2000) 45–51.
3. D. M. Buede and P. Girardi, A target identification comparison of Bayesian and Dempster-Shafer multisensor fusion, IEEE Trans. Systems Man and Cybernetics–Part A: Systems and Humans 27(5) (1997) 569–577.
4. A. P. Dempster, Upper and lower probabilities induced by a multi-valued mapping, Ann. Mathematical Statistics 38 (1967) 325–339.
5. T. Denœx, A $k$-nearest neighbour classification rule based on Dempster-Shafer theory, IEEE Trans. Systems Man and Cybernetics 25(5) (1995) 804–813.
6. T. Denœx, A neural network classifier based on Dempster-Shafer theory, IEEE Trans. Systems Man and Cybernetics–Part A: Systems and Humans 30(2) (2000) 131–150.
7. S. Hégarat-Mascle, I. Bloch, and D. Vidal-Madjar, Application of Dempster-Shafer evidence theory to unsupervised classification in multisource remote sensing, IEEE Trans. Geoscience and Remote Sensing 35(4) (1997) 1018–1031.

8. R. Mahler, Combining ambiguous evidence with respect to ambiguous *a priori* knowledge, I: Boolean logic, IEEE Trans. Systems Man and Cybernetics–Part A: Systems and Humans 26(1) (1996) 27–41.
9. R. Mahler, Combining ambiguous evidence with respect to ambiguous *a priori* knowledge, II: Fuzzy logic, Fuzzy Sets and Systems 75 (1995) 319–354.
10. G. Shafer, A Mathematical Theory of Evidence, Princeton University Press, 1976.
11. P. Smets, The combination of evidence in the transferable belief model, IEEE Trans. Pattern Analysis and Machine Intelligence 12(5) (1990) 447–458.
12. P. Smets and R. Kennes, The transferable belief model, Artificial Intelligence 66 (1994) 191–234.
13. M. Sugeno, Fuzzy measures and fuzzy integrals–a survey, Fuzzy Automata and Decision Processes (M. M. Gupta, G. N Saridis, and B. R. Gaines, Eds.), pp. 89–102, North-Holland, Amsterdam, 1977.
14. Y. Tang, S. Sun and Y. Li, The generalized combination of evidence involved *a priori* knowledge, in: Proceedings of IEEE International Conference on Systems, Man & Cybernetics, Vol. 5, 2003, pp. 4998–5003.
15. F. Tupin, I. Bloch, and H. Maître, A first step toward automatic interpretation of sar images using evidential fusion of several structure detectors, IEEE Trans. Geoscience and Remote Sensing 37(3) (1999) 1327–1343.
16. R. R. Yager, Uncertainty representation using fuzzy measures, IEEE Trans. Systems Man and Cybernetics–Part B: Cybernetics 32(1) (2002) 13–20.
17. R. R. Yager and D. P. Filev, Including probabilistic uncertainty in fuzzy logic controller modeling using Dempster-Shafer theory, IEEE Trans. Systems Man and Cybernetics 25(8) (1995) 1221–1230.
18. J. Yang and M. G. Singh, An evidential reasoning approach for multiple-attribute decision making with uncertainty, IEEE Trans. Systems Man and Cybernetics 24(1) (1994) 1–18.
19. J. Yen, Generalizing the Dempster-Shafer theory to fuzzy sets, IEEE Trans. Systems Man and Cybernetics 20(3) (1990) 559–570.
20. L. M. Zouhal and T. Denœx, An evidence-theoretic $k$-NN rule with parameter optimization, IEEE Trans. Systems Man and Cybernetics–Part C: Applications and Reviews 28(2) (1998) 263–271.

# Ant Colony Optimization Applied to Feature Selection in Fuzzy Classifiers

Susana M. Vieira[1], João M.C. Sousa[1], and Thomas A. Runkler[2]

[1] Center of Intelligent Systems, IDMEC
Instituto Superior Técnico, Technical University of Lisbon
Av. Rovisco Pais, 1049-001 Lisbon, Portugal
{susana,j.sousa}@dem.ist.utl.pt
[2] Siemens AG, Corporate Technology
Information and Communications CT IC 4
81730 Munich - Germany
Thomas.Runkler@siemens.com

**Abstract.** In practice, classifiers are often build based on data or heuristic information. The number of potential features is usually large. One of the most important tasks in classification systems is to identify the most relevant features, because less relevant features can be interpreted as noise that reduces the classification accuracy, even for fuzzy classifiers which are somehow robust to noise. This paper proposes an ant colony optimization (ACO) algorithm for the feature selection problem. The goal is to find the set of features that reveals the best classification accuracy for a fuzzy classifier. The performance of the method is compared to other features selection methods based on tree search methods.

## 1 Introduction

Real-world data analysis, data mining, classification and modeling problems usually involve a large number of candidate inputs or features. Sometimes the number of features is too large, making the problem computationally unfeasible or simply uncomprehensible. Feature selection has been an active research area in data mining, pattern recognition and statistics communities for many years [9]. The main idea of feature selection is to choose a subset of input variables by eliminating features that contribute with little or no information. The methods found in the literature can generally be divided into two main groups: model-free and model-based methods. Model–free methods use the available data only and are based on statistical tests, properties of functions, etc. These methods do not need to develop models to find significant inputs. The methods discussed in this paper belong to the group of model-based methods. Models with different sets of features are compared and the model that minimizes the model output error is selected. Often exhaustive methods are used where all subsets of variables must be tested. Decision tree search methods, with the proper branch conditions, limit the search space to the best performed branches, but do not guarantee to find the global best solution [10].

P. Melin et al. (Eds.): IFSA 2007, LNAI 4529, pp. 778–788, 2007.
© Springer-Verlag Berlin Heidelberg 2007

Nature inspired algorithms like ant colony optimization have been successfully applied to a large number of difficult combinatorial problems like quadratic assignment, traveling salesman problems, routing in telecommunication networks, or scheduling, [6]. Ant colony optimization is particularly attractive for feature selection since no reliable heuristic is available for finding the optimal feature subset, so it is expectable that the ants discover good feature combinations as they proceed through the search space. Recently, nature inspired algorithms have been used to select features [1,7,12].

This paper proposes an ant based feature selection approach for fuzzy classifiers. The method is compared to other feature selection methods, namely two decision tree search approaches: top–down and bottom–up [10]. Our goal is to obtain simpler and more comprehensible fuzzy models for classification. The paper is organized as follows. Fuzzy classification is briefly described in Section 2. Section 3 presents the procedure of structure identification based on decision tree methods. The ant feature selection algorithm is described in section 4. A brief description of the application example, the experiments, and their respective results are presented and commented in Section 5. Finally, some conclusions are drawn in Section 6.

## 2 Fuzzy Classification

We use a fuzzy classifier, more precisely a fuzzy rule based classifier, as it provides a transparent model and a linguistic interpretation in the form of rules[13]. The fuzzy rule based models used in this paper are Takagi-Sugeno (TS) fuzzy models,which are presented in the next section.

### 2.1 Takagi-Sugeno Fuzzy Models

Takagi-Sugeno (TS) fuzzy models [15], consist of fuzzy rules where each rule describes a local input-output relation, typically in an affine form. Usually TS fuzzy models are represented by multi–input single–output (MISO) models. However, when multi–input multi–output (MIMO) models are necessary (like in the present work), they can be obtained as a collection of MISO models without lack of generality [13]. The affine form of a TS MISO model is given by:

$$R_i : \textbf{If } x_1 \text{ is } A_{i1} \textbf{and } \ldots \textbf{and } x_n \text{ is } A_{in} \textbf{then } y_i = a_{i1}x_1 + \ldots + a_{in}x_n + b_i \,, \quad (1)$$

where $i = 1, \ldots, K$, $K$ denotes the number of rules in the rule base, $R_i$ is the $i^{th}$ rule, $\mathbf{x} = [x_1, \ldots, x_n]^T$ is the antecedent vector, $n$ is the number of states, $A_{i1}, \ldots, A_{in}$ are fuzzy sets defined in the antecedent space, $y_i$ is the output variable for rule $i$, $\mathbf{a}_i$ is a parameter vector and $b_i$ is a scalar offset. The consequents of the affine TS model are hyperplanes in the product space of the inputs and the output. The model output, $y$, can then be computed by aggregating the individual rule contributions: $y = \sum_{i=1}^{K} \beta_i y_i / \sum_{i=1}^{K} \beta_i$, where $\beta_i$ is the degree of activation of the $i$th rule, which is defined as: $\beta_i = \prod_{j=1}^{n} \mu_{A_{ij}}(x_j)$, and $\mu_{A_{ij}}(x_j) : \mathbb{R} \to [0, 1]$ is the membership function of the fuzzy set $A_{ij}$ in the antecedent of $R_i$.

### 2.2 Identification

Firstly, the structure of the model must be identified. In this step, the significant features $\mathbf{x}$ of the model must be chosen. This is a very important step, especially for real-world problems. This task can be performed using the algorithms described in Section 3 and in Section 4. The number of variables must be small enough for the sake of simplicity, but with the sufficient number of variables to achieve the desired model accuracy. To identify the model, the regression matrix $\mathbf{X}$ and an output vector $\mathbf{y}$ are constructed from the available data: $\mathbf{X}^T = [\mathbf{x}_1, \ldots, \mathbf{x}_N]$, $\mathbf{y}^T = [y_1, \ldots, y_N]$. Here $N \gg n$ is the number of samples used for identification. The number of rules $K$, the antecedent fuzzy sets $A_{ij}$, and the consequent parameters $\mathbf{a}_i$ and $b_i$ are determined by means of fuzzy clustering in the space of the input and output variables. Hence, the data set $\mathbf{Z}$ to be clustered is composed from $\mathbf{X}$ and $\mathbf{y}$:

$$\mathbf{Z} = [\mathbf{X}, \mathbf{y}]^T . \tag{2}$$

Given the data $\mathbf{Z}$ and the number of clusters $K$, several fuzzy clustering algorithms can be used. This paper uses the fuzzy c-means (FCM) [3] clustering algorithm to compute the fuzzy partition matrix $\mathbf{U}$. The fuzzy sets in the antecedent of the rules are obtained from the partition matrix $\mathbf{U}$, whose $ik$th element $\mu_{ik} \in [0, 1]$ is the membership degree of the data object $\mathbf{z}_k$ in cluster $i$. One-dimensional fuzzy sets $A_{ij}$ are obtained from the multidimensional fuzzy sets defined point-wise in the $i$th row of the partition matrix by projections onto the space of the input variables $x_j$:

$$\mu_{A_{ij}}(x_{jk}) = \text{proj}_j^{\mathbb{N}_{n+1}}(\mu_{ik}), \tag{3}$$

where proj is the point-wise projection operator [8]. The point-wise defined fuzzy sets $A_{ij}$ are approximated by suitable parametric functions in order to compute $\mu_{A_{ij}}(x_j)$ for any value of $x_j$. The consequent parameters for each rule are obtained as a weighted ordinary least-square estimate. Let $\theta_i^T = \left[\mathbf{a}_i^T; b_i\right]$, let $\mathbf{X}_e$ denote the matrix $[\mathbf{X}; \mathbf{1}]$ and let $\mathbf{W}_i$ denote a diagonal matrix in having the degree of activation, $\beta_i(\mathbf{x}_k)$, as its $k$th diagonal element. Assuming that the columns of $\mathbf{X}_e$ are linearly independent and $\beta_i(\mathbf{x}_k) > 0$ for $1 \leq k \leq N$, the weighted least-squares solution of $\mathbf{y} = \mathbf{X}_e\theta + \varepsilon$ becomes

$$\theta_i = \left[\mathbf{X}_e^T \mathbf{W}_i \mathbf{X}_e\right]^{-1} \mathbf{X}_e^T \mathbf{W}_i \mathbf{y} . \tag{4}$$

The second step to identify a model consists of the estimation of the parameters of the model. The number of rules $K$, the antecedent fuzzy sets $A_{ij}$, and the consequent parameters $\mathbf{a}_i$ and $b_i$ are determined in this step, by means of fuzzy clustering in the product space of the input and output variables [2].

The number of fuzzy rules (or clusters) that best suits the data must be determined for classification. For that purpose the following criterion, as proposed in [14], is used to determine the number of clusters:

$$S(c) = \sum_{k=1}^{N} \sum_{i=1}^{c} (\mu_{ik})^m (\| \mathbf{x}_k - v_i \|^2 - \| v_i - \bar{\mathbf{x}} \|^2), \tag{5}$$

where $N$ is the number of data to be clustered, $c$ is the number of clusters ($c \geq 2$), $\mathbf{x}_k$ is the $k^{th}$ data point (usually vector), $\bar{\mathbf{x}}$ is the mean value for the inputs, $v_i$ the center of the $i^{th}$ cluster, $\mu_{ik}$ is the grade of the $k^{th}$ data point belonging to $i^{th}$ cluster and $m$ is an adjustable weight. The parameter $m$ has a great importance in this criterion. The bigger the $m$ the bigger the optimum number of clusters. Therefore, this value is normally around 2. The number of clusters $c$ is increased from two up to the number that gives the minimum value for $S(c)$. Note that this minimum can be local. However, this procedure diminishes the number of rules and consequently the complexity of the fuzzy model. At each iteration, the number of clusters are determined using the fuzzy c-means algorithm and the process stops when $S(c)$ increases from one iteration to the next one. The first term of the right-hand side of (5) is the variance of the data in a cluster and the second term is the variance of the clusters themselves. The optimal clustering achieved is the one that minimizes the variance in each cluster and maximizes the variance between clusters.

The performance criterion used to evaluate the fuzzy rule based classification model is based on misclassifications:

$$MSp = \frac{(n - mis)}{n} \times 100\%, \tag{6}$$

where $n$ is the number of used samples and $mis$ the number of misclassifications.

## 3 Decision Tree Methods

### 3.1 Bottom-Up Approach

The bottom-up approach described in this paper follows the principle of the regularity criterion (RC) approach [14], which is also a bottom-up approach. However, a more recent algorithm that minimizes the computational time with similar performance is used here [10]. The bottom-up approach starts with the most relevant feature(s) and successively adds the most relevant and removes the most irrelevant feature(s).

By using two groups of data, $A$ and $B$, two fuzzy models are built, one for each group, starting with only one feature. At this stage, a fuzzy model is built for each of the $n$ features in consideration. The models are evaluated using the RC performance criterion. The criterion is computed for each model at this stage, and the feature that minimizes the performance criterion is selected as the best one. The one(s) that maximizes the criterion is rejected and is not included in the next stage. At the next stage, the feature already selected is fixed, *i.e.*, it belongs to the model structure. The other feature candidates, excluding the rejected feature(s) in the prior stage, are added to the previous fuzzy model one at a time. When this second stage finishes, the fuzzy model has two features. The second feature is chosen as the one that minimizes the value of the chosen performance criterion, and as before, the feature(s) that maximizes the value of the criterion is rejected. This procedure is repeated until the value of the

**Algorithm 1.** Bottom–up approach

*Cluster the data using fuzzy c-means with two initial clusters;*
*Increase the number of clusters until $S(c)$ in (5) reach its minimum;*
*Divide the data set into two groups $A$ and $B$;*
*For each input in the input vector that does not belong to the inputs of the model:*
**repeat**
  *Build two models, one using data group $A$ and other using data group $B$;*
  *Compute the PC;*
  *Select the input with the lowest value of PC as a new input of the model;*
  *Discard the input with the largest PC;*
**until** *PC increases or the end of the input vector is reached.*
*Select the final inputs;*
*Using the number of clusters given from (5) and the inputs selected using the proposed approach, build a fuzzy model using a fuzzy clustering algorithm.*

performance criterion increases. At this stage, one should have all the relevant features for the considered classification output. In a generic case, using the RC as proposed in [14], the maximum number of iterations is $n \times (n+1)/2$, where $n$ is the number of possible features. The number of iterations using the bottom-up approach decreases. For an odd number of features the maximum number of iterations is $(n+1)^2/4$ and for an even number of features the maximum number of iterations is $n \times (n+2)/4$. Thus, the number of iterations reduces significantly, and then the computational time is also reduced. Assuming that input and output data are collected from a given system, the selection of inputs using this methodology generally entails the algorithm described in Algorithm 1.

Summarizing, the bottom–up approach presented in Algorithm 1 differs from the RC algorithm proposed in [14] because it is possible to exclude one or more variables. This is an advantage, as it allows the reduction of the number of iterations per stage. In some cases, it allows even the reduction of the number of stages, reducing also the computational time.

### 3.2 Top-Down Approach

Another approach proposed to select the input variables is the top-down (TD) approach. This approach begins with all the input variables, and removes the one(s) with the worst performance at each stage. This approach was proposed in [10]. The identification data is divided into two groups, $A$ and $B$, as in the bottom–up approach.

Again, one model is built for each group $A$ and $B$ using all the variables. The proposed approach begins, at stage 0, by using all the variables. The performance criterion (PC) is computed. This is considered as the value to decrease at the following stages. Then, at stage 1, $n$ fuzzy models are obtained, where each one of them is identified without one of the variables used at stage 0. The values of the chosen PC, for each of the $n$ models, are compared to the value obtained

**Algorithm 2.** Top–down approach

*Cluster the data using fuzzy c-means with two initial clusters;*
*Increase the number of clusters until* $S(c)$ *in (5) reach its minimum;*
*Divide the data set into two groups* $A$ *and* $B$*;*
$i = 0$, *where* $i$ *is the stage number;*
**repeat**
  **if** *Stage is zero* **then**
    *Build a model using all the input variables;*
    $m = n$; *where* $n$ *is the number of initial inputs*
  **else**
    $i = i + 2$;
    *Build a model using the input variables not discarded at the previous stage;*
  **end if**
  *Compute* $PC_i$*;*
  **for** $j = 1$ *to* $m$ **do**
    *Build two models, for groups* $A$ *and* $B$*, using all the inputs except input* $j$*;*
    *Compute* $PC_{i+1,j}$
    **if** $PC_{i+1,j} < PC_i$ **then**
      *Discard the input* $j$ *not used in modeling;*
    **end if**
  **end for**
  $m = m - p$, *where* $p$ *is the number of discarded inputs;*
**until** *(no input is discarded) OR (model has only one input) OR* $(PC_i > PC_{i-2})$
*Using the number of clusters given from (5) and the inputs selected by the proposed approach, build a fuzzy model using a fuzzy clustering algorithm.*

at stage 0. For each new value that is smaller, the corresponding input $x_i$ is removed from the vector of inputs. At the next stage, a fuzzy model is identified using only the inputs that have not been discarded at stage 1. The value of the chosen performance criterion is computed, and is used as reference for the next stage. The fuzzy model obtained at stage 2 has $n - p$ inputs, where $n$ is the number of initial inputs and $p$ is the number of inputs removed at stage 1. The presented procedure is repeated until the value of the performance criterion is not decreased by excluding any input. Thus, the inputs considered at stage 2 are the ones that are used in the final model. The top-down approach proposed in this paper is described in Algorithm 2.

This algorithm differs from the bottom–up approach, as it obtains at each stage multivariable fuzzy models, begins with the full feature vector, and discards one or more inputs at each stage. This is a clear advantage, which allows the reduction of the number of iterations per stage. Further, in some cases, it can even reduce the number of stages, and consequently the computational time can be reduced when compared to the BU approach. On the other hand, as the TD approach uses much more inputs to build each model from the beginning, and the identification of each model can be computationally intensive. This is especially critical when the number of inputs is large.

## 4 Ant Feature Selection

Ant algorithms were first proposed by Dorigo *et al.* [5] as a multi-agent approach to difficult combinatorial optimization problems like the traveling salesman problem and the quadratic assignment problem. There is currently a lot of ongoing activity in the scientific community to extend/apply ant-based algorithms to many different discrete optimization problems [4]. Recent applications cover problems like vehicle routing, sequential ordering, graph coloring, routing in communications networks, and so on. Ant algorithms were inspired by the observation of real ant colonies. Ants are social insects, that is, insects that live in colonies and whose behavior is directed more to the survival of the colony as a whole than to that of a single individual component of the colony.

The Ant Colony Optimization (ACO) methodology [6] is an optimization method suited to find minimum cost paths in optimization problems described by graphs. Consider a problem with $n$ nodes and a colony of $g$ ants. Initially, the $g$ ants are randomly placed in $g$ different nodes. The probability that an ant $k$ in node $i$ chooses node $j$ as the next node to visit is given by

$$p_{ij}^k(t) = \begin{cases} \frac{{\tau_{ij}}^{\alpha} \cdot {\eta_{ij}}^{\beta}}{\sum\limits_{r \notin \Gamma}^{n} {\tau_{ir}}^{\alpha} \cdot {\eta_{ir}}^{\beta}}, & \text{if } j \notin \Gamma \\ 0, & \text{otherwise} \end{cases} \tag{7}$$

where $\tau_{ij}$ and $\eta_{ij}$ are the entries of the pheromone concentration matrix $\tau$ and heuristic function matrix $\eta$ respectively, for the path $(i, j)$. The pheromone matrix values are limited to $[\tau_{min}, \tau_{max}]$, with $\tau_{min} = 0$ and $\tau_{max} = 1$. $\Gamma$ is the *tabu list*, which acts as the memory of the ants and contains all the trails that the ants have already passed and cannot be chosen again. The parameters $\alpha$ and $\beta$ measure the relative importance of trail pheromone and heuristic knowledge, respectively.

After a complete tour, when all the $g$ ants have visited all the $n$ nodes, the pheromone concentration in the trails is updated by

$$\tau_{ij}(t+1) = \tau_{ij}(t) \times (1 - \rho) + \Delta\tau_{ij}^q \tag{8}$$

where $\rho \in [0, 1]$ is the pheromone evaporation coefficient and $\Delta\tau_{ij}^q$ are pheromones deposited on the trails $(i, j)$ followed by ant $q$ that found the best solution $f^q(s)$ for this tour:

$$\Delta\tau_{ij}^q = \begin{cases} \frac{1}{f^q(s)} & \text{if arc } (i, j) \text{ is used by the ant } q \\ 0 & \text{otherwise} \end{cases} \tag{9}$$

The algorithm runs $t_{max}$ times.

In this paper, an Ant Feature Selection (AFS) algorithm is proposed. Our goal is to assign $n$ features to a subset of the total set of available features. The main objective is to have the best possible classification accuracy, i.e., to minimize the classification error:

$$E_{min} = |y_{est} - y| \tag{10}$$

**Algorithm 3.** Ant Feature Selection

```
/*Initialization*/
n dimension of the subset of features
for every feature i do
  τ_i(0) = τ_0
end for
for k = 1 to m do
  Place ant k on a randomly chosen feature
end for
Let L^+ be the best feature set found from beginning and E^+ its error;
/*Main Loop*/
for t = 1 to t_max do
  for k = 1 to m do
    Build feature set L^k(t) by applying n − 1 times the following step:
    Choose the next feature j with probability
    p^k_ij(t) = [τ_ij(t)]^α·[η_ij]^β / Σ_{l∈J^k_i} [τ_il(t)]^α·[η_il]^β
    Where i is the current feature
  end for
  for k = 1 to m do
    Compute the fuzzy model using the feature set L^k(t) produced by ant k
    Compute the error E^k(t)
  end for
  if an improved feature set is found then
    update L^+ and E^+
  end if
  for every feature i do
    Update pheromone trails by applying the rule:
    τ_ij(t) ← (1 − ρ) · τ_ij(t) + Δτ_ij(t)
  end for
end for
```

where $y_{est}$ is the classification result. After completion of an iteration, each ant $k$ lays a quantity of pheromone $\Delta\tau_{ij}^k(t)$ on each used feature. The value $\Delta\tau_{ij}^k(t)$ depends on how well the ant has performed. At iteration $t$, the deposited pheromone is given by:

$$\Delta\tau_{ij}^k(t) = \begin{cases} Q/E^k(t), & \text{if feature } i \in L^k(t) \\ 0, & \text{if feature } i \notin L^k(t) \end{cases} \tag{11}$$

where $L^k(t)$ is the set of features produced by ant $k$ at iteration $t$, $E^k(t)$ is the error of the feature set, and $Q$ is a parameter. The pheromone concentration in the features is updated by

$$\tau_{ij}(t+1) = \tau_{ij}(t) \times (1-\rho) + \Delta\tau_{ij}(t) \tag{12}$$

where $\Delta\tau_{ij}(t) = \sum_{k=1}^{m} \Delta\tau_{ij}^k(t)$ and $m$ is the number of ants. The transition rule, that is, the probability for ant $k$ to use feature $i$ while building its $t^{th}$ feature set, is given by

$$p_{ij}^k(t) = \frac{[\tau_{ij}(t)]^\alpha \cdot [\eta_{ij}]^\beta}{\sum_{l \in J_i^k} [\tau_{il}(t)]^\alpha \cdot [\eta_{il}]^\beta} \tag{13}$$

where $\alpha$ and $\beta$ control the relative weight of each feature between the pheromone concentration $\tau_{ij}(t)$, and the heuristic $\eta_{ij} = 1/E_{ij}$.

**Table 1.** Values of parameters used in the experiments

| $\alpha$ | $\beta$ | $\rho$ | $m$ | $Q$ | $\tau$ | $t_{max}$ | $n$ |
|---|---|---|---|---|---|---|---|
| 2 | 1 | 0.1 | 2 | 10 | 0.5 | 100 | 2, 4 or 11 |

## 5 Application

The proposed ant feature selection (AFS) algorithm is applied to a wine classification data set, which is obtained from the repository of University of California at Irvine [11]. The results are compared to decision tree methods for feature selection as described in Section 3. The classification data used in this paper contains the chemical analysis of 178 wines grown in the same region in Italy but derived from three different cultivars. Thirteen continuous attributes are available for classification: alcohol, malic acid, ash, alcalinity of ash, magnesium, total phenols, flavanoids, nonflavanoids phenols, proanthocyanism, color intensity, hue, OD280/OD315 of dilluted wines and proline.

The features are selected using decision tree search methods or the AFS algorithm. The parameters used in the AFS algorithm are given in Table 1. The selected features are used to build fuzzy rule based models for classification.

First, both top-down and bottom-up approaches were applied to the database. The best number of features using these approaches are 11 and 4 features, respectively. Afterwards, AFS was applied using the same number of features. The results are shown in Table 2. This table shows also the results using AFS with only 2 features.

The features selected by the AFS algorithm are similar to the ones selected with both decision tree search approaches. Even so, the AFS has a smaller variability in results than the bottom-up approach, which means that the bottom-up approach is much more dependent on the performance criterion than AFS.

The main advantage of the ant based feature selection algorithm is the search in a much wider space of features subset. In the bottom-up approach, after choosing the best first feature, the following features subsets will always include this feature. This can be a disadvantage, because when this feature is combined with other features, it may turn out not to be the best feature, while in the AFS algorithm this is never the case. Further, AFS can achieve good classification rates even with a small number of features, see Table 2 when AFS is applied with only two features. This can be a very important characteristic for classification problems in very large data sets.

**Table 2.** Correct classification percentage of the wine classification using the bottom-up, top-down and ant feature selection approaches

| | #11 features | | #4 features | | #2 features |
|---|---|---|---|---|---|
| Methods | Top–down | AFS | Bottom–up | AFS | AFS |
| Best | 100 | 100 | 100 | 100 | 100 |
| Average | 99.9 | 99.8 | 96.7 | 99.3 | 97.7 |
| Worst | 99.4 | 97.7 | 92.7 | 97.7 | 89.8 |

## 6 Conclusions

This paper proposed an ant feature selection algorithm and compared it with tree search methods for feature selection. All three algorithms were used to select a subset of features that was then used as inputs of a Takagi–Sugeno fuzzy rule based classifier. We compared the performance of the three feature selection algorithms, when applied to the wine classification data set. The ant based feature selection algorithm yielded the best classification rate for low number of features. The top-down approach method was able to produced slightly better results only when a very high number of features was used.

In the near future we are planing to develop an enhanced algorithm to determine automatically the optimal number of features, and apply the proposed feature selection algorithm to classification problems in very large data sets.

## Acknowledgements

This work is supported by the FCT grant SFRH/BD/25381/2005, Ministério do Ensino Superior da Ciência e Tecnologia, Portugal and by the Programa Operacional Ciência e Inovação (POCI) 2010, FCT, Portugal.

## References

1. Ahmed Al-Ani. Feature subset selection using ant colony optimization. *International Journal of Computational Intelligence*, 2(1):53–58, 2005.
2. R. Babuška. *Fuzzy Modeling for Control.* Boston, 1998.
3. J. C. Bezdek. *Pattern Recognition With Fuzzy Objective Functions.* Plenum Press, New York, 1981.
4. D. Corne, M. Dorigo, and F. Glover. *New Methods in Optimisation.* McGraw-Hill, 1999.
5. M. Dorigo, V. Maniezzo, and A. Colorni. The ant system: Optimization by a colony of cooperating agents. *IEEE Transactions on Systems, Man, and Cybernetics–Part B*, 26(1):1–13, 1996.
6. Marco Dorigo and Thomas Stützle. *Ant Colony Optimization.* The MIT Press, 2004.
7. Richard Jensen and Qiang Shen. Fuzzy-rough data reduction with ant colony optimization. *Fuzzy Sets and Systems*, 149:5–20, 2005.

8. G. J. Klir and B. Yuan. *Fuzzy Sets and Fuzzy Logic: theory and applications.* Upper Saddle River,Prentice Hall, 1995.
9. Huan Liu and Lei Yu. Toward integrating feature selection algorithms for classification and clustering. *IEEE Transactions on Knowledge and Data Engineering*, 17(4):491–502, April 2005.
10. L. F. Mendonça, S. M. Vieira, and J. M. C. Sousa. Decision tree search methods in fuzzy modeling and classification. *International Journal of Approximate Reasoning*, 44(2):106–123, 2007.
11. D. J. Newman, S. Hettich, C. L. Blake, and C. J. Merz. UCI Repository of machine learning databases, 1998. http://www.ics.uci.edu/~mlearn/MLRepository.html.
12. Rahul Sivagaminathan and Sreeram Ramakrishnan. A hybrid approach for feature selection using neural networks and ant colony optimization. *Expert Systems with Applications*, 33:49–60, 2007.
13. J.M. Sousa and U. Kaymak. *Fuzzy Decision Making in Modeling and Control.* World Scientific and Imperial College, Singapore and UK, 2002.
14. M. Sugeno and T. Yasukawa. A fuzzy-logic-based approach to qualitative modeling. *IEEE Transactions on Fuzzy Systems*, 1(1):7–31, 1993.
15. T. Takagi and M. Sugeno. Fuzzy identification of systems and its applications to modelling and control. *IEEE Transactions on Systems, Man and Cybernetics*, 15(1):116–132, 1985.

# Artificial Bee Colony (ABC) Optimization Algorithm for Solving Constrained Optimization Problems

Dervis Karaboga and Bahriye Basturk

Erciyes University, Engineering Faculty, The Department of Computer Engineering
karaboga@erciyes.edu.tr, bahriye@erciyes.edu.tr

**Abstract.** This paper presents the comparison results on the performance of the Artificial Bee Colony (ABC) algorithm for constrained optimization problems. The ABC algorithm has been firstly proposed for unconstrained optimization problems and showed that it has superior performance on these kind of problems. In this paper, the ABC algorithm has been extended for solving constrained optimization problems and applied to a set of constrained problems .

## 1 Introduction

Constrained Optimization problems are encountered in numerous applications. Structural optimization, engineering design, VLSI design, economics, allocation and location problems are just a few of the scientific fields in which CO problems are frequently met [1]. The considered problem is reformulated so as to take the form of optimizing two functions, the objective function and the constraint violation function [2]. General constrained optimization problem is to find $\boldsymbol{x}$ so as to

$$\text{minimize } f(x), \qquad x = (x_1, \ldots, x_n) \in \mathbb{R}^n$$

where $\boldsymbol{x} \in \mathbb{F} \in \mathbb{S}$. The objective function $f$ is defined on the search space $\mathbb{S} \subseteq \mathbb{R}^n$ and the set $\mathbb{F} \subseteq \mathbb{S}$ defines the feasible region. Usually, the search space $\mathbb{S}$ is defined as a $n$-dimensional rectangle in $\Re^n$ (domains of variables defined by their lower and upper bounds):

$$l(i) \leq x(i) \leq u(i), \quad 1 \leq i \leq n$$

whereas the feasible region $\mathbb{F} \subseteq \mathbb{S}$ is defined by a set of $m$ additional constraints ($m \geq 0$):

$$g_j(\boldsymbol{x}) \leq 0, \text{ for } j = 1, \ldots, q$$

$$h_j(\boldsymbol{x}) = 0, \text{ for } j = q+1, \ldots, m.$$

P. Melin et al. (Eds.): IFSA 2007, LNAI 4529, pp. 789–798, 2007.
© Springer-Verlag Berlin Heidelberg 2007

At any point $\boldsymbol{x} \in \mathbb{F}$, the constraints $g_k$ that satisfy $g_k(\boldsymbol{x}) = 0$ are called the active constraints at $\boldsymbol{x}$. By extension, equality constraints $h_j$ are also called active at all points of $\mathbb{S}$ [3].

Different deterministic as well as stochastic algorithms have been developed for tackling constrained optimization problems. Deterministic approaches such as Feasible Direction and Generalized Gradient Descent make strong assumptions on the continuity and differentiability of the objective function [4,5]. Therefore their applicability is limited since these characteristics are rarely met in problems that arise in real life applications. On the other hand, stochastic optimization algorithms such as Genetic Algorithms, Evolution Strategies, Evolutionary Programming and Particle Swarm Optimization (PSO) do not make such assumptions and they have been successfully applied for tackling constrained optimization problems during the past few years [6,7,8,9,16].

Karaboga has described an Artificial Bee Colony (ABC) algorithm based on the foraging behaviour of honey bees for numerical optimization problems [11]. Karaboga and Basturk have compared the performance of the ABC algorithm with those of other well-known modern heuristic algorithms such as Genetic Algorithm (GA), Differential Evolution (DE), Particle Swarm Optimization (PSO) on unconstrained problems [12]. In this work, ABC algorithm is extended for solving constrained optimization (CO) problems. Extension of the algorithm depends on replacing the selection mechanism of the simple ABC algorithm with Deb's [13] selection mechanism in order to cope with the constraints. The performance of the algorithm has been tested on 13 well-known constrained optimization problems taken from the literature and compared with Particle Swarm Optimization (PSO) and Differential Evolution (DE) [14]. The Particle Swarm Optimization (PSO) algorithm was introduced by Eberhart and Kennedy in 1995 [15]. PSO is a population based stochastic optimization technique and well adapted to the optimization of nonlinear functions in multidimensional space. It models the social behaviour of bird flocking or fish schooling. The DE algorithm is also a population based algorithm using crossover, mutation and selection operators. Although DE uses crossover and mutation operators as in GA, the main operation is based on the differences of randomly sampled pairs of solutions in the population. Paper is organized as follows. In Section II, the ABC algorithm and the ABC algorithm adapted for solving constrained optimization problems are introduced. In Section III, a benchmark of 13 constrained functions are tested. Results of the comparison with the PSO and DE algorithms are presented and discussed. Finally, a conclusion is provided.

## 2 Artificial Bee Colony Algorithm

### 2.1 The ABC Algorithm Used for Unconstrained Optimization Problems

In ABC algorithm [11,12], the colony of artificial bees consists of three groups of bees: employed bees, onlookers and scouts. First half of the colony consists of the employed artificial bees and the second half includes the onlookers. For every food source, there is only one employed bee. In other words, the number of employed bees is equal to the number of food sources around the hive. The employed bee whose the food source has been abandoned by the bees becomes a scout.

In ABC algorithm, the position of a food source represents a possible solution to the optimization problem and the nectar amount of a food source corresponds to the quality (fitness) of the associated solution. The number of the employed bees or the onlooker bees is equal to the number of solutions in the population. At the first step, the ABC generates a randomly distributed initial population $P(G = 0)$ of $SN$ solutions (food source positions), where $SN$ denotes the size of population. Each solution $x_i$ $(i = 1, 2, ..., SN)$ is a $D$-dimensional vector. Here, D is the number of optimization parameters. After initialization, the population of the positions (solutions) is subjected to repeated cycles, $C = 1, 2, ..., MCN$, of the search processes of the employed bees, the onlooker bees and scout bees. An employed bee produces a modification on the position (solution) in her memory depending on the local information (visual information) and tests the nectar amount (fitness value) of the new source (new solution). Provided that the nectar amount of the new one is higher than that of the previous one, the bee memorizes the new position and forgets the old one. Otherwise she keeps the position of the previous one in her memory. After all employed bees complete the search process, they share the nectar information of the food sources and their position information with the onlooker bees on the dance area. An onlooker bee evaluates the nectar information taken from all employed bees and chooses a food source with a probability related to its nectar amount. As in the case of the employed bee, she produces a modification on the position in her memory and checks the nectar amount of the candidate source. Providing that its nectar is higher than that of the previous one, the bee memorizes the new position and forgets the old one.

An artificial onlooker bee chooses a food source depending on the probability value associated with that food source, $p_i$ , calculated by the following expression (1):

$$p_i = \frac{fit_i}{\sum\limits_{n=1}^{SN} fit_n} \tag{1}$$

where $fit_i$ is the fitness value of the solution $i$ which is proportional to the nectar amount of the food source in the position $i$ and $SN$ is the number of food sources which is equal to the number of employed bees ($BN$).

In order to produce a candidate food position from the old one in memory, the ABC uses the following expression (2):

$$v_{ij} = x_{ij} + \phi_{ij}(x_{ij} - x_{kj}) \tag{2}$$

where $k \in \{1,2,...,SN\}$ and $j \in \{1,2,...,D\}$ are randomly chosen indexes. Although $k$ is determined randomly, it has to be different from $i$. $\phi_{i,j}$ is a random number between [-1, 1]. It controls the production of neighbour food sources around $x_{i,j}$ and represents the comparison of two food positions visually by a bee. As can be seen from (2), as the difference between the parameters of the $x_{i,j}$ and $x_{k,j}$ decreases, the perturbation on the position $x_{i,j}$ gets decrease, too. Thus, as the search approaches to the optimum solution in the search space, the step length is adaptively reduced.

If a parameter value produced by this operation exceeds its predetermined limit, the parameter can be set to an acceptable value. In this work, the value of the parameter exceeding its limit is set to its limit value.

The food source of which the nectar is abandoned by the bees is replaced with a new food source by the scouts. In ABC, this is simulated by producing a position randomly and replacing it with the abandoned one. In ABC, providing that a position can not be improved further through a predetermined number of cycles, then that food source is assumed to be abandoned. The value of predetermined number of cycles is an important control parameter of the ABC algorithm, which is called "*limit*" for abandonment. Assume that the abandoned source is $x_i$ and $j \in \{1,2,...,D\}$ , then the scout discovers a new food source to be replaced with $x_i$. This operation can be defined as in (3)

$$x_i^j = x_{\mathrm{min}}^j + \mathrm{rand}(0,1)(x_{\mathrm{max}}^j - x_{\mathrm{min}}^j) \tag{3}$$

After each candidate source position $v_{i,j}$ is produced and then evaluated by the artificial bee, its performance is compared with that of its old one. If the new food has an equal or better nectar than the old source, it is replaced with the old one in the memory. Otherwise, the old one is retained in the memory. In other words, a greedy selection mechanism is employed as the selection operation between the old and the candidate one.

It is clear from the above explanation that there are four control parameters used in the ABC: The number of food sources which is equal to the number of employed or onlooker bees ($SN$), the value of *limit*, the maximum cycle number ($MCN$).

Detailed pseudo-code of the ABC algorithm is given below:

Initialize the population of solutions $x_{i,j}, i = 1 \ldots SN, j = 1 \ldots D$
Evaluate the population
cycle=1
**repeat**
    Produce new solutions $v_{i,j}$ for the employed bees by using (2) and evaluate them
    Apply the greedy selection process
    Calculate the probability values $P_{i,j}$ for the solutions $x_{i,j}$ by (1)
    Produce the new solutions $v_{i,j}$ for the onlookers from the solutions $x_{i,j}$ selected depending on $P_{i,j}$ and evaluate them
    Apply the greedy selection process

```
    Determine the abandoned solution for the scout, if exists, and replace it
    with a new randomly produced solution x_{i,j} by (3)
    Memorize the best solution achieved so far
    cycle=cycle+1
until cycle=MCN
```

## 2.2 The ABC Algorithm Used for Constrained Optimization Problems

In order to adapt the ABC algorithm for solving constrained optimization problems, we adopted Deb's constrained handling method [13] instead of the selection process (greedy selection) of the ABC algorithm described in the previous section since Deb's method consists of very simple three heuristic rules. Deb's method uses a tournament selection operator, where two solutions are compared at a time, and the following criteria are always enforced: **1)** Any feasible solution is preferred to any infeasible solution, **2)** Among two feasible solutions, the one having better objective function value is preferred, **3)** Among two infeasible solutions, the one having smaller constraint violation is preferred.

Because initialization with feasible solutions is very time consuming process and in some cases it is impossible to produce a feasible solution randomly, the ABC algorithm does not consider the initial population to be feasible. Structure of the algorithm already directs the solutions to feasible region in running process due to the Deb's rules employed instead of greedy selection. Scout production process of the algorithm provides a diversity mechanism that allows new and probably infeasible individuals to be in the population.

In order to produce a candidate food position from the old one in memory, the adapted ABC algorithm uses the following expression:

$$v_j = \begin{cases} x_{ij} + \phi_{ij}(x_{ij} - x_{kj}) & , if \;\; R_j < MR \\ x_{ij} & , \text{otherwise} \end{cases} \tag{4}$$

where $k \in \{1, 2, ..., SN\}$ is randomly chosen index. Although $k$ is determined randomly, it has to be different from $i$. $R_j$ is randomly chosen real number in the range [0,1] and $j \in \{1, 2, ..., D\}$. MR, modification rate, is a control parameter that controls whether the parameter $x_{ij}$ will be modified or not. In the version of the ABC algorithm proposed for constrained optimization problems, artificial scouts are produced at a predetermined period of cycles for discovering new food sources randomly. This period is another control parameter called scout production period ($SPP$) of the algorithm. At each $SPP$ cycle, it is controlled if there is an abandoned food source or not. If there is, a scout production process is carried out.

Pseudo-code of the ABC algorithm proposed for solving constrained problems is given below:

```
Initialize the population of solutions x_{i,j}, i = 1...SN, j = 1...D
Evaluate the population
cycle=1
```

4: **repeat**
5: Produce new solutions $\upsilon_{i,j}$ for the employed bees by using (4) and evaluate them
6: Apply selection process based on Deb's method
7: Calculate the probability values $P_{i,j}$ for the solutions $x_{i,j}$ by (1)
8: Produce the new solutions $\upsilon_{i,j}$ for the onlookers from the solutions $x_{i,j}$ selected depending on $P_{i,j}$ and evaluate them
9: Apply selection process based on Deb's method
10: Determine the abandoned solution for the scout, if exists, and replace it with a new randomly produced solution $x_{i,j}$ by (3)
11: Memorize the best solution achieved so far
12: cycle=cycle+1
13: **until** cycle=MCN

# 3 Experimental Study and Discussion

In order to evaluate the performance of the ABC algorithm, we used a set of 13 benchmark problems can be found in [16]. This set includes various forms of objective function such as linear, nonlinear and quadratic. The performance of the ABC algorithm is compared with that of the differential evolution (DE) and particle swarm optimization (PSO) algorithms.

## 3.1 Settings

PSO employs Deb's rules for constraint handling. The swarm size is 50 and the generation number is 7000. Hence, PSO performs 350 000 objective function evaluations. Cognitive and social components are both set to 1. Inertia weight is uniform random real number in the range [0.5,1]. All equality constraints are converted into inequality constraints, $|h_j| \leq \epsilon$ with $\epsilon$=0.001 [16].

In DE, F is a real constant which affects the differential variation between two solutions and set to 0.5 in our experiments. Value of crossover rate, which controls the change of the diversity of the population, is chosen to be 0.9 as recommended in [17]. Population size is 40 , maximum generation number is 6000 and it uses Deb's rules.

In ABC, the value of modification rate ($MR$) is 0.8, colony size ($2*SN$) is 40 and the maximum cycle number ($MCN$) is 6000. So, the total objective function evaluation number is 240 000 as in DE. The value of limit" is equal to $SN$ x $D$ where $D$ is the dimension of the problem and $SPP$ is also $SN$x$D$. Experiments were repeated 30 times each starting from a random population with different seeds.

## 3.2 Results and Discussion

The results of the experiments for the ABC algorithm are given in Table 1. Comparative results of the best, mean and worst solutions of the investigated algorithms are presented in Table 2, Table 4 and Table 3, respectively.

**Table 1.** Statistical Results Obtained by the ABC algorithm for 13 test functions over 30 independent runs using 240.000 objective function evaluations

| Problem | Optimal | Best | Mean | Worst | Std. Dev. |
|---|---|---|---|---|---|
| **g01** | -15.000 | -15.000 | -15.000 | -15.000 | 0.000 |
| **g02** | 0.803619 | 0.803598 | 0.792412 | 0.749797 | 0.012 |
| **g03** | 1.000 | 1.000 | 1.000 | 1.000 | 0.000 |
| **g04** | -30665.539 | -30665.539 | -30665.539 | -30665.539 | 0.000 |
| **g05** | 5126.498 | 5126.484 | 5185.714 | 5438.387 | 75.358 |
| **g06** | -6961.814 | -6961.814 | -6961.813 | -6961.805 | 0.002 |
| **g07** | 24.306 | 24.330 | 24.473 | 25.190 | 0.186 |
| **g08** | 0.095825 | 0.095825 | 0.095825 | 0.095825 | 0.000 |
| **g09** | 680.63 | 680.634 | 680.640 | 680.653 | 0.004 |
| **g10** | 7049.25 | 7053.904 | 7224.407 | 7604.132 | 133.870 |
| **g11** | 0.75 | 0.750 | 0.750 | 0.750 | 0.000 |
| **g12** | 1.000 | 1.000 | 1.000 | 1.000 | 0.000 |
| **g13** | 0.053950 | 0.760 | 0.968 | 1.000 | 0.055 |

**Table 2.** The Best Solutions Obtained by DE, PSO and ABC algorithms for 13 test functions over 30 independent runs. – Means That No Feasible Solutions Were Found. Na = Not Available.

| P | Optimal | PSO [16] | DE | ABC |
|---|---|---|---|---|
| **g01** | -15.000 | 15.000 | -15.000 | -15.000 |
| **g02** | 0.803619 | 0.669158 | 0.472 | 0.803598 |
| **g03** | 1.000 | 0.993930 | 1.000 | 1.000 |
| **g04** | -30665.539 | -30665.539 | -30665.539 | -30665.539 |
| **g05** | 5126.498 | 5126.484 | 5126.484 | 5126.484 |
| **g06** | -6961.814 | -6161.814 | -6954.434 | -6961.814 |
| **g07** | 24.306 | 24.370153 | 24.306 | 24.330 |
| **g08** | 0.095825 | 0.095825 | 0.095825 | 0.095825 |
| **g09** | 680.63 | 680.630 | 680.630 | 680.634 |
| **g10** | 7049.25 | 7049.381 | 7049.248 | 7053.904 |
| **g11** | 0.75 | 0.749 | 0.752 | 0.750 |
| **g12** | 1.000 | 1.000 | 1.00 | 1.000 |
| **g13** | 0.053950 | 0.085655 | 0.385 | 0.760 |

As seen from Table 2, the ABC algorithm has found the global minimum of the seven of thirteen problems (g01, g03, g04, g06, g08, g11, g12) through 240 000 cycles. On five functions (g02, g04, g05, g07, g10), the ABC algorithm produced results quite close to the global optima. On one problem, g13, the ABC algorithm could not find the optima in the specified maximum number of cycles.

As seen from Table 2, PSO algorithm is better than ABC on three problems (g09,g10,g13) while the ABC algorithm shows better performance than PSO on four problems (g02, g03, g07, g12). Compared to DE, it is better than ABC on

**Table 3.** The Worst Solutions Obtained by DE, PSO and ABC algorithms for 13 test functions over 30 runs. – Means That No Feasible Solutions Were Found. Na = Not Available.

| P | Optimal | PSO [16] | DE | ABC |
|---|---|---|---|---|
| **g01** | -15.000 | -13.000 | -11.828 | -15.000 |
| **g02** | 0.803619 | 0.299426 | 0.472 | 0.749797 |
| **g03** | 1.000 | 0.464 | 1.000 | 1.000 |
| **g04** | -30665.539 | -30665.539 | -30665.539 | -30665.539 |
| **g05** | 5126.498 | 5249.825 | 5534.610 | 5438.387 |
| **g06** | -6961.814 | -6961.814 | -6954.434 | -6961.805 |
| **g07** | 24.306 | 56.055 | 24.330 | 25.190 |
| **g08** | 0.095825 | 0.095825 | 0.095825 | 0.095825 |
| **g09** | 680.63 | 680.631 | 680.631 | 680.653 |
| **g10** | 7049.25 | 7894.812 | 9264.886 | 7604.132 |
| **g11** | 0.75 | 0.749 | 1 | 0.750 |
| **g12** | 1.000 | 0.994 | 1.000 | 1.000 |
| **g13** | 0.053950 | 1.793361 | 0.990 | 1.000 |

**Table 4.** The Mean Solutions Results Obtained by DE, PSO and ABC algorithms for 13 test functions over 30 independent runs and total success numbers of algorithms. A Result In Boldface Indicates A Better Result Or That The Global Optimum (Or Best Known Solution) Was Reached. – Means That No Feasible Solutions Were Found.

| P | Optimal | PSO [16] | DE | ABC |
|---|---|---|---|---|
| **g01** | -15.000 | -14.710 | -14.555 | **-15.000** |
| **g02** | 0.803619 | 0.419960 | 0.665 | **0.792412** |
| **g03** | 1.000 | 0.764813 | **1.000** | **1.000** |
| **g04** | -30665.539 | **-30665.539** | **-30665.539** | **-30665.539** |
| **g05** | 5126.498 | **5135.973** | 5264.270 | 5185.714 |
| **g06** | -6961.814 | **-6961.814** | – | -6961.813 |
| **g07** | 24.306 | 32.407 | **24.310** | 24.473 |
| **g08** | 0.095825 | **0.095825** | **0.095825** | **0.095825** |
| **g09** | 680.63 | **680.630** | **680.630** | 680.640 |
| **g10** | 7049.25 | 7205.5 | **7147.334** | 7224.407 |
| **g11** | 0.75 | 0.749 | 0.901 | **0.750** |
| **g12** | 1.000 | 0.998875 | **1.000** | **1.000** |
| **g13** | 0.053950 | **0.569358** | 0.872 | 0.968 |

four functions (g07, g09, g10, g13) as the ABC algorithm is better than DE on three problems (g02, g06, g11) with respect to the best results.

From the worst results given in Table 3, PSO is better than ABC on three problems (g05, g06, g09) while ABC outperforms PSO on eight problems (g01, g02, g03, g07, g10, g11, g12, g13). DE show better performance on two problems with respect to the ABC algorithm on three problems (g07, g09,g13) while ABC is better on six problems (g01, g02, g05, g06, g10, g11).

Similarly, with respect to the mean solutions in Table 4, PSO shows better performance with respect to the ABC algorithm on five problems (g05, g06, g09, g10, g13) and ABC algorithm is better than PSO on six problems (g01, g02, g03, g07, g11, g12). DE has better performance than ABC on four problems (g07, g09, g10, g13) while ABC is better than DE on five problems (g01, g02, g05, g06, g11).

From the mean results presented in Table 4, it can be concluded that the ABC algorithm performs better than DE and PSO.

Consequently, the ABC algorithm using Deb's rules can not find the optimum solution for g05, g10 and g13 for each run. g05 and g13 are nonlinear problems and the value of $\rho$ ($\rho = |\mathbb{F}| / |\mathbb{S}|$, F:Feasible Space, S:Search Space) for g05 and g13 is %0.000. Also, these problems have nonlinear equality constraints. g10 is linear and the value of $\rho$ is %0.0020 for this problem. However, g10 has no linear equality and nonlinear equality constraints. Therefore, it is not possible to make any generalization for the ABC algorithm such that it is better or not for a specific set of problems. In other words, it is not clear what characteristics of the test problems make it difficult for ABC.

## 4 Conclusion

A modified version of the ABC algorithm for constrained optimization problems has been introduced and its performance has been compared with that of the state-of-art algorithms. It has been concluded that the ABC algorithm can be efficiently used for solving constrained optimization problems. The performance of the ABC algorithm can be also tested for real engineering problems existing in the literature and compared with that of other algorithms. Also, the effect of constraint handling methods on the performance of the ABC algorithm can be investigated in future works.

## References

1. K. E. Parsopoulos, M. N. Vrahatis, *Particle Swarm Optimization Method for Constrained Optimization Problems*, Intelligent Technologies - Theory and Applications: New Trends in Intelligent Technologies,pp. 214–220. IOS Press, 2002.
2. A. R. Hedar, M. Fukushima, *Derivative-Free Filter Simulated Annealing Method for Constrained Continuous Global Optimization*
3. Z. Michalewicz, M. Schoenauer, *Evolutionary Algorithms for Constrained Parameter Optimization Problems*, Evolutionary Computation, 4:1 (1995), pp. 1–32.
4. C. A. Floudas, P. M. Pardalos, *A collection of test problems for constrained global optimization algorithms*, In: LNCS. Vol. 455. Springer-Verlag (1987).
5. D. M. Himmelblau, *Applied Nonlinear Programming*, McGrawHill (1972).
6. J. A. Joines, C. R. Houck, *On the use of nonstationary penalty functions to solve nonlinear constrained optimization problems with gas*, In: Proc. IEEE Int. Conf. Evol. Comp. (1994) 579-585.
7. X. Hu and R. C. Eberhart,*Solving constrained nonlinear optimization problems with particle swarm optimization.* In Proceedings of the Sixth World Multiconference on Systemics, Cybernetics and Informatics 2002.

8. X. Hu, R. C. Eberhart, Y. H. Shi, *Engineering optimization with particle swarm*, IEEE Swarm Intelligence Symposium, 2003: 53-57.
9. K. E. Parsopoulos, , M. N. Vrahatis,A Unified Particle Swarm Optimization for solving constrained engineering optimization problems, dvances in Natural Computation, Pt. 3 , pages 582-591, 2005. Lecture Notes in Computer Science Vol. 3612.
10. A. E. M. Zavala, A. H. Aguirre, E. R. V. Diharce, *Constrained optimization via particle evolutionary swarm optimization algorithm (PESO)* , In Proceedings of the 2005 conference on Genetic and evolutionary computation (GECCO'05), pp. 209–216.
11. D. Karaboga, *An Idea Based On Honey Bee Swarm For Numerical Optimization*, Technical Report-TR06, Erciyes University, Engineering Faculty, Computer Engineering Department, 2005.
12. B. Basturk, D.Karaboga, *An Artificial Bee Colony (ABC) Algorithm for Numeric function Optimization*, IEEE Swarm Intelligence Symposium 2006, May 12-14, 2006, Indianapolis, Indiana, USA.
13. D.E. Goldberg, K. Deb, A comparison of selection schemes used in genetic algorithms Foundations of Genetic Algorithms, edited by G. J. E. Rawlins, pp. 69-93,1991.
14. R. Storn, K. Price, , *Differential evolution – a simple and efficient heuristic for global optimization over continuous spaces.*, Journal of Global Optimization, 11(1997), pp. 341–359.
15. J. Kennedy, R. C. Eberhart,*Particle swarm optimization*,1995 IEEE International Conference on Neural Networks, 4 (1995), 1942–1948.
16. A. E. M. Zavala, A. H. Aguirre, E. R. V. Diharce, *Constrained optimization via particle evolutionary swarm optimization algorithm (PESO)* , In Proceedings of the 2005 conference on Genetic and evolutionary computation (GECCO'05), pp. 209–216.
17. D. Corne, M. Dorigo, F. Glover.,(Editors), *New Ideas in Optimization*, McGraw-Hill, 1999.

# Beam-ACO Distributed Optimization Applied to Supply-Chain Management

João Caldeira, Ricardo Azevedo, Carlos A. Silva and João M.C. Sousa*

Center of Intelligent Systems, IDMEC
Instituto Superior Técnico, Technical University of Lisbon
Av. Rovisco Pais, 1049-001 Lisbon, Portugal
jmsousa@ist.utl.pt

**Abstract.** The distributed optimization paradigm based on Ant Colony Optimization (ACO) is a new management technique that uses the pheromone matrix to exchange information between the different subsystems to be optimized in the supply-chain. This paper proposes the use of the hybrid algorithm Beam-ACO, that fuses Beam-Search and ACO, to implement the same management concept. The Beam-ACO algorithm is used here to optimize the supplying, the distributer and the logistic agents of the supply-chain. Further, this paper implements the concept in a software platform that allows the pheromone matrix exchange through the different agents, using the TCP/IP protocol and data base systems. The results show that the distributed optimization paradigm can still be applied on supply chains where the different agents are optimized by different algorithms and that the use of the Beam-ACO in the supplying agent improves the local and the global results of the supply chain.

## 1 Introduction

In order to keep their competitiveness in the global market, companies are changing their organization into supply-chains, allowing them to react faster on demand changes and to increase their flexibility. Supply-chains are complex and organized structures of independent companies, decentralized and concentrated on their own interests, working collectively in order to satisfy the end customer. Supply-chain management is the way to supervise the flow of products and information as they move in the supply-chain. Its goal is to optimize the supply-chain in order to increase the profit margin of the end product, by giving exactly what the costumer wants, and to keep low production and distribution costs. Due to the stochastic nature and high level of complexity of a supply-chain, its management is a very difficult and complex process that can only be achieved by advanced planning systems. These systems use mostly deterministic algorithms and are based on centralized optimization, in which each agent optimizes its

* This work is supported by the FCT project POCI/EME/59191/2004, co-sponsored by FEDER, Ministério do Ensino Superior da Ciência e Tecnologia, Portugal and by the Programa Operacional Ciência e Inovação (POCI) 2010, FCT, Portugal.

P. Melin et al. (Eds.): IFSA 2007, LNAI 4529, pp. 799–809, 2007.
© Springer-Verlag Berlin Heidelberg 2007

own problem without considering the other agents involved. A new concept for distributed optimization, where each of the different companies within the supply chain exchange information in order to achieve the global optimum of the supply-chain, was introduced in [9,7]. This is obtained by using the ant colony optimization (ACO) algorithm to optimize each intervenient in the supply-chain and by exchanging the pheromone matrix between them. The exchange of the pheromone matrix can be done in a synchronous or asynchronous manner.

In this paper, this concept is extended by the implementation of the recently proposed optimization algorithm called Beam-ACO [1]. This algorithm has proven to be the state-of-the-art optimization algorithm for several job-shop problems. Therefore, the use of this algorithm instead of standard ACO can improve the global performance of the supply chain. Moreover, it is also important to test whether the distributed optimization paradigm could still be applied through the use of different optimization algorithms by the different partners. The paper also proposes the software platform for supply-chain management to be deployed on the different agents. Each system is an agent of the supply chain and the platform uses the TCP/IP protocol to exchange the information between the several agents involved in the supply-chain. Each agent is equipped with a database that contains all the information necessary for the optimization process. The platform framework is implemented in C#.

The outline of the paper is as follows. Section 2 introduces the supply chain model and describes the optimization problems that each agent has to solve. ACO and Beam-ACO implementations for each agent in the supply chain is presented in Section 3. In Section 4, the developed software platform to implement the distributed optimization concept is described. Section 5 presents the simulation results and some final remarks and future research directions are drawn in Section 6.

## 2 Supply-Chain Management

A supply chain can be defined as a coordinated system of organizations, people, activities, information and resources involved in moving a product or service in physical or virtual manner from supplier to customer [10]. The supply-chain model defined in this work is composed by a logistic, a distribution and two suppliers agents [8]. The management of such a system involves the control of the materials' flow and information through the several stages of production, transportation and distribution along a supply-chain, with the purpose of maintaining low stocks without lowering the level of service guaranteed to the customer, which is defined in this case to deliver the products at the correct date.

### 2.1 Logistic System

The logistic system receives every day new orders requested by different clients, where an order $o_j$ is a set of different types of components in certain quantities, with a certain due date $d_j$. The different components and their quantities are purchased from external suppliers, that deliver the components to the cross-docking centers after a certain period of time. The logistic process task is a scheduling problem that consists of observing the list of $n$ orders and the list of components, and decides which orders are delivered at completion date $C_j$.

The difference between the completion date and the due date is called the *tardiness* $T_j = C_j - d_j$. The objective is to match the completion date with the due date, i.e. to have for all orders $T_j = 0$. This decision step is done once per day. Two disturbances may influence the system: the fact that suppliers service may not be respected; and the fact that some clients ask for desired delivery dates not compatible to supplier services. The optimization objective is to minimize the cost function given by

$$f_L = \frac{\sum_{j \in O} T_j + |O^d_{ND}|}{|O^0_D| + \epsilon} \tag{1}$$

where $\sum_{j \in O} T_j$ accounts for the minimization of the tardiness of the set of released orders $O$; $|O^d_{ND}|$ is the cardinality of subset $O^d_{ND}$ and refers to the minimization of the number of orders that are not delivered and are already delayed; $|O^0_D| + \epsilon$ is the cardinality of subset $O^0_D$ and accounts for the maximization of the number of orders delivered at the correct date. $\epsilon$ is a small constant that avoids the division by zero when no orders are delivered at the correct date. This problem can be formally described by a disjunctive graph $G = \{V, A\}$, where the vertices $V$ represent the $n$ orders waiting to be released.

## 2.2 Supplying System

The supplying sub-system is a network of $p$ different manufacturers, each one producing independently its own subset of components $C_p \subset C$, requested by the logistic sub-system. Each supplier is independent and therefore it optimizes its own problem called the *local supplier* problem. However, for the logistic system this one single entity. From this point of view, the optimization to solve is different and is called *global supplying* problem.

**Local Supplier Problem.** The optimization problem of each $p$ supplier can be modeled as a single machine scheduling problem [4]. There is one machine that has to produce all the components $c$ on the waiting list, minimizing the total tardiness, which is defined as in the logistic system:

$$f_{Sp} = \sum_{c \in C_p} T_c \tag{2}$$

**Global Supplying Problem.** The global supplying problem is modeled as an open job shop problem (OSP) [4]. There are $p$ number of machines, where a machine corresponds to a supplier. In this problem, each machine can only produce certain types of jobs $C_p$. The objective is to minimize the total tardiness:

$$f_S = \sum_{c \in C = C_1 \cup, \ldots, \cup C_p} T_c \tag{3}$$

### 2.3 Distribution System

After the scheduling method has decided which orders will be delivered, a distribution company will pick-up the assigned components and deliver them to the different clients. There is a direct correspondence between clients and orders, but clients are described in this case by their geographical location. In general, a distribution problem consists of determining how many trucks are necessary to transport the orders and which sequence should be followed in order to minimize the transportation costs. We consider here two constraints: the maximum load capacity $Q$ and the maximum travel distance $R$ of each truck. This distribution problem can be modeled as a *Vehicle Routing Problem* (VRP)[2]. In this case, the cost function to be minimized is the distance traveled by all the vehicles

$$f_D = \sum_{i=0}^{m} \sum_{j=0}^{m} \sum_{l=1}^{v} d_{ij} x_{ijl} \tag{4}$$

where $x_{ijl} = \{0, 1\}$ indicates if the vehicle $l$ traveled the distance $d_{i,j}$ from client $i$ to $j$: if yes, $x_{ijl} = 1$; if not $x_{ijl} = 0$. The problem can be represented by a disjunctive graph $G = \{V, A\}$, where the vertices $V$ represent the location of the clients and the arcs $A$ are associated with the traveling distance $d_{ij}$ between the vertices.

## 3 Optimization Algorithms

The ant colony optimization algorithm [3] is an optimization method fitted to find the minimum cost in graph optimization problems, especially when these costs are dynamic. This meta-heuristic is inspired on ant colonies behavior while foraging for food, in particular their ability to find the closest path between the food and the nest. In nature, this behavior is explained by pheromone concentration on the traveled paths, which expresses the colonies experience on finding the shortest path. In this section, we introduce the standard ACO and the Beam-ACO algorithms and its implementation to optimize the different agents of the supply chain.

### 3.1 Ant Colony Optimization

Consider a problem of $n$ nodes and a colony of $g$ ants. Initially the ants depart from the initial node, considered to be the ant colony nest, and move to another node by choosing in a probabilistic way. The probability that a $k$ ant in node $i$ chooses node $j$ is given by

$$p_{ij} = \begin{cases} \frac{\tau_{ij}^{\alpha} \cdot \eta_{ij}^{\beta}}{\sum_{r \notin \Gamma}^{n} \tau_{ij}^{\alpha} \cdot \eta_{ij}^{\beta}} & \text{, if } j \notin \Gamma \\ 0 & \text{, otherwise} \end{cases} \tag{5}$$

where $\tau_{ij}$ and $\eta_{ij}$ are the pheromone concentration matrix $\tau$ and the heuristic function matrix $\eta$ respectively, for the path $(i, j)$. The values of the pheromone matrix are limited to the interval $[\tau_{min}, \tau_{max}]$, where $\tau_{min}$ and $\tau_{max}$ correspond to zero and one respectively. $\Gamma$ is the tabu list and it acts has the ants memory, keeping the list of the nodes that the ants have already passed and can not be chosen again. The $\alpha$ and $\beta$ parameters measure the relative importance of the pheromone trail and the heuristic function respectively. Matrices $\tau_{ij}$ and $\eta_{ij}$ are normalized to the interval [0,1], which allows to better judge their relative weight in choosing the next node through parameters $\alpha$ and $\beta$. If $\alpha < \beta$, then the pheromone matrix will have a greater relative weight than the heuristic. After a complete tour, when all the $g$ ants have visited all the $n$ nodes, the pheromone concentration in the trails is updated by

$$\tau_{ij}(t + g \times n) = \tau_{ij}(t) \times (1 - \rho) + \Delta\tau_{ij}^{q} \tag{6}$$

where $\rho \in [0, 1]$ simulates the evaporation phenomenon of the pheromone, and $\Delta\tau_{ij}^{q}$ are pheromone deposited by ant $q$ that found the best solution, along trail $i, j$, in a given iteration.

$$\Delta\tau_{ij}^{q} = \begin{cases} \frac{1}{f^q(s)} & \text{, if arc } (i,j) \text{ is used by ant } q \\ 0 & \text{, otherwise} \end{cases} \tag{7}$$

**ACO in the Logistic Problem.** In the scheduling problem of the logistic system, the orders waiting to be delivered are the nodes of the graph, and the role of the ants is to find the minimum cost path connecting the orders that should be delivered. We consider that each ant is traveling with a bag with the available stocks and is distributing the stocks between the orders that it is visiting. It only visits orders whose components it is able to deliver. In this way, the ACO only builds feasible solutions. When the stocks' bag is empty or the remaining components are not enough to deliver any missing order, the search for this ant is finished. In this case, the number of visited nodes may not be the same from one ant to another, while for the VRP the number of nodes to visit is fixed and equal to the number of clients to visit [2]. The heuristic function $\eta$ is the order's lateness, as proposed in [6]: if an order has already a positive lateness, the ant will feel a stronger attraction to visit it, because the order is already delayed. We define the heuristic function as an exponential function in the interval $[0, 1]$ where the value 0 is for the order that has the minimum lateness $L_{min}$ and 1 is for the most delayed order $L_{max}$ [6]. The objective is that the orders already delayed attract ants much more than the orders not yet delayed:

$$\eta_j = \frac{e^{\frac{L_j - L_{min}}{L_{max} - L_{min}}} - 1}{e - 1} \tag{8}$$

Notice that in this case the heuristic information is only order dependent, therefore $\eta_j = \eta_{ij}$. The pheromone trails $\tau_{ij}$ are also restricted to the interval $[0, 1]$, therefore $\alpha < \beta$ will indicate a higher relative weight of the pheromones trail. The Tabu list is the list of orders already delivered by the ant and also the orders which is not possible to visit, due to lack of stocks. The objective function to minimize by each ant $k$ is $f_L^k$ defined in (1).

### 3.2 Beam-ACO

Beam-ACO optimization method was introduced in [1]. This method hybridizes the deterministic method Beam Search (BS) with ACO meta-heuristic. BS is a tree search method initially used to solve scheduling problems but was adapted to other problems of combinatorial optimization. This algorithm derives from branch and bound algorithms [4]. The central idea of BS is to extend a set of partial solutions up to $k_{ext}$, called *beam extension*, from a given set $B$, called *Beam.* Each set of obtained solutions is then kept in the set $B_{ext}$, in case they are incomplete solutions and can yet be further extended, or in the set $B_c$, in case they are complete solutions and cannot be further extended. At the end of each step, the algorithm selects up to $k_{bw}$ solutions, called *beam width*, from the set $B_{ext}$ creating a new set $B$. This selection is made through the use of a lower bound, which determines the minimum objective function value for any complete solution that can be built from a partial solution. Set $B$ contains an empty partial solution at the beginning of the algorithm. The node choice made by BS is deterministic and is based on a cost function.

**Beam-ACO in the Local Supplying Problem.** To apply the Beam-ACO to the local supplier problem, which is a single machine problem, is necessary to define a lower bound. In this case, the chosen lower bound is:

$$LB(s_t^p) = \max\{\sum_{c_i \in s_t^p} T_{c_i} + T_{c_j \in N(s_t^p):c_j \text{last operation}}\} \tag{9}$$

where $c_i$ corresponds to a component that belongs to the partial solution $s_t^p$, $c_j$ corresponds to a component in the neighborhood $N(s_t^p)$ and is ordered so that is the last one to be produced from the neighborhood.

**Beam-ACO in the Global Supplying Problem.** The global supplying problem is an open-shop problem [5]. Since the Beam-ACO was originally applied to this type of problems, [1]. The chosen lower bound was the one proposed in that paper and is described as:

$$LB(s_t^p) = \max\{X, Y\} \tag{10}$$

where

$$X = \max_{J_j \in J}\{t_{ec}(o^{J_j}, s_t^p) + \sum_{o \in J_{jt}^+} p(o)\} \tag{11}$$

$$Y = \max_{M_i \in M}\{t_{ec}(o^{M_i}, s_t^p) + \sum_{o \in M_{it}^+} p(o)\} \tag{12}$$

$J_j^+$ is the set of operations still to be processed from job $J_j$, and $M_i^+$ is the set of operations from machine $M_i$ still to be processed. The variable $t_{ec}$ (*earliest completion time*), is the time that takes for machine $M_i$ or job $J_j$ of the partial solution to be completed.

**Beam-ACO in the Distribution Problem.** The distribution problem is modeled as a VRP problem. Two lower bounds were defined to apply the Beam-ACO to the distribution problem. The first bound is defined as:

$$LB_D(s_t^p) = w_{s_t^p} + \sum_{i:i \in s_t^p} \sum_{j:j \in N(s_t^p)} D_{ij} \tag{13}$$

where $w_{s_t^p}$ corresponds to the distance already traveled by the partial solution and $\sum_{i:i \in s_t^p} \sum_{j:j \in N(s_t^p)} D_{ij}$ corresponds to the sum of the distances from the current node to the feasible neighborhood and the distances between the nodes of the feasible neighborhood. The second bound is defined as:

$$LB_\eta(s_t^p) = w_{s_t^p} + \sum_{i:i \in s_t^p} \sum_{j:j \in N(s_t^p)} 1/\eta_{ij} \tag{14}$$

where $\sum_{i:i \in s_t^p} \sum_{j:j \in N(s_t^p)} 1/\eta_{ij}$ corresponds to the sum of the inverse heuristic from the current node to the feasible neighborhood and between the nodes of the feasible neighborhood.

### 3.3 Distributed Optimization

The distributed optimization paradigm considers that there are several problems running in parallel, instead of only one problem. However, if each problem is being optimized by an ACO algorithm, the same principle can be applied to exchange information. This section introduces the implementation of this concept to the distributed optimization paradigm for supply chain management. Each subsystem of the supply chain can be optimized through a pheromone matrix, a matrix that indicates the weights on arcs connecting different nodes. However, the different optimization problems can be described in similar graphs, and therefore, different entities from different problems may be represented by the same nodes and arcs. In this way, it is very easy to exchange the pheromone matrix between different problems. Each colony is solving its problem autonomously taking into consideration relevant information of the colonies that are solving different problems. This is valid for ACO or Beam-ACO algorithms.

For the presented supply chain model, management is done at two distinct moments in time. First, the suppliers decide about their own scheduling policy, that may be influenced or not by the logistic system before it receives the new components at the cross-docking center - this is called the supplying-logistic problem. Then, the logistic process has to decide which orders to deliver after it receives the new components, under the influence or not of the distribution system - this is called the logistic-distribution problem. For more details see [7,5].

## 4 Software Platform

The development of an optimization software applied to supply-chain management introduces many challenges. The most important is the information management, that deals with the information that will be exchanged between the

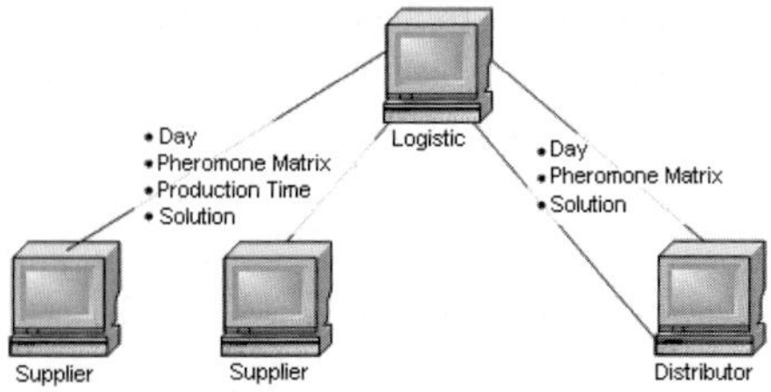


**Fig. 1.** Information exchange

several systems of the supply-chain and how each of them handle that information. The platform is divided into three major parts: the database, the optimization method and the communications section. The optimization method can be implemented in any language. The platform itself was implemented in C#, and manages the information between the optimization algorithm, the database and the communications. Through the use of a database, each system can obtain or modify the required information it needs. There are three ways to establish a link to a database: ODBC, OLeDB and SQL. From these three methods, the OLeDB was chosen. As the information that each agent requires for its optimization is different and has different relations, it is necessary to create one data base for each type of agent. For communication between computers, the TCP/IP protocol is used. This protocol enable communication via internet, allowing the agents to be in different locations. The information necessary to transport between each of the agents can be seen on Fig. 1.

## 5 Results

This section compares the centralized optimization and the distributed optimization methods. The results of the Beam-ACO approach are compared to the traditional ACO using two different supply-chain instances. The instances are composed by one logistic, one distributor and two suppliers. Each instance is described as $(A, B, C, D)$, where A is the average number of orders that enter the system every day; B is the number of different types of components produced by the suppliers; C is the average number of different types of components existing in each order; and D is the maximum quantity of each type of component per order, as described in [7]. Instances $(5, 5, 3, 10)$ and $(10, 10, 2, 30)$ were used. The following tests were executed 20 times for the period between day 1 and day 60. These results were obtained on PC computers using Intel Pentium 4 processors at 3.0 GHz and 1 GB of RAM, running Windows XP.

The parameters used in ACO are: $\alpha = 0.5$, $\beta = 0.2$, $\rho = 0.2$, number of ants is equal to half of the number of nodes and the total number of iterations is equal to 200. The parameters for Beam-ACO are the following: maximum number of iterations is 20, the number of ants is 1, $k_{bw} = 10$ and $k_{ext}$ is equal to the number of nodes that the ant can still travel for a given iteration: $t \leq max\{1, |O|/20\}$, where $|O|$ is the number of nodes. The remaining parameters for Beam-ACO are the same as for ACO. Recall that the instance used for simulation is composed by one logistic, one distributor and two supplier systems. The simulation is running for 60 days.

### 5.1 Results for the Logistic System Using ACO

Table 1 presents the results of the cost function obtained for the tested 60 days in the logistic system. The centralized, distributed synchronous and distributed asynchronous optimizations are compared. The first four lines contain the average, standard deviation, minimum and maximum values of the cost function. The improvement obtained using the distributed optimization is presented in the last two lines. It can be observed that the improvement is remarkable for this system; in fact the cost function is almost reduced to half of the value obtained using centralized optimization. This means that much more orders are getting to the clients at the desired date.

**Table 1.** Cost function results of the logistic for the three variations of supply-chain optimization

| Optimization | centralized | synchronous | asynchronous |
|---|---|---|---|
| Average | 10.09 | 5.38 | 5.26 |
| Standard deviation | 0.00 | 0.02 | 0.16 |
| Minimum | 10.09 | 5.38 | 5.04 |
| Maximum | 10.09 | 5.43 | 5.43 |
| Improvement from centralized | | 46.7% | 47.9% |
| Improvement from synchronous | | | 2.4% |

Similar simulations were implemented in the distribution and in both supplier systems. The improvement in the suppliers were very minor (about 0.012% at most). The distribution system had a slight increase in the cost function of 0.15%. This means that the large improvement of about 48% in the logistic system results in a very slight decrease of performance in the distribution system. Thus, it is clear that the distributed optimization improves significantly the performance of the management of the supply chain.

### 5.2 Results for Beam-ACO

The distributed optimization using the synchronous method was also tested using the Beam-ACO algorithm. The conditions of the simulations are the same as the ones used for ACO. The obtained results are presented in Table 2, where $\mu$ is the average of the 20 runs, $\sigma$ is the standard deviation and $t$ is the computational time in seconds. The results are presented for the four separated systems: Supplier 1, Supplier 2, Logistic and Distribution. By analyzing Table 2 it becomes clear that the Beam-ACO is always slightly worse (less than 1%) than

**Table 2.** Comparison of optimization results using the algorithms Beam-ACO and ACO

| | | Supplier #1 | | Supplier #2 | | Logistic | | Distribution | |
|---|---|---|---|---|---|---|---|---|---|
| | | ACO | Beam-ACO | ACO | Beam-ACO | ACO | Beam-ACO | ACO | Beam-ACO |
| (5,5,3,10) | $\mu$ | 2.13 | 2.15 | 1.94 | 1.94 | 6.63 | 6.66 | 286.5 | 299.1 |
| | $\sigma$ | 0 | 0 | 0 | 0 | 0.002 | 0.000 | 2.257 | 0.074 |
| | $t$(s) | 0.009 | 0.005 | 0.009 | 0.005 | 13.493 | 1.491 | 1.005 | 0.109 |
| (10,10,2,30) | $\mu$ | 3.36 | 3.38 | 4.02 | 4.12 | 11.70 | 11.73 | 514.2 | 536.3 |
| | $\sigma$ | 0 | 0 | 0.004 | 0.000 | 0.000 | 0.003 | 2.52 | 0.341 |
| | $t$(s) | 0.038 | 0.003 | 0.049 | 0.004 | 2.29 | 0.14 | 14.45 | 0.341 |

ACO in terms of cost function. This is a very small value, and can be largely compensated with the computational costs. Note that for the logistic system, which is the more complex system, the Beam-ACO is 9 times faster than ACO. Therefore, in real systems where the complex can be very big, the computational efficiency of Beam-ACO can be an added value.

## 6 Conclusions

This paper proposes the use of the hybrid algorithm Beam-ACO, that fuses Beam-Search and ACO, to optimize a supply-chain in a distributed way. The proposed concept was implemented in a distributed software platform, which allows to exchange the pheromone matrix between the different agents, using the TCP/IP protocol and data base systems. The obtained results show that the use of the Beam-ACO in the supply chain decreases slightly the performance (less than 1%) but reduces drastically the computational time needed to compute a solution. Future work will implement the proposed system in more complex instances, which will be closer to real life situations.

## References

1. C. Blum. Beam-ACO - hybridizing ant colony optimization with beam search: an application to open shop scheduling. *Computers and Operations Research*, 32: 1565–1591, 2005.
2. B. Bullnheimer, R.R. Hartl, and C. Strauss. Applying the ant system to the vehicle routing problem. In I.H. Osman, S. Voß, S. Martello, and C. Roucairol, editors, *Meta-heuristics: Advances and Trends in local search paradigms for optimization*, pages 109–120. Kluwer Academics, 1998.
3. Marco Dorigo and Thomas Stützle. *Ant Colony Optimization.* The MIT Press, 2004.
4. Michael Pinedo. *Scheduling Theory, Algorithms, and Systems.* Prentice Hall, 2002.
5. C. A. Silva. *Distributed Supply Chain Management using Ant Colony Optimization.* Ph.d. thesis, Instituto Superior Técnico, Technical University of Lisbon, 2005.
6. C. A. Silva, T. A. Runkler, J. M. Sousa, and J. M. Sá da Costa. Optimization of logistic processes in supply–chains using meta–heuristics. In *Proceedings of 11th Portuguese Conference on Artificial Intelligence*, pages 9–23. Springer Verlag, 2003.

7. C. A. Silva, T. A. Runkler, J. M. C. Sousa, and J. Sá da Costa. Distributed optimisation of a logistic system and its suppliers using ant colonies. *International Journal of Systems Science*, 37(8):503–512, June 2006.
8. C. A. Silva, J. M. C. Sousa, T. A. Runkler, and R. Palm. Soft computing optimization methods applied to logistic processes. *International Journal of Approximate Reasoning*, 40(3):280–301, November 2005.
9. C.A. Silva, J.M.C. Sousa, T.A. Runkler, and J. Sá da Costa. A multi-agent approach for supply chain management using ant colony optimization. In *Proc. IEEE International Conference on Systems, Man and Cybernetics, IEEE–SMC 2004*, pages 1938–1943, The Hague, The Netherlands, October 2004.
10. N. Viswanadham. The past, present, and future of supply-chain automation. *IEEE Robotics & Automation Magazine*, 9(2):48–56, 2002.

# A Cultural Algorithm with Operator Parameters Control for Solving Timetabling Problems*

Carlos Soza[1], Ricardo Landa[2], María Cristina Riff[1], and Carlos Coello[2]

[1] Universidad Federico Santa María, Departamento de Informática
Av. España No. 1680, Valparaíso, Chile
{csoza,mcriff}@inf.utfsm.cl
[2] CINVESTAV-IPN (Evolutionary Computation Group)
Departamento de Ingeniería Eléctrica, Sección de Computación
Av. IPN No. 2508 Col. San Pedro Zacatenco, México D.F. 07300, Mexico
{rlanda,ccoello}@cs.cinvestav.mx

**Abstract.** A cultural algorithm, together with a set of new operators for the timetabling problem(TP), is proposed in this paper. The new operators extract information about the problem during the evolutionary process, and they are combined with some previously proposed operators, in order to improve the performance of the algorithm. The proposed algorithm is tested with a benchmark of 20 instances, and compared with respect to three other algorithms: two evolutionary algorithms and a simulated annealing algorithm which won an international competition on TP.

## 1 Introduction

The timetabling problem (TP) is a combinatorial problem that can be viewed as an optimzation task. It consists of assigning schedules to several workers or students, which also require some resources. In order to make a feasible timetable, a set of hard constraints must be satisfied(most of them technical constraints); moreover, a good timetable must satisfy some soft constraints (frequently, comfort-related constraints), and if all soft constraints are met, we can consider the solution as optimal. This NP-hard problem presents several variants, such as the employee, exam and university timetabling problems. In 2002, the Metaheuristics Network organized a competition on the University Course Timetabling Problem (UCTP), and published a set of instances of the problem, in order to make easier the comparisons of different algorithms. Cultural algorithms [11] are a particular class of evolutionary algorithm that use domain knowledge extracted during the evolutionary process in order to improve the performance of the search engine (i.e. the evolutionary algorithm) adopted. What we explore in this paper is the use of a combination of knowledge extracted during the evolutionary search with some knowledge that is inserted *a priori* because

* Partially supported by CONACYT/CONICYT Collaboration Project México-Chile No. J110.331/2005 and Fondecyt Project 106377.

P. Melin et al. (Eds.): IFSA 2007, LNAI 4529, pp. 810–819, 2007.
© Springer-Verlag Berlin Heidelberg 2007

it is normally known to be useful when solving combinatorial problems. The main hypothesis in this regard was that the incorporation of knowledge into an evolutionary algorithm would increase its performance as to make it competitive with other approaches whose computational cost is significantly higher. Several heuristics have been used for different types of timetabling problems [9], [13], [3], [8], [15], [10]. Note however, that this paper presents the first attempt (to the authors' best knowledge) to use cultural algorithms to solve TPs.

The proposed approach is compared with respect to an evolutionary algorithm with specialized crossover operators [9], a recently published memetic algorithm [13], and a simulated annealing approach [8] that won the competition of the Metaheuristics Network, in all the test cases proposed for that competition. The obtained results indicate that the proposed approach is a viable alternative for solving, efficiently TPs.

The remainder of this paper is organized as follows: in Section 2 a brief description of the statement of the problem is provided. Section 3 contains an introduction to cultural algorithms which includes a description of their main components and the main motivation to use them. Section 4 contains the details of the proposed approach to solve university course timetabling problems using a cultural algorithm. As part of this section, a description of the representation of solutions adopted in this work is included, as well as the mechanisms implemented to add domain knowledge to the evolutionary algorithm both before and during the search process. Section 5 provides a comparative study. Finally, Section 6 presents the general conclusions and some possible paths for future research.

## 2 Problem Statement

The variant of the problem tackled here was proposed by Ben Paechter for the International Timetabling Competition organized by the Metaheuristics Network [10]. It is referred to in the following as the University Course Timetabling Problem (UCTP). Lecture must be scheduled in 45 timeslots (5 days of 9 hours each) and a number of rooms, with varying facilities and student capacities, so that the following hard constraints are satisfied:

- *H1*: lectures having students in common cannot take place at the same time;
- *H2*: lectures must take place in a room suitable for them in terms of facilities and student capacity; and
- *H3*: no two lectures can take place at the same time in the same room.

We consider as well the following soft constraints:

- *S1*: students should not have to attend lectures in the last timeslot of the day;
- *S2*: they should not attend more than two lectures in a row; and
- *S3*: they should not have a single lecture in any given day.

A timetable in which all lectures have been assigned to a timeslot and a room so that no hard constraints are violated, is said to be feasible. The aim of the problem is to find a feasible solution with minimal soft constraint violations. The corresponding mathematical model is detailed in [16].

## 3 Cultural Algorithms

Cultural algorithms were developed by Reynolds [11] as a complement to the metaphor used by evolutionary algorithms, which had focused mainly on genetic and natural selection concepts. Cultural algorithms are based on some theories originated in sociology and archaeology which try to model cultural evolution (see for example [4]). Such theories indicate that cultural evolution can be seen as an inheritance process operating at two levels: (1) a micro-evolutionary level, which consists of the genetic material that an offspring inherits from its parents, and (2) a macro-evolutionary level, which consists of the knowledge acquired by individuals through generations. This knowledge, once encoded and stored, is used to guide the behavior of the individuals that belong to a certain population.

Culture can be seen as a set of ideological phenomena shared by a population [12]. Through these phenomena, an individual can interpret its experiences and decide its behavior. In these models, it can be clearly appreciated the part of the system that is shared by the population: the knowledge, acquired by members of a society, but encoded in such a way that such knowledge can be accessed by every other member of the society. And then there is an individual part, which consists of the interpretation of such knowledge encoded in the form of symbols. This interpretation will produce new behaviors as a consequence of the assimilation of the corresponding knowledge acquired, combined with the information encoded in the ancestors' genes. Reynolds [11] attempts to capture this double inheritance phenomenon through his proposal of cultural algorithms. The main goal of such algorithms is to increase the learning or convergence rates of an evolutionary algorithm such that the system can respond better to a wide variety of problems [6]. Cultural algorithms operate in two spaces. First, there is the population space, which consists of (as in all evolutionary algorithms) a set of individuals. Each individual has a set of independent features that are used to determine its fitness. Through time, such individuals can be replaced by some of their descendants, which are obtained through the application of a set of operators from the population. The second space is the belief space, which is where the knowledge, acquired by individuals through generations, is stored. The information contained in this space must be accessible to each individual, so that they can use it to modify their behavior. In order to join the two spaces, it is necessary to provide a communication link, which dictates the rules regarding the type of information that must be exchanged between the two spaces. Most of the steps of a cultural algorithm correspond with the steps of a traditional evolutionary algorithm. The main difference lies in the fact that cultural algorithms use a belief space. At each generation the belief space is updated and it incorporates the individual experiences of a select group of members of the population by applying an *accept* function. On the order hand, the variation operators (such as recombination or mutation) are modified by an *influence* function. This function applies some pressure such that the children resulting from the variation operators can exibit behaviors closer to the diserable ones and father away from the undiserable ones, according to the information stored in the belief space. These two functions (*accept* and *influence*) constitute the communication link between the population space and the belief space. The implementation details for these functions in the current proposal are given in the next section. In [11], it is proposed the use of genetic algorithms to model the micro-evolutionary process, and Version Spaces to model the macro-evolutionary process of a cultural algorithm. This sort of algorithm was called the *Version Space guided Genetic Algorithm* (VGA). The main idea behind this approach is to preserve beliefs that are socially accepted and discard (or prune) unacceptable beliefs. Therefore, if a cultural algorithm for global optimization is applied, the acceptable beliefs can be seen as constraints that direct the population at the micro-evolutionary level. In genetic algorithms' theory, there is an expression, called *schema theorem* that represents a bound on the speed at which the best schemata of the population are propagated. Reynolds[11] provided a brief discussion regarding how the belief space could affect the schema theorem. His conclusion was that, by adding a belief space to an evolutionary algorithm, the performance of such algorithm can be improved by increasing its convergence rate. That constitutes the main motivation to use cultural algorithms. Despite the lack of a formal mathematical proof of this efficiency improvement, there is empirical evidence of such performance gains reported in the literature (see for example [1,2]).

## 4 Proposed Approach

The approach proposed in this paper uses, in its population space, a population based on the evolutionary algorithm originally proposed in [9]. A pseudo-code with the main steps of the proposed cultural algorithm is shown in Algorithm 1.. In our algorithm, we have considered three types of knowledge: situational, normative and domain knowledge. Also, we are using five variation operators: two of them use the cultural knowledge (cultural mutation and repair) while the other three are designed to add the exploration component of the algorithm (interchange, sequencing and simple mutation). It is worth mentioning that only one of the exploration operators is applied to each individual.

**Representation:** The representation adopted to encode the solutions plays a very important role when applying an evolutionary computation technique [14]. In this case, a matrix representation was adopted, where columns represent slots of time, and rows represent rooms for the events. This encoding was chosen because it can represent any feasible timetable, and is easier to analyze the violation of some hard constraints, considering only one column at a time.

**Exploration Operators:** The exploration operators are those that allow to maintain diversity of the population. They are listed next. The sequencing operator is similar to the one in [8], and its intention is to generate a large change in the individual since it interchanges two timeslots (this operator is the most destructive one used here). The interchange operator of [9], interchanges two events, and its purpose is to modify the individuals when the problems have in their feasible solutions the same number of places available and events to assign. The simple mutation operator changes the place of an event, and it is useful when the problems have more places availabe in their feasible solutions (without considering the last periods of

**Algorithm 1.** Pseudo-code of the cultural algorithm adopted

```
Generate s random schedules (initial population)
Compute the fitness of each individual in the initial population
Initialize the belief space (copying the best individual to the situational belief space
and create the normative matrix)
repeat
  for each individual in current population do
      Apply cultural mutation operator
      switch (operator)
          case Interchange: Apply Interchange Operator
          case Sequencing: Apply Sequencing Operator
          case SimpleMutation: Apply Simple Mutation Operator
      end switch
      Apply repair operator (with domain knowledge)
  end for
  Selection proccess
  Update the belief space (with the individuals accepted)
until the end condition is satisfied
```

the day) than events to assign. The last two operators make use of the matching algorithm [7] to increase their rate of success.

*Parameter Control for the Application of Exploration Operators*

The parameter control is a process, concurrent to the search of solutions, that allows values of the parameters to change during this process [5]. We use a mechanism of parameter control in order to select the exploration operator (interchange, sequencing or simple mutation) to apply during the mutation process, using a roulette wheel and based on the success rate of each operator. This mechanism consists of updating the probability of each operator to be applied, following some simple rules. If the application of the operator number $i$ results on an improvement of the fitness of the generated individual (with respect to his parent) ($f_{cur} < f_{prev}$), the update of the probabilities is made as follows:

$$operator[i] = operator[i] + \Delta variation$$

where $operator[]$ is the array that contains the probabilities of the operators to be applied, $\Delta variation = \frac{f_{prev} - f_{cur}}{f_{prev} + f_{cur}}$, and $\forall j \in \{1, \ldots, NumOper\}$ and $i \neq j$, $operator[j] = operator[j] - \frac{\Delta variation}{NumOper - 1}$, with $NumOper = 3$ in this case, because we have three operators.

When an operator $i$ is applied and the present solution gets worse ($f_{cur} > f_{prev}$); the updating of the probabilities is made as follows:

$$operator[i] = operator[i] - \Delta variation * \alpha$$

where $\alpha = \frac{PresentTime}{TotalTime}$ , and $\forall j \in \{1, \ldots, NumOper\}$ and $i \neq j$, $operator[j] = operator[j] + \frac{\Delta variation * \alpha}{NumOper - 1}$.

The goal of incorporating the $\alpha$ factor is to mantain controlled the level of decrement, with the objective of not dusturbing those operators whose decreasing ranks are much greater, like the sequencing operator. Initially, the 3 operators in competition start with the same probability of being chosen: $\forall i \in \{1, \ldots, NumOper\}, operator[i] = 1/CantOper$. In order to assure that all operators always have a probability $\neq 0$ of being chosen, all values in $operator[]$ remain between $MinProb = 0.1$ and $MaxProb = 0.8$.

**Mutation Operators with Cultural Influence:** The operator begins selecting an event $E$ and a position $(r, t)$ to move it. This is done through different types of cultural influence.

*Situational Influence:* With the situational influence each individual tries to follow a leader. Such a leader is the best individual found, and is stored in the situational belief space. The key idea is that the individual to be mutated becomes more similar to the leader after the mutation process. The mutation operator randomly selects an event $E$ from the leader, and tries to inherit its position $(r, t)$ to the individual.

The situational belief space is updated at each generation. If the best individual of the current generation is better than the leader in the situational belief space, then the leader is replaced by that individual.

*Normative Influence:* This type of influence is more complex. At each generation, the above average individuals are selected. The idea is to influence the individual to be mutated to inherit some of their characteristics. Before describing the procedure, we need the following definitions:

**Definition 4.1** *We define a* ranking of events *as the set of all the events ordered by the number of events with shared students among them. Thus, the event most connected with other events is the first in the ranking.*

**Definition 4.2** *Given a population $P(g)$ of the generation $g$ and the set $S_g$ composed by the best $s$ individuals of the generation $g$, we define $M$, where each element $M_{ij}$ is the timeslot assigned to the event $i$ in the individual $j$ which belongs to $S_g$.*

The operator proceeds as follows. The room $r$ is fixed. The event is chosen from the ranking of events using a roulette wheel procedure which is biased to the most interconnected events. The new timeslot in the same room $r$ is randomly selected from the matrix $M$, thus the most common timeslot $t$ of the event $E$ in $M$ has the biggest probability of being selected. The hardest event to be assigned, from the constraints point of view, is the event that shares students with the largest number of events. The matrix $M$ is updated at each generation $g$, after the selection of the set $S_g$ (the above average individuals). Once an event $E$ and the position $(r, t)$ have been selected (by any of the cultural influences mentioned), the process of mutation continues as shown in Algorithm 2.. First of all, the operator identifies the current position $(r_E, t_E)$ of the event $E$ in the individual to be mutated. If the new position selected $(r, t)$ is empty and if it is feasible to place $E$ there (from the hard constraints point of view), the current

**Algorithm 2.** mutation($E$, $(r,t)$) procedure, which implements mutation after the influence of cultural selection

$mutation\_finished = FALSE$
identify the position $(r_E, t_E)$ of the event $E$ in the chromosome
**while** $mutation\_finished \neq TRUE$ or $maxtries < 1000$ **do**
  **if** the position $(r,t)$ of the chromosome is empty **then**
    try to move the event $E$ from $(r_E, t_E)$ to $(r,t)$, satisfying the hard constraints
  **else**
    try a swapping move of the event $E_m$ in $(r,t)$
  **end if**
  **if** the position of $E$ was changed **then**
    $mutation\_finished = TRUE$
  **end if**
**end while**

position of event $E$ is modified to $(r,t)$. In case another event $E_m$ is in $(r,t)$, the operator makes swapping moves to change $E_m$ to another position, in order to release $(r,t)$.

**Domain Knowledge:** Our algorithm makes a post-processing procedure which uses the domain knowledge to modify individuals. In the timetabling problem, it is known that the best solution does not include events in the last timeslots of each day, thus the purpose of the repair operator is to try to move the events located in the last timeslots to the earliest ones, always satisfying the hard constraints.

## 5 Comparison of Results

The Cultural Algorithm (CA) is compared with respect to 3 different approaches: a Simulated Annealing (SA) that was the winner of the competition [8], a recent version of a Memetic Algorithm (MA) [13] and the Evolutionary Algorithm (EA) in which this work is based [9]. These references were chosen because they are representative of the state-of-the-art and very competitive on the timetabling problem. The comparison with another EA shows the improvement obtained with the incorporation of culture. The SA approach still presents the best results, but we compare results with it even when it is not an evolutionary algorithm. The benchmark adopted to make the tests and comparisons are the 20 instances of UCTP from the timetabling competition [10]. Those problems are characterized for being of varied difficulty, they consider the individual satisfaction of the students (which allows to consider them individually, not in classes nor groups), and have at least one solution that fulfills both types of restrictions. The proposed approach was implemented in the C++ programming language and was compiled using the GNU g++ compiler in the operating system Debian 3.1. Also, the matching algorithm found in the LEDA library [7] was used.

The cultural algorithm was executed 360 s, as was indicated for the bechmarking program of the competition, for our system configuration.

### 5.1 Cultural Algorithm and Evolutionary Algorithm

The graphs of Figure 1 show the best (right) and the worst case (left) of improvement of CA with respect to EA, in the 20 instances considered. The worst case and the best behavior consider a significant improvement in the first stages which is reflected directly in the final result, in which the cultural algorithm has better results. These graphs show that the incorporation of culture tends to accelerate the convergence of the algorithm and to improve the results.

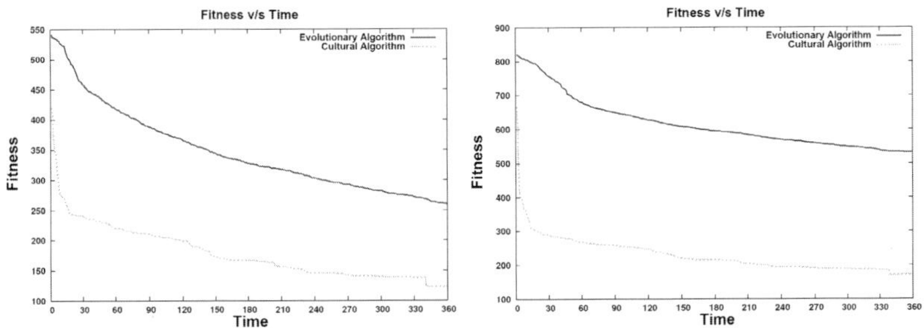


**Fig. 1.** Comparison in time: left the worst case, right the best one

### 5.2 Cultural Algorithm and Other Algorithms

Table 1 shows the results obtained by each algorithm in the 20 problem instances(PI). Table 2 shows a summary of the obtained results emphasizing that the CA improves all the results of EA. The results of the CA are very close in quality from those of MA. Finally, SA is still the most robust approach to solve timetabling problems.

### 5.3 Adaptation on Operators Application Rate

The incorporation of a mechanism to control the parameters of the cultural algorithm, during the selection of the operator to use, resulted on an improvement on the performance of every instance of the benchmark. The graphs of Figure 2 show two representative instances of UCTP. One of them is the instance number 20 (left) where 350 events in 400 places are considered; in such a case the simple mutation operator resulted useful because an important factor was the number

**Table 1.** Comparison of results

| PI | 1 | 2 | 3 | 4 | 5 | 6 | 7 | 8 | 9 | 10 | 11 | 12 | 13 | 14 | 15 | 16 | 17 | 18 | 19 | 20 |
|---|---|---|---|---|---|---|---|---|---|---|---|---|---|---|---|---|---|---|---|---|
| EA | 288 | 260 | 322 | 679 | 557 | 532 | 430 | 305 | 283 | 311 | 328 | 350 | 420 | 469 | 400 | 302 | 521 | 254 | 550 | 424 |
| CA | 140 | 123 | 149 | 330 | 306 | 171 | 159 | 133 | 101 | 147 | 120 | 187 | 233 | 267 | 204 | 102 | 311 | 100 | 296 | 159 |
| MA | 104 | 91 | 126 | 189 | 212 | 90 | 127 | 94 | 78 | 113 | 90 | 138 | 185 | 187 | 120 | 74 | 182 | 75 | 224 | 60 |
| SA | 45 | 25 | 65 | 115 | 102 | 13 | 44 | 29 | 17 | 61 | 44 | 107 | 78 | 52 | 24 | 22 | 86 | 31 | 44 | 7 |

**Table 2.** Summary of results for all instances

| Algorithm | EA | CA | MA | SA |
|---|---|---|---|---|
| Average | 399,25 | 186,9 | 127,95 | 50,55 |
| Std. Desviation | 119,46 | 76,58 | 50,72 | 32,39 |

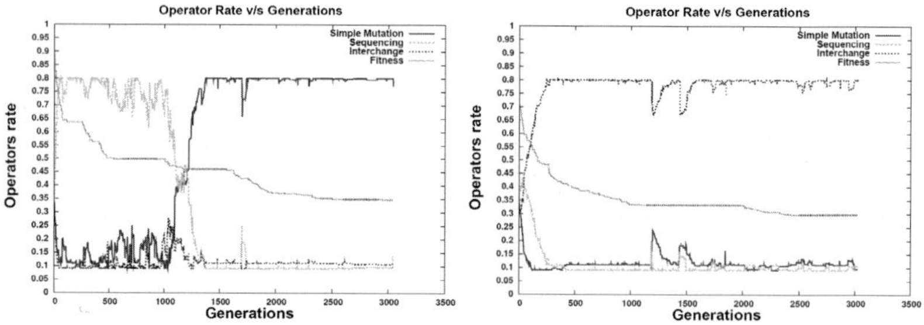


**Fig. 2.** Operators' rate of application: left instance 20, right instance 09

of free places to assign events. On the other hand, instance number 09 (right) has less options to schedule an event, while it has 440 events and just 440 places; in this case, the interchange operator was more useful.

## 6 Conclusions and Future Work

In this paper, we propose the use of domain knowledge, both *a priori* and extracted during the search, to improve the performance of an evolutionary algorithm when solving timetabling problems. The executed experiments provided very encouraging results.

As a future work it would be very interesting to analyze the mechanisms of the simulated annealing method, in order to incorporate them in an evolutionary algorithm or a cultural algorithm. Also, the development of a classification of instances, is a very interesting topic to research, mainly to better understand the performance of different algorithms on different instances.

**Acknowledgments.** The second author acknowledges support from CONACyT through a scholarship to pursue graduate studies at CINVESTAV-IPN. The fourth author acknowledges support from CONACyT through project 45683-Y.

## References

1. Chan-Jin Chung and Robert G. Reynolds. CAEP: An Evolution-based Tool for Real-Valued Function Optimization using Cultural Algorithms. *Journal on Artificial Intelligence Tools*, 7(3):239–292, 1998.

2. Carlos A. Coello and Ricardo Landa. Adding knowledge and efficient data structures to evolutionary programming: A cultural algorithm for constrained optimization. In *In: Proceedings GECCO*, pages 201–209. MK Publishers, 2002.
3. Jean Franois Cordeau, Brigitte Jaumard, and Rodrigo Morales. Efficient timetabling solution with tabu search. International Timetabling Competition, 2003.
4. W. H. Durham. *Co-evolution: Genes, Culture, and Human Diversity.* Stanford University Press, Stanford, California, 1994.
5. A. E. Eiben, R. Hinterding, and Z. Michalewicz. Parameter Control in Evolutionary Algorithms. *IEEE Transactions on Evolutionary Computation*, 3(2):124–141, July 2000.
6. Benjamin Franklin and Marcel Bergerman. Cultural algorithms: Concepts and experiments. In *Proceedings of the 2000 Congress on Evolutionary Computation*, pages 1245–1251, Piscataway, New Jersey, 2000. IEEE Service Center.
7. Algorithmic Solutions Software GmbH. Leda 5.1. `http://www.algorithmic-solutions.com`, 2006. `http://www.algorithmic-solutions.com, 2006.`
8. Philipp Kostuch. Timetabling competition - sa-based heuristic. International Timetabling Competition, 2003.
9. R. Lewis and B. Paechter. New crossover operators for timetabling with evolutionary algorithms. In *5th International Conference on Recent Advances in Soft Computing (RASC 2004)*, LNCS (2004), pages 189–195.
10. Metaheuristics Network. International timetabling competition. `http://www. idsia.ch/Files/ttcomp2002/, 2003.` Last access: December 2006.
11. Robert G. Reynolds. An Introduction to Cultural Algorithms. In A. V. Sebald and L. J. Fogel, editors, *Proceedings of the Third Annual Conference on Evolutionary Programming*, pages 131–139. World Scientific, River Edge, New Jersey, 1994.
12. Peter J. Richerson and Robert Boyd. *Not By Genes Alone: How Culture Transformed Human Evolution.* University Of Chicago Press, December 2004.
13. Olivia Rossi-Doria and Ben Paechter. A memetic algorithm for university course timetabling. In *Proceedings of Combinatorial Optimisation (CO'2004)*, School of Computing, Napier University, Scotland, 2004.
14. Franz Rothlauf. *Representations for Genetic and Evolutionary Algorithms.* Physica-Verlag, New York, 2002.
15. Krzysztof Socha, Michael Sampels, and Max Manfrin. Ant algorithms for the university course timetabling problem with regard to the state-of-the-art. In *Proceedings of EvoCOP 2003*, volume 2611 of *LNCS*, pages 334–345. Springer-Verlag, April 2003.
16. C. Soza, M.C. Riff, and R. Landa. Towards a cultural algorithm to solve timetabling problems. *The Fifth ALIO/EURO conference Paris*, October 2005.

# On Control for Agents Formation*

Rafael Kelly[1], Eusebio Bugarin[1], and Carmen Monroy[2]

[1] CICESE. Apdo. Postal 2615, Adm. 1, Carretera Tijuana–Ensenada Km. 107
Ensenada, B.C., 22800, Mexico
rkelly@cicese.mx

[2] Instituto Tecnológico de la Laguna. Apdo. Postal 49, Adm. 1
Torreón, Coahuila, 27001, Mexico
kmonroy2000@yahoo.com

**Abstract.** Agents control has a broad spectrum of applications in computer science, communications and robotics. This paper focuses on formation of mobile agents, that is, configuration of points in the plane without kinematic restrictions of motion. Several goal formation strategies may be of interest. This paper summarizes control systems for achieving three basic formation structures, namely, absolute positioning with order, absolute positioning without order, and relative positioning with order. Mainly, the paper is devoted to describe each of the above schemes as well as control systems to deal with. Two of the control systems have been already reported in the literature and the remaining one is an original contribution of the paper. The control law proposed is based in the concept of associative memories. Also, simulations are presented to validate the expected behavior.

**Keywords.** Agents, Control Systems, Stability.

## 1 Introduction

Although the word *agent* has a number of meanings, in the modern technology jargon, the underlying concept deals with an entity having the following features: autonomy, adaptability and mobility [1]. *Software agents* differ from conventional software in that they are autonomous and learn from interaction; software agents act as an assistant to the user rather than a tool. *Hardware agents* are abstract points or physical bodies with autonomy of motion and sensory capabilities to learn from the interaction with the environment. Because the real motion of the hardware agents, they are also called mobile agents.

Hardware agents appear in several applications in robotics where the physical configuration or formation of the agents is of interest. The control of collaborative robotic agents to reach and maintain a desired formation is a recent issue in the control and robotics communities [2,3,4].

For the purpose of this paper, an agent is a point in the Cartesian plane with no kinematic constraints of motion [5]. Consider $n$ agents denoted by $\boldsymbol{z}_1, \cdots, \boldsymbol{z}_n$.

* Work partially supported by CONACyT grant 45826 and CUDI.

P. Melin et al. (Eds.): IFSA 2007, LNAI 4529, pp. 820–828, 2007.
© Springer-Verlag Berlin Heidelberg 2007

The configuration of agent $i$ is described by $\boldsymbol{z}_i = [x_i \;\; y_i]^T \in \Re^2$ and the corresponding dynamics is $\dot{\boldsymbol{z}}_i = \boldsymbol{u}_i$ where

$$\boldsymbol{u}_i = \begin{bmatrix} u_{xi} \\ u_{yi} \end{bmatrix}$$

is the control input whose meaning is the agent velocity vector in the Cartesian plane of motion.

The position of all agents gives the group configuration denoted by

$$\boldsymbol{z} = \begin{bmatrix} \boldsymbol{z}_1 \\ \boldsymbol{z}_2 \\ \vdots \\ \boldsymbol{z}_n \end{bmatrix} \in \Re^{2n}$$

and the corresponding model of $n$ agents can be written as

$$\dot{\boldsymbol{z}} = \boldsymbol{u} \tag{1}$$

where $\boldsymbol{u}$ is the control input vector:

$$\boldsymbol{u} = \begin{bmatrix} \boldsymbol{u}_1 \\ \boldsymbol{u}_2 \\ \vdots \\ \boldsymbol{u}_n \end{bmatrix} \in \Re^{2n}.$$

In the context of modern control theory, (1) represents the state equation, thus in order to complete the system description it remains to specify the output equation —in general, a nonlinear function of the state vector $\boldsymbol{z}$—. For the agents model, the output equation arises from the control problem statement and may depend on assumptions about the sensory capabilities of the agents, for instance, to determine the relative displacement or absolute placement of some or all of them. These constraints may derive in a hard control problem to resolve.

Formation of agents may be classified in three basic schemes: absolute positioning with order, absolute positioning without order, and relative positioning with order. The remaining of this paper is devoted to describe each of the above schemes as well as control systems to deal with. Two of the control systems have been already reported in the literature and the remaining one is an original contribution of this work.

## 2 Absolute Positioning with Order

The formation of $n$ agents in absolute positioning with order requires the specification of $n$ desired position $\boldsymbol{z}_1^*, \cdots, \boldsymbol{z}_n^*$ stacked in the vector $\boldsymbol{z}^*$ as

$$\boldsymbol{z}^* = \begin{bmatrix} \boldsymbol{z}_1^* \\ \boldsymbol{z}_2^* \\ \vdots \\ \boldsymbol{z}_n^* \end{bmatrix} \in \Re^{2n}.$$

The formation objective is that each agent, says $\boldsymbol{z}_i$, reaches asymptotically the desired absolute position $\boldsymbol{z}_i^*$. In other terms, the formation objective is

$$\lim_{t\to\infty} \boldsymbol{z}(t) = \boldsymbol{z}^*(t).$$

In general, $\boldsymbol{z}^*$ may be time varying. Without limitation in sensory and communication capability among the agents, that is, each agent may known its absolute location, this is an easy control problem. For this situation, let us define the positioning error vector as $\tilde{\boldsymbol{z}} = \boldsymbol{z}^* - \boldsymbol{z}$. A simple control system is given by the control law [6]

$$\boldsymbol{u} = \dot{\boldsymbol{z}}^* + \boldsymbol{k}(\tilde{\boldsymbol{z}}) \tag{2}$$

where $\boldsymbol{k}(\tilde{\boldsymbol{z}})$ is any function such $\boldsymbol{k}(\boldsymbol{0}) = \boldsymbol{0}$ and $\tilde{\boldsymbol{z}}^T\boldsymbol{k}(\tilde{\boldsymbol{z}})$ is a globally positive definite function.

The closed–loop dynamics is obtained by substituting the control law (2) into the agents model (1). This leads to

$$\dot{\boldsymbol{z}} = \dot{\boldsymbol{z}}^* + \boldsymbol{k}(\tilde{\boldsymbol{z}})$$

which can be rewritten as

$$\dot{\tilde{\boldsymbol{z}}} = -\boldsymbol{k}(\tilde{\boldsymbol{z}}). \tag{3}$$

Since by assumption $\boldsymbol{k}(\tilde{\boldsymbol{z}})$ satisfies $\boldsymbol{k}(\boldsymbol{0}) = \boldsymbol{0}$ and $\tilde{\boldsymbol{z}}^T\boldsymbol{k}(\tilde{\boldsymbol{z}})$ is a globally positive definite function, then $\tilde{\boldsymbol{z}} = \boldsymbol{0}$ is the unique equilibrium of the closed–loop system. The stability of this equilibrium can be analyzed by the following Lyapunov function candidate

$$V(\tilde{\boldsymbol{z}}) = \frac{1}{2}\tilde{\boldsymbol{z}}^T\tilde{\boldsymbol{z}}$$

whose time derivative along the trajectories of the closed–loop system (3) is

$$\dot{V}(\tilde{\boldsymbol{z}}) = -\tilde{\boldsymbol{z}}^T\boldsymbol{k}(\tilde{\boldsymbol{z}}).$$

In virtue that $\dot{V}(\tilde{\boldsymbol{z}})$ is a globally negative definite function, then the Lyapunov's direct method, see e.g. [7], ensures that the equilibrium $\tilde{\boldsymbol{z}} = \boldsymbol{0}$ is globally asymptotically stable. Hence, $\tilde{\boldsymbol{z}}(t) \to \boldsymbol{0}$ as $t \to \infty$ which means that the agents formation is achieved.

## 3 Absolute Positioning Without Order

As in the absolute positioning with order a desired position vector $\boldsymbol{z}^*$ is given. However, no specific desired position is associated to any agent. The formation objective is that the agents reach desired positions regardless the order.

More formally, define the set $\Psi$ as the permutations of $n$ agents in $n$ at a time. This set has $n!$ elements of dimension $2n$. The formation objective is to achieve

$$\lim_{t\to\infty} \mathrm{dist}\left(\boldsymbol{z}(t), \Psi\right) = 0 \tag{4}$$

where $\text{dist}\,(\boldsymbol{z}(t), \Psi)$ denotes the smallest distance from $\boldsymbol{z}$ to any point in the set $\Psi$.

As a contribution of this paper, a control law is proposed based in the concept of associative memories [8,9]. The rational behind this idea is that each element of the set $\Psi$ be seen as a pattern stored into the associative memory. Once a disturbed pattern —initial agents formation— is presented at the associative memory, the corresponding true pattern is recovered —the agents reach one of the formation options in $\Psi$—. This means that the formation objective (4) is achieved.

Based in above arguments, this paper proposes the control law

$$\boldsymbol{u} = -A\boldsymbol{z} + W\boldsymbol{f}(\boldsymbol{z}) + \boldsymbol{b} \tag{5}$$

where $A, W \in \Re^{2n\times 2n}$ with $A > 0$, $\boldsymbol{b} \in \Re^{2n}$, and $\boldsymbol{f}(\boldsymbol{z}) \in \Re^{2n}$ is defined componentwisely by

$$\boldsymbol{f}(\boldsymbol{z}) = \mathbf{tanh}(\boldsymbol{z}) = \begin{bmatrix} \tanh(x_1) \\ \tanh(y_1) \\ \tanh(x_2) \\ \tanh(y_2) \\ \\ \vdots \\ \tanh(x_n) \\ \tanh(y_n) \end{bmatrix}.$$

The first step in the design is to select the matrices $A$ and $W$ and vector $\boldsymbol{b}$ in such a way that

$$-A\boldsymbol{x} + W\boldsymbol{f}(\boldsymbol{x}) + \boldsymbol{b} = \mathbf{0} \qquad \forall\, \boldsymbol{x} \in \Psi. \tag{6}$$

In general the design is a hard algebraic problem to be solved analytically. However several iterative learning procedures —training or learning rules— are available [8,9]. This issue is out of the scope of this paper. Assuming that the design was performed according to (6), then the closed–loop system dynamics obtained by substituting the control law (5) into the agents model (1) yields

$$\dot{\boldsymbol{z}} = -A\boldsymbol{z} + W\boldsymbol{f}(\boldsymbol{z}) + \boldsymbol{b}. \tag{7}$$

This is a nonlinear autonomous system which matches the structure of the well–known Hopfield's neural network [8].

It is worth noticing that in virtue of (6), the elements of $\Psi$ are equilibria of (7). However other spurious equilibria may exist, some of them may be asymptotically stable. Therefore, a sufficient condition to ensure the formation control objective (4) in a local sense is by proving that all equilibria in the set $\Psi$ are asymptotically stable. This implies that the agents will reach a desired formation provided that the initial configuration is sufficiently close of the desired one.

One approach to address the stability analysis is by invoking the Lyapunov's first method [7]. A sufficient condition reported by [10] to determine weather an equilibrium $\boldsymbol{z}^* \in \Psi$ is locally exponentially stable is by testing the positive definiteness of the matrix

$$A - WG(\boldsymbol{z}^*) > 0 \tag{8}$$

where

$$G(\boldsymbol{z}^*) = \begin{bmatrix} \operatorname{sech}^2(x_1^*) & 0 & \cdots & 0 \\ 0 & \operatorname{sech}^2(y_1^*) & \cdots & 0 \\ \vdots & \vdots & \ddots & \vdots \\ 0 & 0 & \cdots & \operatorname{sech}^2(y_n^*) \end{bmatrix}.$$

Thus, a simple test to show that an equilibrium $\boldsymbol{z}^* \in \Psi$ is asymptotically stable is by corroborating that the $2n$ eigenvalues of the symmetric part of $A - WG(\boldsymbol{z}^*)$ are positives.

## 4 Relative Positioning with Order

The formation objective of relative positioning with order is that regardless of the absolute position of any agent, the set of agents pursues a desired geometric form. It is assumed that each agent knows the relative displacement of the next agent in the order, i.e. $\boldsymbol{z}_i - \boldsymbol{z}_{i+1}$ for $i = 1, \cdots, n-1$ and $\boldsymbol{z}_n - \boldsymbol{z}_1$, but the absolute location of each agent $\boldsymbol{z}_i$ cannot be neither measured nor computed.

A geometric shape may be specified by formation functions or formation constraints [11,12] in terms of the relative displacement or distances between each agent. Let us define the desired relative location between agents $i$ and $i+1$ by $\Delta_i \in \Re^2$. For analysis purposes and without loss of generality, choose an arbitrary set of $n$ points $\boldsymbol{z}_1^*, \boldsymbol{z}_2^*, \cdots, \boldsymbol{z}_n^* \in \Re^2$ satisfying the formation constraint, that is

$$\Delta_i = \boldsymbol{z}_{i+1}^* - \boldsymbol{z}_i^*$$

for $i = 1, \cdots, n-1$ and $\Delta_n = \boldsymbol{z}_1^* - \boldsymbol{z}_n^*$.

The formation control of agents under relative positioning with order keeping invariance to translation attempts to guide the agents to reach a desired geometric form with arbitrary unspecified translation. More specifically, the formation objective is to achieve

$$\lim_{t\to\infty} [\boldsymbol{z}(t) - \boldsymbol{z}^* - T\boldsymbol{\xi}] = \mathbf{0} \tag{9}$$

for any $\boldsymbol{\xi} \in \Re^2$ where matrix $T \in \Re^{2n\times 2}$ is given by

$$T = \begin{bmatrix} 1 & 0 \\ 0 & 1 \\ 1 & 0 \\ 0 & 1 \\ \vdots & \vdots \\ 1 & 0 \\ 0 & 1 \end{bmatrix}. \tag{10}$$

A solution to this formation control of agents is the cyclic pursuit method proposed in [5]. The corresponding control law can be written as

$$\boldsymbol{u} = kA\left[\boldsymbol{z} - \boldsymbol{z}^*\right] \tag{11}$$

where $k > 0$ is a control gain, and $A \in \Re^{2n \times 2n}$ is defined by

$$A = \begin{bmatrix} -1 & 0 & 1\,0\,0\,0 \cdots & 0 & 0 \\ 0 & -1 & 0\,1\,0\,0 \cdots & 0 & 0 \\ \vdots & \vdots & \vdots\,\vdots\,\vdots\,\vdots \cdots & \vdots & \vdots \\ 1 & 0 & 0\,0\,0\,0 \cdots & -1 & 0 \\ 0 & 1 & 0\,0\,0\,0 \cdots & 0 & -1 \end{bmatrix}.$$

It is important to observe that for implementation purpose, the control law (11) produces the control actions

$$\begin{aligned} \boldsymbol{u}_i &= k\left[\boldsymbol{z}_{i+1} - \boldsymbol{z}_i\right] - k\left[\boldsymbol{z}^*_{i+1} - \boldsymbol{z}^*_i\right], \\ &= k\left[\boldsymbol{z}_{i+1} - \boldsymbol{z}_i\right] - k\Delta_i, \end{aligned}$$

for $i = 1, \cdots, n-1$ and $\boldsymbol{u}_n = \boldsymbol{z}_1 - \boldsymbol{z}_n - \Delta_n$. We emphasize that implementation of the control law needs only information about the relative location between each agent and the next one in the order, as well as the formation constraints $\Delta_i$, but information about absolute —actual or desired— configuration of the agents is unnecessary.

It is worth noticing that matrix $A$ is negative semidefinite but a singular one whose null space $\mathcal{N}(A)$ is given by

$$\mathcal{N}(A) = \left\{T\boldsymbol{\xi}, \ \ \forall\, \boldsymbol{\xi} \in \Re^2\right\}.$$

The closed–loop dynamics is obtained by plugging the control law (11) into the agents model (1); this leads to

$$\dot{\boldsymbol{z}} = kA\left[\boldsymbol{z} - \boldsymbol{z}^*\right]. \tag{12}$$

Since $A$ is singular, then above equation has an infinity number of equilibria given by

$$\left\{\boldsymbol{z}^* + T\boldsymbol{\xi}, \ \ \forall\, \boldsymbol{\xi} \in \Re^2\right\}. \tag{13}$$

The behavior of the closed–loop system (12), which is an autonomous one, can be studied by invoking the LaSalle's invariance principle [7]. To this end, consider the following nonnegative differentiable function

$$V(\boldsymbol{z}) = \frac{1}{2}\left\|\boldsymbol{z} - \boldsymbol{z}^*\right\|^2.$$

Its time derivative along the trajectories of the closed–loop system (12) yields

$$\dot{V}(\boldsymbol{z}) = k\left[\boldsymbol{z} - \boldsymbol{z}^*\right]^T A\left[\boldsymbol{z} - \boldsymbol{z}^*\right] \tag{14}$$

which satisfies $\dot{V}(\boldsymbol{z}) \leq 0$ for all $\boldsymbol{z} \in \Re^{2n}$ in virtue that $k > 0$ and $A$ is a negative semidefinite matrix. This is a first requirement of the LaSalle's invariance principle. Then, it follows that the system trajectory $\boldsymbol{z}(t)$ tend to the largest invariant set contained in the following domain:

$$\left\{ \boldsymbol{z} \in \Re^{2n} : \dot{V}(\boldsymbol{z}) = 0 \right\}.$$

But from (14) it results that this domain is equivalent to

$$\left\{ \boldsymbol{z} \in \Re^{2n} : A \left[ \boldsymbol{z} - \boldsymbol{z}^* \right] = 0 \right\}$$

which is exactly the equilibria set (13). This is an invariant set, and therefore the largest invariant one; thus, the LaSalle's invariance principle ensures that the system trajectories tend to the equilibria set, so the control objective (9) is attained.

## 5 Simulations

This section presents numerical simulations to demonstrate the control system performance for the "absolute positioning without order" formation structure. Three agents are considered and there is no numbering of the desired locations for each one. The control law has the structure (5) with $A = I \in \Re^{6\times 6}$ and the desired absolute locations for the agents are

$$\boldsymbol{z}_1^* = \begin{bmatrix} -10 \\ 10 \end{bmatrix}; \quad \boldsymbol{z}_2^* = \begin{bmatrix} 10 \\ 10 \end{bmatrix}; \quad \boldsymbol{z}_3^* = \begin{bmatrix} 10 \\ -5 \end{bmatrix}.$$

The set $\Psi$ of permutations has $n! = 6$ elements of dimension $2n = 6$, that is

$$\Psi = \left\{ \begin{bmatrix} \boldsymbol{z}_1^* \\ \boldsymbol{z}_2^* \\ \boldsymbol{z}_3^* \end{bmatrix}, \begin{bmatrix} \boldsymbol{z}_1^* \\ \boldsymbol{z}_3^* \\ \boldsymbol{z}_2^* \end{bmatrix}, \begin{bmatrix} \boldsymbol{z}_2^* \\ \boldsymbol{z}_1^* \\ \boldsymbol{z}_3^* \end{bmatrix}, \begin{bmatrix} \boldsymbol{z}_2^* \\ \boldsymbol{z}_3^* \\ \boldsymbol{z}_1^* \end{bmatrix}, \begin{bmatrix} \boldsymbol{z}_3^* \\ \boldsymbol{z}_1^* \\ \boldsymbol{z}_2^* \end{bmatrix}, \begin{bmatrix} \boldsymbol{z}_3^* \\ \boldsymbol{z}_2^* \\ \boldsymbol{z}_1^* \end{bmatrix} \right\},$$

$$= \left\{ \begin{bmatrix} -10 \\ 10 \\ 10 \\ 10 \\ 10 \\ -5 \end{bmatrix}, \begin{bmatrix} -10 \\ 10 \\ 10 \\ -5 \\ 10 \\ 10 \end{bmatrix}, \begin{bmatrix} 10 \\ 10 \\ -10 \\ 10 \\ 10 \\ -5 \end{bmatrix}, \begin{bmatrix} 10 \\ 10 \\ 10 \\ -5 \\ -10 \\ 10 \end{bmatrix}, \begin{bmatrix} 10 \\ -5 \\ -10 \\ 10 \\ 10 \\ 10 \end{bmatrix}, \begin{bmatrix} 10 \\ -5 \\ 10 \\ 10 \\ -10 \\ 10 \end{bmatrix} \right\}.$$

After tedious but straightforward substitutions, it can be shown that the elements of $\Psi$ satisfy the equilibria matching condition (7) and the asymptotic stability condition (8) with the following selection of $W$ and $b$:

$$W = \begin{bmatrix} -991.0 & 1.0 & -1001.0 & 1.0 & -1001.0 & 1.0 \\ -998.5 & 8.5 & -998.5 & 1.0 & -998.5 & 1.0 \\ -1001.0 & 1.0 & -991.0 & 1.0 & -1001.0 & 1.0 \\ -998.5 & 1.0 & -998.5 & 8.5 & -998.5 & 1.0 \\ -1001.0 & 1.0 & -1001.0 & 1.0 & -991.0 & 1.0 \\ -991.0 & -6.5 & -991.0 & -6.5 & -991.0 & 1.0 \end{bmatrix}; \quad \boldsymbol{b} = \begin{bmatrix} 1000 \\ 1000 \\ 1000 \\ 1000 \\ 1000 \\ 1000 \end{bmatrix}.$$

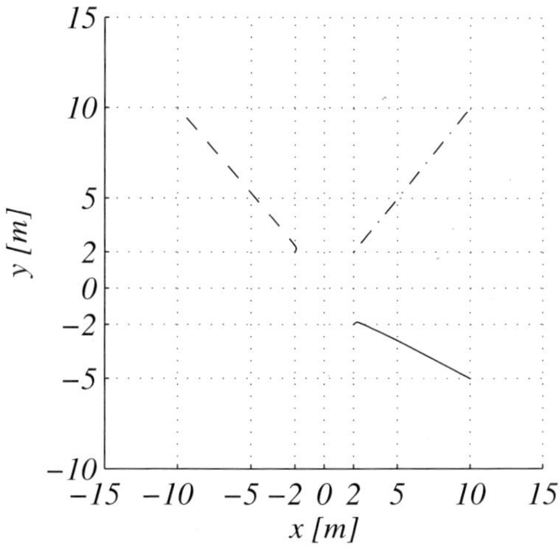


**Fig. 1.** Simulation where $z_1(0) = [2 \ -2]^T$, $z_2(0) = [-2 \ 2]^T$, $z_3(0) = [2 \ 2]^T$

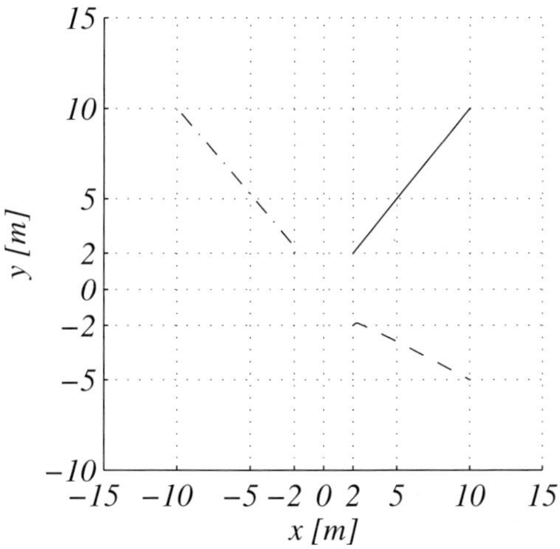


**Fig. 2.** Simulation where $z_1(0) = [2 \ 2]^T$, $z_2(0) = [2 \ -2]^T$, $z_3(0) = [-2 \ 2]^T$

Figures 1 and 2 illustrate the system behavior, each figure represents simulations for the system where the agents start form two different initial configurations and reaches asymptotically the desired formation regardless the order. In both cases, the control objective is achieved, that is, the agents tend to the desired formation.

## 6 Conclusions

The control for agents formation has applications in several fields such as in communications, robotics, and computer sciences. Several agent formations objective can be defined, each one leading to a specific formulation of a control problem. This paper summarized three basic agent formations and introduced a control law inspired from associative memories for dealing with the absolute positioning without order.

In order to illustrate the expected behavior for the proposed controller, simulations are presented showing that the control objective is achieved.

## Acknowledgment

The first author acknowledges the valuable discussions on coordinated control of robots with Dann de la Torre from CICESE, Juan M. Ibarra and Eduardo Aranda from CINVESTAV, and Ricardo Carelli from INAUT. E. Bugarin would like to thank the Instituto Tecnológico de Ensenada for its financial support.

## References

1. M. Wooldridge, *An Introduction to Multiagent Systems*, John Wiley & Sons, (2002).
2. R. Fierro, P. Song, A. Das and V. Kumar, Cooperative control of robot formation, in Cooperative Control and Optimization, R. Murphey & P. Pardalos (Eds.), *Applied Optimization*, vol. 66, Chapter 5, 73–93, (2002).
3. T. Arai, E. Pagello and L. E. Parker, Editorial: Advances in multi–robot systems, *IEEE Transactions on Robotics and Automation*, Vol. 18, No. 5, October, 655–661, (2002).
4. J. P. Desai, A graph theoretic approach for modeling mobile robot team formations, *Journal of Robotic Systems*, Vol. 19, No. 11, 511-525, (2002).
5. Z. Lin, M. Broucke and B. Francis, Local control strategies for group of mobile autonomous agents, *IEEE Trans. on Automatic Control*, Vol. 49, No. 4, 622–629, (2004).
6. R. Kelly, R. Carelli, J. M. Ibarra and C. Monroy, Control de una pandilla de robots móviles para el seguimiento de una constelación de puntos objetivos, *VI Congreso Mexicano de Robótica*, Torreón, México, (2004).
7. H. Khalil, *Nonlinear Systems Analysis*, Second. Edition, Prentice Hall, (1996).
8. S. Haykin, *Neural networks: A comprehensive foundation*, Macmillan College Publishing Company, (1994).
9. M. T. Hagan, H. B. Demuth and M. Beale, *Neural network design*, PWS Publishing Company, (1996).
10. S. I. Sudharsanan and M. K. Sundareshan, Equilibrium characterization of dynamical neural networks and a systematic synthesis procedure for associative memories, *IEEE Tranns. on Neural Networks*, Vol. 2, No. 5, (1991).
11. J. Fredslund and M. J. Mataric, A general algorithm for robot formation using local sensing and minimal communication, *IEEE Transactions on Robotics and Automation*, Vol. 18, No. 5, October, 837–846, (2002).
12. V. Gazi, Formation control of a multi-agent system using nonlinear servomechanism, *International Journal of Control*, Vol. 78, No. 8, 554–565, May, (2005).

# Author Index

Printing: Mercedes-Druck, Berlin
Binding: Stein+Lehmann, Berlin

# Lecture Notes in Artificial Intelligence (LNAI)

Vol. 4539: N.H. Bshouty, C. Gentile (Eds.), Learning Theory. XI, 634 pages. 2007.

Vol. 4529: P. Melin, O. Castillo, L.T. Aguilar, J. Kacprzyk, W. Pedrycz (Eds.), Foundations of Fuzzy Logic and Soft Computing. XIX, 830 pages. 2007.

Vol. 4509: Z. Kobti, D. Wu (Eds.), Advances in Artificial Intelligence. XII, 552 pages. 2007.

Vol. 4496: N.T. Nguyen, A. Grzech, R.J. Howlett, L.C. Jain (Eds.), Agent and Multi-Agent Systems: Technologies and Applications. XXI, 1046 pages. 2007.

Vol. 4483: C. Baral, G. Brewka, J. Schlipf (Eds.), Logic Programming and Nonmonotonic Reasoning. IX, 327 pages. 2007.

Vol. 4482: A. An, J. Stefanowski, S. Ramanna, C.J. Butz, W. Pedrycz, G. Wang (Eds.), Rough Sets, Fuzzy Sets, Data Mining and Granular Computing. XIV, 585 pages. 2007.

Vol. 4481: J. Yao, P. Lingras, W.-Z. Wu, M. Szczuka, N.J. Cercone, D. Ślęzak (Eds.), Rough Sets and Knowledge Technology. XIV, 576 pages. 2007.

Vol. 4476: V. Gorodetsky, C. Zhang, V.A. Skormin, L. Cao (Eds.), Autonomous Intelligent Systems: Multi-Agents and Data Mining. XIII, 323 pages. 2007.

Vol. 4452: M. Fasli, O. Shehory (Eds.), Agent-Mediated Electronic Commerce. VIII, 249 pages. 2007.

Vol. 4451: T.S. Huang, A. Nijholt, M. Pantic, A. Pentland (Eds.), Artifical Intelligence for Human Computing. XVI, 359 pages. 2007.

Vol. 4438: L. Maicher, A. Sigel, L.M. Garshol (Eds.), Leveraging the Semantics of Topic Maps. X, 257 pages. 2007.

Vol. 4429: R. Lu, J.H. Siekmann, C. Ullrich (Eds.), Cognitive Systems. X, 161 pages. 2007.

Vol. 4426: Z.-H. Zhou, H. Li, Q. Yang (Eds.), Advances in Knowledge Discovery and Data Mining. XXV, 1161 pages. 2007.

Vol. 4411: R.H. Bordini, M. Dastani, J. Dix, A.E.F. Seghrouchni (Eds.), Programming Multi-Agent Systems. XIV, 249 pages. 2007.

Vol. 4410: A. Branco (Ed.), Anaphora: Analysis, Algorithms and Applications. X, 191 pages. 2007.

Vol. 4399: T. Kovacs, X. Llorà, K. Takadama, P.L. Lanzi, W. Stolzmann, S.W. Wilson (Eds.), Learning Classifier Systems. XII, 345 pages. 2007.

Vol. 4390: S.O. Kuznetsov, S. Schmidt (Eds.), Formal Concept Analysis. X, 329 pages. 2007.

Vol. 4389: D. Weyns, H.V.D. Parunak, F. Michel (Eds.), Environments for Multi-Agent Systems III. X, 273 pages. 2007.

Vol. 4384: T. Washio, K. Satoh, H. Takeda, A. Inokuchi (Eds.), New Frontiers in Artificial Intelligence. IX, 401 pages. 2007.

Vol. 4371: K. Inoue, K. Satoh, F. Toni (Eds.), Computational Logic in Multi-Agent Systems. X, 315 pages. 2007.

Vol. 4369: M. Umeda, A. Wolf, O. Bartenstein, U. Geske, D. Seipel, O. Takata (Eds.), Declarative Programming for Knowledge Management. X, 229 pages. 2006.

Vol. 4342: H. de Swart, E. Orłowska, G. Schmidt, M. Roubens (Eds.), Theory and Applications of Relational Structures as Knowledge Instruments II. X, 373 pages. 2006.

Vol. 4335: S.A. Brueckner, S. Hassas, M. Jelasity, D. Yamins (Eds.), Engineering Self-Organising Systems. XII, 212 pages. 2007.

Vol. 4334: B. Beckert, R. Hähnle, P.H. Schmitt (Eds.), Verification of Object-Oriented Software. XXIX, 658 pages. 2007.

Vol. 4333: U. Reimer, D. Karagiannis (Eds.), Practical Aspects of Knowledge Management. XII, 338 pages. 2006.

Vol. 4327: M. Baldoni, U. Endriss (Eds.), Declarative Agent Languages and Technologies IV. VIII, 257 pages. 2006.

Vol. 4314: C. Freksa, M. Kohlhase, K. Schill (Eds.), KI 2006: Advances in Artificial Intelligence. XII, 458 pages. 2007.

Vol. 4304: A. Sattar, B.-H. Kang (Eds.), AI 2006: Advances in Artificial Intelligence. XXVII, 1303 pages. 2006.

Vol. 4303: A. Hoffmann, B.-H. Kang, D. Richards, S. Tsumoto (Eds.), Advances in Knowledge Acquisition and Management. XI, 259 pages. 2006.

Vol. 4293: A. Gelbukh, C.A. Reyes-Garcia (Eds.), MICAI 2006: Advances in Artificial Intelligence. XXVIII, 1232 pages. 2006.

Vol. 4289: M. Ackermann, B. Berendt, M. Grobelnik, A. Hotho, D. Mladenič, G. Semeraro, M. Spiliopoulou, G. Stumme, V. Svátek, M. van Someren (Eds.), Semantics, Web and Mining. X, 197 pages. 2006.

Vol. 4285: Y. Matsumoto, R.W. Sproat, K.-F. Wong, M. Zhang (Eds.), Computer Processing of Oriental Languages. XVII, 544 pages. 2006.

Vol. 4274: Q. Huo, B. Ma, E.-S. Chng, H. Li (Eds.), Chinese Spoken Language Processing. XXIV, 805 pages. 2006.

Vol. 4265: L. Todorovski, N. Lavrač, K.P. Jantke (Eds.), Discovery Science. XIV, 384 pages. 2006.

Vol. 4264: J.L. Balcázar, P.M. Long, F. Stephan (Eds.), Algorithmic Learning Theory. XIII, 393 pages. 2006.

Vol. 4259: S. Greco, Y. Hata, S. Hirano, M. Inuiguchi, S. Miyamoto, H.S. Nguyen, R. Słowiński (Eds.), Rough Sets and Current Trends in Computing. XXII, 951 pages. 2006.

Vol. 4253: B. Gabrys, R.J. Howlett, L.C. Jain (Eds.), Knowledge-Based Intelligent Information and Engineering Systems, Part III. XXXII, 1301 pages. 2006.

Vol. 4252: B. Gabrys, R.J. Howlett, L.C. Jain (Eds.), Knowledge-Based Intelligent Information and Engineering Systems, Part II. XXXIII, 1335 pages. 2006.

Vol. 4251: B. Gabrys, R.J. Howlett, L.C. Jain (Eds.), Knowledge-Based Intelligent Information and Engineering Systems, Part I. LXVI, 1297 pages. 2006.

Vol. 4248: S. Staab, V. Svátek (Eds.), Managing Knowledge in a World of Networks. XIV, 400 pages. 2006.

Vol. 4246: M. Hermann, A. Voronkov (Eds.), Logic for Programming, Artificial Intelligence, and Reasoning. XIII, 588 pages. 2006.

Vol. 4223: L. Wang, L. Jiao, G. Shi, X. Li, J. Liu (Eds.), Fuzzy Systems and Knowledge Discovery. XXVIII, 1335 pages. 2006.

Vol. 4213: J. Fürnkranz, T. Scheffer, M. Spiliopoulou (Eds.), Knowledge Discovery in Databases: PKDD 2006. XXII, 660 pages. 2006.

Vol. 4212: J. Fürnkranz, T. Scheffer, M. Spiliopoulou (Eds.), Machine Learning: ECML 2006. XXIII, 851 pages. 2006.

Vol. 4211: P. Vogt, Y. Sugita, E. Tuci, C.L. Nehaniv (Eds.), Symbol Grounding and Beyond. VIII, 237 pages. 2006.

Vol. 4203: F. Esposito, Z.W. Raś, D. Malerba, G. Semeraro (Eds.), Foundations of Intelligent Systems. XVIII, 767 pages. 2006.

Vol. 4201: Y. Sakakibara, S. Kobayashi, K. Sato, T. Nishino, E. Tomita (Eds.), Grammatical Inference: Algorithms and Applications. XII, 359 pages. 2006.

Vol. 4200: I.F.C. Smith (Ed.), Intelligent Computing in Engineering and Architecture. XIII, 692 pages. 2006.

Vol. 4198: O. Nasraoui, O. Zaïane, M. Spiliopoulou, B. Mobasher, B. Masand, P.S. Yu (Eds.), Advances in Web Mining and Web Usage Analysis. IX, 177 pages. 2006.

Vol. 4196: K. Fischer, I.J. Timm, E. André, N. Zhong (Eds.), Multiagent System Technologies. X, 185 pages. 2006.

Vol. 4188: P. Sojka, I. Kopeček, K. Pala (Eds.), Text, Speech and Dialogue. XV, 721 pages. 2006.

Vol. 418[illegible]: J. Euzenat, J. Domingue (Eds.), Artificial Intelligence: Methodology, Systems, and Applications. XIII, 29[illegible] pages. 2006.

Vol. 4180: M. Kohlhase, OMDoc – An Open Markup Format for Mathematical Documents [version 1.2]. XIX, 428 pages. 2006.

Vol. 4177: R. Marín, E. Onaindía, A. Bugarín, J. Santos (Eds.), Current Topics in Artificial Intelligence. XV, 482 pages. 2006.

Vol. 4160: M. Fisher, W. van der Hoek, B. Konev, A. Lisitsa (Eds.), Logics in Artificial Intelligence. XII, 516 pages. 2006.

Vol. 4155: O. Stock, M. Schaerf (Eds.), Reasoning, Action and Interaction in AI Theories and Systems. XVIII, 343 pages. 2006.

Vol. 4149: M. Klusch, M. Rovatsos, T.R. Payne (Eds.), Cooperative Information Agents X. XII, 477 pages. 2006.

Vol. 4140: J.S. Sichman, H. Coelho, S.O. Rezende (Eds.), Advances in Artificial Intelligence - IBERAMIA-SBIA 2006. XXIII, 635 pages. 2006.

Vol. 4139: T. Salakoski, F. Ginter, S. Pyysalo, T. Pahikkala (Eds.), Advances in Natural Language Processing. XVI, 771 pages. 2006.

Vol. 4133: J. Gratch, M. Young, R. Aylett, D. Ballin, P. Olivier (Eds.), Intelligent Virtual Agents. XIV, 472 pages. 2006.

Vol. 4130: U. Furbach, N. Shankar (Eds.), Automated Reasoning. XV, 680 pages. 2006.

Vol. 4120: J. Calmet, T. Ida, D. Wang (Eds.), Artificial Intelligence and Symbolic Computation. XIII, 269 pages. 2006.

Vol. 4118: Z. Despotovic, S. Joseph, C. Sartori (Eds.), Agents and Peer-to-Peer Computing. XIV, 173 pages. 2006.

Vol. 4114: D.-S. Huang, K. Li, G.W. Irwin (Eds.), Computational Intelligence, Part II. XXVII, 1337 pages. 2006.

Vol. 4108: J.M. Borwein, W.M. Farmer (Eds.), Mathematical Knowledge Management. VIII, 295 pages. 2006.

Vol. 4106: T.R. Roth-Berghofer, M.H. Göker, H.A. Güvenir (Eds.), Advances in Case-Based Reasoning. XIV, 566 pages. 2006.

Vol. 4099: Q. Yang, G. Webb (Eds.), PRICAI 2006: Trends in Artificial Intelligence. XXVIII, 1263 pages. 2006.

Vol. 4095: S. Nolfi, G. Baldassarre, R. Calabretta, J.C.T. Hallam, D. Marocco, J.-A. Meyer, O. Miglino, D. Parisi (Eds.), From Animals to Animats 9. XV, 869 pages. 2006.

Vol. 4093: X. Li, O.R. Zaïane, Z. Li (Eds.), Advanced Data Mining and Applications. XXI, 1110 pages. 2006.

Vol. 4092: J. Lang, F. Lin, J. Wang (Eds.), Knowledge Science, Engineering and Management. XV, 664 pages. 2006.

Vol. 4088: Z.-Z. Shi, R. Sadananda (Eds.), Agent Computing and Multi-Agent Systems. XVII, 827 pages. 2006.

Vol. 4087: F. Schwenker, S. Marinai (Eds.), Artificial Neural Networks in Pattern Recognition. IX, 299 pages. 2006.

Vol. 4068: H. Schärfe, P. Hitzler, P. Øhrstrøm (Eds.), Conceptual Structures: Inspiration and Application. XI, 455 pages. 2006.

Vol. 4065: P. Perner (Ed.), Advances in Data Mining. XI, 592 pages. 2006.

Vol. 4062: G.-Y. Wang, J.F. Peters, A. Skowron, Y. Yao (Eds.), Rough Sets and Knowledge Technology. XX, 810 pages. 2006.